Conversion Factors

Conversions that are exact are signified by an asterisk*. Fc factors are written in dimensionless form. To use them, n 60 min/h means that there are 60 minutes in one hour. A quar of these factors changes only the units of the quantity. For example,

$$3.5\text{h} = (3.5\cancel{\text{h}})\left(\frac{60\text{ min}}{\cancel{\text{h}}}\right) = 210\text{ min}.$$

and

$$25\text{ min} = (25\text{ min})\left(\frac{1}{60\text{ min/h}}\right) = (25\,\cancel{\text{min}})\left(\frac{\text{h}}{60\,\cancel{\text{min}}}\right) = 0.42\text{ h}$$

Length
*2.54 cm/in.
*0.3048 m/ft
*1.609344 km/mile
*100 cm/m
12 in./ft
5280 ft/mile

Speed
*0.3048 (m/sec)/(ft/sec)
1.47 (ft/sec)/(miles/h)
0.447 (m/sec)/(miles/h)
1.609 (km/h)/(miles/h)

Power
746 W/hp
550 (ft·lb/sec)/hp

Electrical
1.6022×10^{-19} C/electron charge
96487 C/faraday
*10^4 gauss/tesla

Force
*10^5 dynes/N
4.45 N/lb
0.225 lb/N
1 kg weighs 2.21 lb at $g = 9.80\text{ m/sec}^2$

Time
*86400 sec/day
3.16×10^7 sec/year

Mass
1.6606×10^{-27} kg/u
44.6 kg/slug

Pressure
*1 Pa/(N/m^2)
*1.01325×10^5 (N/m^2)/atm
*1.01325 bar/atm
*0.10 (N/m^2)/(dyne/cm^2)
6.895×10^3 (N/m^2)/psi
133.32 (N/m^2)/mm of Hg at 0°C
76 cm of Hg/atm
14.7 psi/atm
1 torr/mm of Hg

Work and Energy
*10^7 erg/J
*4.184 J/cal
*3.60×10^6 J/kWh
*1,054 J/Btu
1.6022×10^{-19} J/eV
6.242×10^{18} eV/J
0.239 cal/J
0.738 ft·lb/J
1 u → 931.5 MeV

Technical Physics

TECHNICAL PHYSICS

Third Edition

Frederick Bueche

HARPER & ROW, PUBLISHERS, New York
Cambridge, Philadelphia, San Francisco,
London, Mexico City, São Paulo, Singapore, Sydney

Sponsoring Editor: Heidi Udell
Project Editor: David Nickol
Text Design: T. R. Funderburk
Cover Photo: A. Blakesley, Shostal Associates
Text Art: Vantage Art, Inc.
Photo Research: Mira Schachne
Production: Debi Forrest-Bochner
Compositor: York Graphic Services, Inc.
Printer and Binder: R. R. Donnelley & Sons

Technical Physics, Third Edition

Library of Congress Cataloging in Publication Data

Bueche, F. (Frederick), 1923–
Technical physics.

Includes index.
1. Physics. 2. Technology. I. Title.
QC23.B8497 1985 530 84-15659
ISBN 0-06-041036-1

84 85 86 87 9 8 7 6 5 4 3 2 1

CONTENTS IN BRIEF

CONTENTS IN DETAIL

3 MOTION 50

NEWTON'S LAWS 80

FRICTION 106

WORK, POWER, AND ENERGY 119

10 DYNAMICS OF ROTATION 198

11 PROPERTIES OF MATERIALS 218

12 TEMPERATURE AND MATTER 250

13 HEAT ENERGY AND ITS EFFECTS 271

14 INTRODUCTION TO THERMODYNAMICS 287

15 HEAT TRANSFER AND AIR CONDITIONING 309

16 VIBRATORY MOTION 330

REFLECTION AND REFRACTION 419

LENSES AND OPTICAL INSTRUMENTS 445

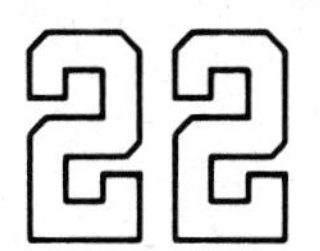

INTERFERENCE AND DIFFRACTION OF WAVES 466

ELECTROSTATICS 486

CIRCUIT ELEMENTS 511

DIRECT CURRENT CIRCUITS 537

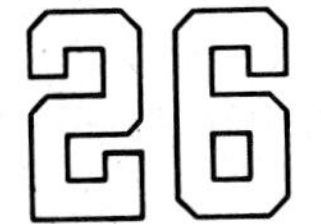

MAGNETISM 559

INDUCED EMF'S 585

TIME-VARYING CURRENTS AND FIELDS 610

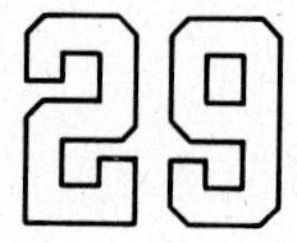

PHYSICS OF THE ATOM 647

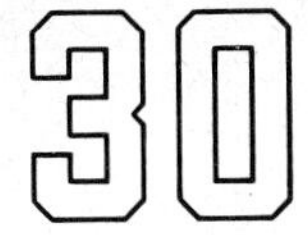

SOLID STATE ELECTRONICS 679

THE ATOMIC NUCLEUS: NUCLEAR ENERGY 698

APPENDIXES

PREFACE

This text has been written for those people who will apply the principles of physics in the world of technology. These people are an important element in our highly industrialized society. It is they who transform the abstractions of basic science into the practical applications that benefit us all. They are the ones who construct, use, and maintain the tools of technology. The precise measurements so necessary to our highly sophisticated industries and laboratories are often the result of their labor. It is to these people, who keep our technological society running, that this text is addressed.

As you look through this book, you will see that it is designed to please its intended users. The principles of physics are presented with an eye to their application in the world about us. We expect the student who uses this text to become a user of the principles. For that reason the principles are illustrated by many worked-out examples. The discussions and explanations have a practical orientation while still teaching the basic laws of physics. Abstractions are made usable by applications related to them. In addition, an effort has been made to keep sentences short and simple so that all students can read the text with ease.

This is a student-oriented text. Words are accompanied by diagrams and photos whenever a visual representation might aid an explanation. Important principles and statements are emphasized visually. Performance goals for each chapter give the student a way to check his or her mastery of the subjects. Detailed chapter summaries help the student to draw together the material of the chapter. The nearly 400 thought-provoking questions at the ends of the chapters are designed to be both informative and searching. They will be useful as focal points for class discussion. More than 1100 problems are provided for student practice. These problems have been carefully prepared to cover the subject matter and to range from easy to rather difficult.

Many students have had insufficient mathematics training. Considerable care is taken to aid the student in this respect. We assume very little mathematical preparation. For the first several chapters most computations are carried out in detail. Power of ten notation and the trigonometry used in the text are taught as the need arises.

We have designed this text for a full-year course. It is organized in such a way that it can be used conveniently in two semesters (for example, Chapters 1–15 and Chapters 16-31) or in three quarters (for example, Chapters 1–10, Chapters 11–22, and Chapters 23–31). The chapters on sound and light waves are grouped together. However, the material on light can be postponed without difficulty if the instructor wishes to do so.

Both the SI and British systems of units are used. However, the British system is relegated to a decidedly secondary position. We still use it because technology in the United States has not yet changed completely to the metric system. But it is absolutely necessary that future technologists be equally familiar with the SI system. Of course, no use is made of such obsolete units as the poundal and the pound mass.

This new edition of *Technical Physics* incorporates many changes suggested by users of previous editions. Although the basic nature of the text remains the same, numerous modifications have been made to increase its teachability. Several chapters have been rearranged and many topics have been rewritten. The problems are now greater in both number and variety. I am deeply indebted to all those who provided suggestions for these and many other improvements we have made. I will appreciate further suggestions for possible changes in the next edition.

Frederick Bueche

Technical Physics

Rogers, Monkmeyer

1 VECTORS

Performance Goals[1]

When you finish this chapter, you should be able to

1. Define vector quantity and differentiate it from a scalar quantity.
2. Determine whether a given quantity is a vector or scalar quantity.
3. Given two displacements, draw a vector diagram to scale showing these displacements and determine the resultant of the two displacements.
4. Draw a vector diagram showing several given displacements. Use

As people progressed from the world of the cave to our modern world of technology, we learned to use symbols to represent objects. We devised a written language which we could use for communication. We devised mathematical symbols and rules which aided us in problem solving. Today in technology we use symbols and language that would have had no meaning to our ancient ancestors. If we are to be a part of that technology, we must learn how to use its language, its concepts, and its symbols. In this chapter we learn how to use symbols called vectors to represent displacements, forces, and motion.

[1] All students should be able to achieve these goals, and the better student should not stop at this level of achievement.

the diagram to determine their resultant.

5. State the Pythagorean theorem and use it to find one side of a right triangle when two sides are given.
6. Given a described vector, draw it and show its x and y components on your diagram.
7. Define the sine, cosine, and tangent of angle θ in terms of the sides of a right triangle. Find θ when any two of the three sides of the triangle are given. Find two sides of the triangle when θ and the hypotenuse are given or when θ and any one of the two sides are given.
8. Given several vectors described to you, make a table showing their components. Find the resultant (magnitude and direction) of these vectors and draw the resultant on a diagram.
9. Repeat objectives 3, 4, 6, and 8 in the case of force vectors.
10. Using your hands, show the approximate magnitude of the following lengths: meter, centimeter, millimeter. Give the abbreviations for each of these units.
11. Convert metric units of length to British units and vice versa.
12. State the unit of force used in the metric system and describe how it is related to the pound.
13. List the metric prefixes, their symbols, and their meanings.
14. Transform a number in scientific notation to regular notation and vice versa.

1.1 Displacement Vectors

Suppose that Bill's Grocery Store is three blocks east of Joe's Gas Station on an east-west street as shown in Figure 1.1. If you move your car from the gas station to the grocery, you have moved it three blocks east. We say that the car has been *displaced* a distance of three blocks to the east. How can we show this *displacement* on the picture? One way is by use of the arrow shown in Figure 1.1. Notice that the arrow starts at the gas station (the initial position of the car) and ends at the grocery (the final position of the car).

The arrow we have drawn represents the displacement. We shall refer to it as the *displacement vector.*[2] The vector (arrow) points in the direction of the displacement. Its length shows how far the car was displaced. We see from this that a displacement vector contains two important pieces of information. The direction in which it points tells us the *direction* of the displacement. Its length tells us the *magnitude* of the displacement (how large the displacement was). Now let us look at a more complex situation.

If you go three blocks east and then two blocks north, the two displacements can be represented by two vectors. This is done in Figure 1.2(a). Here, too, the arrow from the starting point to the ending point represents the resultant displacement. We call it the *resultant displacement vector.* Notice that it shows both the direction and magnitude of the displacement. How far did the two displacements, three blocks east and two blocks north, displace you from the starting point?

To answer this question, we should first notice that the answer IS NOT $2 + 3 = 5$ blocks. This result is clear from the figure. The displacement (i.e., the straight-line distance) from A to C is less than the sum of the distances from A to B and B to C. We see from this example that displacement vectors do not add like simple numbers. In the next section we shall see how to add vector displacements.

1.2 Addition of Vectors: Graphical Method

Since Figure 1.2(a) shows what happens when you go three blocks east and two blocks north, we can use it to find the displacement from A to C. Notice that the displacement vectors have been drawn to scale. This means that we have represented the distance three blocks by a length of 3 units in the figure. Similarly, the two-block displacement is represented by a vector 2 units long. The resultant displacement vector can be measured; it is about 3.6 units long. We therefore conclude that the straight-line distance from A to B is 3.6 blocks. This is the distance displaced. It is the magnitude of the resultant displacement vector.

But the displacement vector tells us direction as well as magnitude. We see in Figure 1.2(a) that C is northeast of A. To tell its direction precisely, we can measure the angle θ (Greek letter theta) shown in the figure. If you use a protractor, you will find that $\theta = 34°$. The complete answer to the question,

[2] In mathematics the word "vector" is given a more abstract meaning.

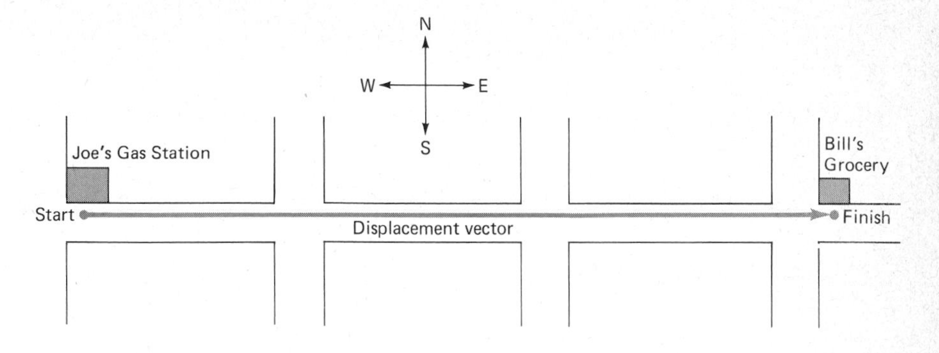

Figure 1.1 We picture the three-block eastward displacement by means of the displacement vector. Note that the arrow (vector) points from the original position to the final position.

"What is the resultant of a 3-block-east displacement and a 2-block-north displacement?" is as follows: The resultant displacement is 3.6 blocks at an angle of 34° north of east.

In the present case we could have found the magnitude of the resultant in another simple way. We recall from geometry the Pythagorean theorem: "In a right triangle the square of the hypotenuse is equal to the sum of the squares of the sides." This theorem is shown in part (b) of Figure 1.2. From Figure 1.2 we see that this means that $h^2 = a^2 + o^2$ or

$$\begin{aligned}(\text{Distance } \overline{AC})^2 &= (3 \text{ blocks})^2 + (2 \text{ blocks})^2 \\ &= 9 + 4 \text{ blocks}^2 \\ &= 13 \text{ blocks}^2\end{aligned}$$

from which

$$\text{Distance } \overline{AC} = \sqrt{13} \text{ blocks} = 3.61 \text{ blocks}$$

This result agrees with the answer of 3.6 blocks that we found by measurement on the figure.

You might ask, "What would happen if you reversed the order of the displacements?" Suppose that you go two blocks north and then three blocks east instead of vice versa. The resultant is still the same. This is shown in Figure 1.3. We see that which displacement is made first does not change the result. Either way the resultant displacement is 3.6 blocks at 34° north of east. In general it can be stated that *the order in which vector displacements are added together causes no effect on the result.*

Let us now go to a much more complicated set of vector displacements. They are shown in Figure 1.4. We are to find their resultant. Notice that we have already drawn the displacements to scale, with 1 unit representing 10 paces. The resultant of these displacements can be measured directly on the figure. It is 5.4 units long, and so the magnitude of the resultant displacement is 54 paces. Its angle as measured with a protractor is 25° north of east. Therefore, the resultant displacement is 54 paces at 25° north of east.

If we are given several displacement vectors, we can always find their resultant by this method. It is called the *graphical method* because we use a scale

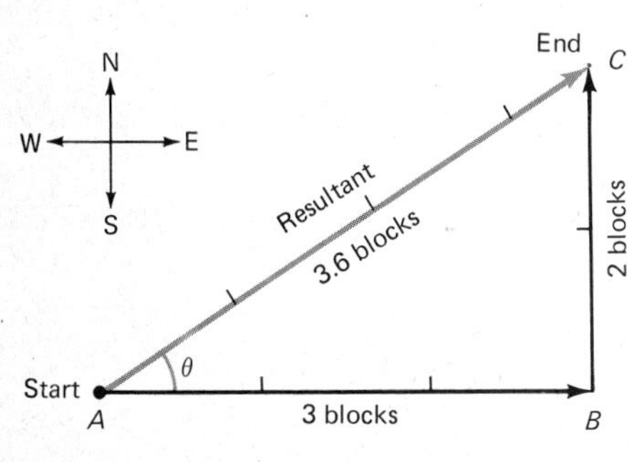

(a) The physical situation

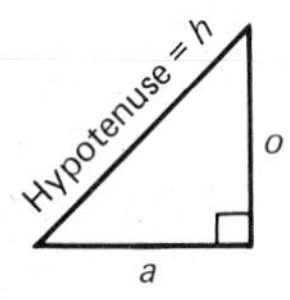

(b) The Pythagorean theorem: $h^2 = a^2 + o^2$

Figure 1.2 The diagram tells us to go 3 blocks east and 2 blocks north. By use of the Pythagorean theorem, the magnitude of the resultant displacement (the distance $\overline{AC}$) is $\sqrt{13}$ blocks, or 3.6 blocks.

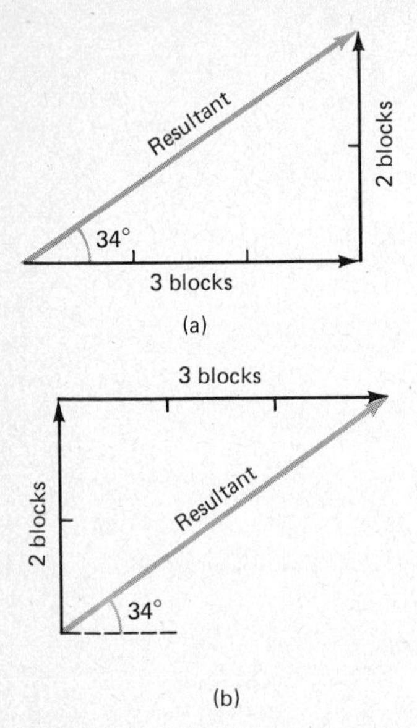

Figure 1.3 Going east first or north first does not change the resultant displacement. The order in which displacement vectors are taken does not change the resultant.

drawing to do it. There are two major disadvantages of the method: First, it is a clumsy method because we must carefully draw a scale diagram; second, it is an inaccurate method because it depends on the accuracy of measurement on a drawing. We shall learn another, usually better, method in a few moments.

■ **EXAMPLE 1.1** Add the following displacements graphically.[3]

45 m east
23 m north
50 m at 60° east of north
30 m south

Solution As we have just seen, the order in which we draw the vectors does not matter. We draw them in the order given, as shown in Figure 1.5. Notice that 1 unit corresponds to 10 m. After drawing in the resultant from the starting to the end point, we measure the vector resultant to be 91 m at an angle of 12° north of east. Notice that the resultant points from the beginning toward the end, as it should since it represents the displacement. ■■

1.3 The Component Method

There is a more convenient and accurate way in which we can add vector displacements. To see what it is, let us add the displacements shown in part (a) of Figure 1.6. This is a special case because all displacements are along either the east-west (x) direction or the north-south (y) direction. Their resultant can be found graphically in the usual way, as we see in (a). But if we look closely, we can see that there is a simpler way.

If you look at the displacements, you will see that we can easily find the total east (or x) distance we have displaced. In the first displacement we went 2 units east. There was no eastward movement in the second displacement. During the third we moved 3 more units east. There was no eastward movement during the

[3]The length unit used here is the meter (m). A meter is about 3 feet (ft) long; precisely, 1 m = 3.28 ft = 1.09 yards (yd).

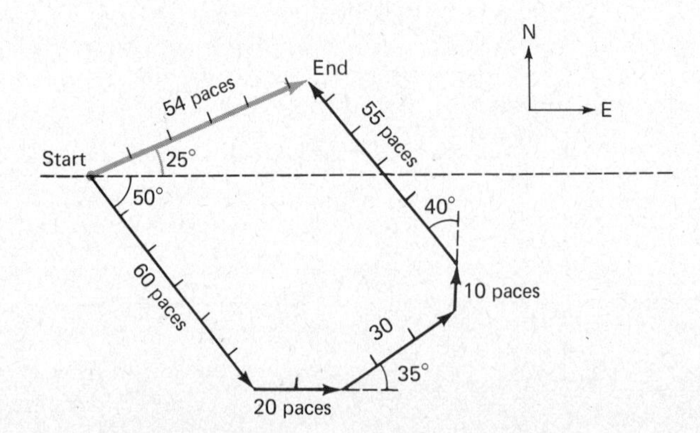

Figure 1.4 The resultant of the five displacements shown is 54 paces at an angle of 25° north of east. Would the answer be changed if the first two displacements were 20 paces east and then 60 paces at 50° south of east?

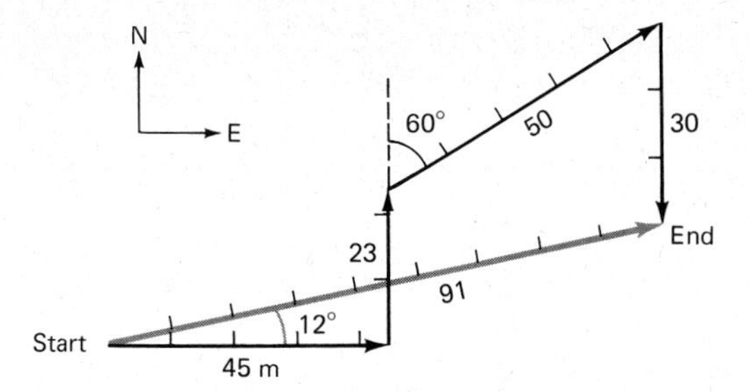

Figure 1.5 The resultant points from the starting point to the end point.

next. The 4-unit displacement west is equivalent to losing 4 units of eastward displacement. So we must count it as −4 east. Finally, the last displacement had no movement eastward. To summarize, the total eastward (or x) displacement $= 2 + 0 + 3 + 0 - 4 + 0 = 1$ unit. In the same way we can compute the total northward (or y) displacement as $0 - 3 + 0 + 5 + 0 + 1 = 3$ units. Examine the figure to see why these numbers are correct.

This example tells us that the effect of all the displacements shown in part (a) has been to move us 1 unit east and 3 units north from the starting point. Therefore, we could just as well replace the diagram in (a) by the diagram in (b). In either case the resultant displacement will be the same. But the diagram in (b) has a great advantage. Using the Pythagorean theorem, we can see at once that

$$(\text{Resultant})^2 = (1)^2 + (3)^2 \text{ units}^2$$

from which

$$\text{Resultant} = \sqrt{1 + 9} = \sqrt{10} = 3.16 \text{ units}$$

Soon we shall see how the angle θ can be found without using a protractor.

Notice the important feature of this method. If we know how far north and how far east each displacement moves us, then the resultant of several displacements is easily found. We simply add all the eastward displacements to find the resultant eastward displacement. The resultant northward displacement is found similarly. Then, since the eastward and northward displacements are the legs of a right triangle, the magnitude of the resultant can be found by use of the Pythagorean theorem.

The displacements in Figure 1.6 were easy to use because each was either northward or eastward. What do we do if a displacement is at an angle to these directions, as in Figure 1.7(a)? As shown in part (b), we wish to know the east and north parts of the displacement. We can easily find them by use of simple trigonometry, as we shall learn in the next section.

1.4 Finding Rectangular Components

You will find that the trigonometry we use in this course is very easy. Most applied scientists and technicians seldom have to use more trigonometry than

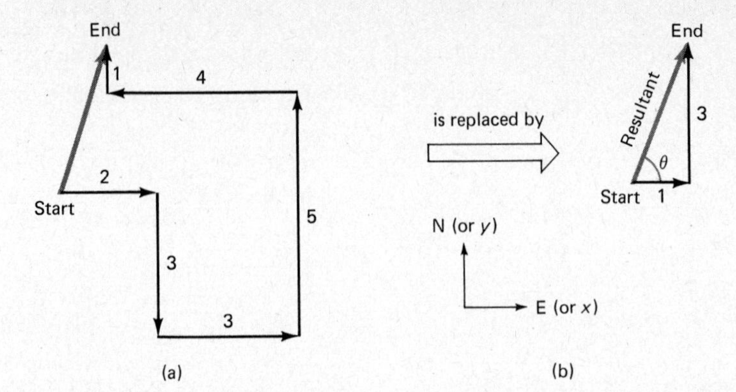

Figure 1.6 The many displacements shown in (a) cause the same effect as the two shown in (b).

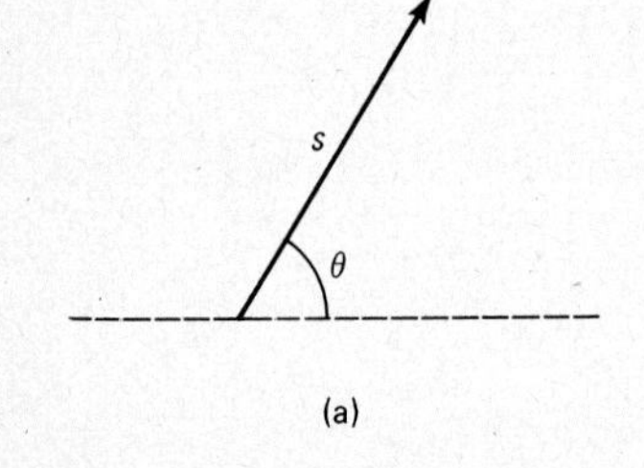

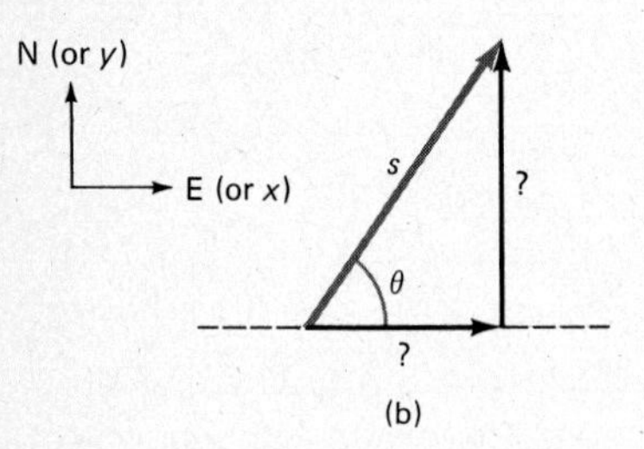

Figure 1.7 How far east and how far north did the displacement shown in (a) move us?

we shall use here. We shall need to be able to find the two sides of a right triangle in terms of the hypotenuse and included angle. To see how we do this, refer to Figure 1.8, where sine, cosine, and tangent are defined. You must learn these definitions. Notice that they apply to right triangles only.

As shown in the figure, the sine of the angle θ is defined to be equal to the length of the side opposite to it (o in this case) divided by the length of the hypotenuse h.

$$\sin\theta = \frac{\text{opposite side}}{\text{hypotenuse}} = \frac{o}{h}$$

Similarly, the cosine is defined as

$$\cos\theta = \frac{\text{adjacent side}}{\text{hypotenuse}} = \frac{a}{h}$$

and the tangent is

$$\tan\theta = \frac{\text{opposite side}}{\text{adjacent side}} = \frac{o}{a}$$

Notice that the trigonometric functions are just ratios. They are pure numbers and have no units. Often it is convenient to remember these definitions in the following form:

$$\text{Opposite side} = (\text{hypotenuse}) \cdot \sin\theta$$
$$\text{Adjacent side} = (\text{hypotenuse}) \cdot \cos\theta$$

Let us now show how we use these definitions.

Trigonometry is useful to us because someone else has done almost all of our computation for us and listed the results in trig tables.[4] A portion of a trig table is shown in Table 1.1. (A complete table is found in Appendix 2.) Notice that the

[4] A trig table is not needed if you have a calculator that provides these data.

values of sin θ, cos θ, and tan θ have been listed for all angles. We need not compute them. This work has already been done for us. Using these tabulated values, we can easily compute many things about right triangles.

Table 1.1 PORTION OF A TRIG TABLE

Angle	Sine	Cosine	Tangent
56°	0.829	0.559	1.483
57°	0.839	0.545	1.540
58°	0.848	0.530	1.600
59°	0.857	0.515	1.664
60°	0.866	0.500	1.732
61°	0.875	0.485	1.804
62°	0.883	0.470	1.881

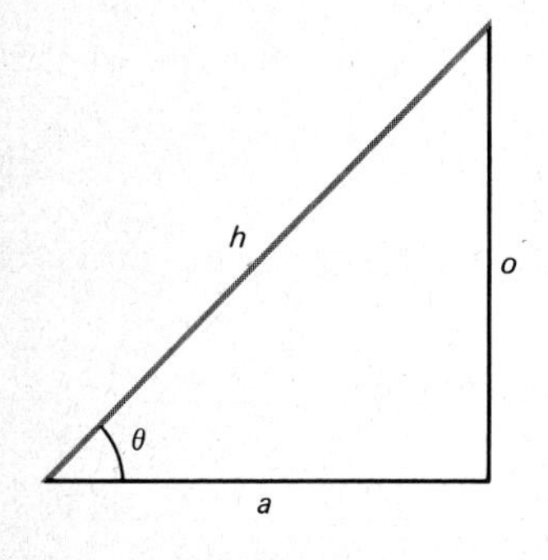

$$\sin\theta \equiv \frac{\text{opposite side}}{\text{hypotenuse}} = \frac{o}{h}$$

$$\cos\theta \equiv \frac{\text{adjacent side}}{\text{hypotenuse}} = \frac{a}{h}$$

$$\tan\theta \equiv \frac{\text{opposite}}{\text{adjacent}} = \frac{o}{a}$$

Figure 1.8 Notice that $a = h\cos\theta$ and $o = h\sin\theta$.

For example, suppose that we are given the data shown in Figure 1.9(a). The hypotenuse of the triangle is 20 units and the angle is 58°. We wish to find the lengths of sides a and o. By our definitions,

$$a = \text{adjacent side} = (\text{hypotenuse}) \cdot \cos\theta$$

and

$$o = \text{opposite side} = (\text{hypotenuse}) \cdot \sin\theta$$

Placing in the values

$$a = 20\cos 58° \qquad \text{and} \qquad o = 20\sin 58°$$

Now we simply refer to the trig table, where we see that cos 58° = 0.530 and sin 58° = 0.848. Therefore,

$$a = (20)(0.530) = 10.6 \text{ units}$$

and

$$o = (20)(0.848) = 17.0 \text{ units}$$

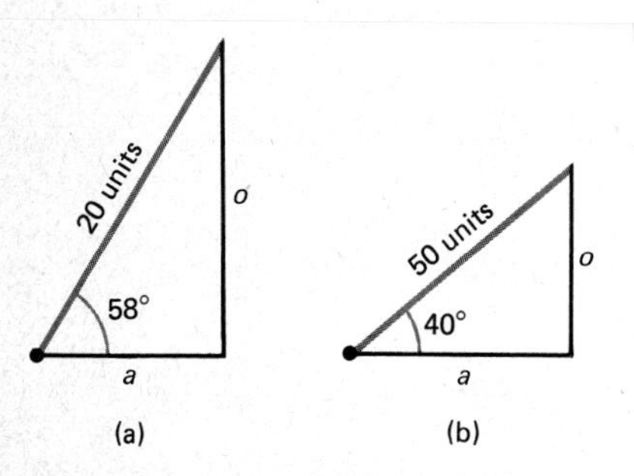

Figure 1.9 You should be able to find a and o for each triangle from the data given in the figure.

Clearly trigonometry provides us with a simple way to find the sides of a right triangle.

It would be a good idea for you to check your understanding by finding a and o in the triangle of Figure 1.9(b). If you do not have a calculator, you will need to refer to the full trig table in Appendix 2. The correct answers are a = 38 and o = 32 units.

You will recall that all this discussion of trigonometry started when we wanted to find the north and east parts of a displacement. Now that we know how to use the sine and cosine, it is simple to do this. A typical situation is

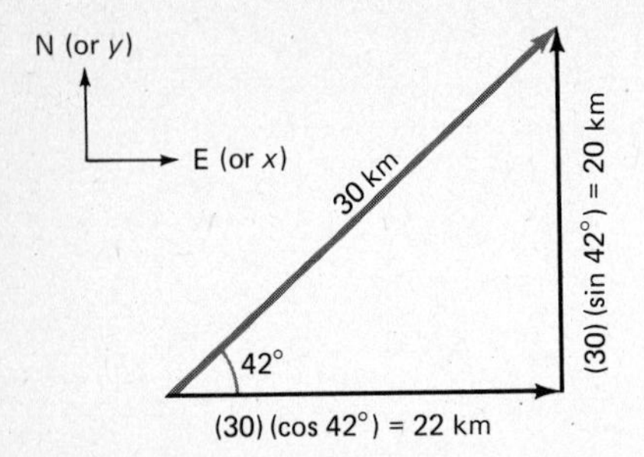

Figure 1.10 We say that the rectangular components of the 30-km displacement are 22 km east and 20 km north.

shown in Figure 1.10, where we have a 30-km displacement[5] at 42° north of east. To find the northward and eastward parts of this displacement, we must find the two sides of the right triangle indicated. We now know how to do this using trigonometry. Check through the calculation given in the figure so that you are sure you see how it was done. We made use of the trig table in Appendix 2 to find the sine and cosine of 42°.

When we find the north and east parts of a vector in this way, we say that we have found the *components* of the vector. The north and east parts are at right angles to each other, and so the name *rectangular components* is also given to these component vectors. To make sure that you know what we are doing, find the rectangular components of the vectors given in Figure 1.11 and check the answers given there. Notice how minus signs are used to show that a vector component points in a direction opposite to north or east. (Notice also the remark about common sense in the figure legend.)

Before leaving our discussion of vector components, let us look at another situation that sometimes arises. In Figure 1.12 we see a vector whose components are known, but the vector and its angle are unknown. We want to find s and ϕ. (The symbol ϕ is often used for angles. It is the Greek letter phi.) To do this we remember that

$$\tan \phi = \frac{\text{opposite side}}{\text{adjacent side}}$$

So we have

$$\tan \phi = \frac{6.1 \text{ m}}{3.8 \text{ m}} = 1.60$$

If we now refer to the trig table shown in Table 1.1, we see under the tangent column that the circled number comes closest to 1.60. The corresponding angle is 58°. Therefore, $\phi = 58°$.

To find s we could use the Pythagorean theorem to give

$$s = \sqrt{(6.1 \text{ m})^2 + (3.8 \text{ m})^2} = 7.2 \text{ m}$$

Or we could use trigonometry in the following way. Since

$$\sin \phi = \frac{\text{opposite side}}{\text{hypotenuse}}$$

we have

$$\sin 58° = \frac{6.1 \text{ m}}{s}$$

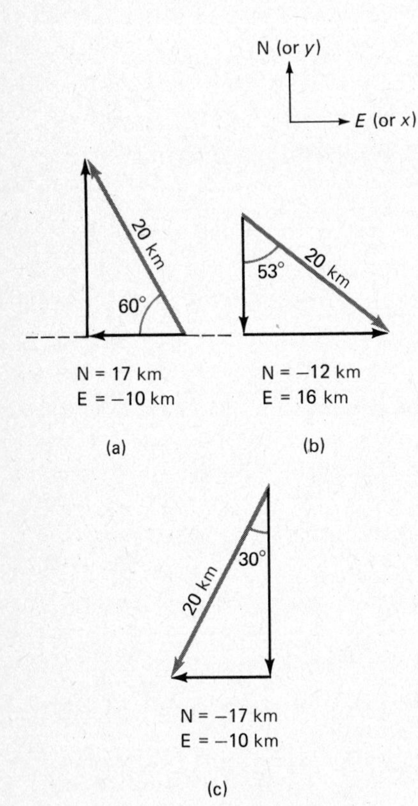

Figure 1.11 Notice that the components have the same general directions as the displacement. Use common sense to avoid errors in direction.

[5]The abbreviation km stands for kilometer. It takes 1000 m to equal 1 km. In addition, 1.6093 km equals 1 mile.

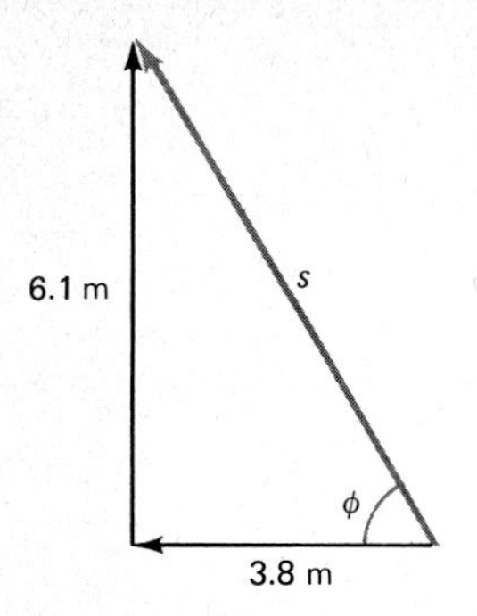

Figure 1.12 You should be able to find both s and ϕ from the data given.

From the table, sin 58° = 0.848. Then

$$0.848s = 6.1 \text{ m}$$

from which

$$s = 7.2 \text{ m}$$

as we found previously by use of the Pythagorean theorem.

1.5 Vector Addition Using Components

Now that we understand how to find the rectangular components of a vector, we can add vectors easily. Our method is simply this:

1. Sketch each vector and show its components.
2. Find the east and north (or x and y) components of each vector.
3. Find the sum of the east (or x) components.
4. Find the sum of the north (or y) components.
5. Use the components of the resultant found in 3 and 4 to find the resultant and its angle.
6. Use a rough sketch to add the vectors graphically. This addition should check your result in 5.

Let us apply this method to a few examples so we are sure we understand how to use it.

EXAMPLE 1.2 A bug crawls along a tabletop in the way shown in Figure 1.13(a). Its displacements are given in centimeters (cm), where 2.54 cm = 1 inch (in.). What is the bug's total displacement?

Solution We apply each step of our method.

1. The vectors and their components are sketched in parts (a) and (b) of the figure. Notice how each component of the 3.0-cm vector is assigned a direction. The original vector points upward and toward the right. Its y component must show this by pointing upward. Its x component must show this by pointing toward the right.
2. The components of the 2.0-cm and 2.5-cm vectors are obvious. Those for the 3.0-cm vector are found as in (b). The x component is 3 cos 40°, or 2.30 cm. The y component is 3 sin 40°, or 1.93 cm.

3. & 4. In order to add the vector components, we prepare a table as follows.

Vector	x	y
2.0 cm	2.00 cm	0 cm
3.0	2.30	1.93
2.5	0	2.50
Resultant	4.30	4.43

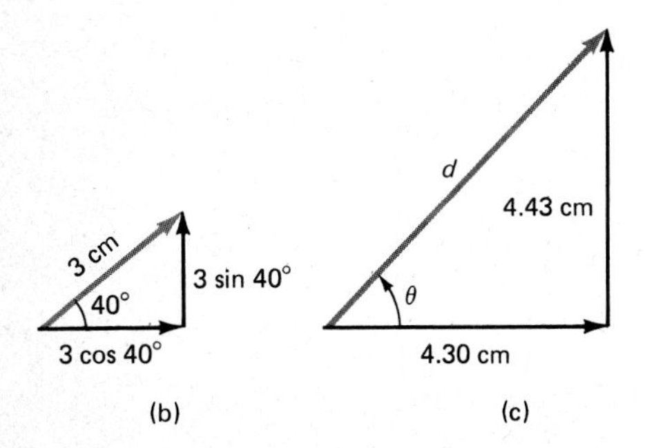

Figure 1.13 A bug makes the three displacements shown in (a). We wish to find out where it is at the end.

5. The components of the resultant lead to the sketch shown in (c) of the figure. Using the Pythagorean theorem, we find the resultant to be

$$d = \sqrt{(4.30)^2 + (4.43)^2} \text{ cm}$$
$$= \sqrt{38.1} \text{ cm} = 6.2 \text{ cm}$$

To find θ we notice that

$$\tan \theta = \frac{o}{a} = \frac{4.43}{4.30} = 1.03$$

The angle that has this tangent is 46°. Therefore, $\theta = 46°$.

6. The rough sketch in (c) checks the resultant found by the sketch in (a).

Before leaving this example, let us point out another way that we could have found d. We could first have found θ, because we already knew that $a = 4.30$ cm and $o =$ 4.43 cm. Then we could write

$$\sin \theta = \frac{o}{h}$$

which becomes in our case

$$\sin 46° = \frac{4.43 \text{ cm}}{d}$$

Multiplying through by d, and using the fact that sin 46° is 0.718, we find that

$$0.718d = 4.43 \text{ cm}$$

From this

$$d = 6.2 \text{ cm}$$

as we found previously. ▮▮

▮ **EXAMPLE 1.3** Find the resultant of the four displacements shown in Figure 1.14(a). The displacements are expressed in meters.

Solution We follow the steps of the component method as shown in the figure. The 30-m vector has a y component of zero whereas its x component is +30 m. The 70-m vector also has a zero y component. But its x component points in the negative x direction. Its x component is therefore −70 m.

The components of the other vectors are found in (b) and (c). Notice how the arrows are placed on the components. The 20-m vector points to the left and upward. Its components, then, must point to the left and upward. This vector therefore has a negative x component and a positive y component. The components are given in the table in part (d). Notice the signs of the components.

If you wished, you might replace the table by the following:

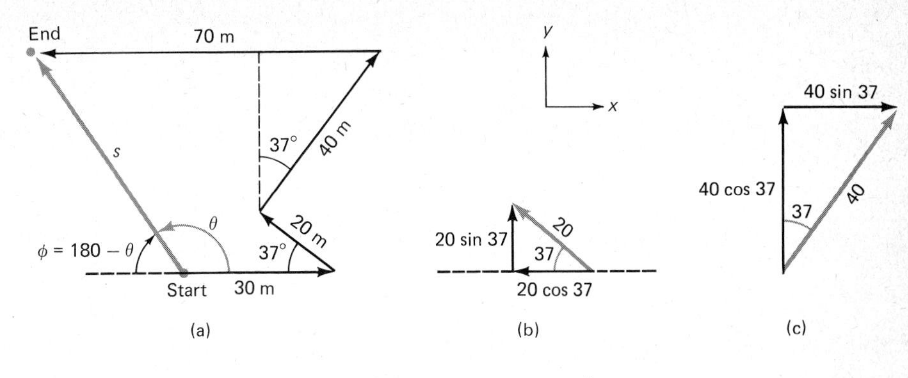

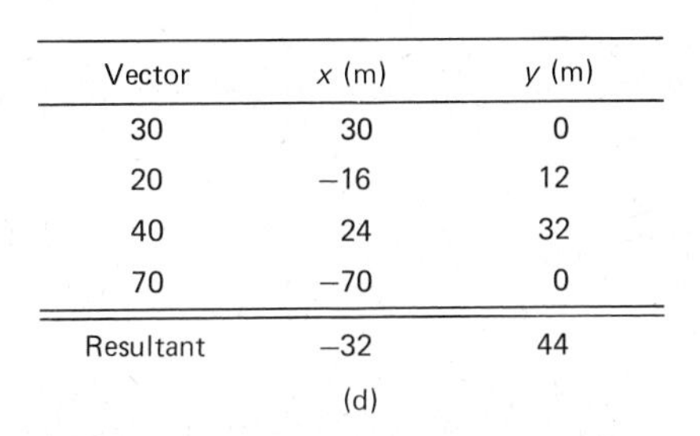

Vector	x (m)	y (m)
30	30	0
20	−16	12
40	24	32
70	−70	0
Resultant	−32	44

(d)

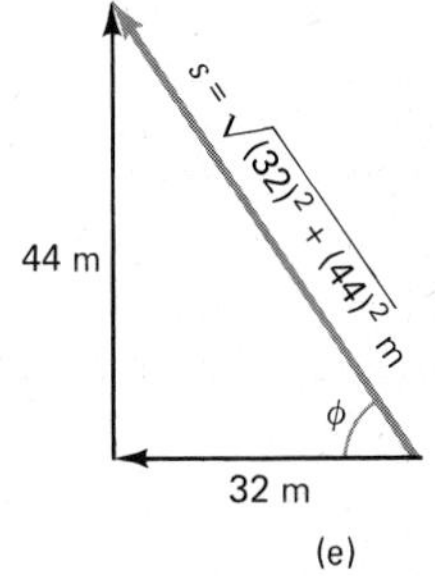

Figure 1.14 The vectors in (a) are added as shown in (b), (c), (d), and (e). Notice in particular that the directions of the components in (b) and (c) agree with common sense. This is also true for the resultant in (e).

$$\text{Resultant's } x \text{ component} = 30 - 16 + 24 - 70 \text{ m} = 54 - 86 \text{ m} = -32 \text{ m}$$

$$\text{Resultant's } y \text{ component} = 0 + 12 + 32 + 0 = 44 \text{ m}$$

We now know that the resultant has an x component of -32 m and a y component of 44 m. The resultant can then be sketched as shown in part (e) of the figure. Notice that the resultant must point toward the left and upward, because that is the way the components point. As we see, the result in (e) checks the sketch in (a).

To find the resultant's magnitude, we can use the Pythagorean theorem to give

$$s = \sqrt{(-32 \text{ m})^2 + (44 \text{ m})^2} = \sqrt{1024 + 1936} \text{ m}$$

Recall that a negative number squared is positive. Then

$$s = 54 \text{ m}$$

The angle of the resultant can be found from the triangle in (e), because

$$\tan \phi = \frac{44}{32} = 1.375$$

The trig tables tell us that this angle is 54°. So we have

$$\phi = 54°$$

But from part (a) of the figure, $\theta + \phi = 180°$, and so $\theta = 180° - \phi$, which gives

$$\theta = 126°$$

You may wonder why we bother to compute θ, the angle shown in (a). It would be more convenient just to compute ϕ and leave the answer in that form. However, most people measure angles counterclockwise (↶) from the x axis. That is, they usually give the angle θ instead of ϕ. We computed θ to conform with this notation. ■■

■ **EXAMPLE 1.4** Call the center point of a football field the origin for an x-y coordinate system. Take the x axis parallel to the sideline and the y axis parallel to the goal line. From a start at the center point, the following displacements are taken:

20 yd at 0° (relative to x axis)
50 yd at 37° (counterclockwise from the x axis is understood)
60 yd at 180°
40 yd at 210°

Find the position of the end point from the center of the field.

Solution A diagram showing these displacements is shown in Figure 1.15(a). Check it to make sure you understand why it is drawn as it is. We wish to find the resultant R and the angle θ shown in (b).

After taking components of each vector, we can prepare the following table.

Vector	x	y
20 yd	20 yd	0 yd
50	40	30
60	−60	0
40	−35	−20
Resultant	−35	10

Now that we know the components of the resultant, we can draw the diagram in Figure 1.15(b).

Using the Pythagorean theorem, we have

$$R = \sqrt{(35)^2 + (10)^2}\ \text{yd} = 36\ \text{yd}$$

To find the angle θ, let us first find the angle labeled ϕ in the diagram. We have

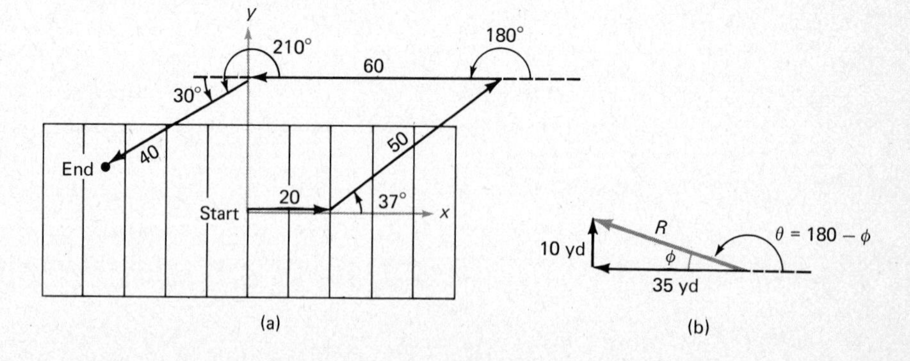

Figure 1.15 We wish to find the displacement of the end point from the center of the football field. The answer is $R = 36$ yd and $\theta = 164°$.

$$\tan \phi = \frac{10}{35} = 0.286$$

from which the trig tables (or calculator) tell us $\phi = 16°$. Since $\theta + \phi = 180°$, we have $\theta = 180° - \phi = 164°$. The displacement from the center is therefore 36 yd at 164° where, as usual, the angle is measured counterclockwise from the x axis. ∎∎

1.6 Vector Quantities Versus Scalar Quantities

As we have seen, displacements can be represented by arrows, which we have called vectors. The vector that represents a displacement shows two things: (1) the length (or magnitude) of the displacement and (2) the direction of the displacement. We might guess that any quantity that has direction as well as magnitude can be represented by a vector in this way. This is true in nearly all cases, as we shall see.

Some things do not have direction and cannot be represented by an arrow or vector. For example, the number of people in the world has a value (magnitude), but this number has no direction associated with it. The same is true for a dozen eggs in a box, the money in your pocket, or the number of birds in a flock. These quantities have magnitude but not direction. Quantities that have magnitude but not direction are called *scalar* quantities.

As we have seen, a displacement has a direction as well as a magnitude. Quantities that have direction as well as magnitude, and that add in the way displacements do, are called *vector* quantities. The man in Figure 1.16(a) pushes on the wall with a force of 30 pounds (lb). Since the push has direction (toward the wall) and magnitude (30 lb), the push can be represented by a vector as shown. In part (b) the rope pulls on the ring with a force of 20 lb. It, too, can be represented by a vector. Pushes and pulls—indeed, any force–can be represented by a vector. Forces are vector quantities.

Forces are measured in several units. In the United States the common unit is the pound. But in science and technology the preferred unit of force is the newton (N). A force of 1 lb is equal to a force of 4.45 N. We shall further discuss the relationship between these two units in Chapter 4. For the present it is sufficient if you know that a push or pull of 44.5 N is the same as a force of 10 lb.

Suppose that a truck is traveling eastward at 60 km/h (h is the abbreviation

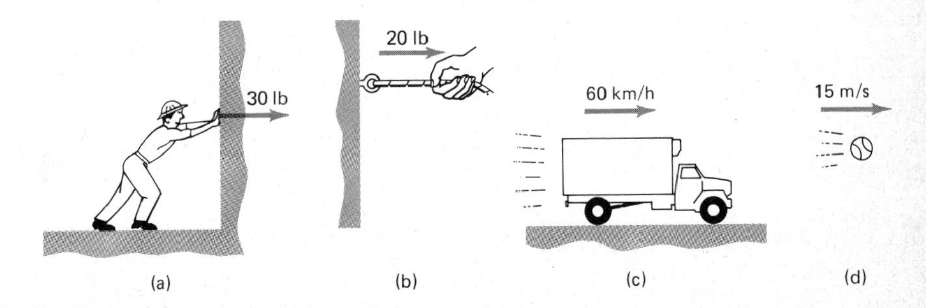

Figure 1.16 Forces and velocities are vector quantities.

for hour). Its velocity (60 km/h eastward) has both magnitude (60 km/h) and direction (eastward). We can therefore represent it by an arrow as shown in part (c) of Figure 1.16. The truck's velocity is a vector quantity. Similarly, the baseball shown in part (d) is flying along the straight line from pitcher to batter. Its velocity is 15 m/s (s is the abbreviation for second) along this line. The velocity has both magnitude and direction. Velocities are vector quantities.

There are other quantities that can be represented by vectors. We shall point them out when we need them later in this text. For now, however, we shall be concerned with three vector quantities: displacements, forces, and velocities. Because we shall sometimes wish to emphasize that a quantity is a vector, we need a quick way of showing this on paper. For example, suppose that we wish to discuss a force of magnitude F with which a man pulls on a rope (F might equal 100 N, for example). Sometimes we are not interested in the direction of the force, but only in its magnitude. We then simply write its magnitude as F. We then treat it like a scalar quantity.

But if we want you to pay attention to the direction of the force as well as its magnitude, we set it in boldface type, **F**. Or on paper or the blackboard, we write it as $\vec{F}$ or $\underset{\sim}{F}$. This representation should be used when we are writing a symbol for a vector quantity. If the representation is not used, we consider the quantity to be a scalar and to have no direction—at least for the discussion at hand.

1.7 Vector Addition of Forces

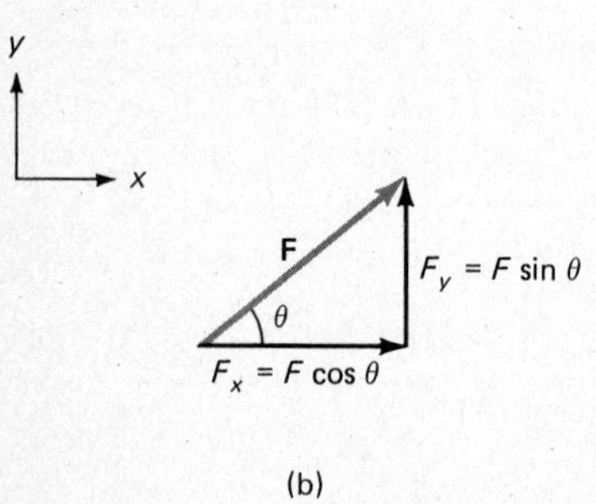

Figure 1.17 The y component of **F** tries to lift the dog, while the x component pulls it along.

Suppose that you pull with a rope on a stubborn dog as shown in Figure 1.17(a). Let us discuss the force **F** with which you pull (notice that the vector symbolism tells us to pay attention to direction as well as magnitude). The vector force can be thought of in terms of its two rectangular components. These are shown in part (b) of the figure. The horizontal or x component of magnitude F_x tends to pull the dog along the ground. But we see from the figure that the y component of **F**, labeled F_y, is doing something completely different than F_x. If **F** is large enough, the upward component F_y will actually lift the dog off the ground. It is important in many cases to consider the components of a force if we are to understand what the force is doing. Notice in the diagram that $F_x = F \cos \theta$ and $F_y = F \sin \theta$. Given both the magnitude (F) and direction (θ) of **F**, we can easily find its components in this way.

We shall soon need to add force vectors together to understand various situations. The addition of force vectors is easy to do by the component method. To show this, consider the three men shown in Figure 1.18(a) pulling on a stuck car. We wish to find the resultant effect of their three pulls. In other words, what is the resultant of these three forces?

In dealing with this situation, it is best to set up x and y axes so we can discuss the situation easily. For convenience, take the axes as shown in (b). The three forces are redrawn in part (b) of Figure 1.18. We shall find their resultant by adding their x components together to find the x component of the resultant. In the same way we can find the resultant's y component.

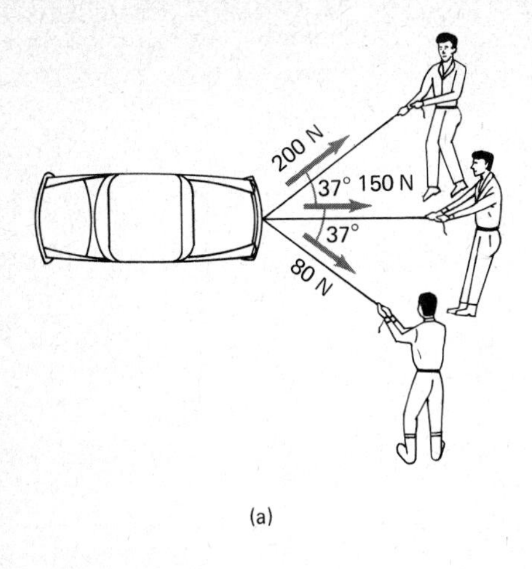

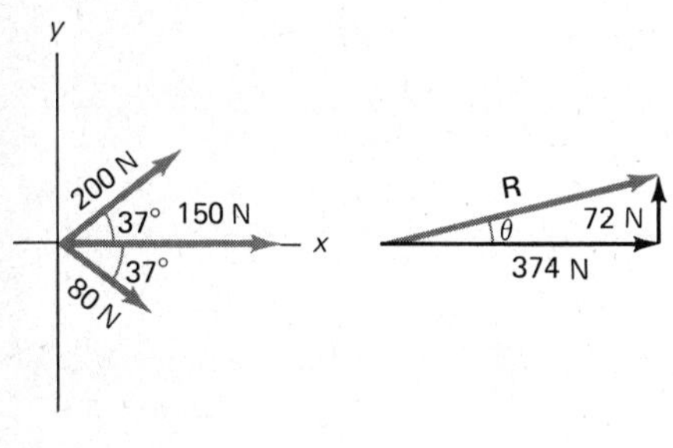

Figure 1.18 We wish to find the resultant force due to the three men pulling on the car.

To carry out this procedure, we find the x and y components of each of the vectors as shown in the following table.

Vector	x	y
150 N	150 N	0 N
200	160	120
80	64	−48
Resultant	374	72

You should check our computation of these values.

From the table we see that the resultant force **R** has x and y components of 374 and 72 N, respectively. The resultant is shown in part (c) of the figure. To find its magnitude we have

$$R = \sqrt{(374)^2 + (72)^2}\text{ N} = 381\text{ N}$$

Its direction θ is found by noting that

$$\tan\theta = \frac{72}{374} = 0.193$$

The trig tables then tell us that $\theta = 11°$.

As we see from this example, the addition of force vectors is simple if we use the component method. Later we shall see many other situations where vectors can be dealt with easily if we break (or *resolve*) the vector into its components. It is important for our future work that you learn well how to resolve vectors into their components and add them. Many of the problems at the end of this chapter will provide you with the necessary practice. Be sure you understand the following two examples.

EXAMPLE 1.5 The following displacements are made when using a jig mounted on a lathe:

2.04 cm at 0°
0.63 cm at 70°

Find the resultant displacement.

Solution A sketch of the displacements is shown in Figure 1.19(a). We find the components of the 0.63-cm displacement as shown in (b). Then we have

$$\begin{aligned}\text{Total } x \text{ displacement} &= 2.04\text{ cm} + 0.63\cos 70°\text{ cm}\\ &= 2.04\text{ cm} + 0.22\text{ cm}\\ &= 2.26\text{ cm}\\ \text{Total } y \text{ displacement} &= 0.63\sin 70°\text{ cm}\\ &= 0.59\text{ cm}\end{aligned}$$

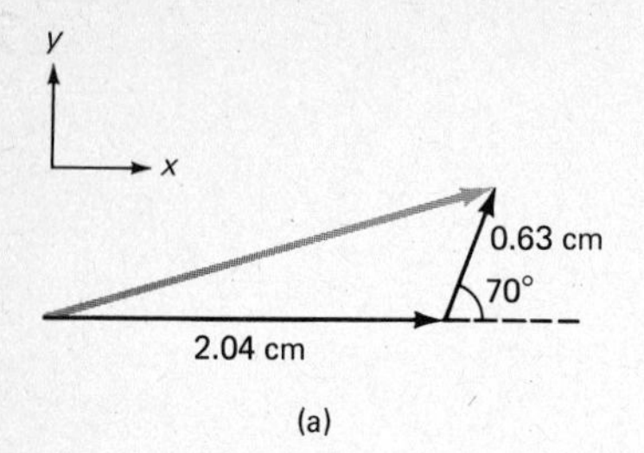

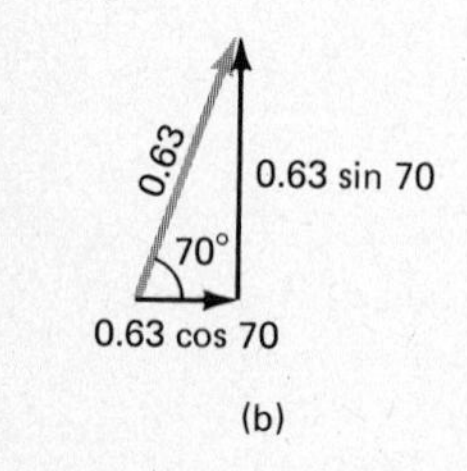

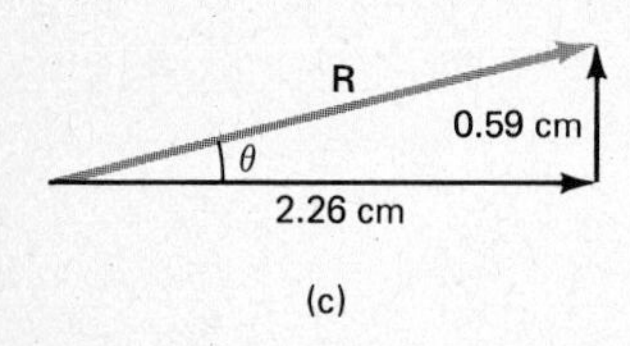

Figure 1.19 Finding the displacement of a jig.

These components, together with the resultant, are sketched in part (c) of the figure.

We could find the magnitude of the resultant displacement by use of the Pythagorean theorem. But for variety, let us do it another way. First find θ from

$$\tan\theta = \frac{o}{a} = \frac{0.59}{2.26} = 0.261$$

This then gives $\theta = 14.6°$.

Now let us use the definition of $\sin\theta$ to find R. From part (c) of the figure we have

$$\sin\theta = \frac{o}{h} = \frac{0.59\ \text{cm}}{R}$$

But $\theta = 14.6°$, so this becomes

$$\sin 14.6° = \frac{0.59\ \text{cm}}{R}$$

The table gives sin 14.6° to be 0.253 and so this becomes

$$0.253 = \frac{0.59\ \text{cm}}{R}$$

Then, after multiplying through the equation by R,

$$0.253R = 0.59\ \text{cm}$$

Dividing through the equation by 0.253 gives

$$R = 2.33\ \text{cm}$$

The resultant is therefore 2.33 cm at 14.6°. ■■

■ **EXAMPLE 1.6** A large box is being pushed across the floor as shown in Figure 1.20(a). The man is pushing with a force of 200 N at an angle 25° below the horizontal. How large a horizontal force does this cause on the box? How large a vertical force?

Solution The 200-N force vector can be resolved into rectangular components as shown in (b). The horizontal component is

$$F_x = 200 \cos 25°\ \text{N} = (200)(0.906)\ \text{N}$$
$$= 181\ \text{N}$$

Similarly, the vertical component is

$$F_y = -200 \sin 25°\ \text{N} = -(200)(0.423)\ \text{N}$$
$$= -85\ \text{N}$$

Notice the minus sign for F_y. It tells us that F_y is a force in the $-y$ direction. ■■

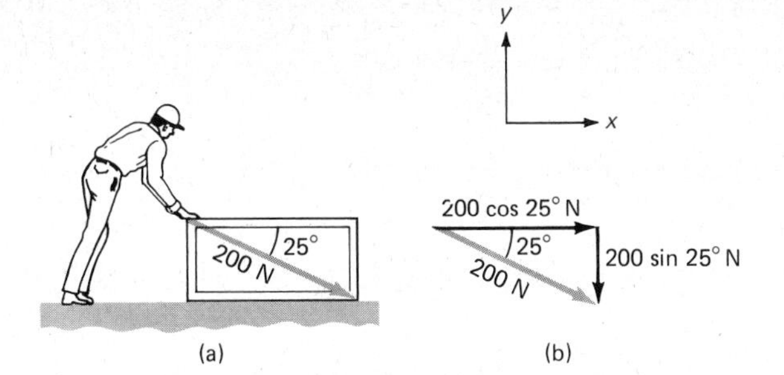

Figure 1.20 The push exerted on the box can be thought of as two component forces.

1.8 Vector Subtraction

Until now we have been discussing vector addition. You might well ask, "How does one subtract vectors?" Before answering this question, let us see where we might expect to use vector subtraction.

Suppose that you go 20 km in the $+x$ direction. Then suppose that you go backwards a distance of 12 km in the $-x$ direction. Your net displacement will be

$$\begin{aligned}\text{Displacement} &= +20\text{ km} - 12\text{ km}\\ &= +8\text{ km}\end{aligned}$$

There are two ways we can think of this displacement. They are shown in Figure 1.21. Method 1 treats the situation as addition. We simply add a vector of -12 km in the x direction to a vector of $+20$ km in the x direction. Method 2 treats the situation as a problem in subtraction. To obtain the same result as in Method 1, we must reverse the direction of the vector and add it.

We define the process of vector subtraction to be the following.

To subtract a vector, reverse its direction and add it.

Most situations that involve vector subtraction can be recast as a situation involving vector addition. As we saw in the present case, the addition and subtraction views are equivalent. We shall almost always use the vector addition approach.

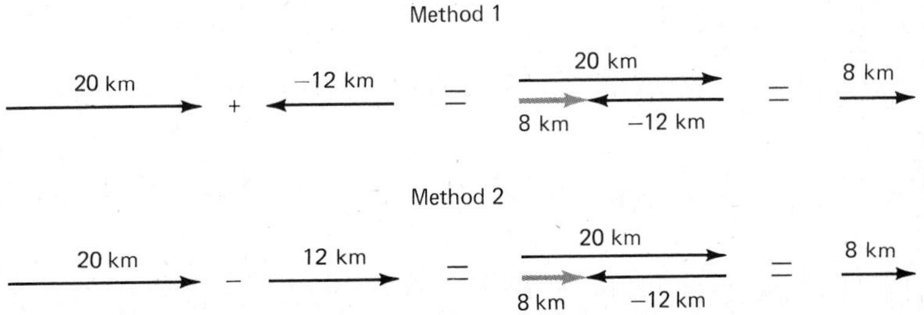

Method 1

20 km + −12 km = 20 km, 8 km, −12 km = 8 km

Method 2

20 km − 12 km = 20 km, 8 km, −12 km = 8 km

Figure 1.21 The two methods are equivalent. In method 2 we subtract the vector. Method 1 shows us that this is the same as reversing the vector and adding it.

1.9 Scientific Notation and the Metric Prefixes

Unless you have a good memory, it is often difficult to recall how many ounces there are in a pound or how many teaspoons in a tablespoon. Most of us know that there are 12 inches in a foot, but how many feet are there in a rod or in a mile? If we use the metric system, difficulties such as these do not occur.

The metric system uses units that are related through factors of ten. For example, there are 100 centimeters in a meter, there are 1000 grams in a kilogram, and one millionth of a gram is a microgram. As you see, factors of ten relate the units. But equally important, the names of the units tell us what the appropriate factor of ten is. Before considering these names, however, let us look at *scientific notation* (often called power of ten notation).

Very large and very small numbers are often inconvenient to handle unless we write them in a shorthand way. For example, the diameter of the nucleus of an atom is about 0.000000000000003 m. The age of the earth is about 4,000,000,000 years. These are clumsy numbers to write. In science and technology we use scientific notation to get rid of this difficulty. In scientific notation, the diameter of a nucleus is about 3×10^{-15} m while the earth's age is about 4×10^9 years. Let us now see the basis for this notation.

You know that $10 \cdot 10 = 10^2$ and $10 \cdot 10 \cdot 10 \cdot 10 \cdot 10 = 10^5$. Keeping this in mind, we can develop a consistent system of power of ten notation. Typical examples are

$$10^0 = 1$$
$$10^1 = 10$$
$$10^2 = 10 \cdot 10 = 100$$
$$10^3 = 10 \cdot 10 \cdot 10 = 1000$$
$$10^4 = 10 \cdot 10 \cdot 10 \cdot 10 = 10{,}000$$

etc.

Therefore, when we write 4×10^9 years for the age of the earth, we mean the following: multiply 4 by 10 nine times. Or, put another way, the factor 10^9 means to move the decimal point nine places to the right. The number 0.0351×10^7 becomes 351,000 when we move the decimal point seven places to the right. Test your understanding by filling in the blanks below.

$$0.73 \times 10^4 = \_\_\_\_\_\_\_\_$$
$$54{,}000{,}000 = 5.4 \times \_\_\_\_\_\_\_\_$$

Your answers should be 7300 and 10^7.

Extending this line of reasoning to numbers smaller than unity, we have

$$10^{-1} = 1/10 = 0.10$$
$$10^{-2} = 1/100 = 0.010$$
$$10^{-3} = 1/1000 = 0.0010$$

etc.

and

$$3 \times 10^{-2} = 3/100 = 0.030$$
$$70 \times 10^{-3} = 70/1000 = 0.070$$
$$0.5 \times 10^{-2} = 0.5/100 = 0.005$$
etc.

A factor 10^{-n} means to divide by 10^n, which means we move the decimal point n places to the left. For our nuclear diameter, 3×10^{-15} m, we would move the decimal point fifteen places to the left. As other examples, 823.7×10^{-3} is 0.8237 and 0.156×10^{-2} is 0.00156. The following will test your understanding.

$$0.173 \times 10^{-3} = \_\_\_\_\_\_$$
$$0.0625 \times 10^{-4} = 6.25 \times \_\_\_\_\_\_$$

Your answers should be 0.000,173 and 10^{-6}.

Now that we know how to use power of ten notation, let us return to the metric system.

An important feature of the metric system is its use of what are called *prefixes.* The metric prefixes are listed in the left-hand column of Table 1.2. Each prefix refers to the power of ten given in the right-hand column. The symbol for each prefix is given in the center column. (The symbol μ is the Greek letter mu.) Using these symbols and their meanings, we can write that a kilogram is 1000 grams (1 kg = 1000 g). As another example, a micrometer is 1×10^{-6} m. The prefixes are used beyond the metric system to give such terms as a kiloton (1000 tons) and a megabuck ($\$1 \times 10^6$). The metric prefixes are widely used and you should be familiar with them.

Table 1.2 METRIC PREFIXES

Prefix	Symbol	Power of Ten Notation
tera	T	10^{12}
giga	G	10^9
mega	M	10^6
kilo	k	10^3
hecto	h	10^2
deka	da	10^1
deci	d	10^{-1}
centi	c	10^{-2}
milli	m	10^{-3}
micro	μ	10^{-6}
nano	n	10^{-9}
pico	p	10^{-12}
femto	f	10^{-15}
atto	a	10^{-18}

■ **EXAMPLE 1.7** In standard scientific notation one digit exists to the left of the decimal point. For example, the number 5280 would most often be written as 5.280×10^3 rather than as 52.80×10^2. Other examples are:

$$872{,}000 = 8.72 \times 10^5$$
$$0.00215 = 2.15 \times 10^{-3}$$
$$0.040510 = 4.0510 \times 10^{-2}$$
$$1000 = 1 \times 10^3 = 10^3$$
$$0.0001 = 1 \times 10^{-4} = 10^{-4}$$

Notice in the last two examples that unity is often omitted when used as a multiplier in the power of ten notation. Therefore 10^4 means 1×10^4, which is 10,000. ■■

1.10 Algebra of Scientific Notation

Algebraic operations involving numbers expressed in scientific notation involve a few simple rules.

Addition and Subtraction When you add numbers such as 625.60 and 0.13, you must align the decimal points.

$$\begin{array}{r} 625.60 \\ +\quad 0.13 \\ \hline 625.73 \end{array}$$

This restriction for numbers in scientific notation means that the numbers must be expressed as the same power of ten. For example,

$$(6.2560 \times 10^2) + (1.3 \times 10^{-1})$$

must be written as

$$(6.2560 \times 10^2) + (0.0013 \times 10^2)$$

so that the powers of ten are the same. They can then be added to give 6.2573×10^2. Other examples are

$$(7.1 \times 10^5) + (8.3 \times 10^5) = 15.4 \times 10^5$$
$$(4.28 \times 10^3) + (340) = (4.28 \times 10^3) + (0.34 \times 10^3) = 4.62 \times 10^3$$
$$(0.016) + (2 \times 10^{-3}) = (16 \times 10^{-3}) + (2 \times 10^{-3}) = 18 \times 10^{-3}$$

Similarly for subtraction:

$$(8.0 \times 10^{-3}) - (0.50 \times 10^{-2}) = (8.0 \times 10^{-3}) - (5.0 \times 10^{-3}) = 3.0 \times 10^{-3}$$

For addition and subtraction, the powers of ten must be the same.

Multiplication We make use of the algebraic rule for exponents that says

$$10^a \times 10^b = 10^{a+b}$$

Therefore

$$(3 \times 10^4) \times (5 \times 10^6) = 3 \times 5 \times 10^4 \times 10^6 = 15 \times 10^{10}$$

Similarly,

$$(8 \times 10^{-2}) \times (4 \times 10^5) = 32 \times 10^{(-2)+(5)} = 32 \times 10^3$$

Notice that, unlike in addition and subtraction, the powers of ten in multiplication can be different.

Division Because $1/10^a = 10^{-a}$, we have

$$\frac{5 \times 10^6}{2 \times 10^4} = \left(\frac{5}{2}\right)\left(\frac{10^6}{10^4}\right) = (2.5)(10^6 \times 10^{-4}) = 2.5 \times 10^2$$

Similarly

$$\frac{4 \times 10^6}{16 \times 10^{-2}} = \left(\frac{1}{4}\right)(10^6 \times 10^2) = 0.25 \times 10^8 = 2.5 \times 10^7$$

Some examples are:

$(25 \times 10^3) \times (3 \times 10^{-3}) = 75 \times 10^0 = 75$
$(8 \times 10^6) \times (0.20) = 1.6 \times 10^6$
$(7 \times 10^{-5}) \times (8 \times 10^{-3}) = 56 \times 10^{-8} = 5.6 \times 10^{-7}$
$(6 \times 10^{-3}) \div (3 \times 10^{-4}) = 2 \times 10^1 = 20$
$(8 \times 10^{-2}) \times (3 \times 10^5) \div (6 \times 10^4) = (24 \times 10^3) \div (6 \times 10^4) = 4 \times 10^{-1}$
$(5 \times 10^4)^3 = (5 \times 10^4) \times (5 \times 10^4) \times (5 \times 10^4) = 125 \times 10^{12} = 1.25 \times 10^{14}$

Summary

Vector quantities possess both direction and magnitude; scalar quantities have no specified direction. Typical vector quantities are displacements, forces, and velocities.

A vector quantity can be represented on a scale diagram by an arrow. The arrow's length represents the magnitude of the quantity, whereas its direction represents the quantity's direction.

To add several vectors graphically on a scale diagram, lay out the vectors (arrows) in series. The head (or tip) of the first coincides with the tail of the second, and so on. The resultant is a vector (arrow) drawn with its tail at the tail of the first vector and with its head at the head of the last vector.

In a right triangle, by definition,

$$\sin\theta = \frac{\text{opposite side}}{\text{hypotenuse}} \quad \text{or} \quad o = h\cdot\sin\theta$$

$$\cos\theta = \frac{\text{adjacent side}}{\text{hypotenuse}} \quad \text{or} \quad a = h\cdot\cos\theta$$

$$\tan\theta = \frac{\text{opposite side}}{\text{adjacent side}} \quad \text{or} \quad o = a\cdot\tan\theta$$

A vector can be resolved (split) into rectangular components. The resultant of several vectors can be found by adding the components of the vectors.

To subtract a vector, change its sign (or reverse its direction in a diagram) and add it.

The basic unit of length in the metric system is the meter (m). A kilometer (km) is 1000 m, whereas a centimeter (cm) is 0.010 m. Also, 1 in. = 2.54 cm and 1 m = 3.28 ft = 39.36 in. Another convenient relation is 1 km = 0.62 mile. The unit of force in the metric system is the newton (N). One newton is equivalent to 0.225 lb.

Very large and very small numbers can be represented conveniently using scientific notation. Powers of ten are employed to show the position of the decimal point.

Questions and Exercises

1. Draw a vector diagram showing a displacement vector of 20 m at an angle of 30° north of east. Use a scale such that 5 cm is equivalent to 10 m.
2. On a diagram for which 1 m is represented by 0.5 cm, draw the appropriate vector diagram to find the resultant of the following displacements.

   8 m east
   5 m at 60°
   7 m west
   10 m at 210°

   All angles are measured counterclockwise from the east axis.
3. A man walks around a block and ends up at his starting point. If each of the four sides of the block is 140 m long, how large is his resultant displacement? How far did he walk?
4. You are given two displacements. One is 25 m north, and the other is 13 m in a direction you can choose. What direction should you choose for it to give the maximum combined displacement? The minimum combined displacement? How large is the displacement in each case?
5. Two displacement vectors are to be added: 25 m east and 10 m at 30° north of east. Why can't we use the Pythagorean theorem to give a resultant equal to $\sqrt{(25)^2 + (10)^2}$ m?
6. Three vector quantities of equal magnitude but whose directions can be changed are to be added. How must they be oriented if their sum is to be zero? Could this be done if the vectors were unequal in magnitude? Can the sum of two unequal vectors be zero?
7. A force in the x direction is to be combined with an equal force in the y direction. What is the direction of the resultant force?
8. Make a list of several quantities that are scalar quantities. Repeat for vector quantities.
9. A certain city has 50,000 people in it. Represent each person by a "person vector." Its point is at the person's nose and its tail is at the person's big toe on the left foot. About how large is the resultant of these "person vectors" at 12 noon? At 12 midnight?

10. Repeat the previous question. But now represent each person by the following "person vector." It is a force vector equal to the force the person exerts on whatever supports the person—the floor, the earth, a chair, and so forth.
11. Devise three different situations in which vector subtraction might be used. Point out in each case why the situation could be considered a problem involving vector addition rather than subtraction. (Most often, the addition method is used in such situations.)

Problems

1. A bug crawls 3.0 mm east, 4.0 mm north, and then 5.0 mm at 45° north of east. Draw a diagram showing its displacements and determine its resultant displacement vector by use of the diagram.
2. Starting from San Francisco, an airplane flies 500 km east, then 200 km at 270°, and then 100 km at 135°. Use a scale drawing to find its resultant displacement from San Francisco. All angles are measured in the way outlined in the text.
3. Add the following displacement vectors by direct measurement on a scale drawing.

   37 km at 20°
   23 km at 150°
   16 km at 270°

4. Using a scale drawing, find the resultant of the following displacement vectors.

   500 km at 60°
   300 km at 120°
   700 km at 300°

*5. City B is 50 km directly north of city A, while city C is 40 km directly west of A. (a) Using a scale drawing, find the distance from city B to city C. (b) What angle does the displacement vector from C to B make with the east?
*6. The displacement from point A to point B is 50 km at an angle of 40°. The displacement from A to point C is 20 km at 90°. Use a scale drawing to find the displacement from (a) point C to point B, (b) point B to point C.
7. Using a scale drawing, find the x and y components of a 20-m displacement at an angle of 120°. Call them s_x and s_y.
8. Using a scale diagram, find the vector (and its angle) that has components $s_x = 30$ ft and $s_y = -20$ ft.
9. Using a scale drawing, find the x and y components of the following vectors: (a) 110 units at 60°, (b) 70 units at 300°, (c) 200 units at 150°.
10. Using the graphical method, find the x and y components of the following displacements: (a) 20 m at 110°, (b) 30 m at −37°, (c) 40 m at 220°.
11. The right triangle shown in Figure P1.1 has side $A = 2$ m and side $B = 3$ m. (a) How large is side C? (b) Evaluate $\sin\theta$. (c) Evaluate $\cos\theta$. (d) Evaluate $\tan\theta$.

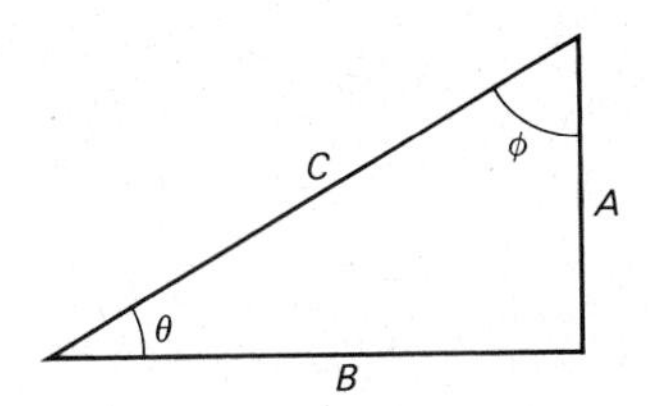

Figure P1.1

12. In the right triangle of Figure P1.1, $C = 7$ m and $A = 4$ m. Find (a) side B, (b) $\sin\theta$, (c) $\cos\theta$, (d) $\tan\theta$.
13. In the right triangle of Figure P1.1, $C = 8$ m and $\theta = 30°$. Find (a) side A and (b) side B.
14. In the right triangle of Figure P1.1, $C = 5$ m and $\phi = 50°$. Find (a) side A and (b) side B.
*15. In the right triangle of Figure P1.1, $\theta = 35°$ and $B = 2$ m. Find (a) side C and (b) side A.
*16. In the right triangle of Figure P1.1, $\phi = 50°$ and $A = 15$ m. Find (a) side B and (b) side C.
*17. In the right triangle of Figure P1.1, $\phi = 50°$ and $B = 3$ m. Find (a) side C and (b) side A.
*18. In the right triangle of Figure P1.1, $\theta = 35°$ and $A = 7.0$ m. Find (a) side B and (b) side C.
19. Using trigonometry, find the x and y components of a 50-m displacement at an angle of 37°.
20. Using trigonometry, find the x and y components of an 80-m displacement at an angle of 210°.
21. Using trigonometry, find the x and y components of the following vectors: (a) 70 m at 110°, (b) 30 m at 240°, (c) 6.0 m at 300°.
22. Using trigonometry, find the x and y components of the following vectors: (a) 20 m at 70°, (b) 40 m at 210°, (c) 5.0 m at −30°.
23. Given a vector whose components are $s_x = -50$ m and $s_y = 30$ m, use trigonometry to find the vector and its angle.

*Problems marked with an asterisk are not as easy as the unmarked ones.

24. The x and y components of a displacement are $s_x = -70$ m and $s_y = -50$ m. Find the displacement and its angle by use of trigonometry.

*25. A displacement vector at an angle of 37° has an x component equal to 6.0 m. Find the magnitude of the displacement vector.

*26. A force pulls at an angle of 53° and has a y component of 200 N. Find the magnitude of the force.

*27. The ratio of the two legs of a right triangle is 0.50. Find the two acute angles of the triangle.

**28. A 1.60-m tall girl standing 25 m from the base of a tree measures its height in the following way. She sights along a meter stick pointed toward the top of the tree. The angle between stick and earth is then 40°. About how tall is the tree?

29. Add the following force vectors by the trigonometric component method.

50 N at 0°
100 N at 37°
70 N at 180°

Check your answer with a rough scale drawing.

30. Using the trigonometric component method, find the resultant of the following forces. Check the direction with a rough scale drawing.

200 N at 70°
50 N at 180°
100 N at 210°

31. Solve Problem 3 by use of trigonometry.

32. Solve Problem 4 by use of trigonometry.

*33. A boat leaves shore and travels as follows on a large quiet lake: 2.00 km at 180°, then 0.50 km at 150°, and finally 1.00 km at 30°, with all angles measured counterclockwise from east. (a) How far is it from its starting point? (b) At what angle must it go if it is to follow a straight-line path back to the starting point? Use the trigonometric component method.

*34. To get from one office to another in the Pentagon (an office building in Washington), one travels as follows (with all angles being counterclockwise relative to east). One must go

8 paces at 180°
40 paces at 90°
50 paces at 162°

Find the straight-line distance between the offices and the direction of a vector pointing from the second to the first.

35. Three forces have the following components, all in newtons: ($F_x = -5$, $F_y = -7$), ($F_x = 0$, $F_y = 30$), ($F_x = 20$, $F_y = -16$). Find their resultant.

36. Three forces have the following components, all in newtons: ($F_x = -6$, $F_y = 0$), ($F_x = 14$, $F_y = -12$), ($F_x = -15$, $F_y = 3$). Find their resultant.

*37. An airplane is taking off from an airport with a velocity of 200 km/h at an angle of 30° above horizontal. (a) How fast is it rising? (b) How fast is it moving parallel to the earth? (Hint: We are asking for the two velocity components.)

*38. A rifle is aimed at an angle of 53° above the horizontal and a bullet is fired from it with a speed of 200 m/s. Just as the bullet leaves the gun, how fast is it rising? How fast is it moving parallel to the earth?

*39. You are given two vectors, **A** = 10 m directed east and **B** = 15 m directed east. Find (a) **A** + **B**, (b) **A** − **B**,(c) 2**A** − **B**.

*40. You are given two vectors, **A** = 5 m in the $+y$ direction and **B** = 3 m in the $-y$ direction. Find (a) **A** + **B**, (b) **A** − **B**, (c) **B** − **A**, (d) **A** − 2**B**.

41. You are given two vectors, **A = 20 m in the $+x$ direction and **B** = 20 m in the $+y$ direction. Find (a) **A** + **B** and (b) **A** − **B**.

*42. A bug is crawling in the $+x$ direction on a board with a velocity of 2.0 cm/s relative to the board. At the same time, the board is being moved in the $+x$ direction with velocity 30 cm/s relative to the floor. What is the velocity of the bug relative to the floor? Repeat if the bug is crawling in the $-x$ direction.

*43. An airplane is flying due east with a velocity of 400 km/h relative to the surrounding air. However, the air itself is part of a strong wind that is blowing straight west with a velocity of 80 km/h relative to the ground. Find the velocity of the airplane relative to the ground. (Hint: See Problem 42 for a similar case.)

**44. A bug is crawling with velocity components $v_x = 2.0$ and $v_y = -5.0$ cm/s relative to a horizontal board. But the board has an x-directed velocity 10.0 cm/s relative to the ground. What are the x and y components of the velocity of the bug relative to the ground?

**45. An airplane is flying straight east with a velocity of 300 km/h relative to the surrounding air. But the air itself is moving southwest (that is, 45° south of west) relative to the earth with a velocity of 100 km/h due to a strong wind. What is the velocity of the plane relative to the earth?

46. Write the following in standard scientific notation (one digit to the left of the decimal): (a) 873, (b) 55,008, (c) 0.00621, (d) 0.000375.

47. Write the following in standard scientific notation: (a) 0.000579, (b) 0.0036, (c) 7490, (d) 20,001.

**Problems marked with a double asterisk are somewhat more difficult than the average.

48. Convert the following to regular notation: (a) 5.7×10^4, (b) 83.1×10^3, (c) 0.91×10^{-3}, (d) 1.67×10^{-4}.
49. Convert the following to regular notation: (a) 672.3×10^3, (b) 2.42×10^2, (c) 0.036×10^{-2}, (d) 36.5×10^{-4}.
50. Find the sum for each of the following sets of two numbers: (a) 3×10^3 and 2.5×10^4, (b) 0.013 and 6×10^{-3}, (c) 3.160×10^{-2} and 2.1×10^{-4}, (d) 200×10^{-4} and 8.21×10^{-2}.
51. Find the sum for each of the following sets of two numbers: (a) 5.1×10^{-2} and 0.123, (b) 74.87×10^4 and 60×10^2, (c) 1.33×10^{-7} and 6.2×10^{-8}, (d) 2.71×10^{-7} and 0.160×10^{-5}.
52. For each of the sets of numbers in Problem 51, multiply the two numbers.
53. For each of the sets of numbers in Problem 50, multiply the two numbers.
54. For each of the sets of numbers in Problem 50, divide the first number by the second.
55. For each of the sets of numbers in Problem 51, divide the first number by the second.

The Port Authority of New York

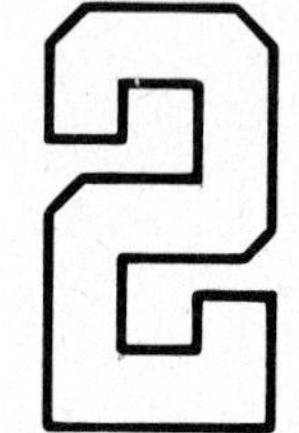

STATICS

Performance Goals

When you finish this chapter, you should be able to

1. Determine whether or not an object subject to simple forces is in static equilibrium.
2. Draw a free-body diagram for an object that is acted upon by known forces.
3. Use the first condition for equilibrium to obtain equations that apply to an object at equilibrium.
4. State whether or not a point object is in equilibrium from a

Statics is the study of objects and systems that do not move. We shall learn in this chapter what must be true if an object is to remain at rest. In doing so, we shall discuss the way in which forces can be balanced. The knowledge of vectors that we acquired in the last chapter will be of great importance to us in all of this study.

consideration of the forces acting upon it.
5. Find the torque due to a given force about any specified pivot point.
6. Use the second condition for equilibrium to tell whether or not an object will be caused to rotate by the forces applied to it.
7. Apply the two conditions for equilibrium to situations such as those given in the examples of the text.
8. Determine the position of the center of gravity of an object by means of an experiment.
9. State whether the equilibrium of a known simple object is stable, unstable, or neutral.

2.1 Static Equilibrium

Suppose that two strings are fastened to a tiny ring as shown in Figure 2.1. Suppose further that the strings pull on the ring in opposite directions as shown. If the ring is not to move, then you know that the two opposite pulls due to the strings must be equal. In other words, the force to the left, $\mathbf{F}_1$, caused by one of the strings must balance the force to the right, $\mathbf{F}_2$, caused by the other string. We see that the ring will not move if there is as much force pulling it to the right as to the left. The force to the left must balance the force to the right.

Now let us look at Figure 2.2. Here we see a heavy ball held up by a string. Because the ball does not fall down, we know that the string must be pulling up on it. Call this upward pull of the string **F.** But we know that the ball would fall down if the string broke. There must be some force pulling down on the ball. It is the force due to gravity.

This force, the force due to gravity, is caused by the earth. Every object on the earth is pulled toward the earth's center by gravity. We call this pull due to gravity on an object the *weight* of that object. When a woman says that she weighs 100 lb, she is telling us that the earth is pulling down on her with this large a force. You, too, are pulled toward the center of the earth by a force equal to your weight. Incidentally, you will recall that 1 N = 0.225 lb. So a 100-lb woman weighs 445 N.

Returning to Figure 2.2, we see that the earth pulls downward on the ball. This force is the weight of the ball, **W.** Why doesn't the ball fall down because of this force? It doesn't fall because the string holds it up. The upward pull of the string, **F,** balances the downward pull of gravity, **W.** For the ball to remain at rest, the upward force **F** must equal the downward force **W.**

Let us consider yet another situation, the one shown in Figure 2.3. Here we see a book lying on a table. The earth pulls down on the book with a force **W,** the weight of the book. If the table were not there, the book would fall. But the table does not let the book fall. It pushes upward on the book just hard enough to hold it at rest. This upward push of the table, **P,** must be just large enough to balance the pull of gravity, **W.** We see that **P** is equal and opposite to **W.**

In each of these examples we see objects that do not move, objects that remain at rest. We notice that in each case the forces on the object are balanced. This is always true. If an object is to remain at rest, all forces pushing and pulling on the object must be balanced. We say that an object that is at rest and remains at rest is in *static equilibrium*. For an object to be in static equilibrium, all forces acting on it must be balanced.

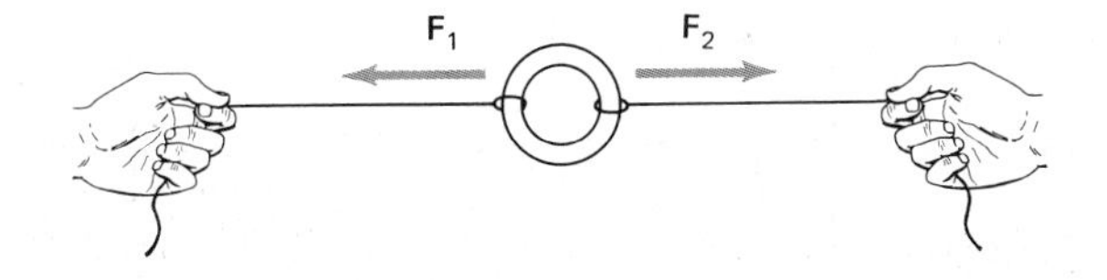

Figure 2.1 What must be true about $\mathbf{F}_1$ and $\mathbf{F}_2$ if the ring is to remain at rest?

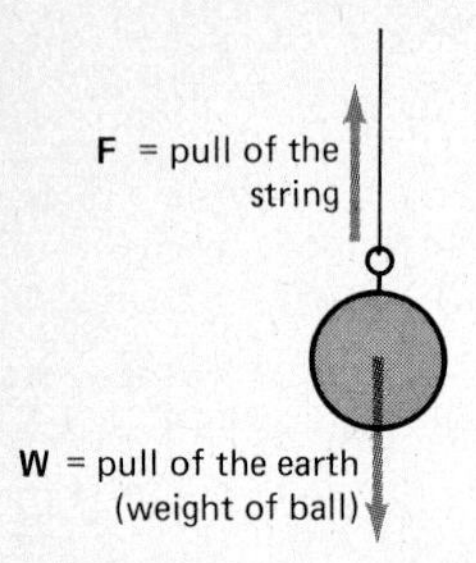

Figure 2.2 How large must **F** be if the ball is to remain motionless?

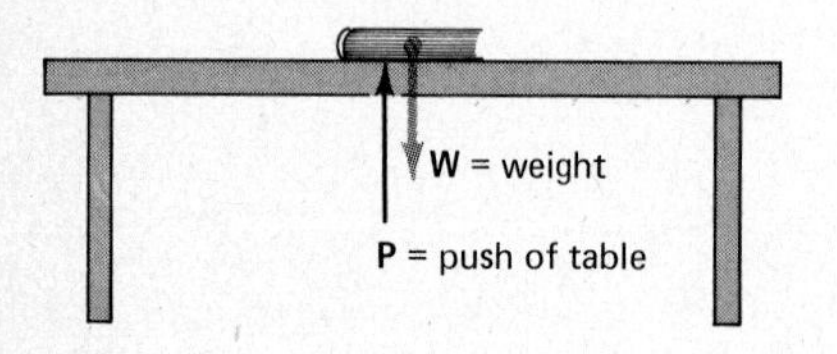

Figure 2.3 What must be true about **P** and **W** if the book is to remain at rest?

2.2 First Condition for Equilibrium

Let us examine more closely what we mean by balanced forces. Suppose that four strings pull on a ring as shown in Figure 2.4(a). We see at once that $\mathbf{F}_1$ must balance $\mathbf{F}_2$ if the ring is not to move in the *x* direction. Similarly, if the ring is not to move in the *y* direction, then $\mathbf{F}_3$ must balance $\mathbf{F}_4$. For the ring to be in static equilibrium, the *x*-directed forces must balance. So, too, must the *y*-directed forces balance.

In part (b) of Figure 2.4 we see a different situation. Here the upward-directed force, $\mathbf{F}_5$, must balance the downward pulls of $\mathbf{F}_6$ and $\mathbf{F}_7$. In addition, $\mathbf{F}_6$ pulls the ring to the left while $\mathbf{F}_7$ pulls it to the right. If the ring is to remain motionless, these two sideways forces must also balance.

We can best understand the situation in Figure 2.4(b) by resolving the forces into *x* and *y* components. This is done in part (c) of Figure 2.4. Clearly, the horizontal force components F_{6x} and F_{7x} must balance. In the vertical direction, $F_{6y} + F_{7y}$ must balance $\mathbf{F}_5$. Only then will the ring be in static equilibrium.

When we state that the forces on an object balance, we are really saying the following: The resultant of all the forces on the object is zero. In other words, when we add up all the *x*-directed forces acting on the object, they must add to zero. The same is true for the *y*-directed forces.

These statements can be written in shorthand. To do so, we use the symbol ΣF_x to represent the words "the sum of all the force components in the *x* direction." Likewise, the symbol ΣF_y means "the sum of all the force components in the *y* direction." Using these symbols, an object will remain at rest if[1]

$$\Sigma F_x = 0 \qquad \text{and} \qquad \Sigma F_y = 0$$

In writing these equations we must be careful of algebraic signs. Forces in the $+x$ or $+y$ direction are given positive signs. Forces in the negative *x* and *y* directions are given negative signs. For example, let us write these two equations for the forces shown in Figure 2.4(c). We have

$$\Sigma F_x = 0 \rightarrow F_{7x} - F_{6x} = 0 \rightarrow F_{7x} = F_{6x}$$

and

$$\Sigma F_y = 0 \rightarrow F_5 - F_{6y} - F_{7y} = 0 \rightarrow F_5 = F_{6y} + F_{7y}$$

Let us restate the result we have found. It is called the *first condition for equilibrium*.

If an object is to remain at rest, then the following must be true for the forces that push or pull on the object:

$$\Sigma F_x = 0 \qquad \Sigma F_y = 0 \qquad \Sigma F_z = 0 \tag{2.1}$$

[1] In three dimensions we must also have $\Sigma F_z = 0$.

Be careful to notice that these are forces that act on the object. Other forces that do not push or pull on the object are omitted.

2.3 Equilibrium of a Point Object; Concurrent Forces

The first condition for equilibrium helps us to solve many problems. We shall now spend some time solving problems by use of it. The situations we shall consider are all alike in one way. The forces acting on the object are *concurrent,* that is, the lines along which the forces act all intersect at a point. Such forces do not cause objects to rotate. Nonconcurrent forces are discussed later.

We shall assume that all forces have only x- and y-directed components. We say that the forces are *coplanar.* No z-directed forces will exist in these cases. Our discussion will be limited to concurrent, coplanar forces.

First let us look at the situation shown in Figure 2.5(a). We see there a 30-N object hanging at rest at the end of the rope.[2] We wish to find the force T with which the rope pulls upward. The situation is extremely simple. We can notice at once that the pull of gravity, W, must be balanced by the upward pull of the rope, T. Therefore $T = W = 30$ N. But, for practice, let us apply Equations 2.1 to this case. Because the object is at equilibrium, Equations 2.1 must apply. We shall follow several definite steps when using these equations.

1. Sketch the situation as done in Figure 2.5(a).
2. Because the equations describe only those forces that act on an object, we must decide what object we are going to talk about. We say that "we isolate an object" for discussion. Other forces that do not push or pull on the object are omitted. Let us consider the oblong object in Figure 2.5(a) as our "isolated" object.
3. It will be necessary to know the forces that act on the object. So that we can see them clearly, we draw and label them on a special diagram called a "free-body diagram." This is the diagram shown in Figure 2.5(b). The forces acting on the object, T and W in this case, are drawn on x-y axes as indicated. Notice: Only the forces that act *on* the object we have isolated are included in the free-body diagram.
4. Our equations describe the x and y components of the forces. We therefore need to find the components of the forces shown in the free-body diagram. In the present case all forces are y-directed and so the components are obvious.
5. Now we can write Equations 2.1. We have, since there are no x forces in Figure 2.5(b),

$$\Sigma F_x = 0 \qquad \text{becomes} \qquad 0 = 0$$

Also,

$$\Sigma F_y = 0 \qquad \text{becomes} \qquad T - 30\text{ N} = 0$$

[2] An object that weighs 30 N weighs about 6.7 lb.

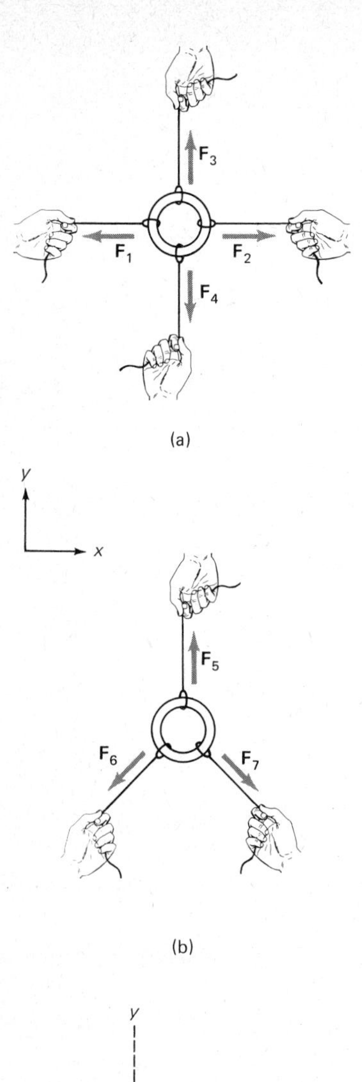

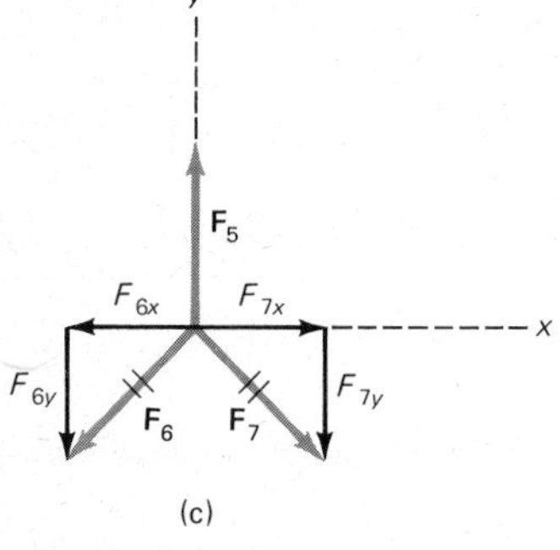

Figure 2.4 If the ring is to remain at rest, the x and y forces must balance. We place slash marks on the $\mathbf{F}_6$ and $\mathbf{F}_7$ vectors to show that we have replaced them by their components.

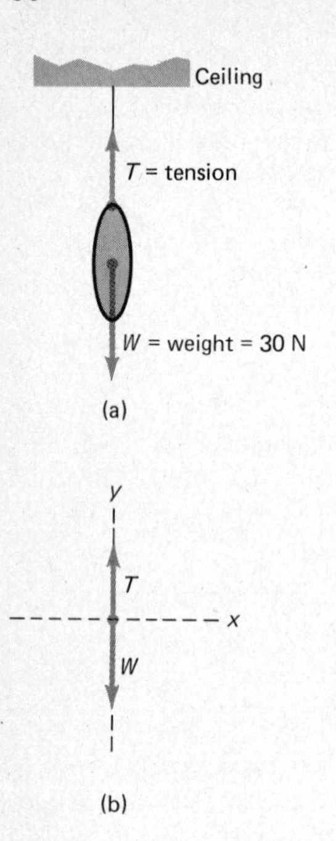

Figure 2.5 The relation $\Sigma F_y = 0$ tells us that $T = W$. What is meant by the *tension* in the rope?

6. Using the equations found in part 5, we can solve for unknown quantities. In this case we solve for T and find

$$T = 30 \text{ N}$$

This completes the solution of the problem.

(Before going further, we should point out that T is called the *tension* in the rope. The force with which a rope or string pulls is called the tension in the rope or string.)

In solving this very simple problem, we have numbered each step in the solution. We did this to point out the method we shall always use. You may think now that this is like killing an ant with a shotgun. Our method was indeed more powerful than the problem required. However, it is a method we shall want to use in most problems in statics, and so we summarize its steps here.

1. Sketch the physical situation.
2. Isolate an object for discussion.
3. Show and label the forces on the object in a free-body diagram.
4. Find the x and y components of the forces in step 3.
5. Write the equations for equilibrium.
6. Solve for the unknowns.

We shall now apply this method to a few examples. Study each one carefully so that you can apply the method to still other problems.

EXAMPLE 2.1 The ropes shown in Figure 2.6(a) support a 200-lb (about 900 N) man. Find the tension in each rope.

Solution Clearly, the tension in the lowest rope supporting the man is 200 lb, equal to the man's weight. We need to find the tensions in ropes 1 and 2. The first decision we must make concerns the object to isolate. If we isolate the man, we have only two forces

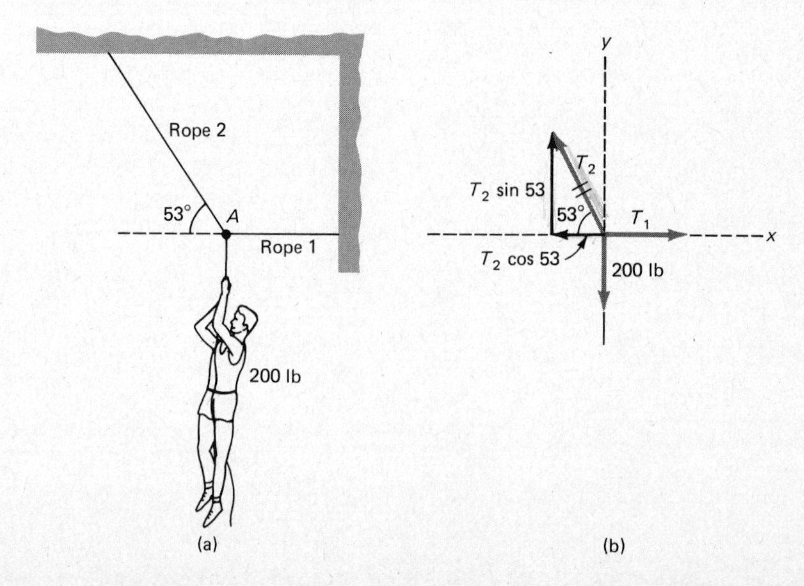

Figure 2.6 Find the tension in each of the three ropes.

acting on him, the pull of gravity (200 lb down) and the pull upward of the rope. This simply tells us the tension in the rope is 200 lb, a fact we already knew.

In situations such as this it is necessary to select the knot at A as the object. We then draw the three forces pulling on it in the free-body diagram shown in (b). Notice that T_1 is the pull of rope 1 on the knot while rope 2 pulls upward with a force T_2.

Both T_1 and the 200-lb force are along the axes, so only T_2 needs to be split into components. This is done on the diagram. Check the values of the components. To show that T_2 is replaced by its components, we place slash marks on it as shown.

Next we write the equilibrium equations.

$$\Sigma F_x = 0 \rightarrow T_1 - T_2 \cos 53° = 0 \rightarrow T_1 - 0.60T_2 = 0$$
$$\Sigma F_y = 0 \rightarrow T_2 \sin 53° - 200 \text{ lb} = 0 \rightarrow 0.80T_2 - 200 \text{ lb} = 0$$

We now have two equations to solve together for T_1 and T_2. They are

$$T_1 - 0.60T_2 = 0 \quad \text{and} \quad 0.80T_2 - 200 \text{ lb} = 0$$

Solving the second one for T_2 gives $T_2 = 250$ lb. Then from the first equation we find $T_1 = 0.60T_2 = 150$ lb. The solution is now complete. ■■

■ **EXAMPLE 2.2** For the equilibrium situation shown in Figure 2.7(a), find the tensions in ropes 1 and 2 as well as the weight of the object. Notice that the tension in one rope is given. It is 40 N.

Solution The tension in rope 1 is equal to the weight it supports, and so $T_1 = W$. We isolate the knot as object. The free-body diagram is shown in (b). After finding the components, we can write the equilibrium equations to give

$$\Sigma F_x = 0 \rightarrow (40 \text{ N}) \cos 37° - T_2 \sin 37° = 0$$
$$\Sigma F_y = 0 \rightarrow T_2 \cos 37° - (40 \text{ N}) \sin 37° - W = 0$$

Using the trig tables, these equations become

$$(40 \text{ N})(0.80) - 0.60T_2 = 0$$

and

$$0.80T_2 - (40 \text{ N})(0.60) - W = 0$$

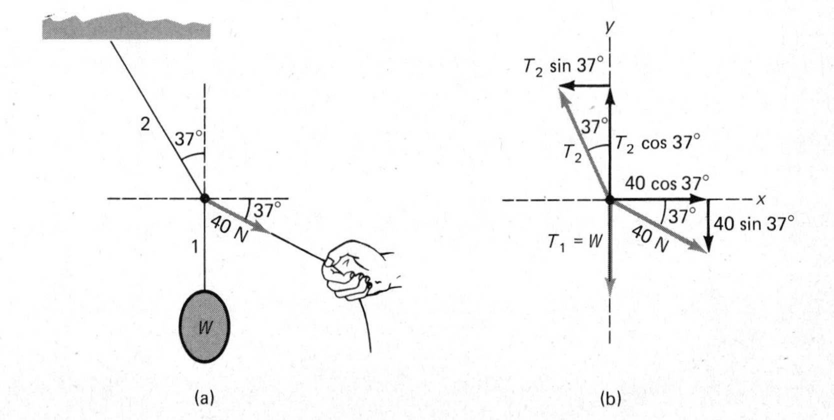

Figure 2.7 Find T_1, T_2, and W.

Solving the first equation for T_2 gives $T_2 = 53$ N. Placing this value in the second equation gives $W = 18.7$ N. Since $T_1 = W$, we see that T_1 is also 18.7 N. ∎∎

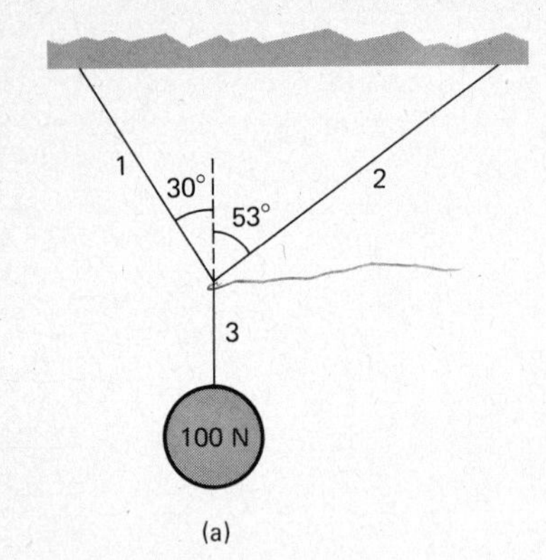

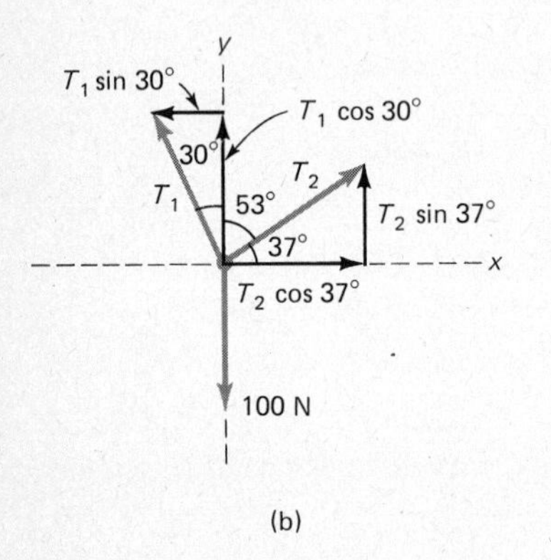

Figure 2.8 Find T_1 and T_2. Why is $T_3 = 100$ N?

▮ EXAMPLE 2.3 Find the tensions in the ropes in Figure 2.8.

Solution T_3 is clearly 100 N, because the object weighs 100 N. We isolate the knot as object and proceed as in part (b) of the figure. Then we write the equilibrium equations.

$$\Sigma F_x = 0 \rightarrow T_2 \cos 37° - T_1 \sin 30° = 0$$
$$\Sigma F_y = 0 \rightarrow T_1 \cos 30° + T_2 \sin 37° - 100 \text{ N} = 0$$

Placing in the trig functions gives

$$0.80T_2 - 0.50T_1 = 0 \rightarrow T_2 = 0.625T_1$$

and

$$0.87T_1 + 0.60T_2 - 100 \text{ N} = 0$$

We can solve for T_2 from the first equation ($T_2 = 0.625T_1$) and substitute it in the second to obtain

$$0.87T_1 + (0.60)(0.625)T_1 = 100 \text{ N}$$

from which

$$T_1 = 80 \text{ N}$$

Then, since $T_2 = 0.625T_1$, we have

$$T_2 = 50 \text{ N}$$ ∎∎

▮ EXAMPLE 2.4 Using the data given in Figure 2.9(a), find the value of the weight W.

Solution We would like to isolate a knot as object. Which shall we choose, A or B? We know nothing about the tensions in the ropes pulling on B, so we shall avoid that knot for now. Instead, isolate knot A as object. Its free-body diagram is shown in Figure 2.9(b). After taking components we can write

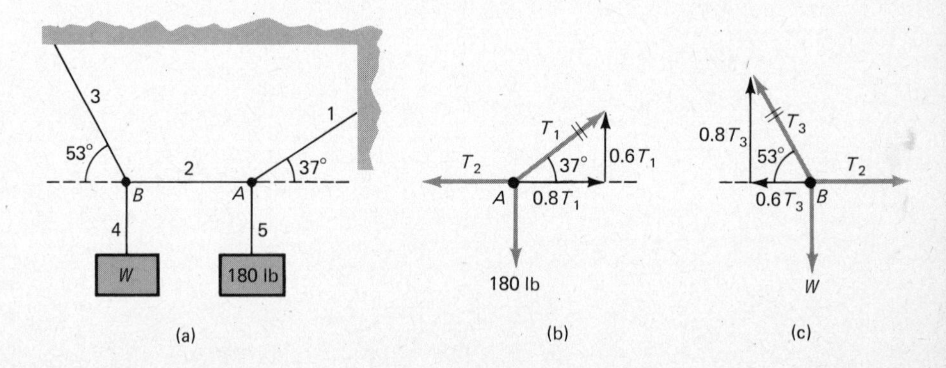

Figure 2.9 To find W we must first deal with point A as object. Then we use point B as a new object.

$$\Sigma F_x = 0 \rightarrow -T_2 + 0.80T_1 = 0 \rightarrow T_2 = 0.80T_1$$

and

$$\Sigma F_y = 0 \rightarrow 0.60T_1 - 180 \text{ lb} = 0 \rightarrow T_1 = 300 \text{ lb}$$

We can substitute this value for T_1 in the first relation to obtain $T_2 = 240$ lb.

Now we know the tension in the horizontal rope $T_2 = 240$ lb. This tells us that the horizontal rope pulls to the right on knot B with a force of 240 lb. Let us now isolate knot B as our object. Its free-body diagram is shown in Figure 2.9(c). After taking components we can write

$$\Sigma F_x = 0 \rightarrow T_2 - 0.60T_3 = 0 \rightarrow T_3 = \frac{240}{0.60} = 400 \text{ lb}$$

and

$$\Sigma F_y = 0 \rightarrow 0.80T_3 - W = 0 \rightarrow W = 0.80T_3 = 320 \text{ lb}$$

Therefore the value of the weight W is 320 lb. ∎∎

2.4 Turning Effects: Torques

Until now we have been talking about concurrent forces. These do not cause rotation of objects. To see how forces cause rotation, let us consider the pivoted meter stick shown in Figure 2.10(a). If you push down on one of its ends, it will rotate like a propeller. In this section we shall find out what causes rotations. We shall learn how forces give rise to turning effects.

Consider first the uniform stick shown in Figure 2.10. It is pivoted at its center, the 50-cm mark. In (a) we see it simply hanging on the pivot. You know from past experience that it will remain at rest at any angle you leave it. (This assumes, of course, that the pivot is exactly at the center.)

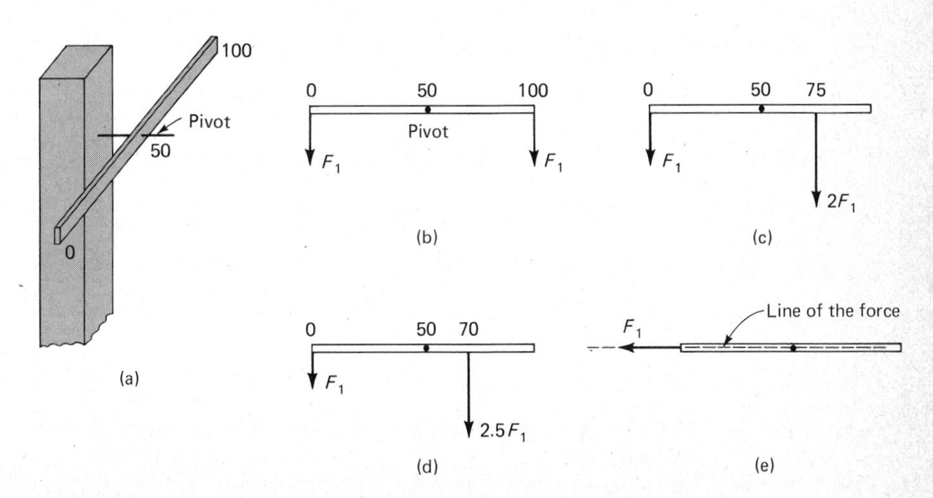

Figure 2.10 The pivoted meter stick does not rotate when positioned as shown.

Suppose now that a weight is hung at the zero mark at one end of the stick. It exerts a force F_1 as shown in part (b) of the figure. We can keep the stick from turning by pulling down on the other end with an equal force F_1 as shown. The turning effect of the force at one end is balanced by an equal force at the other end. You probably know this already.

But this is not the only way we can balance out the turning effect of F_1 at the zero mark. As shown in part (c), it can be balanced by a force $2F_1$ at the 75-cm mark. Or, as in part (d), it can be balanced by a force $2.5F_1$ at the 70-cm mark. Many more balancing combinations are possible.

We see from this that the turning effect of a force depends in some way on its distance from the pivot. So turning effect depends on both the magnitude of the force and the position at which the force is applied.

In addition, the turning effect depends on the direction of the force. An example to show this is given in part (e) of Figure 2.10. We note there that the force F_1 does not cause the stick to rotate. No balancing force is needed. The situation shown there leads us to conclude that

A force whose line goes through the pivot point causes no turning effect.

We shall often refer to the line along which the force vector lies as the *line of the force.* It is shown as the dotted line in Figure 2.10(e).

It should be clear from these examples that the turning action of a force about a pivot point depends on two things. The magnitude of the force is important, of course. In addition, the turning effect depends on the direction and position of the force. How can we summarize these results? That is what we shall now do.

First, we shall agree to describe turning effect in terms of a quantity called *torque* (pronounced "tork"). Before defining the quantity torque, we must first define a certain length called the *lever arm.* (This is also often called the *moment arm.*) By definition,

The lever arm is the length of the perpendicular drawn from the pivot to the line of the force.

To see what we mean by this, refer to Figure 2.11. Notice that the force vector $\mathbf{F}_2$ acts along the dotted line. We have called this line the line of the force. We now start from the pivot point and drop (or draw) a line perpendicular to the line of the force. It is labeled as the lever arm in the figure. Other examples of

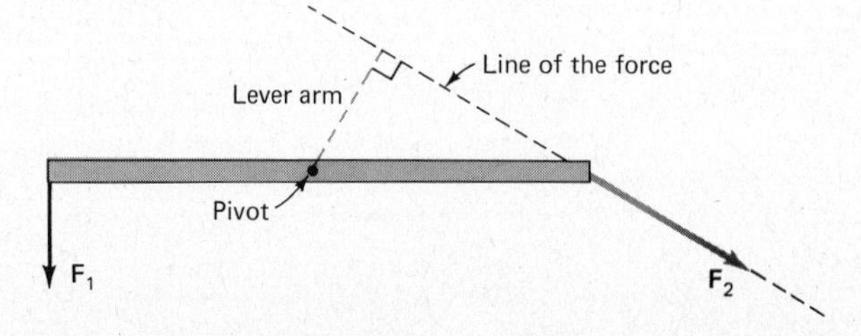

Figure 2.11 The turning effect (or torque) due to $\mathbf{F}_2$ about the pivot point (or axis) is $\mathbf{F}_2$ multiplied by the length of the lever arm.

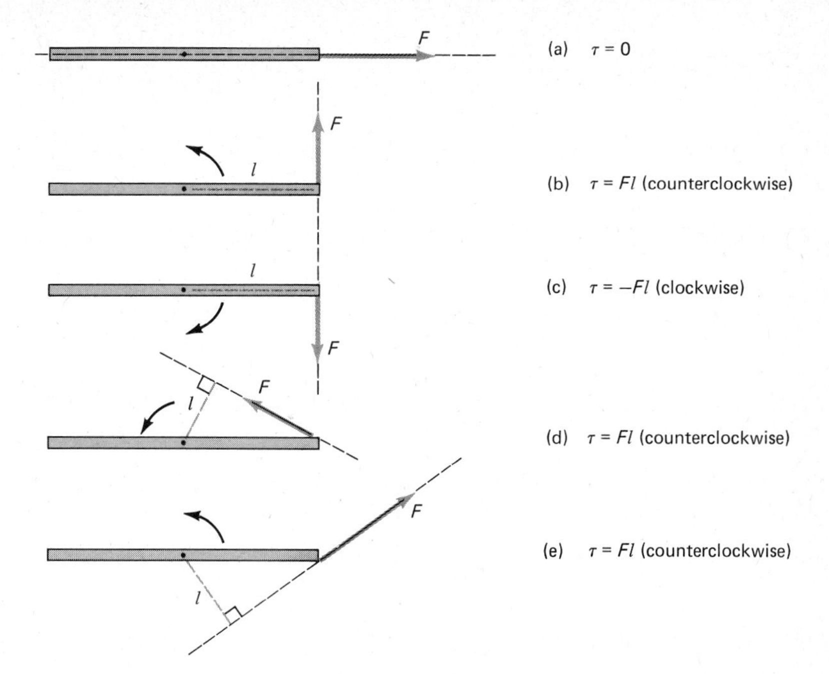

Figure 2.12 Be sure you understand the torques given here. The line of the force (indicated by the dashed line) and the lever arm (l) are shown in each case.

lever arms are shown in Figure 2.12, where each is labeled l. You should examine each carefully so that you can find the lever arm in other, different situations.

Now that we understand what is meant by the lever arm, we are able to define torque. The turning effect due to a force is called a torque. The magnitude of the torque is taken to be the product of the force times the lever arm. Using the Greek symbol tau, τ, to represent torque, we have

$$\tau = (\text{force})(\text{lever arm}) \tag{2.2}$$

Further, we agree to give the torque a positive or negative sign as follows. Torques that try to cause a counterclockwise rotation (c.c.w.) about the pivot are to be considered positive. Torques in the clockwise direction (c.w.) are to be given negative signs. Notice that the units of torque are those of force times distance. Typically this is newton-meters (N·m) or pound-feet (lb·ft).

Let us clarify this definition by reference to Figure 2.12. In part (a) we see a force that causes no turning effect. [It is the case we encountered earlier in Figure 2.10(e).] Notice that the line of the force passes right through the pivot point. The perpendicular distance from the pivot to the line of the force is zero, and so the lever arm is zero. As a result, in this case we have

$$\text{Torque} = (\text{force})(\text{lever arm}) = (F)(0) = 0$$

In part (b) of Figure 2.12 we see a force that tries to turn the bar (looked at as a clock hand) from 3 o'clock back toward 2 o'clock. It is a counterclockwise turning force, and so its torque will be taken positive. The lever arm, l, is the perpendicular from the pivot to the line of the force as shown. In this case

$$\tau = Fl$$

You should examine carefully the other parts of the figure. Try to answer the following questions during your examination. Why is τ negative in (c)? Where is the line of force in each case? Is l really the perpendicular from the pivot to the line of the force in each case?

2.5 Second Condition for Equilibrium

If an object is to remain at rest, then we know that

$$\Sigma F_x = 0 \qquad \Sigma F_y = 0 \qquad \Sigma F_z = 0$$

This we called the first condition for equilibrium. But that is only half of the story. Even if the forces do add to zero, the object might still rotate. To see that this is so, look at the situation shown in Figure 2.13.

There we see a stick resting on a table. Its two ends are pushed by equal and opposite forces. The forces are therefore balanced, and so

$$\Sigma F_x = 0 \qquad \Sigma F_y = 0 \qquad \Sigma F_z = 0$$

In spite of this we know that the stick will move. (Try it with your pencil on the desk top.) The two forces cause the stick to rotate. (This situation did not arise in our earlier discussion of equilibrium. We were concerned there with *concurrent forces*, forces that pass through a single point. Typically that point was a knot. Only *nonconcurrent forces,* forces that do not pass through a single point, can cause rotation.)

We need yet a second condition for equilibrium. It must tell us when no rotation will be caused by the forces. As you might expect, if the torques acting on an object balance, then they will cause no rotation. For coplanar forces this means that the clockwise torques must balance the counterclockwise torques. Or, if we take c.w. torques as negative and c.c.w. torques as positive, we can state the following.

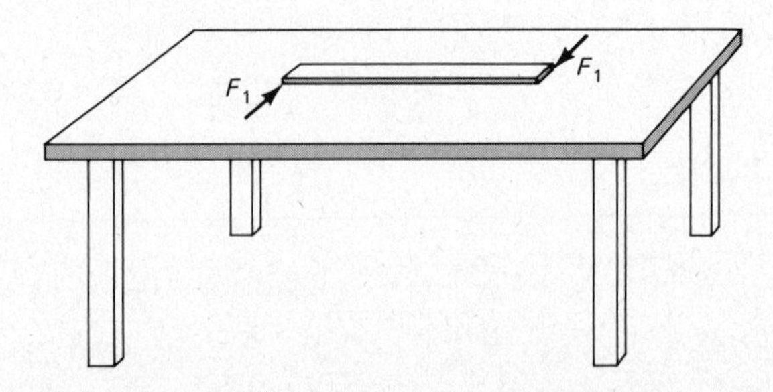

Figure 2.13 Even though the forces balance, the stick will not remain at rest.

For an object to be at equilibrium, the sum of the torques acting on it must be zero.

This may be expressed in shorthand as

$$\Sigma \tau = 0$$

for equilibrium. Remember, c.c.w. torques are taken to be positive, whereas c.w. torques are negative.

This is now the complete story. For an object to be in equilibrium, the following conditions must be true:

$$\Sigma F_x = 0 \qquad \Sigma F_y = 0 \qquad \Sigma F_z = 0 \qquad \Sigma \tau = 0 \tag{2.3}$$

We shall soon use these conditions in the solution of problems. Before doing so, we must point out two facts concerning the last of these equations.

We have been assuming that all rotations can be described as clockwise or counterclockwise. This is most often true. A wheel can rotate only c.w. or c.c.w. on its axle, for example. But sometimes an object can rotate in very complicated ways. For example, a football can tumble end over end while spinning. In these cases a three-dimensional torque equation must be used. We shall not encounter such complex problems in this text.

The second fact is more pleasant. To compute torques, we need a pivot point. (Remember, the lever arm is dropped from the pivot point.) In many cases the pivot point is obvious. It might be the axle point of a wheel or the shaft of a propeller blade. But in two dimensions the important fact is the following: *If an object is in equilibrium, then the sum of the torques acting on the object is zero for any and all points we choose as pivot points.* We can therefore choose the pivot point (or *axis,* as some call it) any place we wish. The importance of this fact will become clear as we proceed. Notice that *the pivot point can be taken anywhere.* No actual pivot or axle need be there.

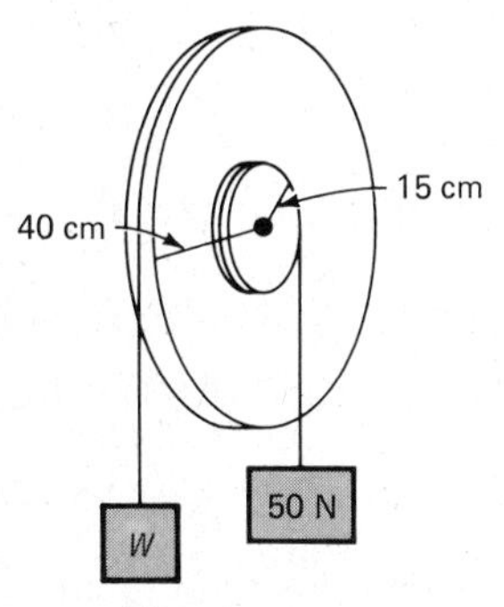

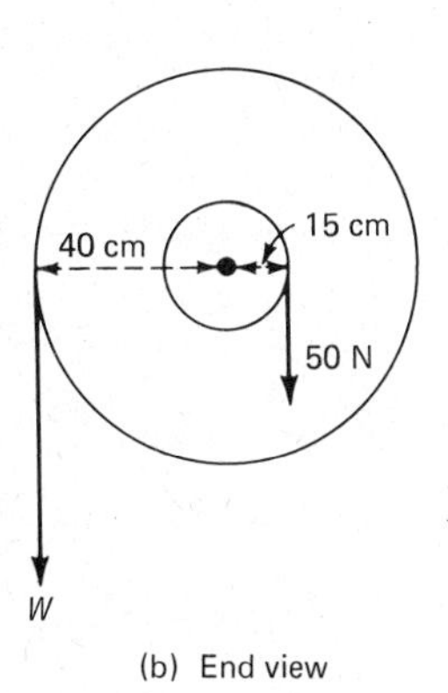

Figure 2.14 How heavy must the weight W be if it is to balance the 50-N weight?

▮ EXAMPLE 2.5 The two wheels shown in Figure 2.14(a) are fastened together and turn on the same axle. How heavy a weight, W, on the outer wheel is needed to balance the 50-N weight?

Solution The 50-N weight causes a c.w. torque, whereas the other weight causes a c.c.w. torque. For equilibrium (that is, for balance), we must have

$$\Sigma \tau = 0$$

Taking torques about the axle as pivot and noting that the lever arms are 40 cm and 15.0 cm gives

$$(W)(0.40 \text{ m}) - (50 \text{ N})(0.150 \text{ m}) = 0$$

(Why do we use the negative sign?) Solving for W gives $W = 18.8$ N. This is a practical and often-used method for lifting heavy loads (or obtaining large force) by use of a small force. ▮▮

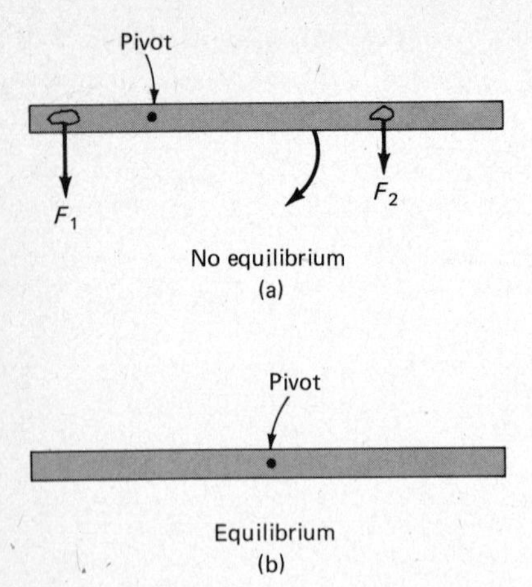

Figure 2.15 When supported at its center of gravity, the rod is at equilibrium.

2.6 Center of Gravity

Very often the weight of an object can cause the object to rotate. For example, the rod shown in Figure 2.15(a) will rotate about the pivot shown when released. A gravitational force pulls down on each little piece of the rod as indicated for the two tiny bits shown. Because there is more downward pull to the right of the pivot, the rod rotates as shown.

But in part (b) of the figure, a uniform rod would not rotate. For this new pivot, the torque due to the weight of the right end of the rod balances that due to the weight of the left. We see that there is a point (or pivot) about which the torques on the object due to gravity cancel out. The object can be balanced by a force equal and opposite to the weight applied at this point.

Moreover, you know that you can balance a meter stick in two ways on your finger tip. You can place your finger under the center of the stick to balance it. But you can also balance the stick on its end by holding it vertical with your finger under its lower end. Apparently, the line of the force must go through the center point of the stick, a very special point, if the force is to balance the weight of the object successfully. We call this special point for an object the *center of gravity* of the object.

It is important to notice the meaning of the center of gravity. When discussing the torques on an object due to the pull of gravity, the following is true.

The torques due to the gravitational force on each tiny piece of an object may be replaced by the torque due to the object's weight acting at the center of gravity.

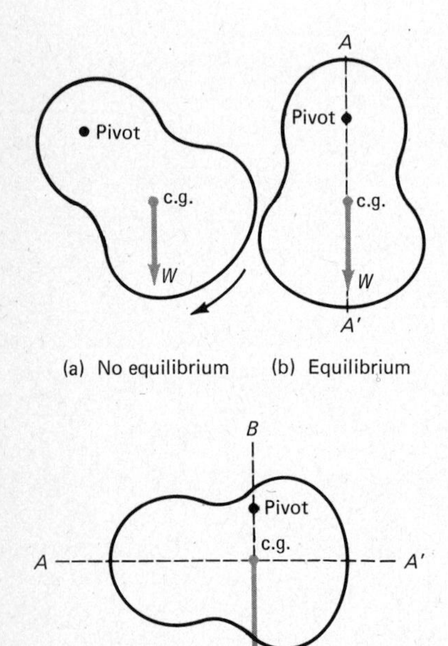

Figure 2.16 Use this method to find the center of gravity of a metal coat hanger. What is amazing about your result?

This allows us to ignore the complicated nature of the gravitational pull. We simply replace it by a single force (the weight) pulling on the center of gravity.

It is easy to see the position of the center of gravity for many objects. The center of gravity of "nice objects" such as uniform spheres, cylinders, rods, and so on, is at their geometrical center. (This is strictly true only in a uniform gravitational field. Scarcely anyone ever worries about this technicality, however.)

In the case of more complicated objects, the center of gravity can be found by calculation (in some cases) or by experiment (in most cases). The experimental method can be understood by reference to Figure 2.16.

If you try to hang the object shown in Figure 2.16 on a pivot, it swings until it hangs straight down. In (a) the weight acts at the center of gravity and exerts a c.w. torque about the pivot. The object rotates as shown because of this torque. But if the center of gravity is straight under the pivot as in (b), then the weight exerts no torque. (The lever arm is zero in this case.) The object therefore hangs at rest. Only if the center of gravity is directly under the pivot will the object be stable and remain at rest. This fact allows us to locate the center of gravity.

We simply hang the object from several pivots in succession. When it hangs from the pivot in (b), we know that the center of gravity is along line AA' somewhere. But by hanging it from a new pivot as in (c), we know that the center of gravity must lie along BB'. Since it must be on both AA' and BB', the center of gravity must be at their point of intersection. A third pivot point will also give rise to a line that intersects this same point. Once this point is found, the center of gravity is known. The object should balance on a support (or pivot) placed at the center of gravity, because gravity will not cause it to tip.

Because the force of gravity causes no rotation about the center of gravity, axles for rotation are usually placed through it. The object will spin without its weight causing wobbling about such an axle. A properly balanced wheel or other rotating device has been adjusted so that its axis of rotation is centered on the center of gravity. If it is not adjusted properly, the torque due to gravity causes the wheel to vibrate. You might wish to compare the object in Figure 2.16(a) to a very poorly balanced car wheel. What effect does its rotation have on the axle of the car? Why is it preferable to have the pivot at the center of gravity?[3]

2.7 Static Equilibrium: General Case

We now possess the tools needed to solve statics problems involving both balanced forces and torques. Our method of approach is not changed from that outlined in Section 2.3. You will recall that the steps were as follows.

1. Sketch the physical situation.
2. Isolate an object for discussion.
3. Draw a free-body diagram with forces.
4. Split the forces into components.
5. Write the equations for equilibrium.
6. Solve for the unknowns.

The only new feature in the application of the method is in part 5. We now must write $\Sigma \tau = 0$ as well as the force component equations. Let us now examine several examples.

EXAMPLE 2.6 A 1000-N sign painter stands on a uniform 250-N board suspended by two ropes, as shown in Figure 2.17. If each of the ropes can hold only 900 N, will it be safe for him to stand at the position shown?

Solution The situation is sketched in Figure 2.17. We want to see if either T_A or T_B must be larger than 900 N. Let us isolate the board as our object. The free-body diagram is shown with the sketch. Notice that the weight of the uniform board, 250 N, is taken to act at the center of the board. The 1000-N man is supported by the board, and so he pushes down on it with a force of 1000 N as shown. All forces are y-directed, and so the components are known.

Let us first write the torque equation. There is no natural pivot point in this problem. But that makes no difference. We know we can choose a pivot for calculations anywhere. Let us choose it at A. Then the c.c.w. torque is caused by T_B, while the 250-N

A B
0.5 m 4 m 1 m 0.5 m

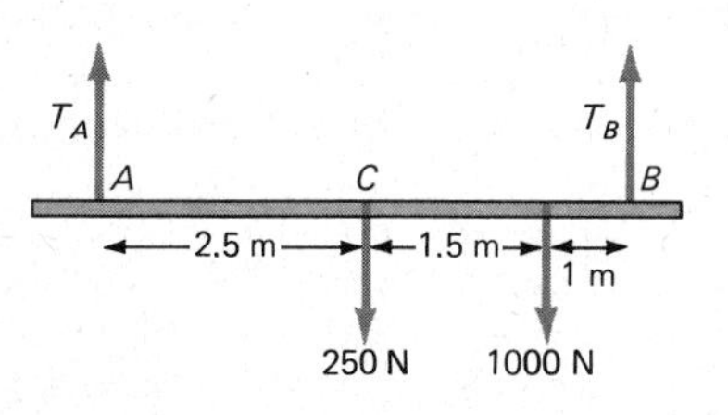

Figure 2.17 If we take point A as the pivot point, then T_A will be missing from the torque equation.

[3] In more advanced texts you would find these questions discussed in terms of the center of mass rather than the center of gravity. For our purposes they can be assumed to coincide.

and 1000-N forces cause c.w. torques. T_A, of course, causes no torque, because its line passes through the axis we have chosen. Look at the diagram and find the lever arm for each force. Did you find them to be 2.5, 4.0, and 5.0 m, respectively, for 250 N, 1000 N, and T_B? The torque equation is then

$$\Sigma \tau = 0 \rightarrow -(250 \text{ N})(2.5 \text{ m}) - (1000 \text{ N})(4.0 \text{ m}) + (T_B)(5.0 \text{ m}) = 0$$

Simplifying gives

$$-625 \text{ N}\cdot\text{m} - 4000 \text{ N}\cdot\text{m} + (5.0 \text{ m})T_B = 0$$

Solving gives

$$T_B = \frac{4625 \text{ N}\cdot\cancel{\text{m}}}{5.0 \cancel{\text{m}}} = 925 \text{ N}$$

(Notice how the units are used just like numbers.) Since rope B can only hold 900 N, it will break.

Just for practice, let us solve for T_A as well. To do so, we need another equation. We have already used $\Sigma \tau = 0$. There remains yet $\Sigma F_y = 0$. Let us write it. We have

$$T_A - 250 \text{ N} - 1000 \text{ N} + T_B = 0$$

But we have already found that $T_B = 925$ N. Using this value for it and solving for T_A gives

$$T_A = 325 \text{ N}$$

Clearly, rope A will not break.

Before leaving this example, you should recognize that we did it the easy way. If we had taken the pivot at the center of the board, the torque equation would have been

$$-(T_A)(2.5 \text{ m}) - (1000 \text{ N})(1.5 \text{ m}) + (T_B)(2.5 \text{ m}) = 0$$

Notice that this equation has two unknowns. We would have had to solve it simultaneously with the $\Sigma F_y = 0$ equation. It pays to take the pivot so that an unknown force's line goes through it. Then that force is absent from the torque equation. ▮▮

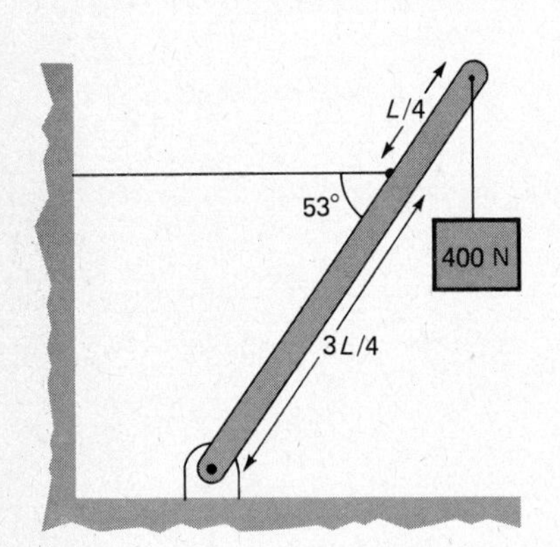

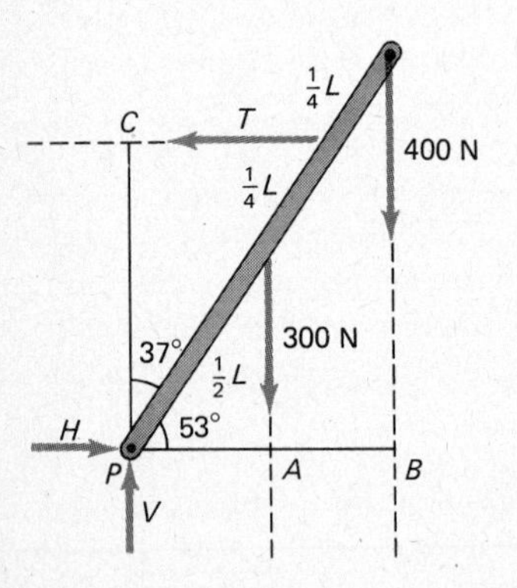

Figure 2.18 With pivot at P, the lever arms are $\overline{PA}$ for 300 N, $\overline{PB}$ for 400 N, and $\overline{PC}$ for T.

▮ **EXAMPLE 2.7** A uniform 300-N boom is pivoted on a pin at the floor as shown in Figure 2.18. It supports a 400-N load. How strong must the horizontal tie rope be to hold the system at this angle?

Solution We want to find the tension T in the tie rope. Let us isolate the boom as our object. Its weight acts at its center. If we assume that the boom is free to rotate on the pin, we can represent the action of the pin by a force that has only a horizontal and a vertical component. In the free-body diagram of Figure 2.18 we therefore show unknown force components H and V exerted on the boom by the pin.

All forces on the free-body diagram are already in component form. We are now ready to write the equilibrium equations.

$$\Sigma F_x = 0 \qquad \Sigma F_y = 0 \qquad \Sigma \tau = 0$$

Let us start with the torque equation. If we take the pivot point at the pin, the unknown forces H and V will not appear in the equation. (Their lever arms are zero.) Be sure that you understand why the lever arms are as given in the figure legend. Notice that the only c.c.w. torque is caused by T. The torque equation is

$$-(300\text{ N})(\overline{PA}) - (400\text{ N})(\overline{PB}) + (T)(\overline{PC}) = 0$$

Now from the geometry of the figure,

$$\overline{PA} = \tfrac{1}{2}L\cos 53° = 0.30L$$
$$\overline{PB} = L\cos 53° = 0.60L$$
$$\overline{PC} = \tfrac{3}{4}L\cos 37° = 0.60L$$

Placing these values in the equation gives

$$-(300\text{ N})(0.30L) - (400\text{ N})(0.60L) + (T)(0.60L) = 0$$

Notice that L can be canceled from the equation (divide through by L). Solving for T, we find that

$$T = 550\text{ N}$$

If we wanted to find H and V as well, we could do so as follows.

$$\Sigma F_x = 0 \rightarrow H - T = 0 \rightarrow H = T = 550\text{ N}$$
$$\Sigma F_y = 0 \rightarrow V - 300\text{ N} - 400\text{ N} = 0 \rightarrow V = 700\text{ N}$$

▮▮

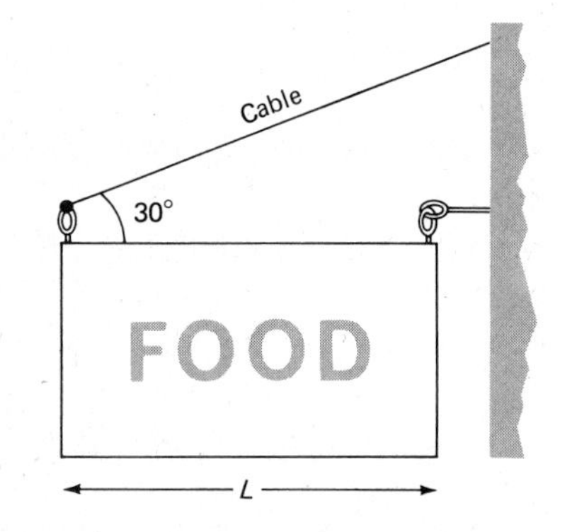

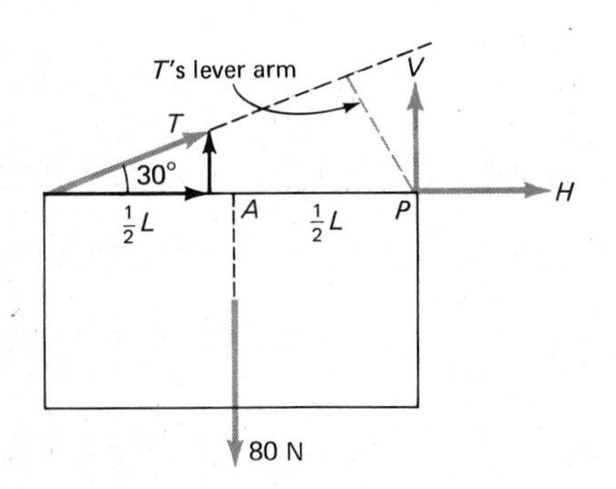

Figure 2.19 Did we draw H in the wrong direction? How will we find out?

▮ **EXAMPLE 2.8** The sign shown in Figure 2.19 is of uniform thickness and weighs 80 N. How strong must the cable be to hold it?

Solution Isolate the sign as object. The unknown force at the wall hinge point is represented by its components H and V. Let us take the pivot point at P so as to eliminate H and V from the torque equation. Notice the lever arm for T. It is $L \sin 30° = (L)(0.50)$. Notice also that the lever arm for the 80-N force is $\overline{AP} = 0.50L$. The torque equation becomes

$$(80\text{ N})(0.50L) - (T)(0.50L) = 0$$

This gives $T = 80$ N.

If we wanted to find H and V, we would write

$$\Sigma F_y = 0 \rightarrow V + T\sin 30° - 80\text{ N} = 0 \rightarrow V = 40\text{ N}$$

and

$$\Sigma F_x = 0 \rightarrow T\cos 30° + H = 0 \rightarrow H = -69\text{ N}$$

The negative sign on H tells us that we drew it in the wrong direction. It really pushes on the sign.

There is another way to deal with the torque due to force T. We could use its components rather than T itself. Its horizontal component ($T\cos 30°$) has its line passing through the pivot, so it exerts no torque. The vertical component ($T\sin 30°$) has a lever arm L because it is applied at the end of the sign. It therefore exerts a torque $-(0.50T)(L)$. In the torque equation the effect of T is simply $-(0.50T)(L)$, which is the same as the term $-(T)(0.50L)$ we used previously. Sometimes it is more convenient to write the torque equation using the components of a force. ■■

2.8 Stable, Unstable, and Neutral Equilibrium

We have been discussing objects in stable equilibrium. They obey the following rule: If they are displaced slightly from their equilibrium position, they tend to come back to that position. For example, if the left end of the sign in Figure 2.19 is lifted a little, it will come back to the position shown when released. As the word "stable" implies, objects in stable equilibrium tend to remain that way even when disturbed slightly. If you examine any such situation, you will see that the original forces acting on it try to pull it back into place when it is disturbed.

A good example of unstable equilibrium is a pencil balanced on its tip. (If you cheat a little by flattening the tip, you may be able to balance it.) Even if it is displaced ever so slightly, it will fall over. The torque on it due to gravity tends to tip it over rather than to urge it back into place.

Neutral equilibrium occurs when the supporting forces neither aid nor stop displacement of the object. A perfectly cylindrical pencil laid on its side on a table is a good example. It lays as well in one position as another. If you roll it slightly and release it, it will remain where you released it. The forces acting on it (gravity down and the table up) do not cause the pencil to roll—or stop it from rolling.

In Figure 2.20 we show an object that can exist in all three forms of equilibrium. Notice in each case what its weight tries to do to it if you tip the object slightly. You might be interested in a general rule for such situations. If the disturbance (1) lifts the center of gravity, the object is in stable equilibrium; (2) lowers the center of gravity, it is unstable; (3) does not raise or lower the center of gravity, it is in neutral equilibrium. We shall understand the reason for this rule after studying about energy in Chapter 6.

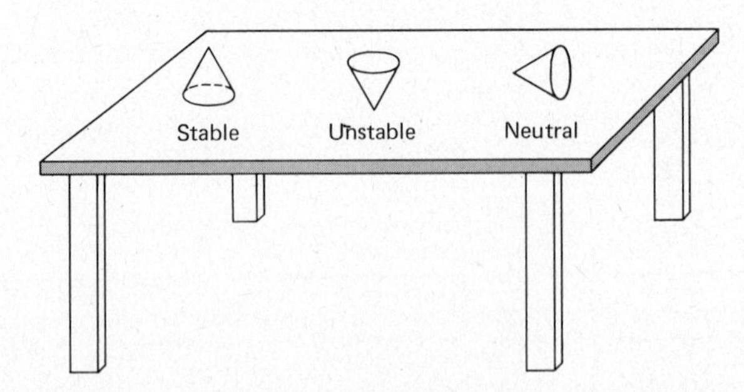

Figure 2.20 What does gravity try to do to each of these cones when the cone is displaced slightly?

2.9 Laboratory Work in Statics

Before ending this chapter, we should take time out for the following information. When you do laboratory experiments in statics, you will probably use weights stamped in kilograms (kg) and grams (g). These stampings give the masses of the weights. We shall discuss the concept of mass in Chapter 4. Until you study that chapter, it will be sufficient if you know the following fact.

A weight stamped 1 kg weighs 9.8 N on earth.

As a result, if you have weights that total 725 g, then the weight is $0.725 \times 9.8 = 7.10$ N. Some laboratory instructors take pity on their less experienced students and allow them to use the factor 10 in place of 9.8 in these experiments. Even so, you must remember that the correct factor is 9.8 N for the weight of 1 kg on earth.

Summary

An object that is at rest and remains at rest is in static equilibrium.

The force with which gravity pulls on an object is called the weight of the object.

The tension in a rope or string is the force with which the rope or string pulls on its support.

To be termed concurrent, the lines of two or more forces must pass through the same point.

The lever arm (or moment arm) of a force is a distance. It is the length of a perpendicular dropped from the pivot to the line of the force.

The torque due to a force about a pivot point is defined to be the force multiplied by its lever arm. Torques are positive if they try to cause counterclockwise rotation about the pivot; negative if they try to cause clockwise rotation. A force whose line passes through the pivot exerts no torque about that pivot.

For an object to be in equilibrium, the following must be true for the forces acting on it: $\Sigma F_x = 0$; $\Sigma F_y = 0$; $\Sigma F_z = 0$; $\Sigma \tau = 0$. These are called the equilibrium equations.

The first condition for equilibrium of an object states that all the forces acting on the object must add to zero (that is, they must balance): $\Sigma F_x = 0$; $\Sigma F_y = 0$; $\Sigma F_z = 0$.

The second condition for equilibrium of an object states that the torques on the object must equal zero, that is, they must balance: $\Sigma \tau = 0$.

At equilibrium the torques about any pivot we choose must add to zero.

In solving problems in statics, we follow the steps outlined on page 39.

The center of gravity of an object is the point at which the weight of an object may be taken to act when torques are being computed. When supported at its center of gravity, the object's weight will not cause it to rotate. The torque due to the gravitational pull on the object is zero when the pivot point is taken at the center of gravity.

There are three types of static equilibrium: stable, neutral, and unstable. They are defined in Section 2.8.

Questions and Exercises

1. Draw a free-body diagram for a man as object in each of the following cases: (a) man standing on the floor, (b) man standing on a stool on the floor, (c) man doing a handstand, (d) man hanging by his feet from a trapeze.
2. Two forces pull on a ring. A force F_1 pulls in the $+x$ direction, while a force F_2 pulls in the $+y$ direction. What can be said about the force F needed to balance these two forces?
3. A water pipe is stuck in a coupling. To unscrew the pipe, why is a long-handled pipe wrench more effective than a short-handled one? Why is a long-handled wrench more effective in loosening a stuck bolt than a short-handled one?
4. A woman attempts to tighten the clothes line so that it won't sag when she hangs a heavy laundry bag at its center. Will she ever be able to remove all the sag? Why?
5. A good way to ruin the hinge on a door is to place a stick near the hinge to prop it open against the wind. Explain why the hinge is often bent or broken in this way.
6. Give several examples of solid objects whose center of gravity is at a point outside the material of the solid.
7. Low-slung, wide-track cars are less easily overturned than others. Explain why this is true. (Hint: Consider the torque due to its weight when the car is tipped on two wheels.)
8. Keeping your body straight, lean forward as far as you can without falling. Why are people with big feet at an advantage in this respect? What determines how far sideways you can lean?
9. Each of the following tools is designed to make use of torques. Describe the torques which exist in each case: wire cutters, nutcracker, nail puller of a claw hammer, wheelbarrow, crescent wrench, bottle opener.
10. Hold a weight in your hand when your arm is straight and horizontal. Compare this to the case when your arm is bent 90° at the elbow. Why is the first method much more difficult? (Hint: Your forearm acts like a boom supported by a muscle attached above and below the elbow. How does the tension in this muscle change?)

Problems

1. A force with components $F_x = 30$ N and $F_y = -20$ N pulls on a ring. What are the components of the force needed to balance this force?
2. Two forces pull on a ring. One is in the $+y$ direction and has a magnitude of 6.0 N. The other pulls in the $-x$ direction with a force of 7.0 N. What are the x and y components of the force needed to balance these forces?
3. A ring remains at rest with three strings pulling on it. One string pulls with a force of 25 N at an angle of 90°. The other pulls with a force of 70 N at an angle of 270°. What is the magnitude and direction of the force caused by the third string?
4. Rope A hangs from the ceiling with a 50-N weight at its end. Rope B hangs from this weight and has a 70-N weight at its lower end. Find the tensions in the two ropes.
5. A 130-lb boy stands on the gym floor pulling down with a force of 60 lb on a rope hanging from the ceiling. One third of the way from the top is a 90-lb boy who is resting while climbing the rope. Find the tension in the two parts of the rope.
6. Two objects of equal weight W hang from a rope that passes over a pulley, as shown in Figure P2.1. If the pulley's weight is negligible in comparison to W, find T_1, T_2, and T_3, the tension in the three cords.

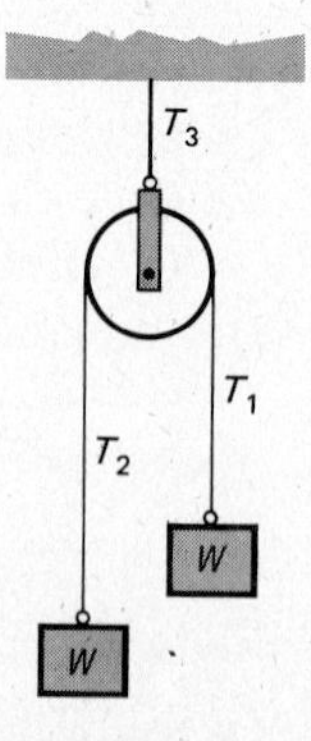

Figure P2.1

*7. The pulleys shown in Figure P2.2 are frictionless and have negligible weight. At equilibrium how large is W_2 if $W_1 =$ 300 N? Evaluate T_1, T_2, T_3, and T_4.

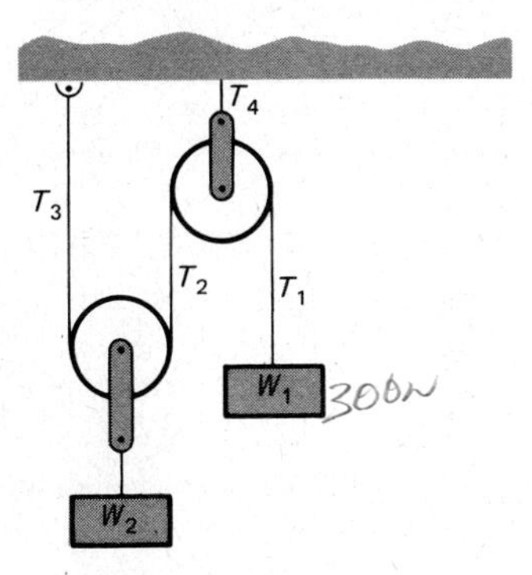

Figure P2.2

**8. The system in Figure P2.2 is at equilibrium. Consider the pulleys to have negligible weight and friction. What is the ratio of W_1 to W_2? Find T_1, T_2, T_3, and T_4 in terms of W_1.

*9. As a gymnastic stunt, two girls each of weight W_g stand on the shoulders of a man who weighs W_m, as shown in Figure P2.3. (a) How large a force must the floor exert upward on the man's feet? (b) How large is the total vertical force exerted on the man by the girls? (c) What must be true about the horizontal forces exerted on the man by the girls?

Figure P2.3

10. A ring remains at rest with three strings pulling on it. One string pulls at 0° with a force of 20 N. Another pulls at 120° with a force of 12 N. Find the magnitude and direction of the force supplied by the third string.

11. A ring is held at rest by three forces: a force of 40 N in the $+x$ direction, a 70-N force in the $+y$ direction, and an unknown force F. (a) What are the x and y components of F? (b) How large is F? (c) At what angle does F pull?

12. How large a horizontal force is needed to pull a 4.0-N pendulum ball aside until the pendulum cord makes an angle of 30° to the vertical? See Figure P2.4.

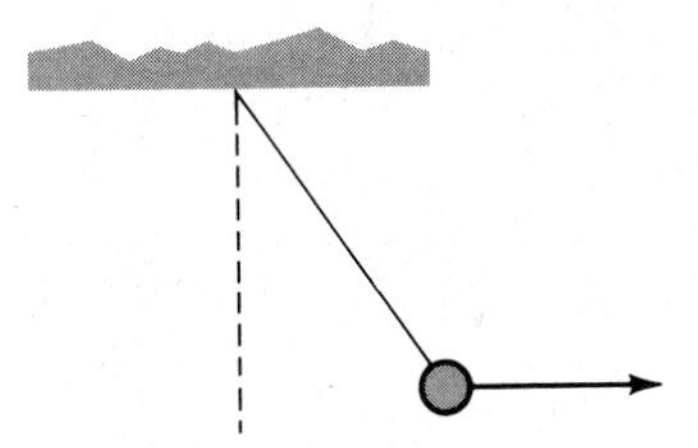

Figure P2.4

13. In Figure P2.4, the horizontal force that holds the pendulum aside is 20 N. If the cord makes an angle of 37° to the vertical, find the tension in the cord and the weight of the ball.

14. As shown in Figure P2.5, an object of weight W is hung from two cords. Find the tension in each cord.

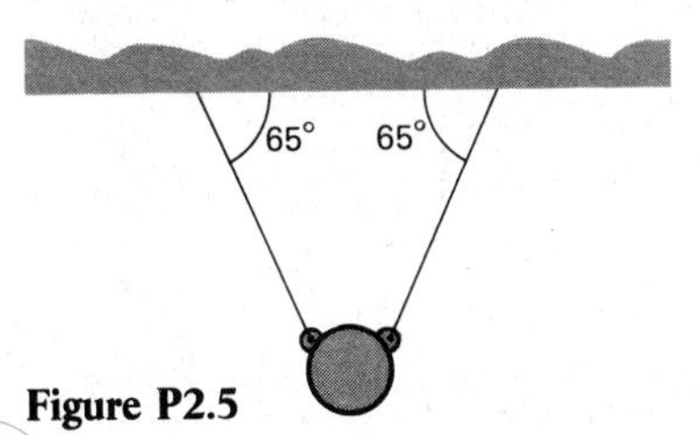

Figure P2.5

*15. Find the tensions T_1 and T_2 in the two cords shown in Figure P2.6 if the object weighs 50 N.

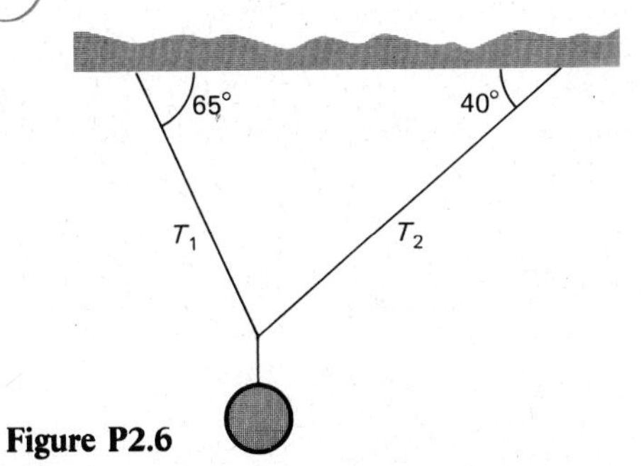

Figure P2.6

*Problems marked with an asterisk are not as easy as the unmarked ones.
**Problems marked with a double asterisk are somewhat more difficult than the average.

16. A 120-lb tightrope walker does not want the rope to sag from the horizontal by an angle of more than 2.0° when she stands at its center. What must be the minimum tension in the rope?

*17. An 800-N street lamp is to hang from the center of a cable stretched between two poles. If the poles are 15 m apart, how large must the tension in the cable be if the rope is to sag by only 75 cm?

*18. For the situation in Figure P2.7, how large must the tension labeled P be if the system is to hang as shown?

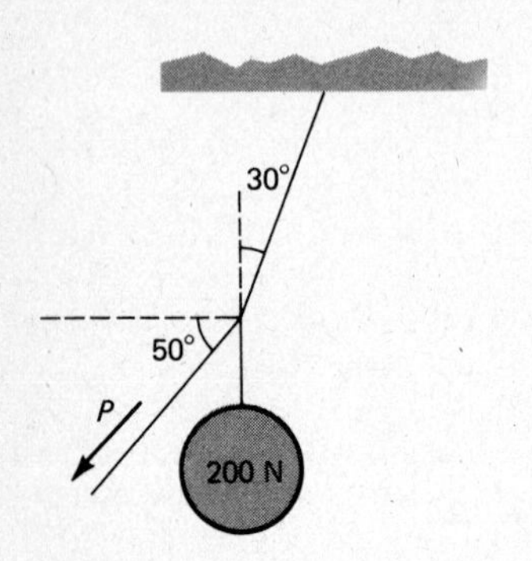

Figure P2.7

**19. Referring to Figure P2.8, what is the tension in the horizontal rope $\overline{AB}$?

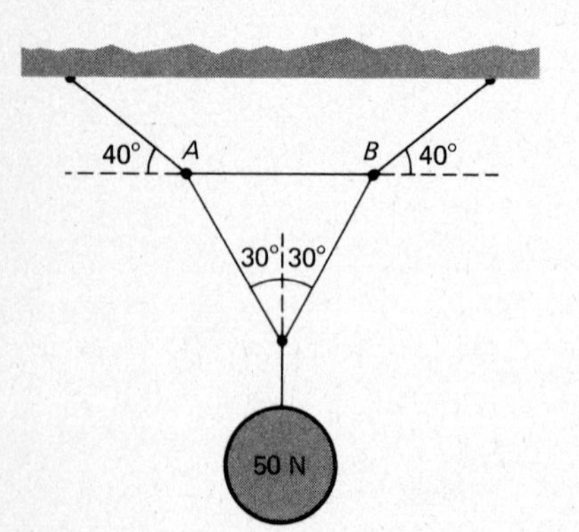

Figure P2.8

20. For the situation shown in Figure P2.9, find the torque about the pivot point due to (a) F_1, (b) F_2, (c) F_3, (d) F_4. Assume $F_1 = F_2 = F_3 = F_4 = 80$ N and $L = 3$ m.

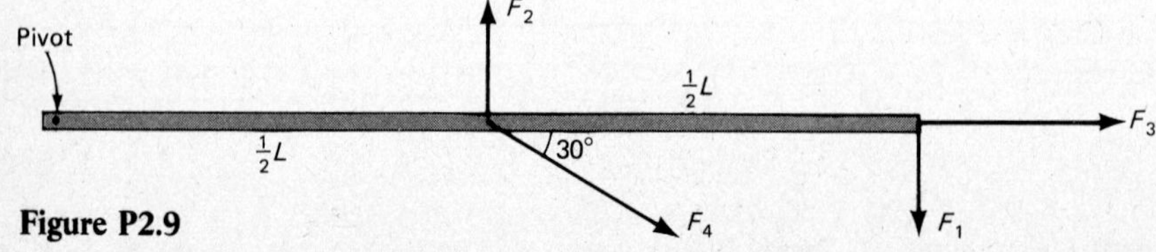

Figure P2.9

21. Assume in Figure P2.9 that $F_4 = 0$ and $F_1 = 400$ N. Assuming the weight of the rod to be negligible, how large must F_2 be if the rod is to be in equilibrium?

22. Assume in Figure P2.9 that $F_1 = 0$ and $F_4 = 60$ N. How large must F_2 be if the rod (of negligible weight) is to be in equilibrium?

23. The rod in Figure P2.10 is uniform and is pivoted at its center. Find the torque about the pivot point due to (a) F_1, (b) F_2, (c) F_3. (d) What must be the relation of F_2 to F_1 if the rod is to be in equilibrium?

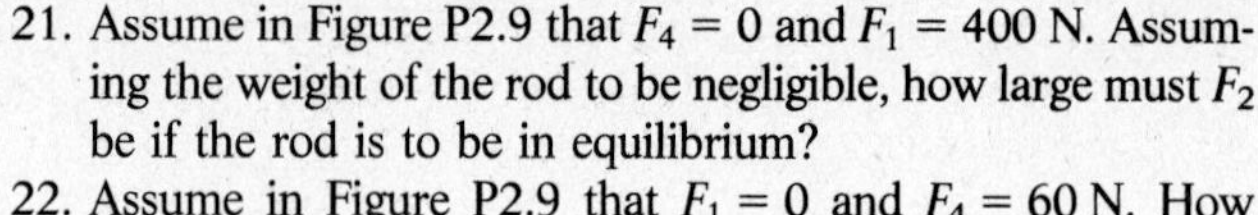

Figure P2.10

24. As shown in Figure P2.11, a small pulley on a motor uses a belt to drive a large wheel. Assuming that the tension in the lower part of the belt is negligible, what is the ratio of the torque experienced by the large wheel to the torque on the pulley of the motor?

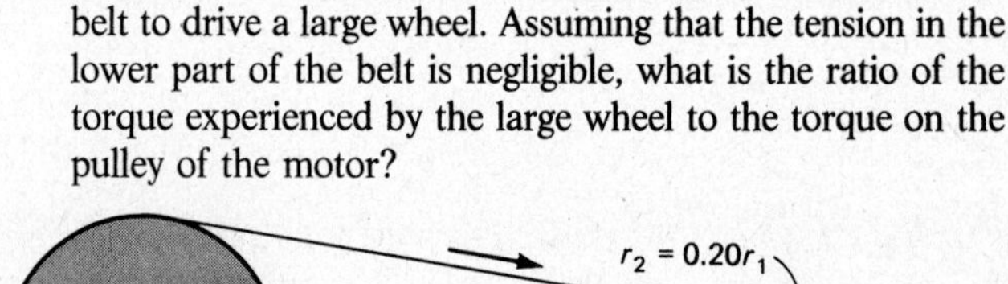

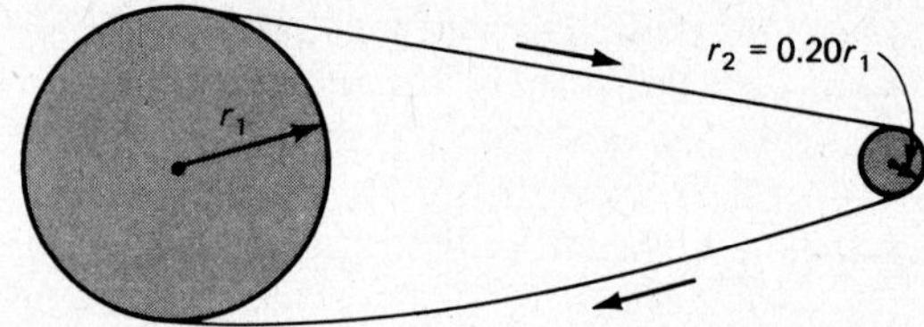

Figure P2.11

25. Consider the nutcracker shown in Figure P2.12. If the user exerts a force F as shown, how large a force is exerted on the nut? (Assume that the nut does not crack.)

Figure P2.12

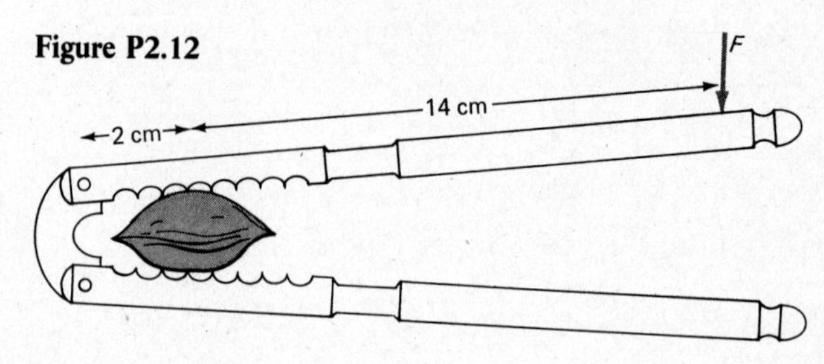

26. Two men are carrying a horizontal pole that is 8 m long and weighs 400 N. They are at its two ends. If the center of gravity of the pole is 3 m from one end, find the load carried by each man.
27. A man and his wife are moving a 200-lb sofa by lifting at its two ends. If their 60-lb child sits one-fourth of the length from one end, with what force must each lift?
28. A uniform 400-N plank supported by two posts (shown in Figure P2.13) serves as a diving board. Find how hard each of the two posts must push up on the board when a 700-N man stands at its end.

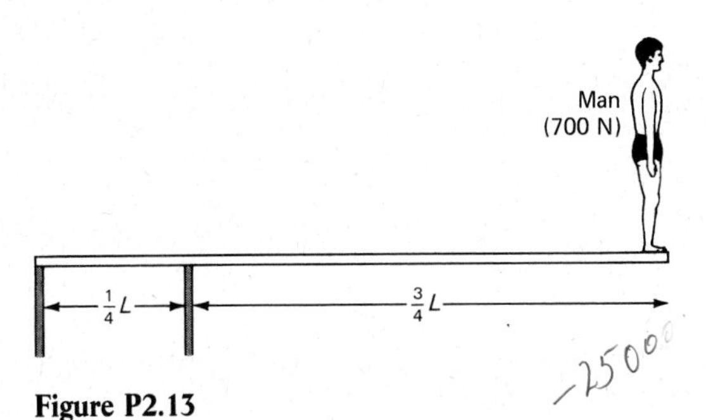

Figure P2.13

29. The following experiment is done in order to locate the center of gravity of a 150-cm-high refrigerator. It is laid on its back supported at its two ends by bathroom scales. The scale at end A reads 800 N and the one at the other end reads 600 N. How far horizontally from A is the center of gravity?
*30. In Figure P2.14 how far from A must you place a pivot in order to balance this device? Assume the connecting rod to have negligible weight.
**31. In Figure P2.14 the connecting rod is uniform and weighs 3.0 N. Where should a pivot be placed if the device is to balance on it?

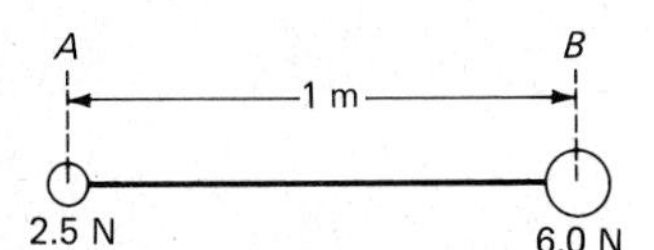

Figure P2.14

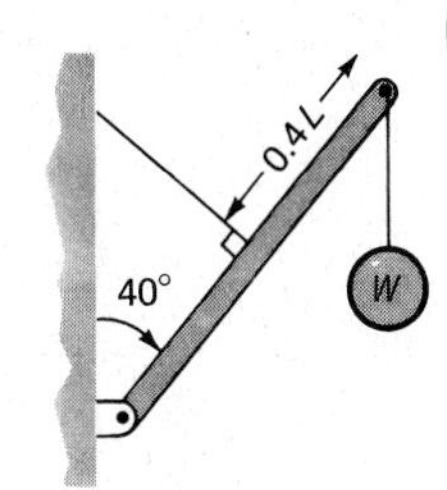

Figure P2.15

*32. The uniform boom in Figure P2.15 weighs 80 N and has a length L. If the tie rope can hold a maximum tension of 400 N, how heavy can the weight W be before the rope will break?
*33. The uniform boom of length L shown in Figure P2.15 weighs 80 N and supports a weight $W = 200$ N. Find the tension in the tie rope and the horizontal and vertical components of the force exerted by the pin at the lower end.
*34. The uniform 60-N boom of length L shown in Figure P2.16 carries a load $W = 140$ N. How large is the tension in the tie rope, and what are the horizontal and vertical components of the force exerted on the boom by the pin?

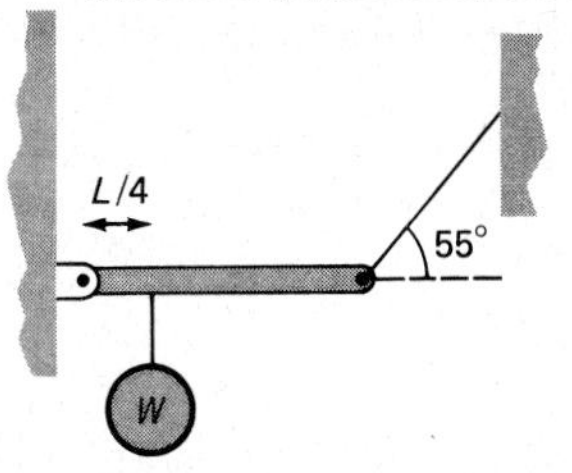

Figure P2.16

*35. A designer wishes to construct the mobile shown in Figure P2.17. The ornament at the right is to weigh 3.0 N. She wants the various lengths to be as shown. What must be the weights W_2 and W_3 for the system to be balanced?

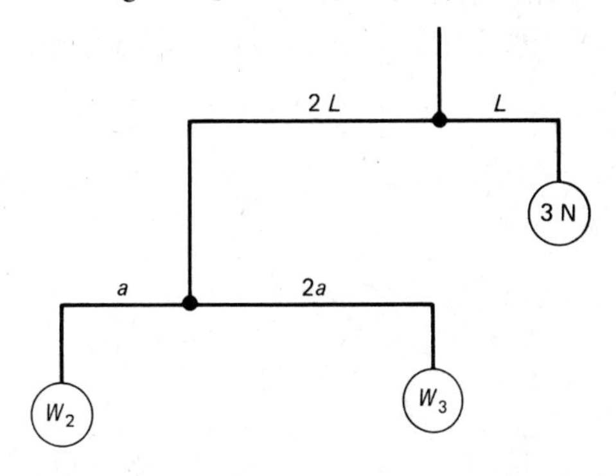

Figure P2.17

*36. A uniform ladder that weighs 200 N leans against a smooth wall as shown in Figure P2.18. At a smooth surface such as

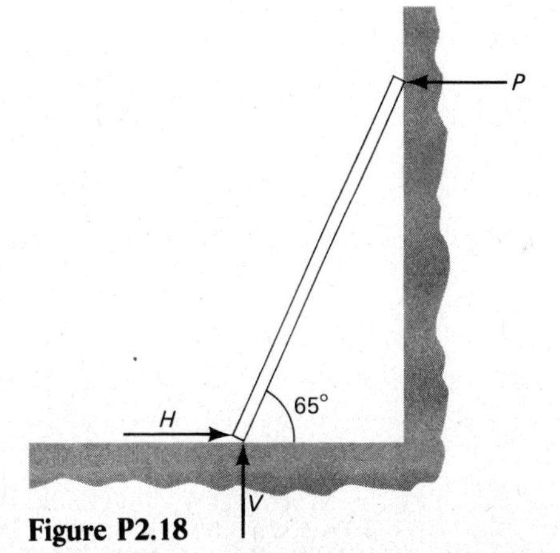

Figure P2.18

the wall, the surface exerts a force P on the ladder which is directed perpendicular to the wall. How large must the horizontal force H on the foot of the ladder be if the ladder is not to slip?

*37. The uniform ladder in Figure P2.18 weighs 200 N. Its top rests against a smooth wall. Because the wall is smooth, it can exert only a force such as P on the ladder. The force P is perpendicular to the wall. If a 140-N boy climbs $\frac{1}{4}$ of the way up the ladder, what then will be the magnitudes of forces P, H, and V?

*38. A stunt man plans to walk up the side of a building using a rope as shown in Figure P2.19. He weighs 800 N. When standing, his center of gravity is 1.00 m above the ground, while his shoulders are 1.40 m from the ground. Find the tension in the rope when he is at the position shown. How large a friction force must act upward on his feet if he is not to slip?

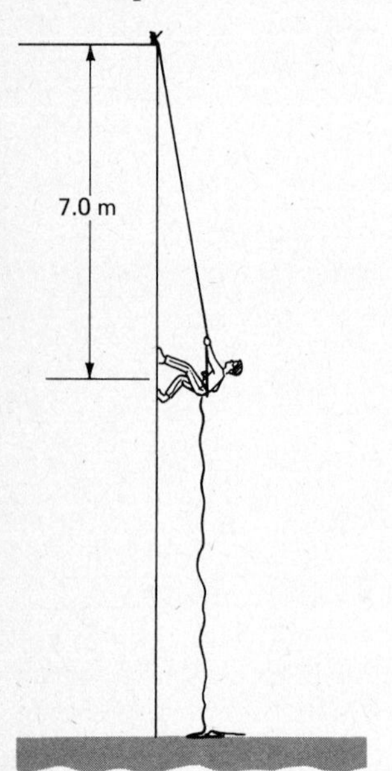

Figure P2.19

**39. A 20-N uniform picture hangs by means of a cord from a nail in a wall as shown in Figure P2.20. (a) Show that the center of gravity must necessarily be directly below the nail as indicated. (b) Can the tensions in the two parts of the cord be equal if the picture hangs as shown? Assume the picture does not touch the wall.

*40. In Figure P2.21 the struts are pivoted at their junction and at each end. Neglecting the weights of the struts, what is the compressive force in the strut due to the weight W? (Hint: Each strut exerts a force on the junction that is in line with

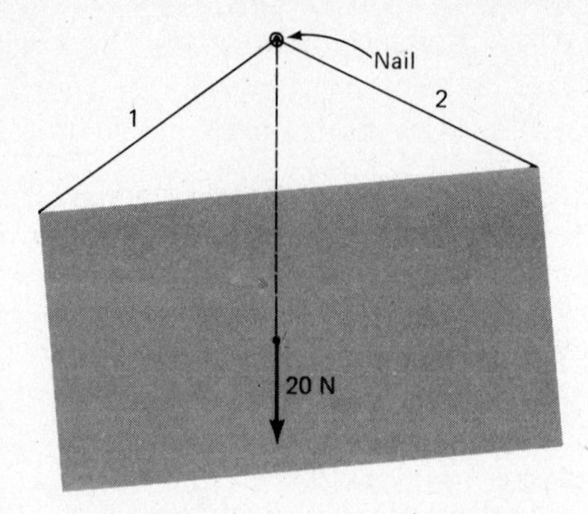

Figure P2.20

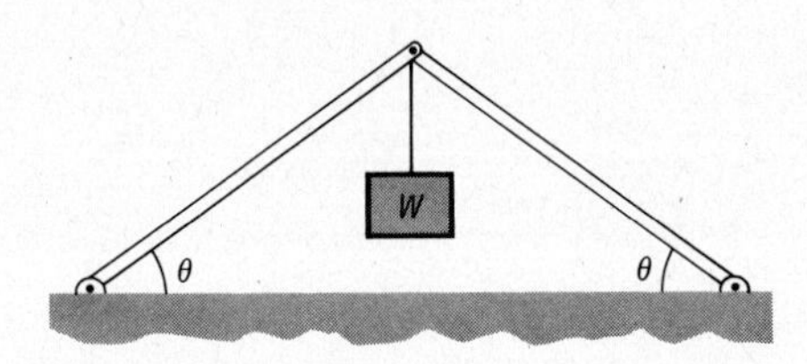

Figure P2.21

the strut. This force is equal in magnitude to the compression in the strut.)

*41. A bowling ball that weighs 60 N rests on the bottom of a groove as shown in Figure P2.22. How large are the forces that the walls of the groove exert on the ball? Assume the walls to be smooth so that the force each exerts is perpendicular to the wall.

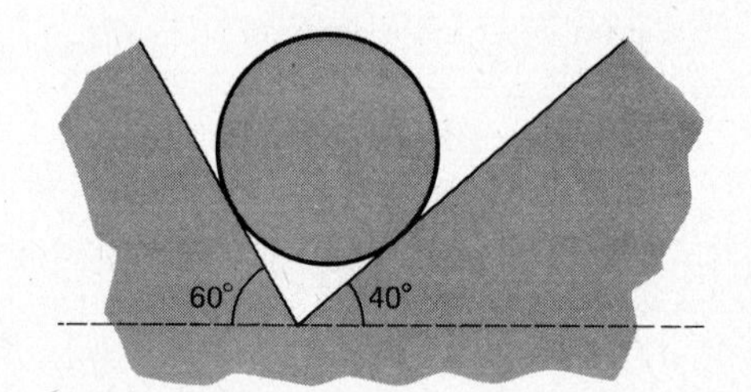

Figure P2.22

**42. Two equal and opposite forces (i.e., equal antiparallel forces) are called a *couple*. If the magnitude of each force is F and the separation of their lines of action is d, then they exert a torque with magnitude Fd about any pivot in their plane. Prove this fact.

**43. In Figure P2.23 the boom is uniform and weighs 50 N. Find the weight of the object W if $T_1 = 100$ N. Also find T_3.

**44. In Figure P2.23 the boom is uniform and weighs 70 N. How

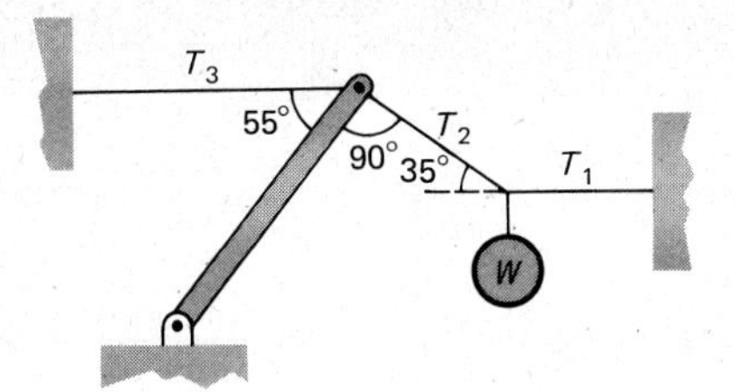

Figure P2.23

large can W be if the tension T_3 is not to exceed 200 N? Also, find the horizontal and vertical components of the force exerted by the boom on the hinge at the floor.

*45. A sealed oil drum is completely filled with oil. It is 1.40 m tall and 0.80 m in diameter. If you tilt the drum far enough, it will tip over when released. What is the largest angle to the vertical that you can tilt it before it will topple?

*46. A riding lawnmower has a wheelbase (side to side to outer edge of its tires) of 1.20 m. When a certain person rides it, the device and rider have a center of gravity that is in its middle and 90 cm above the ground. Consider a situation when the mower is moving on a horizontal line along the side of a hill. What is the maximum angle the hill can make with the horizontal before the mower will tip over?

Harold E. Edgerton, M.I.T., Cambridge

3 MOTION

Performance Goals

When you finish this chapter, you should be able to

1. Recall the equation $s = \bar{v}t$ and use it in simple situations.
2. Change the speed or velocity from one set of units to another set of units.
3. Distinguish between speed and velocity; give an example in which their magnitudes differ.
4. Use tabulated data to plot a graph x versus t. Using the graph, (a) describe the motion in words, (b) tell if the velocity is constant, (c) compute the average velocity for a certain portion of the motion, (d)

In the previous chapter we discussed objects that were at rest. We now begin a discussion of bodies in motion. The simplest type of motion takes place along a straight line. In this chapter we shall learn what is meant by speed, velocity, and acceleration. The way in which objects fall under the pull of gravity will also be examined. In the following chapter we shall see how accelerations are caused by forces.

compute the instantaneous velocity at a point on the graph.
5. Tell what is meant by the slope of a graph. Explain how the slope of the x versus t graph is related to average and instantaneous velocity.
6. Convert a quantity in a given set of units to new units, provided the necessary conversion factors are given.
7. Distinguish between average velocity and instantaneous velocity and state two cases where they are the same.
8. Define acceleration both in words and by equation. Give an example of negative acceleration. Use the change in velocity and the time taken for the change to compute acceleration. Select possible units for acceleration.
9. Write the five equations applicable to uniformly accelerated motion. Define each symbol in them. Point out which are valid even if a is not constant.
10. Use the five motion equations to solve simple problems involving uniform acceleration.
11. State the magnitude and direction of the acceleration experienced by a freely moving object near the surface of the earth. Define the symbol g in this situation.
12. Use the five motion equations to solve problems involving the vertical motion of a freely moving object.
13. Given the initial velocity of a frictionless projectile, find (a) how high it goes, (b) how far it would go along level land, (c) where it would be after a given time, (d) where it would strike a wall a known distance away.

3.1 Speed and Its Units

When you say that a car is moving at a speed of 50 kilometers per hour (50 km/h), we know that you mean to tell us the following: the car will go 50 km in 1 h if it maintains this speed. We can also infer that, at this speed, the car will go 100 km in 2 h. Or in 3 h the car will go $50 \times 3 = 150$ km. Reasoning in this way we conclude that the distance the car will go at this speed is given by

$$\text{Distance traveled} = (\text{speed}) \cdot (\text{time taken})$$

Or, after solving for the speed,

$$\text{Speed} = \frac{\text{distance traveled}}{\text{time taken}} \tag{3.1}$$

Notice that speed has no direction. It is a scalar. The car's speedometer measures how fast the car is moving, not its direction. It makes no difference if the car is going in a circle or along a straight line. The speed is still the distance traveled divided by the time the trip required.

Of course, the speed given by Equation 3.1 is an average speed. When you say it took you 3.0 h to travel a distance of 120 km, your average speed is

$$\text{Speed} = \frac{120\ \text{km}}{3.0\ \text{h}} = 40\ \text{km/h}$$

But this does not mean that your car was always moving at this speed. More likely you sometimes went faster and sometimes slower than this. The average of all your speeds is the number we have found, 40 km/h.

As shown by Equation 3.1, speed is always a distance divided by a time. For that reason its units are a distance unit divided by a time unit. A car goes 90 km/h. We walk 140 ft/min (feet per minute). A snail crawls 0.30 cm/h. Light travels with a speed of 300,000 km/s or 186,000 miles/s. These are all possible units of speed.

3.2 Units and Their Conversion

Since speeds can be measured in many different units, we often need to convert (or change) from one unit to another. For example, when a car is going 60 km/h, you might like to know its speed in feet per second. A simple method can be used to change units. Let us learn how to use it.

We first must know how units are related. All of us know that 1000 m = 1 km and that 1 min = 60 s. A few of these units are listed in Table 3.1. For a more complete listing, refer to the inside cover of the book.

Our method for changing units is based on the following fact: If you multiply

Table 3.1 UNITS[a]

1 inch (in.) = 0.0254 meter (m)
1 foot (ft) = 12 in.
1 mile = 5280 ft = 1610 m
1 m = 100 centimeters (cm)
1 kilometer (km) = 1000 m = 0.62 mile
1 minute (min) = 60 seconds (s)
1 hour (h) = 60 min = 3600 s
1 day = 24 h = 86,400 s

[a]A more complete table will be found on the inside cover of the book.

a quantity by unity, the quantity is not changed. The trick is to obtain the proper value for unity! To see how it is done, suppose that we want to carry out the conversion of days to hours. We know that 1 day = 24 h. To obtain unity from this equation we can divide each side by 1 day. (Notice that we use units just like numbers.) Then we have

$$1 = \frac{24\ \text{h}}{1\ \text{day}}$$

Or we could have divided through the equation by 24 h to obtain unity:

$$\frac{1\ \text{day}}{24\ \text{h}} = 1$$

In either case we obtain unity. If we multiply any quantity by either of these factors, the quantity will not be changed. Let us take a concrete example.

Suppose that we want to find how many days are equivalent to 2100 h. We have 1 day = 24 h, and so

$$\frac{1\ \text{day}}{24\ \text{h}} = 1$$

Multiplying the 2100 h by 1 does not change its value, so

$$2100\ \text{h} = 2100\ \cancel{\text{h}} \times \frac{1\ \text{day}}{24\ \cancel{\text{h}}} = 87.5\ \text{days}$$

Notice how the units cancel. Of course, 1 day = 24 h also gives

$$\frac{24\ \text{h}}{1\ \text{day}} = 1$$

We did not multiply by this because

$$2100\ \text{h} = 2100\ \text{h} \times \frac{24\ \text{h}}{\text{day}} = 50{,}400\ \text{h}^2/\text{day}$$

This is a true statement, but it does not tell us what we wanted to know, the time in days.

As another example, let us find a distance in inches if its value is 300 m. From Table 3.1 we see that

$$1 \text{ in.} = 0.0254 \text{ m}$$

We want to determine

$$300 \text{ m} = ? \text{ in.}$$

To cancel meters we need

$$\frac{1 \text{ in.}}{0.0254 \text{ m}} = 1$$

as our factor. Then

$$300 \text{ m} = 300 \cancel{\text{m}} \times \frac{1 \text{ in.}}{0.0254 \cancel{\text{m}}} = 11{,}810 \text{ in.}$$

Here is another example. Change 1440 lb/ft^2 to pounds per square inch, where 1 ft = 12 in.

$$1440 \text{ lb/ft}^2 = 1440 \frac{\text{lb}}{\cancel{\text{ft}^2}} \times \frac{1 \cancel{\text{ft}}}{12 \text{ in.}} \times \frac{1 \cancel{\text{ft}}}{12 \text{ in.}} = 10 \text{ lb/in.}^2$$

A speed unit conversion is often a little more complicated. For example, suppose we want to change 60 miles/h to a speed in feet per second. We know that 1 mile = 5280 ft and so

$$60 \text{ miles/h} = 60 \frac{\cancel{\text{miles}}}{\text{h}} \times \frac{5280 \text{ ft}}{1 \cancel{\text{mile}}} = 316{,}800 \text{ ft/h}$$

But we also know that 1 h = 3600 s, and so

$$60 \text{ miles/h} = 316{,}800 \frac{\text{ft}}{\text{h}} = 316{,}800 \frac{\text{ft}}{\cancel{\text{h}}} \times \frac{1}{3600} \frac{\cancel{\text{h}}}{\text{s}}$$
$$= 88 \text{ ft/sec}$$

3.3 Velocity Is a Vector

The speedometer on a car is named correctly. It measures speed, a scalar quantity. If we use a compass in addition to the speedometer, we can say "the car is going 40 km/h east." Now we are giving the *velocity* of the car. Notice that velocity has direction as well as magnitude. It is a vector quantity.

If an object moves from point A to B in time t, we define the average velocity of the object to be

$$\text{Average vector velocity} = \frac{\text{displacement vector from } A \text{ to } B}{\text{time taken}}$$

In symbols,

$$\bar{\mathbf{v}} = \frac{\mathbf{s}}{t} \quad \text{or} \quad \mathbf{s} = \bar{\mathbf{v}}t \tag{3.2a}$$

The bar over the "vee" is used to tell the reader that this is an average velocity.

For motion along a straight line, we can use + and − signs to indicate direction. If a car is going in the $+x$ direction at 30 km/h, its velocity is +30 km/h. But if it is going in the reverse direction, its velocity is −30 km/h. In cases where the direction of the velocity can be indicated by its sign, we shall write Equation 3.2a in the form

$$x = \bar{v}t \quad \text{or} \quad y = \bar{v}t \tag{3.2b}$$

If the motion is in either the negative x or y direction, $\bar{v}$ will be a negative number.

It is easy to show a case where the magnitudes of average velocity and average speed are not the same. Consider the case shown in Figure 3.1. The boy throws a ball up to a height of 4.9 m and catches it as it falls. The time taken is 2.0 s. Note that the total distance traveled by the ball from start to finish is 4.9 m up plus 4.9 m down, which equals 9.8 m. We then have

$$\text{Average speed} = \frac{\text{distance}}{\text{time}} = \frac{9.8\text{ m}}{2.0\text{ s}} = 4.9\text{ m/s}$$

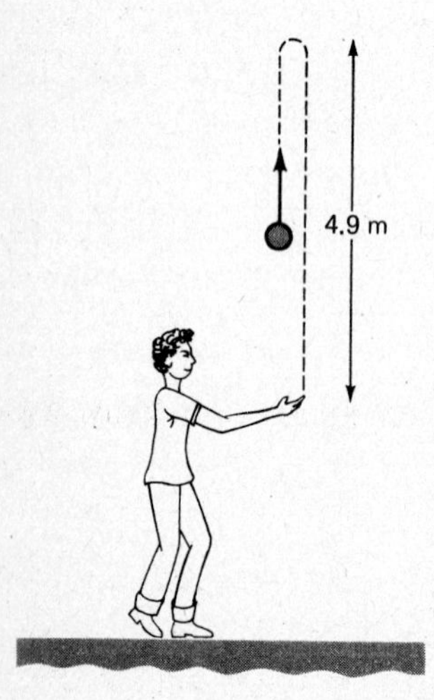

Figure 3.1 The boy catches the ball at the same point from which he threw it. The average velocity of the ball is zero for the trip even though the average speed is 4.9 m/s.

But the average velocity is quite different. To find it, notice that the boy catches the ball at the same point from which he threw it. The displacement vector from starting point (the boy's hand) to end point (the boy's hand) has zero length. The two points are the same. Therefore

$$\text{Average velocity} = \frac{\text{displacement vector}}{\text{time}} = \frac{0\text{ m}}{2.0\text{ s}} = 0\text{ m/s}$$

This is a very special case, of course. But it does show clearly that velocity and speed need not be equal in magnitude. The average velocity was zero in this case because the velocity was positive (when going up) and negative (when coming

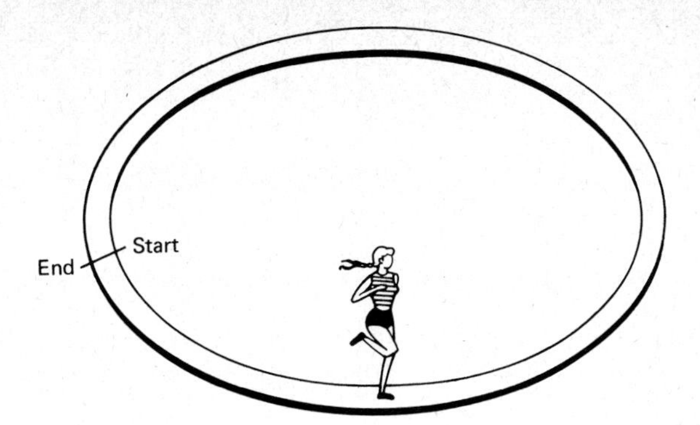

Figure 3.2 How can the girl's average speed be found? Her average velocity?

down). We obtain zero when we take the average of these equal positive and negative velocities.

■ **EXAMPLE 3.1** The runner shown in Figure 3.2 takes a time of 28 s to circle the 200-m track. What was her average speed? Her average velocity?

Solution The total distance she traveled was 200 m. She did this in a time of 28 s. Therefore, from Equation 3.1,

$$\text{Average speed} = \frac{\text{distance traveled}}{\text{time taken}}$$

$$= \frac{200\text{ m}}{28\text{ s}} = 7.1\text{ m/s}$$

To find the average velocity, we need the displacement vector. The runner circled the track, so her start and end point were the same. Therefore, the displacement (the vector from beginning point to end point) has zero magnitude. We have

$$\text{Average velocity} = \frac{\text{displacement vector}}{\text{time taken}}$$

$$= \frac{0\text{ m}}{28\text{ s}} = 0\text{ m/s}$$

Here, too, we see a big difference between the magnitude of speed and velocity. ■■

3.4 Showing Motion by Graphs

Often we would like to draw pictures to show motion. By means of them we can make the situation more clear. But realistic pictures are hard to draw. For this reason technicians draw graphs to show many situations. In many ways graphs are more valuable than even the most realistic picture. Let us see how motion is represented by graphs.

Suppose that a car is going along a straight road as shown in Figure 3.3(a). If we had the proper equipment, we could record its motion. We might measure its

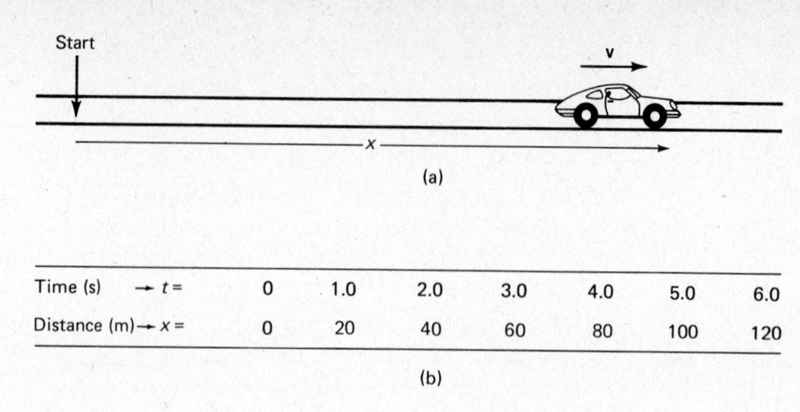

Time (s) → t =	0	1.0	2.0	3.0	4.0	5.0	6.0
Distance (m) → x =	0	20	40	60	80	100	120

(b)

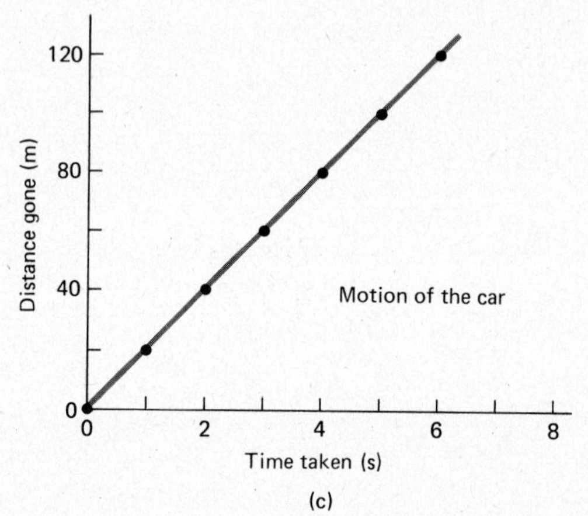

Figure 3.3 The data for the motion of the car in (a) are recorded in the table in (b) and plotted in the graph in (c).

distance x from a certain point and the time since the car passed the point. Our data might be that given in the table in part (b) of the figure. Notice that the timing clock was started as the car passed the starting point. The distance x of the car from this point is then recorded at each second after the clock was started. (Fairly complex equipment might be needed for such measurements.)

If you look at the data in the table, you will see that the car goes 20 m in each second. We know from this that the car's speed is 20 m/s. Because the motion is straight along a line, this is also the magnitude of the car's velocity. Remember, we can show the direction of the velocity by plus and minus signs for motion along a straight line.

The data in the table can be plotted on a graph. This is done in part (c) of Figure 3.3. Each point on the graph represents one x, t pair of data from the table. (To check your understanding, locate on the graph the point that uses the data $x = 80$ m when $t = 4.0$ s from the table.) We can learn much from a glance at such a graph.

The graph of Figure 3.3 is redrawn in Figure 3.4. Notice the two little triangles we have drawn there (each formed by the graph line and two solid, double-pointed arrows). Each tells us that in 1 s the car moves a distance of 20 m. In other words, these triangles tell us the velocity of the car. It is 20 m/s. Clearly, the velocity of the car is constant during the time pictured here. During each

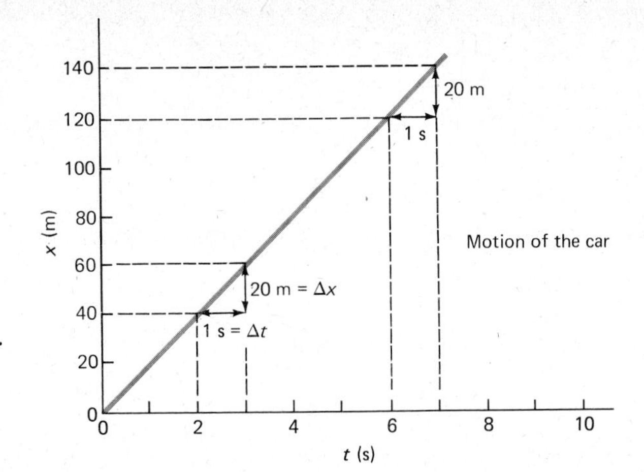

Figure 3.4 The car moves 20 m in 1 s. Its speed is constant at 20 m/s. Note that the distance versus time graph is a straight line if the speed is constant. What is the slope of this line?

second shown the car moves a distance of 20 m. Whenever the x versus t graph is a straight line, the distance traveled is the same each second. So a straight line x versus t graph means that the velocity is constant.[1]

We often describe a graph by what we call its *slope.* The slope measures how fast the graph line is *rising.* (If the graph line is *falling,* its slope is negative.) Precisely, we define the slope in the following way. We draw a triangle on the graph, such as either of the small ones shown. Notice the labels Δt and Δx (read "delta tee" and "delta ex"). The vertical side of the triangle (Δx, or 20 m in this case) is called the *rise* of the graph. The horizontal side of the triangle (Δt, 1.0 s in this case) is called the *run* of the graph. Then, by definition,

$$\text{Slope} = \frac{\text{rise}}{\text{run}} \tag{3.3}$$

In the present case the slope of the graph line is

$$\text{Slope} = \frac{\Delta x}{\Delta t} = \frac{20\text{ m}}{1.0\text{ s}} = 20\text{ m/s}$$

But this is the velocity of the car. We therefore find that the slope of the x versus t graph line is the magnitude of the velocity. This is true, of course, only in the case of motion along a straight line. Let us now make use of this fact in an example.

■ **EXAMPLE 3.2** The motion of a car is shown by the graph of Figure 3.5. Explain what the car was doing during this time.

[1] Those of you who know enough math will recognize that the equation of this straight line is $x = vt$.

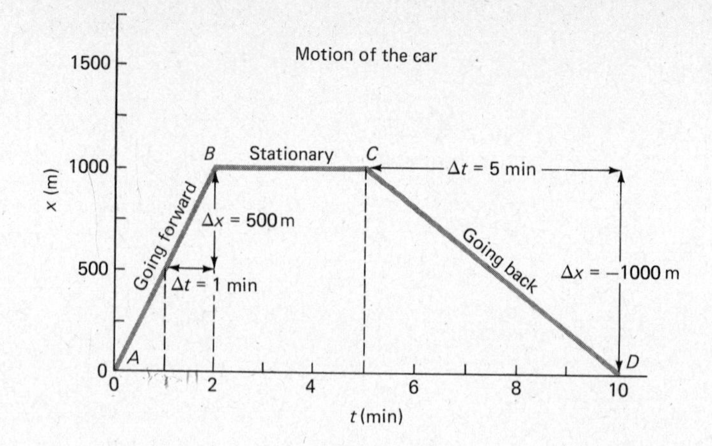

Figure 3.5 The car moved 1000 m in the $+x$ direction, sat there for 3 min, then went back to the starting point. What was its velocity during each portion of the trip?

Solution Read the legend for Figure 3.5. By examining the graph, we can see that what it says is true. During the first 2.0 min the car went 1000 m, as we see from portion *AB* of the graph. Because portion *AB* is a straight line, we know that the car moved with constant velocity. We can find this velocity from

$$\bar{v}_{AB} = \frac{\text{displacement}}{\text{time}} = \frac{1000\text{ m}}{2.0\text{ min}} = 500\text{ m/min}$$

Or we could use the fact that

$$\bar{v}_{AB} = \text{slope} = \frac{\text{rise}}{\text{run}} = \frac{500\text{ m}}{1.0\text{ min}} = 500\text{ m/min}$$

Both methods give the same answer.

For the next 3 min, portion *BC* of the graph, the x value (or position) of the car did not change. The car must be standing still. Its velocity is zero. Or, using the slope, we have

$$\bar{v}_{BC} = \text{slope} = \frac{\text{rise}}{\text{run}} = \frac{0\text{ m}}{3.0\text{ min}} = 0$$

Again both methods give the same result.

Finally, in the last 5 min, the car returned to $x = 0$, the starting point. Its *velocity* is negative, because it is now going in the $-x$ direction. The slope shows this clearly, because the "rise" is now really a drop of 1000 m. We have for portion *CD* of the trip

$$\bar{v}_{CD} = \text{slope} = \frac{\text{rise}}{\text{run}} = \frac{-1000\text{ m}}{5.0\text{ min}} = -200\text{ m/min}$$

Because the motion is in the $-x$ direction, the velocity is negative. The fact that we have a straight line in portion *CD* of the graph tells us that the velocity was constant at this value. Be sure that you understand our interpretation of this graph. ■■

3.5 Instantaneous Velocity

Until now we have been dealing mostly with objects that move with constant velocity. Usually, though, velocities are not constant. A car seldom travels at constant velocity for long. Let us now examine a case where the velocity is changing.

Suppose that a car starts from rest at $x = 0$. It speeds up until it reaches the speed limit. It then travels on with constant speed. Its distance versus time graph is shown in Figure 3.6.

If we recall that the car's velocity is equal to the slope of the curve, we can easily interpret the graph. During the first few seconds the slope is small. But as time goes on, the slope gradually increases. This tells us that the car gradually speeds up. After about 9 s the slope is constant. The curve becomes a straight line. We conclude that the car's velocity is then constant.

To find the car's constant velocity at these longer times, we find the slope of the final straight line. From the upper triangle we see

$$\bar{v}(\text{at long times}) = \frac{\Delta x}{\Delta t} = \frac{5.0\ \text{m}}{1.00\ \text{s}} = 5.0\ \text{m/s}$$

Earlier, however, the car was going more slowly.

To find its average velocity during the time interval from $t = 5$ to $t = 6$ s, we use the triangle shown at that time. We have for the time interval $5 \leq t \leq 6$ s,

$$\bar{v} = \frac{\Delta x}{\Delta t} = \frac{3.0\ \text{m}}{1.00\ \text{s}} = 3.0\ \text{m/s}$$

(Read the symbol $\leq$ as "less than or equal to.") In this same way, we could compute the velocity in any region along the curve.

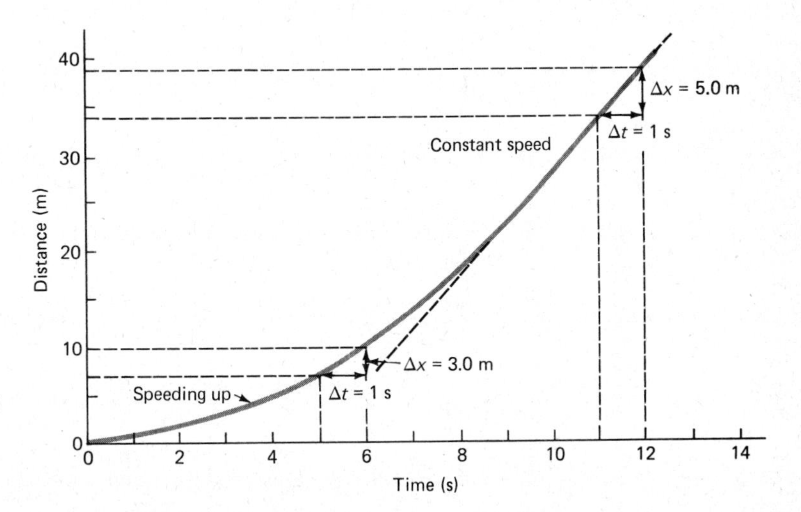

Figure 3.6 The average velocity during the interval $5 \leqslant t \leqslant 6$ was 3.0 m/s. How could we find the instantaneous velocity at $t = 5.10$ s?

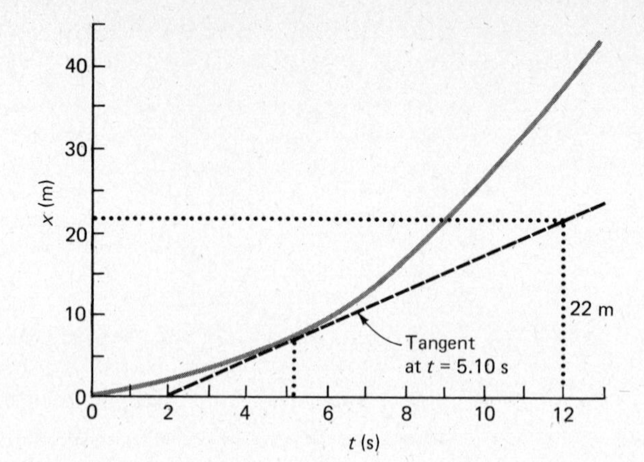

Figure 3.7 The instantaneous velocity at $t = 5.10$ s is the slope of the dashed line.

But still we are computing only an average velocity. If we are asked what the velocity was at $t = 5.10$ s, what could we say? Because the average velocity between $t = 5$ and $t = 6$ s was 3.0 m/s, the velocity at $t = 5.10$ s would be close to this value. To obtain a more accurate value, we should find the slope of the curve exactly at $t = 5.10$ s. That is to say, we should find the slope of a straight line that is tangent to the curve at $t = 5.10$ s. This line, drawn tangent to the curve, is shown in Figure 3.7, where the curve of Figure 3.6 is redrawn. The slope of the tangent line is (22 m) ÷ (10.0 s) = 2.2 m/s. The car's velocity at 5.10 s was 2.2 m/s in magnitude.

As we see, if the velocity of an object is continuously changing, then its x versus t graph is a curve. If we are to find the object's velocity at a particular instant, then we need the slope of the curve at that instant. Its velocity at time $= t$ is the slope of a straight line drawn tangent to the curve at that particular value of t. We call the velocity of the object at a particular instant its *instantaneous velocity.* Of course, if the object is traveling with constant velocity, its average and instantaneous velocities are the same.

3.6 Acceleration

When a car speeds up, we say that it accelerates. We shall now learn the technical definition of acceleration and how to use it. Acceleration is a measure of how fast an object's velocity is changing. If a car is increasing its velocity, it is accelerating. For the present time we shall be dealing with straight-line motion. As before, we shall take care of the vector nature of velocity by using plus and minus signs.

We define the average acceleration of an object in the following way.

$$\text{Acceleration} = \frac{\text{change in velocity}}{\text{time taken for the change}} \tag{3.4}$$

For example, suppose a car is going eastward at 30 km/h and speeds up to 40 km/h in a time of 5.0 s. Then its acceleration is

$$\text{Acceleration} = \frac{\text{change in velocity}}{\text{time}} = \frac{10 \text{ km/h eastward}}{5.0 \text{ s}}$$

$$= 2.0 \frac{\text{km/h}}{\text{s}} \text{ eastward}$$

Its eastward acceleration is 2.0 km/h per second. The car speeds up by 2.0 km/h each second.

We can write the definition of acceleration, **a**, in symbols as follows. The change in velocity is the final velocity, $\mathbf{v}_f$, minus the original velocity, $\mathbf{v}_0$.

$$\mathbf{a} = \frac{\mathbf{v}_f - \mathbf{v}_0}{t} \quad \text{or} \quad \mathbf{v}_f - \mathbf{v}_0 = \mathbf{a}t \tag{3.5}$$

As we see from the second form, the change in velocity is simply the acceleration multiplied by time.

This definition is quite easy to use. Suppose that a car starts from rest. (This means that $v_0 = 0$.) It accelerates along a straight road and 10 s later is going 15 m/s. (This means that $t = 10$ s and $v_f = 15$ m/s.) Our definition of acceleration tells us that

$$a = \frac{15 \text{ m/s} - 0}{10 \text{ s}} = 1.5 \text{ m/s per second}$$

$$= 1.5 \text{ m/s}^2$$

We usually do not write 1.5 m/s per second (or 1.5 m/s/s), but instead write 1.5 m/s^2. But the first form tells us at once what we mean by a. It tells us that the velocity changes by 1.5 m/s each second. That is what we mean by acceleration. It is how much the velocity changes in unit time; it is 1.5 m/s each second in the present case. In this situation the car's velocity was zero at $t = 0$ s. At $t = 1$ s the velocity was $0 + 1.5 = 1.5$ m/s. At $t = 2$ s it was $1.5 + 1.5 = 3.0$ m/s. At $t = 3$ s it was $3.0 + 1.5 = 4.5$ m/s, and so on. The car speeds up 1.5 m/s each second.

What is the acceleration of a car that is slowing down? You might say that the car is not accelerating, but decelerating. But our technical definition is still used. If a car going 30 m/s ($v_0 = 30$ m/s) slows and stops ($v_f = 0$) in a time of 5 s, what is its acceleration? According to our definition it is

$$a = \frac{v_f - v_0}{t} = \frac{0 - 30 \text{ m/s}}{5 \text{ s}} = -6 \text{ m/s}^2$$

Notice the minus sign. In the present case it means that the car is slowing down. The velocity is decreasing by 6 m/s each second. We shall soon see that algebraic signs are very important when dealing with motion along a straight line.

The units used to measure accelerations are important. Because acceleration is change in velocity divided by time, its units are those of velocity divided by time. Typical examples are

$$\begin{array}{ll} \text{m/s} \div \text{s} & \rightarrow \text{m/s}^2 \\ \text{ft/s} \div \text{s} & \rightarrow \text{ft/s}^2 \\ \text{km/h} \div \text{s} & \rightarrow \text{km/h/s} \\ \text{cm/day} \div \text{year} & \rightarrow \text{cm/day/year} \end{array}$$

The last example might be used for the acceleration of a glacier. If it moves 5 cm/day one year and 9 cm/day two years later, its acceleration is

$$a = \frac{(9 \text{ cm/day}) - (5 \text{ cm/day})}{2 \text{ years}} = 2 \text{ cm/day/year}$$

Its velocity increased by 2 cm/day each year. Although certain units are preferred over others (for reasons we shall see later), there are many possible units for acceleration. All are a velocity unit divided by a time unit.

3.7 Uniformly Accelerated Motion

A very simple type of motion results if the acceleration of an object is constant. We then say that the motion is *uniformly accelerated*. The average velocity of the object $\bar{v}$ is quite simple in such a case. Suppose that an object starts with an initial velocity v_0. If the velocity changes in a steady way to a new final value v_f, then its average velocity $\bar{v}$ during this time is easily found. To see what it is, refer to Figure 3.8.

In Figure 3.8 we see a graph that shows how the velocity changes with time. Because it is a straight line, the velocity changes by the same amount each second. This tells us the acceleration is constant (that is, uniform). But the average velocity during the time shown is simply the midpoint velocity. You can

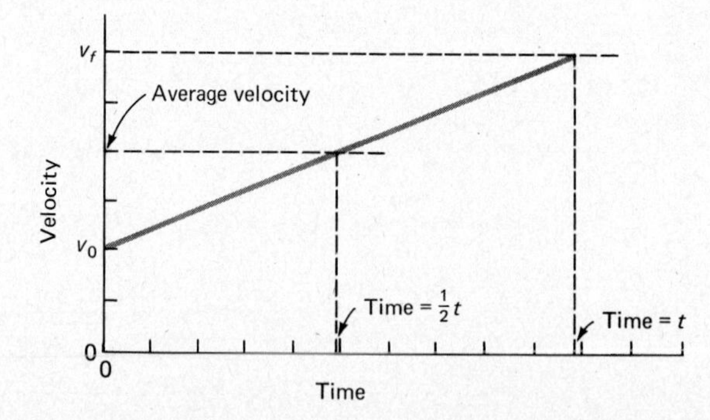

Figure 3.8 If the acceleration is constant (i.e., if the velocity increases the same amount each second), then the average velocity is the midpoint velocity: $\bar{v} = (v_0 + v_f)/2$.

easily convince yourself that the midpoint velocity is the initial plus the final divided by two. (Try it using a few values.) Therefore the average velocity is

$$\bar{v} = \frac{v_0 + v_f}{2} \qquad (3.6)$$

Of course, this is true only if the v versus t graph is a straight line. It is true only if the acceleration is constant.

Let us now summarize the equations that apply to motion along a straight line. We had, from Equation 3.2b, that

$$x = \bar{v}t \qquad (3.2b)$$

From the definition of acceleration, $a = (v_f - v_0)/t$, we had

$$v_f - v_0 = at \qquad (3.5)$$

And now, for uniformly accelerated motion, we have

$$\bar{v} = \tfrac{1}{2}(v_0 + v_f) \qquad (3.6)$$

These three equations make it possible for us to solve a large number of problems involving motion.

Even though these three equations are sufficient for solving problems, two others are convenient to have. They are obtained by combining these three. They contain nothing new, but they aid us in carrying out the arithmetic. To obtain them, we proceed as follows.

Start with Equation 3.2b, $x = \bar{v}t$. We can replace t by solving for t in Equation 3.5:

$$v_f - v_0 = at \rightarrow t = \frac{v_f - v_0}{a}$$

We can also replace $\bar{v}$ by $\frac{1}{2}(v_0 + v_f)$. Doing this, we find that

$$x = \frac{(v_0 + v_f)(v_f - v_0)}{2a} = \frac{v_0 v_f - v_0^2 + v_f^2 - v_f v_0}{2a}$$

After simplifying, we have

$$v_f^2 - v_0^2 = 2ax \qquad (3.7)$$

The second equation is also obtained by starting with $x = \bar{v}t$. We replace $\bar{v}$ by $\frac{1}{2}(v_0 + v_f)$ and then replace v_f by $v_0 + at$ from Equation 3.5. We obtain

$$\begin{aligned} x &= \bar{v}t \\ &= \tfrac{1}{2}(v_0 + v_f)t \\ &= \tfrac{1}{2}(v_0 + v_0 + at)t \end{aligned}$$

and finally

$$x = v_0 t + \tfrac{1}{2}at^2 \tag{3.8}$$

Let us now list these very important equations. The last three of them apply only to uniformly accelerated motion. We have

$$\begin{aligned} &\text{(a) } x = \bar{v}t \\ &\text{(b) } v_f = v_0 + at \\ &\text{(c) } \bar{v} = \tfrac{1}{2}(v_0 + v_f) \\ &\text{(d) } v_f^2 - v_0^2 = 2ax \\ &\text{(e) } x = v_0 t + \tfrac{1}{2}at^2 \end{aligned} \tag{3.9}$$

In the next section we shall see how to use these five equations. Because they are of widespread use, you should memorize them.

3.8 Solution of Motion Problems

You will find that the solution of most problems involving uniformly accelerated motion along a line is easy *if* you are systematic. It is necessary to follow a definite plan. We give you here a method that many people use.

1. Sketch the situation.
2. Locate the starting point and the end point for the motion you wish to deal with.
3. Decide which direction along the line you will take as positive. Displacements, velocities, and accelerations will be positive in that direction.
4. Write down which of the following quantities you know: x, a, t, $\bar{v}$, v_0, v_f.
5. Select the appropriate equation(s) from Equations 3.9 to find the unknowns required.
6. Solve for the unknowns.

Let us now apply this stepwise method to a few examples.

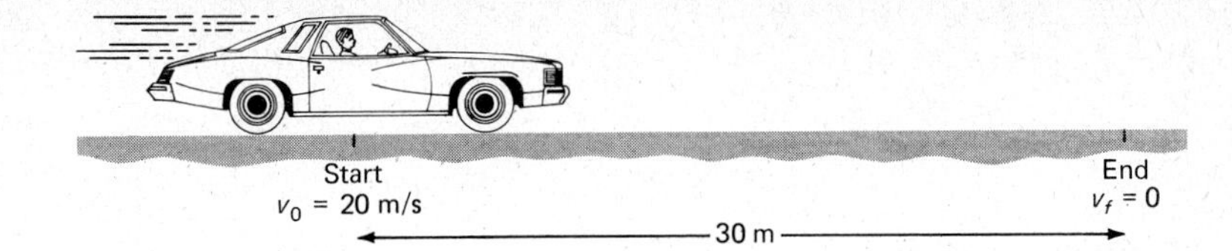

Figure 3.9 Sketch for Example 3.3.

EXAMPLE 3.3 A car going 20 m/s skids to a stop in a distance of 30 m. Assume uniform deceleration. Find the time taken to stop and the acceleration.

Solution We sketch the situation in Figure 3.9. The starting point is the position of the car at the start of the skid. The end point is where it stops. We take the direction of motion as the positive direction. The known quantities are

$$\begin{aligned} v_0 &= 20 \text{ m/s} \\ v_f &= 0 \\ x &= 30 \text{ m} \\ t &= ? \\ a &= ? \end{aligned}$$

We wish to find t and a.

Whenever both v_0 and v_f are known, it is wise to find $\bar{v}$ at once. We have

$$\bar{v} = \tfrac{1}{2}(v_0 + v_f) = \tfrac{1}{2}(20 \text{ m/s}) = 10 \text{ m/s}$$

To find t, we look for the simplest of the Equations 3.9 that contains t as the only unknown. In this case we can use (a).

$$x = \bar{v}t \rightarrow 30 \text{ m} = (10 \text{ m/s})\, t$$

Dividing each side of the equation by 10 m/s gives

$$t = \frac{30 \text{ m}}{10 \text{ m/s}} = 3.0 \frac{\text{m}}{\text{m/s}} = 3.0\, \cancel{\text{m}} \cdot \frac{\text{s}}{\cancel{\text{m}}} = 3.0 \text{ s}$$

as the time taken to skid to a stop. (Notice what we did with the fraction m/s in the denominator. Recall that the rule to invert the denominator and multiply is actually the result of multiplying numerator and denominator by the inverse, s/m in this case.)

We can now find a from either Equation (b), (d), or (e). Of course we use (b) because it is the simplest. Then

$$v_f = v_0 + at$$

gives

$$0 = (20 \text{ m/s}) + a(3.0 \text{ s})$$

Solving gives

$$a = -\frac{20 \text{ m/s}}{3.0 \text{ s}} = -6.7 \left(\frac{\text{m}}{\text{s}}\right)\left(\frac{1}{\text{s}}\right)$$

or

$$a = -6.7 \text{ m/s}^2$$

What does the negative sign tell us? ■■

■ **EXAMPLE 3.4** A car is timed as it passes two posts that are 30 m apart. It takes 2.0 s. If the car's speed is 40 km/h at the first post, what is its speed at the second? Assume uniform acceleration.

Solution The situation is sketched in Figure 3.10. Take the posts as start and end points. Use the direction of motion as positive. Then the knowns are

$$v_0 = \left(40\,\frac{\text{km}}{\text{h}}\right)\left(\frac{1}{3600}\,\frac{\text{h}}{\text{s}}\right)\left(1000\frac{\text{m}}{\text{km}}\right) = 11.1 \text{ m/s}$$
$$t = 2.0 \text{ s}$$
$$x = 30 \text{ m}$$
$$v_f = ?$$

BACK INJURY FROM LIFTING

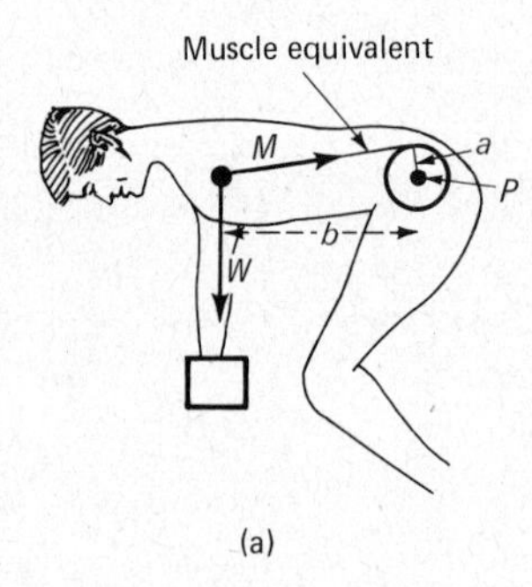

(a)

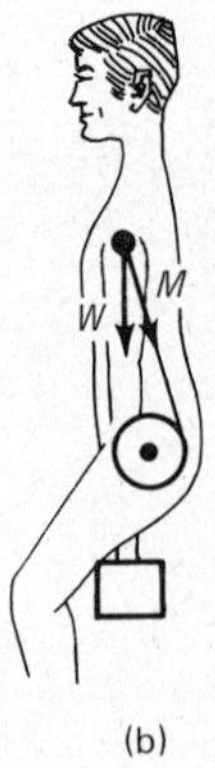

(b)

You have probably been warned that there is a right and a wrong way to lift a weight. We can understand this warning by use of the principles of the previous chapter. In the figure we show the two lifting methods. It is easy to see why method (a) is very bad. To do so, we need only apply the laws of equilibrium to the portion of the body above the hip joint. In this simplified discussion we shall assume the person to be thin enough so the weight of the person's body can be neglected.

Three basic forces act on the upper part of the body. The weight W of the object being lifted is one. In addition, a complicated set of muscles supports the body above the hip. To a crude approximation they can be replaced by the muscle equivalent shown. It pulls with a force M. A third force acts on the spine at point P at the hip. Because we shall take torques about point P, this third force has zero lever arm. It therefore exerts no torque.

If we take torques about P, the force W has a lever arm b as shown. The lever arm for M is a. Because the system is near equilibrium, the torque equation gives

$$Wb = Ma$$

We therefore find the tension in the back muscle M to be given by

$$M = W\left(\frac{b}{a}\right)$$

In a typical case, the ratio b/a might be about 20. As a result, the tension in the muscle will be about $20W$. This tension pulls nearly lengthwise along the spine. It tries to compress the spine between the shoulder and hip. As a result, if a 150-N weight is being lifted, the spine is subjected to a force of about 3000 N. Such a large force could cause injury to the spine.

However, the situation is much different when one uses the lifting method (b). There the lever arms for M and W are about equal. As a result, the compression in the spine is about $2W$. This is about ten times less than in method (a). It is clear from this example why method (a) leads to back injury much more frequently than does method (b).

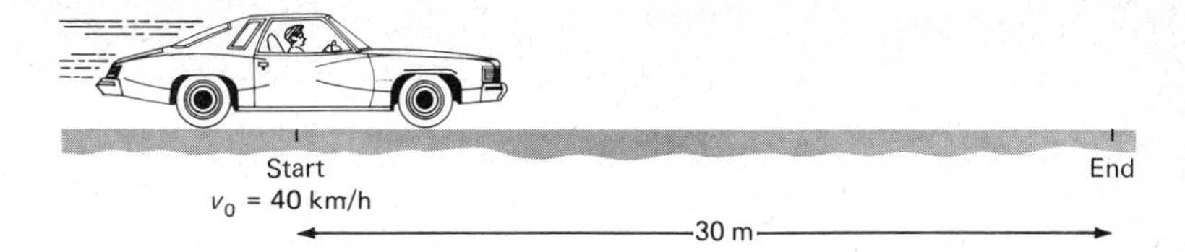

Figure 3.10 Sketch for Example 3.4

(Notice that we changed kilometers per hour to meters per second because t is in seconds and x is in meters.) We wish to find v_f.

None of the equations in Equations 3.9 gives v_f in terms of v_0, t, and x. We therefore look for some other quantity we can calculate. From Equation (a) we can obtain

$$\bar{v} = \frac{x}{t} = \frac{30 \text{ m}}{2.0 \text{ s}} = 15 \text{ m/s}$$

Now we can use (c) to find $\bar{v} = \frac{1}{2}(v_0 + v_f)$ or $2\bar{v} = v_0 + v_f$. Then

$$v_f = 2\bar{v} - v_0 = 30 \text{ m/s} - 11.1 \text{ m/s} = 18.9 \text{ m/s}$$

Often we find it necessary (or easier) to calculate another quantity before we find the one we want. ■■

3.9 Acceleration Due to Gravity in Free-Fall

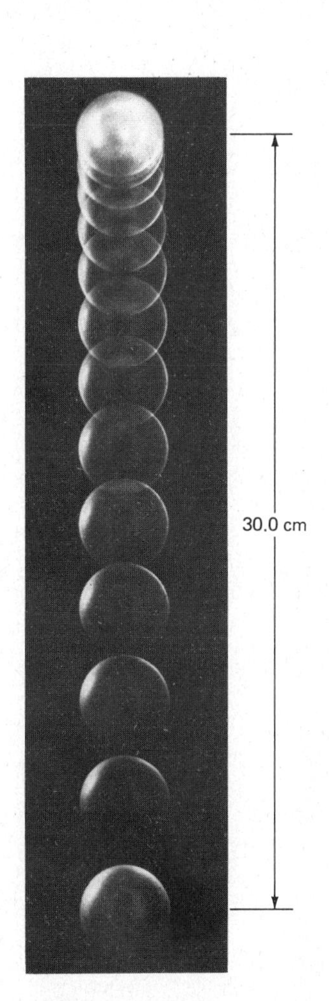

Figure 3.11 The position of a falling ball is shown at equal time intervals. A strobe light flashes each 0.0177 s to show the ball's position. (Courtesy of E.I. du Pont de Nemours and Co., Inc.)

The most common example of uniformly accelerated motion is that of a free-falling object. In Figure 3.11 we see the positions of a ball at equal times after it has been dropped. As we notice, the ball has zero velocity at the instant it is released. But it speeds up as it falls. We can measure its velocity as a function of time. It is found that the ball speeds up by 9.8 m/s each second. In other words, its acceleration is 9.8 m/s^2. (Of course, the direction of the acceleration is downward.) In the British system this acceleration is 32.2 ft/s^2.

Galileo, who lived from 1564 to 1642, was the first to measure how objects fall on the earth. He concluded that all freely falling objects on the earth accelerate downward with $a = 9.8 \text{ m/s}^2 = 32 \text{ ft/s}^2$. Notice that he was talking about *freely* falling objects. This does not apply to a feather falling through the air. Air resistance is strong enough for such a light object that it cannot be considered to fall freely. The acceleration we are talking about applies only to objects for which forces other than gravity are negligible.

Careful measurements have confirmed Galileo's discovery. Freely falling objects have a downward acceleration of about 9.8 m/s^2 (32.2 ft/s^2) on the earth's surface. We call this acceleration the acceleration due to gravity. It is represented by the symbol g. The value of the gravitational acceleration is different on the moon than on earth. There it is 1.6 m/s^2. Even on earth it varies from place to place, ranging from 9.787 to 9.808 m/s^2. The reason for these variations will be given in Chapter 4.

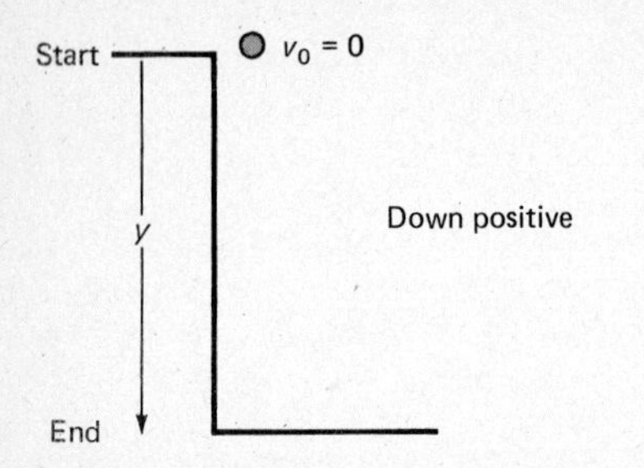

Figure 3.12 Sketch for Example 3.5.

EXAMPLE 3.5 A stone is dropped from a bridge to the river below. It takes 3.0 s for the stone to strike the water. How high is the bridge above the river?

Solution The situation is sketched in Figure 3.12. Let us take the downward direction as positive. Because the displacement vector is directed downward, it will be positive. Also, the acceleration due to gravity is downward. It will therefore be positive in this case. We have

$$v_0 = 0$$
$$a = g = 9.8 \text{ m/s}^2$$
$$t = 3.0 \text{ s}$$
$$y = ?$$

Notice that we call the displacement y instead of x. We do this because the y coordinate, not the x coordinate, is usually taken vertically.

We can find y by use of Equation 3.9e. It is

$$y = v_0 t + \tfrac{1}{2}at^2$$

Substitution gives

$$y = 0 + \tfrac{1}{2}(9.8 \text{ m/s}^2)(3.0 \text{ s})^2$$

We then find

$$y = (\tfrac{1}{2})(9.8)(9.0) \text{ m} = 44 \text{ m}$$

EXAMPLE 3.6 A ball is thrown straight up with a speed of 20 m/s. How high does it go? The problem is sketched in Figure 3.13(a).

Solution The starting point is where the ball leaves the thrower's hand. As end point we take the highest point it reaches, point B in Figure 3.13(b). At that point the ball has stopped and is about to fall back down. Let us take up as positive. Then the initial velocity and the displacement vector (from start to end) will be positive, because they are up. But the acceleration due to gravity is downward, and so it will be negative. We know that

$$v_0 = 20 \text{ m/s}$$
$$v_f = 0 \qquad \text{(because it stops at the top)}$$
$$a = g = -9.8 \text{ m/s}^2$$

We want y, the displacement. To find it, we can use Equation 3.9d with x replaced by y. Then

$$v_f^2 - v_0^2 = 2ay$$

gives

$$0 - 400 \text{ m}^2/\text{s}^2 = -2(9.8 \text{ m/s}^2)y$$

and so

$$y = 20.4 \text{ m}$$

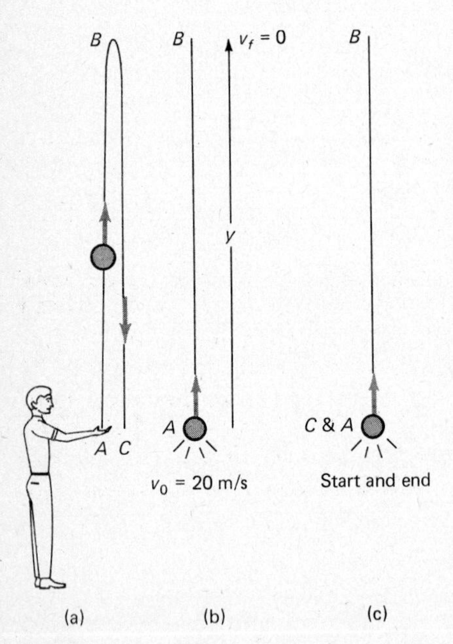

Figure 3.13 Points A and C should actually coincide. Because the ball stops at point B, its velocity is zero there.

In this problem we chose up as the positive direction. You might wish to do it again by taking down as positive. You should obtain $y = -20.4$ m. Why should y be a negative number in this case? ■■

■ **EXAMPLE 3.7** How fast must a ball be thrown straight upward if it is to return to the thrower in 3.0 s? How fast is it going just before it is caught?

Solution The situation is shown in Figure 3.13(c). There are two ways to do this problem depending on where we take the start and end points. Let us take the start point at the position where the ball is thrown (A) and the end point at where it is caught (C). These two points are really the same and so the *displacement from start point to end point* is zero.[2] Take the upward direction as positive. We have as knowns

$$\begin{aligned} y &= 0 \\ a &= g = -9.8 \text{ m/s}^2 \qquad \text{(notice the negative sign)} \\ t &= 3.0 \text{ s} \\ v_0 &= ? \end{aligned}$$

We wish to find v_0. This can be done by use of Equation 3.9e:

$$y = v_0 t + \tfrac{1}{2}at^2$$

which is

$$0 = v_0(3.0 \text{ s}) + \tfrac{1}{2}(-9.8 \text{ m/s}^2)(3.0 \text{ s})^2$$

This gives

$$-3\, v_0 \text{ s} = -\tfrac{1}{2}(9.8)(9.0) \text{ m}$$

from which

$$v_0 = 14.7 \text{ m/s}$$

Its final velocity can be found from

$$v_f = v_0 + at = 14.7 \text{ m/s} + (-9.8 \text{ m/s}^2)(3.0 \text{ s}) = -14.7 \text{ m/s}$$

Notice that its speed at a given point on the way down is the same as its speed at the same point on the way up. ■■

■ **EXAMPLE 3.8** A ball is thrown upward from the ground with a velocity of 30 ft/s. How long will it take to reach a point 10.0 ft above the ground *on its way down*?

Solution The situation is sketched in Figure 3.14. Let us take up as positive with the start point at the ground and the end point 10.0 ft above the ground. Then we know

$$\begin{aligned} y &= 10.0 \text{ ft} \\ a &= g = -32 \text{ ft/s}^2 \qquad \text{(since up is taken as positive)} \\ v_0 &= 30 \text{ ft/s} \\ t &= ? \end{aligned}$$

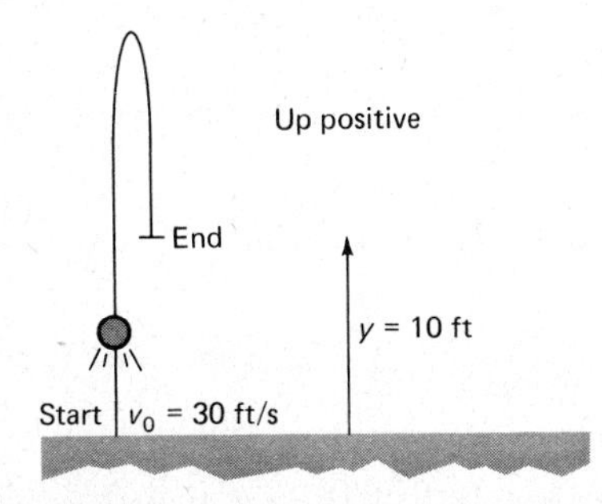

Figure 3.14 Sketch for Example 3.8.

[2] Notice that the end point is taken just before the ball is caught. Once the ball touches the catcher's hand, its acceleration will no longer be g. Our equations will no longer apply to it.

We could find t by use of the equation $y = v_0t + \frac{1}{2}at^2$. But this would require use of the quadratic formula. Instead, let us first use $v_f^2 - v_0^2 = 2ay$ to find v_f. Then

$$v_f^2 = (30 \text{ ft/s})^2 + 2(-32 \text{ ft/s}^2)(10.0 \text{ ft})$$

which gives

$$v_f^2 = 900 \text{ ft}^2/\text{s}^2 - 640 \text{ ft}^2/\text{s}^2$$

or

$$v_f = \sqrt{260} = \pm 16.1 \text{ ft/s}$$

Remember, when you take a square root, the answer can be either plus or minus. Because our final velocity is downward (in the negative direction), we must take $v_f = -16.1$ ft/s. (The positive sign represents the ball on its way up.)

We can now find t by use of $v_f = v_0 + at$. This gives

$$-16.1 \text{ ft/s} = 30 \text{ ft/s} - (32 \text{ ft/s}^2)t$$

(Notice how important the negative signs are here.) Solving for t, we find

$$t = 1.44 \text{ s}$$

■■

3.10 Projectile Motion

In previous sections we have been discussing the straight up-and-down motion of an object. The pull of gravity on it caused the object to accelerate downward with free-fall acceleration g. What happens if an object is shot at an angle to the vertical?

Once the object is free, it is subject to only one force, the force of gravity. (We are ignoring air friction for the present.) The object therefore accelerates downward with acceleration g under this force. Let us see what this means in a simple situation.

Refer to Figure 3.15. Suppose that a ball is thrown parallel to the earth as shown. Its horizontal velocity is v_0. Once it is free, and before it hits the ground, no horizontal force acts upon it. Therefore its horizontal velocity will remain unchanged. Its x component (horizontal) velocity will remain v_0 until the ball hits the ground. This is indicated in Figure 3.15.

When the ball is first thrown, it has zero vertical velocity. But the vertical pull of gravity will accelerate it downward at a rate of 9.8 m/s every second. The vertical (or y component) velocity will increase by 9.8 m/s each second. As shown in the figure, the ball falls downward with increasing vertical velocity. At the end of 1 s, its vertical velocity is 9.8 m/s. At the end of 2, 3, and 4 s, the vertical velocity has increased to 2, 3, and 4 times 9.8 m/s, respectively. All this happens while the ball moves horizontally with *unchanging* velocity v_0.

We see from this that a projectile in flight is doing two things at once: (1) It is flying horizontally with constant speed, and (2) it is moving up or down with

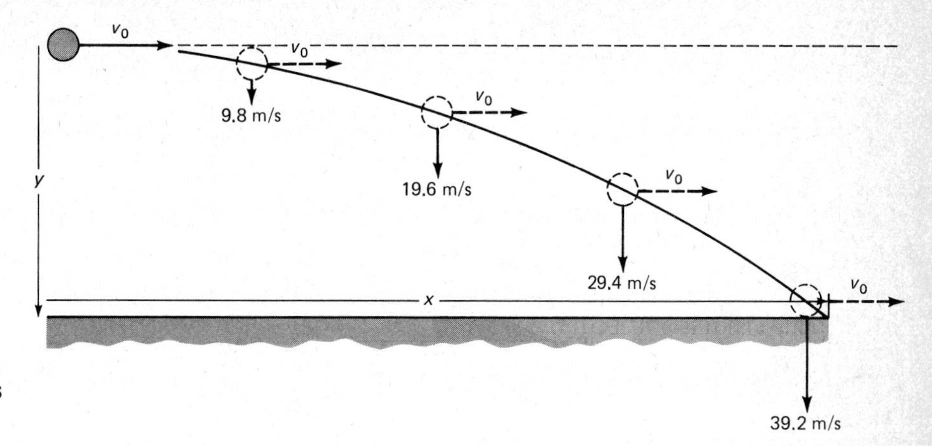

Figure 3.15 The ball follows the solid line. Its horizontal and vertical velocities are given at 1-s intervals.

acceleration g. Once we recognize this, solution of projectile problems is easy. We simply split each problem into two problems. One involves horizontal motion at constant velocity. For it $v_{0x} = \overline{v}_{fx} = \overline{v}_x$ and $a_x = 0$. Hence the motion equation of importance for the horizontal problem is $x = \overline{v}t$.

The vertical part of the motion is also nothing new to us. It is exactly the same as the free-fall motion we have been discussing. We describe it by use of the usual five motion equations. Let us illustrate this by some examples.

■ **EXAMPLE 3.9** Suppose that the ball in Figure 3.15 is thrown horizontally with speed of 30 m/s. It is 1.50 m above the level ground when released. How far will it go before it hits the ground?

Solution We immediately divide the problem into two portions, vertical and horizontal. The start point of the problem is when the ball leaves the thrower's hand. The end point is just before the ball hits the ground.

Vertical: Notice that at the beginning the ball was moving horizontally. Its initial velocity component in the y direction was zero. Therefore, $v_{0y} = 0$. Taking down as positive, the knowns are

$$v_{0y} = 0$$
$$a_y = g = 9.8 \text{ m/s}^2$$
$$y = 1.50 \text{ m}$$

Notice that the start point is 1.50 m above the end point. Therefore, the y displacement is 1.50 m downward.

We can find how long (t) the ball took to reach the ground. To do this we can use $y = v_{0y}t + \frac{1}{2}a_yt^2$. Putting in the known values gives

$$1.50 \text{ m} = 0 + \tfrac{1}{2}(9.8 \text{ m/s}^2)t^2$$

Solving for t^2, we find

$$t^2 = \frac{1.50 \text{ m}}{4.9 \text{ m/s}^2} = 0.306 \text{ s}^2$$

From which

$$t = 0.55 \text{ s}$$

The ball is in the air 0.55 s. Let us now go to the horizontal part of the problem.

Horizontal: At the start, the ball was traveling horizontally with speed $v_{0x} = 30$ m/s. But no horizontal force exists on the ball if we ignore friction. Therefore its horizontal speed will remain unchanged. We therefore know $v_{0x} = v_{fx} = \bar{v}_x = 30$ m/s. Also, because of this, $a_x = 0$. In addition, the vertical part of the problem told us that the ball is in the air for 0.55 s. This same time applies to the horizontal problem, of course. Therefore the knowns are

$$\begin{aligned} v_{0x} &= v_{fx} = \bar{v}_x = 30 \text{ m/s} \\ a_x &= 0 \\ t &= 0.55 \text{ s} \\ x &= ? \end{aligned}$$

To find the horizontal distance x, we can use $x = \bar{v}_x t$. Then

$$\begin{aligned} x &= (30 \text{ m/s})(0.55 \text{ s}) \\ &= 16.6 \text{ m} \end{aligned}$$

This is the distance x shown in the diagram. ■■

■ **EXAMPLE 3.10** A ball is thrown with velocity 20 m/s at an angle of 37° above the horizontal. How high does it go? How far has it moved horizontally at that time?

Solution The situation is shown in Figure 3.16. Notice in the figure that we have already found the x and y components of the velocity. We need them to solve the two different motions. They are $v_{0x} = v_0 \cos 37°$ and $v_{0y} = v_0 \sin 37°$. We then have

$$\begin{aligned} \textit{Horizontal: } v_{0x} &= v_{fx} = \bar{v}_x = 16 \text{ m/s} \\ a_x &= 0 \end{aligned}$$

If we knew t, the time taken to get to the top of the path, we could use $x = \bar{v}_x t$ to find the desired horizontal distance. We need to examine the vertical part of the problem to find t.

$$\begin{aligned} \textit{Vertical: } v_{0y} &= 12 \text{ m/s} &&\text{(up is taken as positive)} \\ a_y &= -9.8 \text{ m/s}^2 &&\text{(}g\text{ is always down)} \\ v_{fy} &= 0 &&\text{(since the vertical velocity decreases to zero at the top)} \end{aligned}$$

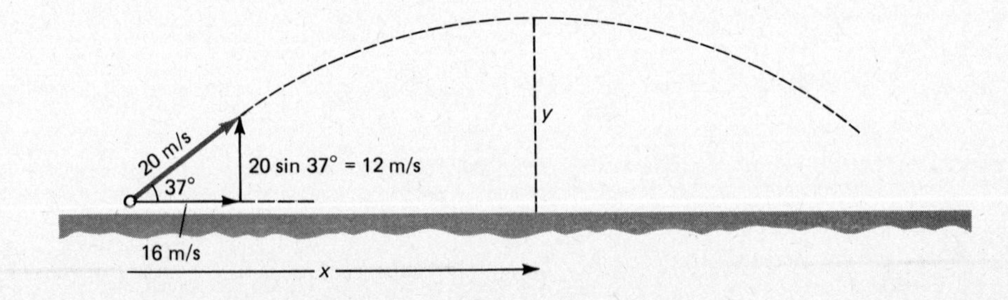

Figure 3.16 Find x and y.

We can find $\overline{v}$ at once from

$$\overline{v}_y = \tfrac{1}{2}(v_{0y} + v_{fy})$$

Then

$$\overline{v}_y = 6.0 \text{ m/s}$$

We can use $v_{fy} = v_{0y} + a_y t$ to find t, the time taken to get to the top. We have

$$0 = (12 \text{ m/s}) + (-9.8 \text{ m/s}^2)t$$

which gives $t = 1.22$ s. Before going back to the horizontal problem, we can find y from $y = \overline{v}_y t = (6.0 \text{ m/s})(1.22 \text{ s}) = 7.3$ m. This is how high the ball goes.

To find how large x (shown in Figure 6.5) is, we can now go back to the horizontal part of the problem. We now know $t = 1.22$ s, and so

$$x = \overline{v}_x t = (16 \text{ m/s})(1.22 \text{ s}) = 20 \text{ m}$$

Notice how simple the problem is *if* you keep its two parts separate. ■■

■ **EXAMPLE 3.11** The motorcycle rider shown in Figure 3.17 wishes to know how fast he must be going to make the jump shown. If air friction can be ignored, what is the necessary speed?

Solution Again we split the problem into two parts.

Horizontal: $v_{0x} = v_{fx} = \overline{v}_x = v_0 \cos 20°$
$a_x = 0$
$x = 20$ m

Then $x = \overline{v}_x t$ tells us that

$$20 \text{ m} = (v_0 \cos 20°)t \qquad \text{(a)}$$

We have two unknowns, v_0 and t, so we need to go on to the vertical problem.

Vertical: $y = 0$ (he starts and lands at same height)
$a_y = -9.8 \text{ m/s}^2$ (up is taken as positive)
$v_{0y} = v_0 \sin 20°$

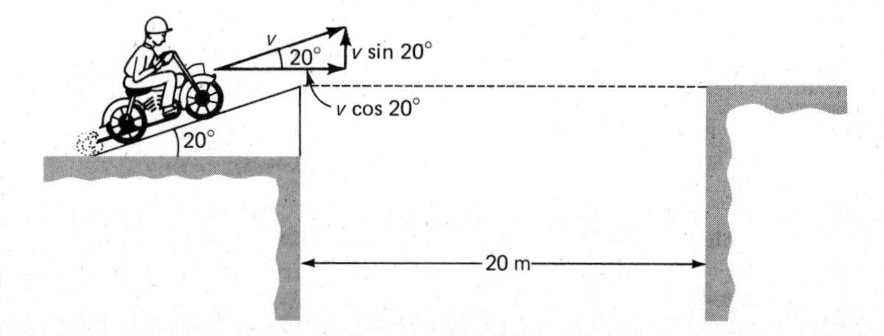

Figure 3.17 To be successful, how fast must the motorcycle be moving?

If we write $y = v_{0y}t + \frac{1}{2}a_yt^2$, we have

$$0 = (v_0 \sin 20°)t + (\tfrac{1}{2})(-9.8\ \text{m/s}^2)t^2$$

Dividing the equation by t and transposing gives

$$(4.9\ \text{m/s}^2)t = v_0 \sin 20°$$

Solving this for t, we have

$$t = \frac{v_0 \sin 20°}{(4.9\ \text{m/s}^2)} \qquad \text{(b)}$$

We can now combine the results of the horizontal and vertical problems. Substituting (b) for t in (a) we have

$$20\ \text{m} = \frac{v_0^2 \cos 20° \sin 20°}{4.9\ \text{m/s}^2}$$

From the trig tables, $\cos 20° = 0.94$ and $\sin 20° = 0.34$. Making use of these values, we find

$$v_0^2 \frac{(0.94)(0.34)}{4.9\ \text{m/s}^2} = 20\ \text{m}$$

From this,

$$v_0 = 17.5\ \text{m/s}$$

This is the minimum speed the motorcycle must have if it is to complete the jump.
This example is more complicated than most, because of the algebra. Neither portion of the problem alone was able to furnish us with an answer. We had to solve two equations simultaneously. In spite of this difficulty, we followed the same procedure. The horizontal and vertical portions of the problem were dealt with separately. ■■

■ **EXAMPLE 3.12** Find the angle at which a projectile should be shot for maximum range.

Solution Calling the initial velocity v_0 and the angle of projection θ, we have

VERTICAL	HORIZONTAL
$v_{0y} = v_0 \sin \theta$	$v_{0x} = v_0 \cos \theta$
$a = -g$	

In the vertical problem the projectile rises and then descends to its original level. Its final speed is the same as its initial speed, but the direction of motion is opposite. Therefore $v_{fy} = -v_0 \sin \theta$. The time the projectile is in the air is given by $v_{fy} = v_{0y} + at$ to be

$$t = \frac{v_f - v_0}{a} = \frac{(-v_0 \sin \theta) - (v_0 \sin \theta)}{-g} = \frac{2v_0 \sin \theta}{g}$$

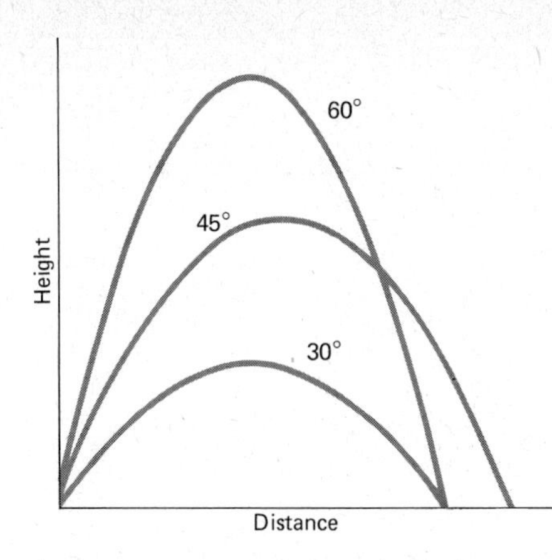

Figure 3.18 Range depends on the angle of projection for a projectile without air friction.

Then the range R is given by

$$R = \overline{v}_x t = v_0 \cos\theta \left(\frac{2v_0 \sin\theta}{g}\right) = \left(\frac{{v_0}^2}{g}\right)(2 \sin\theta \cos\theta)$$

You may remember from trigonometry that $2 \sin\theta \cos\theta = \sin(2\theta)$. Making use of this we find

$$R = \left(\frac{{v_0}^2}{g}\right)\sin(2\theta)$$

This tells us the range is largest when $\sin(2\theta)$ takes on its maximum value, namely 1.00. This occurs when $2\theta = 90°$. Therefore, for maximum range $\theta = 45°$. The projectile goes farthest when shot at a 45° angle above the horizontal. What happens at other angles is shown in Figure 3.18. ■■

Summary

The average speed of an object is defined as the distance traveled divided by the time taken. Its units are a distance unit divided by a time unit. Speed is a scalar quantity.

The average velocity of an object during a displacement is the displacement vector divided by the time taken. It is a vector. In equation form, $\overline{\mathbf{v}} = \mathbf{s}/t$.

The velocity of an object moving along a straight line is equal to the slope of the x versus t graph. When this graph is a straight line, the velocity is constant. The slope of a line is equal to the "rise" divided by the "run."

As the name indicates, instantaneous velocity is the velocity at a particular instant. It is equal to the slope of the tangent line for the x versus t curve at that particular time.

Average acceleration is the change in the velocity vector divided by the time taken for the change. In symbols, $\mathbf{a} = (\mathbf{v}_f - \mathbf{v}_0)/t$. It is a vector quantity. The units of acceleration are a velocity unit divided by a time unit.

For uniformly accelerated motion, $\overline{v} = \frac{1}{2}(v_0 + v_f)$.

Problems involving uniformly accelerated motion can be solved by the six-step method outlined in Section 3.8. The method uses the following five motion equations. The last three of them are true only if a is constant.

$$\begin{aligned}
&\text{(a) } x = \overline{v}t \\
&\text{(b) } v_f = v_0 + at \\
&\text{(c) } \overline{v} = \tfrac{1}{2}(v_0 + v_f) \\
&\text{(d) } {v_f}^2 - {v_0}^2 = 2ax \\
&\text{(e) } x = v_0 t + \tfrac{1}{2}at^2
\end{aligned} \tag{3.9}$$

The acceleration of a freely falling object is called the acceleration due to gravity. We represent it by g. On the earth $g = 9.8\ \text{m/s}^2 = 32.2\ \text{ft/s}^2$. Its direction is downward.

Projectile motion can best be dealt with as two independent motions: horizontal motion at constant speed and vertical motion with acceleration g.

Questions and Exercises

1. Estimate the speeds of the following: a fast-walking man, a fast-moving ant, a high school track star running the 100-m dash, an Olympic miler running the mile.
2. The 400-m dash at a certain school is run by making two laps around a 200-m oval track. A runner takes 60 s to run 400 m. What is her average speed? Average velocity?
3. Figure P3.1 shows the y versus t graph for a ball thrown straight up into the air. Describe in words what the ball is doing at each point, A, B, C, and D. In each case point out how the slope confirms what you say.

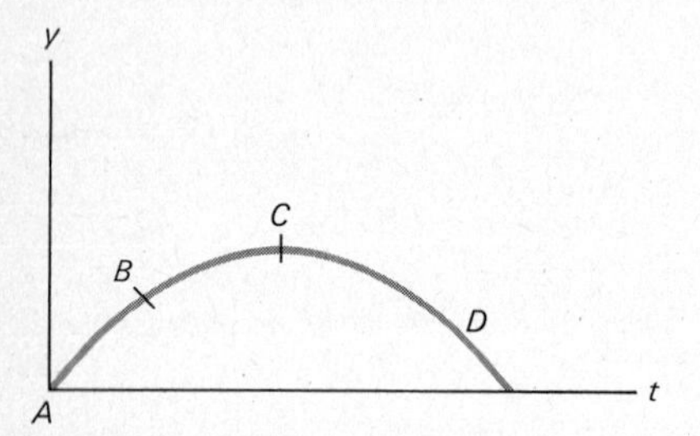

Figure P3.1

4. A billiard ball rolls across a table in a straight line, hits the edge, and bounces straight back. Sketch an x versus t graph for the ball, assuming its speed to be constant. Repeat for the case where the ball slows and stops.
5. A ball is thrown straight up in the air and it then returns to its starting point. Sketch the following graphs for the ball: (a) y versus t; (b) v versus t; (c) a versus t. Restrict the graphs to the time when the ball is moving freely.
6. Can an object have an acceleration greater than zero even though its velocity is zero? Explain your answer.
7. On the moon the acceleration of free-fall is only about 1.6 m/s^2 and is directed toward the center of the moon. About how high should a boy be able to throw a ball there if he can throw it 10 m high on earth?
8. Devise a method for determining the maximum acceleration of a car as it accelerates from rest. Would you expect to obtain the same acceleration for the first 10 s as for the first 5 s?
9. A car going at 5 m/s hits a tree head-on and stops. What data would you need if you wanted to compute its average deceleration?
10. The strobe light for Figure 3.11 flashes every 0.0177 s. Use the picture to determine the acceleration due to gravity.
11. A plane is flying horizontally high above the earth. To bomb a target on the earth, should it release the bomb when it is right above its target? Explain.
12. A smart and quick monkey hangs by his hands from a tree limb. He sees a distant hunter, using an old gun, sight along the barrel of the gun and fire. At the instant he sees the flash of fire from the gun, he drops from the limb to escape the approaching bullet. In spite of his efforts, he is hit by the bullet even though the hunter aimed well. Explain why his plan did not work.
13. If you want to hit a distant object with a rifle, do not aim the gun with the object in line with the gun's barrel. How should the gun be aimed? In practice, how is this taken care of?
14. A railroad boxcar is moving at constant speed along a straight horizontal track when a screw falls from the ceiling of the car. Where will the screw hit the floor as compared to a point right below its original position?
15. If air friction is negligible, where during its flight does a projectile have its minimum speed? Its maximum speed?
16. What factors influence how far a broad jumper will jump? Which of these factors can be influenced by the way the jumper leaves the ground?

Problems

1. The distance an airplane must go to travel from Chicago to London is 6360 km. How long would the trip take at 800 km/h?
2. From Moscow to Cairo is 2900 airline km. What speed must an airplane have to make the trip in 3.0 h?
3. The distance from earth to moon is 385,000 km. A typical flight to the moon takes about 64 h. What is the average speed of the spaceship?
4. The orbital distance around the earth for a spaceship is about 42,000 km. The time taken for the orbiting ship to make one pass around the earth is about 90 min. What is the average speed of the ship during one pass? Its average velocity? (Give answers in kilometers per hour.)
5. In 1970 the Indianapolis 500 was won by Al Unser. He traveled the 500 miles in 3 h 13 min. What was his average speed? What was his average velocity?
6. Using the facts that 1 yd = 36 in. and 1 in. = 2.54 cm, find the number of centimeters in 10 yd.
7. Using the facts that 1 mile = 5280 ft, 1 ft = 12 in., and 1 in. = 2.54 cm, find the number of centimeters in 10 miles.

8. Using the facts that 1 km = 0.62 mile, 5280 ft = 1 mile, and 1 h = 3600 s, find the speed in feet per second that is equivalent to a speed of 100 km/h.
9. Knowing that 1 mile = 5280 ft, 1 km = 0.62 mile, and 3600 s = 1 h, find the speed in kilometers per hour equivalent to a speed of 90 ft/s.
*10. Using the conversion factors found inside the book cover, make the following conversions: (a) 30 km/h to feet per second, (b) 2 cm/day to meter per second, (c) 3 in./year to centimeters per hour, (d) 32 ft/s^2 to meters per second per second, (e) 9.8 m/s^2 to inches per hour per hour.
*11. The following data were obtained for the position of a ball as a function of time as it rolls along a straight track.

t (s) →	0.60	2.10	3.30	7.20	10.00
x (cm) →	2.90	7.40	11.00	22.70	31.10

Plot these data and determine the average velocity of the ball during this time interval. Was the velocity constant?
*12. A wire supports a weight at its end. Because the wire is highly strained, it elongates slowly with time. The position of its end as a function of time is given in the following table:

t (h) →	0	1.60	2.70	3.70	4.50
x (cm) →	4.032	4.076	4.113	4.143	4.167

Using a graph, find the average velocity of the end during this time interval. Was the velocity constant?
13. The motion of an object is graphed in Figure P3.2. Find the velocity of the object at points A, B, and C.

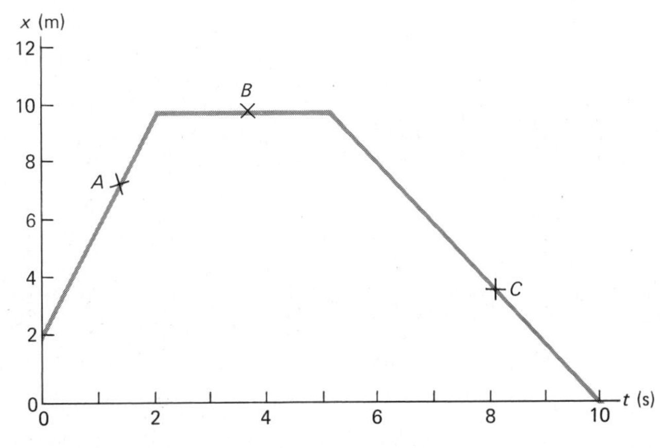

Figure P3.2

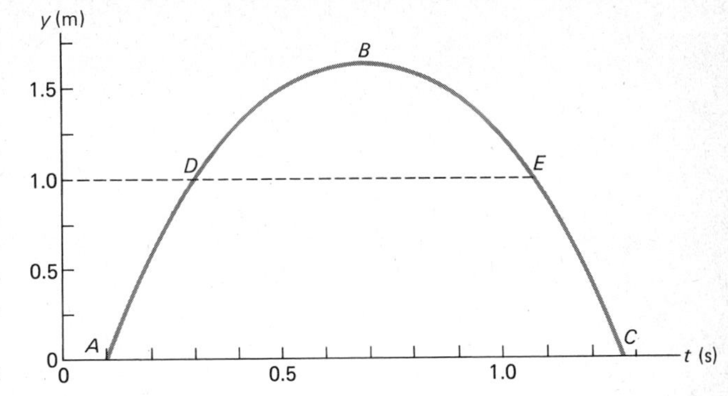

Figure P3.3

*14. Find the velocity at points A, B, and C for the object whose motion graph is shown in Figure P3.3. This is the motion of an object thrown straight upward.
*15. A ball is thrown straight upward and its motion is shown in the graph of Figure P3.3. Find the velocity of the ball at points B, D, and E.
*16. Find the average velocity for the object and time interval shown in Figure P3.3. In Figure P3.2 the velocity at A is 4.0 m/s, whereas it is −2.0 m/s at C. Why can't we say that the average velocity is $\frac{1}{2}[4 + (-2)] = 1$ m/s?
*17. For the motion shown in Figure P3.2, what was (a) the average speed during the 10 s shown and (b) the average velocity during the 10 s?
*18. For the motion shown in Figure P3.3, what was (a) the average velocity during the interval from D to E and (b) the average speed for the interval from A to C?
*19. A car is moving at a speed of 12.0 m/s eastward on a straight road. Inside the car a bug is flying eastward at a speed of 2.0 m/s relative to the car. (a) How fast is the bug moving relative to the road? (b) Repeat for the case in which the bug is flying westward at a speed of 2.0 m/s relative to the car.
*20. Repeat Problem 19 for the case in which the bug is flying at a speed of 3.0 m/s relative to the interior of the car, but it is moving parallel to the axle of the car.
**21. A car is going along a straight road at 70 km/h when it begins to pass a train on a track parallel to the road. The train is going 30 km/h in the same direction as the car. If the train is 120 m long, how long, in seconds, does it take the car to pass the train?
**22. Repeat Problem 21 for the case in which the car and train are going in opposite directions.

*Problems marked with an asterisk are not as easy as the unmarked ones.
**Problems marked with a double asterisk are somewhat more difficult than the average.

23. It takes 8.0 s for a certain car to accelerate from rest to a speed of 20 m/s. What is its average acceleration?
24. A spaceship is moving along a straight-line path at 5000 km/h. It is then accelerated to a speed of 6500 km/h by a 3-s burst from its rockets. What was its average acceleration?
25. A car moving at a speed of 20 m/s slows uniformly to a stop in a time of 5.0 s. What was its average acceleration?
26. A bullet moving at 200 m/s hits a bag of sand and comes to a stop in a time of 0.0030 s. Assuming the acceleration to be uniform, how large was it?
27. The maximum deceleration a car can have on dry pavement is about 6 m/s^2. (a) What is the minimum time a car would take to stop if it is going at 20 m/s? Ignore the driver's reaction time. (b) Repeat for $a = -20$ ft/s^2 and $v_0 = 70$ ft/s.
28. No matter how powerful a car's motor, it can accelerate no faster than the friction force between pavement and wheels allows. For this reason a typical car's maximum acceleration is about 8 m/s^2. (a) What is the shortest time a car would require to accelerate from rest to a speed of 20 m/s? (b) Repeat for $a = 25$ ft/s^2 and $v = 60$ ft/s.
29. A car moving at 20 m/s has its brakes slammed on and stops in a distance of 40 m. (a) What was the car's deceleration if it is assumed to be uniform? (b) How long did it take to stop?
30. A bullet moving at 400 m/s hits a tree and penetrates 0.150 m before stopping. Assuming uniform deceleration, what was its average deceleration? How long did it take to stop?
31. A car accelerating uniformly is going 6.0 m/s when it passes one checkpoint. It passes a second checkpoint 100 m away 10.0 s later. (a) What is its acceleration? What is its speed at the second checkpoint? (b) Repeat for a speed of 20 ft/s and a distance of 300 ft between checkpoints.

*32. A car moves along a straight road. Its velocity versus time graph is shown in Figure P3.4. (a) What was its acceleration in interval *AB*? (b) What was its acceleration in interval *CD*? (c) How far did it go during the time from *A* to *B*? (d) How far did it go during the time from *C* to *D*?

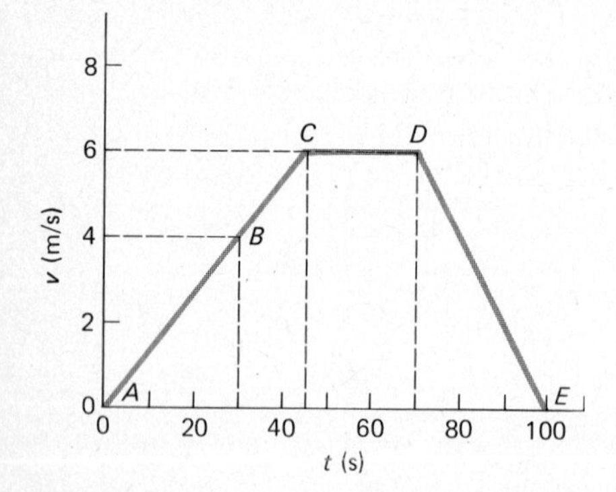

Figure P3.4

*33. Figure P3.4 shows the velocity of a car as it moves in the $+x$ direction. (a) What was the car's acceleration during interval *BC*? (b) During *DE*? (c) How far did the car go during the interval *A* to *E*?
34. A driver of a car going 20 m/s notices a dangerous intersection 100 m ahead. He immediately begins to slow the car uniformly and reaches the intersection 6.0 s later. How large was his deceleration and how fast was he going at the intersection?

**35. A car is going down the road at a speed of 90 km/h in a 50-km/h zone. As the car passes a corner, a police cruiser sitting there notices it. Five seconds later, the cruiser begins accelerating at 5.0 m/s^2. Assuming that he is able to maintain this acceleration, how far from the corner will the police officer catch the car? How fast will the cruiser then be going? Is this realistic?

*36. Two trains are approaching each other on the same track. Both are going 20 m/s. When they are 200 m apart, the brakeman on one sees the other and applies its brakes while the other continues along. If the deceleration of the braking train is 2.0 m/s^2, how fast will it be going when they collide?
37. (a) How long does it take for a stone to fall from a bridge to the water 13.0 m below? How fast is the stone going just before it hits the water? (b) Repeat for a bridge 40 ft above the water.
38. A reluctant swimmer is hanging by his fingertips from a high diving board 10 m above the water. A friend pries him loose. (a) How long does it take him to reach the water? How fast is he going when he hits? (b) Repeat for a distance of 30 ft.
39. (a) How fast must a ball be thrown if it is to rise to a height of 20 m? (b) How long will it take to reach that height?
40. A stone is thrown down from a bridge that is 30 m above a river. It takes 2.0 s for the stone to hit the water. How fast was the stone thrown?
41. A boy throws a ball straight upward and the ball returns to the boy's hands after a time of 3.0 s. (a) How fast was the ball thrown? (b) How high did it rise?
42. Using a slingshot, a girl shoots a pebble straight upward with a speed of 18.0 m/s. (a) How long does it take the pebble to return to her? (b) How high did the pebble go?

*43. A stone is thrown up from a bridge that is 30 m above a river. It takes 3.0 s for the stone to hit the water. How fast was the stone thrown?

*44. While standing on a bridge that is 20 m above the river below, a boy throws a ball upward with a speed of 15 m/s. How long will it take for the ball to reach the water (assuming that it misses the bridge on the way down)? How fast will the ball be going just before it hits?

*45. A ball is thrown in a vertical direction with such a velocity that it is 40 m below its starting point after a time of 8 s has elapsed. Find the direction and magnitude of the ball's initial velocity.

**46. A boy throws a ball straight up alongside a building with speed 20 m/s. As the ball is coming back down, a girl reaches out a window 5.0 m above the boy and catches it. How long a time went by between the throwing and catching of the ball? How high did the ball go?

*47. A balloonist is 30 m above the ground and rising at a speed of 6.0 m/s when she drops a coin from the balloon. (a) How fast is the coin moving as it reaches the ground? (b) How long does it take for the coin to reach the ground? (c) What was the velocity of the coin relative to the ground 0.60 s after being released?

*48. Repeat Problem 47 for a case in which the coin is tossed upward from the balloon with an initial speed of 3.0 m/s relative to the balloon.

49. A ball is thrown horizontally with a speed of 15 m/s at a point 2.0 m above the level ground. How far will it go before it hits the ground?

50. A ball is thrown horizontally from the top of a 20-m-high building. It lands 40 m from the base of the building. How long was it in the air, and how fast was it thrown?

51. What is the range of a projectile fired across level ground with a velocity of 100 m/s at an angle of 30° above the horizontal?

52. When it is 1000 m above the level ground, a dive bomber is following a path at an angle of 37° below the horizontal and has a speed of 200 m/s. It then releases a bomb. Where does the bomb land in relation to the point on the ground directly below the original position of the plane?

53. A boy standing on the ground throws a stone with velocity 20 m/s at an angle of 30° above the horizontal. It hits the wall of a nearby building at a position 3.0 m above the level at which it was thrown. How far is the wall from the boy? (Two answers are possible. Give both. One represents the stone on its way up, the other on the way down.)

**54. An object is to be shot from the level ground at an angle of 37° above the horizontal. How fast must it be shot if it is to remain in the air 4.0 s? How far from its starting point will it be when it hits the ground?

**55. The range of a projectile along flat ground is 20 m when the projectile is shot at an angle of 53° above horizontal. What is the initial speed of the projectile?

*56. A woman holds a hose 2.0 m above the ground such that the water shoots out horizontally. The water strikes the ground at a point 3.0 m away. What is the speed with which the water leaves the hose?

**57. A sprinkler system lies on the ground and shoots a stream of water at an angle of 37° above the horizontal. The water strikes the ground 5.0 m away from the sprinkler. With what speed does the water leave the sprinkler?

NASA

4 NEWTON'S LAWS

Unbalanced forces are the cause of motion. Just how forces are related to motion was first discovered by Isaac Newton. He summarized his discoveries in three laws, which are called Newton's laws of motion. In this chapter we shall see what these laws are and how we use them. You will find that they give us powerful tools for analyzing many physical situations. We will also learn about another famous law discovered by Newton, the law of universal gravitation.

Performance Goals

When you finish this chapter, you should be able to

1. State Newton's first law and apply it to simple situations.
2. Explain the meaning of the term "inertia" and illustrate its meaning by giving examples where it is important.
3. State Newton's third law and give several examples of situations involving action-reaction pairs. Locate the reaction force in a simple situation when the action force is given.

4. State Newton's second law in words and also as an equation. Give the units of the quantities that appear in the equation.
5. Distinguish between the mass and the weight of an object. Give the units in which each is measured. State which of these quantities changes greatly as one goes from the earth to the moon.
6. Find the mass of an object if its weight is given and vice versa. Convert the weight of an object in pounds to its weight in newtons.
7. Write $\Sigma F_x = ma_x$ and $\Sigma F_y = ma_y$ for an object in a simple situation. Explain in words what is meant by ΣF_x.
8. Compute the acceleration of an object of known mass when you are given the forces on the object.
9. State the direction in which the friction force acts on a sliding object.
10. Find the resultant force acting on an object from the acceleration and mass of the object.
11. Compute the tension in the connecting cord and the accelerations of two masses connected in a simple way. Typical situations are found in Section 4.7.
12. Write Newton's law of gravitation and define each symbol in it.
13. Compute the gravitational force one sphere exerts on another when the masses of the spheres and their separation are given. The value of G is also given.
14. Explain in your own words why the weight of an object varies

4.1 Newton's First Law: Inertia

We begin our discussion of motion and forces by considering the most simple situation possible. Consider the case where no unbalanced forces exist on an object. Everyone will agree that if the object is at rest, it will remain at rest. A stone does not begin to move unless something pushes or pulls on it. An apple on the table does not begin to roll unless some force causes it to begin moving. No one will be surprised by the following statement of the way things behave.

An object at rest will remain at rest unless some unbalanced force causes it to do otherwise.

Isaac Newton stated this as a law of nature. It is part of his first law of motion.

But there is another part to the law. It is not as obvious as the one just stated. It concerns an object that is already moving. From his observations, Newton concluded as follows.

An object in motion will continue in motion in a straight line with constant velocity unless some unbalanced force causes it to do otherwise.

This means that a moving object, if left to itself, will move in the same direction forever. It will not slow down. No force is required to keep it moving.

It is often hard for us to accept this part of the law. Our experience does not seem to agree with it. All objects slow and come to a stop. A ball rolling on a table slows down. Any sliding object soon slows to a stop. Nothing keeps going in a straight line forever.

The reason that these examples do not disprove Newton is quite simple. All moving objects are slowed by friction forces. A rolling ball and a sliding object have an unbalanced force holding them back. It is the force of friction. This unbalanced force slows their motion.

Newton had the genius to go beyond such examples. He saw that a sliding block goes farther if the friction forces are decreased. Many other examples led him to believe that, without friction, moving objects would not slow. In the absence of any unbalanced force, an object would move forever with the same velocity. It would not deflect. It would not slow.

Let us now summarize Newton's first law of motion.

(a) An object at rest will remain at rest and (b) an object in motion will remain in motion with the same velocity *provided* no external unbalanced force acts on the object.

Notice that we are using the word *velocity* here. Because velocity has direction as well as magnitude, the law tells us that the object moves along a straight line.

This first law of Newton's is often called the *law of inertia*. You have heard the word "inertia" before. When a person is slow to get started, you say he has a large inertia. In technical language inertia means much the same. Objects do not easily change their state of rest or motion. An object at rest tends to remain at

from place to place. Explain what is meant by the weight of an object on the earth and on the moon.

15. For an object hanging from a cord, find the tension in the cord if the object is (a) at rest, (b) moving with constant speed, (c) has an acceleration upward, (d) has an acceleration downward.

16. Explain how the apparent weight of an object can be measured and point out why the apparent weight can differ from the object's weight.

17. Explain why and under what condition an object may appear weightless.

rest. It has inertia. For example, the large rock shown in Figure 4.1(a) tends to remain at rest. It has large inertia. The car will find this out when it tries to set it in motion.

Inertia also concerns moving objects. An object in motion tends to remain in motion. It has inertia. The moving baseball shown in Figure 4.1(b) requires a force to stop it. This force must be supplied by the man's head when the man stops the ball. Of course, a baseball has less inertia than a bowling ball moving with the same speed. We shall soon see that the more massive an object is, the more inertia it has.

Let us summarize the meaning of the word "inertia."

An object at rest tends to remain at rest; an object in motion tends to remain in motion. "Inertia" is the word used to summarize these properties of an object.

As you see, Newton's first law tells us how inertia effects the behavior of an object free from unbalanced forces. Apply it to the situation shown in Figure 4.2. Why does it tell us that one driver needs a seat harness whereas the other needs a headrest?

4.2 Newton's Third Law: Action and Reaction

Before discussing Newton's second law, let us look at his third law of motion. This law tells us that forces always come in pairs. Newton noticed the following. If one object (call it A) pushes or pulls on a second object (call it B), then object B exerts an equal but oppositely directed force on object A. For example, suppose that you push down on the table with your finger. The table pushes up on your finger with an equal force. These equal, but opposite, forces are called the *action* and *reaction* forces.

Please be careful about these forces. There are always two of them, the action and reaction forces. They are always equal. Their directions are always opposite. And, be careful about this, they act on different objects.

Look at the examples given in Figure 4.3. Be sure that you understand the two forces in each case. Notice that even though the action and reaction forces are equal and opposite, they do not act on the same object. Therefore they do not cancel each other.

We can state Newton's third law as follows.

If object A exerts a force on object B, then object B exerts an oppositely directed force of equal magnitude on object A.

Figure 4.1 In both cases the man is going to learn about the effects of inertia.

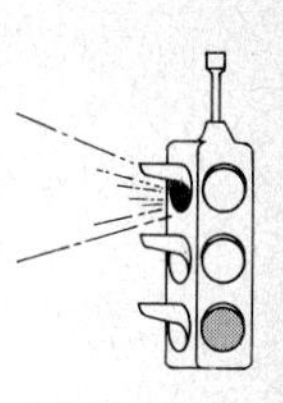

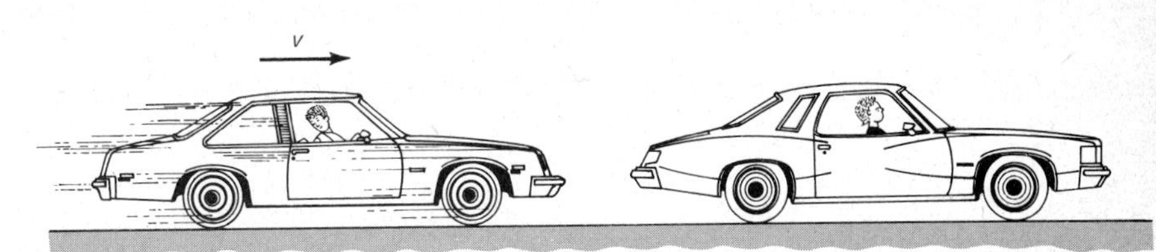

Figure 4.2 Newton's law of inertia aids us in understanding the injuries that will result from this collision.

This law will be helpful to us in many difficult situations. For example, we know that the earth pulls on the moon and holds it in orbit around the earth. Newton's third law tells us that the moon pulls on the earth with an equal, opposite force. We shall encounter other examples later.

4.3 Newton's Second Law: F = *m*a

If you want to speed an object up or slow it down, you have to apply a force to it. Newton's second law tells us how much an object will be speeded up or slowed down by an unbalanced force. It is an equation that relates the unbalanced force **F** acting on an object to the acceleration **a** of the object. We can easily imagine an experiment to determine the relation between **F** and **a.** It is illustrated in Figure 4.4.

Suppose that a cart has tiny, nearly frictionless wheels as shown in the figure. We push it with a force $\mathbf{F}_1$ and find that it speeds up with an acceleration $\mathbf{a}_1$ as indicated. In parts (b) and (c) of the figure we show what happens for larger forces. When a force $2\mathbf{F}_1$ is used, the acceleration increases to $2\mathbf{a}_1$. A force $3\mathbf{F}_1$ causes an acceleration $3\mathbf{a}_1$. We conclude from this experiment (as we might have guessed) that the acceleration is proportional to the unbalanced force pushing the object. In symbols we can write

$$\mathbf{a} \sim \mathbf{F}$$

But the force and acceleration also depend on something else. This we can see by comparing the experiments in parts (a), (d), and (e) of Figure 4.4. As we would expect, the more massive the object, the larger the force must be to accelerate it. The figure shows what must be done to maintain the same acceleration. When the mass is doubled in (d), the force must be doubled. As shown in (e), tripling the mass requires that the force be tripled if the acceleration is to still be the same. If we use the symbol m to represent the massiveness (or *mass*) of the object, then in symbols,

$$\mathbf{F} \sim m \qquad \text{(for constant acceleration)}$$

Push of car on tree = Push of tree on car

Push of hand on jaw = Push of jaw on hand

Force of foot on ball = Force of ball on foot

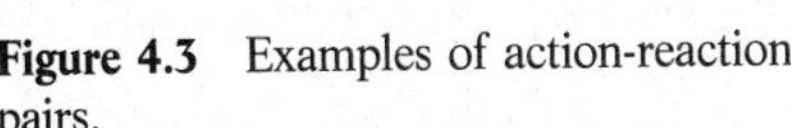

Figure 4.3 Examples of action-reaction pairs.

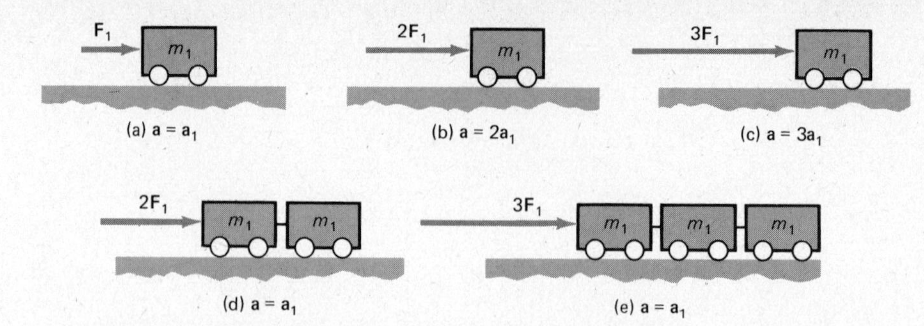

Figure 4.4 The experiments shown here show how F, m, and a are related.

We can combine these two proportionalities into one. It is

$$\mathbf{F} \sim m\mathbf{a}$$

This can be made into an equation by use of a proportionality constant c. We can then write

$$\mathbf{F} = cm\mathbf{a}$$

But now we must find the value of the proportionality constant c. Because it is a constant, once we find it, it will always have that same value.

To find the numerical value of the proportionality constant, we need only measure how large a force **F** is required to give a certain mass m an acceleration **a.** If we substitute these measured values for **F,** m, and **a,** we can obtain a numerical value for c, and it will forever after have this value. But here we have a difficulty. We know that we can measure **a** in meters per second per second (or, perhaps, feet per second per second) and the force **F** in newtons (or pounds). However, we have not yet decided on a unit in which to measure the mass of the object, m. This is fortunate because we can do as we please about it. Let us measure m in such a unit that the proportionality constant c is simply one, $c = 1$. Then we can write

$$\mathbf{F} = m\mathbf{a} \tag{4.1}$$

It is understood that we still have to see what units m should be measured in.

Equation 4.1 is a mathematical statement of Newton's second law. It summarizes the following experimental facts.

1. When a given object is accelerated, the acceleration is proportional to the unbalanced force on the object.
2. The force required to provide a given acceleration is proportional to the mass of the object being accelerated.

Of course, the acceleration has the same direction as the unbalanced force. In the next section we shall learn more about the quantity m, the mass of the object.

4.4 Mass and Weight

The mass of an object is not the same as its weight. Even though the two are related (as we shall see), they have different meanings. We defined the weight of an object to be a force. It is the force with which gravity pulls it down. When we weigh an object on a scale, we are measuring the force that pulls it toward the center of the earth. Weight is a force, the force of gravity. We represent it by the symbol W.

The quantity m, the mass of an object, measures its inertia. If the object has a large mass, it requires a large force to set it in motion. Or if it is moving, it requires a large force to stop it. Mass is a measure of inertia.

But mass and weight are certainly related. We all know that a massive object has a larger weight than a less massive one. It is a simple matter to relate W and m. We do it by applying Newton's law, $F = ma$, to a familiar situation, the acceleration of a freely falling object.

Why does the object in Figure 4.5 fall? It falls because the earth pulls it down with a force W, the weight of the object. When it falls freely under this force W, what is its acceleration? It is the free-fall acceleration g, the acceleration due to gravity. We see, then, that we know two of the quantities in $F = ma$ for the free-fall experiment shown in Figure 4.5. F is simply the weight W. The acceleration a is g, the free-fall acceleration. Therefore

$$F = ma$$

becomes

$$W = mg$$

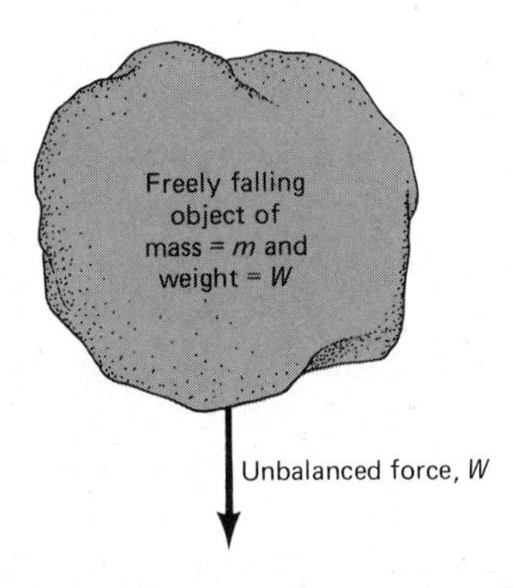

Figure 4.5 The acceleration of the freely falling object is g. It is accelerating because of the unbalanced pull of gravity, its weight W. Hence $F = ma$ gives $W = mg$.

This tells us the relation between mass and weight.

$$W = mg \qquad \text{or} \qquad m = \frac{W}{g} \tag{4.2}$$

Even though weight and mass are different quantities, they are proportional. The proportionality constant is g, the acceleration due to gravity at the place where the weight is measured. Notice that this relation applies on the moon as well as on the earth. There objects weigh about one-sixth what they do on the earth. As a result, g, the free-fall acceleration, is only one-sixth as large on the moon as on earth. Therefore, in the quantity $m = W/g$, the factors of $\frac{1}{6}$ cancel and so m is the same on the earth and the moon. Indeed, the mass of an object is the same everywhere in the universe. Unlike weight, the mass (or inertia) of an object is a property of the object alone. It does not matter where the object happens to be.

4.5 SI Units System

We now understand how an object of mass m is accelerated by an unbalanced force $\mathbf{F}$ that acts on the object; the object obeys $\mathbf{F} = m\mathbf{a}$. But to make use of this relation, which we shall call Newton's second law, we must know what units should be used to measure force, mass, and acceleration. The most important set of units is based on the metric system. It is the set of units we shall prefer to use.

The basic time unit in the SI is the familiar second (s). Its precise definition is as follows: 1 s is the time taken for 9,192,637,770 vibrations of a certain type associated with cesium atoms. These vibrations can be measured to extremely high precision, better than 1 part in 10^{12}. This is equivalent to 1 s in about 30,000 years.

In 1983 the SI unit of length, the meter (m), was redefined in terms of the velocity of light. Its definition is as follows: 1 m is the distance light travels in vacuum in a time of 1/299,792,458 of a second. Because speed is distance divided by time, this definition also *defines the speed of light* in vacuum to be 299,792,458 m/s.

The third basic unit defined in this system is not a force unit. Instead, it is the unit for mass. An object like the one shown in Figure 4.6 is kept near Paris. Its mass is defined to be exactly 1 kilogram (kg). Any object that has the same inertia as this object has a mass of 1 kg. If an object is twice as hard to accelerate, its inertia (or mass) is also twice as large and so its mass is 2 kg. By comparing inertias of objects to the inertia of the standard kilogram, the masses of the objects can be found.

Figure 4.6 The platinum-iridium cylinder shown here is preserved at the U.S. National Bureau of Standards. It is a copy of the standard kilogram kept near Paris. (Courtesy of U.S. National Bureau of Standards)

But there is an easier way to determine the mass of an object. Because $W = mg$, weight and mass are proportional. If an object weighs the same as the standard kilogram does, then its mass is also the same. As you see, masses can be compared by weighing. And this is the way we commonly measure masses. We compare the weight of an object with the weight of objects of known mass. When the weights are the same, then the masses are also the same. Of course, this assumes both weighings are done at the same place. Then g will be the same for the two weighings so that $W = mg$ will allow us to compare masses and weights directly.

But let us give a word of caution: *Mass and weight are not the same.* Mass is a measure of inertia; weight is the force of gravity that acts on an object. The mass and weight are related through $W = mg$, but W and m are completely different quantities.

We see that our scientific set of units is based on the meter, the kilogram, and the second. It is a metric system and is sometimes referred to as the mks system (for meter, kilogram, second). But these units are only a portion of a worldwide set of scientific units. We call the complete system the SI units system (for "Système Internationale"). The SI system is the preferred system of units in scientific and technical work. We shall point out how it measures various other quantities as we proceed with our studies.

Now that we have defined the meter, the kilogram, and the second, we can proceed with our study of $F = ma$. In this equation we know that mass m is measured in kilograms; the acceleration a is measured in meters per second per second. These are the SI units for mass and acceleration. Let us now see how forces are measured in the SI system.

We define the unit of force through Newton's second law, $F = ma$. We must take the units for F that this equation gives. The units will be those of ma. They are

$$(\text{kg})(\text{m/s}^2)$$

This is the unit for force in the SI system. We give it the name "newton" (N). Then

$$1\ \text{N} = 1\ \text{kg}\cdot\text{m/s}^2$$

It is the unit of force in the SI system.

An unbalanced force of 1 N gives a mass of 1 kg an acceleration of 1 m/s^2.

The use of $F = ma$ in defining the unit of force has an important consequence. When we use $F = ma$, we must always use the units we have defined for it. If the force is to be in newtons, then m *must* be in kilograms and a *must* be in meters per second per second. No other units should be used. As we shall see, many other relations we shall encounter are based on $F = ma$, and so they will be subject to this same restriction. To be safe, we should try always to use length in meters, time in seconds, and mass in kilograms when we do calculations.

Frequently we shall need to know how much an object of known mass weighs. For example, the earth pulls on the standard 1-kg mass kept in France. How much does this 1-kg mass weigh? To find out, we recall that weight and mass are related through $W = mg$. In our case the object's mass is 1 kg. The acceleration due to gravity on earth is about 9.8 m/s^2. Therefore, we have

$$\text{Weight} = W = mg$$

which becomes

$$\text{Weight} = (1\ \text{kg})(9.8\ \text{m/s}^2) = 9.8\ \text{N}$$

The 1-kg standard mass weighs 9.8 N on earth. Similarly, the weight of a 5-kg sack of flour is

$$W = (5\ \text{kg})(9.8\ \text{m/s}^2) = 49\ \text{N}$$

Another point we should mention has to do with what happens when no unbalanced force acts on the object we are considering. Then, from $F = ma$ with $F = 0$, the acceleration of the object must be zero. When the resultant force acting on an object is zero, the object will continue with unchanging velocity. We say in a situation such as this that the object is in *dynamic equilibrium*. This is similar in one way to the situation we have called static equilibrium in previous chapters. In both cases the object has no resultant force acting on it. We combine both cases into one by stating an object is in equilibrium if (1) it is at rest and remains at rest or (2) it is in motion and remains in motion with constant velocity.

Although we have done quite a bit in this section, the end result is not as complicated as it might seem. To summarize our results, we have found the following.

1. Weight and mass are related through

$$W = mg$$

where $g = 9.8 \text{ m/s}^2$ on earth.

2. Newton's second law, $F = ma$, is to be used with the following units only:

QUANTITY	UNIT
F	newton (N)
m	kilogram (kg)
a	m/s^2

Never use units other than these.

3. The kilogram is the basic unit of mass in the SI system. Using $W = mg$, we can say that, *on earth*, 1 kg weighs 9.8 N.

In Section 4.7 we shall learn how to use $F = ma$ in practical situations.

4.6 British Engineering Units[1]

Another set of units is still used in the United States and a few other places. Although it is becoming obsolete even there, we shall not ignore it. The British engineering system uses the foot, second, slug, and pound as its units of length, time, mass, and force, respectively. These units are analogous to the meter, second, kilogram, and newton. They are defined in terms of the analogous SI units.

$$\begin{aligned} 1 \text{ ft} &= 12 \times 0.0254 \text{ m} = 0.03048 \text{ m} \\ 1 \text{ slug} &= 14.6 \text{ kg} \\ 1 \text{ lb} &= 4.45 \text{ N} \end{aligned}$$

[1] This section may be omitted if your instructor wishes you to use only SI units.

Because of the way in which these units are defined, they are correct units for use in $F = ma$.

An unbalanced force of 1 lb gives a mass of 1 slug an acceleration of 1 ft/s^2.

When using $F = ma$ in the British system, forces will be in pounds if mass is measured in slugs and acceleration is expressed in feet per second per second. When making computations in the British system, always use these units in fundamental equations.

Many people who use the British system avoid the slug unit. They succeed in this by use of $W = mg$. They always replace m by W/g. As a result, $F = ma$ becomes

$$F = (W/g)a$$

They therefore eliminate mass, and its unit, from the calculations. However, they must be careful. Because both the weight W of an object and g vary from place to place, they must use the value of W and g for the same place. They cannot use the weight of an object on the moon along with $g = 32.2$ ft/s^2, the value appropriate for the earth.

To summarize, those working in the British system can use $F = ma$ provided F is in pounds, m is in slugs, and a is in feet per second per second. If so desired, m can be replaced by W/g, where W is the weight in pounds of the object at a place where the acceleration due to gravity is g ft/s^2.

4.7 Using $F = ma$

Let us review the meaning of Newton's second law. It tells us that an unbalanced force on an object causes the object to accelerate (or decelerate). The direction of the acceleration is the same as the direction of the unbalanced force. Usually it is convenient to work with each force component separately. Then we can write

$$\Sigma F_x = ma_x \qquad \text{and} \qquad \Sigma F_y = ma_y \tag{4.3}$$

where the resultant force (i.e., the unbalanced force) on the object in the x direction is written as ΣF_x. We shall see later how we choose the x and y directions.

In using $F = ma$ we recognize that we are talking about a single object. The procedure we use in applying it is as follows.

1. Isolate an object for discussion.
2. Sketch a diagram showing the forces acting on the object.
3. Find the components of the forces acting on the object.
4. Write Equations 4.3 and solve.

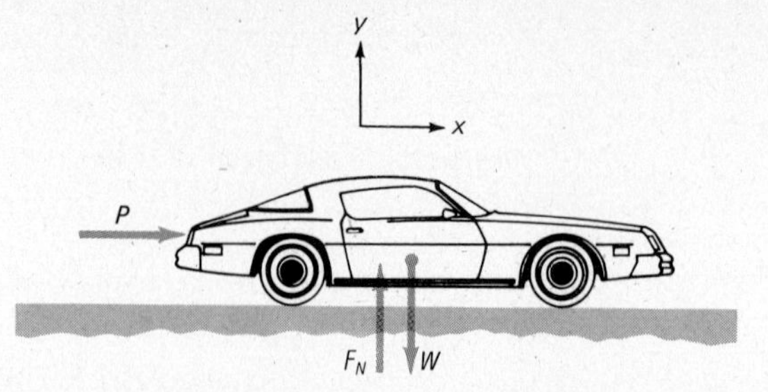

Figure 4.7 The downward pull of the earth on the car, W, is balanced by the upward push of the road, F_N.

The following examples show common uses of $F = ma$. Study them carefully.

■ **EXAMPLE 4.1** How large a horizontal force must be applied to an 800-kg car to give it an acceleration of 3.0 m/s^2 on a level road?

Solution The situation is shown in Figure 4.7. We have shown the y forces as well as those in the x direction. The y forces must balance exactly or the car would accelerate up or down; it does not do this. Therefore $\Sigma F_y = 0$. There is only one x force if we ignore friction. It pushes the car forward. (Actually the horizontal force is applied to the rotating wheels by the pavement.) Let us call this force P as shown in the figure. We shall now write

$$\Sigma F_x = ma_x$$

Since the only x force is P, this becomes

$$P = ma_x$$

But a_x is to be 3.0 m/s^2 and m is 800 kg, so this gives

$$P = (800 \text{ kg})(3.0 \text{ m/s}^2) = 2400 \text{ kg} \cdot \text{m/s}^2 = 2400 \text{ N}$$

In writing this result, we have used the fact that a newton is a kilogram-meter per second per second. We thus find that a force of 2400 N is required to give the car an acceleration of 3 m/s^2. ■■

■ **EXAMPLE 4.2** A 2.0-kg block is sliding along a level floor as shown in Figure 4.8. If its deceleration is 0.60 m/s^2, find the friction force that the floor exerts on it.

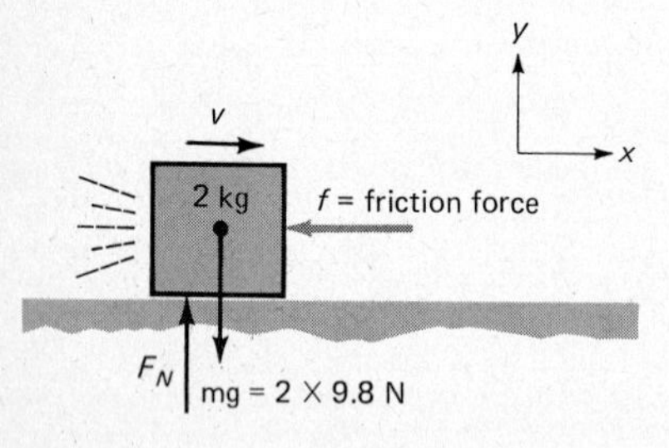

Figure 4.8 The sliding block is slowed by the friction force.

Solution As shown in the figure, the force exerted on the block by the floor consists of two parts, a component F_N perpendicular to the floor and a component f parallel to the floor. The force F_N balances the pull of gravity, mg. The force f, which we call the friction force, causes the block to slow to a stop. It is this force that we wish to find. We wish to write

$$\Sigma F_x = ma_x$$

The only x-directed force is $-f$. Then we have

$$-f = ma_x$$

We were told that the deceleration was 0.60 m/s^2. This means that a_x must be given a negative sign because it is in the negative x direction. (The block is not speeding up in the x direction. It is doing just the opposite.) Hence $a_x = -0.60$ m/s^2. Also, the mass of the block is 2.0 kg. (Remember, kilogram is a unit of mass.) Substituting these values gives

$$-f = (2.0 \text{ kg})(-0.60 \text{ m/s}^2)$$

and so

$$f = 1.20 \text{ kg}\cdot\text{m/s}^2 = 1.20 \text{ N}$$

because 1 kg · m/s^2 = 1 N. The friction force stopping the motion is 1.20 N. ■■

■ **EXAMPLE 4.3**[2] In order to make the 40-lb block in Figure 4.9 move along the floor at constant speed, one must pull as shown with a force of 30 lb. (a) Find the friction force that holds back the motion. (b) If the pull is increased to 50 lb, the acceleration of the block is 23 ft/s^2. How large is the friction force now?

Solution The situation is sketched in part (a) of Figure 4.9. As usual, the y forces must cancel. We are told that the forces shown in (a) apply if the speed is constant. This means that $a = 0$. Then we have

$$\Sigma F_x = ma_x = 0$$

As we see in Figure 4.9(a),

$$\Sigma F_x = 24 \text{ lb} - f$$

where f is the friction force that tends to stop the motion. Therefore

$$\Sigma F_x = 0 \rightarrow 24 \text{ lb} - f = 0 \rightarrow f = 24 \text{ lb}$$

The friction force is 24 lb.

The situation for part (b) of the example is shown in Figure 4.9(b). Because a = 23 ft/s^2, we have

[2]This example may be omitted if you skipped Section 4.6.

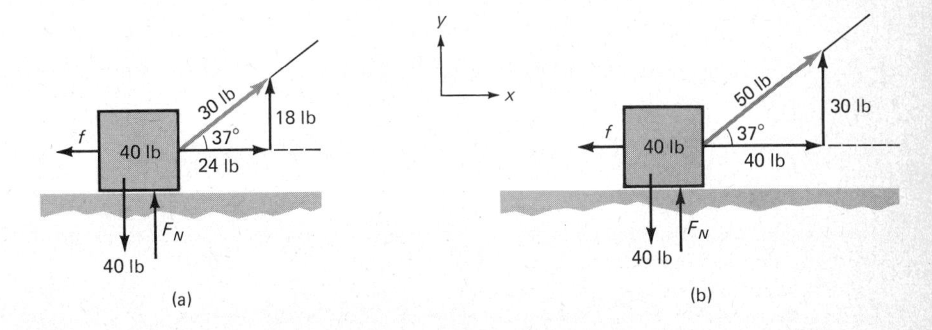

Figure 4.9 In (a) the block moves with constant speed. How large is the force that accelerates it in (b)?

$$\Sigma F_x = ma_x \rightarrow 40 \text{ lb} - f = \left(\frac{40 \text{ lb}}{32 \text{ ft/s}^2}\right)(23 \text{ ft/s}^2)$$

Notice that we have set $m = W/g$. Solving for the friction force gives $f = 11.3$ lb. In the next chapter we shall see why the friction force is less in (b) than in (a). ▮▮

▮ **EXAMPLE 4.4** As a rule of thumb, the maximum friction force a concrete road can exert on the tires of a skidding car is about 0.7 times the car's weight. Find the distance it would require for a car to stop on a concrete road if its initial speed is 30 m/s. Ignore the driver's reaction time.

Solution This is typical of problems that combine a $F = ma$ problem with a motion problem. First let us find the maximum deceleration we could expect for the car. We see in Figure 4.10 that the only horizontal force acting on the car is the friction force f.

Let us take the direction of motion as the positive direction. (We shall usually do this when using $F = ma$.) Then the friction force is a negative force on the car. We then have for $\Sigma F_x = ma_x$,

$$-f = ma_x$$

We were told that the maximum friction force f is $0.7W$, where W is the car's weight. Further, $m = W/g$. Substituting these values gives

$$-0.7W = \left(\frac{W}{g}\right)a_x$$

We can eliminate W by dividing through the whole equation by W. Then

$$-0.7 = \frac{a_x}{g} \qquad \text{or} \qquad a_x = -0.7g$$

Notice that the weight of the car is not important. All cars can be expected to decelerate at about the same rate. Because $g = 9.8 \text{ m/s}^2$, we find that

$$a_x = -0.7(9.8 \text{ m/s}^2) = -6.9 \text{ m/s}^2$$

The negative sign tells us that the acceleration is in the negative x-direction; the car is slowing down, a fact we already knew.

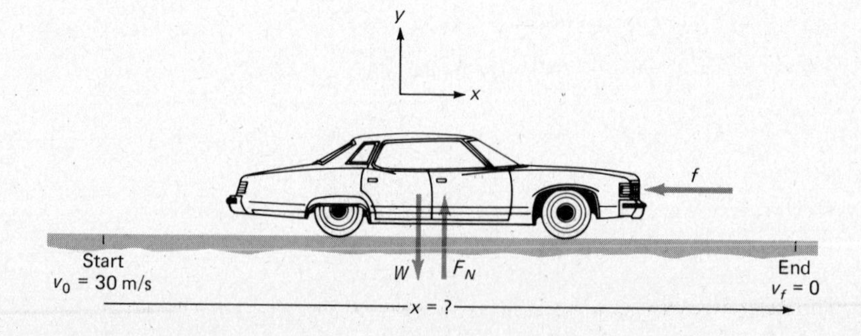

Figure 4.10 The friction force f between the skidding tires and the pavement causes the car to stop.

Now we are prepared to work a motion problem for the car. The following are known (see Figure 4.10).

$$v_0 = 30 \text{ m/s}$$
$$v_f = 0$$
$$a = -6.9 \text{ m/s}^2$$
$$x = ?$$

To find x we can use

$$2ax = v_f^2 - v_0^2$$

Substitution of the values gives

$$2(-6.9 \text{ m/s}^2)x = 0 - (30 \text{ m/s})^2$$

(Notice in passing that x is proportional to the square of the velocity. Therefore the stopping distance increases greatly with speed.) Solving, we find that

$$x = 65 \text{ m}$$

■ ■

■ **EXAMPLE 4.5** The skid marks for a certain car show that the car skidded 20 m on dry concrete before coming to rest. Its driver says that the car was going 15 m/s. Was the driver telling the truth?

Solution Let us use the rule of thumb from the previous example. We know from it that the friction force stopping the skidding car was about $0.7W$, where W is the weight

ACCELEROMETERS

Acceleration can be measured using a ruler and stopwatch. Such a procedure is often not convenient in technical work. Instead, one wants a device that can record accelerations directly. There are many devices designed for this purpose. They are called *accelerometers*.

All accelerometers make use of the fact that an unbalanced force is required to cause a mass to accelerate. A very simple accelerometer is shown in the figure. In (a) the mass m is not accelerating. The two horizontal springs furnish equal and opposite forces to the mass. Because the unbalanced force on it is zero, $\mathbf{a} = 0$ for the mass.

But in (b) the accelerometer (and mass) is accelerating to the left. Notice how the distorted springs exert an unbalanced force to the left on the mass. This unbalanced force is proportional to both the acceleration and the distortion of the springs. As a result, the distance d shown is a direct measure of the acceleration $\mathbf{a}$. This distortion can be used to cause an electrical signal. The signal is then fed to a recording device such as a chart recorder. In this way a record is made of the acceleration of the accelerometer.

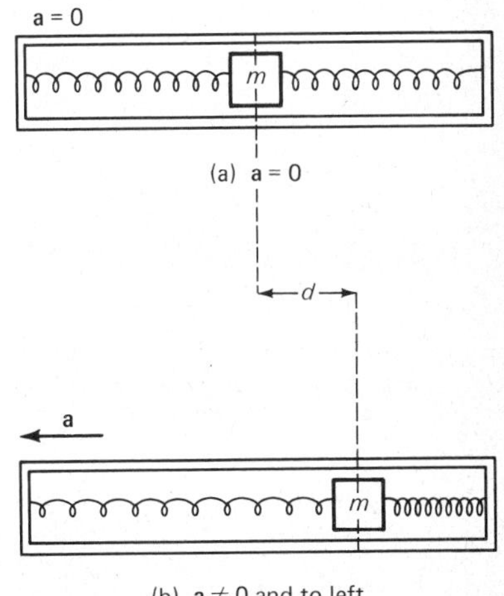

(a) a = 0

(b) a ≠ 0 and to left

of the car. The deceleration of the car can be found from $\Sigma F_x = ma_x$. In the present case this is

$$-0.7W = ma_x$$

The weight of the car is given by $W = mg$ with $g = 9.8 \text{ m/s}^2$. Therefore, the equation becomes

$$(-0.7)(m)(9.8 \text{ m/s}^2) = ma_x \rightarrow a_x = -6.9 \text{ m/s}^2$$

We now have enough data to work a motion problem for the car during the skid. We know that $v_f = 0$, $x = 20$ m, and $a_x = -6.9 \text{ m/s}^2$. To find the initial velocity of the car, we can use $v_f^2 - v_0^2 = 2ax$. Substitution of the known values in it gives $v_0 = 17$ m/s. The driver was not underestimating her speed too much. ■■

■ **EXAMPLE 4.6** As shown in Figure 4.11, two blocks are being pulled along a surface by a force of 5.0 N. If the friction force on the 2.0-kg block is 3.0 N and the friction force on the 1.0-kg block is 1.5 N, find the accelerations of the blocks and the tension in the connecting string.

Solution Let us first apply $\Sigma F_x = ma_x$ to the object composed of the combination of the blocks shown in part (a) of the figure. (The y forces are balanced, so we shall not worry about them.) All the x forces are shown in part (a) of the figure. The only x-directed forces that act on the two-block object are 5 N, −1.5 N, and −3 N. (Notice that the forces T and $-T$ cancel. They are so-called internal forces, forces inside the object we are considering. Such internal forces always cancel because of the law of action and reaction. The forces that do act on the object are outside forces.) We can write

$$\Sigma F_x = ma_x$$

as

$$5.0 \text{ N} - 1.5 \text{ N} - 3.0 \text{ N} = (1.0 \text{ kg} + 2.0 \text{ kg})a_x$$

The mass of our object, the two blocks, is 3 kg. This becomes

$$(3.0 \text{ kg})a_x = 0.50 \text{ N}$$

from which

$$a_x = 0.167 \text{ m/s}^2$$

To find the tension in the connecting cord, we must consider one block alone. Let us isolate the 2-kg block as a new object. The object and its x forces are shown in part (b) of Figure 4.11. Writing $\Sigma F_x = ma_x$ for it gives

$$5.0 \text{ N} - 3.0 \text{ N} - T = (2.0 \text{ kg})(0.167 \text{ m/s}^2)$$

Notice that we have used the value for a_x found previously. We now solve this equation to find $T = 1.67$ N. ■■

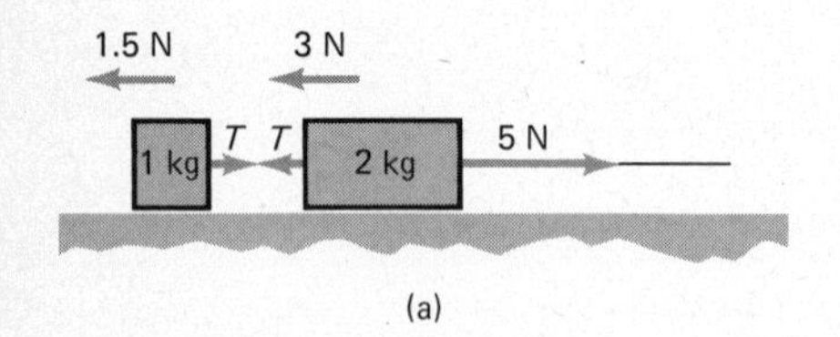

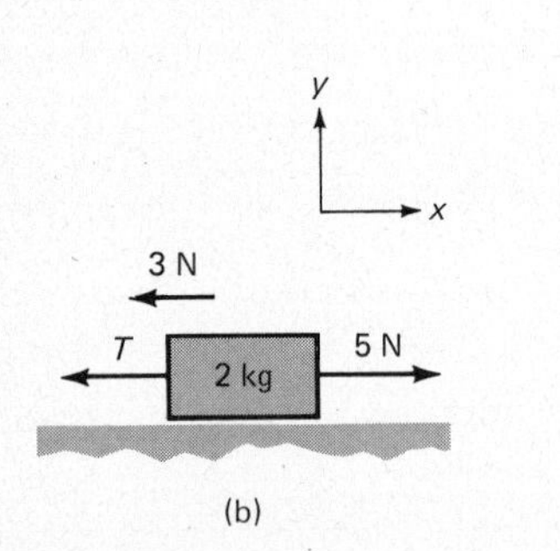

Figure 4.11 What are the y forces in each case? Why are we not concerned with them? The two blocks are connected by a cord. The tension in it is T.

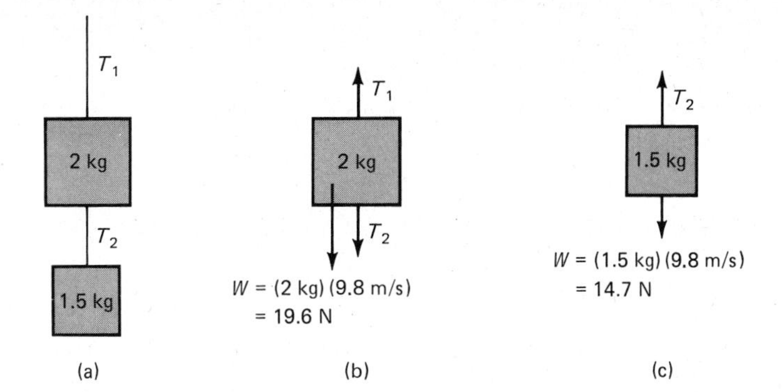

Figure 4.12 We isolate each object in turn. $F = ma$ then gives two simultaneous equations to solve.

EXAMPLE 4.7 The two blocks in Figure 4.12 are to be accelerated upward at $0.20\ \mathrm{m/s^2}$. What must be the tension in the upper rope? In the connecting rope?

Solution We could proceed as in the previous example. Instead, let us first isolate the 2-kg object. The forces on it are drawn in Figure 4.12(b). Notice in particular that the pull of gravity on the object is *mg*. It is

$$mg = (2\ \mathrm{kg})(9.8\ \mathrm{m/s^2}) = 19.6\ \mathrm{N}$$

The tensions T_1 and T_2 in the ropes also pull on it.

Let us take the upward direction as positive because this is the direction of motion. Then we have for $\Sigma F_y = ma_y$, when applied to the upper block as object,

$$T_1 - T_2 - 19.6\ \mathrm{N} = (2\ \mathrm{kg})(0.20\ \mathrm{m/s^2})$$

This simplifies to

$$T_1 - T_2 = 20.0\ \mathrm{N}$$

Because we have two unknowns, we need yet another equation.

We now isolate the other object as shown in (c). Notice that its weight, *mg*, is 14.7 N. Then, for it, $\Sigma F_y = ma_y$ becomes

$$T_2 - 14.7\ \mathrm{N} = (1.5\ \mathrm{kg})(0.20\ \mathrm{m/s^2})$$

This gives

$$T_2 = 15.0\ \mathrm{N}$$

We can now substitute this in the previous equation ($T_1 - T_2 = 20\ \mathrm{N}$) to find

$$T_1 - 15.0\ \mathrm{N} = 20\ \mathrm{N}$$

or

$$T_1 = 35\ \mathrm{N}$$

Notice that T_1 is a little larger than the total weight of the blocks ($3.5 \times 9.8 = 34$ N). This is reasonable because an unbalanced force is needed to cause the acceleration. ∎∎

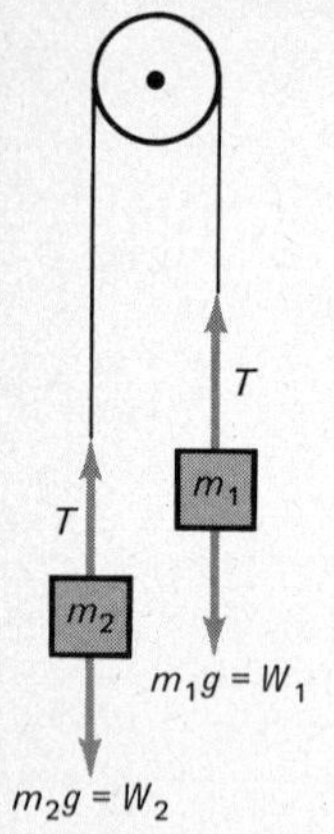

Figure 4.13 Atwood's machine.

∎ **EXAMPLE 4.8** The system shown in Figure 4.13 is called *Atwood's machine.* It consists of two known weights hanging over a small, nearly frictionless pulley. We want to find the acceleration of either mass and the tension in the connecting cord.

Solution Because the masses m_1 and m_2 are only given as symbols, we don't know which is larger. We therefore don't know which way the system will move. In cases such as this, we make a guess as to the direction of motion. If our guess is wrong, our answer for the acceleration will be negative. Let us assume that $m_2 < m_1$, so that m_1 falls and m_2 rises. As usual, we call accelerations and forces in this assumed direction of motion positive.

First we apply $\Sigma F = ma$ to object m_1. We obtain

$$m_1 g - T = m_1 a$$

Notice the signs. T is taken as negative because it is opposite to the assumed direction of motion.

For m_2 we also write $\Sigma F = ma$. This gives

$$T - m_2 g = m_2 a$$

Notice that now T is taken as positive because for this mass the connecting cord pulls in the assumed direction of motion.

We can solve the first of these equations for T to give

$$T = m_1 g - m_1 a$$

This value can then be substituted in the second equation. We then obtain

$$m_1 g - m_1 a - m_2 g = m_2 a$$

Transposing gives

$$m_1 g - m_2 g = m_1 a + m_2 a$$

or

$$a(m_1 + m_2) = g(m_1 - m_2)$$

Dividing through by $(m_1 + m_2)$ gives finally

$$a = g\left(\frac{m_1 - m_2}{m_1 + m_2}\right)$$

It is interesting to notice that this equation says that $a = 0$ when $m_1 = m_2$. This must be true, of course. When m_1 and m_2 are equal, they just balance each other on the two sides of the pulley. What does the answer tell you for $m_2 > m_1$?

To find T we need only substitute back in the equation

$$T = m_1 g - m_1 a$$

Then we have

$$T = m_1 g - m_1 g\left(\frac{m_1 - m_2}{m_1 + m_2}\right)$$

Notice that, for m_1 larger than m_2, the tension in the cord is less than $m_1 g$, the weight of mass 1. The mass will therefore fall, because the upward force on it is less than its weight.

Also, we notice that $T = m_1 g$ if $m_1 = m_2$. Can you explain why this should be? ▪▪

4.8 Newton's Law of Gravitation

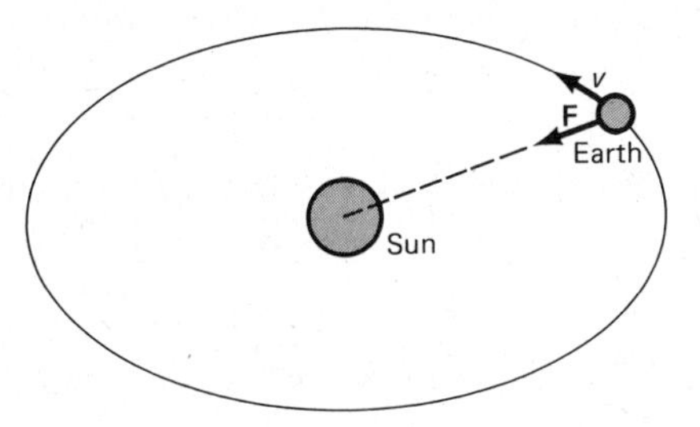

Figure 4.14 The earth orbits the sun.

Newton discovered his law while trying to explain the motion of the earth and the planets around the sun. Consider the earth circling the sun as shown in Figure 4.14. Notice that it does not follow a straight-line path. According to Newton's first law, the earth should move along a straight line unless a force acts on it to cause it to do otherwise. Newton noticed other similar situations: The other planets circle the sun; the moon circles the earth. In each case some sort of force must be pulling the object out of its normal straight-line path.

To explain this behavior, Newton postulated that the sun exerts an attractive force on the earth. This is shown by the force **F** in Figure 4.14. He was able to determine the nature of this force. He did so by examining the motion of the earth and the planets. After lengthy calculations (during which he invented the branch of mathematics called calculus), Newton arrived at the following result. It is called Newton's *law of universal gravitation.*

Two uniform spheres with masses m_1 and m_2 whose centers are a distance r apart attract each other with a force

$$F = G\frac{m_1 m_2}{r^2} \tag{4.4}$$

Newton was unable to determine the value of the proportionality constant G, but we now know it to have the value of $G = 6.67 \times 10^{-11}\ \mathrm{N \cdot m^2/kg^2}$.

We show what the law means in Figure 4.15. There we see two masses m_1 and m_2. The forces $\mathbf{F}_1$ and $\mathbf{F}_2$ are the gravitational forces. Notice that they are attractive and radial. Either force is given by Equation 4.4. Of course, the equation will give the same value for the two forces. This is yet another example of the law of action and reaction. Mass m_1 exerts the action force $\mathbf{F}_1$ on mass m_2; but an equal and opposite reaction force $\mathbf{F}_2$ is exerted by m_2 on m_1.

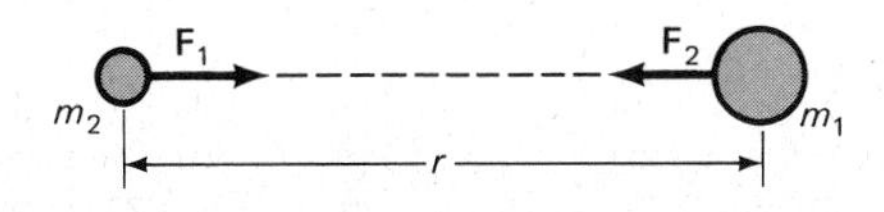

Figure 4.15 According to the law of action and reaction, $\mathbf{F}_1 = -\mathbf{F}_2$.

Newton extended this law to all objects. He concluded that any two objects attract each other with equal forces that are opposite in direction. Moreover, the equation for the force, Equation 4.4, applies even to nonspherical objects as long as r is much larger than the dimensions of the objects. Let us now work out an example to see how large this attractive force is.

EXAMPLE 4.9 Find the attraction force a 30-g marble exerts on a 20-g marble if the distance between their centers is 10 cm.

Solution In using Equation 4.4 it makes no difference which mass we call m_1. We have $m_1 m_2 = (0.020 \text{ kg})(0.030 \text{ kg})$ and $r = 0.10$ m. Substituting in Equation 4.4 gives for the force on either marble

$$F = (6.67 \times 10^{-11})\frac{0.00060}{0.010} \text{ N} = 4 \times 10^{-12} \text{ N}$$

Because a 20-g marble would weigh $(0.020 \text{ kg})(9.8 \text{ m/s}^2) = 0.2$ N, we see that the attractive force is very, very small. It is only about 10^{-11} as large as the weight of the marbles. As a result of its very small magnitude for ordinary-size objects, the gravitational attraction is very difficult to measure. Direct measurements of the attractive force between laboratory-size objects were first made in 1798 by Cavendish. He was then able to measure the gravitational constant G.

4.9 Variation of Weight

Although the attraction force due to ordinary-size objects is very small, the earth is not an ordinary object. Its mass is so huge that the product $m_1 m_2$ is very large even for a small object on the earth's surface. As a result, the earth exerts an easily measured force on an object at the earth's surface. We call this force the weight of the object.

But Equation 4.4, Newton's law of gravitation, tells us a peculiar thing about weight. It tells us that the force of the earth's attraction varies inversely as the square of the distance from the center of the earth. As an object is moved from sea level to the top of a high mountain, its distance from the earth's center, r, increases. The pull of the earth on it must therefore decrease. We see that the weight of an object varies from place to place on the earth.

This variation of weight with distance from the center of the earth is shown in Figure 4.16. (Equation 4.4 applies only if the objects are outside each other. Therefore the graph does not extend to r values less than the radius of the earth.) Notice that even at an r value of 30,000 km (that is, 24,000 km above the earth) the pull of the earth is still 5 percent of the pull at the earth's surface. Indeed the pull of the earth's gravity never goes to zero. No matter how large r may be in Equation 4.4, the force F still has a value greater than zero.

You might well be puzzled at this point. It is commonly known that objects *appear* weightless in a spaceship orbiting the earth. Such a ship may be only a few hundred kilometers above the earth. The difficulty occurs because of what we mean by the words "appears to be weightless." The weight as we have

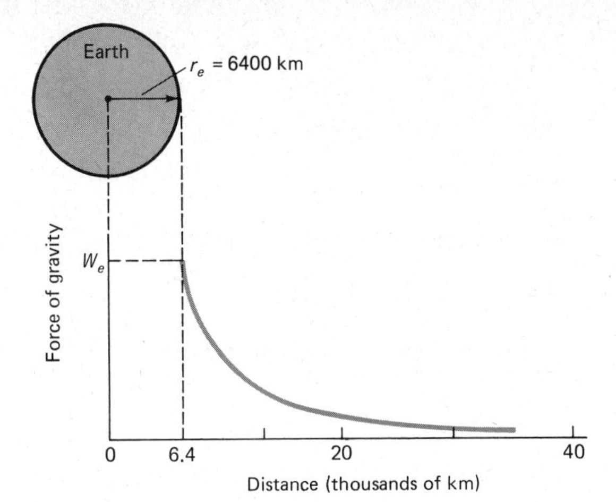

Figure 4.16 At a height of 24,000 km above the earth, an object's weight is about 5 percent as large as at the earth's surface.

defined it is certainly not zero in orbit. Why an object appears to be weightless is discussed in Sections 4.10 and 9.10.

If an object is on the moon, its distance from the earth is large. It is about 60 earth radii from the earth. Since the gravitational force varies as $1/r^2$, the earth's pull on the object is only 1/3600 times as large as it was on earth. Suppose that the object weighs W_e on the earth. When the object is on the moon, the earth's pull on it would be only $W_e/3600$.

But when the object is on the moon, the moon itself pulls strongly on it. Indeed, the gravitational pull of the moon on it is about $W_e/6$. Because this is much larger than $W_e/3600$, the earth's pull, we usually ignore the effect of the earth in such situations. The pull of the moon on the object is called the weight of the object on the moon.

■ **EXAMPLE 4.10** If an object weighs $W_e/6$ on the moon, what is the acceleration due to gravity on the moon? The weight and gravitational acceleration on earth are W_e and $g_e = 9.8\ \text{m/s}^2$, respectively.

Solution On the moon the free-fall acceleration will be due to the pull of gravity $W_e/6$ on the object. So F in $F = ma$ is simply $W_e/6$. The mass of the object is $m = W_e/g_e$. Then $F = ma$ becomes, in the case of free-fall on the moon,

$$\left(\frac{W_e}{6}\right) = \frac{W_e}{g_e}a$$

or

$$a = \frac{g_e}{6}$$

Therefore, the acceleration due to gravity on the moon is only one-sixth as large as on earth. It is about $1.6\ \text{m/s}^2$ or $5.3\ \text{ft/s}^2$.

4.10 Apparent Weightlessness During Free-Fall

Let us now turn our attention to the interesting topic referred to as weightlessness. In this age of space travel, we all know that objects in spaceships often appear to be weightless. We now wish to examine when and why this condition exists.

To begin, we shall consider a very simple case. Suppose that an object of weight W hangs by a string from a spring scale in an elevator. The situation is shown in Figure 4.17(a). What forces act on the weight? There are only two. Gravity pulls down on the object with force W, the weight of the object. The supporting string pulls up on the object with tension T. Notice that the scale reads the pull of the string upon it. The scale reads T, not W. Of course, at equilibrium T balances W and so the scale reading is equal to the weight. This is one common way to measure the weight of an object.

But the scale does not always read W even though it always reads T. There are situations where T does not equal W. We shall examine three of them.

Elevator Accelerating Upward Suppose that the elevator and everything in it are accelerating upward with acceleration a. Each object in the elevator must have an unbalanced force acting upward on it to cause this acceleration. As shown in part (b) of Figure 4.17, the object we are concerned with has forces T up and W down acting on it. Therefore, $F = ma$ for it becomes

$$T - W = ma$$

or

$$T = W + ma$$

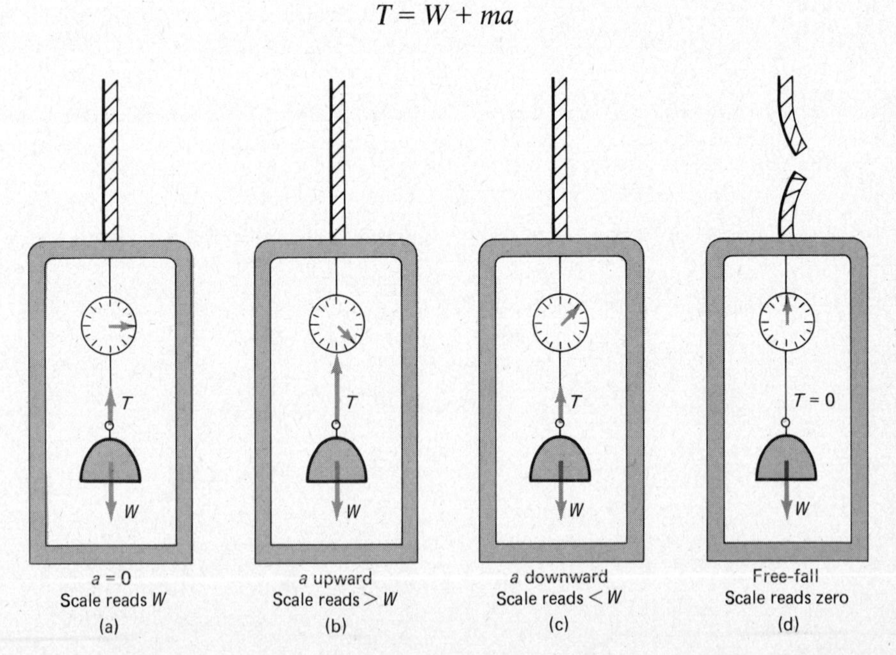

Figure 4.17 The apparent weight of an object in an elevator need not equal its true weight.

The tension in the cord is greater than the weight of the object. Because the scale reads T, it will read an apparent weight greater than W. (You can easily test this. Support a 2-kg mass by a weak string. Now try to accelerate the weight upward. If the acceleration is very large, the tension in the string becomes large enough to break the string.)

Elevator Accelerating Downward If the object is accelerating downward, as shown in Figure 4.17(c), then T must be less than W. An unbalanced downward force must act on the object. Writing $F = ma$ for it (taking down as the positive direction in this case),

$$W - T = ma$$

from which

$$T = W - ma$$

The scale will read this value of T. The apparent weight of the object will be less than its true weight. (You can also test this easily. Support a weight by a string. Now suddenly lower your hand which holds the string. Why does the string go limp?)

Elevator Freely Falling This is simply a special case of the situation described in case 2. Now the object and elevator are accelerating downward with the free-fall acceleration. See Figure 4.17(d). Of course, for the object to be falling freely, the string must not be pulling up on it. So T must be zero. We can also see this by replacing a by g in the equation of case 2. Then

$$T = W - mg$$

But $mg = W$ and so $T = 0$. The object *appears* to be weightless because the scale reads zero.

Now let us return to the case of the spaceship. When it is coasting through space with its motors off, only the pull of gravity acts on it. The ship and everything in it are falling freely through space. We have a situation very much like the freely falling elevator. Because everything is freely falling, no object in the spaceship requires a supporting force. The scale reads zero. No force need be exerted on a book by your hand to hold it—it is already freely falling. Every object in the freely falling spaceship appears to be weightless. We will see in Section 9.11 that this same basic idea gives rise to weightlessness in a ship orbiting the earth.

Notice carefully. Gravity still pulls on these objects. But because they are freely falling through space, no supporting force is needed. If we measure the required supporting force by a spring scale or by "hefting" it with our hand, the necessary supporting force is zero. The object appears to be weightless. We say that its apparent weight is zero.

Summary

Newton's first law is called the law of inertia. It states the following facts concerning an object that has no resultant (unbalanced) force acting on it. If the object is at rest, it will remain at rest. If the object is in motion, it will remain in motion with constant velocity.

Inertia is the tendency of an object to preserve its state of rest or motion. An object at rest tends to remain at rest. An object in motion tends to remain in motion.

The mass of an object is a measure of the object's inertia. It is a property of the object itself and does not depend on its location.

Although mass and weight are related through $W = mg$, they are not the same. The weight of an object is the force of gravitation that pulls the object down. Mass is a measure of inertia.

In the SI system of units, the basic units are those of length (meter), mass (kilogram), and time (second). Forces are measured in newtons (N) in this system.

In the British system of units, force is measured in pounds and mass is measured in slugs. The mass is often replaced in equations by $m = W/g$ when the British system is used.

Newton's second law is summarized in equation form as $\Sigma F_x = ma_x$, $\Sigma F_y = ma_y$, and $\Sigma F_z = ma_z$. Care must be taken in the use of units. In the SI system, the units of force, mass, and acceleration are newtons, kilograms, and meters per second per second, respectively. In the British system, pounds, slugs, and feet per second per second must be used.

Newton's third law is the law of action and reaction. It states that if object A exerts a force on object B, then object B exerts a force on object A that is equal in magnitude but opposite in direction.

Newton's law of gravitation states that two spheres of masses m_1 and m_2 whose distance between centers is r attract each other with a force given by

$$F = G\frac{m_1 m_2}{r^2}$$

where $G = 6.67 \times 10^{-11}\ \mathrm{N \cdot m^2/kg^2}$.

The gravitational attraction force of the earth on an object is called the weight of the object. It varies inversely as the square of the distance from the center of the earth for positions above the earth.

When an object is on the moon or some similar body, its weight is defined as the gravitational force exerted on it by that body. An object weighs about one-sixth as much on the moon as on the earth.

The apparent weight of an object equals its actual weight only if the object is not accelerating. If the object is accelerating, its apparent weight may be larger or smaller than its actual weight. The apparent weight of a freely falling object is zero. Therefore, objects appear to be weightless in spaceships falling freely through space.

Questions and Exercises

1. Explain in what way each of the following is an example of Newton's first law. (a) As a car comes to a stop, a box slides off the car seat. (b) As a car accelerates rapidly forward at a stoplight, a kitten sitting on the top of the driver's seat falls into the back seat. (c) A boxer breaks his hand as he in turn breaks the jaw of his opponent.
2. (a) An automobile moving at a high speed strikes a solid wall head-on. Discuss what sort of injuries the person in the seat next to the driver would sustain, and explain how they are consequences of Newton's laws. Why are seat belts important in such a case?

   (b) A car at rest is hit from behind by another car. Explain what sort of injuries the driver of the car originally at rest will sustain and how they are consequences of Newton's laws. Why are seat belts and headrests important in cases such as this?

   (c) Explain why the driver of a large truck is in far less danger than a driver of a small car in accidents such as those outlined in (a) and (b).
3. An almost legendary party trick is to pull the tablecloth out from under the dishes on a table. Which of Newton's laws is fundamental to the successful completion of it?
4. It is possible to pound a nail into a small block of wood while the block is resting on one's head, provided a heavy block of metal is used as a cushion between the wooden block and one's head. Explain why the metal block serves as a cushion.
5. In ancient times when fathers still spanked their children, a father was likely to tell his child, "This hurts me as much as it does you." Is there any scientific basis for such a statement?
6. A 5-kg mass is suspended from the ceiling by a thin cord. Hanging from the bottom of the mass is a second similar cord. If you pull *slowly* on the lower cord, the upper cord breaks. But if you jerk the lower cord, the lower cord breaks. Explain this difference.
7. Discuss the action and reaction forces in the following situations: (a) a lecturer thumps the table, (b) a hammer hits a nail, (c) a child slides across the floor, (d) a high jumper leaves the ground, (e) a ball hits a window.
8. Newton gave the following example of his third law: "If a horse pulls on a stone tied to a rope, the horse will be equally pulled back toward the stone." Because these two forces are equal and opposite, why should either the horse or stone move?
9. Three outside forces act on a boy as he slides across the ice on a smooth pond. Draw a diagram showing them.
10. A girl stands at rest on a slippery floor. Draw a diagram showing the two outside forces acting on her. She now tries to walk forward. Draw a diagram showing a third outside force that then acts on her. If the floor is completely frictionless, why will she not be able to move?
11. In most countries in the world, groceries are sold by the kilogram or gram rather than by the pound or ounce. If you wanted to buy about 5 lb of sugar, how many kilograms should you ask for? (Use round numbers only, because sugar is usually packaged.) If an Italian visiting the United States wants 500 g of nuts, how many pounds should he ask for?
12. Suppose that a joint Soviet-American space team is exploring the moon. They each have a metal sphere. One sphere is marked 5 lb whereas the other is marked 5 kg. Both were marked correctly on earth, but one label is now wrong. Explain.
13. A woman from Poland is trying to describe her husband to a woman from the United States. She says that he is 160 cm tall and weighs 95 kg. How would you describe him qualitatively in the foot-pound system?
14. One of Newton's three laws is actually a portion of one of the other of the three laws. Explain.
15. An Atwood's machine has as its two weights two identical monkeys. Assume the Atwood machine rope to be light and the pulley to be frictionless. What will happen as either monkey tries to climb the rope on its side? What if both try to climb at the same time? Which will strike the floor first if one lets loose?
16. Objects weigh only one-sixth as much on the moon as on the earth. Yet the mass of the moon is only about one-eightieth that of the earth. Why isn't the weight of an object on the moon only one-eightieth the weight on earth?
17. What is the approximate gravitational force with which a 50-kg woman attracts a 75-kg man when they are 10 m apart? Does the woman find the man equally attractive?
18. The earth's radius is about 6400 km. Knowing that, together with the values of G and g, how could you calculate the mass of the earth? Cavendish was the first to do this.
19. The planet Jupiter is much more massive than earth. The acceleration due to gravity is about 26 m/s^2 there. If a person from earth lived in an enclosure on Jupiter, how would this different value of g influence him?
20. The cable supporting an elevator breaks. As the elevator falls, a passenger in it releases a book. Describe the behavior of the book.
21. When a roller coaster starts its rapid descent down a hill, the people in it have a strange feeling in the pit of the stomach. A similar sensation occurs when an elevator starts down rapidly. Why does this occur? What do you guess from this about one of the problems in adjusting to space flight?

Problems

1. (a) What is the weight of a 3.0-kg object? (b) What is the mass of an object that weighs 600 N? Assume all measurements to be made on earth.
2. (a) A certain car is said to have a mass of 900 kg. How large a force must a repair shop lift exert on it to hold it above the floor for servicing? (b) A bridge can hold a maximum load of 40,000 N. What is the largest mass truck that can cross the bridge safely?
3. How large a resultant force applied to a 500-g object will give it an acceleration of 25 cm/s^2?
4. A resultant force of 0.60 N is applied to a 300-g object. How large an acceleration does it give to the object?
5. A resultant force of 30 N gives an object an acceleration of 5.0 m/s^2. (a) What is the mass of the object? (b) How large an acceleration would a force of 40 N give it?
6. A net force of 50 N acts on an object and gives it an acceleration of 2.0 m/s^2. (a) What is the mass of the object? (b) How large a force would be required to give it an acceleration of 3.0 m/s^2?
7. (a) An unbalanced force of 15 N acts on a 3.0-kg mass. What acceleration does it give it? (b) Repeat for a 7.0-lb force pushing on a 24-lb object.
8. How large an unbalanced force is needed to give a 5.0-kg mass an acceleration of 0.70 m/s^2? Repeat for a 20-lb object whose acceleration is to be 0.30 ft/s^2.
9. (a) How much does a 2.0-kg object weigh on earth? (b) On the moon, where $g = 1.6$ m/s^2?
10. The acceleration due to gravity on Mars is 3.9 m/s^2. If an object has a mass of 6.0 kg on earth, (a) what will be its mass on Mars and (b) how much will it weigh on Mars?
11. (a) What is the mass of an object that weighs 40 N? (b) What is the weight of a 7.0-kg object? Repeat for (c) a 50-lb object and (d) a 20-slug object.
12. A net force of 30 N gives an object an acceleration of 20 cm/s^2. (a) What is the mass of the object? (b) How much does the object weigh? (c) Find the mass and weight of an object that is given an acceleration of 2.0 ft/s^2 by a net force of 15 lb.
13. (a) How large a net force is required to give a 3.0-kg object an acceleration of 2.0 m/s^2? (b) Repeat for a 4.0-lb object being given an acceleration of 5.0 ft/s^2.
14. (a) A net force of 30 N acts on a 6.0-kg object. How large an acceleration does it give the object? (b) Repeat for a force of 25 lb acting on a 12-lb object.

*15. A resultant force of 7.0 N gives a certain mass an acceleration of 0.50 m/s^2. (a) How large a force is needed to give it an acceleration of 2.0 m/s^2? (b) If the mass is cut in thirds, how large a force is needed to give one-third an acceleration of 0.50 m/s^2?

*16. When a resultant force of 200 N is applied to a certain mass, its acceleration is 0.20 m/s^2. (a) How large an acceleration would 120 N give it? (b) If the mass is tripled, how large a resultant force would be needed to give it an acceleration of 0.20 m/s^2?

17. A 700-g block is pushed along the floor by a horizontal force of 20 N. A friction force of 3.0 N opposes the motion. What is the acceleration given to the block?
18. How large a horizontal force is needed to give a 5.0-kg mass an acceleration of 7.0 m/s^2 across a floor if the friction force opposing the motion is 20 N?

*19. A 900-kg car moving at 10 m/s slows uniformly to a stop in a time of 5.0 s. (a) Find the deceleration of the car. (b) How large a force was needed to slow it in this way? (c) Repeat for a 2800-lb car moving at 30 ft/s.

*20. (a) How large a force is needed to accelerate a 1000-kg car from rest to a speed of 20 m/s in a time of 10 s? What was its acceleration? (b) Repeat for a 4000-lb car and 60 ft/s.

*21. A particular gun accelerates a 20-g bullet to a speed of 100 m/s as it passes through the 50-cm-long barrel of the gun. Assuming the acceleration to be uniform (not really true), find the acceleration and the accelerating force.

*22. In a certain car collision the car is decelerated to rest from a speed of 15 m/s. The time taken is 0.20 s. Find the average deceleration of the car. About how large a force must the seat harness exert on a 50-kg passenger to hold her in place during the collision?

*23. (a) A 1200-kg car going 20 m/s collides head-on with a tree and stops in a distance of 2.0 m. About how large was the average stopping force exerted by the tree on the car? (b) Repeat for a 4000-lb car going at 60 ft/s that stops in 6.0 ft.

*24. A car is to tow a 2000-kg car with a rope. How strong must the rope be if they wish to accelerate from rest to 3.0 m/s in a time of 15 s?

*25. (a) The maximum friction force between a 30-kg box and the floor of a truck is 280 N. How fast can the truck accelerate before the box will slip on the floor? (b) Repeat for a 25-lb box if the maximum force is 16 lb.

**26. A 15.0-g bullet going 200 m/s shoots through a piece of wood that is 0.70 cm thick. After leaving the wood, its speed is 80 m/s. What average force did the bullet exert on the wood?

*27. The two blocks shown in Figure P4.1 each have masses of

*Problems marked with an asterisk are not as easy as the unmarked ones.

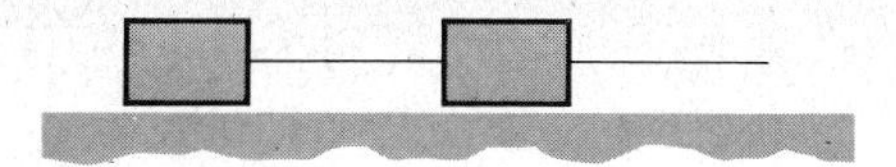

Figure P4.1

0.70 kg. The tension in the cord at the right is 3.0 N. Find (a) the acceleration of the blocks and (b) the tension in the cord connecting them. Ignore friction.

*28. Repeat the previous problem for the case in which the mass on the right is 1.2 kg and the mass on the left is 0.70 kg.

*29. Consider the situation shown in Figure P4.2. If the pull $P =$ 20 N, $m_1 = m_2 = 3.0$ kg, and friction forces are negligible, find the acceleration of the blocks. Also, what is the tension in the connecting cord?

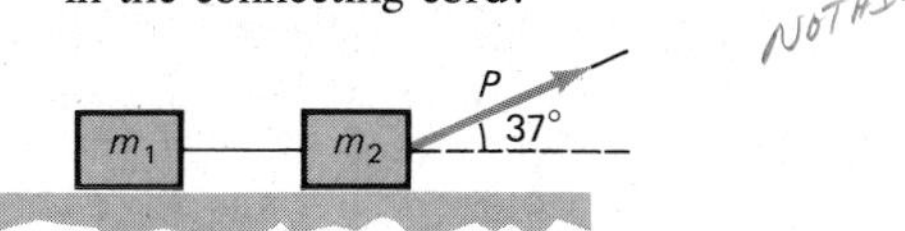

Figure P4.2

**30. In Figure P4.2, $m_1 = m_2 = 5.0$ kg. Each experiences a friction force of 20 N. How large must P be if the acceleration of the blocks is 0.70 m/s²? What will be the tension in the connecting cord then?

**31. Two 5.0-kg masses hang on each side of an Atwood's machine. A 50-g mass is added to one of these and the combined mass falls to the floor 80 cm below. If friction can be ignored, how large is its acceleration? How long does it take to reach the floor?

**32. An Atwood's machine can be used to measure g, the acceleration due to gravity. Assume the pulley to have negligible mass and friction. Its two masses are each 15 kg. When a 50-g mass is placed on one side, the combined mass takes a time of 16.0 s to fall to the floor, 2.0 m below. From these data, determine g.

33. Two 6-kg bowling balls are 80 cm apart (center to center). (a) How large is the attractive force one exerts on the other? (b) Find the ratio of the weight of one of the balls to this force.

34. An 8000-kg satellite is at a height of 9 earth radii from the surface of the earth. How large is the earth's gravitational pull on it? The earth's radius is 6400 km and its mass is 6.0×10^{24} kg.

35. A 20,000-kg spaceship orbits the earth at a height of 500 km. Find the force the earth exerts on it in newtons and pounds. Compare your answer to the weight the spaceship would have on earth. $M_e = 6.0 \times 10^{24}$ kg; $r_e = 6400$ km.

36. The earth has a mass of 6.0×10^{24} kg, whereas the mass of the moon is 7.4×10^{22} kg. Find the force that the earth exerts on the moon. The moon is 3.84×10^8 m from the earth's center.

*37. The radius of the planet Venus is nearly the same as that of the earth. But its mass is only 0.80 that of the earth. (a) If an object weighs W_e on the earth, what do you predict will be its weight on Venus? (b) What do you predict is the acceleration due to gravity on Venus?

*38. The radius of the planet Neptune is about four times larger than that of the earth. Its mass is about 16 times larger than the mass of the earth. If an object weighs W_e on the earth, about how much will it weigh on Neptune? About what is the acceleration due to gravity on Neptune?

*39. How large must the tension in a rope be if the rope is lifting a 20-kg object with (a) a constant speed of 0.50 m/s and (b) a constant upward acceleration of 0.70 m/s²?

*40. How large must the tension in a rope be if the rope is supporting a 15-kg object that is (a) falling at a constant speed of 2.0 m/s, (b) falling with a downward acceleration of 4.0 m/s², (c) falling with a downward acceleration of 9.8 m/s²?

*41. (a) How large a force must a string be able to hold if it is to be used to accelerate a 5-kg mass upward at a rate of 30 cm/s²? (b) What will be the tension in the string if the mass is accelerating downward at 30 cm/s²?

**42. The acceleration due to gravity on the moon is 1.6 m/s². How much does a 5-kg object weigh there? Repeat Problem 31 assuming that the experiment was done on the moon.

**43. For the situation shown in Figure P4.3, each mass is 500 g. How large must the tension P in the upper cord be to move the masses upward at (a) a constant speed of 2.0 m/s, (b) an upward acceleration of 1.20 m/s², (c) a downward acceleration of 1.20 m/s²? (d) What is the tension in the connecting cord in each of these cases?

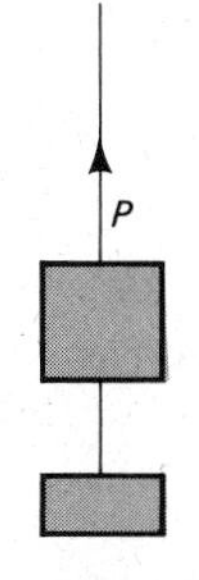

Figure P4.3

**44. Repeat the previous problem for the case in which the lower mass is 200 g and the upper one is 800 g.

**Problems marked with a double asterisk are somewhat more difficult than the average.

American, Frederic Lewis

FRICTION

Performance Goals

When you finish this chapter, you should be able to

1. Explain qualitatively why the following are true for one surface sliding across another: Friction force opposes the motion; maximum static friction force is larger than dynamic friction force; friction force increases with increasing normal force; lubricant decreases friction force.
2. Find the normal force, F_N, for a block of known weight resting on a surface, inclined or flat.
3. Given two of the following three quantities, F_N, μ, f, find the third.

Whenever objects move, they feel the effects of friction. A block sliding across a table is slowed by the friction force on it. A bullet speeding through the air loses speed because of air friction. Even a spaceship in orbit is not completely free from friction. It is slowed by the few gas molecules it must push out of its path. In this chapter we shall learn about this very important force.

4. Find the acceleration of an object on an incline of known angle. Assume that the coefficient of friction between it and the incline is given.
5. Find how far an object of known mass will slide on a level surface. Assume that the initial speed of the object and its coefficient of friction are given.
6. Compute the acceleration of two objects connected by a string. One is on an incline and the other hangs freely. Assume that the coefficient of friction is known.
7. State qualitatively how the friction force varies with speed for an object moving through air or water.
8. Explain what is meant by terminal speed. Tell why an object falling large distances through air or liquid is likely to attain such a speed.

5.1 Measuring the Friction Force

Lay a book on a table and push on it gently with your finger as shown in Figure 5.1. Notice that the book does not move if the pushing force of your finger is small. Why is the push of your finger unable to start the book moving? The reason, of course, is that the table exerts a friction force f on the book. This force exactly balances the pushing force F of your finger. The resultant force on the book is therefore zero. No motion occurs.

But if you push harder, the pushing force of your finger can exceed the force that friction can supply. Then the book begins to move under the action of your finger. Although a friction force f still opposes the motion, it is overpowered by the larger pushing force F.

We can make this clearer by means of the graph in Figure 5.2. There we show how the friction force f compares to the push F of your finger. At small values of F, the friction force can balance it. Therefore f and F vary in exactly the same way; f equals F. But the friction force cannot exceed a certain maximum value f_{max}. When F exceeds this value, it overpowers the friction force and the book begins to move.

But as soon as the book begins to move, a strange thing happens. Try it and see. Once the book begins to move, a somewhat smaller force will keep it moving. This is shown in the graph where f is seen to drop once motion has started. As you see, there are two distinct values of importance for the friction force. One is f_{max}, the maximum value the friction force can have. It occurs when the book is just on the verge of slipping. The other is f_k, the friction force that resists motion of the book once the book begins to slide.

Let us now list the facts that experiments such as the one in Figure 5.1 point out to us.

1. The friction force always opposes the sliding motion. It is therefore tangential to the sliding surface and opposite in direction to the force causing the block to slide.
2. The book will not slide until the pushing force is larger than a certain value; for sliding to occur, F must be larger than f_{max}.
3. Once the book starts to slide, a smaller force is required to keep it sliding. This means that the friction force that opposes the motion is smaller than f_{max} when the object is moving. We call this smaller friction force the *kinetic* (or *dynamic*) friction force f_k. (The word "kinetic" means motional.)
4. The kinetic friction force f_k is often nearly independent of how fast the book is sliding, provided the speed is low.
5. Both f_{max} and f_k are nearly independent of the area of contact between the book and the table.

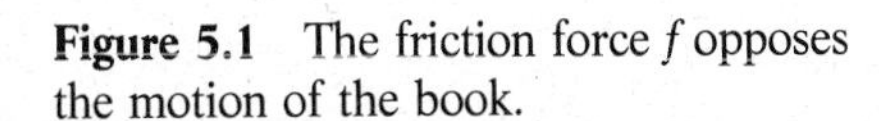

Figure 5.1 The friction force f opposes the motion of the book.

6. Both f_{max} and f_k increase in proportion to the so-called normal force. We shall discuss this force in the next section.

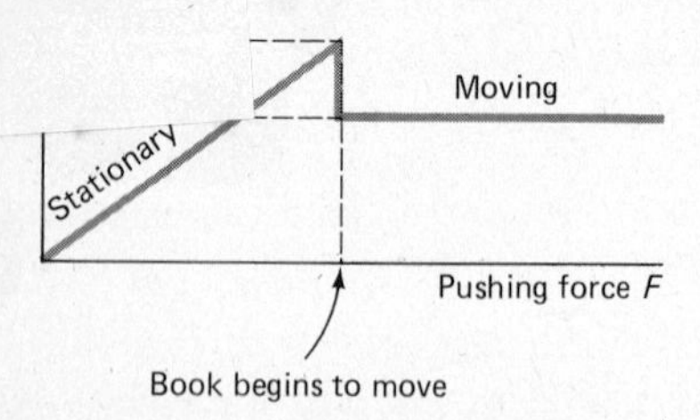

Figure 5.2 The graph shows how the friction force in Figure 5.1 changes as the pushing force is slowly increased.

Let us now look at some of the reasons why the friction force behaves in the way we have found.

5.2 Nature of the Friction Force

If you look at any surface under a high-power microscope, you will see that it is rough. When two surfaces are in contact, the situation is as shown in Figure 5.3(a). The small, rough grooves on the surfaces make it difficult to slide one surface over the other. The two surfaces more or less lock together. We can see several features about friction from this picture.

First, it takes a larger force to start the slipping process than to continue it. Once the surfaces are moving past each other, no time exists for the rough spots to fall together and mesh. The highest points on each surface collide and oppose the motion. But the effect is less than if the surfaces were allowed to settle down close on each other. We therefore recognize that the friction force opposing the motion of an object already sliding is less than that just before the object begins to slide. Dynamic (moving) friction forces are less than those that can occur in static (motionless) situations.

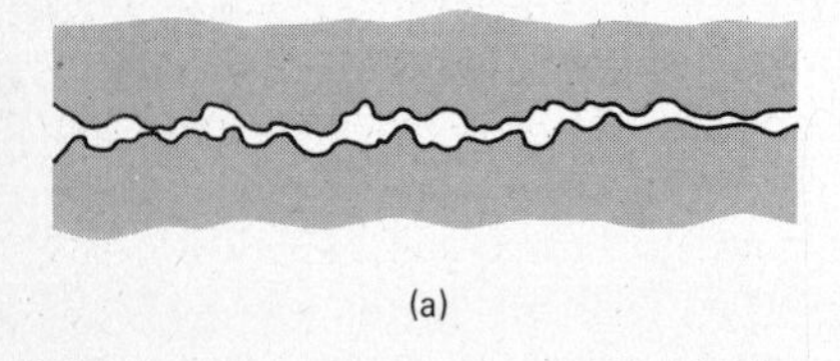

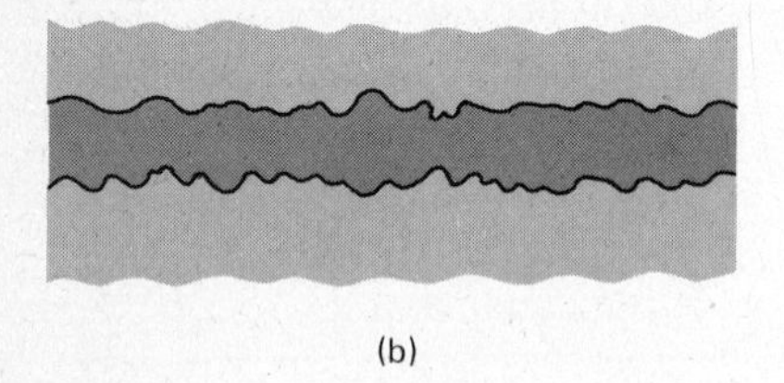

Figure 5.3 The soft layer of oil between the surfaces in (b) makes it easier for the surfaces to slide over each other.

Second, a lubricant can greatly decrease friction forces. In part (b) of Figure 5.3 we see two surfaces with a film of oil between them. The oil layer prevents the rough surfaces from rubbing against each other. As a result, the force needed to cause slipping will be less when a lubricant is present.[1]

Third, the friction force depends on what is called the *normal force*. Notice in Figure 5.3 what would happen if you pressed down heavily on the upper surface. You would force the rough spots to gouge even deeper into each other. The two surfaces would be caused to lock more tightly together. Experiments show that an approximate relation exists between the force pushing the surfaces together and the friction force. To see what the relation is, we must define what we mean by the normal force.

The word "normal" in geometry means perpendicular. This is the meaning we use here. In Figure 5.4 we see several examples of surfaces sitting on each

[1] Even air molecules act as a lubricant. In very high vacuum smooth metal surfaces sometimes weld together.

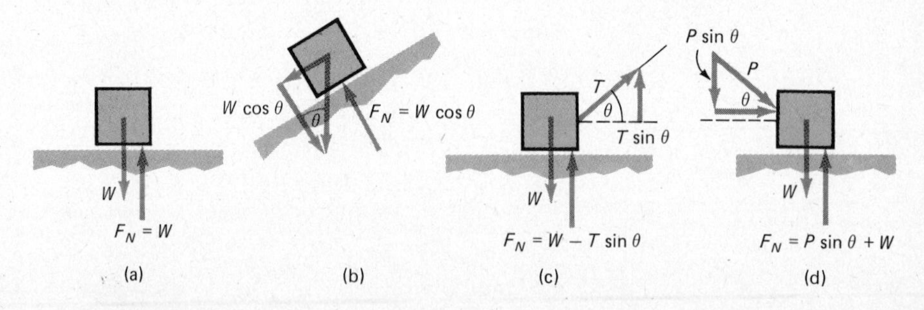

Figure 5.4 The normal force, F_N, depends on the particular situation. It is equal to the perpendicular force the surface must exert to support the object.

other. The normal force is represented by F_N in each case. Notice that the normal force is the perpendicular force the supporting surface exerts on the sliding object. It balances the equal and opposite force the object exerts on the support. As you see, F_N is equal to the force with which the surfaces are pushed together. Be sure you understand the value given for F_N in each case shown.

5.3 Friction Coefficients

Experiment shows that both f_{max} and f_k are proportional to the normal force F_N. If another identical book is set on top of the one shown in Figure 5.1, then the normal force would be doubled. As a result, both f_{max} and f_k would double. For two books stacked on the first, F_N would be three times as large and so would f_{max} and f_k. In other words, both

$$f_{max} \sim F_N \qquad \text{and} \qquad f_k \sim F_N$$

To change these proportions to equations, we use a proportionality constant. We represent the unitless constant by the Greek letter mu, μ. Then

$$f_{max} = \mu_s F_N \tag{5.1}$$

$$f_k = \mu_k F_N \tag{5.2}$$

where μ_s applies to the static (motionless) case and μ_k applies to the dynamic case. They are called the static and dynamic (or kinetic) *coefficients of friction*, respectively. Typical values for them are given in Table 5.1. Let us now see how to use these coefficients.

Table 5.1 COEFFICIENTS OF FRICTION

Surfaces in Contact	Dynamic (μ_k)	Static (μ_s)
Rubber tire on concrete (dry)	0.7–0.9[a]	1.0[a]
Rubber tire on concrete (wet)	0.5–0.7	0.8
Metal on metal (lubricated)	0.07	0.1
Waxed wood on snow	0.05	0.07
Wood on wood	0.3	0.5
Steel on Teflon	0.04	0.04

[a] All values are approximate.

EXAMPLE 5.1 Suppose in Figure 5.1 that the book is replaced by a block of wood that has a mass of 4.0 kg. It slides on a wooden table. Assume the coefficients in Table 5.1 are correct. How large a force F is required to (a) start the block moving and (b) keep it moving?

Solution The table pushes up on the block with a force just large enough to balance the weight of the block. Therefore F_N = weight. Because its weight is

$$\text{Weight} = mg = (4.0\ \text{kg})(9.8\ \text{m/s}^2) = 39\ \text{N}$$

we have that $F_N = 39$ N.

The force needed to start the block moving will be just slightly larger than f_{max}. Therefore, in part (a) F is approximately equal to f_{max} and so

$$F = \mu_s F_N = (0.5)(39\ \text{N}) = 20\ \text{N}$$

When the block is moving, the friction force holding it back is

$$f_k = \mu_k F_N = (0.3)(39\ \text{N}) = 12\ \text{N}$$

This large a force is required to keep the block moving. ∎∎

∎ **EXAMPLE 5.2** In Figure 5.4, $W = 40$ N, $P = T = 10$ N, and θ (where labeled) is 37°. Find F_N for each case.

Solution In part (a) $W = F_N$ and so $F_N = 40$ N.

In part (b) the component of the weight perpendicular to the surface is $W \cos\theta$. Therefore the surface must push back with a force $W \cos\theta$. So $F_N = W \cos 37° = 32$ N.

In part (c) the weight is supported by two forces, both F_N and the vertical component of **T**. As a result,

$$F_N + T \sin 37° = W$$

Placing in the values gives $F_N = 40\ \text{N} - 6\ \text{N} = 34$ N.

In part (d) the surface must support both the weight and the vertical component of the push $P \sin\theta$. Therefore

$$F_N = P \sin 37° + W$$

from which $F_N = 46$ N.

Notice that because $f = \mu F_N$, the friction force in part (d) will be greater than in parts (a) and (c).[2] ∎∎

∎ **EXAMPLE 5.3** Assuming $\mu = 0.90$ between a car's tires and the pavement, how far will a car skid if its initial speed is 28 m/s (about 60 miles/h)? Why can a car stop in a shorter distance if its brakes do not lock?

Solution The friction force acting on the car is μF_N. But F_N is the weight of the car. Hence

$$f = \mu \times (\text{car's weight}) = \mu mg$$

where m is the car's mass. Using $F = ma$ with $-f$ being the unbalanced force, we have

$$-\mu mg = ma$$

[2] Where no ambiguity results, the subscripts on both f and μ will be omitted.

from which the car's acceleration is

$$a = -\mu g$$

(What assumption are we making about the direction that is positive?)

We can now solve a motion problem to find the stopping distance. Known are $a = -\mu g$, $v_0 = 28$ m/s, $v_f = 0$. Using $2ax = v_f^2 - v_0^2$ gives

$$x = \frac{v_f^2 - v_0^2}{2a} = \frac{-(28\text{ m/s})^2}{-2\mu g}$$
$$= \frac{(28\text{ m/s})^2}{2\,(0.90)\,(9.8\text{ m/s}^2)} = 44\text{ m}$$

Compare this with measured stopping distances as given, for example, in the magazine *Consumer Reports*.

Because the coefficient of sliding friction is smaller than the static coefficient, and since the static coefficient applies to a rolling tire on the verge of sliding, a sliding tire exerts less of a stopping force than is possible with a well-controlled rolling tire. ■■

■ **EXAMPLE 5.4** How does the friction force stopping a skidding car depend on the tread width of the car's tires?

Solution The friction force f is equal to $\mu_k F_N$. But no matter what type of tires the car has, F_N must always equal the weight of the car. (This assumes a level road, of course.) Because $f = \mu_k F_N$ and $F_N = W$, we have $f = \mu_k W$. We see that, to this approximation, the friction force does not depend on the area of contact between tire and pavement. In practice, however, this conclusion is not correct. A discrepancy arises because of the abrading action the rough pavement causes on the rubber tire. Because the amount of rubber abraded depends on the contact area, μ becomes a function of F_N, and this complicates the situation. ■■

■ **EXAMPLE 5.5** As shown in Figure 5.5, a rope pulls on a block with a force of 70 N. Find the acceleration of the block if the coefficient of friction is 0.30 between the surfaces.

Solution Notice that the rope does two things. It pulls the block forward with a force of 56 N. But it also lifts the block with a force of 42 N, thereby decreasing the normal force. To find the normal force F_N, we make use of the fact that the vertical forces must balance. Otherwise the block would move up or down. Therefore

$$\Sigma F_y = 0 \rightarrow F_N + 42\text{ N} - (5.0 \times 9.8)\text{ N} = 0$$

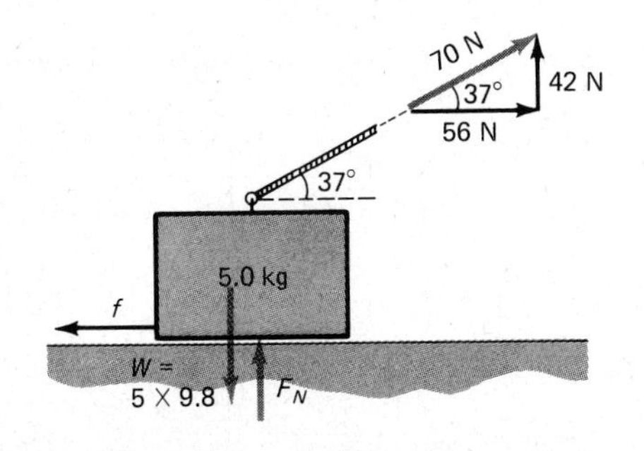

Figure 5.5 If $\mu = 0.30$, what will be the acceleration of the block?

Solving gives $F_N = 7$ N.

We can now find the friction force because

$$f = \mu F_N$$
$$= (0.30) \times (7\text{ N}) = 2.1\text{ N}$$

Now that we know all of the horizontal forces acting on the block, we can write $F = ma$ for it. We have that

$$56\text{ N} - 2.1\text{ N} = (5.0\text{ kg})a$$

Solving, the acceleration is found to be

$$a = \frac{54\ \text{N}}{5.0\ \text{kg}} = 10.8\ \text{m/s}^2$$

5.4 Motion on an Incline

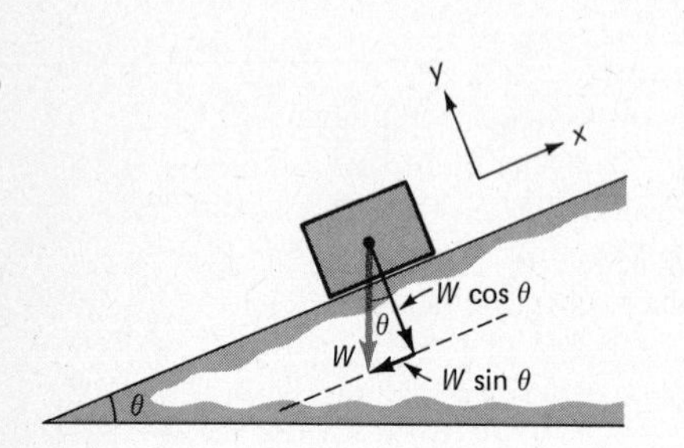

Figure 5.6 The two components of **W** serve different functions. Can you prove from geometry that the two angles labeled θ are equal?

Quite frequently we are interested in objects moving on an incline or hill. We shall examine what to do in such situations. A typical case is shown in Figure 5.6. The object of weight W sits on an incline. Notice the angle labeled θ in the diagram. This angle is called the angle of the incline or the *angle of inclination*.

Because the object will move parallel to the incline, we take our axes as shown in the figure. The x axis is taken along the incline. The y axis is taken perpendicular to it. We shall be concerned with motion along the x direction.

The weight W of the object does two things to the object. As shown in the figure, its y component is $W \cos \theta$. This component of the weight simply pushes the object tightly against the surface of the incline. As you can see from the figure, the x component of W is $W \sin \theta$. It pulls the object down along the incline. It is this part of the weight that makes the object slide down the incline. Always we shall replace W by its two components in situations such as this.

Notice carefully: One component of the weight is perpendicular to the incline's surface; the other component is parallel to the surface. These two components have completely different effects. One holds the object tight on the incline. The other causes the object to slide down the incline. We shall see the importance of this distinction in the examples that follow.

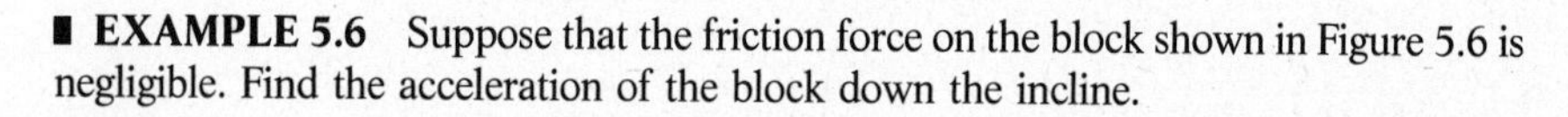

EXAMPLE 5.6 Suppose that the friction force on the block shown in Figure 5.6 is negligible. Find the acceleration of the block down the incline.

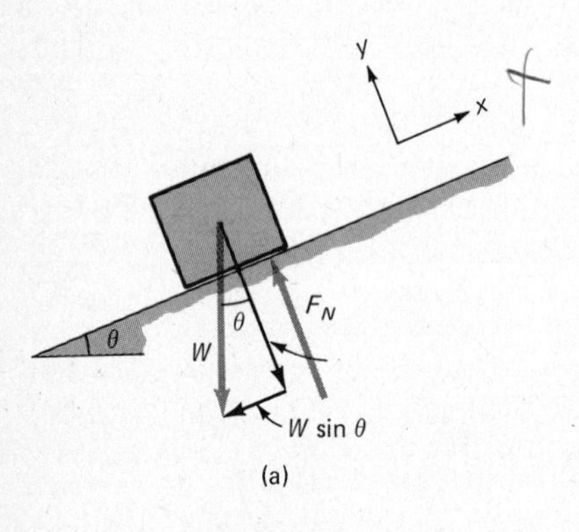

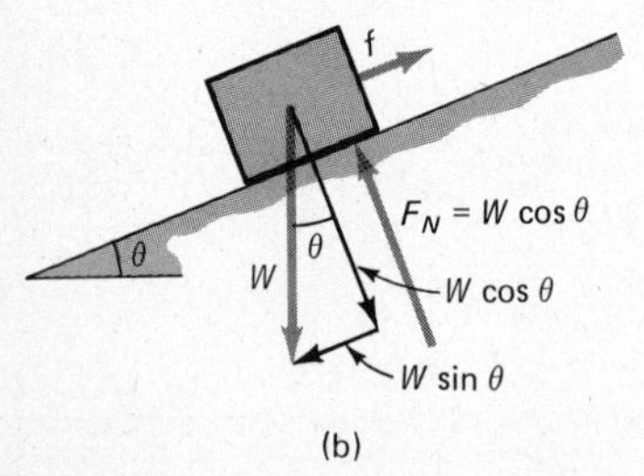

Figure 5.7 How large is the force pulling the block down the incline?

Solution The situation is diagramed in Figure 5.7(a). Because the object does not crash down and break the incline, the y-directed forces must balance. This means that

$$F_N = W \cos \theta$$

Our main concern is with the x-directed forces, because motion occurs in that direction. Only one x-directed force acts on the object. It is $W \sin \theta$. Applying $\Sigma F_x = ma_x$ to this situation, we have (taking x positive up along the incline)

$$-W \sin \theta = ma_x$$

But $W = mg$, and so we have

$$-mg \sin \theta = ma_x$$

Solving for a_x gives

$$a_x = -g \sin \theta$$

This is the acceleration of the object. It is directed down the incline.

Notice what a_x becomes in the following cases. If $\theta = 0$, the incline becomes flat. The object would not slide, of course. Our answer tells us this because $\sin 0° = 0$ and so $a_x = 0$. But if the incline were straight up and down ($\theta = 90°$), the object would fall straight down. In this case our answer tells us that

$$a_x = -g \sin (90°) = -g$$

The object has the free-fall acceleration due to gravity, as it should. ■■

■ **EXAMPLE 5.7** Repeat the previous example supposing the friction forces are not negligible. Assume $\mu_k = \mu$ to be known.

Solution In this case the friction force f shown in part (b) of Figure 5.7 must be included. Because $f = \mu F_N$, and because $F_N = W \cos \theta$, we find that

$$f = \mu W \cos \theta$$

For motion in the x direction, $\Sigma F_x = ma_x$ becomes

$$-W \sin \theta + \mu W \cos \theta = ma_x$$

As before, $W = mg$. After substituting and canceling m from all terms of the equation, we find that

$$a_x = -g \sin \theta + \mu g \cos \theta$$

From this

$$a_x = g(-\sin \theta + \mu \cos \theta)$$

Notice a peculiar fact about this result. To make sense, a_x must be negative. The object certainly will not accelerate up the incline—the positive x direction. To be negative, our result tells us that $\sin \theta$ must be larger than $\mu \cos \theta$. If the friction coefficient μ is too large, $a_x = 0$ or might even be positive. Both results tell us in that case that the block will not slide. The friction force holds it from sliding. ■■

■ **EXAMPLE 5.8** The dynamic coefficient of friction between the 30-N box and incline shown in Figure 5.8 is 0.30. Find the acceleration of the block up the incline caused by the horizontal force of 80 N pushing on it.

Solution First assure yourself that the three angles labeled 37° are all equal. Now notice that the 80-N push does two things. Its y component, 48 N, pushes the block against the surface. The x component, 64 N, pushes the block up the incline.

To find the friction force f, we need to know F_N. Because the y forces are balanced, we can equate them to zero. Therefore

$$-48 \text{ N} - W \cos 37° + F_N = 0$$

From this

$$F_N = 48 \text{ N} + W \cos 37° = 48 \text{ N} + 24 \text{ N}$$
$$= 72 \text{ N}$$

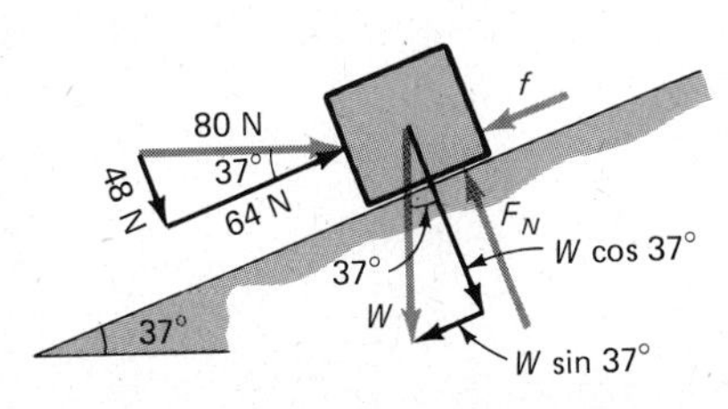

Figure 5.8 Notice that the 80-N push must be balanced partly by F_N.

Now we can use $f = \mu_k F_N$ to find the friction force. It is

$$f = (0.30)(72\text{ N}) = 22\text{ N}$$

To find the acceleration of the block along the incline, we must write $\Sigma F_x = ma_x$. In writing this we recall that $m = W/g = 30\text{ N}/9.8\text{ m/s}^2 = 3.1\text{ kg}$. Then, using the x forces shown in the figure, $\Sigma F_x = ma_x$ becomes

$$64\text{ N} - W \sin 37° - 22\text{ N} = (3.1\text{ kg})a_x$$

Substituting for W its value, 30 N, gives

$$a_x = 7.7\text{ m/s}^2$$

EXAMPLE 5.9 For the situation shown in Figure 5.9, find the acceleration of the blocks and the tension in the connecting cord. Assume that $\mu = 0.40$ between the upper block and the surface.

Solution If the system moves at all, the 5-kg block will fall. The friction force f will be in such a direction as to oppose this motion. That is why its direction is as shown. We take quantities to be positive if they are in the direction of motion.

As in the last chapter, we isolate each block in turn and write $F = ma$ for each. Notice that the weight of the 5-kg block is 5×9.8 N. We have drawn the forces on it. We see from them that, for the 5-kg block, we have for $F = ma$.

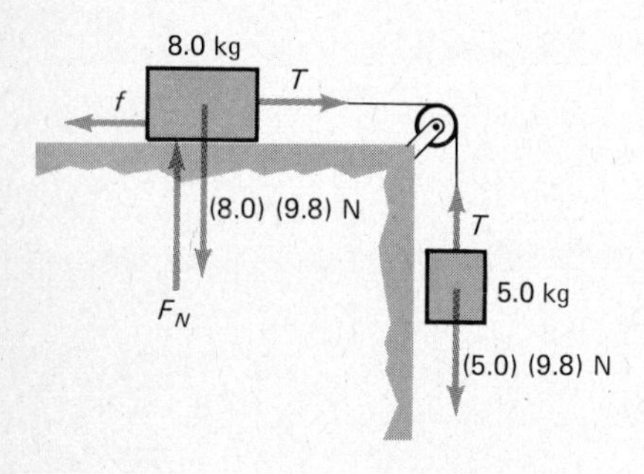

Figure 5.9 We take the direction of motion as the positive direction.

$$(5.0)(9.8)\text{ N} - T = (5.0\text{ kg})a \qquad \text{(a)}$$

To write a similar equation for the 8-kg block we must first find the friction force f. Because the vertical forces on it must balance, we see that $F_N = (8.0)(9.8)$ N. The friction force is

$$f = \mu F_N = (0.40)(8.0)(9.8)\text{ N} = 31\text{ N}$$

Then $F = ma$ for this block becomes

$$T - 31\text{ N} = (8.0\text{ kg})a \qquad \text{(b)}$$

If we add Equation (b) to Equation (a), T will drop out. We then find that

$$49\text{ N} - 31\text{ N} = (13.0\text{ kg})a$$

from which

$$a = 1.38\text{ m/s}^2$$

To find T we can substitute in either Equation (a) or Equation (b). We then find that $T = 42$ N. As a check on this, we notice that the weight of the 5-kg block is about 50 N. Because T is smaller than this, the block should fall—as it does.

EXAMPLE 5.10 The system shown in Figure 5.10 accelerates at 0.50 m/s² when released. How large a friction force resists the motion?

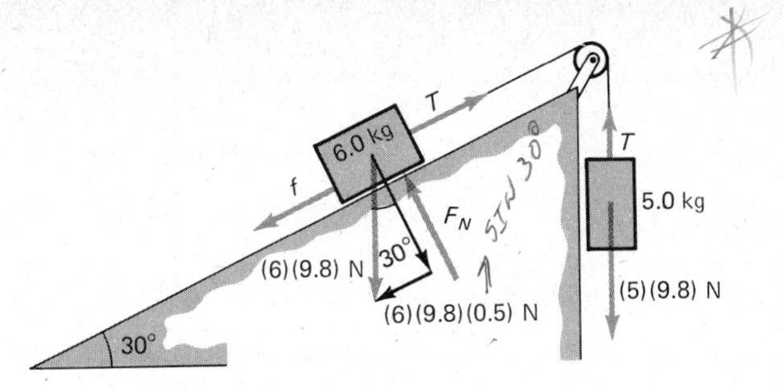

Figure 5.10 How do we decide which way to draw the friction force in a case such as this?

Solution The forces acting on each object are shown in the figure. Notice that the weight of the 5-kg block (about 50 N) pulls on that end. The component of the weight of the 6-kg block pulling it down the incline is about 30 N (that is, $6 \times 9.8 \times 0.5$). As a result, the 5-kg block will fall and the 6-kg block will rise. Because friction always opposes the motion, f must act in the direction shown.

We now write $F = ma$ for each block separately. For the 5-kg block we have

$$(5)(9.8)\text{ N} - T = (5\text{ kg})a$$

Notice that the motion direction is called positive.

For the 6-kg block we have

$$T - f - (6)(9.8)(0.5)\text{ N} = (6\text{ kg})a$$

Solving the two equations simultaneously gives[3]

$$-f + 19.6\text{ N} = (11\text{ kg})a$$

But we were told that $a = 0.50\text{ m/s}^2$, and so we find that

$$f = 14\text{ N}$$

5.5 Terminal Velocity (or Speed)

In the previous sections we have been dealing with friction forces that do not depend seriously on speed. Most friction forces are of this type. However, the friction forces due to wind or water are not of this type. You know that the faster a car goes, the greater is the resistance of the air through which it moves. This simply means that the friction force exerted by the air on the car is largest at high speeds. If you hold your arm out the window of a car, this effect is easily seen. At low speeds your arm is hardly disturbed by the air. But at high speeds your arm is pushed back strongly by the air's friction forces.

Similarly, when a boat moves through water, the boat is held back by friction forces due to the water. In this case, too, the faster the boat moves, the greater

[3] The easiest procedure is to add the two equations together. Then T drops out. Or you can solve one equation for T. Then substitute its value in the other equation.

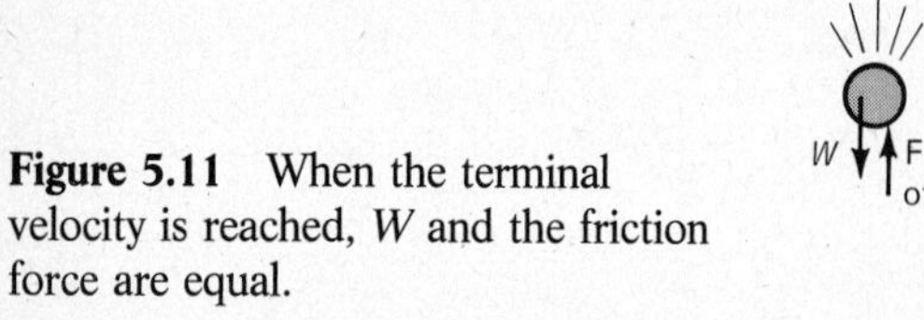

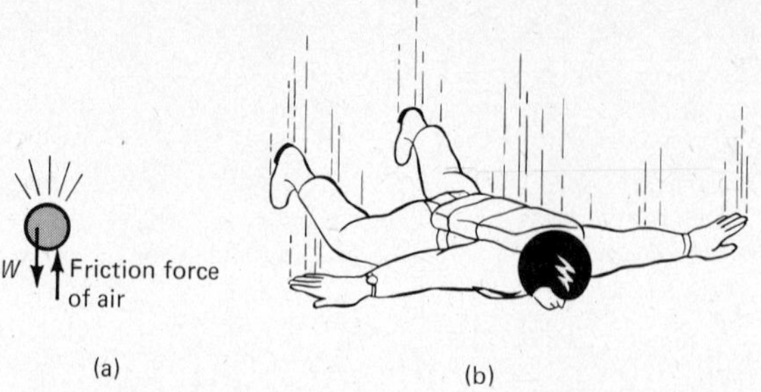

Figure 5.11 When the terminal velocity is reached, W and the friction force are equal.

are the friction forces. The friction force on an object moving through a liquid increases with speed.

Suppose that an object is falling through the air. At first its speed is low and the friction force due to the air rushing past it is small. But as the object speeds up, the friction force opposing its motion increases. Soon the upward friction of the air nearly equals the pull of gravity on the object. [See Figure 5.11(a).] At that time, the vertical forces on the object essentially balance out to zero. The weight of the object is balanced by the friction force of the air rushing past it.

When the speed of the falling object becomes large enough for this to happen, there is no longer an unbalanced force on the object. The object will no longer accelerate. Its speed of fall will remain constant. We call this constant speed the *terminal speed* (or *terminal velocity*) of the object. Typical values are the following for an object falling through air: raindrop, 8 m/s; smoke particle, 0.1 cm/s; man, 80 m/s. Of course, these values depend on the shape and orientation of the falling object. See Figure 5.11(b) for a drawing of a person falling with terminal velocity. Why is the person's orientation important?

Summary

Friction forces oppose the motion of sliding objects. More force is required to start an object sliding than to keep it sliding. Static friction forces are larger than kinetic friction forces when solids slide across solids.

The coefficient of static friction, μ_s, is defined as the ratio of the friction force to the normal force when the object is just on the verge of sliding. The coefficient of dynamic friction, μ_k, is the ratio of the friction force to the normal force when the object is in motion. Coefficients of static friction are larger than the coefficients for dynamic friction.

When dealing with motion along an incline, the x axis is taken parallel to the incline. The y axis is taken perpendicular to it. All forces are split into components along these axes.

Friction forces on objects moving through air or liquids (such as water) depend on speed. The faster the object moves, the larger is the friction force.

An object falling through the air (or a liquid) may eventually move fast enough so that the upward friction force balances the downward pull of gravity. The object then falls with constant speed, its terminal speed.

Questions and Exercises

1. The force you must apply to a push-type broom to push it along the floor depends on the broom's angle. Explain why. Why can't you make the broom move if the handle is too close to the vertical?
2. If given the choice, it is usually better to pull a short but heavy box along the floor than to push it. Explain why.
3. To increase the traction of a car or truck, it is common to load the back end with heavy objects. What is the reasoning behind this procedure?
4. Is there any reason why the static coefficient of friction between two surfaces cannot exceed unity?
5. Why is it that pavement is usually much more slippery when it first becomes wet than after it has rained for a few hours? This effect is most noticeable after a long dry spell. (Hint: Consider the nature of the surface.)
6. In which case can a car accelerate faster from rest: (a) when it is on clean pavement or (b) when it is on a gravel road?
7. Suppose you want to stop a car in the shortest distance possible. Which of the following is the best method: (a) lock the brakes and skid to a stop, (b) apply the brakes just enough so the car will not skid, (c) pump the brakes?
8. It is claimed that cars get better gasoline mileage at 80 km/h than at 100 km/h. Why should this be?
9. What factors influence whether or not a book will slide off the seat of a car as the car is suddenly brought to a stop?
10. A feather has a much lower terminal velocity than a tightly wadded ball of paper of the same weight. Why?

Problems

1. A horizontal force of 300 N is needed to push a crate along the floor at constant speed. What is the friction force on the crate?
2. A box sitting on the floor is pulled by a rope inclined 40° to the horizontal. The box slides with constant speed when the tension in the rope is 500 N. What is the friction force on the box?
3. Find the normal force on each of the blocks shown in Figure P5.1 if (a) the block has a mass of 10 kg and $P = 50$ N, (b) the block weighs 40 lb and $P = 30$ lb.
4. Repeat Problem 3 for the situations shown in Figure P5.2.
5. A 50-kg box sits on a 37° incline. Find the normal force that acts on it. How large a force must friction supply if the box is not to slide?
6. A 20-kg box sitting on a level floor is pulled with a force of 80 N by a rope inclined at 37° above the horizontal. Find the normal force on the box. How large must the friction force be if the box is not to move?
7. A horizontal force of 30 N is needed to start an 8.0-kg box moving along the floor. After the box is in motion, a force of 25 N is required to keep it moving at constant speed. Find μ_s and μ_k for this situation.

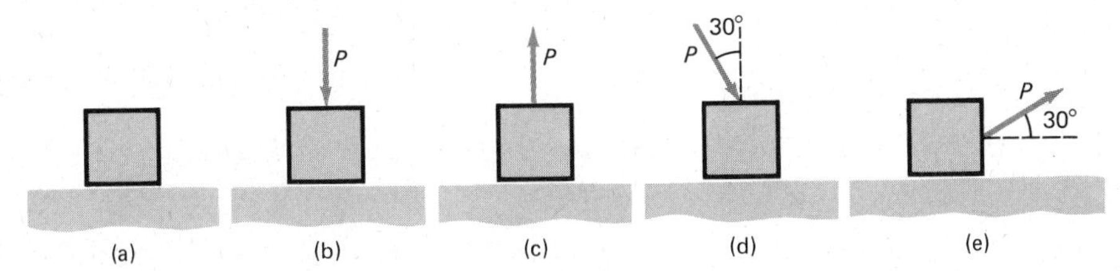

Figure P5.1

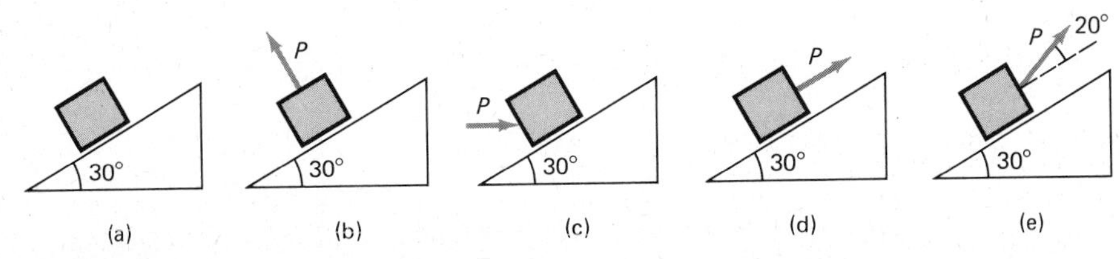

Figure P5.2

8. A 30-kg crate is to be pulled along level cement by a horizontal rope. It is found that a force of 220 N is needed to start it moving, but only 90 N is needed to keep it moving. Find the coefficients of static and dynamic friction in this situation.
9. A box weighing 800 N is to be pushed across the floor by a horizontal force. If the coefficient of static friction is 0.80 and that of dynamic friction is 0.60, how large a force is required to start the box moving? To keep it moving?
10. A 12-kg box sits on a floor for which $\mu_s = 0.70$ and $\mu_k = 0.60$. How large a horizontal force is needed to start it moving? To keep it moving at constant speed?

**11. Refer to Figure P5.1(d). (a) If μ_k is 0.20 and the box has a mass of 5.0 kg, how large must P be to keep the box moving with constant speed? (b) Repeat for a box that weighs 40 lb.

**12. Repeat Problem 11 for the situation shown in Figure P5.1(e).

*13. One way to measure the coefficient of friction is as follows. A box sits on a variable-angle incline. It is found that the box will start to move when the angle is increased to 50° or larger. Once the object is moving, it will continue to move if the angle of the incline is larger than 40°. From these data, find the static and dynamic coefficients of friction.

*14. A 200-N box sits on a level floor. It is pushed by a force P directed downward at an angle of 37° below the horizontal. If $P = 140$ N to start it moving and 100 N to keep it moving, find the static and dynamic friction coefficients.

**15. The static and dynamic coefficients of friction for a 150-kg box on a level floor are 0.80 and 0.70. How large a force, pushing down at an angle of 37° to the horizontal, is needed to start the box moving? To keep it moving?

*16. In Figure P5.2(d) the 5.0-kg box moves up the incline with constant speed when $P = 40$ N. Find μ_k.

*17. In Figure P5.2(c) the 2.0-kg box moves up the incline at constant speed when $P = 30$ N. Find μ_k.

*18. In Figure P5.2(e) the 10-kg box moves up the incline at constant speed when $P = 90$ N. Find μ_k.

*19. Refer to Figure P5.2(d). How large is P if the box moves up the incline at constant speed? The box has a mass of 5.0 kg and $\mu_k = 0.40$.

**20. Refer to Figure P5.2(c). How large is P if the 2.0-kg box moves up the incline at constant speed? $\mu_k = 0.30$.

**21. Refer to Figure P5.2(e). How large is P if the 10-kg box moves up the incline at constant speed? $\mu_k = 0.10$.

22. The friction force between a 5.0-kg box and the floor is 20 N. How large a horizontal force is needed to give the box an acceleration of 3.0 m/s²?
23. The dynamic coefficient of friction between a 4.0-kg box and the floor is 0.60. How large a horizontal force is needed to give the box an acceleration of 2.0 m/s²?

**24. Refer to Figure P2.19 (page 48). The man shown there is climbing a wall. He weighs 800 N and his center of gravity is 1.00 m from the wall. His armpits are 1.40 m from the wall. (a) How large a friction force must the wall supply to his shoes? (b) What must be the coefficient of friction between the shoes and the wall?

*25. A 500-g block moving at 80 cm/s slides 60 cm across a table before stopping. (a) What was its deceleration? (b) How large a friction force stopped it? (c) What was μ_k in this case?

*26. A 1500-kg car going 20 m/s skids to a stop in a distance of 70 m. (a) What was the car's deceleration? (b) What was the coefficient of friction for the tires sliding on the road?

**27. Refer to Figure P5.3. (a) If $m_1 = 3.0$ kg, $m_2 = 2.0$ kg, and the coefficient of friction between block m_1 and the surface is 0.40, find the acceleration of m_1 and the tension in the cord. (b) Repeat for a friction coefficient of 0.80.

**28. How large a force must pull to the left on m_1 in Figure P5.3 to give the system an acceleration of 0.30 m/s² (a) to the left and (b) to the right? Take $m_1 = 3.0$ kg, $m_2 = 2.0$ kg, and $\mu = 0.40$ for block m_1.

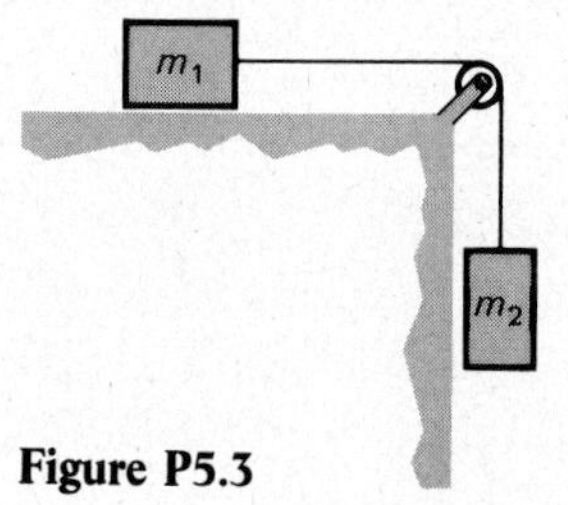

Figure P5.3

**29. In Figure P5.4 the masses are $m_1 = 10.0$ kg, $m_2 = 3.0$ kg. The coefficient of friction for each block is 0.20. Find the acceleration of the blocks and the tension in the connecting cord.

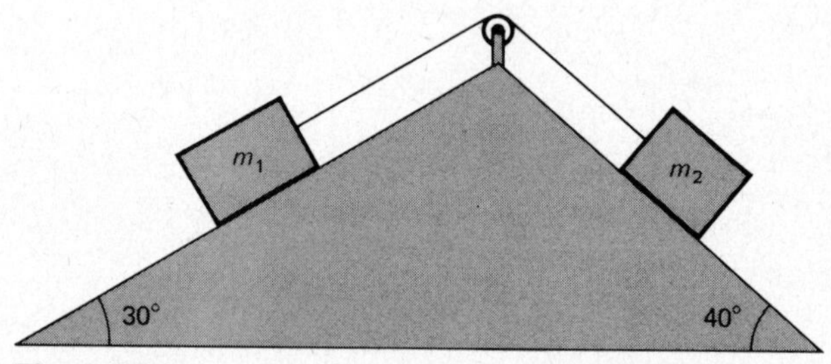

Figure P5.4

30. A tiny metal sphere is timed as it falls through a column of oil. The following data are taken for it. What is the terminal speed of the particle?

Time (s)	→ 0	20	40	60	80	100	120
Depth (mm)	→ 1.67	2.38	3.18	4.01	4.86	5.71	6.56

*Problems marked with an asterisk are not as easy as the unmarked ones.
**Problems marked with a double asterisk are somewhat more difficult than the average.

Griffiths, DPI

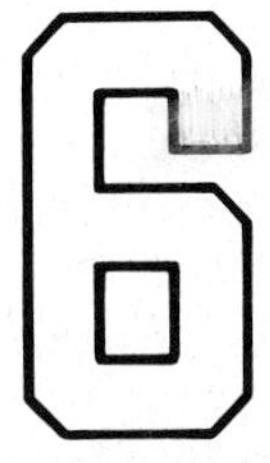

WORK, POWER, AND ENERGY

Performance Goals

When you finish this chapter, you should be able to

1. Give the technical definition of work together with its defining equation(s). Define each quantity in the equation(s).
2. Define joule (and foot-pound).
3. Compute the work done by a force on an object during a displacement. Assume that the force and displacement are given.
4. Give and explain several

Our modern industrial society is dependent on the availability of vast amounts of energy to be able to do the many kinds of work that our civilization requires. Long ago each person supplied the energy for the work that he or she did. Later, animals were domesticated and aided people in doing various chores. With the invention of engines and complex machines, people greatly increased their ability to do work. Today energy is in short supply. Progress of our civilization is being retarded because of the lack of energy to carry out numerous tasks. Work, energy, and our world are closely linked. In this chapter you will learn the meaning of work and energy so you will be able to recognize and use them in technology.

situations where a force on an object does no work on the object. Some of these should include objects that are in motion.

5. Tell whether the work done by a force is zero, positive, or negative in a given situation.
6. Compute the work done against gravity in carrying a mass *m* from one given point to another.
7. Give the technical definition of power and the units in which power is measured.
8. Given power in one of the following units (watts, foot-pounds per second, or horsepower), find its value in the other two units systems.
9. Find the unknown quantity in Equations 6.3 and 6.4 for power if all but one quantity are given.
10. Compute how much work a machine can do in a given time if its power output is known. Find how large a force the machine can exert as it moves an object at a known speed.
11. Explain what is meant by energy and give its units. Given several examples of objects that possess energy, select those objects that have (a) kinetic energy (KE), (b) gravitational potential energy (GPE), and (c) both KE and GPE.
12. Find the GPE of an object of known mass relative to any specified reference level.
13. Compute the KE of an object if you are given its mass and velocity.
14. Give three illustrations of each of the following: an object gains KE from work done on it; an object loses KE in doing work; an object's GPE decreases while

6.1 Work: Its Technical Meaning

The word "work" means different things to different people. We say "we go to work," "we work in school," "a machine or device works," and so on. In science and technology, though, we must define exactly what we mean by work. As you will see, the technical meaning is not the same as the common meaning.

We define work in the following way. Suppose, as shown in Figure 6.1, that a force **F** pulls an object through a displacement **s**. If F_s is the component of the force in the direction of **s**, then the work done by the force is

$$\text{Work} \equiv F_s s \qquad (6.1)$$

We can express this equation in words. It says that

The work done by a force is the product of two quantities: (1) the magnitude of the displacement and (2) the magnitude of the force component in the direction of the displacement.

There are two important points we should notice about this definition. First, no matter how strongly a force pushes or pulls, it does no work if it causes no displacement. You can push as hard and long as you wish against the wall of a room. If the wall doesn't move, you have done no work on the wall in a technical sense. Suppose that you try to lift a huge rock. If you are unable to move it, our definition tells us that you have done no work on it.[1]

But it is not enough that the object moves. The force must have a component in the direction of the motion. To see this, refer to Figure 6.2. A vertical force **F** is required to hold the pail. This force may or may not do work on the pail. If the pail is moving horizontally, not vertically, the force does no work on the pail because there is no force component in the direction of the displacement. *For work to be done by a force, the force must have a component in the direction of the displacement.*

The defining equation for work is often written in a form different from Equation 6.1. Notice in Figure 6.1 that $F_s = F \cos \theta$; θ is the angle between the force vector **F** and the displacement vector **s**. Substituting this value for F_s in Equation 6.1 gives

$$\text{Work} = Fs \cos \theta \qquad (6.2)$$

Equation 6.2 is often used in place of Equation 6.1 to define work. The two are equivalent, of course. Equation 6.2 can be stated in words as follows.

[1] In these examples the person supplying the force may become tired. Various parts of the person's body are doing work. The heart is pumping blood, for example. Even so, the force we are talking about is doing no work.

its KE increases by an equal amount; an object gains GPE from work done on it.

15. State the law of conservation of energy in your own words.
16. Write the work-energy theorem in words as well as in symbols. Explain carefully what is meant by each term in it. Pay particular attention to when a term is positive and when negative.
17. Write the work-energy theorem equation for a given simple physical situation that involves the loss of KE in doing friction work.
18. Write the work-energy theorem equation for a given simple physical situation that involves a change of KE and GPE. Assume that friction work is also done.

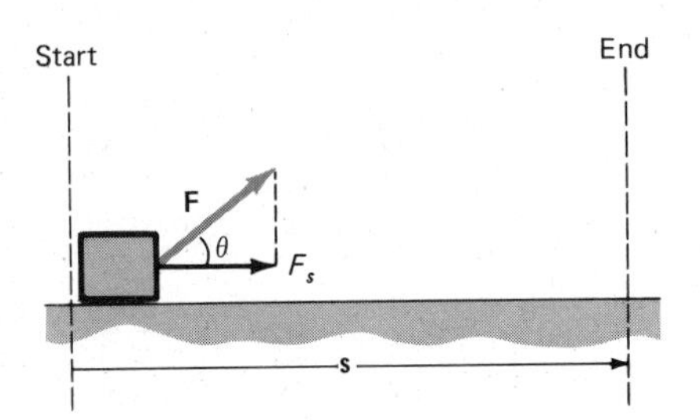

Figure 6.1 In moving the object from the start position to the end position, the work done by **F** is $F_s s$. This is the same as $(F \cos \theta)s$.

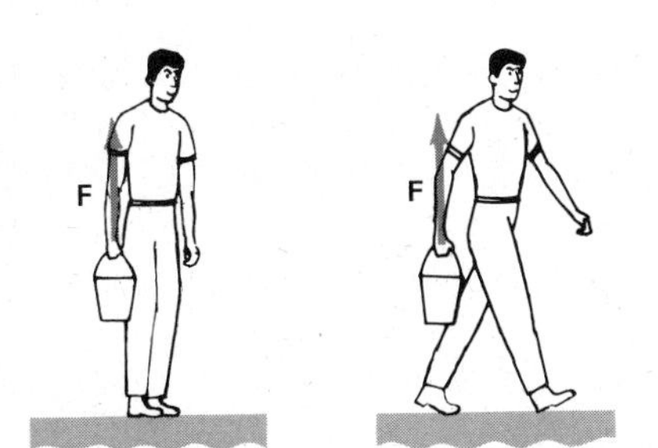

Figure 6.2 Why is the force **F** doing no work in each of these cases?

The work done by a force F during a displacement s is equal to the product of three quantities: (1) the force magnitude, (2) the magnitude of the displacement, and (3) the cosine of the angle between them.

You should now go back and apply this alternate form to the previous examples. Why is no work done on a motionless wall by a person pushing on it? On a rock that refuses to budge? On a pail held stationary? By the supporting force on a pail moved horizontally?

6.2 Units of Work

We need units in which to measure work. The units are easily defined in terms of Equation 6.1. Because work $= F_s s$, its units are a unit of force multiplied by a unit of length.

The SI unit of work is a force unit (newton) multiplied by a length unit (meter). It is therefore a newton-meter. But, more commonly, this unit is called the joule[2] (rhymes with *rule*) and its abbreviation is J. We have $1\ \text{J} = 1\ \text{N}\cdot\text{m}$. In the British system forces are measured in pounds and distances in feet. The unit of work is the foot-pound (ft·lb). (It is customary to call it a foot-pound and not a pound-foot.)

$$1\ \text{J} = 0.738\ \text{ft}\cdot\text{lb}$$

There is a third unit of work often used in electrical applications. It is called the kilowatt-hour (kW·h). We shall see why it is useful in a later section. For now we state simply that

$$1\ \text{kW}\cdot\text{h} = 3.6 \times 10^6\ \text{J}$$

You may already have heard of this unit. It is the unit of work that the power company uses when selling electricity to you.

Those people who still make use of the dyne as a force unit often use the erg as a unit of work. For us it is sufficient to know that $1\ \text{J} = 10^7$ erg.

In atomic physics the electron volt (eV) is used for work and energy. It is given by

$$1\ \text{eV} = 1.602 \times 10^{-19}\ \text{J}$$

where the conversion factor is the charge on the electron.

▌EXAMPLE 6.1 How much work is done in slowly lifting a 2.0-kg book from the floor to a height of 1.2 m above the floor?

[2] After James Joule, who did early experiments to show the relation between work, heat energy, and electrical energy.

Solution The lifting force will be almost exactly[3] equal to the weight of the book, (2.0)(9.8) N = 19.6 N. It is directed in the same direction as the displacement and so

$$\text{Work} = F_s s = (19.6 \text{ N})(1.2 \text{ m}) = 23.5 \text{ N}\cdot\text{m} = 23.5 \text{ J}$$

■ **EXAMPLE 6.2** How much work is done in slowly lifting a 2.0-lb textbook from the floor to a height of 4.0 ft above the floor?

Solution The lifting force will be almost exactly 2.0 lb. It is directed in the same direction as the 4.0-ft displacement. Therefore

$$\text{Work} = F_s s = (2.0 \text{ lb})(4.0 \text{ ft}) = 8.0 \text{ ft}\cdot\text{lb}$$

■ **EXAMPLE 6.3** Suppose in Figure 6.1, a 600-N force pulls a 50-kg box a distance of 2.0 m along the floor. If $\theta = 37°$, how much work does the 600-N force do?

Solution We have no need for the mass of the box. We only need $F_s = 600 \cos 37°$ N and $s = 2.0$ m. Then

$$\text{Work} = F_s s = (600 \cdot 0.80 \text{ N})(2.0 \text{ m}) = 960 \text{ N}\cdot\text{m} = 960 \text{ J}$$

The alternative expression for work, $F_s \cos\theta$, would give the same result.

6.3 Negative and Positive Work

Work is not a vector quantity. It is a scalar. Even though force and displacement are vectors, we do not give direction to their product, work. However, the work done by a force can be either positive or negative, as we shall now show.

In Figure 6.3 we see a force **F** doing work on an object. But the value of the work done in (a) is quite different from the value in (b). In (a) the force is in the direction of the motion, so we have that

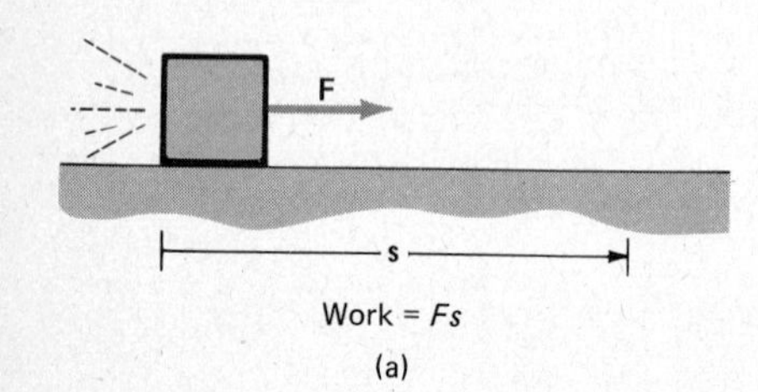

$$\text{Work} = F_s s = Fs$$

Or, in terms of Equation 6.2, because the angle between **F** and **s** is zero, we have

$$\text{Work} = Fs\cos\theta = Fs\cos 0° = Fs$$

In part (b) of the figure the results are not the same. We have there a moving object being stopped by a negatively directed force. Then

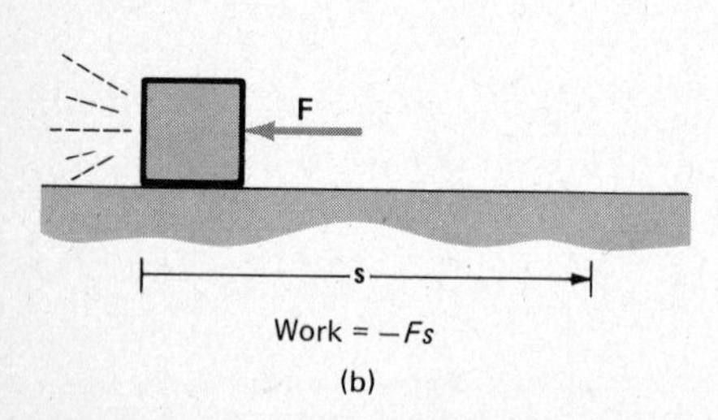

Figure 6.3 When the force and displacement are in opposite directions, the work done by the force is negative.

$$\text{Work} = F_s s = (-F)s = -Fs$$

or, in terms of Equation 6.2,

$$\text{Work} = Fs\cos\theta = Fs\cos 180° = Fs(-1) = -Fs$$

[3] When the lifting is first started, a slight acceleration is given to the book. A force slightly larger than 19.6 N is needed to produce this very small acceleration.

As we see, the force does negative work in Figure 6.3(b). Apparently, the work will always be negative when the force and the displacement have opposite directions.

We are concerned about this difference in sign because positive work and negative work have very different effects. For example, when you push on a car in the direction of its motion, you speed the car up. The work done is positive. But when you push on a car in a direction opposite to its motion, you slow the car down to a stop. *Negative work tends to stop the motion.*

6.4 Work Against Gravity

Let us now look at the situation shown in Figure 6.4. In (a) a force is lifting an object; in (b) a force is lowering the same object. If the object is not accelerating appreciably, then the supporting force is essentially mg, the weight of the object. To lift or lower slowly an object of mass m, a force $F = mg$ is required.

The work done by this force in *lifting* an object through a height h is

$$\text{Lifting work} = Fs\cos\theta = (mg)(h)(1) = mgh$$

But in *lowering* the object, the supporting force does negative work. This follows because

$$\text{Lowering work} = Fs\cos\theta = (mg)(h)(-1) = -mgh$$

We see, then, that there is a simple rule for the work done by a supporting force.[4] If the object is lifted, the work done is mgh. If the object is lowered, the work done by the supporting force is $-mgh$. Because the supporting force balances the pull of gravity, we often call this work "work done against gravity."

We now ask if this same expression for work is correct even if the object is lifted along a nonlinear path. The answer to this question is simple: The work done against gravity is *always* mgh, where h is the distance the object is lifted (assuming the weight of the object is constant throughout the distance h).

To prove this assertion, suppose that an object is lifted from A to B in Figure 6.5(a) by moving it over the path shown. Notice that during the movement the supporting force is mg upward. No appreciable force opposes motion in the horizontal direction. For that reason only a negligible horizontal force is needed to move the object from A to B.

To understand what is going on here, look at part (b) of Figure 6.5. Suppose that instead of following the original path, we lift the object along the jagged path shown by the arrows. As discussed before, the lifting force does no work during the horizontal displacements. No work is done on the object except during the up-and-down motions. The total work done in moving the object from A to B along the jagged path is

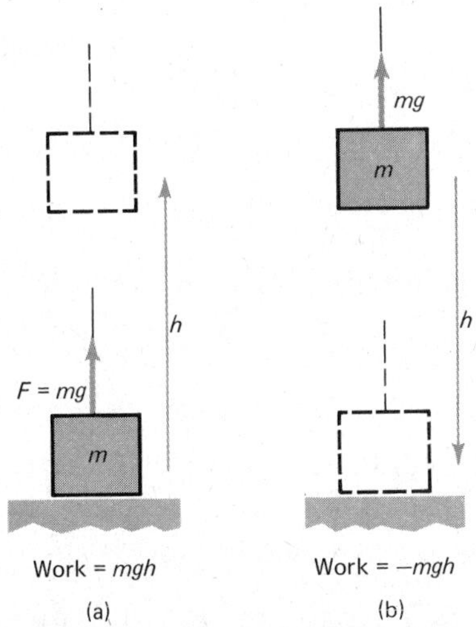

F = mg
m
h
Work = mgh
(a)
mg
m
h
Work = −mgh
(b)

Figure 6.4 Work done against gravity is mgh or $-mgh$.

[4] This assertion is strictly true only if the height is small enough so the weight of the object does not change appreciably.

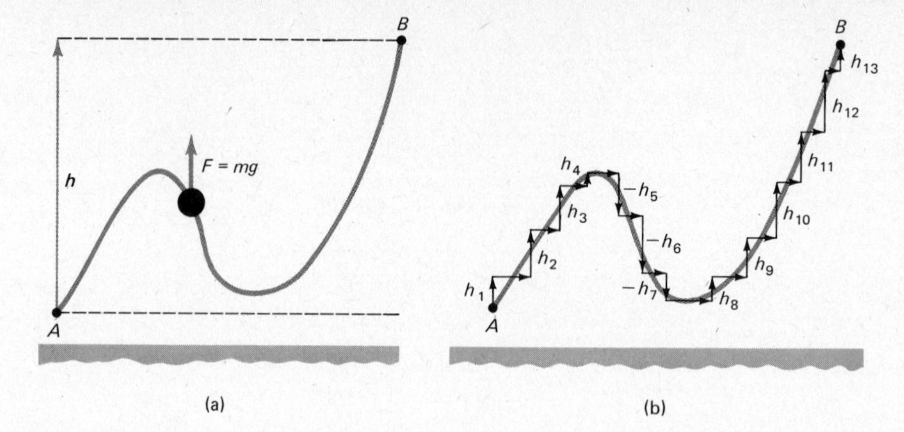

Figure 6.5 The work done against gravity in moving a mass m from A to B by any path is mgh.

$$\text{Work} = mgh_1 + mgh_2 + mgh_3 + mgh_4 - mgh_5 - mgh_6 - mgh_7 + mgh_8 + \cdots + mgh_{13}$$

Or, after factoring out mg,

$$\text{Work} = mg(h_1 + h_2 + h_3 + h_4 - h_5 - h_6 - h_7 + h_8 + h_9 + h_{10} + h_{11} + h_{12} + h_{13})$$

Notice that the h's that lower the object contribute negative work.

But now comes the important point. The sum of all these plus and minus h's must be h, the upward displacement of B from A. Therefore, the work done by carrying the object along the jagged path is mgh. This is the same as the work done in simply lifting the object straight up through a distance h.

We would obtain the same result for any similar path. If we wished, we could make the jags extremely tiny, so there would be perhaps a billion of them from A to B. The jagged path would then be so close to the original curved path that no one could tell the difference. Even so, the work done on this path from A to B would still be mgh.

It therefore makes sense to conclude our original assertion. Let us now state it again.

The work done against gravity on an object of mass *m* in lifting it through a height *h* by any path is *mgh*.

If the object is lowered instead of lifted, then the work done is $-mgh$.

The work done in moving an object from place to place under the action of gravity does not depend on the path by which the object was moved. Only the displacement is important in computing the work done, as we have just seen. Because of this property, we say that the gravitational force is, by definition, a *conservative force.* In our study of electricity later in this text, we shall encounter another important type of conservative force. It is the electrostatic force.

6.5 Power

In many applications involving work, the rate at which work is done has importance. For example, a man with a shovel can move the same amount of dirt as a woman with a bulldozer. But the man will require a much longer time. Often the man's rate of doing the work would be impractical. Similarly, a 1.0 liter engine can accelerate even a large car to a high speed. But a 3.0 liter engine can do it much more quickly. These are examples involving the rate at which work is done. We call this quantity *power.*

Power is a measure of how fast work is being done. If a certain amount of work is done in a time, then we define

$$\text{Power} \equiv \frac{\text{work done}}{\text{time}} \tag{6.3}$$

From this definition we see that the units of power are those of work divided by time. In the SI system the unit of power is the joule per second, which is named the watt (W).[5] In the British system the unit of power is the foot-pound per second (ft·lb/s). It has no other name.

Another widely used unit is the horsepower (hp). The relation between it and our standard units is:

$$1 \text{ hp} = 746 \text{ W}$$
$$1 \text{ hp} = 550 \text{ ft}\cdot\text{lb/s}$$

We should remember, however, that horsepower should not be used in Equation 6.3. Only watts (and foot-pounds per second) are to be used in it.

Now that we have defined power and its unit, the watt, we can understand one of the work units discussed earlier. It is the kilowatt-hour. If we rearrange Equation 6.3 we have

$$\text{Work done} = \text{power} \times \text{time}$$

We see that the unit of work could be taken to be the product of a power unit with a time unit. Sometimes it is convenient to measure power in kilowatts (that is, 1000 watts) and time in hours. The product of these two quantities gives the work done (or energy used) in units of kilowatt-hours (kW·h). This unit is widely used when discussing practical applications of electricity.

Before leaving the subject of power, we should point out another expression for power that is sometimes convenient. We can obtain it by rearranging the power equation.

[5] After James Watt, who developed the steam engine.

$$\text{Power} = \frac{\text{Work}}{\text{time}}$$

$$P = \frac{F_s s}{t} = F_s \frac{s}{t}$$

But s/t is simply the velocity. Therefore, the equation becomes

$$P = F_s v \tag{6.4}$$

where v is the velocity at which the force works.

■ **EXAMPLE 6.3** An 80,000-kg truck follows a winding path up a mountain. It rises 700 m in elevation in 40 min. What average horsepower must the truck's engine produce to carry out this lifting work? Of course, the engine must produce more power than this to overcome friction, etc.

Solution The lifting work done by the truck is

$$\begin{aligned}\text{Lifting work} = mgh &= (80{,}000\ \text{kg})(9.8\ \text{m/s}^2)(700\ \text{m}) \\ &= 5.5 \times 10^8\ \text{J}\end{aligned}$$

It took the following time to do this work:

$$\text{Time} = (40\ \text{min})(60\ \text{s/min}) = 2400\ \text{s}$$

Using the definition of power, we have

$$P = \frac{\text{work}}{\text{time}} = \frac{5.5 \times 10^8\ \text{J}}{2.4 \times 10^3\ \text{s}} = 2.3 \times 10^5\ \text{W}$$

Because 746 W = 1 hp, this becomes

$$P = (2.3 \times 10^5\ \text{W})(1\ \text{hp}/746\ \text{W}) = 310\ \text{hp}$$

To accomplish the task in the time allowed, the output power of the truck's engine must be at least 310 hp. ■■

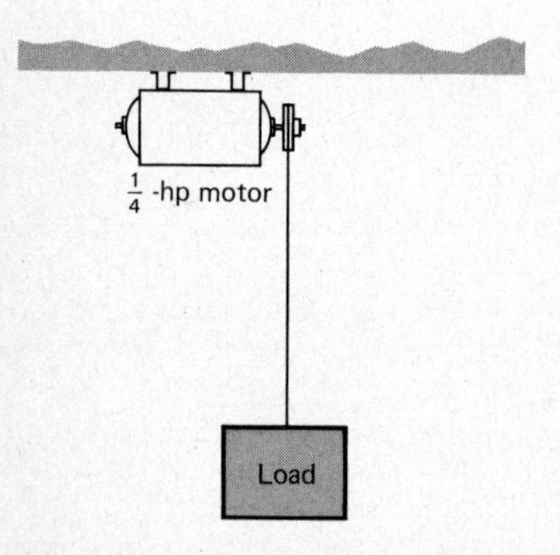

Figure 6.6 The load the motor can lift depends on the speed with which the load is lifted.

■ **EXAMPLE 6.4** Suppose the $\frac{1}{4}$-hp motor shown in Figure 6.6 lifts its load at a speed of 0.20 m/s. How large a load (maximum) could it lift at this speed?

Solution Ideally a $\frac{1}{4}$-hp motor would be able to do work at the rate of

$$(\tfrac{1}{4})(746)\ \text{J/s}$$

each second because 1 hp = 746 J/s. But in lifting the load, the motor exerts a force F to pull the load with a speed of 0.2 m/s. It therefore must do work at the rate

$$F_s v = F(0.20\ \text{m/s})$$

We now equate these two quantities: the work the motor can do each second and the work done each second in lifting the load. This gives

$$(\tfrac{1}{4})(746)\ \text{J/s} = F(0.20\ \text{m/s})$$

Solving, we find that $F = 930$ J/m $= 930$ N. This is the maximum load the motor could lift at this speed. In practice, because of friction and other energy losses, the maximum load would be less than this. ■■

6.6 Energy

Energy is the ability to do work. If an object has energy, it is capable of doing work by itself. For example, a car coasting along the street has energy. It can do work because it is moving. Before it comes to rest, it can push against and displace almost any object with which it collides. All moving objects possess energy because they can do work as they are being stopped. We call this kind of energy *kinetic energy,* KE. (The word "kinetic" comes from the Greek word for motion, *kinetikos.*) Kinetic energy is the ability to do work that an object has because it is moving.

An object that can fall also has energy. It has the ability to do work because of its position. For example, a pile driver consists of a heavy object that falls onto the post (or pile) to be driven down. It pushes the pile down and thereby does work. When an object is positioned so that it can fall and do work, we say that it has *gravitational potential energy,* GPE. The term "potential" indicates that the object possesses the ability to do work, but that the work cannot be done unless the object falls.

The hot gas in the piston of a motor also can do work. It pushes the piston out and this motion of the piston causes the motor to operate. We say that the gas has *thermal energy.* As we shall see later in this text, thermal energy and kinetic energy are closely related.

There are other forms of energy, too. An explosive charge in a gun can throw a bullet from the gun when detonated. The explosive has the ability to do work on the bullet because of its chemical structure. This form of ability to do work is called chemical energy. Electrical energy and nuclear energy are certainly known to you. All these forms of energy have one thing in common: Energy is the ability to do work.

We need a quantitative way to describe energy. Because work and energy are so closely related, it is convenient to measure energy in the following way. If we could design and build perfect equipment, we could obtain the maximum work possible from an object's energy. We define the energy of an object to be the maximum work the object can do if all its energy could be utilized. Because of this definition, the units of work and energy are the same. If an object has the ability to do 20 J of work, it has an energy of 20 J. An object that has 300 J of kinetic energy (KE) is capable of doing 300 J of work because of its motion. Let us now find out how gravitational potential energy (GPE) is related to the height through which an object can fall.

6.7 Gravitational Potential Energy

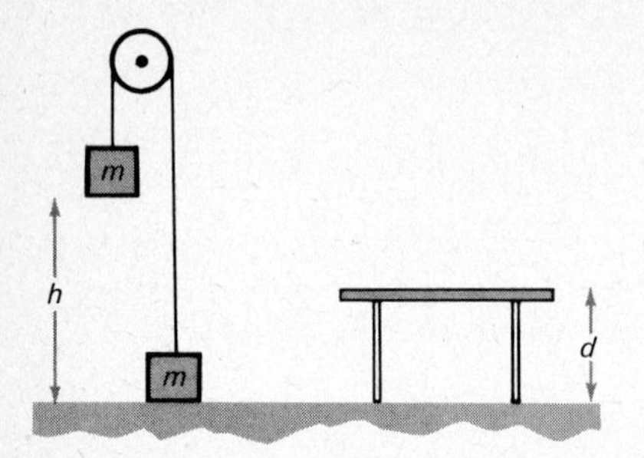

Figure 6.7 The upper mass has a potential energy *mgh*. Is it wrong to say that the potential energy is $mg(h - d)$?

Suppose that an object is at a height h above the floor. How much work can it do as it falls to the floor? This will be its gravitational potential energy.

Let us imagine the experiment illustrated in Figure 6.7. Because the experiment is imaginary, we shall state that the pulley is frictionless. The masses are identical. If we give the upper mass a very slight push, it will start to fall slowly. It will fall with constant speed because both masses experience equal pulls due to gravity. They balance each other.

Clearly the upper mass does work as it falls. It lifts the other mass. We already know how much work is done in lifting a mass m a distance h. It is mgh. This is how much work the upper mass can do as it falls to the floor. Therefore, the upper mass must have had an amount mgh of gravitational potential energy in the beginning. It is able to do an amount mgh of work because of its height.

From simple considerations such as this we arrive at the following.

An object of mass *m* that can fall through a height *h* has an amount *mgh* of gravitational potential energy. The symbol GPE is used for gravitational potential energy.

The units of potential energy are joules and foot-pounds. Energy has the same units as work.

Sometimes people give different values for the potential energy of an object. This is because the value of h is more or less arbitrary. For example, if a table was under the falling mass of Figure 6.7, another person might not measure h to be the distance to the floor. It also makes good sense to measure h to the tabletop. Fortunately, it makes no real difference which reference point (the tabletop or the floor) we use to measure h. This fact will become apparent as we use potential energy. For that reason we insist only that, once a reference level is chosen, the reference level must not be changed during a discussion. Usually we take the lowest level to be the reference level.

■ EXAMPLE 6.5 Niagara Falls has a drop of 51 m. If all its energy could be used, how much work could 1000 kg of water do as it drops over the falls?

Solution Let us take the zero level (or reference level) at the bottom of the falls. At the top of the falls, the water has a potential energy *mgh*. Ideally it could do this much work as it drops. Therefore,

$$\begin{aligned} \text{Work} &= mgh \\ &= (1000\ \text{kg})(9.8\ \text{m/s}^2)(51\ \text{m}) = 500\ \text{kJ} \end{aligned}$$

Electric energy is generated from this potential energy at the falls. The water turns turbine generators as it drops down the 51 m.

6.8 Kinetic Energy

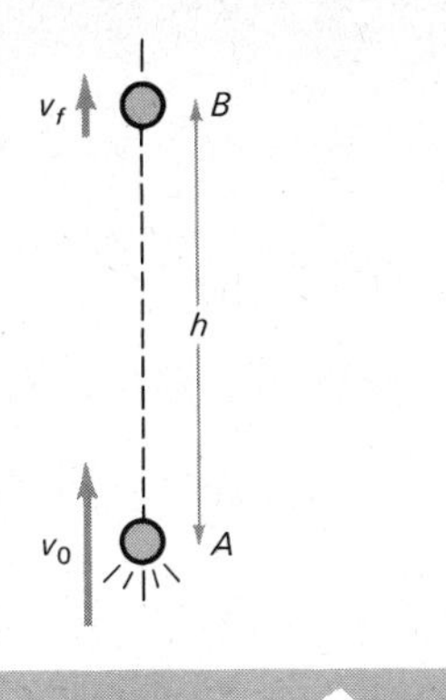

Figure 6.8 As the ball of mass m does work mgh in rising, it loses motional energy. We find that $mgh = \frac{1}{2}mv_0^2 - \frac{1}{2}mv_f^2$.

Let us next find an expression for the kinetic energy of a moving object. Because it is the result of the object's motion, it should in some way be related to v, the velocity of the object. We need some way to relate the motion of an object to the work it can do. Let us consider the simple experiment shown in Figure 6.8.

We see there a ball of mass m shooting upward. As it rises, it lifts itself. But the work it does in lifting itself causes the ball to slow down. It loses kinetic energy in the process. It loses some of its ability to do work. We shall compute how its loss in motion is related to the work it does in lifting itself.

The easiest way to approach the situation in Figure 6.8 is to use the motion equations. Use points A and B as start and end points. We know that the equation

$$v_f^2 - v_0^2 = 2ay$$

applies. In the present case, $y = h$ and $a = -g$. Therefore

$$v_f^2 - v_0^2 = -2gh$$

Or, after rearranging,

$$\tfrac{1}{2}v_0^2 - \tfrac{1}{2}v_f^2 = gh$$

But we want to relate the motion of the ball to work. We already know that the work the ball does in lifting itself from A to B is mgh. To obtain this quantity, we multiply the above equation by m to find

$$mgh = \tfrac{1}{2}mv_0^2 - \tfrac{1}{2}mv_f^2$$

But because mgh is the work the ball does in lifting itself from A to B, we conclude the following.

A moving object of mass m and initial speed v_0 can do an amount of work $\frac{1}{2}mv_0^2 - \frac{1}{2}mv_f^2$ as it slows from v_0 to v_f.

A special case of this result occurs when $v_f = 0$ and the object stops. Then the object has lost all its original kinetic energy. Our result says that it can do work equal to $\frac{1}{2}mv_0^2$ in coming to a stop. For this reason we conclude that

The kinetic energy (KE) of a moving object (mass = m, speed = v) is $\frac{1}{2}mv^2$. It can do this amount of work in coming to rest.

$$\mathbf{KE} = \tfrac{1}{2}mv^2 \qquad (6.5)$$

Although these results were obtained in a special case, work done against gravity, they can be shown to be true for all types of work. The work an object can do in slowing from speed v_0 to speed v_f is $\frac{1}{2}mv_0^2 - \frac{1}{2}mv_f^2$.

6.9 Interchange of Work, Potential Energy, and Kinetic Energy

In the last two sections we showed the following facts:

1. When an object slows from v_0 to v_f, it can do work equal to $\frac{1}{2}mv_0^2 - \frac{1}{2}mv_f^2$. The work it can do equals its loss in kinetic energy.
2. When an object falls through a height h, it can do work equal to mgh. The work it can do equals its loss in potential energy.

Now we shall go one step further. We shall find how much work one must do *on* an object in order to *increase* its potential and kinetic energy.

When one lifts an object of mass m through a height h, the work done on the object is mgh. This fact was shown in Section 6.5. But in the lifting process, the object is given potential energy. Its potential energy is increased by mgh. We see from this that work and potential energy are interchangeable. When work mgh is done in lifting an object, its potential energy increases by mgh. When an object falls and loses an amount mgh of potential energy, it can do mgh amount of work.

A similar situation exists for kinetic energy. We saw that the work an object can do while slowing down is equal to the kinetic energy it loses. But the reverse is also true. Suppose that you push an object forward so as to increase its speed. Then the work done on the object can cause an equal increase in kinetic energy of the object. Work and kinetic energy are interchangeable. When an object loses kinetic energy, it can do work equal to the energy lost. When work is done to accelerate an object, the object can gain kinetic energy equal to the work done.

Finally, we should notice that potential energy and kinetic energy are interchangeable. We saw this in our discussion of Figure 6.8. When the ball rises, it loses kinetic energy, but in the process it gains an equal potential energy. Similarly, when the ball falls, it loses potential energy, but in this process it gains an equal kinetic energy.

6.10 Conservation of Energy

Great progress is made in science and technology when an important unifying principle is found. We have seen that work, potential energy, and kinetic energy are all related. These relations can all be summarized in a basic principle or law of nature. Once we understand the law, we acquire a powerful tool for analyzing physical systems.

The important law we are about to present is called the law of conservation of energy. As we indicated earlier in this chapter, there are many forms of energy. The list of them would include gravitational potential energy, kinetic energy, electrical energy, chemical energy, thermal energy, and nuclear energy. There are still others, some of which we shall encounter later in this text.

The law of conservation of energy states what happens to the energy of an isolated system. An isolated system is a group of objects that neither receives from nor gives energy to objects outside the system. The whole universe is an isolated system. Because there is nothing outside it, the universe cannot share energy with things outside it. To a good approximation, the hot coffee in a stationary thermos jug is an isolated system. Although the molecules within the jug are bouncing against one another, the insulated jug protects them from losing (or gaining) energy from outside.

If we are given an isolated system, the *law of conservation of energy* tells us an important fact about the energy of the system. It states:

The total energy of an isolated system is constant.

This simple statement has far-reaching effects. For example, suppose that an object in such a system falls and loses potential energy. Then the law tells us that the system must gain an equal amount of some other kind of energy to compensate for this loss. The total amount of energy does not change.

As another example, suppose that a projectile explodes in flight. If we neglect the effects of gravity and air friction during the short time of the explosion, then the projectile approximates an isolated system. Before explosion it had kinetic, potential, chemical, and perhaps other kinds of energy. After explosion, its total energy must be unchanged. This is true in spite of the fact that chemical energy was lost during the explosion. The lost chemical energy was transformed, partly at least, into increased kinetic energy of the fragments. No matter where the energy went, we know from the law that energy was not lost by the system.

The law of conservation of energy applies to all forms of energy. It is more general than we need in the study of mechanics. After all, we shall not usually be concerned with chemical and nuclear energy. We therefore look for a way to make the law more useful in mechanics. In the next section we shall learn how to apply the law in mechanics.

6.11 Work-Energy Theorem of Mechanics

Energy and work are closely related, as we have seen. Indeed, they are so closely related that often one can be replaced by the other. For example, suppose that you lift an object through a height h. You can describe what you did in two different ways. You can say (1) that you increased the potential energy of the object by mgh or (2) that you did work in the amount mgh on the object. Either description is acceptable.

We can see that there are two ways of describing an object or system. We can speak in terms of its energy or we can speak in terms of work. It proves convenient in mechanics to use both methods.

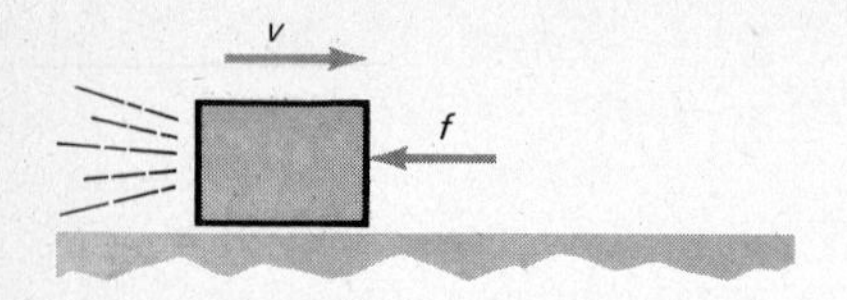

Figure 6.9 What happens to the block's KE as the friction force slows it to a stop?

As another example, consider what happens to the kinetic energy of the block shown in Figure 6.9. Friction forces eventually cause the block to stop. Clearly the block has lost all its original kinetic energy. But where did the kinetic energy go? According to the law of conservation of energy, it must have simply changed its form.

In this case the kinetic energy was changed to thermal energy. You know of many examples where friction forces lead to heat. The American Indians rubbed sticks together to start a fire. A poorly lubricated bearing burns out. These are simple examples to show that friction work results in thermal energy. In fact, as you might have guessed, the friction work done is equal to the thermal energy produced. Here, too, we can describe the mechanical effect in two ways. We can speak about work done against friction or we can speak about the equal amount of thermal energy generated.

From these considerations we can conclude that we have a choice. If we wish, we can replace a particular kind of energy in a discussion by the work that is equivalent to the energy. It is convenient in mechanics to use only potential and kinetic energy. All other types of energy, such as thermal energy, will be replaced by their equivalents in terms of work. The most important case has to do with friction work and thermal energy. We shall describe most situations in terms of the friction work done rather than the thermal energy generated.

Let us now see how we apply the law of conservation of energy to mechanical systems. The law tells us that:

$$\begin{pmatrix}\text{Change in KE}\\ \text{of object}\end{pmatrix} + \begin{pmatrix}\text{change in GPE}\\ \text{of object}\end{pmatrix} = \begin{pmatrix}\text{outside energy}\\ \text{added to object}\end{pmatrix}$$

In writing this we agree that increases in energy are positive, while decreases are negative.

But how can energy be added to an object? In mechanics we shall be concerned with accelerating (and decelerating) forces acting on the object. These add (or subtract) energy equal to the work the force does on the object. We can therefore write that

$$\begin{pmatrix}\text{Change in KE}\\ \text{of object}\end{pmatrix} + \begin{pmatrix}\text{change in GPE}\\ \text{of object}\end{pmatrix} = \begin{pmatrix}\text{work done on object}\\ \text{by outside forces}\end{pmatrix}$$

Thus we have a relation between the object's energy and the external work done on the object.

There is one important feature you must remember about the work term. It represents energy added to the system. Accelerating forces thereby do positive work on the object. But friction forces, because they cause a loss of KE, contribute negative work to the object.[6]

The work done on an object by friction forces is negative.

[6] To see this mathematically, use work = $Fs \cos \theta$. The friction force acting on the object is opposite in direction to **s**. Therefore, $\theta = 180°$ and so $Fs \cos \theta = -Fs$.

We can summarize what we have been saying in terms of an equation. Recall that the symbol delta (Δ) means "change in." Our result can be written as

$$\Delta\mathbf{KE} + \Delta\mathbf{GPE} = \textbf{work done on object} \qquad \textbf{(6.6)}$$

We refer to this equation as the *work-energy theorem.* In using it, we must remember that the work term includes the work done on the object by all external forces except gravity. Forces that add energy (such as accelerating forces) give rise to positive work. Those that cause the object to lose energy (such as friction) give rise to negative work. Let us now see, by example, how we make use of this powerful theorem.

EXAMPLE 6.6 A ball falls to the earth from a height h. What is its speed just before it hits the earth?

Solution We could solve this using the motion equations. Instead, let us use the work-energy theorem. The ball is our system. Its change in potential energy as it falls is $-mgh$. The negative sign is used because the change is a loss. As it falls, the ball gains kinetic energy. The amount gained is simply its final kinetic energy $\frac{1}{2}mv^2$ because the original kinetic energy was zero. If we assume friction to be negligible, then the work term is zero. We have, from the work-energy theorem,

$$\Delta\text{KE} + \Delta\text{GPE} = \text{work}$$

Using the values just given, this becomes

$$\tfrac{1}{2}mv^2 + (-mgh) = 0$$

Solving for v gives

$$v = \sqrt{2gh}$$

EXAMPLE 6.7 A 800-kg car moving on a level road at 20 m/s is to stop in a distance of 70 m. How large a stopping force is needed?

Solution The car is the system. Its height does not change, so $\Delta\text{GPE} = 0$. It loses all its original KE, and so

$$\Delta\text{KE} = -\tfrac{1}{2}m{v_0}^2$$

where the negative sign is needed because this is a loss in energy. Therefore

$$\begin{aligned}\Delta\text{KE} &= -\tfrac{1}{2}(800\ \text{kg})(20\ \text{m/s})^2\\ &= -16 \times 10^4\ \text{J}\end{aligned}$$

Notice the minus sign. Kinetic energy was lost.

The work done on the car by stopping forces (friction forces in this case) is $F_s s$, where F_s is the stopping force and $s = 70$ m. We then have, from the work-energy theorem,

$$\Delta\text{KE} + \Delta\text{GPE} = \text{work}$$

so that

$$-16 \times 10^4 \text{ J} + 0 = F_s(70 \text{ m})$$

This then gives a stopping force

$$F_s = -2290 \text{ N}$$

It is negative because it is a stopping force and contributes a negative term. ∎∎

∎ **EXAMPLE 6.8** A pendulum with ball of mass 50 g swings back and forth. At its highest position the ball is 20 cm higher than at its lowest. How fast is the ball going as it passes through the lowest position?

Solution Consider the ball as it goes from the lowest to the highest position. We have $\Delta\text{GPE} = mgh = (0.050 \text{ kg})(9.8 \text{ m/s}^2)(0.20 \text{ m}) = 0.098$ J. Because the ball has KE = 0 at the top, we have $\Delta\text{KE} = -\frac{1}{2}mv^2$, where v is the speed at the bottom. The minus sign is needed because the ball loses kinetic energy. We shall ignore the slight work done against friction. Then the work-energy theorem

$$\Delta\text{GPE} + \Delta\text{KE} = \text{work}$$

becomes

$$mgh - \tfrac{1}{2}mv^2 = 0$$

and gives

$$0.098 \text{ J} - (\tfrac{1}{2})(0.050 \text{ kg})v^2 = 0$$

Solving gives

$$v = 1.98 \text{ m/s}$$ ∎∎

∎ **EXAMPLE 6.9** Refer to Figure 6.10. The 1.7-kg block has an initial velocity of 0.50 m/s at the instant shown. If a friction force of 7.0 N acts to slow it, how fast is it going just before it reaches the bottom of the incline?

Solution In going from top to bottom the block loses *mgh* potential energy. Therefore

$$\begin{aligned}\Delta\text{GPE} &= -mgh = -(1.7 \text{ kg})(9.8 \text{ m/s}^2)(1.5 \text{ m})\\ &= -25.0 \text{ J}\end{aligned}$$

The work done by friction is $-(f)(s)$, where $f = 7.0$ N and $s = 3.0$ m. Therefore

$$\text{Work} = -(7.0 \text{ N})(3.0 \text{ m}) = -21.0 \text{ J}$$

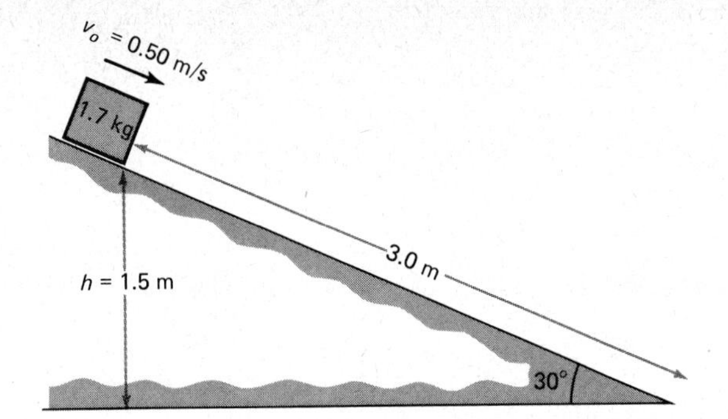

Figure 6.10 If the friction force retarding the motion is known, find the speed of the block at the bottom of the incline.

Note the negative sign. Friction forces slow the object and thus do negative work on it. The kinetic energy of the block at the start was $\frac{1}{2}mv_0^2$, where $m = 1.7$ kg and $v_0 =$ 0.50 m/s. We then have (because the change in a quantity is always the final value minus the initial value)

$$\begin{aligned}\Delta\text{KE} &= (\text{final KE}) - (\text{initial KE}) \\ &= \tfrac{1}{2}mv_f^2 - \tfrac{1}{2}mv_0^2 \\ &= \tfrac{1}{2}(1.7)(v_f^2 - 0.25)\ \text{J}\end{aligned}$$

Using the work-energy theorem,

$$\Delta\text{GPE} + \Delta\text{KE} = \text{work}$$

gives

$$-25.0\ \text{J} + \tfrac{1}{2}(1.7)(v_f^2 - 0.25)\ \text{J} = -21.0\ \text{J}$$

and so

$$v_f = 2.2\ \text{m/s}$$

▮▮

▮ **EXAMPLE 6.10** As the 400-kg roller coaster of Figure 6.11 leaves the top of the hill at *A*, its speed is 2.0 m/s. (a) What will be its speed as it passes over the hill at *B*? Solve assuming friction to be negligible. (b) If its speed at *B* is actually 3 m/s and the distance from *A* to *B* along the track is 200 m, how large was the average friction force that opposed its motion?

Solution In going from *A* to *B*, the coaster lost potential energy.

$$\Delta\text{GPE} = mgh_B - mgh_A = -mg(h_A - h_B)$$

Notice that the distance "fallen" from *A* to *B*, $h_A - h_B$, is important. But the reference level for measuring the heights of *A* and *B* is of no importance.

The change in kinetic energy of the coaster is

$$\Delta\text{KE} = \tfrac{1}{2}mv_B^2 - \tfrac{1}{2}mv_A^2$$

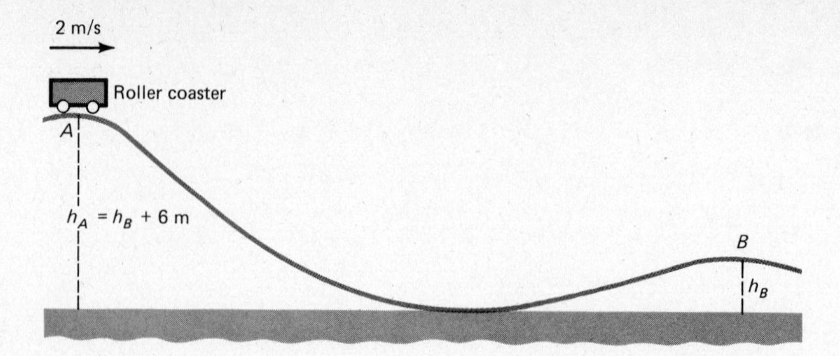

Figure 6.11 In finding the roller coaster's speed at B, h_A and h_B are not needed. Only the difference in heights, 6 m, is required.

In part (a) the friction work done is zero. The work-energy theorem

$$\Delta\text{KE} + \Delta\text{GPE} = \text{work}$$

becomes

$$\tfrac{1}{2}mv_B^2 - \tfrac{1}{2}mv_A^2 - mg(h_A - h_B) = 0$$

We can divide through the equation by m, so the mass of the coaster cancels out. Putting in the known values, $v_A = 2.0$ m/s and $h_A - h_B = 6$ m, gives

$$v_B^2 = (2\text{ m/s})^2 + (2)(9.8\text{ m/s}^2)(6\text{ m})$$

from which

$$v_B = 11\text{ m/s}$$

In part (b) we must use the work term because friction is not negligible. This term is $-(f)\cdot$(distance along track) or $-(f)\cdot(200\text{ m})$. The work-energy theorem

$$\Delta\text{KE} + \Delta\text{GPE} = \text{work}$$

becomes

$$\tfrac{1}{2}mv_B^2 - \tfrac{1}{2}mv_A^2 - mg(h_A - h_B) = -(f)(200\text{ m})$$

We know that $m = 400$ kg, $v_A = 2$ m/s, $v_B = 3$ m/s, and $h_A - h_B = 6$ m. Placing these values in the equation and solving for f gives

$$f = 113\text{ N}$$

Summary

The work done by a force **F** in giving an object a displacement **s** is defined as follows: work = (component of **F** in the direction of the displacement) · (magnitude of the displacement). This is stated in two equivalent equations.

$$\text{Work} = F_s s$$
$$\text{Work} = Fs \cos \theta$$

where θ is the angle between the force vector and the displacement vector. Work is a scalar quantity.

No work is done by a force if it causes no displacement. No work is done by a force if the force has no component in the direction of the displacement.

The units of work are the joule (J) in the SI system and foot-pound (ft · lb) in the British system.

When a supporting force raises an object, positive work is done. When it lowers the object, negative work is done. The work done in lifting (lowering) a mass m through a height h is mgh ($-mgh$).

The gravitational force is a conservative force. That is, the work done against gravity in lifting a mass from one point to another is not dependent on the path along which the mass is lifted.

Power is the rate of doing work. Its defining equation is

$$P = \frac{\text{work done}}{t}$$

where t is the time taken to do the work. The units of power are the watt (or joule per second) and foot-pound per second. Another common unit, the horsepower, is related to these by

$$1 \text{ hp} = 746 \text{ W} = 550 \text{ ft} \cdot \text{lb/s}$$

Energy is the ability to do work. A moving object can do work because of its motion. It is said to have kinetic energy. An object that can fall is capable of doing work while falling. It is said to have gravitational potential energy.

An object of mass m and speed v has kinetic energy KE $= \frac{1}{2}mv^2$. In slowing from speed v_0 to speed v_f, it can do work equal to $\frac{1}{2}mv_0^2 - \frac{1}{2}mv_f^2$.

An object of mass m at a height h has gravitational potential energy GPE $= mgh$. In falling through a height h, it can do an amount mgh of work.

Many other forms of energy exist. A few of them are thermal energy, electrical energy, nuclear energy, and chemical energy.

The law of conservation of energy concerns an isolated system, a system that cannot give or take energy from its surroundings. The law states that the total of all forms of energy within the system is constant.

The following is a statement of the work-energy theorem of mechanics. For an object,

$$\Delta\text{GPE} + \Delta\text{KE} = \text{external work done on object}$$

In computing the work done, the work of gravity is not included. It is already contained in the term ΔGPE.

Questions and Exercises

1. A man holds a child on his shoulders while watching a parade. According to our definition of work, he has done no work on the child. Try to reconcile this with the fact that the man feels tired after holding the child.
2. A pendulum swings back and forth. Does the tension in the string do work on the ball of the pendulum? Discuss the energy possessed by the pendulum at each instant as it swings back and forth and eventually stops. Where, finally, does all its energy go?
3. A man wishes to lift a large, heavy crate up onto two blocks. But it is too heavy for him to lift. He therefore tips up one end and slides one block under it. He repeats for the other end. Compare the work he did on the crate with that which two men would do in lifting it directly onto the blocks.
4. What data would you need in order to estimate the power output of a boy climbing a rope?
5. Estimate the maximum power output of a human in the following way. Run up a flight of stairs as fast as you can. Compute how much work you did in lifting your body. Knowing the time it took, compute the power output in watts and in horsepower.
6. When an object loses kinetic energy, the energy can go (a) to potential energy, (b) to doing work on another object, and (c) into work against friction forces. Give an example of each case.
7. Give an example for each of the following ways in which potential energy can be lost: (a) lost to work against friction; (b) changed to kinetic energy; (c) lost to doing useful work.
8. While doing a lab experiment, a girl decides to take the top of the lab table as the zero level of potential energy. What will be the potential energy of a mass m at a height y above the tabletop? At a distance y below the tabletop? Should she be criticized for her choice of reference level?
9. You have probably been warned that a car going 60 km/h can do four times as much damage as one going half as fast. What is the basis for this warning? How is this related to the fact that the length of the skid marks of a strongly braked car varies as the square of the car's speed?
10. Estimate the time it takes to accelerate an auto from 0 to 15 m/s. Knowing the change in kinetic energy of the car, how much work was done on it during the acceleration? Find the power output (in horsepower) needed to do this in the allotted time. Compare to the car's horsepower rating.

Problems

1. Determine how much work is done by the force in each of the following cases. (a) A 50-N vertical force lifts an object 30 cm. (b) A 50-N friction force stops a book that slides 20 cm across a tabletop. (c) A child is held 80 cm above the floor by a 150-N force. (d) A car jack exerts a force of 5000 N on a car and lifts it 20 cm. (e) A car jack exerts a force of 5000 N on a car as the car is lowered 20 cm.
2. Determine how much work must be done by an outside force to (a) push an object 20 cm with constant speed against a 30-N friction force, (b) slowly lift a 3-kg object 40 cm, (c) lift a 7-kg child from the floor to a 50-cm-high stool, (d) lower a 7-kg child from a 50-cm-high stool to the floor, (e) hold a 7-kg child in your arms at a height of 120 cm above the floor.
3. (a) A car is towing another car along a straight road with a rope that can stand a maximum tension of 1000 N. How much work does the car do in towing the other car 20 m at this maximum tension? (b) Repeat for a 300-lb tension and a distance of 50 ft.
4. (a) The skid marks of a car on the pavement are 20 m long. If the friction force of the pavement on the car during the skid was 8000 N, how much work was required during the skid to bring the car to rest? (b) Repeat for a 3000-lb car that skids 60 ft.
5. A 20-kg box sits on a level floor. The coefficient of kinetic friction between floor and box is 0.60. How large a friction force opposes the motion of the box while moving? How much work must be done against friction to move the box 5.0 m?
6. In order to drag a 30-kg box across a level floor at constant speed, a horizontal force of 90 N is required. How large a friction force opposes the motion? What is the coefficient of friction between the box and the floor? How much work is done against friction in moving the box 7.0 m?
7. (a) A force of 90 N pulling at an angle of 37° above the horizontal moves a crate across the level floor. How much work does the force do in moving the crate 8 m? (b) Repeat for a 200-lb crate moved 20 ft by a force of 50 lb directed at 37°.
8. A large box is pushed along the level floor by a force F directed 53° below the horizontal. How much work does the force do in pushing the box a distance D? (a) Use $F = 90$ N and $D = 300$ cm. (b) Repeat for $F = 60$ lb and $D = 40$ ft.
9. (a) An 8-kg child is lifted from the floor and placed on a

1.0-m-high table. The child is then lifted from the table and placed in a crib that is 70 cm above the floor. How much work was done in each of these displacements? (b) Repeat for a 15-lb child with a 40-in.-high table and a 30-in.-high crib.

10. (a) A 400-kg elevator is lifted from the basement to the tenth floor, a distance of 50 m. It is then lowered to the eighth floor, a distance of 10 m. How much work was done on the elevator in each of these displacements? (b) Repeat for a 300-lb elevator with distances of 150 ft and 30 ft.

11. How much does the GPE of a 20-kg object change as the object is (a) lifted 3 m, (b) lowered 3 m, (c) moved 3 m horizontally?

12. Determine how much the GPE of a 900-kg car changes as it (a) goes from the bottom to the top of a 30-m-high parking garage, (b) travels 20 m horizontally on the top floor, (c) goes from the top to the bottom of the garage.

13. How much work is needed to lift 500 cm^3 of water through a height of 80 cm? (Hint: 1 cm^3 of water has a mass of 1 g.)

14. How much GPE does 1 m^3 of water lose as it falls a distance of 40 m? (Hint: 1 m^3 of water has a mass of 1000 kg.)

*15. A horizontal force of 240 N is needed to push a 31-kg crate up a 37° incline. (a) How much work is done by the force as the box moves 2 m along the incline? (b) How high is the box lifted during this 2-m displacement? (c) How much does the GPE for the box change because of the displacement? (d) What can you conclude about the friction force acting on the crate?

*16. In order to slide a 90-kg crate slowly down a 37° incline, a horizontal force of 600 N must push against it to retard its motion. (a) How much work is done by the force as the crate slides 3 m along the incline? (b) How much is the box lowered as it slides 3 m? (c) What is its change in GPE as it slides 3 m? (d) What can be said about the friction force retarding the crate?

17. (a) A 200-kg load of bricks is to be lifted from the ground to the top of a 30-m-high building in ½ min. What minimum horsepower motor would be needed to do this? Assume 100 percent efficiency. (b) Repeat for a 500-lb load and a 60-ft-high building.

18. (a) A crane lifts 1500 kg of junk metal through a distance of 6 m in a time of 15 s. How much horsepower is delivered to the load? (b) Repeat for 4000 lb lifted 20 ft.

*19. A 2-hp motor is being used to lift a 100-kg load. What is the greatest speed at which the load can be lifted with this motor?

*20. A 90-W electric motor is to lift a load at a speed of 2 cm/s. How large a load could it lift at this speed?

**21. A 900-kg car has a motor that can deliver 45 hp to move the car forward. The maximum steady speed of the car on a level road is 100 km/h. How large a retarding force acts on the car at that speed?

**22. A 900-kg car has a motor that can deliver 45 hp to move the car forward. At what maximum steady speed could the car go up a 20° incline (a) in the absence of a retarding friction force and (b) if the resultant retarding friction force on the car is 2000 N?

23. What is the kinetic energy of a 2000-kg car moving at 20 m/s? Of a 4000-lb car moving at 60 ft/s?

24. What is the kinetic energy of a 20-g bullet moving with speed 150 m/s? A 2-oz bullet moving at 400 ft/s? (1 lb = 16 oz.)

25. How much work is needed to stop a 20-g bullet moving with a speed of 150 m/s? A 2-oz bullet moving at 400 ft/s? (1 lb = 16 oz.)

26. How much work is needed to stop a 2000-kg car moving at 20 m/s? A 4000-lb car moving at 60 ft/s?

27. How large an average force is needed to give a 20-g bullet a speed of 150 m/s as it passes down the 40-cm-long barrel of a gun? A 2-oz bullet at 400 ft/s in a 20-in.-long barrel? Use the energy approach.

28. A 1500-kg car is coasting along a level road. Its speed at point A is 15 m/s, whereas at point B it is 10 m/s. If the distance between A and B is 50 m, how large is the average friction force opposing the car's motion?

29. An egg is released at a height of 70 cm above the floor. How fast is it moving just before it strikes the floor? Use energy methods.

30. A ball is moving freely with an upward velocity of 20 m/s. How high will it rise? Use energy methods.

31. A block sits on a nearly frictionless incline. It is released. When it reaches a point 30 cm below its original height, what is its speed?

32. A child, playing in an auto, releases the parking brake. The car coasts backward down the drive into the street. If the street is 150 cm below the car's original level, how fast is the car moving? Assume negligible friction.

*33. The speed of a bowling ball is 2.0 m/s as it falls past a window of a tall building. How fast is it going as it passes another window 3.0 m below?

*34. A ball is thrown straight up from the ground with speed 30 m/s. How high will it be when its speed is 10 m/s?

*35. Refer to Problem 32. If the car has a speed of 2.0 m/s as it rolls onto the street, how large a friction force opposed its motion? The car's mass is 1500 kg and the length of the drive along which it rolled was 12 m.

*36. Refer to Problem 31. If the speed of the block at the point in

*Problems marked with an asterisk are not as easy as the unmarked ones.
**Problems marked with a double asterisk are somewhat more difficult than the average.

question is 1.50 m/s, how large a friction force opposed its motion? The block has a mass of 2.0 kg and it slides 300 cm along the incline.

*37. A pendulum is 80 cm long. It is pulled aside to an angle of 37° with the vertical. Find the speed of the pendulum ball as the ball goes through its lowest position. How high above its lowest position will the pendulum swing on the other side? Ignore the small friction forces present.

38. In an amusement park a 500-kg car swings in a vertical circle at the end of a relatively light 15-m-long bar. How fast must the car be going at the bottom of its swing if it is to be able to coast over the top of the 30-m-diameter circle?

*39. A 2.0-kg block slides down an incline and out onto the level floor. It starts from a height 1.50 m above the floor. If it slides a total distance of 30 m before stopping, how large was the average friction force that opposed its motion?

*40. In Figure P6.1 is shown a wire along which a 20-g bead slides. If friction forces can be ignored, and the bead starts from rest at A, how fast will it be going at points C and D?

*41. Refer to Figure P6.1. The 20-g bead has a speed of 0.80 m/s as it passes point A. It finally coasts to a stop at point E. If the distance along the wire from A to E is 6.0 m, how large is the average friction force that slows its motion along the wire?

*42. If there were no friction forces, what power (in horsepower) would be needed to accelerate a 1500-kg car from rest to 15.0 m/s in a time of 8.0 s?

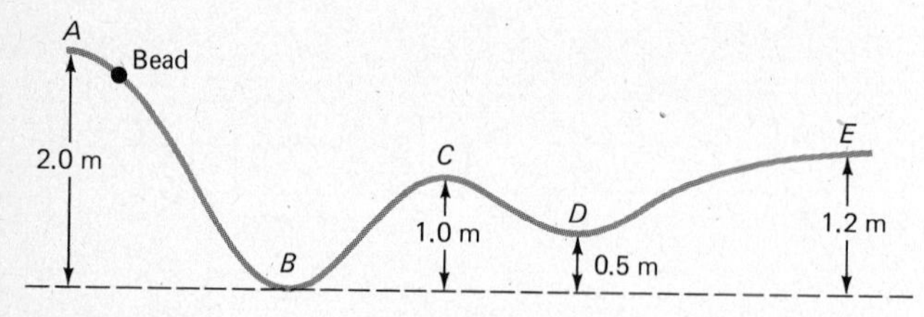

Figure P6.1

*43. (a) Suppose that a 1500-kg car has a 200-hp motor. Suppose further that the power could be used effectively to accelerate the car. If the car starts from rest, how long would it take to accelerate the car to 20 m/s along a level road? (b) Repeat for a 4000-lb car accelerated to 60 ft/s.

**44. A pump is to lift 0.100 kg of water from a 30-m-deep well each second. If conditions were ideal, what minimum horsepower motor would the pump have to have? Assume the well pipe to be large enough so that the water is given negligible kinetic energy. (Question: Analyze the situation and see why the pipe size should be important.)

**45. A pump is to fill a 1000-gal (gal = gallon) tank with water by pumping the water from a lake. The pump outlet to the tank is 5.0 m above lake level. What is the shortest conceivable time a 0.50-hp pump could take to do this? One gallon of water weighs 37.1 N. In practice, why would a pump not nearly achieve this performance?

**46. According to the magazine *Consumer Reports*, a certain 1200-kg car takes a minimum distance of 52 m to decelerate to rest from a speed of 27 m/s. What is the coefficient of friction between the car's tires and the pavement if other friction forces can be ignored?

**47. The car mentioned in Problem 46 is capable of accelerating from rest to a speed of 26 m/s in a time of 4.9 s. (a) What average horsepower does the car put out to do this? (b) If the limiting factor is friction at the wheels, what is the coefficient of friction between tires and pavement?

*48. A car manufacturer states that its car expends only 15 hp when traveling at a speed of 22 m/s along a level roadway. If this is true, how large a friction force opposes the motion of the car?

T. Chitra, DPI

7

SIMPLE MACHINES

Performance Goals

When you finish this chapter, you should be able to

1. Draw diagrams showing the essentials of the three basic machine types.
2. State which basic type of machine is used in simple devices such as a claw hammer, a chisel, scissors, and others.
3. Find the AMA, IMA, and efficiency of a machine if s_o, s_i, F_o, and F_i are given.
4. Write an equation that relates output work, input work, and friction work for a machine. Using it, together with the definition of efficiency, explain why the efficiency of a machine cannot exceed unity.

Nearly everything that we do involves the use of machines. Even a screwdriver and a knife are machines in technical terminology. Machines vary widely in complexity. Nevertheless, all machines have certain features in common. We devote this chapter to a study of the principles that apply to all machines.

5. Draw diagrams showing the essentials of the following machines: pliers, nutcracker, wheelbarrow, screw jack, single-pulley system, block and tackle with IMA = 3, belt-pulley system with IMA = 4, wheel and axle, spur-gear system with IMA = 4, belt-pulley system with IMA = 0.5.
6. Find the IMA, AMA, and efficiency for simple machines such as those listed in 5, provided sufficient data are given.
7. Design a block-and-tackle system having a specified IMA for any value less than 8. Repeat for a belt-pulley system and a spur gear system.

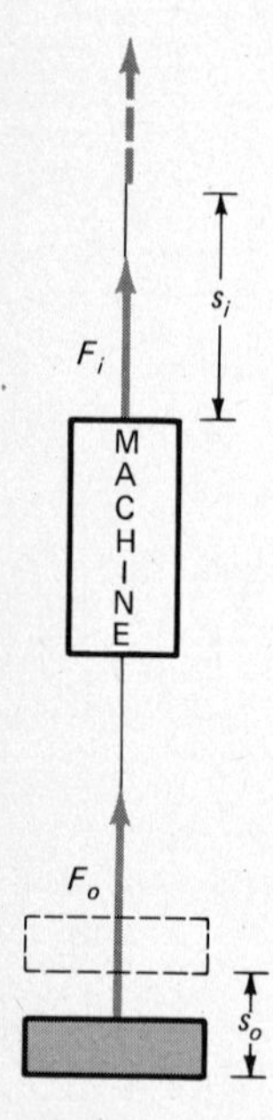

Figure 7.1 The input force F_i works through a distance s_i when the output force F_o works through a distance s_o.

7.1 The Black Box Machine

Any device that makes use of an input force to do work is called a machine. Because machines vary widely in design, it is important to know the principles that apply to all of them. We therefore begin our study by discussing a so-called black box machine, shown in Figure 7.1. Its workings are hidden from view so that we will not be led astray by details. Let us now see what can be said about this machine.

Notice in Figure 7.1 that an input force F_i gives rise to an output force F_o. The output force lifts the load. Very often machines are used to help us lift heavy loads. We therefore are interested in the ratio F_o/F_i. This ratio tells us the load that can be lifted by one unit of input force. The ratio is given a special name, the *actual mechanical advantage* of the machine.

$$\text{Actual mechanical advantage} = \text{AMA} \equiv \frac{F_o}{F_i} \tag{7.1}$$

Typical values for AMA might be 30 for a car jack and 100 for a chain hoist. For an input force F_i applied to the handle of a car jack, the jack can lift a load of $F_o = 30\,F_i$.

The actual mechanical advantage is easy to determine because F_i and F_o are easily measured. Even so, the AMA is a complicated quantity. It depends not only on the design of the machine, but also on the friction and other energy losses within the machine. To illustrate this point and arrive at an important result, let us now consider a friction-free machine.

Suppose we have a machine that has no friction or other energy losses within it. This ideal machine would convert all the energy given to it by the input force to usable output work. We could therefore write that

$$\text{Output work} = \text{input work} \qquad \text{(ideal machine)}$$

The input work is simply F_is_i, while the output work is F_os_o. The above equation thus becomes

$$F_os_o = F_is_i \qquad \text{(ideal machine)}$$

Dividing through the equation by F_is_o gives

$$\frac{F_o}{F_i} = \frac{s_i}{s_o} \qquad \text{(ideal machine)} \tag{7.2}$$

This expression gives the ratio of the output force to the input force in the case of an ideal machine. We call it the *ideal* (or *theoretical*) mechanical *advantage*.

$$\text{Ideal mechanical advantage} = \text{IMA} \equiv \frac{s_i}{s_o} \quad (7.3)$$

In any real machine the AMA is less than the IMA; the output of the machine is decreased by energy losses within the machine. Typically these losses are friction losses. The car jack we discussed previously often has an AMA less than half its IMA.

7.2 Efficiency

One of the most important quantities used to describe machines (and processes) is *efficiency*. This quantity tells us how effectively we make use of energy. We define it as follows.

$$\text{Efficiency} \equiv \frac{\text{output work}}{\text{input work}} = \frac{\text{AMA}}{\text{IMA}}$$

$$\text{Percent efficiency} \equiv 100 \times \frac{\text{output work}}{\text{input work}} \quad (7.4)$$

When we say that a process or machine is 70 percent efficient, we mean that its efficiency is 0.70; only 0.70 of the input energy is utilized effectively.

We have just defined efficiency as (output work) ÷ (input work). Recall that work and power are related by the equation work = (power) · (time). Combining the two definitions, we have

$$\text{Efficiency} = \frac{\text{output work}}{\text{input work}} = \frac{(\text{output power}) \cdot (\text{time})}{(\text{input power}) \cdot (\text{time})}$$

Simplifying, we have

$$\text{Efficiency} = \frac{\text{output power}}{\text{input power}} \quad (7.5)$$

7.3 Work and Energy in Machines

Machines do not create energy. The work that they do comes from the input work. Even in an ideal machine, the output work is no larger than the input work. Usually the output work is considerably less than the input work. Part of the input work is wasted doing work against friction forces.

The great advantage of machines has to do with the way energy is used. We do work on a machine and thereby give it energy. This energy is then used by the machine to do work. For example, consider the car jack. It allows us to lift a load by use of a force too small to lift the load directly. Because of this feature, it is a useful machine. Even so, because of friction forces, the work we get out of the machine is considerably less than what we put in. The machine is less than 100 percent efficient.

All machines that multiply the input force to obtain a larger output force have one feature in common. In all of them the input force must act through a larger distance than the output force. This is certainly true for the car jack. The load is lifted much less than the distance moved by your hand.

This is a general feature of all machines of this type. To see why, consider the energy equation for the ideal machine. We have

$$\text{Input work} = \text{output work} \qquad \text{(ideal)}$$

This becomes

$$F_i s_i = F_o s_o$$

or

$$s_i = s_o\left(\frac{F_o}{F_i}\right)$$

Therefore, if the output force is larger than the input force, F_o/F_i is larger than unity. In this case the input distance, s_i, must always be larger than the distance moved by the load, s_o. We also see that this is a direct result of the fact that energy is conserved.

EXAMPLE 7.1 By experiment it is found that a certain pulley system can lift a 20-kg load by means of an input force of 65 N. The load is lifted 30 cm when the input force pulls the pulley rope 150 cm. Find the IMA, AMA, and efficiency.

Solution By use of the equations we have stated

$$\text{IMA} = \frac{s_i}{s_o} = \frac{150\text{ cm}}{30\text{ cm}} = 5.0$$

$$\text{AMA} = \frac{F_o}{F_i} = \frac{20 \times 9.8\text{ N}}{65\text{ N}} = 3.0$$

$$\text{Efficiency} = \frac{\text{AMA}}{\text{IMA}} = \frac{3.0}{5.0} = 0.60 = 60\%$$

EXAMPLE 7.2 A certain electric motor uses 200 W of power while its output is $\frac{1}{4}$ hp. What is the motor's efficiency?

Solution By definition,

$$\text{Efficiency} = \frac{\text{output power}}{\text{input power}}$$

Substituting our numbers gives

$$\text{Efficiency} = \frac{(\frac{1}{4}\ \text{hp})(746\ \text{W/hp})}{200\ \text{W}} = 0.93$$

The electric motor is 93 percent efficient. Electric motors are usually very efficient.

7.4 Inclined Plane

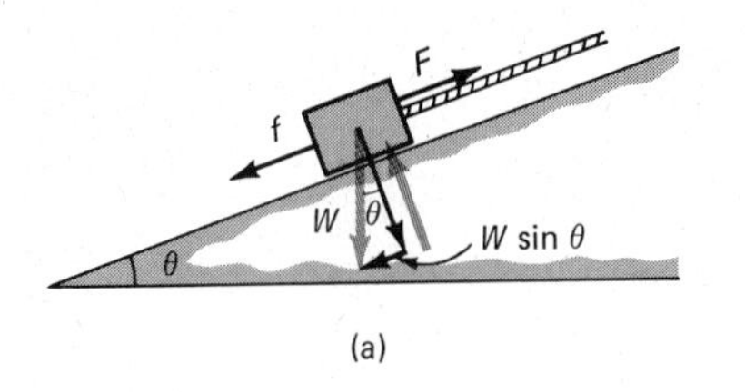

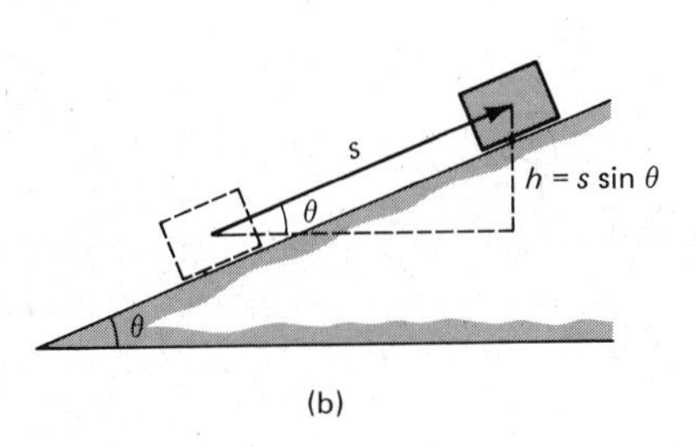

Figure 7.2 If $f = 0$, what would be the efficiency of this machine?

Three simple machines form the basis for many complex machines. They are the inclined plane, the lever, and the hydraulic press. We will discuss each in turn.

An inclined plane is shown in use in Figure 7.2. We can see from part (a) why the machine is of great use. A force equal to the object's weight W would be needed to lift the object directly. By use of the incline, a force equal to $W \sin \theta$ can pull the object up the incline if friction is negligible. As a result, a load W can be lifted by a force $W \sin \theta$, a much smaller force if the incline angle is small. In practice, however, friction forces are usually important.

The IMA of the inclined plane can be obtained from Figure 7.2(b). We see that for an input distance s, the load is lifted a distance $s \sin \theta$. Therefore, from Equation 7.3,

$$\text{IMA} = \frac{s_i}{s_o} = \frac{s}{s \sin \theta} = \frac{1}{\sin \theta}$$

If the angle of the incline is 10°, we have $\sin \theta = 0.174$ and IMA = 5.8. Let us now look at a few common machines that make use of the inclined plane.

The Wedge As we see from Figure 7.3(a), the wedge is a combination of two inclines. When the wedge is driven forward by a force, a much larger force is exerted to each side by the wedge surfaces. As the wedge is driven in a distance s_i, the surfaces being separated are moved apart through a distance s_o as shown. Calling the length of the wedge L and its end thickness t, we have

$$\text{IMA (wedge)} = \frac{s_i}{s_o} = \frac{L}{t} \tag{7.6}$$

We see from this that a long, thin wedge has a higher mechanical advantage than a short, thick one. Typical wedges are found in the ax, chisel, knife, and other cutting tools.

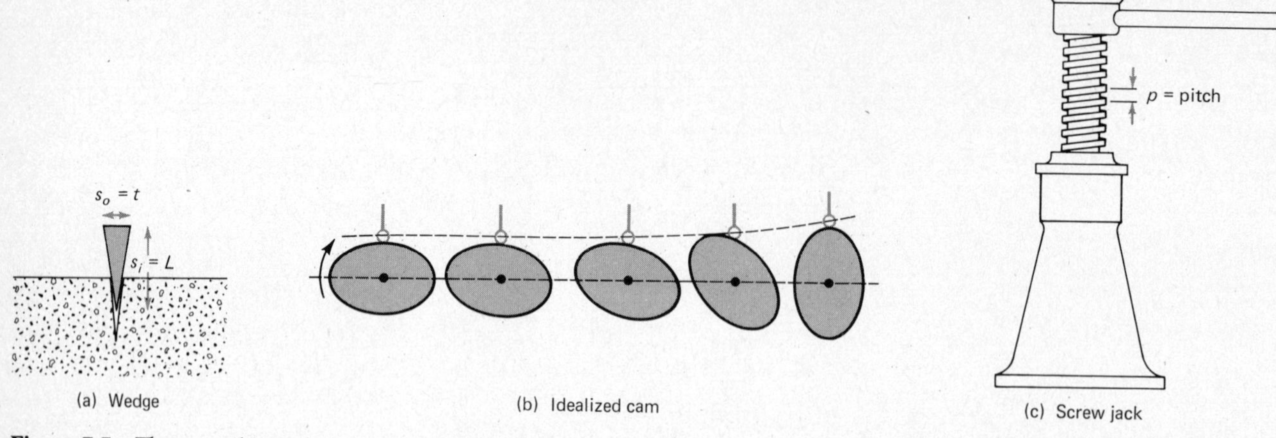

Figure 7.3 Three machines that use inclined planes.

The Cam There are many types of cams. In all of them a rotating wheel is caused alternately to lift and lower a rod as shown in Figure 7.3(b). The wheel must be noncircular or mounted off center. As you can see in the figure, the rod is lifted as the incline moves beneath it. One of the most well-known uses of a cam is to move the valve lifters in engines of automobiles. They have wide use elsewhere as well.

The Screw There are many different types of screw devices. The simple wood screw is usually only a portion of a machine that includes a screw driver. A purer use of the screw is the screw jack shown in part (c) of Figure 7.3. The device is operated by a force at the end of the crank. As the crank moves around once, the input force acts through the circumference of a circle of radius r. Therefore, $s_i = 2\pi r$. As the screw turns, the top of the jack is lifted up the incline of the screw.

For each turn of the screw, the screw lifts a distance equal to its pitch, p. We therefore have $s_o = p$. Then, from Equation 7.3, we find

$$\text{IMA (screw jack)} = \frac{s_i}{s_o} = \frac{2\pi r}{p} \tag{7.7}$$

As you can see, if r is perhaps 60 cm and $p = 0.5$ cm, then IMA = 750. Although jacks of this type have very poor efficiency, they are still effective force multipliers.

7.5 The Lever

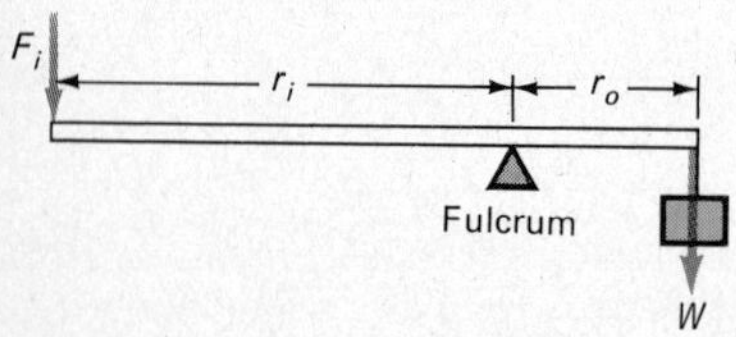

The second type of simple machine, the *lever*, is shown in Figure 7.4. To slowly lift the load W, the counterclockwise torque furnished by F_i must balance the torque due to W. Taking torques about the fulcrum, we then have

$$F_i r_i = W r_o$$

Figure 7.4 The IMA for a lever is r_i/r_o.

where energy losses within the machine are assumed negligible. Rearranging gives

$$\frac{W}{F_i} = \frac{r_i}{r_o}$$

But in this ideal case W/F_i equals the IMA. Therefore,

$$\text{IMA (lever)} = \frac{r_i}{r_o} \tag{7.8}$$

A few machines that use the lever in a simple way are shown in Figure 7.5. As you see, the lever arms are always measured from the fulcrum. Notice in all these cases that the balance of torques about a pivot point is basic to the device. In (a) through (e) the load is given a lever arm shorter than the lever arm of the input force. As a result, the input force can be much smaller than the load it is to lift. In part (f) the input force must be larger than the output force. A machine such as this last one has utility even though F_i is larger than F_o. Try to think of other machines like the one in (f). Why are they constructed in this way?

Figure 7.5 The levers shown in (a), (b), and (c) are called levers of the first kind. The ones in (d) and (e) are levers of the second kind. The one in (f) is a lever of the third kind. They are classified according to the position of the fulcrum relative to F_i and F_o.

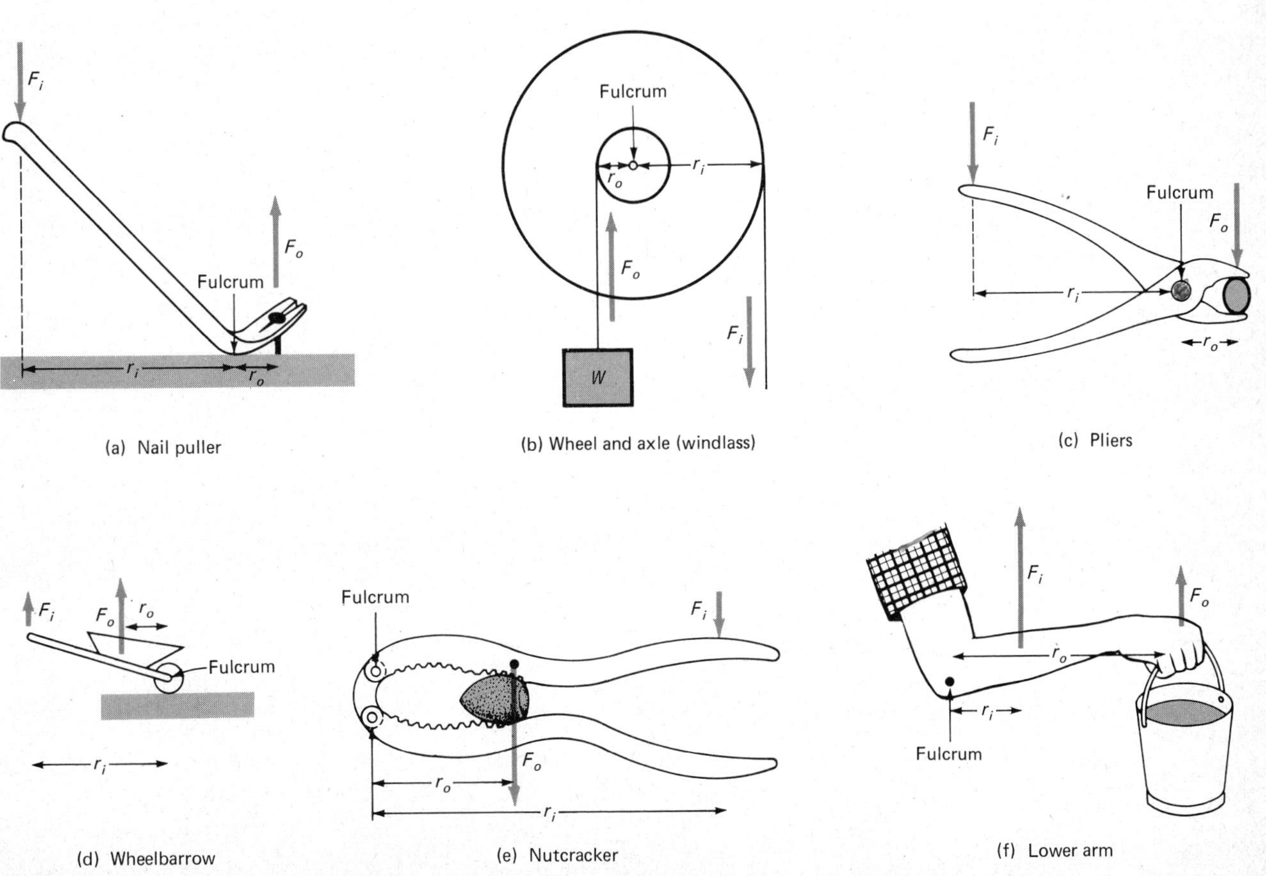

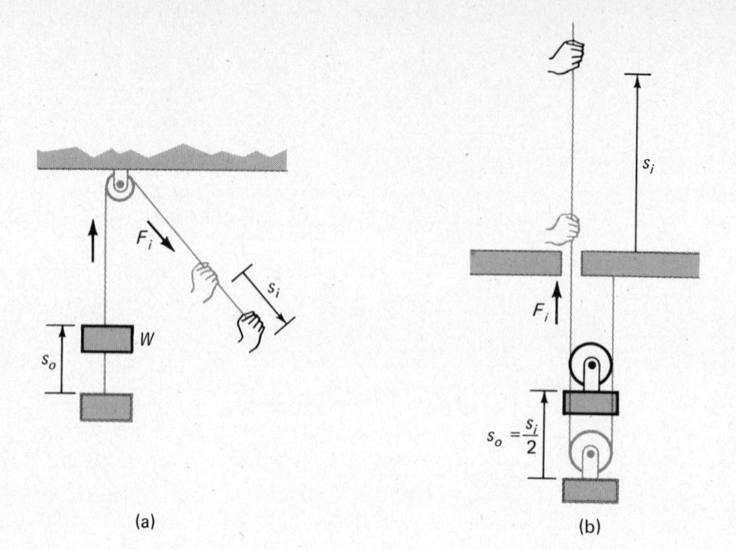

Figure 7.6 The system in (a) has an IMA = 1, whereas the one in (b) has an IMA = 2.

7.6 Pulley Systems

The wheel and axle of Figure 7.5(b) is an example of a wide class of machines called pulley systems. One of the simplest devices of this sort is shown in Figure 7.6(a). Its IMA is unity because $s_o = s_i$. Use is made of a single pulley like this to change the direction in which a force is to be exerted.

Another way in which a single pulley can be used is shown in Figure 7.6(b). We call this a block-and-tackle system. Notice that, for each centimeter that the load moves up, the force F_i must pull the rope up 2 cm. This is the result of the following fact: When the pulley moves 1 cm closer to the ceiling, both the right and left ropes to the ceiling will be 1 cm shorter. This means that 2 cm of rope must be pulled up through the hole to shorten the support by 1 cm.

For this system, then, $s_i = 2s_o$, and so

$$\text{IMA} = \frac{s_i}{s_o} = 2.00$$

It is clear that, if the system were not moving, both ropes would be pulling up on the pulley and weight. Hence if the pulley were weightless, the tension in each rope would be $W/2$, so that their combined upward pull on the pulley would be W. In that case F_i would be equal to $W/2$. Actually, of course, F_i must be large enough so that the ropes also support the weight of the pulley and overcome the action of friction as well. Because of this, it would be impossible for this machine to have an AMA = 2.00. If the pulley weighed nearly as much as W, the machine would have a very low efficiency, even though the friction in the pulley itself was very small. Nevertheless, it would be an improvement over a single rope in some respects.

More complicated pulley systems are shown in Figure 7.7. Using the same reasoning as for the previous case, we find that the IMA for systems (a), (b), and

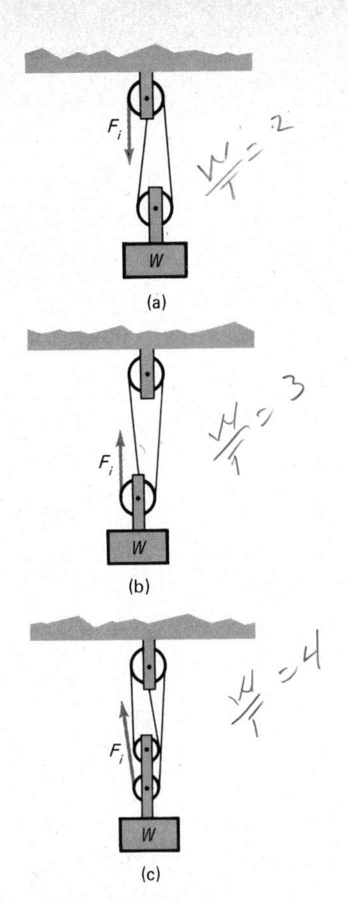

Figure 7.7 The IMAs for the systems shown are 2, 3, and 4. Notice that these are also the number of ropes pulling up on the load-bearing pulley.

(c) are 2, 3, and 4, respectively. Notice that in this case as well as in Figure 7.6, the IMA is numerically equal to the number of ropes pulling up on the free pulley. This fact provides us with a simple rule of thumb for determining the IMA of some block-and-tackle systems.

The IMA of a block and tackle that makes use of one continuous rope is equal to the number of strands that support the movable block.

EXAMPLE 7.3 Consider the block-and-tackle system shown in Figure 7.8. It is found that a force of 40 N is needed to lift 100 N by use of this device. Find the IMA, AMA, and efficiency for this device.

Solution Because four strands support the movable block (i.e., the load), the IMA of this system is 4. Its AMA can be found from the definition

$$\text{AMA} = \frac{F_o}{F_i} = \frac{100}{40} = 2.5$$

To find its efficiency, we recall from Equation 7.4 that

$$\text{Eff} = \frac{\text{AMA}}{\text{IMA}} = \frac{2.5}{4.0} = 0.625$$

The efficiency is 62.5 percent.

7.7 The Hydraulic Press

The third type of simple machine we wish to discuss is the hydraulic press, shown in Figure 7.9. A fluid such as oil is confined to two cylinders by two movable pistons. Using this device, a small input force F_i gives rise to a huge output force F_o. You have probably seen hydraulic car jacks that are based on this principle. When you pump the handle of the jack, you apply a force to the small piston of end area A_i. Your car is lifted by the large piston which has an area A_o.

To see the principle upon which this device is based, we can proceed as follows. Suppose the input force pushes the piston on the left down a distance s_i. The work it does is $F_i s_i$. Because the fluid in the cylinders is not compressible, the piston on the right must rise to make way for the fluid displaced by the piston on the left. The work done by this piston is $F_o s_o$. In the ideal case losses are negligible, the input and output work must be equal. Therefore,

$$F_o s_o = F_i s_i$$

which gives

$$F_o = \frac{F_i s_i}{s_o}$$

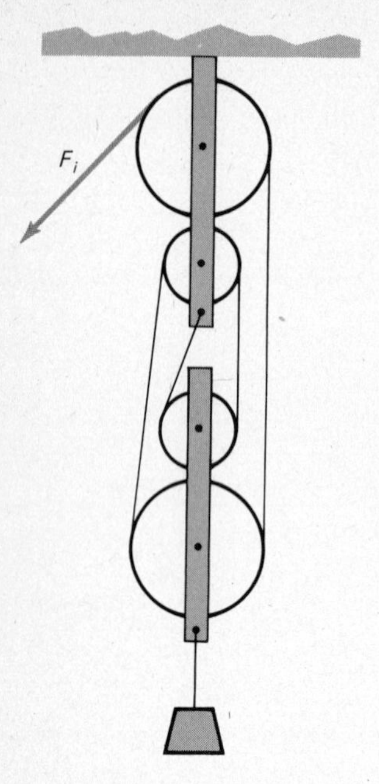

Figure 7.8 What is the IMA for this block-and-tackle system?

The ratio s_i/s_o depends on the design of the system as we shall now show.

We can obtain the ratio s_i/s_o by considering the volumes displaced by the two pistons as they move. When the input piston moves a distance s_i, it displaces a cylindrical volume of fluid with base area A_i and height s_i. Because the volume of a cylinder is (base) × (height), the volume displaced by the input piston is $A_i s_i$. Similarly, the volume displaced by the output piston is $A_o s_o$. Because the fluid is not compressible, these two volumes must be equal. Therefore,

$$A_i s_i = A_o s_o \qquad \text{or} \qquad \frac{s_i}{s_o} = \frac{A_o}{A_i}$$

which yields

$$F_o = F_i \frac{A_o}{A_i} \tag{7.9}$$

The output force is a factor A_o/A_i larger than the input force. It is easy to construct a hydraulic press for which this area ratio is several hundred. At a service station, the input force needed to lift a 20,000 N car might be only 50 N. Hydraulic presses are widely used in industry to stamp out metal parts from flat metal sheets. They are the major way in which industry obtains huge forces.

7.8 Devices That Transmit Torque

A motor is frequently used to drive some machine with the two devices connected by a belt. For example, the small wheel in Figure 7.10 might be that of the motor driving a large wheel on a band saw. In cases such as this the torque exerted by the motor can be increased or decreased by a suitable choice of pulleys (belt wheels).

For the case shown in Figure 7.10, the upper portion of the belt is tight whereas the lower portion is quite loose. The torque at the motor is r_i multiplied by the tension in the upper belt, T. The torque exerted on the large wheel is $r_o T$. We see that the torque advantage in this ideal system is

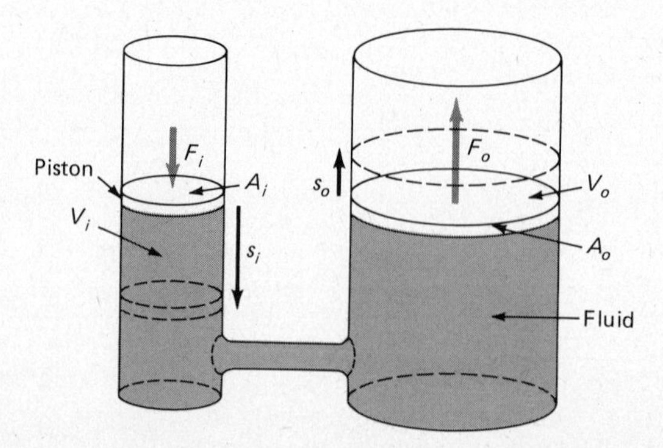

Figure 7.9 A hydraulic press.

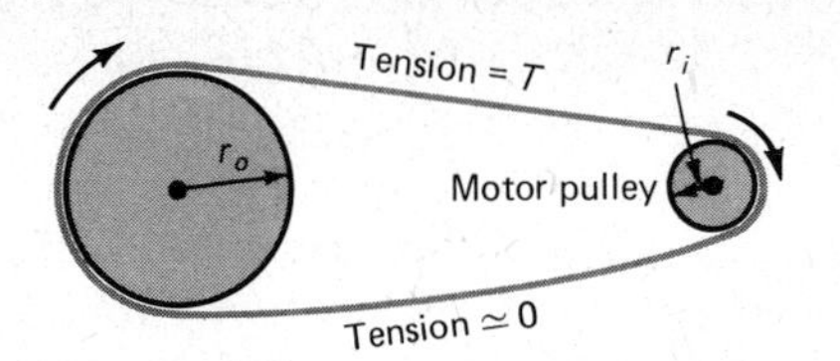

Figure 7.10 If the tension in the upper part of the belt is T, the torque on the large wheel is r_oT, and the torque on the small driving wheel is r_iT. The torque advantage is r_o/r_i.

$$\text{IMA (belt drive)} = \frac{\text{output torque}}{\text{input torque}} = \frac{r_oT}{r_iT} = \frac{r_o}{r_i}$$

In practice the tension in the lower belt is not zero. The actual mechanical advantage is therefore smaller than the ratio of the radii.

It is often of interest to know the relative speeds of the two pulleys. As the large wheel goes around once, a length of belt equal to its circumference ($2\pi r_o$) must pass over it. But the small wheel turns once each time a length of belt $2\pi r_i$ passes over it. So in one revolution of the large wheel, the small wheel has to turn the following number of times:

$$\frac{2\pi r_o}{2\pi r_i} = \frac{r_o}{r_i}$$

In other words, the input wheel turns r_o/r_i times faster than the output wheel. As we see, this ratio has the same value as the IMA.

Torques can be transmitted by gears as well as by belts. For example, the spur gears shown in Figure 7.11(a) are typical. They differ from the pulleys of Figure 7.10 in that the torque is transmitted by direct contact of the wheels instead of

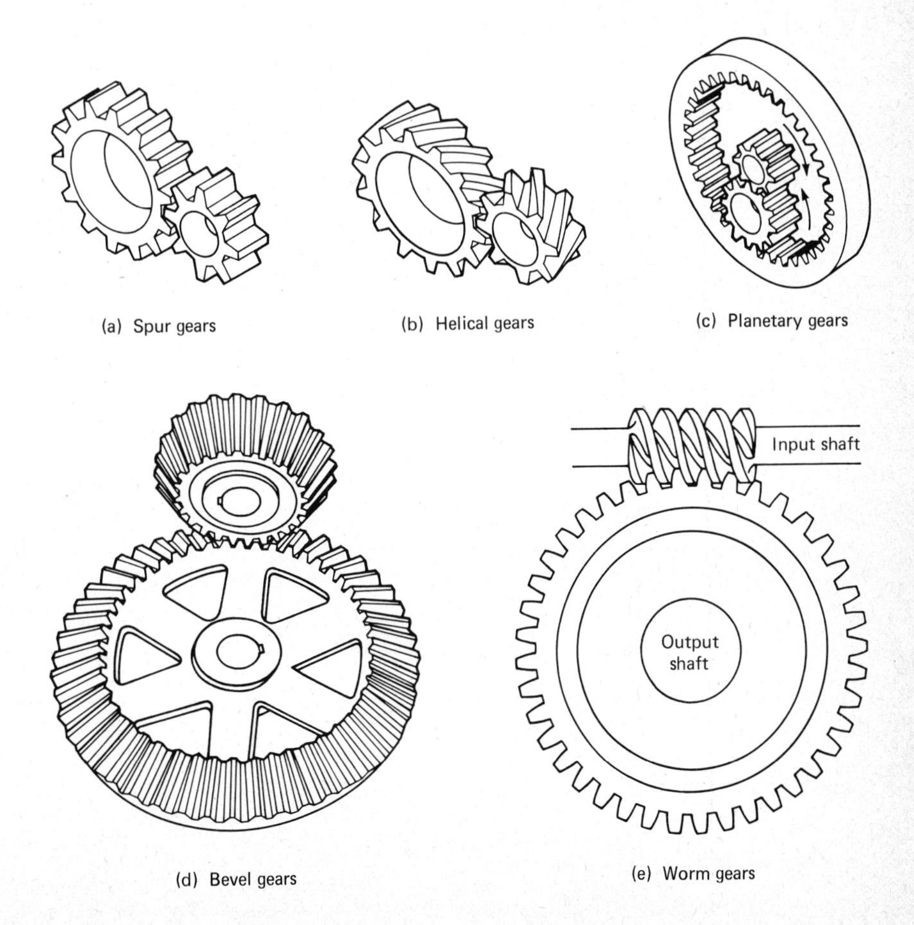

Figure 7.11 Several types of gear systems.

by a belt. It is still true that the IMA is equal to the ratio of the rotation rates of the two geared wheels. If the input gear has N_i teeth and the output gear has N_o teeth, then the ratio of their rotation rates is obviously N_o/N_i. Therefore,

$$\text{IMA (spur gears)} = \frac{N_o}{N_i}$$

Also, as before,

$$\text{IMA} = \frac{\text{input rotation rate}}{\text{output rotation rate}}$$

Other types of gears are also shown in Figure 7.11. Their IMA can be reasoned out in the same way we have done for the spur gears.

Summary

A machine is any device that transforms an input force into an output force that does work. There are many different kinds of machines. Many of them are composed of one or more of three basic types. The three basic types of machine are the inclined plane, the lever, and the hydraulic press.

Two types of mechanical advantage are used to describe machines. The ideal (or theoretical) mechanical advantage (IMA) is defined in terms of the distance s_i the input force must move when the output force moves a distance s_o. It is

$$\text{IMA} = \frac{s_i}{s_o}$$

The actual mechanical advantage (AMA) is defined in terms of the input force F_i and the output force F_o. It is

$$\text{AMA} = \frac{F_o}{F_i}$$

If the machine had no friction or other energy losses, then IMA = AMA.

All machines have friction. The output work is equal to the input work less the friction work and internal energy losses. We express the efficiency of a machine by

$$\text{Eff} = \frac{\text{output work}}{\text{input work}}$$

The percent efficiency is found by multiplying this ratio by 100. It also is true that

$$\text{Eff} = \frac{\text{AMA}}{\text{IMA}}$$

The efficiency is always less than unity (or 100 percent). This result follows from the fact that the output work is always less than the input work. All machines have friction and/or internal energy losses.

Questions and Exercises

1. The moving element of a simple mousetrap (see Figure P7.1) is a lever of the third kind [see Figure 7.5(f)]. Why is it important to have the input force applied close to the fulcrum (axis) in this device?

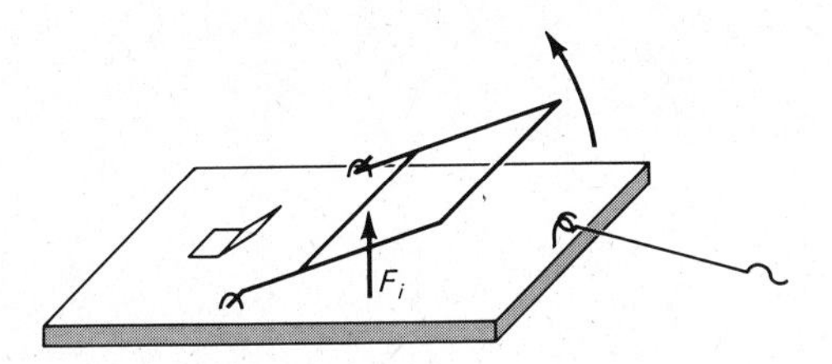

Figure P7.1 Mousetrap executioner mechanism.

2. Which of the following statements are true? If untrue, point out the untruth.
   a. All real machines have efficiencies less than unity.
   b. The IMA of a machine is always larger than the AMA.
   c. The IMA of a machine can be larger, equal to, or smaller than unity.
   d. No machine can continuously do more work than the input force does on it.
   e. Some machines have F_i larger than F_o. For them, s_i is less than s_o.
3. Many proposed perpetual motion machines have the following feature. The machine produces enough energy (or does enough work) to keep itself operating and, in addition, does work outside the machine. Why, without further investigation of the machine, can we say it will not work? Would the machine be possible if it didn't do work outside the machine?
4. Design a block-and-tackle system that has an IMA of 7.
5. Design a spur-gear system that has an IMA of $\frac{1}{5}$. What possible use could such a gear system have?
6. By watching the motions of the hands on a sweep-second-hand watch, one can see that it probably contains gear systems with IMAs in the ratio 1 to 12 and 1 to 60. Explain.
7. In the typical use of a flyswatter, the user's arm and swatter together form a lever of the third kind [see Figure 7.5(f)]. What property of this type of lever is critical for its successful use?
8. Refer to Figure 7.5 and relate it to the following statement: Levers can be classed as being of the first, second, or third kind depending on the order in which the fulcrum (f), F_i, and F_o are applied along the lever. These classes are

   First kind: F_o, f, F_i
   Second kind: f, F_o, F_i
   Third kind: f, F_i, F_o

   What advantageous features does each class have?
9. A three-speed bike has advantages and disadvantages in each gear setting. Explain these by reference to the IMA and F_i for each setting.
10. Often the steering gear systems for cars with power steering have different gear ratios than for those without power steering. Explain why. Is there any similarity between this situation and the fact that cars are designed to shift gears at various speeds?

Problems

1. In a particular machine the input force moves 18 cm while the load is lifted 0.50 cm. Experiment shows that the machine can lift a load of 500 N with a force of 30 N. What are the IMA, AMA, and efficiency of this machine?
2. A 600-N object is to be lifted by a machine using a force of 40 N. The machine found suitable for this lifts the load $\frac{1}{2}$ m when the applied force moves 10 m. Find the IMA, AMA, and efficiency of the machine.
3. For a mechanical car jack it is noticed that a force of 50 N must push down the handle a distance of 1.50 m for each

0.042 m the jack raises. What is the IMA of the jack? Assuming the jack to be 40 percent efficient, how large a load can the applied 50 N lift?

4. For a particular socket wrench the operator's hand is applied at 8.0 cm from the center of the socket. What is the IMA of the wrench when it is turning a nut with an effective radius of 0.40 cm?
5. Figure P7.2 shows a load being lifted in three different ways by the three possible classes of lever. Assume that the machines are 100 percent efficient so that AMA = IMA. How large must F_i be in each case? What is the IMA in each case?

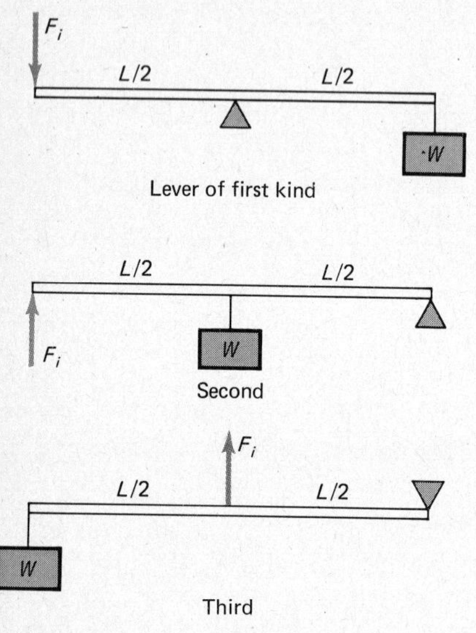

Figure P7.2 The three classes of levers.

6. Give the IMA, AMA, and efficiency for each of the levers shown in Figure 7.5.

*7. A certain electric motor uses electricity at the rate of 140 W. It has an efficiency of 85 percent when lifting a weight. How large a weight can it lift at a speed of 8.0 cm/s?

*8. Two screwdrivers are identical except for the radii of the handles. One is 1.50 cm, while the other is 2.50 cm. The torque you can apply to a screw is larger for the large-handled screwdriver. About how many times larger?

*9. A mechanic notes he cannot budge a nut using a 20-cm-long wrench. But when he inserts the wrench handle in the end of a 2.00-m length of pipe, the nut loosens easily. By what factor does the pipe increase the torque he could apply?

10. The lever of the third kind shown in Figure P7.2 is used to increase the speed of lifting. Suppose in such a lever that F_i is 5 cm from the fulcrum and W is 30 cm from the fulcrum. How fast would the load be moving if F_i was moving at 10 cm/s?
11. A 1000-N load on a wheelbarrow has its center of gravity 0.40 m from the axle. How much force must be exerted on the handles, 1.60 m from the axle, to lift the load?
12. In order to loosen a horizontal pipe, an 80-cm-long pipe wrench is clamped on it. When a 70-kg man stands on the end of the wrench, how much torque is applied to the pipe?
13. In a certain wheel-and-axle arrangement, the wheel has a radius of 50 cm, whereas the axle has a 6.0-cm radius. What minimum force on the wheel could lift a 20-kg mass hanging from the axle?
14. A typical pair of wire-cutting pliers is pushed together by one's hand placed about 8 cm from the pivot. If the wire to be cut is 0.8 cm from the pivot, how large a force is exerted on the wire when 30 N of force is applied by the hand?
15. For a certain pulley system a load of 500 N can be lifted 3 cm by an input force of 100 N pulling through a distance of 24 cm. What are the IMA, AMA, and efficiency for the system?
16. A certain pulley system can lift a load of 2000 N by use of a 500-N force. If its efficiency is 80 percent, what is the IMA of the system?
17. Draw the following pulley systems: (a) two fixed pulleys and one movable one with IMA = 3 and (b) two fixed and three movable pulleys with IMA = 6.
18. Refer to the pulley system shown in Figure P7.3. What is its IMA? If the efficiency of the system is 0.80, how large a load can be lifted by an input force of 400 N?

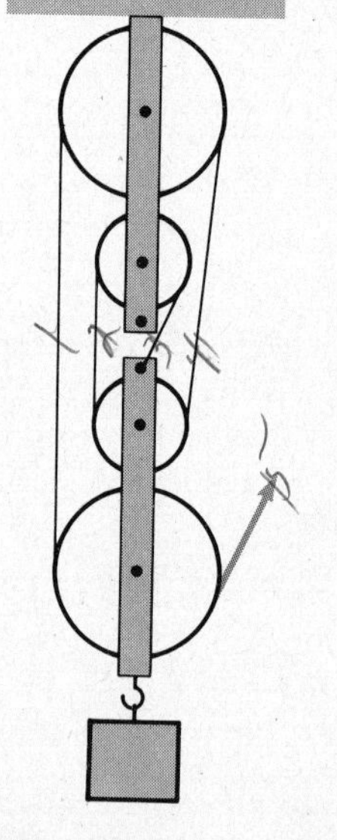

Figure P7.3

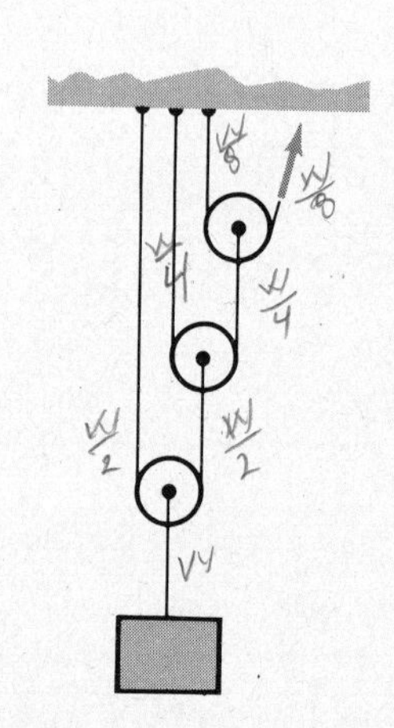

Figure P7.4

*Problems marked with an asterisk are not as easy as the unmarked ones.

**19. Refer to the pulley system shown in Figure P7.4. This is not a conventional block and tackle. Find its IMA by considering the distances s_i and s_o. What value would you obtain by simply counting the strands holding the load?

20. A 20-kg object is pulled up an incline by a force of 80 N parallel to the incline. The incline rises 15 cm for each 1 m along the incline. What are the IMA, AMA, and efficiency for the incline?

21. A knife blade is ground so that the blade is wedgelike. At a distance of 1 mm back from the cutting edge, the blade thickness is 0.15 mm. What is the IMA for this cutting device? Repeat for an ax blade that has a blade width of 2 mm at a distance of 0.8 cm from the cutting edge.

22. A certain hydraulic press has an output piston of 5.0-cm diameter. The input piston has a diameter of 0.625 cm. How large a force must be supplied to the input piston to provide an output force of 70,000 N?

23. Compressed air furnishes a force of 60 N to the 0.80 cm^2 area of an oil line that furnishes the oil used to lift the piston in a car lift at a gas station. What must be the cross-sectional area of the lift piston if the device is to be able to lift a 1100-kg vehicle?

24. In a certain system of gears the input shaft rotates 60 times for each rotation of the output shaft. Ideally how many times larger would the output torque be than the input torque?

*25. The pedals on a certain child's tricycle are 17 cm from the pedal axle. The pedaled wheel of the tricycle is 60 cm in diameter. How large a maximum force pushes the tricycle forward when a 20-kg girl exerts her full weight on the pedal?

*26. The belt-and-gear system of a lathe uses a driving motor whose rotation speed is 3450 revolutions per minute (rev/min). When the output chuck of the lathe rotates at 2.0 rev/s, by what factor does the gear system multiply the torque? Make the (poor) assumption that friction is negligible.

*27. Many belt-driven machines vary their speeds by means of two sets of pulleys (step pulleys), as shown in Figure P7.5. In the belt position shown the output rotation speed is much higher than the input speed. By shifting the belt to the left-hand-side pulleys, the speed ratio can be reversed. Suppose that the input is driven by a 5200-rev/min motor and that the pulleys have diameters of 4, 6, and 12 cm. What are the three possible output speeds?

**28. Figure P7.6 shows a differential chain hoist. The chain turns the pulleys without slipping. Show that the section of chain below A and B shortens by $2\pi(R - r)$ as the upper wheel turns through one revolution. Using this fact, prove that the IMA of the hoist is $2R/(R - r)$.

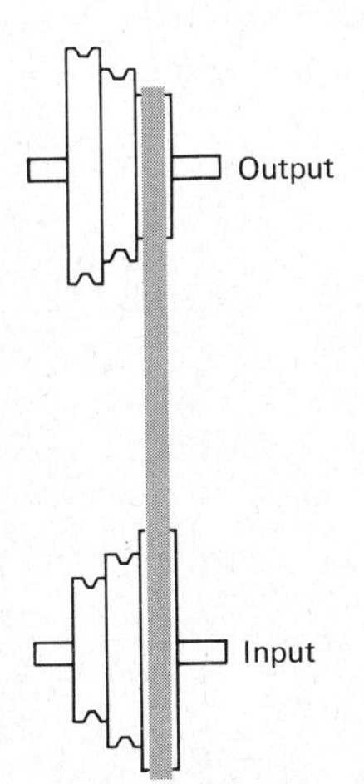

Figure P7.5

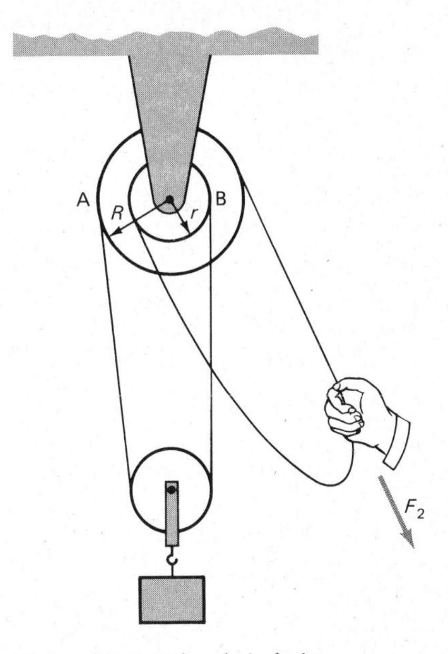

Figure P7.6 The chain hoist.

*29. A certain 0.25-hp motor rotates at 3450 rev/min. It pulls a belt on its pulley. The radius of the pulley is 3.0 cm. (a) What length of belt passes over the pulley each second? (b) If the motor were 100 percent efficient, how much work could it do each second? (c) What would be the tension in the belt under this optimum condition?

**30. A circular saw (diameter = 18 cm) is driven by a 5200-rev/min, 1.25-hp motor. The motor and blade shafts are coupled by spur gears. The one on the blade shaft has eight times as many teeth as the one on the motor shaft. (a) What is the

**Problems marked with a double asterisk are somewhat more difficult than the average.

rotation speed of the blade? (b) How far does a tooth on the blade travel in 1 s? (c) If the saw-motor combination is 40 percent efficient, how much useful work can the saw do in a second? (d) About what force does each tooth of the saw exert on the wood it is sawing?

*31. Figure P7.7 shows a device called a turnbuckle, which is used to tighten wires and cables. As it is twisted about the wire as axis, the two threaded rods move lengthwise toward the center. The radius of the center section is r as shown, and the pitch of the screw is p. Using energy input = energy output for an ideal machine, show that the ideal mechanical advantage of this device is $2\pi r/p$. (Assume that the driving force is applied as indicated.)

**32. Consider a worm-gear system such as the one shown in Figure 7.11(e). As the worm shaft rotates once, the cogged wheel moves by one cog. Suppose that the cogged wheel has n cogs. If the input and output shafts have equal-diameter pulleys on them, what is the IMA of the device? Repeat for an output shaft with a pulley three times larger in diameter than the input shaft pulley.

**33. For the compound machine shown in Figure P7.8, find (a) the IMA of the inclined plane, (b) the IMA of the block and tackle, (c) the IMA of the combined system.

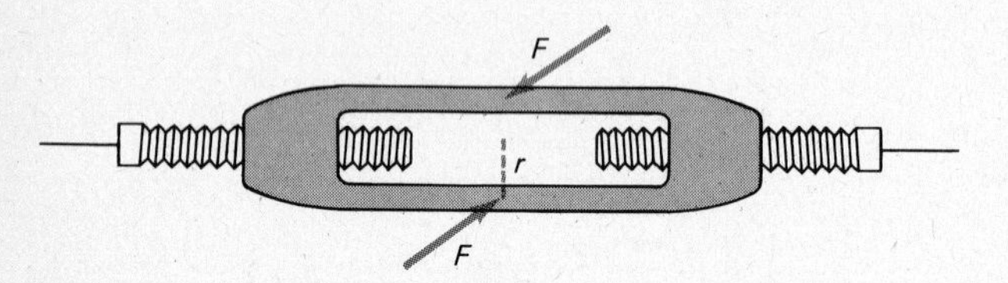

Figure P7.7 The turnbuckle.

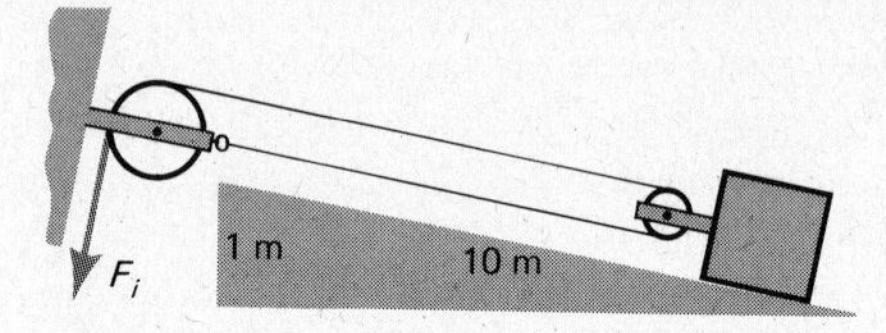

Figure P7.8 A compound machine.

Golf ball being struck by a golf club.

Harold E. Edgerton, M.I.T., Cambridge

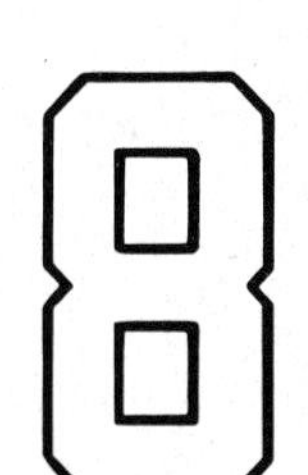

IMPUSE AND MOMENTUM

Performance Goals

When you finish this chapter, you should be able to

1. Find one of the following quantities if the other two are given: average force, time during which it acts, and impulse due to the force.
2. Describe qualitatively what will happen to a moving object when it is subject to a given impulse. Describe what will happen as the average force and time of impact change.
3. Find the change in momentum of an object when subjected to a known impulse (one-dimensional motion only).
4. Find the impulse that acted on an object if its initial and final

As we have seen, the concepts of work and energy are of great aid to us. In this chapter we present two other useful quantities. They are impulse and momentum. We shall see that momentum obeys a conservation law. By use of it, we shall be able to analyze easily many complex situations. It is of particular use when discussing collisions and explosions.

momentum are known (one-dimensional motion only).

5. Explain the law of conservation of linear momentum. In your explanation pay particular attention to: the meaning of "system of particles," restrictions on the system, internal versus external forces, the importance of the impulse equation, the importance of the law of action and reaction.
6. State whether or not kinetic energy is conserved in the collision of two objects. Assume that the objects move on the same straight line and stick together after collision. Justify and write the (momentum before) = (momentum after) equation for the collision. Solve for any unknown.
7. Find the velocity of one piece of an object that has just exploded. Assume that the object was originally hanging at rest from a weak thread. It exploded into two pieces and the velocity of the other piece is known. Both masses are known.
8. Explain the basis for the operation of rockets and jet engines.
9. State under what conditions it is allowable to write the law of conservation of linear momentum for a collision. Repeat for the equation for the conservation of kinetic energy.
10. Find two of the following quantities provided the others are given: m_1, m_2, v_{10}, v_{20}, v_{1f}, v_{2f}. Assume that two masses m_1 and m_2 are moving on a straight line and undergo a perfectly elastic collision.

8.1 Impulse

Impulse is usually associated with collisions. When a hammer strikes a nail, the nail experiences an impulse. When a fighter hits his opponent's jaw, the jaw may be broken by the impulse. These are examples of the common meaning given to this word.

In science and technology we make this meaning quantitative. Suppose that there is an impact of one object with another. If the time of contact is t and the average vector force during this time is $\overline{\mathbf{F}}$, then by definition

$$\text{Vector impulse} = \overline{\mathbf{F}}t \tag{8.1}$$

Notice that impulse is a vector. It has the same direction as $\overline{\mathbf{F}}$.

We see the effects of impulse in many ways each day. Each time we strike something, we exert an impulse on the object struck. Most situations are complex. For example, when a bat hits a ball, the force acting on the ball varies widely. The impulse involves the average of this variable force. In addition, the time of contact between ball and bat is difficult to measure. So the impulse $\overline{\mathbf{F}}t$ is not simple. But we shall see in the next section that the effect of an impulse can be treated easily if we look at another quantity, the momentum.

8.2 Linear Momentum

Let us examine a simple situation so that we can easily see the effect of an impulse. Suppose that a golf club strikes a golf ball as shown in Figure 8.1. What can we do with such a complex situation? Common sense tells us that the club's effect on the ball will involve at least two factors, the force and the duration of the force. To begin, let us see what we can say about the average force on the ball, $\overline{F}$. According to Newton's law, we can write $\mathbf{F} = m\mathbf{a}$ for the ball. In this case $\mathbf{F} = \overline{\mathbf{F}}$ and m is the mass of the ball. Then

$$\overline{\mathbf{F}} = m\overline{\mathbf{a}}$$

where $\overline{\mathbf{a}}$ is the average acceleration of the ball. But from the definition of $\overline{\mathbf{a}}$, we have

$$\overline{\mathbf{a}} = \frac{\mathbf{v}_f - \mathbf{v}_0}{t}$$

(Of course, in the present case $v_0 = 0$. But let us carry v_0 along so as to make the discussion more general.) Substitution of this value for $\overline{\mathbf{a}}$ gives

$$\overline{\mathbf{F}} = \frac{m(\mathbf{v}_f - \mathbf{v}_0)}{t}$$

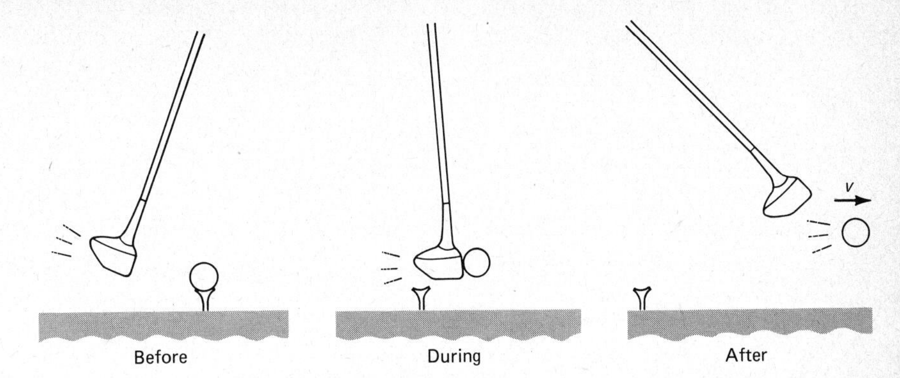

Figure 8.1 During the impulse of duration t, the club exerts an average force $\bar{F}$.

Or, after clearing fractions,

$$\bar{\mathbf{F}}t = m\mathbf{v}_f - m\mathbf{v}_0 \qquad (8.2)$$

This is an important equation for two reasons. It tells us how the impulse changes the motion of the ball, that is, how the impulse causes the quantity $m\mathbf{v}$ to change. But even more important, this relation tells us that $m\mathbf{v}$ is a simple quantity we can consider in order to discuss a collision.

We call the product of the mass and velocity of an object, *m*v, the linear momentum of the object. It is designated by *p*.

Notice that momentum is a vector. It has direction, the direction of **v**.

We can now state Equation 8.2 in words.

When an impulse is exerted on an object, the change in momentum of the object equals the impulse.

The following examples will show some uses of these concepts. In the next section we shall see that momentum—not impulse—is the more useful of the two quantities.

■ **EXAMPLE 8.1** Suppose that you are running straight ahead in the dark. You run head-on into an object that stops you. Why will you be hurt less if the object is a person than if it is a tree?

Solution Upon collision the object exerts a force on your body. The larger the force, the larger the damage to your body may be. In both cases your momentum was changed by the same amount. You were stopped. Therefore, both objects must have exerted the same impulse on your body. In other words, $\bar{F}t$ was the same in each case.

But t was greatly different in the two cases. By t we mean the time during the collision. When a tree is hit, the tree stops you almost at once. The tree does not "give." But when you hit another person, the person moves away as the collision occurs. The

(a) Impulse = $F \times t$

(b) Impulse = $F \times t$

Figure 8.2 Although the impulse required to stop the car is the same in both cases, the damage to the car is much less in (b). Why?

person "gives." As a result, the stopping process is spread over a longer time. The quantity t is much larger in the case of a person than in the case of a tree.

Because $\bar{F}t$ is the same in each case, $\bar{F}$ is larger when t is smaller. Therefore, the object that "gives" most will cause the smaller force and less injury. See Figure 8.2 for another example of this. This same idea is basic to the operation of most impact-absorbing devices. ■■

■ **EXAMPLE 8.2** A 10-g bullet is shot with a speed of 250 m/s from a gun. It took 6 ms (milliseconds) to accelerate the bullet in the gun. Find the average force that accelerated the bullet.

Solution We make use of the impulse equation:

$$\bar{F}t = mv_f - mv_0$$

with $t = 6 \times 10^{-3}$ s, $m = 0.010$ kg, $v_f = 250$ m/s, and $v_0 = 0$. Substitution of these values gives

$$\bar{F} = 420 \text{ N}$$

■■

■ **EXAMPLE 8.3** An 0.050-kg ball moving at 2.0 m/s toward a wall rebounds straight back with the same speed. If the time of contact with the wall is 0.010 s, find the average force exerted by the wall upon the ball.

Solution This example shows clearly the vector nature of momentum. If it were a scalar, then the ball's momentum would not be changed. It has the same *speed* before and after collision. Actually, though, its *velocity* does change. Originally it was 2.0 m/s and finally it is −2.0 m/s. Using the impulse equation, Equation 8.2, we have

$$\bar{F}t = mv_f - mv_0$$

which is

$$\begin{aligned}\bar{F}(0.010 \text{ s}) &= (0.050 \text{ kg})(-2.0 \text{ m/s}) - (0.050 \text{ kg})(2.0 \text{ m/s}) \\ &= -0.20 \text{ kg} \cdot \text{m/s}\end{aligned}$$

Solving gives

$$\bar{F} = -20 \text{ N}$$

The negative sign tells us that $\bar{F}$ is in the direction opposite to the direction we have taken as positive. ■■

8.3 Conservation of Linear Momentum

The law of action and reaction leads us to a powerful fact concerning momentum. Suppose that we have several objects (balls on a billiard table, for example). Taken together, their total momentum at any instant will be

$$m_1\mathbf{v}_1 + m_2\mathbf{v}_2 + \cdots + m_N\mathbf{v}_N$$

if there are N objects.

Let us now call the objects a *system of objects*. Two kinds of forces act on the objects. When one of the objects of the system strikes another, they exert forces on each other. We call these *internal forces*. (For example, the collision forces between two of the billiard balls would be internal forces.) Outside forces (or *external forces*) may also act on the system. The billiard ball system has gravity pulling down on the balls, but since the table's upward push on each ball balances out the gravitational pull, there is no resultant outside force acting on the system. An unbalanced external force is produced when a ball is hit by a cue or strikes the edge of the table.

The important fact we are about to learn deals with an isolated system of objects, that is, a system in which the resultant of all outside forces acting on the system is zero (or at least negligible). In such a system the only unbalanced forces on the objects are due to internal forces. For example, if two of the billiard balls collide, then each will exert an unbalanced force on the other. Let us see what happens in a case such as this.

Suppose that two balls collide as in Figure 8.3. There are no unbalanced forces on the balls until they hit each other. During the collision they exert forces on each other. The law of action and reaction tells us that the forces are equal and oppositely directed. The change of momentum of one ball will be $\overline{\mathbf{F}}t$, where t is the contact time and $\overline{\mathbf{F}}$ is the force exerted on it by the other ball. But the force on the other ball is $-\overline{\mathbf{F}}$. So its change in momentum is $-\overline{\mathbf{F}}t$. The balls together have zero momentum change because $\overline{\mathbf{F}}t + (-\overline{\mathbf{F}}t) = 0$. The momentum change of one exactly cancels the change of the other.

We conclude from this that collisions within a system of objects do not change the total momentum of the system. Therefore, any change in the total momentum of a system must come from unbalanced external forces. In the absence of a resultant external force, the total momentum of a system remains constant. This fact is summarized as the *law of conservation of linear momentum*.

When the vector sum of the external forces acting on a system is zero, then the total linear momentum of the system remains unchanged.

This law has widespread application. We shall show its use in the following examples.

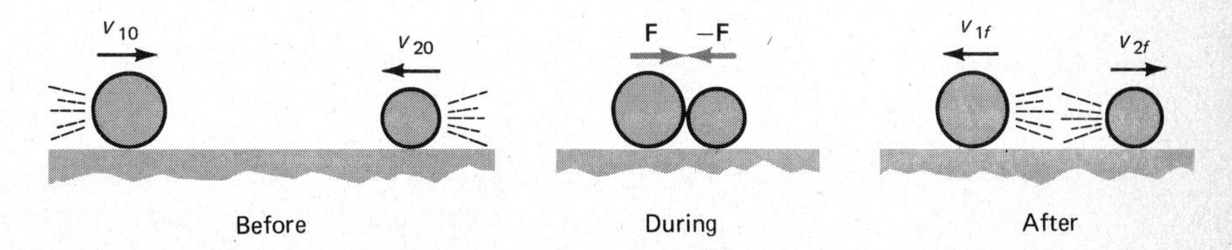

Figure 8.3 The law of action and reaction tells us that the collision forces are equal but opposite. We then conclude that the combined change in momentum of the two balls is zero.

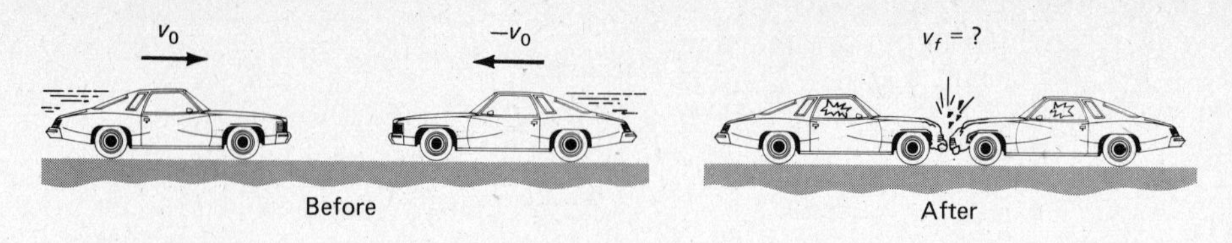

Figure 8.4 What happens if the cars stick together after the collision?

■ **EXAMPLE 8.4** Two identical cars going in opposite directions with equal speeds collide head-on. After collision, they stick together. Find how fast they are moving just after the collision.

Solution The situation is shown in Figure 8.4. Notice that one car has a velocity v_0 whereas the velocity of the other is $-v_0$. No large unbalanced external forces act on the cars. During collision huge forces are exerted on the cars, but these are internal forces.

Because unbalanced external forces are absent, the momentum of the two-car system does not change. We can write the following word equation:

$$\text{Momentum before collision} = \text{momentum after}$$

But the momentum of one car was mv_0 and that of the other was $m(-v_0)$ before collision. After collision they stick together, so their velocity is v_f. Placing these quantities into our word equation gives

$$mv_0 + m(-v_0) = mv_f + mv_f$$

This becomes

$$0 = 2mv_f$$

from which

$$v_f = 0$$

In other words, the cars come to a dead stop upon colliding head-on. This is true only if the cars have equal but opposite momentum before collision. Their combined momentum before collision was zero. It must also be zero after collision. Therefore, $v_f = 0$. ■■

Figure 8.5 What happens to the log as the man jumps from it?

■ **EXAMPLE 8.5** An 80-kg man is standing on a 60-kg log floating on a lake. He jumps horizontally from the log with a speed of 5 m/s relative to the water. This causes the log to shoot backwards. (See Figure 8.5.) Find the recoil velocity of the log just after he jumps.

Solution Strictly speaking, the friction force of the water on the recoiling log is not balanced. However, that force acts to slow the motion of the log. If we are interested in the speed of the log just after the man jumps, we can ignore this relatively small unbalanced force.

Before the man jumps, the man and the log both have zero momentum. After he jumps, the man's momentum is (80 kg)(5 m/s). We are calling the man's motion direction positive. The log's final momentum is (60 kg)(v_f). The law of conservation of momentum allows us to write

$$\text{Momentum before} = \text{momentum after}$$

which is

$$0 = (80 \text{ kg})(5 \text{ m/s}) + (60 \text{ kg})\, v_f$$

Solving gives

$$v_f = -6.7 \text{ m/s}$$

The negative sign tells us that the log's velocity is in the negative direction. The log recoils. ■■

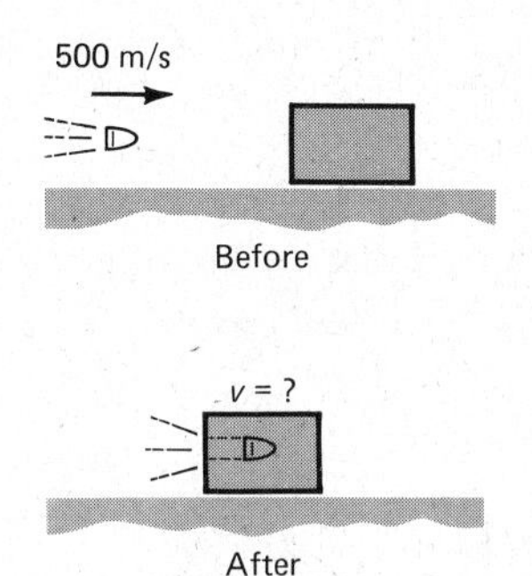

Figure 8.6 Apply the law of momentum conservation to this collision.

■ **EXAMPLE 8.6** An 8.0-g bullet moving at 500 m/s is fired into a 6-kg block of wood resting on a table, as shown in Figure 8.6. The bullet sticks in the wood. How fast is the block sliding across the table just after collision?

Solution Here, too, we can ignore the friction between table and block because we are interested in conditions just after collision. The law of momentum conservation allows us to write

$$\text{Momentum before} = \text{momentum after}$$
$$(0.0080 \text{ kg})(500 \text{ m/s}) + 0 = (6.008 \text{ kg})v$$

where v is the velocity of the combined block and bullet after collision. Solving gives $v = 0.67$ m/s. ■■

8.4 Rockets and Jet Propulsion

In Example 8.5 we saw a situation involving recoil. As the man jumped one way, the log moved in the opposite direction. This is a simple example of an important phenomenon. Another example is shown in Figure 8.7. Those who use guns are aware that they "kick" backwards, that is to say, they recoil. Exploding gunpowder in a gun forces the bullet forward in the barrel (to the right in the illustration). At the same time the explosion exerts a leftward (backward) force on the gun. This causes the gun to recoil. Can you show, using the law of momentum conservation, that the recoil velocity of the gun is given by

$$V_{gun} = v_{bullet} \frac{m_{bullet}}{M_{gun}}$$

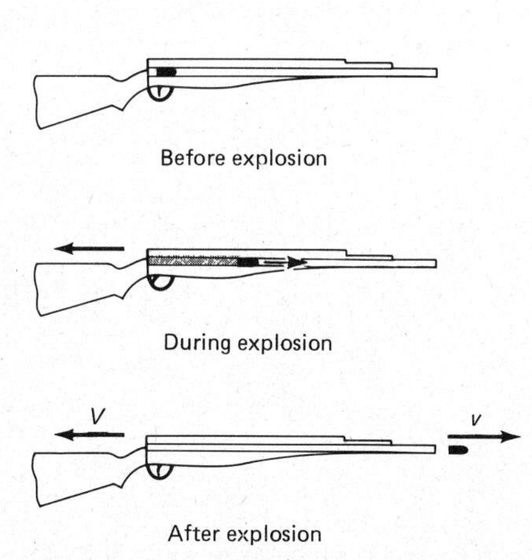

Figure 8.7 The exploding gunpowder pushes the bullet forward and the gun backward.

Notice that the recoil has nothing to do with the environment of the gun. It could be anywhere, even in outer space where there is no air. Recoil action does not depend on the air or any other surroundings. We emphasize this because rockets and jet engines make use of this same recoil action.

Newton was one of the first to recognize that recoil action could be used for propulsion. He described a jet-propelled cart something like the one shown in Figure 8.8. Steam from the boiler ejects to the rear. The escaping steam is like

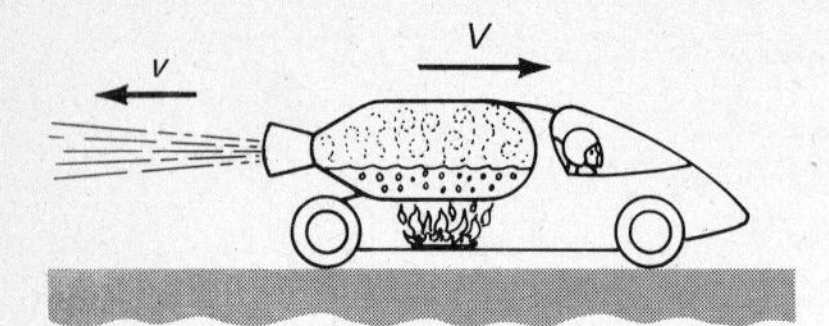

Figure 8.8 A jet-propelled cart.

the bullet shooting from the gun. Just as the gun recoils, so too does the cart recoil. Hence the cart recoils forward (toward the right) as the steam shoots backward (toward the left).

Rockets and jet engines cause propulsion in the same way as Newton's steam boiler. Hot gas is generated by burning fuel within the rocket or jet engine. The hot gases eject with high velocity toward the rear, and the rocket recoils forward. Forces generated within the rocket by the burning fuel throw the reaction gases one way and the rocket the opposite way.

Notice that this whole phenomenon is best summarized by the law of conservation of momentum. Since the system's momentum cannot be changed if there are no unbalanced forces acting on it, the rearward momentum given to the exhaust gases must equal the forward momentum given to the rocket. Such a system operates best in the absence of air because air friction will retard the motion given to the rocket by recoil action.

EXAMPLE 8.7 A 20,000-kg space ship is coasting through space at a speed of 50 m/s. It turns on its rocket engine and ejects 30 kg of vaporized fuel to the rear with a velocity of 15,000 m/s relative to the rocket. Find the new velocity of the rocket. (Assume negligible gravitational force acts on the ship.)

Solution From the momentum conservation law, assuming the effect of external forces to be negligible,

$$\text{Momentum before} = \text{momentum after}$$
$$(20{,}000\ \text{kg})(50\ \text{m/s}) = (20{,}000 - 30\ \text{kg})(v) + (30\ \text{kg})(50 - 15{,}000\ \text{m/s})$$

Solving for v gives the new velocity of the rocket to be 72 m/s. ■■

EXAMPLE 8.8 The Polaris missile experiences a thrust of 450,000 N under full power. That is to say, the force pushing the rocket forward is 450,000 N. This thrust is caused by ejection of hot fuel gases with a speed of 1500 m/s. How much fuel does the rocket exhaust each second?

Solution By the law of action and reaction the force on the missile is equal and opposite to the force on the exhaust gas. Applying the impulse equation to the exhaust gas (mass m) shot out in 1 s, we have from Equation 8.2

$$(\text{Force on gas})(1\ \text{s}) = (\text{final momentum of gas}) - (\text{initial momentum})$$
$$(450{,}000\ \text{N})(1\ \text{s}) = m(1500\ \text{m/s}) - 0$$

from which $m = 300$ kg. The rocket burns 300 kg of fuel each second. ■■

8.5 Perfectly Elastic Collisions

Consider the straight-line collision shown in Figure 8.9. There are only two objects involved. Suppose we know their masses and original velocities. What can we say about the motion after collision? Unless we know more than the masses and initial velocities, there is nothing more that we can say about the collision. Let us see why not.

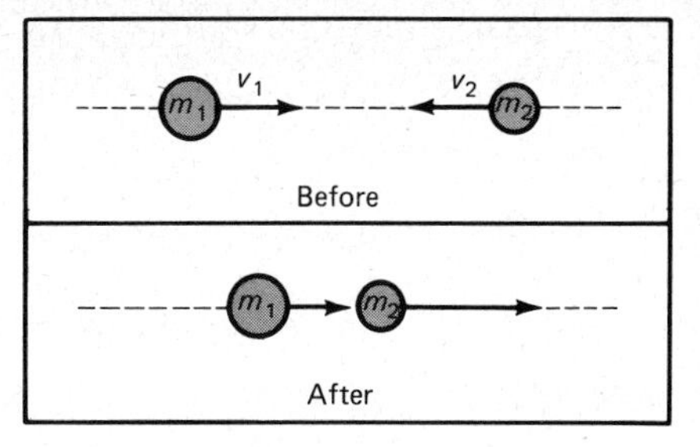

Figure 8.9 If you know v_1 and v_2 in this head-on collision, what can be said about the final velocities?

The only equation we can write for the collision is:

$$\text{Momentum before} = \text{momentum after}$$

We know only the momenta of the two balls before collision. The two final velocities are unknown. But we have only one equation, the momentum equation, to solve for these two unknowns. As you know from algebra, you need two independent equations to solve for two unknowns. We therefore need another equation if we are to find the motion after collision.

How did we get around this difficulty in earlier sections of this chapter? The situations we encountered gave us data equivalent to another equation. For example, when we said that the objects stuck together, we provided the follow-

DAMAGE FROM MECHANICAL SHOCK

In designing almost any commercial device, one must consider the effects of mechanical impact or shock upon it. We all know that considerable work has been done to reduce automobile impact damage. Proper design of a car's front end can reduce damage in a front-end collision. But the problem is not confined to autos. For example, what will happen to a portable TV when it is dropped? This problem is of great concern in the design of a quality product. There are many other examples one could give where good design must consider the effects of impact.

How does one design in protection against impact? One way would be to house the device in a massive shell. But an auto built like a military tank is not a reasonable solution. In most situations the massive-shell approach is not practical. Fortunately, there is another alternative.

As we learned in this chapter, the duration of the shock or impact is important. Because impulse is $\bar{F}t$, the impact force can be reduced by increasing the time for the impact. It is for this reason that car front ends are now designed to give. The car is not brought to a rigid stop. Instead, it is brought to rest more slowly as the front end gives. The time of impact is lengthened. As a result, the impact force (and damage) is reduced.

In a similar way electronic components in a TV can be nonrigidly suspended. Should the TV be dropped, a spring cushion is provided for its interior parts. In many situations flexible housings are used. They give upon impact so as to reduce damage.

Even on a molecular scale, use is made of this. High-impact plastics, for example, are designed to give on a molecular scale. In the photo are shown common examples of impact-resistant containers, which give on impact. As we see, the factor t in $\bar{F}t$, the impulse expression, is of great practical importance.

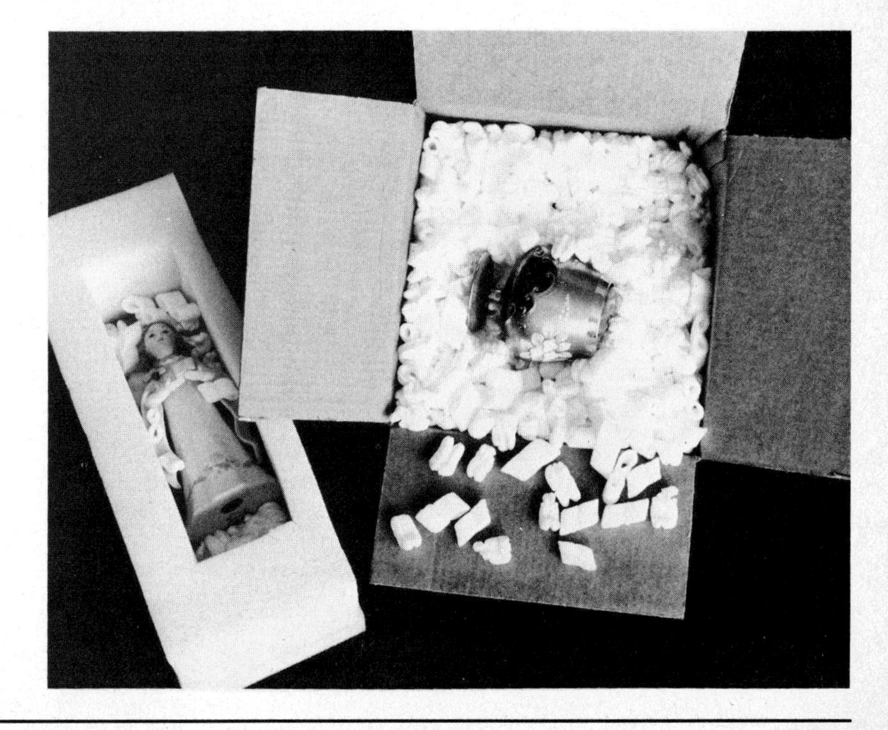

ing equation: The two final velocities are equal. Or in another case we gave one of the final velocities. This left only one unknown. These were practical and useful situations. But there are other important situations where the second equation must be found in other ways.

We can gain useful data for a collision if we know something about the kinetic energy of the colliding objects. In Example 8.4 we discussed the head-on collision of two identical cars with equal speeds. They came to a dead stop. In this case the kinetic energy before collision was $2(\frac{1}{2}mv_0^2)$. After collision it was zero. All the kinetic energy was lost to other energy forms during collision.

The opposite type of collision, in which no kinetic energy is lost, is also important. We call this a *perfectly elastic* collision.

A perfectly elastic collision is one in which the total kinetic energy is not changed by the collision.

Two examples of collisions that are often very nearly of this type are the collisions of billiard balls or of molecules and atoms. Let us see what we can conclude for perfectly elastic collisions.

For any isolated system of two objects we can always write the momentum collision equation. Call the masses of the objects m_1 and m_2 and their velocities v_1 and v_2. Then *if the motion is along a straight line,*

$$\text{Momentum before} = \text{momentum after}$$

becomes

$$m_1v_{10} + m_2v_{20} = m_1v_{1f} + m_2v_{2f}$$

If the collision is perfectly elastic, then we can also write

$$\text{KE before} = \text{KE after}$$

which is

$$\tfrac{1}{2}m_1v_{10}^2 + \tfrac{1}{2}m_2v_{20}^2 = \tfrac{1}{2}m_1v_{1f}^2 + \tfrac{1}{2}m_2v_{2f}^2$$

This provides us with enough equations to find v_{1f} and v_{2f} if all the other quantities are known.

We can combine these two equations to find an interesting result. After some simple algebra, the two equations become

$$m_1(v_{10} - v_{1f}) = m_2(v_{2f} - v_{20}) \qquad \text{(a)}$$

and

$$m_1(v_{10}^2 - v_{1f}^2) = m_2(v_{2f}^2 - v_{20}^2)$$

This latter equation can be factored to give

$$m_1(v_{10} - v_{1f})(v_{10} + v_{1f}) = m_2(v_{2f} - v_{20})(v_{2f} + v_{20}) \tag{b}$$

If we now divide equation (b) by (a), we find

$$v_{10} + v_{1f} = v_{2f} + v_{20}$$

This can be put in the following form by transposing:

$$v_{10} - v_{20} = -(v_{1f} - v_{2f}) \tag{9.3}$$

Notice what this equation says. The quantity $v_{10} - v_{20}$ is the speed with which the particles approach each other. For example, if they are heading toward each other with speeds v_0, then $v_{10} = v_0$ and $v_{20} = -v_0$ so that their speed of approach is $2v_0$. If they rebound with speeds v_0, then $v_{1f} = -v_0$ and $v_{2f} = v_0$. Thus

$$-(v_{1f} - v_{2f}) = 2v_0$$

This is their speed of separation after collision.

We can summarize our result in the following way. Notice that there are restrictions on the result.

For (a) a perfectly elastic collision between two objects (b) moving along a straight line, the relative velocity before collision ($v_{10} - v_{20}$) equals the negative of the relative velocity after collision $-(v_{1f} - v_{2f})$.

This is a convenient fact to remember when dealing with perfectly elastic collisions along a straight line. Let us illustrate its use.

■ **EXAMPLE 8.9** In a nuclear reactor a neutron going in the $+x$ direction with speed v_0 strikes a stationary hydrogen atom head-on, as shown in Figure 8.10. Assuming the collision to be perfectly elastic, find the speeds of the two particles after collision. The masses of the two particles are very nearly the same.

Solution The collision is along a straight line (head-on). We can write

$$\begin{gathered}\text{Momentum before} = \text{momentum after}\\ mv_0 + 0 = mv_n + mv_H\end{gathered} \tag{a}$$

We call the final velocities of the neutron and hydrogen v_n and v_H, respectively. Notice that m cancels from the equation.

Because the collision is perfectly elastic, we can write Equation 9.3.

$$v_0 - 0 = -(v_n - v_H)$$

Let us solve this equation for v_n. Then

$$v_n = v_H - v_0 \tag{b}$$

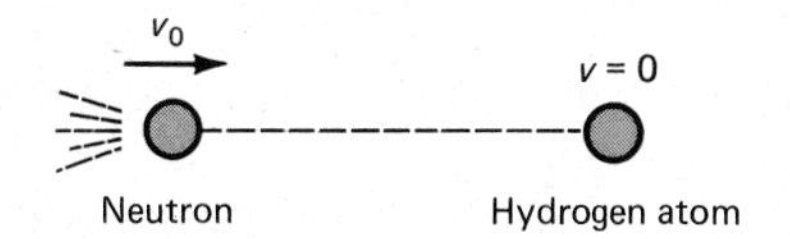

Figure 8.10 If the masses are identical and the collision is perfectly elastic, what can be said about the final velocities of these particles?

We then substitute (b) in (a) to find

$$v_0 = v_H - v_0 + v_H$$

which gives

$$v_H = v_0$$

Substituting this in (b) gives

$$v_n = 0$$

Notice what happened. The neutron came to a dead stop. The hydrogen atom went off with the same velocity as the original neutron. This is an important result worth remembering. But before stating it, let's look at a more general case. ▮▮

▮ **EXAMPLE 8.10** Two identical objects with initial velocities v_1 and v_2 collide head-on. Find their final velocities if the collision is perfectly elastic.

Solution As before, conservation of momentum gives

$$mv_1 - mv_2 = mv_{1f} + mv_{2f} \quad \text{(a)}$$

Because the collision is perfectly elastic, we can use Equation 9.3,

$$v_1 - (-v_2) = -(v_{1f} - v_{2f}) \quad \text{(b)}$$

Dividing through equation (a) by m and adding equations (a) and (b) gives

$$2v_1 = +2v_{2f}$$

from which

$$v_{2f} = v_1$$

Placing this result in (a) gives

$$v_{1f} = -v_2$$

In other words, the particles just exchange velocities! ▮▮

Let us now summarize the result of these two examples. Notice the restrictions.

> **Given (a) two identical particles that collide (b) perfectly elastically in (c) straight-line motion, the two particles simply interchange their velocities.**

This is an easy result to remember. But be careful. It applies only to this particular situation.

■ EXAMPLE 8.11 A hydrogen atom ($m = 1$ u) with velocity v_0 collides head-on with a stationary carbon atom ($m = 12$ u). The collision is perfectly elastic. Find their speeds after collision.

Solution The unit of mass used here is the atomic mass unit (1 u = 1.66×10^{-27} kg). The momentum equation is

$$(1\text{ u})v_0 + 0 = (1\text{ u})v_H + (12\text{ u})v_C \qquad \text{(a)}$$

Notice that the units (u) cancel from the equation. The mass unit used is not important because it cancels.

Equation 9.3 is also applicable and gives

$$(v_0 - 0) = -(v_H - v_C)$$

Solving for v_H gives

$$v_H = v_C - v_0 \qquad \text{(b)}$$

Placing this in (a) yields

$$v_C = \left(\frac{2}{13}\right)v_0$$

Now using this in (b) gives

$$v_H = -\left(\frac{11}{13}\right)v_0$$

■■

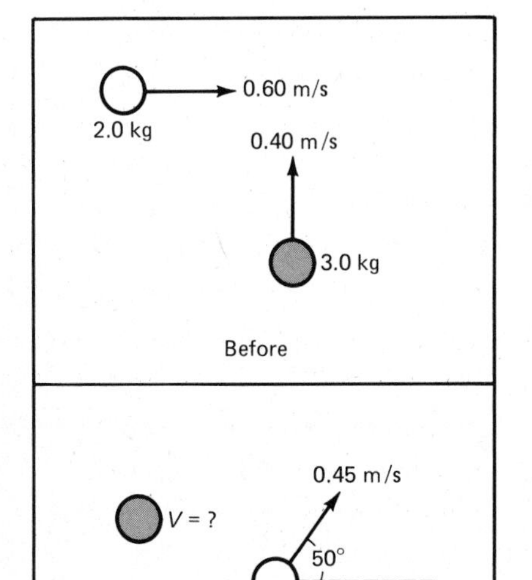

Figure 8.11 What equations do the momentum components obey?

8.6 Collisions in Two Dimensions

Until now we have considered collisions between objects moving along a straight line. Collisions in two and three dimensions are only slightly more difficult to deal with. For them we need only split the momentum vector into its components. We then note that the direction as well as the magnitude of the momentum obeys the conservation law. We can therefore write our equations in component form. The law of conservation of linear momentum applies to each component of the total momentum.

Referring to Figure 8.11, we see a two-dimensional collision. The law of conservation of momentum tells us that

x-direction momentum before = *x*-direction momentum after
y-direction momentum before = *y*-direction momentum after

If the collision had involved three dimensions, there would be a similar equation for the *z* direction.

■ EXAMPLE 8.12 A 2.0-kg ball moving at 0.60 m/s in the $+x$ direction collides with a 3.0-kg ball moving at 0.40 m/s in the $+y$ direction as shown in Figure 8.11. After collision the 2.0-kg ball is moving at 0.45 m/s at 50° to the x axis. What is the 3.0-kg ball doing after the collision?

Solution Momentum must be conserved in both the x and y directions. We can therefore write

$$(\text{Momentum before})_x = (\text{momentum after})_x$$
$$(2.0\ \text{kg})(0.60\ \text{m/s}) + 0 = (2.0\ \text{kg})(0.45 \cos 50^\circ\ \text{m/s}) + (3.0\ \text{kg})v_x$$

and

$$(\text{Momentum before})_y = (\text{momentum after})_y$$
$$(3.0\ \text{kg})(0.40\ \text{m/s}) + 0 = (2.0\ \text{kg})(0.45 \sin 50^\circ\ \text{m/s}) + (3.0\ \text{kg})v_y$$

We can solve these two equations for the components of the 3.0-kg ball's velocity. The result is

$$v_x = 0.207\ \text{m/s} \qquad \text{and} \qquad v_y = 0.170\ \text{m/s}$$

from which

$$v = \sqrt{v_x^2 + v_y^2} = 0.27\ \text{m/s}$$

Can you show that the angle the velocity makes with the $+x$ axis is 39°? Is kinetic energy conserved in this collision? ■■

Summary

The impulse exerted by an average force $\overline{\mathbf{F}}$ acting for a time t is $\overline{\mathbf{F}}t$. It is a vector and has the direction of $\overline{\mathbf{F}}$.

An impulse $\overline{\mathbf{F}}t$ acting on an object of mass m causes the velocity of the object to change from $\mathbf{v}_0$ to $\mathbf{v}_f$, where

$$\overline{\mathbf{F}}t = m\mathbf{v}_f - m\mathbf{v}_0$$

The linear momentum $\mathbf{p}$ of an object of mass m and velocity $\mathbf{v}$ is $m\mathbf{v}$. It is a vector and has the same direction as $\mathbf{v}$.

When the vector sum of the external forces acting on a system is zero, then the total linear momentum of the system remains constant. This is a statement of the law of conservation of linear momentum.

In the absence of a sizable resultant external force, one can write the following equation for the objects involved in a collision or explosion:

$$\text{Momentum before} = \text{momentum after}$$

A perfectly elastic collision is one in which the total kinetic energy of the system is unchanged by the collision. In a perfectly inelastic collision the colliding objects stick together after collision.

Only in the case of a perfect elastic collision can one write an equation stating that the KE before the collision equals the KE after the collision.

For a perfectly elastic collision of two objects moving along a straight line, one can equate the relative velocity of approach to the negative of the relative velocity of recession:

$$v_{10} - v_{20} = -(v_{1f} - v_{2f})$$

When two *identical* particles that move on a *straight line* collide *perfectly elastically,* the two particles interchange velocities.

When dealing with a situation involving motion in two or three dimensions, the law of conservation of linear momentum yields a component equation for each dimension. In the absence of a sizable resultant external force, one can write

$$(\text{Momentum before})_x = (\text{momentum after})_x$$

plus similar equations for the y and z coordinates.

Questions and Exercises

1. Reasoning from the impulse equation, explain why it is much more fun to dive into a pool when it is full of water than when it is empty.
2. They say that when you jump down onto the floor or ground, you should not hold your legs rigid. You should let them give. Explain, in terms of the impulse equation, why this is true.
3. An astronaut is trying to fix a mechanism with a wrench on the outside of her crippled spaceship. She loses her tie rope and finds herself floating 3 m from the ship. Can she get back to the ship by moving her body? What can she do to get back without aid from the ship?
4. All large military guns are mounted in a frame that allows them to recoil. Why are they not fastened rigidly to the support?
5. Explain the principle behind the operation of impact-absorbing bumpers on autos. Make reference to the impulse equation in your explanation.
6. A self-propelled rocket accelerates best in space where there is no air. If there is no air, what does the rocket push against to accelerate itself? How is this question related to Question 3?
7. Compare the momentum of a 120-kg lineman with a 60-kg quarterback if they both have the same speed. How does their kinetic energy compare?
8. A car is to be stopped by a stationary cement wall. Joe claims that the stopping force must double if the car's speed is doubled. Mary says that the force must be four times as large. Is either one of them correct? Who is nearest to being right?
9. After studying this chapter all day, a student dreamed the following dream. He was locked in a frictionless railroad car on a straight, level track. To get the car moving, he started at one end and ran to the other. The car recoiled backward. He slowly walked back to the original end and repeated the run. After doing this many times, the car was going so fast that he was killed as the car collided with a stationary car on the same track. Should he have studied more?
10. An 8-kg baby is dropped from a burning building into the arms of a fireman 10 m below. Estimate the average force he must exert in catching the child.
11. Estimate the force a man's head exerts on his neck if his stationary car is hit from the rear by a huge truck moving at 10 m/s. Why does this type of accident lead to the so-called whiplash injury?

Problems

1. (a) What is the momentum (in kilogram-meters per second) of a 500-g ball that is traveling in the x direction with a speed of 60 cm/s? (b) What is the momentum of a 3-lb ball moving in the y direction with a speed of 0.5 ft/s?
2. (a) What is the momentum (in kilogram-meters per second) of a 900-kg car going east at 20 km/h? (b) What is the momentum of a 2000-lb car going north at a speed of 30 ft/s?
3. Find the momentum of: (a) a 1500-kg car moving at 30 m/s, (b) a 20-g bullet moving at 400 m/s, (c) a 3400-lb car moving at 60 miles/h.
4. What is the momentum of: (a) a 20-g nut falling with a speed of 18 m/s, (b) a 0.25-kg ball moving at 0.70 m/s, (c) a 240-lb man running the 100-yd dash in 20 s?

*5. A 20-g ball falls freely from rest through a height of 80 cm, strikes a hard floor, and rebounds 60 cm. What is its momentum (a) just before it strikes the floor and (b) just after it leaves the floor during rebound? (c) How much was the change in momentum caused by the impact?

*6. A girl throws a 500-g ball straight up and it rises to a height of 12.0 m. What was the momentum of the ball (a) just as it left her hand and (b) just before it reaches her hand on the way down? (c) How much did the ball's momentum change during its flight?

7. A stationary ball is hit by an average force of 30 N for a time of 0.020 s. How large is the impulse applied to the ball? What is the momentum of the ball after the impulse?
8. A boy kicks a stationary can with his toe. If he exerts a 30-N average force for a time of 0.50 s, how large is the impulse applied to the can? What is the momentum of the can after the impulse?
9. How large an impulse is given to a 30-g bullet if it leaves a gun at a speed of 500 m/s?
10. How large an impulse does a 40-g bullet moving at 500 m/s exert on a tree as it strikes and comes to rest in the tree?
11. A 20-g ball going 3.0 m/s in the $+x$ direction is hit in such a way that it ends up going 2.0 m/s in the $-x$ direction. How large an impulse acted on it? What was the direction of the impulse?
12. A 5.0-g ball is dropped onto a floor. Just before hitting, its speed is 4.0 m/s. It rebounds with a speed of 3.0 m/s. How large an impulse does the floor exert on the ball? What is the direction of this impulse?
13. It is claimed that a 1100-kg car can accelerate from rest to a speed of 25 m/s in a time of 8 s. How large an average force is needed to cause this acceleration? Use the impulse method.
14. (a) A 1500-kg car going 20 m/s stops in a time of 4.0 s. Using the impulse method, find the average stopping force that acted on the car. (b) Repeat for a 3000-lb car going 50 ft/s.
15. (a) A 20-g bullet moving at 400 m/s hits a bag of sand and comes to rest in 0.00110 s. Using the impulse method, find the average force that stopped the bullet. (b) Repeat for a 0.04-lb bullet at 1200 ft/s.

*16. A 900-kg car moving at 20 m/s applies its brakes and skids to rest in a distance of 40 m. (a) How long did it take to come to rest if its deceleration was constant? (b) Use the impulse method to find the force that caused the deceleration.

*17. A 20-g bullet moving at 400 m/s strikes a tree and comes to rest after going 8 cm into the tree. (a) How long did it take to stop if its deceleration was constant? (b) Using the impulse method, find the average force that stopped it.

**18. A 60-kg woman drops 2.50 m to the ground from a second story window. Upon reaching the ground, she lets her legs "give" for a distance of 20 cm to cushion the fall. What average force must her legs withstand to stop her fall? Assume she decelerates uniformly. Do not include the force her legs must supply simply to support her while standing.

19. A 20,000-kg truck moving 6.0 m/s hits the rear end of a 1200-kg car standing at rest. What will be the speed with which the truck pushes the car along the street just after the collision? Assume that they move together.
20. A 2.0-kg melon sits on top of a post. It is hit by a 500-g arrow moving with a speed of 20 m/s. The arrow sticks in the melon. What will be the speed of the melon just after impact?
21. A 70-kg boy on roller skates is moving at 4.0 m/s when he runs squarely into the back of a 50-kg girl standing on roller skates. He holds onto her and they move off together. What is their speed after collision?
22. A 2.5-kg rifle lies on a smooth table when it suddenly discharges. It fires a 0.015-kg bullet with a speed of 600 m/s. Find the recoil speed of the gun.

*23. Radium atoms are radioactive. A stationary radium atom (m = 226 u) shoots out an alpha particle (m = 4.0 u) with a speed of 1.50×10^7 m/s. Find the recoil speed of the remaining atom.

*24. A 90-kg soldier in the Arctic is coasting on his ice skates at a speed of 3.0 m/s. He fires 25 bullets directly forward from his 8.0-kg machine gun. Each bullet has a mass of 50 g and a speed of 400 m/s. What is his final velocity?

*25. A Saturn rocket uses five jet engines each producing a thrust

*Problems marked with an asterisk are not as easy as the unmarked ones.
**Problems marked with a double asterisk are somewhat more difficult than the average.

of 7.7×10^6 N. It shoots out gas at a velocity of 2000 m/s. How much gas, in kilograms, does each engine expel each second?

*26. A 900-kg car is waiting at a stoplight when a 1100-kg car coasting at 5.0 m/s strikes it from the rear. The cars do not stick together. The front car moves ahead at 3.0 m/s after the collision. What is the speed of the rear car after collision?

*27. A 2.0-kg ball is going west at 3.0 m/s when it strikes head-on a 1.5-kg ball that is going east at 4.0 m/s. After collision the 2.0-kg ball is going east at 0.50 m/s. What is the 1.5-kg ball doing?

*28. A billiard ball (*A*) with speed 2.0 m/s collides head-on with an identical billiard ball (*B*) that is at rest. After collision *A* is still going forward with speed 0.30 m/s. Find the velocity of *B* after collision. Was this collision perfectly elastic?

*29. Two identical billiard balls are approaching each other head-on. One (*A*) has a velocity 3.0 m/s, whereas the other (*B*) has a velocity $-$ 2.0 m/s. After collision *A*'s velocity is -1.5 m/s. What is *B*'s velocity? Was this collision perfectly elastic?

**30. Suppose that the collision described in Problem 29 were perfectly elastic. What would have been the final velocity of ball *A*? Ball *B*?

**31. Balls *A* and *B* collide head-on in a perfectly elastic collision. It is known that $m_A = 2m_B$ and that the initial velocities are $+3.0$ m/s for *A* and -2.0 m/s for *B*. Find their velocities after the collision.

**32. A nitrogen molecule (m = 28 u) is moving in the x direction with a velocity of 800 m/s. It collides head-on with an oxygen molecule (m = 32 u). After collision the nitrogen molecule comes to a dead stop. What were the initial and final velocities of the oxygen molecule if the collision was perfectly elastic?

*33. (a) A 3-kg block of wood sits on the floor. A 25-g bullet with a speed of 300 m/s is fired down into the block at an angle of 37° below horizontal. Find the speed of the block along the floor just after the bullet lodges in the block. (b) Repeat for a 6-lb block and 0.050-lb bullet going 1000 ft/s.

**34. A military gun is mounted on a railroad car. It fires its projectile of mass m along the train tracks but at an angle θ above the horizontal. The speed of the projectile is v and the combined mass of the gun and car is M. Show that the unbraked car would recoil with speed $(m/M)v \cos \theta$.

*35. Two streets intersect at right angles. A 900-kg car moving at 12 km/h along one (the $+x$ direction) collides with a 1100-kg car moving at 20 km/h in the $+y$ direction. They stick together after collision. Find the x and y components of their velocity after collision. Give your answer in kilometers per hour.

*36. A 1.5-kg ball is moving at 3.0 m/s in the $+x$ direction. It strikes a stationary 2-kg ball and bounces off so that it is going 0.20 m/s in the $+y$ direction. What are the x and y components of the 2-kg ball's velocity after collision?

*37. A 2.0-kg ball is moving in the $+x$ direction with a speed of 3.0 m/s when it collides with a 1.5-kg ball moving at 4.0 m/s in the $+y$ direction. After collision the 2.0-kg ball has v_x = 1.0 m/s and v_y = 2.0 m/s. What are the components of the velocity for the other ball?

**38. Refer to Figure P8.1. Balls *A* and *B* are identical. Ball *A* is released from the position shown. If the collision with *B* is perfectly elastic, how high will *B* swing after impact? What happens to *A*?

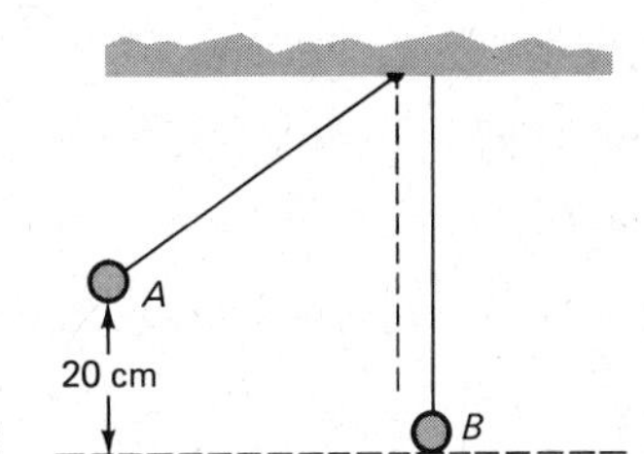

Figure P8.1

**39. Refer to Figure P8.1. Suppose that *A* has a mass half as large as *B*. If the system is released from the position shown, how high will *B* swing after collision? Assume the collision to be perfectly elastic.

**40. Repeat Problem 38 if *A* and *B* stick together. The collision is no longer perfectly elastic, of course.

*41. The ballistic pendulum shown in Figure P8.2 is sometimes used to find the speed of a bullet. The bullet lodges in the pendulum and one measures h, the height to which the pendulum rises. (a) Show that the velocity of the pendulum just after impact is $\sqrt{2gh}$. (b) If the bullet's mass is m and the pendulum mass is M, show that the bullet's speed was $(m + M) \cdot \sqrt{2gh}/m$.

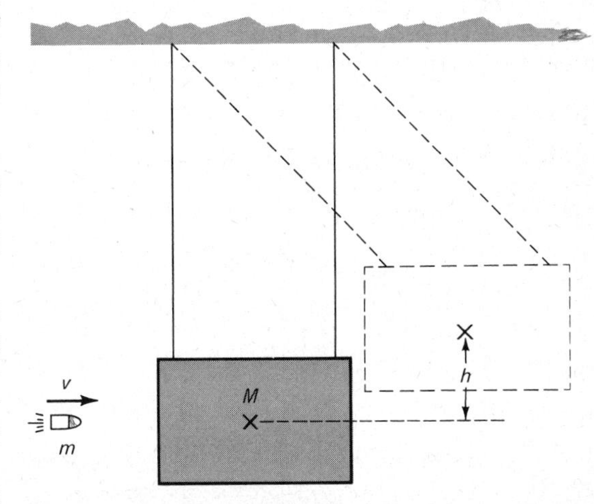

Figure P8.2 The ballistic pendulum.

Kings Island, Kings Island, Ohio

9 ROTATIONAL MOTION

Performance Goals

When you finish this chapter, you should be able to

1. State the value of an angle in degrees, radians, and revolutions if its value is given in one of these units.
2. Write the defining equations for average angular velocity and angular acceleration.
3. Write the five linear motion equations and their angular counterparts. Identify the angular symbols corresponding

Most of the machinery used in our modern world makes use of rotation. A complex machine without a gear, wheel, or rotating shaft is not often found. In this chapter we learn how to describe rotations. We shall find out what is meant by angular distances, velocities, and accelerations. They will be found to obey equations similar to their counterparts in linear motion. Also in this chapter we shall learn about centripetal force and acceleration. The orbital motion of earth satellites will also be discussed.

to x, v, and a. Give the restriction on three of these equations.

4. Given the angular velocity or angular acceleration in one set of units, convert it to any other set.
5. Use the five angular motion equations to find the quantities θ, $\bar{\omega}$, ω_f, ω_0, α, and t if three of the quantities are given.
6. Define the tangential quantities s, v_T, and a_T in three ways: (a) in terms of the rotation of a wheel on its axis, (b) in terms of a string or belt wound on a wheel, (c) in terms of the distance a wheel rolls on a surface.
7. Give the relations between θ and s, ω and v_T, α and a_T. Point out the restriction on their use.
8. Given the angular motion of a wheel, be able to find the corresponding tangential motion and vice versa.
9. Define by an equation the centripetal acceleration. Find the centripetal acceleration for an object traveling in a circle with known radius and speed.
10. Explain how Newton's first law tells us that a force is required if an object is to move in a circular path. Give the name of the force and its direction. Give the equation by which it is related to the object's mass, the tangential velocity, and the radius of the circle.
11. An object is moving in a horizontal circle. You have three of the quantities F_c, m, v_T, and r. Find the fourth.
12. Find the force, other than gravity, that must act on an object (a) at the top of the circle

9.1 Angular Measurements

Let us review how we describe the rotation of a wheel. Consider the wheel shown in Figure 9.1. It is rotating counterclockwise (c.c.w.) as indicated.[1] Point P was originally on the x axis. At the instant shown it has reached the point where it is in the figure. The wheel has rotated through the angle θ as labeled. We call θ the *angular displacement.*

There are three common ways of measuring θ. We could use a protractor and obtain its value in degrees. As drawn, $\theta = 45$ degrees (deg).[2] Or we could describe the angle θ in terms of what fraction of a revolution it is. Because

$$1 \text{ rev} = 360 \text{ deg}$$

we have as our conversion factor

$$\frac{1 \text{ rev}}{360 \text{ deg}} = 1$$

Therefore, using our usual conversion method,

$$45 \text{ deg} = (45 \cancel{\text{deg}}) \times \frac{1 \text{ rev}}{360 \cancel{\text{deg}}} = \frac{45}{360} \text{ rev} = 0.125 \text{ rev}$$

We can therefore say that $\theta = 0.125$ rev.

Both degrees and revolutions are sometimes convenient units. But other times there is a third, more convenient unit. It is called the *radian* (abbreviated *rad*). It is defined in the following way. If you refer to Figure 9.2 you will see that point P moves a distance s as an angle θ is swept out. The arc length s can be used to measure θ. However, to remove the effect of the radius of the circle, we use the ratio s/r as the measure of θ. We then define

$$\theta \text{ (in radians)} \equiv \frac{s}{r} \qquad (9.1)$$

Notice that θ in radians is simply the ratio of two lengths. Therefore, the angle has no units. We carry along the term "radians" with θ to remind ourselves how θ is being measured. The radian is not a unit in the ordinary sense.

We can easily relate the radian to the degree. When θ is 360 deg, then $s = 2\pi r$. Therefore, 360 deg corresponds to

$$\theta = \frac{s}{r} = \frac{2\pi r}{r} = 2\pi \text{ rad}$$

[1] The positive direction for most rotations in science is taken to be counterclockwise.

[2] To show how units cancel we will use the abbreviation deg for degrees throughout this chapter, instead of the degree sign (°).

and (b) at the bottom of the circle if it is to move in a vertical circle. Assume that m and v_T are known.

13. Describe what happens to an object shot at various speeds parallel to the earth but high above it. Discuss when the object will (a) hit the earth, (b) orbit the earth, (c) escape from the earth.

14. Explain why it is correct to say that an orbiting satellite is freely falling. Using this fact, explain why objects appear weightless in an orbiting spaceship.

From this we find that

$$2\pi \text{ rad} = 360 \text{ deg} \qquad \text{or} \qquad 1 \text{ rad} = 57.296 \text{ deg}$$

The radian unit will be found in later sections to have some useful properties. Until then you may consider it to be just another unit in which to measure angles.

Let us now summarize our angular units. An angle can be measured in three different units. They are the degree (deg), revolution (rev), and radian (rad). We can convert between them using the following equations:

$$1 \text{ rev} = 360 \text{ deg}$$
$$1 \text{ rev} = 2\pi \text{ rad} = 360 \text{ deg}$$
$$1 \text{ rad} = 57.30 \text{ deg}$$

EXAMPLE 9.1 Change the following angles to the remaining two units: 20 deg, 6.4 rad, 0.39 rev.

Solution We make use of the conversion equations just given to find conversion factors. Then

$$20 \text{ deg} = (20 \cancel{\text{deg}}) \times \frac{1 \text{ rev}}{360 \cancel{\text{deg}}} = 0.0556 \text{ rev}$$

$$20 \text{ deg} = (20 \cancel{\text{deg}}) \times \frac{1 \text{ rad}}{57.3 \cancel{\text{deg}}} = 0.349 \text{ rad}$$

$$6.4 \text{ rad} = (6.4 \cancel{\text{rad}}) \times \frac{57.3 \text{ deg}}{1 \cancel{\text{rad}}} = 366.7 \text{ deg}$$

$$6.4 \text{ rad} = (6.4 \cancel{\text{rad}}) \times \frac{1 \text{ rev}}{2\pi \cancel{\text{rad}}} = 1.02 \text{ rev}$$

$$0.39 \text{ rev} = (0.39 \cancel{\text{rev}}) \times \frac{360 \text{ deg}}{1 \cancel{\text{rev}}} = 140.4 \text{ deg}$$

$$0.39 \text{ rev} = (0.39 \cancel{\text{rev}}) \times \frac{2\pi \text{ rad}}{1 \cancel{\text{rev}}} = 2.45 \text{ rad}$$

9.2 Angular Velocity

The *angular velocity* of a wheel is a measure of how fast the wheel turns. If a wheel makes 3000 revolutions in one minute, we say its angular velocity is 3000 rev/min. This simple example tells us that angular velocity is the angular displacement divided by the time taken.[3] It is customary to represent the angular velocity by the symbol ω (Greek letter omega). Then by definition,

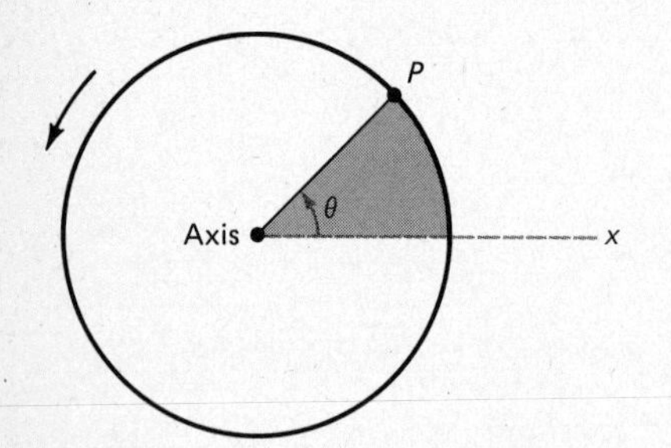

Figure 9.1 As the wheel rotates counterclockwise, the angle θ increases.

[3] When the angular velocity is expressed in rev/s or rev/min, it is sometimes referred to as the *angular frequency* or *rotation frequency f*.

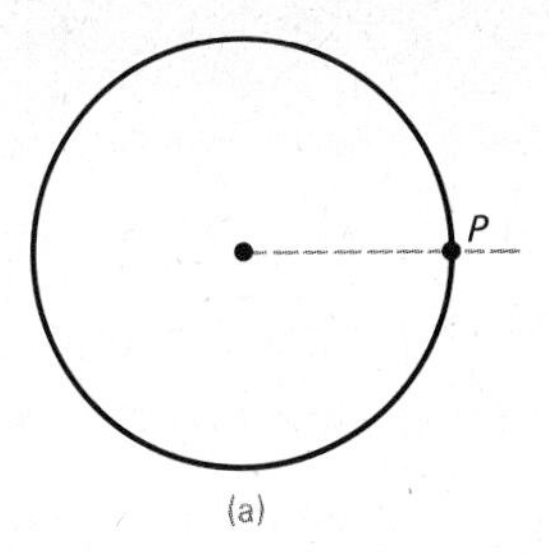

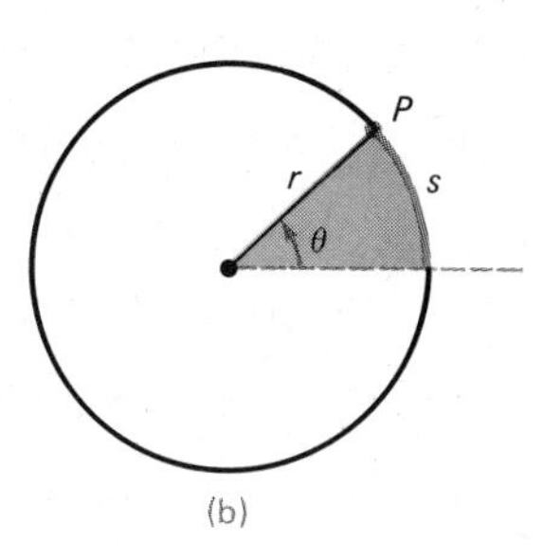

Figure 9.2 As the wheel turns an angular distance θ, the point P moves through a tangential distance s. If θ is measured in radians, then $s = r\theta$.

$$\text{Average angular velocity} = \frac{\text{angular displacement}}{\text{time taken}}$$

or

$$\bar{\omega} = \frac{\theta}{t} \qquad (9.2)$$

This is the average velocity for the time interval t.

The units for angular velocity are seen at once from its defining equation. They are an angle unit divided by a time unit. Typical examples are degrees per second, revolutions per second, radians per second, revolutions per minute, and so on. We can change from one unit to the other by use of the conversion factors given in the previous section.

Before looking at some examples that make use of Equation 9.2, we must mention that, as the name implies, angular velocity is a vector. However, for all but a few complex technical applications, the vector part of the definition is not important. In this chapter we shall not complicate the discussion by assigning direction to ω. Instead we shall use the terms angular speed and angular velocity interchangeably. The vector nature of angular velocity is important in discussing the motion of gyroscopes and similar devices. These topics are treated in advanced courses in mechanics.

■ **EXAMPLE 9.2** The shaft of a motor rotates 4000 rev in 0.75 min. Find its angular speed in revolutions per minute, revolutions per second, radians per second, and degrees per second.

Solution From the data given, we compute

$$\bar{\omega} = \frac{\theta}{t} = \frac{4000 \text{ rev}}{0.75 \text{ min}}$$
$$= 5330 \text{ rev/min}$$

By use of the usual conversion factors,

$$\bar{\omega} = \left(5330 \frac{\text{rev}}{\text{min}}\right)\left(\frac{1 \text{ min}}{60 \text{ s}}\right) = 88.9 \text{ rev/s}$$

$$\bar{\omega} = \left(88.9 \frac{\text{rev}}{\text{s}}\right)\left(\frac{2\pi \text{ rad}}{1 \text{ rev}}\right) = 558 \text{ rad/s}$$

$$\bar{\omega} = \left(88.9 \frac{\text{rev}}{\text{s}}\right)\left(\frac{360 \text{ deg}}{1 \text{ rev}}\right) = 32{,}000 \text{ deg/s}$$

■■

9.3 Angular Acceleration

When we say that a wheel's rotation is speeding up, we are saying that the wheel has an *angular acceleration.* Angular acceleration measures how much the rotation velocity is increasing each second. To be precise, angular acceleration is the change in angular velocity divided by the time taken for the change. We represent it by the Greek symbol α (alpha). The defining equation is

$$\text{Angular acceleration} = \frac{\text{change in angular velocity}}{\text{time taken}}$$

In symbols,

$$\alpha = \frac{\omega_f - \omega_0}{t} \quad \text{or} \quad \omega_f = \omega_0 + \alpha t \tag{9.3}$$

As an example, suppose that a wheel is rotating at 5 rev/s and 40 s later is rotating at 20 rev/s. Its (average) angular acceleration during this time is, from Equation 9.3,

$$\alpha = \frac{(20 - 5)\ \text{rev/s}}{40\ \text{s}} = 0.375\ \text{rev/s}^2$$

The angular speed increases by 0.375 rev/s every second. Angular acceleration has units of angular distance divided by time squared. (Notice the similarity to linear acceleration which has units of linear distance divided by time squared.)

■ **EXAMPLE 9.3** A motor's shaft is turning at 4000 rev/min when the motor is turned off. It stops in 20 s. Find the deceleration in revolutions per second per second, radians per second per second, and degrees per second per second.

Solution Because all the quantities we wish to find are in terms of seconds, let us make the following conversion.

$$4000\ \text{rev/min} = \left(4000\ \frac{\text{rev}}{\text{min}}\right)\left(\frac{1\ \text{min}}{60\ \text{s}}\right) = 66.7\ \text{rev/s}$$

We know that $\omega_0 = 66.7$ rev/s, $\omega_f = 0$, and $t = 20$ s. Therefore,

$$\alpha = \frac{\omega_f - \omega_0}{t} = \frac{-66.7\ \text{rev/s}}{20\ \text{s}} = -3.33\ \text{rev/s}^2$$

The negative sign tells us that this is a deceleration. Converting in the usual way gives

$$\alpha = -\left(3.33\ \frac{\text{rev}}{\text{s}^2}\right)\left(\frac{2\pi\ \text{rad}}{1\ \text{rev}}\right) = -20.9\ \text{rad/s}^2$$

and

$$\alpha = -\left(3.33\ \frac{\text{rev}}{\text{s}^2}\right)\left(\frac{360\ \text{deg}}{1\ \text{rev}}\right) = -1200\ \text{deg/s}^2$$

9.4 Analogies Between Linear and Angular Motion

Our discussion concerning angular motion may seem slightly familiar to you. Our equations are similar to those we had for linear motion. The angular displacement θ replaces the linear distance x. The angular velocity ω replaces the linear velocity v. And the angular acceleration α replaces the linear acceleration a. In both cases we have similar defining equations.

$$\bar{v} = \frac{x}{t} \quad \rightarrow \bar{\omega} = \frac{\theta}{t}$$

$$a = \frac{v_f - v_0}{t} \rightarrow \alpha = \frac{\omega_f - \omega_0}{t}$$

Similarly, for uniformly accelerated motion we can find an average velocity.

$$\bar{v} = \frac{v_0 + v_f}{2} \rightarrow \bar{\omega} = \frac{\omega_0 + \omega_f}{2}$$

If you look back to Chapter 3, you will see that these equations were used to find the other two linear motion equations. We can do the same for the angular equations.

$$v_f^2 - v_0^2 = 2ax \rightarrow \omega_f^2 - \omega_0^2 = 2\alpha\theta$$
$$x = v_0 t + \tfrac{1}{2}at^2 \rightarrow \theta = \omega_0 t + \tfrac{1}{2}\alpha t^2$$

As we see, the five linear motion equations have analogs in angular motion. We already know how to deal with linear motion. Now we find that the equations for angular motion are the same except for changed symbols. We can therefore use these equations for angular motion just as we did those for linear motion. The analogy between the two types of motion is summarized as follows.

	LINEAR	ANGULAR	
	x	θ	
	v	ω	
	a	α	
	$x = \bar{v}t$	$\theta = \bar{\omega}t$	(9.4a)
	$v_f = v_0 + at$	$\omega_f = \omega_0 + \alpha t$	(9.4b)
Uniform acceleration	$\bar{v} = \frac{1}{2}(v_0 + v_f)$	$\bar{\omega} = \frac{1}{2}(\omega_0 + \omega_f)$	(9.4c)
Uniform acceleration	$v_f^2 - v_0^2 = 2ax$	$\omega_f^2 - \omega_0^2 = 2\alpha\theta$	(9.4d)
Uniform acceleration	$x = v_0 t + \frac{1}{2}at^2$	$\theta = \omega_0 t + \frac{1}{2}\alpha t^2$	(9.4e)

Let us now use these equations in a few examples.

EXAMPLE 9.4 A wheel turning at 2000 rev/min coasts to a stop in 20 s. Assuming uniform deceleration, through how many revolutions did it turn in this time?

Solution We know that $\omega_0 = 2000$ rev/min, $\omega_f = 0$, and $t = 20\text{ s} = \frac{1}{3}$ min. (Notice that we don't mix time units.) We want to find θ. Because ω_0 and ω_f are known, we can find $\bar{\omega}$.

$$\bar{\omega} = \tfrac{1}{2}(\omega_0 + \omega_f) = 1000 \text{ rev/min}$$

Then

$$\theta = \bar{\omega}t = \frac{1000}{3} \text{ rev}$$

■■

EXAMPLE 9.5 A wheel starts from rest and in 5.0 s reaches a speed of 300 rev/s. Find its angular acceleration during this time.

Solution We know that $\omega_0 = 0$, $\omega_f = 300$ rev/s, and $t = 5.0$ s. Then, because

$$\omega_f = \omega_0 + \alpha t$$

we have

$$300 \text{ rev/s} = 0 + \alpha(5 \text{ s})$$

from which

$$\alpha = 60 \text{ rev/s}^2$$

■■

9.5 Tangential Displacement

Look again at the wheel in Figure 9.2. In going from part (a) to (b) the wheel has turned through an angle θ. No matter what portion of the wheel we choose to talk about, it has turned this same angle θ about the axis. Sometimes, though, we are interested in how far a point on the rim of the wheel has moved. In the figure point P moved along an arc of length s. The arc of length s is tangent to the circle. Because of this, s is called the *tangential displacement* of point P.

We already know a relation between s, r, and θ *provided that* θ *is measured in radians.* From the definition of the radian measure for angles we had Equation 9.1, which was

$$\theta = \frac{s}{r} \qquad (\theta \text{ measured in radians})$$

Or in the form usually remembered, we have

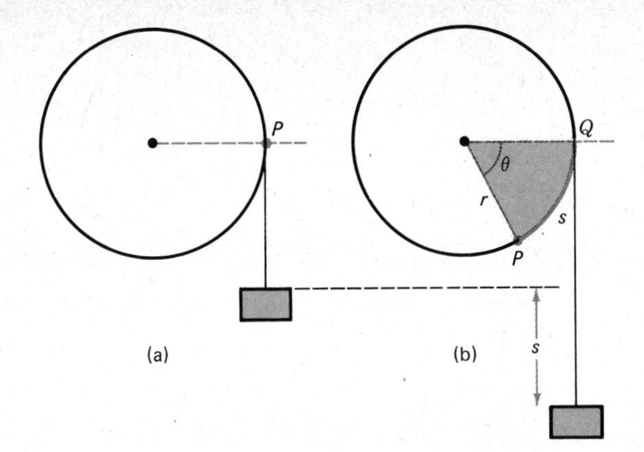

Figure 9.3 As the wheel turns, the point P moves through a tangential distance s. A length s of the rope unwinds from the wheel also.

$$s = r\theta \qquad (\theta \text{ in radians}) \tag{9.4}$$

This equation between arc length and angle is of great usefulness. For example, refer to Figure 9.3. How far does the weight drop as the wheel turns through the angle θ?

If you examine part (b) of Figure 9.3, you will see that as the wheel turns, P moves through the arc s shown. But this arc length is equal to the arc distance between P and Q. There was a length s of string wound on the wheel between these two points. As the wheel turned, this length s of string unwound. In the process the weight dropped a distance s. We know from Equation 9.4 that $s = r\theta$.

This is a general result. We can summarize it as follows.

When a wheel of radius r turns through an angle θ, a length $s = r\theta$ will be wound or unwound from its rim. θ must be measured in radians.

This can be expanded to the case of a rolling wheel. By referring to Figure 9.4, you will see:

When a wheel turns through an angle θ as it rolls without slipping on a flat surface, the wheel moves a distance $s = r\theta$ across the surface. θ must be in radians.

Let us now make use of these facts in two examples.

▮ **EXAMPLE 9.6** As a car travels 1 km, how many revolutions does one of its 60-cm-diameter wheels make?

Solution We know that the tangential distance traveled is s = 1.00 km. Because r = 0.3 m, we find from $s = r\theta$ that

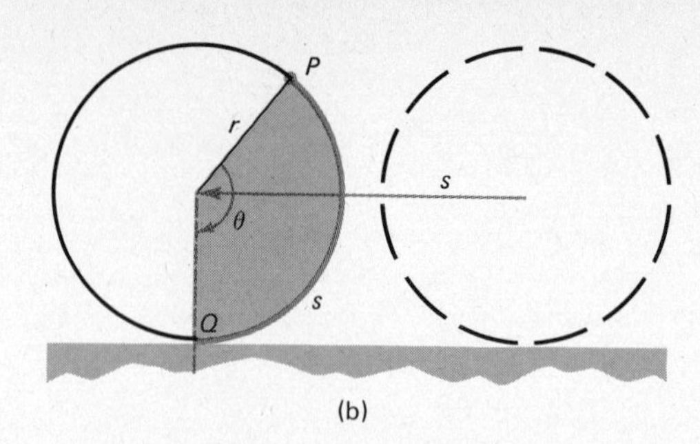

Figure 9.4 As the wheel rolls a distance s, it turns through an angle θ. The relation is $s = r\theta$.

$$\theta = \frac{1000 \text{ m}}{0.30 \text{ m}} = 3300 \text{ rad}$$

Notice the unit of θ. We used Equation 9.1 in obtaining θ. This equation always gives θ in radians. But we want θ in revolutions. Therefore,

$$\theta = (3300 \text{ rad})\left(\frac{1 \text{ rev}}{2\pi \text{ rad}}\right) = 530 \text{ rev}$$

■ **EXAMPLE 9.7** A belt runs on a pulley with diameter 5.0 cm. The pulley turns through 2000 rev in a minute. What length of belt passes over the pulley each minute?

Solution We are told that $\theta = 2000$ rev $= 4000\pi$ rad. Because $r = 2.5$ cm, we have from $s = r\theta$ that

$$\begin{aligned} s &= (2.5 \text{ cm})(4000\pi) \\ &= 314 \text{ m} \end{aligned}$$

This is the length of belt that passed over the pulley. Notice that we used θ in radians because $s = r\theta$ demands this. Notice also that the angle units are not carried because the angle units were dropped when deriving $s = r\theta$.

9.6 Tangential Velocity

If you look at Figure 9.5, you can see that the point P has a very definite velocity. Although the direction of the velocity changes as the wheel rotates, the velocity vector is always tangent to the wheel. We call this the *tangential velocity*, v_T, of the point P.

We can easily obtain a relation between v_T and ω, the angular velocity of the wheel. From our usual motion equations and Figure 9.5 we have

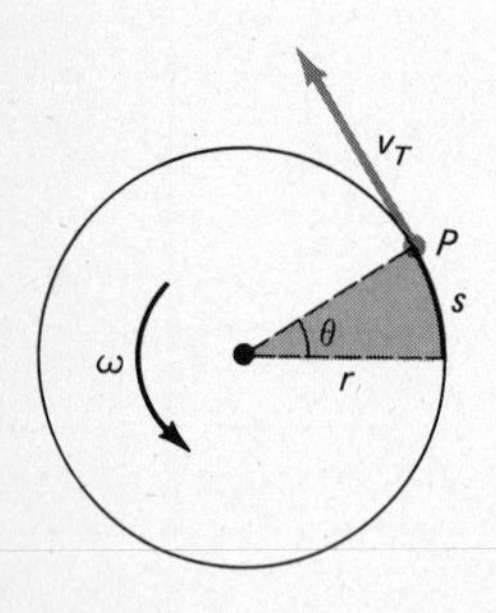

Figure 9.5 The point P moves through a tangential distance s in time t. Its tangential velocity is $v_T = s/t$.

$$v_T = \frac{s}{t}$$

where s is the distance P moves in time t. We can replace s by use of Equation 9.1 which tells us that $s = r\theta$. Making this substitution for s gives

$$v_T = r\frac{\theta}{t}$$

Now θ/t is simply the angular speed of the wheel, ω. Therefore,

$$v_T = r\omega \qquad \text{(angles in radians)} \qquad (9.5)$$

The restriction on this relation is that radian measure must be used. This is because we used $s = r\theta$ to obtain it. As we have pointed out, $s = r\theta$ is true only if θ is in radians. Therefore, in using Equation 9.5, we must always express ω in radians per unit time.

Equation 9.5 also applies to a rolling wheel. If ω is the speed at which the wheel is rotating, then the linear speed of the rolling wheel is simply $r\omega$. Further, if a rotating wheel is unwinding rope, the speed of the rope is also $r\omega$. As we see, Equation 9.5 applies to a number of situations.

■ **EXAMPLE 9.8** A speck of dust is 9.0 cm from the center of a phonograph record as the record turns at 33 rev/min. What is the speed of the speck?

Solution We wish to find v_T for a point at a radius of 0.090 m. To make use of $v_T = r\omega$, we must have ω in radian measure. We have

$$\omega = \left(33\,\frac{\text{rev}}{\text{min}}\right)\left(\frac{2\pi\ \text{rad}}{1\ \text{rev}}\right)\left(\frac{1\ \text{min}}{60\ \text{s}}\right) = 3.5\ \text{rad/s}$$

Now we can use Equation 9.5 to give

$$v_T = r\omega = (0.090\ \text{m})(3.5\ \text{rad/s}) = 0.31\ \text{m/s}$$

as the speed of the speck. Notice how we drop the nonunit radians as angular measure disappears from our equations. ■■

■ **EXAMPLE 9.9** A 60-cm-diameter car wheel is rotating at a speed of 3.0 rev/s. How fast is the car going?

Solution The angular speed of the wheel, ω, must be changed to radians per second. It is converted in the usual way to give $\omega = 18.8$ rad/s. We want v_T, the tangential speed at which the wheel rolls along the road. It can be found from

$$\begin{aligned} v_T &= r\omega \\ &= (0.30\ \text{m})\left(18.8\,\frac{\text{rad}}{\text{s}}\right) = 5.6\,\frac{\text{m}}{\text{s}}\cdot\text{rad} = 5.6\ \text{m/s} \end{aligned}$$

This is the linear speed of the rolling wheel and of the car. ■■

■ **EXAMPLE 9.10** The small pulley in Figure 9.6 has a radius of 2.0 cm and rotates at 3600 rev/min. (a) How fast does the belt move in centimeters per second? (b) How fast does the large wheel, radius = 40 cm, rotate?

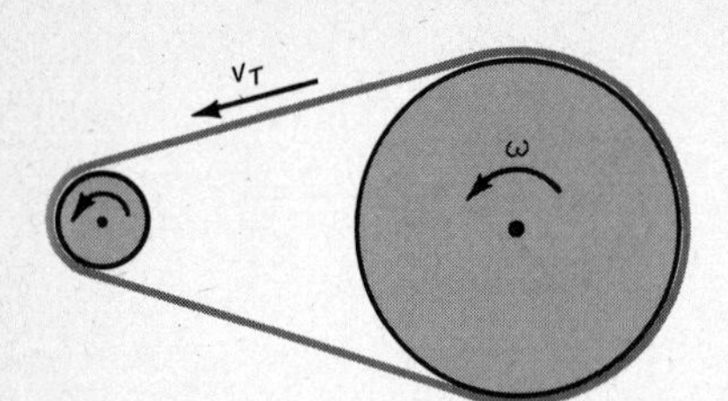

Figure 9.6 The angular velocities of the pulleys are related to the tangential velocity of the belt.

Solution If we are to use $v_T = r\omega$, we need ω in radian measure. Can you show it is $\omega = 377$ rad/s? The belt speed is v_T and is

$$v_T = \omega r = (377 \text{ rad/s})(2.0 \text{ cm}) = 754 \text{ cm/s}$$

Now we can use the same equation for the large wheel. Its angular velocity ω' is

$$\omega' = \frac{v_T}{r'} = \frac{754 \text{ cm/s}}{40 \text{ cm}} = 18.8 \text{ rad/s}$$

If we want this in revolutions per second, we have

$$\omega' = \left(18.8 \frac{\text{rad}}{\text{s}}\right)\left(\frac{1 \text{ rev}}{2\pi \text{ rad}}\right) = 3.0 \text{ rev/s}$$

■■

9.7 Tangential Acceleration

If we refer again to the rotating wheel shown in Figure 9.5, we know that the tangential speed of point P is v_T. If the wheel's rotation rate changes, then the tangential speed of P will also change. As a result, P will undergo a *tangential acceleration, a_T*. Let us define this tangential acceleration in the following way.

Suppose a mass hangs by a cord from a wheel as shown in Figure 9.7. When the wheel is released, it turns and the mass falls. As we saw in the last section, the speed with which the mass falls is equal to the tangential speed of a point on the edge of the wheel. Suppose the object is falling with speed v_{T1} at one instant and, at a time t later, has a speed of v_{T2}. Its acceleration, which we call the tangential acceleration, is then

$$a_T = \frac{v_{T2} - v_{T1}}{t}$$

Since $v_T = \omega r$, this becomes

$$a_T = r\left(\frac{\omega_2 - \omega_1}{t}\right)$$

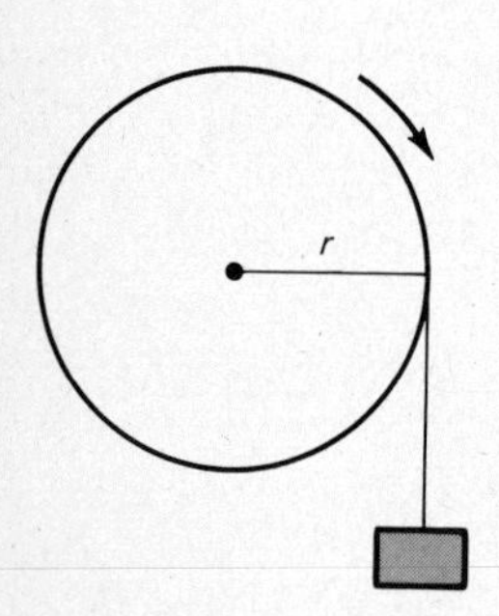

Figure 9.7 The weight is accelerating downward.

We recognize the quantity in parenthesis to be α, the angular acceleration. Therefore, tangential acceleration is related to angular acceleration through

$$a_T = r\alpha \qquad \text{(angles in radians)} \qquad (9.6)$$

This relation is important because it allows us to relate the rotational acceleration of a wheel to the linear motion of the wheel or a rope wound on it.

EXAMPLE 9.11 In Figure 9.7 the weight is falling with an acceleration of 4.8 m/s^2. What is the angular acceleration of the wheel? The radius of the wheel is 30 cm.

Solution We are told that $a_T = 4.8\ m/s^2$. To find α we make use of $a_T = r\alpha$. Then

$$\alpha = \frac{a_T}{r} = \frac{4.8\ m/s^2}{0.30\ m} = 16\ rad/s^2$$

Notice that here, too, the angle units do not carry through the equations. We must supply them at the end. Of course, they are radians because of the restriction on $a_T = r\alpha$.

9.8 Centripetal Acceleration

You may have wondered why we did not define the tangential acceleration in the usual way, simply as the rate of change of tangential velocity. The reason we did not do this has to do with the vector nature of velocity and acceleration. Even when a wheel is turning at constant speed, the tangential velocity of a point on the rim of the wheel is changing direction. The point is therefore accelerating. But the tangential acceleration we defined in the previous section is zero. As you see, the tangential acceleration is not the same as the total acceleration of a point on the wheel's rim. For that reason, we must discuss further the acceleration of such a point.

Consider the object in Figure 9.8 that is moving on a circular path with *constant* speed v. Although its *speed* is constant, its *velocity* is changing. As you see in the figure, its velocity has a different direction at point 1 than at point 2. Although the magnitude of the velocity does not change, its direction does. Hence there is a change in the velocity and the object is therefore accelerating. The acceleration is entirely the result of the change in direction of the velocity.

To find the acceleration, we start from the definition

$$\text{Acceleration} = \frac{\text{change in velocity}}{\text{time taken}}$$

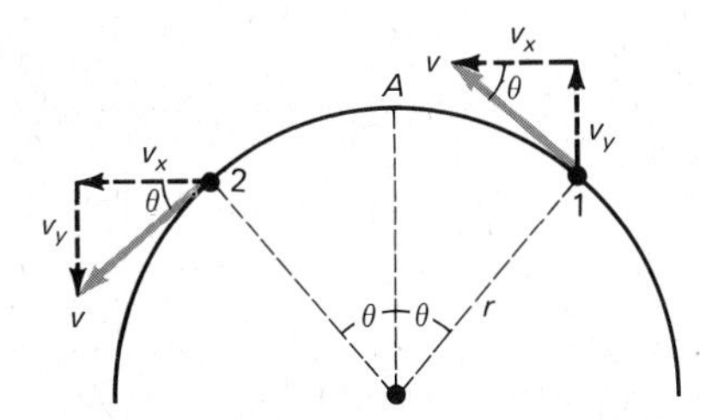

Figure 9.8 The object is accelerating even though it is traveling at constant speed.

Let us consider the change as the object goes from point 1 to 2 in Figure 9.8. If the speed of the object is v, the time taken to move from 1 to 2 is

$$\text{time} = \frac{\text{distance}}{\text{speed}}$$

$$t = \frac{r(2\theta)}{v}$$

We have used the fact that the arc length from 1 to 2 is $r(2\theta)$ as you can see from the figure.

Next, we must find the change in velocity. The x component, v_x, did not change. But the y component did. It changed from v_y to $-v_y$, a total change of $2v_y$. Since $v_y = v \sin\theta$, as you can see from the figure, we have

$$\text{Change in velocity} = 2v_y = 2v \sin\theta$$

Substituting these values for the velocity change and time in the acceleration equation gives

$$\text{Acceleration} = \frac{2v \sin\theta}{2r\theta/v} = \frac{v^2 \sin\theta}{r\theta}$$

This is the average acceleration as the object moves between points 1 and 2. We, however, are interested in the instantaneous acceleration of the object, the acceleration at a point.

To obtain the instantaneous acceleration, we simply take θ as approaching zero. Then points 1 and 2 nearly coincide with point A. In this case we can simplify the acceleration equation by recalling from trigonometry that for very small angles in radians

$$\sin\theta \cong \theta$$

Making this replacement, we have

$$\text{Acceleration} = \frac{v^2\theta}{r\theta} = \frac{v^2}{r}$$

Let us now find the direction of this acceleration.

Look again at Figure 9.8. When θ is small, v_y is nearly in line with the vertical radius of the circle. In going from 1 to 2, the change in v is due entirely to the change in v_y. Because v_y changes from radially outward to radially inward, the change in v_y is directed radially inward. This is also the direction of the acceleration we have computed. It is directed radially inward toward the center of the circle. We call this the *centripetal acceleration* of the object moving in a circular path. Let us summarize.

An object moving with constant speed v along a circular path of radius r experiences a centripetal acceleration toward the center of the circle.

$$\textbf{Centripetal acceleration} = a_c = \frac{v^2}{r} \qquad \textbf{(9.7)}$$

EXAMPLE 9.12 A space satellite circles the earth at a height of 130 km, that is, in a circular path with $r = 6500$ km. It takes about 86 minutes (5160 s) for one orbit.[4] Find its centripetal acceleration.

Solution Using

$$r = 6.5 \times 10^6 \text{ m}$$

$$v = 2\pi r/t = 2\pi(6.5 \times 10^6 \text{ m})/5160 \text{ s} = 7.9 \times 10^3 \text{ m/s}$$

we have

$$a_c = \frac{v^2}{r} = \frac{(7.9 \times 10^3 \text{ m/s})^2}{6.5 \times 10^6 \text{ m}} = 9.6 \text{ m/s}^2$$

This acceleration is directed toward the center of the earth. Surprising, isn't it, that this is about what you would guess for the gravitational acceleration at this altitude. We will soon see why this is an exact correspondence.

9.9 Centripetal Force

As we have just seen, an object traveling on a circular path is accelerating toward the center of the circle. Because forces are required to cause accelerations, the centripetal acceleration must be accompanied by a centripetal force. We can easily see that this is true by referring to the experiment shown in Figure 9.9.

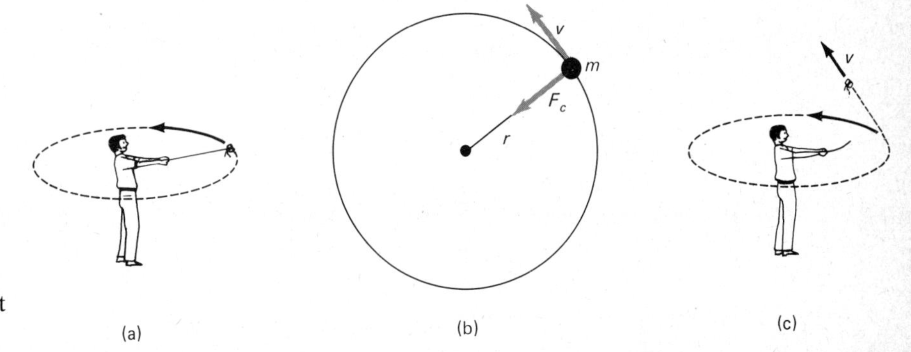

Figure 9.9 A force is needed to pull the ball out of its normal straight-line path. The centripetal force F_c causes it to follow a circular path.

As shown there, a ball at the end of a string is swung in a circular path. The string pulls radially inward on the ball so as to furnish the required centripetal force. Part (b) diagrams the situation. Notice that the ball of mass m and speed v is pulled toward the center of the circle by the unbalanced force F_c furnished by the string. This unbalanced force gives the ball the centripetal acceleration needed if it is to travel along a circle. Newton's second law, $F = ma$, tells us the

[4]The time taken for one complete orbit or rotation is often referred to as the *period* of the motion.

relation between the centripetal force F_c and the centripetal acceleration it produces. It is $F_c = ma_c$, or, because $a_c = v^2/r$,

$$\text{Centripetal force} = F_c = \frac{mv_T^2}{r} \qquad (9.8)$$

It is interesting to consider what would happen to the ball in Figure 9.9 if the string broke. The centripetal force would then be zero and so the ball would no longer be accelerated. It would simply fly off along a straight line tangent to the circle as shown in Figure 9.9(c). In order to be pulled from its normal straight-line path, the ball requires an unbalanced force, the centripetal force, to cause it to follow a circular path.

All objects that travel in a circle (or an arc of a circle) require a centripetal force. The earth is pulled toward the sun by gravitational attraction. This pull furnishes the needed centripetal force and causes the earth to circle the sun. Similarly, the moon (and earth satellites) circle the earth because of the earth's gravitational pull on them. This pull furnishes the needed centripetal force. We shall see more examples of centripetal forces later.

It is of interest to notice that no work is done by the needed centripetal force. To do work, a force must have a component in the direction of motion. But the centripetal force is directed along the radius of a circle, whereas the motion occurs tangent to the circle. Because the tangent is perpendicular to the radius, the centripetal force has no component in the direction of motion. It therefore does no work. It simply changes the direction of the motion.

EXAMPLE 9.13 A 500-g ball is to be swung on a string in a horizontal circle of radius 120 cm at a speed of 0.80 rev/s. How large a tension must the string be able to hold? Neglect the effect of gravity.

Solution The tension in the string furnishes the needed F_c and must be that large. To find F_c we need m (= 0.50 kg), r (= 1.20 m), and v. We are given $\omega = 0.80$ rev/s. To find v we can use $v = \omega r$ provided that ω is in radians per second. We have

$$\omega = \left(0.80 \frac{\text{rev}}{\text{s}}\right)\left(\frac{2\pi \text{ rad}}{1 \text{ rev}}\right) = 5.0 \text{ rad/s}$$

From this,

$$v = \omega r = (5.0 \text{ rad/s})(1.20 \text{ m}) = 6.0 \text{ m/s}$$

Then we can use

$$F_c = \frac{mv^2}{r}$$

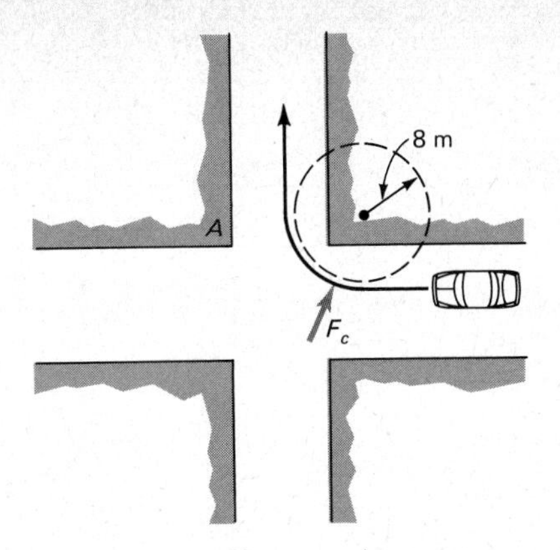

Figure 9.10 How fast can the car be going and still be able to make the turn?

to find

$$F_c = 15 \text{ N}$$

Notice that the gravitational force (about 4.9 N) is not really negligible in this situation. How could the problem be solved without its neglect? ■■

■ **EXAMPLE 9.14** An 800-kg car is to round a corner as shown in Figure 9.10. How large a friction force will be needed if the car's speed is 15 m/s? Would such a friction force usually be possible?

Solution The friction force between tires and road must provide the needed force F_c shown in the figure. We know that

$$F_c = \frac{mv^2}{r}$$

In the present case $m = 800$ kg, $v = 15$ m/s, and $r = 8$ m, as seen in the figure. Therefore, we find after substitution

$$F_c = 22{,}500 \text{ N}$$

This large a friction force would be required.

To see if this large a force is possible, recall that $f = \mu F_N$. If the road is flat, then F_N is equal to the weight of the car, (9.8)(800) = 7840 N. Because the required value of $f = 22{,}500$ N, we find that

$$22{,}500 = \mu(7840)$$

from which the coefficient of friction would need to be

$$\mu = 2.87$$

In practice μ would usually be less than unity. Therefore, it is clear that the car would not be able to make the turn at this speed. It would skid along a nearly straight-line path and strike the side of the road near the corner labeled *A*. Often roadways are banked along curves that cars take at high speed. When the road is banked, the normal force F_N has a component in the direction of F_c. It helps the friction force hold the car on a circular path. (See Problem 40 at the end of this chapter.) ■■

■ **EXAMPLE 9.15** In Figure 9.11 we see a plane going around a vertical loop. Its speeds at the two points indicated are v_1 and v_2. Find the force that the seat exerts on the pilot as he passes by each of the points in the loop. The pilot has a mass *m*.

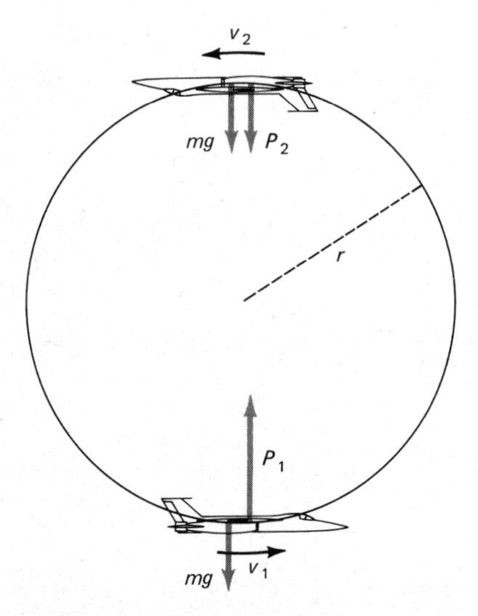

Figure 9.11 Find the force *P* that the seat exerts on the pilot.

Solution Two forces act on the pilot. The force of gravity, *mg*, and the push of the seat on him, *P*. Together these two forces must supply the required centripetal force, $F_c = mv^2/r$. Let us consider each point in turn.

Point 1: The seat pushes up with a force P_1, while gravity pulls down. Together these two forces must supply F_c, which is directed upward. Therefore,

$$P_1 - mg = F_c$$

Because

$$F_c = \frac{mv_1^2}{r}$$

we have

$$P_1 = mg + m\left(\frac{v_1^2}{r}\right)$$

As the result shows, the push of the seat must do two things at point 1. It must support the pilot's weight, mg. It must also furnish the required centripetal force, mv_1^2/r.

An interesting feature is seen if we rewrite the equation for P_1 as

$$P_1 = m\left(g + \frac{v_1^2}{r}\right)$$

The centripetal acceleration v_1^2/r simply adds to the gravitational acceleration. Because of this people sometimes talk about the pilot experiencing an acceleration of so many g's. For example, if v_1^2/r was $3g$, then the pilot would be said to experience an acceleration of 4g's.

Point 2: As we see from Figure 9.11, the pull of gravity and the push of the seat, P_2, are now in the same direction. They together supply the needed centripetal force. Therefore,

$$P_2 + mg = F_c$$

Placing in the value for F_c and solving for P_2 gives

$$P_2 = m\left(\frac{v_2^2}{r}\right) - mg$$

The seat supplies less than the required centripetal force. The pull of gravity is helping the seat hold the pilot in a circular path.

It is interesting to notice that P_2 can become zero. This occurs if the plane is moving slow enough so that the centripetal acceleration v_2^2/r just equals g, the gravitational acceleration. At that instant the pilot is falling freely. What would happen if the plane was going slower than this? ■■

9.10 Satellite Motion

The first satellite to orbit the earth was launched billions of years ago. It is the moon. Today there are many artificial satellites. All earth satellites have one feature in common. They are held in an approximately circular path by the earth's gravitational pull. It is this force that supplies the needed centripetal force.

Suppose that a mass m circles the earth in an orbit of radius r as shown in Figure 9.12. To keep it in its orbit, the earth must pull on it to supply the needed centripetal force mv^2/r, where v is the tangential speed of the mass in the orbit.

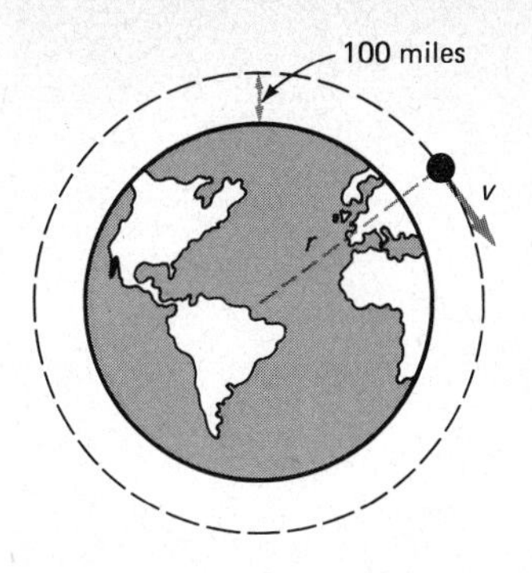

Figure 9.12 All satellites circling the earth in the same orbit must have the same speed.

This force must be supplied by the gravitational pull of the earth. If we call the mass of the earth M, then Equation 4.4, Newton's law of gravitation, gives the earth's pull to be

$$G\frac{Mm}{r^2}$$

Of course, $G = 6.67 \times 10^{-11}\ \mathrm{N \cdot m^2/kg^2}$, the gravitational constant.

Equating these two values for the same force gives

$$\frac{mv^2}{r} = G\frac{mM}{r^2}$$

From this we find that the satellite speed v is given by

$$v = \sqrt{\frac{GM}{r}} \tag{9.9}$$

Notice that the mass of the satellite does not enter. All satellites in orbit with radius r must have the same speed.

In a practical case a satellite might be 161 km (100 miles) above the earth. Because the earth's radius is about 6370 km, we have $r = 6530 \times 10^3$ m. The mass of the earth is $M = 6.0 \times 10^{24}$ kg. Using these values in Equation 9.9 gives

$$v = 7830 \text{ m/s}$$

This corresponds to a speed of 17,500 miles/h. The satellite must have this speed if it is to orbit the earth at a height of 161 km.

It is of interest to compute how long it takes the satellite to make one trip around its orbit. The length of the orbit is the circumference of a circle, $2\pi r$. In the present case $r = 6.53 \times 10^6$ m. We have from $x = \bar{v}t$ that $2\pi r = \bar{v}t$, or

$$2\pi(6.53 \times 10^6 \text{ m}) = (7830 \text{ m/s})t$$

Solving for t, we find

$$t = 5240 \text{ s} = 87 \text{ min}$$

Satellites at other heights will take somewhat different times to complete an orbit.

EXAMPLE 9.16 Show that the centripetal acceleration for a ship in orbit equals the gravitational acceleration at that altitude.

Solution In orbit the gravitational force on the ship supplies the centripetal force. We wrote this force as $G(Mm/r^2)$ in the previous discussion. But this is also the same as the

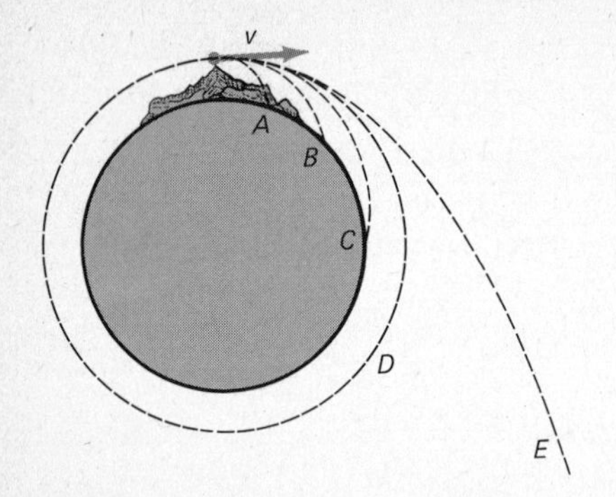

Figure 9.13 The fate of an object thrown parallel to the earth depends on its speed. Only D has the proper speed to orbit the earth.

weight of the ship, mg, where g is the acceleration due to gravity at that altitude. Equating gravitational and centripetal forces gives, because $F_c = ma_c$,

$$mg = ma_c \qquad \text{or} \qquad a_c = g$$

The centripetal acceleration equals the gravitational acceleration at that altitude. ▮▮

9.11 Weightlessness in Orbit

You will recall that we discussed apparent weightlessness in Section 4.10. We saw there that a freely falling elevator has a peculiar property. All objects in it appear to be weightless. This resulted from the fact that no supporting force is needed if an object is to fall freely. We shall now see that objects in orbit appear weightless for the same reason.

Newton was probably the first to discuss orbital motion clearly. In his famous book the *Principia,* he shows a figure much like Figure 9.13. It illustrates what happens to objects shot at various speeds from a mountaintop. He idealized the situation and ignored air friction.

We know that a projectile shot parallel to the earth will fall to the earth. It is pulled by gravity toward the earth's center. But if it is shot fast enough, a peculiar thing happens. As we see in Figure 10.12, the faster the object is shot, the farther away it will land. Due to the curvature of the earth, an object that is shot fast enough will never strike the earth.

For example, consider orbit D in Figure 9.13. Gravity is constantly pulling the object toward the center of the earth, but the curvature of the earth causes the earth's surface to fall with the same curvature as the object's path. Even though the object is freely falling, it never strikes the earth! The gravitational pull of the earth simply bends its orbit into a circle about the earth.

A satellite ship in orbit about the earth follows a path like D. The ship is freely falling at all times. It does not strike the earth because its initial horizontal velocity causes it to miss the earth's surface. But the main point for now is that the ship is freely falling under gravity. Everything stationary within the ship is also freely falling. Objects within the ship need no support because they must fall freely to remain in place in the ship. As a result, objects within the ship appear to be weightless, and travelers experience weightlessness (see Figure 9.14). We should mention that a spaceship coasting anywhere in space without firing its engines is freely falling also. The previous considerations apply to it as well.

Figure 9.14 Astronauts Pogue and Carr demonstrate weightlessness in orbit during their journey in Skylab 4. (Courtesy of NASA)

Summary

The rotation of a wheel is measured by its angular displacement, θ. The units used for this measurement are revolutions, degrees, and radians. To convert from one unit to another we use

$$1 = \frac{1 \text{ rev}}{360 \text{ deg}} = \frac{1 \text{ rad}}{57.30 \text{ deg}} = \frac{2\pi \text{ rad}}{1 \text{ rev}} = \frac{360 \text{ deg}}{2\pi \text{ rad}}$$

The rate of rotation of a wheel is measured in terms of its angular velocity, ω. If an angular displacement θ occurs in a time t, then the average angular velocity is given by

$$\bar{\omega} = \frac{\theta}{t}$$

As a wheel speeds up, it undergoes an angular acceleration, α. If t is the time taken for the angular velocity to change from ω_0 to ω_f, then

$$\alpha = \frac{\omega_f - \omega_0}{t}$$

There are five angular motion equations. They are identical to those used for linear motion except that x, v, and a are replaced by θ, ω, and α. See Section 9.4 for a list of them.

As a wheel of radius r turns, a point on its rim undergoes a tangential displacement, s. The tangential and angular displacements are related by

$$s = r\theta$$

In this relation θ must be measured in radians.

When a wheel turns through an angle θ, it can wind or unwind a length of belt on its rim given by $s = r\theta$. This is also the distance a wheel would roll without slipping.

The tangential velocity, v_T, of a point on the rim of a wheel is

$$v_T = r\omega$$

where radians must be used as angular measure. This is also the speed at which the belt is wound on the wheel's rim.

The tangential acceleration, a_T, of a point on the rim of a wheel is

$$a_T = r\alpha$$

Again, radian measure must be used. This is also the linear acceleration of a point on a string or belt wound on the wheel.

An object traveling in a circle with constant tangential speed v is accelerated toward the circle's center. This centripetal acceleration is given by v^2/r. It is caused by the centripetal force.

If an object of mass m is to move in a circular path with tangential velocity v, then a centripetal force is needed. Its magnitude is

$$F_c = \frac{mv^2}{r}$$

and it must pull the object toward the center of the circle. The needed centripetal force does no work on the object.

Satellites circling the earth are freely falling. As a result, objects in the satellite appear to be weightless.

Questions and Exercises

1. Identify each of the following as θ, ω, or α. (a) The earth rotates on its axis once each 24 h. (b) The nameplate of a motor lists 3200 rev/min. (c) Some phonograph records are listed as $33\frac{1}{3}$ and others as 45. (d) A child's swing has an angle of 30° to the vertical.
2. Two cars are going around two different circular paths at the same angular velocity. The speed of one car is 60 km/h on a track of radius R. What is the speed of the other car if its track has a radius $\frac{1}{2}R$?
3. Discuss the principle of operation of the spin-dry cycle in a washing machine. If you were to redesign a machine to increase its drying power, how would you change (a) its rotation speed and (b) its drum diameter? What other factors affect its drying power?
4. What furnishes the required centripetal force in each of the following: (a) the earth circles the sun, (b) a fly rides on a rotating phonograph record, (c) a car turns a corner on a level road, (d) a track star runs high on a banked turn of an oval track, (e) an airplane pulls out of a dive?
5. Figure P9.1 shows an amusement park device. As the rotation speed of the central shaft is increased, the angle θ increases. (a) Explain what furnishes the centripetal force to the cars in which the riders sit. (b) Why must θ increase as the rotation speed increases?

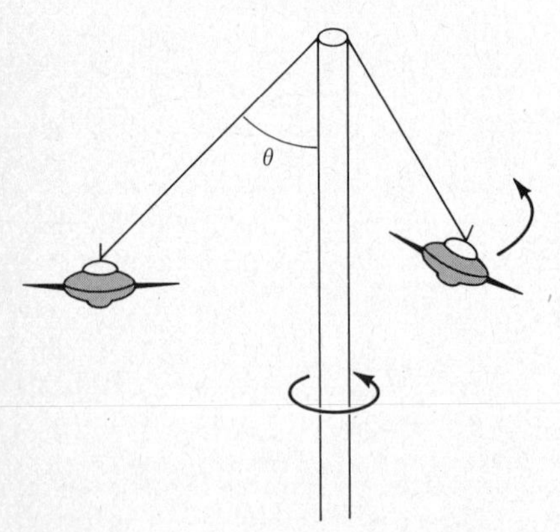

Figure P9.1 A whirling amusement park ride.

6. One common type of angular speed control device (a *governor*) is shown in Figure P9.2. The cylinders at B and B' slide in or out as the rotation speed is changed. Their movement is designed to actuate appropriate controls. Explain how the system shown operates. In particular, what furnishes the required centripetal force to the massive cylinders?

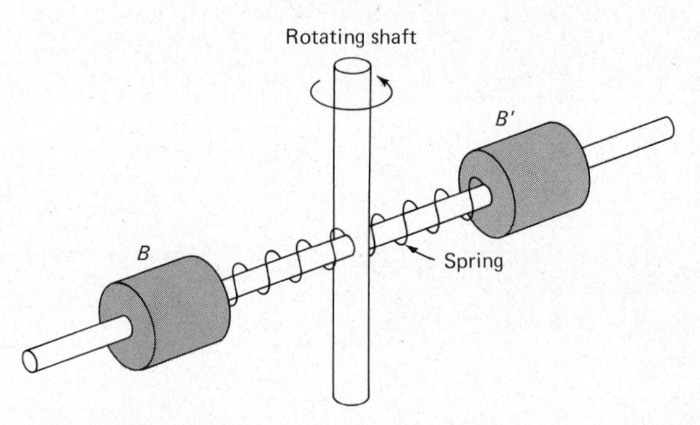

Figure P9.2 A spring-type governor.

7. Another type of governor is shown in Figure P9.3. In this

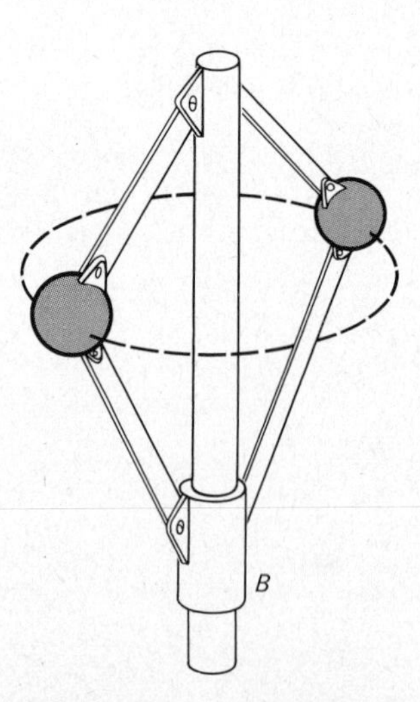

Figure P9.3 A governor.

case the collar at B moves up or down depending on speed. Explain how the device operates. In particular, what furnishes the needed centripetal force to the two massive balls?

8. Many industries use what is called a cyclone-type dust-and-soot remover to clean exhaust gases. As shown in Figure P9.4, the air to be cleaned is shot at high speed around a curved path. The soot and dust particles are trapped and collected at T by a collecting pan and/or a water spray. Explain the principle of operation for such a device.

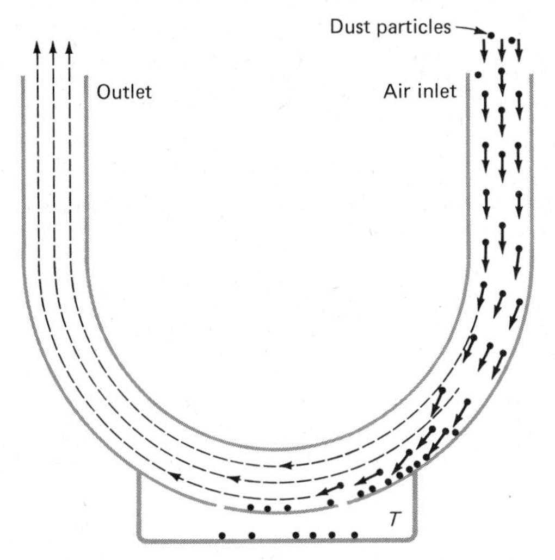

Figure P9.4 A cyclone-type dust remover.

9. If an airplane pilot pulls out of a dive too fast, he may pass out. Explain why he may faint.
10. The earth is flattened at the poles and bulges at the equator. (Polar radius = 6.357×10^6 m; equatorial radius = 6.378×10^6 m.) Relate this to the rotation of the earth about its axis. What effect should a gradual speeding up of the earth's rotation rate have?
11. Many people insist that a force exists which pushes them radially outward as they try to round a curve. They cite the fact that a box sitting on a car seat is thrown to the side of a car as the car rounds a sharp bend. According to them, some force threw the box to the side. Give the proper interpretation of the effect. How is it related to the fact that one is thrown forward when a car suddenly stops?
12. A tiny insect sits about halfway to the edge of a phonograph record. It holds on for dear life when the speed is $33\frac{1}{3}$ rpm. Describe the insect's motion as the speed is increased to 45 rpm and the insect begins to slip.
13. In 1970 the spaceship Apollo 13 encountered trouble when it was about halfway to the moon. It was known at that point that the ship could not carry out its mission. In spite of this the ship did not turn around immediately. Instead it continued on to the moon, passed behind it, and then returned to the earth. Supposedly, this maneuver was possible even though immediate return would have been impossible because of lack of fuel. Explain.

Problems

1. Express the following angles in degrees, revolutions, and radians: (a) 36 deg, (b) 0.47 rad, (c) 2.18 rev.
2. Express the following angles in degrees, revolutions, and radians: (a) 208 deg, (b) 5.19 rad, (c) 0.23 rev.
3. Express the following values for ω in degrees per second, revolutions per second, and radians per second: (a) 3000 rev/min, (b) 260 deg/h, (c) 47 rad/min.
4. Express the following values for ω in degrees per second, revolutions per second, and radians per second: (a) 1 rev/day, (b) 6 deg/min, (c) 400 rad/h.
5. Express the following values for α in degrees per second per second, revolutions per second per second, and radians per second per second: (a) 3600 rad/min^2, (b) 4 rev/s^2, (c) 2×10^5 deg/h^2.
6. Express the following values for α in degrees per second per second, revolutions per second per second, and radians per second per second: (a) 2.3 deg/s^2, (b) 0.170 rad/s^2, (c) 1.74 rev/min^2.
7. A shaft is timed to rotate 87 rev in 0.50 min. What is its angular speed in radians per second?
8. The earth rotates one revolution in one day. Find its angular speed in radians per second.
9. A large fan takes 12.0 s to reach its operating speed of 2.0 rev/s after it is turned on. Find its average acceleration in radians per second per second.
10. A large gambling wheel turning at a speed of 2.0 rev/s coasts to rest in an agonizing time of 14 s. Find its deceleration in radians per second per second.
11. After the power is turned off to a motor whose shaft speed is 3000 rev/min, the motor takes 5.0 s to stop. Through how many revolutions did the shaft turn while the motor came to rest? Assume uniform acceleration.
12. If a wheel accelerates uniformly from rest to a speed of 20 rev/s in a time of 0.50 min, through how many revolutions did it turn while accelerating?
13. A wheel has an acceleration of 1.70 rad/s^2. It starts from rest. How fast is it turning (radians per second) after it has completed 20 rev?
14. Due to friction, a rotating wheel has a deceleration of 0.80 rad/s^2 as it coasts to rest. If its initial speed was

3000 rev/min, through how many revolutions does it turn as it comes to a stop?

15. A wheel rotating at 3 rev/s slows down and stops after turning through 45 rev. Assuming uniform deceleration, (a) how long did it take to stop and (b) what was its deceleration?

*16. As a wheel speeds up from rest with uniform acceleration, it is found to have a speed of 2 rad/s at a certain time. At a time 5 s later, its speed is 8 rad/s. (a) What was the acceleration of the wheel? (b) How long before it had a speed of 2 rad/s did it start to move? (c) Through how many revolutions did it turn in the first 3 s after starting?

17. A phonograph record rotates at a rate of 45 rpm. How fast is a dust speck on the record moving (in meters per second) if it is 5.0 cm from the axis?

18. A fan blade is rotating at a speed of 300 rev/min and the radius of the blade is 12.0 cm. How fast (in meters per second) is the outer tip of the blade moving?

19. A 25-cm-diameter bowling ball rolls a distance of 30 m without slipping. How many times did it turn around?

20. A car with 80-cm-diameter wheels is going at a speed of 20 m/s. How fast, in radians per second, are its wheels turning?

21. If a bicycle is traveling at 15 km/h, how fast are its 50-cm-diameter wheels turning? (Give answer in revolutions per second.)

22. In a factory thread is being wound on a 2.5-cm-diameter spool as the spool rotates at 2000 rev/min. How fast is the thread being wound? How long does it take to wind 1000 m of thread on the spool?

23. A cord is wound on the rim of a 30-cm-radius wheel. A weight hung from the end of the cord accelerates downward at 0.8 m/s^2. (a) What is the angular acceleration of the wheel? (b) Through how many revolutions does the wheel turn as the weight falls 2 m from rest?

24. A disklike wheel has a radius of 20 cm. It is turning at a rate of 3.0 rev/s when its power source is cut off. The wheel coasts to rest in 8.0 s. (a) What is the deceleration of the wheel in radians per second squared? (b) What length of belt (in meters) passes over the rim of the wheel as the wheel coasts to rest?

25. How strong must a string be (i.e., what load can it hold) if a 250-g mass is to be swung from it in a horizontal circle of 80 cm radius at a rate of 3 m/s?

26. A string holds a 2.0-kg object in a horizontal circle of 75-cm radius while the object moves at a rate of 1.50 rev/s. (a) What is the tension in the string? (b) Repeat for a 2-lb object in a 3-ft-radius circle.

*27. A coin sits on a phonograph record at a distance of 5 cm from the center. How large must the coefficient of friction be between coin and record if the coin is not to slip when the rotation rate is $33\frac{1}{3}$ rpm?

*28. The radius of the earth's orbit around the sun is 1.5×10^{11} m and it goes around the orbit in a time of 365 days. (a) What is the angular speed (in radians per second) of the earth in the orbit? (b) How large a force must the sun exert on the earth to hold it in orbit? $M_e = 6 \times 10^{24}$ kg.

*29. How large must be the friction coefficient between road and tires of a car if it is to be able to follow a circular arc of 15-m radius on level road while going 20 m/s?

*30. The maximum friction coefficient between a 1500-kg car and a certain level roadway is 0.90. How fast can the car be moving to negotiate safely a curve of 20-m radius of curvature?

*31. A 2-kg object is swung in a vertical circle at the end of a string 120 cm long. If the object is to move at constant speed of 3 rev/s, how strong must the string be?

*32. A 500-g object at the end of a string is swung with a speed of 4 rev/s in a vertical circle of radius 80 cm. What is the tension in the string at (a) the bottom of the circle and (b) the top of the circle?

*33. A 0.80-kg object is swung in a vertical circle at the end of a 1.50-m-long cord. If the object has a constant speed of 4.0 m/s, what is the tension in the cord at (a) the bottom of its path and (b) at the top?

**34. A pendulum consisting of a 2.0-kg ball at the end of a 1.50-m-long cord is pulled aside to an angle of 37° to the vertical and released. Find the speed of the ball at the lowest position. Find the tension in the cord when it is vertical.

*35. Refer to the amusement park ride shown in the photo on page 174. Assuming the curved path has a radius of 8.0 m, how fast must the cart be moving if the passengers in the cart are to be just on the verge of falling out?

*36. A common trick is to swing a pail of water in a vertical circle and show that the water does not fall out even when the pail is upside down. What must be the angular speed of the pail at the top of the path if the trick is to succeed?

**37. A spaceship is to orbit the moon at a height of 800 km above the moon's surface. How fast must the ship be moving? (For the moon $r = 1740$ km, $M = 7.3 \times 10^{22}$ kg.)

**38. How high above the earth must a satellite be to complete its orbit once each day? Such a satellite, orbiting above the equator, is called a *synchronous satellite*. It remains motionless above a given point on the earth because it turns just as fast as the earth does. Synchronous satellites are used for worldwide communications. (See the tables in the front of the book for the necessary data.)

*Problems marked with an asterisk are not as easy as the unmarked ones.
**Problems marked with a double asterisk are somewhat more difficult than the average.

**39. The governor of Figure P9.3 is much like the conical pendulum shown in Figure P9.5. The tension in the pendulum cord balances mg of the bob and also furnishes the centripetal force. Show that the pendulum will make an angle θ to the vertical given by $\tan\theta = v^2/gR$ as the bob goes around the circle of radius R with speed v.

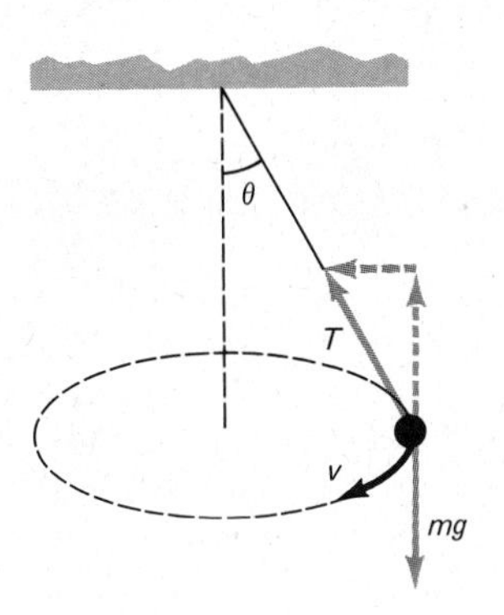

Figure P9.5 The conical pendulum.

**40. Roadways are often banked on curves as shown in Figure P9.6. When the speed of the car is exactly right, the horizontal component of the normal force exactly supplies the required centripetal force. The vertical component of $\mathbf{F}_N$ exactly balances the weight of the car. Therefore, no friction force between car and roadway is needed to negotiate the curve. Show that the relation between the banking angle θ and the car's speed v is given by $\tan\theta = v^2/gR$ where R is the radius of the curve.

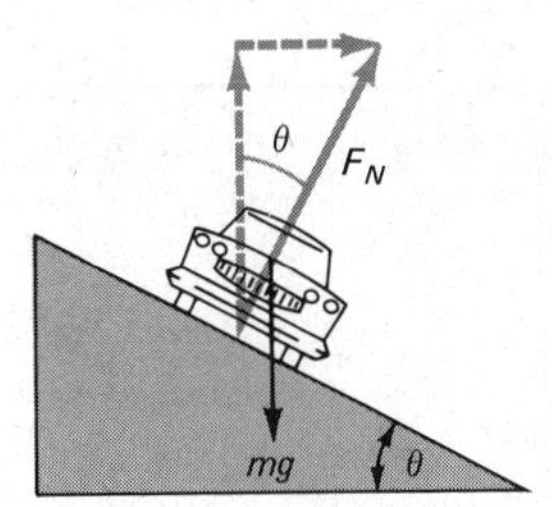

Figure P9.6 For proper banking the normal force balances mg and supplies the centripetal force as well.

Hurricane as seen by an infrared camera.

Westinghouse

10 DYNAMICS OF ROTATION

Performance Goals

When you finish this chapter, you should be able to

1. Give three examples that illustrate rotational inertia. Each example should show the effect of torque on rotation. The force and lever arm should be pointed out in each case.
2. Point out which of three objects has the greatest (or least) rotational inertia about a specified rotation axis. The objects are given, have equal

In the last chapter we saw how to describe rotational motion. Except for the concept of centripetal force, little attention was paid to the forces that cause rotation. In this chapter we shall see that unbalanced torques result in angular accelerations. It will be shown that each object possesses inertia which causes it to resist changes in its angular motion. The concept of kinetic energy will be extended to include the rotational motion of objects. In addition, we shall see that rotating objects possess angular momentum. A conservation law for this type of momentum will be stated and used.

masses, but have obviously different mass distributions.
3. Point out why the moment of inertia of an object is different for different choices of the axis.
4. Find τ, I, or α for a given wheel if two of these three quantities are known. Furnish proper units for each quantity.
5. Find I, M, or k for an object if two of these are known. Furnish proper units for each quantity. Give the qualitative meaning of k. Evaluate k for a hoop with an axis that goes through the center of the hoop.
6. Apply $F = ma$ and $\tau = I\alpha$ to the following system in order to compute a, α, I, m, or τ as required. The system consists of a mass m hanging from the rim of a wheel that has a radius r.
7. Give the formula for KE_r and show how it applies in three simple situations.
8. Give the total kinetic energy in terms of I, ω, m, and r for an object that is rolling without slipping on a surface.
9. Apply the work-energy theorem to the following situations: (a) an object rolls up or down an incline, (b) a wheel lifts or lowers a mass hanging from it.
10. Give the rotational analogs for F, m, a, v.
11. State the equation that relates torque to the work done by the torque. Repeat for the equation that relates torque to power.
12. Find the magnitude and direction of the angular momentum of an object when I and ω are given.
13. State the law of conservation of angular momentum. Give four examples that illustrate its use.

10.1 Rotational Inertia and Energy

You have certainly noticed that it is difficult to stop the spin of a rotating wheel. No one would knowingly use a finger to stop a rapidly rotating fan blade even if it were coasting to rest. The blade is not easily stopped, although air friction eventually stops it. This is but one example of the fact that objects that can rotate have rotational inertia. They resist being set into rotation; when already rotating, they resist attempts to stop the rotation. Let us now investigate the cause of this type of inertia.

Consider the wheel of radius a shown in Figure 10.1. It can rotate about an axle through its center as shown. To start its rotation, we wind a string around it and pull on the string with a steady force F. Let us first compute the work we do on the wheel as the wheel is turned through an angular distance θ.

When the wheel turns through an angle θ, the tangential distance a point on its rim moves is $a\theta$. This is also the length of string unwound. It is also the distance through which the force F works. Therefore

$$\begin{pmatrix}\text{Work done by } F \text{ in} \\ \text{turning through } \theta\end{pmatrix} = Fa\theta$$

We can put this in a more suitable form by noticing that Fa is force times lever arm. It is the torque applied to the wheel by F. Hence

$$\text{Work} = \tau\theta \qquad (10.1)$$

This relation gives the work done on a wheel by a torque τ as the wheel turns through θ. Why must θ be in radians in this equation?

What happens to this work? Assuming frictionless conditions, it must appear as kinetic energy of the wheel. It is easy to see that a rotating wheel has KE. The wheel is composed of many tiny masses; a few are shown schematically in Figure 10.1. Each piece has a velocity, such as v_1 shown for m_1. As a result, m_1 has a KE of $\frac{1}{2}m_1v_1^2$ and similarly for the other masses composing the wheel. If the wheel consists of N such masses (much smaller than shown), then the total KE of the wheel is the sum of all these individual KE's. We have

$$\text{KE of wheel} = \tfrac{1}{2}m_1v_1^2 + \tfrac{1}{2}m_2v_2^2 + \tfrac{1}{2}m_3v_3^2 + \cdots + \tfrac{1}{2}m_Nv_N^2$$

The ellipsis (three dots) between the two plus signs mean there are many other similar terms we should have written, one for each bit of mass in the wheel.

We can simplify this expression greatly if we notice that the v's are the tangential velocities of the various masses. Therefore, because $v_T = r\omega$, we can write $v_1 = r_1\omega_1$, $v_2 = r_2\omega_2$, and so on. But the rotation rate of the wheel is ω, and this is also the rotation rate of each tiny piece of mass. Therefore, $\omega = \omega_1 = \omega_2 = \omega_3 = \omega_4$, etc., and so we have

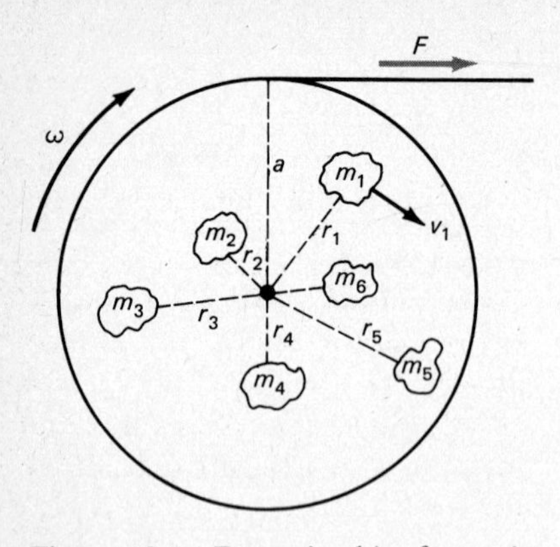

Figure 10.1 Every tiny bit of mass in the rotating wheel has the same angular velocity ω.

$$\text{KE of wheel} = \tfrac{1}{2}m_1 r_1{}^2\omega^2 + \tfrac{1}{2}m_2 r_2{}^2\omega^2 + \tfrac{1}{2}m_3 r_3{}^2\omega^2 + \cdots + \tfrac{1}{2}m_N r_N{}^2\omega^2$$

Or, after factoring out the common factor $\frac{1}{2}\omega^2$,

$$\text{KE of wheel} = \tfrac{1}{2}\omega^2(m_1 r_1{}^2 + m_2 r_2{}^2 + m_3 r_3{}^2 + \cdots + m_N r_N{}^2)$$

As you might expect, the KE of a rotating object is an important quantity. But this expression for it is too long for our use. We therefore agree to write the quantity in parenthesis as I and call it the *moment of inertia* of the wheel. Thus

$$\text{Moment of inertia} \equiv I \equiv m_1 r_1{}^2 + m_2 r_2{}^2 + \cdots + m_N r_N{}^2 \quad (10.2)$$

We will discuss the moment of inertia in more detail in a moment.

Our expression for the KE of the rotating wheel can now be rewritten in terms of I. It is

$$\text{KE}_r \equiv \text{rotational KE} = \tfrac{1}{2}I\omega^2 \quad (10.3)$$

In other words, the kinetic energy possessed by the moving pieces of a rotating wheel is given by Equation 10.3. We call this KE *rotational kinetic energy* because it arises from the rotation of the wheel about its axis. A wheel on a car moving down the road would be translating at the same time, that is, it would have translational (straight-line motion) KE as well.

Where did this rotational KE come from? It was supplied to the wheel in Figure 10.1 by the work done by the force F. We have already computed this work in Equation 10.1. The work-energy theorem tells us

$$\text{Work done on wheel} = \text{KE of wheel}$$

or

$$\tau\theta = \tfrac{1}{2}I\omega^2$$

This equation is inconvenient because it contains both θ and ω. But there is a relation between the two. Our angular motion equations give us

$$2\alpha\theta = \omega^2 - \omega_0{}^2$$

In our case the wheel started from rest, so $\omega_0 = 0$. We therefore have

$$2\alpha\theta = \omega^2$$

Substituting this value for ω^2 gives

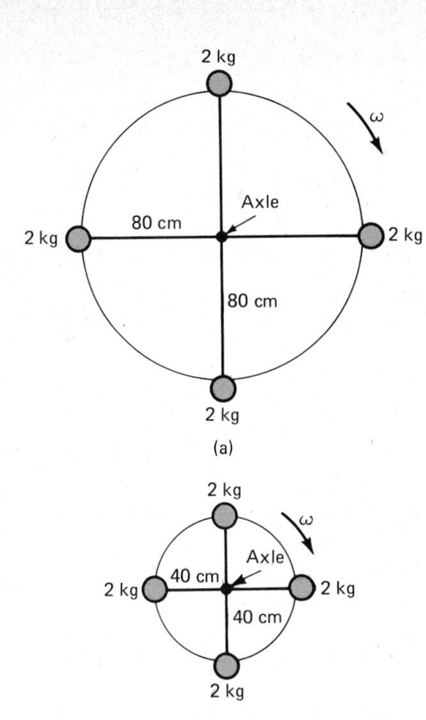

Figure 10.2 Even though the objects in (a) and (b) have identical mass, one shows more rotational inertia than the other. Which has the larger moment of inertia?

$$\tau\theta = \tfrac{1}{2}I(2\alpha\theta)$$

or

$$\tau = I\alpha \qquad (10.4)$$

We have therefore arrived at a relation between the way in which a wheel's rotation speeds up and the torque applied to the wheel. This is an important relation. Notice that we used $s = r\theta$ in deriving Equation 10.4. Therefore, α must be measured in radians per second squared when using it.

10.2 Meaning of Our Equations

Let us first examine Equation 10.4. Notice that it is the rotational analog to Newton's second law, $F = ma$. The linear form of the law, $F = ma$, tells us a great deal about the translational inertia of an object. Solving for the object's mass m gives

$$m = \frac{F}{a}$$

So the mass of an object (its measure of translational inertia) is a measure of the force needed to give it unit acceleration. Massive objects require large forces to accelerate them.

If we solve Equation 10.4 for I (its measure of rotational inertia), we find it to be

$$I = \frac{\tau}{\alpha}$$

So the moment of inertia tells us how large a torque is required to produce unit angular acceleration. Clearly, I is a measure of an object's rotational inertia. If an object has a large moment of inertia, it will be difficult to set into rotation.

Now let us examine the defining equation for I in view of our previous discussion. Consider the two objects shown in Figure 10.2. Both objects have the same mass, but the mass is distributed differently in the two cases. (We consider the wires connecting the masses to have negligible mass. The spheres are to be much smaller than shown.) Suppose we want to accelerate each object with the same angular acceleration α. To do so, we must apply a torque to each. But if we want to give each the same angular acceleration, we discover by experiment that the object in (a) requires four times the torque we need for the object in (b). Why is this?

The reason is quite simple. Suppose each wheel is rotating with $\omega = 1$ rev/s.

Although the wheels have the same angular speed, their masses are moving with far different speeds. The masses in (a) travel *twice* as far per revolution as the masses in (b). Hence the speed of the masses in (a) is *twice* that in (b). This makes the KE's of the wheels different even though they are rotating with the same angular speed. Because KE is $\frac{1}{2}mv^2$ for each mass, the masses in (a) have *four* times the KE of those in (b). This means that four times more work must be done to accelerate the wheel in (a) to 1 rev/s compared to that in (b). The torque needed to provide the same acceleration to (a) as to (b), is four times larger in (a) than in (b).

We can also see this difference in rotational inertia for the two wheels in Figure 10.2 by looking at the defining equation for I, the moment of inertia. In the present case Equation 10.2 becomes

$$I_a = (2\ \text{kg})(.80\ \text{m})^2 + (2\ \text{kg})(.80\ \text{m})^2 + (2\ \text{kg})(.80\ \text{m})^2 + (2\ \text{kg})(.80\ \text{m})^2$$
$$= 5.12\ \text{kg}\cdot\text{m}^2$$

and

$$I_b = (2\ \text{kg})(.40\ \text{m})^2 + (2\ \text{kg})(.40\ \text{m})^2 + (2\ \text{kg})(.40\ \text{m})^2 + (2\ \text{kg})(.40\ \text{m})^2$$
$$= 1.28\ \text{kg}\cdot\text{m}^2$$

As we see, the rotational inertia for the wheel in (a) is indeed four times that in (b). Notice that the difference is the result of the different distribution of mass in the two objects.

■ **EXAMPLE 10.1** How large a torque is required to accelerate the wheel in Figure 10.2(b) from rest to a speed of 5.0 rev/s in a time of 30 s?

Solution We will make use of $\tau = I\alpha$ to find τ. The moment of inertia I for this wheel was found in the previous discussion: $I = 1.28\ \text{kg}\cdot\text{m}^2$. Now let us find α from an angular motion problem. Of course, α must be in rad/s^2.

We know that $\omega_0 = 0$, $\omega_f = 5.0$ rev/s $= 10\pi$ rad/s, and $t = 30$ s. Using $\omega_f - \omega_0 = \alpha t$, we have

$$\alpha = \frac{\omega_f - \omega_0}{t} = \frac{31.4\ \text{rad/s}}{30\ \text{s}} = 1.05\ \text{rad/s}^2$$

Substituting in $\tau = I\alpha$ gives

$$\tau = (1.28\ \text{kg}\cdot\text{m}^2)(1.05\ \text{rad/s}^2) = 1.34\ \text{N}\cdot\text{m}$$

as the required torque. ■■

10.3 More About Moment of Inertia

The defining equation for moment of inertia is Equation 10.2,

$$I = m_1 r_1^2 + m_2 r_2^2 + m_3 r_3^2 + \cdots + m_N r_N^2$$

We picture the object to be composed of N tiny masses. The distance from the rotation axis of mass 2, for example, is r_2. As we see, I is dependent on both the mass of the object and how far the mass is from the axis.

The moment of inertia of a car wheel, for example, is not easily computed. The positions of the masses composing the wheel are not easily described. In cases such as this, it is customary to express I as a product of the mass of the wheel, M, and a type of average distance, k, for the masses that compose it. Then we have that

$$I = Mk^2 \tag{10.5}$$

where M is the mass of the wheel and k is called the *radius of gyration* of the wheel. We will see soon how it can be measured. In a crude sense, k is an average distance from the axle to the various pieces of mass from which the wheel is made.

The moments of inertia of several simple objects are shown in Figure 10.3. Let us look at a few of them. Notice that for the hoop all the mass is at a distance b from the axis. Therefore, the radius of gyration for the hoop must be b and so $I = Mb^2$.

But the uniform solid disk is not so simple. All the mass is at a radius less than b. Hence the radius of gyration will be considerably smaller than b. The figure shows us that $k = b/\sqrt{2}$. And so we have for a disk

$$I = Mk^2 = M\left(\frac{b}{\sqrt{2}}\right)^2 = \tfrac{1}{2}Mb^2$$

If you look at the other examples in Figure 10.3, you will see that they are also reasonable.

One further point becomes clear if we look at the case of the thin rod in Figure 10.3. In one situation the rod is rotated about an axis through its center. The radius of gyration is obviously less than $\frac{1}{2}L$ and the value given is $\frac{1}{2}L/\sqrt{3}$. In the other situation the rod is rotated about an axis through its end. Now k is obviously less than L and probably greater than $\frac{1}{2}L$. The figure tells us k is $L/\sqrt{3}$ in this case, twice as large as before. Clearly, the position of the rotation axis is important when we are giving the moment of inertia.

▮ EXAMPLE 10.2 A string is wound on a 40-kg wheel as shown in Figure 10.4. When the tension in the string is 9 N, the wheel accelerates from rest to 3.0 rev/s in 15 s. Find both I and k for the wheel.

Solution We can find I from $\tau = I\alpha$. The torque is (9 N)(0.20 m). To find α, we must work a motion problem. We know that $\omega_f = 3.0$ rev/s $= 6\pi$ rad/s, whereas $\omega_0 = 0$ and $t = 15$ s. Therefore

$$\alpha = \frac{\omega_f - \omega_0}{t}$$

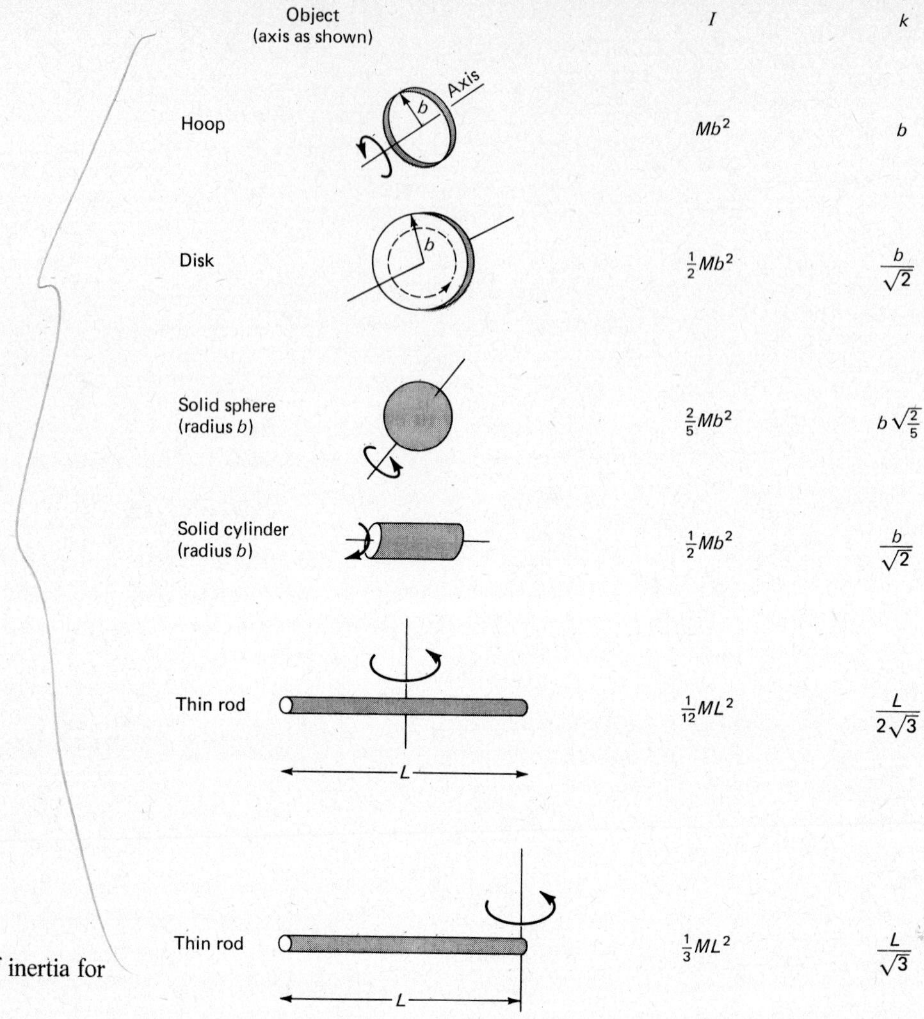

Figure 10.3 Moments of inertia for several common objects.

gives

$$\alpha = \frac{6\pi}{15} \text{ rad/s}^2 = 1.26 \text{ rad/s}^2$$

We recall that α must be expressed in radians per second per second when using $\tau = I\alpha$. We then have

$$I = \frac{\tau}{\alpha} = \frac{(9 \text{ N})(0.20 \text{ m})}{1.26 \text{ rad/s}^2}$$

and so

$$I = 1.43 \text{ N}\cdot\text{m}\cdot\text{s}^2 = 1.43 \text{ kg}\cdot\text{m}^2$$

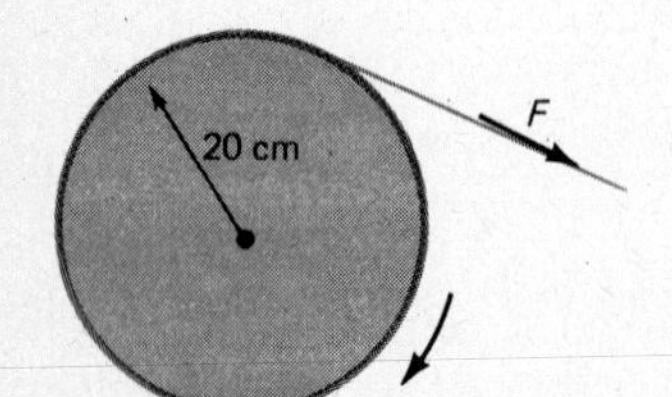

Figure 10.4 If F and a are known, how can I be found?

FURNISHING GRAVITY TO SPACE COLONIES

It has been suggested that humans, if they wished, could establish colonies in space.* Huge, sealed cylindrical tubes perhaps 4 miles in diameter and 20 miles long would be used. They would float freely in space or might orbit the earth. It is estimated that such a tube could be self-sufficient for about a million people. The cross-section view of the cylinder might be much like that in the figure.

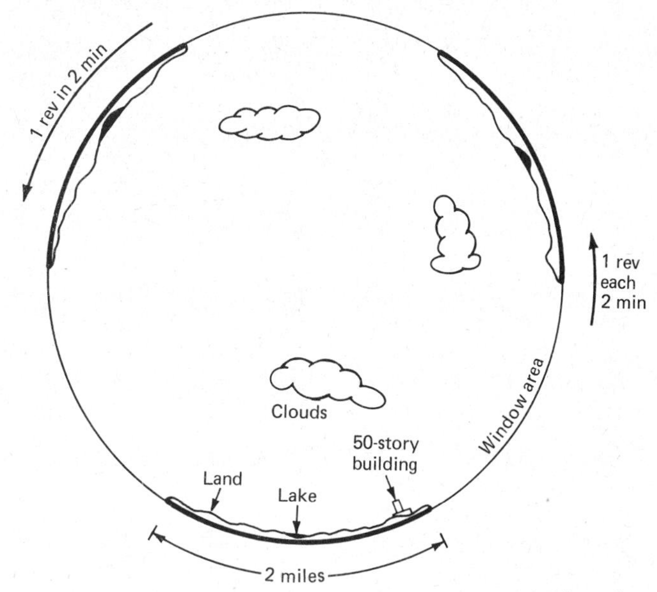

The people living within the tube would use its inside surface as "earth." In a spaceship such as this, everything appears weightless. Some way must be found to provide weight to objects in it. Otherwise, liquids would not pour and objects would not fall. To provide the effects of weight on earth, the cylinder would be rotated on its axis. A rotation rate of about 0.5 rev/min would be needed.

Everything within the cylinder will rotate with the cylinder. A man on the "earth" would then move in a circular path. To keep him in a circular path, a centripetal force is needed. As he stands on the "earth," the cylinder must push inward on him with the centripetal force. The stated rotation speed is chosen so as to make the required centripetal force nearly equal to each person's weight on earth. (Can you show that the rate is the same independent of the person?)

As a result of the rotation, everything within the cylinder must be supported. A force about equal to the object's weight must push "upward" (that is, radially inward) in order to support the object. If the object is not supported, it will fall to "earth." Liquids will settle to the bottom (that is, "earth" side) of a container as they normally do on earth. In this way everything within the space colony would experience its usual effects of weight.

There are many fascinating aspects of a space colony such as this. It appears that current technology is far enough advanced to make such a colony feasible. The reference given in the footnote presents a readable discussion of the technical problems that arise.

*G. K. O'Neill, *The High Frontier,* Bantam Books, 1977.

Because $I = Mk^2$, its units are clearly a mass unit multiplied by a length unit squared, and kilogram·meter squared is commonly used. To find k, if $I = Mk^2$, $M = 40$ kg, and $I = 1.43\ \text{kg}\cdot\text{m}^2$, we have

$$1.43\ \text{kg}\cdot\text{m}^2 = (40\ \text{kg})k^2$$

Solving for k^2 and taking the square root gives

$$k = 0.190\ \text{m}$$

It appears that most of the mass is at the rim of the wheel. ■■

EXAMPLE 10.3 A wheel for which $I = 0.050\ \text{kg}\cdot\text{m}^2$ is spinning freely at 20 rev/s. It coasts to rest in 40 s. How large is the friction torque that slows its motion?

Solution We wish to use $\tau = I\alpha$, so we must first find α. In a motion problem, we know that $t = 40$ s, $\omega_f = 0$, and $\omega_0 = 20$ rev/s. Then we can use

$$\omega_f - \omega_0 = \alpha t \rightarrow 0 - 20\ \text{rev/s} = \alpha(40\ \text{s})$$

to find

$$\alpha = -0.50\ \text{rev/s}^2$$

To convert α to radians per second per second, we have

$$\alpha = -(0.50\ \text{rev/s}^2)(2\pi\ \text{rad/rev}) = -3.14\ \text{rad/s}^2$$

Substituting this and the given value for I in $\tau = I\alpha$ gives

$$\begin{aligned}\tau &= (0.050\ \text{kg}\cdot\text{m}^2)(-3.14\ \text{rad/s}^2)\\ &= -0.157\ \text{kg}\cdot\text{m}^2/\text{s}^2 = -0.157\ \text{N}\cdot\text{m}\end{aligned}$$

The second set of units follows because a newton is a kilogram-meter per second per second. ■■

EXAMPLE 10.4 A 50-g mass hangs from a wheel as shown in Figure 10.5. The moment of inertia of the wheel is $0.20\ \text{kg}\cdot\text{m}^2$. Find the acceleration with which the mass falls. How far will it fall in the first 5.0 s after its release?

Solution In situations such as this we write two equations. For the mass we can write $F = ma$. Let us take the direction of motion as the positive direction. From the figure F is $mg - T$, and so $F = ma$ becomes

$$(0.050)(9.8)\ \text{N} - T = (0.050\ \text{kg})a \tag{a}$$

Our second equation is $\tau = I\alpha$ for the wheel. Because $\tau = T(0.08\ \text{m})$,

$$(T)(0.08\ \text{m}) = (0.20\ \text{kg}\cdot\text{m}^2)\alpha \tag{b}$$

Since α and a are related by $a = \alpha r$, we substitute for α in equation (b).

$$(0.08\ \text{m})T = (0.20\ \text{kg}\cdot\text{m}^2)\left(\frac{a}{0.08\ \text{m}}\right)$$

from which we find that

$$T = (31\ \text{kg})a$$

Substituting this value in equation (a) and solving for a gives

$$a = 0.0158\ \text{m/s}^2$$

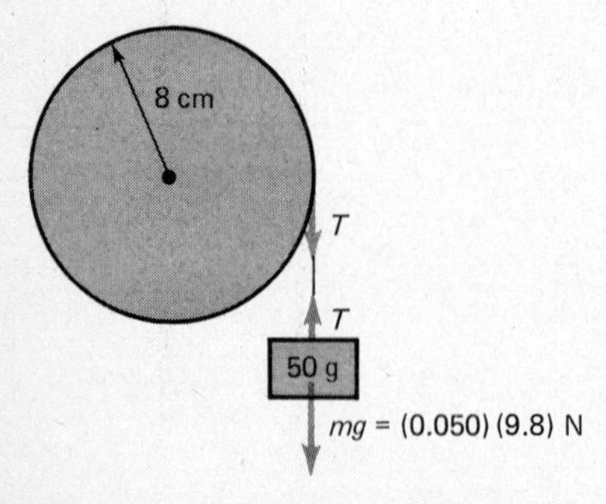

Figure 10.5 If $I = 20\ \text{kg}\cdot\text{m}^2$ for the wheel, what will be the acceleration of the mass?

In order to find the distance fallen, we solve a motion problem. We know that $t = 5.0$ s, $v_0 = 0$, and $a = 0.0158\ \text{m/s}^2$. Using

$$y = v_0 t + \tfrac{1}{2}at^2$$

gives

$$y = 0.198\ \text{m}$$

■■

10.4 Work-Energy Theorem Extended

In Section 6.11 we stated the work-energy theorem of mechanics. It relates the potential and kinetic energy of a system to the work done on the system. It says

$$\begin{pmatrix}\text{Change in KE}\\ \text{of object}\end{pmatrix} + \begin{pmatrix}\text{change in GPE}\\ \text{of object}\end{pmatrix} = \begin{pmatrix}\text{work done on object}\\ \text{by outside forces}\end{pmatrix}$$

The work done by accelerating forces is positive, while the work done by stopping forces, such as friction, is negative.

At that time we discussed only objects that have translational kinetic energy. However, the theorem is still true for rotating objects. For them the kinetic energy term must contain both types of energy, translational and rotational. Let us examine a typical case where both forms exist.

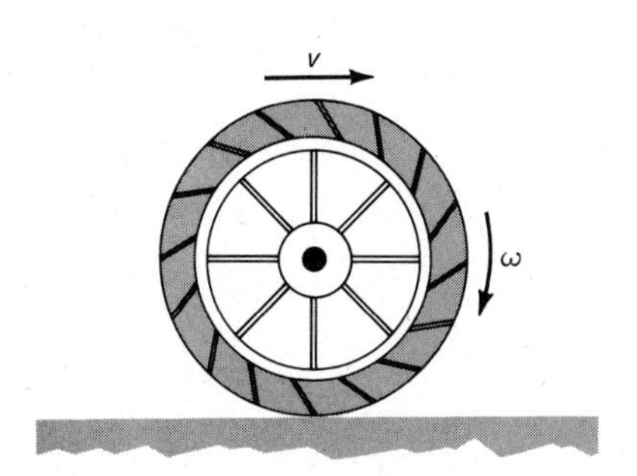

Figure 10.6 The wheel has $\text{KE}_t = \frac{1}{2}mv^2$ and $\text{KE}_r = \frac{1}{2}I\omega^2$. If the wheel does not slip, $v = \omega r$. In these expressions ω must be in radians per second.

In Figure 10.6 we see a wheel rolling along level ground. The wheel is rotating about its center of mass with angular speed ω. It therefore has rotational kinetic energy, $\text{KE}_r = \frac{1}{2}I\omega^2$. Because we are dealing with rotation about the mass center, I must be the moment of inertia about an axis through the mass center of the wheel. This is the moment of inertia usually given for a wheel.

At the same time the mass center of the wheel is translating with speed v. It therefore has translational kinetic energy $\text{KE}_t = \frac{1}{2}mv^2$. The total kinetic energy of the wheel is therefore

$$\text{KE} = \text{KE}_t + \text{KE}_r = \tfrac{1}{2}mv^2 + \tfrac{1}{2}I\omega^2$$

If the wheel does not slip on the surface, then we also know that $v = \omega r$. We shall see that this is a convenient equation in many cases. Let us now see how the work-energy theorem is used when rotations are involved.

■ **EXAMPLE 10.5** In Figure 10.7 a disk of radius b and mass m rolls down an incline. Assuming that it started from rest at point A, find its translational speed when it reaches B. Ignore friction work, because it is usually small for rolling objects.

Solution The work-energy theorem tells us (because no work is done) that

$$\text{Change in KE} + \text{change in GPE} = 0$$

Therefore

$$[(\text{KE}_t + \text{KE}_r)_B - (\text{KE}_t + \text{KE}_r)_A] + [mgh_B - mgh_A] = 0$$

Figure 10.7 How fast is the disk moving at B?

But $KE_t = \frac{1}{2}mv^2$ and $KE_r = \frac{1}{2}I\omega^2$, and so this becomes

$$[(\tfrac{1}{2}mv_B{}^2 + \tfrac{1}{2}I\omega_B{}^2) - 0] + [0 - mgh] = 0$$

Now I for a disk is $\frac{1}{2}mb^2$. Also, $\omega_B = v_B/b$. The equation above then becomes

$$\left[\tfrac{1}{2}mv_B{}^2 + \tfrac{1}{2}(\tfrac{1}{2}mb^2)\left(\frac{v_B}{b}\right)^2\right] - mgh = 0$$

Notice that m can be canceled from each term of the equation. Then we have

$$\tfrac{1}{2}v_B{}^2 + \tfrac{1}{4}v_B{}^2 - gh = 0 \rightarrow \tfrac{3}{4}v_B{}^2 = gh$$

Solving for the speed of the disk, we find that

$$v_B = \sqrt{\frac{4gh}{3}}$$

Notice that v_B is the same for disks of any mass and radius. It is of interest to recall that the speed of an object that slides without friction down such an incline is $\sqrt{2gh}$. This is larger than the speed of the rolling disk. The rolling object is moving less fast because some of the original potential energy is changed into kinetic energy of rotation. Its translational kinetic energy is less than it would be without rotation. ■■

■ **EXAMPLE 10.6** The mass m in Figure 10.8 is released from rest. As it falls, it causes the wheel (moment of inertia I) to turn. Ignoring friction, how fast will the wheel be turning after the mass has fallen a distance h?

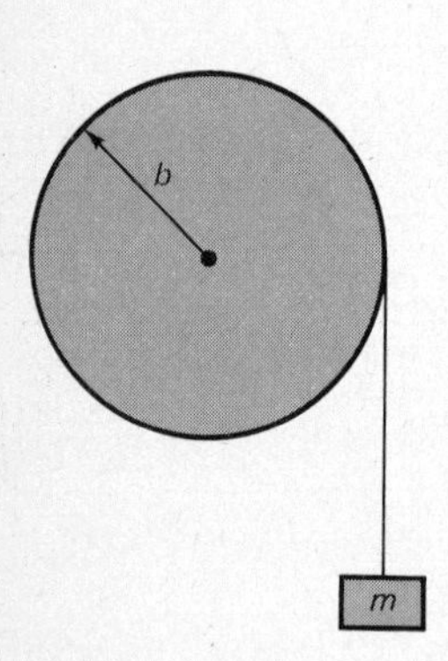

Figure 10.8 How fast is the wheel rotating after the mass has fallen a distance h?

Solution We shall make use of the work-energy theorem. Because no outside work is done, we have for the wheel and mass (using subscripts 0 and f for the initial and final situations, respectively)

$$[(KE_t + KE_r)_f - (KE_t + KE_r)_0] + [mgh_f - mgh_0] = 0$$

where KE_t is for the falling mass and KE_r is for the wheel. Placing in the appropriate values gives

$$[(\tfrac{1}{2}mv^2 + \tfrac{1}{2}I\omega^2) - 0] + [0 - mgh] = 0$$

Since $v = \omega r$, and we want to find ω, we have

$$\tfrac{1}{2}m(\omega b)^2 + \tfrac{1}{2}I\omega^2 = mgh \rightarrow \tfrac{1}{2}\omega^2(mb^2 + I) = mgh$$

Solving for ω^2 gives

$$\omega^2 = \frac{2mgh}{mb^2 + I}$$

from which

$$\omega = \sqrt{\frac{2mgh}{mb^2 + I}}$$

■■

10.5 Analogies Between Linear and Rotational Motion

In Chapter 9 we pointed out several analogies between linear and angular motion. For example, the five uniform motion equations carry over to angular motion with x replaced by θ, v by ω, and a by α. Once we recognize these correspondences, it becomes much easier to remember the equations for angular, or rotational, motion.

We have learned in this chapter that there are several other correspondences.

LINEAR	ROTATION
Work = $F_s s$	Work = $\tau\theta$
$F = ma$	$\tau = I\alpha$
$\text{KE}_t = \frac{1}{2}mv^2$	$\text{KE}_r = \frac{1}{2}I\omega^2$

m corresponds to I
F corresponds to τ

Recognizing these correspondences helps us to remember the rotational equations.

These correspondences can be carried over to other equations we learned for linear motion. For example, the relation for power in terms of speed is

$$\text{Power} = F_x v_x \qquad\qquad \text{Power} = \tau\omega$$

In the next section we will see yet another correspondence between linear and rotational motion.

■ **EXAMPLE 10.7** A belt passes over a 4.0-cm-radius pulley as shown in Figure 10.9. The tension is 200 N in the tight section of the belt and is negligible in the other section. If the pulley is driven at a rate of 180 rev/min by the belt, what is the average power delivered to the pulley?

Solution We shall make use of the relation

$$\text{Power} = \tau\omega$$

The torque the belt exerts on the pulley is

$$\tau = (200\ \text{N})(0.040\ \text{m}) = 8.0\ \text{N}\cdot\text{m}$$

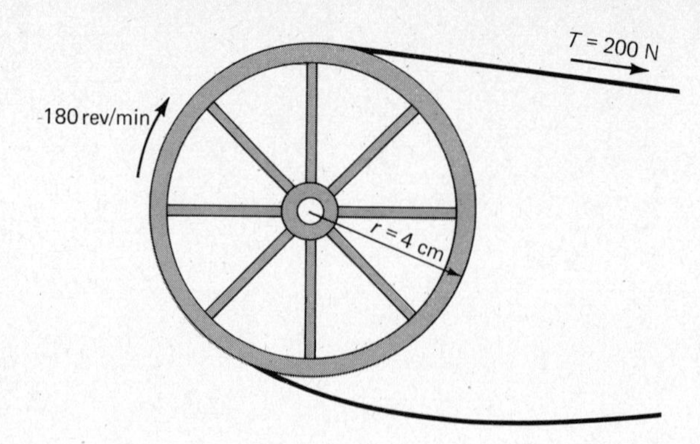

Figure 10.9 The pulley is being driven at 180 rev/min by the belt.

The angular speed in radians per second is

$$\omega = (180 \text{ rev/min})\left(\frac{2\pi \text{ rad}}{1 \text{ rev}}\right)\left(\frac{1 \text{ min}}{60 \text{ s}}\right)$$
$$= 18.8 \text{ rad/s}$$

Therefore we find that

$$\text{Power} = (8 \text{ N} \cdot \text{m})(18.8/\text{s}) = 150 \text{ N} \cdot \text{m/s} = 150 \text{ W}$$

Because 1 hp = 746 W, this is equivalent to a power of 0.20 hp. ■■

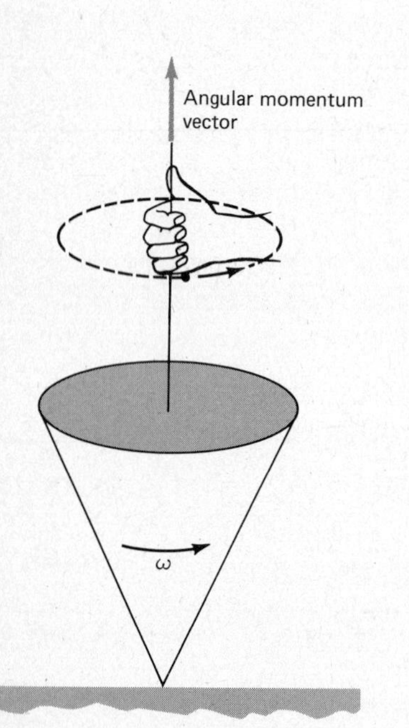

Figure 10.10 The right-hand rule: Grasp the axis with the right hand with the fingers circling in the direction of rotation; the thumb then points along the axis in the direction of the vector.

10.6 Angular Momentum

We define angular momentum in exactly the way you might expect. You will recall that linear momentum was simply mv. As usual, we replace m by I and v by ω to have

$$\text{Angular momentum} = I\omega \qquad (10.6)$$

Although we ignored that fact in the last chapter, like linear momentum, angular momentum is given direction. It is a vector. The direction of the vector is taken parallel to the axis of rotation. This is shown in Figure 10.10.

One of the most important properties of angular momentum is that it obeys a conservation law. The law may be stated as follows.

Consider a system (or object) upon which no external unbalanced torque acts. The angular momentum of the system (or object) will not change either in magnitude or direction.

This is the law of conservation of angular momentum. We can use the law to explain several interesting features of rotating objects.

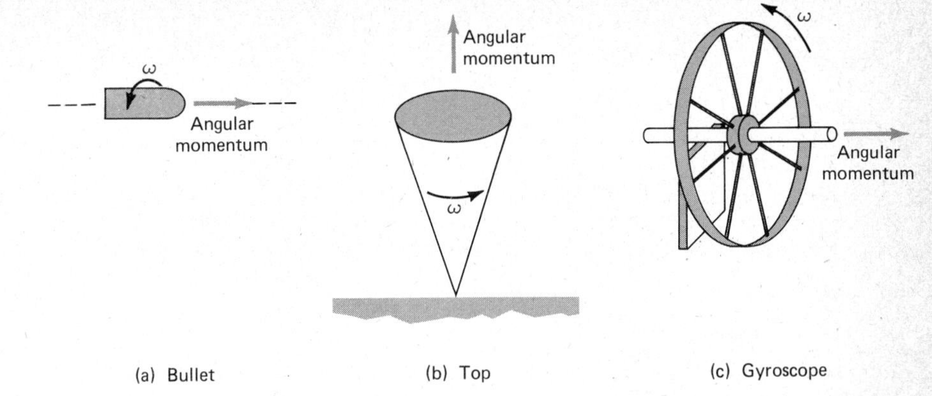

Figure 10.11 The rapid spinning of these objects causes them to remain oriented as shown.

For example, a rapidly spinning projectile will resist wobble. The bullet shown in Figure 10.11(a) spins rapidly as indicated. If it were to wobble, the direction of the angular momentum vector would change. But because no appreciable torque acts on the bullet to reorient the angular momentum, the bullet must remain oriented as shown. A spinning football also behaves in this way.

In part (b) of the figure we see a rapidly spinning top. It will not wobble and fall until its spin rate has become small. The angular momentum vector resists being reoriented.

Part (c) of Figure 10.11 shows a simple gyroscope. If the wheel were not spinning rapidly, the wheel would topple over from the support. Instead, it remains with its axis horizontal, although gravity does cause it to disorient slowly. The exact motion of a gyroscope (and most spinning objects) is not usually simple to describe. Even so, the tendency for the angular momentum vector to resist changes in orientation is obvious in most cases.

■ EXAMPLE 10.8 Stars such as our sun slowly contract over billions of years as their internal nuclear fuel is depleted. Gravity is responsible for the contraction. Suppose that a uniform sphere of mass M originally rotates with angular speed ω_0 on its axis. As time goes on, its radius decreases from r_0 to r_f. Find its new angular speed, ω_f.

Solution We know that angular momentum, $I\omega$, for the star must be conserved. Therefore, the final and initial angular momenta must be equal. We have

$$I_0\omega_0 = I_f\omega_f$$

But I for a uniform sphere is $2mr^2/5$ (see Figure 10.3). So this equation becomes

$$\left(\frac{2Mr_0^2}{5}\right)\omega_0 = \left(\frac{2Mr_f^2}{5}\right)\omega_f$$

From this we find

$$\omega_f = \omega_0\left(\frac{r_0}{r_f}\right)^2$$

The star rotates faster as it collapses. ■■

Figure 10.12 Why does the skater spin faster in (b) than in (a)?

■ **EXAMPLE 10.9** Ice skaters often do the trick shown in Figure 10.12. As the skater draws her body into a thin object close to the rotation axis, she spins faster. If her moment of inertia is originally I_0 and finally I_f, find ω_f in terms of ω_0.

Solution Only a negligible torque is applied to the girl by the ice skate. Therefore, angular momentum is conserved. We have

$$I_0\omega_0 = I_f\omega_f$$

Solving, we find the desired result

$$\omega_f = \omega_0\left(\frac{I_0}{I_f}\right)$$

Notice that the average radius for the mass of her body is less in (b) than in (a). Therefore, I_f is smaller than I_0. Because I_f is much smaller than I_0, the final rotation speed is much larger than ω_0. ■■

Summary

Objects that can rotate have rotational inertia. They require a torque to set them in rotation. They require a torque to slow their rotation.

Rotational inertia depends on two factors: the mass of the object and the distance of the mass from the axis of rotation. It is measured by the moment of inertia of the object I. Suppose the object to be composed of tiny masses, m_1, $m_2, \ldots, m_N$. The distance of each mass from the rotation axis is $r_1, r_2, \ldots, r_N$. Then

$$I = m_1r_1^2 + m_2r_2^2 + \cdots + m_Nr_N^2$$

The moment of inertia can be written as

$$I = Mk^2$$

where M is the total mass of the object. The radius of gyration k is a measure of the average radius of the mass from the axis.

The unbalanced torque τ needed to give an angular acceleration α to an object of moment of inertia I is

$$\tau = I\alpha$$

In using this, α must be in radians per second per second. Notice the similarity between this expression and its linear analogy $F = ma$.

An object rotating about an axis with angular speed ω has rotational kinetic energy, KE_r.

$$KE_r = \tfrac{1}{2}I\omega^2$$

where ω must be in radians per second. Notice the analogy between this and translational kinetic energy, $KE_t = \frac{1}{2}mv^2$.

An object of mass M that is rotating and translating at the same time has both KE_r and KE_t. If KE_r is taken for an axis through the mass center, then the total kinetic energy is $KE_r + \frac{1}{2}Mv^2$, where v is the speed of the mass center.

The work-energy theorem of mechanics can be applied to rotating objects. In so doing, the kinetic energy term must include both the rotational and translational kinetic energies.

The work done by a torque τ in giving an angular displacement θ is

$$\text{Work} = \tau\theta$$

where θ must be in radians.

The magnitude of the angular momentum of an object is given by $I\omega$. Angular momentum is a vector. It is directed along the axis of rotation in the direction given by the right-hand rule.

The law of conservation of angular momentum is as follows: Consider a system (or object) upon which no external unbalanced torque acts. The angular momentum of the system (or object) will not change either in magnitude or direction.

Questions and Exercises

1. The following all have the same mass and same radius from the axis to the outermost point. Compare their moments of inertia about an axis through the center of each: hoop, solid sphere, hollow sphere, solid thin cylindrical rod, disk.
2. Large, massive flywheels are sometimes used to smooth the motion imparted by one- or two-cylinder engines. Explain the purpose of the flywheel.
3. One proposal for storing energy is by means of a massive,

fast-rotating flywheel. Discuss its pros and cons as (a) an automobile power source and (b) a means for the power company to equalize its power production during various hours of the day.

4. Suppose that you wish to design a rotating device that is as close to freely rotating as possible. What considerations will influence your choice of axle and support of the axle? Why is an axle tapered to a pinpoint at the support sometimes used? If a cylindrical bearing is to be used, what effect will shaft diameter have?
5. A mass m is hung from a string wound on the rim of a wheel. As the mass falls, the tension in the string is *not* mg. Why not? If the wheel has a very large I, the tension will be close to mg. Why?
6. Two spheres have the same mass and radius. One is solid wood, whereas the other is aluminum and hollow. If they are both spinning with the same speed on an axis through their centers, compare their rotational kinetic energies. Which would roll down an incline fastest?
7. The spool shown in Figure P10.1 takes off with greatly increased speed when it reaches the bottom of the incline. Why? From where does its increased KE_t come?

Figure P10.1

8. Hamsters are often placed in exercise cages such as the one shown in Figure P10.2. Assume that the bearing at the rotation axis has negligible friction. Analyze the motion of the hamster and cage from the standpoints of (a) action and reaction and (b) conservation of angular momentum. Does gravity have any importance in this situation?

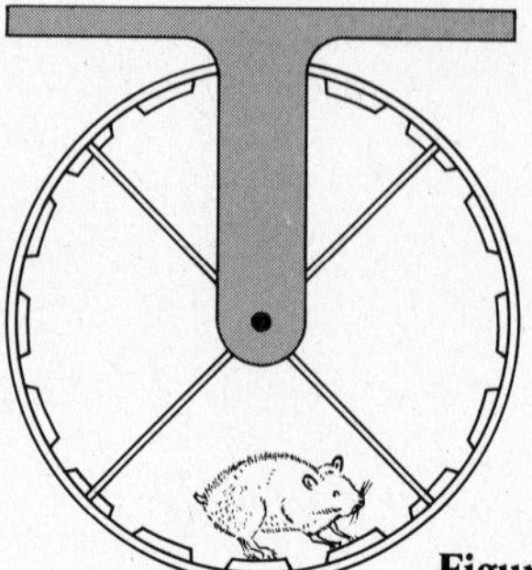

Figure P10.2

9. An earth satellite circles the earth in a highly elliptic orbit. At its closest point (called *perigee*) it is moving faster than at its farthest point (*apogee*). Use the law of conservation of angular momentum to explain this behavior. What is I for the satellite about an axis through the earth's center? Compare ω at perigee and apogee.
10. A do-it-yourselfer builds a helicopter with a single propeller on a vertical axis. In its maiden flight, the operator becomes sick because the whole helicopter tends to spin about a vertical axis. What went wrong? How do more experienced designers get around this problem?
11. Any one of us can influence the motion of the earth. Do you speed up or slow down the rotation of the earth as you begin to run east on it? Can you give an example of a situation in which an effect much like this is of more importance?
12. In Figure P10.3 we see a young man on a stool that rotates about a vertical axis. He has weights in his hands. With his arms outstretched he rotates slowly. When he pulls his arms in, he rotates much faster. Explain.

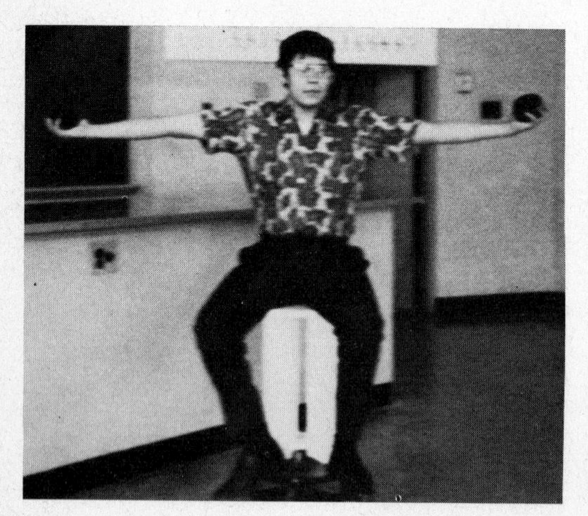

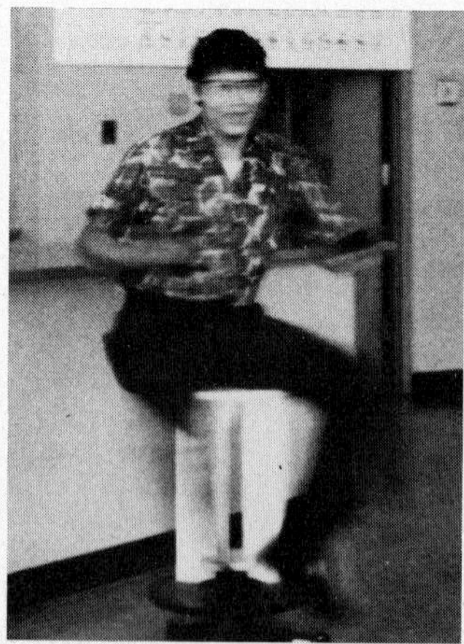

Figure P10.3

Problems

1. A 500-g hoop has a diameter of 60 cm. (a) What is its moment of inertia about an axis through its center? (b) What is its radius of gyration? (c) Repeat for a 1.5-lb hoop with a 2-ft diameter.
2. A uniform, solid 800-g sphere has a radius of 5.0 cm. (a) What is its moment of inertia about an axis through its center? (b) What is its radius of gyration? (c) Repeat for a 0.8-lb sphere with a 2-in. radius.
3. Find the moment of inertia of the device in Figure P10.4 about the axis shown. (a) Assume a = 20 cm and each ball has a mass of 50 g. (b) Repeat for a = 0.80 ft with each ball weighing 0.25 lb.

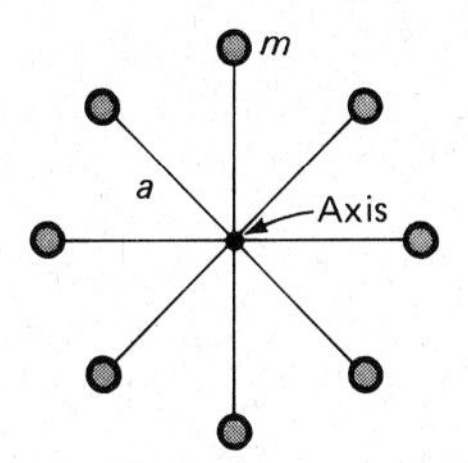

Figure P10.4

4. Find the moment of inertia of the device in Figure P10.5 about the axis shown. The rods connecting the balls to the axis have negligible mass. (a) Assume m = 500 g and a = 40 cm. (b) Repeat for a device in which each ball weighs 0.40 lb and a = 24 in.

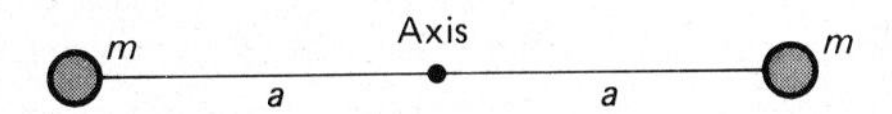

Figure P10.5

**5. A solid, disklike wheel has a moment of inertia of I_d. A rim of radius 80 cm and mass 7.0 kg is slipped onto the wheel. Show that the moment of inertia of the combination is $(I_d + 4.5)\ \text{kg}\cdot\text{m}^2$.

**6. Two wheels have moments of inertia I_1 and I_2. They are fastened together on the same axis so as to make a single wheel. Starting from the definition of I, prove that the moment of inertia of the combined wheel is $I_1 + I_2$.

7. A small merry-go-round has a moment of inertia of 10,000 $\text{kg}\cdot\text{m}^2$. How large a torque is required to give it an acceleration of (a) 0.8 rad/s^2 and (b) 0.50 rev/s^2?
8. (a) A grinding wheel has $I = 50 \times 10^{-4}\ \text{kg}\cdot\text{m}^2$. It coasts to rest with a deceleration of 20 rev/s^2. How large a friction torque acts to stop it?

*9. (a) A 10.0-N force is applied tangent to the rim of a 20-cm-radius wheel. How large a torque is exerted about an axis through the center of the wheel? (b) If the wheel accelerates from rest to a speed of 7.0 rev/s in 16.0 s, what is the moment of inertia for the wheel?

*10. The friction torque between a bearing and the 1.50-cm-radius shaft, on which the wheel turns, slows the wheel. It takes the wheel 25 s to slow to rest from an initial speed of 3.5 rev/s. The moment of inertia of the wheel is 0.80 $\text{kg}\cdot\text{m}^2$. Find the value of (a) the friction torque and (b) the friction force that acts at the bearing.

*11. A small clock motor has a maximum output torque of $4.0 \times 10^{-3}\ \text{N}\cdot\text{m}$. It is used to accelerate a uniform disk (m = 2 kg, r = 15 cm). (a) How large an angular acceleration can it give the disk? (b) How long would it take to bring the disk up to its operating speed of 1 rev/min?

*12. A phonograph turntable coasts from $33\frac{1}{3}$ rpm to rest in 12 s after the motor is turned off. If $I = 1200\ \text{kg}\cdot\text{cm}^2$ for the turntable, how large is the friction torque that slows it?

*13. An 80-kg wheel with a 75-cm radius is to be accelerated from rest to a speed of 150 rev/min in 25 s by a belt on its rim. If the wheel has a radius of gyration of 50 cm, how large must be the net force furnished to the wheel by the belt?

*14. The 0.80-kg disk on a grinding wheel has a radius of 15 cm. It is mounted on the axle of a motor whose operating speed is 2100 rev/min. Assume the wheel to be a uniform disk. (a) What is its moment of inertia? (b) What is its radius of gyration? (c) How large a torque is needed to get it up to speed in 3.0 s?

*15. A 150-g meter stick is pivoted at one end. It is held horizontal and released. (a) Find its angular acceleration in radians per second per second just after its release. (b) Repeat for when it passes through its lowest position.

*16. A pendulum 50 cm long consists of a small ball of mass 20 g at the end of a string. (a) What is the moment of inertia of the pendulum about its support? (b) The pendulum is pulled aside until it makes a 37° angle to the vertical. It is then released. What is its angular acceleration in radians per second per second just after release?

*17. A 50-g mass hangs from a cord wound on the rim of an 80-cm-diameter wheel and causes the wheel to rotate. If the mass drops with an acceleration of 400 cm/s^2, what is (a) the

*Problems marked with an asterisk are not as easy as the unmarked ones.
**Problems marked with a double asterisk are somewhat more difficult than the average.

moment of inertia of the wheel and (b) the tension in the cord?

*18. A solid metal disk has a mass of 30 kg and a radius of 25 cm. It is to be rotated on an axle by a mass m hanging from a string wound on its rim. How large should the mass be if it is to accelerate the wheel at 0.40 rev/s^2?

**19. In Figure P10.6 $m_1 = 800$ g, $m_2 = 900$ g, and $a = 12$ cm. The pulley has a moment of inertia of 0.020 kg · m^2 and rotates on a frictionless axle. When the system is released, with what acceleration will the larger mass fall? (Assume that the cord does not slip on the pulley.)

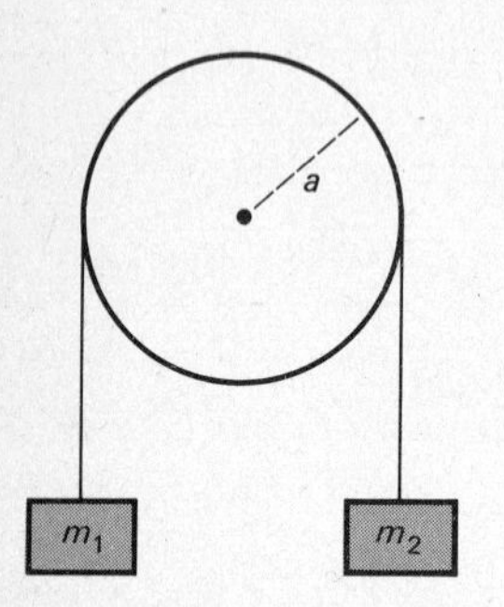

Figure P10.6

20. What is the rotational KE for a wheel that is rotating at 2.0 rev/s if I for the wheel is 0.50 kg · m^2?

21. What is the rotational KE of a uniform disk that is rotating at 800 rev/min if the disk has a radius of 6.0 cm and $m =$ 400 g?

22. A 50-cm-diameter hoop with 2-kg mass is rolling along the floor at 3 m/s. (a) How fast is it rotating? (b) What is its moment of inertia? (c) How much rotational KE does it have? (d) How much translational KE does it have?

23. Find the rotational, translational, and total kinetic energy of a 7.0-kg bowling ball rolling along the floor at 8.0 m/s. The ball has a diameter of 18 cm.

24. A 90-kg drum full of oil is rolling along the floor at 60 cm/s. The diameter of the drum is 70 cm. Find the rotational, translational, and total kinetic energy of the drum.

*25. A bowling ball is rolling along the floor at 3.0 m/s. It is to roll up an incline and stop just as it reaches the top of the incline. How high should the top be above the floor?

*26. A hoop rolls down an incline without slipping. What is its translational speed when it is a vertical distance of 2.0 m below the point where it started from rest?

*27. Repeat Problem 26 for a uniform solid sphere.

*28. A heavy drum of oil is to roll along the floor onto a ramp and up to a platform 70 cm above the floor. How fast (in meters per second) must the drum be rolling if it is to stop just as it arrives on the platform?

*29. A 10-kg mass hangs from the rim of a wheel with radius of gyration of 20 cm and 14.0-kg mass. The radius of the wheel is 30 cm. Find the rate at which the wheel is turning (in revolutions per second) after the weight has fallen 1.2 m. Solution is quickest using the energy method.

**30. Suppose in Figure P10.6 that the wheel has $I =$ 0.020 kg · m^2 and $a = 12$ cm. If $m_1 = 800$ g and $m_2 =$ 900 g, how fast will m_2 be moving when it has fallen 2 m from the position where the system started from rest? Use energy methods.

*31. The rotating part of a certain motor (the rotor) has a mass of 0.90 kg and has a radius of gyration of 2.2 cm. When running without load, the motor has a speed of 3200 rev/min. After being turned off, it stops after rotating through 400 rev. How large a friction torque opposes its motion? Use the work-energy method.

*32. A certain $\frac{1}{4}$-hp motor (based on output) normally operates at 2100 rev/min under load. (a) How large a torque (in newton-meters) does the motor develop? (b) If it drives a belt on an 8.0-cm-diameter pulley, what is the difference in tensions of the two portions of the belt when the motor is exerting this torque?

*33. A certain motor is capable of an output power of 150 W at a rotation rate of 20 rev/s. When operating in this way, with what maximum force can it pull on a belt that runs over a 2-cm-radius pulley on its shaft?

*34. By use of a braking device, it is found that a certain motor develops a torque of 0.90 N · m. It operates at a speed of 3200 rev/min. (a) What is the output horsepower of the motor under these conditions? (b) If the electrical input power to the motor is 500 W, what is the motor's efficiency?

35. What is the angular momentum of a uniform disk that is rotating with a speed of 20 rev/s if $r = 5.0$ cm and $m =$ 1.5 kg?

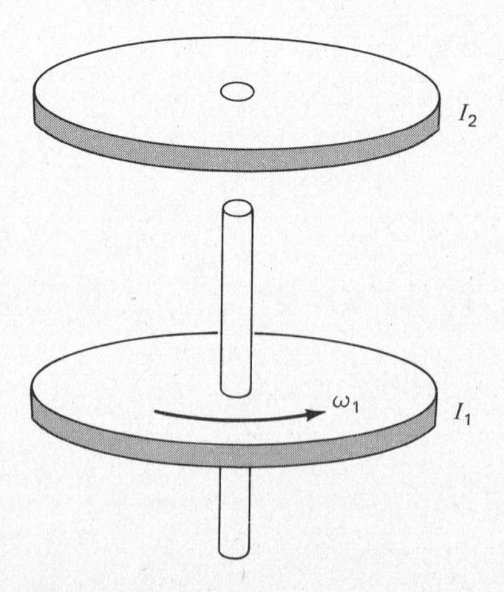

Figure P10.7

*36. Figure P10.7 shows a disk and shaft (moment of inertia = I_1) coasting with angular velocity ω_1. A nonrotating disk of moment of inertia I_2 is dropped onto the first disk. (a) Find the angular velocity of the combination after the disks couple together. (b) Repeat for the case in which the dropped disk has an angular speed ω_2 in the same direction as ω_1. (c) Repeat for the case in which the rotation directions are opposite. (See Problem 6.)

37. Suppose the earth suddenly contracted so that its moment of inertia became one-third its present value. How long would the present 24-h day become?

**38. The 30-g block shown in Figure P10.8 revolves in a circle on a frictionless table. It is held by a cord that passes through a tiny hole at the center of the circle. The angular speed of the block is 2.00 rev/s, while $r = 60$ cm. (a) How large is the force F? (b) If the string is pulled down a distance of 15 cm, what will be the new angular speed of the block? (c) How much work did the force F do to shorten the circle radius to 45 cm? Assume the block is quite small compared to the radius of the circle.

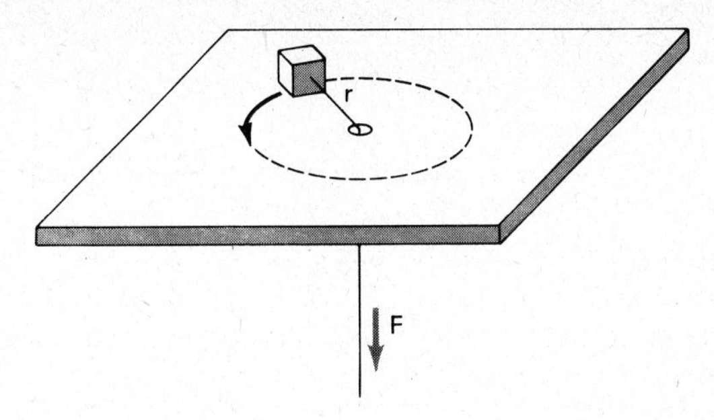

Figure P10.8

Diamonds made in the laboratory.

General Electric Company

11 PROPERTIES OF MATERIALS

Performance Goals

When you finish this chapter, you should be able to

1. From a description of several materials, tell which are solids, liquids, or gases. Distinguish between crystalline and amorphous solids in terms of atom arrangement. State the difference between the terms "fluid" and "liquid."
2. Given a graph showing distortion versus distorting force, state in which region of the

Materials can be divided into three basic groups: solids, liquids, and gases. In this chapter we shall learn some of the basic properties of each of these groups. We shall see how to measure and characterize the way materials behave. Hooke's law and the effects of distorting forces will be discussed. In addition we shall learn about the concepts of pressure and buoyancy.

graph the distortion obeys Hooke's law.

3. State Hooke's law in words. Write an equation involving stress and strain that is a statement of Hooke's law.
4. Do the following for tensile, shear, and bulk compression types of distortion. Draw a picture of a typical experiment; define the stress in relation to the picture; define the strain in relation to the picture; define the modulus.
5. Define the following terms in relation to a stress-strain experiment: elastic, elastic limit, ultimate strength.
6. State how stress, strain, and modulus are related. Explain how one could measure the modulus in tensile elongation (or compression), simple shear, bulk compression.
7. Find the compressibility of a material if its bulk modulus is given.
8. Compute the pressure exerted on a given area by a known force. Knowing the pressure in a fluid at rest, compute the force exerted by the fluid on a flat surface placed in the fluid.
9. Given two of the following quantities for an object, compute the third: mass, volume, mass density. Repeat for weight, volume, weight density. Given mass density or weight density, find the other.
10. Compute the pressure due to a fluid at a given depth in terms of both the weight and mass density.
11. Given a problem involving the balancing of a column of liquid in a U tube by another column

11.1 Classification of Materials

It is customary to divide materials into three groups: solids, liquids, and gases. All of us know in a general way what is meant by each of these. A solid object maintains its shape when sitting on a table. A liquid settles to the bottom of its container. Its shape is determined by the container. A gas fills the whole container in which it is placed. Unlike a liquid, a gas does not settle to the container bottom.

In spite of these statements, the divisions between these groups are not always clear. Some solid materials actually flow slowly over a period of many years. Glass and many plastics are of this type. For example, glass windows in ancient buildings often show evidence of flow. The glass is usually thicker at the bottom of the window than at the top. Because of this uncertainty in classification, the terms "solid," "liquid," and "gas" are not completely precise.

We usually divide solids into two classes. They are crystalline and noncrystalline solids. In *crystalline solids* the atoms are precisely arranged. For example, Figure 11.1(a) shows the arrangement of the sodium and chloride atoms in NaCl (sodium chloride, common table salt). Notice that the atom centers are arranged in a cubical pattern. The geometrical pattern in which the atoms are arranged is called a *crystal lattice.* Sodium chloride crystallizes in a cubic lattice. Other compounds and elements crystallize in different lattices. Often the crystals of a substance give a strong hint as to the crystal's lattice. For example, the calcite crystal in Figure 11.1(c) gives us a clue concerning its lattice. The actual lattice is illustrated in part (b) of the figure. See also the regular shape of the diamond crystals pictured on the previous page.

Glass and many plastics (polystyrene and Lucite, for example) are typical noncrystalline solids. We say that they are *amorphous solids.* In these materials no long-range pattern exists for the atom positions. This fact influences many mechanical and thermal properties of the solid. It is often important in technology to know whether a solid is crystalline or amorphous. We shall see later that X rays can be used to determine whether or not a solid is crystalline.

Both gases and liquids are able to flow. For this reason they are often grouped together and called *fluids.* Because fluids do not have a rigid shape, they respond to forces much differently than solids do. We shall examine the effect of forces on fluids later in this chapter.

11.2 Distortion of Solids: Hooke's Law

Suppose that a stretching force F is applied to a rod as shown in Figure 11.2. We call a stretching force such as this a *tensile* force. Because of it, the rod elongates an amount ΔL. A graph showing ΔL as a function of F is shown in Figure 11.3. It is often found that the stretch, ΔL, is proportional to the stretching force, F. This behavior is shown as the solid portion of the curve. In this range of forces the rod will retract to its original length when the force is removed. The material of the rod is said to be *elastic.* An elastic material will return to its original condition when a distorting force is removed.

of a different liquid, find the ratio of the two heights in terms of the densities of the liquids.

12. Draw a diagram of a manometer measuring a pressure unequal to atmospheric pressure. Explain how the manometer readings can be used to find the gauge pressure. State what further data must be known to find the absolute pressure.
13. Draw a diagram of a mercury barometer and explain how it is used to obtain atmospheric pressure. State which it measures, gauge pressure or absolute pressure.
14. Given a pressure in any one of the following units, transform it to the other units listed: newtons per square meter, pascals, centimeters of Hg, millimeters of Hg, atmospheres, torrs, bars. Repeat for pounds per square inch, pounds per square foot, atmospheres.
15. Draw a diagram and use it to explain the basic idea of the aneroid-type gauge. Repeat for the Bourdon-type gauge.
16. State Pascal's principle in your own words. Explain its meaning in relation to an experiment of your own choice.
17. State Archimedes' principle in your own words. Given an object of known volume totally submerged in a liquid of known density, compute the buoyant force on the object.
18. Find the volume of an object in terms of its apparent weight in a liquid of known density. Assume that the object does not float.

If you stretch a material too far, the material will yield or perhaps break. The dotted portion of the curve in Figure 11.3 shows this type of behavior. Notice what happens if forces larger than that at the elastic limit point are applied to the rod. The graph in Figure 11.3 shows that the material stretches more easily beyond this point. When the rod is stretched beyond the elastic limit, it usually will not return to its original length after the force is removed.

The behavior shown by the solid line in Figure 11.3 is of great importance. As we noted, the stretch (or distortion) is proportional to the stretching (or distorting) force. If the force is doubled, the distortion doubles. If the force is tripled, the distortion triples. And so on. This is a common type of behavior.

One of the first people to examine the effects of distorting forces was Robert Hooke. He noticed that most things distort in proportion to the applied force. Springs, for example, behave in this way. Even a hair from your head stretches in proportion to the stretching force. Hooke went even further. He found that the same rule applies to the bending and twisting of rods. If a rod is distorted by bending it or twisting it, the distortion is proportional to the distorting force.

Hooke was led to state what is now called *Hooke's law.* It can be stated in words as follows.

When an object is distorted, the distortion is proportional to the distorting force.

We know, however, that Hooke's law is not always obeyed. As we saw in Figure 11.3, a rod does not always stretch in proportion to the force. If the force and elongation are too large, then the F versus ΔL graph is no longer linear. This is typical of most types of distortion. We can say:

Hooke's law is usually obeyed only at small distortions.

In Section 11.4 we shall see how Hooke's law is stated in mathematical form.

11.3 Stress and Strain

Distortions and the distorting forces that cause them are important in many practical applications. They are of primary concern whenever heavy loads must be held. In technical applications we must place numerical values on distortions and the forces that cause them. For this purpose we define and use the two quantities, *stress* and *strain.*

We shall always define stress in the following way. Stress is a measure of force in relation to the area over which it is applied. If a distorting force F is applied to an area A, then

$$\text{Stress} = \frac{\text{distorting force}}{\text{area}} = \frac{F}{A} \qquad (11.1)$$

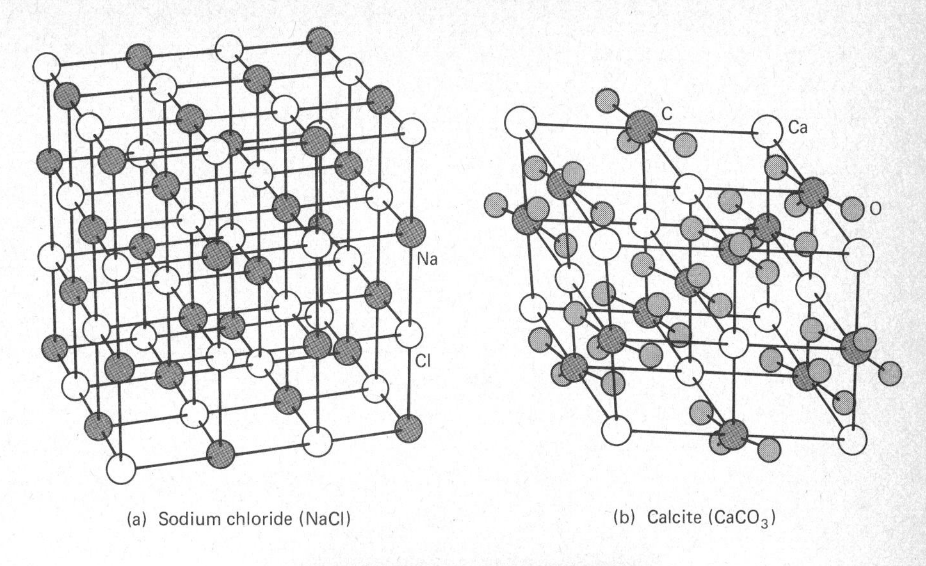

Figure 11.1 Crystalline solids have atoms arranged in a lattice.

For example, in Figure 11.4 the stress applied to the rod is F/A. From its definition, the units of stress are always those of force divided by area. The SI unit for stress is newtons per square meter (N/m^2) and this unit is called the pascal (Pa).

$$1 \text{ Pa} = 1 \text{ N/m}^2$$
$$1 \text{ kPa (kilopascal)} = 10^3 \text{ Pa}$$
$$1 \text{ GPa (gigapascal)} = 10^9 \text{ Pa}$$

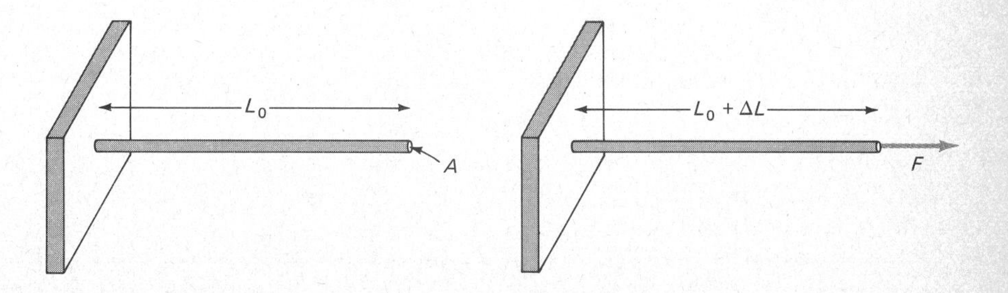

Figure 11.2 The force F causes the rod to stretch an amount ΔL.

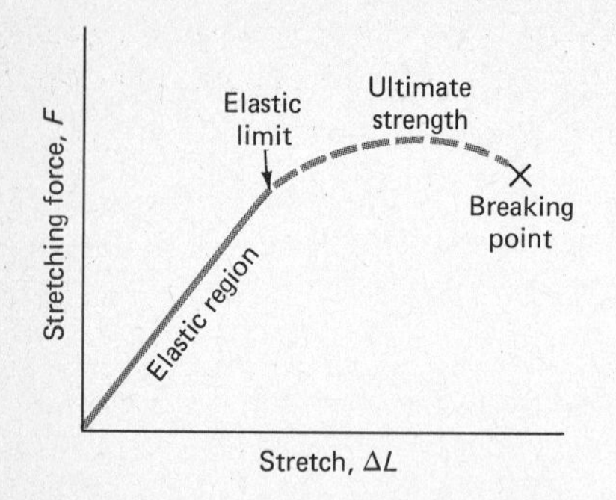

Figure 11.3 At low stretching forces, $F =$ (constant) ΔL. The material obeys Hooke's law.

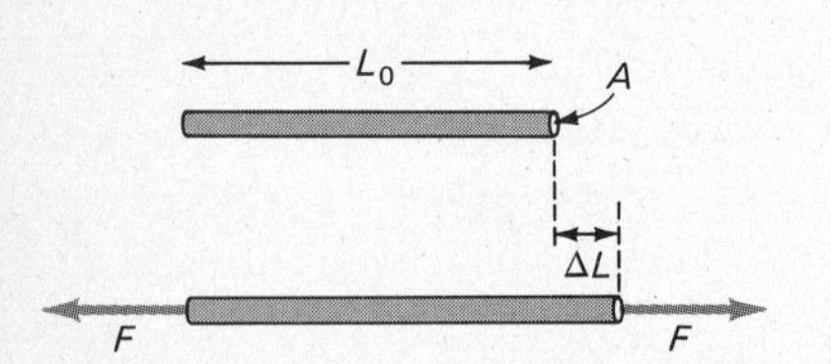

Figure 11.4 The stress is F/A, whereas the strain is $\Delta L/L_0$. What are the units of each?

Strain is a measure of deformation. We do not take it equal to ΔL in Figure 11.4. The reason for this is that ΔL depends on the original length L_0 of the bar. A bar twice as long would stretch twice as much under the same load. To eliminate this effect, we take the strain equal to $\Delta L/L_0$ in this case. In the more general case, we define strain in the following way.

$$\text{Strain} = \frac{\text{distortion}}{\text{undistorted size}} \tag{11.2a}$$

As we stated, for the rod shown in Figure 11.4, we have

$$\text{Strain} = \frac{\Delta L}{L_0} \tag{11.2b}$$

From its definition, we see that strain has no units. It is simply a ratio.

11.4 Hooke's Law and Moduli of Elasticity

If a material obeys Hooke's law, then the distortion is proportional to the distorting force. Let us describe the distortion by the strain and the distorting force by the stress. Then Hooke's law can be written as

Stress is proportional to strain.

$$\textbf{Stress} \sim \textbf{strain} \tag{11.3}$$

This can be made into an equation by using a proportionality constant. The constant will depend on the experiment we are performing and the material being used. We call this constant the *modulus of elasticity.* In terms of it, Hooke's law is

$$\text{Stress} = (\text{modulus of elasticity}) \cdot (\text{strain}) \tag{11.4a}$$

Or,

$$\text{Modulus of elasticity} = \frac{\text{stress}}{\text{strain}} \tag{11.4b}$$

The modulus measures how large a stress is required to produce one unit of strain. Because strain has no units, the modulus has the same units as stress. Its units are those of force per unit area. Let us now discuss several types of moduli.

Tensile (or Young's) Modulus, Y This modulus of elasticity applies when materials are being stretched along a straight line. The stretching experiment shown in Figure 11.4 is typical. In that type of distortion we have

$$\text{Strain} = \frac{\Delta L}{L_0}$$

and

$$\text{Stress} = \frac{F}{A}$$

Notice that A is the cross-sectional area of the bar. The modulus in this type of distortion is

$$\text{Young's modulus} = Y = \frac{\text{stress}}{\text{strain}}$$

or

$$Y = \frac{(F/A)}{(\Delta L/L_0)} = \frac{FL_0}{A\,\Delta L} \tag{11.5}$$

This equation also applies to small compressions. In that case the direction of F in Figure 11.4 is reversed. Typical values for Young's modulus are given in Table 11.1.

Shear Modulus, S (Modulus of Rigidity) The distortion referred to as *shear* is shown in Figure 11.5(a). Notice that this is quite different from a tensile distortion. In computing the stress we use F/A, where A is shown in the figure. Notice that F is parallel to the surface in this case. The distorting force tries to slide the area A along. This is unlike the tensile case, where the force tries to pull the area out.

Because the strain is a measure of distortion, we take it to be equal to $\Delta L/L_0$. These quantities are also shown in the figure. If the distortion is small, then it is shown in trigonometry that $\Delta L/L_0 = \phi$, where ϕ is the angle shown measured in radians. We therefore have, for the shear modulus,

$$\text{Shear modulus} = S = \frac{\text{stress}}{\text{strain}}$$

or

$$S = \frac{F/A}{\Delta L/L_0} = \frac{F/A}{\phi} \tag{11.6a}$$

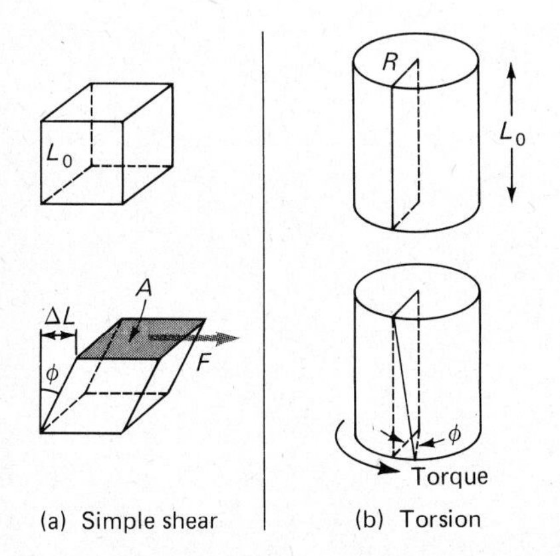

Figure 11.5 The shear modulus $S = F/A\phi$. The distortion shown in (b) also involves shear, but not in a simple way.

which gives

$$S = \frac{F}{A\phi} \tag{11.6b}$$

Of course, ϕ must be measured in radians. Typical shear moduli are given in Table 11.1.

A torsion-type distortion is shown in part (b) of Figure 11.5. It is not easy to relate stress due to the twisting torque τ and the strain in this case. However, by use of calculus one can show that

$$S = \frac{2\tau L_0}{\pi R^4 \phi} \tag{11.7}$$

It is usually easier to measure the shear modulus in torsion than in pure shear.

Bulk Modulus, *B* (Compressibility, *k*) The third important type of modulus applies to both solids and liquids. It involves the compression of a volume. Suppose that we have a cube of material as shown in Figure 11.6(a). The area of each face is A. The volume of the cube will be taken to be V_0. If a force F is applied to each face, then the volume will be compressed an amount ΔV. But ΔV is negative because the cube shrinks. We therefore take the strain to be $-(\Delta V/V_0)$. As usual, the stress is F/A. We then have

$$\text{Bulk modulus} = B = \frac{\text{stress}}{\text{strain}}$$

or

$$B = -\frac{(F/A)}{(\Delta V/V_0)} \tag{11.8}$$

The bulk modulus is a measure of how large a force is needed to compress a substance. Typical values are given in Table 11.1. Notice that liquids are easier

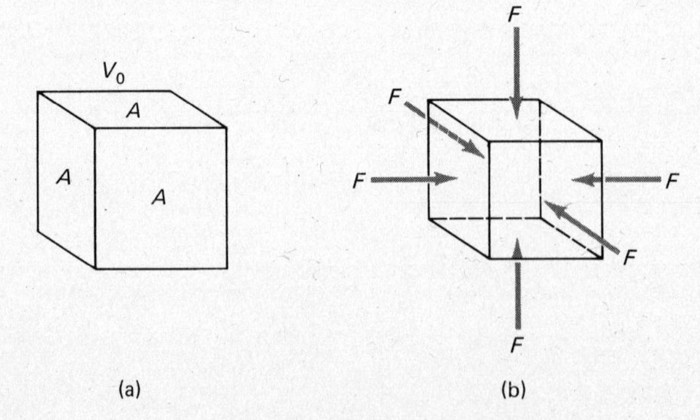

Figure 11.6 The bulk modulus B is a measure of force required to compress a material; $B = -(F/A)/(\Delta V/V_0)$.

Table 11.1 APPROXIMATE ELASTIC PROPERTIES (in GPa)[a]

Material	Young's Modulus	Shear Modulus	Bulk Modulus	Elastic Limit	Tensile Strength
Aluminum	70	23	70	0.13	0.14
Brass	90	36	60	0.35	0.45
Copper	110	42	140	0.16	
Glass	55	23	37		
Iron (wrought)	90	70	100	0.17	0.32
Lead (rolled)	16	6	8		0.02
Polystyrene	1.4	0.5	5		0.05
Rubber	0.004	0.001	3		0.03
Steel	200	80	160	0.24	0.48
Tungsten	350	120	20		0.41
Benzene			1.0		
Mercury			28		
Water			2.2		

[a]For conversion purposes, 1 GPa = 10^9 Pa and 1 Pa = 1 N/m^2 = 1.45×10^{-4} lb/in.^{2}.

to compress than solids. What would you guess about the bulk modulus for foam rubber?

Sometimes people describe a material in terms of its compressibility, k. The compressibility measures how much a given force will compress a substance. It is defined as $1/B$. We have

$$\text{Bulk compressibility} = k = \frac{1}{B} \tag{11.9}$$

We do not list values of k, because they are easily found from the bulk modulus.

Elastic Limit and Tensile Strength Also given in Table 11.1 is the elastic limit for a rod under a tensile stress. It is the stress beyond which the material will no longer return to its original length when released from a stress. Hooke's law does not apply in a rod if the stress exceeds this limit. Therefore, you should be very careful to confirm that the stress is below the elastic limit before applying Hooke's law in a problem.

The last column in Table 11.1 gives the tensile strength of the material. A rod will break when the tensile stress in it exceeds this value.

■ **EXAMPLE 11.1** How much will a 50-cm length of brass wire stretch when a 2-kg mass is hung from its end? The wire has a diameter of 0.10 cm.

Solution This is a tensile-type strain, as shown in Figure 11.7. So we are concerned with Young's modulus. For brass, Table 11.1 tells us that Y is 9×10^{10} N/m^2. We are told that $L_0 = 0.50$ m. The applied force is the weight of 2 kg, namely, mg. Therefore,

$$F = (2\text{ kg})(9.8\text{ m/s}^2) = 19.6\text{ N}$$

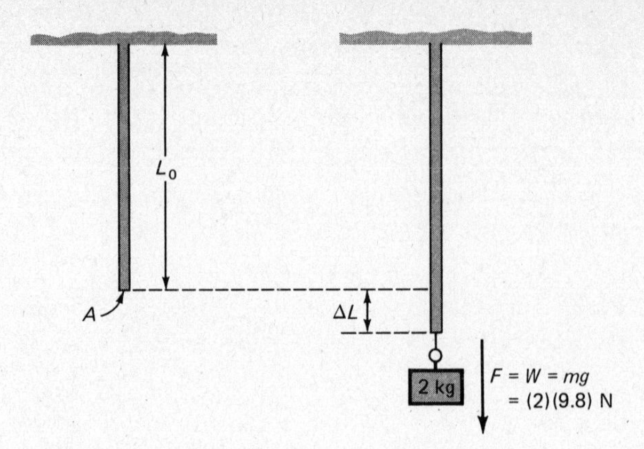

Figure 11.7 We wish to find how far the wire stretches under the 2-kg load.

The cross-sectional area is $\pi d^2/4$ and so

$$A = \frac{\pi(1 \times 10^{-3}\text{ m})^2}{4} = 7.9 \times 10^{-7}\text{ m}^2$$

The stress is therefore

$$\text{Stress} = \frac{F}{A}$$

$$= \frac{19.6\text{ N}}{7.9 \times 10^{-7}\text{ m}^2} = 2.5 \times 10^7\text{ Pa}$$

Looking at Table 11.1, we see that this stress is well below the elastic limit for brass. So we are still in the Hooke's law range. In that range

$$\text{Modulus} = \frac{\text{stress}}{\text{strain}}$$

and so we have

$$\text{Strain} = \frac{\text{stress}}{\text{modulus}}$$

$$= \frac{2.5 \times 10^7\text{ Pa}}{Y}$$

For brass, $Y = 9 \times 10^{10}$ Pa and so

$$\text{Strain} = \frac{2.5 \times 10^7\text{ Pa}}{9 \times 10^{10}\text{ Pa}} = 2.8 \times 10^{-4}$$

But the strain is $\Delta L/L_0$ with $L_0 = 0.50$ m in this case. We therefore find that

$$\Delta L = (L_0)(\text{strain}) = (0.50\text{ m})(2.8 \times 10^{-4}) = 0.00014\text{ m}$$

This tells us that the wire will stretch 0.014 cm. ■■

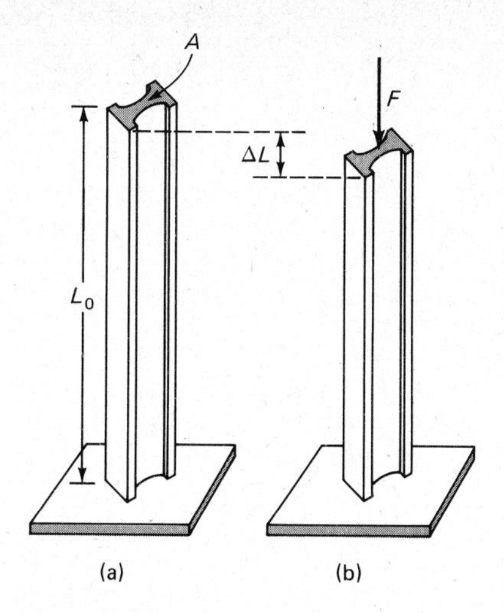

Figure 11.8 We wish to find the compression of the beam.

▮ **EXAMPLE 11.2** A 5-m-long vertical steel beam with 20-cm^2 cross section is to support a load of 20,000 N placed on its top end. How much will the beam be compressed by this load? See Figure 11.8.

Solution Young's modulus applies here. From Table 11.1 we have $Y = 20 \times 10^{10}$ Pa. We know that $F = 20{,}000$ N and $A = 20$ cm^2. Because $(F/A)/(\Delta L/L_0) = Y$,

$$\Delta L = \frac{FL_0}{AY}$$

$$= \frac{(2 \times 10^4 \text{ N})(5 \text{ m})}{(20 \times 10^{-4} \text{ m}^2)(20 \times 10^{10} \text{ Pa})}$$

$$= 2.5 \times 10^{-4} \text{ m}$$

▮▮

▮ **EXAMPLE 11.3** A motor is mounted on foam rubber feet. The feet are in the form of cylinders 1.2 cm high and cross-sectional area 5.0 cm^2. How large a sideways pull on the motor will be required to displace it 0.10 cm?

Solution The situation for any one of the feet is shown in Figure 11.9. Comparing this with Figure 11.5(a) tells us what quantities to use in Equation 11.6a. We have $A = 5 \times 10^{-4}$ m^2, $\Delta L = 0.10$ cm, $L_0 = 1.2$ cm. From Equation 11.6a we have

$$F = A\left(\frac{\Delta L}{L_0}\right)S$$

Table 11.1 tells us that $S = 1 \times 10^6$ Pa. Then

$$F = (5 \times 10^{-4} \text{ m}^2)\left(\frac{0.10 \text{ cm}}{1.2 \text{ cm}}\right)(1 \times 10^6 \text{ Pa})$$

or

$$F = 42 \text{ N}$$

But there are four legs under the motor. Therefore, the sideways force would have to be about 4(42 N) or about 170 N.

▮▮

11.5 Density

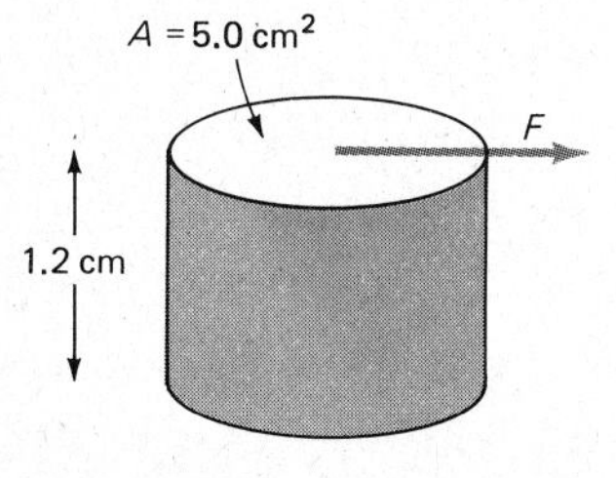

Figure 11.9 The horizontal force is to displace the top of the cyclinder by 0.10 cm.

One of the most important properties of a substance is its density. In the metric system we deal with the *mass density.* It is defined in the following way.

> **If a mass m of a substance occupies a volume V, then the mass density ρ of the substance is the mass per unit volume, or**

$$\rho = \frac{m}{V} \tag{11.10}$$

Table 11.2 DENSITIES[a]

Material	ρ (kg/m^3)	ρ (g/cm^3)	D (lb/ft^3)
Water (3.98°C)	1,000	1.000	62.4
Water	998	0.998	62.3
Air (0°C),(1 atm[b])	1.29	0.00129	0.0805
Air (1 atm)	1.20	0.00120	0.0750
Aluminum	2,700	2.70	168
Brass	8,700	8.70	540
Copper	8,890	8.89	557
Glass	2,600	2.6	160
Gold	19,300	19.3	1,204
Ice	920	0.92	57
Iron	7,860	7.86	491
Lead	11,340	11.34	708
Mercury (0°C)	13,600	13.6	849
Benzene	879	0.879	56.1
Gasoline	680	0.68	42

[a]At room temperature (20°C) unless specified otherwise.
[b]A pressure of 1 atm (1 atmosphere) is equivalent to 1.01325×10^5 Pa.

The symbol ρ is the Greek letter rho. (Sometimes d is used in place of ρ.)

Typical units for mass density are kilograms per cubic meter (kg/m^3) and grams per cubic centimeter (g/cm^3). You can easily verify that

$$1 \text{ kg/m}^3 = 0.001 \text{ g/cm}^3$$

Although ρ has the units slugs per cubic foot in the British system, this set of units for density is not often used. The values of ρ for various substances are given in Table 11.2. Notice that water has a density close to 1.0 g/cm^3 (1000 kg/m^3).

Those who work in the British system of units usually make use of the *weight density* of substances. It is defined in the following way.

If a volume V of a substance weighs W, then the weight density D of the substance is the weight per unit volume, or

$$D = \frac{W}{V} \tag{11.11}$$

The units of weight density are pounds per cubic foot. The values of D for various substances are given in Table 11.2. This type of density is usually used only in the British system.

There is a simple and convenient relation between the mass density ρ and the weight density D. Because

$$D = \frac{W}{V}$$

and $W = mg$, we see that

$$D = \frac{W}{V} = \frac{m}{V}g \qquad \text{or} \qquad D = \rho g$$

In other words, the weight density is related to mass density in the same way weight is related to mass. Simply multiply the mass density by the acceleration due to gravity to obtain the weight density.

Sometimes people make use of the *specific gravity* of a substance. By definition,

The specific gravity of a substance is the ratio of the density of the substance to the density of water.

Because it is the ratio of two densities, it has no units. It is a pure number. In giving the specific gravity of a substance, one must specify the temperature of both the water and the substance, because densities vary with temperature. However, because it is a unitless ratio, specific gravity is the same in all units systems.

EXAMPLE 11.4 A 100-cm^3 flask "weighs" 23.00 g empty and 97.84 g when full of oil. Find the density of the oil.

Solution Notice the terminology. It is customary to say that one *weigh*s an object. But when the data are given in grams, one is actually giving the object's *mass*. We have enough data to find the mass of 100 cm^3 of oil. There was 100 cm^3 of oil in the flask. We have

$$\begin{aligned}\text{Mass of oil} &= (\text{mass of flask full}) - (\text{mass empty}) \\ &= 74.84 \text{ g}\end{aligned}$$

The volume of the oil was 100 cm^3. Therefore,

$$\begin{aligned}\rho = \frac{m}{V} &= \frac{(74.84 \text{ g})}{(100 \text{ cm}^3)} \\ &= 0.748 \text{ g/cm}^3 = 748 \text{ kg/m}^3\end{aligned}$$

EXAMPLE 11.5 The density of uranium is 18.7 g/cm^3. Find the volume occupied by 1000 lb of uranium.

Solution Notice that the weight is given in the British system and the density in the metric system. We need to change one to the other. We shall change to the metric system. Because

$$1 \text{ N} = 0.225 \text{ lb}$$

we see that

$$1000 \text{ lb} = 4440 \text{ N}$$

Therefore the volume in which we are interested weighs 4440 N. But

$$\text{Weight} = mg$$

and so the mass in this volume

$$m = \frac{\text{weight}}{g} = \frac{4440 \text{ N}}{9.8 \text{ m/s}^2} = 454 \text{ kg}$$

But the density of uranium is given to be 18.7 g/cm^3, which is 18,700 kg/m^3. From the definition of mass density, $\rho = m/V$, we have

$$V = \frac{m}{\rho}$$

In our case, using the values for ρ and m,

$$V = \frac{454 \text{ kg}}{18{,}700 \text{ kg/m}^3} = 0.024 \text{ m}^3$$

This is the volume of a cube that is about 29 cm on an edge, about 1 ft^3. One cubic foot of uranium weighs about half a ton! ■■

11.6 Pressure

Suppose that a book lying on a table exerts a force **F** perpendicular to the tabletop. The force acts on an area A of the table. We define the pressure exerted by the book on the tabletop in the following way:

> **If a force *F* is applied perpendicularly to an area *A*, then the pressure *P* on the area is the force per unit area, or**

$$\boldsymbol{P = \frac{F}{A}} \qquad \textbf{(11.12)}$$

The SI unit for pressure is the pascal (Pa), where 1 Pa = 1 N/m^2. (Why is this unit the same as that for stress?)

Notice the restriction in the definition. The force must be perpendicular to the area. We usually deal with the pressures exerted by fluids. (The air pressure in a tire is the force per unit area exerted by the air on the inner wall of the tire.) It turns out that stationary fluids always obey this restriction. The reason for this

MODIFICATION OF MATERIALS BY HEAT AND PRESSURE

On the earth only 92 different elements exist in quantity. From these are made all the materials we use. Many of these materials are made by living things, plants and animals. They are the organic substances such as wood, flesh and bone. Other substances were made when the earth formed. Among them are granite, iron, and diamonds. But in the past century human beings, through technology, have been able to exceed nature in the preparation of new materials.

For example, we now make use of many manufactured modifications of iron. By alloying it with various elements, steels and alloys of great variety are formed. The same substance treated in different ways takes on different properties. For example, steel ingots are often held at high temperature for several hours. During this heat treatment important changes occur in the steel's mechanical properties.

Perhaps one of the most striking triumphs of technology is the production of diamonds. Diamonds were first formed at the high temperatures and pressures that existed at the birth of the earth. Not until 1955 were science and technology advanced enough to produce diamonds in the laboratory. By 1957 small, imperfect diamonds were being made by the General Electric Company for industrial uses. In 1970 the General Electric research team produced gem-quality diamonds. Typical artificial diamonds are shown in the photo. To produce them, carbon is compressed under huge pressures at high temperatures. A picture of a device used for this purpose is also shown.

Today we are able to make many thousands of different substances. Some of these, like metal alloys, are simply solutions and mixtures of the 92 pure elements. Others, like diamond, are a single element with its atoms arranged in a unique way. But most are composed of new molecules, molecules first made in the laboratory. Among these are the wide variety of plastics, fibers, rubbers, and other synthetics so familiar to all of us.

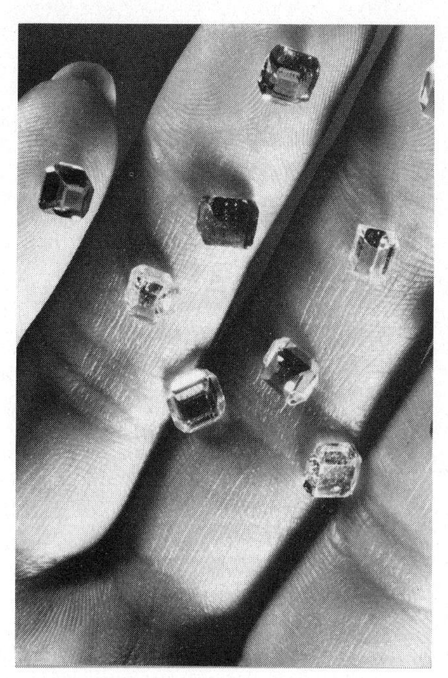

The diamonds in the photo on the left were made in the apparatus shown on the right. (Courtesy of General Electric Co.)

is that nonperpendicular forces on a surface have a component parallel to the surface which would cause the fluid to move; but if the fluid is stationary, there can be no parallel component so any force on the surface must be perpendicular to the surface.

■ **EXAMPLE 11.6** An aluminum sphere of 0.80 cm radius is placed in oil under a pressure of 2×10^9 Pa. What is the fractional change in volume, $\Delta V/V_0$, of the sphere due to this pressure?

Solution From the definition of the bulk modulus (Equation 11.8) we have

$$B = -\frac{(F/A)}{(\Delta V/V_0)}$$

But the force involved in the definition of B is perpendicular to A. Therefore $F/A = P$, the pressure of the oil in this case.

After multiplying through the equation by $\Delta V/V_0$ and dividing by B, we have

$$\frac{\Delta V}{V_0} = \frac{-P}{B}$$

From Table 11.1 we see that B for aluminum is 7×10^{10} N/m^2. Therefore, because $P = 2 \times 10^9$ N/m^2, we have

$$\frac{\Delta V}{V_0} = \frac{2 \times 10^9 \text{ N/m}^2}{7 \times 10^{10} \text{ N/m}^2} = -0.029$$

The volume of the ball decreased by 3 percent under this extremely high pressure. (This pressure is over 30,000 times as large as the pressure of the air in the atmosphere.) ■■

11.7 Pressure in a Fluid

There are many ways to measure the pressure of a fluid. One possible device is shown in Figure 11.10(a). The pressure (shown by the arrow P) causes the piston to compress the spring. It will compress until the spring's force balances the force due to the outside pressure. As a result, the position of the piston can be used to measure the pressure on the end of the piston.

We can use this or some other device to measure the pressure in a fluid. One interesting result is shown in Figure 11.10(b). As indicated, the gauge reading does not depend on its orientation in the liquid. We therefore conclude:

For a fluid at rest, the pressure at a point is the same in all directions.

This fact is convenient to know. If a surface is placed in a fluid, the pressure P that acts on the surface does not depend on the orientation of the surface.

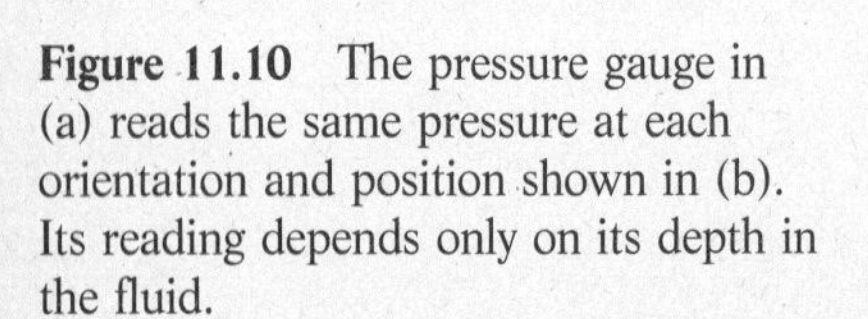

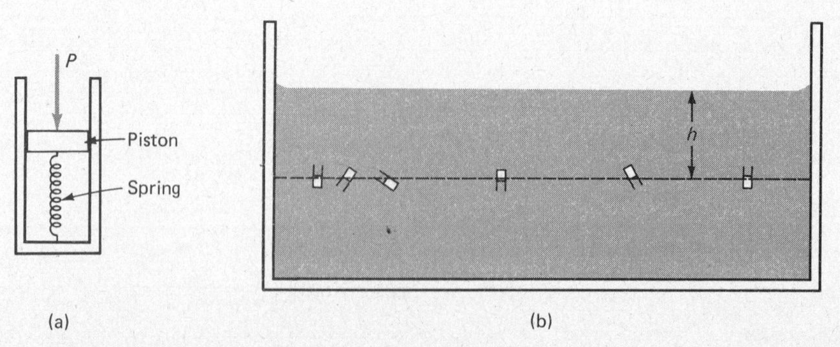

Figure 11.10 The pressure gauge in (a) reads the same pressure at each orientation and position shown in (b). Its reading depends only on its depth in the fluid.

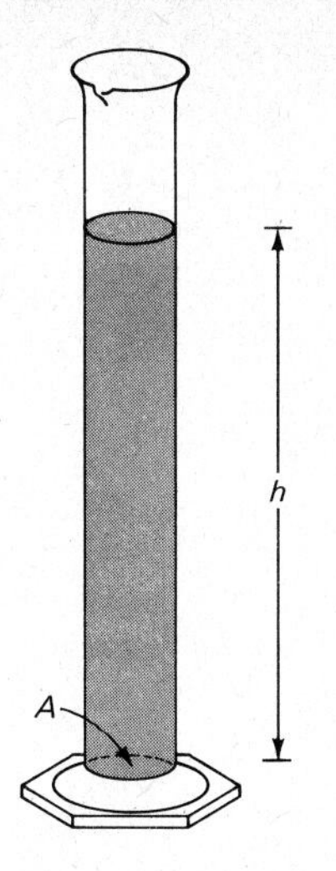

Figure 11.11 We wish to find the pressure due to the fluid at the bottom of the container.

Another point of interest is also shown in Figure 11.10(b). The pressure in a fluid is the same everywhere at the same depth below the fluid's surface. Of course, the deeper we go in the fluid, the greater the pressure becomes. So P does depend on h, the depth. Let us calculate the fluid pressure at a depth h below the surface of a fluid.

To compute the pressure due to a fluid, consider the simple situation shown in Figure 11.11. We wish to find the pressure at the bottom of the fluid. To find it, we first find the force F that the fluid exerts on the bottom of the container. Then because $P = F/A$, we divide this force by the area of the container bottom.

The force exerted by the fluid on the container bottom is easy to find. It is equal to the weight of the fluid, because the bottom supports the fluid. We therefore have

$$F = \text{weight of the fluid}$$

But because weight = mg, we can write this as

$$F = (\text{mass of fluid})(g) \tag{a}$$

We know the volume of the fluid. It is

$$V = (\text{area of base})(\text{height}) = Ah \tag{b}$$

because the volume is a cylinder. To relate the mass to its volume, we recall that the mass density ρ is given by

$$\rho = \frac{m}{V}$$

or

$$m = \rho V \tag{c}$$

If we now substitute (b) in (c), we obtain

$$m = \rho Ah$$

Placing this value for the mass in (a) gives

$$F = \rho Ahg$$

Or after dividing by A, we find that

$$\boxed{P = \frac{F}{A} = \rho gh} \tag{11.13a}$$

This is an important result. It tells us the pressure at a depth h in a fluid of mass density ρ. The pressure due to the fluid at depth h is simply ρgh. We can also write this in terms of the weight density D. Because $D = \rho g$, we have

$$P = Dh \qquad (11.13b)$$

Although this result was derived for a cylinder of fluid, it is true in general. We can therefore state the following.

The pressure due to a fluid of density ρ (or D) is ρgh (or Dh) at a depth h in the fluid.

Let us now use this result in a few examples.

EXAMPLE 11.7 Find the pressure due to a column of mercury that is 76 cm high.

Solution Using Equation 11.13a, we have

$$P = \rho gh$$

In the present case $g = 9.8\ \text{m/s}^2$, the density of mercury from Table 11.2 is 13,600 kg/m^3, and $h = 0.76$ m. This gives

$$P = (13{,}600\ \text{kg/m}^3)(9.8\ \text{m/s}^2)(0.76\ \text{m})$$

from which

$$P = 1.01 \times 10^5\ \text{N/m}^2 = 101\ \text{kPa}$$

As we shall see later, the atmosphere about us also exerts this pressure. ■■

EXAMPLE 11.8 The atmospheric pressure on our bodies is about 100 kPa. Compare this to the pressure due to the water at a depth of 10.2 m in water.

Solution We have

$$P = \rho gh$$

For water, $\rho = 1000\ \text{kg/m}^3$. Therefore,

$$P = (1000\ \text{kg/m}^3)(9.8\ \text{m/s}^2)(10.2\ \text{m}) = 100{,}000\ \text{Pa} = 100\ \text{kPa}$$

But atmospheric pressure also has this same value. We therefore see that the pressure of the atmosphere around us is equivalent to the water pressure at the bottom of a water column that is 10.2 m high. This is really a large pressure. Its effect on a metal can is shown in Figure 11.12. ■■

EXAMPLE 11.9 Find the pressure in pounds per square inch at a depth of 34 ft below the surface of the water in a lake.

Figure 11.12 The air pressure inside the can at the rear balances the pressure on the outside. When a vacuum pump is used to remove the air from the inside of a can, the outside pressure is no longer balanced. The can collapses.

Solution The pressure is given by

$$P = Dh$$

in the British system. From Table 11.2 $D = 62.4\ \text{lb/ft}^3$, and so this becomes

$$P = (62.4\ \text{lb/ft}^3)(34\ \text{ft}) = 2122\ \text{lb/ft}^2$$

We can change this to pounds per square inch (1 ft = 12 in.).

$$P = \left(2122\frac{\text{lb}}{\text{ft}^2}\right)\left(\frac{1}{144}\frac{\text{ft}^2}{\text{in.}^2}\right) = 14.7\ \text{lb/in.}^2$$

This is normal atmospheric pressure. We see from this that atmospheric pressure can support a 34-ft-high column of water. ■■

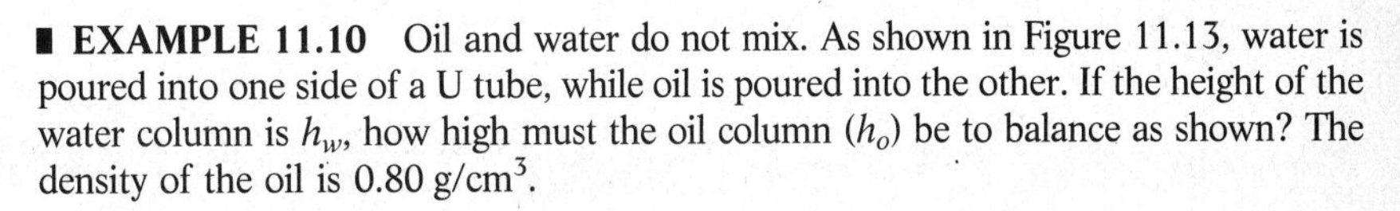

■ **EXAMPLE 11.10** Oil and water do not mix. As shown in Figure 11.13, water is poured into one side of a U tube, while oil is poured into the other. If the height of the water column is h_w, how high must the oil column (h_o) be to balance as shown? The density of the oil is $0.80\ \text{g/cm}^3$.

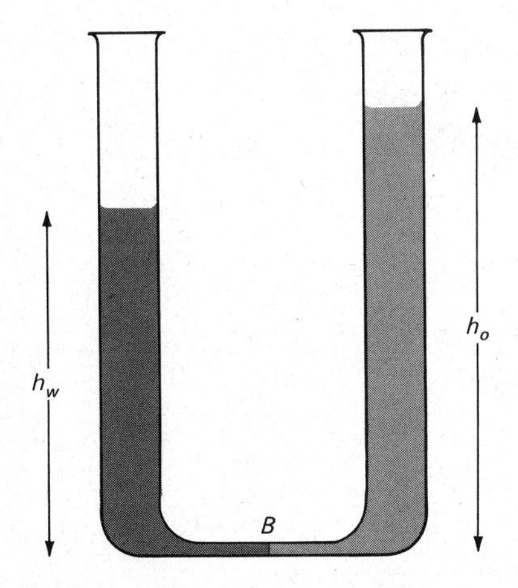

Figure 11.13 The pressure due to the oil at B must be the same as that due to the water if they are to balance.

Solution Notice in Figure 11.13 that the pressure of the water at B must equal the pressure of the oil. If this were not true, their contact point at B would be caused to move one way or the other. The pressure at B due to the water is

$$P_w = \rho_w g h_w$$

whereas that due to the oil is

$$P_o = \rho_o g h_o$$

Because they must be equal for the two columns to balance,

$$\rho_w g h_w = \rho_o g h_o$$

This gives

$$h_o = h_w\left(\frac{\rho_w}{\rho_o}\right)$$

But $\rho_w = 1000\ \text{kg/m}^3$, whereas $\rho_o = 800\ \text{kg/m}^3$. Therefore,

$$h_o = h_w\left(\frac{1000\ \text{kg/m}^3}{800\ \text{kg/m}^3}\right) = 1.25\ h_w$$

11.8 Measurement of Pressures

There are many ways in which pressures of fluids (liquids and gases) are measured. We shall outline several of the most important ways. In doing so, we shall see that the methods fall into two groups. Some devices measure the true pressure, called the *absolute pressure*. Others measure the difference between the true pressure and atmospheric pressure. We call this the *gauge pressure*. This distinction will become apparent as we proceed. Not let us examine several pressure-measuring devices.

The Manometer The manometer, shown in Figure 11.14, is widely used in technical work. A liquid (usually mercury or oil) is placed in a U tube as shown. When the manometer is open to the atmosphere, the liquid stands at the same level in the two arms of the U. As shown in Figure 11.14(a), the atmospheric pressure P_a on its two surfaces balances out.

But suppose, as in Figure 11.14(b), that one arm of the U tube is connected to a gas whose pressure P is to be measured. Suppose further that P is larger than P_a. Then the two columns of the liquid will not experience the same pressure on their surfaces. One has a pressure P and the other has a pressure P_a. For the columns to balance, the liquid heights must differ by just enough to compensate for $P - P_a$, that is, the height difference Δh must furnish a pressure to balance the pressure difference $P - P_a$. A height Δh of liquid with density ρ furnishes a pressure $\rho g\ \Delta h$. Therefore

$$P - P_a = \rho g\ \Delta h$$

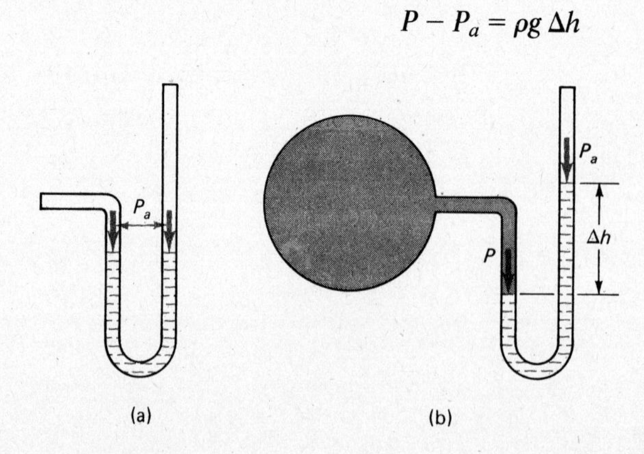

Figure 11.14 The manometer measures the quantity $P - P_a$. It reads the gauge pressure.

Notice that Δh is a measure of $P - P_a$. It measures the difference between the desired pressure and atmospheric pressure. We call this pressure difference the *gauge pressure.*

Oftentimes it is not convenient to evaluate $\rho g \Delta h$ to obtain the gauge pressure. For this reason it is customary to give only Δh. Suppose that a mercury-filled manometer is being used. If Δh is 20 cm, for example, one says that "the pressure difference is 20 cm of mercury." This, of course, is equivalent to a pressure difference of

$$\rho g \Delta h = (13{,}600 \text{ kg/m}^3)(9.8 \text{ m/s}^2)(0.20 \text{ m}) = 26{,}700 \text{ Pa}$$

In obtaining this result we used the fact that ρ for mercury is 13,600 kg/m^3. Obviously it is more convenient to give the pressure in centimeters of Hg (mercury) in this case. One would give the pressure difference in centimeters of oil for an oil-filled manometer.

The Mercury Barometer We show the essentials of a mercury barometer in Figure 11.15. A glass tube is placed in a bowl of mercury as shown. All the air is pumped out of the upper part of the tube. The tube's upper end is then sealed off. As a result, there is no air pressure in the upper, empty, end of the tube.

We know that the pressure in the mercury-filled tube at B must balance the pressure of the atmosphere P_a. Otherwise mercury would be caused to flow into or out of the tube. The pressure at B is entirely due to the mercury column above it. (Remember, the air pressure above it is zero.) This pressure is ρgh. We therefore have, because P_a balances the mercury column's pressure,

$$P_a = \rho gh$$

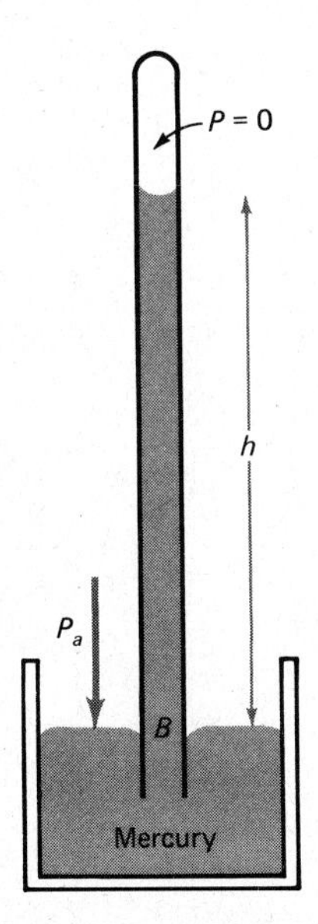

Figure 11.15 P_a, the pressure of the atmosphere, balances the pressure due to the column of height h.

Here, too, we see that it would be convenient to measure P_a in centimeters (or inches) of mercury. Under standard atmospheric conditions h is 76 cm (or 30 in.) in a barometer. We then say that atmospheric pressure is 76 cm (or 30 in.) of mercury. You have no doubt heard the weather forecaster give the daily atmospheric pressure in these units. To convert to pascals, we know that 76 cm of Hg is equivalent to a pressure of $\rho g(0.76 \text{ m})$. This is

$$(13{,}600 \text{ kg/m}^3)(9.8 \text{ m/s}^2)(0.76 \text{ m}) = 1.01 \times 10^5 \text{ Pa}$$

Standard atmospheric pressure is 101 kPa. This pressure is often expressed as a unit called the atmosphere (atm). Various units are used to measure pressure. The most important are

$$1 \text{ atm} = 1.01325 \times 10^5 \text{ Pa} = 14.7 \text{ lb/in.}^2$$
$$1 \text{ torr} = 1 \text{ mm of mercury} = 133.32 \text{ Pa}$$
$$1 \text{ lb/in.}^2 = 6895 \text{ Pa}$$
$$1 \text{ bar} = 1 \times 10^5 \text{ Pa}$$

Mechanical Gauges Both the barometer and manometer are too bulky for many uses. Mechanical pressure gauges are used instead. Two versions are

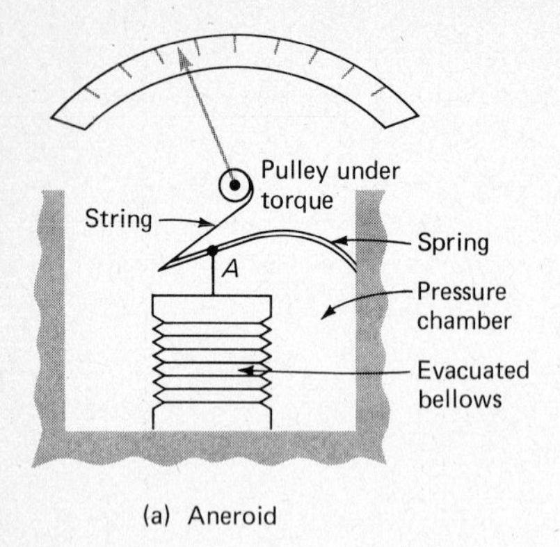

(a) Aneroid

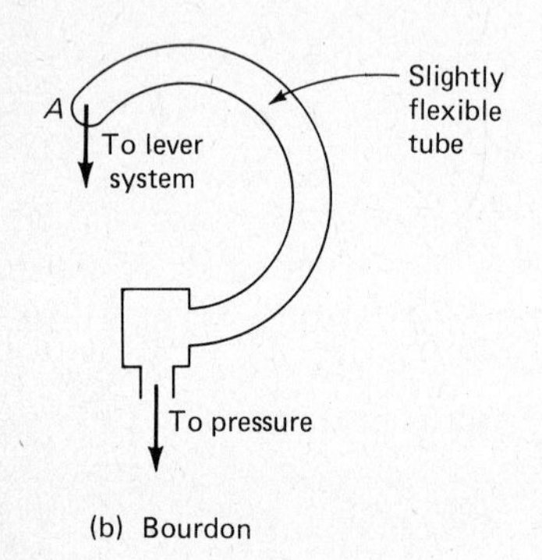

(b) Bourdon

Figure 11.16 The aneroid and Bourdon mechanisms are two types of mechanical gauges used to measure the pressure of gases. In (a) point A lowers as the pressure increases. The opposite happens in (b).

shown schematically in Figure 11.16. Both gauges read zero when open to the atmosphere. They therefore read gauge pressure, not absolute pressure.

In part (a) of Figure 11.16 we see one form of the *aneroid* mechanism. The heart of the device is the bellows. Although the bellows is not always of this shape, it is always empty of air and flexible. In operation the gas to be measured is allowed into the pressure chamber around the bellows. The pressure of the gas compresses the bellows and point A moves down. This motion is then transmitted to a needle. The movement of the needle along a precalibrated scale reads the pressure. As you see, it reads gauge pressure, not absolute pressure. Moreover, it must be calibrated.

Part (b) of Figure 11.16 shows the basic element of a *Bourdon* gauge. The tube is so made that it responds to the gas or liquid pressure within it. As the pressure is increased, the curvature of the tube decreases. Point A rises as the pressure increases. This point is connected through a lever system to a needle on a scale. Usually the gauge reads the difference between the measured pressure and atmospheric. It measures gauge pressure.

Other types of pressure-measuring devices exist. A tire gauge is a well-known example. Gauges based on electronic effects are now beginning to replace mechanical gauges. Nearly all these devices measure gauge pressure. To obtain the true (or absolute) pressure, atmospheric pressure must be added to the reading.

■ **EXAMPLE 11.11** On a certain day the morning newscast says atmospheric pressure is steady at 752 mm of mercury. You measure the pressure in a car tire at the gas station and find it to be 26 lb/in.^{2}. What is the absolute pressure of the air in the tire?

Solution The pressure measured at the gas station is the gauge pressure. We must add atmospheric pressure to it to obtain the true (or absolute) pressure. The barometer reading is 752 mm of mercury. This is equivalent to a pressure of a 0.752-m-high column of mercury. Therefore, using the density of mercury from Table 11.2,

$$\begin{aligned}\text{Atmospheric pressure} &= \rho g h \\ &= (13{,}600\ \text{kg/m}^3)(9.8\ \text{m/s}^2)(0.752\ \text{m}) = 100 \times 10^3\ \text{Pa} \\ &= 100\ \text{kPa}\end{aligned}$$

To convert the pounds per square inch reading to the same units, we have

$$26\ \text{lb/in.}^2\left(\frac{6.895\ \text{kPa}}{\text{lb/in.}^2}\right) = 180\ \text{kPa}$$

The absolute pressure in the tire is therefore

$$\begin{aligned}\text{Absolute pressure} &= P\text{ of atmosphere} + P\text{ of gauge} \\ &= 100\ \text{kPa} + 180\ \text{kPa} \\ &= 280\ \text{kPa}\end{aligned}$$

■■

11.9 Pascal's Principle

Let us refer to the experiment shown in Figure 11.17. A piston supplies a pressure to the liquid in a cylinder. The liquid is at rest. If we now apply more

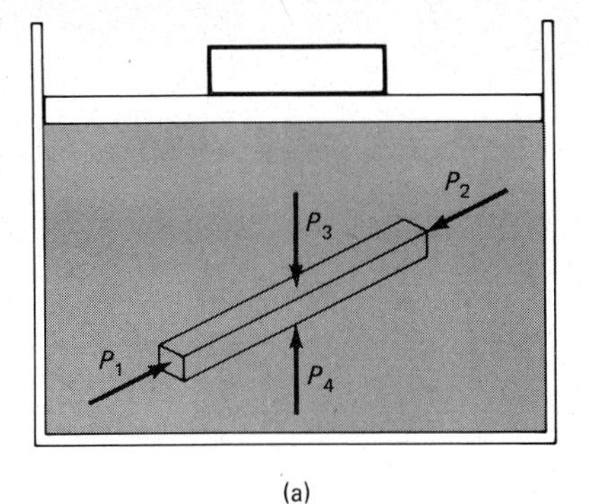

(a)

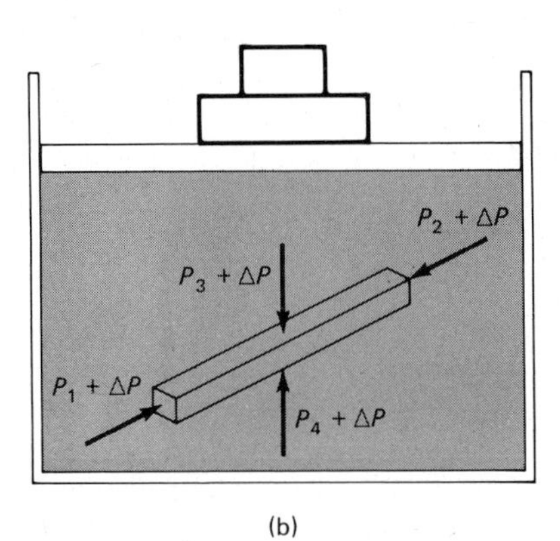

(b)

Figure 11.17 When the pressure of the fluid is increased at one point, the same increase occurs at all points.

weight to the piston, as in (b), the pressure within the liquid will obviously increase. *Pascal's principle* makes a statement about this pressure increase.

If a change in pressure is applied to a confined liquid, the same change in pressure occurs everywhere within the liquid.

To prove Pascal's principle, consider the portion of the liquid within the shaded, imaginary box in Figure 11.17(a). Because the liquid is at rest, we know that the pressures P_1, P_2, etc., that act on the liquid in the boxed region hold the liquid just where it is. The forces acting on the liquid in the boxed region must add to zero.

Suppose, as in (b), that the pressure is increased by putting extra weight on the piston. Because the liquid remains at rest, the forces on the boxed region must still add to zero. This means that pressures P_1 and P_2 must increase by the same amount, ΔP. If they did not increase by the same amount, then an unbalanced force would exist on the boxed region and the liquid would move.

Proceeding in this same way with various imaginary boxes within the liquid, we can show that the pressure increase is the same at any two points within the liquid. Thus we are able to conclude that ΔP is the same for all points. The pressure increase is transmitted unchanged to all points within the confined liquid. Because of this fact, hydraulic systems such as brakes and controls can use a liquid pressure line to operate devices far removed from the point where the operator causes a pressure change. Can you show that this principle is important in the hydraulic press discussed in Chapter 7?

11.10 Buoyancy; Archimedes' Principle

We all know that some objects float in liquids. Even those that do not are buoyed up (that is, partially supported) by liquids in which they are submerged. In this section we shall discuss Archimedes' principle. It tells us how large the buoyant force due to a liquid is. Let us first see why a buoyant force should exist.

Suppose that a cylinder is immersed in a liquid as shown in Figure 11.18. The pressure at the bottom of the cylinder, P_b, will be larger than the pressure at the top of the cylinder, P_t, because it is at a depth h greater in the liquid. The upward force due to the liquid on the bottom of the cylinder is greater than the downward force on its top. The liquid therefore causes a net upward force on the cylinder. We shall now compute this force, the buoyant force.

The forces on the top and bottom of the cylinder are

$$F_t = P_tA \qquad \text{and} \qquad F_b = P_bA$$

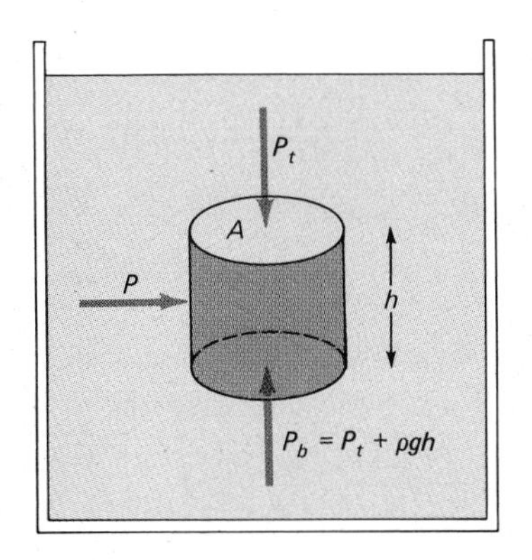

Figure 11.18 Because P_b is larger than P_t, the cylinder experiences a supporting (or buoyant) force $P_bA - P_tA$ due to the liquid.

where A is the area shown in the figure. We know that the increased pressure due to a depth h of liquid with density ρ is ρgh. Therefore,

$$P_b = P_t + \rho gh$$

We are now ready to compute the buoyant force on the cylinder. It is

$$\text{BF (buoyant force)} = F_b - F_t = P_b A - P_t A$$

which is

$$\text{BF} = (P_t + \rho gh)A - P_t A = \rho ghA$$

or

$$\text{BF} = \rho g(hA)$$

But hA is simply the volume V of the cylinder. *It is equal to the volume of liquid displaced by the cylinder.* Therefore

$$\text{BF} = g(\rho V)$$

Finally, we note that ρV is the mass of the liquid displaced. Therefore, $(\rho V)g$ is mg, the weight of the displaced liquid. We therefore have found that

$$\text{BF} = \text{weight of displaced liquid}$$

Although we have derived this for a simple case, it has much wider validity. It applies to an object of any shape. The principle involved was first pointed out by Archimedes. It is known as Archimedes' principle.

An object partly or wholly immersed in a fluid is buoyed up by a force. The force is equal to the weight of the fluid displaced.

Before making use of this principle, we should point out that an object submerged in a fluid appears to weigh less than it does in vacuum. It appears to weigh less because of the buoyant force. For an object of volume V that is completely submerged,

$$\begin{aligned}\text{Apparent weight} &= \text{true weight} - \text{buoyant force} \\ &= mg - \text{weight of displaced fluid} \\ &= (\rho_m V)g - (\rho_f V)g\end{aligned}$$

where ρ_m is the density of the object and ρ_f is the density of the fluid. Collecting terms,

$$\text{Apparent weight} = (\rho_m - \rho_f)Vg \qquad (11.14)$$

for an object that is completely submerged. What does this relation tell us for $\rho_m = \rho_f$? For $\rho_f > \rho_m$?

EXAMPLE 11.12 An object weighs 0.50 N in air and has a volume of 8.00 cm³. Find its mass density.[1]

Solution Because $W = mg$, the mass of the cube is

$$m = \frac{W}{g} = \frac{0.50\ \text{N}}{9.8\ \text{m/s}^2} = 0.051\ \text{kg}$$

Its mass density is

$$\rho = \frac{m}{V} = \frac{0.051\ \text{kg}}{8 \times 10^{-6}\ \text{m}^3} = 6380\ \text{kg/m}^3$$

EXAMPLE 11.13 An object weighs 1.200 N in air and has a volume of 10.0 cm³. How much will it appear to weigh when completely submerged in water?

Solution Its apparent weight in water will be smaller than its weight in air because of the buoyant force.

$$\text{Apparent weight} = \text{true weight} - \text{buoyant force}$$

The object will displace 10 cm³ of water. The mass of this much water is

$$\begin{aligned}\text{Mass of water} &= \rho V = (1000\ \text{kg/m}^3)(10 \times 10^{-6}\ \text{m}^3)\\ &= 10 \times 10^{-3}\ \text{kg}\end{aligned}$$

A mass m has a weight mg, and so the weight of the displaced water is

$$\text{Weight of water} = (10 \times 10^{-3}\ \text{kg})(9.8\ \text{m/s}^2) = 0.098\ \text{N}$$

Because the buoyant force (BF) is equal to the weight of the displaced liquid, BF = 0.098 N. Therefore

$$\text{Apparent weight} = 1.200\ \text{N} - 0.098\ \text{N} = 1.102\ \text{N}$$

EXAMPLE 11.14 An object "weighs" 0.1224 kg in air and has a volume of 10 cm³. How much will it "weigh" when completely submerged in water?

Solution We are actually given the mass of the object. Its weight is

$$W = mg = (0.1224\ \text{kg})(9.8\ \text{N}) = 1.200\ \text{N}$$

The buoyant force is equal to the weight of 10 cm³ of fluid. Therefore,

$$\begin{aligned}\text{BF} &= \rho_f V g = (1000\ \text{kg/m}^3)(10 \times 10^{-6}\ \text{m}^3)(9.8\ \text{m/s}^2)\\ &= 0.098\ \text{N}\end{aligned}$$

[1] In these problems we shall neglect the small buoyant force due to the air.

Therefore,

$$\text{Apparent weight} = 1.200 - 0.098 = 1.102\ \text{N}$$

Since the weight and mass of an object are proportional, it is natural to associate an "apparent mass" with the "apparent weight" of a submerged object, even though we know its mass has not really changed. It is this "apparent mass" we are looking for when we ask how many kilograms it "weighs" when submerged.

$$\text{Apparent mass} = \frac{\text{apparent weight}}{g} = \frac{1.102\ \text{N}}{9.8\ \text{m/s}^2}$$
$$= 0.112\ \text{kg}$$

■■

■ **EXAMPLE 11.15** An object weighs 0.56 lb in air and has a volume of 3.0 in.^{3}. How much will it weigh when totally submerged in water?

Solution We know that its weight $W = 0.56$ lb. Further, we know that its volume

$$V = 3.0\ \text{in.}^3 = (3\ \text{in.}^3)\left(\frac{1\ \text{ft}}{12\ \text{in.}}\right)^3 = 1.74 \times 10^{-3}\ \text{ft}^3$$

When submerged, it displaces 1.74×10^{-3} ft^3 of water. This much water's weight can be found from

$$D_{\text{water}} = \frac{W \text{ of water}}{V \text{ of water}}$$

or

$$W \text{ of water} = (62.4\ \text{lb/ft}^3)(1.74 \times 10^{-3}\ \text{ft}^3)$$
$$= 0.108\ \text{lb}$$

This will equal the buoyant force on the submerged object. Its apparent weight when submerged will therefore be

$$\text{Weight} - \text{BF} = \text{apparent weight}$$

which is

$$0.56 - 0.11\ \text{lb} = \text{apparent weight}$$

The object therefore weighs 0.45 lb when submerged. ■■

■ **EXAMPLE 11.16** A piece of aluminum weighs 5.0 N when weighed in air. How large a supporting force will be required to support it when immersed in oil of density 850 kg/m^3?

Solution The situation is shown in Figure 11.19. We wish to find the supporting force T. It will, of course, equal the apparent weight of the object. Equation 11.14 states that

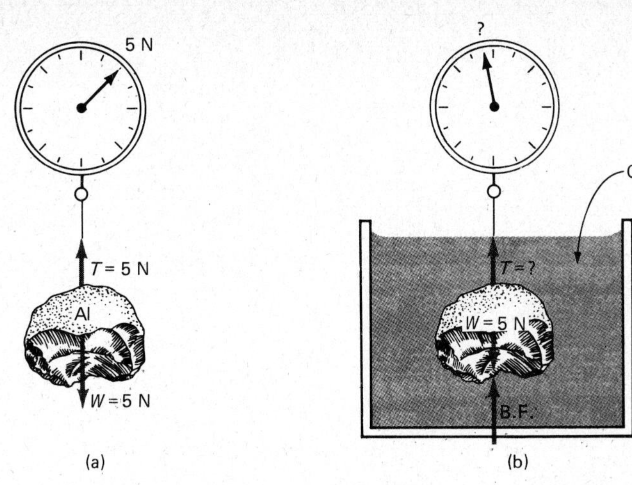

Figure 11.19 An example of Archimedes' principle.

$$\text{Apparent weight} = (\rho - \rho_f)Vg$$

From Table 11.2 we know that ρ for aluminum is 2700 kg/m^3. We are told that $\rho_f =$ 850 kg/m^3.

To find the volume of the object, we make use of $m = \rho V$ and $W = mg$ to give

$$V = \frac{m}{\rho} = \frac{W/g}{\rho} = \frac{(5.0\ \text{N})}{\rho g} = 1.89 \times 10^{-4}\ \text{m}^3$$

Substituting gives

$$\begin{aligned}\text{Apparent weight} &= (2700 - 850\ \text{kg/m}^3)(1.89 \times 10^{-4}\ \text{m}^3)(9.8\ \text{m/s}^2)\\ &= 3.4\ \text{N}\end{aligned}$$

▮ EXAMPLE 11.17 A solid object "weighs" 23.00 g in air and 12.70 g when fully submerged in oil of density 800 kg/m^3. Find the mass density of the material of the object.

Solution Using precise terminology, the weight in air is (0.023)(9.8) N. In oil the apparent weight is (0.0127)(9.8) N. We shall use the definition of the density:

$$\rho = \frac{\text{mass}}{\text{volume}}$$

The mass of the object is 0.0230 kg. We still need its volume.

To find its volume, we shall use Archimedes' principle. The difference in apparent weight of the object is

$$(0.023)(9.8)\ \text{N} - (0.0127)(9.8)\ \text{N} = (0.0103)(9.8)\ \text{N}$$

We know from Archimedes' principle that (0.0103)(9.8) N of oil was displaced. Because $m = W/g$, the mass of oil displaced is 0.0103 kg. The volume displaced can be found from the fact that

$$\rho_{oil} = \frac{\text{mass of oil}}{\text{volume}}$$

In our case, $\rho_{oil} = 800\ \text{kg/m}^3$ and its mass was 0.0103 kg. Substituting in the relation above gives

$$\text{Volume} = \frac{\text{mass}}{\rho} = \frac{0.0103\ \text{kg}}{800\ \text{kg/m}^3}$$
$$= 1.29 \times 10^{-5}\ \text{m}^3$$

This is also the volume of the object.

Now that we know both the mass and volume of the object, we have

$$\rho = \frac{\text{mass}}{\text{volume}} = \frac{0.0230\ \text{kg}}{1.29 \times 10^{-5}\ \text{m}^3} = 1940\ \text{kg/m}^3$$

■■

■ **EXAMPLE 11.18** A solid piece of iron is to float on mercury. What fraction of the volume of the iron will be submerged?

Solution The iron will sink until it displaces a weight of mercury equal to its own weight. From the definition of weight density, we have

$$\text{Weight of iron} = \rho_i V_i g$$
$$\text{Weight of mercury displaced} = \rho_m V_m g$$

At equilibrium, these two forces are equal. Therefore

$$\rho_i V_i = \rho_m V_m$$

But V_m is the volume of displaced mercury. It is equal to the volume of iron submerged. The fraction of the iron submerged is therefore V_m/V_i. We can obtain this by rearranging the equation. Then

$$\frac{V_m}{V_i} = \frac{\rho_i}{\rho_m}$$

Using the values in Table 11.2, we have

$$\frac{V_m}{V_i} = \frac{7860\ \text{kg/m}^3}{13{,}600\ \text{kg/m}^3} = 0.58$$

So 58 percent of the iron object will be submerged.

■■

■ **EXAMPLE 11.19** The man shown in Figure 11.20 is having his blood pressure taken. An inflatable sleeve has been wrapped around his upper arm. The doctor has inflated the sleeve by use of the squeeze bulb shown. The air pressure in the sleeve is measured by the mercury manometer. As air slowly leaks from the sleeve, the doctor listens to the man's pulse with a stethoscope. At first there is no pulse beat. When the

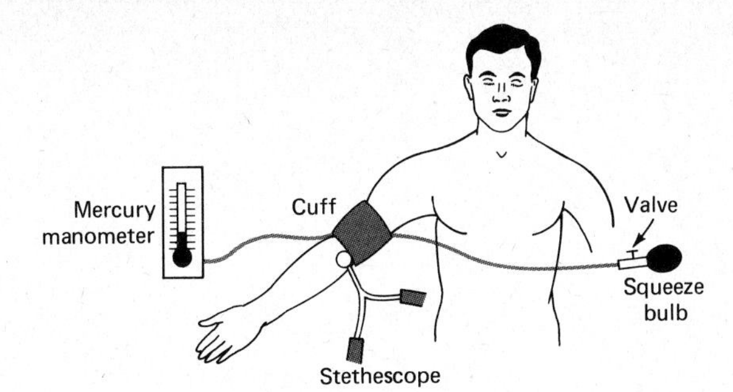

Figure 11.20 By listening to the pulse and reading the manometer, the blood pressure can be determined.

pressure falls to a certain value, the pulse can be heard. This is called the *systolic* pressure. As the pressure in the sleeve continues to drop, the pulse sound becomes fainter. Finally, it disappears. The pressure at which it disappears is called the *diastolic* pressure. How can we interpret these measurements?

Explanation The sleeve applies pressure to the arm. The pressure applied is the same as the pressure measured within the sleeve. When the sleeve pressure is high, it stops blood from flowing in the arm. Reduced pressure in the sleeve begins to let blood flow. It spurts slightly into the arm each time the heart pumps. A pulse can therefore be heard. The systolic pressure is the maximum blood pressure in the arm artery.

When the sleeve pressure is low enough, the spurting action stops. Blood flows smoothly through the artery even at the lowest pressure in the artery. Therefore, the diastolic pressure measures the lowest pressure of the blood in the artery.

During a heartbeat cycle, the blood pressure in the artery goes up and down. Its highest value is the systolic pressure. Its lowest value is the diastolic pressure. When a doctor quotes a blood pressure of 150/80 (or 150 over 80), these two pressures are being given. The units used are millimeters of mercury. ■■

Summary

Materials can be separated into three groups: solids, liquids, and gases. Because liquids and gases flow, they are called fluids. Solids can be either crystalline or amorphous. The atoms of a crystalline solid are placed in a definite geometric pattern called a lattice.

Hooke's law states that the stress applied to an object is proportional to the strain it causes. This law is obeyed at small deformations by elastic materials. An elastic material returns to its original shape upon removal of the stress.

The stress exerted by a force F on an area A is F/A. It causes a strain defined as the ratio of the distortion divided by the undistorted size. Strain has no dimensions.

A modulus of elasticity for a material is the stress divided by the strain. Young's modulus, Y, applies to a tensile stress. The shear modulus, S, applies to a shearing stress. The bulk modulus, B, applies to volume compression. We call the reciprocal of B the compressibility, k.

Pressure exerted on an area A is F/A, where F is the force perpendicular to the area. Pressure is measured in pascals (Pa), where $1\ \text{Pa} = 1\ \text{N/m}^2$.

Forces exerted on walls by fluids at rest are always perpendicular to the surface.

Two types of density can be specified for any material. The mass density ρ is the mass of the material divided by its volume, m/V. The weight density D is the weight of the material divided by its volume, W/V. The two are related in the same way W is related to m ($W = mg$), so $D = \rho g$.

The specific gravity of a material is the ratio of its density to the density of water.

For fluids at rest, the pressure at a point is the same in all directions. The pressure P due to a height h of liquid with density ρ is $P = \rho gh = Dh$.

Typical pressure-measuring devices are the manometer, mercury barometer, aneroid-type gauge, and Bourdon-type gauge. Of these, only the mercury barometer measures the true (or absolute) pressure. The others measure the pressure excess above atmospheric. They are said to read gauge pressure.

Pascal's principle states that if a pressure is applied to a confined liquid, the pressure is transmitted to every point within the liquid.

Archimedes' principle states that if an object is partly or wholly submerged in a fluid, it is buoyed up by a force equal to the weight of the displaced fluid.

Questions and Exercises

1. Gelatin dessert is often cut into cubes perhaps 4 cm × 4 cm × 4 cm. How would you go about measuring the shear modulus of gelatin using a cube like this? Explain why the shear modulus is greatly increased when large sections of peaches or oranges are placed in the gelatin.
2. List as many types of distortion as you can. What does Hooke's law imply for each?
3. As shown in Figure 11.12, air pressure exerts large forces on an empty can. Why do we not notice this huge force on our bodies?
4. The depth to which a submarine can safely dive is determined by the strength of its walls. Why don't they simply increase the air pressure inside the submarine so as to balance the force due to the water?
5. Estimate the pressure your shoe exerts on the floor as you stand on one toe to reach an object high on a shelf.
6. Make a list of the most dense solids you can think of. Repeat for the least dense solids. Can you draw any conclusions from your lists?
7. Some people float better in water than others. Estimate the density of a typical human. What factors influence the floating ability of a person?
8. A large ice cube is placed in a glass. The glass is then filled to the brim with water. Of course, the ice cube now floats with part of it above the top of the glass. Will the glass overflow as the ice cube melts?
9. Refer to Figure P11.1. Experiment shows that at equilibrium the liquid rests as shown. Of course, the pressures at A, B, and C must be the same or the liquid would move to the left or right. Shouldn't the pressures be highest at B and C because more liquid is supported above these points?

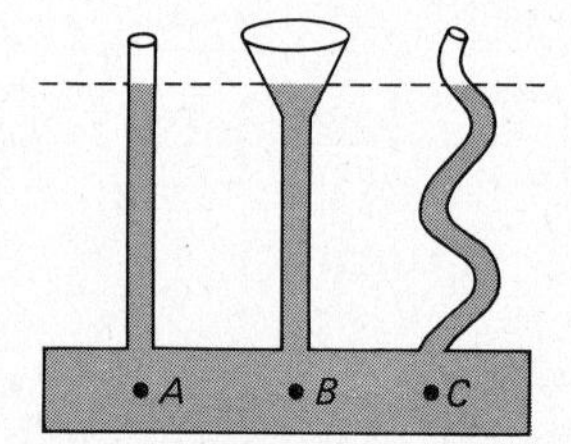

Figure P11.1

10. How does an ordinary tire gauge work?
11. A fish at the bottom of a deep lake gives off a bubble of air. Describe the motion of the bubble as it rises to the surface. What makes it rise? How does its speed vary with time?
12. An aboveground swimming pool is to have water in it 3 m deep. The designer wishes to provide two versions, both rectangular. One is to be 20 m × 20 m. The other is to be 20 m × 40 m. How do the forces each of the walls must support compare in the two cases?
13. A certain scientific measuring instrument uses a metal ball

floating on mercury. Suppose that the device is to be used on the moon. Compare the level at which the ball floats in the device on the earth and moon.

14. It is suspected that a steel ball bearing has a sizable hole at its center. How could you determine whether or not this was true without harming the ball?
15. Explain the principle by which a suction-type water pump works. Repeat for a siphon.

Problems

1. The following data were taken for a rubber band under load. Does the rubber band obey Hooke's law in this range?

Load (g) →	0	50	100	150	200	250
ΔL (cm) →	0	0.25	0.75	1.80	2.15	2.30

2. The following data were taken for the angle of twist in a wire as a function of the applied torque. Over what range of twist angles does Hooke's law apply?

Torque (N·cm) →	0	10	20	30	40
Angle (deg) →	0	23	46	70	94

Torque (N·cm) →	50	60	70	80
Angle (deg) →	117	141	163	185

3. A 10-kg load is hung from the end of a 2-m-long steel wire. The wire has a radius of 0.70 mm. (a) What is the tensile stress in the wire? (b) What is the tensile strain in the wire? (c) How much does the wire stretch under this load? Use the data in Table 11.1.
4. A 40-kg load is hung at the end of a 5-m-long brass rod that has a radius of 4 mm. Find (a) the stress in the rod, (b) the strain in the rod, and (c) the rod's elongation. Use the data in Table 11.1.
5. A 2-mm-diameter wire that is 160 cm long stretches 0.50 mm when subjected to a 30-kg load. What is Young's modulus for the material of the wire?
6. A steel girder has a cross-sectional area of 50 cm^2. How large a compressive force would be required to shorten the girder by 0.01 percent?
7. The tensile strength of aluminum is about 1.4×10^8 N/m^2. (a) How large a load can a 2-mm-diameter aluminum wire hold before breaking? (b) If the wire obeyed Hooke's law up to this load, by what percent would it elongate under it? Use the data in Table 11.1.
8. Steel piano wire has an ultimate strength of about 2.3×10^9 N/m^2. (a) How large a diameter wire would be needed to support a load of 300 kg? (b) If the wire obeyed Hooke's law even at such high stresses, what would be its percent elongation just before breaking? Use the data in Table 11.1.
9. As shown in Figure P11.2, two metal plates are sealed to two faces of a rubber block. When a force of 200 N is applied to the upper plate as shown, the upper plate moves sideways 0.30 cm. Find the shear modulus of the rubber. The block has dimensions 4.0 cm × 4.0 cm × 2.5 cm.

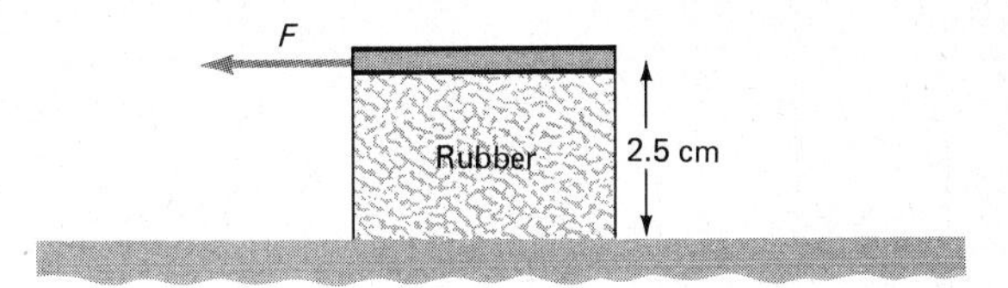

Figure P11.2

*10. A 2000-N shearing force is applied to the upper surface of a brass cube that is 3.0 cm on a side. How large a distortion angle (in degrees) is caused by the force?

**11. A 500-g ball is to be swung in a 40-cm-radius circle at the end of a 0.050-cm diameter steel wire. What is the maximum speed (in rev/s) that the ball can have if the circle is (a) horizontal and (b) vertical?

12. A cube of material is 2.00 cm on each edge. It has a mass of 53 g. What is the density of the material?
13. The density of uranium is 18,700 kg/m^3. (a) What is the mass of a cube of uranium that is 2.00 cm on each edge? (b) How many times heavier is such a cube than a similar-size cube of water?
14. Using the data in Table 11.2, find what volume of air at 20°C will have a mass of 2.00 kg.
15. An empty 100-cm^3 calibrated flask "weighs" 107.30 g. When filled to the mark with a certain liquid, it "weighs" 196.43 g. What is the density of the liquid in kilograms per cubic meter?
16. The density of the human body is close to that of water. Find the approximate volume of a 25-kg boy.

*17. A length L of uniform brass wire is to be suspended from a support far above the earth's surface. How long could the

*Problems marked with an asterisk are not as easy as the unmarked ones.
**Problems marked with a double asterisk are somewhat more difficult than the average.

wire be if it were to be on the verge of breaking under its own weight? Assume $g = 9.8\ \text{m/s}^2$.

*18. Show that the maximum length of a material in the form of a cylinder that can hang under its own weight without breaking is equal to (tensile strength)/ρg.

19. A cube of wood 5 cm on each edge is placed on the floor. What pressure does the block exert on the floor when a 70-kg person stands on the block?

20. A 2-cm-internal-diameter pipe is standing vertically with its lower end plugged. The pipe is filled with water up to a level so that there are 2500 g of water in the pipe. What is the pressure of the water on the plug at the pipe's bottom?

21. How large a pressure is required to compress water by 0.0100 percent?

22. A certain fluid compresses 0.20 percent under a pressure of 2.0 MPa. What is the bulk modulus for this fluid?

23. How large a pressure in pascals would be required to compress water by 10 percent? How high a column of water (in kilometers) would be needed to furnish this pressure?

24. Find the pressure due to the water at a depth of 1.61 km in the ocean. By what percent is the water at this depth compressed? Assume the seawater to behave about the same as pure water.

25. A vertical glass tube is filled with mercury to a height of 50 cm. (a) What is the pressure due to the mercury at the bottom of the tube? (b) If the tube has an inner diameter of 4 mm, how large a force must the plug at the bottom of the tube support?

*26. The U tube shown in Figure P11.3 has water in one side and oil in the other. If a 40-cm-high column of oil is supported by a 35-cm-high column of water, what is the density of the oil?

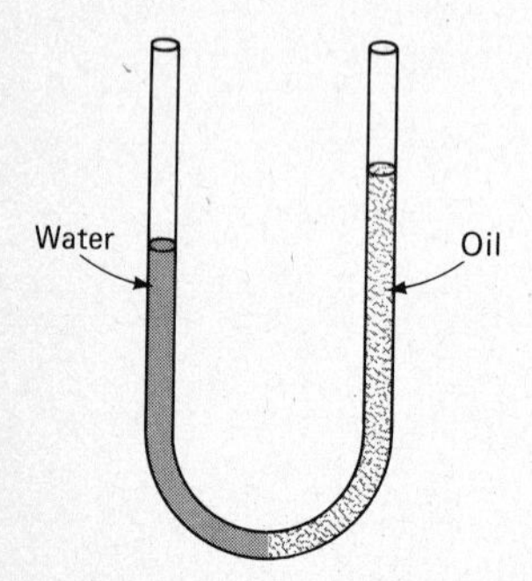

Figure P11.3

*27. In Figure P11.3 a 70-cm-high column of oil is supported by a 60-cm-high column of water. What is the specific gravity of the oil?

*28. A certain flask "weighs" 15.0 g when empty, 74.8 g when filled with oil, and 85.3 g when filled with water. What is the specific gravity of the oil?

29. The water supplied to a tall building comes from a tank on its top. What will be the water pressure in a faucet 70 m below?

30. Standard atmospheric pressure is 101 kPa. How tall must a water barometer be if it is to be able to read this pressure? (Hint: In a barometer the liquid column balances atmospheric pressure.) Note that this is the maximum height to which a simple suction system can lift water.

31. Find the barometric pressure in kilopascals on a day when the barometer reads 74 cm of Hg.

32. On a day when atmospheric pressure is exactly 100 kPa, what is the total pressure 6 m deep in a freshwater lake?

33. A mercury manometer is used to read the gas pressure in a tank. The column attached to the tank is 12 cm higher than the column open to the atmosphere. If atmospheric pressure is 100 kPa, what is the pressure in the tank?

34. A manometer using oil ($\rho_{\text{oil}} = 800\ \text{kg/m}^3$) has a 25-cm difference in heights for its two columns. The lower column is attached to a gas-filled container. If atmospheric pressure is exactly 100 kPa, (a) what is the pressure in the tank and (b) what will an ordinary pressure gauge (calibrated in kilopascals) read the tank pressure to be?

*35. The system in Figure P11.4 is at equilibrium as shown. If a 3.0-kg mass is set on the top piston, which has an area A_t, by what force must F_b be increased if the system is to remain in equilibrium? Express your answer in terms of A_t and A_b, the piston areas.

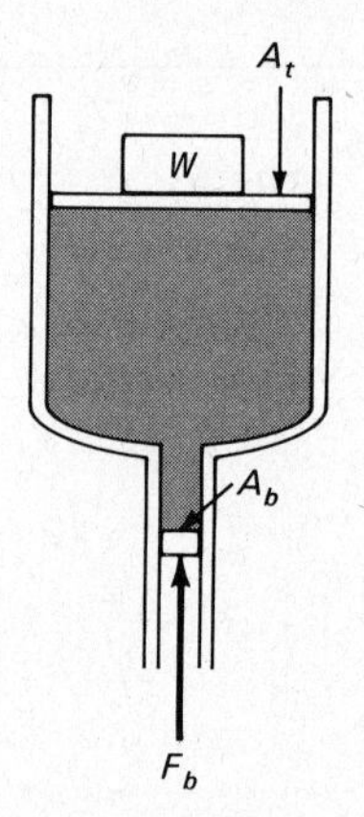

Figure P11.4

36. (a) How large is the buoyant force on a 2.00-cm³ piece of metal completely submerged in benzene? See Table 11.2. (b) Repeat for a 0.030-ft³ piece in benzene.

37. (a) A large, spherical balloon has a radius of 2.50 m. Find the buoyant force on it due to air at 20°C and 1 atm pressure. (b) Repeat for a radius of 8.0 ft.

*38. A certain piece of metal "weighs" 63.2 g in air and 29.1 g when submerged in water at 20°C. What is the mass density of the metal?

*39. A solid piece of aluminum "weighs" 57.6 g in air and 26.5 g when submerged in a liquid. (a) What is the volume of the metal piece? (b) What is the mass density of the liquid? Use Table 11.2.

*40. The density of helium in a balloon is about 0.175 kg/m^3. How large in volume must a balloon be if it is to lift a total load of 700 kg? The 700 kg includes the mass of the balloon but not of the helium in it.

**41. Life preservers are sometimes made of polystyrene foam with density 0.150 g/cm^3. How large a volume must they have if each is to be able to keep a 90-kg person one-third out of the water? Assume the person's density to be 1.10 g/cm^3.

**42. In order to fish on a partly frozen lake, a 90-kg man drives his 2000-kg car onto the ice. Although the ice is 28 cm thick, it begins to break up. How large an area ice sheet must his car be on if it is to remain floating? (ρ for ice is 920 kg/m^3.)

U.S. Air Force

12 TEMPERATURE AND MATTER

Performance Goals

When you finish this chapter, you should be able to

1. Change any given temperature on the Fahrenheit, Celsius (or centigrade), Kelvin, and Rankine scales to all the other scales.
2. Give the value of the ice point and steam point for water on all four temperature scales.
3. Given the molecular (or atomic) weight of a substance, compute the mass of one of its molecules

We begin our study of heat by learning how temperatures are measured. The various temperature scales are defined, and we see how to convert from one to the other. Temperature has an important physical meaning. We learn what it is by examining how the pressure of a gas depends on temperature. The ideal gas law is obtained, and we find out how to use it in practical situations. Finally, we discuss the way in which liquids and solids expand with temperature. The thermal expansion equations are applied to problems of practical importance.

(or atoms). Repeat for the reverse.

4. Sketch the following graphs for an ideal gas: P versus t_C with V constant; V versus t_C with P constant. Repeat for t_C replaced by t_F, T, and T_R. Locate the numerical value of absolute zero on each graph.
5. Write the ideal gas law and define each quantity in it. Show that it predicts the behavior shown in the P versus T graphs of the previous goal. Using it, explain what factors influence the slope and intercepts of these graphs.
6. Using the ideal gas law, find one of the following quantities when all others are supplied: P_1, V_1, T_1, P_2, V_2, T_2.
7. Given all but one of the following quantities, use the gas law to find the unknown quantity: P, V, M, T, m.
8. Select from a group of equations the equation(s) that properly relate T to v for the molecules in an ideal gas. Use it to find v for a known gas at a stated temperature.
9. Sketch the graph that shows the probability versus speed relation for an ideal gas. Explain what it means. Show how it changes as T is changed.
10. Explain the molecular meaning of thermal motion. In terms of your explanation, show why one would usually expect liquids and solids to expand as their temperatures are increased.
11. Given all but one of the following quantities for an object, find the unknown quantity: original size, final size, t_1, t_2, α or β.

12.1 Temperature

Long before children enter school, they know the difference between hot and cold. These are concepts we learn through experience. Even primitive peoples were aware of these sensations. But as time progressed, people saw the need to measure hot and cold. For this purpose thermometers were invented.

There are many types of thermometers in use today. They measure the hotness of objects in terms of a quantity we call temperature. In spite of their wide variety, nearly all thermometers use one of four temperature scales. These scales are named the Celsius (or centigrade) scale, the Kelvin scale, the Fahrenheit scale, and the Rankine scale. Only the Celsius and Kelvin scales are used in modern technical and scientific work.

In principle any quantity that changes with temperature can be used as the basis for a thermometer. The common bulb thermometer, such as the one shown in Figure 12.1, makes use of the fact that liquids expand as they are heated.[1] A liquid partly fills a capillary tube attached to a bulb. Because the liquid expands as the thermometer temperature increases, the height of the liquid in the tube can be used as a measure of temperature. The tube can be calibrated in any temperature scale.

A typical Celsius thermometer is shown in Figure 12.2. Two calibration points can be used to construct such a thermometer. The reading when the thermometer is at the freezing point of water is taken to be 0°C (read "zero degrees Celsius"). When the thermometer is at the temperature of boiling water, its reading is taken to be 100°C. The thermometer scale between these points is then divided into 100 degrees.[2]

The Kelvin thermometer uses the same size degrees as the Celsius scale, but the freezing point of water is taken to be 273.15 K (read 273.15 kelvins).[3] The boiling point of water is taken to be 373.15 K. Its absolute zero point is taken at −273.15°C. As you can see, the Kelvin scale is displaced from the Celsius scale by 273.15 Celsius-size degrees.

Where the British system of units is used (chiefly the United States), temperature is measured on the Fahrenheit scale. It assigns 32°F to the freezing point of water and 212°F to the boiling point. You can convert between the two scales by use of the following relation.

$$\frac{t_F - 32}{t_C} = \frac{9}{5} \qquad \textbf{(12.1)}$$

where, t_F and t_C are the Fahrenheit and Celsius temperatures, respectively.

The Rankine temperature scale uses the same size degrees as the Fahrenheit

[1] There is one very important exception. Water contracts as it is heated from 0°C to 4°C.
[2] Because the freezing and boiling points depend on pressure, the pressure must be standard pressure, 101 kPa.
[3] It is customary to write Kelvin temperatures as 55 K, for example, not 55°K.

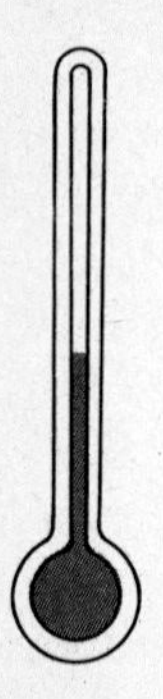

Figure 12.1 As the thermometer bulb is heated, the liquid expands and rises in the tube.

scale, but its absolute zero point is taken at $-460°F$. On it water freezes at 492°R and boils at 672°R. As you see, the Rankine scale is offset from the Fahrenheit scale by 460 Fahrenheit-size degrees.

Figure 12.3 compares these various temperature scales. Notice that the lower limit for each scale is what is called *absolute zero.* We will later learn the meaning of this temperature. For now we will simply state that it is the lowest possible temperature. Both the Kelvin and Rankine scales take their zero at absolute zero. They are therefore called *absolute temperature scales.*

▮ EXAMPLE 12.1 When room temperature is 75°F, what is the temperature on the Celsius scale?

Solution We substitute directly in Equation 12.1.

$$\frac{75 - 32}{t_C} = \frac{9}{5}$$

After cross multiplying, this becomes

$$375 - 160 = 9t_C$$

From this

$$t_C = \frac{215}{9} \approx 24°C$$

▮▮

12.2 Avogadro's Number and the Mole

In the next section we discuss how gases behave. To do that, we use terms that you learned in chemistry and may only faintly remember. We take time out in this section to refresh your memory.

A quantity called the *mole* (mol) has been defined by international agreement. This quantity is a measure of the *number* of objects. It is a very large number, about 6×10^{23}. We define it in terms of the number of carbon atoms in 12 g of one form of carbon, called carbon 12 (read this as "carbon twelve").

One mole is the amount of substance that contains the same number of objects as there are atoms in 12 g of carbon 12.

In keeping with our emphasis on the kilogram, we shall usually use the kilomole (kmol).

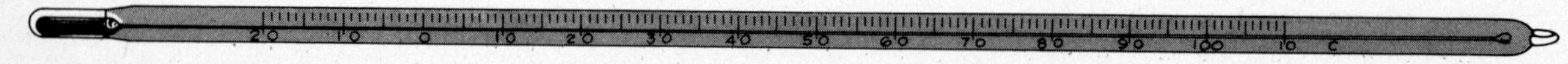

Figure 12.2 A typical Celsius thermometer.

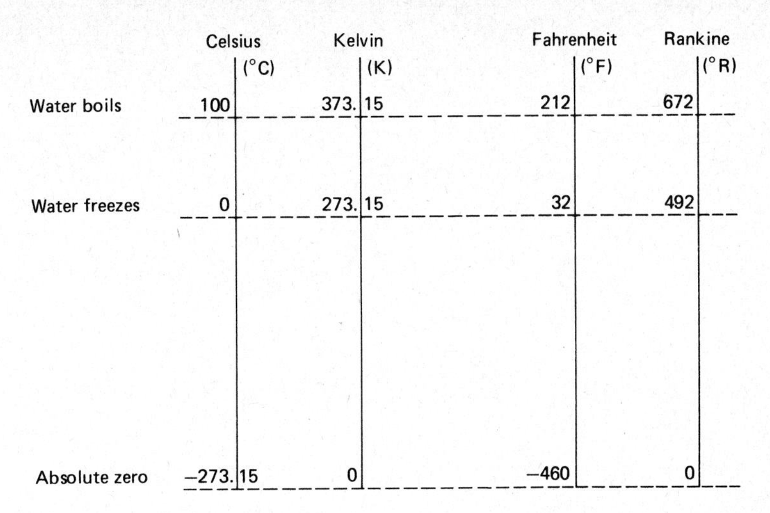

Figure 12.3 The chart compares the temperature scales.

One kilomole is the amount of substance that contains the same number of objects as there are atoms in 12 kg of carbon 12.

Another term that you may have heard is *Avogadro's number,* N_A. It is defined in terms of the mole or kilomole.

Avogadro's number, N_A, is the number of objects in a kilomole.

We can find this number by counting the number of carbon atoms in 12 kg of carbon 12. The experimental results give

$$N_A = 6.02 \times 10^{26}/\text{kmol}$$

Oftentimes people give $N_A = 6.02 \times 10^{23}/\text{mol}$. This is correct, of course, because there are 1000 more objects in a kilomole than in a mole. But we shall find it more convenient to use the value per kilomole.

The third term we wish to recall to your memory is the term *molecular weight, M.*

The molecular weight, *M*, of a substance is the mass in kilograms of one kilomole of the substance. Its units are kilograms per kilomole.[4]

For example, by definition, 1 kmol of carbon 12 has a mass of 12 kg. Therefore, $M = 12$ kg/kmol for carbon 12. (As you see, M is a mass and so the term "molecular weight" is a misnomer. It should be called "molecular mass," and it is sometimes so-termed. However, we follow the more common usage and refer to M as molecular weight.)

[4] Often people measure M in grams per mole (g/mol). The two values are numerically the same. Why?

Other familiar examples are $M = 2$ kg/kmol for hydrogen molecules and $M = 28$ kg/kmol for nitrogen molecules. (Recall that each of these molecules contains two atoms. They are diatomic gases.) Or the molecular weight of hydrogen *atoms* is 1 kg/kmol while for nitrogen *atoms* it is 14 kg/kmol. In the case of single atoms we usually use the term *atomic weight* (or *atomic mass*) in place of molecular weight. We shall follow custom and use these three terms interchangeably.

Now that we know Avogadro's number and the meaning of molecular weight, it is easy to compute the mass of a molecule. To do so, we note that M kilograms of a substance contains N_A molecules (or atoms if M is the atomic weight). Hence the mass of a single molecule (or atom) is given by

$$\text{Molecular mass} = m_0 = \frac{M}{N_A} \qquad (12.2)$$

For example, we know that the molecular weight of an oxygen molecule, O_2, is 32 kg/kmol. The mass of an oxygen molecule is therefore

$$m_0 \text{ for } O_2 \text{ molecule} = \frac{32 \text{ kg/kmol}}{6.02 \times 10^{26}/\text{kmol}} = 5.3 \times 10^{-26} \text{ kg}$$

Similarly, the atomic weight of the most abundant form of uranium is 238 kg/kmol. One of these atoms has a mass given by

$$m_0 = \frac{M}{N_A} = \frac{238 \text{ kg/kmol}}{6.02 \times 10^{26}/\text{kmol}} = 3.95 \times 10^{-25} \text{ kg}$$

12.3 The Ideal Gas Law

In this section we will learn how the pressure of a gas depends on temperature. This has great practical importance in industry, of course. But, equally important, we shall find that the pressure of a gas can be used as a tool to tell us about the molecular meaning of temperature.

Let us begin by considering the experiment shown in Figure 12.4. A gas, such as nitrogen, is held in a container of fixed volume V. The absolute pressure of the gas is measured as a function of its temperature on the Kelvin scale. (We use the Kelvin absolute scale because, as we will see, it is related in a simple way to the behavior of the molecules in the gas.) A typical experiment might produce the data shown in Figure 12.5(a).

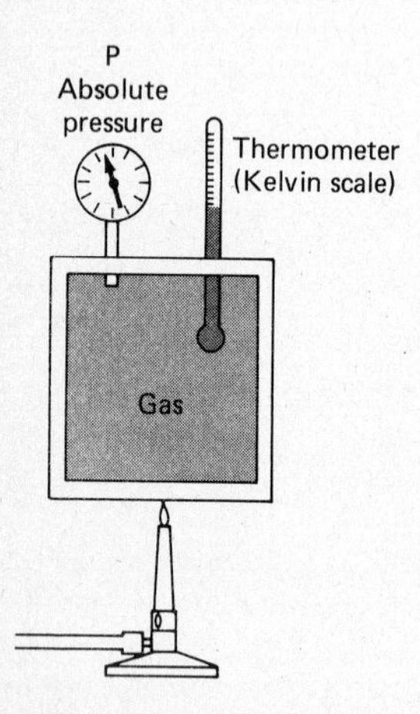

Figure 12.4 The pressure is measured as a function of temperature for a gas of fixed volume.

As we see in the figure, the pressure of the enclosed gas is proportional to the temperature. We can write that

$$P \sim T \qquad \text{(volume constant)}$$

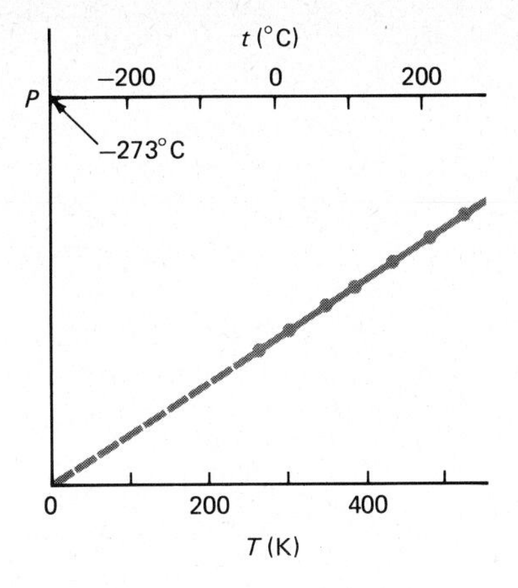

(a) Constant volume

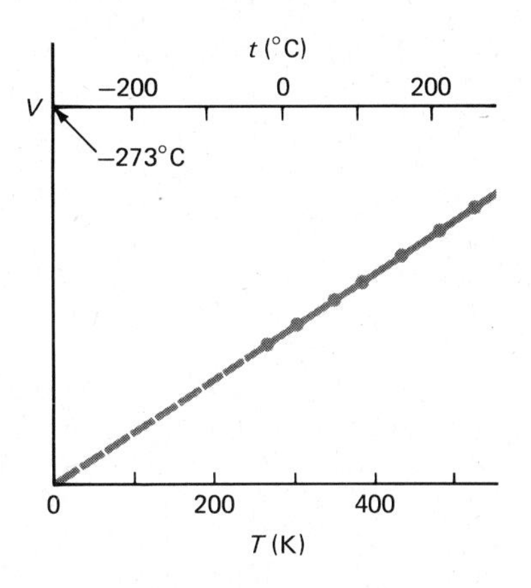

(b) Constant pressure

Figure 12.5 The graphs demonstrate the behavior of an ideal gas. Note that absolute zero (0 K or − 273.15°C) is the temperature at which P and V extrapolate to zero. The Celsius scale is included for comparison. T is used instead of t to represent absolute temperature.

Notice two things: P is the total pressure, not just the gauge pressure; and T is the temperature on the Kelvin scale, an absolute temperature scale.[5] Only on this scale does $P = 0$ when $T = 0$. Clearly, there is something distinctive about the temperature $T = 0$ K, −273.15°C, absolute zero. The pressure of a gas extrapolates to zero at that temperature.

If you repeat this experiment with any gas, you will obtain the same general result *provided* the gas is far above the temperature at which it liquifies. The data fit a straight line that extrapolates to $P = 0$ at $T = 0$. Of course, long before most gases are cooled to −273°C, they liquify. That is why no experimental points are shown at very low temperatures.

A second experiment we can do is shown in Figure 12.6. The gas is confined to a cylinder by a movable piston. The piston supplies a constant pressure to the gas. As the temperature of the gas changes, its volume changes. We show the results of a typical experiment in Figure 12.5(b). Notice that we plot the volume of the gas as a function of its absolute temperature.

The volume increases as the temperature is increased. Here, too, we obtain a straight line. The line tells us that the volume (at constant pressure) is proportional to the absolute temperature.

$$V \sim T \qquad \text{(pressure constant)}$$

Notice that absolute zero (0 K) is again a distinctive temperature. The volume of the gas extrapolates to zero at absolute zero.

Careful experiments such as these lead us to an important conclusion concerning a gas far removed from conditions under which it liquifies. Consider a mass m of gas whose molecular weight is M. The gas is confined to a volume V. Then the pressure of the gas is related to these other variables through the following experimental relation.

$$\boldsymbol{PV = \left(\frac{m}{M}\right)RT} \tag{12.3}$$

where T is the absolute temperature of the gas and R is a constant called the *gas constant.* The value of the gas constant, as determined by experiment, is 8314 J/(kmol)(K) and is the same for all gases. We call Equation 12.3 the *ideal* (or *perfect*) *gas law.* Any gas that obeys it is called an *ideal gas.*

We can see that Equation 12.3 agrees with the data in Figure 12.5(a). If V is constant, then

$$P = \left[\left(\frac{1}{V}\right)\left(\frac{m}{M}\right)R\right]T = (\text{constant})\ T$$

The pressure varies linearly with absolute temperature T, and $P = 0$ when $T =$

[5] Absolute temperatures are usually represented by T rather than by t.

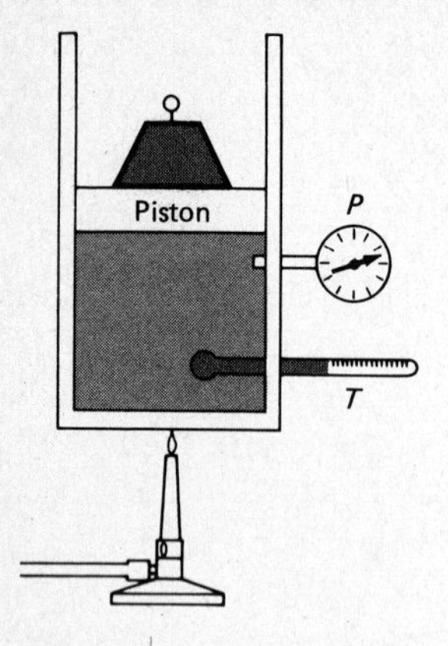

Figure 12.6 The pressure of the gas in this device is constant.

0. Similar considerations will show you that Equation 12.3 also corresponds to the data in Figure 12.5(b). Let us now use the gas law to solve a simple problem.

■ **EXAMPLE 12.2** A 100-cm^3 flask is filled with nitrogen at standard atmospheric pressure ($P = 1.01 \times 10^5$ Pa). Find the mass of nitrogen in the flask. The temperature is 20°C. The molecular weight of nitrogen, N_2, is 28 kg/kmol.

Solution We simply substitute in the ideal gas law, Equation 12.3. The values used are

$$P = 1.01 \times 10^5 \text{ Pa}$$
$$V = 100 \times 10^{-6} \text{ m}^3$$
$$T = 273 + 20 = 293 \text{ K}$$
$$M = 28 \text{ kg/kmol}$$

Substitution gives

$$m = \frac{PVM}{RT} = 1.2 \times 10^{-4} \text{ kg}$$

Notice that we changed all units to SI. In addition, T must be the absolute temperature, not the Celsius temperature. ■■

12.4 Gas Law Problems

Often one wishes to find how a change in P, V, and T affects a given quantity of ideal gas. For example, suppose that originally a gas has pressure = P_1, volume = V_1, and absolute temperature = T_1. After some change has taken place, the new values are P_2, V_2, T_2. But m, the mass of gas in the container, does not change.

In cases such as this we write the gas law, Equation 12.3, twice.

$$P_1V_1 = \left(\frac{m}{M}\right)RT_1 \quad \text{and} \quad P_2V_2 = \left(\frac{m}{M}\right)RT_2$$

If we divide one equation by the other, we obtain

$$\frac{P_1V_1}{P_2V_2} = \frac{T_1}{T_2} \tag{12.4}$$

This equation is of great utility when dealing with perfect gases. Notice that only the ratios of P and V appear. For this reason units frequently cancel from the equation. When units can be canceled, there is no need to change units from British to SI or to any particular SI units. However, T and P must always be absolute temperature and pressure.

■ **EXAMPLE 12.3** A car tire is inflated to a pressure of 24.0 lb/in.^{2} at a temperature of 20°C. Find its pressure at 60°C. Assume that the tire volume does not change. Atmospheric pressure is 14.7 lb/in.^{2}.

Solution We have

$$T_1 = 273 + 20 = 293 \text{ K}$$
$$T_2 = 273 + 60 = 333 \text{ K}$$

The pressure given is the tire gauge reading. Therefore

$$P_1 = 24.0 \text{ lb/in.}^2 + 14.7 \text{ lb/in.}^2 = 38.7 \text{ lb/in.}^2$$

We wish to find P_2.

From Equation 12.4 we have, after cross multiplication,

$$P_2V_2T_1 = P_1V_1T_2$$

Dividing by V_2T_1 gives

$$P_2 = P_1\left(\frac{V_1}{V_2}\right)\left(\frac{T_2}{T_1}\right)$$

Since $V_1 = V_2$,

$$P_2 = (38.7 \text{ lb/in.}^2)\left(\frac{333 \text{ K}}{293 \text{ K}}\right) = 44.0 \text{ lb/in.}^2$$
$$\text{Gauge pressure} = 44.0 \text{ lb/in.}^2 - 14.7 \text{ lb/in.}^2 = 29.3 \text{ lb/in.}^2$$

Notice that because this type of use of the gas law involves only ratios, the units cancel out. There is no need to change the units to the SI system. However, T must be absolute temperature and P must be the absolute, not gauge, pressure. ▮▮

▮ **EXAMPLE 12.4** In a diesel engine air and oil droplets are inserted into the cylinder. A piston compresses the air to about $\frac{1}{16}$ its original volume. The temperature rises rapidly during the compression and ignites the oil. Suppose that air at $P = 1$ atm is compressed to $\frac{1}{16}$ its original volume and a pressure of 50 atm. Find the temperature of the compressed air. Assume that $T_1 = 20°\text{C}$.

Solution From Equation 12.4 we have

$$T_2 = T_1\left(\frac{P_2}{P_1}\right)\left(\frac{V_2}{V_1}\right) = (293 \text{ K})\left(\frac{50}{1}\right)\left(\frac{1}{16}\right)$$
$$= 915 \text{ K}$$
$$= 643°\text{C}$$

This temperature is high enough to ignite the oil. ▮▮

12.5 The Kinetic Theory of Gases

The ideal gas law can be arrived at by considering a simple model of a gas. In so doing, we will learn how the motion of the gas molecules is related to the temperature of the gas. The model which we shall use is a simplified version of a portion of the *kinetic theory of gases.*

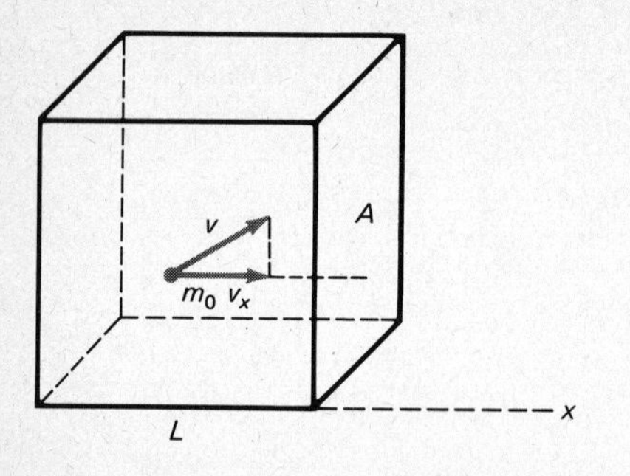

Figure 12.7 There are n molecules per unit volume. Hence in this box of volume $V = LA$, there are $n\ LA$ molecules.

Consider the situation in Figure 12.7. Shown here is one of the many molecules in the container of volume $V = LA$. The molecule has a mass m_0 and its average speed is v. Let us say that there are n molecules in each unit volume of the gas. Then the number of molecules in the box is nV which is the same as nLA.

When a molecule strikes the right-hand end of the box, its velocity component v_x will be reversed. It will therefore give an impulse to the wall of the box. The impulse is given by Equation 8.2.

$$\text{Impulse} = \text{change in momentum} = 2m_0v_x$$

Billions upon billions of such molecules strike the wall each second. Their combined impulses are the pressure of the gas on the wall.

Let us consider an average of the many molecules. This "average" molecule will have an x-directed speed v_x. Between collisions with the right-hand wall, it must travel back and forth across the box, a distance $2L$. Therefore, the time between collisions for it with this wall will be, from $x = v_xt$,

$$\text{Time between collisions} = \frac{2L}{v_x}$$

Let us now use this time to find the average force the molecule exerts on this wall.

From the definition of impulse, we know that

$$\text{Impulse} = \bar{F}t$$

where $\bar{F}$ is the average force during the time interval t. We have already found the impulse due to a molecule to be $2m_0v_x$. And the time interval for which we are finding the average is the time between collisions, $2L/v_x$. Substituting these values and solving for $\bar{F}$, we have

$$\text{Average force per molecule} = \frac{\text{impulse}}{t} = \frac{2m_0v_x}{2L/v_x} = \frac{m_0{v_x}^2}{L}$$

Our container has nLA molecules in it. The total force on the right-hand wall will be nLA times larger than the force due to a single molecule. We then have

$$\text{Force on right-hand wall} = nLA\left(\frac{m_0{v_x}^2}{L}\right) = nm_0{v_x}^2A$$

The pressure on this wall is simply F/A, and so we find

$$P = nm_0{v_x}^2$$

As we see, the pressure depends on the number of molecules per unit volume, the mass of each, and the square of the molecular speed.

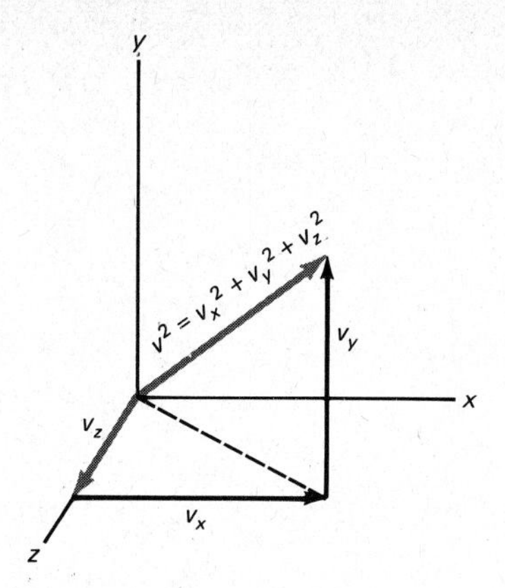

Figure 12.8 For the "average" molecule, $v_x^2 = v_y^2 = v_z^2 = \frac{1}{3}v^2$.

We can simplify this by referring to Figure 12.8. For our "average" molecule, $v_x^2 = v_y^2 = v_z^2$. Then, if you recall from geometry,

$$v^2 = v_x^2 + v_y^2 + v_z^2 = 3v_x^2$$

So we can replace v_x^2 by $v^2/3$. Therefore, the pressure of the gas is given by

$$P = \tfrac{1}{3}nm_0v^2 \tag{12.5}$$

This is to be compared with the ideal gas law, $PV = (m/M)RT$. To make this comparison more obvious, let us replace a few quantities in the gas law. We know from Equation 12.2 that $m_0 = M/N_A$ and so the gas law becomes

$$PV = \frac{m}{m_0N_A}RT$$

But m/m_0 is the number of molecules in the volume V. Therefore, $(m/m_0)/V$ is just n, the number of molecules per unit volume. We thus find that

$$P = n(R/N_A)T = nkT \tag{12.6}$$

where the constant $R/N_A = k$ is often called Boltzmann's constant. Its value is $k = 1.38 \times 10^{-23}$ J/K.

Equation 12.6 gives an experimental value for the pressure, while Equation 12.5 gives the value we have calculated from our bouncing ball model. The two values for P must be the same. So let us equate them.

$$\tfrac{1}{3}nm_0v^2 = nkT$$

Or, after a little algebra,

$$\tfrac{1}{2}m_0v^2 = \tfrac{3}{2}kT \tag{12.7}$$

Look what we have here! The kinetic energy of an average gas molecule, $\frac{1}{2}m_0v^2$, is proportional to the absolute temperature of the gas. The proportionality constant is $(3/2)k$ which is the same as $(3/2)(R/N_A)$. We conclude that the absolute temperature of an ideal gas has a molecular meaning:

The absolute temperature of an ideal gas measures the average translational energy of each gas molecule.

We see, then, that hot gases contain swiftly moving molecules. Cold gases contain molecules that move more slowly.

We have now given a molecular meaning to temperature. Let us use it to learn more about absolute zero. If $T = 0$, then Equation 12.7 tells us that $v = 0$. The gas molecules would no longer be moving. This makes sense. If the molecules were not moving, they would exert zero pressure on the container walls. They would also fall to the bottom of the container and so the volume filled by the gas would be nearly zero. These facts were indicated in the graphs of Figure 12.5. However, we know in practice that all gases condense at temperatures above $T = 0$. Therefore, even though our interpretation of T is correct, we cannot extrapolate the interpretation to $T = 0$.

EXAMPLE 12.5 How fast is the average nitrogen molecule moving in air at 20°C? M for nitrogen gas is 28 kg/kmol.

Solution We will make use of the relation we have found in Equation 12.7. First we calculate the mass of the nitrogen molecule.

$$m_0 = \frac{M}{N_A} = \frac{28 \text{ kg/kmol}}{6.02 \times 10^{26}\text{/kmol}} = 4.7 \times 10^{-26} \text{ kg}$$

Placing this in Equation 12.7 gives

$$v^2 = \frac{3kT}{m_0} = \frac{(3)(1.38 \times 10^{-23} \text{ J/K})(293 \text{ K})}{4.7 \times 10^{-26} \text{ kg}}$$

Evaluation of this gives $v = 510$ m/s. The average speed of nitrogen molecules in the air at 20°C is about 500 m/s. ∎∎

EXAMPLE 12.6 What volume does a kilomole of ideal gas occupy under standard conditions?

Standard temperature and pressure, abbreviated STP, are 1.01×10^5 Pa and 273.15 K.

Solution From Equation 12.3, the gas law, we have, because $m = M$ kg in this case,

$$PV = \left(\frac{M \text{ kg}}{M \text{ kg/kmol}}\right) RT$$

Using the values stated, this becomes

$$(1.01 \times 10^5)V = (8314)(273.15)$$

all in SI units. Solving for V gives

$$V = 22.4 \text{ m}^3$$

In other words, 1 kmol of ideal gas occupies a volume of 22.4 m^3 at STP.

12.6 Speed Distribution in Gases

As we have seen, the temperature of a gas is a measure of the translational kinetic energy of its molecules. We found that $T \sim v^2$, where v is an average molecular speed. Of course, all of the molecules will not have this speed. Some will, for an instant, even be at rest. But as the molecules collide with each other, their speeds will change.

In spite of the complicated way in which molecules move, we can still learn much about them. We have already learned that their average translational kinetic energy is given by Equation 12.7. We can go one step further. It is possible to state the chance that a molecule has a certain speed. In technical terminology we talk about the probability that a molecule has a given speed.

Suppose that one looks at N different molecules and measures the speed of each. Suppose that n of these N molecules have a speed within the range $v \pm \frac{1}{2}$ m/s. Then we say that the probability that a molecule has a speed v is n/N.

For example, suppose that we measure the speed of 1,000,000 molecules. Suppose further that 83 have speeds between 520.5 and 521.5 m/s. Then the probability that a molecule has a speed of $521 \pm \frac{1}{2}$ m/s is $83/10^6 = 0.83 \times 10^{-4}$ or 83×10^{-4} percent. Let us now see what this means for our ideal gas.

James Clerk Maxwell was the first man to compute how the speeds of gas molecules are distributed. (Maxwell also was a pioneer in the field of electromagnetic waves.) Using the ideas of the kinetic theory, he computed the probability that a gas molecule has a speed v. His result for nitrogen gas molecules is shown graphically in Figure 12.9.

As we see in Figure 12.9, most molecules of N_2 gas at 460 K (187°C) have speeds close to 550 m/s. At 1700 K the speeds are closer to 1000 m/s. But the molecular speeds range widely. Some molecules have zero speed. The speeds of

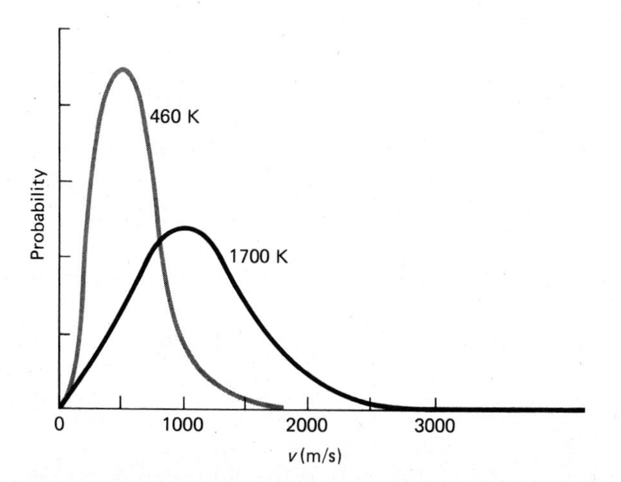

Figure 12.9 The curves describe the probability that nitrogen gas molecules have a speed v at two different temperatures. The graph is called the Maxwell-Boltzmann speed distribution.

others are very high. However, as we would expect, the number of molecules with high speeds increases as the temperature of the gas is raised. We shall see later that this distribution in speeds (and energies) is of great importance in explaining evaporation.

■ **EXAMPLE 12.7** How high above the earth can an average nitrogen molecule rise? Assume that the temperature of the air is 20°C. Neglect the variation in T and g with height.

Solution The average molecule has a translational kinetic energy given by Equation 12.7.

$$\tfrac{1}{2}m_0v^2 = \tfrac{3}{2}kT = (\tfrac{3}{2})(1.38 \times 10^{-23}\ \text{J/K})T$$

In our case $T = 293$ K and so

$$\tfrac{1}{2}m_0v^2 = 6.1 \times 10^{-21}\ \text{J}$$

The average molecule can rise to a height that changes this kinetic energy to potential energy. Therefore,

$$\text{GPE at top} = \text{KE at bottom}$$

or

$$m_0gh = \tfrac{1}{2}m_0v^2$$
$$m_0gh = 6.1 \times 10^{-21}\ \text{J}$$

Solving for h, we have

$$h = \frac{6.1 \times 10^{-21}\ \text{J}}{(9.8\ \text{m/s}^2)m_0}$$

Now the mass of a nitrogen molecule ($M = 28$ kg/kmol) can be found as follows.

$$m_0 = \frac{M}{N_A} = \frac{28\ \text{kg/kmol}}{6.02 \times 10^{26}/\text{kmol}} = 4.65 \times 10^{-26}\ \text{kg}$$

Then, after substituting and evaluating,

$$h = 13{,}400\ \text{m}$$

This is a height of about 8.3 miles.

As we see, the average nitrogen molecule has enough energy to rise about 13 km above the earth. For this reason the earth's atmosphere thins greatly at high altitude. Only a small fraction of the molecules have enough energy to reach great heights. Notice that, because $h \sim 1/m_0$, the lighter gas molecules such as hydrogen can go higher than nitrogen does. The composition of the air therefore changes as we go high above the earth. ■■

12.7 Thermal Expansion

As we have seen, the molecules in a gas speed up as the temperature is increased. In fact, a simple relation exists between T and v for a gas. This was given in Equation 12.7. The situation is not so simple for liquids and solids.

The molecules in liquids and solids also have kinetic energy. This kinetic energy increases as the temperature of the substance is increased. They also possess potential energy which increases with temperature. We call this temperature-dependent energy *thermal energy.*

Thermal energy is dependent on temperature and is the energy that molecules have because their temperature is higher than absolute zero. It causes thermal motion of the molecules.

Because of their thermal energy, the molecules in a liquid or solid vibrate, rotate, and sometimes translate. The molecules are in continual motion. We call this motion *thermal motion.*

At low temperatures the molecules in a liquid or solid do not have much energy. But as the temperature is increased, they vibrate more widely. They have enough energy to push their neighbor molecules out of the way. As a result, their motion causes them to occupy more space than they occupy at lower temperatures. Therefore, a liquid or solid usually expands as the temperature is increased. We call this expansion as the temperature is increased *thermal expansion.*

Linear Thermal Expansion Suppose, as shown in Figure 12.10, that a rod has a length L_0 at temperature t_0. If the temperature is raised to t, the rod's length increases to L. Experiment shows that, when $t - t_0$ is not too large,

$$L - L_0 \sim (t - t_0)L_0$$

This can be written as an equation by defining the coefficient of linear thermal expansion, α. We have

$$\frac{L - L_0}{L_0} = \alpha(t - t_0) \qquad (12.8a)$$

or

$$\Delta L = L_0\alpha \, \Delta t \qquad (12.8b)$$

where $\Delta L = L - L_0$ and $\Delta t = t - t_0$.

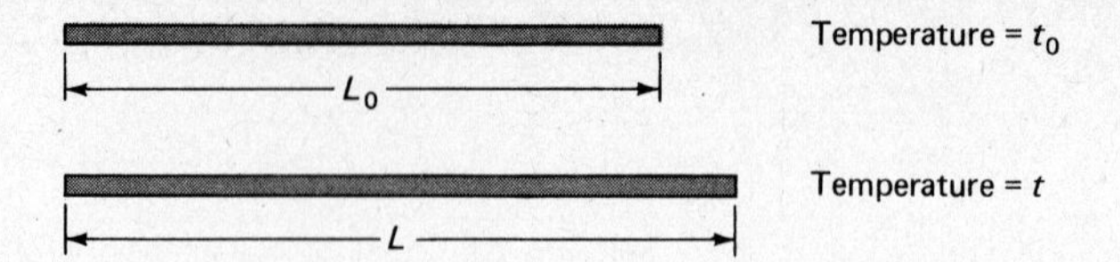

Figure 12.10 The coefficient of linear expansion is $\alpha = (L - L_0)/L_0(t - t_0)$.

Each material expands differently, so α is different for different materials. Because the units cancel in $(L - L_0)/L_0$, α has units of 1/°C or 1/°F. A few typical values for α are given in Table 12.1. Notice that α is a measure of the fractional elongation per each degree of temperature change.

Volume Thermal Expansion As a solid object is heated, each of its linear dimensions will change. This will result in an expansion of the volume of the object. To describe this effect, we define a volume thermal expansion coefficient β (β is the Greek letter beta). Consider an object of volume V_0 that is subjected to a temperature increase Δt. Then the object's volume will change by an amount

$$V - V_0 = \Delta V = V_0 \beta \, \Delta t \tag{12.9}$$

Typical values for β are given in Table 12.1. ∎∎

∎ **EXAMPLE 12.8** A certain piece of brass is 0.5000 cm thick and has a volume of 885.0 cm^3 at 20°C. Find the increase in these quantities when the brass is placed in boiling water at 100°C.

Table 12.1 EXPANSION COEFFICIENTS[a]

Material	Linear α(1/°C)	Volume β(1/°C)
Aluminum	25×10^{-6}	75×10^{-6}
Brick	$\approx 10 \times 10^{-6}$	$\approx 30 \times 10^{-6}$
Brass	19×10^{-6}	57×10^{-6}
Concrete	$\approx 10 \times 10^{-6}$	$\approx 30 \times 10^{-6}$
Diamond	1.2×10^{-6}	3.6×10^{-6}
Glass (Pyrex)	$\approx 3 \times 10^{-6}$	$\approx 9 \times 10^{-6}$
Iron	12×10^{-6}	36×10^{-6}
Lead	30×10^{-6}	90×10^{-6}
Steel	12×10^{-6}	36×10^{-6}
Tungsten	4.4×10^{-6}	13×10^{-6}
Vinyl siding	$\approx 150 \times 10^{-6}$	$\approx 450 \times 10^{-6}$
Acetone	—	1490×10^{-6}
Benzene	—	1240×10^{-6}
Gasoline	—	$\approx 950 \times 10^{-6}$
Mercury	—	182×10^{-6}

[a]These are average values for the range 0–100°C. To change the coefficients to 1/°F, multiply the given value by $\frac{5}{9}$.

Solution The thickness change can be obtained by using Equation 12.8. We have

$$L - L_0 = L_0\alpha(t - t_0)$$

From Table 12.1, $\alpha = 19 \times 10^{-6}/°\text{C}$. So

$$\begin{aligned} L - L_0 &= (0.50\ \text{cm})(19 \times 10^{-6}/°\text{C})(80°\text{C}) \\ &= 6.4 \times 10^{-4}\ \text{cm} \end{aligned}$$

To find the change in volume, we use Equation 12.9.

$$\begin{aligned} V - V_0 &= V_0(\beta)(t - t_0) \\ &= (885\ \text{cm}^3)(57 \times 10^{-6}/°\text{C})(80°\text{C}) \\ &= 4.0\ \text{cm}^3 \end{aligned}$$

■■

■ **EXAMPLE 12.9** A 22-gal gas tank is filled with gasoline at 15°C. After the tank has been standing in the sun, the temperature of the gas rises to 32°C. How much gasoline flows out of the overflow on the tank?

Solution If you look at Table 12.1, you will notice that gasoline expands much more than do metals. Therefore, to a fair approximation, we can ignore the expansion of the gas tank.

The expansion of the gasoline is, from Equation 12.9,

$$V - V_0 = V_0\beta(t - t_0)$$

Therefore

$$\begin{aligned} V - V_0 &= (22\ \text{gal})(950 \times 10^{-6}/°\text{C})(17°\text{C}) \\ &= 0.36\ \text{gal} \end{aligned}$$

This much gas will overflow.

■■

■ **EXAMPLE 12.10** If the gas tank in the previous example had been made of steel, how much would its capacity have increased?

Solution The metal tank will expand in the same way as if it were solid.[6] Then from Equation 12.9,

$$(V - V_0)_{\text{tank}} = V_0\beta(t - t_0)$$

Using the value of β for steel gives

$$\begin{aligned} (V - V_0)_{\text{tank}} &= (22\ \text{gal})(36 \times 10^{-6}/°\text{C})(17°\text{C}) \\ &= 0.013\ \text{gal} \end{aligned}$$

[6] A hole within any uniform solid must expand in the same way as the solid itself expands. The solid around the hole expands in the same way whether or not the hole is there. Therefore the boundary of the hole and the hole itself expand exactly as though the hole were filled with the same solid.

Therefore, only 0.35 gal instead of 0.36 gal would have overflowed in the previous example. ■■

■ **EXAMPLE 12.11** A steel beam has cross-sectional area 200 cm^2 and length 15 m. Its two ends are embedded in concrete pillars. The beam was put in place when the temperature was 0°C. With what force does the beam push against the supports when the temperature is 35°C?

Solution First let us find how much the beam would lengthen if the supports were not there.

$$\begin{aligned} L - L_0 &= L_0\alpha(t - t_0) \\ &= L_0(12 \times 10^{-6}/^\circ\text{C})(35^\circ\text{C}) \\ &= 4.2 \times 10^{-4} L_0 \end{aligned}$$

But the pillars will not allow the beam to lengthen. In effect, they must compress the beam by the amount we have just calculated. From the definition of Young's modulus we have

$$\frac{\Delta L}{L_0} = \frac{(F/A)}{Y}$$

For steel $Y = 20 \times 10^{10}$ N/m^2 from Table 11.1. Also, we were told that $A = 0.0200$ m^2. Because

$$\frac{\Delta L}{L_0} = \frac{L - L_0}{L_0}$$

we have

$$4.2 \times 10^{-4} = \left(\frac{F}{0.0200\ \text{m}^2}\right)\frac{1}{20 \times 10^{10}\ \text{N/m}^2}$$

From this

$$F = 1.68 \times 10^6\ \text{N}$$

Because 1 N = 0.225 lb, this is equivalent to a force of 378,000 lb.

As we see, huge forces are needed to prevent thermal expansions. You have no doubt heard of roadways and bridges that failed because of thermal expansion. In all situations where expansion is of importance, suitable expansion gaps must be provided. ■■

Summary

The Celsius (or centigrade) temperature scale takes the melting point of ice and the boiling point of water (under 1 atm pressure) to be 0°C and 100°C, respectively. These two temperatures are 32°F and 212°F on the Fahrenheit scale. We can convert between these two scales by means of the relation

$$\frac{t_F - 32}{t_C} = \frac{9}{5}$$

We represent a Fahrenheit scale temperature by t_F and a Celsius temperature by t_C.

An absolute temperature scale has its zero at absolute zero. Absolute zero is at -273.15°C or -460°F. Two absolute temperature scales exist. The Kelvin scale uses Celsius-size degrees. We have, for temperatures on the Kelvin scale,

$$T = 273.15 + t_C$$

The Rankine absolute scale uses Fahrenheit-size degrees. For temperatures on it, we have

$$T_R = 460 + t_F$$

All gases that are far removed from their condensation pressure and temperature approximate ideal gases. They obey the ideal (or perfect) gas law:

$$PV = \left(\frac{m}{M}\right)RT$$

In this relation a mass of gas m with molecular weight M is confined to a volume V. Its pressure is P and its absolute temperature is T. The quantity R is the gas constant and has a value 8314 J/(kmol)(K).

The ideal gas law is frequently applied to a fixed mass of gas. In that case the relation between the gas under two different conditions is

$$\frac{P_1V_1}{P_2V_2} = \frac{T_1}{T_2}$$

The kinetic theory of gases shows that the absolute temperature T has the following meaning. In a gas with molecules of mass m_0, the absolute temperature is related to the translational kinetic energy of a molecule by

$$\tfrac{1}{2}m_0v^2 = \tfrac{3}{2}(R/N_A)T = \tfrac{3}{2}\mathrm{k}\mathrm{T}$$

where k is Boltzmann's constant, 1.38×10^{-23} J/K. Absolute temperature measures the average translational KE of the molecules in an ideal gas.

The molecules in a gas have a wide range of speeds. When the temperature of the gas is increased, the distribution of speeds shifts to higher speeds. At room temperature the nitrogen molecules of the air have average speeds close to 500 m/s.

Within a limited temperature range the thermal expansion of many substances is given by

$$\text{Linear} \qquad \Delta \mathrm{L} = L_0\alpha\,\Delta t$$
$$\text{Volume} \qquad \Delta \mathrm{V} = V_0\beta\,\Delta t$$

The quantity α is the coefficient of linear expansion and β is the coefficient of volume expansion.

Questions and Exercises

1. Water contracts as its temperature is raised from 0°C to 4°C. Why does this make it unsuitable for a thermometer liquid?
2. Many thermometers do not use the expansion of a liquid. For example, circular dial thermometers operate on an entirely different principle. List as many different kinds of thermometers as you can and describe the principle basic to the operation of each.
3. Boyle's law for gases states: "The volume of a gas varies inversely to the pressure provided the mass and temperature of the gas are maintained constant." Show that Boyle's law is a special case of the ideal gas law.
4. Charles' law for gases states: "The volume of a gas increases in direct proportion to the absolute temperature provided the pressure and mass of the gas are maintained constant." Show that this is a special case of the ideal gas law.
5. Dalton's law of partial pressure states: "The total pressure of a mixture of gases is equal to the sum of the partial pressures of the gases in the mixture." Using the ideal gas law and our kinetic theory ideas, justify Dalton's law.
6. Tiny smoke, dust, and water particles obey the equations we have found for molecules of an ideal gas. By reference to Equation 12.7, explain why large dust particles settle to the bottom of a closed container, whereas the smallest ones do not.
7. In order to escape from the earth, a rocket must be shot out from it with at least a speed of 11,200 m/s. Explain why only a tiny amount of hydrogen exists in the atmosphere even though billions of years ago there may have been more hydrogen than nitrogen in it.
8. It is common practice to shrink-fit hollow metal cylinders tightly to rods. The outer sleeve has a hole slightly smaller than the rod onto which it is to fit. The heated sleeve is slid over the cool rod. After cooling, the rod is firmly fastened to the sleeve. Explain the physics of this operation. Can you think of any similar applications of this idea?
9. Vinyl house siding has prepunched slots through which the siding is to be nailed. Why are slots, rather than circular holes, used? What precautions must one take when applying this type of siding? Why does this problem not arise with wood and aluminum siding?
10. Chemists and others who work with glass never, never mix pieces of soft glass and heat-resistant (Pyrex) glass. The two have very different coefficients of thermal expansion. Explain why a joint made by fusing the two glasses together almost always breaks when cooled.
11. Water in the range 0°C to 4°C behaves much differently than most substances. It contracts rather than expands. It increases in volume when the liquid freezes. Discuss how the world as we know it would be changed if water behaved like most liquids.

Problems

1. Mercury freezes at -38.9°C and boils at 356.6°C. What are these two temperatures on the Fahrenheit, Kelvin, and Rankine scales?
2. On a certain day the lowest and highest temperatures in the United States were -21°F and 73°F. What are these two temperatures on the Celsius, Kelvin, and Rankine scales?

*3. At what temperature do the Celsius and Fahrenheit thermometers read the same numerical value?

*4. At what temperature do the Celsius and Fahrenheit thermometers read the same numerical value except that the signs of the two values are opposite?

5. How many moles are there in 100 g of water? M for water is 18.0 kg/kmol.
6. A chemist wishes to dissolve 0.150 mol of sodium chloride in a 250-cm^3 flask. How much NaCl should she put in the flask? M for NaCl is 58.44 kg/kmol.
7. What is the mass of a water molecule? M for water is 18.0 kg/kmol.
8. The molecular weight for benzene is 78 kg/kmol. What is the mass of a benzene molecule?

*9. (a) How many atoms are there in 50 g of carbon 12? (b) How many atoms are there in 4 kg of sodium? The atomic weight of sodium is 23 kg/kmol.

*10. (a) How many atoms are there in 4 kg of helium? M_{He} = 4 kg/kmol. (b) How many atoms are there in 5 g of nitrogen? The atomic mass for nitrogen is 14 kg/kmol.

*11. How many kilomoles of substance are there in each of the following: (a) 3 kg iron, (b) 100 kg uranium, (c) 5 g copper? Use the atomic masses given in Appendix 3.

*12. Use the data in Appendix 3 to find the mass of a mole of each of the following substances: (a) sulfur, (b) silicon, (c) diatomic oxygen gas molecules. How many *atoms* are there in a gram of each of these substances?

*Problems marked with an asterisk are not as easy as the unmarked ones.

**13. The molecular weight of drinking alcohol (ethanol) is 46 kg/kmol. How many alcohol molecules are there in 100 cm^3 of ethanol? The density of ethanol is 790 kg/m^3.

14. A student took the following data for air confined to a constant volume. Plot the data and use them to find absolute zero on the Celsius scale. What value is obtained?

P (cm of Hg) →	78.0	84.0	91.1	97.6	103.6
t (°C) →	1.0	23	45	70	91

15. Air is trapped in the sealed-off end of a capillary tube by a mercury column above it. Call the length of the air column h. The temperature of the tube is varied. The following data were taken at constant pressure. Plot the data and use them to find absolute zero on the Fahrenheit scale. What value is obtained?

h (cm) →	21.7	22.2	22.9	23.5	24.3
t (°F) →	33.0	45.0	57	73	91

16. Air at atmospheric pressure (101 kPa) is to be compressed to one-tenth of its original volume. What pressure will the air be at if its temperature is unchanged?

17. A pressure gauge on a 2000-cm^3 tank of oxygen gas reads 600 kPa. How much volume will the oxygen occupy at the pressure of the outside air, 100 kPa?

18. (a) A 90-gal water tank is filled with air at atmospheric pressure, 100 kPa. Water is pumped from a well into the tank without allowing any air to escape. What is the volume of the air when the tank pressure gauge reads 306 kPa? (b) Water is now drained out of the tank until the pressure reading on the gauge drops to 136 kPa. What is the volume of air in the tank then?

*19. While swimming at a depth of 12.0 m in a freshwater lake, a fish emits an air bubble of volume 2.0 mm^3. (a) What is the original pressure on the bubble? (b) What is the volume of the bubble as it reaches the surface? Atmospheric pressure there is 100 kPa.

20. A gas is confined to a cylinder by a piston. Its original conditions are $P = 1$ atm, $t = 27°C$, and $V = V_0$. After compression, $P = 12$ atm and $V = V_0/10$. What is the final temperature of the gas?

21. Gas is confined to a cylinder by a piston. The original temperature of the gas is 20°C and its gauge pressure is 180 lb/in.^{2}. The gas is now allowed to expand until its pressure reaches atmospheric, 14.5 lb/in.^{2}. Its temperature is then 13°C. Find the ratio of its initial to final volume.

22. A 2-m^3-volume tank holds nitrogen gas ($M = 28$ kg/kmol) at an absolute pressure of 500,000 Pa. The temperature is 27°C. Find the mass of nitrogen in the tank.

23. Five kilograms of helium gas ($M = 4$ kg/kmol) are contained in a 0.50-m^3 tank at 20°C. What is the absolute pressure in the tank?

24. How large a volume does 3 kg of oxygen gas ($M = 32$ kg/kmol) occupy under standard conditions? (Standard conditions are $T = 0°C$, $P = 101$ kPa.)

*25. A droplet of liquid nitrogen (0.030 g) is sealed off in a cold tube whose volume is 20 cm^3. Find the nitrogen pressure in the tube when the temperature has risen to 20°C.

*26. A 2.0-liter (2×10^3 cm^3) tank of helium gas ($M = 4.0$ kg/kmol) has a gauge pressure of 610 kPa and is at room temperature. About what mass of helium is there in the tank?

*27. About what mass of air is in a $3 \times 5 \times 6$-m^3 room under STP?

28. The average temperature of outer space is about −270°C. Find the average speed of a helium atom ($M = 4$ kg/kmol) in outer space.

29. It is estimated that the temperature of the sun's interior is 10×10^6 K. What is the average speed of a helium atom ($M = 4$ kg/kmol) in a gas at this temperature?

*30. A 2.0×10^{-14}-g dust particle is at the same temperature as the air (22°C) in a room. Assuming the particle has average thermal energy, how high should it be able to rise in the still air?

31. A steel measuring tape is 10.000 m long at 20°C. How much longer will it be at 30°C?

32. A steel gauge block is calibrated to be correct at 20°C. Find the fractional error in its dimension at 32°C. What will be its percent error?

33. At 20°C an aluminum ball of diameter 2.000 cm just barely passes through a hole in an aluminum plate. How much larger must the hole diameter be if the ball is to just pass through at 100°C? (a) Assume that both plate and hole are at 100°C. (b) Assume that the plate is at 20°C.

34. Back in the stagecoach era, rims were placed on wooden wheels by shrink-fitting. Suppose that the wheel has a diameter of 100.00 cm, whereas the rim has an inside diameter of 99.75 cm at 20°C. How hot must the iron rim be heated if it is to be slid over the wheel?

35. A certain amount of benzene occupies 250 cm^3 at 20°C. What will its volume be at 30°C?

36. A brass cube has a volume of exactly 4 cm^3 at 20°C. What will its volume be at 0°C?

37. A Pyrex flask has an internal volume of exactly 100 cm^3 at 20°C. What will its internal volume be at 30°C?

38. A 100-cm^3 Pyrex flask at 20°C is filled to the mark with

**Problems marked with a double asterisk are somewhat more difficult than the average.

acetone at 30°C. When the acetone cools to 20°C, how much more acetone must be added to fill to the mark?

**39. A chemist is preparing a dilute benzene solution and wishes the concentrations to be correct at 0°C. He is working at 20°C. He fills a 50-cm^3 volumetric flask to the mark. How much more benzene should be added to make the flask full at 0°C?

**40. How many nitrogen molecules exist in 1 cm^3 of nitrogen under STP?

**41. Under standard conditions, a nitrogen molecule in the air moves about 10^{-6} cm between collisions. (a) About how many collisions does it have each second? (b) What would the pressure have to be (in centimeters of Hg) for its distance between collision to be about 1 m?

**42. Show that for a uniform material the volume coefficient of expansion is three times larger than the linear expansion coefficient. Hint: Consider the expansion of a cube and assume the expansion is small.

**43. Use the gas law to show that the bulk modulus of a gas at constant temperature is equal to the pressure of the gas.

Forsyth, Monkmeyer

13 HEAT ENERGY AND ITS EFFECTS

Performance Goals

When you finish this chapter, you should be able to

1. Define heat energy in terms of energy flow between objects at different temperatures. State the direction of energy flow. Explain what happens within an object when it is cooled or heated.
2. List the common units used to measure heat energy. State how they are related to the joule and foot-pound. In terms of them, state how much heat energy is

In the last chapter we found that temperature is a measure of molecular thermal energy. We shall see in this chapter how various substances are affected by the addition of heat energy. Effects such as melting, boiling, and vaporization will be discussed. In order to carry out these discussions, we must first decide on a unit in which to measure heat energy.

needed to heat 1 g of water by 1°C and 1 lb of water by 1°F.

3. In your own words, define specific heat capacity of a substance. Give its value for water.
4. Solve the following type of problem. A mass m (or weight w) of substance is heated ΔT degrees by an amount of heat ΔQ. Given three of the following quantities, find the fourth: m (or w), ΔQ, ΔT, c.
5. Solve the following type of problem. A mass m (or weight w) of substance is melted (or frozen) with a heat gain (or loss) ΔQ. Given two of the following, find the third: m (or w), ΔQ, H_f (heat of fusion).
6. Solve the following type of problem. A mass m (or weight w) of substance is vaporized (or condensed) with a heat gain (or loss) ΔQ. Given two of the following, find the third: m (or w), ΔQ, H_v (heat of vaporization).
7. Solve the following type of problem. A known quantity of substance is mixed with a known quantity of another substance at a different temperature. H_v, H_f, and c (specific heat capacity) are known for both substances. Given the initial temperatures, find the final equilibrium temperature.
8. Explain in your own words what is meant by latent heat of fusion and latent heat of vaporization. In your explanation describe the molecular processes involved and tell what happens to the temperature during melting and boiling.

13.1 The Meaning of Heat

We all know that heat flows by itself from hot to cold, but not from cold to hot. When placed in contact with a cold object, a hot object warms the cold one. The reverse does not occur. Although this fact had been known for many centuries, the meaning of heat flow was unknown until about 1800. Not until the kinetic theory of gases was developed did we learn the real meaning of heat energy. As we saw in the last chapter, the kinetic theory gives us the meaning of temperature. A hot gas has in it molecules with high average kinetic energy. The molecules of a cold gas have lower average kinetic energy. Let us now discuss what happens when a hot gas is placed in contact with a similar cold gas.

Suppose that two containers of nitrogen gas have temperatures T_1 and T_2, as shown in Figure 13.1. Let T_1 be greater than T_2. We know, then, that the nitrogen molecules in container 1 are moving faster than those in container 2. Suppose now that the two containers are connected so that the molecules can mix. The fast molecules from 1 will collide with the slow molecules from 2. Finally, the molecules completely mix. Their average molecular energy is then somewhere in between the energies of the original two gases. The final temperature T_f is in between T_1 and T_2.

Notice what is happening here. The two gases come into contact. They interchange kinetic energies until their average molecular energies are equal. Energy is transferred from the molecules originally in container 1 to those of container 2. We say that heat energy flows from container 1 to container 2.

It is not necessary for the two gases to mix for heat energy to flow. Suppose that a thin metal film is placed between the two containers as shown in part (d) of Figure 13.1. The molecules of the hot gas give energy to the molecules of the film as they strike it. The molecules of the film then give this energy to the molecules of the cold gas as these molecules collide with the film. Molecular kinetic energy is thereby transferred from the hot to the cold gas.

We can extend this reasoning to include a hot solid (or liquid) brought into contact with a cold solid (or liquid). Experiments show that the hot object cools while the cold object warms. Eventually, the objects reach thermal equilibrium. Their temperatures are then equal.

Let us now summarize what we have said.

Two objects with temperatures T_1 and T_2 ($T_1 > T_2$) are placed in thermal contact. Molecular kinetic energy is then transferred from the hot to the cold object. When equilibrium is attained, both objects have the same temperature T with $T_1 > T > T_2$.

Further,

The energy transferred in a situation such as this is called *heat energy*. Heat energy is the energy that is transferred from a hot object to a cooler one as a result of their difference in temperature. It always flows from the hot object to the cold object.

9. Use a set of data for the vapor pressure of a liquid as a function of temperature to give the pressure of the vapor in a closed container partly filled with the liquid. State from the data, if possible, at what temperature the liquid would boil at a given pressure upon the liquid.
10. Explain in your own words why a liquid boils at different temperatures depending upon the pressure.
11. Determine c and H_f from a graph such as Figure 13.3.
12. Knowing the heat of combustion of a process, compute how much heat is generated as a substance is burned.

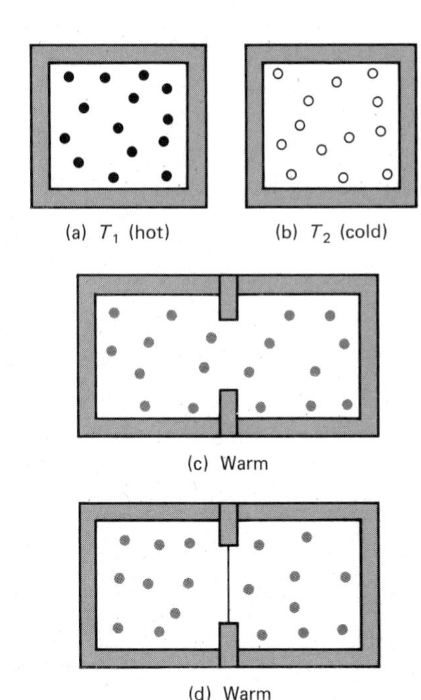

Figure 13.1 When the two containers of gas are brought into thermal contact, the final temperature T is such that $T_1 > T > T_2$.

13.2 Units of Heat Energy

As we have seen, heat energy is simply one form of energy. We should therefore measure it in joules or foot-pounds. However, people were using heat long before they knew its basic nature. They needed units in which to measure it. These units are still widely used today. They were originally defined in the following way.

1 calorie (cal) is the amount of heat energy needed to raise the temperature of 1 g of water from 14.5° to 15.5°C.

1 British thermal unit (Btu) is the amount of heat energy needed to raise the temperature of 1 lb of water[1] from 63° to 64°F.

These two units are related by 1 Btu = 252 cal. The heat needed to change the temperature of water varies slightly with temperature. In the range 0° to 100°C, the two extremes are as follows: 1.008 cal/(g)(°C) at 0°C and 0.997 cal/(g)(°C) at 40°C. We shall usually ignore this slight variation.

In recent years accurate measurements have been made to determine the relation between the calorie and the joule. We now define the calorie and Btu in terms of the joule as follows:

$$1 \text{ cal} = 4.184 \text{ J}$$
$$1 \text{ Btu} = 1054 \text{ J} = 778 \text{ ft}\cdot\text{lb}$$

In other words, 4.184 J of work is equivalent to 1 cal of heat. This conversion is often referred to as the *mechanical equivalent of heat.* As a word equation it becomes

$$\frac{\text{Mechanical energy (in joules)}}{\text{Equivalent heat energy (in calories)}} = \text{mechanical equivalent of heat} = 4.184 \text{ J/cal}$$

It is useful in situations such as that discussed in the following example.

Before taking up that example, we should note that another unit, the kilocalorie (kcal), is also used. Nutritionists call this a Calorie, and so confusion sometimes results. The nutritionist's Calorie is actually a kilocalorie. It is sometimes referred to as a "large calorie."

EXAMPLE 13.1 The work done each second against friction in a certain bearing is 0.40 J/s. How much heat develops each second in the bearing?

Solution The friction work results in heat energy. Therefore, 0.40 J/s of heat is developed each second. Using the conversion factor, we have

[1] It is assumed that the weight 1 lb is measured on the surface of the earth.

$$\text{Heat developed per second} = 0.40\ \text{J/s}$$

$$= \left(0.40\frac{\text{J}}{\text{s}}\right)\left(\frac{1\ \text{cal}}{4.184\ \text{J}}\right) = 0.096\ \text{cal/s}$$

To keep the bearing cool, 0.096 cal of heat must be carried away from the bearing each second. ▮▮

13.3 Specific Heat Capacity

Each substance heats up in its own way. Most substances require less heat than water does to change temperature. Let us see why substances differ in the way they heat up.

When molecular kinetic energy is added to a substance, the energy does not all go to translational energy of the molecules. Energy is also used to increase the rate of rotation and vibration of the molecules. Often energy is also used to break small groups of molecules apart from each other. Substances therefore differ in the heat energy needed to increase their temperature.

Because substances heat differently, we need some way to determine how much heat energy is required to raise the temperature of a given substance. To be able to do this we use the *specific heat capacity* of the substance, defined as follows.

The specific heat capacity, c, of a substance is the quantity of heat needed to raise the temperature of a unit quantity of substance by one degree.

It is customary to call the quantity of heat added (or taken away) ΔQ. If this causes a change in temperature ΔT to a mass m of substance, then

$$c \equiv \frac{\Delta Q}{m\,\Delta T} \qquad \text{or} \qquad \Delta Q = cm\,\Delta T \tag{13.1a}$$

The units for c are J/(kg)(K) in the SI, although the unit cal/(g)(°C) is frequently used. When using ΔQ in British thermal units, we usually express the quantity of substance by its weight, w.

$$c \equiv \frac{\Delta Q}{w\,\Delta T} \qquad \text{or} \qquad \Delta Q = cw\,\Delta T \tag{13.1b}$$

Notice that c is the quantity of heat required to change the temperature of unit mass (or unit weight) of substance by one degree. Because of our choice of the calorie and British thermal unit, c has the same value in calories per gram-degree Celsius as in British thermal units per pound-degree Fahrenheit. Typical values for several substances are given in Table 13.1. These values are not precise since c varies somewhat with temperature.

■ **EXAMPLE 13.2** How much heat in calories is needed to change the temperature of 50 g of copper from 10°C to 70°C?

Solution This is a typical situation in which Equation 13.1a is used. We wish to find ΔQ. We know that $\Delta T = 60°\text{C}$ and $m = 50$ g. From Table 13.1 we see that $c = 0.092$ cal/(g)(°C) for copper. Therefore, from Equation 13.1a,

$$\Delta Q = cm\,\Delta T$$
$$= \left(0.092\frac{\text{cal}}{(\text{g})(°\text{C})}\right)(50\text{ g})(60°\text{C}) = 276\text{ cal}$$

■■

■ **EXAMPLE 13.3** How much water at 20°C is needed to cool 5.0 kg of aluminum from 80°C to 30°C?

Solution We assume that the water will be mixed with the aluminum. The final temperature of both will be 30°C. Conservation of energy tells us that the heat lost by the aluminum equals the heat gained by the water. (This assumes negligible loss to the surroundings.) As an equation,

$$\text{Heat lost by Al} = \text{heat gained by water}$$
$$(mc\,\Delta T)_{\text{Al}} = (mc\,\Delta T)_{\text{water}}$$

Table 13.1 SPECIFIC HEAT CAPACITIES,[a] c

Substance	J/(kg)(K)	cal/(g)(°C) or Btu/(lb)(°F)
Solids		
Aluminum	920	0.21
Concrete	800	0.2
Copper	380	0.092
Glass	400–800	0.1–0.2
Gold	130	0.031
Granite	800	0.19
Human body	3500	0.83
Ice (−10°C)	2100	0.50
Iron and steel	460	0.11
Lead	130	0.031
Rubber	1900	0.45
Silver	230	0.056
Wood	1700	0.4
Liquids		
Alcohol (ethyl)	2400	0.58
Ether	2200	0.53
Mercury	140	0.033
Turpentine	1760	0.42
Water	4184	1.00
Gases (at 1 atm)		
Air	700	0.17
Steam (100°C)	1920	0.46

[a] At about 20°C unless otherwise noted.

That is,

$$(5000\ \text{g})\left(0.21\frac{\text{cal}}{(\text{g})(^\circ\text{C})}\right)(80^\circ\text{C} - 30^\circ\text{C}) = m\left(1\frac{\text{cal}}{(\text{g})(^\circ\text{C})}\right)(10^\circ\text{C})$$

Solving for m gives 5300 g of water. ■■

■ **EXAMPLE 13.4** The normal food intake of a 60-kg person might be 2000 nutritionist's Calories. If this is transformed to heat energy and retained by the body, how much would the person's body warm? Assume the human body to have $c =$ 0.83 cal/(g)(°C). •

Solution The heat added to the person is 2000 kcal. We have

$$\Delta Q = cm\,\Delta T$$

From this,

$$2 \times 10^6\ \text{cal} = \left(0.83\frac{\text{cal}}{(\text{g})(^\circ\text{C})}\right)(60 \times 10^3\ \text{g})\ \Delta T$$

Solving gives

$$\Delta T = 40^\circ\text{C}$$

Such a temperature rise is not observed, of course. How does the body get rid of this heat? ■■

■ **EXAMPLE 13.5** How much heat in British thermal units is needed to heat 20 lb of iron from 70°F to 212°F?

Solution This is a simple application of Equation 13.1b. We have

$$\Delta Q = cw\,\Delta T$$

This becomes

$$\Delta Q = \left(0.11\frac{\text{Btu}}{(\text{lb})(^\circ\text{F})}\right)(20\ \text{lb})(142^\circ\text{F})$$

$$= 312\ \text{Btu}$$ ■■

13.4 Melting and Freezing

Substances that crystallize show the behavior we call melting and freezing. You will recall that in a crystal the atoms are arranged in a pattern called a lattice. Even at low temperatures the atoms have some thermal energy. They therefore vibrate about their equilibrium position in the lattice. Thermal energy, if large enough, can cause the atoms to vibrate hard enough to break free from the lattice.

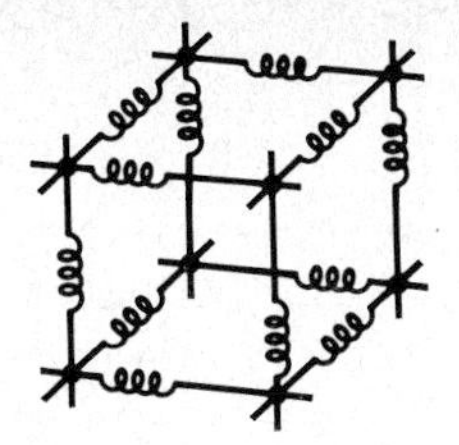

Figure 13.2 The atoms are held in the crystalline lattice by springlike forces. Thermal energy causes them to vibrate about the positions in which they are held. At no time would they all be placed exactly as shown.

In Figure 13.2 we show in a schematic way the situation in a crystalline solid. The atoms (or molecules) are held in position by springlike forces. Thermal energy causes the atoms to vibrate against the restoring forces of these springs. The amplitude of their vibration increases as the temperature is increased.

Each crystalline substance has a definite temperature at which the lattice breaks up. At that temperature the molecular vibration becomes larger than the springs can hold. The springs break and the atoms break free from the lattice. As a result, the atoms lose their regular arrangement. Moreover, the atoms are relatively free to move about. But they still are not energetic enough to escape entirely from the others. At this temperature the crystalline solid changes to a liquid. We say that the crystal melts.

The reverse process is also easy to picture. When a liquid is cooled, the thermal energy of its atoms (or molecules) decreases. At a certain low temperature the forces between the particles are able to hold the particles in place and cause them to arrange in a definite way. They bind together in the geometrical pattern we call their lattice. We say the substance crystallizes.

Let us now discuss what happens to the total energy of the system as the lattice melts. As you know, energy is required to stretch and break a spring. It

CHANGE IN VOLUME UPON MELTING

Most crystalline substances expand as the crystals melt. This is not unexpected. The crystals contain atoms packed carefully together in a definite lattice. Anyone who has packed a suitcase or placed cans in a box knows that more objects can usually be packed in a definite space if they are carefully arranged. This same idea carries over to atoms in crystals and liquids. The crystal is almost always a more dense packing than is the liquid. As a result, most substances contract as they freeze from liquid to crystal.

There is one extremely important exception to this rule. Water expands as it freezes to crystalline ice. An example of this is shown in the photo. The water (mixed with a little milk) in the bottle expanded as it froze. This expansion not only forced ice out of the neck of the bottle, it also exerted such a strong outward force on the bottle that the bottle broke. In the case of water the expansion is about 9 percent.

It is extremely fortunate that water does not behave like other substances. If it did not expand as it freezes, the earth would be greatly changed. Ice would form at the bottom of lakes rather than the top. Icebergs would sink to the bottom of the oceans. The ice in rivers and lakes would not melt in the summer because the water above it would act like an insulating blanket. As a result, our climate would change greatly. There are many other ways in which the world would change if water contracted as it froze. Can you list a few of them?

should be no surprise, then, that energy is required to break up the lattice. At the melting temperature the heat energy added to the crystal is used to break the "springs" that hold the molecules in the lattice. As more energy is added, more of the springs break. The increase in energy furnished to the solid is used to break the springs until all of the springs are broken. As a result, the average temperature of the molecules does not increase until the lattice has been destroyed completely. *Only after the crystal has been melted does the addition of heat cause the temperature to rise.*

We can illustrate this behavior by the graph in Figure 13.3 which shows how ice and water heat up as heat energy is added. The graph is based on 1 g of ice. As we see, each calorie heats the ice 2°C at temperatures from −20°C to 0°C. This value agrees with the value for c given in Table 13.1, 0.5 cal/(g)(°C). When the ice reaches 0°C, it begins to melt. Now the added heat energy is all used to break up the crystal. After 80 cal have been added, the crystal lattice has been destroyed. All the ice is now changed to water. Then, and only then, does the addition of further heat cause the temperature to rise. The temperature now rises 1°C for each calorie added.

A similar graph could be drawn for the freezing of a substance. The substance cools to the same temperature at which it melted. Then as heat is continued to be taken away from it, it crystallizes. During this time its temperature does not change. Finally, after it is all crystalline, further heat removal causes the crystal to cool.

We call the heat needed to melt (or freeze) a unit quantity of a crystal-forming substance the latent heat of fusion of the substance. We represent it by H_f. To melt a mass m

$$\Delta Q = H_f m \tag{13.2}$$

As we see from Figure 13.3, the heat of fusion of water is 80 cal/g. Values for other substances are given in Table 13.2.

■ **EXAMPLE 13.6** How many grams of ice at −14°C are needed to cool 200 cm³ of Coke (essentially water) from 25°C to 10°C?

Solution The heat lost by the Coke is gained by the ice. In doing this, the ice first warms to 0°C, then it melts, and then the resultant water warms to 10°C. Therefore

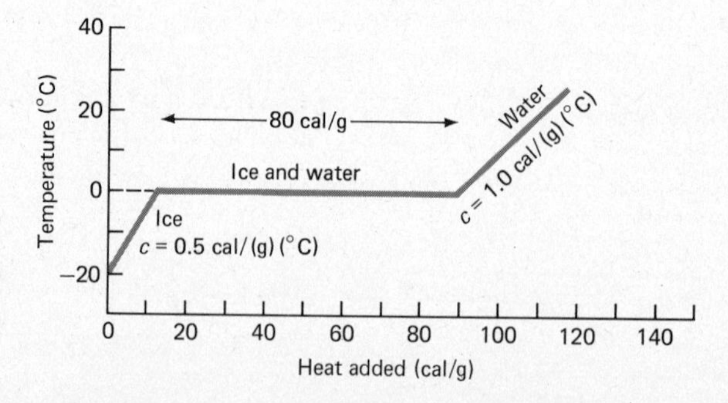

Figure 13.3 This graph can be retraced as heat is removed. The heat of fusion of water is 80 cal/g.

Table 13.2 HEATS OF FUSION AND VAPORIZATION[a]

Substance	Melting Temperature (°C)	Boiling Temperature (°C)	Heat of Fusion (cal/g)	Heat of Vaporization (cal/g)
Alcohol (ethyl)	−114	78.5	24.9	204
Aluminum	658	2057	76.8	—
Copper	1080	2310	42	—
Helium	−273	−268.9	1.2	5
Lead	327	1620	5.9	208
Mercury	−39	358	2.8	71
Nitrogen	−210	−196	6.1	48
Oxygen	−219	−183	3.3	51
Silver	961	1950	21	560
Water	0	100	80	539

[a]At 1 atm pressure.

$$\text{Heat lost by Coke} = \text{heat gained by ice}$$

becomes

$$(cm\,\Delta T)_{\text{Coke}} = (cm\,\Delta T)_{\text{ice}} + mH_f + (cm\,\Delta T)_{\text{water}}$$

Or

$$[1\ \text{cal/(g)(°C)}](200\ \text{g})(15\text{°C}) = [(0.5\ \text{cal/(g)(°C)}](14\text{°C})m + (80\ \text{cal/g})m + [1\ \text{cal/(g)(°C)}](10\text{°C})m$$

Solving for m gives 31 g. About how many ice cubes is this? It is interesting to notice that most of the cooling is the result of the heat of fusion in a situation where ice is used. ■■

13.5 Vaporization and Vapor Pressure

Substances tend to vaporize at all temperatures. This results from the following fact. As you recall from the previous chapter, the thermal energy of a molecule continually changes. As it collides with its neighbors, it gains and loses energy. Sometimes molecules obtain very high energy. Other times they have nearly zero energy. Let us consider what this means for a molecule at the surface of a liquid or solid.

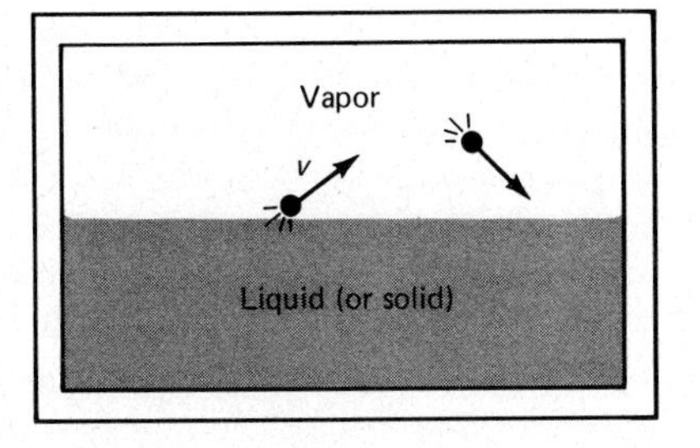

Figure 13.4 Equilibrium results when the number of molecules escaping from the liquid (or solid) equals the number returning from the vapor.

In Figure 13.4 we see a molecule escaping from the surface of a liquid or solid. To escape, it must have a high thermal energy. Only then can it break loose from the attraction forces of its neighbors. But there are always a few molecules with energy very much greater than average. Therefore, a few molecules are always escaping from the surfaces of substances. For solids this number is usually negligible. However, you may know that dry ice (solid carbon dioxide) vaporizes from the solid (*sublimes*) in this way. Snow also sublimes appreciably in the wintertime.

Table 13.3 SATURATED VAPOR PRESSURE OF WATER AND ICE

Temperature (°C)	Pressure (kPa)	Pressure (torr or mm of Hg)
−90	9.31×10^{-6}	0.000070
−50	4.0×10^{-3}	0.030
−10	0.26	1.95
0	0.61	4.58
10	1.22	9.21
30	4.23	31.8
60	19.9	149.4
90	70.0	526
94	81.1	611
99	97.5	733
100	101	760
110	143	1,075
150	475	3,570
200	1550	11,650

In the case of liquids this loss of molecules from the surface is called *evaporation.* Liquids such as ether and alcohol evaporate more easily than mercury or oil. This simply means less energy is required to tear ether and alcohol molecules loose from their neighbors. More of them succeed in escaping from the liquid's surface.

Suppose that the liquid is placed in a closed container, as shown in Figure 13.4. Molecules evaporate into the space above the liquid. When enough molecules exist in the vapor, an equilibrium is established. The same number of molecules evaporate each second from the liquid as return to the liquid from the vapor. Under these conditions the vapor is said to be *saturated.* The pressure of the molecules in the vapor is called the *saturated vapor pressure.*

The saturated vapor pressure varies with temperature. When the liquid is hot, more molecules have enough energy to escape than when it is cold. We list in Table 13.3 the saturated vapor pressure for water as a function of temperature. Notice that the vapor pressure for ice is also given. As we see, ice sublimes (changes from solid to gas) appreciably even though it is a solid.

13.6 Heat of Vaporization; Cooling by Evaporation

In order to escape from a liquid, a molecule must have considerable energy. We can imagine a process in which we tear all of the molecules in a liquid apart from each other. We thereby change the substance from a liquid to a gas. To do this requires work or energy on our part. Usually we furnish this energy in the form of heat.

The heat of vaporization, H_v, of a substance is the energy required to change a unit quantity of the substance from liquid to vapor. This same

amount of heat is released as the vapor condenses to liquid. To vaporize a mass *m*,

$$\Delta Q = H_v m \qquad (13.3)$$

Typical values of H_v are given in Table 13.2. The values of H_v vary slightly with temperature. In the table are listed values at the normal boiling temperature of the liquid.

When molecules evaporate from a liquid, they cool the liquid that remains behind. The reason for this is quite simple. Only the highest-energy molecules can escape from the liquid. When they escape, they carry more than their share of energy away from the surface. As a result, the average thermal energy of the molecules left behind becomes lower as evaporation proceeds. The liquid is therefore cooled by the evaporation process.

■ **EXAMPLE 13.7** How much sweat must evaporate from the surface of a 60-kg human to cool the person 1°C? Assume that $c = 0.83$ cal/(g)(°C) for a human and $H_v = 600$ cal/g for water at body temperature.

Solution Each gram that evaporates carries away an energy 600 cal. Therefore, the heat lost by the person as m grams evaporate is (600 cal)·m. This is ΔQ in the expression $\Delta Q = cm\,\Delta T$. Substitution of the known quantities gives

$$(600\text{ cal})(m) = [(0.83\text{ cal/(g)(°C)}](60 \times 10^3\text{ g})(1\text{°C})$$

Solving for m, we find that $m = 83$ g. In practice, considerable water vapor is expelled as a person breathes. ■■

13.7 Boiling

When visible bubbles form in large numbers within a liquid, we say that the liquid is boiling. If we examine the bubbles, we see that they grow and rise to the liquid surface. The bubbles are filled with vapor from the liquid. Let us now see what must be true for these bubbles to form and grow.

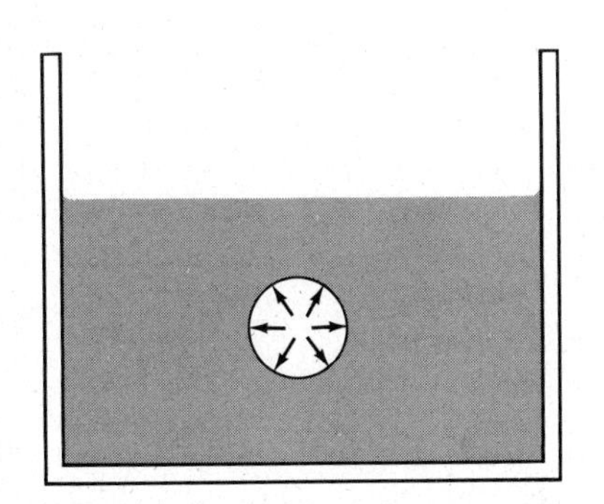

Figure 13.5 If the vapor pressure within the bubble is large enough, the bubble will grow.

Suppose that a tiny bubble of vapor forms within the liquid. A typical one, exaggerated in size, is shown in Figure 13.5. Two possibilities exist. If the vapor pressure in the bubble is less than the pressure of the liquid around it, the bubble will collapse. If the vapor pressure inside the bubble is larger than the pressure of the liquid, the bubble will grow. In this latter case it will rise to the liquid surface and explode.

Tiny bubbles are constantly being formed within liquids. Small groups of a few molecules acquire enough thermal energy to tear a hole within the liquid. This happens most easily near sharp points or pieces of dust in the liquid. When the temperature of the liquid is low, the bubbles soon collapse. The vapor pressure within them is too low to balance the pressure of the liquid on the bubble.

But the vapor pressure of a liquid increases with temperature. If the liquid is heated hot enough, the vapor pressure inside the bubble will equal the pressure of the liquid on the bubble. (Think of the bubble as a balloon, pressed on from all sides by the pressure of the liquid.) When the vapor pressure becomes this large, the bubbles within the liquid will grow. The liquid will boil.

Usually the pressure in the liquid is nearly equal to the pressure of the atmosphere above the liquid. (This would not be true deep under water, of course. But it is certainly true for a pan of water.) We therefore conclude the following:

A liquid will boil when its saturated vapor pressure equals the atmospheric pressure above it.

As you can see from Table 13.3, the vapor pressure of water reaches 101 kPa (760 torr) at 100°C. This is standard atmospheric pressure. We therefore call the boiling temperature of water 100°C.

But water can boil at a temperature less than 100°C. High in the mountains atmospheric pressure is less than 760 mm of Hg. At Denver, for example, atmospheric pressure is usually close to 600 torr. Saturated water vapor has this pressure at about 94°C. Therefore, in Denver water boils at 94°C rather than at 100°C.

This dependence of boiling temperature on pressure has wide importance in technology. Often a liquid needs to be purified by evaporation. But many times the liquid cannot withstand a temperature as high as its boiling temperature. One gets around this by reducing the pressure above the liquid by use of a partial vacuum. Then the liquid can be boiled (and vaporized) at a temperature much below its normal boiling point.

Once a liquid has started boiling, its temperature cannot be increased further. When heat is added to the liquid, the liquid simply boils faster. The temperature can only rise higher if boiling is caused to stop. Only then will the exploding vapor bubbles stop carrying away the heat energy supplied to the liquid. This situation is much like the one that exists when crystals melt. In both cases, melting and vaporization by boiling, the temperature will not rise until the process is completed.

■ **EXAMPLE 13.8** How many grams of hot steam at 115°C are needed to heat 1000 g of water at 20°C to 90°C?

Solution We make use of the law of conservation of energy. In this case

$$\text{Heat lost by steam} = \text{heat gained by water}$$

But the steam loses heat in three steps. It cools to 100°C, condenses to water, and then the water cools to 90°C. Therefore the equation above becomes

$$(cm\,\Delta T)_{\text{steam}} + mH_v + (cm\,\Delta T)_{\text{water}} = (cm'\,\Delta T)_{\text{water}}$$

Placing in the numbers,

$$[0.46\ \text{cal/(g)(°C)}]m(15°\text{C}) + m[539\ \text{cal/(g)(°C)}] + [1\ \text{cal/(g)(°C)}](m)(10°\text{C}) = [1\ \text{cal/(g)(°C)}](1000\ \text{g})(70°\text{C})$$

Simplifying gives

$$m[6.9 + 539 + 10] = 70{,}000\ \text{g}$$

or

$$m = 556\ \text{g}$$

Notice that most of the heating effect of the steam occurs when it releases its heat of vaporization. It is for this reason that live steam can cause such serious burns. ■■

13.8 Heat of Combustion

Most of the heat energy we use is obtained from combustion, the burning of substances. We heat our homes by burning oil or gas. We propel our cars by burning gasoline. Coal is burned to generate electricity and to provide heat for industrial purposes. Let us now investigate the amount of heat we can obtain by burning various materials. The quantity of concern to us is called the *heat of combustion.*

The heat of combustion, H_c, of a substance is the heat given off when a unit mass (or volume) of the substance is completely burned.

It is measured in units of calories per gram, British thermal units per pound, calories per cubic centimeter, and so forth. Typical values are given in Table 13.4.

■ **EXAMPLE 13.9** How much heat is given off as 200 g of fuel oil is burned completely?

Solution According to Table 13.4, 1 g of fuel oil gives off 10,500 cal when burned. Therefore, the heat given off by 200 g is

$$\text{Heat} = (200\ \text{g})(10{,}500\ \text{cal/g}) = 2.1 \times 10^6\ \text{cal}$$

■■

Table 13.4 Approximate Heats of Combustion[a]

Material	cal/g	cal/cm^3
Alcohol	6,500	—
Coal	7,800	—
Fuel oil	10,500	—
Gasoline	11,500	—
Wood	3,300	—
Hydrogen	29,000	2.6
Natural gas	13,000	10
Propane	12,000	23

[a]At STP.

Summary

When two objects having different temperatures are placed in thermal contact, heat energy is transferred from the warmer to the colder object. At equilibrium the two objects have a single temperature. This temperature is intermediate between the original two temperatures.

Heat energy is the energy that is transferred from a hot object to a cooler object as a result of their difference in temperature.

The units of heat energy are the calorie (cal) and the British thermal unit (Btu). Nutritionists use a unit they call a Calorie (or large calorie), which is really a kilocalorie. Conversions are

$$1 \text{ cal} = 4.184 \text{ J}$$
$$1 \text{ Btu} = 1054 \text{ J} = 778 \text{ ft} \cdot \text{lb}$$

These conversions are called the mechanical equivalent of heat.

The specific heat capacity, c, of a substance is the heat energy needed to raise the temperature of a unit quantity of the substance by one degree. From this definition a mass m (or weight w) has its temperature increased by ΔT when ΔQ heat flows into it.

$$\Delta Q = cm\,\Delta T \qquad \text{or} \qquad \Delta Q = cw\,\Delta T$$

The relation also applies to cooling. For water $c = 1.00$ cal/(g)(°C) = 1.00 Btu/(lb)(°F).

When a crystalline substance melts, it must be supplied with its latent heat of fusion, H_f. This is the quantity of heat required to melt a unit mass of the crystalline substance. This heat is given off when the substance crystallizes. Its value for ice is 80 cal/g. Summarizing, $\Delta Q = H_f m$.

For a unit mass of substance to be boiled away, the heat of vaporization, H_v, must be furnished to it. This same quantity of heat is given off as the vapor condenses. For water, H_v is about 540 cal/g. We have that $\Delta Q = H_v m$.

Evaporation causes cooling because the most energetic molecules escape. They carry more than their share of thermal energy away with them.

When a substance vaporizes directly from the solid, it is said to sublime.

Boiling occurs when the vapor pressure of a liquid equals the pressure of the atmosphere on the liquid. The boiling temperature can be altered by changing the external pressure.

Questions and Exercises

1. The king decreed that there be made three massive (but equal mass) coffee mugs, one of gold for him, one of iron for his army chief, and one of aluminum for the court jester. The king always complains that his coffee is too hot, while the jester complains his is too cold. Can you think of an explanation for this? The mugs hold equal volumes.
2. Two solid spheres are heated in boiling water. One sphere is made of copper and the other of aluminum, but they both have a mass of 10.0 g. Compare the sizes of the holes they will melt in a block of ice onto which they are dropped.
3. A bearing that is not lubricated heats up and often burns out. Yet no heat energy flows into the bearing. Explain the mechanism responsible for its temperature rise.
4. It is possible to add heat to a substance without the tempera-

ture of the substance changing. Explain under what conditions this is true.

5. Sometimes it is difficult to tell by using X rays whether or not a substance is partly crystalline. If one measures the temperature of the substance as heat is added at a constant rate, even a small amount of crystallinity can often be detected. Explain how. (One technique that uses this idea is in common use. It is called differential thermal analysis. DTA.)
6. Freeze-drying techniques are widely used in the chemical industry. They all make use of the same basic idea as that used in making freeze-dried coffee. Explain the basic idea of this technique. What technical word presented in this chapter is pertinent to the process?
7. A jar is partly filled with water and then sealed. After equilibrium is attained, the space above the water has water vapor in it. Compare the amount of water vapor you would expect in the following two cases: (a) the jar was sealed in air, (b) the jar was sealed under near-vacuum conditions.
8. Meat and vegetables are composed mainly of large polymer molecules plus water. During the cooking process the molecules are broken down into much smaller molecules. This breakdown occurs by destroying chemical bonds between some of the atoms in the molecules. Explain in terms of molecular concepts why potatoes take much longer to cook in Denver than in San Francisco. Why are pressure cookers often used? Is a floor burn (say, obtained by accident in a gym class) really a burn?
9. In some industrial processes a liquid is heated just enough so it continues to boil slowly. In other processes the liquid is heated so it boils vigorously. What determines which type of boiling should be used?
10. In Table 13.4 the values of H_c for the gases appear to vary oppositely in the two columns. Reconcile the data in the two columns.
11. Discuss the advantages and disadvantages in substituting hydrogen for natural gas as a fuel.
12. By reference to Table 13.4, discuss the advantages and disadvantages in using various fuels for steam generation and for use in automobiles.

Problems

1. (a) How many calories are given off by 50 g of water as it is cooled from 100°C to 20°C? (b) How many British thermal units are needed to heat 1 ft^3 (62.4 lb) of water from 70°F to 212°F?
2. In order to heat 60 g of water from 10°C to 83°C, how much heat is required? How many British thermal units are given off by 3.0 lb of water as it is cooled from 93°F to 62°F?
3. The specific heat capacity of the human body is about 0.83 cal/(g)(°C). (a) How much heat must be added to a 60-kg person if the person's temperature is to be increased 1.5°C? (b) How much heat must be removed from a 160-lb person to lower the person's body temperature from 98.8°F to 97.0°F?
4. A 120-g piece of copper is to be heated from 15°C to 300°C. How much heat will be required? How much heat is given off by 0.25 lb of copper as it is cooled from 70°F to 32°F?
5. An 800-kg car moving at 20 m/s slams on its brakes and skids to a stop. (a) How much friction work was done? (b) How many calories is this equivalent to?
6. A 2000-kg rocket falls straight back to the earth from a height of 150 km. (a) How much potential energy does it lose in the process? (b) After the rocket comes to rest on the earth, about how many calories has it generated in its fall?

*7. On his honeymoon, James Joule (after whom the unit is named) compared the water temperature at the bottom of a waterfall to that at its top in an effort (partly successful) to determine the mechanical equivalent of heat. How high would a waterfall have to be in order to raise the temperature of the water 1°C? Assume no evaporation or other heat loss.

*8. How many nutritionist's Calories of food must one eat in order to supply the energy needed to lift a 50-kg person through a height of 200 m? Assume that the food energy is completely changed to lifting work, although this is a poor assumption.

*9. A steel ball falls from a height of 20 m into a pile of sand. If one-half of its energy ends up as heat in the ball, how much is the ball heated?

*10. A lead bullet is moving at a speed of 150 m/s. It strikes a wood block and stops. If one-half of its original energy ends up heating the bullet, how much does the temperature of the bullet rise?

11. How much cold water (6°C) must be added to 200 g of boiling water (100°C) to cool it to 60°C?
12. How much turpentine at 90°C must be added to 50 g of turpentine at 20°C to give it a final temperature of 30°C?
13. A 90-g piece of hot iron is dropped into a 500-g vat of oil at 20°C. The final temperature of the oil and iron is 30°C. Find the original temperature of the iron. [$c = 0.50$ cal/(g)(°C) for the oil.]
14. Eighty grams of oil at 100°C is mixed with 50 g of turpentine at 10°C. After mixing, the final temperature is 80°C. Find the specific heat capacity of the oil.

*Problems marked with an asterisk are not as easy as the unmarked ones.

*15. A 500-g steel plate at 400°C is laid on a 2000-g aluminum plate at 20°C. The aluminum plate is on asbestos so heat loss from it is negligible. (a) What is the final temperature of the two plates when they reach the same temperature? (b) What is the average temperature of the aluminum plate when the average temperature of the steel plate is 250°C?

16. How much heat is needed to change 30 g of solid lead at 327°C to molten lead at 327°C?

17. How much heat must be removed from 50 g of molten lead at 327°C to change it to solid?

18. It is found that 80 g of a certain substance requires 680 cal to melt it at constant temperature. What is the heat of fusion for this substance?

19. How much heat must be added to 25 g of ice at −12°C in order to change it to water at 20°C?

20. How much heat must be removed from 80 g of water at 30°C to change it to ice at −5°C?

21. How much heat is required to change 50 g of lead at 20°C to molten lead at 327°C?

22. How much heat must be removed from 20 cm^3 of mercury at 18°C in order to just freeze all of it? The density of Hg is 13.5 g/cm^3.

23. How much heat is required to change 20 g of ethyl alcohol from liquid to gas at its normal boiling point?

24. When oxygen is liquified at −183°C, how much heat must be removed from 300 g of the oxygen?

25. How many grams of steam at 100°C must be condensed in 500 g of water at 20°C to raise its temperature to 30°C?

*26. Steam at 100°C is bubbled into 500 g of water originally at 20°C. What will be the temperature of the water after 30 g of steam has condensed?

*27. Thirty grams of liquid nitrogen at −196°C are poured into an open cavity in a 500-g copper block. The block is originally at 20°C. Find the final temperature of the block. Assume that the nitrogen escapes from the block as soon as it vaporizes.

28. A 60-W light bulb is painted black so that it emits no light. Its total power is then released as heat. How long would the bulb take to heat 2000 g of water by 30°C if the bulb is completely submerged?

*29. An electric motor operates in such a way that it is 70 percent efficient when consuming power at the rate of 200 W. The motor is basically copper and has a mass of 0.90 kg. Assuming that all the lost power goes into heating the motor, how much will its temperature rise in 1 min?

*30. How much heat is required to melt a 15-g lead bullet that is originally at 20°C? How fast must the bullet be going if it is to generate this much heat energy upon impact?

31. (a) How many calories are generated when 50 g of alcohol is burned completely? (b) If all the energy could be used, how high could it lift a 50-kg mass?

*32. A gasoline engine is used to lift a 500-kg load of bricks to the top of a 60-m-high building. Assuming the engine to have an efficiency of 25 percent, how many grams of gasoline will be used in the process?

33. Often one finds heats of combustion tabulated as calories per mole. (a) Use the data in Table 13.4 to find the value of H_c for hydrogen in calories per mole. (M = 2 kg/kmol.) (b) Find the value for propane. (M = 44 kg/kmol.)

Bettmann

14 INTRODUCTION TO THERMODYNAMICS

Performance Goals

When you finish this chapter, you should be able to

1. State the first law of thermodynamics both in words and as an equation. Give examples for each of the following cases: $\Delta U = 0$, or $\Delta Q = 0$, or $\Delta W = 0$.
2. Describe what is meant by the internal energy of a gas for a gas consisting of polyatomic molecules.
3. State how much work a system

The science of mechanics is based on Newton's laws together with the laws of conservation of energy and conservation of momentum. The science of heat can also be based on a few fundamental laws. We call these laws the laws of thermodynamics. In this chapter we shall state these laws. They will then be applied to several topics of importance in technology.

does on its surroundings as it undergoes a small expansion ΔV. Given a *P-V* graph showing how a system behaves as it expands or contracts from V_A to V_B, calculate the external work done by the system.

4. Define an isothermal process. Give an example of such a process. Given descriptions of several processes, select the one or more processes that are isothermal.
5. Repeat 4 for an adiabatic process.
6. Repeat 4 for an isochoric process.
7. Compute ΔQ, ΔU, and ΔW for an isothermal process and an adiabatic process given the appropriate *P-V* diagram.
8. Justify why the temperature increases in an adiabatic compression. Also explain why the temperature decreases in an adiabatic expansion. Give an example of a throttling process.
9. Sketch the indicator diagram for a thermodynamic cycle if the cycle is described to you in words. Explain how the indicator diagram can be used to find the net output work for the cycle.
10. Given the indicator diagram for a thermodynamic cycle, compute the net output work for the cycle. Also compute the net heat energy that flowed into the system during the cycle.
11. Compute the maximum efficiency possible for a thermodynamic cycle if the temperatures during the cycle are given. Explain your answer by reference to a Carnot cycle.
12. Draw the indicator diagram for

14.1 The First Law of Thermodynamics

The law of conservation of energy is of fundamental importance in mechanics. It is equally important in other branches of physics. When the energy conservation law is applied to topics involving heat, it is called the *first law of thermodynamics.*

To state the first law of thermodynamics, let us consider a group of molecules that we shall call a system. The system might be the gas in a bottle. Or it might be the gasoline exploding in the cylinder of an engine. Or it might be the molecules in a block of substance. The law we are about to state applies to any system.

The law is concerned with what happens to heat energy that is furnished to a system. Suppose that the system is the gas shown in Figure 14.1. The gas molecules are confined to a cylinder by a piston. These molecules have several kinds of energy. They have kinetic energy of translation and rotation. They might have energy of vibration due to the vibration of atoms within each gas molecule. If the molecules attract each other, there is energy associated with this, too. In addition, the electrons in the atoms have energy. As we see, the energy of the system of molecules is quite complex. We refer to this energy within a system as the system's internal energy, U.

The energy of the molecules in a system is called the internal energy, U, of the system.

When heat energy is added to a system, the added energy, called ΔQ, may be used to increase U by an amount ΔU. But the system shown in Figure 14.1 can use the added heat energy in another way. As the gas receives heat energy, it will try to expand. If the piston can move, the gas will lift it. In so doing, the gas does work on the piston. Let us call the work done by the system ΔW. We shall not restrict ΔW to be just the lifting of a piston. It can be any type of work done by the system. (Notice that this is the work done *by* the gas. When work is done *on* the gas, ΔW is negative.)

If we analyze even the most complex system, we find the heat energy added to the system can be used for only two purposes.

The heat energy ΔQ added to a system can increase the internal energy by an amount ΔU. It can also cause work, ΔW, to be done.

The law of conservation of energy then tells us that

The heat energy added to a system must equal the sum of the system's change in internal energy and the work done by the system.

This statement of the law of conservation of energy as applied to heat is called the *first law of thermodynamics*. As an equation, it can be written

a Carnot cycle. Describe it in words. Explain why this is an important cycle.

13. Explain the operation of a four-stroke internal combustion engine with spark ignition. Sketch its indicator diagram and use it in the explanation.
14. State the second law of thermodynamics in your own words. Explain its major implication for all heat engines. Explain why it leads to the concept of the degradation of energy.

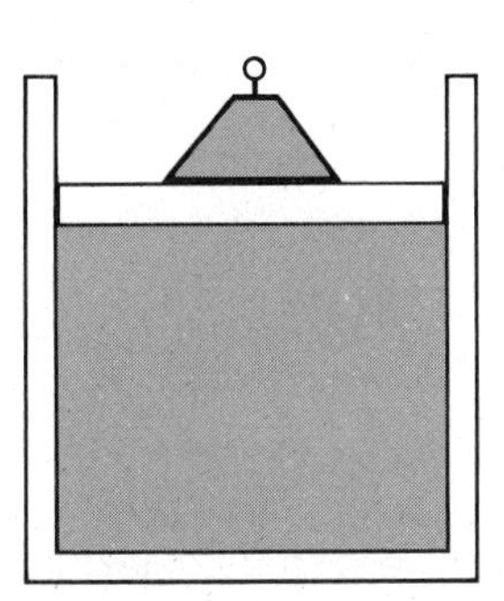

Figure 14.1 When heat energy ΔQ is added to the system of molecules, the internal energy of the system increases by ΔU. The system does work ΔW lifting the piston. From the first law of thermodynamics $\Delta Q = \Delta U + \Delta W$.

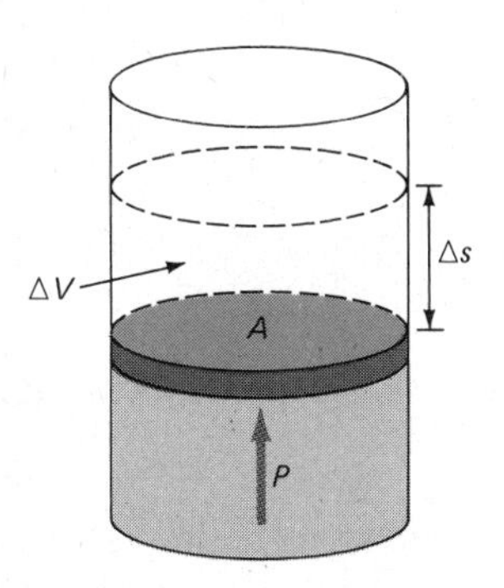

Figure 14.2 The force exerted by the gas on the piston is PA. Work is done as the gas expands the amount $\Delta V = A\,\Delta s$ shown.

$$\Delta Q = \Delta U + \Delta W \qquad (14.1)$$

Notice that ΔQ is the heat *added* to the system, ΔU is the *increase* in its internal energy, and ΔW is the work done *by* the system.

14.2 Expansion Work

In applying the first law of thermodynamics to practical systems, the term ΔW is often of a very simple form. It is quite frequently the work done by the system as its volume changes. For example, when you heat a metal cube, it expands. The cube must do work against the atmosphere in the expansion. This work is the term ΔW in Equation 14.1.

To compute the expansion work in general, let us consider the system shown in Figure 14.2. A gas at pressure P pushes the piston upward a distance Δs. The gas does work on the piston because it exerts a force and causes a displacement in the direction of the force. We wish to calculate the work done.

If Δs were as large as shown, the pressure would decrease during the expansion. But if Δs is very small, then P is essentially constant. In that event the force exerted on the piston of area A is constant and has a value PA. The work done by this force is

$$\text{Work} = F_s s$$
$$\Delta W = (PA)\,\Delta s$$

This can be put in a more useful form by noticing that the quantity $A\,\Delta s$ is the increase in volume of the gas during expansion. Call this volume increase ΔV. Then we find that

When a gas at pressure P expands through a small volume ΔV, the gas does the following amount of work:

$$\Delta W = P\,\Delta V \qquad \text{(constant pressure)} \qquad (14.2)$$

(When the gas is compressed, ΔV and ΔW are negative.)

Let us now consider the case where the change in volume is large. Then P most likely will change during the expansion. Suppose the pressure changes as shown in Figure 14.3 as the volume of the gas changes from V_A to V_B. What can we say in a situation such as this?

To begin with, consider the tiny expansion shown by the shaded, nearly rectangular, area in Figure 14.3(a). During this small expansion, ΔV, the pres-

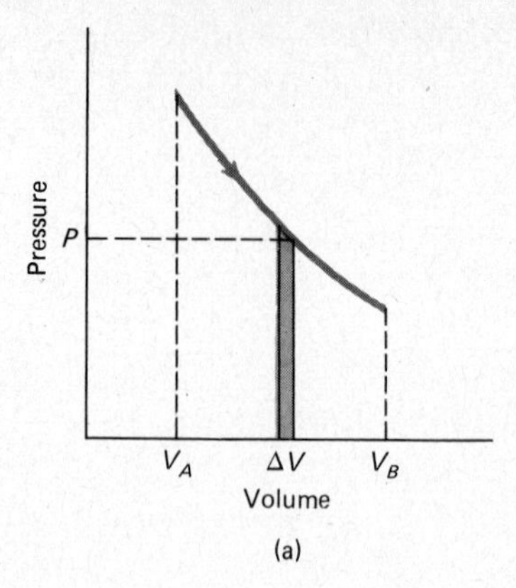

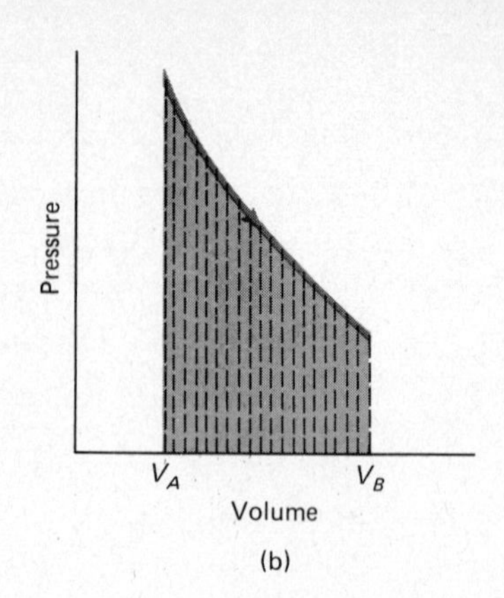

Figure 14.3 Even in complicated situations, the work done by the gas is equal to the area under the P versus V graph for the expansion.

sure P is essentially constant and the work done is, from Equation 14.2, just $P\,\Delta V$. Now notice an important fact. The area of the shaded rectangle is

$$\text{Shaded area} = (\text{height})\cdot(\text{width}) = P\,\Delta V$$

We therefore conclude that this area is equal to the work done during the expansion ΔV. This gives us the tool we need to evaluate the work done as the gas expands from V_A to V_B.

As shown in Figure 14.3(b), the whole expansion can be thought of as the sum of many tiny expansions. Since the work done during each tiny expansion is the area of a tiny vertical rectangle, the total work done is simply the sum of the area of all the rectangles. This sum area is identical to the total shaded area, the area under the curve traced out during the expansion. To summarize,

The work done by a gas during an expansion from V_A to V_B is the area under the P versus V curve between V_A and V_B.

The P versus V curve is often called a P-V diagram. Let us now apply this result to two examples.

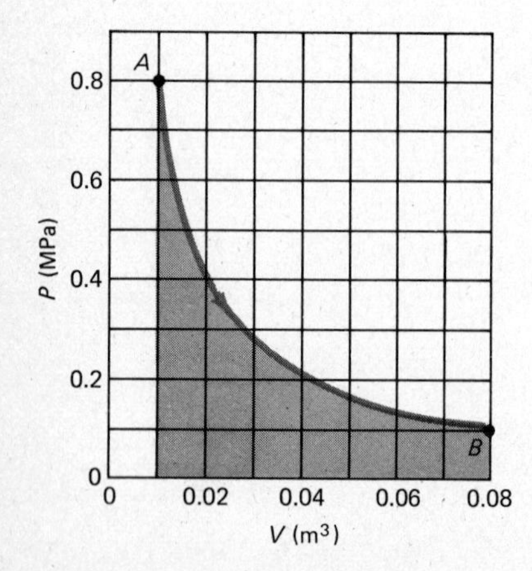

Figure 14.4 How much work did the gas do during this expansion?

■ EXAMPLE 14.1 A cylinder of gas expands against a piston. Its P versus V curve is shown in Figure 14.4. Find the work done by the gas as it expands from A to B.

Solution The work done is the shaded area under the curve. Machines exist for measuring areas. However, we shall find the area by counting squares on the graph. Notice that each square has a height of 0.10×10^6 Pa and a width of $0.01\ \text{m}^3$. Therefore

$$\text{Area of 1 square} = (0.10 \times 10^6\ \text{N/m}^2)(0.01\ \text{m}^3) = 1 \times 10^3\ \text{J}$$

Counting squares, we find the total area under the curve to be about 15.0 squares. Therefore

$$\begin{aligned}\text{Area under curve} &= 15.0\ \text{squares}\\ &= 15.0 \times 10^3\ \text{J}\end{aligned}$$

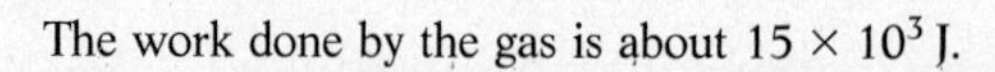

The work done by the gas is about 15×10^3 J. ■■

EXAMPLE 14.2 A 1-cm^3 cube of brass is heated 1°C. How much expansion work does the cube do? How much does its internal energy change? For brass, $c = 0.092\ \text{cal/(g)(°C)} = 0.38\ \text{J/(kg)(°C)}$, $\beta = 5.7 \times 10^{-5}/°\text{C}$, and $\rho = 8.7\ \text{g/cm}^3$. Use 100 kPa as atmospheric pressure.

Solution We will use $\Delta Q = \Delta U + \Delta W$ as applied to the cube. We know

$$\Delta Q = cm\,\Delta T$$

and

$$m = \rho V$$

Therefore, we have

$$\begin{aligned}\Delta Q &= cm\,\Delta T \\ &= [0.092\ \text{cal/(g)(°C)}](8.7\ \text{g})(1°\text{C})\end{aligned}$$

From this,

$$\Delta Q = 0.80\ \text{cal} = 3.35\ \text{J}$$

As the cube is heated, it expands an amount

$$\begin{aligned}\Delta V &= V_0\beta\,\Delta T \\ &= (10^{-6}\ \text{m}^3)(5.7 \times 10^{-5}/°\text{C})(1°\text{C}) = 5.7 \times 10^{-11}\ \text{m}^3\end{aligned}$$

This expansion was carried out against the pressure of the atmosphere. So the expansion work done is

$$\Delta W = P\,\Delta V = (1.00 \times 10^5\ \text{N/m}^2)(5.7 \times 10^{-11}\ \text{m}^3) = 5.7 \times 10^{-6}\ \text{J}$$

Now that we know both ΔQ and ΔW, we can compute ΔU.

$$\begin{aligned}\Delta Q &= \Delta U + \Delta W \\ \Delta U &= \Delta Q - \Delta W = 3.35\ \text{J} - 5.7 \times 10^{-6}\ \text{J} \cong 3.35\ \text{J}\end{aligned}$$

The expansion work is negligible in this case. Essentially all the added heat energy is used to increase the internal energy of the molecules in the brass block. ∎∎

14.3 Isochoric Processes

Processes in which the temperature, pressure, volume, and internal energy of a system all change can be quite complicated. But if only a few variables are allowed to change at one time, some important and informative limiting processes are found to occur. In this and the next two sections we will examine three such processes.

Consider the gas shown in Figure 14.5. The piston is fastened in place so the volume of the system cannot change. Therefore, the work done by the system, ΔW, is zero.

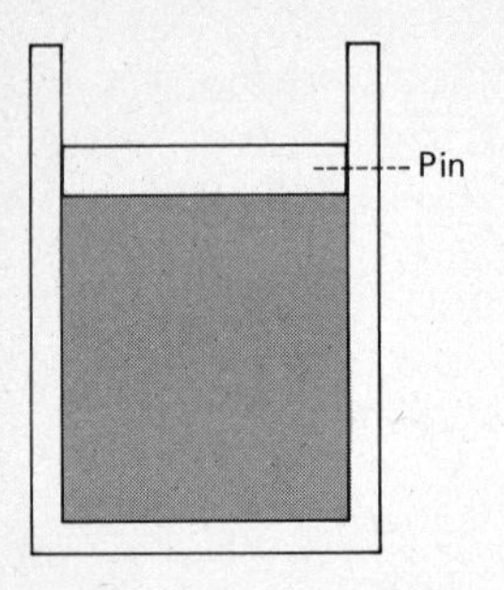

Figure 14.5 In an isochoric process the volume remains constant.

A process in which no work is done either by or on the system is called an *isochoric process*.[1]

Let us see what the first law of thermodynamics tells us about processes that are isochoric. Because $\Delta W = 0$, Equation 14.1 becomes

$$\Delta Q = \Delta U$$

Under the condition of $\Delta W = 0$, the first law tells us that the change in internal energy of a system equals the heat energy added to the system. Two examples of such a change follow.

Situation 1 Suppose that we melt a mass m of crystalline substance. During the melting process we shall not allow the substance to do appreciable work on its surroundings. As we saw in the last chapter, during melting the temperature remains constant. To melt the substance, the heat of fusion H_f must be added to each unit mass of it. Therefore, the heat needed to melt the mass m is

$$\Delta Q = mH_f$$

We have just seen that $\Delta Q = \Delta U$ in this special case where $\Delta W = 0$. We must therefore conclude that the internal energy of the substance increases by mH_f during the melting process.

As we have said, the internal energy of a substance is often quite complex. In this case the internal energy increases because the crystal is broken up. Energy is needed to tear the crystal apart. The molecules of the liquid thus formed receive this energy. It then becomes part of their internal energy.

Situation 2 Suppose that a mass m of substance is heated by an amount ΔT. In the last chapter we saw that, to do this, the heat required is

$$\Delta Q = cm\,\Delta T$$

Assuming the substance does negligible work on its surroundings as it heats up, we have

$$\Delta Q = \Delta U$$

We now have two quantities (ΔU and $cm\,\Delta T$) that are equal to the same quantity ΔQ. They must therefore be equal. Equating them gives

$$cm\,\Delta T = \Delta U$$

This relation should not surprise us. It says that the internal energy of a substance increases as the temperature increases. But we learned that T was a

[1] Also called an isovolumetric process for a gas.

measure of molecular kinetic energy. This relation tells us how the total internal energy (not just kinetic energy) is related to temperature.

■ **EXAMPLE 14.3** By how much does the internal energy of 20 g of water change as it is cooled from 50°C to 10°C?

Solution We can easily find the heat lost by the water by using

$$\Delta Q = cm\,\Delta T$$

In this case we have $c = 1$ cal/(g)(°C), $m = 20$ g, and $\Delta T = (-40°\text{C})$. The minus sign occurs because ΔT is a decrease in temperature. Placing in these values we find that

$$\Delta Q = -800 \text{ cal}$$

But where did this energy come from? According to the first law,

$$\Delta Q = \Delta U + \Delta W$$

Because the water did no work as it cooled, $\Delta W = 0$. Therefore, $\Delta Q = \Delta U$, and we find, for the change in internal energy,

$$\Delta U = -800 \text{ cal} = -3350 \text{ J}$$ ■■

14.4 Isothermal Processes

An isothermal process is a process carried out in such a way that the temperature does not change.

Isothermal means constant temperature. In the case of an ideal gas we have, from the gas law,

$$PV = \left(\frac{m}{M}\right)RT$$

Under isothermal conditions T is a constant. Upon dividing by V,

$$P = \frac{\text{constant}}{V} \qquad \text{(isothermal)} \qquad (14.3)$$

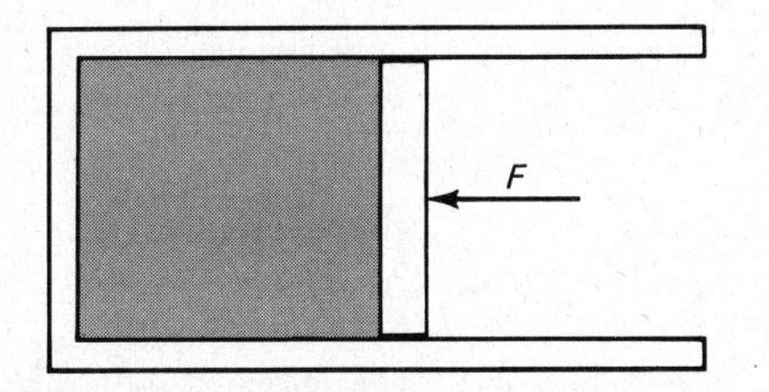

Figure 14.6 An isothermal expansion can be carried out by slowly decreasing F on the movable piston.

We see that P varies inversely with V for an ideal gas under isothermal conditions.

Equation 14.3 is often referred to as *Boyle's law.* It is easy to carry out an isothermal expansion. For example, the piston system shown in Figure 14.6 could be used. Suppose that the pressure within the cylinder is larger than 1 atm (1×10^5 Pa). Then the force F in Figure 14.6 is needed to hold the piston in place. If the force is decreased slowly, the piston will slowly move out. The temperature of the gas can be kept constant if the gas can exchange heat with the surroundings as the slow expansion occurs. Therefore the process will be isothermal.

As we have seen, the gas does work during the isothermal expansion. Where does the energy come from to do this work? According to the first law,

$$\Delta Q = \Delta U + \Delta W$$

Therefore ΔW must come from a change in ΔU or ΔQ or both. For an ideal gas the internal energy U is proportional to T, the temperature of the gas. Because the process is isothermal, T does not change. Therefore, U does not change and so $\Delta U = 0$. We see, then, for an isothermal process that

$$\Delta W = \Delta Q \qquad \text{(isothermal gas)}$$

This equation tells us that heat energy is furnished to the gas as the gas expands. The gas uses this heat energy to do an equivalent amount of work. No change in the internal energy of the gas occurs. As a result, the process is carried out at constant temperature. It is an isothermal process.

We can also use the device of Figure 14.6 to compress the gas. In a compression ΔV is negative. Therefore, the work done by the gas, $P\,\Delta V$, is also negative. This is merely a complicated way of saying that the external force F does work on the gas. What happens to the energy given to the gas as the force F compresses it?

This question can be answered easily if the compression is isothermal. In that case, as we have seen, $\Delta U = 0$. Then once again the first law tells us that

$$\Delta Q = \Delta W$$

But ΔW is negative because work is done by the gas—the gas is compressed. Therefore ΔQ is negative. What does a negative ΔQ mean? We recall that ΔQ is the heat energy that flows *into* the system. When ΔQ is negative, the heat must flow *out* of the system.

We therefore analyze the isothermal compression of the gas of Figure 14.6 in the following way. As the force F compresses the gas, it does work on the gas. It gives energy to the gas. Because the process is isothermal, the internal energy of the gas does not change, that is, the gas must give away this added energy immediately. It gives the energy away by allowing an equivalent amount of heat to flow out of the gas. Overall, this isothermal process amounts to the following: The work done on the gas by the compressing force causes an equal quantity of heat energy to flow from the gas.

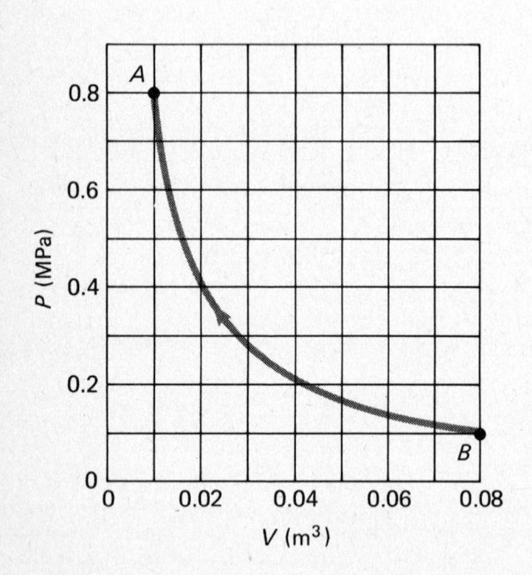

Figure 14.7 How much work did the gas do during this compression?

■ **EXAMPLE 14.4** A cylinder of gas is compressed isothermally by a piston. Its *P-V* diagram is shown in Figure 14.7. How much heat is given off by the gas as it is compressed from *B* to *A*?

Solution You should be able to show that the area under the *P* versus *V* curve is

1.50×10^4 J. This is the work done in compressing the gas from B to A. Because the compression is isothermal, $\Delta U = 0$. Therefore

$$\Delta Q = \Delta W = -1.50 \times 10^4 \text{ J}$$

But 1 cal = 4.184 J, and so

$$\Delta Q = -3.6 \times 10^3 \text{ cal}$$

This much heat flowed from the gas during the compression. ▮▮

14.5 Adiabatic Processes

An adiabatic process is a process carried out in such a way that no heat flows into or out of the system.

For an adiabatic process $\Delta Q = 0$. It is impossible to obtain a perfectly adiabatic process. We can never insulate a system so well from its surroundings that no heat transfer occurs. However, if a process is carried out quickly, very little heat has time to escape. We often find that quick compressions and expansions are nearly adiabatic.

Suppose that the gas in the cylinder of Figure 14.6 is suddenly expanded from V_1 to V_2. During this quick process little heat exchange occurs with the surroundings. Therefore ΔQ for the system is zero. The first law then tells us that

$$0 = \Delta U + \Delta W$$

or

$$\Delta U = -\Delta W \qquad \text{(adiabatic)}$$

In words, the internal energy of the gas decreases by an amount equal to the amount of work done by the gas.

This result is nearly obvious. If the system does work ΔW, the energy for this must come from somewhere. It could only come from heat added to the system, ΔQ, or from the internal energy already in the system. Because the process is adiabatic, $\Delta Q = 0$. Therefore the internal energy must have decreased enough to supply the energy needed to do this work. This means, as we have found, that $\Delta U = -\Delta W$.

But what happens to the gas as it loses internal energy? You will recall that the temperature of a gas measures U. Therefore, because U decreases, the temperature must also decrease. We conclude that

In an adiabatic expansion the temperature of a gas decreases.

As we have seen, this decrease is due to the loss of internal energy as the gas does work.

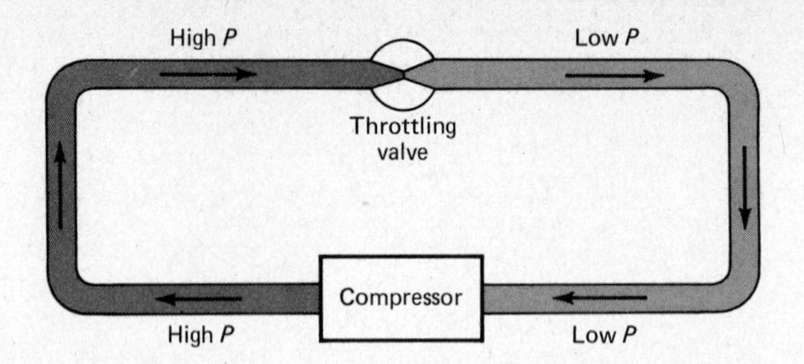

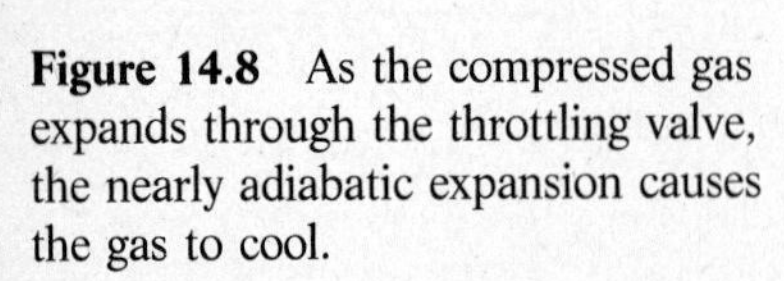
Figure 14.8 As the compressed gas expands through the throttling valve, the nearly adiabatic expansion causes the gas to cool.

The reverse process, adiabatic compression of a gas, is also of interest. In this case the external force doing the compressing does work on the gas. Because no heat flows from the gas, this work must appear as an increase in internal energy. Because T measures the internal energy, the temperature of the gas rises. We therefore conclude that

In an adiabatic compression the temperature of a gas increases.

This fact is used in the diesel engine. The gas in the engine's cylinder is rapidly (nearly adiabatically) compressed. It is thereby heated hot enough to ignite the fuel in the cylinder. No use is made of a spark plug for ignition. In this same connection, why does a tire heat up as it is being inflated?

In the next chapter, when we discuss refrigeration, we shall mention another important adiabatic process. It is referred to as a *throttling process*. In it, a gas is allowed to expand quickly, usually through a small hole or a porous plug. As shown in Figure 14.8 the gas flows from a region of high pressure to one of lower pressure. It expands as it reaches the low-pressure region. During the expansion it does work on the gas that is already there. Because the expansion is rapid, little heat flows into the gas. The work is therefore done at the expense of the internal energy of the gas. As a result, the gas is cooled considerably by the expansion.

There are many ways to show the cooling effect of a throttling process. Many refrigeration systems make use of it. A more direct method is the following. A tank of compressed carbon dioxide (CO_2) has a small valve on it that can act as a throttle. When the valve is opened, the gas shoots through the valve into the atmosphere. If conditions are right, the CO_2 will be cooled enough to condense to a powdery solid. A similar procedure is used to liquify nitrogen. But in this case the gas must be cooled in several stages.

14.6 Heat Engines

Any device that uses heat energy to do work is called a heat engine. Among these are steam engines, gasoline and diesel engines, and even the jet engine shown on page 164. Let us discuss a few of the features that all of these engines have in common.

Heat engines must have a source of heat to drive them. The gasoline engine

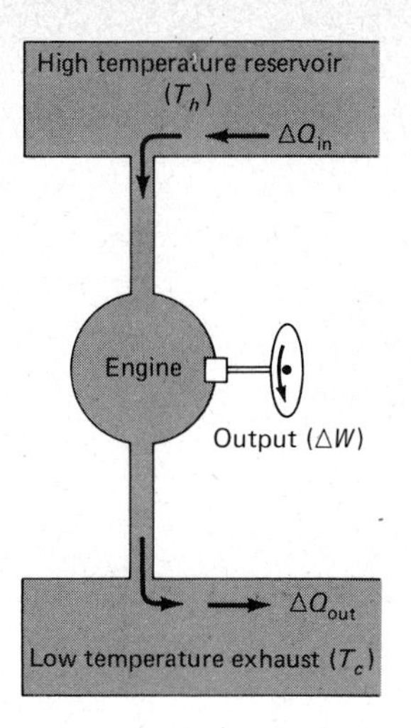

Figure 14.9 For any heat engine the first law tells us that $\Delta Q_{in} - \Delta Q_{out} = \Delta W$.

uses burning gasoline. The heat from the burning gasoline produces a high pressure gas in the engine cylinder. This causes the piston that closes the cylinder to move, thereby allowing the engine to do work. Eventually, the expanded, cooler gas in the cylinder is vented to the outside. This exhaust gas is still warm and so some of the heat is lost from the engine. (The exhaust heat causes the manifold, muffler, and tail pipe on a car to become very hot.) If you examine any heat engine you can see that it also possesses analogous features.

Because of the similarity of all heat engines, we can summarize them in a diagram. Such a diagram is shown in Figure 14.9. It reminds us of the following facts about all engines.

1. A high temperature heat source, called a heat reservoir, at temperature T_h, supplies an amount of heat ΔQ_{in}.
2. The engine does mechanical work ΔW, thereby using part of the input energy.
3. The remainder of the heat energy ΔQ_{out} is discarded to a low temperature exhaust, or cold reservoir, at temperature T_c.

We can conclude at once from the law of conservation of energy that

$$\Delta W = \Delta Q_{in} - \Delta Q_{out}$$

One of the most important features of an engine is its efficiency, that is, how effectively it uses the input energy. As with all machines, we define an engine's efficiency as

$$\text{Efficiency} = \frac{\text{output work}}{\text{input energy}}$$

As Figure 14.9 indicates, the input energy is ΔQ_{in} and the output work is ΔW. Therefore,

$$\text{Efficiency} = \frac{\Delta W}{\Delta Q_{in}}$$

Since $\Delta W = \Delta Q_{in} - \Delta Q_{out}$,

$$\text{Efficiency} = \frac{\Delta Q_{in} - \Delta Q_{out}}{\Delta Q_{in}} = 1 - \frac{\Delta Q_{out}}{\Delta Q_{in}} \qquad (14.4)$$

This equation for efficiency tells us an important fact. A perfectly efficient engine requires ΔQ_{out} to be zero. This is almost obvious even without our equation. The fraction of the input energy thrown away, namely, $\Delta Q_{out}/\Delta Q_{in}$, is the fraction of energy unavailable for doing useful work. Equation 14.4 states

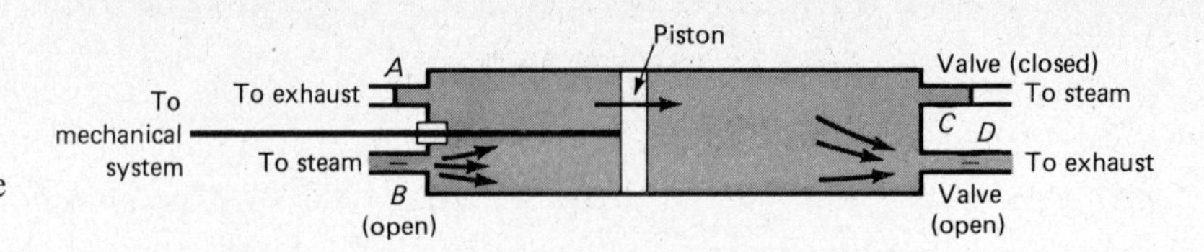

Figure 14.10 At the moment shown steam valve B and exhaust valve D are open. The piston is being forced to the right. In an instant these valves will close and valves A and C will open. The piston will then return to the left.

this fact in a concise way. The most efficient engine is the one that rejects (or throws away) the smallest fraction of its input energy. We shall soon see that ΔQ_{out} can never be made zero for even a perfect engine. Hence an engine can never have an efficiency of 1, or 100 percent.

14.7 Cyclic Processes

All piston and rotor engines cycle the same process over and over again. In order to understand the engine, we must understand its operation cycle. To illustrate some of the features of such an engine, we shall consider the simplified steam engine shown in Figure 14.10. It is called a one-cylinder, double-stroke engine. The engine operates as follows.

At the instant shown hot steam at temperature T_h is entering the left chamber through open valve B. This high-pressure steam pushes the piston to the right. Little resistance to this motion occurs in the right-hand chamber because the exhaust valve D is open there and the steam entrance valve C is closed. The steam in the left chamber is doing work during this time.

After the piston moves to the far right, all of the valves change position simultaneously. Those that were open close and vice versa. The pressure in the left chamber drops swiftly as the exhaust valve A opens and the steam valve B closes. In the right chamber the reverse occurs. C is open and D is closed. Hot, high-pressure steam enters to push the piston to the left. Eventually, the piston nears the left end and the valves reverse again. Thus the piston is pushed back and forth in the cylinder. This back-and-forth motion is changed to rotational motion by suitable mechanical linkages.

The steam engine clearly wastes a lot of heat. Steam, still quite hot, is released to the outside when the exhaust valve opens. As a result, the engine is not very efficient. A normal efficiency is usually no greater than 15 percent.

We can understand this engine in more detail by considering what is called its *indicator diagram.* This is simply a graph of the pressure and volume in one chamber of the engine throughout one cycle. Consider the gas in the left chamber as the cycle proceeds. We show its pressure versus volume graph in Figure 14.11.

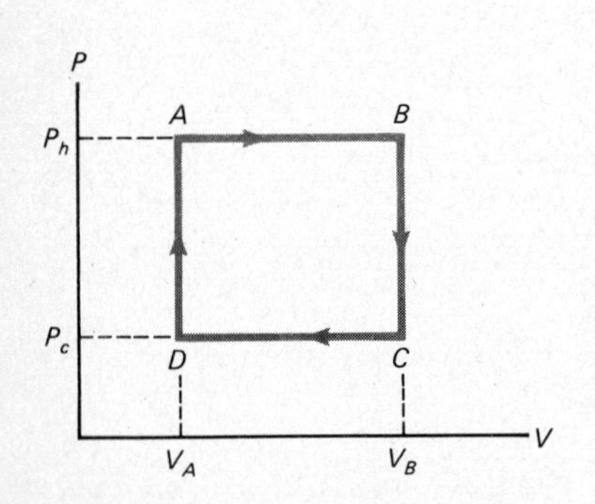

Figure 14.11 This is the indicator diagram for one side of the engine shown in Figure 14.10.

At point A on the graph, the piston is far to the left (small volume) and valve B is open. The chamber is filled with high-pressure, hot steam (pressure = P_h). As the piston moves to the right, the volume expands from V_A to V_B along line AB at the same high pressure. At point B on the graph, the valves reverse and the pressure in the chamber drops sharply to P_c as the gas exhausts. This happens so fast that the volume changes very little. The gas in the chamber follows portion BC of the graph.

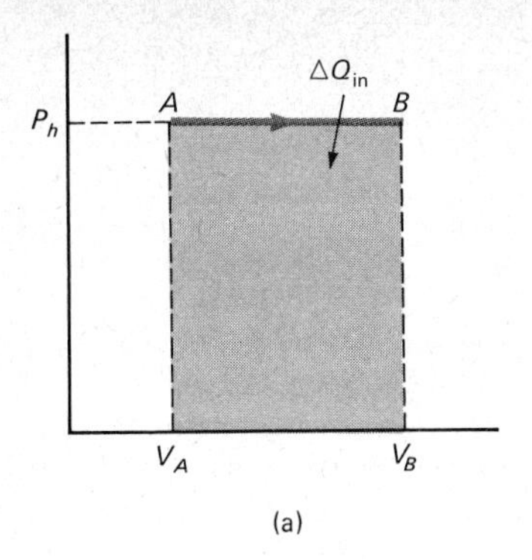

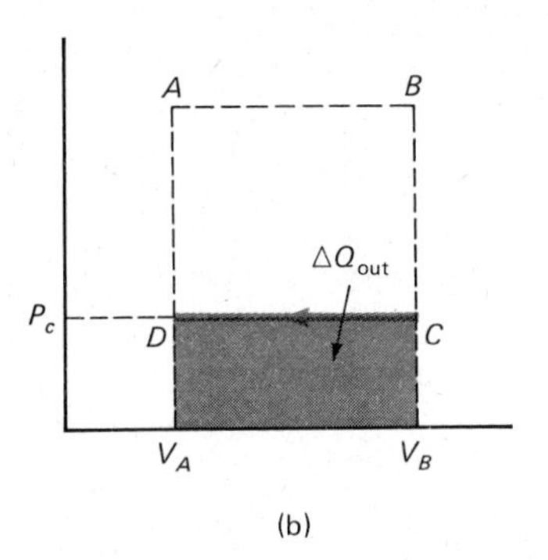

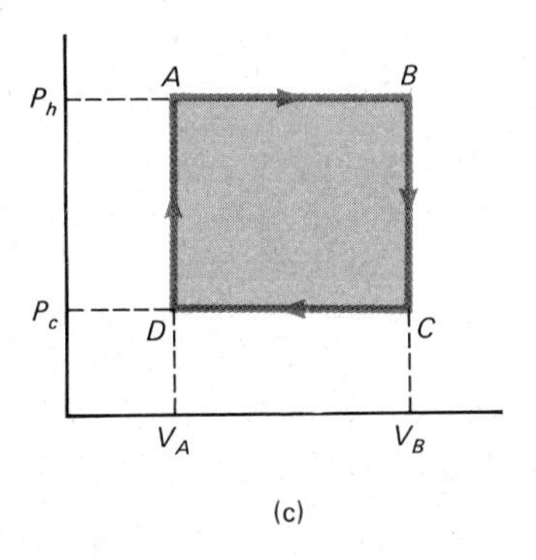

Figure 14.12 The output work during one cycle is the area enclosed by the indicator diagram.

As the piston moves back to the left, the volume reduces from V_B to V_A along the portion CD of the graph. The pressure of the cooler gas in the chamber, now open to the exhaust, is P_c. The piston reaches the left end of its motion and the valves reverse again. Valve B admits hot steam into the chamber and the pressure rises rapidly to P_h. This is portion DA of the cycle.

The cycle has now been completed. The gas in the chamber has returned to its original condition. Its volume is V_A and its pressure is P_h. Moreover, its temperature is that of the hot steam in the hot reservoir, T_h. In effect, in this cycle heat (ΔQ_{in}) enters the engine at a temperature T_h and is exhausted (ΔQ_{out}) at a temperature T_c, the exhaust gas temperature. The difference between ΔQ_{in} and ΔQ_{out} is the output work of this chamber of the engine during one cycle.

14.8 Work per Cycle

From a practical standpoint, we are interested in the output work of an engine per cycle. This work can be found directly from the indicator diagram for the engine. Let us examine the indicator diagram more closely in Figure 14.12.

During portion AB of the cycle, heat ΔQ_{in} flows into the chamber. At the same time the gas pushes the piston to the right, thereby doing work. We know from Section 14.2 that the work done is the area under the P-V curve from A to B. This is the shaded area in Figure 14.12(a).

During portions BC and DA of the cycle, the piston moves only a negligible distance. No work is done during these portions of the cycle.

But during portion CD of the cycle, outside work is done on the gas in our chamber as it is pushed out the exhaust. The heat lost is ΔQ_{out}. Moreover, the work done on the gas is equal to the area under this portion of the P-V curve. This is the shaded area in Figure 14.12(b). Notice, however, that this work is negative because ΔV is a decrease in volume.

The useful work (or net work) done by the engine during the cycle is the algebraic sum of these two values. It is the area in (a) minus the area in (b). The final result for the work done is the shaded area in (c). Notice that it is the area enclosed by the indicator diagram for the cycle. We conclude that

The net output work done by a system during one cycle is equal to the area enclosed by its indicator diagram.

As you see, the indicator diagram for an engine yields a great deal of valuable information.

EXAMPLE 14.5 For the thermodynamic cycle shown in Figure 14.13, how much net output work is done by the engine each cycle? If the engine runs at 4 cycles per second (cps), what is its output horsepower?

Solution We know that the net output work per cycle is equal to the enclosed area of the indicator diagram. Each square of the graph has an area of $(100 \text{ cm}^3)\cdot(1 \times 10^5 \text{ Pa})$. Therefore

$$\begin{aligned}\text{Area/square} &= (1 \times 10^{-4} \text{ m}^3)(1.00 \times 10^5 \text{ N/m}^2)\\ &= 10.0 \text{ J}\end{aligned}$$

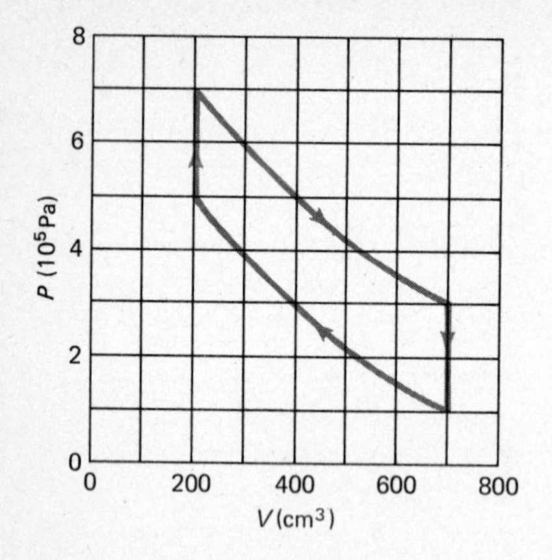

Figure 14.13 Find the net output work per cycle.

The area enclosed by the indicator diagram is about 10.0 squares. Therefore

$$\text{Net work/cycle} = 100\ \text{J}$$

At a rate of 4 cps, this is

$$\begin{aligned}\text{Work/second} = \text{power} &= 400\ \text{W} \\ &= (400\ \text{W})\left(\frac{1\ \text{hp}}{746\ \text{W}}\right) \\ &= 0.54\ \text{hp}\end{aligned}$$

■■

14.9 The Carnot Cycle

Heat-operated engines vary widely in their efficiencies. Diesel engines are usually more efficient than the common gasoline engines. These in turn are more efficient than steam engines. It would be of value to know just how efficient a perfect engine could be. You might at first think that a perfect engine should have an efficiency of unity. But this is not true. Even if no friction and unwanted heat losses are present, a heat engine cannot be 100 percent efficient. This was first pointed out clearly by Sadi Carnot (pronounced *car-no*).

Carnot pictured an ideal heat engine called the *Carnot engine.* He was able to prove from theory that no other engine could exceed its efficiency.[2] One cycle of the engine is shown in Figure 14.14. The indicator diagram for the so-called *Carnot cycle* is also shown there. Let us discuss the operation of this ideal engine.

In portion *AB* of the cycle the gas is expanded slowly. The gas is in thermal contact with a heat reservoir. The expansion is done in such a way that the temperature of the gas remains constant. Portion *AB* of the cycle is therefore isothermal. During it an amount of heat ΔQ_{in} is fed into the gas.

In portion *BC* of the cycle the gas is expanded adiabatically. The gas is insulated thermally, and so no heat exchange occurs. As the gas expands during the portions *AB* and *BC*, the gas does work. During these portions of the cycle, the engine does useful work.

In portion *CD* of the cycle the gas is compressed isothermally. To maintain its temperature, it is in contact with a cold reservoir. During this portion of the cycle ΔQ_{out} heat is rejected to the reservoir.

In portion *DA* of the cycle the gas is compressed adiabatically. No heat transfer occurs. In both *CD* and *DA* an external force must cause the compression. Work is done on the engine during these parts of the cycle.

As we saw in the previous section, the area of the indicator diagram has meaning. It is equal to the net work done by the engine during the cycle. Further, the efficiency of the engine is given by Equation 14.4.

$$\text{Efficiency} = 1 - \frac{\Delta Q_{out}}{\Delta Q_{in}}$$

[2] Because the Carnot engine is an idealized engine, it is impossible to build. Even the best practical engines fall short of its predicted efficiency.

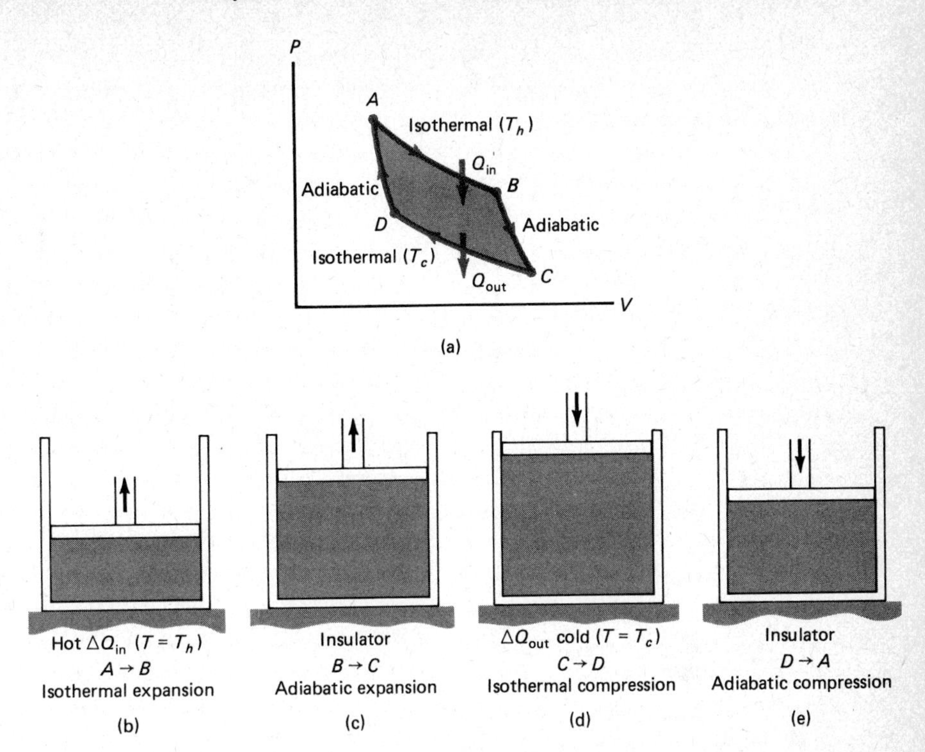

Figure 14.14 The Carnot cycle.

Carnot was able to prove an important fact about the ratio $\Delta Q_{out}/\Delta Q_{in}$. He showed that it was equal to the ratio T_c/T_h. You will see from Figure 14.14 that T_c is the absolute temperature of the cold reservoir. T_h is the hot reservoir's absolute temperature. We therefore have the following equation for the best possible efficiency of a heat engine.

$$\text{Best efficiency} = 1 - \frac{T_c}{T_h} \qquad (14.5)$$

Notice what this relation says. As long as $T_c > 0$, the efficiency of the engine must be less than 100 percent. All practical engines are limited by this condition. Let us discuss the meaning of this.

The Carnot engine is the most efficient heat engine possible. Its efficiency is best if the hot reservoir is very hot (T_h, large) and the cold reservoir is very cold (T_c small). In the case of other engines, this means that the driving gases (or fluid) should be very hot. The exhaust gases (or fluid) should be very cold. Unfortunately, most practical engines operate with the exhaust at about atmospheric temperature. The driving gas will have a temperature determined by engine design. In the steam engine it is the steam temperature. In the gasoline engine it is the temperature of the burning gasoline. As we see, T_h and T_c are not easily changed by large amounts.

■ **EXAMPLE 14.6** Find the maximum efficiency of a steam engine that uses steam at 250°C. Assume that it exhausts to the atmosphere at 100°C.

Solution A typical one-cylinder, double-stroke steam engine is sketched in Figure 14.10. The temperatures T_h and T_c are 523 K and 373 K, respectively. The Carnot engine efficiency would be

$$1 - \frac{373}{523} = 0.29$$

Even the best-designed steam engine operating between these temperatures could do no better than this. Therefore, the maximum efficiency would be 29 percent. In fact, no engine could be as efficient as the Carnot engine. In practice, the efficiency is usually around 10 percent. ■■

14.10 Internal Combustion Engines

A heat engine is any device that converts heat energy into mechanical energy. A heat engine that burns fuel within the engine itself is called an *internal combustion (IC)* engine. The gasoline engine is of this type. Most gasoline engines use four strokes per cycle of the piston. We show these four strokes schematically in Figure 14.15.

In the intake stroke gas and air are injected from the carburetor. This mixture is compressed during the compression stroke. After compression, the gas is ignited by the spark plug. (In the diesel engine the temperature rises high enough during the quick compression to ignite the gas.) The burning gasoline generates heat and gases, which push the piston back out. This is the power stroke. During it the engine does work. Then the exhaust valve opens and the spent gases are ejected. This occurs during the exhaust stroke.

The idealized indicator diagram for this engine is shown in part (e) of the figure. Notice that the ignition process supplies energy to the system. In a typical engine the temperature of the ignited gas might be about 800 K. This is T_h. The exhaust temperature would be T_c. It cannot be lower than about 400 K. From this we see that the best possible efficiency would be

$$\text{Efficiency} = 1 - \frac{400}{800} = 0.50$$

In practice, the gasoline engine has an efficiency approaching 30 percent.

14.11 The Second Law of Thermodynamics

Like the first law, the second law of thermodynamics is a statement of experimental fact. It can be stated in several different, equivalent ways. For our discussion we shall state the second law of thermodynamics as follows.

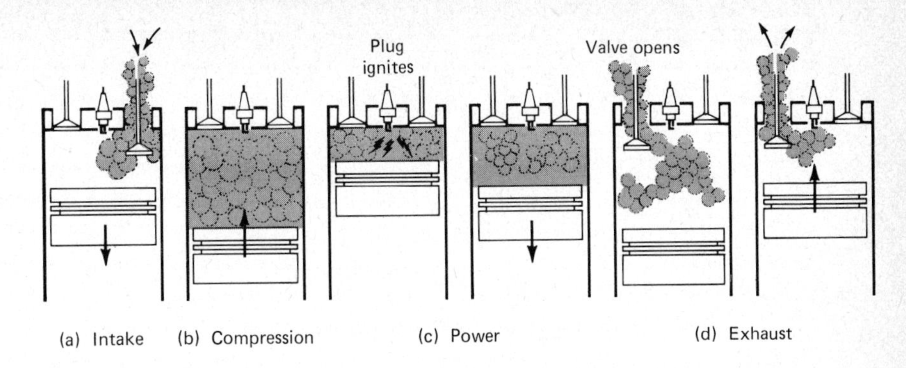

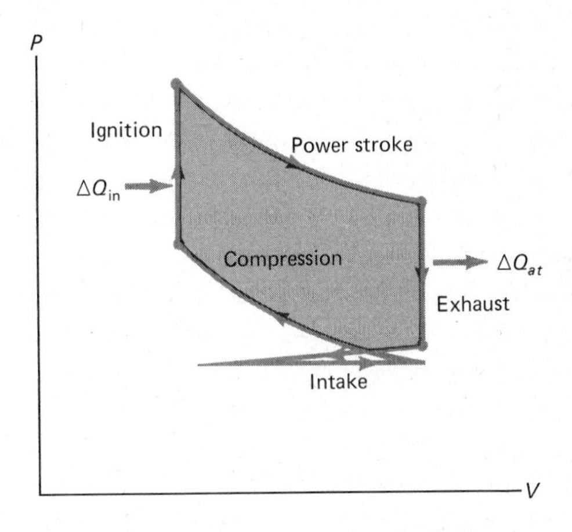

Figure 14.15 The four-stroke gasoline engine. The Otto cycle.

Heat flows by itself from a hotter to a colder object but not vice versa. As a result, a heat engine must have both a heat source and a heat exhaust. The heat source must be at a higher temperature than the heat exhaust.

Also implied by the law is the following: A heat engine must exhaust some of the heat energy it receives. It cannot use all the heat energy it receives to do work.

We can make the law reasonable by reference to our molecular picture of heat. As we have seen many times, heat energy is related to molecular energy. Hot molecules give energy to colder molecules upon collision. For this reason heat energy flows from hot objects to cold ones. This flow continues until the two objects reach the same temperature. Then the flow stops.

Because heat engines use the flow of heat for their operation, this law has importance for them. Heat flow can occur only if we have both a hot and a cold reservoir. This heat is lost from the system. It does not produce useful work. For this reason the efficiency of a heat engine must always be less than 100 percent.

The second law has far-reaching consequences in the universe. It tells us that heat energy always flows from hot to cold when allowed to do so. As a result, hot objects tend to cool and cold objects tend to warm. They therefore become less useful heat sources and heat exhausts. Of course, we can burn fuel to make objects hot again. But the fuel itself is then used up. To make more fuel requires heat from the sun or from some other source of energy. Eventually all these energy sources will be gone. Then the universe will reach a uniform temperature throughout. Although molecular energy will still exist, it will no longer be useful. There will be no cooler objects to which heat energy can flow.

There is a name for this general trend of energy to become less useful. It is called the *degradation of energy*. Even though energy cannot be destroyed, it becomes less usable as time goes on. In the end, when everything is at the same temperature, no process that requires energy can be carried out. This situation is often referred to as the *heat death of the universe*.

Fortunately, this heat death of the universe is billions of years away. In fact, there is good reason to believe that it will never occur. Instead, the big bang theory of the universe indicates that quite a different catastrophe may occur. There is a good possibility that gravitational forces will cause the universe to collapse. The kinetic energy of the objects falling to the center of collapse will be tremendous. Upon collision they will heat the universe to fantastically high temperatures. This extremely hot fireball will explode. As it expands, it will cool. Parts of it will aggregate into suns, stars, planets, and moons. The universe, much like we presently know it, will be formed again.

Summary

The energy of the molecules in a system is called its internal energy, U.

The first law of thermodynamics states: The heat energy added to a system (ΔQ) must equal the sum of the system's change in internal energy (ΔU) and the work done by the system (ΔW). In symbols,

$$\Delta Q = \Delta U + \Delta W$$

It is a statement of the law of conservation of energy.

The work done during expansion through a volume ΔV against a constant pressure P is given by $\Delta W = P\,\Delta V$. It is also given by the area under the P versus V curve.

An isochoric process is one in which the system does no external work. For it the first law becomes $\Delta Q = \Delta U$.

An isothermal change within a system occurs at constant temperature. During it $\Delta U = 0$ provided that no phase change occurs. Therefore the first law becomes $\Delta Q = \Delta W$.

An adiabatic change within a system takes place in such a way that $\Delta Q = 0$. In this case the first law becomes $\Delta U = -\Delta W$.

In an adiabatic expansion the temperature of a gas usually decreases. In an adiabatic compression the temperature of a gas rises.

The *P-V* diagram for a thermodynamic cycle is called an indicator diagram. Its enclosed area equals the net output work done during the cycle.

A heat engine is any device that converts heat energy into mechanical work.

The efficiency of a heat engine is given by

$$\text{Efficiency} = \frac{\text{output work}}{\text{input heat energy}}$$

This is equivalent to

$$1 - \frac{\text{heat rejected}}{\text{heat input}} = 1 - \frac{\Delta Q_{\text{out}}}{\Delta Q_{\text{in}}}$$

For the most efficient heat engine possible, the Carnot engine,

$$\text{Efficiency} = 1 - \frac{T_c}{T_h}$$

where T_c and T_h are the temperatures of the cold and hot reservoirs, respectively. No engine can be 100 percent efficient.

The second law of thermodynamics states: Heat flows by itself from hot to cold but not vice versa. A heat engine must have both a heat source and heat exhaust. The temperature of the exhaust must be lower than that of the source.

Questions and Exercises

1. Suppose a distant cousin of yours comes to you for financing to develop the following invention. To start his machine, a 12 volt battery is needed. But once it is running, its internal electric generator furnishes enough electricity to keep it running and to light the lights in a house. Should you contribute your savings to help him?
2. A well-stirred glass of water has an ice cube melting in it. Apply the first law to the system of water plus ice cube. Describe the values of ΔQ, ΔW, and ΔU as time goes on. Is ΔW negligible in this case? Is ΔW zero?
3. Devise an isothermal system that does no work even though heat energy is being added to it.
4. How can the temperature of a gas be raised without heat flowing into it? How can it be cooled without heat flowing from it? Is it possible to do the same for a metal plate?
5. It is possible for the internal energy of a substance to increase without its temperature increasing. Explain. Can the substance be a gas? A liquid? A solid?
6. The specific heat capacity of a gas maintained at constant volume, c_V, is smaller than when maintained at constant pressure, c_P. Consider a gas confined to a cylinder by a piston. Using the first law, explain why $c_P > c_V$.
7. In order to test a tire for leaks, a gas station attendant filled it to high pressure. He later let some of the gas out of the tire by opening the tire valve. As the air rushed out, frost formed on the valve stem. His fingertip, which was holding the valve open, became very cold. Explain why these effects occurred.
8. An engineer is able to measure the heat energy input and the heat energy discarded in the exhaust gases of a machine. The machine uses heat energy as its energy source. How can she compute the efficiency of the machine from these measurements?
9. A gas is confined to the lower half of an insulated cylinder by a piston. Above the piston is vacuum. Compare what will happen to the gas in the following two ways for it to fill the upper half of the cylinder: (a) The piston springs a large leak; (b) the piston rises rather slowly to the top of the cylinder. Explain your answer in two ways: by reference to the first law and by molecular arguments.
10. What is the basic source for most energy used on earth?
11. The compression ratio of a piston-type gasoline engine is the ratio of the maximum volume of the cylinder to its minimum volume during the cycle. Why might one guess that, all other things being equal, a high-compression engine will have a higher efficiency than one with a lower compression ratio?
12. Hydrogen and oxygen are placed in a strong container of fixed volume. (Such a device is called a "bomb" by chemists.) A tiny spark ignites the gas mixture and the following reaction occurs: $2H_2 + O_2 \rightarrow 2H_2O$. The final products are

in the form of a high-pressure, high-temperature gas. Apply the first law to the reaction and describe each term, ΔQ, ΔU, and ΔW.

13. Comment on the following quotation from a science magazine: "As the Royal Society's committee on nutritional science put it, 'the joule should be adapted as the unit of energy in all nutritional work and the calories should fall into disuse.' The joule, which is the metric unit of energy, is defined as the work done in moving one kilogram of mass through one meter of space in one second of time. A food's energy value in joules can be obtained by multiplying the caloric value (which is usually expressed in kilocalories) by 4.18."

Problems

1. How much does the internal energy (in joules) of a block of metal change (a) when 30 cal of heat is added to it and (b) when 60 cal of heat is taken away from it? Assume the metal's volume to remain constant.
2. A 15-g block of substance with specific heat capacity of 0.20 cal/(g)(°C) is heated 15°C. Assume its volume change to be negligible. How much did its internal energy increase? Give your answer in joules.
3. By how many joules does the internal energy of a 20-g block of aluminum change as it is cooled from 370°C to 100°C? Ignore the small change in volume of the aluminum; $c_{Al} =$ 0.21 cal/(g)(°C).
4. It requires 80 cal to change 1 g of water at 0°C to 1 g of ice at 0°C. By how many joules does the internal energy of 1 g of the substance change in the process? Ignore the small volume change.
5. The latent heat of vaporization of steam is 539 cal/g. By how many joules does the internal energy of 1 g of steam change as it transforms from steam to water all at 100°C? Assume that the steam originally filled a container of fixed volume.

*6. A piece of hot metal is placed in 100 g of 20°C water. By the time the metal reaches the same temperature as the water, the water temperature is 23.5°C. Neglecting heat losses to the surroundings, how much does (a) the internal energy of the water change and (b) the internal energy of the metal change?

**7. An ideal gas is in a closed container at a temperature of 300 K and a pressure of 150 kPa. Its internal energy is 84 J. Suppose 20 cal of heat energy is added to the system. What will be its new internal energy, its new pressure, and its new temperature? Assume the only type of energy the molecules can have is translational KE.

*8. A 30-g steel block is moving at 80 cm/s when it slides across a tabletop and comes to rest. (a) How much friction work was done on the block? (b) How much heat, in calories, was generated in the process? (c) Assuming half the energy went into the block, how much did the internal energy of the block increase? (d) Under the same assumption, how much did the temperature of the block rise?

*9. A 20-g bullet (made of lead) is moving at 400 m/s when it strikes a tree and becomes embedded in it. How much would the temperature of the bullet be raised if half of all the frictional heat went into the bullet?

*10. An electric cake mixer is driven by a 120 watt motor. It mixes 1500 g of cake batter; $c = 0.90$ cal/(g)(°C) for 3.00 min. Assuming 80 percent of the motor's energy goes into friction losses in the batter, how much will the temperature of the batter rise? Neglect heat loss to the mixing bowl.

*11. Figure P14.1 shows the P-V curve for a gas undergoing a certain process. How much work does the gas do during the following portions of the process: (a) B to C, (b) C to D, and (c) D to A?

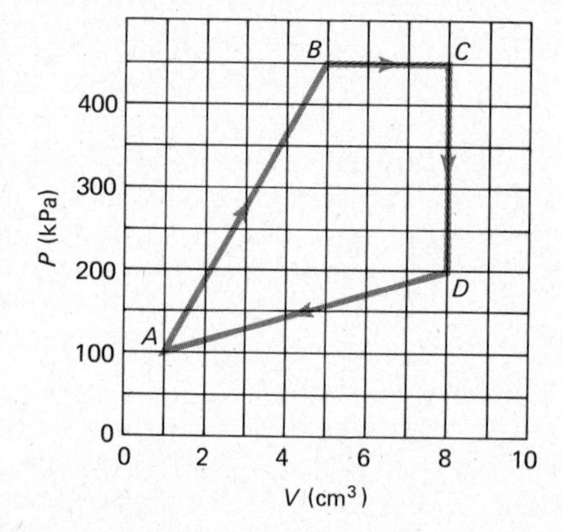

Figure P14.1

*12. A gas follows the P-V curve shown in Figure P14.1. It starts at A and goes to D, C, B, and back to A. Compute the work done by the gas during the following parts of the graph: (a) C to B, (b) B to A, and (c) D to C.

*13. Consider the P-V diagram for a system shown in Figure

*Problems marked with an asterisk are not as easy as the unmarked ones.
**Problems marked with a double asterisk are somewhat more difficult than the average.

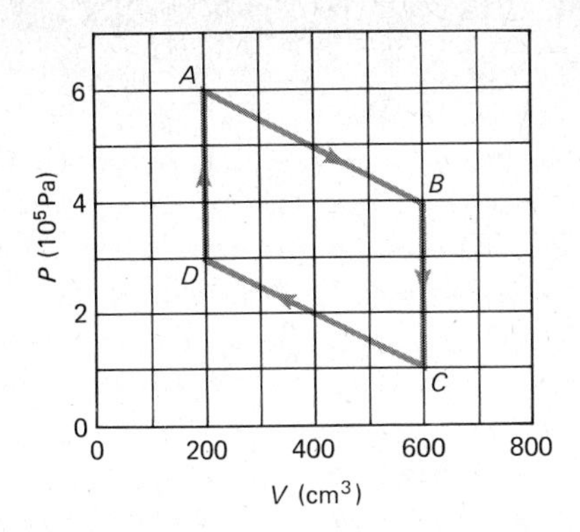

Figure P14.2

P14.2. (a) How much work was done by the system as it expanded from *A* to *B*? (b) How much work was done during portion *BC* of the process?

*14. Refer to the *P-V* diagram for a system shown in Figure P14.2. (a) How much work did the system do during the contraction from *C* to *D*? (b) How much work was done during portion *DA* of the process?

15. An outside force does 900 J of work in an isothermal compression of an ideal gas. (a) How much heat flowed from the gas? (b) How much did the internal energy of the gas change?

16. An outside force does 700 J of work in the adiabatic compression of a gas. (a) How much heat flowed from the gas? (b) How much did the internal energy of the gas change?

17. An insulated cylinder contains 300 g of gas at 30°C. It is suddenly compressed by a piston, which does 38 J of work on the gas. Find the change in internal energy of the gas.

18. The 500 g of nitrogen in an insulated cylinder is at a pressure of 3×10^5 Pa. It is allowed to expand suddenly. During the expansion it does 170 J of work on the confining piston. How much did the internal energy of the gas change in the process?

19. A volume of gas is increased from 40 cm^3 to 60 cm^3 under a constant pressure of 200 kPa. How much work did the gas do?

*20. A gas confined to a piston-cylinder arrangement is compressed by an outside force. The force does 320 J of work in the process. During the process 60 cal of heat flows out of the gas. How much did the internal energy of the gas change during the process?

21. A vertical cylinder is closed by a freely movable piston at its top. The piston has a mass of 5.0 kg and an area of 300 cm^2. Atmospheric pressure is 1.000×10^5 Pa. As the gas is heated, the piston rises 8.0 cm. How much work did the gas do during the expansion?

22. Refer to the cylinder of gas described in the previous problem. A 10-kg mass is placed on the piston and considerable time is allowed to pass. Then the gas is cooled so that the piston falls 7.0 cm. (a) What is the pressure of the gas? (b) How much work did the gas do?

*23. An ideal gas is confined to a vertical cylinder by a 2.0-kg piston. The piston has an area of 12.0 cm^2. When 1.70 cal of heat is added to the gas, the piston lifts 25 cm. Find (a) the increase in internal energy of the gas and (b) the work it does. (c) Do you really need to know the piston's area?

**24. A volume V of water at 100°C is vaporized to steam at 100°C. The steam now has a volume of 1670 V and its pressure is 1 atm (101 kPa). Find the following quantities as 100 g of water is vaporized at 100°C and 1 atm: ΔQ, ΔW, ΔU.

*25. A certain steam engine has an output of 500 hp and its efficiency in use of fuel is 14 percent. It burns fuel oil that has a heat of combustion of 10,500 cal/g. How many kilograms of oil does the engine consume each hour?

*26. Figure P14.2 shows a thermodynamic cycle for an ideal gas confined to a cylinder by a movable piston. During which portions of the cycle (a) was heat added, to the gas, (b) was heat lost from the gas?

*27. Suppose that the thermodynamic cycle shown in Figure P14.2 was run backwards, that is, the process went from *A* to *D* to *C* to *B* to *A*. During which portions of the cycle (a) was work done by the gas, (b) was heat lost from the gas, (c) was heat added to the gas? Assume an ideal gas.

*28. Repeat Problem 26 for the cycle in Figure P14.1.

*29. Repeat Problem 27 for the cycle in Figure P14.1.

**30. For the cycle in Figure P14.1, find the output work per cycle. What would be the horsepower of this engine if it operates at 0.40 cycles per second?

*31. Suppose that the thermodynamic cycle in Figure P14.2 was run backwards, that is, the process went from *A* to *D* to *C* to *B* to *A*. Find the (a) net output work done and (b) net amount of heat energy that flowed into the gas.

*32. For the thermodynamic cycle in Figure P14.2, find the (a) net output work done and (b) net amount of heat energy that flowed into the gas.

**33. When confined to a constant volume, a certain mass of gas requires the loss of 2500 cal to cool it 10°C. Suppose that the gas is originally at 20°C. It is now expanded adiabatically. During the expansion it does 450 J of work. What will be its temperature after the expansion?

**34. The gas in the cylinder of a certain diesel engine requires 0.18 cal to heat it 1°C at constant volume. Suppose that the gas is compressed adiabatically. How much work must be done on the gas to raise its temperature from 20°C to 620°C?

35. What is the efficiency of a Carnot engine that operates between the two temperatures 240°C and 20°C?
36. What is the efficiency of a Carnot engine that exhausts 40 percent of the heat it takes in?
37. A certain heat engine has an efficiency of 25 percent. For each 1000 J of heat energy it takes in, (a) how much output work does it do and (b) how much heat does it exhaust?
38. If a Carnot engine has its cold reservoir at 20°C, what must be the temperature of its hot reservoir if its efficiency is to be 90 percent?

*39. A particular Carnot cycle operates between reservoirs with temperatures at 300°C and 20°C. (a) What is its efficiency? (b) If the net work done per cycle is 500 J, how many calories of heat must flow from the hot reservoir during each cycle? (c) How much heat flows into the cold reservoir each cycle?

*40. A particular Carnot cycle has its exhaust temperature at 20°C. During each cycle it takes in 500 cal from the hot reservoir and exhausts 150 cal to the cold reservoir. Find the (a) net output work per cycle, (b) efficiency, (c) temperature of the hot reservoir.

AP/Wide World

15 HEAT TRANSFER AND AIR CONDITIONING

Performance Goals

When you finish this chapter, you should be able to

1. Give three examples of heat transfer by conduction. Explain the basic mechanism of transfer by conduction.
2. Give three examples of heat transfer by convection. Explain the basic mechanism of convection.

There are many situations in which heat energy must be moved from place to place. Furnaces are used to heat buildings; steam heat is used in industrial processes; heat must be removed from a refrigerator. These are but a few examples. In addition, to conserve energy we must stop the transfer of heat by use of insulation. Heat transfer has many practical applications and we shall examine some of them in this chapter.

3. Give three examples of heat transfer by radiation. How is heat transferred by this mechanism?
4. Use the heat conduction equation to solve simple problems involving it.
5. Explain the meaning of R-value.
6. State the Stefan-Boltzmann law and be able to use it in simple situations.
7. Sketch the general details of a compressor-type refrigeration system. Explain the function of each of its parts. Locate the parts on a real refrigerator.
8. Explain what is meant by the terms "coefficient of performance" and "ton of refrigeration."
9. Explain the operation of a heat pump. Why is it less economical in cold climates than in moderate climates?
10. Define both absolute and relative humidity.
11. Given two of the following four quantities, find the third and fourth: RH, dew point, temperature, m. Assume Table 15.2 is available.

15.1 Heat Transfer by Conduction

There are three basic ways in which heat energy is carried from place to place. They are conduction, convection, and radiation. In this section we shall discuss conduction of heat. This method of heat transfer is shown in Figure 15.1. When one end of a metal rod is placed in a flame, the whole rod becomes hot. What happens is this. The flame heats the one end of the rod. It gives energy to the atoms in that end of the rod. This causes the atoms to vibrate with more and more energy. But by collision with their neighboring atoms, they share this energy with their cooler neighbors. These atoms in turn strike their neighbors and pass energy along to them. In this way thermal energy is transported from atom to atom along the rod. The atoms at the hot end of the rod thereby pass along their energy to the rest of the rod, and so the whole rod becomes warm.

In heat transfer by conduction heat energy is transferred from particle to particle by direct collision of the particles.

Metals are among the best heat conductors. They contain many free electrons. The electrons as well as the atoms undergo collision and thereby aid in transporting the heat. Metals are also good electrical conductors owing to their free electrons. Because of this dual role played by the free electrons, the following rule of thumb applies: Good electrical conductors are usually good conductors of heat.

Everyone knows that metals conduct heat much better than wood or plastic, but in technology we must make this observation quantitative. We do this by defining a quantity called the thermal conductivity. To define it, we consider a slab of material such as the one shown in Figure 15.2. Its thickness is L and its face area is A. If a temperature difference $T_2 - T_1$ exists between its faces, then heat will flow through the slab. We would guess that the heat flow per unit time, $\Delta Q/\Delta t$, through the slab will depend on all these factors. Experiment shows that

$$\frac{\Delta Q}{\Delta t} \sim \frac{A(T_2 - T_1)}{L}$$

The quantity $(T_2 - T_1)/L$ is often called the *thermal gradient*. It is the temperature change per unit length.

We can make this proportion into an equation by use of a proportionality constant, k. Then

$$\frac{\Delta Q}{\Delta t} = k\frac{A(T_2 - T_1)}{L} \tag{15.1}$$

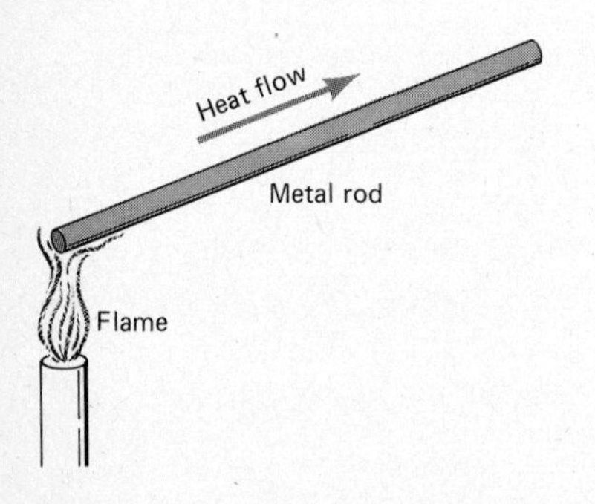

Figure 15.1 In conduction energy is passed along the rod by collisions between particles in the rod.

The constant k varies from substance to substance because the heat flow also depends on the material of the slab. We call k the *thermal* or *heat conductivity*. It is large for good thermal conductors such as metals and small for thermal

Table 15.1 THERMAL CONDUCTIVITY, K, AND APPROXIMATE R-VALUES

Material	k [W/(m)(°C)]	k [cal/(cm)(s)(°C)]	k [(Btu)(in.)/(ft²)(h)(°F)]	R-value (per inch)
GOOD CONDUCTORS				
Aluminum	210	0.50	1390	—
Brass	109	0.26	750	—
Copper	390	0.92	2700	—
Silver	420	1.00	2900	—
INTERMEDIATE				
Asbestos	0.60	14×10^{-4}	4.0	0.25
Brick	0.70	17×10^{-4}	5.0	0.20
Concrete	1.70	40×10^{-4}	12.0	0.08
Glass	≈0.80	$\approx 20 \times 10^{-4}$	≈5.5	0.18
Ice	2.20	52×10^{-4}	15.0	0.07
Water[a]	0.60	14×10^{-4}	4.2	0.24
Wood	0.10	2.4×10^{-4}	0.7	1.40
INSULATORS				
Air[a]	0.022	5.3×10^{-5}	0.16	6.2
Cork	0.042	10.0×10^{-5}	0.30	3.3
Glass wool	0.040	9.3×10^{-5}	0.27	3.7
Kapok	0.035	8.3×10^{-5}	0.24	4.2
Vacuum[a]	0	0	0	∞

[a] For conduction only.

insulators such as cork. Table 15.1 shows typical values of k. Notice the units given for k in the table. The first column gives k when $\Delta Q/\Delta t$ is in watts and with A and L in SI units. The units for the other columns are given at the top of the column. Let us now use these values in a few examples.

EXAMPLE 15.1 A 3-m × 15-m concrete wall is 30 cm thick. (a) On a day when its outer surface is at −10°C and its inner surface is at 20°C, at what rate (in calories per second) does heat flow through it? (b) What thickness of glass wool would act as an equivalent insulator?

Solution (a) We have (in proper units)

$$\frac{\Delta Q}{\Delta t} = k\frac{A(T_2 - T_1)}{L} = \left[40 \times 10^{-4} \frac{\text{cal}}{(\text{cm})(\text{s})(°\text{C})}\right] \frac{(300 \times 1500\ \text{cm}^2)(30°\text{C})}{(30\ \text{cm})}$$

$$= 1800\ \text{cal/s}$$

Notice how important it is to carry along the proper units in such a calculation.

(b) We wish to have

$$\left(\frac{\Delta Q}{\Delta t}\right)_{\text{concrete}} = \left(\frac{\Delta Q}{\Delta t}\right)_{\text{gw}}$$

Then

$$k_{\text{concrete}}\frac{A(T_2 - T_1)}{L_{\text{concrete}}} = k_{\text{gw}}\frac{A(T_2 - T_1)}{L_{\text{gw}}}$$

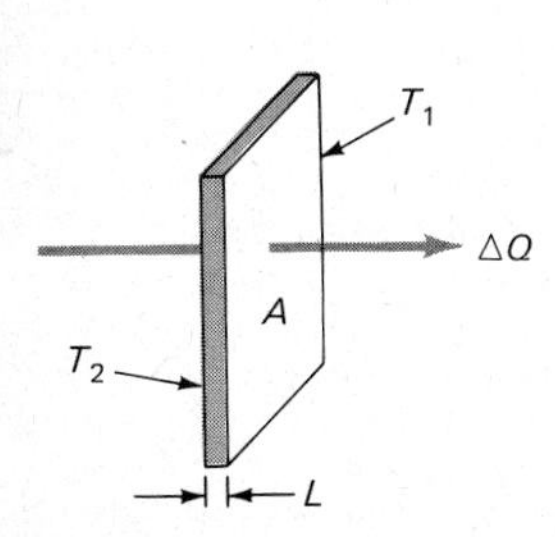

Figure 15.2 The heat ΔQ transported through the slab is proportional to A, to $(T_1 - T_2)$, to $(1/L)$, and to the time Δt for which it flows.

Canceling $A(T_2 - T_1)$ gives, after simplifying,

$$L_{gw} = L_{concrete}(k_{gw}/k_{concrete})$$

Putting in the known values, we have

$$L_{gw} = (30 \text{ cm})\left(\frac{9.3 \times 10^{-5}}{400 \times 10^{-5}}\right) = 0.70 \text{ cm}$$

In other words, 0.70 cm of glass wool is as good an insulator as 30 cm of concrete. ■■

■ EXAMPLE 15.2 Two large metal plates are separated by a glass plate that is 0.25 in. thick. The metal plates have temperatures 32°F and 70°F. How many British thermal units flow through each square foot of the glass in a minute?

Solution From Table 15.1, $k = 5.5$ (Btu)(in.)/(ft^2)(h)(°F). In these units ΔQ is in British thermal units and

$$T_2 - T_1 = 38°\text{F}$$
$$\Delta t = \tfrac{1}{60} \text{ h}$$
$$L = 0.25 \text{ in.}$$
$$A = 1 \text{ ft}^2$$

Therefore, from Equation 15.1,

$$\Delta Q = [5.5(\text{Btu})(\text{in.})/(\text{ft}^2)(\text{h})(°\text{F})]\left(\frac{1}{60}\text{ h}\right)\frac{(38°\text{F})(1 \text{ ft}^2)}{(0.25 \text{ in.})}$$
$$= 14 \text{ Btu}$$ ■■

■ EXAMPLE 15.3 A 30.0-cm-thick concrete house wall is covered by 2.0-cm-thick wood paneling on the inside. The outer surface of the wall is at 10°C and the inner wood surface is at 18°C. Find the temperature of the concrete-wood interface.

Solution The situation is shown in Figure 15.3. Using the symbols shown, for an area A of wall, the heat flowing into the wood from the room is

$$\frac{\Delta Q_{wood}}{\Delta t} = k_{wood}\frac{A(T_1 - T_2)}{L_1}$$

The heat flowing out of the concrete to outdoors is

$$\frac{\Delta Q_{con}}{\Delta t} = k_{con}\frac{A(T_3 - T_2)}{L_2}$$

But, at equilibrium, as much heat must flow out of the concrete as flows into the wood. Equating ΔQ_{wood} and ΔQ_{con} gives

$$k_{wood}\frac{A(T_1 - T_2)}{L_1} = k_{con}\frac{A(T_3 - T_2)}{L_2}$$

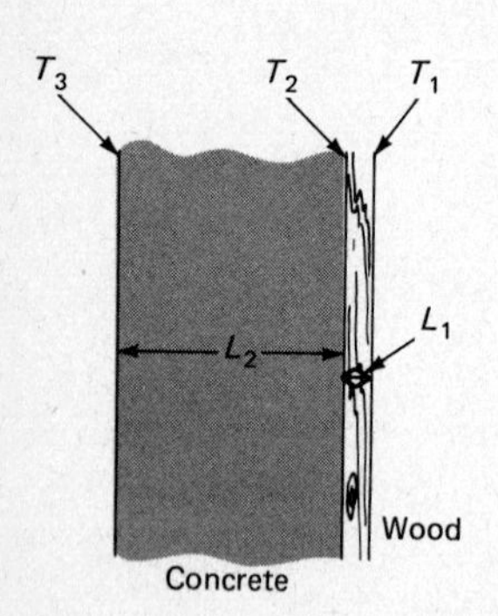

Figure 15.3 $T_1 > T_2 > T_3$.

Using the values for k given in Table 15.1, we have, after canceling A,

$$[0.10\ \mathrm{W/(m)(°C)}]\frac{(18°\mathrm{C} - T_2)}{0.020\ \mathrm{m}} = [1.70\ \mathrm{W/(m)(°C)}]\frac{(T_2 - 10°\mathrm{C})}{0.30\ \mathrm{m}}$$

$$(5.0)(18°\mathrm{C} - T_2) = (5.7)(T_2 - 10°\mathrm{C})$$

Solving for T_2 gives it to be 13.7°C. As we see, 30 cm of concrete is about as effective an insulator as 2 cm of wood. ▮▮

15.2 R-values

In the United States thermal insulation is frequently measured in terms of R-values. The R stands for thermal "resistance." While k, the thermal conductivity, measures the ease with which heat flows through a substance, the R-value measures the difficulty of heat flow. The larger the R-value, the better insulator the layer of insulating material will be. Typical R-values are 19 for a 6-inch-thick layer of fiberglass and 1.2 for 6 inches of brick.

The technical definition of the R-value for a layer of material is as follows.

The R-value is the reciprocal of the heat flow (in Btu) through 1 square foot of the layer per hour per degree Fahrenheit of temperature differential.

If you measure k in (Btu)(in.)/(ft^2)(h)(°F), then the R-value of a layer of the material is given by

$$\text{R-value} = \left(\frac{1}{k}\right)\cdot(\text{thickness in inches}) \qquad (15.2)$$

Some R-values per inch are given in Table 15.1.

The total R-value of several layers of material is quite simple to find. It is the sum of the R-values for the various layers. You will be asked to prove this in Problem 16 at the end of this chapter.

15.3 Heat Transfer by Convection

Molecular kinetic energy, thermal energy, can be moved from place to place by moving the molecules themselves. For example, hot air can be blown into a room to heat it. Hot water can be sent through a pipe to heat an object. Unlike conduction, heat energy is not transported by collisions. It is transported by the movement of the molecules themselves over large distances. We call this type of heat transport *convection*.

In heat transfer by convection hot molecules are moved from place to place. Heat energy is carried along with them.

Notice that convection involves long-range motion of molecules. It is therefore only seen in gases and liquids (i.e., fluids) and never in solids. In most industrial processes the fluid is pumped or blown from place to place. We call this *forced convection*. Typical examples of this are forced-air (or water) home and auto heating systems.

But *natural convection* (as distinguished from forced) is also important. Natural convection is the result of the fact that hot gases and liquids normally rise. They do so because fluids usually expand when heated. Hot air is therefore less dense than cold air. As a result, a region of warm air surrounded by cooler air will experience a buoyant force. The hot-air region will therefore rise through the cooler air so as to float above the denser cool air. Large-scale air circulation on the earth and in homes results from this fact. Typical examples of this situation are shown in Figure 15.4. Similarly, hot water rises above cool water. An example of convection where this phenomenon is important is shown in Figure 15.5. You can no doubt think of other situations where convection is of importance.

When a hot solid is in contact with a fluid, it is observed to lose heat to the fluid at a rate proportional to the area of contact between the solid and fluid as well as to the temperature difference between the hot solid and the fluid. Doubling the area or doubling the temperature difference will double the heat loss rate.

$$\frac{\Delta Q}{\Delta t} \sim A(T_2 - T_1)$$

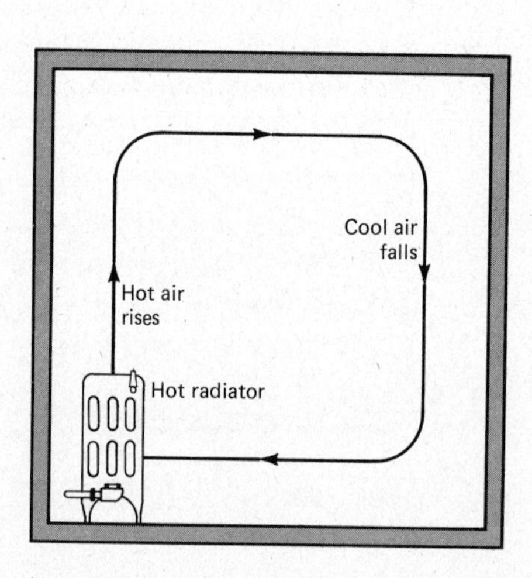

(a) Room heating

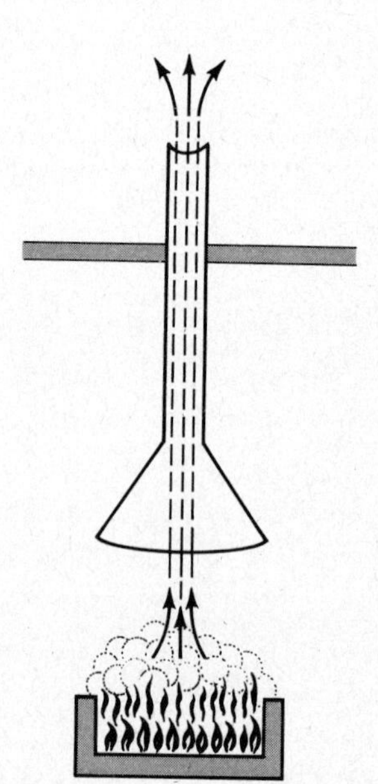
(b) Smoke carried along with the hot air

Figure 15.4 Two common examples of natural convection.

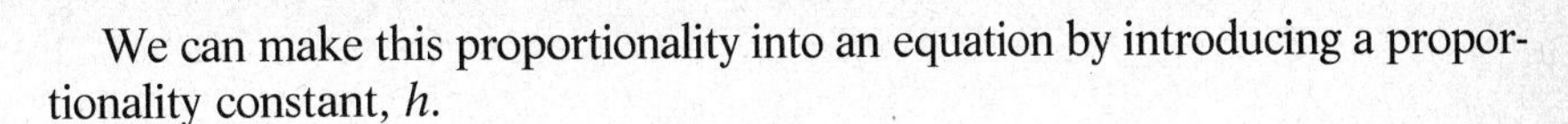

We can make this proportionality into an equation by introducing a proportionality constant, h.

$$\frac{\Delta Q}{\Delta t} = hA(T_2 - T_1) \tag{15.3}$$

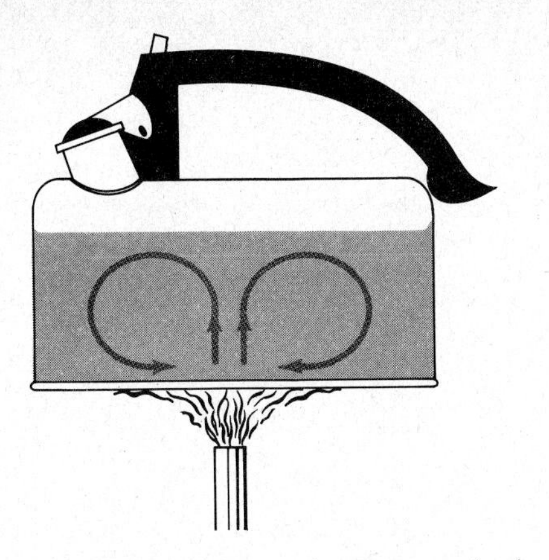

Figure 15.5 Convection currents in the water cause the hot water to mix through the teakettle.

The proportionality constant h is called the *convective surface coefficient*. Its units are $W/(m^2)(°C)$ or $Btu/(h)(ft^2)(°F)$. The value of h depends on the nature of the surface and the properties of the fluid. In most cases it can be described only by empirical equations based on experiments. However, a simple example will show how it is used.

Suppose a glass windowpane has its *outside* surface at 20°C when the air temperature far from the window is 5°C. (Notice we are not concerned here with the temperature between the two panes of the window.) Then $T_2 - T_1$ is 15°C. Suppose further that the wind speed is 5 m/s. Then appropriate tables show that the convection constant in this case is $h = 25\ W/(m^2)(°C)$. Taking the window area to be 1.5 m^2, Equation 15.2 gives

$$\frac{\Delta Q}{\Delta t} = hA(T_2 - T_1) = [25\ W/(m^2)(°C)](1.5\ m^2)(15°C)$$
$$= 563\ W$$

As you see, tabulated values for h are necessary if convective losses of this type are to be computed.

15.4 Heat Transfer by Radiation

The two heat transfer methods we have discussed make use of molecules. In conduction the energy is passed from molecule to molecule by collision. In convection the molecules carry energy with them as they move over large distances. The third method by which energy is carried from hot to cold objects is quite different. It does not use molecules to carry the energy. In this process, called *radiation,* energy moves through empty space, vacuum.

Heat radiation is evident to all of us as the sun sends energy to the earth. Your body is warmed by the sun's rays. More spectacularly, a lens can be used to focus the sun's rays to burn an object as shown in Figure 15.6. Heat energy, as well as light, travels from the sun to the earth. The space between the sun and the earth is a better vacuum than we can produce in large volumes on the earth. Heat energy flows through the vacuum from the sun to the earth in this important example of heat radiation.

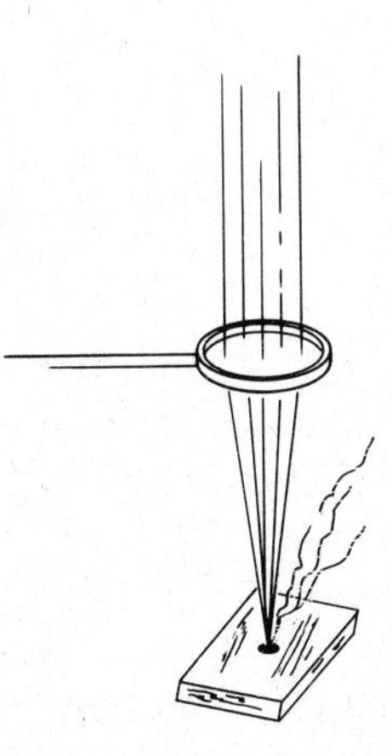

Figure 15.6 Heat energy is carried from the sun to the earth as this burning lens shows.

Another example of heat radiation is seen in Figure 15.4(b). You are warmed as you stand near such a fireplace. But the air close to it moves toward the fire and up the chimney and so air does not carry the warmth to you; the air flows toward the fire. This type of device heats you by radiation. The air in the room actually hinders the flow of heat to you!

Heat radiation is much like light and radio waves. These three—heat radiation, light, and radio waves—are examples of electromagnetic radiation. There are other forms of this radiation. You have heard of some of them, such as X rays, infrared, and ultraviolet light. We shall learn the meaning of this type of wave after we have studied electricity. For now, though, we shall simply state that the sun and all hot objects radiate heat energy. Energy travels by radiation through empty space from hot objects to cool ones.

In heat transfer by radiation electromagnetic waves carry energy from hot objects to cool ones. This type of energy transfer can occur through vacuum.

The law that governs the radiation of heat is called the *Stefan-Boltzmann law*. A surface at absolute temperature T radiates the following amount of heat per second and per unit area:

$$\left(\frac{\Delta Q}{\Delta t}\right)/A = e\sigma T^4 \qquad (15.4)$$

In this expression σ (Greek letter sigma) is a constant of nature, known as the Stefan-Boltzmann constant. Its value is 5.67×10^{-8} W/(m^2)(K^4). We can better understand the units of the Stefan-Boltzmann constant, if we notice that $\Delta Q/\Delta t$ is energy per second, which is power. So the left side of Equation 15.4 has the units watts per square meter.

The quantity e in Equation 15.4 is called the *emissivity* of the surface. Its values range from 0 to 1 depending on the surface involved. Qualitatively, it measures how well the surface absorbs radiation. Dull surfaces have e values near unity while e is near zero for shiny surfaces.

In estimating the value of e, there is a simple rule that is often helpful. A surface that is a good absorber of light is also usually a good absorber of heat. In addition, good heat absorbers are good heat emitters. What this means for the emissivity e can be stated this way. A perfect heat absorber is one that retains all the heat radiation which strikes its surface. It is called a *blackbody* since it usually absorbs all light that strikes it. For a blackbody, $e = 1$. As we see from Equation 15.4, the blackbody surface emits heat well because e has the largest possible value for it. For a shiny surface e will be near zero. It neither absorbs nor emits heat well.

Finally, we should point out that objects at any temperature radiate heat energy. A chair in a room emits heat radiation. The radiation strikes and is absorbed by the wall of the room perhaps. But the wall and all other objects in the room also emit heat radiation. Some of this is absorbed by the chair. At temperature equilibrium the chair emits as much radiation as it absorbs, provided conduction and convection are negligible.

At temperature equilibrium with its surroundings, an isolated object emits as much heat radiation as it absorbs.

We shall learn more about electromagnetic radiation in our study of waves later in this book.

EXAMPLE 15.4 A dirty solid aluminum sphere ($m = 11.3$ g, area $= 12.5\ \text{cm}^2$) has a temperature of 427°C. How much will it cool in 1 s if it loses heat only through radiation? Assume $e = 0.60$ for the sphere.

Solution From Equation 15.4 we have

$$\frac{\Delta Q}{\Delta T} = (A)e\sigma T^4 = (12.5 \times 10^{-4}\ \text{m}^2)(0.60)[5.67 \times 10^{-8}\ \text{W}/(\text{m}^2)(\text{K})^4](700\ \text{K})^4$$
$$= 10.2\ \text{J/s} = 2.44\ \text{cal/s}$$

Notice that we used absolute temperature, not Celsius. Heat will enter the sphere due to radiation from its surroundings, but if they are near room temperature, say 27°C, we would use $T = 300$ K in our equation. Thus the heat from the surroundings will be a factor of about $(300/700)^4 = 0.03$ smaller. We will ignore this much smaller heat flow into the sphere.

The heat lost by the sphere is 2.44 cal per second. Then, from $\Delta Q = cm\,\Delta T$, the temperature change per second of the sphere is

$$\Delta T = \frac{\Delta Q}{cm} = \frac{2.44\ \text{cal}}{[0.21\ \text{cal}/(\text{g})(°\text{C})](11.3\ \text{g})} = 1.0°\text{C}$$

Of course, as the sphere cools, its heat loss per second and rate of temperature change will decrease. ▮▮

15.5 Heat Conservation in Buildings

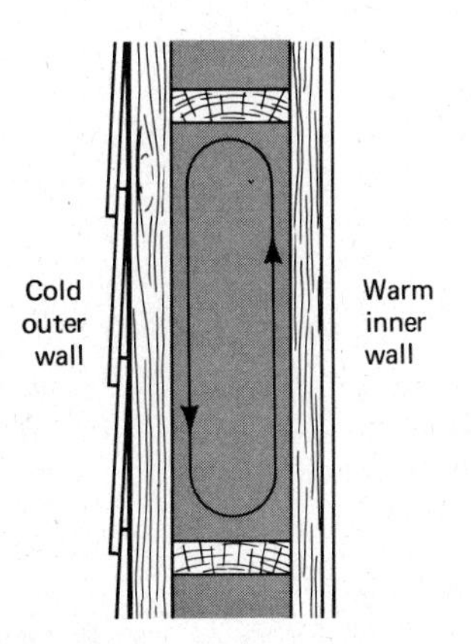

Figure 15.7 Convection in wall spaces decreases the effectiveness of air as an insulator.

Now that we understand the three methods by which heat is transferred, we can discuss how to prevent heat loss in the home. Ideally, the living area of a home should be so well insulated that negligible heat could escape. Looking at Table 15.1 we see that such materials as glass wool, corkboard, and even wood are desirable insulating materials. Other materials such as foamed plastic and shredded paper are also widely used.

One might well ask why an airspace between outer and inner house walls should not be extremely effective. After all, air has a low k value. The difficulty with air as an insulating material has to do with the fact that air can undergo convection. As seen in Figure 15.7, the warm air next to the inner wall rises and changes place with the air at the outer wall. Moreover, any wall opening to the cooler attic air allows the cool attic air to fall and displace the warm air between the walls. Because of this fact, air by itself is not a good insulation material in situations such as this. (In most double-glazed windows the space between the layers of glass is small enough so that air convection is not a great problem.)

Even so, glass wool and similar porous insulating materials owe much of their effectiveness to air. In foam insulation, for example, a large fraction of the insulator is air. But in all these cases the air is unable to undergo convection because of the presence of the glass fibers, plastic, and shredded paper. As a result, they function as much better insulators than would air alone.

Obviously, we should insulate heavily all exterior walls and the ceiling below a cold attic or roof. Windows cannot be covered with insulation, but considerable heat can be saved by use of storm windows or double-pane glass, such as Thermopane. These procedures interfere with heat loss through conduction.

However, windows also allow heat to escape by radiation. Or in summertime the windows allow sun radiation to enter and cause undesired heating. To reduce these effects, builders sometimes use tinted or metalized windows. These windows partly reflect the sun's rays and keep the home cooler in summer. In winter the partly reflecting windows help to keep heat from radiating out of the house interior. But these measures compete with the primary purpose of a window, to let light enter the house.

A much better solution to heat effects due to windows is the proper choice of window locations. Windows on the southern side of a house, if unobstructed by evergreen trees, are almost always a heat source for a home. The sun's rays cause more heat to enter, on the average, than is lost, when averaged over 24 h. This situation is not true for north windows, and so in cold climates windows on that side should be kept to a minimum. The problem of excess heat entrance through the windows in summer can be largely eliminated by overhanging eaves. They shield the window from the nearly vertical rays of the summertime sun. Of course, shades, curtains, shutters, and so on are also valuable as heat insulators for windows.

15.6 Refrigeration

Left to itself, a system will always undergo heat flow from hot to cold. However, if outside work is done on the system, it is possible to cause heat to flow from cold to hot. We will now discuss the refrigeration system, a device designed to force heat to flow from a cold region to a hot region.

A schematic diagram of a typical refrigeration system is shown in Figure 15.8. It causes cooling to occur by means of evaporation of a fluid called the *refrigerant.* (The refrigerant, or working fluid, is typically Freon 12, which has a normal boiling point of $-30°C$.) Let us see how the system operates by tracing the path of the fluid as it is cycled through the system. We begin with the fluid entering the compressor.

1. In the *compressor* the gaseous refrigerant is compressed nearly adiabatically. It emerges as a hot, high-density gas.
2. In the *condenser* the hot fluid is allowed to cool in a coil exposed to flowing air. This unit exhausts heat ΔQ_1 to the surroundings. It accounts for the hot air expelled from a running refrigerator. The

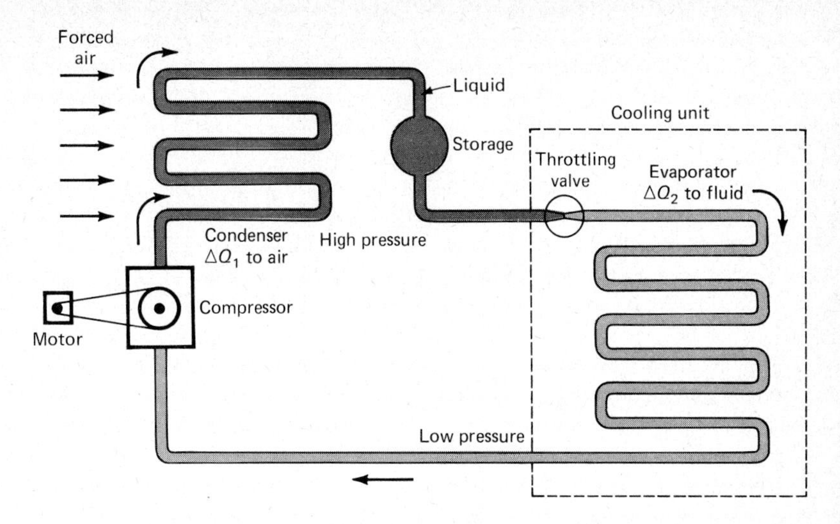

Figure 15.8 A refrigerator system.

working fluid cools enough to condense in this unit. It is now a warm liquid under high pressure.

3. The condensed high-pressure working fluid is stored in the *storage chamber.*
4. The fluid is allowed to escape from the high-pressure side to the low-pressure side through a *throttling valve.* In the process it partly vaporizes and cools.
5. As the cold working fluid moves through the *evaporator,* heat ΔQ_2 flows into the fluid from the chamber around it. The fluid does not warm because of it. Instead, the heat is used to evaporate the remaining liquid portion of the fluid. As we see, the flowing fluid removes heat from the chamber around it. This surrounding chamber therefore becomes cold.
6. The gaseous coolant now returns to the compressor, where it is compressed. The energy it has acquired is removed in the condenser as before. We see that the overall process reduces to the following. Heat ΔQ_2 is removed from the cooling unit to maintain its low temperature. This heat plus the work of compression leaves the system as ΔQ_1 at the condenser. The system pumps heat from the cold chamber to the outside. A refrigeration unit therefore heats the region outside it, while cooling the region close to its evaporator coil.

The operation of a refrigeration system is much like a heat engine run in reverse. The heat engine uses heat energy to produce work and exhaust heat. A refrigerator uses outside work to run the compressor, which causes heat to flow from a cold region to a warm region. These modes of operation are summarized in Figure 15.9.

As we see in Figure 15.9(a), in a heat engine the input energy comes from the high-temperature reservoir in the form of ΔQ_1. Part of this energy is lost to the

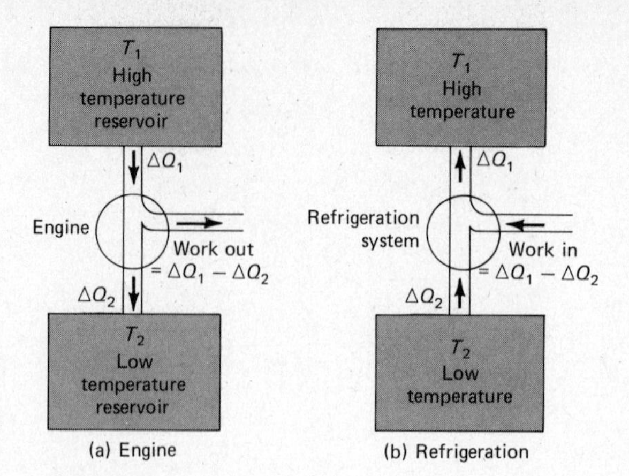

Figure 15.9 The schematic diagrams show the heat flow in an engine and in a refrigerator.

exhaust, the low-temperature reservoir. The remainder is used by the engine to do work.

Figure 15.9(b) shows the heat flow for a refrigeration system. Much of the input energy comes from the work done on the system by the compressor motor. Some input energy, ΔQ_2, is drawn into the system from the cool chamber of the refrigerator. This is done as the working fluid evaporates. Finally, the input work and ΔQ_2 are exhausted to the hot chamber, the region outside the refrigerator. Notice that heat energy flows from cold to hot in a refrigeration system. The outside energy used to run the compressor motor makes this possible.

15.7 Refrigeration Specifications

Refrigeration systems are most often used for refrigerator-freezers and for air conditioning. In both situations the system is of value because it takes heat from a cold region (thereby making it cooler) and transfers it to a warmer region. To measure the system's ability to carry out this chore, we use the coefficient of performance and the cooling capacity of the system.

We define the *coefficient of performance* of a refrigeration system in terms of the input work, the extracted heat ΔQ_2, and the exhausted heat ΔQ_1. These quantities were shown in Figure 15.9(b). It seems reasonable to describe the effectiveness of a refrigeration system in terms of the ratio of extracted heat to input work. The coefficient of performance, η (Greek letter eta), is taken equal to this ratio.

$$\eta = \frac{\text{extracted heat}}{\text{input work}}$$

From Figure 15.9(b), ΔQ_2 is the heat extracted from the cold chamber. Because the law of conservation of energy tells us that

$$\Delta Q_2 + \text{work input} = \Delta Q_1$$

we have

$$\text{Input work} = \Delta Q_1 - \Delta Q_2$$

Therefore

$$\eta = \frac{\Delta Q_2}{\Delta Q_1 - \Delta Q_2} \qquad \text{(ideal refrigerator)}$$

It is possible to express this ratio in terms of the absolute temperatures T_1 and T_2. The end result is

$$\eta = \frac{T_2}{T_1 - T_2} \qquad \text{(ideal refrigerator)} \qquad (15.5)$$

Notice that η becomes infinite if $T_1 = T_2$. In that case the inside of the refrigerator is at the same temperature as the surroundings. It is easy to transfer heat from the inside to outside under these conditions. Therefore the coefficient of performance is large. But when the temperature difference between outside and inside is large, it is difficult to transfer heat. More work must be done. The coefficient is therefore smaller. Of course, Equation 15.5 applies to an ideal refrigerator. In an actual system heat losses and other inefficiencies would cause η to be smaller.

The *cooling capacity* of a refrigerator or air conditioner's refrigeration system is usually expressed in *tons of refrigeration.* One ton of refrigeration is equivalent to the removal of 12,000 Btu in 1 h (i.e., 12.65 MJ/h). Of course, the heat removed by a refrigeration system depends not only on the system itself, but also on the temperature difference through which it must work.

EXAMPLE 15.5 What is the cooling capacity of an ideal $\frac{3}{4}$-hp refrigerator unit operating between 30°C and 18°C?

Solution For this ideal system we have

$$\eta = \frac{T_2}{T_1 - T_2} = \frac{273 + 18}{12} = 24$$

The input power to the unit is $\frac{3}{4}$ hp = ($\frac{3}{4}$ hp)(746 W/hp) = 560 W. Therefore,

$$\eta = \frac{\text{extracted heat per second}}{\text{input work per second}}$$

becomes

$$24 = \frac{\Delta Q \text{ per second}}{560 \text{ W}}$$

from which

$$\Delta Q \text{ per second} = 13{,}400 \text{ J/s}$$

Because there are 3600 s in an hour.

$$\Delta Q \text{ per hour} = (3600) \times (13{,}400) = 48 \times 10^6 \text{ J/h}$$

Since 1 ton of refrigeration is equivalent to 12.65 MJ/h, this system has a cooling capacity of

$$(48 \text{ MJ/h})\left(\frac{1 \text{ ton}}{12.65 \text{ MJ/h}}\right) = 3.8 \text{ tons}$$

The ideal refrigeration system has a capacity of 3.8 tons. ■■

15.8 Heat Pumps

In recent years a device capable of both heating and cooling has come into widespread use. It is called a *heat pump*. Basically, the heat pump is a refrigeration unit. It is designed in such a way that it refrigerates the interior of a building in summer and refrigerates the exterior of the building in the winter. The heat removed from the outside during the winter is exhausted into the building and thereby the building is warmed.

Electric refrigeration units use electric energy to pump heat usually from a cold region to a warmer region. The same is true for a heat pump. Everything we have said about refrigeration units also applies to a heat pump. The basic equation governing the operation of such a device is obtained from the law of conservation of energy. Because energy must be conserved,

$$\begin{pmatrix}\text{Heat output}\\ \text{of unit}\end{pmatrix} = \begin{pmatrix}\text{heat removed}\\ \text{from cold region}\end{pmatrix} + \begin{pmatrix}\text{heat equivalent of}\\ \text{electric energy input}\end{pmatrix}$$

In a heat pump the interior of the building can be used as either the warm or cold region. This operation is shown in Figure 15.10. In winter the unit takes heat from the outside (the cold region) and exhausts it to the building interior. But because the equation above still applies, the electric energy input to the system is also delivered as needed heat to the building. As a result, use is actually made of the electrical energy that would ordinarily be wasted when the heat pump is used for cooling.

Because refrigeration units transport heat most efficiently between small temperature differences, a heat pump works best in a moderate climate. Even so, heat pumps are finding increasing use in colder climates. As the cost of electricity decreases in relation to other fuels, heat pumps will become even more attractive for air-conditioning systems.

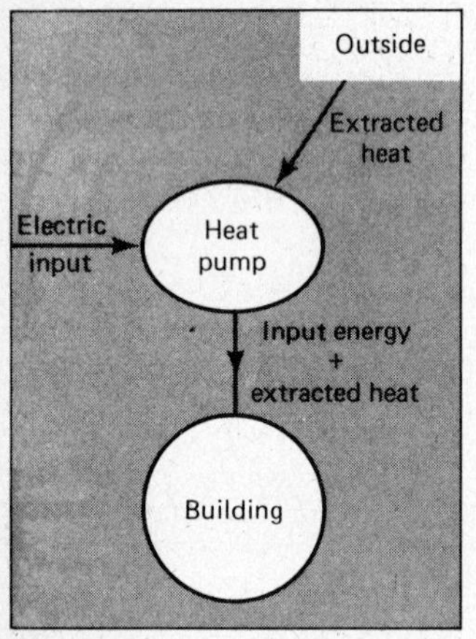

Figure 15.10 The schematic diagram shows the energy flow for a heat pump system.

▮ EXAMPLE 15.6 How does the cost of operation of an ideal heat pump in the heating mode vary with temperature difference?

Solution We need the electric input energy to the refrigeration unit. This can be found from (for an ideal system)

$$\eta = \frac{\text{extracted heat}}{\text{input energy}} = \frac{T_2}{T_1 - T_2}$$

This gives

$$\text{Extracted heat} = (\text{input energy}) \times \left(\frac{T_2}{T_1 - T_2}\right)$$

But the output heat is given by

$$\text{Output heat} = \text{input energy} + \text{extracted heat}$$

which becomes, after substitution,

$$\text{Output} = \text{input} + \left[(\text{input}) \times \left(\frac{T_2}{T_1 - T_2}\right)\right]$$

Therefore, for each unit of input energy the output heat is

$$\frac{\text{Output}}{\text{Input}} = 1 + \frac{T_2}{T_1 - T_2} = \frac{T_1 - T_2 + T_2}{T_1 - T_2} = \frac{T_1}{T_1 - T_2}$$

For each unit of input energy the output energy is

$$\text{Output per unit input} = \frac{T_1}{T_1 - T_2}$$

But we know that

$$\text{Output cost/unit} = \frac{\text{input cost/unit}}{\text{output units per unit of input}}$$

from which

$$\text{Output cost} = (\text{input cost}) \times \left(\frac{T_1 - T_2}{T_1}\right)$$

The cost of heat from an ideal heat pump therefore rises in proportion to the temperature difference. ▮▮

▮ EXAMPLE 15.7 A certain heat pump has a coefficient of performance of 1.5 when the outside temperature is 5°C and the inside temperature is 20°C. Compare its heating cost to the cost of electric resistance heating.

Solution The heat pump furnishes heat given by

$$\text{Output heat} = \text{input energy} + \text{extracted heat}$$

The input energy is equal to the electric energy furnished to the heat pump. It is also equal to the amount of energy an equivalent amount of electricity would furnish in baseboard heating. We therefore need to find (output heat)/(input energy).

We found this in the previous example for an ideal system. It was

$$\frac{\text{Output}}{\text{Input}} = 1 + \frac{T_2}{T_1 - T_2}$$

But for a real system $T_2/(T_1 - T_2)$ must be replaced by η, the actual coefficient of performance. In our case

$$\frac{\text{Output}}{\text{Input}} = 1 + \eta = 1 + 1.5 = 2.5$$

Two and one-half times as much heat is given off by the heat pump as compared to direct electric heating. Therefore, the heat pump should operate at about $1/2.5 = 0.4$ (i.e., 40 percent) the cost of electric resistance heating. ■■

15.9 Humidity

It is well known that a person's sensation of warmth or cold is influenced by the humidity of the air. When the air is too dry in a home during cold weather, the room temperature must be raised to make the room seem more comfortable. During the summer we hear people say, "It's not the heat but the humidity." Let us now investigate what is meant by "humidity" and why it influences how hot or cold we feel.

Humidity is directly concerned with the amount of water vapor in the air. You know that the air in a closed bottle partly filled with water, as in Figure 15.11, is saturated. We say that the humidity is 100 percent for such air. Because the bottle is closed, no net evaporation of water is occurring in the bottle. As many molecules leave the surface of the water as return to it from the water-laden, saturated air above it. On the other hand, water evaporates most readily if the air above it is dry. Many more molecules leave the water surface than return from the dry air above it. Hence evaporation is most rapid when the air contains little water, that is, when the humidity is low.

The technical definition of relative humidity is as follows:

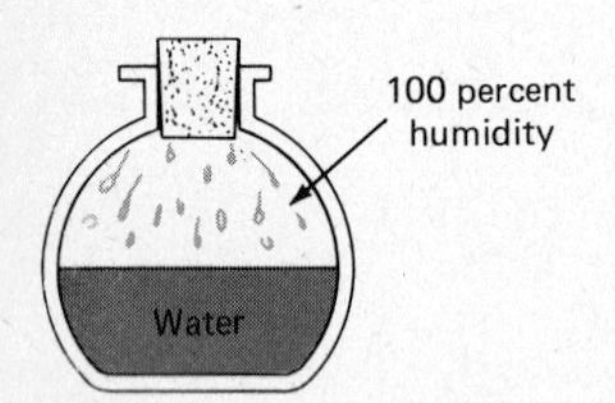

Figure 15.11 The air in the closed bottle is saturated with water vapor.

The *relative humidity* (RH) is the ratio of (the mass of water vapor per unit volume in the air) to (the mass of water vapor per unit volume of saturated air) at the same temperature. The *absolute humidity* is the mass of water vapor per unit volume of air.

In order to use this definition, we must know how much water vapor exists in saturated air. Table 15.2 tabulates these data. For example, we see from it that

Table 15.2 WATER VAPOR PER UNIT VOLUME OF SATURATED AIR

Temperature (°C)	Water Content (g/m^3)
−8	2.74
−4	3.66
0	4.84
4	6.33
8	8.21
12	10.57
16	13.50
20	17.12
24	21.54
28	26.93
32	33.45
36	41.82

1 m^3 of saturated air holds only 2.74 g of water vapor at −8°C. The hotter the air, the more water vapor it contains when saturated.

It is now a simple matter to make use of relative humidity. For example, suppose that the air temperature is 32°C on a particular day. We see from the table that saturated air holds 33.45 g of water per cubic meter of air. Suppose, though, that there was only 15.0 g of water per cubic meter of air on that day. Then the relative humidity would be

$$RH = \frac{\text{amount present}}{\text{amount possible}} = \frac{15.0}{33.45} = 0.45$$

RH is often expressed as a percent; in this case it would be 45 percent. Let us now discuss what happens when humid air is cooled.

Humid air is nearly saturated with water vapor. Suppose that the air temperature is 24°C and the amount of water the air contains is 13.5 g/m^3. According to Table 15.2, saturated air at 24°C contains 21.5 g/m^3. Therefore, the air is *not* saturated. However, suppose that the air is cooled. We see from the table that the air will be saturated when the temperature drops to 16°C. (The air contains 13.5 g/m^3. This is how much water saturated air contains at 16°C.) If the temperature drops below 16°C, the air must release some of its water. At temperatures below 16°C, it cannot hold 13.5 g/m^3. The excess water then falls out of the air as fog, dew, or rain.[1]

In the example just given, 16°C is called the *dew point* of the air. It is the temperature at which the air would be saturated.

The dew point is the temperature at which the air would be saturated if it retained its present amount of water vapor.

[1]Under special conditions the air contains more than this amount of water for a short time. The air is then said to be supersaturated.

At its dew point, air has a humidity of 100 percent. When cooled below the dew point, precipitation must occur.

EXAMPLE 15.8 According to the weather report, the RH on a certain day is 83 percent and the temperature is 24°C. How cold must a water pipe be if it is to sweat?

Solution When the pipe sweats, dew is forming on it. We must compute the dew point of the air. At 24°C saturated air holds 21.54 g of water per cubic meter of air. We have

$$\text{RH} = \frac{\text{amount present}}{\text{amount possible}}$$

$$0.83 = \frac{\text{amount present}}{21.54\ \text{g/m}^3}$$

This gives

$$\text{Amount present} = 17.9\ \text{g/m}^3$$

Looking at Table 15.2 we see that saturated air at 20°C contains 17.1 g/m^3. Therefore, the dew point will be slightly above 20°C. When the water pipe reaches this temperature, it will begin to sweat.

Summary

Heat can be transferred in three ways. In conduction transfer of energy occurs by direct particle interaction. Energy is passed from hotter particles to cooler ones by collisions between the particles. In convection fluid flow occurs. Warm material displaces cooler material and thereby carries heat energy from hot regions to cool regions. In radiation electromagnetic waves carry energy from place to place. Unlike the other two methods, this type of heat transfer can occur through empty space.

For the conservation of energy, homes should be well insulated. Large masses of air are subject to convection, and so air is often a poor insulator. However, most insulation, such as glass wool and foams, owe their superior insulating qualities to trapped air. Properly placed windows can decrease the average heat loss from a building.

A refrigeration system extracts heat from a cold reservoir and exhausts it to a hot reservoir. An outside energy source is required to do this. The coefficient of performance of such a system is

$$\eta = \frac{\text{extracted heat}}{\text{input work}}$$

In the case of the best possible cycle this becomes

$$\eta = \frac{T_2}{T_1 - T_2}$$

T_2 and T_1 are the cold and hot reservoir temperatures, respectively.

Heat pumps can be used to both cool and heat a building. They are basically refrigeration units whose direction of operation can be reversed. In their heating mode, heat from the cooler outside air is pumped into the building. Heat pumps are most efficient when the temperature difference between outside and inside is small.

Relative humidity is the ratio of the mass of water vapor in unit volume of air to the mass present at saturation. At the dew point the RH is 1.0, or 100 percent. Evaporation is much less rapid in humid air than in dry air.

Questions and Exercises

1. A metal rod and a wooden stick have been side by side in a freezer for considerable time. When they are touched by one's hand, the metal rod feels colder than the wooden stick. Why?
2. On a certain day the temperature is 0°C outdoors and 20°C inside. To compute the heat flow through a windowpane consisting of a single sheet of glass, a student uses the heat conduction equation. Even though he uses the area of the pane, its thickness, $\Delta T = 20°C$, and the proper k, he obtains a result for $\Delta Q/\Delta t$ that is far too large. Why? Why are double-pane windows used rather than a single plate twice as thick?
3. The draft on an industrial-type chimney increases with the height of the chimney. Explain why. What other reason is there to use a high chimney?
4. The thermal conductivity for air is much less than for glass. Yet builders insulate homes by filling the partitions with fiberglass and similar substances. Explain why.
5. How is a Thermos jug constructed? Why should its inside surface be metalized?
6. Along the edge slopes of a mountain range, airline pilots are always alert for thermal updrafts and downdrafts. Explain the origin of these air currents. Why do they occur so noticeably near mountains?
7. Even though the outside and inside temperatures are the same on a windy and a windless day, more heat is lost from a building on the windy day. Why?
8. An attic floor has 10 cm of insulation. If 10 cm more insulation is added, by about what factor will heat loss to the attic be changed? Repeat if 20 cm instead of 10 cm is added.
9. The Gulf Stream is an example of what form of heat transfer?
10. A melon can be cooled by wrapping it in a large, wet towel and placing it in the sun. Explain why this works better on some days than on others.
11. Even though the air in a room may be at 20°C, you feel cooler in the room if the walls are cold than if they also are at 20°C. Explain why.
12. Under otherwise equivalent conditions, which will be warmer in the sunshine, a cement pavement or cement covered with asphalt. Which will be warmer at midnight after a hot day?
13. A friend said that an advertisement for a portable electric air conditioner claimed that it could be used in the middle of any room. The air conditioner is simply plugged into an electric outlet and turned on. Why must the claim (or the friend) be wrong? Compare this with the attempt to cool a home using a refrigerator with its door open.
14. A certain heat pump is used between the 0°C outside and 20°C inside. These are not realistic temperatures to be used for the cold and hot temperatures when computing the coefficient of performance of the pump. Why not?
15. On a cold day it is possible to see a person's breath. Explain why it is visible then but not on a warmer day.

Problems

1. How many joules of heat will flow in 1 h through a 40-cm × 100-cm windowpane that is 3.0 mm thick? Assume the outer surface is at 5°C and the inner surface is at 20°C.
2. A 15-cm layer of glass wool is used to insulate a 15-m × 20-m attic floor. How many calories of heat are lost to the attic each minute if the temperatures at the two sides of the insulation are 18°C and −6°C?
3. In order to determine the thermal conductivity of a material, one measures the heat flow through it. It is found that 500 cal flows in 1 min through a 3-cm^2 area that is 2 mm thick. The temperature difference is maintained at 30°C. What is k for this material?
4. A metal pan with boiling water in it is set on a 0.50-cm-thick corkboard. This, in turn, rests on a metal stove top at 20°C. How much heat is conducted out of the bottom of the pan in

1 min? The area of the bottom of the pan is 90 cm^2. Assume that equilibrium has been established at the two faces of the corkboard.

*5. The inner metal wall of an oven is separated from its outer metal wall by an asbestos sheet, $\frac{1}{2}$ in. thick. When the inner wall is at a temperature of 360°F and the outer is 80°F, how much heat (in British thermal units) flows through the oven walls each second? (The wall area is 600 $in.^2$.) How many watts are being lost this way?

6. Assuming convection effects to be negligible, what thickness of air will insulate as well as a 3-mm-thick glass windowpane?

7. What thickness of brick will have the same insulation effectiveness as a 5-cm-thick layer of cork?

*8. A flat sheet is made by a 2-cm-thick layer of cork on a 10-cm-thick layer of brick. The heat flow through the sheet is 5 cal/min for each 100 cm^2 of area. What is the temperature difference across (a) the cork, (b) the brick, and (c) the entire sheet?

*9. A certain double-glazed window has a 2-mm air gap between its two 3-mm-thick glass panes. The temperatures on the two sides of the inside pane are 20.0°C inside the house and 18.0°C at the air gap side. Find the following: (a) the heat flow in calories per second through a 1-cm^2 area of the inner pane, (b) the heat flow through the air gap, (c) the temperature at the cool side of the gap, (d) the temperature at the outside of the window. Assume equilibrium temperature conditions. Ignore convection.

**10. A 2-mm-thick layer of asbestos covers the surface of a 3-cm-thick brass plate. Find the heat flow (in calories per second) through the composite sheet if the temperature difference across it is 200°C. The plate area is 400 cm^2.

**11. A brass rod has a cross-sectional area of 4.0 cm^2 and is 8.0 cm long. One end is brazed to a brass plate maintained at 5.0°C. The other end is sealed to a large plate of glass 0.20 cm thick. The other side of the glass plate is sealed to a brass plate maintained at 20°C. Find the approximate temperature of the rod at the point where it is sealed to the glass. (Hint: The heat flow down the rod must be the same as through the glass.)

*12. A flat-bottomed glass beaker sits squarely on a brass plate maintained at 200°C. The beaker contains 150 cm^3 of water originally at 20°C. About how long does it take the water to heat to 40°C? Assume the bottom of the beaker has an area of 50 cm^2 and is 0.20 cm thick. The water is stirred during the process.

*13. An ice chest is maintained at 0°C by ice placed in it. The chest is cubical with each inside wall having an area of 1200 cm^2 and a thickness of 4.0 cm. Each wall is made of foam plastic with $k = 1.30 \times 10^{-4}$ cal/(cm)(s)(°C). How much ice will melt in the chest in an hour when the outside temperature is 24°C?

*14. A 6.0-ft^2 sheet of material allows 70 Btu to flow through it in 30 min when the temperature difference across it is 25°F. What is the R-value for the sheet?

*15. How much heat (in Btu) flows through 25 ft^2 of R-10 insulation in 24 h if the temperature difference across the sheet is 40°F?

**16. Three sheets of thermal insulation having R-values of R_1, R_2, and R_3 are laid one on top of the other. Show that their combined R-value is equal to $R_1 + R_2 + R_3$.

17. A hot metal sphere is allowed to cool by radiation. Find the ratio of its cooling rate at 500°C to its rate at 400°C. Assume the surroundings are quite cold.

*18. The planet Mercury has a diameter about 0.40 that of the earth. Its average surface temperature is about 300°C compared to the earth's average temperature of about 20°C. Find the approximate ratio of the heat radiated per second by Mercury to that of the earth. Assume for this rough computation that the two planets have equal emissivities.

*19. How much heat (in calories) will radiate each hour from a sphere with a radius of 2.0 cm if the temperature of the sphere is 500°C and its emissivity is 0.35?

*20. The temperature of the sun's surface is about 6000 K. Its radius is about 7.0×10^8 m. Assuming it to have an emissivity of 0.90, how much heat (in calories) does it radiate each minute?

**21. The earth receives radiation from the sun at a rate of 0.134 $J/cm^2/s$. The earth is 1.50×10^{11} m from the sun. Compute the temperature of the sun's surface assuming its radius is 7.0×10^8 m and its emissivity is 1.00.

*22. An ideal refrigerator has its cooling unit at 17°F and its exhaust unit at 98°F. What is the value of the coefficient of performance for this ideal refrigerator? An actual refrigerator operating between these two temperatures has a $\frac{1}{2}$-hp, completely efficient compressor. It is capable of removing 140 cal/s from the cold chamber. Find its coefficient of performance.

*23. A $\frac{3}{4}$-hp window air conditioner has a coefficient of performance of 3.2 when taking air at 90°F and cooling it to 70°F. (a) What would be the coefficient of performance of an ideal refrigerator operating between these two temperatures? (b) How many calories per second does the actual refrigerator remove from the warm air? Assume the compressors to be 100 percent efficient.

24. How long would it take a 3-ton-cooling-capacity refrigera-

*Problems marked with an asterisk are not as easy as the unmarked ones.
**Problems marked with a double asterisk are somewhat more difficult than the average.

tion system to remove 500,000 cal of heat from a large vat of liquid.

25. How many tons of refrigeration would be needed to change 0.10 m^3 of water to ice (all at 0°C) in a time of 1 h?

*26. What is the cooling capacity of a 0.50-hp refrigeration unit if its coefficient of performance is 4.0?

*27. How much power (in horsepower) is required to operate a 3.5-ton-cooling-capacity refrigerator if its coefficient of performance is 4.5?

28. A certain heat pump extracts 350 cal/s from outside and delivers 500 cal/s to the inside of a building. What is its coefficient of performance?

*29. A certain heat pump uses 1000 W and delivers heat at a rate of 600 cal/s. (a) How many calories of heat does the pump extract each second from the outside? (b) What is its coefficient of performance?

30. Commonly, the relative humidity in a house is about 40 percent and the temperature is 20°C. How much water is there in a room $5 \times 8 \times 3$ m^3?

31. The internal volume of a certain house is 900 m^3. Its temperature is maintained at 20°C. How much water must a dehumidifier remove from the house to decrease the RH from 90 percent to 40 percent?

32. On a cold day the outdoor temperature is −4°C and RH is 60 percent. If this air is taken into a house and heated to 20°C, what will be its relative humidity?

33. What is the relative humidity on a day when the temperature is 24°C and the dew point is 12°C?

34. As the sun goes down one windless day, the temperature is 32°C and the humidity is 80.5 percent. How much must the air cool before fog will form?

Travers, © Peter Arnold

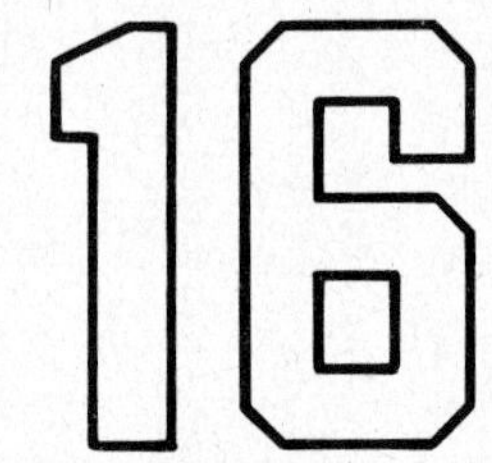

16 VIBRATORY MOTION

Vibration is one of the most important types of motion. Often we wish to eliminate vibrations. For example, we don't want the front end of a car to shimmy. We don't want a refrigerator to vibrate as though it were falling apart when its motor runs. In other instances vibration is desired. Most things driven by springs depend on vibration for their operation. This chapter is devoted to oscillating motion. In it we shall learn the basic principles that govern vibration.

Performance Goals

When you finish this chapter, you should be able to

1. Measure the amplitude, frequency, and period of motion for a slowly vibrating system. State the mathematical relation between period and frequency.
2. Sketch a graph showing the displacement of a vibrating system as a function of time. On the graph, identify the amplitude and period of the motion.
3. Given a simple vibrating system, identify the restoring force or torque. Explain how the

restoring force causes the system to oscillate.

4. State the force law that the restoring force or torque must obey if a system is to undergo simple harmonic motion.
5. Measure the force constant of a spring. Knowing the force constant and mass, compute the frequency of vibration. Check the computation by direct measurement.
6. State the equation for the energy stored in a spring in terms of k and x.
7. Describe how energy is divided between kinetic and potential as a mass vibrates at the end of a spring. Repeat for a pendulum. Repeat for a torsion pendulum.
8. Given k, x_0, and m for a spring-mass system, find v for the mass at a given value of x.
9. Starting from $F = ma$, compute the relation between a, m, k, and x for a mass vibrating at the end of a spring. For such a system, if k, m, and x are given, compute the acceleration of the mass.
10. Describe qualitatively how the motion of a point P on the reference circle is related to simple harmonic motion.
11. Define the "natural vibration frequency" of an oscillator. Give the equation by which it can be computed for a spring-mass system and for a torsion pendulum.
12. Compute the period of a pendulum of known length. Point out the restriction on the relation you use.
13. Discuss qualitatively how the amplitude of a driven oscillator varies with frequency of the

16.1 Periodic Motion

The systems shown in Figure 16.1 will vibrate when disturbed and released. Motion such as this, that repeats itself over and over again, is called *periodic motion.* The *period, T,* of the motion is the time the system takes to complete one complete cycle. For example, one complete cycle for the systems shown in Figure 16.1 is motion from A to B to C back to B and finally to A again. The period T of the motion is the total time taken for the system to move from A to B to C and back to A.

Another term we use to describe periodic motion is *frequency.* It is the number of cycles the system goes through per unit time. We represent it by f; the Greek symbol ν (nu) is also used. For example, if the pendulum swings back and forth through five complete cycles each second (a fast pendulum, to be sure), then its frequency is $f = 5$ cycles per second. The unit cycles per second is given a special name in the SI. It is called the Hertz (Hz).[1]

$$1 \text{ Hz} = 1 \text{ cycle/s}$$

The frequency of our 5-cycle/s pendulum is 5 Hz.

There is a simple relation between frequency and period. If an oscillator has a frequency of 10 cycles/s, then it takes $\frac{1}{10}$ s to complete 1 cycle. Its period, that is, the time for one oscillation, is $\frac{1}{10}$ s. In equation form

$$\text{Period} = \frac{1}{\text{frequency}} \quad \text{or} \quad T = \frac{1}{f} \tag{16.1}$$

As an another example, if a pendulum makes 2.0 cycles/s its period is

$$\text{Period} = \frac{1}{2.0 \text{ cycles/s}} = 0.50 \text{ s}$$

Notice that "cycles" is just a descriptive word. It is not a unit. Therefore it does not carry through in equations.

To describe how far a system swings during its vibration, we use the term *amplitude of vibration.* It is the farthest distance from its central position that the system reaches in its cycle. In Figure 16.1 the amplitude of the vibration is the distance from A to B in each case. Notice that the amplitude is only half of the total swing.

Often it is convenient to represent oscillations by a picture or graph. One way of doing this is shown in Figure 16.2. As the oscillating mass moves up and down, it traces out a graph of its motion. The coordinates of the graph in this case are y (= vertical displacement from equilibrium) and time t. This type of

[1] After Heinrich Hertz, a pioneer in the field of electromagnetic waves.

driving force. Sketch a graph showing the variation of amplitude for two different amounts of friction. Locate the resonance frequency on the graph.

curve is called a *sinusoidal curve* by mathematicians. For this reason vibratory motion is often called sinusoidal motion. Notice in the figure that $\overline{DE}$ is the amplitude of motion. The time from A to E is the period.

16.2 Energy of Vibration

The systems we showed in Figure 16.1 obviously have energy when they are vibrating. When the pendulum bob and the mass at the end of the spring pass through point B, they are in motion. Clearly, the systems then have kinetic energy. But the pendulum bob and the mass on the spring come to rest at point C. What happens to the KE they had at B?

In the case of the pendulum, it gains gravitational PE as it swings from B to C. Apparently, all the KE it has at B is changed to PE at C. Therefore, as the pendulum swings back and forth, it exchanges its PE at A to KE at B and then back to PE at C. If there were no friction, the back and forth vibration would continue forever. Continuous interchange of PE and KE would occur.

The vibration of a mass at the end of a spring is similar. To show this most simply, let us look at the cart vibrating back and forth at the end of a spring in Figure 16.3. We pull the cart to position A in (a) and then release it. At that instant the cart is motionless. But the spring is stretched and has energy stored in it. We call this stored energy *spring potential energy,* SPE. The stretched spring pulls the system to the left and accelerates the cart. In the process the spring relaxes so SPE is changed to KE.

When the cart is at B, as in part (c), the spring is no longer stretched. All of the SPE has been changed to KE of the cart. But the inertia of the cart carries it beyond B and causes it to compress the spring. In so doing, KE is changed to SPE. When the cart reaches C, as in part (e), it has lost all its KE to SPE. The compressed spring starts the cart into motion toward the right, and the SPE is once again changed to KE.

As we see, the vibrating system shows a continual interchange of energy between SPE and KE. At the ends of its swing, at A and C, the system has only SPE. But as the system goes through its equilibrium position, its energy is all kinetic. Notice how similar this is to the energy interchange for a vibrating pendulum.

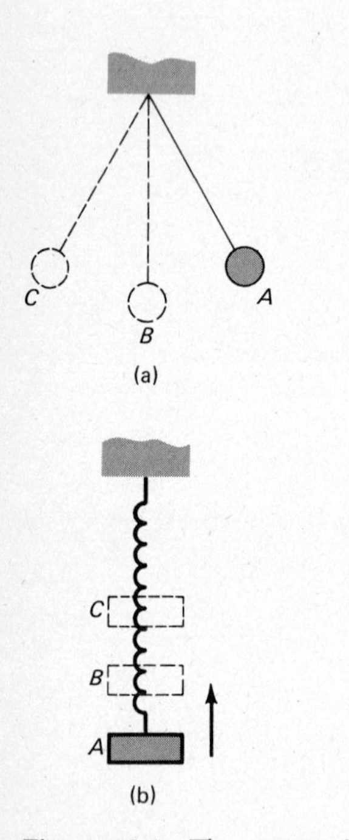

Figure 16.1 The systems in (a) and (b) both complete one cycle by vibrating from A to B to C to B and back to A.

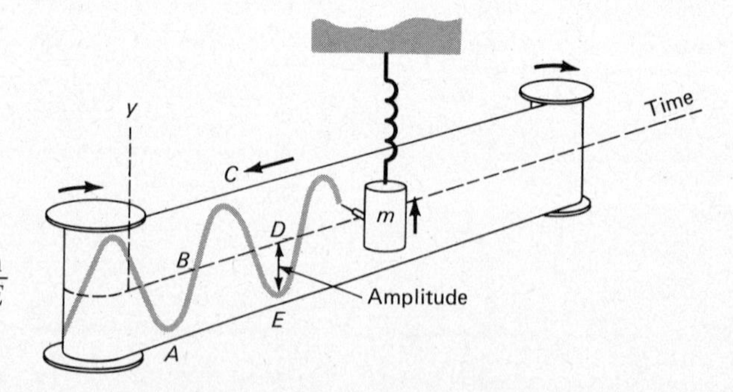

Figure 16.2 A record of the oscillation is left on the moving sheet of paper. $\overline{DE}$ is the amplitude. The time from A to E is the period.

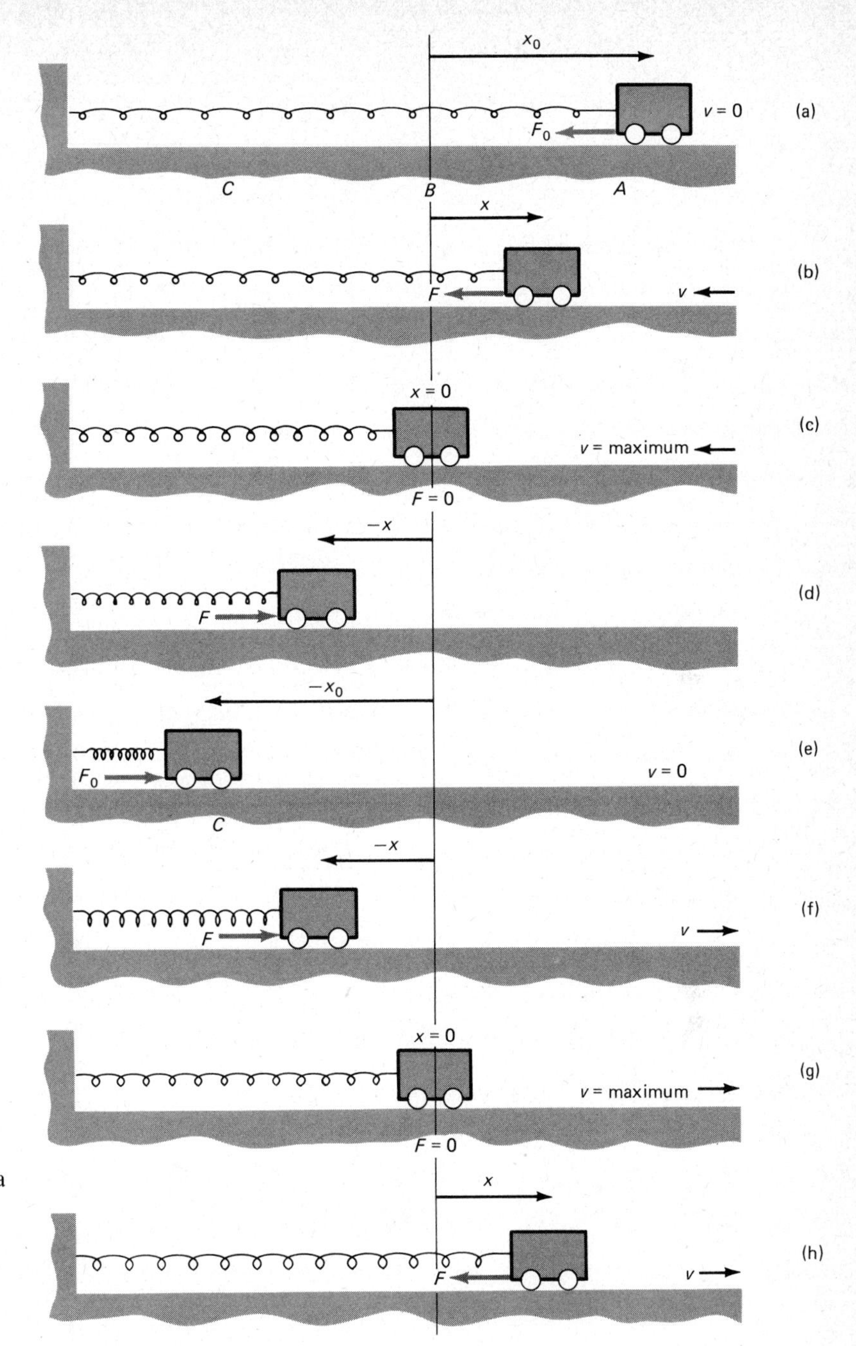

Figure 16.3 The spring's force first speeds up the cart. Then it slows it to a stop and reverses its motion. From (a) through (e) and back to (a) is one complete cycle of vibration. One complete cycle requires a time T, the period of vibration.

In any vibrating system a similar energy interchange occurs. At the end of its swing, when the system is momentarily at rest, the energy is stored in a spring or in some other form of PE; But when the system passes through its equilibrium position, its energy is all kinetic.

EXAMPLE 16.1 In Figure 16.3 the SPE in (a) is 25 J. If the cart has a mass of 5.0 kg, how fast will it be moving when it reaches position B?

Solution The SPE the system has at A will all be changed to KE when the system reaches B. We can therefore write

$$\text{SPE at } A = \text{KE at } B$$

which becomes

$$25 \text{ J} = \tfrac{1}{2}mv_B^2$$

Using $m = 5.0$ kg, we can solve for v_B. The speed of the cart at B is 3.2 m/s.

16.3 Energy Stored in Springs

Because springs (and other elastic devices) are so important for vibrations, we will now take time out to discuss them. The springs we shall be discussing obey Hooke's law. Recall Hooke's law from Section 11.2: The distortion is proportional to the distorting force. Most springs obey this law for small deformations. The experiment shown in Figure 16.4 illustrates this behavior.

Figure 16.4 shows a spring being stretched by a force F. The experimental data lead to the graph shown in (b). The linear relation shown tells us the stretching force F is proportional to the stretch x. This can be summarized in the Hooke's law equation:

$$F = kx \qquad \text{(Hooke's law)} \qquad (16.2)$$

The proportionality constant k is called the *spring* (or *force*) *constant*. Solving for it, we find

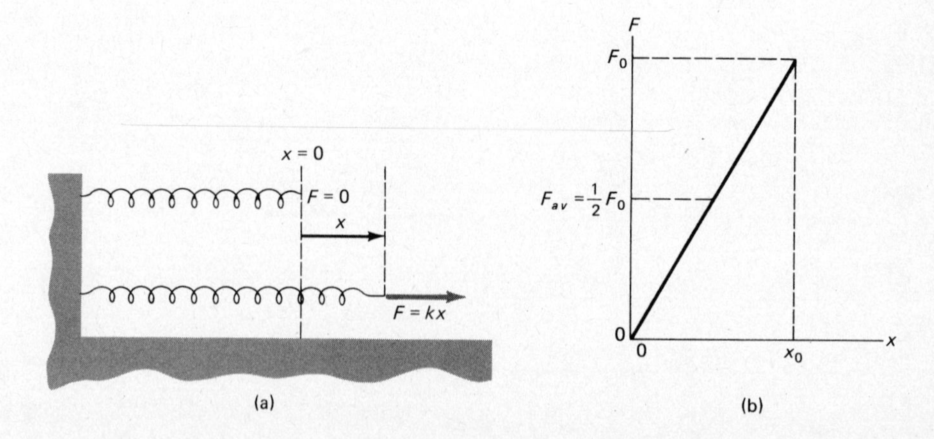

Figure 16.4 The stretching force F stores energy in the spring as it stretches the spring. The energy stored in the spring is $\frac{1}{2}kx^2$.

$$\text{Spring constant} = k = \frac{F}{x}$$

which tells us that k measures the force needed to stretch the spring unit distance. Its units in the SI are newtons per meter. The force constant is large for very stiff springs.

It is easy to measure the force constant of a spring. Suppose you hang a 200-g mass from the end of a spring and the spring stretches 5.0 cm. The force you have applied to the spring is

$$F = mg = (0.20\text{ kg})(9.8\text{ m/s}^2) = 19.6\text{ N}$$

This force caused the spring to stretch 0.050 m. Then, because $F = kx$, we have

$$k = \frac{F}{x} = \frac{19.6\text{ N}}{0.050\text{ m}} = 390\text{ N/m}$$

The force constant is thus found to be 390 N/m.

To find the energy stored in a stretched spring, we need only compute the work done in stretching it. This work must equal the stored energy given to the spring. We can evaluate this work by reference to the graph in Figure 16.4. It shows the behavior of a spring being stretched an amount x_0. Notice that the stretching force varies linearly from zero to F_0. Since the variation is linear, the average stretching force is $\frac{1}{2}F_0$. Then we have

$$\text{Work} = F_s s = F_{av} x_0 = \tfrac{1}{2}F_0 x_0$$

The stretching work is $\frac{1}{2}F_0 x_0$.

We know that the spring obeys Hooke's law, $F = kx$. In our case, this tells us that $F_0 = kx_0$. Substituting this value for F_0 gives

$$\text{Work} = \tfrac{1}{2}kx_0^2$$

for the work needed to stretch a spring a distance x_0.

The work the force does on the spring results in an equal amount of energy stored in the spring. We could find the same result by considering the compression of the spring. Therefore, we conclude that:

The energy stored in a Hooke's law spring stretched (or compressed) a distance x is

$$\textbf{Spring PE} = \textbf{SPE} = \tfrac{1}{2}kx^2 \qquad \textbf{(16.3)}$$

This relation applies to any springlike device provided that it obeys Hooke's law. It is not restricted to common-type springs.

To find the acceleration of a mass at the end of a Hooke's law spring, we can make direct use of $F = ma$. The only unbalanced force acting on the mass is the force the spring exerts on it. We call this force the *restoring force* because it tries to restore the system to its equilibrium position. The restoring force is given by Hooke's law to be $F = kx$ when the spring is stretched (or compressed) a distance x.

Newton's law gives $a = F/m$ and we have found that $F = kx$. Substituting this value for F gives

$$a = \frac{kx}{m} \tag{16.4}$$

for the acceleration of a mass at the end of a Hooke's law spring. We see at once that the acceleration is greatest when the spring is stretched the most because it is at that point that the restoring force is largest. The direction of the acceleration is the same as the direction of the restoring force. Thus the acceleration is always directed toward the equilibrium position.

■ **EXAMPLE 16.2** The 500-g block shown in Figure 16.5 strikes the end of the spring head-on. It stays attached to the spring and compresses it. How far does the spring compress if its constant is 2.0 N/m?

Solution We have to be careful in collisions. Ordinarily, kinetic energy is not conserved. Only momentum is conserved. In this case, though, we shall assume that the mass is slowed in such a way that we can say that the kinetic energy of the mass is all lost in compressing the spring. As an equation,

Kinetic energy lost by mass = final energy stored in spring

Or

$$\tfrac{1}{2}mv_0^2 = \tfrac{1}{2}kx^2$$

Solving for x gives

$$x = v_0\sqrt{\frac{m}{k}}$$

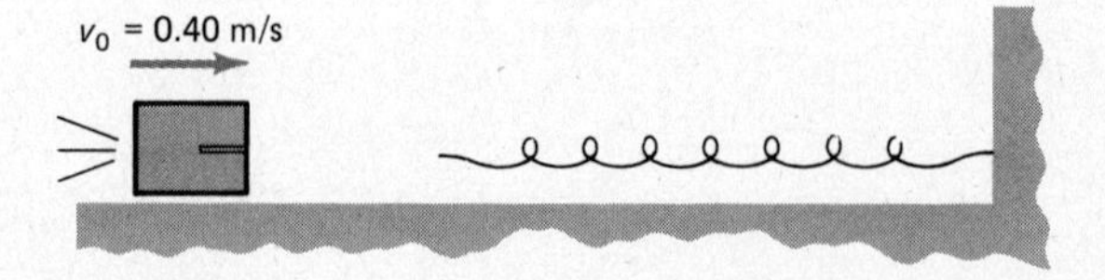

Figure 16.5 We wish to find the compression of the spring caused by the mass.

In our case $k = 2.0$ N/m, $m = 0.50$ kg, and $v_0 = 0.040$ m/s. Substituting these values gives $x = 0.20$ m. ■■

■ EXAMPLE 16.3 A spring whose constant is 5.0 N/m has a 200-g mass at its end. The mass is displaced 12 cm and released. Find the speed of the mass when it is at the following positions: (a) $x = 12$ cm, (b) $x = 0$, (c) $x = 4$ cm.

Solution This is much like the situation we showed in Figure 16.3 with 12 cm as the distance $\overline{AB}$ and 200 g as the mass of the cart. The mass oscillates back and forth between $x = 12$ cm and $x = -12$ cm.

(a) At $x = 12$ cm, the mass is at the end of its swing, where its speed is zero.

(b) At $x = 0$ there is no energy stored in the spring, but its speed is largest. Because all the SPE the system had at $x = 12$ cm is changed to KE when the mass is at $x = 0$, we have

$$(\text{KE at } x = 0) = (\text{SPE at } x = 12 \text{ cm})$$

or

$$\tfrac{1}{2}mv^2 = \tfrac{1}{2}kx_0{}^2$$

where $x_0 = 12$ cm. Placing in the known values gives

$$\tfrac{1}{2}(0.20 \text{ kg})v^2 = \tfrac{1}{2}(5.0 \text{ N/m})(0.12 \text{ m})^2$$

which gives 0.60 m/s for the speed of the mass at the center point. This is the maximum speed of the mass.

(c) At $x = 4$ cm, the mass has KE and the spring has SPE. We have

$$(\text{KE} + \text{SPE at } x = 4 \text{ cm}) = (\text{SPE at } x = 12 \text{ cm})$$
$$\tfrac{1}{2}mv^2 + \tfrac{1}{2}k(0.04 \text{ m})^2 = \tfrac{1}{2}k(0.12 \text{ m})^2$$

Since $k = 5.0$ N/m and $m = 0.20$ kg, $v = 0.57$ m/s. ■■

■ EXAMPLE 16.4 For the situation described in Example 16.3, find the acceleration of the mass at (a) $x = 12$ cm, (b) $x = 0$, and (c) $x = 4$ cm.

Solution (a) Equation 16.4 gives us

$$a = \frac{kx}{m} = \frac{(5.0 \text{ N/m})(0.12 \text{ m})}{(0.20 \text{ kg})} = 3.0 \text{ m/s}^2$$

when $x = 12$ cm.

(b) When $x = 0$, we see at once that $a = 0$. At that point the spring is not stretched and so there is no restoring force to accelerate the mass.

(c) At $x = 4$ cm,

$$a = \frac{kx}{m} = \frac{(5.0 \text{ N/m})(0.04 \text{ m})}{0.20 \text{ kg}} = 1.0 \text{ m/s}^2$$

■■

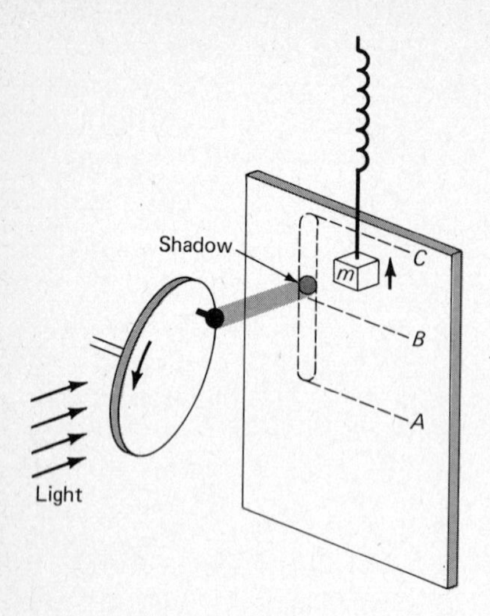

Figure 16.6 The shadow of the object on the rotating wheel duplicates the motion of the mass.

16.4 Simple Harmonic Motion and the Reference Circle

The particular spring system that we have been discussing is but one example of many systems that vibrate. Any system that has a Hooke's law restoring force will vibrate in the same general way. We showed a graph of the motion, called sinusoidal motion, in Figure 16.2. The mathematical form of this vibration is quite simple; such motion is called *simple harmonic motion,* SHM.

Vibrations in which the restoring force obeys Hooke's law are simple harmonic motion.

Reasoning from the energy of such systems, we have already learned a great deal about them. We saw how to find the speed of the vibrating object as a function of its position. We also learned that the acceleration of the object is kx/m when the object is a distance x from the equilibrium position and that the acceleration is always directed toward the equilibrium position.

We still need an equation that tells us the frequency and period of the motion. Calculus could give us all the details of the motion, but another method gives the same result. It involves use of what we call the *reference circle.* This circle is suggested by the experiment shown in Figure 16.6.

The experiment is a comparison of the motion of the shadow of an object fixed to a rotating wheel and the motion of a vibrating mass. If the wheel turns at just the proper speed, the shadow and the mass will move together. The shadow's motion is therefore SHM. If we can describe the motion of the shadow, this will be equivalent to describing the motion of the vibrating object. It turns out that this is quite simple to do.

Let us consider the two motions shown in Figure 16.7. At the bottom a vibrator oscillates back and forth between the limits A and C. It keeps step with the shadow of the point that travels around the circle with tangential speed v_T. We call this point traveling around the circle the *reference point.* Notice that the radius of the reference circle is taken as equal to the amplitude of the vibratory motion.

We can easily find the period for the motion of the reference point. It is simply the time taken for one trip around the circle.

$$\text{Period} = T = \frac{2\pi r}{v_T} = \frac{2\pi x_0}{v_T}$$

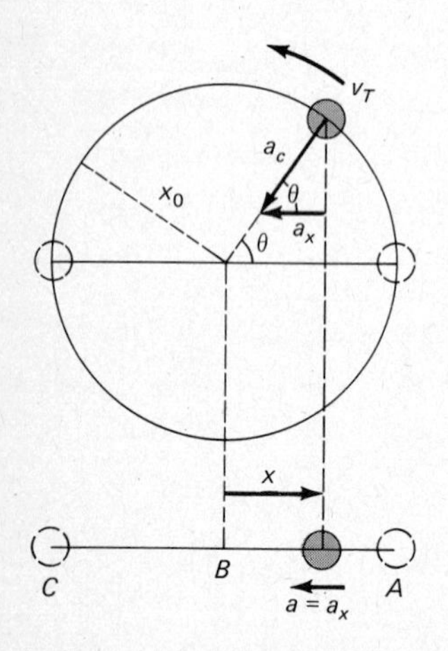

Figure 16.7 As the ball moves around the circle with speed v_T, its shadow keeps pace with the vibrator moving back and forth between A and C.

This is also the period of vibration for the vibrator. Unfortunately, we do not know v_T so this is not too useful a result. But we can obtain more information by computing the acceleration of the vibrator.

Even though the reference point travels around the reference circle at a constant speed, v_T, it is accelerating since it has a centripetal acceleration toward the center of the circle. You will recall from Section 9.8 that centripetal acceleration is given by

$$a_c = \frac{v_T^2}{r} = \frac{v_T^2}{x_0}$$

We know that the vibrator is also accelerating. Its acceleration must be the same as the x-component of the reference point's acceleration, namely, a_x. We see from the figure that

$$a_x = a_c \cos \theta$$

From the triangle at the center of the circle, we see that

$$\cos \theta = \frac{x}{x_0}$$

We therefore find that

$$a_x = a_c\left(\frac{x}{x_0}\right) = \left(\frac{v_T^2}{x_0}\right)\left(\frac{x}{x_0}\right)$$

from which

$$v_T^2 = a_x\left(\frac{x_0^2}{x}\right)$$

The acceleration of the oscillator, a_x, was found in Equation 16.4 to be

$$a_x = \frac{kx}{m}$$

Substituting this value yields

$$v_T^2 = \left(\frac{kx}{m}\right)\left(\frac{x_0^2}{x}\right) = \frac{kx_0^2}{m}$$

from which

$$v_T = x_0\sqrt{\frac{k}{m}} \tag{16.5}$$

We can now substitute this in the expression for the period and obtain

$$T = 2\pi\sqrt{\frac{m}{k}}$$

The frequency of the motion is simply $1/T$. We therefore conclude the following for a Hooke's law oscillator.

$$T = 2\pi\sqrt{\frac{m}{k}} \quad \text{and} \quad f = \frac{1}{2\pi}\sqrt{\frac{k}{m}} \tag{16.6}$$

EXAMPLE 16.5 When a 50-g mass is hung at the end of a spring, the spring stretches 20 cm. Find the force constant of the spring in newtons per meter.

Solution We wish to find k, where

$$k = \frac{F}{x}$$

The force is equal to mg, or

$$(0.050\ \text{kg})(9.8\ \text{m/s}^2) = 0.49\ \text{N}$$

It causes an elongation $x = 0.20$ m. Therefore

$$k = \frac{0.49\ \text{N}}{0.20\ \text{m}} = 2.45\ \text{N/m}$$

EXAMPLE 16.6 When a 50-g mass is hung at the end of a spring, the spring stretches 20 cm. Find the frequency and period with which the spring will oscillate when an 80-g mass is hung at its end.

Solution This is the same spring we encountered in Example 16.5. We found there that $k = 2.45$ N/m. In this case a 0.080-kg mass hangs at its end. Its frequency of vibration will be

$$f = \frac{1}{2\pi}\sqrt{\frac{k}{m}} = \frac{1}{2\pi}\sqrt{\frac{2.45\ \text{N/m}}{0.080\ \text{kg}}}$$
$$= 0.88\ \text{Hz}$$

Its period of oscillation will be

$$T = \frac{1}{f} = \frac{1}{0.88\ \text{Hz}} = 1.13\ \text{s}$$

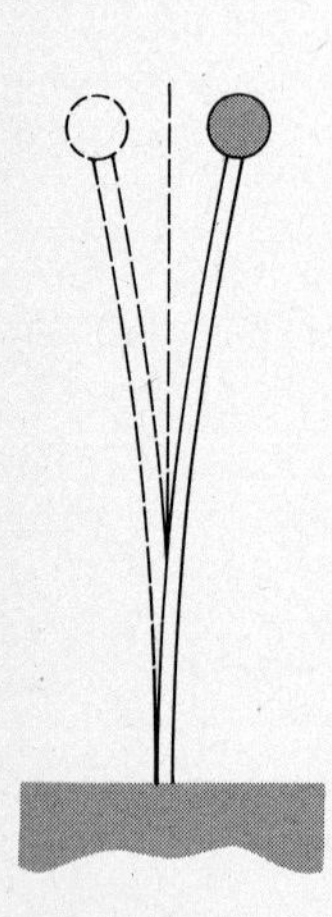

Figure 16.8 For small displacements the ball vibrates back and forth with simple harmonic motion.

EXAMPLE 16.7 A 2.0-kg ball is held at the end of a spring steel rod as shown in Figure 16.8. A force of 32 N is required to displace the ball 4 cm to the left or right from its equilibrium position. Find the frequency and period with which the ball will oscillate. (Ignore the small mass of the rod.)

Solution The spring constant k of the rod is

$$k = \frac{F}{x} = \frac{32 \text{ N}}{0.04 \text{ m}} = 800 \text{ N/m}$$

From Equation 16.6, the frequency is

$$f = \frac{1}{2\pi}\sqrt{\frac{k}{m}} = \frac{1}{2\pi}\sqrt{\frac{800 \text{ N/m}}{2 \text{ kg}}}$$
$$= 10/\pi \text{ Hz} = 3.18 \text{ cycles/s}$$

To find the period of vibration we use

$$T = \frac{1}{f} = \frac{1}{3.18 \text{ cycles/s}} = 0.314 \text{ s}$$

■■

16.5 Rotational Oscillations

The results we have obtained in this chapter were for an oscillator moving on a straight line. It is not difficult to carry over the results to oscillators that rotate. A typical example, the torsion pendulum, is shown in Figure 16.9.

The torsion pendulum obeys Hooke's law. The restoring torque τ is proportional to the distortion angle θ. We therefore have

$$\tau = k'\theta \qquad (16.7)$$

The proportionality constant k' is the spring constant for torsion. It measures how large a torque is required to produce a distortion of 1 rad. We often call k' the *torsion constant* in a situation such as this.

As we see, in rotational oscillations force is replaced by torque. Linear displacement is replaced by angular displacement. If you guess that all of our usual correspondences apply here, you are correct. Mass is replaced by moment of inertia. Linear acceleration and velocity are replaced by α and ω.

Because of this correspondence, we can describe the behavior of the torsion pendulum at once. The most important result is the frequency of vibration. We had for a linear oscillator

$$f = \frac{1}{2\pi}\sqrt{\frac{k}{m}}$$

The frequency of the torsion pendulum is

$$f = \frac{1}{2\pi}\sqrt{\frac{k'}{I}} \qquad (16.8)$$

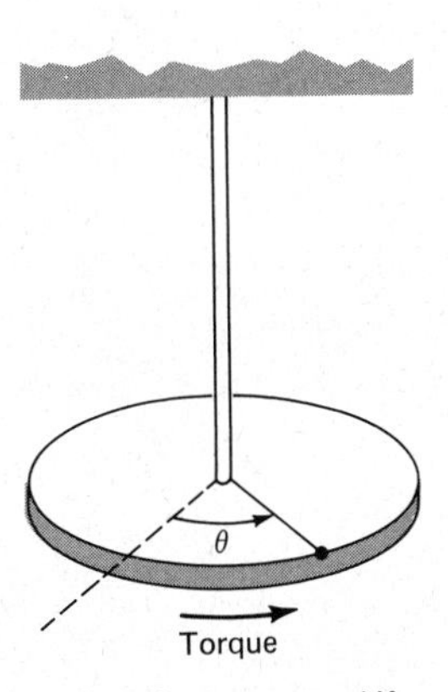

Figure 16.9 The torsion pendulum oscillates with simple harmonic motion.

In this expression I is the moment of inertia of the disk about the torsion wire as axis.

EXAMPLE 16.8 In Figure 16.9 the disk is uniform and has a mass of 4.0 kg and an outer radius of 20 cm. The torsion wire has a constant of 0.30 N · m/rad. Find its frequency of oscillation.

Solution We need only substitute in Equation 16.8. We know that $k' =$ 0.30 N · m/rad. To find I we recall that, for a uniform disk,

$$I = \tfrac{1}{2}Mr^2 = \tfrac{1}{2}(4\text{ kg})(0.20\text{ m})^2 = 0.080\text{ kg}\cdot\text{m}^2$$

Substitution gives

$$f = \frac{1}{2\pi}\sqrt{\frac{0.30\text{ N}\cdot\text{m}}{0.080\text{ kg}\cdot\text{m}^2}} = 0.31\text{ Hz}$$

It takes about 3 s for this torsion pendulum to complete one oscillation.

16.6 The Simple Pendulum

One of the most common oscillating devices is the pendulum. It is of interest to find its frequency of oscillation. We can do this easily in the following way.

Consider the simple pendulum shown in Figure 16.10. The gravitational pull, mg, on the mass causes a restoring torque. To find the torque about the suspension point, we resolve the gravitational force into components as shown. One component has its line of force passing through the pivot (or suspension) point. It has a lever arm of zero and therefore its torque is zero. The other component, $mg \sin \theta$, has a lever arm L. (Notice that the force acts at the position of m.) It causes a torque of

$$\tau = (mg \sin \theta)L = mgL \sin \theta$$

This equation is *not* of Hooke's law form. The Hooke's law form is

$$\tau \sim \theta$$

We therefore conclude that a pendulum does not obey Hooke's law in general. It therefore does not always undergo simple harmonic motion.

But a simplification exists. We know from trigonometry that, if θ is *small*, then

$$\sin \theta \approx \theta$$

To that approximation we have, for the pendulum,

$$\tau = mgL\theta$$

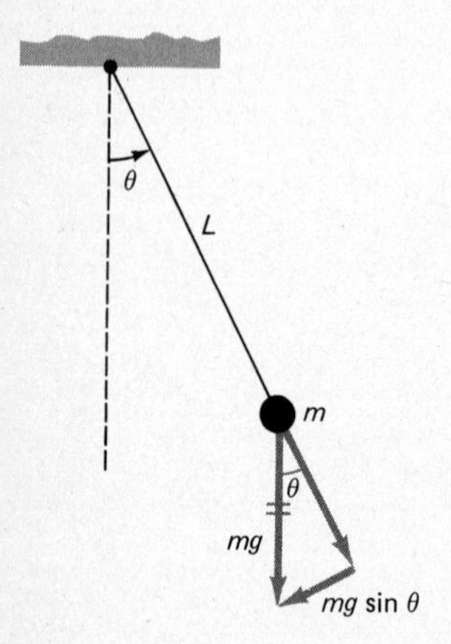

Figure 16.10 The torque on the pendulum is $(mg \sin \theta)L$. If θ is small, then $\sin \theta \cong \theta$ and the motion is simple harmonic.

This is of Hooke's law form, $\tau = k'\theta$, with $k' = mgL$.

We found in the previous section that

$$f = \frac{1}{2\pi}\sqrt{\frac{k'}{I}}$$

for a system that rotates. As we have just seen, $k' = mgL$ for the pendulum. We still need its moment of inertia about the suspension point. From the definition of I we have

$$I = \Sigma\, mr^2$$

In the present case there is only one mass and its $r = L$. Therefore, for the simple pendulum,

$$I = mL^2$$

If we substitute this into the equation for f, we have

$$f = \frac{1}{2\pi}\sqrt{\frac{g}{L}} \qquad (16.9)$$

Recall that this is correct only if θ is small. In practice, even for $\theta = 30°$ the approximation is not too bad. For small to moderate amplitudes of swing, a simple pendulum undergoes simple harmonic motion. Its natural frequency is given by Equation 16.9. For small and moderate amplitudes the frequency of a pendulum is independent of amplitude.

It is interesting to notice that a pendulum can be used to find g, the acceleration due to gravity. If the frequency of the pendulum and its length are known, then g can be calculated from Equation 16.9. This method can be made very accurate.

16.7 Driven Oscillations and Resonance

Friction forces cause oscillations to die down and eventually stop. A pendulum, a child in a swing, a mass on a spring—all these oscillations eventually stop. To keep an oscillator in motion, a periodic push or pull from outside is required. The external force that furnishes energy to an oscillator is called the *driving force.* The oscillator is said to be driven.

Consider the situation shown in Figure 16.11. The mass m held by the springs with combined constant k has a natural frequency f_0 given by

$$f_0 = \frac{1}{2\pi}\sqrt{\frac{k}{m}}$$

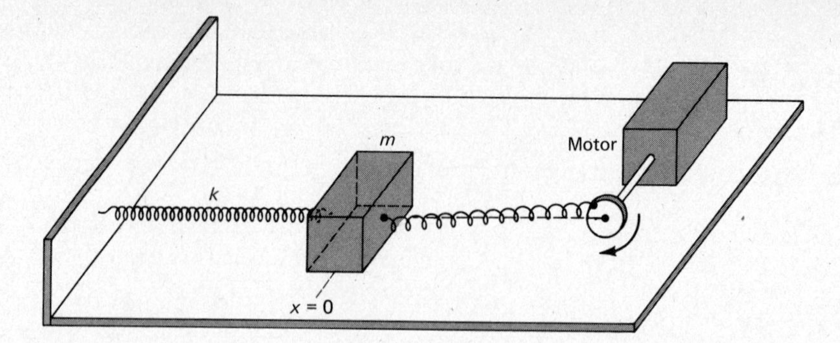

Figure 16.11 As the motor shaft rotates, it applies an oscillating force to the mass. The frequency of the force can be changed by changing the speed of the motor.

A variable-frequency driving force can be applied to it by a variable-speed motor as shown. We are interested in the amplitude of the block's motion as a function of the frequency of the driving force.

The behavior of the mass is shown in Figure 16.12. Notice what happens if the driving frequency f is much larger or much smaller than f_0, the natural frequency. In such cases the block remains nearly motionless. But if the driving frequency is close to the natural frequency, then the block vibrates with large amplitude. Its amplitude of motion reaches a maximum when $f \approx f_0$. This frequency that causes maximum response is called the *resonance frequency* of the system. For negligible damping the resonance frequency is exactly f_0. It departs from f_0 for large damping.

This resonance behavior of vibrating systems is well known to all of us. We know that, to swing high, we must push on a swing at just the right times. We always push on it so as to aid the swinging process. Only then do we add energy to the system. Only then will its oscillation energy become large.

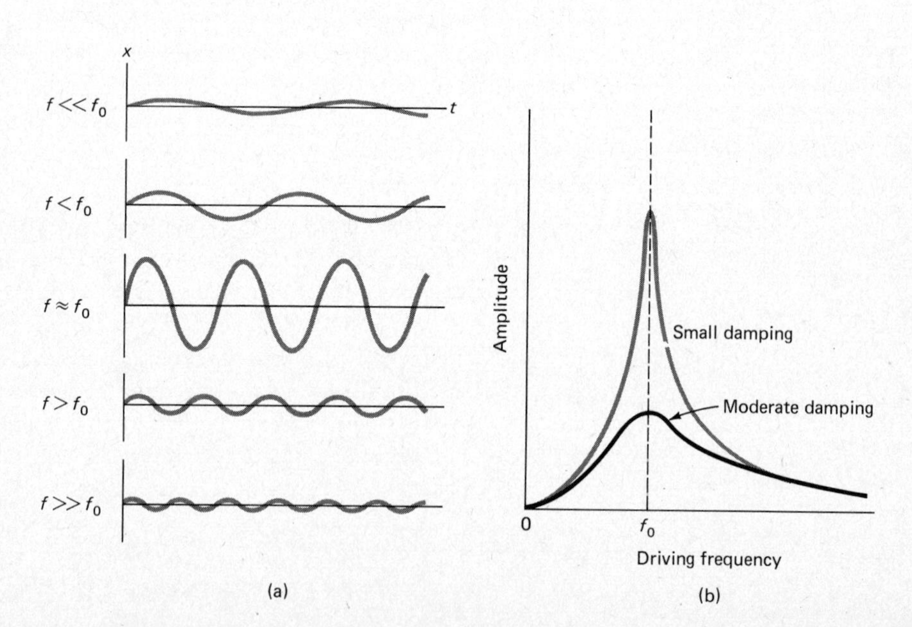

Figure 16.12 The response of the mechanical system in Figure 16.11 depends on the difference between the driving frequency and the natural frequency.

Another familiar example of resonance is observed for cars with defective tires or wheels. Suppose that one tire has a slight bump on it. Each time the wheel rolls over the bump, the car is given a slight upward push. The frequency of the push depends on the speed of the car. If the frequency is just right, it will match some resonance frequency of the car. That portion of the car will then vibrate strongly. It is for this reason that some cars show very bad properties at certain speeds. In fact, most rattles of machinery of all types can be traced to an undesired driving force and resonance.

Summary

In order to vibrate, a system must possess a restoring force or torque. By means of it, the system is pulled or pushed toward its equilibrium position.

The maximum displacement of an oscillator from its equilibrium position is its amplitude of vibration. The time taken for one complete vibration is the period of vibration T. The number of complete vibrations per unit time is the frequency f. Frequencies are commonly measured in the unit hertz (Hz), which is equivalent to 1 cycle/s. Frequency and period are related through $T = 1/f$.

If the restoring force for an oscillator obeys Hooke's law, then the oscillation will be simple harmonic.

A Hooke's law spring stretched (or compressed) a distance x has

$$\text{Spring potential energy (SPE)} = \tfrac{1}{2}kx^2$$

stored in it. During oscillation the energy of a mass-spring system continually interchanges from kinetic to potential energy. At any instant the law of conservation of energy tells us that

$$\tfrac{1}{2}kx_0^{\,2} = \tfrac{1}{2}kx^2 + \tfrac{1}{2}mv^2$$

If a spring requires a force F to give it an elongation x, then the spring constant $k = F/x$. Its units are those of force per unit elongation. Springs usually obey Hooke's law at small to moderate extensions and compressions.

For a mass m undergoing oscillatory motion at the end of a spring with constant k, the following may be said. Its natural frequency of oscillation is

$$f = \frac{1}{2\pi}\sqrt{\frac{k}{m}}$$

Its acceleration at any instant is

$$a = \left(\frac{k}{m}\right)x$$

where x is the displacement of the mass.

The vertical (or horizontal) component of the uniform motion of a point around a reference circle is simple harmonic motion. The frequency of rotation equals the oscillation frequency. The amplitude of oscillation equals the radius of the reference circle.

Rotational oscillators obey Hooke's law if $\tau = k'\theta$. If the moment of inertia of the rotator about the axis is I, then its natural frequency is

$$f = \frac{1}{2\pi}\sqrt{\frac{k'}{I}}$$

The natural frequency of vibration of a pendulum of length L is

$$f = \frac{1}{2\pi}\sqrt{\frac{g}{L}}$$

where g is the acceleration due to gravity.

A driven oscillator responds best if the driving force has a frequency close to the natural frequency. The frequency of maximum response is called the resonance frequency of the oscillator.

Questions and Exercises

1. A pendulum swings back and forth with constant amplitude. Where is it in each of the following cases? (a) Speed is maximum. (b) Speed is zero. (c) Acceleration is zero. (d) Acceleration is maximum. (e) Energy is all potential. (f) Energy is all kinetic. (g) Energy is half kinetic, half potential.
2. List several systems that you think undergo SHM. How could you test your assumption in each case?
3. Discuss the following statement: "A spring-mass system vibrates in the same way whether it is vertical or horizontal on a frictionless table. This is because the restoring force the mass experiences when disturbed is the same in either case. If the spring did not obey Hooke's law, this might not be true."
4. If you have a suitable spring of known force constant, a stop watch, and a meter stick, you can measure the mass of an unknown object. How?
5. Compare the period of a pendulum as measured on the earth and on the moon. Repeat for a spring-mass system.
6. Would a pendulum clock tick out time properly on the moon if it was adjusted properly on the earth? How would it behave in an earth satellite ship?
7. A piece of 2 × 4 wood floats in a large pan of water. The block is pushed down a little and released. If friction between it and the water did not slow it, would you expect its vibration to be simple harmonic?
8. How could you compute the up-and-down resonance frequency of a car using data obtained from the lowering of the car with increased load? Estimate this frequency for an automobile. When might it be important?
9. During the spin-dry portion of a washer's cycle, the washer sometimes vibrates strongly. Why? Is the unbalance of the load the whole story? What should a washer designer do to minimize this problem?
10. People sometimes notice that something in the kitchen (perhaps a pan in a cupboard) vibrates strongly while the refrigerator is running. Explain the origin of the vibration.
11. One common way to free a stuck car is to rock it back and forth, either by pushing or by changing gears from low to reverse back and forth. Discuss the physical basis for this method. Pay particular attention to the potential and kinetic energy of the car.

Problems

1. Refer to Figure P16.1. Find the amplitude, period, and frequency for the vibration shown there.
2. A sinusoidal vibration has a frequency of 5.0 Hz and an amplitude of 3.0 cm. Sketch an x versus t graph for it. Place appropriate numbers on the axes.
3. A mass at the end of a spring undergoes 100 complete oscil-

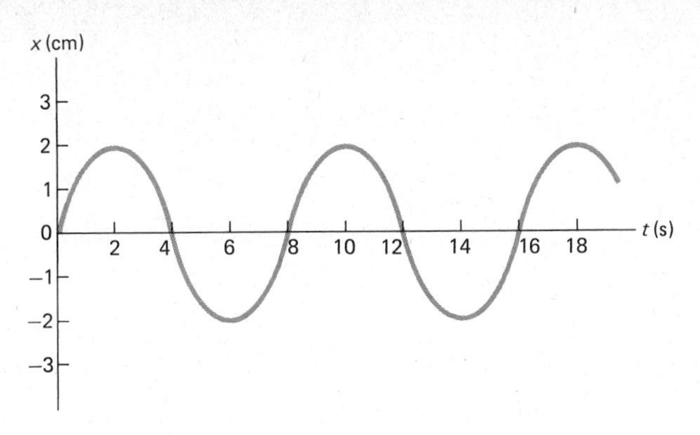

Figure P16.1

lations in 243 s. What is its (a) period and (b) frequency of oscillation?

4. When a 20-g mass is added to the end of a vertical spring, the spring stretches 4.0 cm. What is the spring constant for the spring?
5. The spring constant for a certain spring is 250 N/m. When a 50-g load is added to the top of the vertical spring, how much will the spring compress?
6. A 50-g mass vibrates horizontally at the end of a spring for which $k = 12$ N/m. What is the acceleration of the mass when its displacement from equilibrium is (a) zero, (b) 2 cm, (c) 5 cm?

*7. When a 50-g mass is hung from a certain spring, it stretches 2.0 cm. Suppose the mass is pulled down an additional 3.0 cm, and released. What is the acceleration of the mass when it is (a) at the end of its path, (b) at the center of its path, (c) at 2.0 cm from the center?

*8. A 200-g mass vibrates with amplitude 15 cm at the end of a spring whose constant is 0.75 N/cm. Find the acceleration of the mass when its displacement from the equilibrium position is (a) 15 cm, (b) 0 cm, (c) 5 cm.

*9. A 0.50-kg mass vibrates with amplitude 4.0 cm at the end of a spring whose constant is 25 N/m. Find the acceleration of the mass when its displacement from equilibrium is (a) 0, (b) 4.0 cm, (c) 3.0 cm.

10. A certain spring stretches 2 cm under a 50-g load. How much energy is stored in the spring when (a) it is stretched 5 cm and (b) it is compressed 5 cm?
11. A certain slingshot obeys Hooke's law. To pull it back 6.0 cm requires a force of 8.0 N. (a) What is its spring constant? (b) How much energy is stored in it when it is pulled back 5.0 cm?
12. In order to cock a certain popgun, a force of 100 N is required to compress the spring 5 cm to its operating position. (a) What is the spring constant? (b) How much energy is stored in the cocked gun?

*13. For the situation described in Problem 9, find the speed of the mass in each case.

*14. For the situation described in Problem 8, find the speed of the mass in each case.

**15. Prove that the speed of a mass undergoing SHM with frequency f is, at a displacement x,

$$v = 2\pi f \sqrt{(x_0^2 - x^2)}$$

*16. A bow with spring constant 1500 N/m is used to shoot a 75-g arrow. If the bow is pulled back 20 cm, what will be the speed of the arrow as it leaves the bow?

*17. A washer (5.0 g) sits at the end of a horizontal spring ($k =$ 100 N/cm) which is slipped over a large bolt. The washer is pressed back against the spring so as to compress it 2.0 cm. How fast will the washer shoot off the bolt if the washer is released?

18. When a 2.0-kg mass is hung from the end of a spring, the spring stretches 0.50 cm. The mass is then pulled down an additional 1.50 cm and released. Find the spring constant and the amplitude, frequency, and period of the resulting vibration.
19. A 40-g mass hangs from the end of a rubber band. The mass is pulled down 3 cm and released. It vibrates with a frequency of 3.0 Hz. Find the period and amplitude of the motion as well as the spring constant for the rubber band.
20. By pushing up and down on a certain auto, it is found to have a natural period of oscillation equal to 1.5 s. Find the spring constant for the car's suspension system. Assume that the mass held by the suspension is 2000 kg. How far will the car sink when an 80-kg man enters the middle of it?

*21. A 14-kg motor is mounted on rubber feet. The motor can be pulled aside a distance of 0.50 cm by a force of 200 N. (a) Find the spring constant for the sideways motion of the motor. (b) What is the natural frequency for sideways vibration of the motor? (c) The motor is to rotate the drum of an electric dryer by means of a belt connected to a pulley. If the driven pulley is to rotate at 9 rev/s, is trouble likely to result? Why?

*22. A 20-g mass undergoes simple harmonic motion at the end of a spring. The period of the motion is 1.20 s. The mass oscillates back and forth through a total distance of 15 cm. Find the maximum speed and acceleration of the mass.

**23. The piston in a car's engine undergoes approximate simple harmonic motion of amplitude A and frequency f. Find the maximum speed and acceleration of the piston.

*Problems marked with an asterisk are not as easy as the unmarked ones.
**Problems marked with a double asterisk are somewhat more difficult than the average.

24. How long (in centimeters) should a pendulum be if its period is to be exactly 1 s on earth?
25. The acceleration due to gravity on the moon is about 1.61 m/s^2. What will be the period of a 200-cm-long pendulum on the moon?
*26. A 15.0-kg wheel is suspended as the disk of a torsion pendulum. It takes a torque of 9.0 N·m to rotate the wheel through an angle of 180°. The natural frequency of rotation of the device is 0.30 Hz. What is the moment of inertia of the wheel?
**27. One spring has spring constant k_1 and another has constant k_2. Find their combined spring constant when connected as shown in Figure P16.2(a). Repeat for the situation shown in part (b) of the figure.

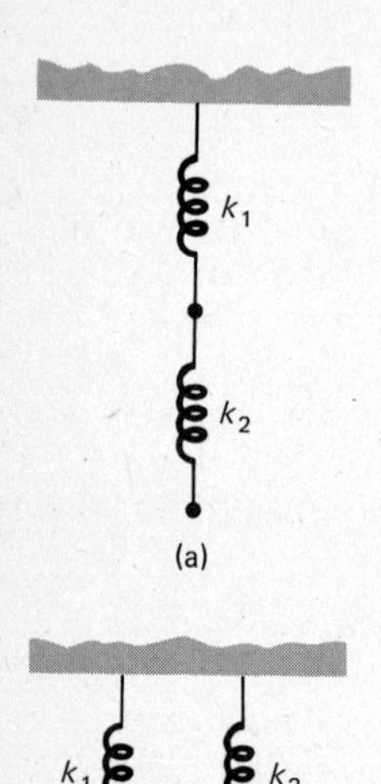

Figure P16.2

**28. Show that the maximum speed of a pendulum bob is given by

$$v = \sqrt{2gL(1 - \cos\theta)}$$

where θ is the angle from which the pendulum is released.

**29. A piston vibrates up and down with sinusoidal motion. Its amplitude is 5 cm. A small metal washer sits on top of the piston. As the frequency of motion is increased, a frequency is reached at which the washer doesn't fall as fast as the piston at the start of the piston's downward motion. At what frequency does this occur?
**30. Suppose that a hole was bored straight through the center of the earth. It can be shown that the weight of an object within the hole at a distance r from the earth's center is

$$W = W_0\left(\frac{r}{R_0}\right)$$

where W_0 is the weight of the object on the surface of the earth and R_0 is the earth's radius. Why will an object dropped into the hole oscillate with simple harmonic motion if friction effects can be ignored? Show that the frequency of the motion is

$$\left(\frac{1}{2\pi}\right)\sqrt{\frac{g}{R_0}}$$

This corresponds to a period of about 85 min. It is of interest to notice that this is also the period for a satellite orbiting the earth.
**31. As shown in Figure P16.3, a 20-g bullet is shot with speed 300 m/s into a 1.50-kg block at the end of a spring (k = 70 N/m). The bullet sticks in the block. Find the resultant amplitude of motion of the block on the nearly frictionless table.

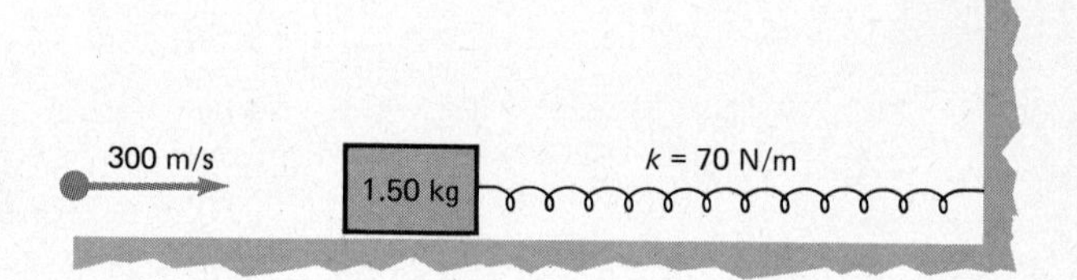

Figure P16.3

Beckwith Studios

17 WAVES

Performance Goals

When you finish this chapter, you should be able to

1. Define the following quantities by reference to a sketch of a periodic wave: wavelength, amplitude, crest, trough, line or direction of propagation, direction of energy flow.
2. Find f for a wave disturbance when T is given and vice versa.
3. Find λ, v, and T when two of these quantities are given.
4. Do the following for two waves traveling on a string at the same time. If the two waves are given

There are many types of waves. Water waves, waves on a rope or string, sound waves, and light waves are familiar to all of us. There are other types as well. All waves behave alike in many ways, even though they differ in other ways. In this chapter we shall examine some of the ways in which they behave alike. Later we shall learn about the more specialized properties of each wave type.

at a certain instant, find the string's displacement at that instant. Explain your procedure in terms of the superposition principle.

5. Do the following for a string fixed at both ends. Draw the standing wave patterns for the first few resonances of the string. Point out the positions of nodes and antinodes. Locate the segments of the string. Use them to compute λ for the wave in terms of the string length.
6. Explain qualitatively why a string will resonate to only certain frequencies of vibration.
7. Find either v or f for the wave if one of these two quantities is given together with the length of the string and its standing wave pattern.
8. State if a given wave is transverse or longitudinal. Give two examples of each type of wave.
9. Sketch graphs for the first several standing wave patterns for a spring of known length. For each find the resonance frequency if v is known. Find v if f is known.
10. Sketch the first few resonance modes of longitudinal motion for a rod (a) fastened at one end, (b) free at both ends, (c) fastened at the center, (d) fastened one-quarter of the way along the rod.
11. Given two of the quantities L, v, and f_n for the rod of the previous goal, find the third quantity.
12. Sketch the first few sound resonance modes for a tube (a) closed at both ends, (b) open at both ends, (c) closed at one end and open at the other.

17.1 Waves on a String

Suppose that you hold one end of a tight string or rope as shown in Figure 17.1. If you quickly raise and lower your hand, a pulse will be sent down the string. The pulse moves with a definite speed down the string. Because the pulse shown in the figure moves 20 cm in 0.10 s, its speed is, from $x = \bar{v}t$,

$$\bar{v} = \frac{x}{t} = \frac{20\text{ cm}}{0.10\text{ s}} = 200\text{ cm/s}$$

The speed of the pulse on the string depends on two things. The larger the tension in the string, the faster the pulse moves. The more massive the string, the slower the pulse moves.[1]

A repeating set of pulses sent down a string is called a *wave.* If the pulses are all identical, then the wave is said to be *periodic.* An important type of periodic wave is shown in Figure 17.2(a). It is sent down the string by vibrating the end of the string sinusoidally. You can approximate such a wave by moving your hand up and down. As the figure shows, the pulses traveling down the string make a sinusoidal pattern. We call such a wave a *sinusoidal wave.*

Other types of wave patterns are also possible. In part (b) of Figure 17.2 we see a *square wave.* The wave in part (c) is called a *ramp* (or *sawtooth*) *wave.* The type of wave sent down the string depends on how you vibrate the end of the string.

Let us now learn the terms we use to describe these waves. Refer to Figure 17.2 for examples.

The *wavelength* of a wave is the distance between identical points along the repeating wave. This distance is always represented by λ (Greek letter lambda).

The *amplitude* of a wave is the maximum displacement of the wave from its equilibrium position. As in the case of vibration, this is only half of the distance moved.

Now look at Figure 17.3. There we see a wave moving to the right with speed v. Let us ask what will happen as the wave moves through point P. The string there will vibrate up and down as the wave moves past. As one wavelength moves past, the string will make one complete up-and-down motion. It vibrates through one complete vibration.

The *frequency* of a wave is the number of complete vibrations that a point such as P undergoes in 1 s. We represent the frequency by f.

The frequency of a wave is identical to the frequency of the source that sent it out. If our hand vibrates up and down two times each second, then the fre-

[1] As a formula, $v = \sqrt{F/(m/L)}$ where F is the tension in the string and (m/L) is the string's mass per unit length.

13. Given two of the quantities L, v, and f_n for the tube of the previous goal, find the third quantity.
14. Describe qualitatively what is meant by the Fourier analysis of a wave or vibration.

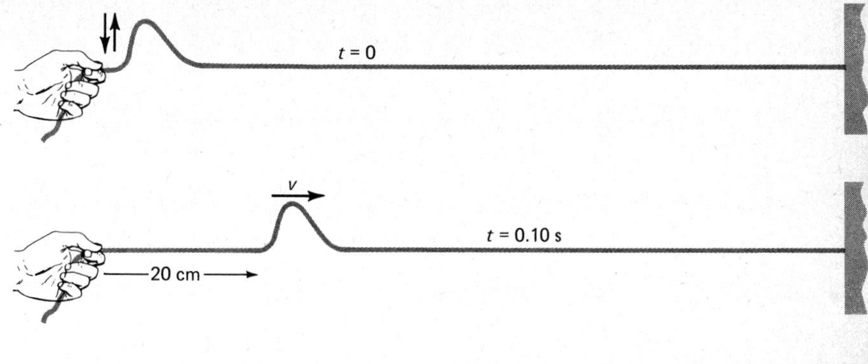

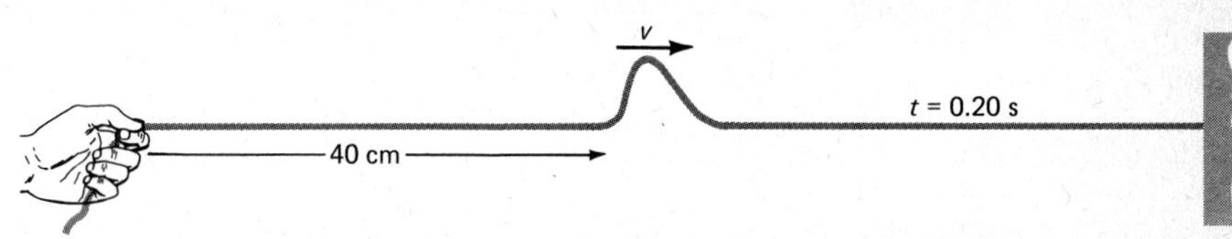

Figure 17.1 A pulse can be sent down a string by moving your hand up and down.

quency of the wave is 2 cycles per second (cycles/s), or 2 hertz (Hz). You will recall that

$$1 \text{ cycle/s} = 1 \text{ Hz}$$

We might also ask how long it takes for one complete wave to pass point P in Figure 17.3.

The *period* of a wave is the time taken for a point such as P to undergo one complete vibration. The period is represented by T.

As in the case of vibrations, the period and frequency are related through

$$f = \frac{1}{T} \tag{17.1}$$

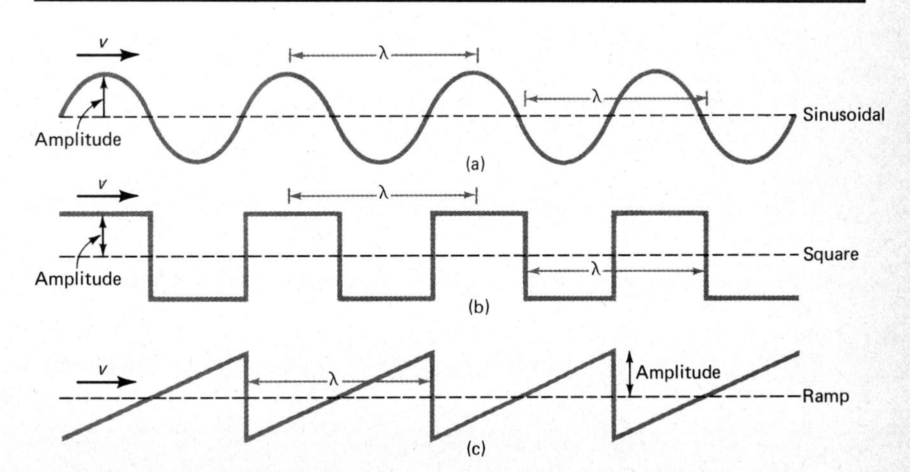

Figure 17.2 Various forms of waves can move along a string. (In practice, one cannot obtain the sharp edges shown.)

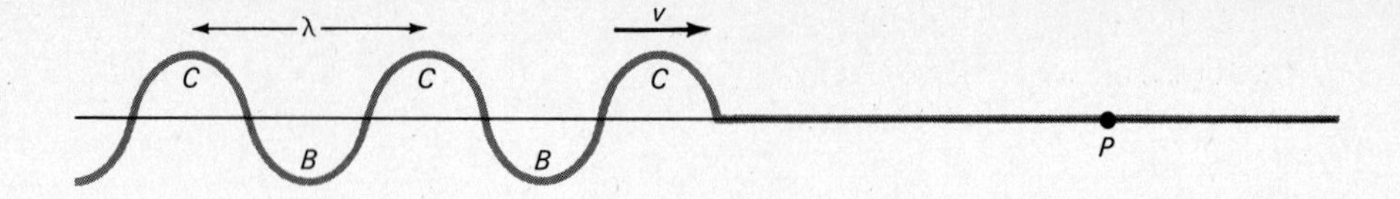

Figure 17.3 As the wave passes through point P, the string at P moves up and down with a frequency f and period T.

There is an important relation between the period, wavelength, and speed of all waves. To see what it is, consider the wave as it moves through point P in Figure 17.3. Let us apply $x = \bar{v}t$ to it. How long does it take for one complete wavelength to move past point P? During that time a length λ of the wave will have moved past P. The point P will have undergone one complete vibration during that time. We call that time the period T of the vibration. Therefore, $x = \lambda$ and $t = T$, so $x = \bar{v}t$ becomes

$$\lambda = vT \tag{17.2a}$$

where v is the speed of the wave down the string. This relation is true for all waves, and we shall use it over and over again. Notice that it is simply $x = \bar{v}t$ written for this special case. Equation 17.2a can be written in another form by replacing T by $1/f$. We then have

$$\lambda = \frac{v}{f} \quad \text{or} \quad v = \lambda f \tag{17.2b}$$

Two more terms are used in connection with waves. The top of a wave, indicated as C in Figure 17.3, is called the *crest* of the wave. The bottom of the wave, indicated as B, is called the *trough* of the wave.

One of the most important features of a wave is the energy it carries. In Figure 17.4 the oscillator at the end of the string acts as an energy source for the wave. It does work on the string as it pulls the string up and down. This energy travels down the string with speed v, the speed of the wave. We call the direction in which the wave travels the *direction of propagation* of the wave.

■ **EXAMPLE 17.1** The oscillator shown in Figure 17.4 vibrates at 40 Hz. It sends out the wave shown. What is the period of the wave? Its wavelength? Its amplitude? Its speed along the string?

Solution For any wave $T = 1/f$. So in this case

$$T = \frac{1}{40\ \text{Hz}} = 0.025\ \text{s}$$

Remember that 1 Hz = 1 cycle/s.

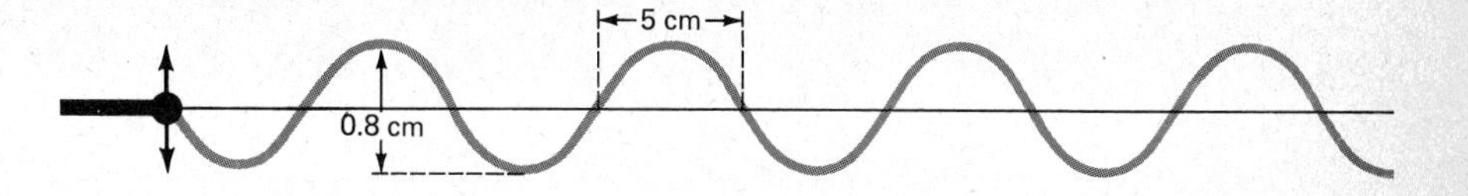

Figure 17.4 The frequency of the vibrator at the end of the string is 40 Hz. Find λ, T, and v.

The wavelength is the distance between neighboring crests of the wave. From Figure 17.4 we see that this distance is 10 cm. Therefore $\lambda = 10$ cm.

The amplitude of the wave is the distance of a crest (or trough) from the centerline of the wave. In our case the amplitude is half of 0.80 cm or 0.40 cm.

To find the speed of the wave, we can use $x = \bar{v}t$ in its wave case, namely, $\lambda = vT$. We then have

$$v = \frac{\lambda}{T} = \frac{10 \text{ cm}}{0.025 \text{ s}} = 400 \text{ cm/s}$$

■■

17.2 Wave Fronts and Rays

Suppose that a water wave travels out from its source as shown in Figure 17.5(a). The wave crests form circles around the source. The crests travel out along the surface with a definite speed v. We can represent the waves in a shorthand way as shown in part (b). The circles represent the positions of the

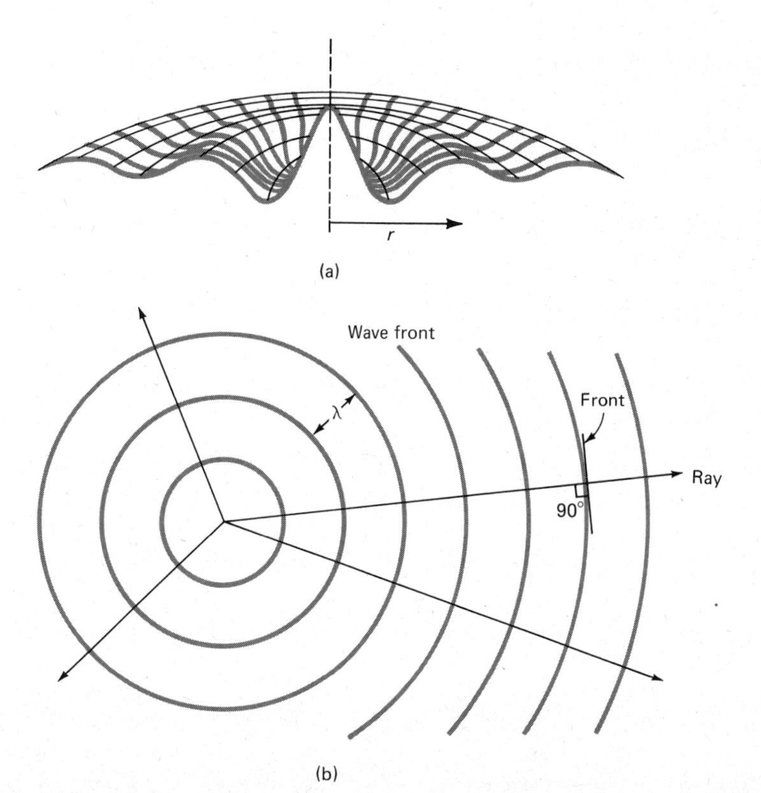

Figure 17.5 At large distances from the source a wave has little curvature.

crests of the waves. They are called the *wave fronts* of the waves. The distance between them is λ, the wavelength of the wave.

Notice how the curvature of the waves changes with distance. As the circle increases in size, its curvature decreases. Clearly, when the circle has become very large, the wave front has very little curvature at all. As we see, the wave front far from the source is nearly straight. We call a wave that has essentially no curvature a *plane wave.*

Also shown in Figure 17.5(b) are straight lines called *rays.*

Rays show the direction of propagation or travel of the wave. They are always perpendicular to the wave fronts.

Energy from the source is carried out along the rays by the wave. Be careful to notice that the rays and wave fronts are mutually perpendicular.

Another feature of waves is illustrated in Figure 17.5. In part (a) the wave crests decrease in height as the wave moves away from the source. The height of the crest is a measure of the energy carried by each unit length of the wave front. As the circular wave front becomes larger, the energy is spread over a larger length. Therefore, the wave has less energy per unit length of wave front as the wave moves away from the source. This is one major reason that a wave dies down as it spreads out into space.

17.3 Reflection and Superposition of Waves

Let us now return to the subject of waves on a string. In our discussion of waves on a string we assumed that the string had one end fastened to something, but we ignored the fact that a pulse traveling down the string will hit the end. What happens then?

Of course, the pulse carries energy down the string. It gives kinetic energy to the portion of the string through which it is passing. This energy moves down the string with the pulse. When the pulse hits the end of the string, the energy must be accounted for. In some cases the support exerts friction forces on the string. Part or all of the pulse energy is then lost to heat as it strikes the end. Very often, though, the support simply reverses the direction of motion of the pulse. This situation is shown in Figure 17.6. We say that the pulse is *reflected* by the support.

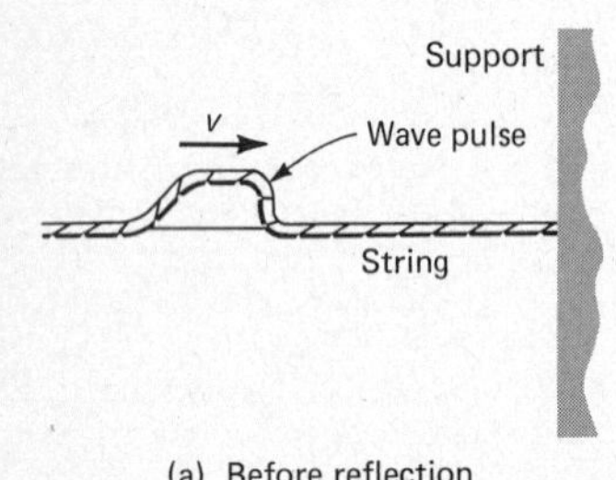

(a) Before reflection

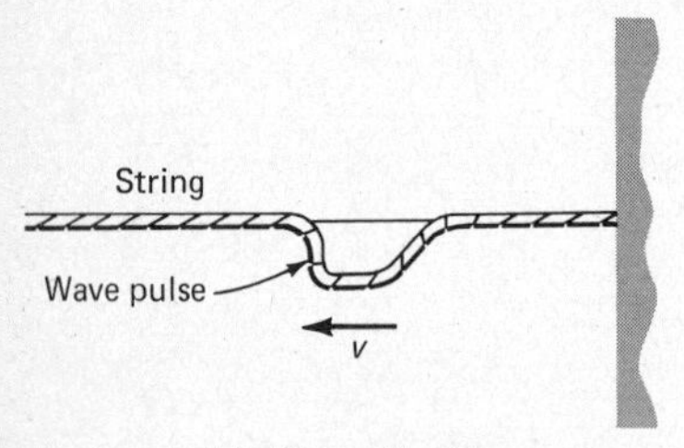

(b) After reflection

Figure 17.6 A rigid support causes the pulse to invert (flip over) as it is reflected.

An interesting feature of the reflection is shown in Figure 17.6. Notice that the pulse is turned upside down by the reflection. We say that the pulse has been *inverted.* When the support is rigid, a pulse on a string is always inverted by the reflection process. However, if the string is able to move freely up and down at its end, inversion does not occur. This is an unusual case for strings. However, it will be of importance to us later when we discuss other types of waves.

Suppose that a series of pulses, a sinusoidal wave, is sent down the string. When it hits the end of the string, the wave will be reflected. This is shown in Figure 17.7. Now there will be two waves on the string at the same time. The *incident wave* from the vibrating source is still traveling to the right on the

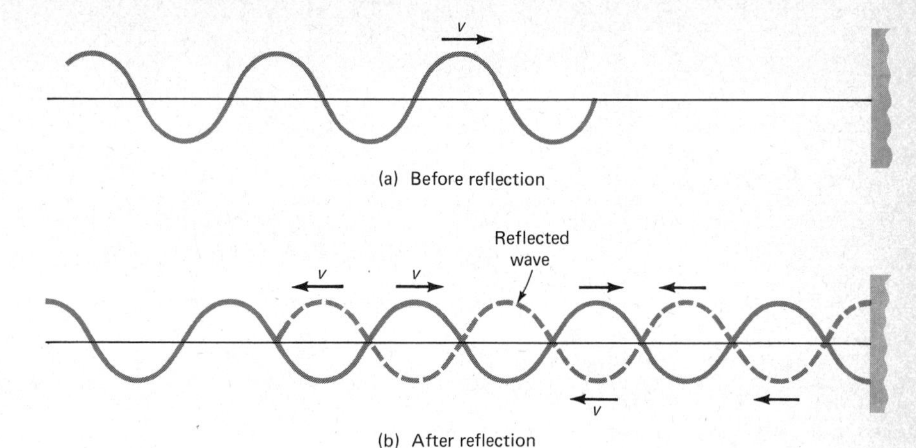

Figure 17.7 The reflected wave is moving along the string and so is the incident wave. What do they cause the string to do at the instant shown?

string. The *reflected wave* is traveling to the left. Both want the string to move in their own special way. What will the string do?

One of the nice things about nature is that waves usually combine in a simple way. They obey the following rule.

The *superposition principle* states that two or more waves passing through the same point at the same time add vectorially.

In other words, if one wave would cause the string to rise 5.0 mm and the other would cause it to drop 2.0 mm, the string will actually rise

$$5.00 \text{ mm} - 2.0 \text{ mm} = 3.0 \text{ mm}$$

Note that we take up as positive and down as negative.

Let us apply this principle to the situation shown in Figure 17.7(b). This situation will last only for an instant. The waves will soon move out of the position shown. But at that instant the waves exactly cancel each other. Everywhere that the incoming wave wants the string to rise, the reflected wave wants it to drop by the same amount. Their vector sum is therefore zero all along the string. At the instant shown, the string will position itself along the horizontal line. Let us see what happens as time goes on.

In Figure 17.8 we see the wave positions at four different times. We also show their vector sum. The string will behave in the way the sum of the two waves behaves. As time goes on, various portions of the string move up and down. From Figure 17.8 we see that the points labeled A move farthest from equilibrium (horizontal). These points are called the *antinodes* of the combined wave. Other points on the string, labeled N, never move at all. These points are called the *nodes* of the combined wave. The portion of the string between two neighboring nodes is called a vibrating *segment* of the string.

We see from this discussion that

If two identical waves travel in opposite directions through the same points, they give rise to a vibration with nodes and antinodes.

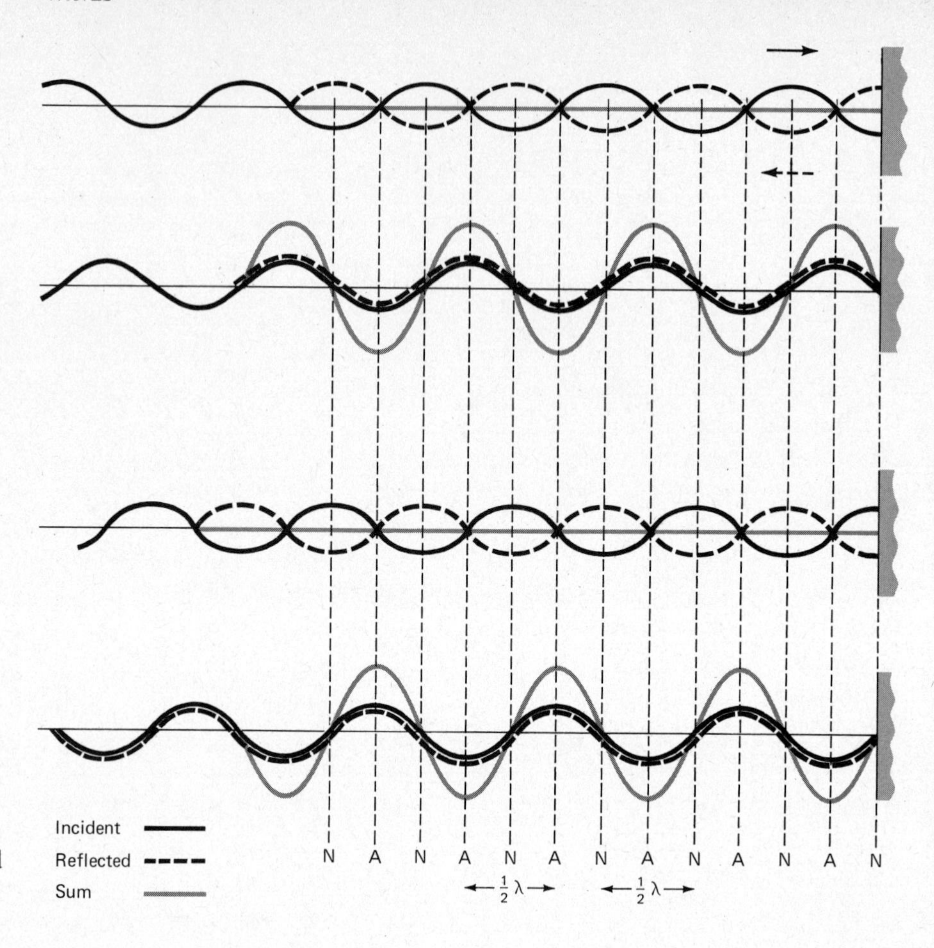

Figure 17.8 The incident and reflected waves give rise to nodes and antinodes.

Figure 17.8 shows that

> **The antinodes are separated by a distance of a half wavelength, $\lambda/2$. The nodes are also $\lambda/2$ apart.**

We shall soon see that these facts give us a powerful tool for studying waves.

17.4 Resonance of Waves

When we studied vibrations in Chapter 16, we learned about mechanical resonance. If we push on a swing or some other vibrator just right, it will resonate. Its motion will become very large. To have resonance, the pushing force must be timed so as to aid, not stop, the motion.

A similar situation exists with waves on a string. We can build up a huge vibration on the string provided that we vibrate it in just the proper way. What we want to do is add more and more energy to the string. The vibrator at its end must be timed so that it always helps the string to move. Let us see what must be true for this to happen.

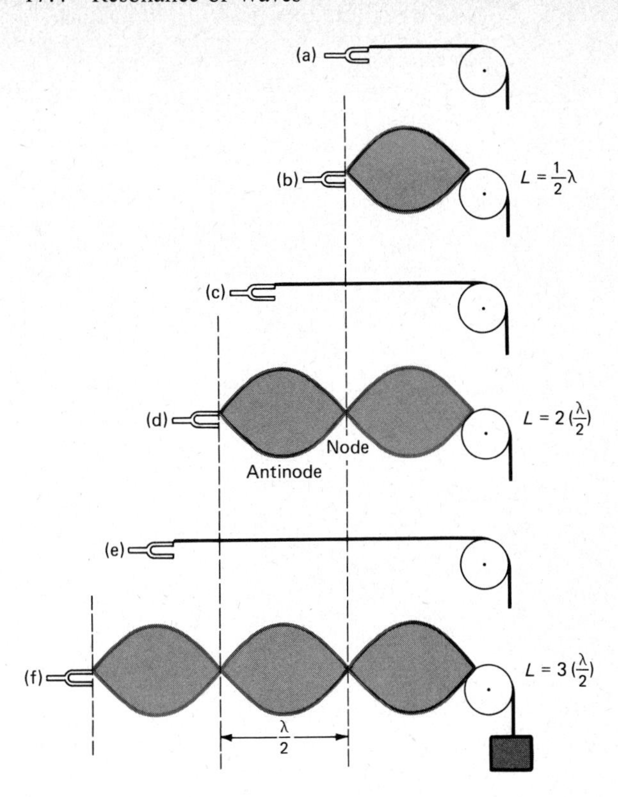

Figure 17.9 The string resonates only when it is a whole number of half wavelengths long, that is, when $L = n(\frac{1}{2}\lambda)$, where n is an integer.

Suppose we perform the experiment shown in Figure 17.9. The vibrating tuning fork sends waves down the string and they are reflected by the pulley at its end. Under nearly all conditions, the string simply quivers a little, so little that its motion can hardly be seen, as in (a). But when the string is lengthened, as in (b), it vibrates widely. Apparently, at this length the wave sent down the string and reflected back reaches the tuning fork at just the right time. It is re-reflected along with an additional wave just being sent out by the fork. The reflected wave is in resonance with each new wave being sent down the string.

When the string is lengthened a little more, as in (c), the resonance condition is lost. Not until the string reaches the length in (d) is resonance again observed. Another resonance is observed when the string reaches the length in (f). Further resonances exist at still larger lengths. Because the distance between nodes is always $\lambda/2$, we see from the figure that

A string held firmly at its two ends will resonate if it is $n(\frac{1}{2}\lambda)$ long, where n is an integer ($n = 1, 2, 3, 4$, and so on).

The pattern in which a wave resonates is called a *standing wave.* Each resonance frequency corresponds to its own standing wave. We shall often draw these standing wave patterns and use them for our computations.

▮ EXAMPLE 17.2 The resonating string shown in Figure 17.10 is 80 cm long and is driven by a 120-Hz vibrator. Find the speed of the wave on the string.

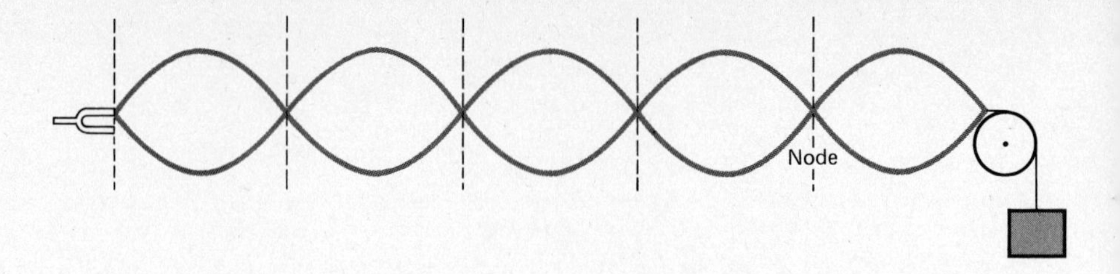

Figure 17.10 If the tuning fork has a frequency of 120 Hz and the string is 80 cm long, what is the speed of the wave?

Solution The distance between nodes is $\frac{1}{2}\lambda$. We see on the figure that the string is $5(\frac{1}{2}\lambda)$ long. We can equate this to the known length of the string.

$$5(\tfrac{1}{2}\lambda) = 80 \text{ cm}$$

Solving for λ gives $\lambda = 32$ cm.

We now know $\lambda = 32$ cm and we were told that $f = 120$ Hz. Because

$$\lambda = vT = v\left(\frac{1}{f}\right)$$

we have

$$\begin{aligned} v = \lambda f &= (32 \text{ cm})(120 \text{ Hz}) \\ &= 3840 \text{ cm/s} \end{aligned}$$

The speed of the wave on the string is 38.4 m/s. ■■

■ **EXAMPLE 17.3** A certain guitar string is 50 cm long and is tightened until it resonates to a frequency of 264 Hz. If this is the lowest of its resonance frequencies, to what other frequencies will it resonate? (We call the lowest resonance frequency the *fundamental* frequency of the string, in this case, 264 Hz. Higher resonance frequencies are called *overtones.*)

Solution Let us first find the speed of the wave on the string. We know that the more nodes on a string, the faster it is vibrating. Therefore, the lowest frequency vibration must be the one with the least nodes on the string. In this case the string vibrates with a frequency of 264 Hz when it vibrates as a single segment. This is the situation shown in Figure 17.9(b). Because the distance between nodes is $\frac{1}{2}$ wavelength, the wavelength in this case is (2)(50 cm) = 100 cm. Let us represent this wavelength, the fundamental wavelength, by λ_1. Then

$$\lambda_1 = 100 \text{ cm} \qquad \text{and} \qquad f_1 = 264 \text{ Hz}$$

We can now use

$$\lambda = vT = v\left(\frac{1}{f}\right) = \frac{v}{f}$$

to find

$$v = \lambda_1 f_1 = 26{,}400 \text{ cm/s}$$

But the string will also resonate whenever $n(\frac{1}{2}\lambda) = 50$ cm. It will therefore resonate to wavelengths λ_n, given by

$$n(\tfrac{1}{2}\lambda_n) = 50 \text{ cm}$$

which can be written as

$$n\lambda_n = 100 \text{ cm}$$

from which

$$\lambda_n = \frac{100 \text{ cm}}{n}$$

Therefore

$$\lambda_2 = \frac{100 \text{ cm}}{2}$$

$$\lambda_3 = \frac{100 \text{ cm}}{3}$$

and, in general,

$$\lambda_n = \frac{100 \text{ cm}}{n}$$

We know, though, that $\lambda = v/f$, and so

$$\lambda_n = \frac{26{,}400 \text{ cm/s}}{f_n}$$

Placing in the value we have found for λ_n gives

$$\frac{100 \text{ cm}}{n} = \frac{26{,}400 \text{ cm/s}}{f_n}$$

After cross multiplying and solving for f_n, we find that

$$f_n = 264n \text{ Hz}$$

where $n = 1, 2, 3, \ldots$. The string will therefore resonate to all these frequencies. ▮▮

17.5 Transverse Versus Longitudinal Waves

The waves we have been discussing move along a string. Each pulse or wave crest moves (or propagates) in the direction of the string. It carries energy down the string as it moves along the string.

The direction in which the energy of a wave moves is called the direction (or line) of propagation of the wave.

In the case of a string the line of propagation is the string itself.

When a pulse moves down a string, the pieces of the string do not oscillate along the direction of propagation. Instead, they move perpendicular to the string. We call this type of wave a *transverse wave* (from the Latin *trans*, "across," and *vertere*, "to turn").

In a transverse wave the direction of oscillation is perpendicular to the direction of propagation.

The vibration direction is across the direction of propagation.

Water waves are nearly transverse. Consider a water wave that travels out across a puddle. Although the wave travels along the surface of the puddle, the water itself does not. If you watch a bug or dust particle sitting on the surface of the water, it simply moves up and down as the wave travels past it, as shown in Figure 17.11(a). The direction of oscillation in the wave is up and down. The direction of wave propagation is horizontal. Therefore, the wave is transverse.

We shall see later that light waves and radio waves are also transverse. They are part of a group called electromagnetic waves. As their name indicates, their oscillation involves electric and magnetic quantities. The quantities that oscillate will be discussed in our study of electricity. For now we shall simply state that the oscillation is perpendicular to the direction of propagation, as shown in Figure 17.11(b). Therefore, electromagnetic waves are transverse.

Not all waves are transverse. There is another type of wave in which the oscillation direction is the same as the direction of propagation. We call this a *longitudinal* (or *compressional*) *wave*.

In a longitudinal (or compressional) wave the direction of oscillation is parallel to the direction of propagation.

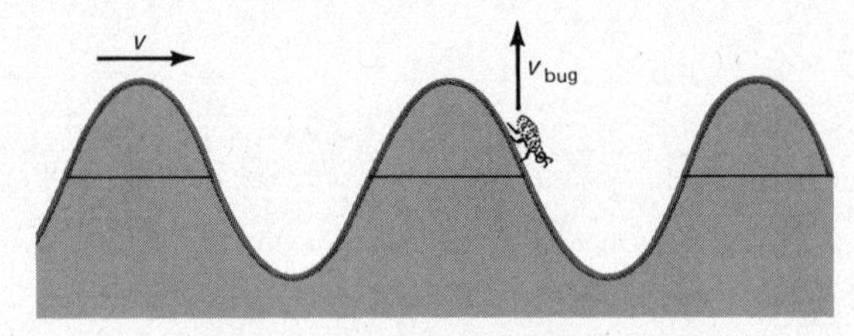

(a) The bug moves up and down with the water as the wave travels past.

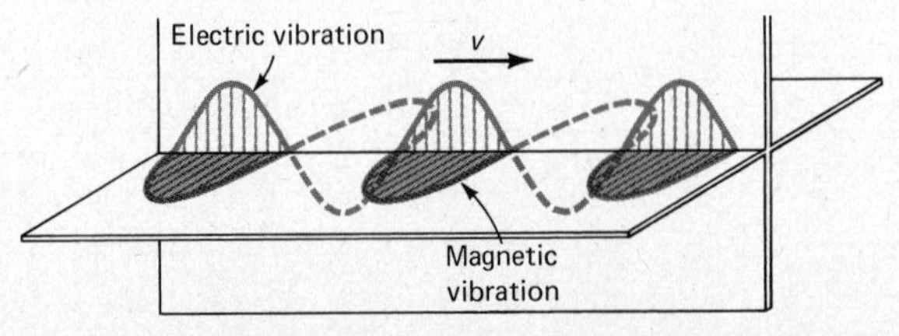

(b) Radio and light waves are of this type.

Figure 17.11 Examples of transverse waves.

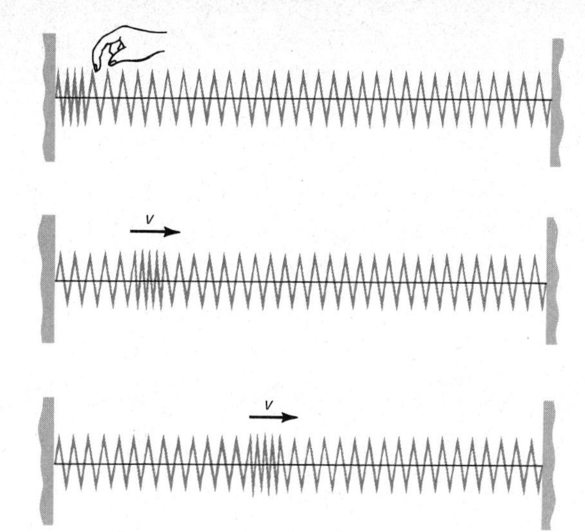

Figure 17.12 When the spring is released, the compression pulse moves with speed v down the spring. As it moves past the wires of the spring, it causes them to oscillate back and forth (not up and down). Because the oscillation direction is the same as the propagation direction, pulses such as this give rise to longitudinal waves.

An example of a longitudinal wave pulse is shown in Figure 17.12. We see there a compression pulse moving down a coil spring. One end of the spring is compressed. When the compression is released, it travels along the spring as shown. Until the pulse reaches it, each coil of the spring remains motionless. But as the pulse passes through a point, the coils near the point move horizontally. They move closer together so as to become the compressed region. As the pulse passes, the coils return to their original position. As we see, the coils move back and forth along the direction of propagation. For this reason pulses such as this cause longitudinal (compressional) waves.

It is also possible to send a stretching-type pulse down a spring. For example, the end of the spring in Figure 17.13 is given a sudden jerk, and the end region is stretched. This stretched region then travels as a pulse down the spring. A stretching-type pulse, such as this, is called a *rarefaction.* It is the opposite of a compression. It, too, gives rise to a longitudinal wave.

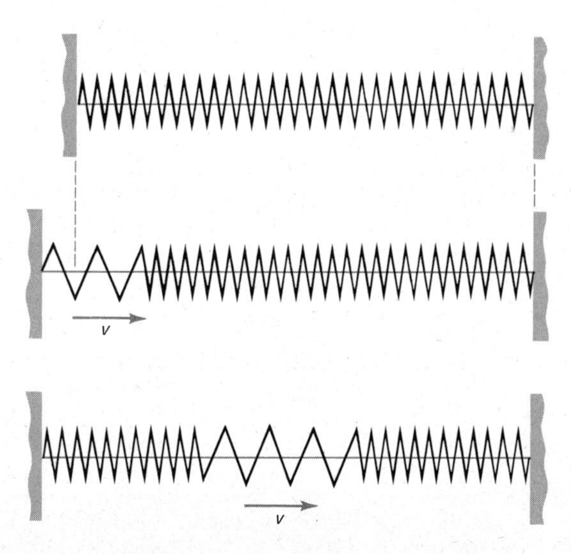

Figure 17.13 The rarefaction (or stretching) pulse moves down the spring.

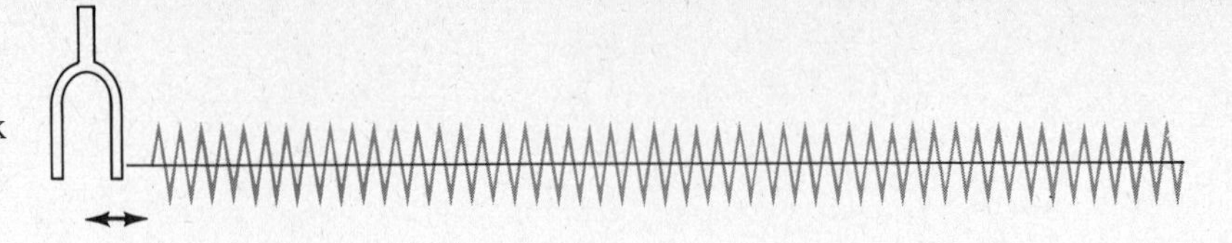

Figure 17.14 The vibrating tuning fork can send a series of compressions and rarefactions down the spring.

Suppose that one end of the spring is attached to a tuning fork prong as shown in Figure 17.14. As the prong vibrates back and forth, it will alternately compress and stretch the spring. Alternate compressions and rarefactions will then move down the spring. The resultant wave on the spring at a certain instant is shown in the upper part of Figure 17.15. Of course, the tension in the spring varies as we go from compression to rarefaction. At a compression the spring is compressed, not stretched. The tension is negative. But at a rarefaction the spring is stretched, and the tension is positive. The tension along the spring is plotted in the lower part of Figure 17.15.

As we see, the tension along the spring varies in a sinusoidal way. The tuning fork sends a sinusoidal type of compression wave down the spring. This is a longitudinal wave because the oscillation direction is along the line of propagation. As usual, the wavelength is the distance between crests (or troughs) of the wave. The relation $\lambda = vT = v(1/f)$ found for a wave on a string applies here, too. We shall now see that, like a wave on a string, a wave on a spring can also resonate.

Suppose that the spring is fastened solidly at one end. A vibrator such as a tuning fork sends compressions and rarefactions down the spring, as in part (a) of Figure 17.16. The fixed end must be at a motion node because no motion can occur there. The end where the tuning fork is placed cannot move much either. Therefore that end, too, must be very close to a motion node. From this we see at once that the spring will resonate with nodes at its two ends.

To show the resonances of the spring, we draw diagrams such as those shown in parts (b), (c), and (d) of Figure 17.16. Unlike the diagrams for a string, these

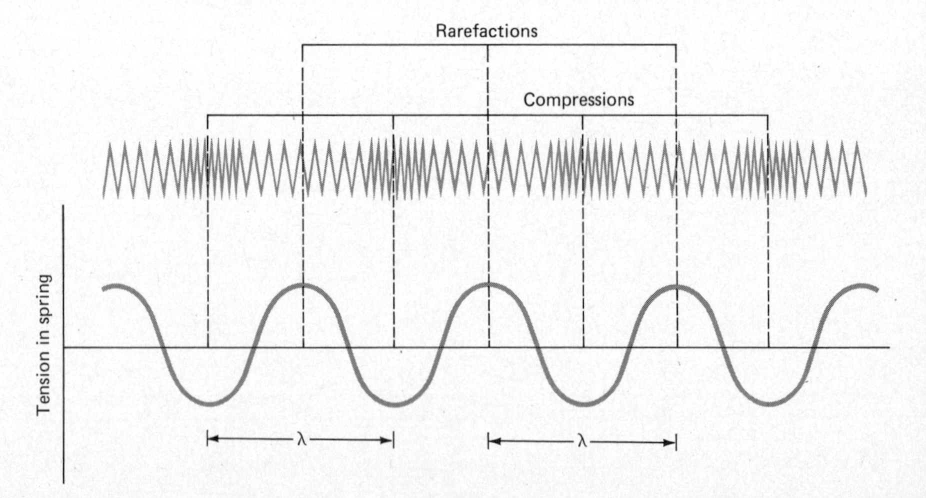

Figure 17.15 A compressional wave is sent down a spring by a vibrator at its end. The tension in the spring varies as shown in the lower portion of the figure. Notice that λ is the distance from tension crest to tension crest.

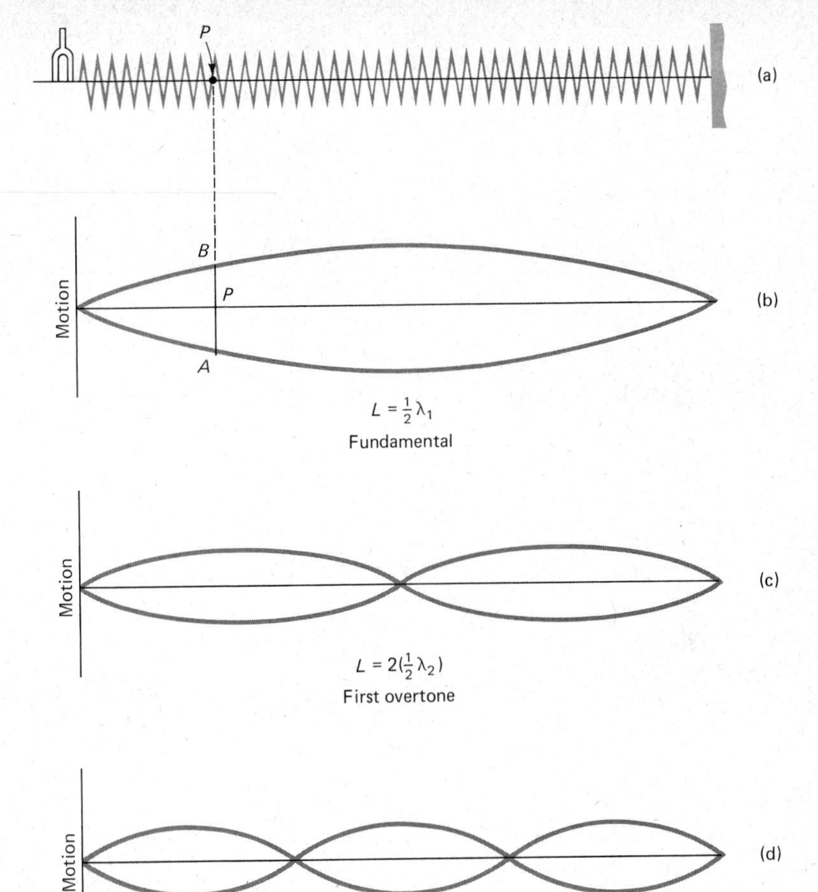

Figure 17.16 Because the spring is held at its two ends, only slight motion occurs there. They are nodes. For resonance, the tuning fork frequency must be chosen such that $L = n(\frac{1}{2}\lambda_n)$, where n is an integer.

do not represent transverse motion. Instead, for this longitudinal wave they represent the longitudinal vibration. For example, notice the line AB at point P in part (b). It tells us that point P on the spring vibrates back and forth (not up and down) through a distance proportional to $\overline{AB}$.

The first few resonances for the spring are shown in parts (b), (c), and (d) of Figure 17.16. Recall that each segment is $\frac{1}{2}\lambda$ long. It is clear that the relation between spring length L and resonance wavelength λ_n is

$$L = n(\tfrac{1}{2}\lambda_n) \rightarrow 2L = n\lambda_n \qquad n = 1, 2, 3, \text{etc.}$$

from which

$$\lambda_n = \frac{2L}{n}$$

The spring will resonate to those wavelengths λ_n given by

$$\lambda_n = \frac{2L}{n} \qquad n = 1, 2, 3, \text{ etc.}$$

Because $\lambda = vT = v(1/f)$, this corresponds to tuning fork frequencies of

$$f_n = \frac{v}{\lambda_n} = \frac{v}{2L/n}$$

which becomes, after inverting the denominator and multiplying,

$$f_n = n\left(\frac{v}{2L}\right)$$

The vibrator must have one of these frequencies if the spring is to resonate.

The resonance frequencies of a spring, string, or any other system are called *characteristic* (or *eigen*) *frequencies* of the system. Their resonance patterns are referred to as *resonance modes.*

■ **EXAMPLE 17.4** A spring 80 cm long and held at its two ends is vibrated longitudinally. It resonates in its second overtone when the driving frequency is 5.0 Hz. Find the speed of a compressional pulse along the spring.

Solution We want the speed v of the wave on the spring. The spring is vibrating like the one shown in part (d) of Figure 17.16. Therefore $L = 3(\frac{1}{2}\lambda)$ in this case. Because $L = 80$ cm, we have

$$\lambda = \frac{2L}{3} = 53.3 \text{ cm}$$

Using $\lambda = vT = v(1/f)$, we find that

$$\begin{aligned} v &= \lambda f = (53.3 \text{ cm})(5.0 \text{ Hz}) \\ &= 267 \text{ cm/s} \end{aligned}$$

This is the speed with which a pulse will move along the spring. ■■

17.6 Compression Waves in Air: Sound

Let us now look at a more important type of compression wave, sound waves. We shall introduce this type of wave by a simple example of it. Suppose, as in Figure 17.17, that a piston is placed at one end of a tube filled with a gas such as air. By means of a motor, the piston can be vibrated back and forth. As the piston moves to the right, a compression is sent down the tube through the gas. When the piston moves to the left, a rarefaction is sent down the tube. This motion results in a longitudinal wave moving down the tube as indicated.

If you were to listen at the right-hand end of the tube, these alternate com-

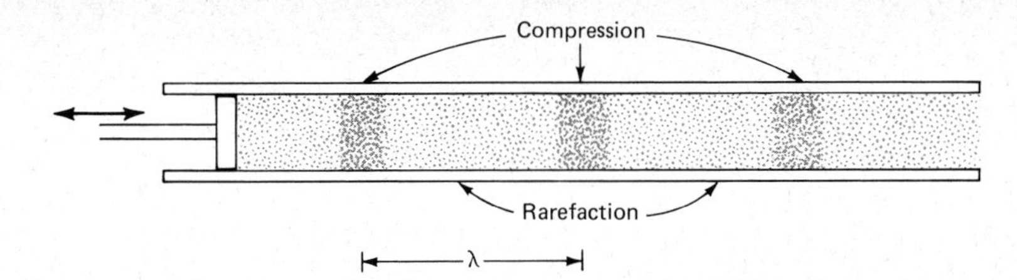

Figure 17.17 A compression-rarefaction longitudinal wave is sent through the air by the vibrating piston.

pressions and rarefactions would strike your ear. They would cause your eardrum to vibrate back and forth. If the frequency f of the wave was in the range of about $30 \leq f \leq 17{,}000$ Hz, you would hear a sound.[2] Longitudinal waves in gases (and other materials) are therefore called sound waves.

Sound waves are longitudinal vibrations in gases and other materials. The normal ear can hear sound waves with frequency in the approximate range $30 \leq f \leq 17{,}000$ Hz.

Waves with frequencies higher than about 17,000 Hz are called *ultrasonic waves.*

Suppose that the tube in Figure 17.17 is closed at the right-hand end. Suppose further that the piston does not vibrate very much. We then would like to know under what conditions the sound wave would resonate in the tube. Clearly, both ends of the tube (shown in Figure 17.18) must be nearly motion

[2] Remember that ≤ means "less than or equal to."

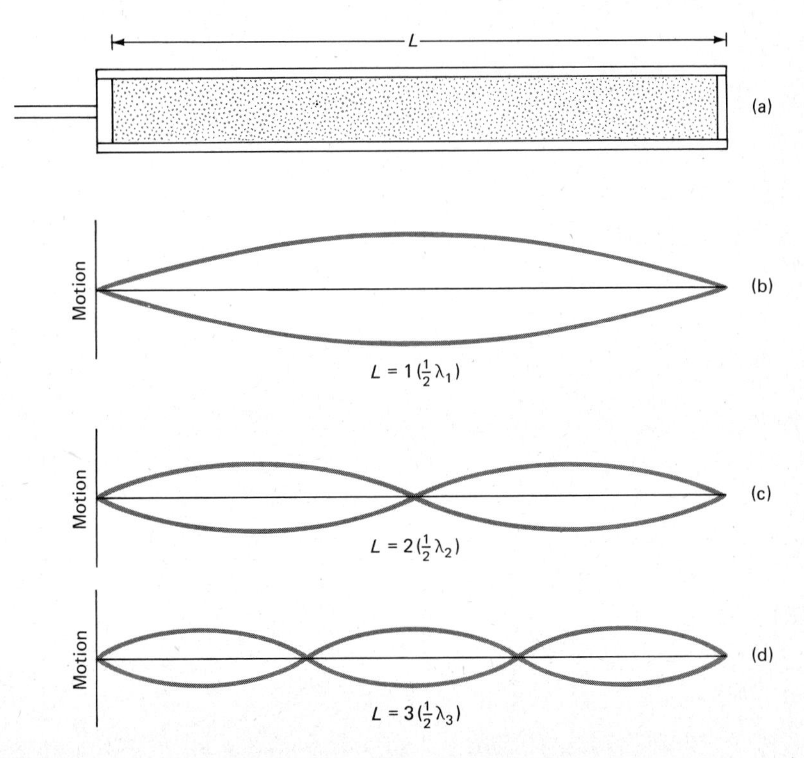

Figure 17.18 If the piston does not move much, it will be near a node. For a tube closed at both ends, resonance occurs when $L = n(\frac{1}{2}\lambda_n)$, where n is an integer.

nodes. The air vibrates scarcely at all there, because the two ends are nearly motionless. This fact tells us at once that the resonance modes of the tube will be as shown in parts (b), (c), and (d) of the figure. For resonance to occur the length of the tube L must be related to the wavelength by the equation

$$L = n(\tfrac{1}{2}\lambda_n) \qquad n = 1, 2, 3, \text{etc.}$$

Only for these wavelengths will the wave fit in the tube properly.

A new and different situation exists if the tube is open at its right-hand end. This is shown in Figure 17.19(a). When the wave pulses reach the end of the tube, much of the energy is reflected. But now, unlike the closed-end case, the air molecules are not prevented from moving there. Instead, they can move most easily there. As a result, the open end of a tube is a position of maximum motion. If resonance is to occur, it must be close to the position of an antinode.[3]

In order for the tube open at one end to resonate, the open end must be an antinode. The closed end must be a node. Therefore, the resonances of the tube must be as shown in parts (a), (b), (c), and (d) of Figure 17.19. As usual, the distance between adjacent nodes (or antinodes) is $\frac{1}{2}\lambda$. We see, therefore, that the tube will resonate only if λ has one of the following values:

$$L = 1(\tfrac{1}{4}\lambda_1)$$
$$L = 3(\tfrac{1}{4}\lambda_2)$$

[3] Advanced work shows that the open end is slightly short of the antinode position. If λ is large compared to the diameter of the pipe, this effect is negligible.

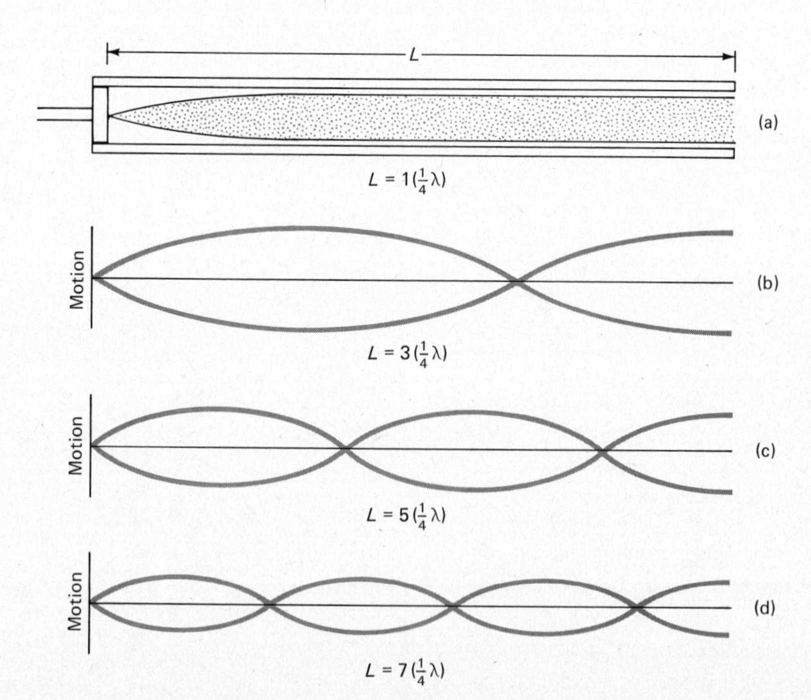

Figure 17.19 A tube open at one end and closed at the other will resonate when $L = n(\frac{1}{4}\lambda_n)$, where $n = 1, 3, 5, \ldots$

$$L = 5(\tfrac{1}{4}\lambda_3)$$

and so on. In general, $L = n(\frac{1}{4}\lambda_n)$, where $n = 1, 3, 5$, etc., an odd integer.

EXAMPLE 17.5 The speed of sound waves in air is about 340 m/s. To what frequencies of sound will a tube 60 cm long closed at both ends resonate?

Solution This is the situation shown in Figure 17.18 with $L = 60$ cm. We had for resonance that $L = n(\frac{1}{2}\lambda_n)$. Using our values, we have

$$\lambda_n = \frac{1.20 \text{ m}}{n}$$

where λ is in meters. But $\lambda = vT = v/f$ tells us that the resonance frequencies are

$$f_n = \frac{v}{\lambda_n}$$

Because $v = 340$ m/s and $\lambda_n = (1.20 \text{ m})/n$, we find that

$$\begin{aligned} f_n &= \frac{340 \text{ m/s}}{(1.20 \text{ m})/n} = \frac{340n}{1.20}\,\frac{1}{\text{s}} \\ &= 283n \text{ s}^{-1} \\ &= 283n \text{ Hz} \end{aligned}$$

The fundamental resonance frequency is obtained by setting $n = 1$. When $n = 2$ we obtain the first overtone, and so on.

$$\begin{aligned} f_1 &= 283 \text{ Hz} \\ f_2 &= 566 \text{ Hz} \\ f_3 &= 850 \text{ Hz} \end{aligned}$$

and so on.

EXAMPLE 17.6 A piece of glass tubing is 1.50 m long and open at both ends. Find the two lowest frequencies of sound to which it will resonate. The speed of sound in air is about 340 m/s.

Solution Because both ends are open, both ends will be antinodes. The two lowest resonance modes are shown in Figure 17.20. For them we have

$$\begin{aligned} L &= 2(\tfrac{1}{4}\lambda_1) \\ L &= 4(\tfrac{1}{4}\lambda_2) \end{aligned}$$

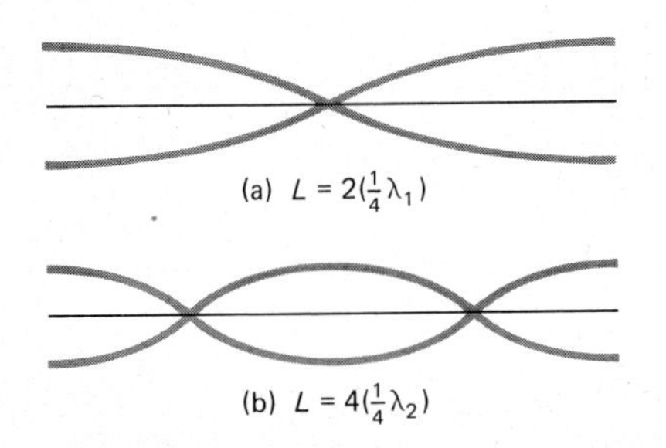

Figure 17.20 The first two resonances of a tube open at both ends.

Because $L = 1.50$ m, we have $\lambda_1 = 3.0$ m and $\lambda_2 = 1.50$ m.

The corresponding frequencies can be found from $\lambda = v/f$ to be

$$f = \frac{340 \text{ m/s}}{\lambda}$$

Therefore, $f_1 = 113$ Hz and $f_2 = 227$ Hz.

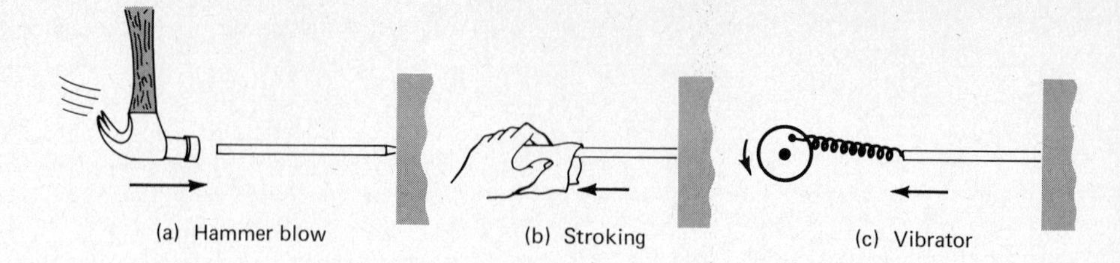

Figure 17.21 Three ways to send a longitudinal wave down a rod.

17.7 Longitudinal Resonance of Rods

Compressional waves can be sent down a rod easily. Three simple ways for doing this are shown in Figure 17.21. In (a) the rod is struck lengthwise at its end. The wave sent down the rod by such a blow is quite complicated. In (b), the rod is stroked by pulling a cloth dusted with rosin over it. The cloth alternately sticks and slips. This gives rise to compressions and rarefactions within the rod. In (c) the end of the rod is vibrated lengthwise by a vibrating mechanical device.

The rod will resonate to certain frequencies. To find out what they are, we need to know where the nodes and antinodes are along the rod. If the rod is firmly held at one end, as in Figure 17.21, this end must be a node. The free end of the rod should be a place of greatest vibration. An antinode is therefore located near there. Let us consider such a rod in some detail.

The rod is shown in Figure 17.22. When vibrated lengthwise, a longitudinal wave is sent down the rod. To resonate, the wavelength must allow a node at the right-hand end and an antinode at the left-hand end. The first three resonances are shown in the figure. Remember that the vibration occurs lengthwise of the rod. The diagrams are simply graphs of this motion.

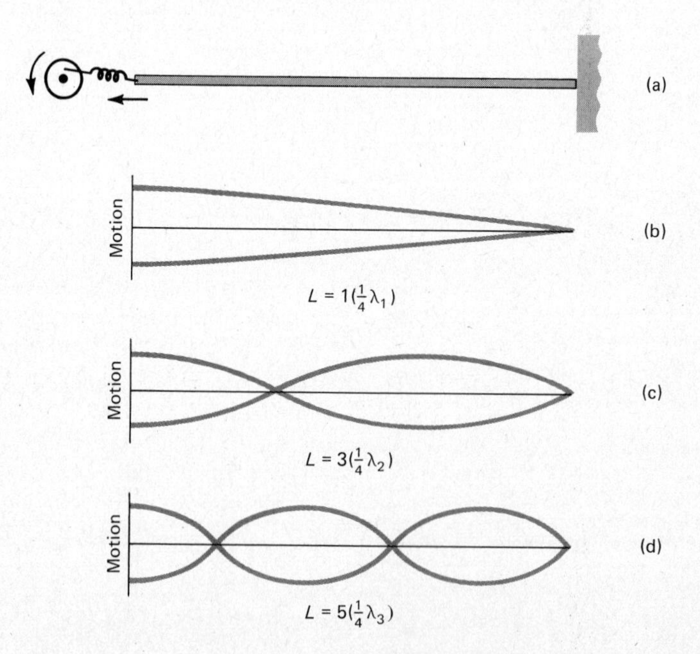

Figure 17.22 The first three longitudinal resonances for the bar shown in part (a).

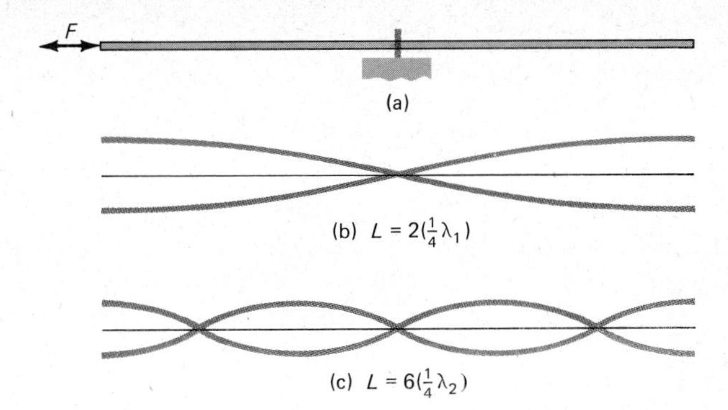

Figure 17.23 Longitudinal resonances for a rod fastened at its center.

EXAMPLE 17.7 The speed of sound waves in steel is 5000 m/s. A steel rod 90 cm long is held firmly at its center as shown in Figure 17.23(a). Find the two lowest frequencies to which it will resonate when excited longitudinally.

Solution The center of the rod must be a node, whereas its two ends are antinodes. Its resonant motions are shown in parts (b) and (c) of the figure. We see at once that

$$L = 2(\tfrac{1}{4}\lambda_1)$$
$$L = 6(\tfrac{1}{4}\lambda_2)$$

Because $L = 0.90$ m, this gives $\lambda_1 = 1.80$ m and $\lambda_2 = 0.60$ m. Then, because $v =$ 5000 m/s, we can use $\lambda = vT = v/f$ to find the resonant frequencies. Substitution gives

$$f_1 = \frac{5000 \text{ m/s}}{1.80 \text{ m}} = 2778 \text{ Hz}$$

and

$$f_2 = \frac{5000 \text{ m/s}}{0.60 \text{ m}} = 8333 \text{ Hz}$$

17.8 Complicated Waves; Fourier Analysis

All the periodic waves we have so far discussed are "nice" waves. Usually we have made use of sinusoidal-type waves. But frequently waves are much more complicated than this. For example, when a guitar string is plucked, the resulting wave is very complex. Similarly, when a violin bow is drawn across a violin string, the wave on the string is far from sinusoidal. The hammer that strikes the string of a piano also sends a complex wave down the string. How can we treat such complicated situations?

There is a simple answer to this question. It is given by Fourier's principle,[4] which can be summarized as follows:

[4] Pronounced "foor-yea's" principle.

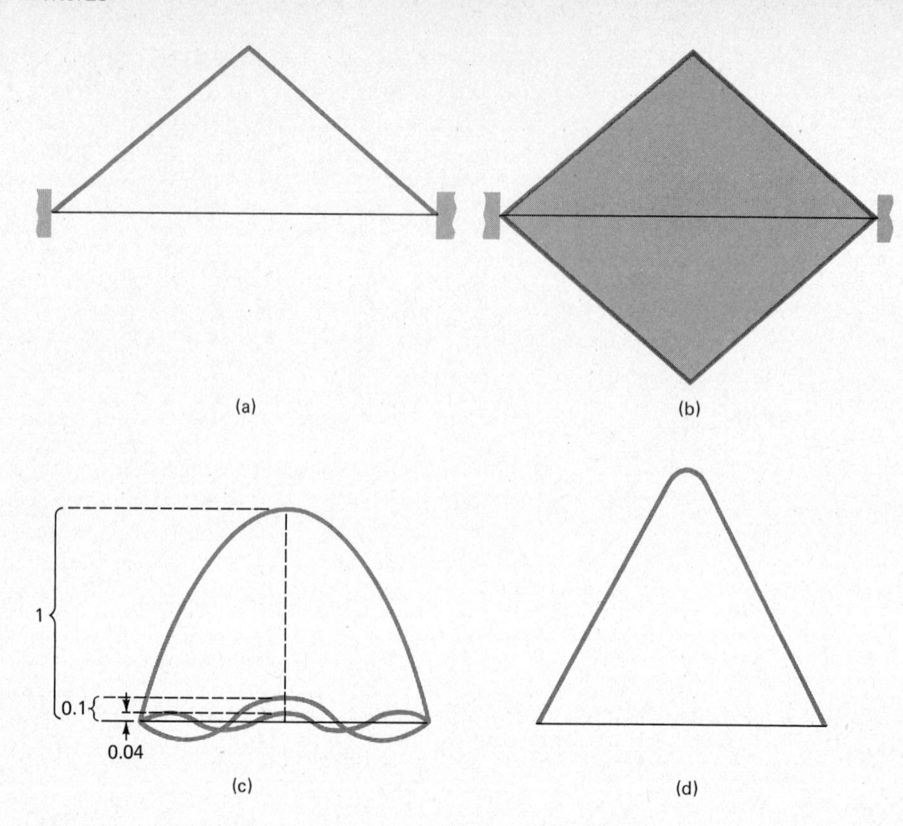

Figure 17.24 The plucked string in (a) vibrates through the limits shown in (b). It vibrates as though it had a number of sinusoidal waves resonating on it. The three waves in (c) add to give the form shown in (d).

Any wave, no matter how complicated, can be thought of as being the sum of sinusoidal waves.

Let us see the meaning of this statement in the case of a plucked string.

Suppose that a guitar string is pulled aside as shown in Figure 17.24(a). When released, the string will vibrate back and forth between the limits shown in (b). This is a vibration quite unlike the simple resonances we have been discussing. In spite of this, Fourier's principle tells us that this vibration can be thought of as a sum. It is the sum of many sinusoidal waves resonating on the string.

To see this more clearly, refer to Figure 17.24(c). Let us add the three sinusoidal resonance waves shown in part (c). Their sum is shown in (d) of the figure. Notice how these three sinusoidal waves added together nearly form the wave shown in (a). By adding still other waves of higher frequency and smaller amplitude, the correspondence can be made nearly perfect. This can be done for any resonating wave, no matter how complicated. Let us review what we have found.

Any resonating wave can be duplicated by a sum of sinusoidal resonating waves.

This type of analysis of complex waves, that is, the process of separating a complex wave into its sinusoidal components, is called *Fourier analysis.*

Technical people make much use of Fourier analysis of vibrations and waves. For example, most sound and electromagnetic waves are not simple sinusoidal waves. For the interpretation of their effects, the waves are analyzed into their sinusoidal building blocks. But the effects of the simple sinusoidal waves are usually easy to deal with. Viewed in this way, a problem involving a complex wave is reduced to one that is more easily handled. We shall use this technique when we discuss the sounds emitted by musical instruments in the next chapter.

Summary

The wavelength of a wave, λ, is the distance between adjacent corresponding points along the wave. The maximum displacement from equilibrium caused by the wave is the amplitude of the wave. The line along which the wave advances is called the direction of propagation of the wave.

Waves carry energy away from the vibration source. The energy moves out along the line (or lines) of propagation.

The time taken for a point through which a wave is passing to undergo one complete vibration is called the period T of the wave. Its inverse, $1/T$, is the frequency of the wave, f. This is the number of complete vibrations that pass through the point in unit time. Both f and T are identical for the wave and its vibration source.

Applying $x = \overline{v}t$ to a wave yields the fundamental relation $\lambda = vT$. Because $f = 1/T$, this may be rewritten as $\lambda = v/f$.

According to the superposition principle, for two or more waves passing through the same point at the same time, the wave disturbances add vectorially.

When identical waves can be reflected back and forth upon each other, a large wave motion can be built up. This condition is called resonance. In order for resonance to happen, the vibration frequency must be chosen properly. These frequencies are called the resonance frequencies for the wave.

At resonance a wave gives rise to a standing wave pattern. In this pattern the nodes are points where no motion occurs. The antinodes are points of maximum disturbance. The distance between adjacent nodes (or antinodes) is $\frac{1}{2}\lambda$.

In a transverse wave the oscillation direction is perpendicular to the direction of propagation. Waves on a string are transverse. In a longitudinal (or compressional) wave the oscillation direction is along the line of propagation. Sound waves are longitudinal.

Sound waves consist of alternate compressions and rarefactions sent through a material. The normal ear can hear sound waves with frequencies in the approximate range $30 \leq f \leq 17{,}000$ Hz. Waves of frequency higher than this are called ultrasonic waves.

Any complex wave (or vibration) can be duplicated by a sum of sinusoidal waves. Fourier analysis is the process of separating a complex wave into its sinusoidal component waves.

Questions and Exercises

1. A wave pulse moves down a stretched string. Initially a point on the string has zero energy. It gains energy as the pulse strikes it. Then it loses the energy as the pulse moves away. Explain how it gains energy; how it loses the energy it gained. (Hint: Note the work done on it by the nearby portion of string.)
2. A pebble is dropped into a large pool of water. Wave crests spread out as circles from the point of impact. Explain why these crests die down as they move away from the center.
3. Suppose that the waves shown in Figure 17.8 are square waves rather than sinusoidal. How will the string vibrate in this case? Will nodes still exist?
4. One way to eliminate standing waves is to place an energy absorber at the end opposite the vibrator. Explain why this will get rid of standing waves. What sort of absorber might one use for sound waves? For waves on a string?
5. A wave on a string always has the same frequency as the vibrator. But the wavelength and speed depend on other factors. What are they? Can the frequency of a sound wave change without changing the frequency of the sound source?
6. A wave with frequency f_0 is sent down a string in which its speed is v_0. Tied to the end of this string is another one in which the speed is v. The wave continues on into this attached string. What will be the values of f and λ for this new wave in terms of f_0, v_0, and v?
7. Two narrow brass pipes have equal lengths. One pipe is open at both ends, while the other is open only at one end. Do they have any resonance frequencies in common? Which has the largest number of resonance frequencies at low frequencies?
8. A brass rod is dropped end first onto a concrete floor. What sort of vibratory motion does it undergo because of the impact? What can you say about its vibratory motion? Why do different length rods dropped in this way give off different sounds?
9. A xylophone consists of a set of bars of different lengths. Each bar emits its own distinctive tone when hit in the middle. Discuss the physics of this device. Where should the bars be supported?
10. When you blow across the top of a pop bottle, it resonates to one particular sound. Where is the vibrator that causes the wave? The vibrator gives off waves of many frequencies. (It is a complex wave with many Fourier components.) Explain why the bottle selects only one wave. How could you carry out a crude Fourier analysis of a complex sound wave by use of a series of bottles?

Problems

1. Microwaves have wavelengths of a few centimeters and travel with the speed of light, 3×10^8 m/s. What must be the frequency of a microwave transmitter that sends out 20-cm-wavelength microwaves?
2. The laser light beam from a helium laser has a wavelength of 633 nm. Laser light, like all light waves, travels with a speed of 3×10^8 m/s. What is the frequency of this type of laser wave?
3. Radio station KDKA in Pittsburgh sends out radio waves with frequency 1020 kHz. The speed of radio waves is 3.0×10^8 m/s. Find the wavelength of the waves sent out by this station.
4. Radio station KILT of Houston operates at a frequency of 610 kHz. Like all radio waves, the radio waves it sends out have a speed of 3.0×10^8 m/s. Find the wavelength of the waves sent out by this station.
5. Refer to the wave shown in Figure 17.4. What is its (a) amplitude and (b) wavelength? (c) If the speed of the wave on the string is 2 m/s, what is the frequency of the wave source?
6. Refer to Figure P17.1, which shows a wave on a string at a certain instant. The oscillator that sends out the wave is vibrating at 120 Hz. Find the following for the wave: (a) wavelength, (b) amplitude, (c) period, (d) speed.

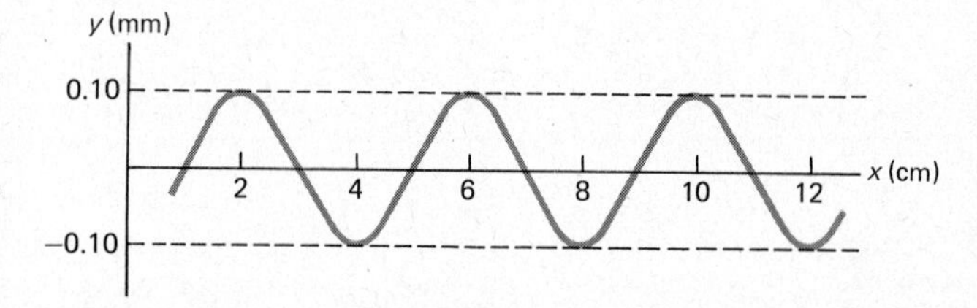

Figure P17.1

7. Suppose that the speed of the wave in Figure P17.1 is 5.0 m/s. (a) What is its frequency? (b) Its amplitude? (c) If the frequency of the vibrator that sends it out is now doubled, what will be the wavelength of the new wave?
8. When a 30-cm-long banjo string resonates in one segment, it vibrates with a frequency of 200 Hz. (a) What is the wavelength of the wave on the string? (b) What is the speed of waves on the string?
9. A 180-cm-long string resonates in three segments when it is

vibrated at a frequency of 40 Hz. (a) What is the wavelength of the wave on the string? (b) What is the speed of the wave?

10. The speed of waves on a 90-cm-long string is 20 m/s. With what frequency must the string be vibrated if it is to resonate in (a) one segment, (b) two segments, and (c) three segments?
11. What are the four lowest frequencies to which a 70-cm-long string will resonate if the speed of waves on the string is 30 m/s?
12. A wire is strung between two poles that are 10 m apart. When the wind blows on the wire, it hums with a tone of 60 Hz. Assuming the wire to be vibrating in its fundamental, (a) what is the wavelength of the wave on the wire? (b) What is the speed of the wave? (c) What are the next three higher frequencies to which the wire will resonate?

*13. A heavy weight hangs at the end of a 140-m cable. A worker at the top of the cable notices that when he strikes the top of the cable sideways, the pulse hits the weight after a time of 4.0 s. (a) With about what frequency must he vibrate the top end if the rope is to resonate in its fundamental? (b) In its first overtone? (c) Why is your answer only approximate?

**14. A string is found to resonate to 135 Hz and 162 Hz. It does not resonate to any frequency in between. (a) What is the fundamental resonance frequency for the string? (b) In how many segments does it vibrate at 135 Hz?

15. A stretched coil spring is held firmly at its two ends that are 7.0 m apart. The speed of longitudinal waves on the spring is 3.0 m/s. When vibrated longitudinally, what are its lowest three resonance frequencies?
16. A coil spring is stretched between two points 5.0 m apart. When a compressive pulse is sent along the coil, it takes 1.50 s to make one complete trip down and back. Find the three lowest frequency compressional waves to which the spring will vibrate.
17. Refer to Figure 17.16. The spring is 3.0 m long. It vibrates in three segments to a frequency of 200 Hz. (a) What is the wavelength of the wave in that case? (b) What is the speed of waves on the spring? (c) What is the fundamental resonance frequency?
18. A coil spring is stretched to a length of 2.8 m. The spring vibrates longitudinally under the action of a vibrator at one end. It is found that the spring resonates in four segments when the vibrator frequency is 1.50 Hz. (a) What is the speed of these waves along the spring? (b) What will be its fundamental resonance frequency?

*19. A coil spring clamped at its end resonates in its fundamental mode to longitudinal waves of frequency 40 Hz. A clamp is now placed one-third of the way from one end so that this point will be a node. (a) To what fundamental frequency will its short end resonate? (b) Its long end?

20. A 50-cm length of straight pipe is attached to a car's muffler. This tail pipe acts as a tube open at both ends. Find its three lowest resonance frequencies. Take the speed of sound to be 360 m/s for the hot gases in the tail pipe.
21. Repeat Problem 20 assuming that the tube acts as though it is closed at one end and open at the other.
22. What are the four lowest resonance frequencies for sound waves in a 2-m-long tube closed at both ends? Use $v =$ 340 m/s.
23. What are the four lowest resonance frequencies for sound waves in a 2-m-long tube open at one end and closed at the other? Use $v = 340$ m/s.
24. What are the four lowest resonance frequencies for sound waves in a 3-m-long tube open at both ends? Use $v =$ 340 m/s.

*25. The device shown in Figure P17.2 is sometimes used to measure the speed of sound in air. By raising and lowering the can, the water level in the tube at the right can be changed. Using a 2500-Hz tuning fork, resonances are found

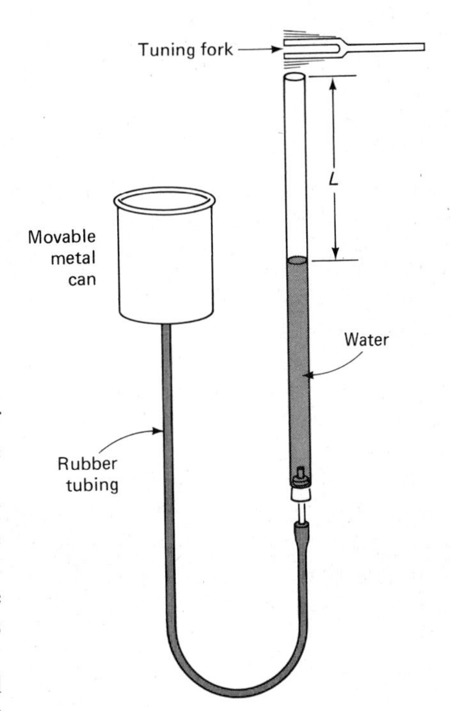

Figure P17.2

*Problems marked with an asterisk are not as easy as the unmarked ones.
**Problems marked with a double asterisk are somewhat more difficult than the average.

at two values for L separated by 6.7 cm. No resonances exist for water levels in between. What is the speed of sound in air according to these results?

*26. In order to determine the speed of sound in helium gas, the following experiment is performed. A helium-filled tube is closed at one end by a movable piston. A tiny loudspeaker is mounted at the other closed end. Resonances are obtained by moving the piston. When the loudspeaker frequency is 1000 Hz, resonance positions of the piston are 27.6 cm, 76.8 cm, and 125.9 cm measured from an arbitrary reference mark. Find the speed of sound in helium.

*27. The experiment described in Problem 26 is used to calibrate the frequency of a loudspeaker. Air is used as the gas. The speed of sound in it is 343 m/s at the temperature used. Resonances are found at piston positions of 39.3 cm, 51.7 cm, and 64.0 cm. Find the loudspeaker frequency.

28. Theory shows that the speed of transverse waves on a string or cable is given by

$$v = \sqrt{\frac{T}{m/L}}$$

In this expression T is the tension and m is the mass of length L of the string. Two meters of a particular string "weigh" 0.80 g. If the tension in the string is 20 N, what is the speed of waves on the string?

29. Using the formula given in Problem 28, find the speed of waves on a string that is under a tension of 150 N. The string has a mass of 0.20 g per meter length.

*30. A string 70 cm long has a mass per unit length of 0.40 g/m. What must be the velocity and tension in the string if its fundamental resonance frequency is to be 440 Hz? (See Problem 28.)

*31. A 6.0-kg mass hangs at the end of an 80-cm-long wire. This length of wire weighs 2.0 g. (a) Find the speed of transverse waves in the wire. (b) To what frequencies will the wire resonate? See Problem 28 for the relation between speed and tension.

32. A metal rod 60 cm long is clamped firmly at its midpoint. When stroked lengthwise, it gives off a tone of frequency 4100 Hz. Find the speed of compressional waves (sound) in the metal.

33. A metal rod 70 cm long is clamped firmly at one end. When stroked lengthwise, it gives off a 1400-Hz tone. Find the speed of compressional waves (sound) in the metal.

34. The speed of compressional waves in a certain metal rod is 4000 m/s. What are the three lowest resonance frequencies for the rod when it is clamped at one end? The rod is 180 cm long.

35. Find the lowest resonance frequency for the rod in Problem 34 if it is clamped at its center point.

*36. Find the two lowest resonance frequencies for a 60-cm-long rod clamped one-quarter of the way from one end. Compressional waves with speed 4800 m/s are being used.

**37. Find the lowest frequency to which both parts of the rod in Problem 34 would resonate simultaneously if the rod is clamped one-sixth the length (that is, 30 cm) from one end.

**38. A 90-cm-long metal bar is clamped firmly at a point 15 cm from one end. The speed of compressional waves in the bar is 5000 m/s. Find the lowest resonance frequency for the bar.

39. Theory shows that the speed of compressional waves in a rod is given by

$$v = \sqrt{\frac{Y}{\rho}}$$

In this expression Y is Young's modulus of the material and ρ is its density. A 70-cm-long metal bar clamped at its center has a fundamental resonance frequency of 3500 Hz for compressional waves. (a) Find the speed of sound in the rod. (b) If the density of the material is 8.3 g/cm^3, what is Young's modulus for the material?

*40. A ramp-type wave can be Fourier-analyzed into an infinite sum of sine waves. Suppose that the wavelength and amplitude of the first wave are λ_1 and A_1. Then the second wave has wavelength $\lambda_1/2$. Its amplitude is $A_1/2$, but its sign is reversed. The third wave has wavelength $\lambda_1/3$ and amplitude $A_1/3$ and is positive. The fourth has wavelength $\lambda_1/4$ and amplitude $A_1/4$ and is negative. And so on. Sketch the first three waves to scale over a length equal to $2\lambda_1$. Add them and see how closely they approximate a ramp wave.

Ellis Herwig, Stock, Boston

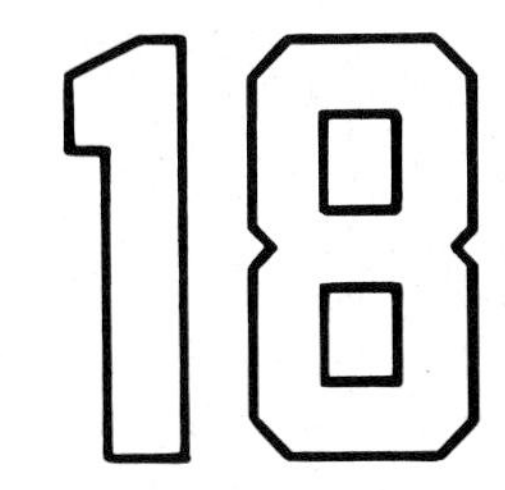

18 SOUND

Performance Goals

When you finish this chapter, you should be able to

1. Describe the nature of sound waves in gases, liquids, and solids. In the description explain the meaning of the graph of such a wave. Distinguish among audible, ultrasonic, and infrasonic sound waves. Explain how sound waves are sent out by a loudspeaker and by a violin string.
2. Distinguish between wave fronts and rays. Explain the relation of one to the other. What does

We give waves in the air that we can hear the name "sound." But sound waves are more than this. They are longitudinal vibrations in any material and at any frequency. In this chapter we shall learn about their nature and the ways in which they are measured. We shall also learn what factors influence the audibility and quality of the sounds we hear.

each show? What is a plane wave, and how can it be obtained?
3. Give the speed of sound in air and tell how it changes with temperature and pressure. Give the approximate (within a factor of 2) speed of sound in water and iron.
4. Define what is meant by sound intensity. Use the definition to find the energy carried through a given area in a stated length of time by a sound of known intensity. Given the power radiated by a sound source and the area through which it radiates, find the intensity of the sound.
5. Assign approximate intensity values to the following: least audible sound; pain-producing sound.
6. Describe the decibel sound level scale and state why it is used. Convert from sound level in decibels to intensities in watts per square meter and vice versa.
7. Sketch a rough graph that illustrates the variation with frequency of the sensitivity of the normal ear. Point out on the graph where the ear is most sensitive and least sensitive. Define hearing threshold.
8. Give several examples of sounds whose quality differs. In reference to them, explain what is meant by the quality of a sound. Explain what a Fourier analysis of a sound is and what it would show for sounds of differing quality.
9. Use the superposition principle to find the wave for a sound if the sinusoidal sound waves that make it up are given.

18.1 Nature of Sound Waves

In Chapter 17 we discussed compressional waves in general terms. It was stated that

> **Sound waves are longitudinal vibrations in gases and other materials. The normal ear can hear sound waves with frequency in the approximate range $30 < f < 17{,}000$ Hz.**

We are most familiar with sound waves in the air.

One way for producing sound waves in air is shown in Figure 18.1. The piston vibrates back and forth. In so doing, it sends alternate compressions and rarefactions through the air in the tube. If you place your ear at the end of the tube, the disturbance in the air will be carried to your eardrum. The alternate compressions and rarefactions cause your eardrum to move back and forth. These movements are communicated to your brain and you thereby hear the sound.

The ear cannot detect sound waves of all frequencies. We shall return to that subject later, but for now we simply point out that sound waves of too high or too low a frequency are not registered by the ear. We call sound waves with frequencies above about 20,000 Hz *ultrasonic waves* and those with frequencies less than about 20 Hz *infrasonic waves.*

Longitudinal-type waves in liquids and solids (as well as in gases) are also called sound waves. They may be sent out by any vibrator that makes contact with the liquid or solid. Sound generators for use under water often consist of a small vibrating plate (or crystal) immersed in the water. Electrical methods are used to set the crystal into vibration.

Sound waves in air are produced in many ways. A vibrating guitar string causes the air around it to vibrate. The alternate compressions and rarefactions thus caused result in sound waves through the air. The vibrating vocal cords in a person's throat cause sound waves to be generated. A knife dropped on the floor vibrates and sends out a sound into the air. All the sounds we hear can be traced to a vibrating source of some kind. Often the source vibrates in a complex way. The sound wave it gives off is also complex. For now, though, we shall consider only sinusoidal waves. Later we shall use Fourier's principle to see how complex waves can be understood.

A loudspeaker such as that shown in Figure 18.2 is often used to send out a nearly pure sinusoidal sound wave. An electric voltage generator (an oscillator) operates a transducer, a device for changing electrical signals to mechanical motion. The transducer causes the loudspeaker diaphragm to oscillate sinusoidally. This causes a sound wave to move out from the loudspeaker. At the points of compression (C) the pressure is higher than atmospheric. At the rarefactions (R) the pressure is lower than atmospheric. These results can be shown by plotting the gauge pressure, because gauge pressure measures the difference between actual and atmospheric pressure. We do this in the lower part of Figure 18.2.

Of course, the wave shown in Figure 18.2 moves out with a speed v. The

10. Do the following for a source that sends circular, sinusoidal waves out on the surface of a pool of water. Draw the wave fronts in three cases: (a) stationary source, (b) moving source with speed less than wave speed, (c) moving source with speed greater than wave speed. Use the diagrams to explain the following terms: bow wave, shock wave, Doppler effect.
11. Explain qualitatively how the Doppler effect arises in each of the following cases: approaching sound source; receding sound source; stationary source with listener moving.
12. Pick the Doppler effect equation from a list of given equations. Use it to compute the frequency shift in the situations outlined in Goal 11.
13. Compute the beat frequency to be expected in a typical situation.

figure is simply a picture of its position at a certain instant. In addition, the wave spreads out to the sides as well. Each loudspeaker has its own directional properties. Some speakers send the wave out primarily in the forward direction. Others spread the wave energy out over a wide angle. We can show this effect better by drawing the wave in another way. This is done in Figure 18.3.

In Figure 18.3 we see a wave sent out from a loudspeaker. The circular arcs represent the crests of the wave. They are also called wave fronts. The energy given to the wave by the sound source flows out along the straight lines, called rays. As pointed out in the last chapter, the wave fronts and rays are perpendicular to one another.

The wave fronts should really be viewed in three dimensions. Figure 18.3 shows only their cross sections in the plane of the page. The fronts are actually portions of spherical surfaces centered on the loudspeaker. They are surfaces rather than lines. At large distances from the source the spheres become very large. They are scarcely curved at all except over very large distances. In the case of such large spheres the curvature is often negligible. Wave fronts from an infinitely faraway source appear much like flat planes. We call such a wave a *plane wave.*

The wave fronts of a plane wave are flat planes. Waves from a distant source are nearly plane waves.

There are other ways to obtain plane waves. We shall learn about them soon in our study of optics.

18.2 Speed of Sound Waves

The speed of sound waves depends on the material in which they are traveling. Usually the less compressible a material is, the faster sound travels through it. This result is not unreasonable. After all, we would expect tightly packed molecules to pass a compression along faster than loosely packed ones. A metal rod, hit on the end, should transmit the impulse more swiftly than would the gas in a tube. Table 18.1 gives the speed of sound in various materials. Notice that two values are given for solids. As we see from the table, sound travels more slowly down rods than it does through large volumes of the material.

Theory shows that the speed of sound in ideal gases is given by

$$v = \sqrt{\frac{\gamma P}{\rho}} \quad \text{(gases)} \qquad (18.1a)$$

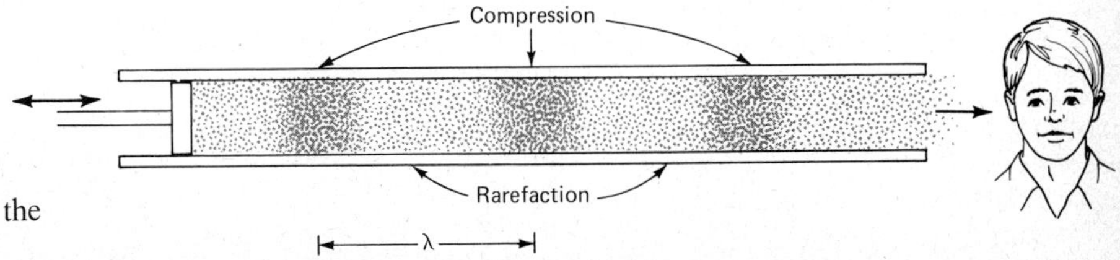

Compression

Rarefaction

λ

Figure 18.1 The oscillating piston generates sound waves that travel to the listener's ear.

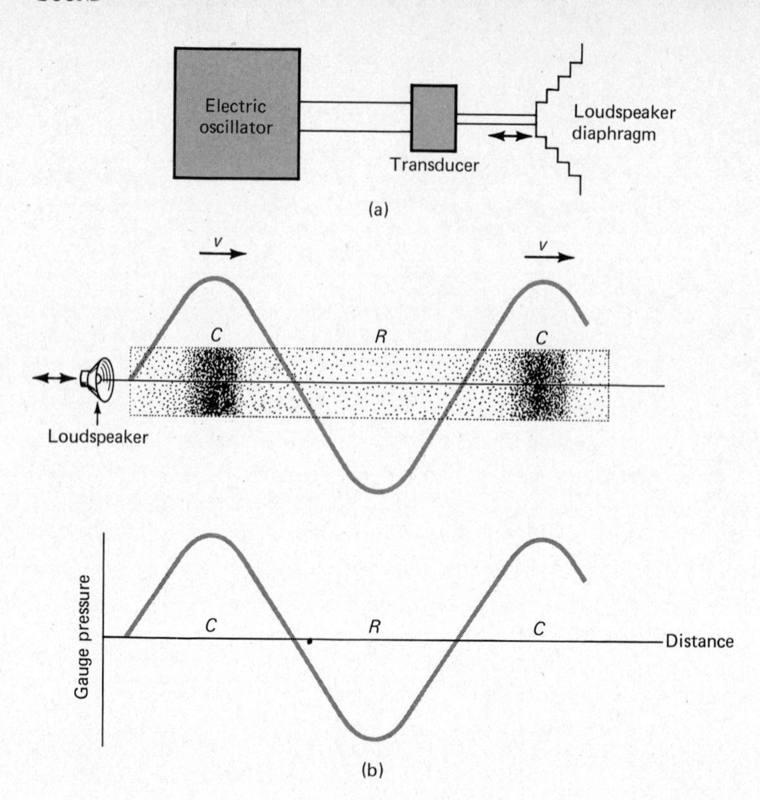

Figure 18.2 The transducer in (a) is enclosed in the loudspeaker housing. A sinusoidal electric signal causes a good loudspeaker to send out a sinusoidal sound wave.

P is the pressure of the gas and ρ is its density. The quantity γ (Greek letter gamma) is related to the nature of the gas. It is about 1.67 for monatomic gases such as He. For O_2 and N_2 and other diatomic gases, it is about 1.40. More complex gases such as CO_2, H_2O (steam), and CH_4 have values close to 1.30.

This equation can be put in a more enlightening form by use of the gas law, $PV = (m/M)RT$. Substituting $\rho = m/V$ gives

$$v = \sqrt{\frac{\gamma RT}{M}} \tag{18.1b}$$

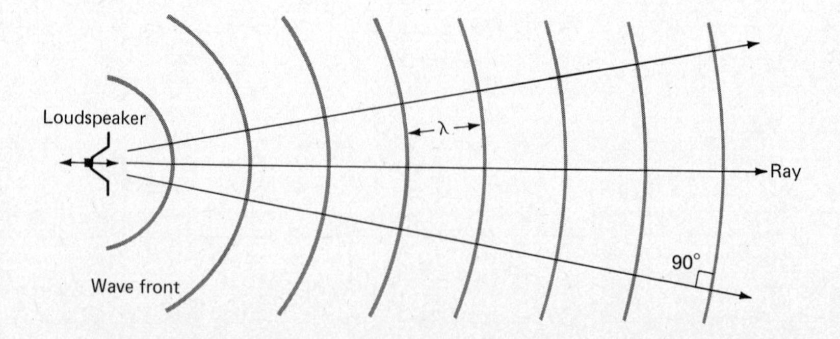

Figure 18.3 The wave fronts are portions of spherical surfaces centered on the loudspeaker. Notice how they become nearly planar at large distances. The energy of the wave flows out along the rays.

Table 18.1 SPEED OF COMPRESSIONAL (SOUND) WAVES

Material	v (m/s)	
Gases (0°C)		
Air		331
Argon		319
Helium		965
Nitrogen		334
Oxygen		316
Steam (134°C)		494
Liquids (25°C)		
Acetone		1,174
Ethanol		1,207
Kerosene		1,324
Turpentine		1,255
Water (pure)		1,498
Water (sea)		1,531
	Bulk	*Rod*
Solids		
Aluminum	6,420	5,000
Beryllium	12,890	12,870
Brass	4,760	3,810
Glass (Pyrex)	5,640	5,170
Iron	5,950	5,120
Lucite	2,680	1,840
Polystyrene	2,350	2,240
Steel	5,850	5,070
Rubber (butyl)	1,830	—
Titanium	6,070	5,080

This equation shows clearly that the speed of sound in an ideal gas depends on temperature but not on pressure.

In the case of air, evaluation of Equation 18.1 gives

$$\begin{aligned} v_{\text{air}} &\simeq 331.5 + 0.61 t_{\text{C}} \quad \text{m/s} \\ &\simeq 1087 + 1.1(t_{\text{F}} - 32) \quad \text{ft/s} \end{aligned} \tag{18.2}$$

where t_{C} and t_{F} are the Celsius and Fahrenheit temperatures. The speed of sound in air increases about 0.6 m/s for each 1°C of temperature rise. For example, at 20°C the speed of sound in air is

$$v = 331.5 + 12.2 = 343.7 \text{ m/s}$$

The speed of sound in large volumes of fluids and simple solids is given by

$$v = \sqrt{\frac{B}{\rho}} \qquad \text{(extended fluids and solids)} \tag{18.3}$$

where ρ is the density of the material and B is the bulk modulus (discussed in Chapter 11).

A somewhat different relation applies to sound waves in rods. There the material undergoes simple extension and compression. As a result, the bulk modulus is replaced by Young's modulus. We have

$$v = \sqrt{\frac{Y}{\rho}} \qquad \text{(rods)} \tag{18.4}$$

That is why two values of v are given for solids in Table 18.1.

EXAMPLE 18.1 A depth sounder on a boat sends a sound pulse straight down toward the bottom of the lake. The reflected pulse reaches the boat 0.013 s after it was sent out. How deep is the lake there?

Solution From Table 18.1 we take the speed of the sound pulse in the lake to be 1500 m/s. It takes a time 0.013 s to travel the distance down to the bottom and back, $2h$. Therefore, $x = \bar{v}t$ gives

$$2h = (1500 \text{ m/s})(0.013 \text{ s})$$

from which the depth

$$h = 9.8 \text{ m}$$

18.3 Sound Intensity

The human ear judges the loudness of sounds qualitatively. Often in technical applications of sound, we must place a quantitative value on loudness. We do this by specifying the intensity of sound. As we shall see, intensity and loudness are related, but the relation between them is not a linear one. Let us now see how sound intensity is defined.

Suppose that a sound wave is moving along the x direction as shown in Figure 18.4. Consider a unit area erected perpendicular to the x axis as shown. The direction of propagation of the wave is perpendicular to this surface. As the wave moves through this unit area, it carries energy through the area. We define the *sound intensity, I,* to be the sound energy that passes through the area each second.

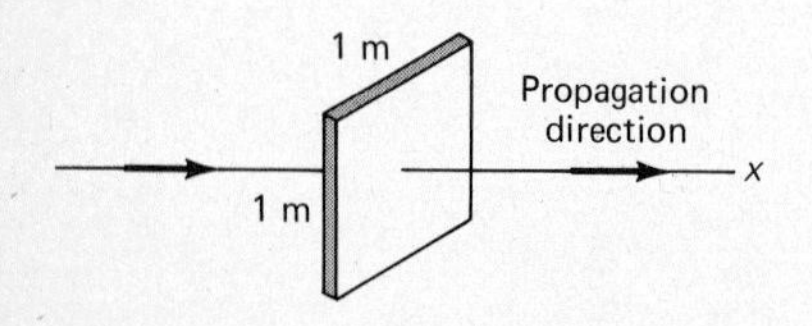

Figure 18.4 Sound intensity is the energy flowing through unit area each second. The area must be perpendicular to the direction of propagation.

Sound intensity, *I*, is the sound energy carried each second through a unit area perpendicular to the direction of propagation.

Notice that intensity is really power per unit area. Its units will normally be watts per square meter (W/m^2). Typical sound intensities are shown in Table 18.2.

Table 18.2 TYPICAL SOUNDS

Situation	Intensity (W/m^2)	Sound Level (dB)
Hearing threshold	10^{-12}	0
Whisper	10^{-10}	20
Average home	10^{-8}	40
Loud party	10^{-6}	60
Noisy traffic	10^{-4}	80
Power mower	10^{-2}	100
Amplified band	10^{-1}	110
Jet engine	10^{2}	140

The ear can hear a tremendous range of intensities. The least audible sound is about 10^{-12} W/m^2. The loudest safe sound is about 10^2 W/m^2. If you think about the sounds you hear, you will realize that the ear is not a linear device. By this we mean that the loudness we hear is not directly proportional to the sound intensity. We do not hear the sound of a rock band to be 10 million times louder than the sounds in an average home. Yet the intensities, according to Table 18.2, do differ by this large a factor. Clearly, the loudness of a sound does not vary in linear proportion to the sound intensity.

Experiment shows that the ear judges loudness in the following way. Suppose that a person is asked to listen to the sounds listed in Table 18.2. She is told to assign a value zero to the lowest intensity sound that she can hear. The sound of a power mower is to be rated as 100 on an arbitrary scale. She is then told to assign loudness ratings to the other sounds using this scale. Her ratings will be close to those given in the "sound level" column of Table 18.2. The scale shown there is therefore a good way to represent sound levels (i.e., the loudness of sound).

The sound level scale shown in the last column of Table 18.2 is called the *decibel scale.* Notice that it increases by 10 units (decibel or dB units) for each factor of 10 that the intensity increases. As an equation, the decibel sound level scale is defined by

$$\text{Sound loudness level in decibels} = 10 \log\left(\frac{I}{I_0}\right) \qquad (18.5a)$$

where I is the sound intensity and I_0 is a reference sound intensity, taken to be 1×10^{-12} W/m^2. This equation can be rewritten as

$$\text{Sound loudness level} = 10(\log I - \log I_0)$$

Because $\log(1 \times 10^{-12}) = -12$, this becomes

$$\text{Sound loudness level in decibels} = 10 \log I + 120 \qquad (18.5b)$$

where I must be in watts per square meter.

Notice that when $I = I_0$, the sound level is zero decibels. The value of I_0 was chosen as 10^{-12} W/m^2, the lower limit of audible sound. Hence a loudness level of 0 dB corresponds to zero audible sound.

As an example of the use of Equation 18.5, consider the sound level close to a power mower. We see from Table 18.2 that I is about 10^{-2} W/m^2 there. Therefore, from Equation 18.5,

$$\begin{aligned}\text{Sound level} &= 10 \log(10^{-2}) + 120 \text{ dB} \\ &= -20 + 120 = 100 \text{ dB}\end{aligned}$$

As we see, the ear responds to the logarithm of the sound intensity. The ear is therefore called a *logarithmic* detector of sound.

Let us now summarize what we have learned about the decibel scale.

The decibel scale assigns a value of zero to the least audible sound. A very loud sound has a value near 100 dB. The sound level in decibels is related to the sound intensity in watts per square meter by Equation 18.5. When the loudness level increases by 10 dB, the sound intensity increases by a factor of 10.

■ **EXAMPLE 18.2** The loudness levels of two sounds are 30 dB and 70 dB. How do their intensities compare?

Solution An increase of 10 dB means that the intensity has increased by a factor of 10. In this case the increase is 40 dB. Hence the intensity increased by a factor $10 \times 10 \times 10 \times 10$. The intensity of one sound is 10^4 times larger than that of the other. ■■

■ **EXAMPLE 18.3** A sound level meter reads a sound to be 67 dB. Find the sound intensity in watts per square meter.

Solution From Equation 18.5 we have

$$\text{Level in decibels} = 10 \log I + 120$$

This becomes

$$67 = 10 \log I + 120$$

from which

$$\log I = -5.3$$

Therefore

$$I = 10^{-5.3} = \frac{1}{10^{5.3}} = \frac{1}{10^5} \times \frac{1}{10^{0.3}}$$

But $10^{0.3}$ is the number whose logarithm is 0.3. Looking for it in the log tables, we find that

$$10^{0.3} = 2.0$$

Therefore

$$I = \frac{1}{2 \times 10^5} = 0.5 \times 10^{-5}$$

As stated in the definition given for the decibel scale, I is measured in watts per square meter. Therefore, $I = 0.5 \times 10^{-5}$ W/m^2. ▮▮

18.4 Response of the Ear

In the last section we learned that the ear is a logarithmic sound detector. Let us now see how it responds to different frequencies of sound. We can generate pure sine wave sounds by use of an electric oscillator and loudspeaker (see Figure 18.2). Set the generator to the frequency you want to test. Then the intensity of the sound is increased until the listener can just barely hear it. We call this limiting sound level the *threshold of hearing* for that frequency.

The sound level for a frequency that is just barely audible is called the *threshold of hearing.*

Each person has his or her own individual threshold of hearing. Approximately 1 percent of the people in the United States have their hearing threshold below the bottom curve in Figure 18.5. About 50 percent have their hearing

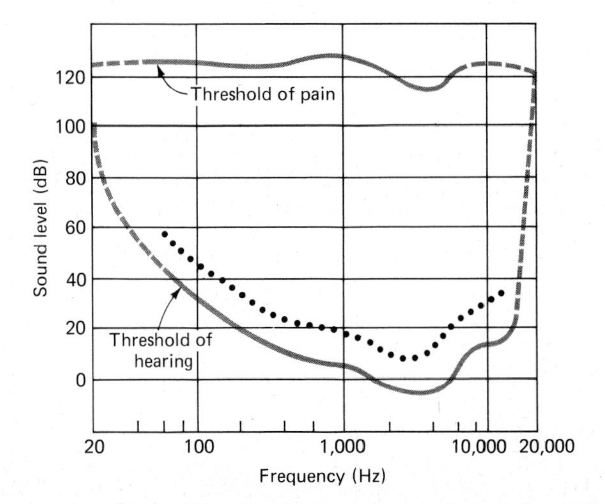

Figure 18.5 About 1 percent of the population can hear sounds at the level indicated by the lower solid curve. The dotted curve gives the same data for 50 percent of the population.

threshold below the dotted curve. As we see, the ear is most sensitive to sounds near 3000 Hz. At very low and high frequencies, the sound must be very loud to be audible.

When a sound is very loud, it is felt as a painful sensation. Most people respond in this way to the sound levels given by the upper curve in Figure 18.5. If a sound is much louder than this, the ear may be damaged by it. In fact, continued exposure to sounds above about 100 dB can cause serious injury to the ear. The hearing ability of those subjected to long exposure to intense sounds is greatly decreased. Sound levels in factories are now controlled by law in many places. However, in this age of the electric sound amplifier, many situations exist where ear damage can, and does, occur.

As people grow older, their hearing sensitivity decreases. Usually their response to high frequencies decreases first. Some people suffer ear injury during childhood diseases. Frequently these people cannot hear frequencies above about 6000 Hz. Because most sounds are a mixture of both low and high frequencies, these people can still hear most sounds. For this reason they often do not realize they have a hearing problem until a frequency response test is given to them. It would be of interest to test the students in your class for this impairment. The whole class can be tested at once by use of a variable-frequency oscillator connected to a good loudspeaker.

18.5 Quality of Sounds

We all know that people's voices differ. Even if they sing the same note, their sounds are not the same. This is also true for musical instruments. A clarinet and a trumpet sound different even when they play the same note. We have no difficulty in distinguishing between a violin and a trombone when they play the same musical tone. Let us see why this difference in sound occurs.

The sounds we hear are due to the sound waves that strike the ear. To analyze these sounds, we can change them to electrical signals. These signals can then be displayed as graphs. In that way we can obtain a graph of a sound wave. One way for doing this is shown in Figure 18.6.

A nearly pure sinusoidal sound can be obtained from a loudspeaker. If the loudspeaker is driven by an electrical oscillator, its sound wave is as shown in part (a) of Figure 18.7. Waves from other sound sources are shown in parts (b), (c), and (d). Most musical instruments give waves somewhat more complicated than the one shown for the piano. In fact, the waveform one obtains from a piano varies depending on many factors.

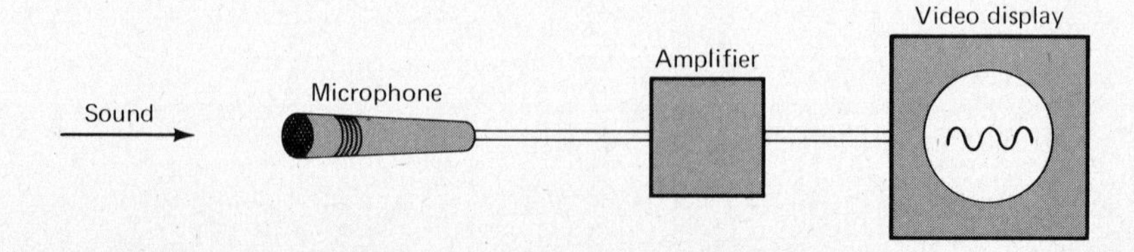

Figure 18.6 Sound waves can be displayed on a screen by electrical means.

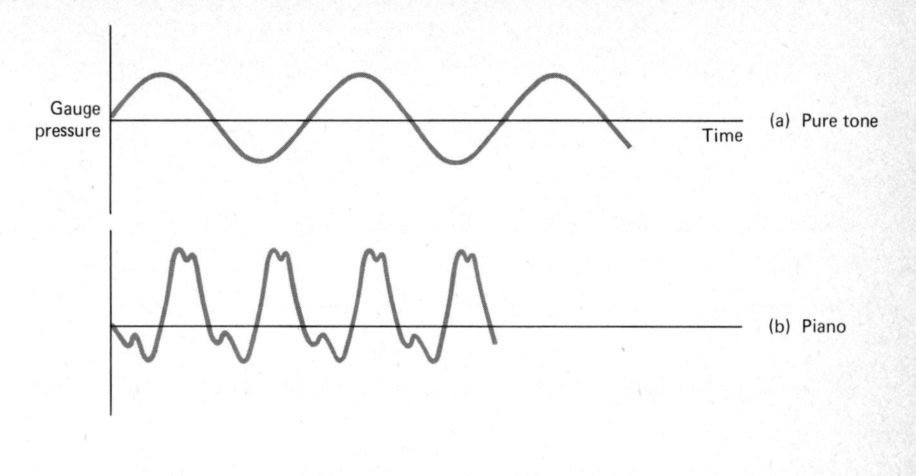

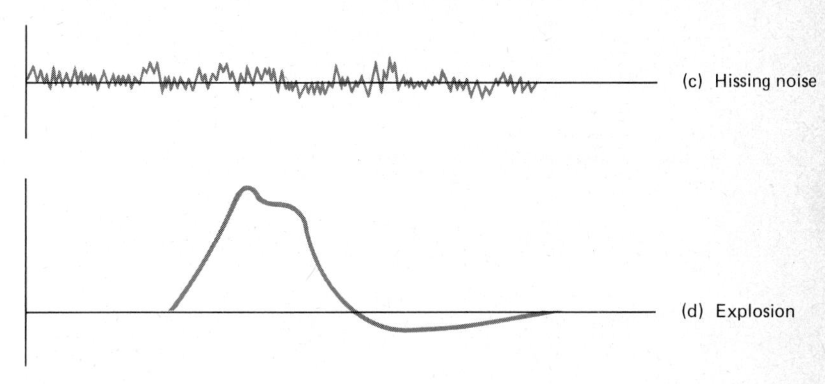

Figure 18.7 Each type of sound produces a different graph. Both (a) and (b) give rise to periodic vibrations of the eardrum and are therefore heard as definite tones. The eardrum simply quivers erratically under the action of the hissing noise. An explosion causes a sudden pulse to hit the eardrum, and we hear something quite different from the other sounds.

In the last chapter we pointed out that a complex wave can be thought of as the sum of sinusoidal waves. This forms the basis for analyzing sounds. For example, consider the wave shown in Figure 18.8(a). It is the sound wave given off by a certain violin string when bowed smoothly. This complex wave can be thought of as being the sum of several waves. It can be duplicated by adding together five pure sine waves. These waves have frequencies in the ratio of 1:2:3:4:5. Their amplitudes are shown in Figure 18.8(b). The first frequency in part (b) is that of the fundamental vibration of the string. The others are the

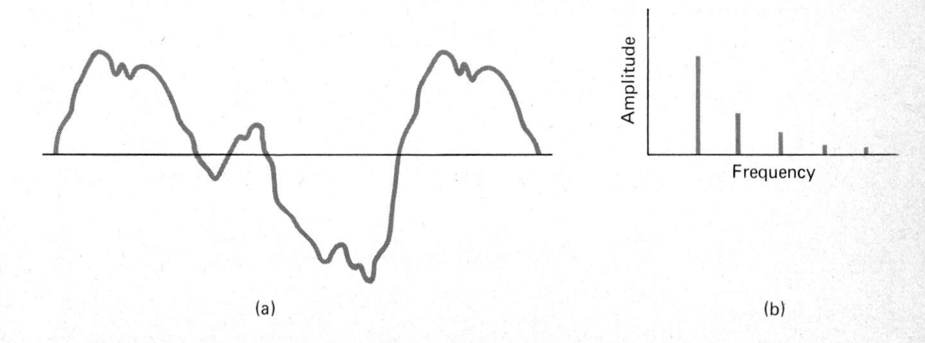

Figure 18.8 Part (a) shows the sound wave from a violin string. It consists of the frequencies shown in (b).

first, second, third, and fourth overtone vibrations. We conclude the following from the sound given off by the string: The string vibrates in five modes at one time; the sound wave it sends out is the sum of the sine waves sent out by these five simultaneous vibrations.

Each different sound we hear can be Fourier-analyzed in this way. Identical sounds must have identical component waves. Different sounds are composed of different component waves. Let us take a specific case.

Suppose that a piano and a violin sound the same note, middle C. The fundamental frequency of vibration of the string in both instruments is 264 Hz. But the two instruments differ in the sound waves they emit [compare Figures 18.7(b) and 18.8]. This means that the two strings vibrate differently in regard to their overtones. Either the amplitudes or the frequencies of the overtones are not in the same ratio in the two cases. Or perhaps both relative amplitudes and frequencies differ. The net result is that the two tones are different sounding.

As we see, the composition of a sound wave determines how it will sound. These differences in the way identical notes sound are called differences in *quality* of the sound. When a beginner bows a violin string, the quality of the sound may be quite different from the quality an expert would obtain. The beginner's sound might be rasping and unpleasant. Not so for the expert. Although the fundamental frequency is the same in the two cases, the overtones differ markedly. We therefore conclude that

The quality of a sound is determined by the number and relative amplitude of the overtones present.

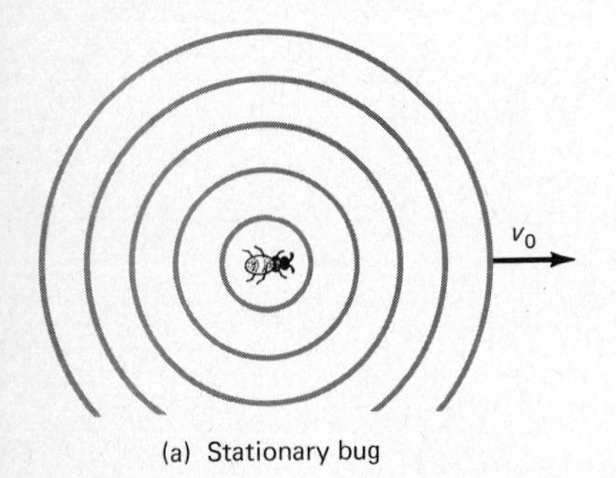

(a) Stationary bug

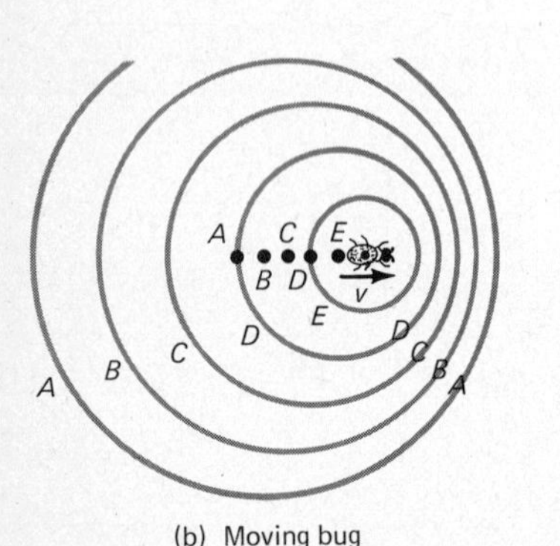

(b) Moving bug

Figure 18.9 The water waves sent out by a stationary bug are as shown in (a). When the bug is moving, the wave crests are as shown in (b). The wavelength of the wave depends on direction.

18.6 Doppler Effect

Suppose that a water bug sits on the smooth surface of a pond as shown in Figure 18.9(a). To while away the time, it keeps bobbing its head up and down with simple harmonic motion. This motion causes circular wave crests to move out across the water as shown. Let us say that the bug is bobbing its head with period T_0 and frequency f_0. The speed of the waves on the water surface is v_0. We can easily find the wavelength of the waves.

$$\lambda_0 = v_0 T_0$$

Now suppose that the bug starts to swim to the right with speed v as shown in part (b). It is still bobbing its head with frequency f_0. As before, it sends out circular wave crests. But now they each have a different center. Each comes from where the bug was when the crest was sent out. Notice that now the crests are not the same distance apart as they were in (a). Because the distance between crests is the wavelength, we see that the wavelength has changed. Even though the frequency of the source is still f_0 and the speed of the waves is still v_0, the wavelength is no longer λ_0.

We saw in the previous chapter that the relation $\lambda = vT = v/f$ is true for all waves. It applies to these waves as they pass by a leaf floating on the water. The frequency with which the leaf will bob up and down is, from $\lambda = v/f$,

$$f = \frac{v}{\lambda}$$

Therefore the leaf will experience a different frequency depending on where it is on the water surface. If it is to the right of the bug in Figure 18.9(b), waves with small λ will strike it. It will therefore bob up and down with a high frequency. But if it is to the left, waves with large λ will strike it, and it will bob up and down with low frequency.

These same ideas apply to all sources of waves. For example, they apply to a car horn. If the horn is blown while the car is stationary, the ear hears waves with frequency f_0. But if the car is moving toward the listener [see Figure 18.10(a)], the listener hears a higher frequency sound from the horn. If the car is moving straight toward the listener with speed v, then the frequency heard will be (see Problem 41)

$$f = f_0\left(\frac{v_0}{v_0 - v}\right) \qquad \text{(moving source approaching)} \qquad (18.6a)$$

where v_0 is the velocity of the waves.

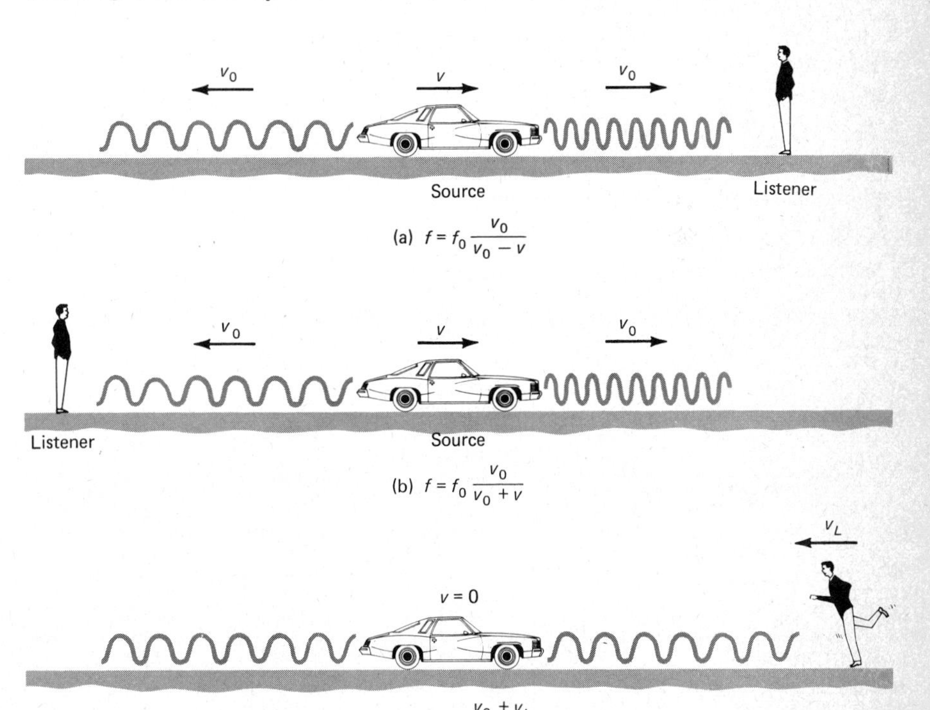

Figure 18.10 The observed frequency of a sound depends on the relative motion of the source and the observer.

If the car horn is moving away from the listener, as in Figure 18.10(b), fewer waves will strike the ear each second. The frequency heard will be

$$f = f_0\left(\frac{v_0}{v_0 + v}\right) \qquad \text{(moving source receding)} \qquad (18.6b)$$

Notice that the frequency heard will be less than that of the stationary horn.

A similar situation exists if the sound source is stationary but the listener is moving. For example, in Figure 18.10(c) the listener is running toward the stationary horn with speed v_L. His ears will therefore encounter more wave crests each second than they would if he stood still. He will hear a sound with frequency higher than that of the horn. It is given by

$$f = f_0\left(\frac{v_0 + v_L}{v_0}\right) \qquad \text{(moving listener approaching)} \qquad (18.6c)$$

Similarly, if the listener is running away from the horn, he will hear a lower frequency. It is given by

$$f = f_0\left(\frac{v_0 - v_L}{v_0}\right) \qquad \text{(moving listener receding)} \qquad (18.6d)$$

These four relations can be combined into one that includes all cases. It is

$$f = f_0\left(\frac{v_0 \pm v_L}{v_0 \mp v}\right) \qquad (18.6)$$

In this expression the speed of the waves is v_0, the source speed is v, and the listener's speed is v_L. The upper sign in each case is to be used if the velocity is one of approach. If it is one of recession, then the lower sign is to be used.

This general phenomenon is called the Doppler effect. We can summarize it qualitatively as follows.

The *Doppler effect* is a shift in the frequency of a wave received caused by relative motion of the wave source and the wave receiver. Approaching motion causes a rise in frequency. Receding motion causes a drop in frequency.

A similar situation occurs with light waves. By use of it, we are able to measure the recession speed of distant stars.

An interesting limiting case occurs when the speed of the bug in Figure 18.9 is the same as the speed of the waves. Then the waves at the right all pile up on top of each other. This causes a large amount of wave energy to be concentrated in the area just ahead of the wave source, the bug in this case. We call such a concentration of waves a *shock wave.*

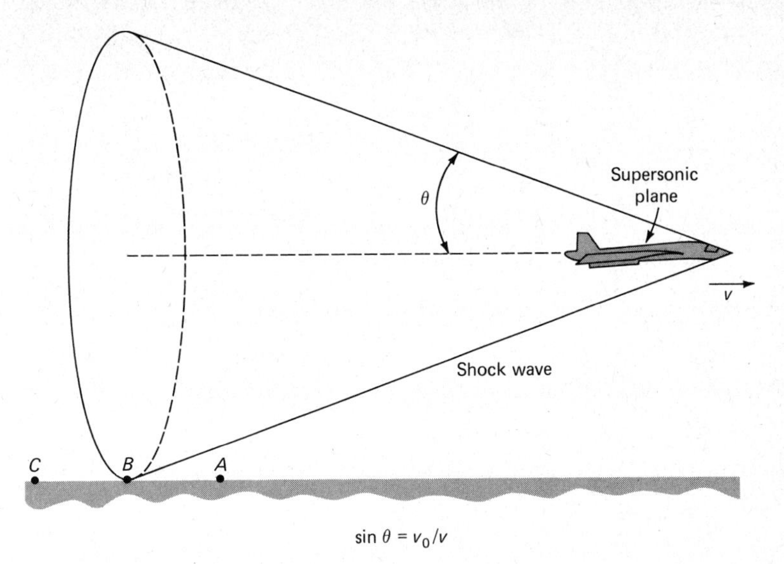

Figure 18.11 A shock wave trails a supersonic craft.

Fast-moving aircraft traveling at or above the speed of sound give rise to shock waves in the air. The shock wave of a supersonic plane forms a conical surface with the plane at its apex as shown in Figure 18.11. When the shock wave passes by us, we hear what is called a *sonic boom.* In the figure the sonic boom has just been heard at *C,* it is now occurring at *B,* and soon those at *A* will experience it.

■ **EXAMPLE 18.4** A bat locates its prey by sending out ultrasonic waves and noticing how they are reflected. Suppose that a bat is moving east at a speed of 20 m/s while sending a sound wave of frequency 50,000 Hz ahead of it. Another bat, coming toward it with a speed of 15 m/s, hears the wave. What frequency does the second bat hear?

Solution The general formula for the Doppler effect is Equation 18.6:

$$f = f_0\left(\frac{v_0 \pm v_L}{v_0 \mp v}\right)$$

In our case $v_L = 15$ m/s and $v = 20$ m/s. We shall take the speed of sound to be 340 m/s. Because the listener is approaching the source, we use the upper (+) sign for v_L. Because the source is moving toward the listener, use the upper (−) sign for v. Then, since $f_0 = 50{,}000$ Hz, we have

$$f = 50{,}000\left(\frac{340 + 15}{340 - 20}\right)\text{Hz} = 55{,}500 \text{ Hz}$$

■■

■ **EXAMPLE 18.5** The Doppler formula applies to electromagnetic waves such as radar provided the object's speed is far less than the speed of light. A police off[illegible] sitting along a roadside is tracking an oncoming car with radar. If the speed of th[illegible] 30 m/s, what is the ratio of the frequency for the radar waves reflected from[illegible] the frequency of the waves sent out by the officer?

Solution We first calculate the frequency of the waves as observed by the driver of the moving car. We have, with the speed of radar waves being 3×10^8 m/s,

$$f_{car} = f_0\left(\frac{v_0 + v_L}{v_0}\right) = f_0\left(\frac{3 \times 10^8 + 30}{3 \times 10^8}\right) = f_0(1 + 1 \times 10^{-7})$$

This is the frequency of the waves reflected to the officer. These reflected waves act as though they come from a moving source, the car. The frequency received by the officer is

$$f = f_{car}\left(\frac{v_0}{v_0 - v}\right) = f_{car}\left(\frac{3 \times 10^8}{3 \times 10^8 - 30}\right) = f_{car}\left(\frac{1}{1 - 1 \times 10^{-7}}\right)$$

Substituting for f_{car} gives

$$\frac{f}{f_0} = \frac{1 + 1 \times 10^{-7}}{1 - 1 \times 10^{-7}} \cong (1 + 2 \times 10^{-7}) = 1.0000002$$

where use has been made of the mathematical fact that $1/(1 - x) \cong 1 + x$ for $x <<< 1$.

Notice that the frequency is shifted by only 0.2 parts in 1 million. Even though this is a very small shift, it is easy to detect by the use of beats between f_0 and f. We shall discuss beats in Section 18.7. ∎∎

18.7 Beats Between Waves

As we have seen, waves obey the superposition principle. This allows us to analyze a sound wave into its various sinusoidal waves when we describe the quality of sound. Now we will proceed one step further and discuss the interference of waves from two separate sound sources.

Suppose, as shown in Figure 18.12, two loudspeakers send identical sinusoidal waves to a microphone. Or, if you prefer, the microphone could be replaced by your ear. The waves will travel to the microphone and meet there. Notice that in the case shown a crest of the wave on the left will meet a trough of the wave on the right. Indeed, the two waves will continue to meet at the microphone with crest on trough and trough on crest. Their combined effect is to cancel each other. So, in this case, no sound will be heard at the microphone. We call such a situation *complete destructive interference.* We say that the waves are half a wavelength (or 180°) *out of phase.*

Of course, if the loudspeaker on the left is moved $\frac{1}{2}\lambda$ closer to the microphone, then crests of the two waves will arrive at the microphone together. Then

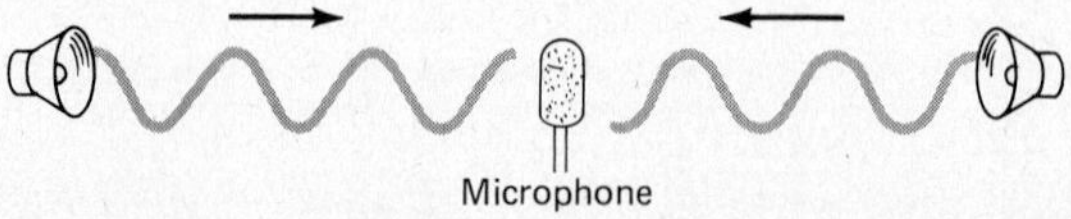

Figure 18.12 The two sets of waves cancel each other at the microphone.

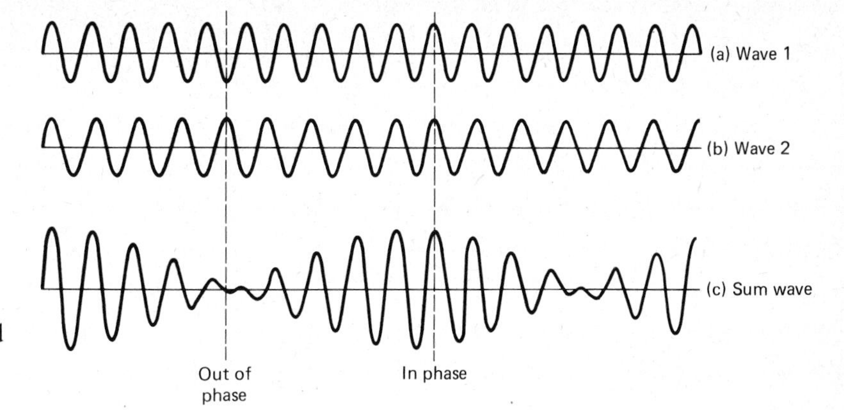

Figure 18.13 Two waves of slightly different frequency combine to give beats. The number of beats each second equals the frequency difference of the waves.

crest will fall on crest and trough will fall on trough. The two waves will reinforce each other and a loud sound will be heard at the microphone. We call this situation *constructive interference* and we say that the two waves are *in phase.* More will be said about this terminology when we study the interference of light.

Let us now suppose that the two sources in Figure 18.12 do not have quite the same frequency. Their waves are no longer of the same length. As a result, the waves cannot continuously reinforce each other at the microphone. If at some instant they meet crest on crest, this reinforcement will soon be lost. This happens because the next crest of the shorter wavelength wave takes less time to reach the microphone than the next crest from the longer wave.

This feature is shown more clearly in Figure 18.13. The two waves joining at the microphone are shown in (a) and (b). Their sum is shown in (c). Notice how sometimes the waves reinforce and sometimes they cancel. When you listen to these two sounds together, you will hear the sum wave. This sum wave will be alternately loud and soft and you will hear what are called *beats.*

An analysis of this situation shows that the number of beats per second, the beat frequency, equals the difference in frequencies between the two waves. For example, suppose one loudspeaker has a frequency of 600 Hz and the other's is 603 Hz. Then their combined sound will beat with a frequency of 3 Hz. Beats provide an extremely sensitive way to compare frequencies. Musicians tune their instruments so as to reduce the beat frequency to zero.

■ **EXAMPLE 18.6** In Example 18.5 we found that $f/f_0 = 1.0000002$ for the original and reflected radar beams. If $f_0 = 1 \times 10^{10}$ Hz, find the beat frequency between the two beams.

Solution We have

$$f = 1.0000002\, f_0 = 10{,}000{,}002{,}000 \text{ Hz}$$

Thus $f - f_0 = 2000$ Hz. The beat frequency is equal to the difference in frequency between the two waves. We therefore see that the beat frequency is 2000 Hz. Police radar measures the beat frequency to determine the speed of the oncoming car. ■■

Summary

Sound waves are longitudinal vibrations in gases and other materials. The normal ear can hear sound waves with frequencies in the approximate range $30 < f < 17{,}000$ Hz. Ultrasonic waves have frequencies higher than the audible range.

The location of the crest of a wave is called a wave front. The wave fronts from a distant source are nearly flat planes. Such a wave is called a plane wave.

Energy in a wave travels out from the source along rays. The rays are always perpendicular to the wave fronts.

The speed of sound waves is usually low in highly compressible materials. It is higher in materials that are less compressible. In air at 0°C the speed of sound is 331.5 m/s. It increases by 0.61 m/s for each 1°C temperature rise.

Sound intensity is a measure of the energy flow through a unit area in unit time. The intensity I is the sound energy flowing per second through a unit area erected perpendicular to the propagation direction. It is measured in watts per square meter. The weakest sound normally audible is about 10^{-12} W/m^2. Sound intensities above about 10^2 W/m^2 cause the ear to feel pain.

The ear is a logarithmic device. For example, two sounds whose intensities differ by a factor 10 do not appear to differ in loudness by this amount. Sounds of intensities 10^{-12}, 10^{-11}, 10^{-10}, and 10^{-9} W/m^2 are heard to have loudness in approximate ratio to 0, 10, 20, and 30, respectively.

A scale that approximates the loudness of sounds is the sound level scale. It uses decibels (dB) as the unit of measure. If I is the intensity of a sound in watts per square meter, then the

$$\text{Sound level in decibels} = 10 \log I + 120$$

On this scale the least audible sound is at about 0 dB, whereas a pain-producing sound is near 120 dB.

The average human ear is most sensitive to sounds near 3000 Hz. Its sensitivity drops rapidly toward zero at frequencies near 30 Hz and 17,000 Hz. The sound level that is just barely audible is called the hearing threshold.

Sounds above 100 dB can cause damage to the ear. The damage increases as the time of exposure increases.

Even though two sound sources may be giving off tones with the same fundamental frequency, they may sound different. The quality of a sound is determined by the number and relative amplitude of the overtones present.

Relative motion of a wave source and wave receiver causes a shift in the frequency received. This result is called the Doppler effect. Approaching motion causes a rise in frequency. Recession causes the frequency to be lowered. If the source frequency is f_0, the source speed is v, the receiver speed is v_L, and the speed of sound is v_0, then the received frequency is

$$f = f_0\left(\frac{v_0 \pm v_L}{v_0 \mp v}\right)$$

The upper sign is used if the velocity is one of approach. If it is one of recession, the lower sign is used.

An object moving through air at a speed faster than sound causes a shock wave to form. Trailing behind the object is a cone. The surface of the cone is the position of the shock wave. The larger the speed of the object, the smaller is the apex angle of the cone. When the shock wave passes by a point, it gives rise to a sonic boom.

When combined, two similar waves of slightly different frequency give rise to beats. The beat frequency is equal to the difference in frequency of the two waves.

Questions and Exercises

1. Ultrasonic devices are widely used to clean intricate objects. The intense ultrasonic waves are sent through a cleaning solution in which the object is immersed. Explain how the cleaning action occurs.
2. The sound from a vibrating tuning fork can be made much louder by holding the handle of the fork tightly against the top of a table with the prongs away from the table. Explain why the sound is so greatly increased.
3. Ships and submarines use sonar to "see" under water. Sound pulses are sent out and the "seen" object reflects them back. Bats also use a form of sonar to see in pitch-black caves. Explain how such methods work.
4. People who live near air terminals are victims of noise pollution. Why would this not be a problem for astronauts living close to a landing site on the moon?
5. An underwater sound source vibrates with a fixed frequency of 2000 Hz. The sound is intense and can be heard by a person standing in the air nearby. Compare the frequency heard to the frequency of the source.
6. A famous opera singer was said to be able to shatter a fine glass goblet with his voice. He sang the note with which the glass vibrated when gently tapped. What features of his voice and the glass made this trick possible?
7. The sound of your voice is greatly influenced by resonances within cavities in your throat, mouth, and nose. Make the sound "ooo" with your voice and change it to "ahhh" by opening your mouth more. What happened to cause this change in sound? Did you change the frequencies of the sound? The quality?
8. Electronic sound simulators can duplicate the sound of nearly any musical instrument. Suppose that you were given five variable-frequency oscillators that could simultaneously drive a single loudspeaker. How could you synthesize different sounds?
9. Refer to the bug in Figure 18.9(b). Suppose there is a wall at the right edge of the pond. Compare the frequencies with which the waves hit various points along the wall.
10. In a shock wave the wave crests are piled on top of each other. The distance between crests, λ, is small. Therefore, because $f = v/\lambda$, the frequency is high, too high to be audible. Therefore, a shock wave cannot be heard. Where did we go wrong in this reasoning?
11. Common ways to decrease the sound level in a noisy room are as follows: Open the windows; hang curtains or drapes on the walls; cover the ceiling with a porous material; put a rug on the floor; add more (but quiet) people to the room. Explain why each method helps.
12. The reverberation time of an auditorium is (roughly) the time taken for the sound of a sharp handclap to die out. Why is the reverberation time for an empty auditorium longer than for a fully occupied one? What factors influence the reverberation time? Compare the effects of reverberation time on the suitability of a room for a concert hall and a lecture hall.
13. Beatlike behavior can occur for many vibrating systems. Suppose two pendulums have natural frequencies of 10 and 12 Hz. They are started in unison. Explain why they will be vibrating again in unison after $\frac{1}{2}$ s. What is their beat frequency? Extend this reasoning to pendulums with frequencies of 50 and 58 Hz.

Problems

1. Sound waves cover the frequency range from about 20 Hz to 20 kHz. To what range in wavelengths does this correspond? Assume air at 0°C.
2. What is the wavelength in centimeters of a 10,000-Hz sound wave in air at 0°C? In water?
3. Ultrasonic cleaning baths use frequencies near 50 MHz.

What is the wavelength of these waves in water? Notice that the wavelength is comparable to the size of the dirt particles that are being removed.

*4. An underwater sound source has a frequency of 8000 Hz. What is the wavelength of its sound waves in water? The sound leaves the water and enters the air. What is its wavelength in 0°C air?

5. Geologists learn about the structure of the earth by timing sound pulses (one form of seismic waves) sent into the earth. It is found that the pulse from an explosion at the earth's surface returns to the explosion site after 5.2 s. How far below the surface is the layer that reflected it? Assume the speed of the pulse to be 5600 m/s.

6. Seven seconds after a lightning flash, thunder is heard. How far away was the lightning flash? Assume the speed of sound to be 340 m/s and the speed of light to be nearly infinite.

7. What is the speed of sound in air at 30°C?

8. A tube filled with air is open at one end and closed at the other. Its fundamental resonance frequency is 660 Hz when the air is at 20°C. What will be its resonance frequency when the air temperature is 30°C?

* 9. An organ pipe resonates to a sound of frequency 264.0 Hz at 20°C. If the pipe is used on a cold day when the air temperature in it is 12°C, what will be its resonance frequency?

10. Use Equation 18.1 to find the speed of sound in hydrogen gas at 0°C. For hydrogen gas $M = 2$ kg/kmol and the gas is diatomic.

11. Argon gas is monatomic and has $M = 40$ kg/kmol. Find the speed of sound in argon at 0°C.

12. Find the speed of compressional waves through mercury. For mercury the bulk modulus is 2.8×10^{10} N/m^2.

13. Using the fact that the speed of sound in water is 1500 m/s, find the bulk modulus of water.

14. The speed of sound in a certain type of brass rod is 3800 m/s. Make use of this to find Young's modulus for the brass.

15. A mountaineer notices that a distant mountain face echoes her handclap 5 s after she claps her hands. How far away is the mountain face?

16. The lecture platform in a large auditorium is 50 m from the rear row of seats. The lecturer's voice reaches these seats by two means, a loudspeaker mounted there and the usual path through the air. A simple loudspeaker system responds almost instantaneously to the speaker's voice. Find the time difference between the sounds from the two paths. (Use $v = 340$ m/s.) This delay is usually compensated for electrically in good audio systems.

17. A plane sound wave with intensity 2.5×10^{-7} W/m^2 has its propagation direction perpendicular to a window. The window is open and has an area of 3500 cm^2. How much sound energy is lost out the window in a minute?

18. The intensity of the sound at a certain loudspeaker is 0.021 W/m^2. How much energy is carried from the speaker each second by the sound waves? The loudspeaker is circular and has a 4-cm radius.

19. The sound intensity in a certain factory is 0.13 W/m^2. How much energy strikes a person's ear each second? Assume the area of the person's ear is 12 cm^2.

**20. A loudspeaker sends out 15 W of power in the form of sound waves. The sound goes out uniformly into a hemisphere with the speaker at its center. Find the sound intensity at a distance of 2.0 m from the speaker. What is the sound level there in decibels?

**21. For the sound source described in Problem 20, show that the sound intensity at a radius r_1 from the source is related to that at r_2 by $I_1/I_2 = (r_2/r_1)^2$.

*22. A record player that uses 25 W furnishes sound to a single loudspeaker that has an opening area of 60 cm^2. When the speaker is blaring out sound with an average intensity of 0.5 W/m^2 at its opening, how efficient is the phonograph system in producing sound energy?

**23. Suppose you were to talk loudly and continuously for 5 h. Let us say that the average sound intensity coming from your 6-cm^2 mouth opening is 0.01 W/m^2. (a) How much sound energy would you pour out in this time? (b) How many nutritionist's Calories of food energy must you eat to supply this much energy?

24. From Figure 18.5, about what is the intensity of (a) a 200-Hz sound at the threshold of hearing and (b) a 2000-Hz sound at the threshold of pain?

25. What are the sound levels in decibels for the following intensities: (a) 10^{-7} W/m^2, (b) 10^3 W/m^2, (c) 10^{-14} W/m^2?

26. In order to increase the sound level of a hi-fi set from 90 to 100 dB, by what factor must the intensity of the sound be increased? To increase it from 90 to 110 dB?

*27. What is the sound level in decibels of a sound whose intensity is 6×10^{-7} W/m^2?

*28. What is the sound level in decibels of a sound with intensity 3.7×10^{-5} W/m^2?

*29. In a certain factory the sound level is read to be 83 dB. Find the sound intensity in watts per square meter.

*30. Just outside the loudspeaker of a hi-fi system, a sound level meter reads 94 dB. What is the sound intensity in watts per square meter?

**31. The sound level at a distance of 3.0 m from a loudspeaker is 95 dB. Find the sound intensity there. If this same sound

*Problems marked with an asterisk are not as easy as the unmarked ones.
**Problems marked with a double asterisk are somewhat more difficult than the average.

level existed in all directions from the speaker, how much sound power is the loudspeaker putting out?

*32. A 45-rpm phonograph record is mistakenly played at 33 rpm. When a singer sings middle C (264 Hz) on the record, what frequency is given off by the phonograph?

33. A car horn is sounding at 400 Hz as it drives down a straight road at a speed of 20 m/s. What frequency does a hitchhiker hear as the car approaches him? Recedes from him? Assume v of sound to be 340 m/s.

*34. About how fast must a car be approaching a stationary listener if the sound of its horn is to be inaudible because its frequency is too high? How fast must it be receding for the sound frequency to be too low? The horn's frequency is 400 Hz. Use 340 m/s as the speed of sound.

*35. Two motorcyclists race toward each other at speeds of 25 m/s. Their identical horns blare out a 900-Hz tone. What frequency tone do they hear from each other when approaching? When receding from each other? Take the speed of sound to be 340 m/s.

**36. A car is moving toward a wall at a speed of 20 m/s. It is blowing its horn, which has a frequency of 1000 Hz. What frequency sound reflected from the wall is heard by (a) a listener standing some distance from the wall close to the path of the car and (b) the driver of the car? Assume the speed of sound to be 340 m/s.

37. A 1000.0-Hz tuning fork is sounded simultaneously with a fork of unknown frequency. A beat is heard every 0.70 s. What are the possible frequencies of the unknown fork?

*38. A violinist is tuning her violin to 440.00 Hz by comparison with a piano. She notices one beat each three seconds. She can decrease the beat frequency by tightening the violin string. What was the original frequency of the string?

*39. Two identical organ pipes have fundamental frequencies of 240.00 Hz at 20°C. If they are sounded together when one is at 20°C and the other at 15°C, what will be the beat frequency between their two sounds?

**40. The speed of a receding baseball is being measured by 1×10^{10}-Hz radar. If the beat frequency between the reflected and original wave is 1400 Hz, what is the speed of the ball?

**41. To derive the Doppler equation for an approaching sound source, show that the wavelength in front of the source in Figure 18.10 is $\lambda = \lambda_0 - (vT)$ where T is the period of the source. Replace the λ's in terms of frequencies and find the Doppler result.

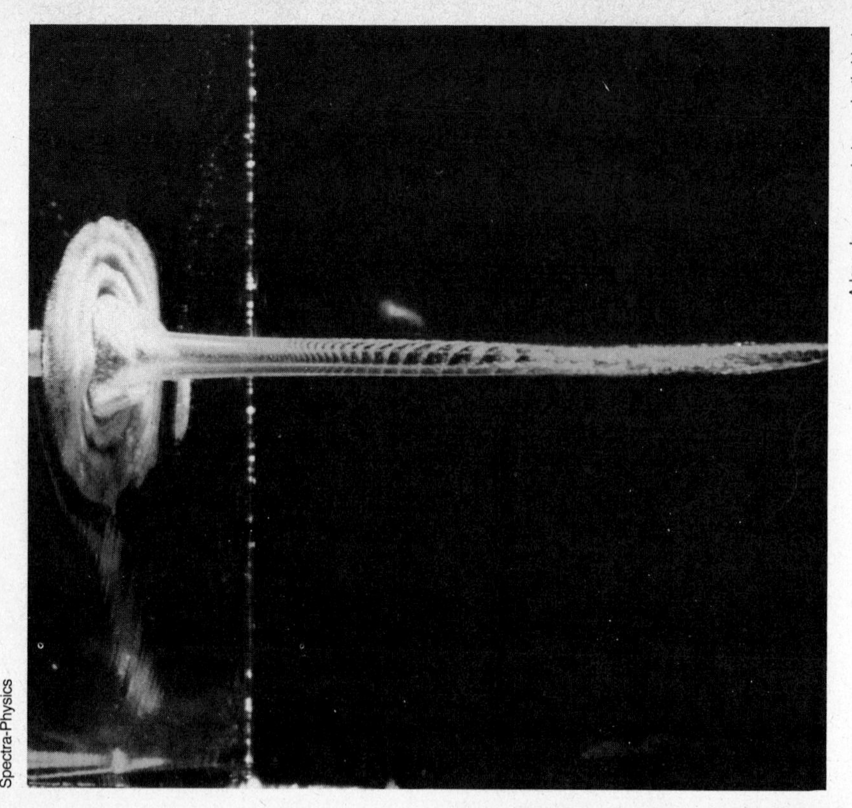

A laser beam penetrates a plastic block.

Spectra-Physics

19 LIGHT WAVES

Performance Goals

When you finish this chapter, you should be able to

1. Select those waves that are em waves from a list of wave types.
2. List a given series of em wave types in order of increasing or decreasing wavelengths. Match this series with a list of wavelength ranges.
3. Give an instantaneous graph of an em wave and the speed of an em wave in a vacuum. State whether the wave is longitudinal or transverse, what is vibrating,

In this chapter we introduce the subject of electromagnetic waves. Typical waves of this type are light and radio waves. We shall see that there are many others. Our primary concern in this chapter will be with light. Topics such as wavelength, speed, color, polarization, and types of light sources will be discussed. In the following chapters we shall learn how lenses, mirrors, prisms, and diffraction gratings are used in technical application of waves.*

*The chapters on light may be postponed until after the study of electricity.

and what is meant by the **E** vector.

4. Give the wavelength limits of the visual spectrum in the following units: meters, nanometers, angstroms, millimicrons.
5. Distinguish between continuous and line sources of light. List two examples of each. State what is meant by a white light source.
6. Discuss qualitatively what one sees as the temperature of an object is changed from about 500°C to 3000°C. In connection with the discussion sketch a rough graph showing the intensity of radiation emitted by a hot source as a function of wavelength.
7. Tell qualitatively how emission spectroscopy can be used to analyze a vaporized material.
8. Make a sketch that shows the shadow of an object on a screen. Label both the umbra and the penumbra. Point out the difference in shadows from a point and an extended source.
9. Sketch a light beam in such a way as to indicate a ray and a wave front on it.
10. Give the approximate color the eye sees when exposed to any single wavelength of light. Explain why the color of an object cannot usually be specified by a single wavelength.
11. Define the following: standard luminosity curve, luminous intensity, flux, illuminance, luminous efficiency. Where applicable, give the appropriate SI unit.
12. Explain the difference between

19.1 Nature of Electromagnetic Waves

Water waves travel through water. Waves on a string move along the string. Sound waves travel through air or some other material. All the waves we have discussed so far consist of vibrations of a material of some sort. The waves cause atoms and molecules in their path to vibrate back and forth or up and down. All these waves need some form of material in which to travel. Obviously, a water wave cannot exist if there is no water, a wave on a string requires a string, and sound waves need air or some other material. These waves *cannot* travel through vacuum, because no material exists in a vacuum.

Electromagnetic waves (em waves) are not like these other waves. Electromagnetic waves require no material in which to vibrate. They travel best in the absence of all material, in vacuum. Clearly, the vibration in em waves is *not* a vibration of atoms or molecules. It is a vibration of a force vector, not a material. Let us now see what this vibrating force is.

In Figure 19.1 you see an experiment you may have done in grade school. It illustrates the force we shall be concerned with. The experiment shows the effects of like charges on each other. You will recall, perhaps, that "like charges repel" each other. If a glass rod is rubbed with silk, the rod acquires a charge which we call positive charge. Some of this charge can be rubbed off onto a tiny ball suspended from a thread. We now find that the ball is repelled by the rod as shown in Figure 19.1. The positive charge on the rod repels the like charge on the ball. The strength of this repelling force is characterized by the *electric field strength* and is represented by the symbol **E.**

Experiment shows that this repulsion force exists even if the experiment is done in vacuum. The force exists even though there is no material between the rod and ball. Electric forces do not require material in the region where they act. These forces can pass through vacuum as well as through air. The electric field force is quite unlike the tension in a string or the pressure in a gas.

How can we tell if there is an electric field force at some point? In principle, all we need to do is perform the experiment shown in Figure 19.2. We place a tiny, positively charged ball at the point. If because of its charge the ball experiences a force, then an electric field force exists there. We say that there is an *electric field* at that point.

An electric field exists at a point if a stationary positive test charge placed there experiences a force because of its charge. The strength and direction of the electric field force is represented by the electric field vector E.

Electric fields exist close to all charged objects. We shall learn more about them in our study of electricity. For now it is sufficient if we know what is meant by the electric field vector **E**. It represents an electric force. As we have seen, this electric force can exist in vacuum as well as in matter. It is this force that vibrates in electromagnetic waves.

If a charge is oscillated back and forth, a changing electric field can be sent

radiation beam intensity and luminous intensity.

13. Explain the meaning and use of the following relations: $F = I\Omega$, $F = 4\pi I$, $E = F/A$, $E = I/r^2$, $E = (I \cos \theta)/r^2$, $E_1/E_2 = (r_2/r_1)^2$.

14. Give the ratio of the illuminations due to an isotropic source at two specified distances from the source.

15. Using a diagram, show the difference between plane-polarized and nonpolarized light. Explain one method for obtaining polarized from nonpolarized light. Describe a method by which one could determine if a light beam was plane-polarized.

out into space.[1] As shown in Figure 19.3, the situation is much like a vibrator sending a wave down a string. The vibrator at the end of the string sends a disturbance out along the string. The oscillating charge also sends out a disturbance. This disturbance consists of an electric field that continually reverses its direction. Notice in the figure that the electric field force vector points alternately upward and downward as the charge oscillates up and down. This oscillating electric field then travels out through space much like the disturbance on the string travels down the string. Notice that the force vector vibrates perpendicularly to the direction of propagation. Therefore these waves are transverse in nature.

In Figure 19.4 we see a more realistic situation. At a radio station charge is oscillated up and down in the transmitting antenna. This causes an electric field oscillation to move out from the station. At a certain instant the electric field wave along the earth might be as shown. At that instant the electric field is zero at points such as *B, D, F,* and *H.* But at points such as *A* and *E* along the earth, the **E** field is directed upward and is very strong. Also at this same instant the field is strong—but directed downward—at points such as *C* and *G.*

As time goes on, the electric field wave shown in Figure 19.4 travels to the right at high speeds. A person at point *P* notices a changing electric field as the wave passes by the point. When a crest reaches *P,* the electric field is strong and upward. When a trough reaches *P,* the field is strong but directed downward. As you see, the oscillating charge on the radio antenna sends an electric field wave out across the earth. We call this wave an electromagnetic wave. In the present case the em wave is simply a radio wave.

We see from this discussion that oscillating charges cause electric field waves, em waves. These waves have a definite frequency and wavelength. The frequency is the same as that of the original charge oscillation. As usual, the speed of the wave v is related to its frequency f and wavelength λ through $v = f\lambda$. Unlike the situation for other waves we have studied, nothing material vibrates in an em wave. Only the electric field (or force) vector **E** vibrates back and forth. Because electric forces exist even in vacuum, em waves can travel through

[1] A changing magnetic field is sent out with the electric field. Hence the name electromagnetic. We shall not be concerned with the magnetic part at present.

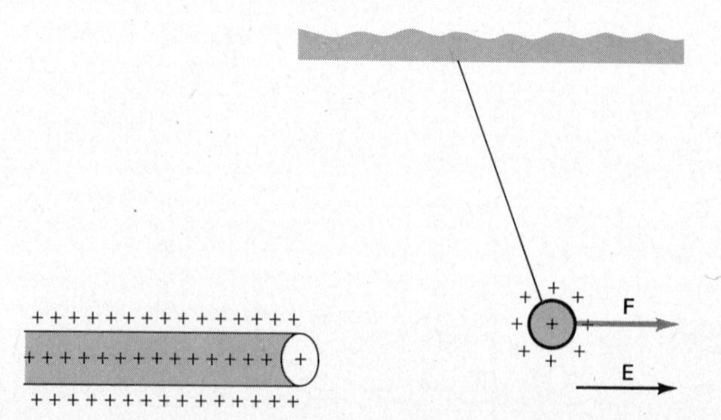

Figure 19.1 The positively charged glass rod repels the positively charged ball. (Like charges repel.) This force is a measure of the electric field **E** caused by the rod at the position of the ball.

empty space. In the next section we shall see how fast em waves move through vacuum.

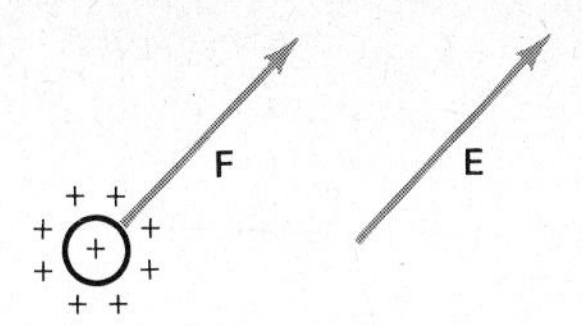

Figure 19.2 The stationary charge experiences a force **F** as shown because of its charge. We therefore know that an electric field exists at this point. It is represented by the electric field vector **E**, which points in the same direction as **F**.

19.2 Types of em Waves

Long before radio waves had been produced, it was predicted (in 1860) that they should be possible. James Clerk Maxwell, who lived from 1831 to 1879, derived a theory for the effects of oscillating electric charges. He predicted that electromagnetic waves should exist. Of course, radios had not yet been invented at that time. So his prediction remained untested for many years thereafter.

But Maxwell predicted more than just this. He was able to predict the speed of em waves. The speed he calculated was 2.998×10^8 m/s in vacuum. This is a very high speed. No object on earth has ever moved that fast.[2] But this was well known to be the measured speed of light. Was this simply coincidence? Or are em waves and light actually the same thing? Maxwell proposed that light waves *are* em waves. As time went on, further experimental facts were found to support Maxwell's proposal.

[2] We shall see in Chapter 29 that no object *can* move faster than this speed.

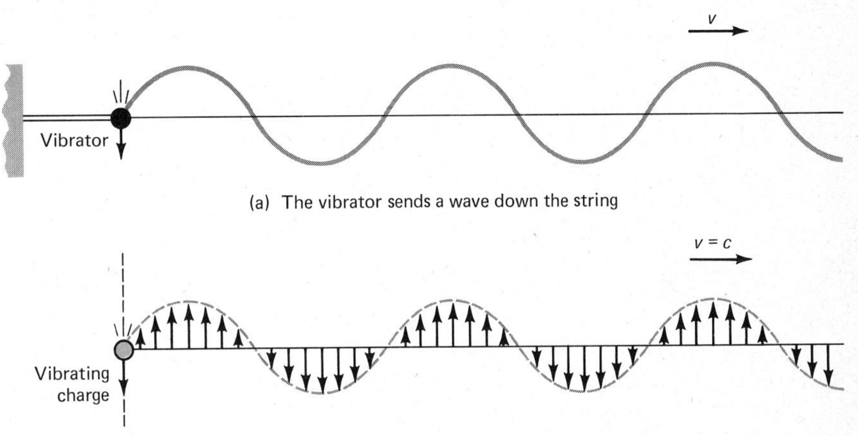

Figure 19.3 The electric field wave, unlike the wave on a string, travels out from the vibrator through vacuum.

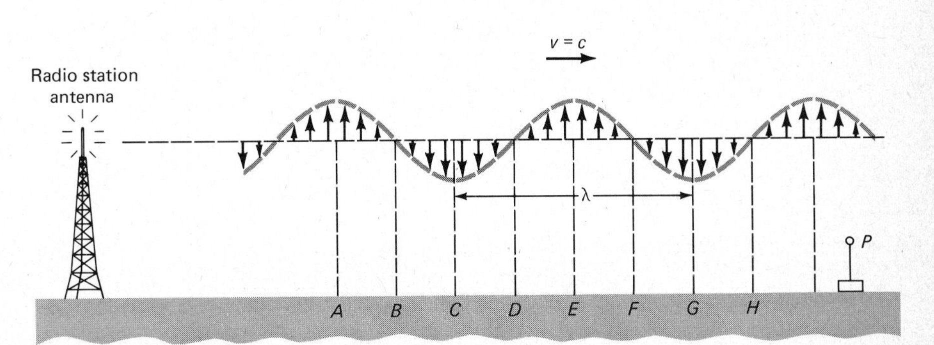

Figure 19.4 The radio station sends out an em wave. Its frequency is the same as the frequency with which the station oscillates charge on the antenna. Its wavelength λ is shown.

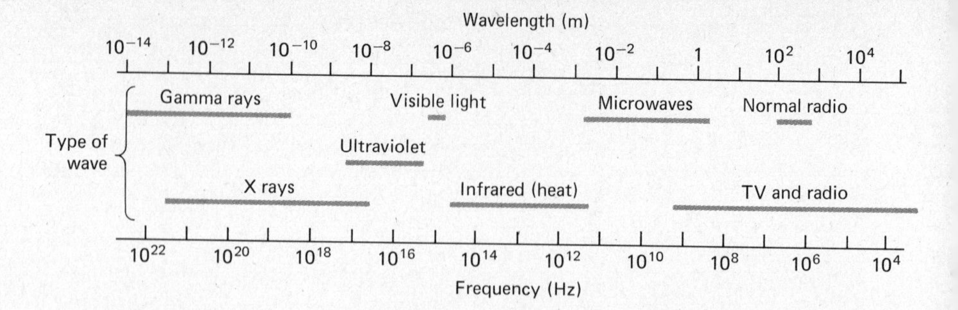

Figure 19.5 Types of electromagnetic waves.

Today we know that Maxwell was correct. There are many kinds of electromagnetic waves. Among them are radio, radar, infrared, light, ultraviolet, X-ray, and gamma-ray waves. All consist of oscillating electric field vectors, **E**. They all have the same speed. They differ only in frequency and wavelength. We conclude the following.

Electromagnetic waves travel through vacuum with speed 2.998×10^8 m/s = 186,000 miles/s. We designate this speed, the speed of light in vacuum, by the symbol *c*.

Because all waves obey the relation $v = f\lambda$, we have, for electromagnetic waves in vacuum,

$$c = f\lambda \quad \text{with} \quad c = 2.998 \times 10^8 \text{ m/s} \qquad (19.1)$$

The wavelengths and frequencies of the various types of em waves are given in Figure 19.5.

Notice from Figure 19.5 that each type of wave covers a range of frequencies and wavelengths. The limits on these ranges are not sharp. As we see, long-wavelength X rays can also be called ultraviolet light. Similarly, microwaves are the same as long-wavelength infrared light. Heat radiation and infrared light are basically the same. We shall learn more about these types of em radiation as our studies progress. In this chapter, though, our primary concern will be with the small region designated visible light.

Figure 19.6 (*Facing page*) A beam of light can be separated into its various colors (its spectrum) as shown in the top portion of the figure. Depending on the light source, the spectrum has different forms. In 1 is shown the spectrum of a white-hot object, such as the white-hot filament in an incandescent light bulb. This is called a continuous spectrum. The light reaching the earth from the sun is of this type, but certain colors are missing from the sun's spectrum, as shown in 2. These wavelengths are absorbed as the light passes through the atoms in the space between the earth and the sun. The other spectra shown are the line spectra given off by the ionized gas atoms indicated: sodium, Na; hydrogen, H; calcium, Ca; mercury, Hg; and neon, Ne. (From *General College Chemistry*, 6th ed., by Charles W. Keenan, Donald C. Kleinfelter, and Jesse H. Wood. Copyright © 1980 by Harper & Row, Publishers. By permission of the publishers.)

19.3 Visible Light

The eye can see only a very limited range of em wavelengths. Visible light wavelengths extend from 4×10^{-7} m to 7×10^{-7} m. As you can see in Figure 19.5, this is only a small region of the em wavelengths. But all the colors of the rainbow are within it. Violet light has the shortest wavelength and red the longest. In between these are all the other colors. We show their approximate wavelengths in line 1 of Figure 19.6. Be sure to read the figure legend.

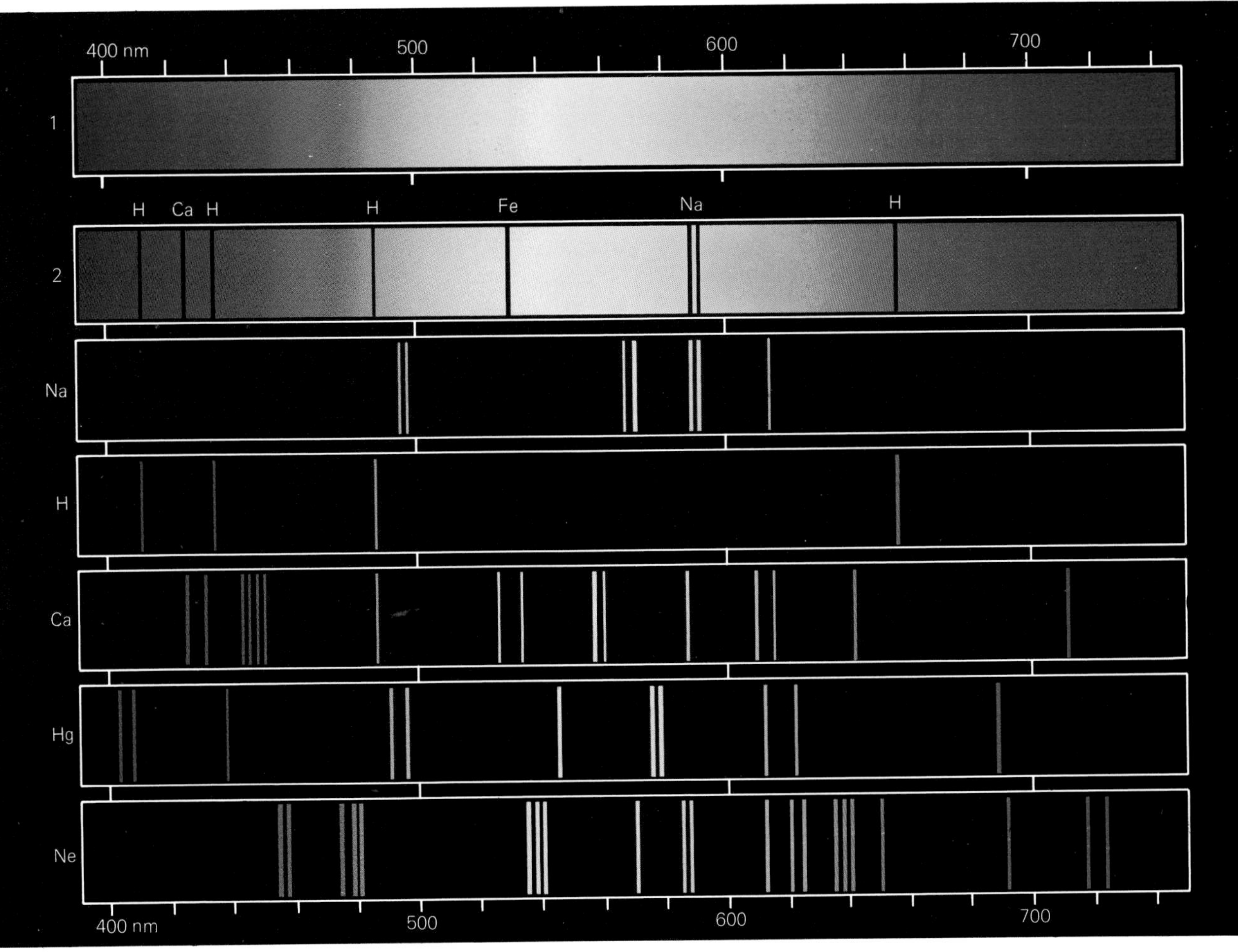

400 nm
500
600
700
1
H
Ca
H
H
Fe
Na
H
2
Na
H
Ca
Hg
Ne
400 nm
500
600
700

There are several units commonly used in measuring wavelengths of light. They are as follows.

NAME	SIZE	RANGE OF VISIBLE SPECTRUM
nanometer (nm)	1 nm = 10^{-9} m	400–700 nm
millimicron (mμ)	1 mμ = 10^{-9} m	400–700 mμ
angstrom (Å)	1 Å = 10^{-10} m	4000–7000 Å

The angstrom unit is notable because atoms have radii of about 1 Å. You should be familiar with all of these units because they are widely used in technology. However, the nanometer is the preferred unit.

Most light sources give off waves of many different wavelengths. The range of wavelengths given off by a light source is called a *spectrum.* Sunlight contains all the colors of the rainbow. It is what we call a *continuous spectrum* because all wavelengths are present in it.

As you might guess, the eye is not equally sensitive to all colors. We see best those colors near the center of the visible range. The eye is least sensitive to violet and deep red, the colors at the two ends of the visible spectrum. In the next section we discuss the sources of light and the colors they give off.

19.4 Spectra of Light Sources

The light from the sun consists of all colors. These colors can be separated from each other by use of a prism (a topic to be discussed later). When the colors of sunlight are separated, they give rise to the continuous spectrum we see in line 1 of Figure 19.6. Notice that the sun's spectrum contains all colors. It is what we call a *continuous spectrum.*

A continuous spectrum contains an unbroken range of wavelengths.

The sun is not the only continuous spectrum source. All other glowing, hot, solid and liquid bodies give off a continuous spectrum. We call solid objects that are hot enough to give off light *incandescent* light sources. An incandescent light bulb is a source of this type. The white-hot filament within the bulb gives off a continuous spectrum. All solid objects that are red- and white-hot act as incandescent light sources. They all give off a continuous spectrum.

You probably know that a red-hot object is not as hot as a white-hot object. The reason for this is that the radiation given off by an incandescent source varies with temperature. This variation is shown in Figure 19.7. Notice that much of the radiation is not visible. Most of the radiation is beyond the red, in the infrared. Some of the radiation is below the violet, in the ultraviolet. But we see that the radiation shifts more toward the visible as the temperature increases. The 3400-K curve would be characteristic of a white-hot object. At 2880 K the radiation is shifted more toward the infrared. The shift is much greater for a 1000-K object (not shown). It glows a dull red.

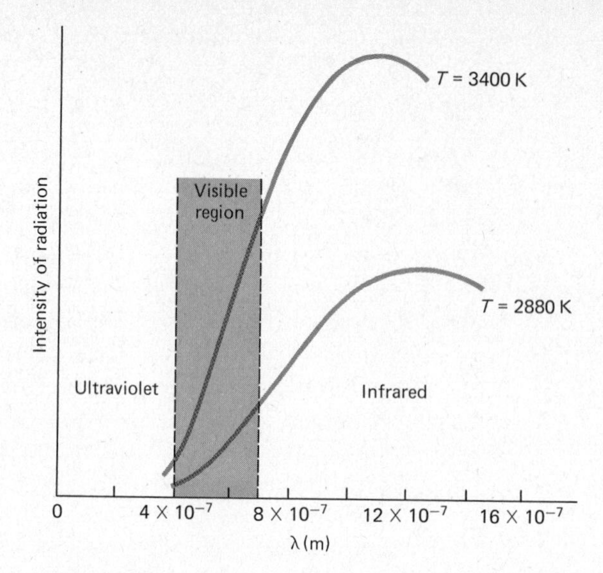

Figure 19.7 The hotter the incandescent object, the more the radiation increases and shifts to the blue.

In Figure 19.6 line 1 shows the spectrum emitted by an incandescent solid such as the sun. But the spectrum reaching the earth from the sun is as shown in line 2. We see certain dark lines in it. This is called a *dark-line absorption spectrum.* Some of the light coming to the earth from the sun is absorbed by atoms in the space between the earth and the sun. For example, hydrogen atoms in space absorb at the dark-band wavelengths that are labeled H. These wavelengths, even though they are emitted by the sun, are absorbed by hydrogen atoms before the light reaches the earth. Similarly, the lines labeled Na are due to absorption by sodium atoms in the space between earth and sun. We shall now see that these same atoms, when heated to high temperature, emit the same wavelengths they absorb.

When sodium atoms are vaporized and strongly heated, they emit the wavelengths shown as the spectrum labeled Na. Hot hydrogen atoms emit the spectrum shown in line H. Similar spectra are emitted by hot atoms in vaporized calcium (Ca), mercury (Hg), and neon (Ne). Indeed, for reasons we shall discuss in Chapter 29, all hot atoms in a vapor emit spectra such as these. They are called *bright-line emission spectra.*

A line spectrum consists of a discrete set of emitted wavelengths.

A neon sign consists of electrically excited neon atoms in a gas. Its color is red because its spectrum consists of intense lines in the red wavelength region. A sodium vapor lamp emits a yellow light. The vaporized, electrically excited, sodium atoms in it emit lines in the yellow. Mercury arc lights use electrically excited mercury vapor atoms. They emit strong lines whose wavelengths are 4.0×10^{-7} m, 4.4×10^{-7} m, 5.5×10^{-7} m, and 5.8×10^{-7} m plus other weaker lines. Because it has strong lines in several portions of the spectrum, its light is more nearly white than that of neon and sodium. Lasers emit a single wavelength of light. They also are line sources.

Each different type of atom emits its own characteristic wavelengths. Use is made of this fact to identify atoms in unknown materials. For example, if an alloy is vaporized in an electric arc, it gives off light. The wavelengths it gives off are used to identify the atoms in it. This type of analysis, *emission spectroscopy,* is widely used in industry.

There are other sources of light in addition to incandescent and arc lights. Most common is the fluorescent light source. The spectrum emitted by fluorescent bulbs is usually a combination of a continuous and a line spectrum. In them a gas inside the bulb emits a line spectrum. This radiation strikes the fluorescent material, which is coated on the inside of the bulb. The bulb then glows as the fluorescent material gives off light. By using just the right mixtures of fluorescent materials, the emitted light can be made to appear approximately white.

19.5 Color

The human eye sends signals to the brain, and the brain interprets these signals as color. It is important to note that mixtures of wavelengths are interpreted as a single color. For example, suppose that light with wavelength in the red is added to light with wavelength in the green. The combined light is then viewed by the eye. We do not see both red light and green light at once. Instead, we see this mixture as being yellow light.

Most of the colors we see about us are actually mixtures of wavelengths. We usually see them in reflected light. The sun or a light bulb furnishes the light with which we see. This light is reflected off the object we are viewing. If all of the light is reflected equally, the light entering the eye will contain all wavelengths. The object will appear white. The paper of this book reflects the white light that shines on it. The page is therefore white. But the letters on it reflect no light. They therefore appear black.

As another example, why do the leaves of a tree appear green? The white sunlight is shining on them, but only certain wavelengths are strongly reflected by the leaves. These reflected wavelengths combine in our eyes and we assign the color green to them. Even so, leaves vary in their shade of green. This tells us that they reflect different mixtures of wavelengths. Each mixture appears as a slightly different color to the eye.

Sometimes we view the light transmitted, rather than reflected, by an object. A glass of beer looks yellow when white light shines through it. The original white light contains all wavelengths. But the beer absorbs (or stops) many wavelengths that try to pass through it. Wavelengths in the yellow portion of the spectrum are absorbed least. As a result, the transmitted light is rich in wavelengths in the yellow portion of the spectrum. The eye sees the transmitted light to have a yellow tinge to it.

As we see, color is a complex quantity. It depends on the wavelengths that reach the eye. It also depends on the way the eye adds different wavelengths. The colors and hues that surround us depend on the kind of light we use to view them. For example, suppose that a striped yellow and red cloth is viewed in the light from a sodium vapor lamp. This lamp gives off yellow light. When viewed in this yellow light, the yellow stripes on the cloth will be seen as yellow. They reflect the yellow light of the lamp. But the red stripes do not reflect yellow light. No light reaches the eye from them. They therefore appear black when viewed in the yellow sodium light.

As another example, the light from a mercury arc contains several wavelengths. Therefore, objects viewed in its light do not look the same as in sunlight or in sodium light. Even many fluorescent lights do not duplicate sunlight exactly. For this reason a fabric may appear different in sunlight from the way it appears inside a store with fluorescent lights. As you can see, the subject of color has many fascinating aspects. If you wish to study further about it, most libraries have readable books on the subject.

19.6 Beams of Light and Shadows

For many purposes light may be considered to follow a straight-line path. The simple experiment in Figure 19.8 is an easy way to show this. We see there a light bulb enclosed in a box. Light from it can escape through a small hole in its side. The light coming out of the hole is narrowed into a cylindrical beam by a second hole. As shown, the beam of light then travels in a straight line away from the source. You have no doubt seen a situation somewhat like this when a spotlight shines through a slightly smokey auditorium or theater. The path of the beam can be seen clearly as it passes through the smokey air. A straight line that follows the path of a light beam is called a *ray*.

Because it follows a straight line, light can be used to produce shadows. For example, the point light source shown in Figure 19.9(a) casts a clear shadow of an object placed in its path. All the rays that would ordinarily strike the screen between *A* and *B* are stopped by the obstacle. The situation is somewhat different for an extended (or nonpoint) light source as in part (b). As you can see, the rays from the top of the source cast a shadow between *C* and *B*. The rays from the bottom of the source cast a shadow between *A* and *D*. The region *CD* is special. No rays from any portion of the source can reach this region and therefore it is a region of complete darkness. We call this the *umbra*.

The umbra is the portion of the shadow that is completely dark.

In regions *AC* and *DB* part, but not all, of the rays are stopped by the obstacle. For this reason these regions are not completely dark. We call this portion of the shadow the *penumbra*.

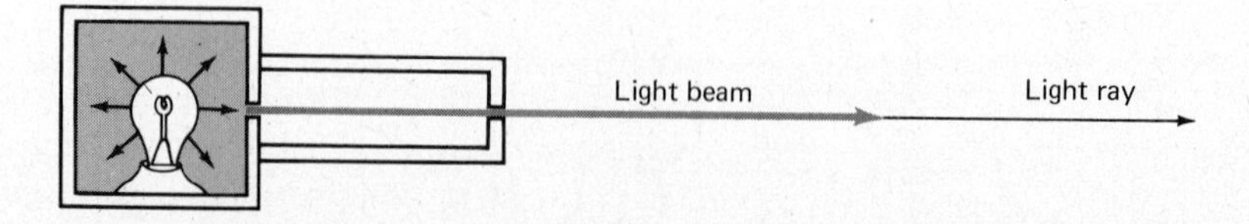

Figure 19.8 A series of small holes can be used to form a beam of light.

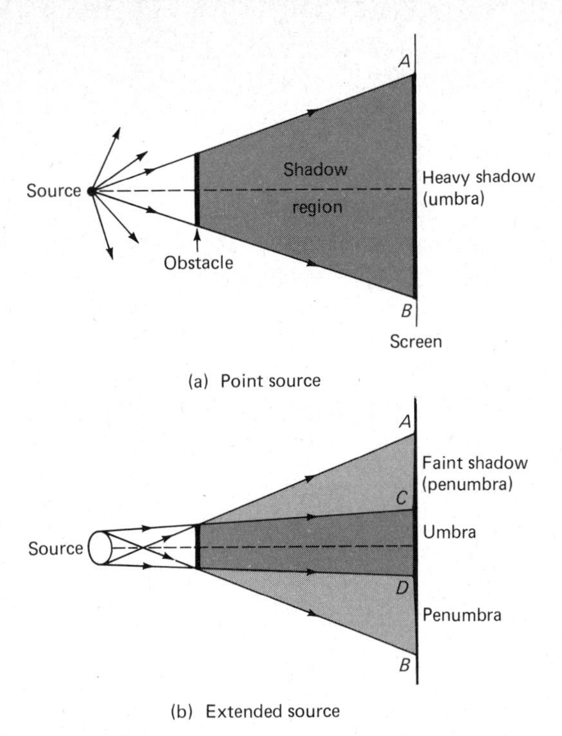

Figure 19.9 Rays from the source are stopped by the obstacle. No rays reach the umbra. Only a portion of the rays reach the penumbra.

The penumbra is the portion of the shadow that is partly exposed to light rays from the source.

This portion of the shadow is of importance only for shadows cast by extended sources.

19.7 Energy in em Waves; Light Intensity

Consider the em wave beam shown in Figure 19.10. It might be a radio wave at a large distance from the station or, as indicated, it might be a light beam. The beam carries energy with it. In the situation shown in the figure, energy moves along the beam with the speed of light.

Energy is carried along the rays of an electromagnetic wave beam. It moves with the speed of light.

We define the intensity of a light (or em) wave beam much as we did for sound waves. As shown in Figure 19.10, we erect an area perpendicular to the beam.

Beam intensity is the energy carried each second through a unit area perpendicular to the direction of propagation.

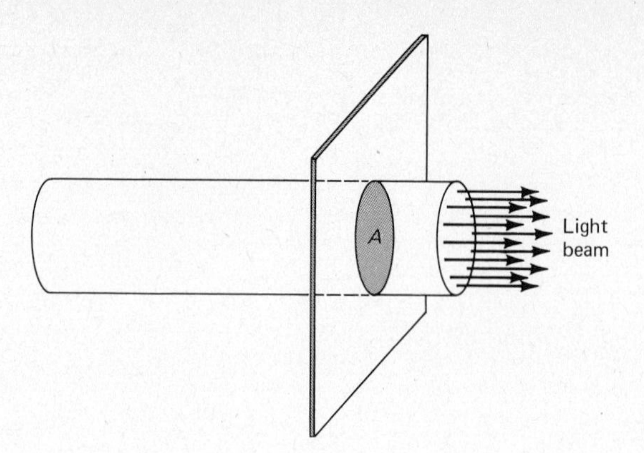

Figure 19.10 The intensity of the em wave beam is the (energy flowing through A each second) $\div A$. Its units are watts per square meter.

Notice that the area must be perpendicular to the rays of the beam. Typical units for intensity are watts per square meter.

The definition for intensity can be written in equation form. It is

$$\text{Intensity} = \frac{\text{(energy/second) through the area}}{\text{area}}$$

Or, because energy per second is simply power,

$$\text{Intensity} = \frac{\text{power flowing through the area}}{\text{area}} \qquad (19.2)$$

Equation 19.2 shows clearly that the units of intensity are watts per square meter.

19.8 Measurement of Light

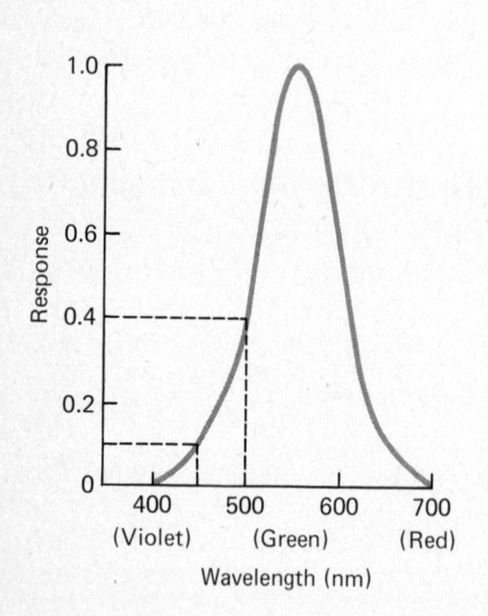

Figure 19.11 Normal response curve of the eye.

The beam intensity we have been discussing is often not a convenient measure for light. You can see why from the following example. Ultraviolet light (or black light) sources are nearly invisible to the human eye. Yet they emit intense beams of radiation. Our eyes are not sensitive to ultraviolet light and so we cannot see the beam even though its intensity might be high. We need a better measure of the brightness of a beam as seen by our eyes.

In order to understand how light is measured, we must know how the eye responds to various wavelengths of em radiation. The normal eye can detect wavelengths within the approximate range 400 nm to 700 nm. It is most sensitive near 550 nm and least sensitive near the ends of the wavelength range.

We show the response of the normal eye in Figure 19.11. This curve is called the *standard luminosity curve.* It is assumed for the purposes of the graph that a light beam of constant intensity but variable wavelength enters the eye. The observer is then asked to judge the brightness of the beam for various wavelengths. Even though the intensity of the beam is the same for all wavelengths, the beam appears to be brightest for wavelengths near 550 nm. Since the eye cannot detect wavelengths shorter than 400 nm or longer than 700 nm, the curve drops to zero at these wavelengths.

Photocell systems can be built that have a response similar to that of the eye. Their response as a function of wavelength is the same as the standard luminosity curve. We can use such a photocell for light measurements. Common light meters used by lighting engineers and photographers are examples. Let us now see how such a photocell can be used to define the quantities in terms of which we measure light.

19.9 Source Luminous Intensity

A 40-watt light bulb gives off far less light than a 120-W bulb does. We describe this feature of a light source in terms of the *luminous intensity,* I_L, of the source. Our discussion will be restricted to sources that are small enough so that we can approximate them by pinpoint size sources called *point sources.* For many purposes an ordinary incandescent bulb acts like a point source if you are not too close to it.

The SI unit system uses white hot molten platinum as a standard source. (An approximation to the standard source is shown at the bottom of Figure 19.12.) A pool of molten, white-hot platinum at 2046 K is used. The platinum is shielded from view except for an opening 1/60 cm^2 in area. This small, white-hot area is the standard source. We define its luminous intensity, I_L to be one candela (1 cd).

The standard light source has a luminous intensity I_L = 1 cd.

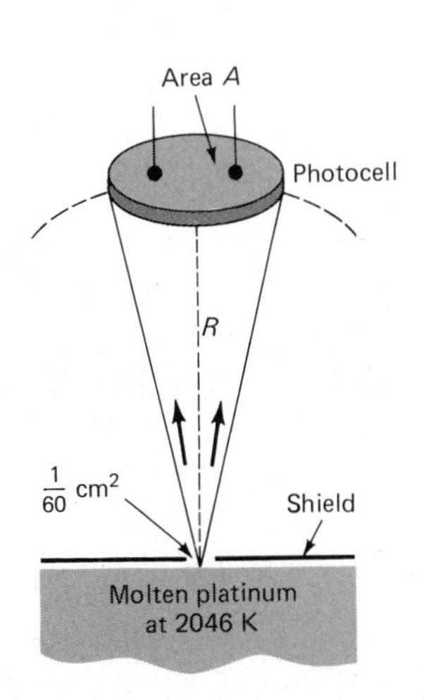

Figure 19.12 The SI standard light source uses white-hot platinum.

The luminous intensity of any other point source is found by comparison. We measure the response of a light meter type photocell to the standard source by placing it above the standard source, as shown in Figure 19.12. Then the standard source is replaced by the unknown source. If the photocell response is twice as large, we know the luminous intensity of the unknown is 2 cd. As you see, the unknown is measured by direct comparison.

Prior to the use of the SI standard, other units were used for source intensity. They are related to the candela by

$$1 \text{ candle} = 1 \text{ candlepower} = 0.981 \text{ cd}$$

Most light sources have different luminous intensities in different directions. For example, our standard photocell aimed at the end of a light bulb will read differently than when aimed at the side of the bulb. Similarly, a searchlight has a high luminous intensity only when viewed along the beam. Sometimes the average of the luminous intensity in all directions is used. It is called the *mean spherical luminous intensity.*

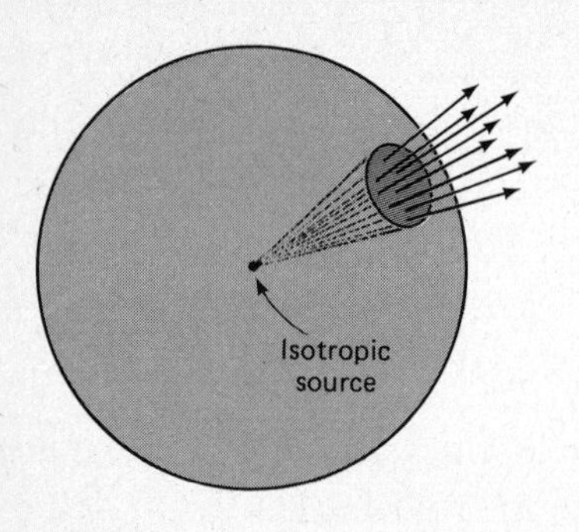

Figure 19.13 The isotropic source of luminous intensity I_L sends a total flux $4\pi I_L$ out through the surface of the sphere that encloses it.

An idealized light source that emits light equally in all directions is called an *isotropic source.* Our standard photocell would read the same no matter what side it was viewing.

19.10 Luminous Flux

We often draw rays of light coming from a light source as shown in Figure 19.13. Such a diagram is meant to represent the flow of light outward from the source. To make the concept of light flow from a source quantitative, we define a quantity called *luminous flux, F.*

Let us define luminous flux in terms of the light that comes from a point source whose intensity is I_L. You will recall that an isotropic source emits light equally in all directions. Consider the isotropic source in Fig. 19.13. We define the amount of flux coming from this source to be

$$\text{Luminous flux from isotropic point source} = F_{\text{total}} = 4\pi I_L \quad (19.3)$$

The unit for flux is the *lumen* (lm). As an example, the total luminous flux coming from a 10 cd isotropic source is $4\pi I_L$, which is 40π lm.

It is instructive to consider the factor 4π in this definition of flux. It arises because of the way we define solid angles. To see how this comes about, refer to Figure 19.14. Take C to be the center of a sphere. The area A is a portion of the sphere's surface. This area subtends the solid angle Ω (Greek letter capital omega). The unit for solid angles is the steradian (sr). From the definition of a solid angle, the angle in steradians is

$$\text{Solid angle} = \frac{\text{subtended area of sphere's surface}}{\text{square of sphere's radius}}$$

or

$$\Omega(\text{in sr}) = \frac{A}{R^2} \qquad (19.4)$$

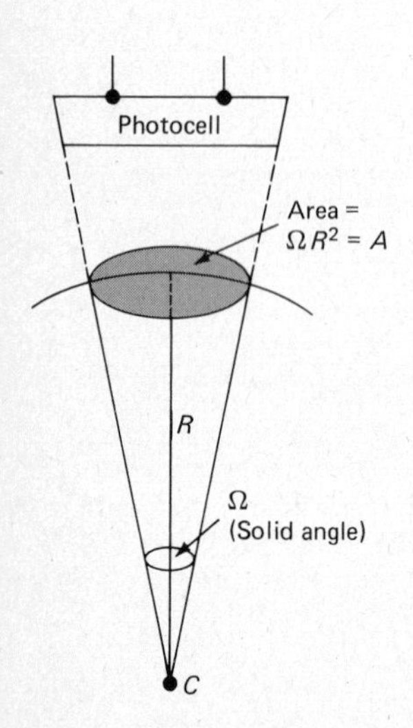

Figure 19.14 A source at C sends luminous flux through the solid angle Ω. The flux is $F = \Omega I_L$. Since the photocell subtends this same angle, it intercepts this same flux.

Since the complete surface area of a sphere is $4\pi R^2$, the total solid angle about point C is

$$\Omega_{\text{total}} = \frac{4\pi R^2}{R^2} = 4\pi$$

This is the reason for the factor 4π. The total flux from an isotropic point source is $4\pi I_L$. The flux through each steradian of solid angle is

$$\text{Flux per unit solid angle} = F/\Omega = \frac{4\pi I_L}{4\pi} = I_L \qquad (19.5)$$

Once we know this result, we can easily find the flux due to nonisotropic sources. Suppose a source has a luminous intensity I_L in a particular direction. Then the luminous flux sent out in this direction through a solid angle Ω is simply ΩI_L.

A practical concern is the total luminous flux emitted by a source compared to the electric power used by it. A typical 40-W incandescent bulb emits about 500 lm. We define its *luminous efficiency* to be the lumens emitted per watt of power. In this case it is about $500/40 \cong 12$ lm/W. A typical 40-W fluorescent bulb emits about 2000 lm. Hence its efficiency is about 50 lm/W, much higher than for an incandescent bulb. You need only touch each when operating to see why the incandescent bulb is so inefficient.[3] The luminous efficiency is usually listed on the package of a bulb.

EXAMPLE 19.1 A standard photocell system deflects 100 divisions on its scale when it is aimed at a point source 200 cm away. The source has a luminous intensity of 150 cd. (It might be a 60-W bulb.) The opening through which light strikes the photocell has an area of 3.0 cm^2. To how many lumens does each division on the scale correspond? If the source is isotropic, what would its luminous flux output be?

Solution The situation is sketched in Figure 19.15. Notice that the area of the photocell surface is small compared to the area of the sphere of radius 2 m. Therefore, the portion of the area of the sphere subtended by the photocell is almost exactly 3.0 cm^2. Using Equation 19.4 to find the solid angle subtended by the photocell gives

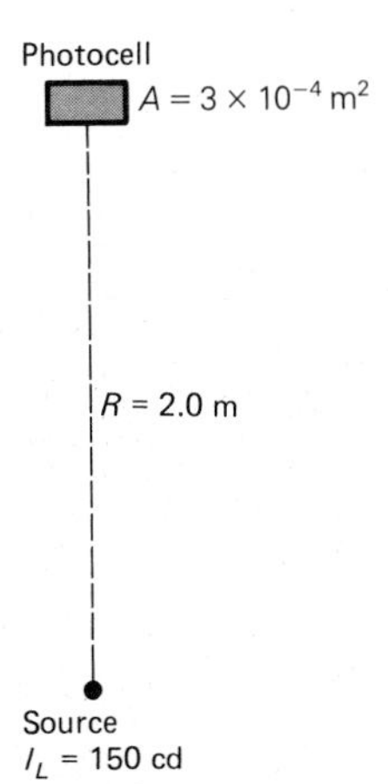

Figure 19.15 The 150 cd source causes a scale deflection of 100 on the photocell meter.

$$\Omega = \frac{3.0 \times 10^{-4}\ \text{m}^2}{4.0\ \text{m}^2} = 7.5 \times 10^{-5}\ \text{sr}$$

The luminous flux sent out into this solid angle by the source is

$$F = \Omega I_L = (7.5 \times 10^{-5}\ \text{sr})(150\ \text{cd}) = 0.0112\ \text{lm}$$

Hence the flux entering the photocell is 0.0112 lm. Because this causes a 100-division deflection of the photocell, each scale division (or unit) corresponds to a flux of 1.12×10^{-4} lm entering the photocell.

If the source is isotropic, its luminous flux output would be

$$F_{\text{total}} = \left(\frac{4\pi}{\Omega}\right)(0.0112\ \text{lm}) = 1880\ \text{lm}$$

[3] The hot incandescent bulb tells you that much of the bulb's energy is dissipated as heat, not light.

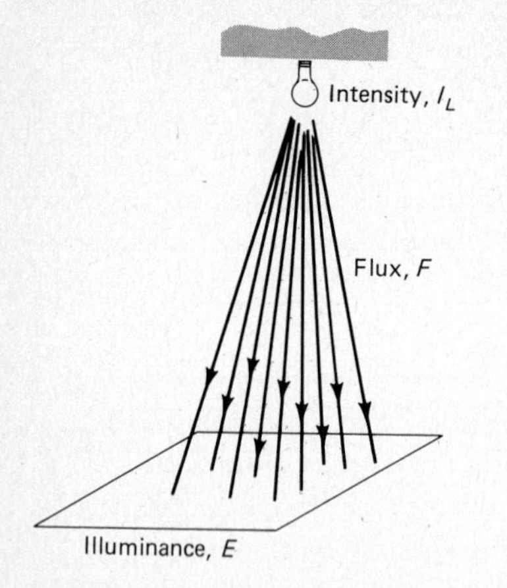

Figure 19.16 The sketch illustrates the relation between luminous intensity, flux, and illuminance.

19.11 Illuminance

When you were a child, did you ever set a piece of paper on fire by using a lens? As the sun's rays passed through the lens, they were brought together at a point by the lens. This concentrated the rays into an extremely bright spot on the paper. Using our present terminology, the luminous flux passing through the lens was directed onto a small area of the paper. As a result, the flux per unit area was very large. And this gave rise to the extreme brightness of the spot.

We see from this example that the quantity of flux is not the only important factor in determining brightness. The degree of concentration is also important. In other words, it is the amount of flux per unit area of a surface that determines how brightly the surface is lighted, or illuminated. To measure this facet of lighting, we define a quantity called *illuminance, E.* If a quantity F of luminous flux strikes an area A of surface, then the illuminance E of the surface is defined to be

$$\text{Illuminance} = E = F/A \tag{19.6}$$

Figure 19.16 shows how source intensity, flux, and illuminance are related.

The unit of illuminance is the lux (lx), or lumen per square meter:

$$1 \text{ lm/m}^2 = 1 \text{ lx}$$

Another unit sometimes used for illuminance is the lumen per square foot, also called a footcandle.

$$1 \text{ lm/ft}^2 = 1 \text{ foot candle} = 10.76 \text{ lx}$$

A photocell can be calibrated to read illuminance. Indeed, that is the purpose of most light meters. For example, in Example 19.1 we found that a flux of 0.0112 lm enters the 3.0×10^{-4} m^2 area of the photocell when its deflection is 100 units (divisions). The illuminance of the photocell surface is

$$E = \frac{F}{A} = \frac{0.0112 \text{ lm}}{3.0 \times 10^{-4} \text{ m}^2} = 37 \text{ lx}$$

Therefore, the meter deflects 100 divisions for an illuminance of 37 lux. Some typical values for illuminance are given in Table 19.1.

Table 19.1 ILLUMINANCE IN TYPICAL SITUATIONS

Situation	*Illuminance* (lx)	(lm/ft^2)
General illumination in home	50	5
Reading areas	500–1000	50–100
Classrooms	500–1000	50–100
Workbench (fine work)	2000	200
Brightly sunlit window area	2000	200

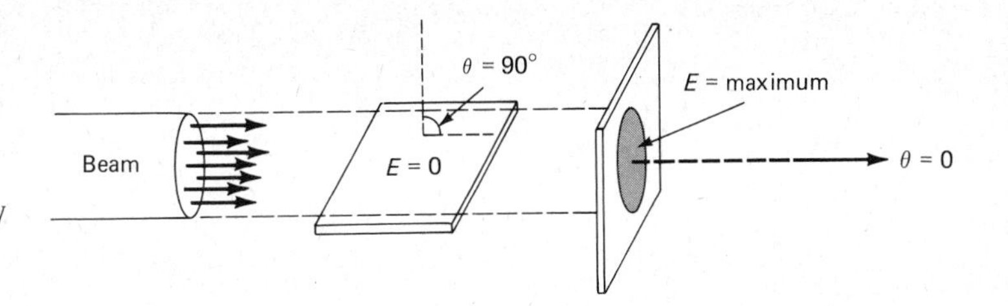

Figure 19.17 The area is most brightly lit if the flux lines are perpendicular to the area.

The orientation of an area upon which light flux falls affects the illuminance of the surface. This is easily seen in Figure 19.17. Although the two surfaces shown there are of equal area and are in the same beam, they will not be equally bright. (You can prove this to yourself by holding a sheet of paper at various angles to a light beam. The paper appears most brightly lit when perpendicular to the beam.) It turns out that the illuminance varies with the cosine of the angle θ shown in Figure 19.17. For the poorly lit area, $\theta = 90°$ and so $\cos\theta = 0$. This area has zero illuminance. But for the other area, $\theta = 0°$ and $\cos\theta = 1$. It receives maximum illumination.

19.12 Point Source Relations

Often a light source can be approximated by a point source. If you are perhaps ten times farther from a source than the source is large, the approximation becomes fairly good. We can then picture the situation as in Figure 19.18. (You must use your imagination a little because this is supposed to represent two concentric spheres.)

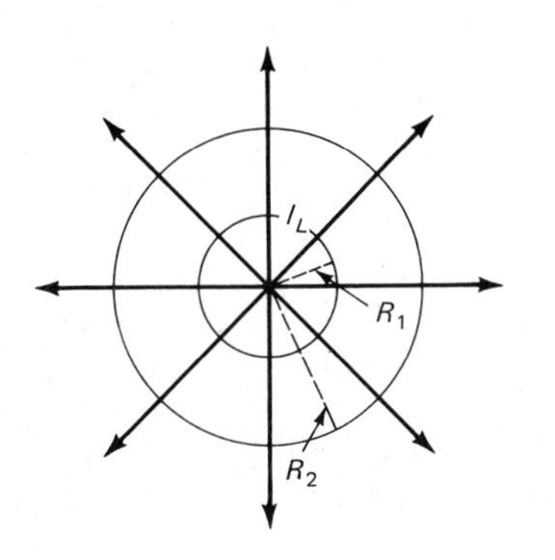

Figure 19.18 The luminous flux through both of the concentric spheres is $4\pi I_L$. The circles represent the surfaces of two concentric spheres.

Because the source emits a quantity of luminous flux given by $F = 4\pi I_L$, the illuminance of the surface of sphere 1 is, from $E = F/A$,

$$E_1 = \frac{4\pi I_L}{4\pi R_1^2} = \frac{I_L}{R_1^2}$$

Similarly, $E_2 = I_L/R_2^2$. We therefore have the following fact.

The illuminance caused by a point source varies inversely as the square of the distance from the source.

$$E = \frac{I_L}{R^2} \tag{19.7}$$

The illumination from a point source obeys an inverse square law.

What we have just stated applies to a surface that is perpendicular to the light rays. (The surface of a sphere is perpendicular to its radii.) We show the more general situation in Figure 19.19. As mentioned earlier, the flux is now spread

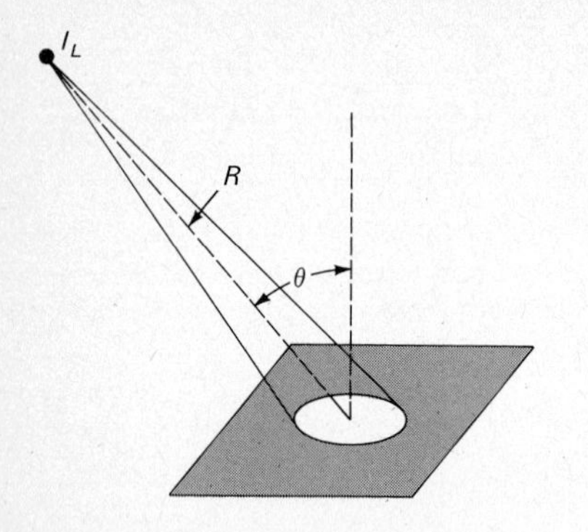

Figure 19.19 The illuminance of the area shown is $E = I_L \cos\theta/R^2$.

over a larger area, so the illuminance is decreased. In this case the illuminance of the surface is given by

$$E = \frac{I_L \cos\theta}{R^2} \tag{19.8}$$

where θ is the angle indicated.

EXAMPLE 19.2 An unshaded light bulb gives $E = 200$ lx at a distance of 50 cm directly below it. What will be the illuminance at 70 cm directly below it?

Solution To a rough approximation we can consider the bulb to be an isotropic point source. From Equation 19.7

$$E_1 = \frac{I_L}{R_1^2} \quad \text{and} \quad E_2 = \frac{I_L}{R_2^2}$$

from which

$$\frac{E_1}{E_2} = \frac{R_2^2}{R_1^2}$$

Solving gives

$$E_2 = E_1\frac{R_1^2}{R_2^2} = (200\ lx)\left(\frac{50\text{ cm}}{70\text{ cm}}\right)^2 = 102\text{ lx}$$

EXAMPLE 19.3 All the light from a spotlight is to be collected by a reflector and focused on a screen. The area lighted is to be 0.20 m² and the illuminance required there is 500 lx. What must be the mean spherical luminous intensity of the light?

Solution From Equation 19.6 the required total flux from the bulb is

$$F_{\text{total}} = EA = (500\text{ lm/m}^2)(0.20\text{ m}^2) = 100\text{ lm}$$

But $F_{\text{total}} = 4\pi I_L$, so we find that

$$I_L = \frac{F_{\text{total}}}{4\pi} = \frac{100\text{ lm}}{4\pi} = 8.0\text{ cd}$$

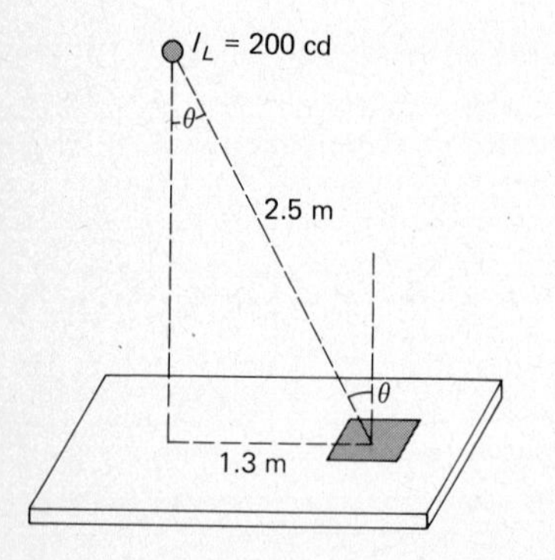

Figure 19.20 What is the illuminance of the sheet of paper?

EXAMPLE 19.4 A sheet of paper on a table is lighted as in Figure 19.20. Find the illuminance at the paper due to the 200-cd lamp bulb.

Solution Considering the bulb to act like a point source, we have

$$E = \frac{I_L \cos\theta}{R^2} = \frac{(200\text{ cd})\cos\theta}{(2.5\text{ m})^2}$$

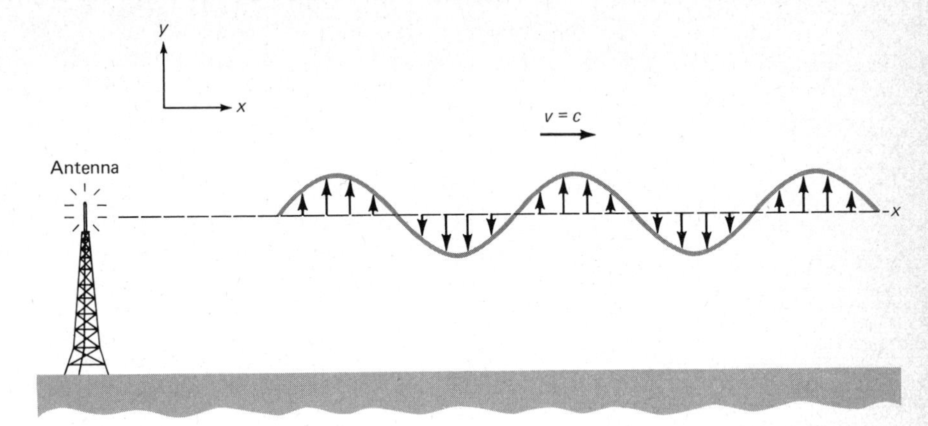

Figure 19.21 The oscillation of charge up and down the vertical antenna causes an em wave as shown. Notice that the **E** vector is always in the y direction.

From the diagram we see that $\sin \theta = (1.3)/(2.5) = 0.52$, so $\theta = 31°$. The cosine of the angle is 0.85 and so

$$E = \frac{(200 \text{ cd})(0.85)}{(2.5 \text{ m})^2} = 27 \text{ lx}$$

■■

19.13 Polarization of Light

Let us now return to the em wave sent out by a radio antenna. The situation is shown in Figure 19.21. The wave travels in the x direction, out along the earth. But the quantity that vibrates, the **E** vector, vibrates in the y direction, up and down. The vibration (up and down) is perpendicular to the direction of propagation. A wave for which this is true is called a transverse wave, as we stated in Chapter 17. Electromagnetic waves are transverse.

Another feature of the wave in Figure 19.21 should also be noticed. The vibration is up and down. As we see, the vibrating **E** vector is always parallel to the same plane, the plane of the page in this case. We say that a wave such as this is *plane-polarized.*

In a transverse wave the vibration is perpendicular to the direction of propagation. If the vibrations are all parallel, the wave is said to be plane-polarized. The vibration and propagation lines determine a plane. This is called the plane of polarization.

In Figure 19.21 the plane of polarization is the plane of the page.

An ordinary light source does not send out a plane-polarized beam. The oscillating charges that send out the light are in atoms, not in a single antenna. These vibrating charges are in billions upon billions of atoms. Each sends out its own em wave. The beam of light from the source is really composed of billions of waves like the one shown in Figure 19.21.

Because the charges oscillate in all directions, not just up and down, the individual waves are not all in the x-y plane. We have made an attempt to show

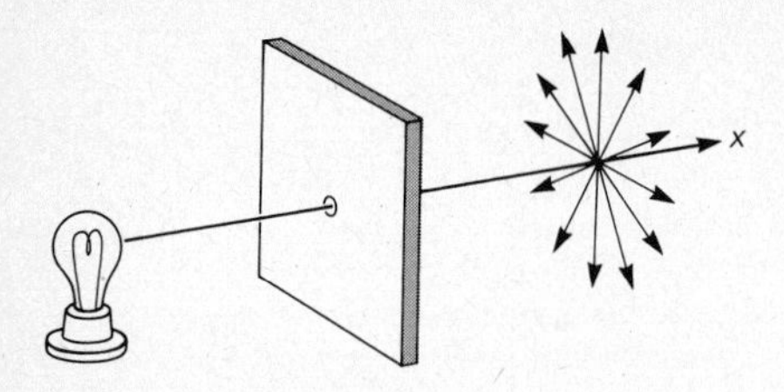

Figure 19.22 The light from a source is usually not polarized. At any point on the beam, such as the one shown, the **E** vectors vibrate in all directions except one. None of the **E** vectors has a component along the line of propagation.

the true situation in Figure 19.22. In the figure the **E** vectors are all perpendicular to the line of propagation. (This is always true for any transverse wave.) But they vibrate in many directions. They do not define a single plane as they did in the polarized wave of Figure 19.21. This type of beam is *nonpolarized.*

A polarized beam of light can be obtained from a nonpolarized beam by using Polaroid sheets. These sheets consist of plastic with needlelike crystals oriented lengthwise, all in one direction. The crystals allow light to pass through only if the **E** vector is in a certain direction. They may therefore be used as shown in Figure 19.23.

Notice that the Polaroid on the left in Figure 19.23 allows only the vertical vibrations to pass through. The unpolarized beam is therefore polarized by the first sheet. In part (a) the transmission direction of the second sheet is parallel to that of the first. It therefore allows the **E** vectors to pass through. In (b), however, the analyzer (second Polaroid) has been rotated through 90°. It now allows only horizontal vibrations to pass through. Because the beam is vertically polarized, it cannot get through. We say in this second case that the polarizer and analyzer are *crossed.*

There are many uses of polarized light in industry and technology. One of the most important is in stress analysis. When a transparent object, under stress, is placed between crossed Polaroids, the second Polaroid does not stop all the light. For example, Figure 19.24 shows the light that passes through crossed Polaroids when a strained plastic model is placed between them. Instead of complete darkness, one sees an informative picture. Where the bands are closest together in it, the stress variation is highest. Such pictures can be used to tell when and how objects will fail under high stresses.

There are some techniques for producing polarized light other than Polaroid sheets. We list a few.

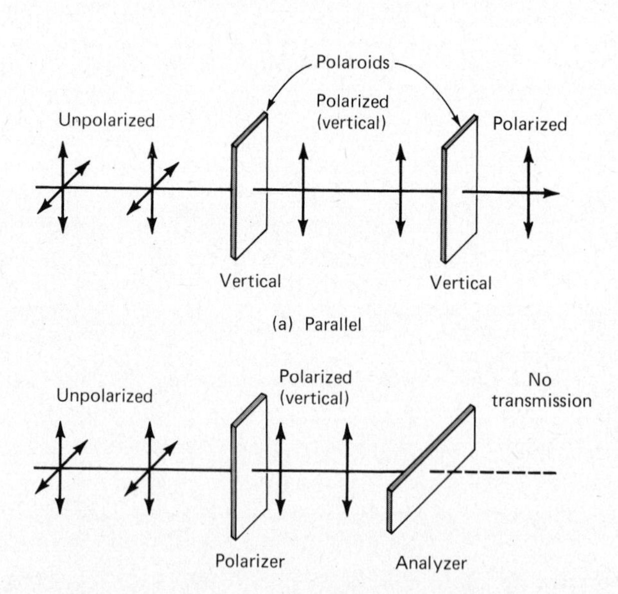

Figure 19.23 Polaroid sheets can be used to produce and analyze polarized light.

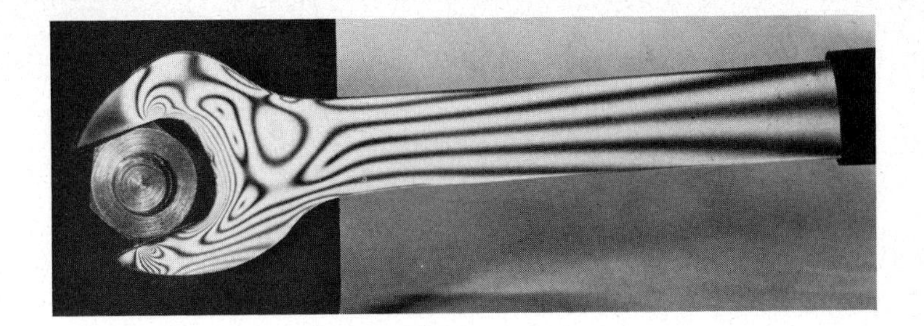

Figure 19.24 Stress variation in a plastic model is seen between crossed Polaroids.

1. *Nicol prism.* This device makes use of a phenomenon called *double refraction.* When a beam of unpolarized light enters certain types of materials, called *doubly refracting materials,* two distinct beams are formed. Each beam is plane-polarized, one at right angles to the other. The nicol prism is so designed that one beam is rejected and the other is passed straight through. As a result, the portion of the beam that is transmitted is plane-polarized.
2. *Reflection.* When light is reflected from a smooth surface, the reflected light is partly polarized. Only in the case of dielectrics, such as glass, plastic, and water, can total polarization occur and even then only at one certain angle of reflection. In such cases the light reflected at this angle is plane-polarized.
3. *Scattering.* We see the daytime sky as bright because sunlight is scattered to our eyes by the molecules and other particles in the atmosphere. This scattered light is partly polarized. Similarly, light scattered from a beam by dust, smoke, and so forth is also partly polarized.

Although these methods for producing polarized light are important in certain applications, Polaroid has largely replaced them in technical applications.

Summary

In electromagnetic (em) waves the vibrating quantity is the electric force vector **E.** Electromagnetic waves can be generated by vibrating electric charges. The frequency of the wave is the same as the vibration frequency of the source. Radio waves are generated by charges oscillating on the transmitting antenna of the station. Light waves are generated by charge motions within atoms.

From longest to shortest wavelength, the em wave spectrum consists of radio, TV, radar and microwaves, infrared and heat, visible, ultraviolet, X rays, and gamma rays. The visible region extends from about 400 nm to 700 nm.

Electromagnetic waves are transverse waves. They need no material in which to travel. Their speed through vacuum is the speed of light, $c = 3 \times 10^8$ m/s = 186,000 miles/s. Like all waves, they obey the relation $\lambda = v/f$. For vacuum $v = c$.

The eye is most sensitive to light with wavelength near the center of the visible range. Light of wavelength 400 nm is violet; that with wavelength = 700 nm is deep red. These are the limits of visible em radiation. The complete visible spectrum in order of increasing wavelength is violet, blue, green, yellow, orange, red.

Incandescent light sources consist of hot solids and liquids. Typical are the sun and the white-hot wire in a light bulb. They give off all wavelengths of light. Their spectrum is called a continuous spectrum. The color of an incandescent source depends on its temperature. At about 1000 K, the source is a dull red. Near 2500 K it is white. The hotter the source, the more radiation it emits at short wavelengths.

Excited gas atoms and molecules emit only certain discrete wavelengths. Their spectrum is called a line spectrum. Typical line spectrum sources are neon signs, mercury vapor lamps, and sodium lamps. Lasers also give a line spectrum.

The eye responds in a complex way to light. A beam of light consisting of a mixture of wavelengths is seen as a single color. Wide ranges of hues and colors can be obtained by mixing various wavelengths of light. When viewed by reflected light, the color of an object depends on both the nature of the light used and the ability of the object to reflect light.

For many purposes light can be considered to travel in straight lines. A beam of light therefore forms shadows. When the light source is a point source, the shadow is sharp and uniformly dark. An extended source forms a shadow that has an umbra and penumbra. The umbra is completely dark. Rays from some portions of the source strike the penumbra region.

The intensity of an em wave (or beam) is defined in terms of an area A perpendicular to the beam. It is the power flowing through A divided by A. Its units are power per unit area, watts per square meter.

The luminous intensity I_L of a light source is specified in candelas. An isotropic source of intensity I_L emits a luminous flux F given by $F = 4\pi I_L$ lumens. Illuminance E measures the light flux striking a unit area of surface, $E = F/A$. Its units are lumens per square meter, called the lux.

An isotropic (or spherical) source emits radiation uniformly in all directions. The illuminance due to the radiation from such a source decreases as $1/R^2$, where R is the distance from the source.

A plane-polarized wave has its vibration direction parallel to a plane. The vibration line and the line of propagation determine a plane. This plane is called the plane of polarization.

Light beams from ordinary light sources are not polarized. The beam can be polarized by sending it through a Polaroid sheet.

Questions and Exercises

1. Suppose that a nuclear bomb was exploded on the moon. We on earth would not be able to hear the explosion even though we could see it. Explain.
2. Galileo tried to measure the speed of light. He and a friend stood a distance D apart. The friend opened a cover on a lantern he held. As soon as Galileo saw the light, he opened the cover of his own lantern. The friend timed how long it took after he opened his lantern until he saw light from Galileo's. If the measured time was t, then the speed of light should be $2D/t$. Why? Unfortunately, all Galileo could conclude was that light travels very fast. Why couldn't he give a better answer?

3. An eclipse of the sun occurs when the moon comes between the earth and sun. On a sketch show the umbra and penumbra regions in the shadow cast on the earth. How are these related to the regions of total and partial eclipse?
4. Light travels fastest through vacuum. Its speed through window glass is about 2×10^8 m/s. A beam of yellow light enters a piece of glass. Because $v = \lambda f$, either λ, f, or both must decrease as it enters the glass. Which is correct? Why?
5. Some people illuminate their yards with high-intensity mercury lights. Explain why people look unnatural in such light.
6. Explain the following statement: Equation 19.7 is a special case of Equation 19.8.
7. The pupil of your eye narrows in bright light. Which of the following is this change in opening designed to control: luminous intensity, luminous efficiency, flux, illuminance?
8. Like the ear, the human eye is an approximately logarithmic measuring device. Explain what is meant by this statement.
9. Polaroid sheets become very hot when they are used to polarize an intense beam of light. Explain why. Would you expect the heating of a Polaroid analyzer to depend on whether it was parallel to or crossed with the polarizer?
10. A solution of sugar in water is optically active. By this we mean the following. When a beam of vertically plane-polarized light is shined through the solution, the plane of polarization is rotated. (For example, a 90° rotation would cause the plane of polarization to be horizontal.) But still the emerging beam is plane-polarized. This fact is used as a basis for a rapid method for determining sugar concentrations in solutions. Devise a method for doing this.

Problems

1. The red light from a helium-neon laser has a wavelength of 633 nm. Find the frequency of this light wave.
2. Cobalt 60 is commonly used in hospitals as a γ-ray source. The wavelength of the γ ray it gives off is 0.061 Å. Find the frequency of these waves.

*3. Light from a sodium vapor arc is yellow and has a wavelength of 589 nm in vacuum. Find the frequency and wavelength of a beam of this light in water. The speed of light in water is 2.25×10^8 m/s.

*4. The wavelength of the red light from a helium-neon laser is 6328 Å in vacuum. Find the frequency and wavelength of a beam of this light as it passes through glass, in which the speed of light is 2.00×10^8 m/s.

*5. A 5.0-cm diameter light bulb hangs 1.50 m above a tabletop. Directly below the bulb is a pinhole through the very thin surface of the table. Find the diameter of the image formed of the bulb on the floor, 80 cm beneath the table.

*6. A point source of light at the ceiling of a room is 2.0 m above the center of a circular table. The diameter of the tabletop is 120 cm. How large is the diameter of the shadow of the table on the floor, 80 cm below?

**7. A vertical, 50-cm-long fluorescent bulb is 2.0 m from a vertical meter stick. The centers of bulb and stick are at the same level. Find the length of the umbra and penumbra for the stick on a wall 3.0 m beyond the stick.

**8. A frosted light bulb is used end-on as a light source. It then acts as a sphere of 3-cm diameter. The bulb is 96 cm from a wall. A ball with a 9-cm diameter is placed midway between the bulb and the wall. Find the approximate diameter of the umbra part of the shadow on the wall. Repeat for the penumbra part.

9. A certain laser sends out a beam that has a total power of 3.0 mW. The diameter of the beam is 2.0 mm. What is the intensity of the beam?
10. The radiation intensity at the earth from the sun is 1340 W/m^2. A 25-m^2 solar panel on a space ship utilizes this energy with an efficiency of 5 percent. How many 60-W bulbs could the panel light?

*11. The beam from a laser strikes a block of metal and is absorbed in it. From the temperature rise of the block, the beam is found to deliver 2.0 cal/s to the block. (a) Find the power in watts furnished to the block. (b) If the cross-sectional area of the beam is 0.15 cm^2, what is the radiation intensity (watts per square meter) in the beam?

*12. Lasers for student use have beams with power equal to about a milliwatt. (a) Find the radiation intensity in a 1-mW beam if the cross section of the beam has an area of 0.050 cm^2. (b) Suppose that a 1-mW beam is absorbed in 100 cm^3 of black coffee. How much will the coffee be heated by it in 60 s?

13. How much flux does a bulb emit if its mean spherical luminous intensity is 60 cd?
14. What is the mean spherical luminous intensity of a bulb that emits 2000 lm of flux?
15. If the diameter of the pupil of your eye is 2.00 mm, how large a spherical angle does it subtend for a bulb that is 1.50 m away?
16. A sheet of paper, 8 cm by 10 cm, lies on a table 1.50 m

*Problems marked with an asterisk are not as easy as the unmarked ones.
**Problems marked with a double asterisk are somewhat more difficult than the average.

directly below a tiny bulb. About how large a solid angle does the paper subtend at the bulb? Why is your answer only approximate?

17. How much luminous flux does a 40-cd isotropic point source emit?
18. A typical 60-W incandescent bulb emits about 800 lm of flux. Assuming it to act like an isotropic point source, what is the luminous intensity of the bulb in candelas?
19. A certain 100-W bulb has a mean spherical luminous intensity of 130 cd. How much total luminous flux does the bulb emit?

*20. The intensity of the sun's radiation at the position of the earth is 1340 W/m^2. How much energy does the sun radiate each minute? The earth-to-sun distance is 1.5×10^{11} m.

*21. Using the data given in Problem 20, find the intensity of the sun's radiation at the planet Mercury. The distance of this planet from the sun is 5.8×10^{10} m.

*22. Repeat Problem 21 for the planet Pluto, which is 5.9×10^{12} m from the sun.

23. A certain 75-W bulb has an average output of 1170 lm. What is its luminous efficiency?
24. A certain 60-W bulb is rated at 14 lm/W luminous efficiency. What is its output in lumens?
25. All the light from a bulb is focused onto a wall. The lighted area is 600 cm^2. The bulb emits a total luminous flux of 3000 lm. What is the illuminance at the wall?
26. A 20-cm × 20-cm sheet of paper lying on a table receives a total flux of 2000 lm. What is the average illuminance at the paper?

*27. A certain searchlight beam provides an illuminance of 20,000 lx close to the searchlight. The beam there has an area of 2.0 m^2. The beam spreads somewhat as it travels away from the light. What illuminance does it provide at a position where its cross section has increased to 8.0 m^2?

*28. A certain isotropic light bulb produces an illuminance of 1000 lx when it is 60 cm directly above a workbench. If it is raised another 30 cm, how large will the illuminance be?

*29. A light meter reads 2000 lx when aimed at a point light source from a distance of 10 cm. What will it read when placed 80 cm from the source and aimed at the source?

*30. A point source of gamma rays (or X rays) is provided by a small piece of radioactive substance. If the intensity of the radiation from the source is I at a distance of 10 cm, what is the intensity at 50 cm from the source?

*31. An X-ray tube portal acts like a point source for X rays. How far from the source must one be for the X-ray intensity to be only 1 percent of its value 5 cm from the source?

**32. A point-like bulb hangs 50 cm above a table. An illumination of 700 lx is wanted on the tabletop. Assuming the bulb has a luminous efficiency of 15 lm/W, what should be the wattage of the bulb?

*33. Two isotropic light sources are 200 cm apart. One has a luminous intensity of 30 cd. A screen is moved along the line connecting the sources. Equal illumination on the two sides of the screen is noticed when the screen is 50 cm from the 30-cd source. What is the luminous intensity of the other source?

**34. A fluorescent bulb is a cylindrical source of light. For points close to the bulb, the cylinder can be considered infinite in length. Using reasoning much like that used for a spherical source, prove that

$$\frac{E_1}{E_2} = \frac{r_2}{r_1}$$

for a cylindrical source.

**35. The helium-neon laser beam has a single wavelength, 6328 Å. Consider one that has an output energy of 2.00 mW. It is to be replaced by another type of laser that has a blue beam (λ = 440 nm). About what must be the output of this laser if its beam is to appear equally bright to the eye? (Use Figure 19.11.)

**36. A certain isotropic source provides an illuminance of 40 lx at a wall that is 2.0 m away. A reflector system is then placed on the source and concentrates all its light on a 0.50-m^2 area of the wall. What now will be the illuminance at the area?

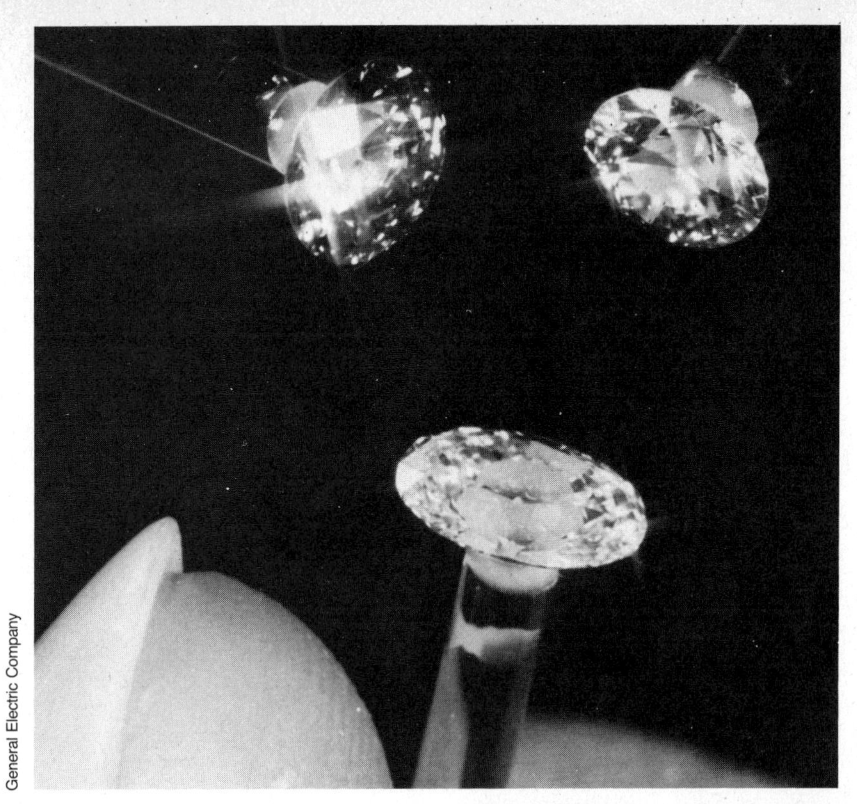

General Electric Company

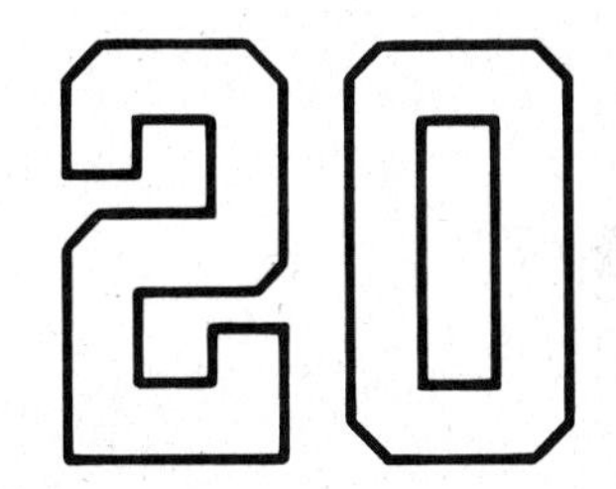

20 REFLECTION AND REFRACTION

Performance Goals

When you finish this chapter, you should be able to

1. State the law of reflection and illustrate its meaning with a diagram.
2. Distinguish between specular and diffuse reflection.
3. Locate the image (or object) for a plane mirror when the object (or image) distance is given. Further, state the relative size and orientation of object and image.

The optical devices used in technology usually involve reflection and refraction of light. As we shall see, mirrors can be used to give images of distant objects. The fact that the paths of waves can be changed by refraction is basic to the use of lenses. In this chapter and the next we shall learn about this behavior of waves and how reflection and refraction are used in various applications.

4. Distinguish between a real and virtual (or imaginary) image.
5. Draw appropriate rays to locate the image of any object placed in front of a concave mirror, including a very distant object. Further, state the character (real or imaginary, erect or inverted, enlarged or diminished) of the image.
6. Repeat Goal 5 for a convex mirror.
7. Given two of the following three quantities (image position, object position, focal length or radius of curvature) for a mirror system, use the mirror equation to obtain the third quantity. In so doing, state the rules governing the signs to be used for p, q, and f.
8. Calculate the magnification and ratio of image height to object height for a mirror system when q and p are given.
9. Show by means of a diagram what is meant by spherical aberration.
10. Find the index of refraction or speed of light for a material when one or the other is given.
11. Find the wavelength and frequency of a light wave in a material of known index of refraction when its wavelength in vacuum is given.
12. Use Snell's law to find what happens to a beam as it travels from one material to another.
13. Two materials of known index of refraction are in contact. Calculate θ_c for total internal reflection and, with the aid of a diagram, show the meaning of θ_c for two materials of known index of refraction.
14. Explain in words the physical basis for the way in which a prism separates colors.

20.1 The Law of Reflection

Light reflects in a simple way from a flat surface. Suppose, as in Figure 20.1(a), that a flashlight beam shines on a flat mirror or other reflecting surface. Experiment tells us that the beam is reflected in the way shown. In terms of the angles we have labeled in the figure the experimental results can be stated as follows.

The angle of incidence equals the angle of reflection.

This is called the *law of reflection.*

We shall often sketch beams of light as shown in part (b) of Figure 20.1. One ray of the beam, with equal angles of incidence and reflection, is used in the sketch. It is assumed that the beam consists of a large number of such rays.

The situation shown in Figure 20.1 is called *specular reflection.*

Specular reflection is the type of reflection that occurs at a mirror surface.

More often reflection is *diffuse.* By this we mean the type of reflection that occurs from a nonmirrorlike surface. An example of diffuse reflection is shown in Figure 20.2. This might represent the reflection from a rough plaster wall. Or it might represent the highly magnified reflection from a sheet of paper. Each ray still obeys the law of reflection, but because the surface is rough, the rays are scattered in many directions.

20.2 Plane Mirror

The flat, or plane, mirror is such a common optical device that we take it for granted. In spite of its simplicity, it illustrates many interesting features. Let us see how such a mirror works.

Suppose you look in a mirror as shown in Figure 20.3. You see your face on the opposite side of the mirror as shown. We call the face you see behind the mirror the *image.* The image is lifelike. If the mirror is perfect, and if you didn't know better, the image could be mistaken for an actual person. The light reaching the viewer's eyes appears to come from the image behind the mirror. Perhaps you have been temporarily fooled by the realistic images formed in plane mirrors in a store. This is because our eyes tell us where the light rays appear to come from. The plane mirror fools the eye by changing the path of the light rays.

An image such as the one in Figure 20.3 is called a *virtual image* (or, sometimes, an *imaginary image*). Notice that the light rays do not go through the image. As a result, a screen or sheet of paper placed at the image position would not show a picture of the image. We therefore make the following distinction.

If a screen placed at the position of an image can show a picture of the image, the image is said to be real. Otherwise the image is called virtual or imaginary.

As we proceed, you will see many examples of this distinction.

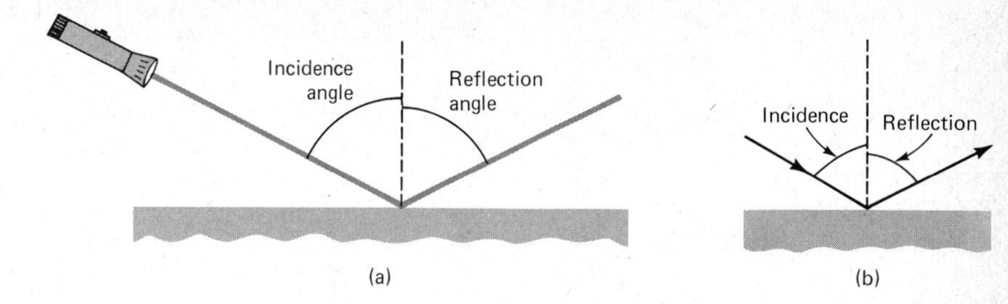

Figure 20.1 Rays of light are reflected in such a way that the angle of incidence equals the angle of reflection.

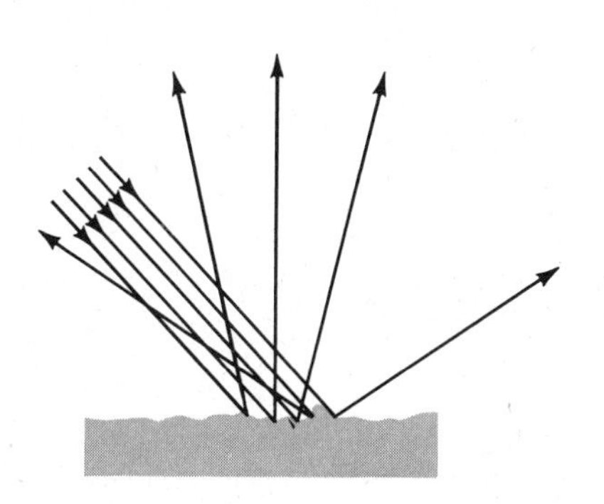

Figure 20.2 Each ray of a beam obeys the law of reflection. When the beam strikes a rough, that is, nonmirror, surface, it is scattered. This is called diffuse reflection.

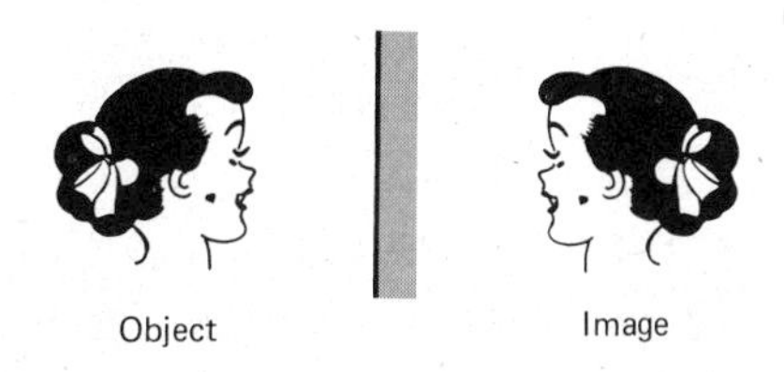

Figure 20.3 The image formed by a plane mirror in this way is a virtual (imaginary) image. A screen placed at the image position would not show the image.

It is of interest to know how the plane mirror causes our eyes to see an image. Suppose that a tiny ball is placed in front of a mirror as shown in Figure 20.4. Light in the room reflects off the ball and so rays leave the ball. In a sense the ball appears to be a light source. It gives off light even though the light it gives off is only reflected room light. Consider two of these rays of light coming from the tiny ball. They hit the mirror. Because they must obey the law of reflection, they reflect back as indicated. They then enter the eye of the viewer. Where does the eye see them coming from?

The eye sees the rays coming from the position located at the intersection of the broken lines. All the rays from the object that reach the eye after reflection also come from this same point. The eye therefore "sees" the object as being at the image position. As far as the eye can tell, the object exists at the position of the image. Of course, the light rays do not come from there. A sheet of paper placed there would not show a picture of the object. The image is virtual.

One other feature of the image should be noticed. As you can see, the image is as far behind the mirror as the object is in front of it. This can be proved by means of similar triangles. (We shall not do it here, but leave it for a problem at the end of this chapter.)

A plane mirror forms a virtual image that is as far behind the mirror as the object is in front of it.

20.3 Focal Point of a Concave Mirror

Let us now turn our attention to curved mirrors. A common type of curved mirror is shown in Figure 20.5(a). It is a portion of a spherical surface. The inner, concave, portion of the sphere is used as the reflecting surface. We therefore refer to this as a concave spherical mirror. Parts (b) and (c) show the ways in which we shall draw such a mirror. The diagram in (c) is more realistic because most spherical mirrors are only slightly curved.

You may have performed the following simple experiment with a concave mirror. (A curved shaving or makeup mirror will work well.) If light from the sun shines on a concave mirror, it forms an image of the sun. As shown in Figure 20.6 it is a real image. When a piece of paper is placed at the image position, the image is often hot enough to burn the paper. Let us refer to Figure 20.6 to see what is happening.

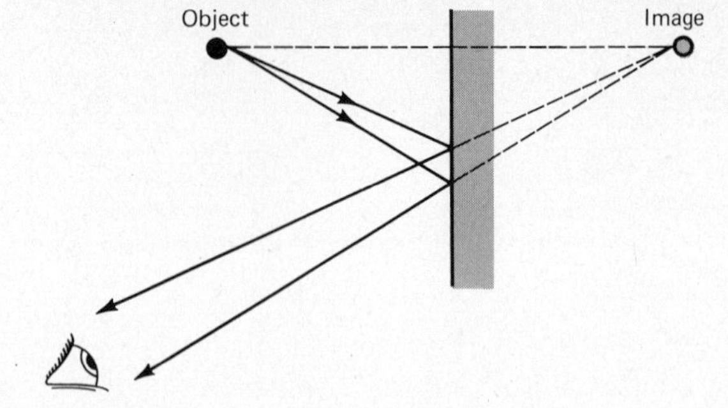

Figure 20.4 The eye sees the rays from the object as though they come from the image. Why is the image virtual?

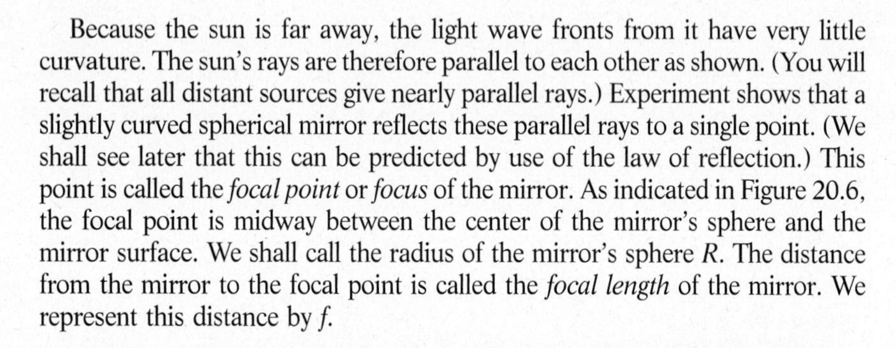

Because the sun is far away, the light wave fronts from it have very little curvature. The sun's rays are therefore parallel to each other as shown. (You will recall that all distant sources give nearly parallel rays.) Experiment shows that a slightly curved spherical mirror reflects these parallel rays to a single point. (We shall see later that this can be predicted by use of the law of reflection.) This point is called the *focal point* or *focus* of the mirror. As indicated in Figure 20.6, the focal point is midway between the center of the mirror's sphere and the mirror surface. We shall call the radius of the mirror's sphere R. The distance from the mirror to the focal point is called the *focal length* of the mirror. We represent this distance by f.

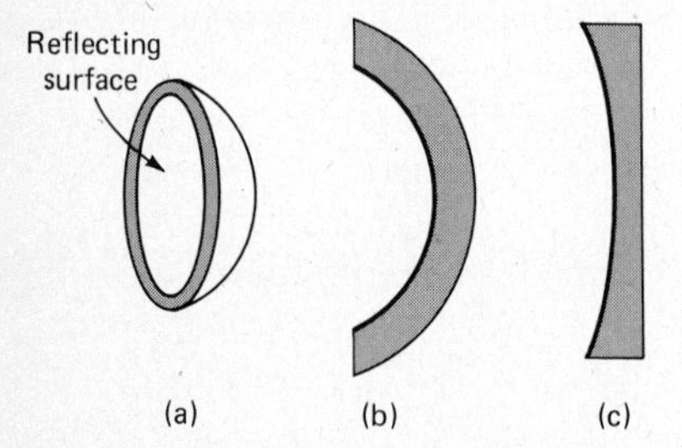

Figure 20.5 The inner side of the spherical surface is used as a reflecting surface. Such a mirror is called a concave spherical mirror.

Parallel rays are focused to the focal point of a concave spherical mirror. The focal length f and the mirror's curvature radius R are related by

$$f = \tfrac{1}{2}R \qquad (20.1)$$

This is true only if the rays are nearly parallel to the axis of the mirror. (See Figure 20.6 for what we mean by the mirror axis.) Of course, rays from any distant source are parallel and will also be focused in this way. Therefore, a concave spherical mirror causes any distant object to be imaged at the focal point.

We should notice another fact from Figure 20.6. If you place your eye at the position labeled E, then the image of a distant source will be seen at the focal

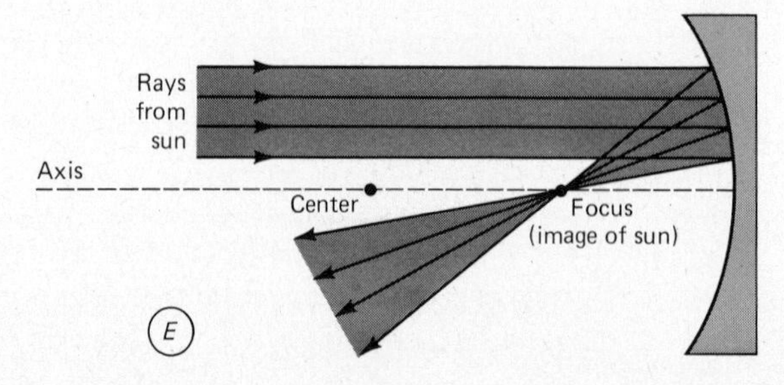

Figure 20.6 Parallel rays are reflected through the focal point (or focus) of the mirror. Notice that the focal point is midway between the mirror surface and its center.

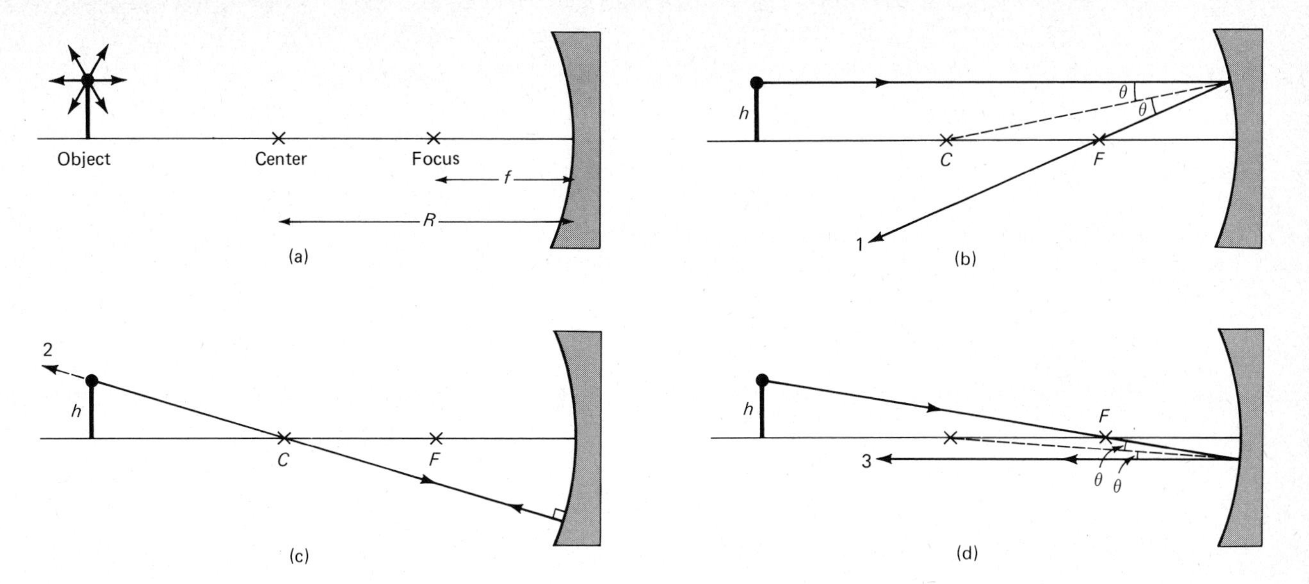

Figure 20.7 The three rays shown in (b), (c), and (d) are easily drawn. You should know how to draw them no matter where the object is placed.

point. (In this case use a distant light bulb instead of the sun. The sun's light is so strong that it may hurt your eye.) The rays coming to the eye come from the focal point. The eye will therefore see the real image that exists there.

20.4 Images Formed by a Concave Mirror

As we just saw, distant objects are imaged at the focal point by a concave mirror. Where is the image of a less distant object formed? To answer this question we shall consider what happens to three rays that leave a point on the object and strike a concave mirror. Rays of light come off a point on an object in all directions, as shown in Figure 20.7(a).

1. Ray 1, shown in Figure 20.7(b), is parallel to the mirror axis. It is reflected through the focus. Notice that the law of reflection causes this ray to behave as it does.
2. Ray 2, shown in part (c), goes through the center of curvature and therefore strikes perpendicular to the mirror surface. (Recall that a radius of a circle is perpendicular to a tangent.) Therefore it reflects straight back on itself as shown.
3. Ray 3, shown in part (d), is much like ray 1. However, it first goes through the focus. Then it reflects parallel to the axis. Here, too, the law of reflection causes the ray to follow the path shown.

We now have three rays that we can draw without measuring angles. They are drawn together in Figure 20.8. Be sure that you see how each ray is drawn in both cases. You should be able to draw all three rays no matter where the object is placed. When these three rays enter the eye , the eye sees them as coming from the image position. These three rays are typical. All other rays give rise to the

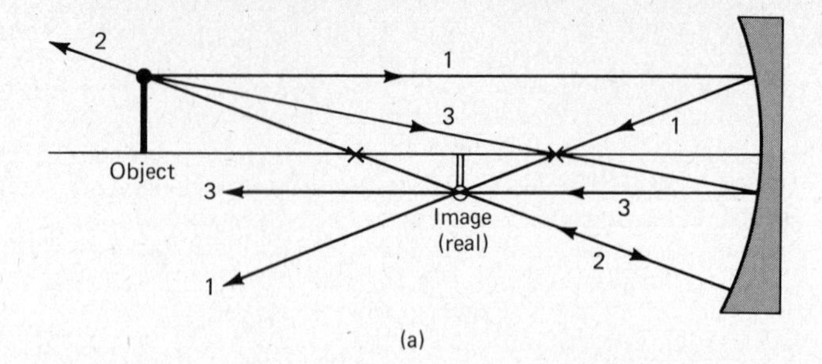

Figure 20.8 When rays 1, 2, and 3 enter the eye, they cause the eye to see the image indicated. Be sure that you understand each ray. Note especially rays 2 and 3 in (b).

same image as these do. Notice that the eye sees a real image in (a), but a virtual image in (b).

▮ **EXAMPLE 20.1** An object is placed at the center of curvature of a spherical mirror. Where is the image formed? Describe the image by stating if it is (a) real or virtual, (b) erect or inverted, and (c) enlarged or diminished.

Solution The situation is shown in Figure 20.9. We draw rays 1 and 3 to locate the image as shown. (Try to draw ray 2. Why don't we show it? Sometimes only two of the three rays are convenient to draw.) The image is real. A screen placed at the image position would be struck by the reflected rays. They would form an image of the object on the screen. The image is inverted (turned upside down) in comparison to the object. In this case, the image is the same size as the object. ▮▮

20.5 The Mirror Equation

It is always possible to find the image position by drawing a ray diagram. However, accurate sketches are often troublesome to make. We would like an algebraic method to find the image position. A rough ray diagram could then be

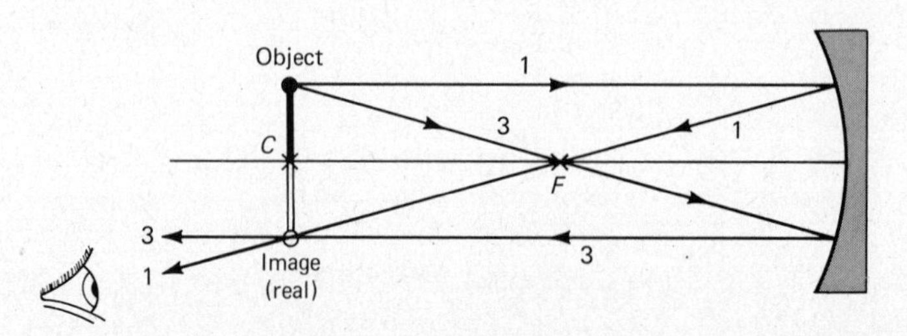

Figure 20.9 The image is real, inverted, and unchanged in size.

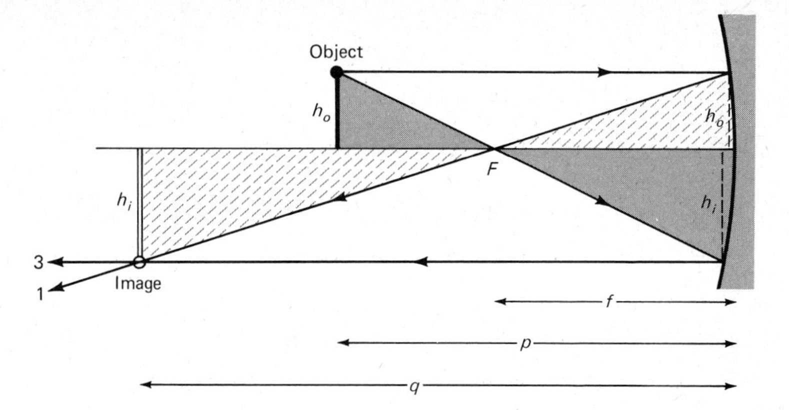

Figure 20.10 If the mirror is only slightly curved, the right-hand sides of the triangles lie nearly on the mirror surface.

used to check the exact arithmetic. In this section we shall find the equation that concave mirrors obey.

To find the mirror equation, let us refer to Figure 20.10. We have drawn rays 1 and 3 on it. Consider the two heavily shaded triangles. They are right triangles with equal acute angles. Therefore, geometry tells us, they are similar triangles. Notice that one side of one triangle is h_o, the height of the object, and the similar side in the other triangle is h_i, the image height.

In similar triangles the ratios of similar sides are equal. Therefore for these two triangles

$$\frac{\text{End of one}}{\text{End of other}} = \frac{\text{side of one}}{\text{side of other}}$$

As you can see from the figure, these ratios are

$$\frac{h_o}{h_i} = \frac{p - f}{f}$$

As shown in the diagram, p is the object distance, that is, it is the distance of the object from the mirror surface. In writing this, we have assumed that the curvature of the mirror is so small that the end of the triangle can be taken to coincide with the mirror surface.

We can obtain another equation for h_o/h_i by considering the triangles shaded with lines. They, too, are similar triangles. The ratios of their sides can be written as

$$\frac{h_o}{h_i} = \frac{f}{q - f}$$

As shown in the diagram, q is the image distance, that is, it is the distance of the image from the mirror surface. (Notice that we again assume the curvature of the mirror to be small.)

If we equate the two expressions for h_o/h_i, we have

$$\frac{p-f}{f} = \frac{f}{q-f}$$

Now, cross multiply to obtain

$$(p-f)(q-f) = f^2$$

or

$$pq - pf - fq + f^2 = f^2$$

Because f^2 occurs on both sides of the equation, subtract it from each side. Then divide through the whole equation by pqf and find that

$$\frac{1}{f} - \frac{1}{q} - \frac{1}{p} = 0$$

After transposing, this gives what is known as the *mirror equation:*

$$\frac{1}{p} + \frac{1}{q} = \frac{1}{f} \qquad (20.2)$$

In using the mirror equation, we must be careful about signs. Objects and images in front of the mirror have positive values for p and q. As we have seen in our ray diagrams, images sometimes exist behind the mirror. Virtual images such as that will have q values that are negative.

EXAMPLE 20.2 A concave mirror has a 30-cm radius of curvature. If an object is placed 10 cm from the mirror, where will the image be found?

Solution We make use of the mirror equation, Equation 20.2. Recall that Equation 20.1 tells us that $f = \frac{1}{2}R$ which is 15 cm in this case. Also, we are told that p, the object distance, is 10 cm. Upon substitution in Equation 20.2, we have

$$\frac{1}{10} + \frac{1}{q} = \frac{1}{15}$$

with all distances in centimeters. We can easily find a least common denominator and write

$$\frac{3}{30} + \frac{1}{q} = \frac{2}{30}$$

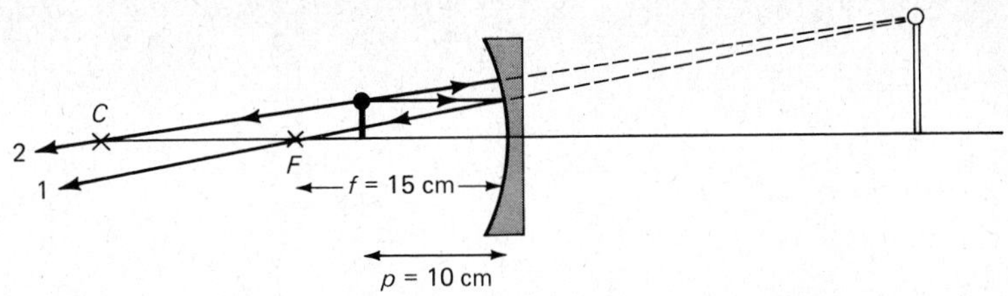

Figure 20.11 A very accurate diagram would be needed to show that the image is exactly at 30 cm.

which gives

$$\frac{1}{q} = -\frac{1}{30}$$

and so

$$q = -30 \text{ cm}$$

Or if you prefer, you can use a calculator to write the equation

$$0.100 + \frac{1}{q} = 0.0667$$

Then you obtain

$$\frac{1}{q} = -0.0333$$

from which

$$q = -30 \text{ cm}$$

The negative image distance tells us that the image is behind the mirror. We should always check our computations with a rough ray diagram. This is done in Figure 20.11. Notice that the image is virtual and erect (right side up). ■■

20.6 Magnification

It is often important to find the ratio of image height to object height. This tells us by what factor the image is magnified. We call the ratio h_i/h_o the *magnification.* A simple equation can be found for this ratio. To find it, consider the special rays shown in Figure 20.12. The law of reflection tells us the angles they make are equal. The image must exist either in front of the mirror as in (a) or behind it as in (b). In either case the shaded triangles are similar. From the ratio of sides we have at once that

$$\frac{h_i}{h_o} = \frac{q}{p}$$

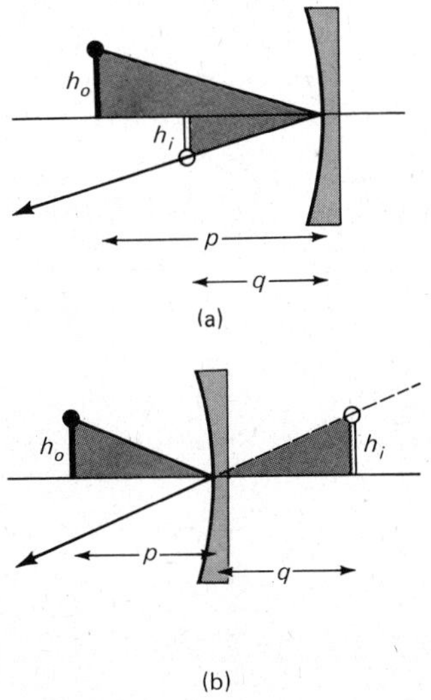

Figure 20.12 To find the magnification, use the similar triangles shown. In each case $h_i/h_o = q/p$.

In other words, the magnification is equal to the ratio of the image to object distance.

$$\text{Magnification} = \frac{h_i}{h_o} = \frac{q}{p} \qquad (20.3)$$

EXAMPLE 20.3 A 20-cm-focal-length mirror is to be used to obtain a real image of an object. The image is to be one-third as large as the object. Where should the object be placed? Where will the image be?

Solution We know that the ratio h_i/h_o is to be $\frac{1}{3}$. Therefore, from Equation 20.3

$$\frac{1}{3} = \frac{q}{p}$$

Cross multiplication gives $p = 3q$.

Because the image is to be real, the signs of both p and q are to be positive. We can use the mirror equation with $f = 20$ cm, $p = 3q$, and $q = q$.

$$\frac{1}{p} + \frac{1}{q} = \frac{1}{f}$$

becomes

$$\frac{1}{3q} + \frac{1}{q} = \frac{1}{20\text{ cm}} \qquad \text{or} \qquad \frac{1}{3q} + \frac{3}{3q} = \frac{1}{20\text{ cm}}$$

Solving this for q yields $q = 26.7$ cm. And since $p = 3q$, $p = 80$ cm.

We should place the object 80 cm from the mirror. The image will then be found at 26.7 cm and it will be one-third as large as the object. You should check this computation with a ray diagram.

20.7 Convex Spherical Mirror

A convex spherical mirror reflects light from the outside of a mirrored sphere. This is shown in Figure 20.13(a). When sunlight is reflected from a convex mirror, the rays are diverged (or fanned out), as shown in Figure 20.13. They all appear to come from a point F, the focal point or focus of this type of mirror. As with concave mirrors, the focal length f is related to the radius of curvature by Equation 20.1, namely, $f = \frac{1}{2}R$.

There are three simple rays we can draw to locate the image for a convex mirror. They are shown in Figure 20.14.

1. Ray 1 is parallel to the axis. In this respect it is similar to ray 1 we drew for a concave mirror. Here, however, the ray diverges from the focus after reflection.

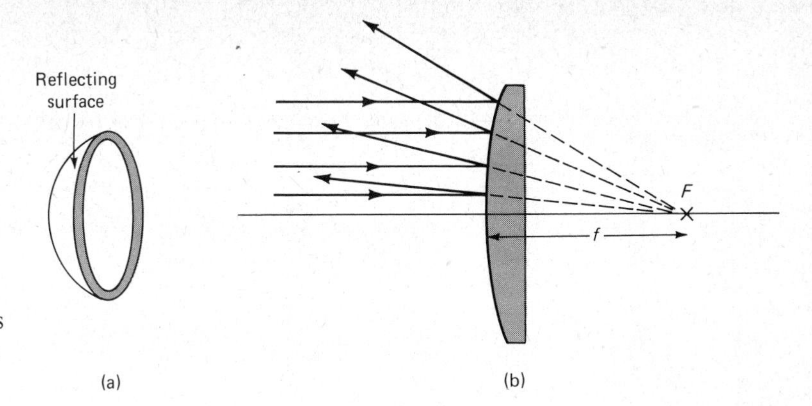

Figure 20.13 A convex mirror diverges parallel rays. They appear to come from the focal point, *F*.

2. Ray 2 strikes the mirror surface perpendicularly. It is reflected straight back upon itself.
3. Ray 3 leaves the object heading for the focus. It is reflected parallel to the mirror axis.

We show in Figure 20.15 how these three rays are used to locate the image. Examine each carefully. If the reflected rays (1, 2, and 3) enter the eye, they are seen as coming from the image behind the mirror. You should sketch a few other similar situations with the object placed differently. Then you will find that the image is always behind the mirror; it is always virtual.

The mirror equation can also be derived for a convex mirror. However, we should notice two facts in Figure 20.15. The image position is behind the mirror. We earlier agreed that q will be a negative number in this case. In addition, unlike the situation for a concave mirror, the focal point F is behind the mirror.

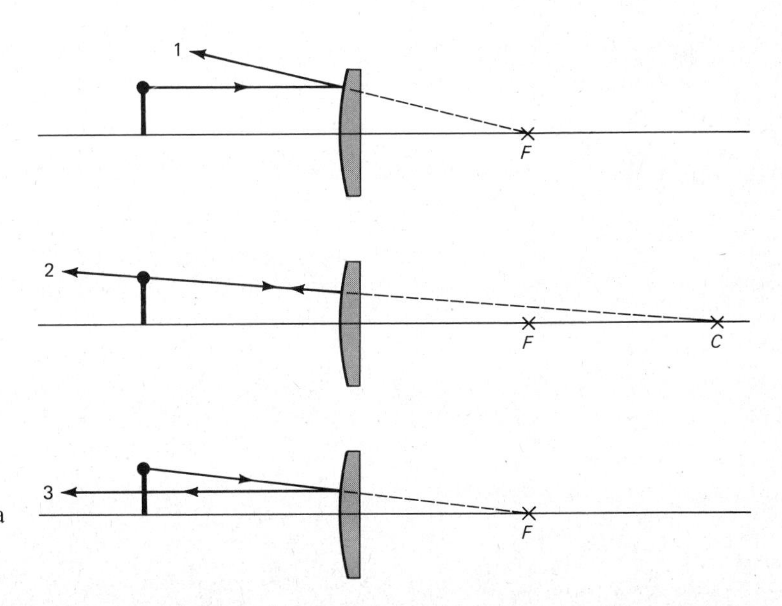

Figure 20.14 The three rays used for a convex mirror. Each is similar to the corresponding ray for a concave mirror.

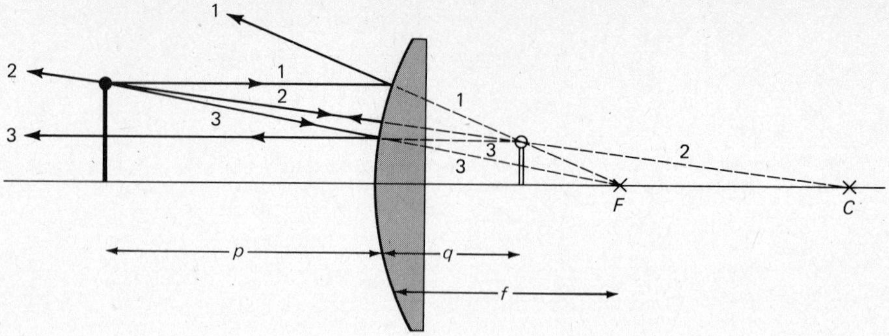

Figure 20.15 The three rays are used to locate the image for a convex mirror. What happens to the image as the object is moved closer to the mirror? As the object is moved farther from the mirror?

As a result, its distance, the focal length f, must be taken negative. It is then found that Equation 20.2, the mirror equation, applies to convex mirrors as well as concave ones. The mirror equation for both mirrors is

$$\frac{1}{p} + \frac{1}{q} = \frac{1}{f} \tag{20.2}$$

In using the mirror equation we must remember the following.

The focal length f is a positive number for concave mirrors and a negative number for convex mirrors. The image distance q is positive if the image is on the lighted side of the mirror; otherwise it is negative. The object distance p is always positive except in one very rare case. If the object is not real, it can exist behind the mirror. In that case its distance p must be taken negative.

Before giving examples involving convex mirrors, let us consider the size of the image formed. In the same way as with a concave mirror, we can derive Equation 20.3. For a convex mirror, too, we find that the ratio $h_i/h_o = q/p$. The magnification relation applies to both types of mirror.

EXAMPLE 20.4 A 2.0-cm object is placed 30 cm in front of a convex mirror. The radius of curvature of the mirror is 80 cm. What is the focal length of the mirror? Where is the image formed. How large is the image? Is it real or virtual? Is it erect or inverted?

Solution We shall solve the problem using the mirror equation. Then we shall check our solution and answer the last two questions by drawing a rough ray diagram.

The focal length has a numerical value equal to $\frac{1}{2}R$, or 40 cm. Because this is a convex mirror, a negative sign must be given to f. Therefore, $f = -40$ cm. Because we were told the object distance is 30 cm, the mirror equation becomes, in centimeters,

$$\frac{1}{30} + \frac{1}{q} = \frac{1}{-40}$$

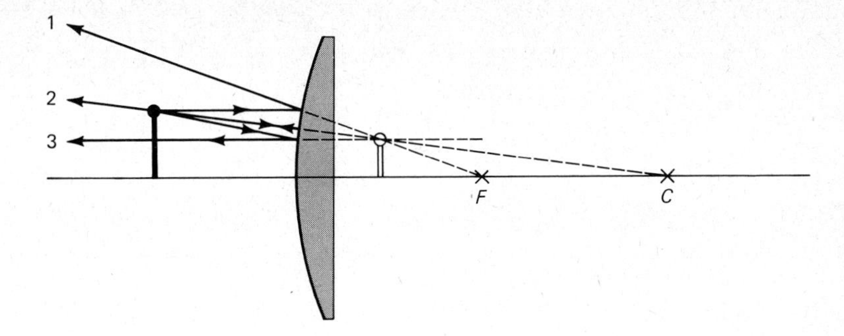

Figure 20.16 The image is virtual, erect, and decreased in size.

From which

$$\frac{1}{q} = -0.0583$$

and

$$q = -17.2 \text{ cm}$$

Notice that q is negative. The image is therefore behind the mirror.

To find the size of the image, we use the magnification relation:

$$\frac{h_i}{h_o} = \frac{q}{p} \quad \text{or} \quad h_i = h_o\left(\frac{q}{p}\right)$$

which gives

$$h_i = (2 \text{ cm})\left(\frac{17.2}{30}\right) = 1.15 \text{ cm}$$

Notice that we do not carry signs along when using the magnification equation. If you do carry the signs, there are rules for using them to predict whether or not the image is inverted. Because we shall obtain this information directly from a ray diagram, we shall not make use of these rules.

We now check our solution by means of the ray diagram shown in Figure 20.16. As we see, the image is virtual and erect. ■■

20.8 Spherical Aberration

Our mirror diagrams show the mirror to be only a small portion of a sphere. The reason for this is shown in Figure 20.17(a). If you examine each ray shown, you will see that it obeys the law of reflection. Clearly, though, the rays do not all pass through the focus F. Only the rays that lie close to the axis are reflected through F. Therefore, all our previous discussions apply only to these rays.

Rays that strike the outer portions of a highly curved spherical mirror do not focus properly. Therefore, to obtain sharp images, only slightly curved spherical

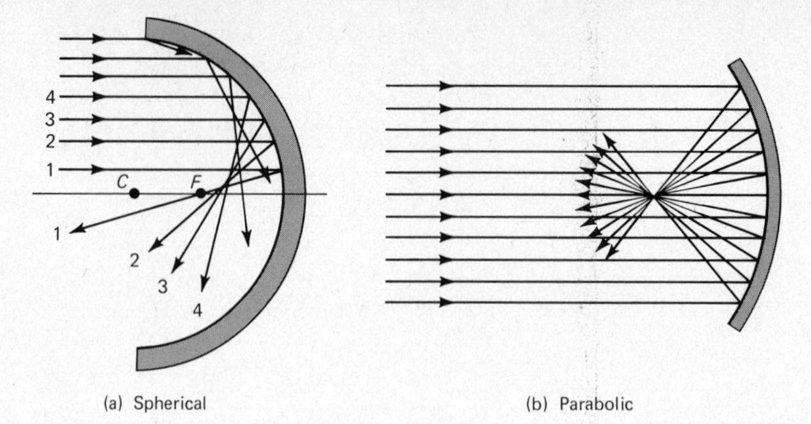

Figure 20.17 Only the central portion of a spherical mirror reflects rays through the focus *F*. A parabolic mirror does not suffer from spherical aberration.

mirrors can be used. We call this defect of spherical mirrors *spherical aberration.* Because of spherical aberration, only a small portion of a sphere should be used for a spherical mirror.

To overcome this defect, the best mirrors are parabolic rather than spherical. Such a mirror is shown in Figure 20.17(b). Notice that the parabolic mirror reflects all the parallel rays to the focus. It therefore forms sharper images than a spherical mirror does. Unfortunately, parabolic mirrors are more difficult to construct than spherical ones are. For this reason spherical mirrors are much more common than are parabolic.

EXAMPLE 20.5 A hot arc light source is to be used with a spherical reflector to make a searchlight. Where should the arc be placed relative to the reflector?

Solution Usually one wants a searchlight to send out a parallel beam. This ensures that the beam will travel over large distances and still remain concentrated. The situation is shown in Figure 20.18.

As we see, the searchlight ray diagram reminds us of the way parallel rays are focused. Parallel rays are reflected to the focus. But the rays must obey the law of reflection no matter which way they are going. Therefore, light rays from a source at the focus will be reflected parallel to the axis, as shown. We therefore conclude that the arc should be placed at the focus of the mirror. Of course, a parabolic reflector would be preferable to the spherical one. (How would you solve this example by formula? Notice that $p = f$.)

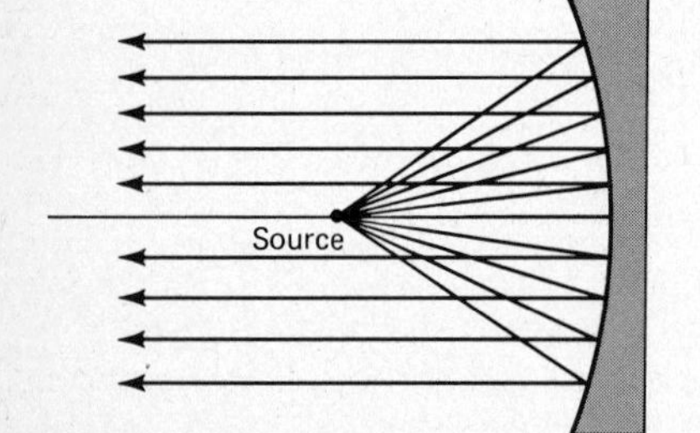

Figure 20.18 Light rays obey the law of reflection. If a mirror reflects parallel rays through the focus, then a light source at the focus of that mirror gives rise to parallel rays.

20.9 Speed of Light; Refractive Index

We learned in Chapter 19 that light travels through vacuum with a speed of 2.998×10^8 m/s. This speed is designated by the symbol c.

$$\text{Speed of light in vacuum} = c = 2.998 \times 10^8 \text{ m/s}$$

Usually this is rounded off to 3.00×10^8 m/s.

Light travels less fast than c in all materials. For example, in water the speed of light is $0.75c$. Even in air the speed of light is somewhat less than in vacuum. It is about $0.9997c$. The speed in a material is designated v.

It is customary and convenient to make use of a quantity called the *refractive index,* μ, of a material. By definition,

$$\text{Absolute refractive index} \equiv \frac{\text{speed of light in vacuum}}{\text{speed of light in material}}$$

Or

$$\mu = \frac{c}{v} \tag{20.4}$$

Because c is always larger than v, the refractive index is always larger than 1.00. Typical values are given in Table 20.1. Of course, μ has no units because it is a ratio.

EXAMPLE 20.6 What is the speed of light in diamond?

Solution From Table 20.1, μ for diamond is 2.42. Therefore, the speed of light in diamond, v, is obtained from

$$\mu = \frac{c}{v} \qquad \text{or} \qquad v = \frac{c}{\mu}$$

to be

$$v = \frac{3.00 \times 10^8\ \text{m/s}}{2.42} = 1.24 \times 10^8\ \text{m/s}$$

Table 20.1 REFRACTIVE INDEXES (λ = 589 nm)

Material	$\mu = c/v$
Normal air	1.0003
Water	1.33
Ethanol	1.36
Acetone	1.36
Fused quartz	1.46
Benzene	1.50
Lucite or Plexiglas	1.51
Crown glass	1.52
Polystyrene	1.59
Carbon disulfide	1.63
Flint glass	1.66
Methylene iodide	1.74
Zircon	1.92
Diamond	2.42

20.10 Wavelength in Materials

Let us now investigate what happens to a set of waves as it moves from one material to another. What we shall find is true for any wave. Therefore, for ease in thinking about it, let us consider a particularly simple system.

Suppose that we consider the marching band (or perhaps soldiers) shown in Figure 20.19. The waves of marchers can be considered the positions of wave crests. At the left of the diagram they are marching on pavement, where their speed is v_1. The distance between waves, the wavelength, is λ_1. They march to the right onto grass, where their speed is less, v_2. Their wavelength there is λ_2.

The marchers move less rapidly on the grass than on the pavement. During the time that the marchers move a distance λ_1 on pavement, they will move a smaller distance λ_2 on the grass. Therefore, the wavelength λ_2 is less than λ_1. The wave crests are closer together in the slow marching region.

To find the exact relation between λ_1 and λ_2, we note the following. The time taken between successive waves striking the edge of the grass is T, the period of the wave. During this time, the waves on the pavement will move a distance (given by $s = \bar{v}T$), which is

$$\lambda_1 = v_1 T$$

During the same time the waves on the grass will move a distance

$$\lambda_2 = v_2 T$$

If we divide one of these equations by the other, we find that

$$\frac{\lambda_1}{\lambda_2} = \frac{v_1}{v_2}$$

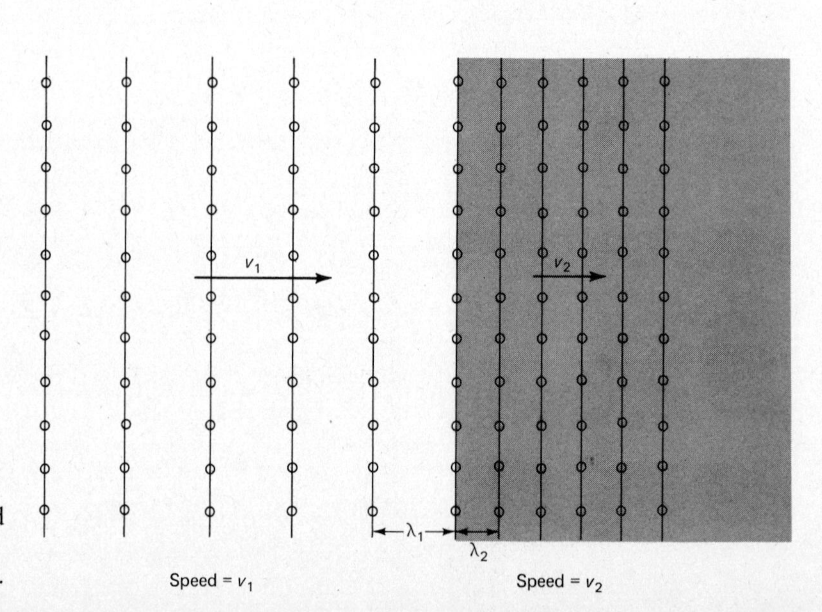

Figure 20.19 As the waves of marchers enter the shaded area (area of grass), they slow from speed v_1 to speed v_2. The distance between waves decreases from λ_1 to λ_2; $\lambda_2 = \lambda_1\,(v_2/v_1)$.

We can therefore state the following.

Waves with wavelength λ_1 in a material where the speed is v_1 will have a wavelength λ_2 given by

$$\lambda_2 = \lambda_1\left(\frac{v_2}{v_1}\right) \tag{20.5}$$

in a material where the speed is v_2.

Because the same number of waves of marchers passes the boundary (or any other point) in a given length of time, the frequency of the wave is the same in the two materials.

We can draw another conclusion from Equation 20.5. Suppose that material 1 is vacuum. Then $v_1 = c$ and the wavelength is λ_v. We have

$$\lambda_2 = \lambda_v\left(\frac{v_2}{c}\right)$$

But c/v_2 is μ_2, the index of refraction of material 2. Therefore we find that

$$\lambda_2 = \frac{\lambda_v}{\mu_2} \tag{20.6}$$

In other words:

If the wavelength of a light beam in vacuum is λ_v, then its wavelength in a material of refractive index μ is λ_v/μ.

Again we emphasize that the frequency of the wave does not change.

20.11 Refraction of Waves: Snell's Law

Consider a flashlight beam shining from air into water as shown in Figure 20.20. Experiment shows that the light beam changes direction (or bends) at the water surface. Notice in Figure 20.20(a) the names we give the angles the beam makes to the perpendicular line. We call this phenomena *refraction.*

The bending of a light beam (or ray) as it passes from one material to another is called refraction.

To see what causes refraction, consider once again our marching wave fronts. Now, however, they strike the boundary between two materials at an angle, as shown in Figure 20.20(b). If v_2 is less than v_1, the distance between waves will shorten as they enter material 2. Notice the wave front that is half in and half out

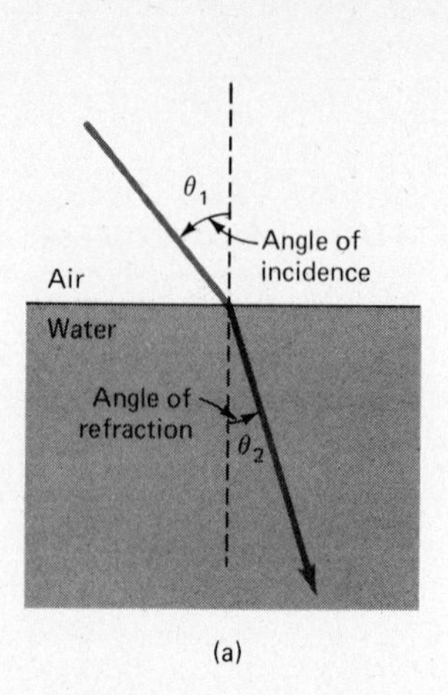

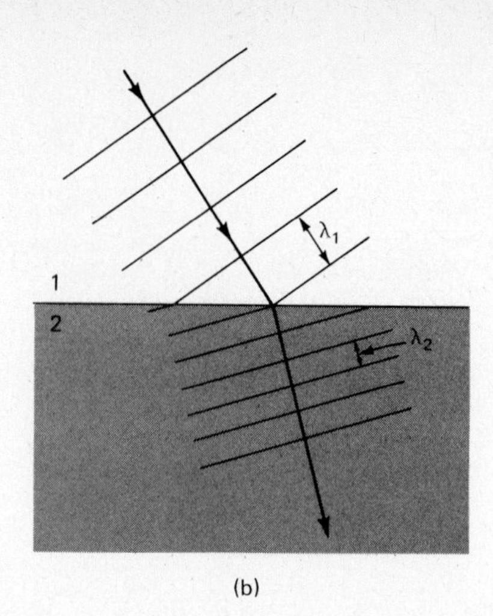

Figure 20.20 The beam refracts (changes direction) because of the difference in wave speed in the two materials. Notice that the rays must always be perpendicular to the wave fronts.

of material 2. The part in 2 has been slowed down. This causes the wave to change its orientation. Because the rays are always perpendicular to the wave fronts, the beam direction is also changed. (The marchers always march in straight lines in a direction perpendicular to the lines of marchers.)

We can obtain an important relation between θ_1, the angle of incidence, and θ_2, the angle of refraction. To do so, consider the two wave fronts shown in Figure 20.21. You should be able, using geometry, to satisfy yourself that the angles are labeled correctly. Notice that the distance between wave fronts is λ_1 in material 1 and λ_2 in material 2.

Consider the two right triangles with common side AB. From them we have

$$\sin\theta_1 = \frac{\lambda_1}{AB} \qquad \text{and} \qquad \sin\theta_2 = \frac{\lambda_2}{AB}$$

Since $\lambda_1 = v_1 T$ and $\lambda_2 = v_2 T$, these become

$$AB = \frac{v_1 T}{\sin\theta_1} \qquad \text{and} \qquad AB = \frac{v_2 T}{\sin\theta_2}$$

We can equate these two values for AB to obtain

$$\frac{v_1 T}{\sin\theta_1} = \frac{v_2 T}{\sin\theta_2}$$

Rearranging and canceling T gives

$$\frac{1}{v_1}\sin\theta_1 = \frac{1}{v_2}\sin\theta_2$$

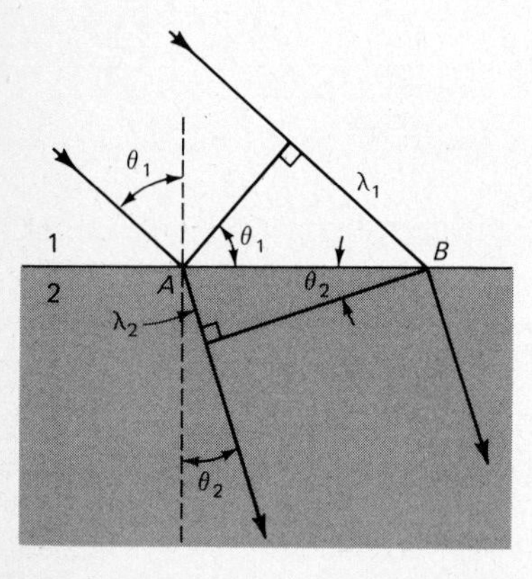

Figure 20.21 The two wave fronts are λ_1 apart in material 1 and λ_2 apart in material 2.

It is customary to write this in terms of the refractive indexes. Multiplying both sides of the equation by c and remembering that $c/v = \mu$, we have

$$\mu_1 \sin \theta_1 = \mu_2 \sin \theta_2$$

What we have found is known as *Snell's law.* It can be summarized as follows.

If a light beam at angle θ_1 in material 1 enters material 2 where its angle is θ_2, the relation between θ_1 and θ_2 is

$$\mu_1 \sin \theta_1 = \mu_2 \sin \theta_2 \qquad (20.7)$$

The quantities μ_1 and μ_2 are the two refractive indexes.

It is convenient to remember when using Snell's law that the index of refraction of vacuum (or, approximately, air) is 1.000.

EXAMPLE 20.7 (a) A layer of oil floats on water. A beam of light enters the oil from air at the angle shown in Figure 20.22. Find the angle the beam makes in the oil and in the water. (μ for the oil is 1.52.) (b) What would the angle in the water be if the oil were not present?

Solution (a) We can use Snell's law at the air-oil surface. Then

$$\mu_a \sin 70° = \mu_o \sin \theta_o$$

But $\mu_a = 1.00$ and $\mu_o = 1.52$, so

$$\sin \theta_o = \frac{1.00}{1.52} \sin 70° = 0.618$$

From this,

$$\theta_o = 38.2°$$

The same procedure can now be used at the oil-water surface. We have

$$\mu_o \sin 38.2° = \mu_w \sin \theta_w$$

But $\mu_o = 1.52$, whereas $\mu_w = 1.33$. We then have

$$\sin \theta_w = \frac{1.52}{1.33} \sin 38.2° = 0.707$$

from which

$$\theta_w = 45°$$

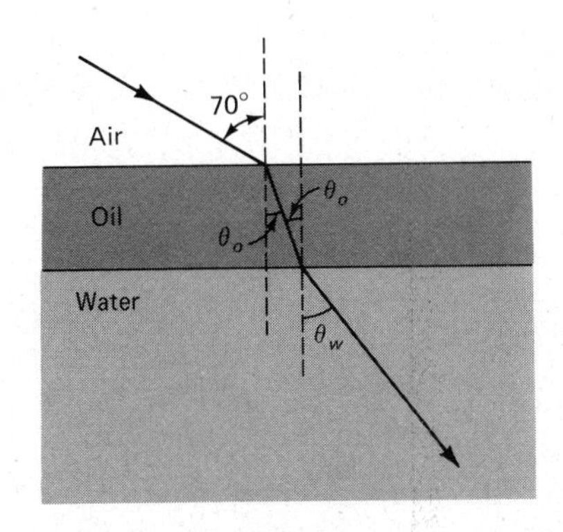

Figure 20.22 As a beam goes from a small-μ material to a larger-μ material, it bends toward the normal. It bends away from the normal in the reverse case.

(b) In this case

$$\mu_a \sin 70° = \mu_w \sin \theta_w$$

Then

$$\sin \theta_w = \frac{\mu_{air}}{\mu_w} \sin 70° = \frac{1.000}{1.33} (\sin 70°) = 0.707$$

from which $\theta_w = 45°$. Notice that this is the same as we obtained in (a). This is an example of the fact that

A parallel layer does not affect the direction of travel of a light ray.

You will be asked to prove this in general in a problem.

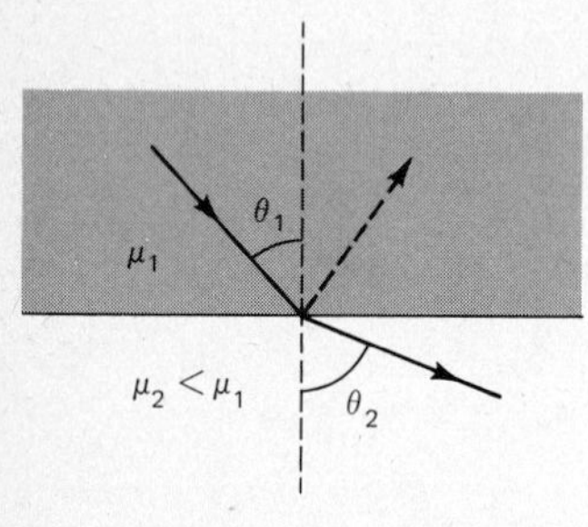

(a) Notice that $\mu_1 > \mu_2$

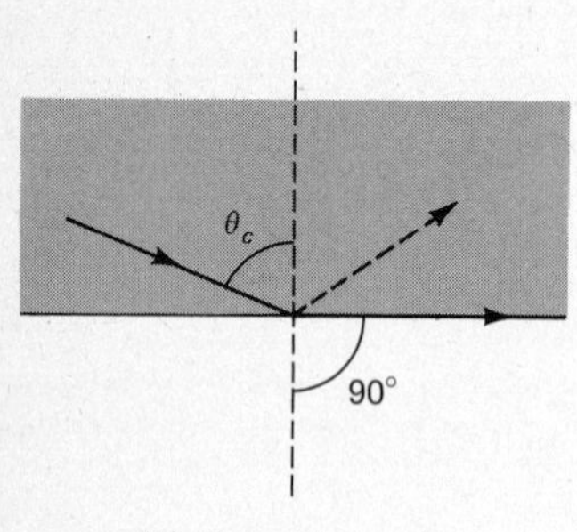

(b) The critical angle

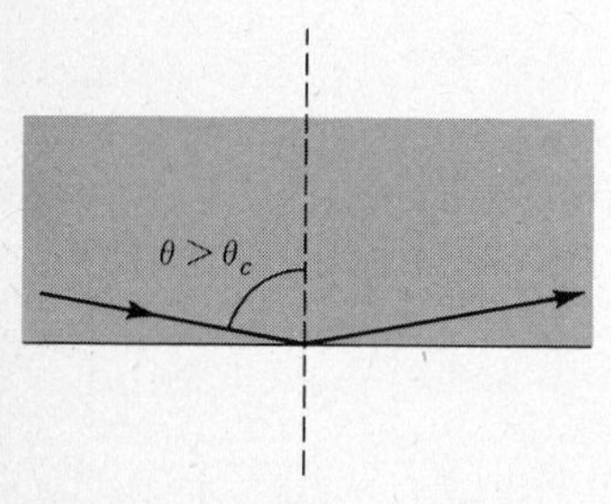

(c) Total internal reflection

Figure 20.23 If θ_1 is greater than the critical angle, the beam cannot enter material 2. It is totally internally reflected.

20.12 Total Internal Reflection

Suppose, as shown in Figure 20.23, that a light beam passes from a material with high index of refraction to one of lower index. (We say that the higher index material is *more dense optically* than the lower index material.) Notice that the beam is deflected away from the normal (or perpendicular) line as it enters the lower material.

An interesting and useful effect can occur. If θ_1 is large enough, the refracted beam skims along the surface. The value of θ_1 at which this occurs is labeled θ_c and is called the *critical angle.* If θ_1 is still larger, as in (c), there is no refraction. All the light is reflected at the surface. None of it enters material 2 as a refracted beam.

To take a specific case, suppose that material 1 is glass and material 2 is air. The beam is going from an optically dense material to a less dense material. It will be refracted as shown in Figure 20.23. Some of the beam is always reflected at the boundary. It is shown as the broken ray in (a) and (b). But when θ_1 becomes larger than θ_c, the light beam cannot escape from the glass. It is totally internally reflected.

Let us now find θ_c. It is the value of θ_1 that makes $\theta_2 = 90°$. From Snell's law,

$$\mu_1 \sin \theta_1 = \mu_2 \sin \theta_2$$

we have in this critical case

$$\mu_1 \sin \theta_c = \mu_2 \sin 90°$$

But sin 90° = 1, and so

$$\sin \theta_c = \frac{\mu_2}{\mu_1} \tag{20.8}$$

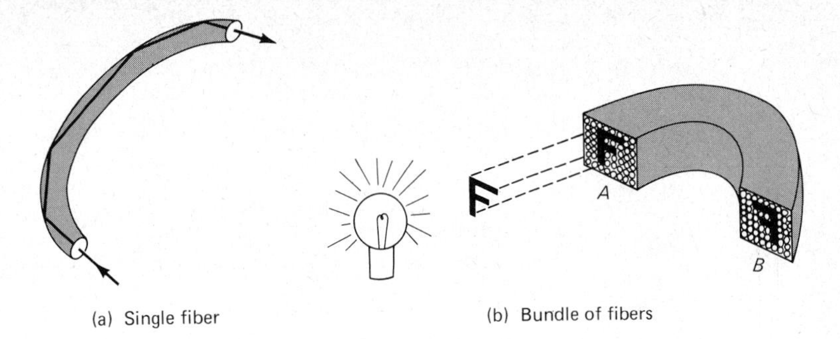

Figure 20.24 The fiber bundle transmits a light pattern.

Notice that this relation makes sense only if $\mu_2 < \mu_1$. The sine of an angle can never be greater than unity. Therefore, total internal reflection (and Equation 20.8) can exist only if $\mu_2 < \mu_1$.

In the particular case of crown glass and air, we have the following value for θ_c:

$$\sin \theta_c = \frac{1.00}{1.52} = 0.658$$

From this, $\theta_c = 41°$. In other words, if θ_1 exceeds 41°, a beam of light in crown glass will be completely reflected at its surface. No light will escape into the air.

Total internal reflection is used in light pipes. Consider the glass fiber shown in Figure 20.24(a). Because of the gentle curvature of the fiber, the light beam is unable to escape from the fiber. The fiber therefore acts like a pipe for the beam.

Bundles of glass fibers can be used to carry designs from place to place. This is shown in part (b) of the figure. Such fiber optics methods can be used in many other ways. For example, the internal walls of the stomach and intestine can be seen by means of fiber optical devices.

20.13 Prisms and Color

Until now we have considered only monochromatic (single-color) beams of light. Many interesting effects and devices make use of the fact that the speed of light depends on the wavelength of light. All wavelengths of electromagnetic radiation have the same speed, *c*, in vacuum. But this is not true in materials such as glass, water, and so on. For example, blue light travels less fast in glass than does red light. This is true for most other materials as well.

Since the index of refraction of a material μ is equal to c/v, we see that the value of μ depends on color. The index of refraction for blue light is usually larger than that for red light. Therefore, in refraction, blue light is usually bent more than red light.

This variation of μ with λ allows a prism to be used to separate colors. As we see in Figure 20.25, a beam of white light (which contains all colors) is sepa-

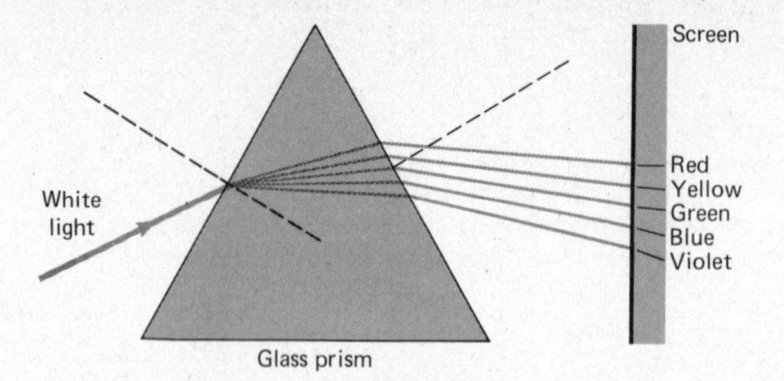

Figure 20.25 The prism disperses the white light into a spectrum of colors. On the screen the white-light spectrum appears as a continuous rainbow gradually shading from red to violet.

rated into its colors by a prism. We say that the beam is *dispersed* into its component colors by the prism. Notice that this effect is due to the beam's refraction at the two surfaces of the prism. Because wavelengths at the blue end of the spectrum are bent more than those near the red, the white light shows its complete spectrum after leaving the prism. This is one way of showing what wavelengths exist in a beam of light. We shall discuss it in more detail in later chapters. (See also the spectra in Figure 19.6.)

POLAROID SUNGLASSES AND GLARE

We pointed out in the previous chapter that ordinary light consists of unpolarized light. The **E** vector in it vibrates in all possible directions perpendicular to the light ray. Consider the **E** vector shown in part (a) of the figure. It can be split into two components as shown. Suppose now that a sheet of Polaroid is placed in such a way that it transmits only the vertical part of the vibration. Then only E_v but not E_h is transmitted. On the average, for all directions of vibration, the Polaroid transmits only half of each **E** vector. Therefore, a Polaroid sheet reduces the amplitude* of the unpolarized light vibration by a factor of $\frac{1}{2}$.

Polaroid sunglasses consist of Polaroid sheets. They are oriented so that only vertical vibrations are transmitted. For unpolarized light they reduce the light amplitude by $\frac{1}{2}$. But they serve another purpose. They reduce the effects of glare.

Sunlight reflected to the eye by water, snow, pavement, a page of a book, and so on, is one important source of glare. It turns out that such light is partly polarized. Most of it consists of **E** vectors that have a horizontal vibration direction. Because Polaroid sunglasses allow only vertical vibrations to pass through, much of the glare is eliminated. This is shown in (b).

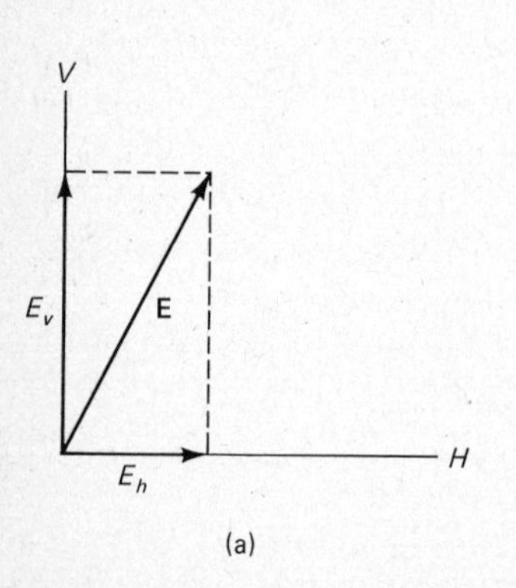

(a)

(b)

Notice that the horizontal eyeglasses reduce the glare much more than the vertical ones do.

*Light intensity turns out to be proportional to the amplitude squared. As a result, the intensity is reduced by a factor of $\frac{1}{4}$.

Summary

Light obeys the law of reflection: The angle of incidence equals the angle of reflection. Even when diffuse reflection occurs from a rough surface, reflection at each individual point of the surface obeys the law.

Plane mirrors form an image as far behind the mirror as the object is in front of it. Such an image is virtual (imaginary), because a screen placed at that position will not show the image. Real images, on the other hand, can be captured on a screen.

A concave mirror focuses parallel light to the focal point of the mirror. The focal point is a distance f from the mirror surface. We call f the focal length of the mirror. It is $\frac{1}{2}R$, where R is the radius of curvature of the mirror surface.

The image formed by a spherical mirror can be located by drawing three basic rays once the position of the mirror, its center of curvature, and the focal point are known.

The mirror equation can be used to locate images. If an object is placed a distance p from the mirror surface, its image will be found a distance q from the surface. These two distances are related to the mirror's focal length by the mirror equation:

$$\frac{1}{p} + \frac{1}{q} = \frac{1}{f}$$

For this equation images on the unlighted side of the mirror have negative q values. The focal length of a concave mirror is taken to be a positive number, whereas f for a convex mirror is negative.

The ratio of the image height h_i to the object height h_o is given by the magnification equation,

$$\frac{h_i}{h_o} = \frac{q}{p}$$

Spherical mirrors give rise to spherical aberrations. Rays that are reflected from a highly curved mirror form blurred images. Only the rays reflected from a small portion of a sphere are focused together. Parabolic mirrors do not suffer from this difficulty.

Light has its highest speed, c, in vacuum. If the speed of light is v in some material, the index of refraction μ of the material is defined by $\mu = c/v$. Because $v < c$, indexes of refraction of materials are always greater than unity.

A beam of light that has a wavelength λ_v in vacuum has a wavelength $\lambda = \lambda_v/\mu$ in a material of refractive index μ. The frequency of a given light wave is the same in all materials. Even though $\lambda = v/f$, only λ and v change from material to material.

As a beam of light passes from one material to a second, its direction of travel may change. This change in direction is called refraction. The angles of the beam in the two materials are related through Snell's law:

$$\mu_1 \sin \theta_1 = \mu_2 \sin \theta_2$$

Total internal reflection can occur if a beam is directed from an optically dense material to a less dense material. If $\mu_2 < \mu_1$, the critical angle $\theta_1 = \theta_c$ occurs when $\theta_2 = 90°$. For angles $\theta_1 > \theta_c$, the beam is unable to enter material 2. It is completely reflected back into material 1. This fact is fundamental to the operation of fiber optics devices.

The speed of light in a material varies according to wavelength. Blue light (short wavelength) usually has a slower speed than does red (long wavelength). As a result, different colors are refracted differently. Use is made of this fact in a prism in order to separate a beam of light into its various component colors.

Questions and Exercises

1. Bill and Joe are identical twins. Each has a mole on his left ear. When Bill looks at his face in a plane mirror, how does his image differ from Joe's face?
2. What is the minimum height a full-length vertical mirror must have if you are to be able to see your whole body in it? At what height from the floor must the mirror be placed?
3. Two plane mirrors are placed together so that they form a 90° angle to each other. An object is placed between them. Describe the three images of the object.
4. Two nearly parallel walls are covered with plane mirror surfaces. Suppose that you stand midway between them and look at your image in one of the walls. What will you see?
5. A concave mirror of radius R has an object placed at a distance p in front of it. Where is the image if (a) $p \to \infty$, (b) $p > R$, (c) $R > p > f$, (d) $f > p > 0$? State the character (real? erect? enlarged?) of the image in each case.
6. Repeat Question 5 for a convex mirror.
7. The two-wheel axle device shown in Figure P20.1 rolls from cement to sand as shown. Because it slips a little in the sand, its speed is less in sand than on cement. Describe the motion of the device. In what way is this similar to refraction of waves? What plays the role of the wave front in this case?
8. Show why the following statement is true: A beam of light does not change its direction upon going through a flat glass plate (such as window glass) but may be displaced by the plate.
9. A semitransparent mirror separates a brightly lit room and a dimly lit room. A person in the dark room can see a person in the bright room but not vice versa. Explain why.
10. A prism can be used to disperse white light into its colors. But this does not happen when a beam of light passes through a windowpane. Why not?

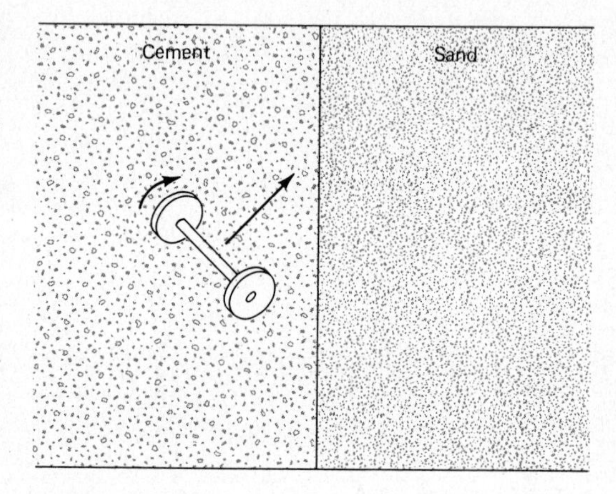

Figure P20.1

11. What is the radius of curvature of a plane mirror? What does the mirror equation say in this case?
12. Which would be more frightening to a kitten as it looks into a nearby mirror—if the mirror is concave or convex?
13. Why can't a transparent object be seen if it is immersed in a liquid with the same index of refraction as the object?
14. A tiny light source is placed at the center of a hollow metal sphere that has a reflecting inside surface. Where is the image of the source? Reason this out; do not use a formula. Then check your answer by formula. Why could the formula not usually be used in such a situation?

Problems

1. A 1.75-m-tall man stands 3.0 m from a plane mirror. (a) How far is he from his image? (b) How tall is his image? (c) If the man moves 50 cm closer to the mirror, how much closer does he get to his image?
2. How long must a plane mirror be for a 2-m-tall person to be able to see both her eyes and her feet in it? Assume the mirror is mounted on a wall and the woman stands upright in front of it. Where should the top of the mirror be placed?

*3. A beam of light strikes a plane mirror perpendicularly so that the reflected beam comes back along the incident beam. If the mirror is now turned through 20°, what will be the angle between the incident and reflected beams?

*4. The surfaces of two plane mirrors make an angle of 60° with each other. A beam of light is sent parallel to the bisector of the angle between the mirrors and is reflected from the mirrors. Through what angle(s) will the final reflected beam be deflected?

*5. The two opposite walls of a long straight hallway are covered with plane mirrors. Locate the images of a person who stands in the middle of the hall. Assume the mirrors are 3 m apart.

6. A large spherical mirror has a radius of curvature of 2 m. It is to be used to focus the sun in a solar furnace. Where, relative to the mirror, should the object to be heated be placed?

7. A 2-cm-high object is placed 120 cm in front of a concave mirror whose radius of curvature is 60 cm. Use a ray diagram to find the position, size, and character of the image. Repeat for object distances of (b) 80, (c) 40, (d) 20 cm.

8. Repeat Problem 7 for a concave mirror with $f = 40$ cm.

9. Repeat Problem 7 for a convex mirror with a focal length of 30 cm.

10. Repeat Problem 7 for a convex mirror with a focal length of 40 cm.

11. Using the mirror equation, find the position, size, and character of the image for an object 3 cm long placed in front of a concave mirror with a 40-cm focal length. Use object distances of (a) 120 cm, (b) 80 cm, (c) 15 cm.

12. Repeat Problem 11 for a concave mirror with a 60-cm radius of curvature.

13. Repeat Problem 11 for a convex mirror with an 80-cm radius of curvature.

14. Repeat Problem 11 for a convex mirror with a 60-cm radius of curvature.

15. A 0.5-cm-high object is placed 25 cm in front of a concave mirror. Its image is real and 200 cm from the mirror. (a) What is the size of the image? (b) Is it upright or inverted?

16. A 0.2-cm-high object is placed 14 cm in front of a convex mirror. Its image is 6 cm from the mirror. (a) What is the size of the image? (b) Is it real or virtual? (c) Is it upright or inverted?

*17. Starting from a ray diagram such as Figure 20.4, prove that the image is as far behind the plane mirror as the object is in front.

*18. It is desired to form a real image five times as large as an object by use of a concave mirror with $R = 70$ cm. Where should the object be placed? Where will the image be found?

*19. A concave mirror ($R = 90$ cm) is to be used to form a virtual image of an object. The image is to be five times as large as the object. (a) Where should the object be placed? (b) Where will the image be found?

*20. Using a convex mirror ($R = 70$ cm), it is desired to obtain an image half as far behind the mirror as the object is in front of it. (a) Where should the object be placed? (b) What will be the mirror's magnification in this case?

*21. A highly polished metal ball has a radius of 15 cm. (a) Where will it image a nose placed 20 cm from its surface? (b) An ear that is 35 cm away? (c) What will be the magnification in each case?

**22. Starting from the ray diagram shown in Figure P20.2, prove that the mirror equation applies to a convex mirror. Use the same sign conventions as we have used in the text. Show also that $h_i/h_o = q/p$.

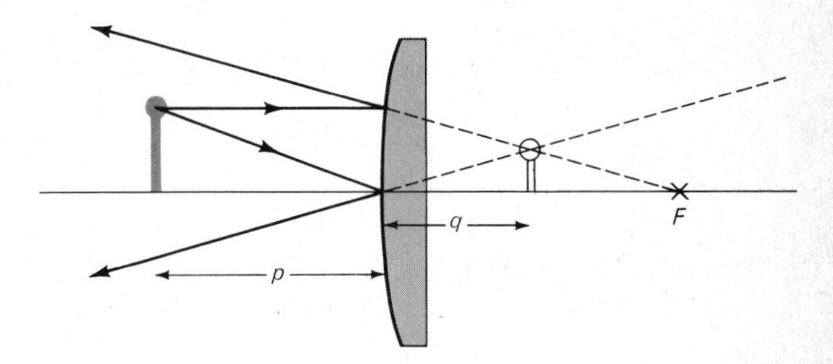

Figure P20.2

23. The index of refraction of a certain type of glass is 1.58. What is the speed of light in this glass?

24. The speed of light in a certain plastic is 2.04×10^8 m/s. What is the index of refraction of the plastic?

25. The yellow light from a sodium vapor lamp has a wavelength of 589 nm in vacuum. When this light is moving through glass of index of refraction 1.55, what is its speed, wavelength, and frequency?

26. The red light from a helium-neon laser has $\lambda = 633$ nm in vacuum. Its speed in a certain oil is 2.06×10^8 m/s. Find (a) the index of refraction of the oil, (b) the wavelength in oil, (c) the frequency in the oil, (d) the frequency in vacuum.

27. A beam of light enters water from air. (a) If its angle of incidence is 70°, what is the angle of refraction? (b) At what angle must it enter the water if its angle of refraction is to be 20°?

28. A beam of light in water enters a plastic block (index of refraction = 1.50). (a) If the angle of incidence is 50°, what is the angle of refraction in the plastic? (b) At what angle

*Problems marked with an asterisk are not as easy as the unmarked ones.

**Problems marked with a double asterisk are somewhat more difficult than the average.

must the beam be incident on the plastic if the refraction angle is to be 30°?

29. A beam of light travels from benzene to water. If the angle of incidence in the benzene is 48°, what will be the angle of refraction in the water?
30. At what angle of incidence must the light beam in the previous problem strike the water if it is to be at the critical angle for total internal reflection?
31. A drop of oil is placed on a glass block. The glass has an index of refraction of 1.55. When light is shined from the glass into the droplet, total internal reflection occurs at a critical angle of 59.0°. Find the index of refraction of the oil. (This method is capable of high accuracy and is used in many refractometers.)
32. A particular glass has the following indexes of refraction: blue light, 1.5273; red light, 1.5147. If white light is incident at an angle of 48.00° on this glass, what will be the angle between the red and blue beams in the glass?

*33. A beam of light strikes a flat glass plate at an angle of incidence of 70°. The glass has an index of refraction 1.55. Find (a) the angle of refraction in the glass and (b) the angle of refraction of the beam in air as it exists from the glass plate into the air.

*34. A light beam enters a piece of window glass ($\mu = 1.54$) at an incidence angle of 62°. Find the angle it makes to the normal as it exits from the other side of the flat window glass. What general rule can you conclude?

**35. A ray of light in a material of index of refraction μ_1 strikes a flat glass plate (refractive index μ_2) at an incidence angle θ_1. It then passes through the plate and enters a third material of refractive index μ_3. Show that its angle θ_3 in material 3 is the same whether or not the glass plate is present.

**36. Suppose a laser beam strikes a prism as shown in Figure 20.25. The top angle of the prism is 60° and the angle of incidence on the left is 55°. The index of refraction of the glass is 1.50. Find the angle between the incident beam and the beam that exits the prism.

*37. A fish is 5.0 ft below the surface of a still pond. At what angle to the vertical must the fish look if it is to see a fisherman sleeping on the shore of the pond?

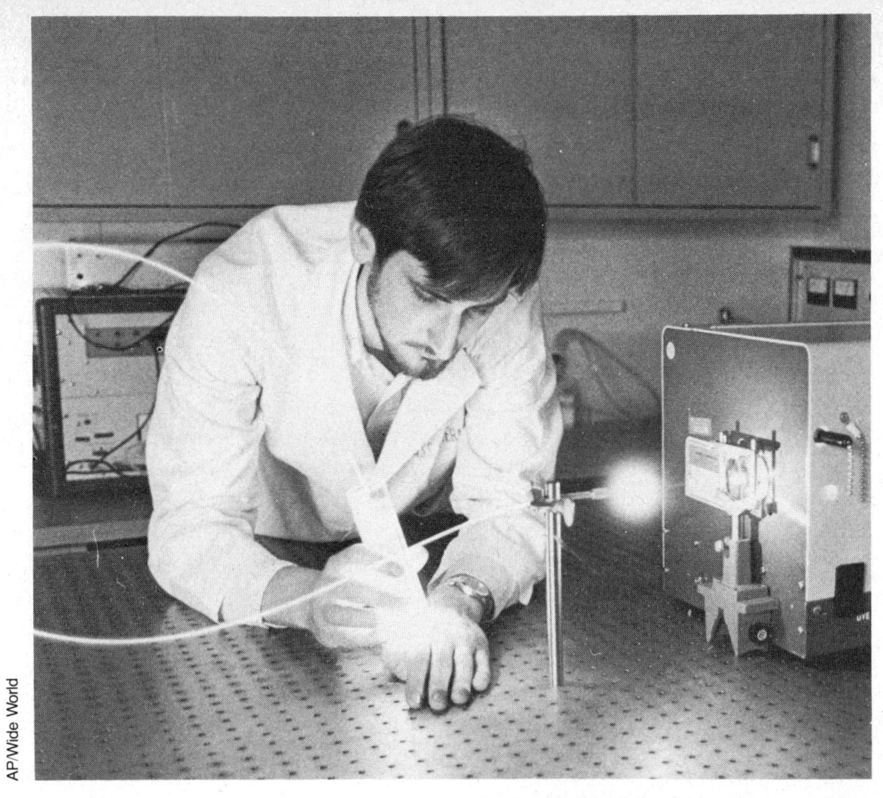

AP/Wide World

21 LENSES AND OPTICAL INSTRUMENTS

Performance Goals

When you finish this chapter, you should be able to

1. Separate several given lenses into two groups, converging and diverging.
2. Show in a diagram what happens to parallel rays when they pass through a given lens.
3. Locate the image of an object by the ray method for converging

In Chapter 20 we learned how light is refracted. We shall see in this chapter what lenses are and how they produce images. Lenses make use of refraction to focus light. We shall show how optical devices use lenses and mirrors. Examples of such devices—the microscope, the telescope, binoculars, and others—will be discussed.

and diverging lenses. State the character of the image.
4. Repeat Goal 3 by the lens equation method. Also, state the sign rules for the equation.
5. Write the lens magnification equation and use it in an example. Also, apply it to a combination of two lenses.
6. By use of a diagram, explain the operation of a simple camera.
7. Repeat Goal 6 for the eye, a magnifier, the microscope, the telescope, and binoculars.
8. Explain what is meant by spherical aberration and chromatic aberration. Point out what should be done to minimize the first effect for a given lens.
9. Sketch a prism spectrometer and explain its operation. Explain what is meant by a spectral line. Describe the spectrum one observes for a hot gas and for a white-hot solid.

21.1 Converging Lens: Focal Point

As we saw in Chapter 20, light travels slower in glass than it travels in air. We can make use of this fact to focus a parallel beam of light. To see how this can be done, consider Figure 21.1. We see there a parallel beam of light (perhaps from the sun). Its wave fronts are flat planes as shown. Remember, the wave fronts are always perpendicular to the rays. What happens as these rays and wave fronts pass through the glass lens shown?

Because light travels slower in glass than in air, the wave fronts will change as they pass through the lens. The portion of the wave front near the edge of the lens travels through less glass than that near the center portion. As a result, the center of the wave front falls behind the outer part. The wave fronts are therefore curved as shown after passing through the lens.

A properly designed lens of this type causes the plane wave fronts to become spherical. They then travel along radii of the sphere toward the point F, the *focal point* of the lens. As you see in the figure, the originally parallel rays converge on the focal point F. The distance f of this point from the center of the lens is the *focal length* of the lens. Lenses that converge parallel light in this way are called *converging lenses.* Notice that the light waves would be held back the same amount even if the lens were turned with its other face forward. Therefore, lenses can be thought of as having two focal points. Each is a distance f from the lens center.

A converging lens focuses parallel rays to the focal point.

Notice that the ray directions in Figure 21.1 could be reversed. A source of light at F would send out rays as shown. These would then be bent by the lens so as to form parallel rays. We therefore see that a light source placed at the focus of a converging lens gives rise to a parallel beam of light. This fact allows us to obtain parallel light from a source even though the source is not far away.

We have not yet stated just how the lens should be shaped to produce the focusing effect outlined above. It must, of course, be thicker in the middle than at the edges. If the lens is very thin, then it will focus properly if its surfaces are portions of spheres. For highly curved lenses spherical surfaces are not correct.

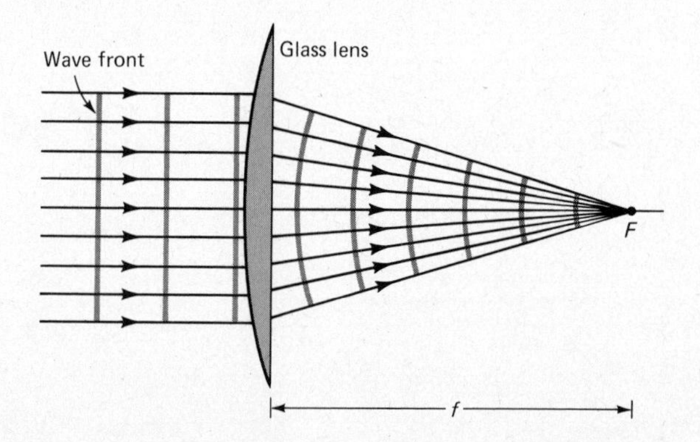

Figure 21.1 Since light travels more slowly in glass than in air, the plane wave fronts are held back more near the center than near the edge of the lens. This is a converging lens.

If the lens surfaces are too curved, the parallel rays do not all focus at the same point. We say that the lens shows *spherical aberration.* The effect is much like that we saw for highly curved spherical mirrors. We shall assume in our discussions that the lens is thin enough so that spherical aberration is not important.

21.2 Converging Lens: Ray Diagrams

In Figure 21.2(a) we show a few of the multitude of rays from an object point that pass through a lens. We can locate the image formed by a converging lens by drawing three special rays. Once the focal points and lens position are known, the rays in Figure 21.2(e) can be drawn. Let us take each ray in turn.

1. Ray 1, shown in part (b), is originally a ray parallel to the lens axis. It is converged through the focal point as we saw in the previous section.
2. Ray 2 is shown in (c). To draw this ray, we notice that the lens surfaces are parallel at the center of the lens. As we saw in Chapter 20, a flat plate does not deflect the beam. Therefore, ray 2 proceeds straight through the lens center as shown.
3. Ray 3 comes from the focal point. We saw in Section 21.1 that rays coming from the focus are bent parallel to the axis by the lens. Therefore, as shown in part (d), this ray proceeds parallel to the axis after leaving the lens.

In Figure 21.2(e) we combine these three rays. As we see, the rays produce an image. The eye gathers the rays and sees them as coming from the image. Because the rays actually do pass through the image, a screen placed there would show the image. This is therefore a real image. You can easily see from the diagram what happens as the object is changed to other positions. As the object is moved out to infinity, the image moves in to the focal point. (Distant objects give parallel light; parallel light is focused at the focal point.)

Suppose, however, that the object is between the focal point and the lens. The correct ray diagram is shown in Figure 21.3. Notice that the image is now seen behind the lens. The rays do not actually come from it. Therefore, a screen placed at the image position would not show the image. We conclude that the image is virtual in this case. You should be able to use a ray diagram to locate the image no matter where the object is in front of the lens.

21.3 The Lens Equation

Ray diagrams allow us to locate an image quickly, but for exact work we need an algebraic method for finding the image. Let us now find the equation that applies to thin lenses.

We shall use the same reasoning we used in deriving the mirror equation. Consider the two gray triangles in Figure 21.4. Since they are similar, the ratio of sides gives

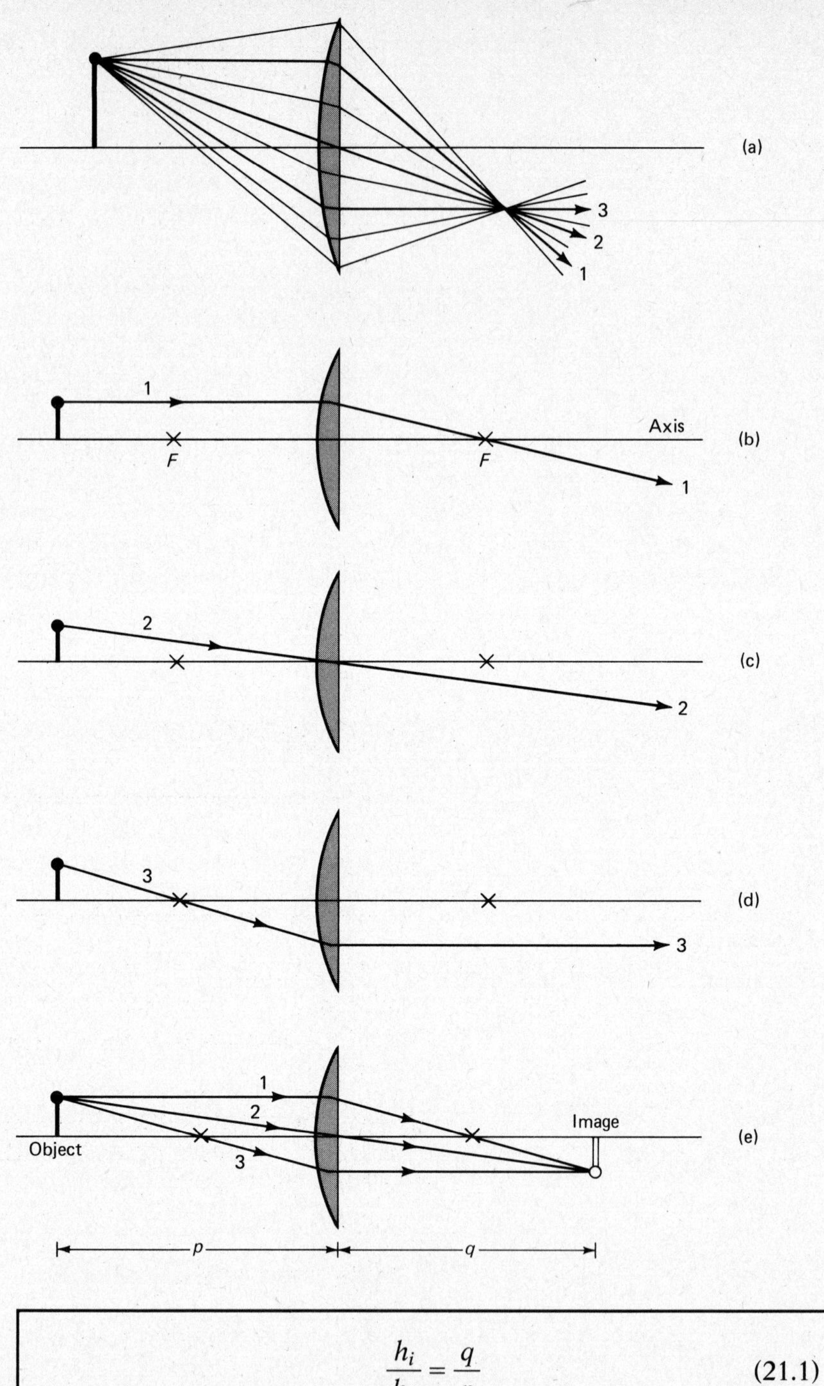

Figure 21.2 Three rays are used to find images made by converging lenses. The lens should be much thinner than shown.

$$\frac{h_i}{h_o} = \frac{q}{p} \tag{21.1}$$

(Notice in passing that this is the same magnification formula as we found for mirrors.)

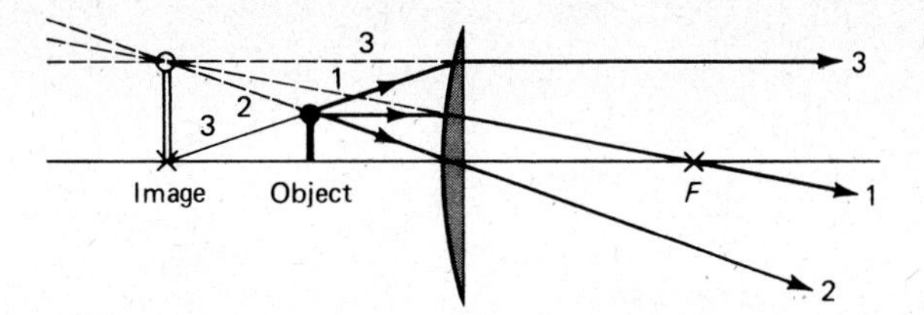

Figure 21.3 If the object is inside the focal point of a converging lens, then the image is virtual.

Now let us look at the triangles filled with broken lines. The ratio of sides in this case gives

$$\frac{h_i}{h_o} = \frac{q - f}{f}$$

We equate these two expressions for h_i/h_o and obtain

$$\frac{q}{p} = \frac{q - f}{f}$$

Cross multiplying gives

$$qf = pq - pf$$

Or, dividing the whole equation by pqf and transposing the last term give

$$\frac{1}{p} + \frac{1}{q} = \frac{1}{f} \qquad (21.2)$$

This is called the *thin-lens equation.* It is identical in form to the mirror equation. Notice, though, that the image distance q is positive when the image is on the side to which light is going as in Figure 21.4. This is unlike the situation for mirrors.

■ **EXAMPLE 21.1** A thin converging lens has a focal length of 20 cm. An object is placed 30 cm from the lens. Find the image distance, the character of the image, and the magnification.

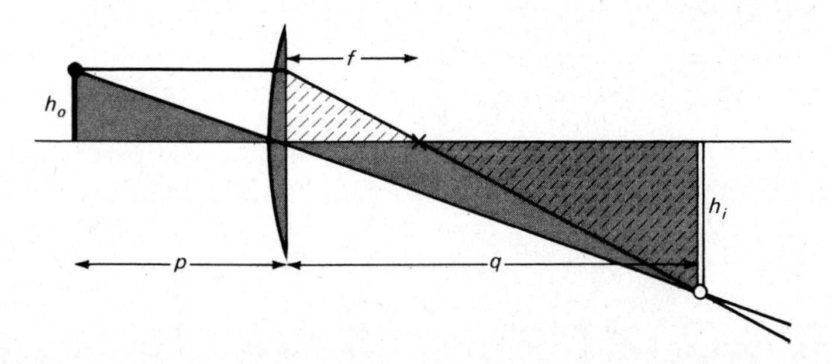

Figure 21.4 In this situation p, q, and f are positive.

Solution We make use of the lens equation with $p = 30$ cm and $f = 20$ cm. We have, in centimeters,

$$\frac{1}{30} + \frac{1}{q} = \frac{1}{20} \quad \text{or} \quad \frac{2}{60} + \frac{1}{q} = \frac{3}{60}$$

which gives

$$\frac{1}{q} = \frac{1}{60}$$

and so

$$q = 60 \text{ cm}$$

We should check this result by using a ray diagram. It will be much like Figure 21.4. Notice that the image is real and inverted.

To find the magnification, we use Equation 21.1.

$$\frac{h_i}{h_o} = \frac{q}{p} = 2.0$$

■ **EXAMPLE 21.2** Repeat Example 21.1 for a case in which $p = 9.0$ cm.

Solution The lens equation becomes

$$\frac{1}{9} + \frac{1}{q} = \frac{1}{20} \quad \text{or} \quad 0.111 + \frac{1}{q} = 0.050$$

from which $q = -16.4$ cm. Because q is negative, the image is on the side of the lens from which the light is coming.

The ray diagram for this case is similar to Figure 21.3. Notice that the image is virtual and erect. Its magnification, given by Equation 21.1, is

$$\frac{h_i}{h_o} = \frac{16.4}{9.0} = 1.82$$

21.4 Diverging Lens

Lenses that are thicker in the center than at the edge converge parallel rays of light in air. Typical converging lenses are given in Figure 21.5(a). We shall now discuss an opposite type of lens.

If a lens is thinner at the center than near its edges, it will diverge a beam of parallel rays. We call a lens such as this a *diverging lens.* Examples are shown in Figure 21.5(b).

To see why such a lens diverges the beam, refer to Figure 21.6. Notice that the wave fronts pass through glass that is thicker at the lens edge than it is at its center. Because light travels slower in glass than in air, the central portion of the

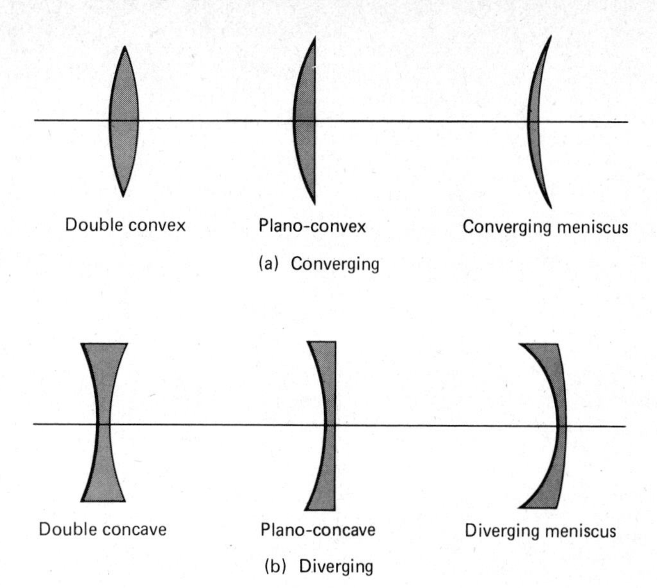

Figure 21.5 Types of lenses.

wave front gets ahead of the edges. By proper shaping of the lens, the plane wave fronts are changed to spherical ones. These spherical wave fronts appear to come from a single point, *F*, as shown. This is the focal point of the diverging lens.

Two focal points exist for a thin diverging lens. They are at equal distances on each side of the lens. Their distance from the lens center is the focal length, *f*.

A diverging lens causes originally parallel rays to appear to diverge from the focal point.

The three rays we can draw to locate the image for a diverging lens are shown in Figure 21.7. Ray 2 is exactly the same as ray 2 for a converging lens. Rays 1 and 3 tell us what happens to rays whose lines pass through the focal point. Notice how they differ from the similar rays for a converging lens. You should be able to draw these rays for any object position.

In the lower part of Figure 21.7 we show an image located by means of these three rays. As you see, rays 1, 2, and 3 appear to come from a point behind the

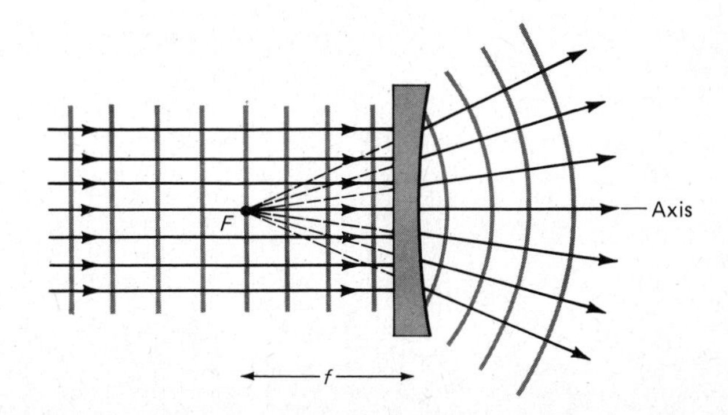

Figure 21.6 Parallel rays appear to diverge from the focal point after passing through a diverging lens.

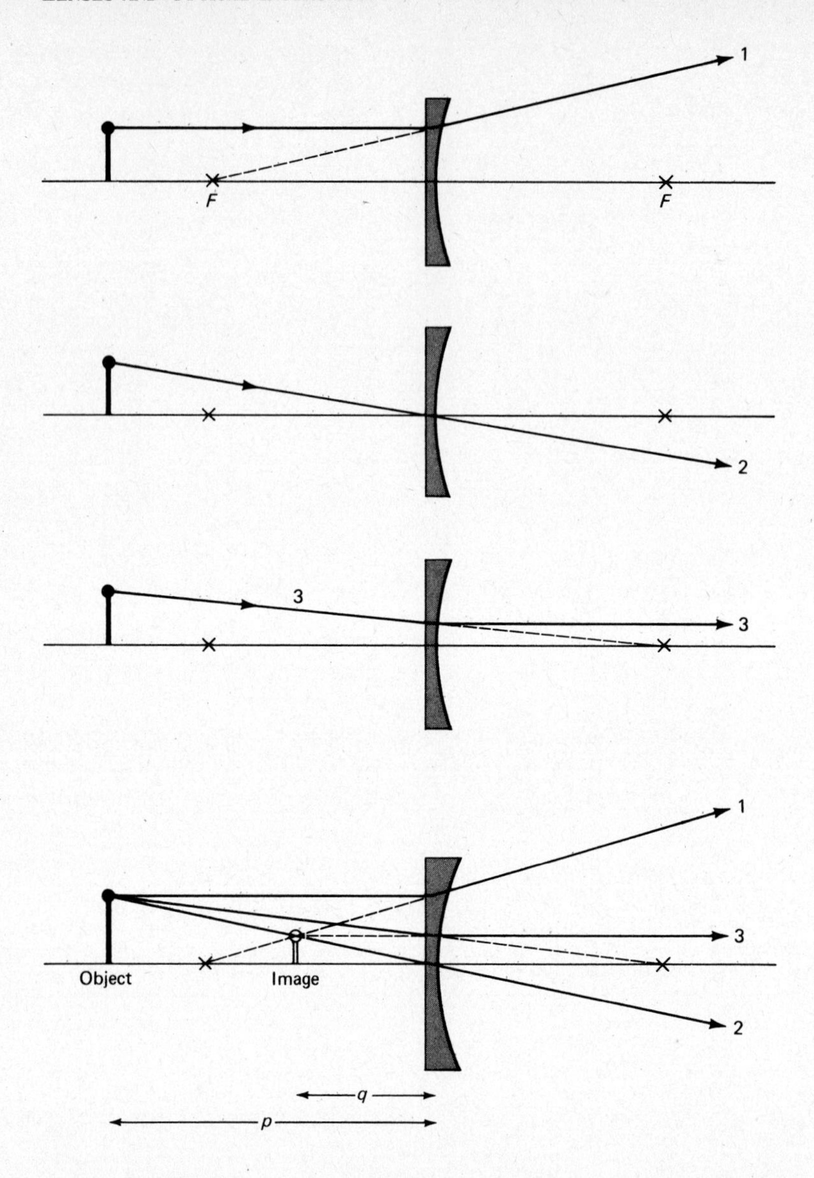

Figure 21.7 What happens to the image as the object position changes?

lens. Because the rays do not actually go through this point, the image is virtual. If you examine the figure, you will see that the image is always virtual for all values of p between zero and infinity.

It is easy to show that the lens equation applies to a diverging lens as well as to the converging one. However, one precaution must be taken. The focal length of the diverging lens must be given a negative sign. As usual, the object distance is positive if the object is on the same side as the incoming light. Image distances are positive if they are on the side of the lens toward which the light is going.

The thin-lens equation (Equation 21.2) applies to both converging and diverging lenses provided (a) f is taken positive for converging lenses

and negative for diverging lenses; (b) p is taken as positive if the object is on the side of the lens from which the light is coming; (c) q is taken positive if the image is on the side to which the light is going.

Equation 21.1, the magnification equation, applies to both types of lenses as well.

■ **EXAMPLE 21.3** An object is placed 30 cm in front of a thin diverging lens whose focal length is 20 cm. Find the image position, character, and magnification.

Solution We use the lens equation. Because this is a diverging lens, we must take f as a negative number, $f = -20$ cm. Also, $p = 30$ cm. We have, in centimeters,

$$\frac{1}{30} + \frac{1}{q} = \frac{1}{-20}$$

Solving gives

$$q = -12 \text{ cm}$$

The negative sign tells us that the image is positioned as in Figure 21.7. In fact, Figure 21.7 is the rough ray diagram for this situation. We see that the image is virtual and upright.

To find the magnification, we use Equation 21.1.

$$\frac{h_i}{h_o} = \frac{12}{30} = 0.40$$

■■

21.5 The Simple Camera

Now that we understand lenses, we shall discuss some devices in which they are used. We begin our discussion with the simple camera. A diagram of one type is shown in Figure 21.8.

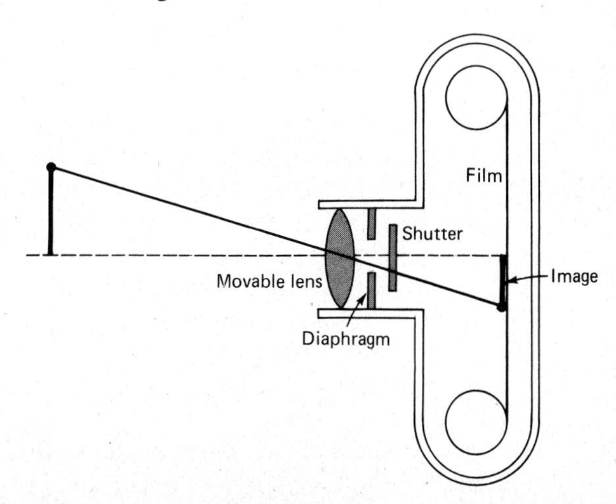

Figure 21.8 A simple camera is focused by moving the lens in or out until the image lies on the film.

The camera lens can be moved back and forth so as to focus the camera. Proper focus occurs when the image to be photographed lies exactly on the film. The shutter is kept in place except during the short instant when the film is exposed. In order to control the amount of light that strikes the film, the diaphragm is used. Its opening can be varied. Under dim light conditions the diaphragm can be fully open. But usually the lens is "stopped down." By this we mean that the diaphragm opening is kept small, so that only the central portion of the lens is used. This reduces the effect of spherical aberration. Another reason for keeping the lens stopped down is that with small diaphragm openings, the image will be fairly sharp even if the camera is not in perfect focus. You will learn more about this effect in Exercise 4 at the end of this chapter.

21.6 The Eye

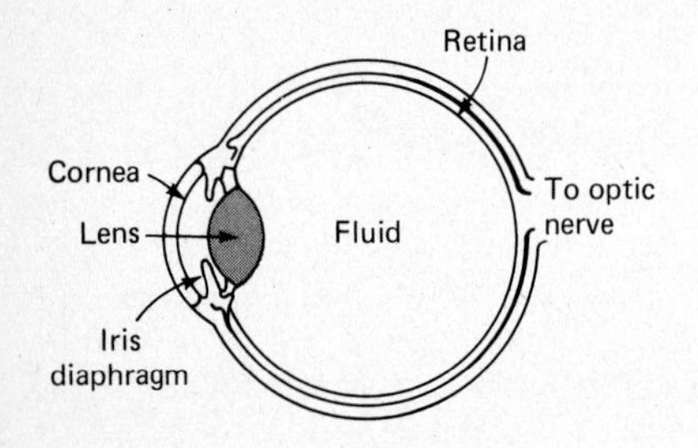

Figure 21.9 The eye.

A schematic diagram of the eye is shown in Figure 21.9. It is like a camera in several respects. In place of the film, the retina records the picture. The picture is then sent to the brain by means of the optic nerve.

The lens of the eye is more or less soft. It is held at its edges by muscles that can change its shape by pulling on it. The muscles can thereby cause the focal length of the lens to change. To focus the image of an object on the retina, we adjust the focal length of the eye lens. We say that the eye is capable of accommodation; its focal length can be varied so as to focus objects properly.

The iris diaphragm controls the quantity of light entering the eye. Its opening is the pupil of the eye. You have no doubt noticed that the pupil is small in bright light and large in dim light. The eye automatically adjusts its opening to compensate for changes in light intensity. The outer protective cover on the eye is called the cornea of the eye.

Defects in accommodation may be corrected by eyeglasses. Two different types of trouble occur, nearsightedness and farsightedness. A nearsighted eye is able to focus near objects properly but not distant ones. As shown in Figure 21.10(a), the eye lens is too converging. It focuses the distant objects too close to the lens. A diverging eyeglass placed in front of the eye can partly compensate

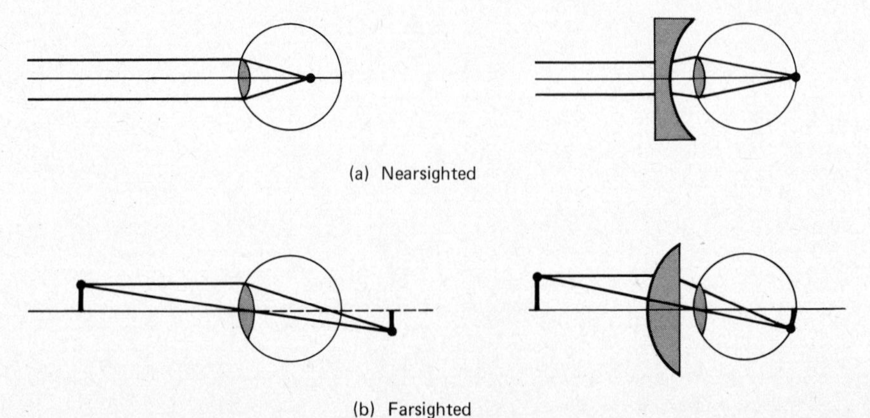

Figure 21.10 A diverging lens is used to correct myopia (nearsightedness), whereas a converging lens is needed to correct hyperopia (farsightedness). The diagrams exaggerate the effect. In practice the lenses are not shaped exactly as shown. Examine a few eyeglasses to see.

for the overconvergence. The image of a distant object then occurs at the retina of the eye.

A farsighted eye can focus distant objects clearly but not near ones. As shown in Figure 21.10(b), the eye lens cannot be made converging enough. We therefore place a converging lens in front of the eye to aid it.[1] As a result, a close object is focused properly.

Even the normal eye cannot see objects clearly if they are too close. The normal eye lens can focus objects no closer than a certain point in front of the eye. This point is called the *near point* of the eye. It is usually assumed to be about 25 cm.

21.7 The Magnifying Glass

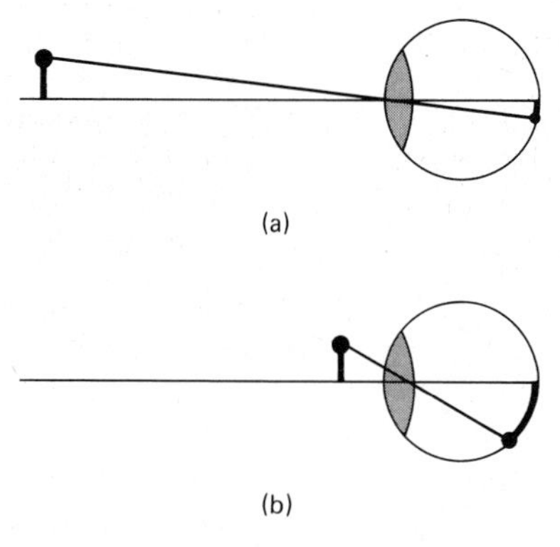

Figure 21.11 The closer the object is to the eye, the larger its image on the retina.

If you look at Figure 21.11, you can see an important feature of the eye. The size of the image on the retina depends on how close the object is to the eye. If we are to see the fine details of an object, the object should be placed as close to the eye as possible. Only then will the image on the retina be the largest possible. Unfortunately, the eye has a near point. It cannot focus on objects closer than about 25 cm.

A magnifying glass allows us to bring the object closer than the near point and still be seen. As shown in Figure 21.12, a magnifying glass is simply a converging lens. Part (a) shows the virtual image it forms of an object placed closer than the focal point. If the eye looks at the light from the lens, the virtual image is seen. The eye can focus on the faraway virtual image even though the object is much too close to be seen clearly. As seen in part (b), the size of the image on the retina is determined by the object. Even though the actual object is very close to the eye, it can still be seen clearly with the aid of the magnifying glass.

[1] Optometrists frequently use the term "power of a lens." The power of a lens (in diopters) is 1 over the focal length of the lens in meters.

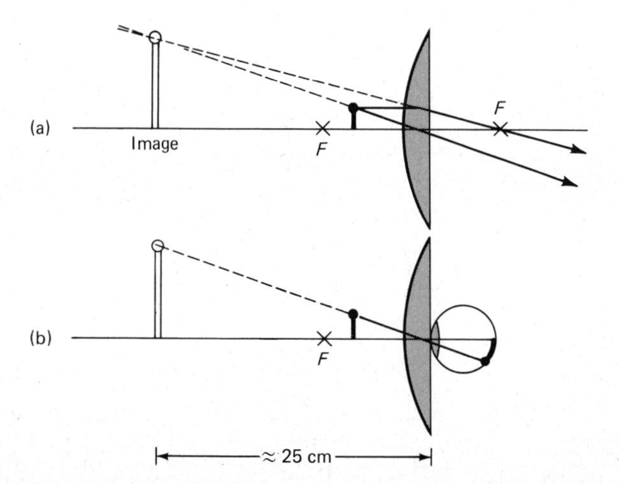

Figure 21.12 The magnifying glass allows one to bring the object closer than the near point of the eye. The eye focuses on the virtual image produced by the magnifier.

The magnification produced by a magnifying glass is the same as for any lens.

$$\text{Magnification} = \frac{q}{p}$$

In the present case the image distance q will be about 25 cm, the near-point distance. The object is close to the focal point, so $p \simeq f$. To this approximation,

$$\text{Magnification} \simeq \frac{25\ \text{cm}}{f}$$

If the magnifier has a focal length of 8 cm, then the magnification it produces is about 3.

21.8 Lens Defects

A thick, short-focal-length lens does not work well as a magnifier. The image is distorted and has colored edges. These effects are most noticeable for simple, highly curved lenses. Let us now see how they arise.

Spherical Aberration We mentioned spherical aberration in our discussion of mirrors and in Section 21.1. Parallel rays are focused to the focal point of a lens only if the rays pass close to the center of the lens. Rays through the highly curved outer portion do not focus properly. This effect, spherical aberration, is always present for spherical lenses. It is worst for highly curved lenses. To minimize it, only the central portion of the lens should be used. Many optical devices use a diaphragm to cover all but the center portion of the lens.

Chromatic Aberration As shown in Figure 21.13, a lens approximates two prisms. Like a prism, the lens refracts various colors differently. As a result, the focal point for one color is not the same as for others. Images formed by the lens therefore show color separation. This effect is called chromatic aberration.

To reduce this defect, compound lenses may be used. These are actually two or more lenses of different types of glass cemented together. Commonly one lens is converging and has a low index of refraction, whereas the other lens is diverging and has a higher index of refraction. The lenses' focal lengths are chosen so that the combination is still converging. But if the two lenses are chosen properly, the color separation caused by one lens is partly canceled by the other. Exact cancellation occurs at only one wavelength. However, even this is a great improvement over a single lens. Lenses corrected for chromatic aberration are called *achromatic lenses*.

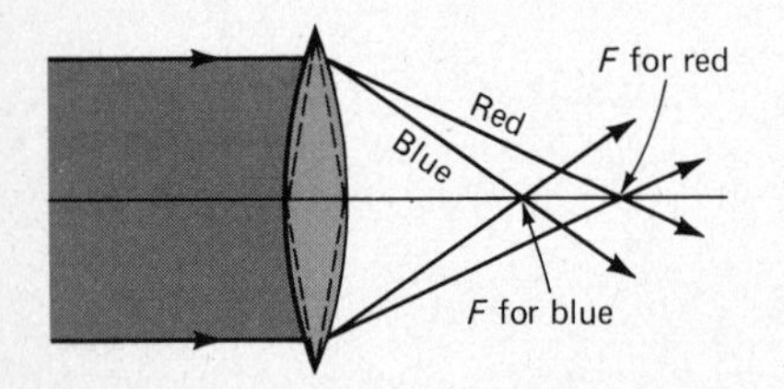

Figure 21.13 Because of its prismlike shape, a lens focuses the various colors to different points. This effect is called chromatic aberration.

There are other lens aberrations in addition to these. To correct for them, several types of glass lenses are sometimes cemented together. These highly corrected lenses are often found in the optical instruments used in science and industry. Even moderate-priced cameras use lenses that are partly corrected for some of these aberrations.

21.9 The Microscope

A microscope consists of two (usually highly corrected) converging lenses as shown in Figure 21.14. The eyepiece lens is used as a magnifying glass. It examines the image of the object produced by the objective lens.

Notice that the object to be seen is placed just outside the focal point of the objective lens. It forms a real image inside the barrel of the microscope. The eyepiece uses this image as object. As we saw for the magnifying glass, the eyepiece magnifier gives rise to a virtual image about 25 cm from the lens. These facts are all shown in Figure 21.14.

We define the magnification of the microscope to be h_2/h_o (see the figure for these quantities). For the first lens we know that the magnification M_1 is given by

$$M_1 = \frac{h_1}{h_o} = \frac{q_o}{p_o}$$

For the second lens we have

$$M_2 = \frac{h_2}{h_1} = \frac{q_e}{p_e}$$

We notice that the product of M_1 and M_2 will yield h_2/h_o, the total magnification M. Therefore,

$$M = M_1M_2 = \frac{h_2}{h_o} = \left(\frac{q_o}{p_o}\right)\left(\frac{q_e}{p_e}\right)$$

We therefore arrive at the following general result.

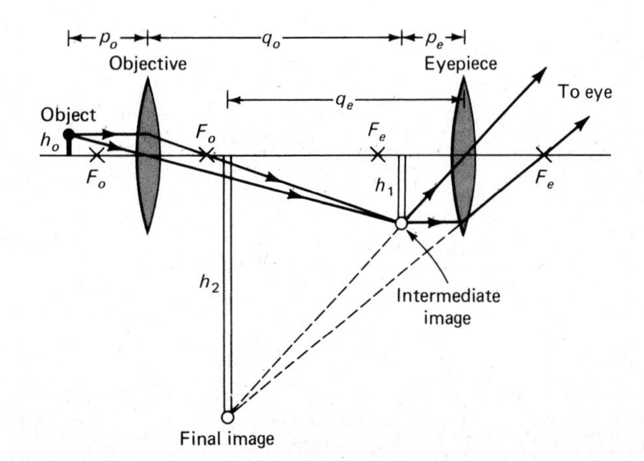

Figure 21.14 In the microscope the eyepiece acts like a magnifier. Be sure you understand how the images are found using our usual rays.

The magnification produced by a combination of lenses is equal to the product of the separate magnifications.

For a typical microscope the object is usually placed close to the focal point of the objective lens. Therefore $p_o \simeq f_o$ is the focal length of the objective lens. The distance q_o might be about 20 cm and q_e would be about 25 cm. Because the intermediate image would be close to the focal point of the eyepiece, we can use $p_e \simeq f_e$, the focal length of the eyepiece. Then the magnification becomes

$$M \simeq \frac{(20)(25)\ \text{cm}^2}{f_o f_e}$$

where f_o and f_e are to be measured in centimeters.

Notice that the shorter the focal lengths are, the higher will be the magnification. Short-focal-length lenses must be highly curved. Therefore to produce a good image, the microscope lenses should be complex, corrected lens combinations.

21.10 The Telescope

A telescope is used for looking at distant objects. In the refracting telescope shown in Figure 21.15, two lenses are used. The objective lens gathers light from the distant object. This light is focused to give an image close to the focal point of the objective. The second lens, the eyepiece, acts like a magnifier, to examine this image. Check through the ray diagram in Figure 21.15.

Our usual definition of magnification is not useful for a telescope. The object distance is nearly infinite, and so the ratio h_i/h_o for the objective lens is near zero. A more useful quantity, the *angular magnification*, is defined for such cases. It is defined in terms of the two rays reaching the eye from the two ends of the distant object. Call the angle between these two rays θ when the object is observed with the naked eye. Call the angle θ' when viewed through the telescope. Then the angular magnification is θ'/θ. These angles are shown in Figure 21.15. For the case shown, this ratio turns out to be f_o/f_e. Therefore, high magnification is obtained if the objective lens' focal length is long. As usual, the magnifier (eyepiece) should have a short focal length.

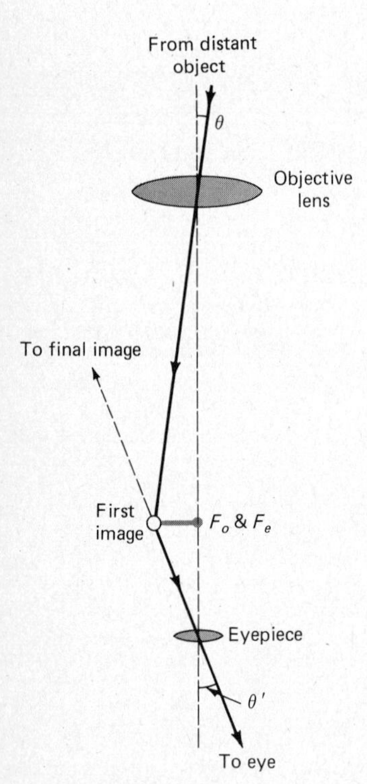

Figure 21.15 Because this telescope produces upside-down images, it is not suitable for viewing objects that are on earth.

In astronomy the objects are distant stars and planets. Because the light from them is extremely faint, the objective lens should be very large so that a large amount of light is gathered. But large lenses (a meter in diameter, for example) are difficult to build. Therefore, the objective lens is often replaced by a concave mirror. This type of telescope is called a reflecting telescope. Typical designs are shown in Figure 21.16.

Radar signals are picked up by radio telescopes. One is shown in Figure 21.17. In these a concave mirror for radar waves is used. Because these waves have long wavelengths (a few centimeters), the "mirror" can actually consist of a parabolic-shaped metal screen. The rays from the distant object are focused on a radar wave detector system at the focal point of the mirror.

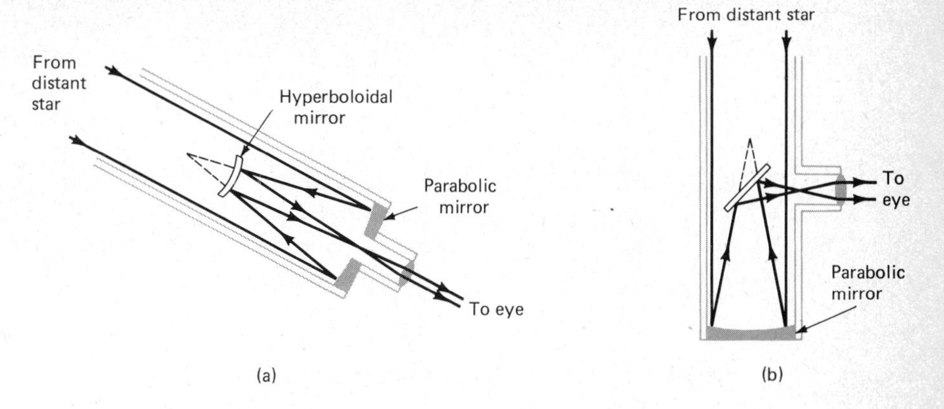

Figure 21.16 The Cassegrainian telescope in (a) uses a small hyperboloid mirror to deflect the rays back through a small hole in the much larger parabolic mirror. In effect, the hyperboloidal mirror increases f_o. The Newtonian telescope in (b) uses a small plane mirror to extract the beam. (The telescopes are not drawn to scale.)

Figure 21.17 This radio telescope is located in an isolated section of Australia, far from sources of electromagnetic noise. (Courtesy of Australian Information Service)

(a)

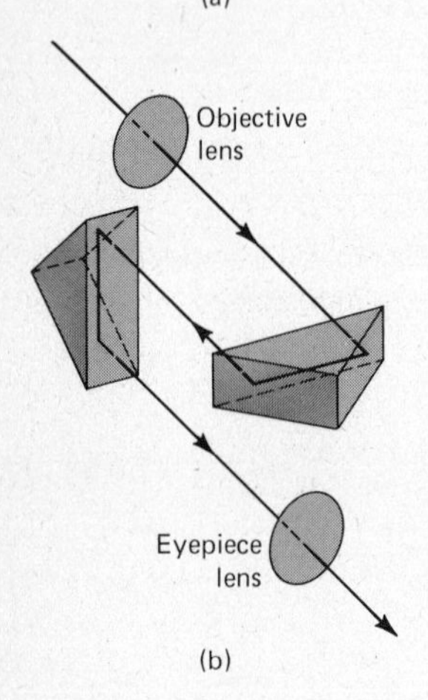

(b)

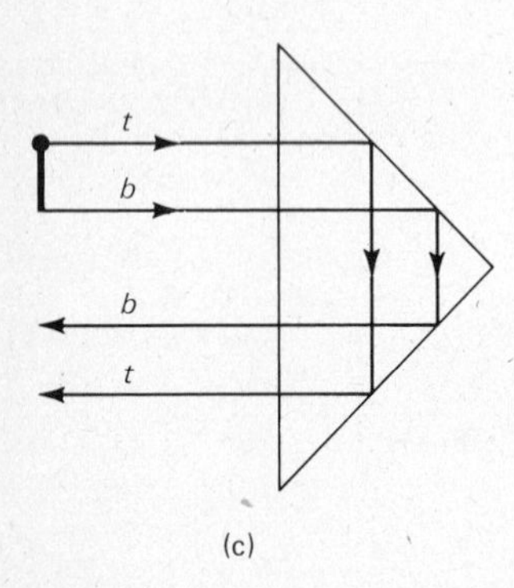

(c)

Figure 21.18 In the binoculars two total internal reflection prisms invert the image and reverse right and left. Part (b) shows the lens arrangement in one side of the binoculars.

21.11 Binoculars

For observations on earth or even in the sky that do not require the high magnification provided by a large telescope, binoculars may be used. This device, shown in Figure 21.18(a), gives us an upright image of distant objects.

In addition to a telescope objective and eyepiece, two prisms are placed in each side of the binoculars, as shown in Figure 21.18(b). The light is totally internally reflected inside the prisms. In the process, as shown in part (c), the image is inverted so that it reaches the eye right side up. Moreover, the right and left sides of the image are reversed to correct for a right-left reversal caused by the objective lens. (You might want to sketch a horizontal object and show that the horizontal image orientation is reversed by the objective lens.) The end result is that the eye sees an image that is upright and whose right and left sides have not been reversed.

21.12 The Prism Spectroscope (or Spectrometer)

In the last chapter we saw that a prism can separate light into its various colors. Let us now see how this is done in a practical way.

Suppose, to begin with, that a light source gives off only one color (or wavelength) of light. A yellow sodium arc light is nearly monochromatic (or single-colored). To examine its light, we use the prism spectroscope shown in Figure 21.19. The light source sends a beam through a slit, and the slit acts like a source for the collimating lens. (To collimate is to make parallel.) Because the slit is placed at the focal point of the collimating lens, parallel light exits from the collimator. These parallel rays are then refracted by the prism. Because we are assuming monochromatic light, all the light is refracted the same. This parallel beam of light now enters the telescope and is focused by the objective lens. The objective lens forms an image of the collimator slit. So if you look through the eyepiece, you will see a very bright image of a slit. This image is simply a bright, vertical line. It is called a *spectral line.* A photographic film could be placed at the image position. The photograph would show the slit image as a line.

Most sources of light give off more than one color or wavelength. Because each color is refracted differently, the slit images for each color are at different positions. For example, a mercury arc lamp produces several slit images, as shown in one part of the color plate, Figure 19.6. We say that mercury light has several spectral lines. Each line represents a separate wavelength of light given off by the mercury vapor arc. Also shown in the color plate are the spectral lines given off by excited hydrogen and neon gases, among others.

Red-hot solids and molten liquids also give off light. Their light contains a continuous band of wavelengths. As a result, a continuous band of color is seen rather than the sharp lines observed for hot gases. The continuous spectrum given off by the hot wire in a light bulb is also shown in the color plate. We shall discuss the causes of these spectra in Chapter 29.

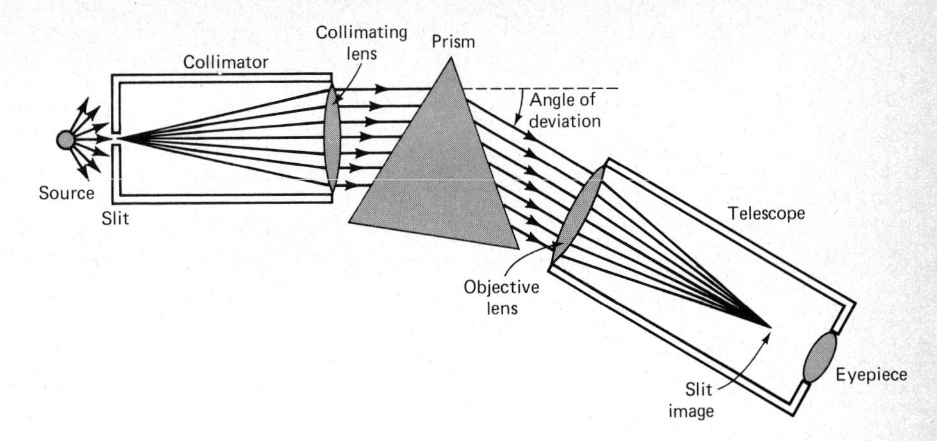

Figure 21.19 If the source is monochromatic, a single image of the slit is formed in the telescope.

21.13 Virtual Objects

Sometimes when two or more lenses (or mirrors) are being used, an object for one may not be real. To see what we mean by this statement, consider the two-lens system shown in Figure 21.20(a). An object is placed as shown in front of lens 1. We wish to find the final image formed by the lens system.

In situations such as this we simply start with the object and work through the system. Lens 1 would form an image of the object as shown in (b) if the

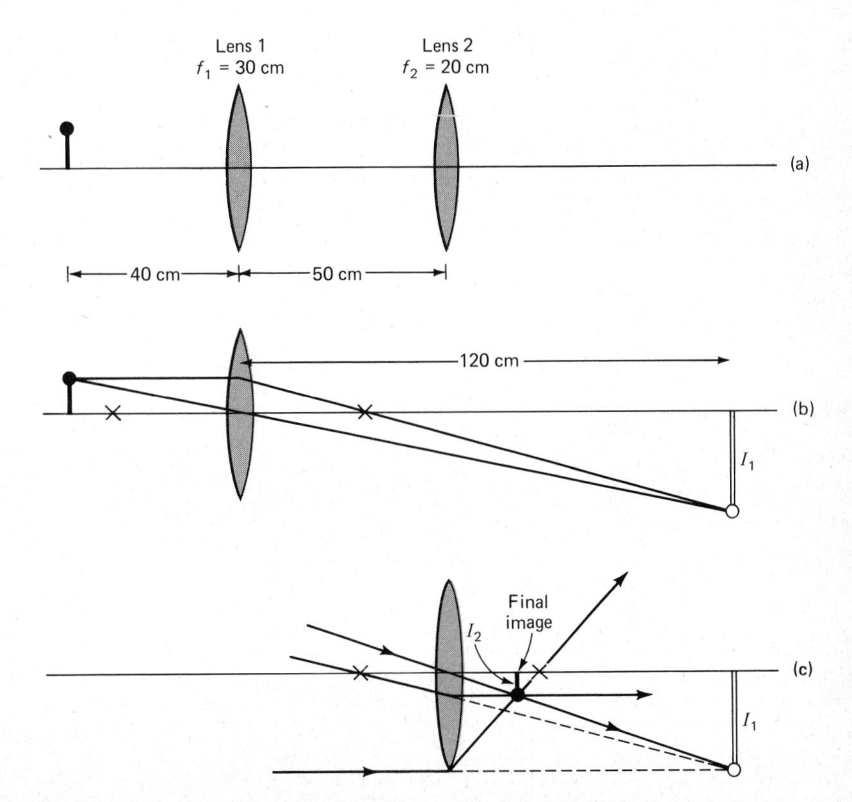

Figure 21.20 In this system a virtual object occurs.

second lens were not there. Let us locate its position using the lens equation. We have, using the values given in the figure,

$$\frac{1}{40} + \frac{1}{q} = \frac{1}{30} \qquad \text{or} \qquad q = 120 \text{ cm}$$

So the image formed by lens 1, I_1, would be as shown in Figure 21.20(b).

We can now use image I_1 as the object for lens 2. But it is not a real object because it never forms. Lens 2 interrupts the light before it can get to I_1. We call this type of object a *virtual object* for lens 2. In this rather strange case the object is on the side of the lens to which light is going. Its distance must therefore be taken as negative.

Because I_1 is $120 - 50 = 70$ cm from lens 2, we have $p = -70$ cm for the second lens. The lens equation gives

$$\left(\frac{1}{-70}\right) + \frac{1}{q} = \frac{1}{20}$$

From this we find that $q = 140/9 = 15.6$ cm. Its position is shown as I_2 in Figure 21.20(c).

We also show in part (c) the rays that can be used to locate I_2. Notice that they are our usual rays heading toward I_1 before hitting lens 2. You should be able to explain why each is drawn as shown.

To find the magnification of the system, we note that

$$M_1 = \frac{120}{40} = 3.00 \qquad \text{and} \qquad M_2 = \frac{15.6}{70} = 0.222$$

Therefore the total magnification is

$$M = M_1 M_2 = 0.67$$

Notice that the final image is inverted and real.

As we see from this example, complex systems can be analyzed simply. We start with the object and find its image due to the first lens (or mirror). This image is used as the object for the second lens (or mirror). And so on through the system until the light exits from it. You will find a few problems of this type at the end of this chapter.

Summary

When used in air, glass lenses that are thicker in the center than at the edges are converging. Those that are thinner at the center are diverging.

Parallel rays are converged to the focal point by a converging lens. A diverging lens causes them to diverge from the focal point. Parallel rays can be made by a light source placed at the focal point of a converging lens.

There are three rays one can draw with a straightedge to locate images. One goes through the center of the lens unchanged. The other two go through the two focal points of the lens. On the other side of the lens they are parallel to the axis.

The lens equation is:

$$\frac{1}{p} + \frac{1}{q} = \frac{1}{f}$$

The object distance p is positive if it is on the side of the lens from which light is coming. The image distance q is positive if it is on the side to which light is going. The focal length f is positive for converging lenses and negative for diverging lenses.

A lens magnifies an object by the following factor, called the magnification:

$$M = \frac{q}{p}$$

The magnification caused by a series of lenses is equal to the product of their individual magnifications.

Highly curved lenses suffer from aberrations. Spherical aberration refers to the fact that only the rays near the axis of a spherical lens focus properly. This effect is minimized by using only the central portion of the lens. Chromatic aberration refers to the fact that the focal point is different for each color. Therefore colored images result. To correct this defect, achromatic lenses, actually combinations of lenses of different indexes of refraction, are used.

A magnifying glass (a converging lens) forms a virtual image of a close object. In effect, it allows one to observe the object at a distance closer than the near point of the eye. For high magnification, the focal length of the lens should be small.

A microscope uses a magnifier (the eyepiece) to examine the image produced by the objective lens. For high magnification both lenses should have short focal lengths.

The simple telescope gives rise to an image that is inverted and has right and left reversed. In the binoculars total internal reflection corrects for these inversions of the image.

A prism spectroscope (or spectrometer) separates light into its component colors. It records each color as a separate image of a slit. These slit images are called spectral lines. Gaseous light sources show sharp spectral lines. Red-hot solids and liquids show a continuous band of color. They are said to emit a continuous spectrum.

Questions and Exercises

1. Suppose that a glass lens ($f = 20$ cm) is submerged in a huge vat of oil which has the same index of refraction as the glass. What will be the focal length of the lens then? (That is, how far from the lens will parallel rays be focused?)
2. A spherical air bubble in a large block of glass acts like a spherical lens. Is it a converging or a diverging lens?
3. From time to time one reads a news story something like the following: "Joe Blow's house burned down yesterday. The

fire was started by sunlight shining through a rounded globe filled with water." Explain how a fire could be started in this way.

4. Draw a small, bright object about 1 mm high at a distance of 10 cm from a 1-cm opening in a large, opaque screen. Show how the bright spot cast by the object on a screen 5 cm behind the opening decreases in size as the opening is made smaller. Show that, in the limit of a pinhole opening, two objects 1 cm apart and both 10 cm from the opening will give rise to well-defined images on the screen. This is the basis for the operation of the pinhole camera. In view of your results can you explain why even expensive cameras give better images when stopped down?
5. Fill in the blanks in the table below for a converging lens.

p/f	q/f	Is Image Real?	Is Image Erect?
∞	______	______	______
2	______	______	______
1	______	______	______
$\frac{1}{2}$	______	______	______
0	______	______	______
$-\frac{1}{2}$	______	______	______

6. Repeat Exercise 5 for a diverging lens after replacing f by $-f$.
7. An object is placed in front of a lens. If the lens is converging, where must the object be placed to obtain a virtual image? If the lens is diverging, can a real image be obtained?
8. A coin is at the bottom of a glass full of water. If you look at the coin from above, it appears to be considerably above the location you know it to have. Explain the effect by considering a wave front as it progresses to your eye from a point on the coin. Remember, rays are perpendicular to wave fronts.
9. A glass sphere has an air bubble at its exact center. Would the bubble appear to be (a) at the center or (b) closer to or (c) farther from the observer than the bubble actually is?
10. Describe a practical method by which the focal length of a lens can be determined to an accuracy of about 5 percent. The focal length is about 30 cm. Consider both a converging and diverging lens.

Problems

Note: Whenever possible, check your answer by a ray diagram.

1. A thin lens focuses the light from the sun at a point 12 cm from the lens. (a) Is the lens converging or diverging? (b) What is the focal length of the lens?
2. An object is placed 25 cm in front of a flat plate of glass. Locate the image seen for the light that passes through the glass by use of (a) a ray diagram and (b) the lens formula. (c) Is the image real or virtual?
3. An object is imaged by a converging lens of 40 cm focal length. Find the image position by means of a ray diagram for the following object distances: (a) 80 cm, (b) 50 cm, (c) 20 cm. In each case is the image real? Erect?
4. A converging lens whose focal length is 50 cm casts an image of an object. Find the image distance by means of a ray diagram for the following object distances: 100 cm, 80 cm, 20 cm. In each case is the image real? Erect?
5. Repeat Problem 3 using the lens equation. If the object is 5 cm high, how large is the image in each case?
6. Repeat Problem 4 using the lens equation. If the object is 10 cm high, how large is the image in each case?
7. A thin converging lens is used to cast an image of a street lamp 25 m away. The image is 60.0 cm from the lens. What is the focal length of the lens?
8. In a certain slide projector the object (the slide) is placed 15 cm from the projection lens. The image appears on the screen 3.0 m away. What is the focal length of the lens? If the slide is 2 cm $\times$ 3 cm, how large will the image be on the screen?
9. A diverging lens with 60-cm focal length forms an image of an object. Use a ray diagram to find the image for the following object distances: (a) 200 cm, (b) 100 cm, (c) 30 cm. Is the image real? Erect?
10. A 40-cm-focal-length diverging lens casts an image of an object. Use the ray diagram method to find the image position for each of the following object positions: (a) 100 cm (b) 70 cm, (c) 30 cm. Is the image real? Erect?
11. Repeat Problem 9 using the lens equation. What is the ratio of image to object height in each case?
12. Repeat Problem 10 using the lens equation. What is the ratio of image to object height in each case?
13. A certain lens forms a real image on a screen 80 cm from the lens when the object is 20 cm from the lens. What is the focal length of the lens?
14. What should the focal length of a lens be if an object's image is to be cast on a screen that is 60 cm from the object? The lens is to be placed 20 cm from the object.

*15. A converging lens with 80-cm focal length is to be used to form a 2-m-long real image of an object that is 50 cm long. (a) What should be the object distance? (b) The image distance?

*16. Repeat Problem 15 if the image is to be virtual.

*17. It is desired to use a 60-cm-focal-length diverging lens to form a virtual image of an object. The image is to be one-third as large as the object. (a) Where should the object be placed relative to the lens? (b) What will be the image distance?

*18. Use the lower ray diagram in Figure 21.7 to derive the lens equation for a diverging lens.

**19. The distance between an object and a screen is 200 cm. Where should a 40-cm-focal-length converging lens be placed to form a real image on the screen? Two answers are possible. Give both.

**20. An object is 80 cm away from a screen. At what two positions could a converging lens with $f = 15$ cm be placed so as to cast an image of the object onto the screen?

**21. An object is a distance D away from a screen. Show that a converging lens of focal length f will image the object on the screen provided the lens is placed such that $p = \frac{1}{2}D[1 \pm \sqrt{1 - (4f/D)}]$.

*22. A magnifying glass with 10.0-cm focal length is used to look at an object. Where should the object be placed if the image is to be 30 cm from the lens? What will the magnification then be?

*23. Two lenses are both converging, and each has a 20-cm focal length. They are placed 40 cm apart. A 5-cm-high object is placed 80 cm in front of the first lens. Find the final image position and its size. Is it real? Is it erect?

24. A camera with a thin converging lens is used to take a picture of a tower that is 30 m away. The focal length of the camera lens is 12.0 cm. When the film is developed, the image of the tower is found to be 2.5 cm high. What is the height of the tower?

25. A certain reflecting telescope uses a concave mirror with a focal length of 5.0 m as its objective. The image cast by the mirror is photographed directly. How large is the diameter of the image of the moon on the photograph? (Moon diameter = 3.5×10^6 m; earth-to-moon distance = 3.8×10^8 m.)

*26. Prove that the angular magnification (θ'/θ) for the telescope of Figure 21.15 is f_o/f_e. You may need to make use of the fact that for small angles $\sin\theta \cong \tan\theta \cong \theta$.

*27. A 20-cm-focal-length converging lens is placed 50 cm in front of a 30-cm-focal-length diverging lens. A 4-cm-high object is placed 60 cm in front of the converging lens. Find the final image and its size. Is it real? Erect?

**28. A system consists of the following along the x axis:

4-cm-high object at $x = 0$
$f = 20$-cm converging lens at $x = 30$ cm
Plane mirror at $x = 100$ cm

Find the position and sizes of all the images formed by this system.

**29. Two converging lenses of focal lengths f_1 and f_2 are placed in contact. Their separation is negligible compared to the focal lengths. Show that the final image position and object position are related by

$$\frac{1}{p} + \frac{1}{q} = \frac{1}{f_1} + \frac{1}{f_2}$$

This tells us that the combination lens has a focal length given by

$$\frac{1}{f} = \frac{1}{f_1} + \frac{1}{f_2}$$

This is true only for a close combination.

**30. Repeat Problem 28 for a system in which the plane mirror is replaced by a convex mirror with a 30-cm focal length.

**31. A Galilean telescope (also called an opera glass) uses a diverging lens as eyepiece ($f = -3.0$ cm) and a converging lens ($f = 12.0$ cm) as objective. The two lenses are 8.5 cm apart. Find the position of the final image it forms of a person who is 30 m away. What is the angular magnification it produces? What advantage does the opera glass have over the ordinary telescope?

*Problems marked with an asterisk are not as easy as the unmarked ones.
**Problems marked with a double asterisk are somewhat more difficult than the average.

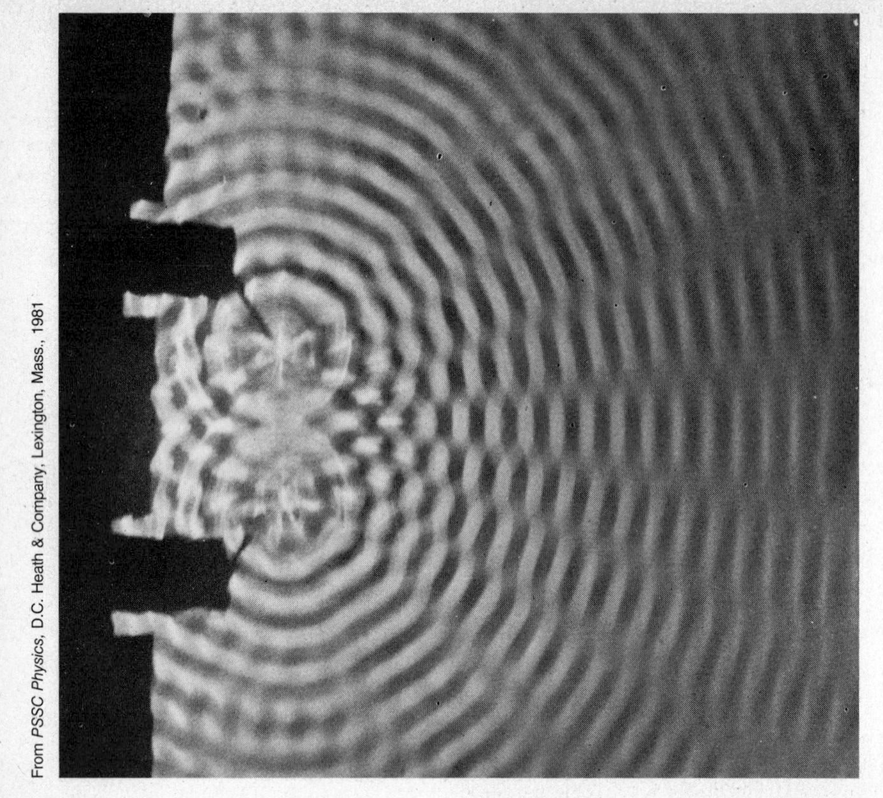

From *PSSC Physics*, D.C. Heath & Company, Lexington, Mass., 1981

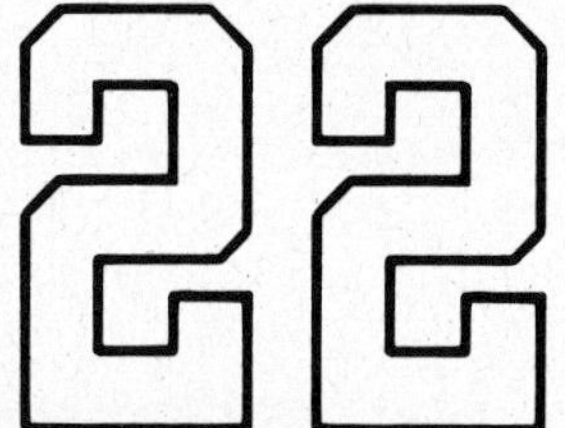

INTERFERENCE AND DIFFRACTION OF WAVES

We have already seen in earlier chapters that waves can interfere with each other. In some situations they reinforce one another; in others they cancel. In this chapter we investigate these effects in detail for light waves. It will be seen that, because of them, light does not always travel in straight lines. Many curious and useful effects occur because of the interference of waves. They occur not only with light but with all types of waves.

Performance Goals

When you finish this chapter, you should be able to

1. Explain how two coherent sources can give rise to interference effects. For example, given two coherent sources sending waves along the same line, find the separations of the

sources that give reinforcement or cancellation.
2. State whether two sources and their waves are coherent when the sources and their waves are described to you.
3. State Huygens' principle and explain how it provides a way to obtain coherent sources.
4. Explain how interference fringes are formed in a Young's double-slit experiment. Point out why maxima (bright fringes) are obtained at certain Δs values and minima (dark fringes) are obtained at others.
5. Sketch a small portion of a diffraction grating and describe an optical arrangement in which it might be used to measure λ. By means of your sketch, show why brightness would be expected at certain special angles. In your explanation point out why the grating equation gives these angles.
6. Use a sketch of a Young's double-slit experiment to point out the meaning of the following terms: central maximum, first-order bright spot, third-order bright fringe, third-order dark fringe, zeroth-order fringe.
7. Explain how spectral lines are formed in a diffraction grating spectrometer and what they are an image of.
8. Give four examples of diffraction and use them to explain the meaning of the term.
9. Give a qualitative explanation for the fact that, using waves of wavelength λ, it is impossible to observe details of an object that have linear dimensions less than λ.

22.1 Interference Between Waves

In Chapter 17 we studied what happens when two waves are brought together. We found that the resultant disturbance is the sum of the disturbances of the two waves. Since the sum is an algebraic one, the waves can cancel each other's effect. Refer to Figure 22.1 to review this fact. We see two identical sound sources that send out wave crests at the same time. The two waves travel toward the right and are joined, or added, at the detector.

In part (a) of the figure the two sources are at the same distance from the detector. Their waves therefore reach the detector in phase. A wave crest from one source reaches the detector at the same time as a crest from the other. As a result, the two waves reinforce each other. The detector registers a loud sound. The waves are *in phase.* Reinforcement is also called *constructive interference.*

But the situation in (b) is quite different. One source is now $\frac{1}{2}\lambda$ (one-half wavelength) closer to the detector than the other. The lengths of the paths from source to detector differ by an amount

$$\text{Path difference} = \Delta s = \tfrac{1}{2}\lambda$$

The waves no longer reach the detector in phase. A crest from one and a trough from the other arrive at the same time. Since the crest cancels the trough, the two waves cancel exactly at the detector. In the case where one wave is $\frac{1}{2}\lambda$ behind the other, we say the waves are 180° or $\frac{1}{2}\lambda$ *out of phase.*

Two identical waves that are 180° (or $\frac{1}{2}\lambda$) out of phase cause complete cancellation.

Complete cancellation is also referred to as *destructive interference.*

Now look at part (c) of Figure 22.1. The path lengths to the detector differ by a distance λ.

$$\text{Path difference} = \Delta s = \lambda$$

Once again the two waves reach the detector in step. Two crests arrive together and, later, two troughs arrive together. The two waves help each other at the detector. Reinforcement again occurs.

We could continue the process. If the separation of sources is increased by another $\frac{1}{2}\lambda$, cancellation will again occur. As we see, cancellation occurs when the path difference is $\frac{1}{2}\lambda$ or $(\lambda + \frac{1}{2}\lambda)$. In fact, a crest from one source will join with a trough from the other when Δs is $\frac{1}{2}\lambda$, $(\lambda + \frac{1}{2}\lambda)$, $(2\lambda + \frac{1}{2}\lambda)$, and so on. Therefore, we can state

Destructive interference (complete cancellation) occurs when the path difference between identical sources is $\Delta s = n\lambda + \frac{1}{2}\lambda$, where $n = 0, 1, 2, \ldots$.

10. Explain how interference fringes arise in the case of a thin film. Compute the difference in thickness of the film between two adjacent fringes.
11. State the Bragg equation and explain how it can be used to find the distance between atomic layers in crystals.

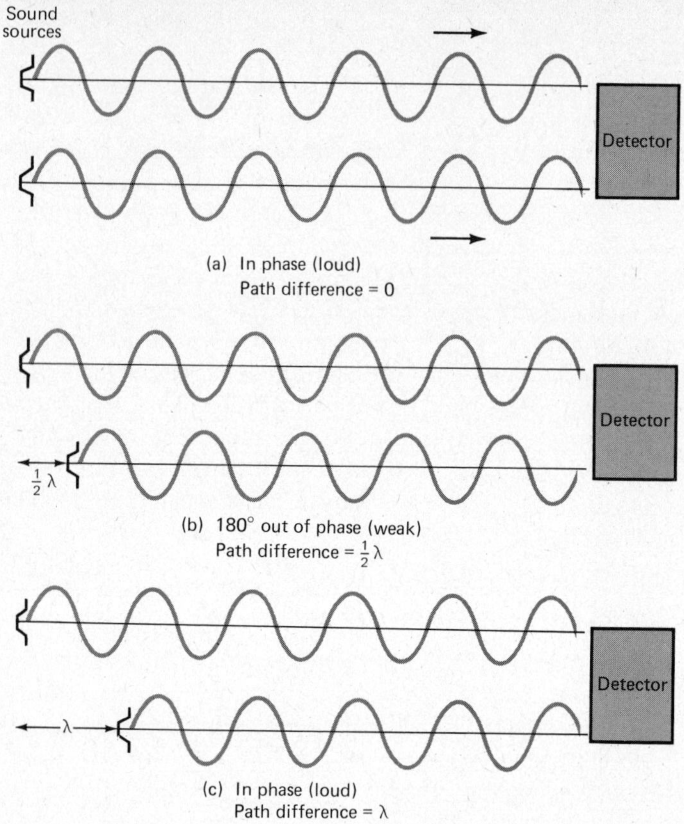

Sound sources

Detector

(a) In phase (loud)
Path difference = 0

Detector

$\frac{1}{2}\lambda$

(b) 180° out of phase (weak)
Path difference = $\frac{1}{2}\lambda$

Detector

λ

(c) In phase (loud)
Path difference = λ

Figure 22.1 The path distance from the two sources to the detector can be varied. Loudness is heard if $\Delta s = 0, \lambda, 2\lambda$, and so on. Cancellation occurs if $\Delta s = \frac{1}{2}\lambda, \lambda + \frac{1}{2}\lambda, 2\lambda + \frac{1}{2}\lambda$, and so on.

You should not try to memorize this fact. If you understand it, it is a fact that should be obvious to you when two identical waves are combined.

Similarly, we see that the two waves reinforce each other when Δs is λ, or 2λ, or 3λ, and so forth. Therefore,

Constructive interference (reinforcement) occurs when the path difference between identical sources is $\Delta s = n\lambda$, where $n = 0, 1, 2, \ldots$.

This, too, should be an obvious fact if you understand the discussion above.

22.2 Coherent Sources; Huygens' Principle

The experiment just described assumed that the wave sources were identical. In order to have easily noticeable interference, the waves from the two sources have to be the same. If their frequencies are unequal, the waves will not remain in step. Even though the waves might cancel at a certain instant, a little later they would reinforce. Our discussion applies only if the two waves have the same frequency and shape.

In this chapter we shall discuss interference between waves of identical frequency and shape. Other types of wave interference occur, but they are very complicated. The waves we shall consider are called coherent waves.

Coherent waves are waves that are identical in shape and frequency. The sources that send them out are called coherent sources.

The waves need not be in phase. However, their phase difference remains constant. Usually we shall assume the waves to be sinusoidal. As pointed out in Chapter 17, complicated waves can be considered to be made up of sinusoidal waves. Using this fact, our results can be applied to more complex cases.

It is easy to obtain coherent sound and water waves. All we need to do is use wave sources that vibrate in identical ways. But coherent light waves are more difficult to obtain. Light is given off by the atoms in the light source. Two light sources will not give out the same shape of waves unless the atoms in them are behaving exactly the same. For this reason the waves given off by two light sources will not, in general, be coherent waves.[1]

There are ways, though, to obtain coherent light waves. They involve breaking a light beam into two or more parts. We shall see several ways of doing this as our discussion proceeds. For now we shall point out only one of them.

Suppose, as shown in Figure 22.2, that a series of water waves hits a wall that has a hole in it. Experiment shows that the hole acts as a new source of water waves. Waves spread out from it in all directions beyond the wall. The schematic diagram in part (b) of the figure represents the physical situation shown in (a). Christian Huygens (1629–1695) extended this idea to all waves. *Huygens' principle* states:

Each point on a wave front acts as a new source of identical waves.

[1] Two identical lasers, however, can closely approximate coherent sources. We shall discuss the laser in Chapter 29.

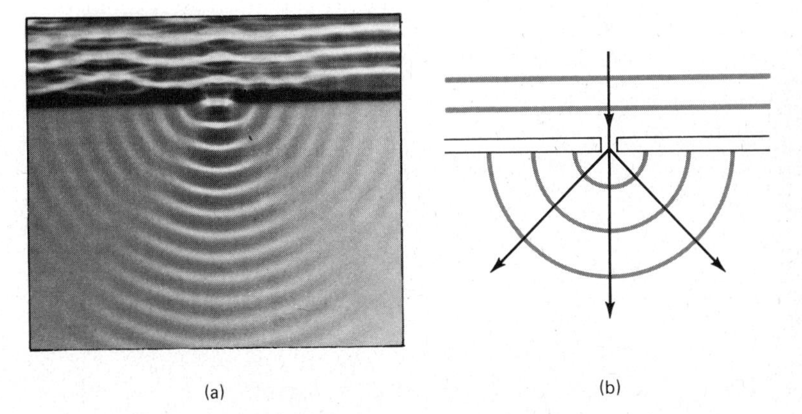

Figure 22.2 The portion of the water wave that strikes the hole in the wall acts as a new source for water waves. (Photo from Education Development Center, Newton, Mass.)

This principle is easily demonstrated for water and sound waves. Its application to electromagnetic waves is less easily justified, but both theory and experiment prove it to be correct. We shall apply Huygens' principle to all types of waves. In the next section we shall use it to show one way of obtaining coherent waves.

22.3 Young's Double-Slit Experiment

Suppose that a series of waves strikes a barrier that has two holes in it, as shown in Figure 22.3(a). These waves could be of any type, perhaps water, sound, or light waves. Huygen's principle tells us that the two holes in the barrier will act as two new wave sources. Because they send out identical waves, these two holes will be coherent wave sources.

Let us now consider what happens at a point such as P in Figure 22.3(b). Clearly, the wave from hole (or slit) A has to travel farther than that from slit B to reach P. If the distances PB and PD are equal, then the extra path length, Δs, is AD as shown.

This reminds us of the situation discussed previously. Identical waves are sent out from identical sources. But one wave has to travel a distance Δs farther than the other before the waves are joined together. If $\Delta s = 0, \lambda, 2\lambda$, and so on, the waves will be in phase with and reinforce each other. Therefore, point P will be a point of strong wave motion if $\Delta s = 0$, or λ, or 2λ, and so on.

But if point P is chosen so that $\Delta s = \frac{1}{2}\lambda$, $(\lambda + \frac{1}{2}\lambda)$, $(2\lambda + \frac{1}{2}\lambda)$, and so on, the two waves will be 180° out of phase and will exactly cancel. In this case no wave motion will occur at P.

Thomas Young was the first to carry out an experiment such as this with

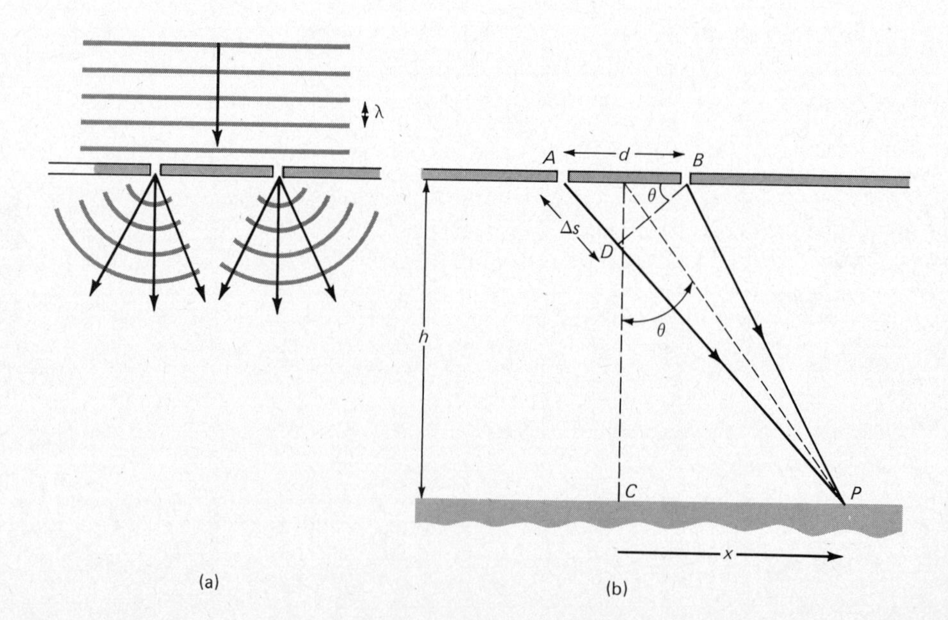

Figure 22.3 If Δs is λ, 2λ, 3λ, and so on, the waves reaching point P will reinforce. If Δs is $\frac{1}{2}\lambda$, $\lambda + \frac{1}{2}\lambda$, $2\lambda + \frac{1}{2}\lambda$, and so on, the waves reaching P will cancel.

light. He followed the procedure shown in Figure 22.4(a). The two coherent sources were two slits as shown. He then noticed that a series of bright and dark lines (or *fringes*) appeared on a screen beyond the slits.[2] Figure 22.4(b) is a schematic drawing of the pattern on the screen.

The bright fringes in Figure 22.4(b) are given the names indicated in the figure. The central bright fringe is also called the *zeroth-order bright fringe.* This region is bright because the light waves travel equal distances from the slits to it. In Figure 22.3(b) it is point C. Clearly, for it $\Delta s = 0$. The waves from the two slits reinforce each other there.

The two bright fringes on each side of the central fringe are called the *first-order bright fringes.* When point P is at either of them, light from one slit must travel λ farther than from the other. As a result, $\Delta s = \lambda$ and the waves from the two slits reinforce each other.

Similarly, at the *second-order bright fringes* light from one slit travels 2λ farther than from the other. Therefore $\Delta s = 2\lambda$ and reinforcement occurs. As you would expect, the *third-order bright fringes* occur where $\Delta s = 3\lambda$. And so on for the fringes farther out in the pattern.

What about the dark fringes? These are positions where the waves from the two slits cancel. At the first dark fringe, the one next to the central maximum, $\Delta s = \frac{1}{2}\lambda$. Complete cancellation occurs there. At the second dark fringe $\Delta s = \lambda + \frac{1}{2}\lambda$ and so darkness occurs there, too. In a similar way we conclude that at the fifth dark fringe $\Delta s = 4\lambda + \frac{1}{2}\lambda$. And so on for all the other dark fringes.

The dimensions in Figures 22.3(b) and 22.4(a) are exaggerated. The slits are very close together, and the fringes are closely spaced. Typical dimensions are given in the legend to Figure 22.4. Notice that $d << h$ and $x << h$. In this case $\sin\theta \cong \tan\theta$ and so one can write for the triangles of Figure 22.3(b) that

[2]The analogous effect for water waves is shown on page 466.

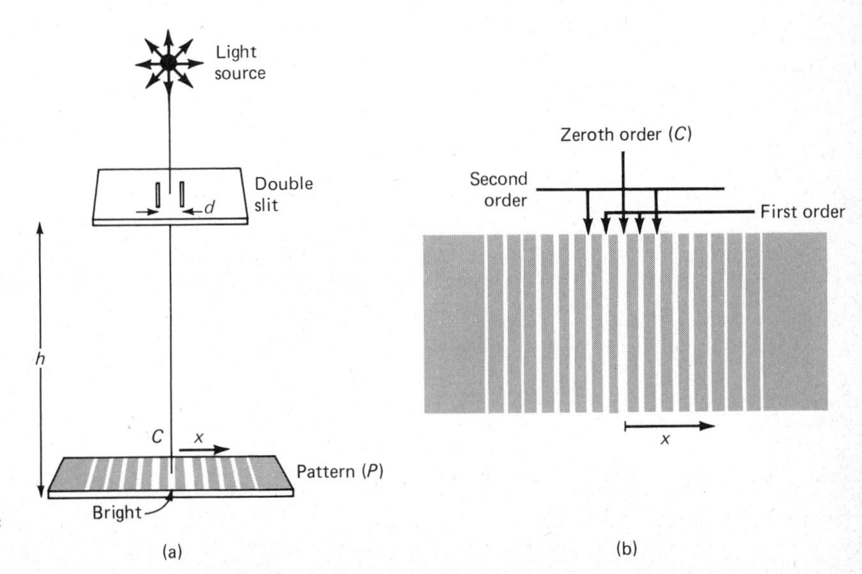

Figure 22.4 A Young's double-slit experiment demonstrates reinforcement and cancellation. Typical dimensions are $h = 100$ cm, $d = 0.05$ cm, $x = 0.4$ cm.

$$\sin\theta = \frac{\Delta s}{d} = \frac{x}{h} \qquad (22.1)$$

This relation allows us to find the wavelength of light, as we shall see in the following example.

■ EXAMPLE 22.1 In a certain Young's double-slit experiment, the slit separation is 0.050 cm. The slit-to-screen distance is 100 cm. When blue light is used, the distance from the central fringe to the fourth-order bright fringe is 0.36 cm. Find the wavelength of the blue light.

Solution The experiment tells us that $d = 0.050$ cm, $h = 100$ cm, and $x = 0.36$ cm. As we found above, at the fourth-order bright fringe we know that $\Delta s = 4\lambda$. Substituting these values in Equation 22.1 gives

$$\frac{4\lambda}{5.0 \times 10^{-4}\ \text{m}} = \frac{3.6 \times 10^{-3}\ \text{m}}{1.00\ \text{m}}$$

Solving for λ gives

$$\lambda = 4.5 \times 10^{-7}\ \text{m} = 450\ \text{nm}$$

Therefore blue light has a wavelength of about 4.5×10^{-7} m. The first measurements of the wavelength of light were made in this way. Today we have a much more accurate way for measuring λ. It is described in the next section. ■■

22.4 Diffraction Grating

The fringes in the Young's double-slit experiment are wide and fuzzy. As a result, their centers cannot be measured easily. This means that the quantity x in Equation 22.1 will never be very accurate. We therefore cannot obtain precise values of λ using the method of Example 22.1.

This difficulty can be gotten around by using not two but thousands of slits. The interference fringes become thinner and sharper as the number of slits is increased. If thousands of slits are used, the fringes become extremely thin and can therefore be located precisely. Let us now see how this fact can be used for accurate wavelength determination.

We call a large number of parallel slits a *diffraction grating.* The distance between slits is usually designated by d. An ordinary grating might be 1 cm × 1 cm and contain 10,000 parallel slits. In that case d would be 10^{-4} cm. A typical experimental arrangement using a diffraction grating is shown in Figure 22.5.

The apparatus in Figure 22.5, called a grating spectrometer, reminds us of the prism spectrometer discussed in the last chapter. In this apparatus the prism is replaced by a diffraction grating. Suppose for the present that the light source is monochromatic. Then only a single wavelength of light need be considered.

It is found that considerable light goes straight through the grating. There-

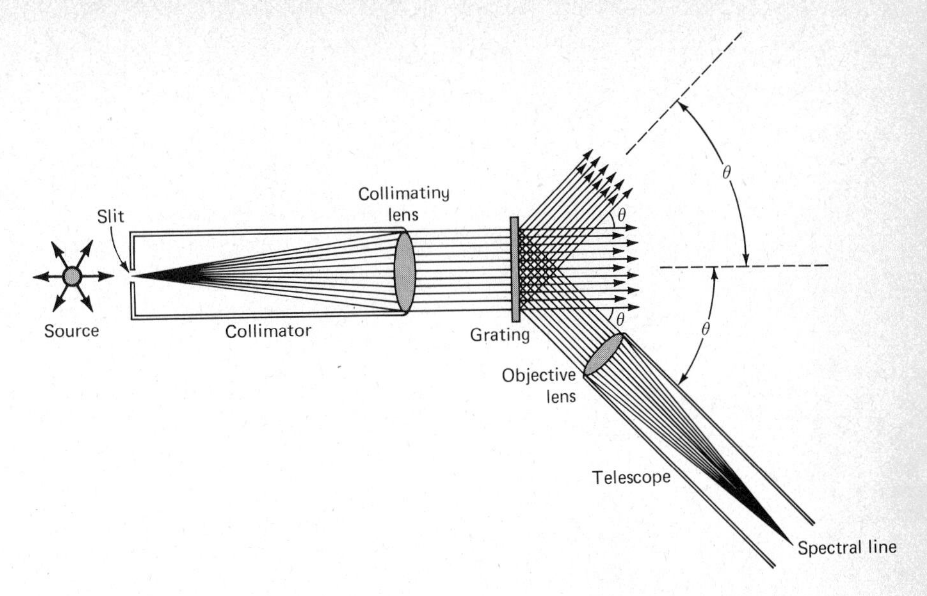

Figure 22.5 The diffraction grating gives brightness at $\theta = 0$ as well as at selected symmetric angles.

fore, when the telescope is rotated until $\theta = 0$, a sharp, bright image of the slit appears. As before, this can be photographed or simply looked at with an eyepiece. But if θ is changed slightly away from $\theta = 0$, no light is seen. Apparently no light comes through the grating at these angles.

However, if θ is increased to perhaps the angle shown in Figure 22.5, a bright line is again seen in the telescope. The grating sends out strong light at this particular angle but at no others between it and zero. We can measure this angle precisely because the image of the slit in the telescope (the spectral line) is extremely sharp. At an equal angle θ on the other side of $\theta = 0$, brightness is also seen. Let us now see why intense light is noticed at this particular angle.

Refer to Figure 22.6. We see there a small portion of the grating. According to Huygens' principle, each slit acts as a source of light. Consider the rays coming away at the angle θ shown. As we see, the light from slit B travels a distance Δs farther than the light from slit A. In fact, the path length from each slit differs from that of the previous slit by this same amount, Δs.

When will the rays from all of these slits reinforce each other? As we have learned, reinforcement (brightness) will occur if the path length difference Δs is λ, 2λ, or 3λ, and so on, or, in shorthand notation,

$$\Delta s_n = n\lambda \qquad n = 1, 2, 3, \ldots$$

for brightness.

We can easily find Δs from the geometry of Figure 22.6. Look at the triangle near slits D and E. We see from it that

$$\Delta s = d \sin \theta$$

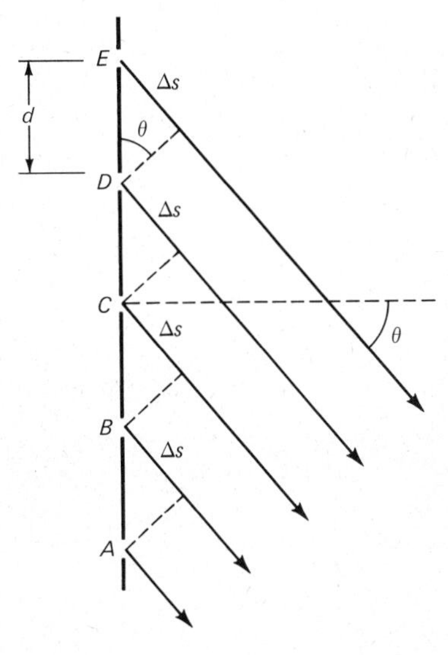

Figure 22.6 If $\Delta s = \lambda$, 2λ, 3λ, and so on, then the rays will all reinforce. Brightness will be seen at angles θ_n such that $\Delta s = n\lambda$.

where d is the grating spacing, the distance between slits. Therefore, brightness will occur at angles θ_n for which

$$d \sin \theta_n = \Delta s_n$$

But $\Delta s_n = n\lambda$, and so we find for brightness

$$n\lambda = d \sin \theta_n$$

This is an important relation.

The grating equation

$$\boldsymbol{n\lambda = d \sin \theta_n} \qquad \textbf{(22.2)}$$

tells us that brightness occurs at the angles specified by it.

Notice that brightness may occur at more than one angle. It can occur at θ_1, where $\Delta s = \lambda$; at θ_2, where $\Delta s = 2\lambda$; at θ_3, where $\Delta s = 3\lambda$; and so on. However, as you see from Figure 22.6, Δs can never be larger than d. As a result, the number of images is limited.

We should also notice that brightness can occur on both sides of the beam that passes directly through. This situation is shown in Figure 22.7. Also shown there are the names we give these bright lines (slit images). You should learn these names, because they are often used.

EXAMPLE 22.2 The diffraction grating gives exceedingly sharp lines. When the wavelength of a particular spectral line is measured, it is found that its second-order image occurs at 72.33°. The grating used has, very accurately, 8000 lines (or slits) per centimeter. Find the wavelength of the light.

Solution For this grating we know that $d = 1 \text{ cm}/8000 = 1.25 \times 10^{-4}$ cm. Because we are observing the second-order image, $n = 2$. Therefore, the grating equation

$$n\lambda = d \sin \theta$$

becomes

$$2\lambda = (1.25 \times 10^{-6} \text{ m}) \sin (72.33°)$$

from which

$$2\lambda = (1.25 \times 10^{-6} \text{ m})(0.953)$$

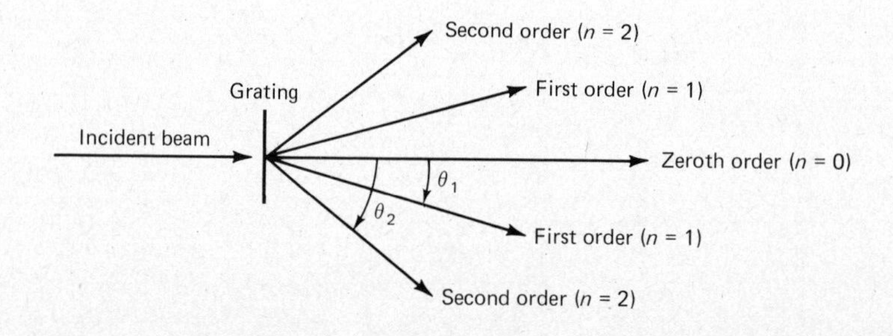

Figure 22.7 Brightness occurs at angles such that $n\lambda = d \sin \theta_n$. These correspond to $\Delta s_n = n\lambda$.

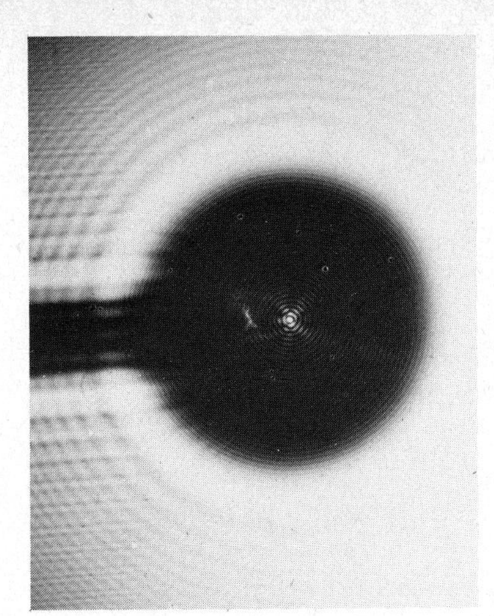

Figure 22.8 The shadow cast by a map tack. (Courtesy of R. C. Nicklin and J. Dinkins)

and so

$$\lambda = 5.95 \times 10^{-7} \text{ m} = 595 \text{ nm}$$

Notice in this case that the third-order image is impossible to obtain. Why?

22.5 Diffraction

The word *diffraction* in the term "diffraction grating" represents an important phenomenon. We have seen examples in this chapter where waves do not cause sharp shadows. Water waves spread out into the whole region beyond a barrier as shown in Figure 22.2. In Figure 22.4 the light coming through two slits does not remain as two beams. It spreads out into the region of shadow. This same type of behavior was seen with the diffraction grating. Indeed, Huygens' principle tells us that waves spread out from every point on the wave front. This fact gives rise to the phenomenon called *diffraction.*

The ability of waves to bend around obstacles in their path is called diffraction.

Interference effects determine the way that waves behave after being diffracted. For example, the double-slit interference pattern results as the diffracted waves from the slits combine. Because of diffraction, no object casts an absolutely sharp shadow. In Figure 22.8 we show the shadow cast by a map tack. If you look closely, you can see that there is a bright spot in the very center of the shadow. The details of the shadow pattern can be calculated in a straightforward way from Huygens' principle and our ideas of interference. As you might guess, the computation in this particular case is not simple.

An important example of diffraction is shown in Figure 22.9. In part (a) we see a plane wave striking a slit of width w. If there was no diffraction, a bright spot with width w should appear on the screen. Instead, one notices a fuzzy bright spot. Actual photos of the bright spot are shown in part (b) of that figure.

Notice that the bright spot becomes larger as the slit width becomes very small. Theory and experiment both show that the width of the bright spot can be given rather simply. As shown in part (a) of Figure 22.9, we define an angle θ_c. It turns out that the width of the bright spot as measured by θ_c is given by

$$\sin \theta_c = \frac{\lambda}{w} \qquad (22.3)$$

In this expression λ is the wavelength of the light and w is the slit width.

Equation 22.3 tells us that the ratio of λ/w is important. If the slit is much wider than one wavelength, then $\lambda/w \rightarrow 0$ and θ_c will be very small. The bright spot will be very narrow. But as w becomes smaller, λ/w becomes larger. The bright spot spreads out. Finally, when $w = \lambda$, θ_c has increased to 90°. The bright spot will be infinitely wide.

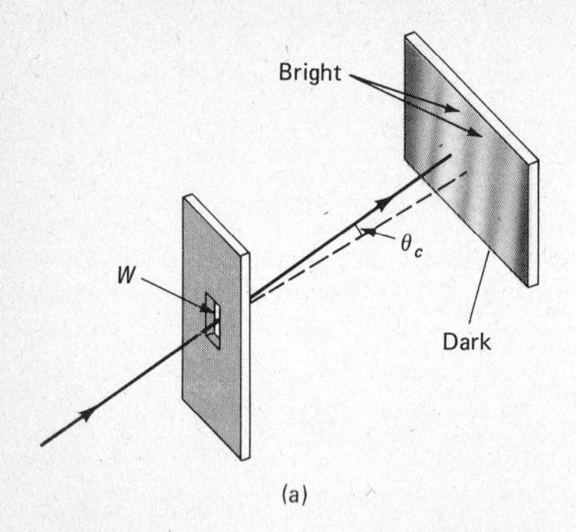

(a)

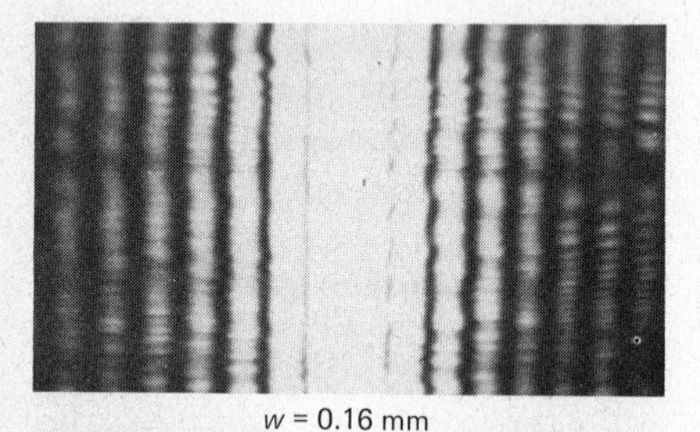

w = 0.16 mm

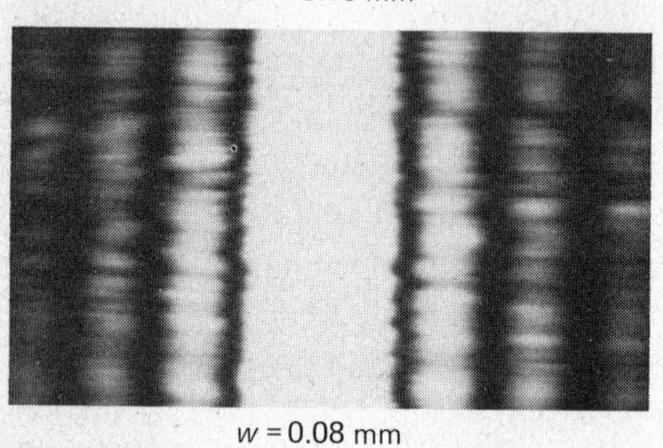

w = 0.08 mm

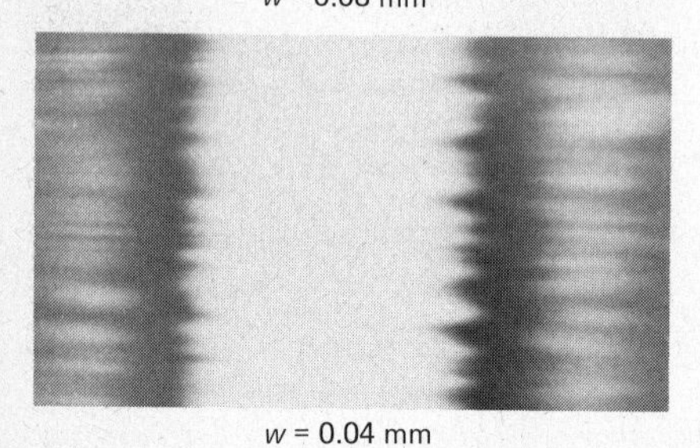

w = 0.04 mm

(b)

Figure 22.9 Notice how the bright spot increases in width as the slit width w is narrowed. (Photos courtesy of R. C. Nicklin and J. Dinkins)

This same sort of behavior occurs when an obstacle casts a shadow. If the obstacle is fairly large compared to λ, the shadow will be sharp. But the shadow becomes more fuzzy the smaller the obstacle is. For obstacles whose linear dimensions are about equal to λ, the shadow becomes so fuzzy that it cannot be seen.

This diffraction behavior of slits and obstacles is of great fundamental importance. Because of it we must conclude that

No object or detail can be seen distinctly if its dimensions are close to or smaller than the wavelength of waves being used to examine it.

In our case the waves are light waves. But this statement is true for all waves. In fact, as we shall see in Chapter 29, even electrons sometimes behave as waves. As a result, this statement places a limit on the detail one can see with an electron microscope as well as with light microscopes.

22.6 Limit of Resolution in Optical Devices

At first you might think the results of the previous section are not meaningful for most optical devices. After all, the lens opening (aperture) for a telescope or microscope or the eye is far, far larger than λ. Even so, diffraction is still important in these cases. To see why, refer to Figure 22.10. (The pattern shown there is only suggestive. It is not accurate in detail.)

We see there two point objects being imaged by a lens. The lens opening D corresponds to the slit width w. Diffraction causes the images to be spread out as shown. (Of course, the drawing exaggerates the situation.) In the case of a circular aperture rather than a slit, it turns out that

$$\sin \theta_c = 1.22\frac{\lambda}{D} \qquad (22.4)$$

So θ_c will be very small because D will certainly be much, much larger than λ.

However, suppose that the two objects are fine details of an insect and that the lens is part of a microscope. Then in order to be seen clearly, the images of these objects must not overlap seriously. Usually one states that the objects can no longer be resolved (seen individually) if the angle shown is less than θ_c.

Two objects cannot be resolved if the angle they subtend at the circular aperture is less than θ_c, where $\sin \theta_c = 1.22\lambda/D$.

This same reasoning applies to any objects and for any optical device. It provides a serious limitation for all tools that use waves to examine detail.

■ **EXAMPLE 22.3** An object is examined at a distance of 0.50 cm in front of a microscope objective. The diameter of the objective is 0.30 cm. How small a detail can the microscope see?

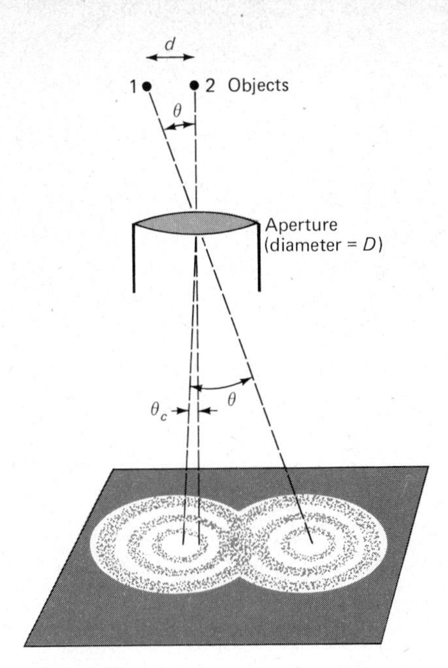

Figure 22.10 If the angle θ subtended by the objects is not larger than θ_c, then the images will overlap too much to resolve the objects.

Solution We must assume that the microscope is perfect so that diffraction is the limiting factor. From Figure 22.10, assuming that the distance from lens to object is 0.50 cm, we see that

$$\sin\theta \simeq \frac{d}{0.50 \text{ cm}}$$

In the limiting case,

$$\sin\theta = \sin\theta_c = 1.22\frac{\lambda}{D}$$

We shall use an average value $\lambda = 5 \times 10^{-5}$ cm, which is green light. Also, we were told that $D = 0.30$ cm. Therefore, when just resolved,

$$\frac{d}{0.50 \text{ cm}} = 1.22\left(\frac{5 \times 10^{-5} \text{ cm}}{0.30 \text{ cm}}\right)$$

Solving for d, we find that

$$d = 1.0 \times 10^{-4} \text{ cm}$$

No matter how perfect, this microscope could not see details closer together than 1×10^{-4} cm. Thus we find that detail smaller than λ cannot be seen. ■■

22.7 Interference in Thin Films

All of us have seen interference effects in thin films. As soapy water drains off a glass or dish, the surface sometimes appears to be highly colored. Soap bubbles may display the rainbow colors. An oil slick on a puddle of water shows interference colors in bright sunlight. Let us now examine this effect.

First let us look at the idealized situation shown in Figure 22.11. A beam of monochromatic light shines nearly perpendicular to two glass plates. The plates are parallel to one another and separated by a distance x. The two surfaces facing each other are lightly metalized so they act as semitransparent mirrors. Most of the light goes straight through the plates, but beams 1 and 2 are reflected by the mirrored surfaces as shown. For simplicity assume that beams 1 and 2 are of equal intensity. Furthermore, assume the angle θ to be so small that the beams can be considered to go straight up and down.

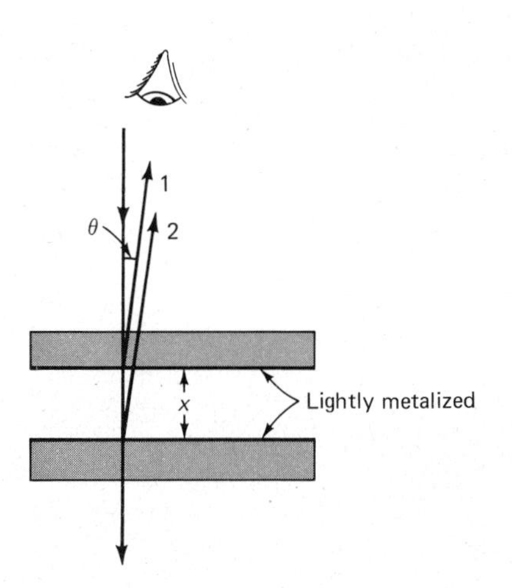

Figure 22.11 Beam 2 travels a distance $2x$ farther than beam 1. If $2x$ equals $n\lambda$, where n is zero or an integer, the eye will see brightness. If $2x$ is $\frac{1}{2}\lambda$ or $n\lambda + \frac{1}{2}\lambda$, then darkness will result. (θ is assumed to be very small.)

We see that beams 1 and 2 are coherent, because they are part of the same original beam. As a result, we can expect them to interfere in a simple way. Notice that beam 2 travels a distance $2x$ farther than beam 1, where x is the separation between the plates (the second beam goes down and back up, a total distance $2x$). When the two reflected rays join together, they will interfere with each other. If $2x$ is equal to λ, beam 2 will be one whole wavelength behind beam 1. It will therefore interfere constructively with 1, and a bright reflected beam will result. This will also be true if $2x$ is equal to 2λ, 3λ, and so on. In general, bright reflection will occur if $2x$ equals $n\lambda$, where n is an integer.

But if $2x$ is only $\frac{1}{2}\lambda$, then beam 2 will be 180° out of phase with beam 1. The two reflected beams will then cancel each other. This cancellation will occur if $2x$ is $\frac{1}{2}\lambda$, $\lambda + \frac{1}{2}\lambda$, $2\lambda + \frac{1}{2}\lambda$, and so on. Whether or not strong reflection occurs depends on the plate spacing x.

Now suppose that the upper plate can be moved up and down. To begin with, suppose that the two plates are nearly in contact. Then $x \simeq 0$. Beam 1 is still reflected from the upper metal film while 2 is reflected from the lower one. But because x is now assumed to be nearly zero, the two beams will have traveled the same distance and will be in phase. Therefore, brightness will be observed for the reflection.

If x is now increased to $\frac{1}{4}\lambda$, then the path difference, $2x$, will be $\frac{1}{2}\lambda$. The two beams now cancel, and darkness is observed. If x is increased still further to $\frac{1}{2}\lambda$, then the path difference is $2x = \lambda$ and so brightness is again observed. This process of alternating dark and bright continues as the separation is increased. As you see, the closely spaced metal surfaces give rise to two coherent beams. Whether or not the beams cancel or reinforce depends on the separation between the surfaces.

If you use white light (sunlight, for example) instead of monochromatic light, you will observe colored reflections. This occurs because only one wavelength of light undergoes perfect reinforcement at a given value of x. If $2x = 450$ nm, then you will see reinforcement of blue light. But if $2x = 550$ nm, then the interference will show a greenish color. It is for this reason that most interference effects we observe show colored bands (or *fringes*) of light.

An interesting situation occurs if the region between the plates is filled with a substance of refractive index μ instead of air. Then light beam 2 will travel with speed c/μ in that region. As a result, it takes a longer time for beam 2 to travel its distance. This causes it to act as though it had traveled further than the distance x it actually travels. In fact, the distance x in air acts like a distance μx when the region is filled with a substance other than air. We say that the equivalent optical path length is μx.

A distance x in a material of refractive index μ is equivalent to an optical path length of μx.

What this means in a practical sense is the following. For constructive interference to occur in Figure 22.11, $2\mu x = \lambda$, or 2λ, or 3λ, etc. In the case we discussed before, $\mu = 1.00$ and so we obtained the results previously given.

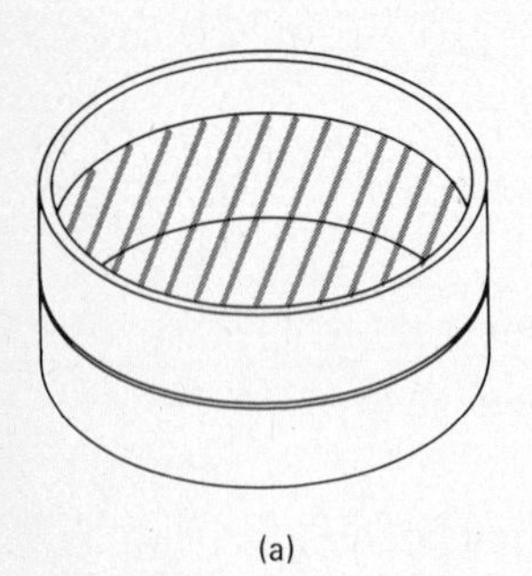

(a)

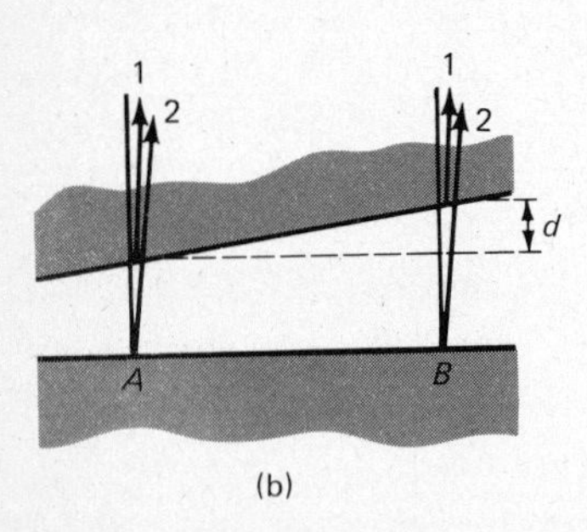

(b)

Figure 22.12 If A and B are at the positions of adjacent dark fringes, then the thickness difference d must be $\frac{1}{2}\lambda$.

■ EXAMPLE 22.4 The optical flat is a device used to measure the perfection of surfaces. The two optical flats in Figure 22.12(a) are illuminated by 580 nm light. A narrow, wedgelike gap between them gives rise to the fringes shown. The distance between fringes is 0.60 cm. Find the angle of the wedge assuming the gap to be filled with (a) air and (b) water.

Solution Consider the dark fringes because they are sharper. The positions of two adjacent ones are shown in (b). Beams 1 and 2 cancel at A and B. Hence the path length must be λ greater at B than at A. This path is longer because beam 2 travels through the distance d twice. If the gap is filled with air, then

$$2d = \lambda$$
$$d = \tfrac{1}{2}\lambda = 290 \text{ nm}$$

If the gap is filled with water, then

$$2\mu d = \lambda \qquad \text{or} \qquad 2(1.33)d = 590 \text{ nm}$$
$$d = 218 \text{ nm}$$

To find the wedge angle, notice that $\overline{AB} = 6.0 \times 10^{-3}$ m. Therefore, for the gap filled with air

$$\tan\theta = \frac{d}{AB} = \frac{290 \times 10^{-9}\text{ m}}{6 \times 10^{-3}\text{ m}} = 4.8 \times 10^{-4}$$

For the gap filled with water we find in the same way

$$\tan\theta = 3.6 \times 10^{-4}$$

For such small angles $\tan\theta$ is essentially equal to the angle in radians. Therefore we find (a) $\theta = 4.8 \times 10^{-4}$ radians and (b) $\theta = 3.6 \times 10^{-4}$ radians. ■■

22.8 Newton's Rings

Figure 22.13 shows an interference experiment performed by Newton. A slightly curved plano-convex lens is placed on an optical flat. The fringes observed are shown in (b). As we see, the fringes are circular. This phenomenon is called *Newton's rings.*

We can explain these fringes easily. At the darkness positions ray 1 and ray 2 must be 180° out of phase and cancellation occurs. The air wedge must increase in thickness by $\frac{1}{2}\lambda$ as we move from one fringe to the next larger one. Why?

But there is a puzzling feature of the pattern in (b). The center point is dark. The rays cancel there even though their path difference is essentially zero. Early

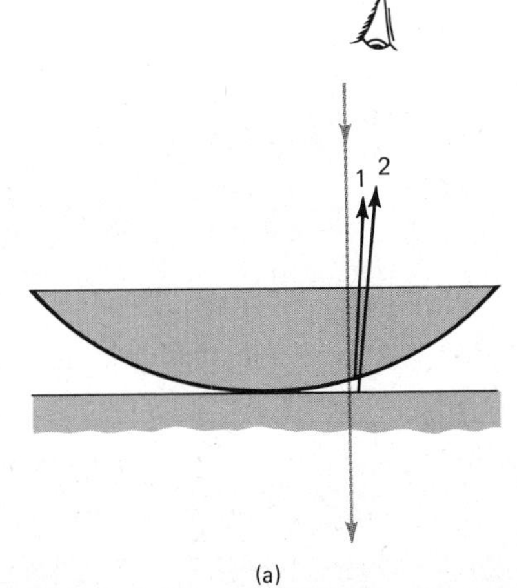

(a)

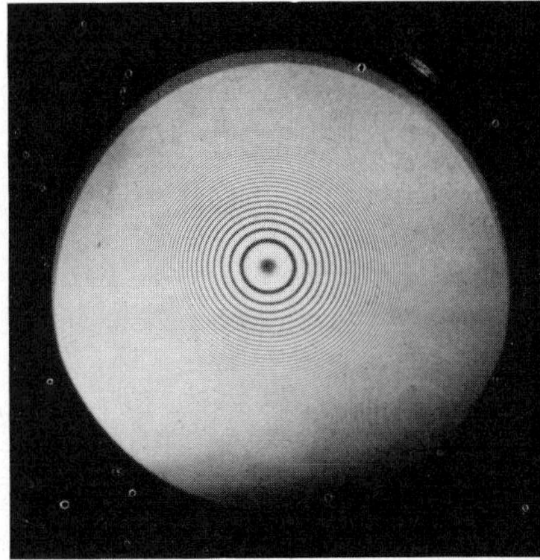
(b) Pattern seen from above

Figure 22.13 Newton's rings. The curvature of the upper surface is much less than shown. Rays 1 and 2 have been displaced for clarity. Notice that the center of the pattern is dark. (Photo courtesy of Bausch & Lomb Analytical Systems Division)

researchers tried in vain to get the surfaces closer together so the central spot would be bright. But the harder they worked at it, the darker the center became.

The explanation for this was found much later. It turns out that a change in phase sometimes occurs when a beam of light is reflected. This occurs only if the reflecting material has a larger index of refraction than the material in which the beam is traveling.

When a wave is reflected by an optically more dense material into an optically less dense material, the wave undergoes a 180° phase change. In effect, the reflection process holds it back through $\frac{1}{2}\lambda$.

In the case of Newton's rings ray 2 is held back by $\frac{1}{2}\lambda$ during the reflection process. Ray 2, but not ray 1, is reflected by an optically more dense material. As a result, even though the path lengths are the same, rays 1 and 2 cancel at the center position.

▮ **EXAMPLE 22.5** The lens of a good camera can be seen to show color. This is the result of interference due to a coating on the lens. To prevent reflection, lenses are coated with a thin film of magnesium fluoride ($\mu = 1.38$). How thick must this film be so that light with $\lambda = 500$ nm is not reflected?

Solution Notice that the light goes from air ($\mu = 1.0$) to coating ($\mu = 1.38$) to glass ($\mu = 1.55$). Therefore, a phase reversal of the reflected beam occurs at both the fluoride and glass surfaces. In order for the reflected beams to cancel, beam 2 in Figure 22.14 must be held back through $\frac{1}{2}\lambda$. Therefore,

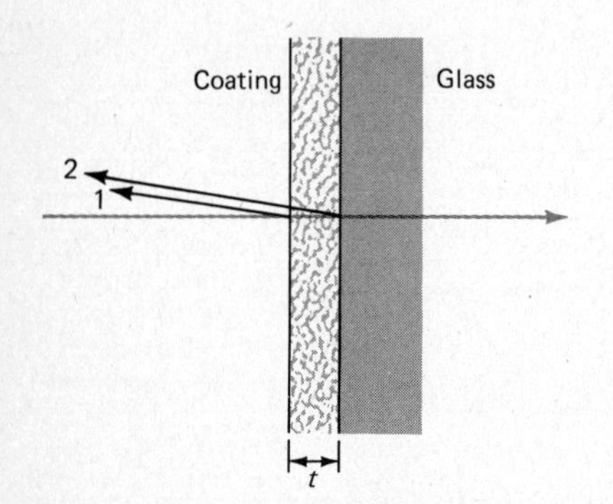

Figure 22.14 The nonreflecting coating on the lens surface causes the two reflected beams to cancel. Do all colors of light cancel?

$$\text{Extra optical path} = \tfrac{1}{2}\lambda = \tfrac{1}{2}(500 \times 10^{-9}\ \text{m})$$

But the coating has a thickness t and is equivalent to an optical path length of μt or $1.38t$. Because the beam travels through this distance twice, we have

$$\text{Extra optical path} = 2(1.38t)$$

Equating these two expressions gives

$$\tfrac{1}{2}(500 \times 10^{-9}\ \text{m}) = 2(1.38t)$$

from which

$$t = 9.1 \times 10^{-8}\ \text{m}$$

The fluoride coating should be this thick. Why does the lens appear colored? ▮▮

22.9 X-Ray Diffraction

Electromagnetic radiation includes X rays as well as light. It is possible, therefore, to imagine interference and diffraction experiments using X rays. However, X rays have wavelengths about 1 nm and less. This means that the slits and film thicknesses used for such experiments are of atomic size. (The diameter of

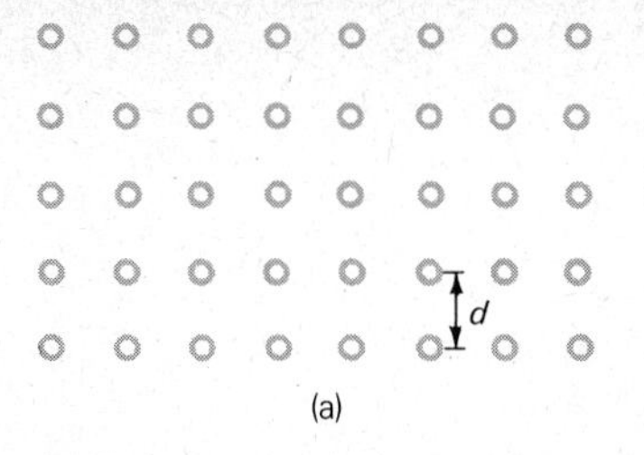

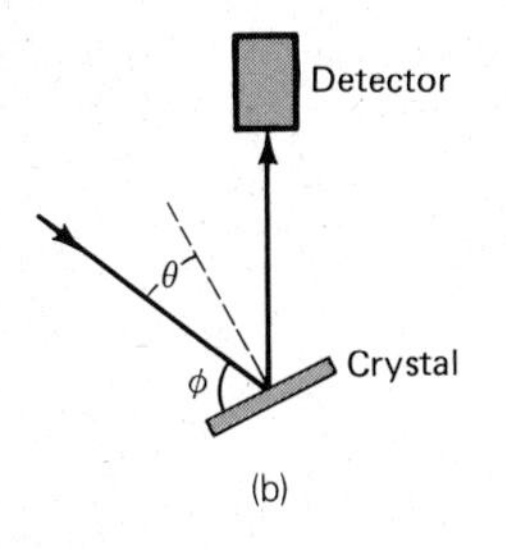

Figure 22.15 Sharp, strong reflections occur at certain angles.

a hydrogen atom is about 0.1 nm.) It turns out that X-ray interference and diffraction provides us with a major tool for examining atoms, molecules, and crystals. We will discuss only the topic of crystal structure determination at this time.

You know that crystals consist of precise geometric arrangements of atoms. Consider the cubic crystal shown in cross section in Figure 22.15(a). We wish to use X rays to determine the spacing d between the layers of atoms. To do so, we reflect a beam of X rays off the crystal as shown in (b). It is observed that little reflection occurs at most angles, but the beam is strongly reflected at certain special angles. At these angles the beam reflections from the parallel atomic layers reinforce each other.

To obtain a relation for these strong reflections, refer to Figure 22.16. Beam 2 travels a distance $2d \sin \phi$ further than beam 1. Reinforcement between these (and all similar) beams will occur if this extra path is λ, or 2λ, or 3λ, and so on. Thus, for reinforcement,

$$n\lambda = 2d \sin \phi \tag{22.5}$$

where n is 1, 2, 3, ... This relation is called the *Bragg equation,* after W. H. Bragg and his son W. L. Bragg who first used it extensively in 1913. Notice in particular how ϕ is defined. It is not the angle of incidence, but 90° − the angle of incidence.

■ **EXAMPLE 22.6** A certain flat crystal strongly reflects a 0.200 nm X-ray beam at an angle of incidence of 30°. Find the possible spacing of the atomic layers causing this reflection.

Solution If the incidence angle is 30°, then ϕ in Figure 22.16 is 60°. Can you show this by use of the figure? The Bragg equation becomes

$$n(2.0 \times 10^{-10}\ \text{m}) = 2d \sin 60°$$

Solving gives

$$d = (1.15 \times 10^{-10}\ \text{m})n$$

where n can be any integer. Hence the atom layer spacing could be 0.115 nm, or 0.230 nm, etc. ■■

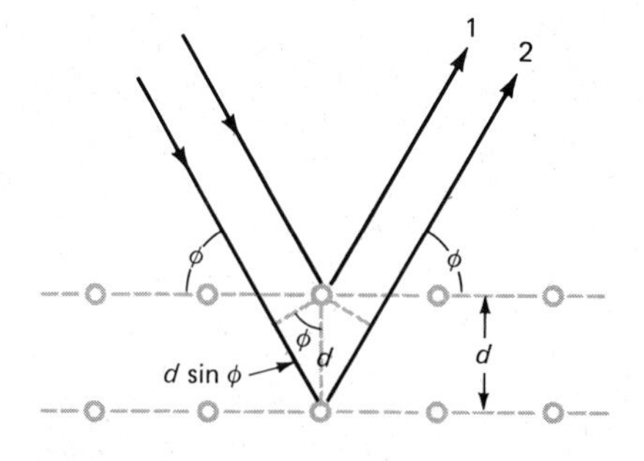

Figure 22.16 The path difference for the two beams is $2d \sin \phi$.

Summary

Coherent waves have identical frequencies and shape. The sources that send them out are called coherent sources.

Two coherent waves will interfere destructively, or cancel, if they add with the crest of one wave on a trough of the other. We say that such waves are 180° or $\frac{1}{2}\lambda$ out of phase. The waves will interfere constructively, or reinforce, if the crest of one falls on the crest of the other. These waves are said to be in phase.

Huygens' principle states that each point on a wave front acts as a new source of identical waves.

In Young's double-slit experiment two coherent sources are obtained by allowing a wave to strike two slits. Beyond the slit the waves from these sources interfere to give bright and dark fringes. The positions of the fringes can be used to calculate the wavelength of the waves.

A diffraction grating consists of thousands of parallel slits a distance d apart. It produces sharp, bright fringes at angles θ_n given by the grating equation:

$$n\lambda = d \sin \theta_n \qquad n = 0, 1, 2, \ldots$$

Use is made of gratings in precise determinations of λ.

The ability of waves to bend around obstacles in their path is called diffraction. When the dimensions of slits and obstacles are comparable to or smaller than λ, distinct beams and shadows are not found. As a result, no object or detail can be seen distinctly if its dimensions are close to or smaller than the wavelength of the waves being used to examine it.

Two objects can be resolved from each other in an optical device of aperture diameter D only if the angle θ_c they subtend at the aperture is greater than that given by

$$\sin \theta_c = 1.22\frac{\lambda}{D}$$

Two coherent beams of light can be obtained by reflecting portions of a single beam from the two surfaces of a thin film. The optical path length difference between the two beams due to an actual path length difference d will be μd, where μ is the index of refraction of the material of the film.

In a wedge-shaped thin film the thickness of the film changes by an amount Δd from dark fringe to next dark fringe (or from bright fringe to bright fringe). The optical path length increase due to Δd is $2\mu\ \Delta d$. The factor of 2 occurs because one beam travels the distance Δd twice. In going from one fringe to the next similar fringe, this added optical path length is equal to λ. Therefore,

$$2\mu\ \Delta d = \lambda$$

Interference fringes obtained with white light show colors. This is a result of the fact that interference effects depend on the relation of path difference to λ. Usually positions of reinforcement for one color are not the same as for others.

When a beam is reflected by a substance more dense optically than the material in which the beam is moving, a 180° phase change occurs upon reflection. It is as though the reflection process holds the beam back by $\frac{1}{2}\lambda$. No such effect is observed if the reflecting substance is less dense optically than the material in which the beam is moving.

When a beam of X rays is reflected from a crystal, strong reflection due to reinforcement is observed at certain angles. These angles are given by the Bragg equation

$$n\lambda = 2d \sin \phi$$

where d is a layer spacing within the crystal and ϕ is 90° − angle of incidence.

Questions and Exercises

1. The two prongs of a tuning fork are two coherent sound sources. If you hold a 10,000-Hz tuning fork vertically at arm's length and rotate the fork slowly on its axis, alternate loud and weak sounds are heard. Explain this effect.
2. In order to perform a double-slit experiment, a student shines light from two headlights of a car onto a wall. Will a pattern be seen on the wall?
3. Two identical loudspeakers are 20 m apart in an open field and are aimed at each other. They send identical sound waves toward each other. What would you expect to hear as you walk along the line from one to the other?
4. In a double-slit experiment would the fringes be closer together using blue or red light? What would you see if white light was used? What would you see if white light was used in a diffraction grating spectrometer? In a prism spectrometer?
5. Both radio and light waves are em waves, but the wavelengths of radio waves are in the range of hundreds of meters. Explain why a tree trunk and a metal pole cast clear shadows for light but not for radio waves.
6. An interesting pattern may be observed when a distant light is seen through a thin, finely woven curtain (often called a "sheer"). A pattern of light in the form of a cross is seen. Explain its origin.
7. Estimate how far away a car could be and still have its headlights resolved by an observer. Consider the limiting factor to be diffraction limitation of the pupil of the eye.
8. Bats can fly in dark caves without striking obstacles. They send out very high-frequency sound waves that, upon reflection, warn them of obstacles in their path. Why would waves of much lower frequency be far less effective?
9. Semitransparent metal films can be deposited on glass plates by evaporating a metal film onto the plate in vacuum. The thickness of the film can be estimated by observing the reflected white light from the film. Describe what should happen to the color as the film slowly increases in thickness.
10. Figure P22.1 shows interference fringes observed when laser light is reflected from the two surfaces of a thin glass plate (a microscope slide). What can you say about the uniformity of thickness of the plate from this picture?

Figure P22.1 Laser interference in a microscope slide. (Courtesy of R. C. Nicklin and J. Dinkins)

11. Many industrial diffraction grating spectrometers use so-called reflection gratings. These consist of thousands of parallel grooves (or lines) cut in a metal mirror surface. Would you expect the grating equation to apply to these gratings? (Often the grating surfaces are curved so that they perform the functions of the collimating and objective lens all in the reflection process.)
12. In Example 22.6, how might taking measurements at different angles enable you to decide which layer spacing is the correct one?

Problems

1. Two tiny loudspeakers are sealed in the same end of a long tube. They send sound waves down the tube. The waves have the same amplitude and frequency. What can you say about the sound in the tube (a) if both speakers send out compressions at the same time, (b) if one sends out a wave crest while the other sends out a trough, (c) if the speakers vibrate in phase, (d) if the speakers vibrate 180° out of phase?
2. Two identical loudspeakers are very close together and send identical waves out along the x axis. The amplitudes of the waves are both A. Their wavelengths are both 130 cm. Assume both speakers send out crests at the same time. (a) When both speakers are together at $x = 0$, what is the amplitude of their combined wave? (b) Repeat for one speaker at $x = 0$ and the other at $x = 65$ cm. (c) Repeat for one at $x = 0$ and the other at $x = -65$ cm. (d) Repeat for one at $x = 65$ cm and the other at $x = -65$ cm.
3. Two loudspeakers send out identical sound waves along the x axis. The wavelength of the waves is 70 cm. One speaker is at $x = 0$. A listener is at $x = 20$ m. The other speaker is at

$x = 0$ but is slowly moved out along the x axis. At what speaker positions between $0 \le x \le 5$ m will the listener hear the loudest sounds?

4. Repeat Problem 3 but give the positions for weakest sound.

*5. Two identical loudspeakers send identical sound waves out along the x axis to a listener at $x = 20$ m. The speakers are at $x = 0$ and $x = 1.40$ m. The frequency of the two waves is slowly increased from a very low value. Give the first four wavelengths at which weak sound will be heard.

*6. Repeat Problem 5 but give the wavelengths for loudest sound.

7. A Young's double-slit arrangement uses green light with $\lambda = 546$ nm, the mercury green line. The slits are 0.070 mm between centers. Fringes are formed on a screen 2 m away. (a) How much further is it from one slit to the zeroth-order fringe than it is from the other slit? (b) The first-order bright fringe? (c) The second-order bright fringe? (d) The third-order bright fringe?

8. For the situation in Problem 7, how far along the screen is (a) the first-order bright fringe from the zeroth-order fringe? (b) The third-order bright fringe from the central maximum?

9. In a Young's double-slit experiment, the slit separation is 0.100 mm and the slit-to-screen distance is 1.50 m. The yellow light from a sodium lamp is used ($\lambda = 589$ nm). Find the distance from the center bright fringe (a) to the third bright fringe and (b) to the third dark fringe.

*10. For the situation in Problem 9, find the separation between any (a) two adjacent bright fringes and (b) two adjacent dark fringes.

11. Using a diffraction grating with 8000 lines/cm, the second-order maximum is found at an angle of 72° on each side of the straight-through beam. (a) What is the wavelength of the light? (b) Will the third-order maximum be visible?

12. Red light from a helium-neon laser ($\lambda = 633$ nm) passes through a diffraction grating. The first-order maximum is found to be 67° on each side of the central maximum. (a) What is the distance between centers of the lines on the grating? (b) How many lines does the grating have per centimeter?

13. For the calibration of a certain diffraction grating, the blue mercury line at 435.8 nm is observed. Its second-order line appears at an angle of 72° to the straight-through beam. (a) What is the grating spacing for this grating? (b) How many lines does the grating have per centimeter?

14. A diffraction grating with 8000 lines/cm is illuminated perpendicularly with light having $\lambda = 546$ nm. (a) At what angle will the first-order maximum be found? (b) The second-order?

*15. A diffraction grating with 10,000 lines/cm is used to examine sodium yellow light. This light is composed of two very close lines (a doublet) with wavelengths 588.995 nm and 589.592 nm. Find the angle between the two lines in the first-order spectrum. (This grating will show the doublet easily.)

*16. A diffraction grating used in an infrared spectrometer has 300 lines/cm. What is the largest wavelength line that can be formed by this grating? In what spectral order is it found?

17. Refer to the single-slit situation shown in Figure 22.9. How large is θ_c if the slit is 0.020 mm wide and 579-nm yellow light is being used?

*18. Suppose in Problem 17 that the light coming through the slit strikes a screen 70 cm away from the slit. How wide will the central bright spot be on the screen?

*19. In a certain grating spectrometer the sodium yellow doublet lines (see Problem 15) are separated by an angle of 0.040°. (a) Can these two lines be resolved if the aperture of the objective lens has a diameter of 2.0 cm? Assume diffraction to be the limiting factor. (b) What is the smallest angle the system could resolve?

*20. The telescope at the Yerkes Observatory in Wisconsin has a lens with a diameter of 102 cm. How far apart must two objects be on the moon if they are to be resolved by this telescope? The earth-moon distance is 3.84×10^8 m. Assume that $\lambda = 500$ nm. (In practice, variations in the earth's atmosphere prevent the telescope from doing this well.)

**21. Suppose diffraction limits the ability of a person's eye to see detail. Suppose further that the pupil of the eye has a 0.25 cm diameter. How far from the eye can a millimeter scale be held and still have the millimeter lines clearly resolved? Assume $\lambda = 500$ nm.

22. Two plates are 0.50 mm apart. What is the equivalent optical path length between the plates when the gap between the plates contains (a) vacuum, (b) water with $\mu = 1.33$, (c) oil with $\mu = 1.52$?

23. Two glass plates are to be $\lambda/2$ apart when red light with $\lambda = 633$ nm is being used. How far apart (in meters) must the plates be if the gap between them is filled with (a) vacuum, (b) water, (c) oil with $\mu = 1.48$?

**24. Laser light with $\lambda = 633$ nm is used in a Young's double-slit experiment. The slits are 0.100 mm apart and the screen is 2 m beyond the slits. If the whole experiment is being done under water, how far will the first-order bright fringe be from the central maximum?

**25. What form will the diffraction grating equation have if the grating is to be used under the surface of a lake?

26. Light waves with wavelength 440 nm (the blue mercury

*Problems marked with an asterisk are not as easy as the unmarked ones.

**Problems marked with a double asterisk are somewhat more difficult than the average.

line) are reflected by two semitransparent metal films as shown in Figure 22.11. The metal films start at zero separation and are slowly moved apart. Give the first four separations for which strongest reflection occurs.

27. Repeat Problem 26 but give the positions of minimum reflection.

*28. The green mercury line ($\lambda = 546$ nm) is reflected from the variable air gap between two glass plates. Starting with zero gap width, give the first four gap widths at which near zero reflection occurs.

*29. Repeat Problem 28 but give the gaps for the strongest reflection.

30. Refer to Figure 22.12. Suppose it is known that the air gap thickness between the two optical flats increases from dark fringe 1 to dark fringe 5. How much thicker is the gap at 5 than it was at 1? Assume the yellow mercury line, $\lambda = 579$ nm, is used.

*31. Figure P22.1 shows the interference pattern for He-Ne laser light ($\lambda = 632.8$ nm) reflected perpendicularly from the two faces of a glass microscope slide (a glass plate). The index of refraction of the glass is 1.56. By how much does the thickness of the plate change as we move from one fringe to the next?

**32. An oil slick on a water puddle shows colored interference fringes in the sunlight. Assume the oil to have an index of refraction of 1.50. The color sequence between two red fringes is red-yellow-green-blue-red. Is the second fringe at a thicker or less thick position? Assuming the red color to have $\lambda = 600$ nm, what is the difference in thickness between the two fringes?

**33. A glass plate ($\mu = 1.55$) has a thin transparent coating ($\mu = 1.36$) on its surface. It is observed that green light with $\lambda = 546$ nm is much more strongly reflected from the surface than are other nearby wavelengths. (a) What is the minimum thickness the coating can have? (b) What is the next larger possible thickness?

*34. Refer to Figure P22.2. Light from a He-Ne laser ($\lambda = 632.8$ nm) is reflected from the two plates as shown. The plate on the right is mounted on a screw. As the screw is turned through one revolution, the mirror moves through a distance equal to the pitch of the screw. Starting at minimum reflection (darkness), 1500 fringes pass as the screw is turned through one revolution. (That is, the reflection goes from dark to bright to dark 1500 times.) What is the pitch of the screw?

**35. Refer to Figure P22.2. Light from a He-Ne laser ($\lambda = 632.8$ nm) is reflected from the two plates as shown. These plates are the two ends of a vacuum cell, and they are 3.00 m apart. Originally the reflection is bright. Now air is slowly let into the tube. As this happens, the reflection goes from bright to dark to bright 2770 times. We say that 2770 fringes pass in the field of view. From these data what is the index of refraction of air?

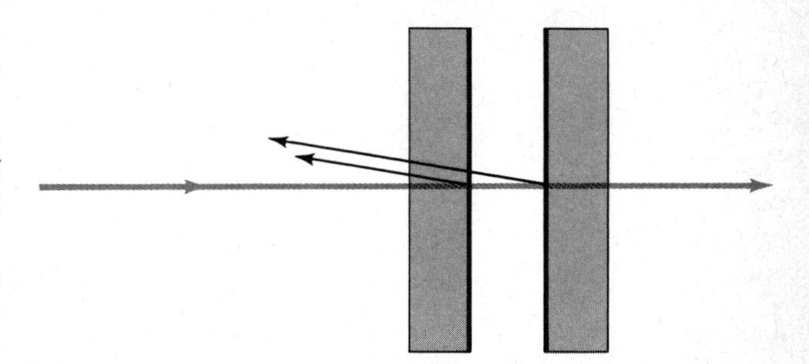

Figure P22.2 The angle of the reflected rays is much exaggerated.

UPI

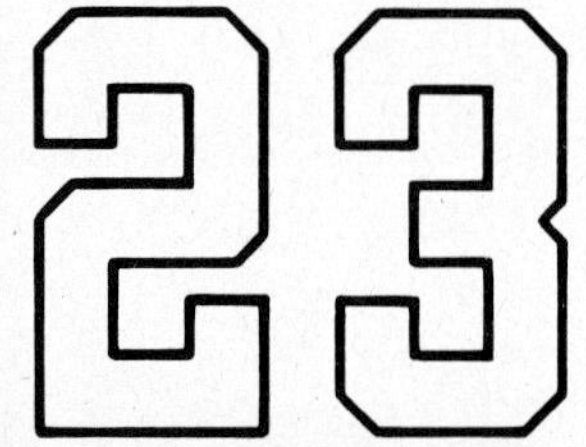

ELECTROSTATICS

Performance Goals

When you finish this chapter, you should be able to

1. Explain the meaning of the following statements.
   a. Two kinds of charge exist in nature.
   b. Likes repel, unlikes attract.
   c. There is a fundamental unit of charge.
   d. The proton and electron have equal, but opposite, charges.
   e. The atom is mostly empty space.
2. Label the following on a sketch of the atom (such as Figure

Electrostatics is the study of electric charges at rest. In this chapter we shall learn how charges are separated and what kind of forces they exert on each other. We shall see how to represent these forces in terms of the electric field strength **E.** The concept of electric field lines will be introduced and used. Also, we shall learn the relation between battery voltage and work. It will be seen that energy methods are extremely useful in understanding the acceleration of charges in electronic devices.

23.2): electron, nucleus, positive charge, location of neutrons, location of protons. Give an order-of-magnitude value for the radius of the atom. Indicate the true size of the electrons and nucleus on a scaled diagram.

3. State the SI unit of charge and give e to two significant figures.
4. Explain the difference between electric conductors and insulators. Give four examples of each.
5. Sketch the induced charges on a metal object when a charge is brought close to the object.
6. State Coulomb's law and the types of charges to which it is restricted. When a situation involving two point charges is given, calculate the force on one due to the other.
7. Show on a sketch with several charges the forces exerted on one of them by all the others. State in words how the resultant force could be found. Relate your statement to the superposition principle.
8. Define the following terms in your own words: electric field, electric field strength or intensity **E**, test charge. In terms of the test charge concept, explain how one can determine **E**.
9. Show the approximate electric field lines on a sketch for the case of two point charges. Using your result, state the direction of **E** at any given point. Also, use the diagram to explain where **E** is strong or weak.
10. Use the relation $\mathbf{F} = q\mathbf{E}$ in simple situations.
11. Sketch the electric field lines in situations similar to (but not the same as) those in Figures 23.11,

23.1 Two Kinds of Charge

Long before people knew about atoms, they were aware that there are two kinds of electric charge. One kind could be produced by rubbing a glass rod with silk. The charge that was produced on the glass rod was called positive charge. A different type of charge could be obtained by rubbing an ebonite rod with fur. (Ebonite is a stiff, hard form of rubber.) This charge was called negative.

It is easy to see that the two charges are different. Figure 23.1 shows an experiment in which small balls are suspended by threads and given some of the charge from either rod so that they are either positive or negative. From the experiment we see the following.

Like charges repel each other. Unlike charges attract each other.

Positive repels positive and negative repels negative. But positive and negative always attract each other.

We know today that these charges are the only types of charges that exist.

There are only two types of charge, positive and negative.

In the next section we shall see how these charges are combined within atoms.

23.2 Charges Within Atoms

In the early 1900s much was learned about the makeup of atoms through experiments in which tiny particles were shot at them.[1] It was found that atoms are not solid balls as some people had thought. Instead, they are mostly empty space.

The picture of the atom that was indicated by these experiments is shown in Figure 23.2. First we should notice that the atom is electrically neutral; it contains exactly as much positive charge as negative charge. In the outer part of the atom are tiny, negatively charged particles called electrons. Each electron carries a charge equal to $-e$. (The numerical value of e will be given soon.) The mass of an electron is about 9.1×10^{-31} kg.

At the center of the atom is a piece of matter called the nucleus. It contains exactly as much positive charge as there is negative charge on all the electrons. The nucleus' positive charge exactly balances the total negative charge on the atom's electrons. The net charge on each atom is zero. Each atom is electrically neutral.

Today we know that the nucleus contains particles called protons and neutrons. Both of these particles have masses many times larger than that of the electron. The proton's mass is 1.6726×10^{-27} kg, and the neutron's mass is

[1] In the most famous of these experiments, E. Rutherford and his associates shot alpha particles (helium atom nuclei) at extremely thin sheets of gold.

23.12, and 23.13. Justify your sketch in terms of the following: induced charge, E in a metal, E at a metal surface, origin and end point of field lines.

12. State how much work is required to carry a charge $\pm Q$ from one terminal of a given battery to the other. State how you obtained your result by use of the definition of potential difference.

13. State (a) which of two objects connected to the two terminals of a battery is at the higher potential, (b) which way a positive or a negative charge between the objects would fall, (c) how much electric potential energy a charge loses as it falls from one object to the other.

14. Calculate the change in KE of a charged particle as it is shot (or let fall) from one plate to another if a given potential difference exists between the plates.

15. Define the unit electronvolt. Using mental calculation only, give the energy in electronvolts acquired by a charge $\pm ne$ as it falls through a given potential difference.

1.6749×10^{-27} kg. For many purposes we can consider these two masses to be the same.

The neutron particle has no net charge. Its charge is zero. However, each proton has a positive charge $+e$. In other words, the magnitude of the proton's charge is the same as that for the electron, but the proton's charge is positive whereas that on the electron is negative. Because the atom is neutral, there must be as many electrons outside the nucleus as there are protons within it.

Finally, we must notice that Figure 23.2 is not at all drawn to scale. The electrons and the nucleus would have to be drawn smaller than pinpricks to relate more accurately to the distance between them. With that in mind, we see that, indeed, the atom is mostly empty space. The diameter of a typical atom is about 2×10^{-10} m (i.e., about 2 Å).

23.3 Unit of Charge

The charge on the proton is $+e$ and on the electron $-e$. To give a numerical value to this charge, we must define a unit of charge. The unit of charge is defined in such a way as to be most useful in practical electricity.

The SI unit of charge is the coulomb (C). It is a large amount of charge when compared to e, the electron and proton charges. The numerical value of e as found from experiment is

$$e = 1.60219 \times 10^{-19}\ \text{C}$$

It would take about 10^{19} electrons to give a charge of -1 C, yet nearly a coulomb of charge flows through a lighted light bulb each second. In practical electricity billions of billions of electrons move through a lighted bulb each second.

Other particles in nature carry a charge, and their charges are always found to be multiples of e. Apparently charge always comes in pieces of magnitude either plus or minus e. All charges have magnitudes that are integer multiples of e, that is, e, $2e$, $3e$, $4e$, and so on.[2]

[2] Current theories postulate particles called *quarks* that carry fractional charge. However, they have not been isolated as yet.

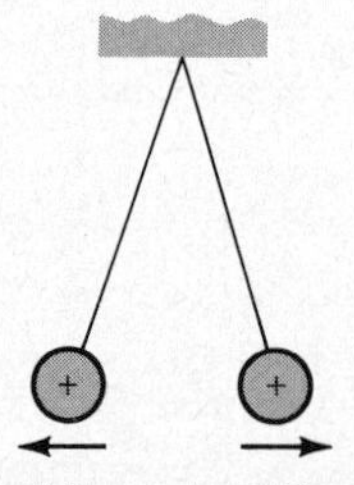

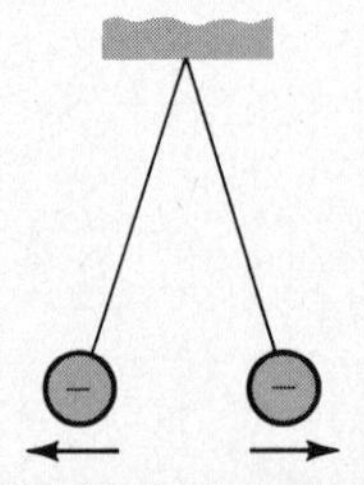

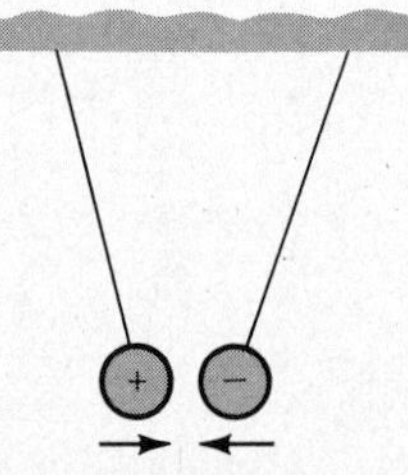

Figure 23.1 Like charges repel each other, whereas unlike charges attract.

23.4 Conductors Versus Insulators

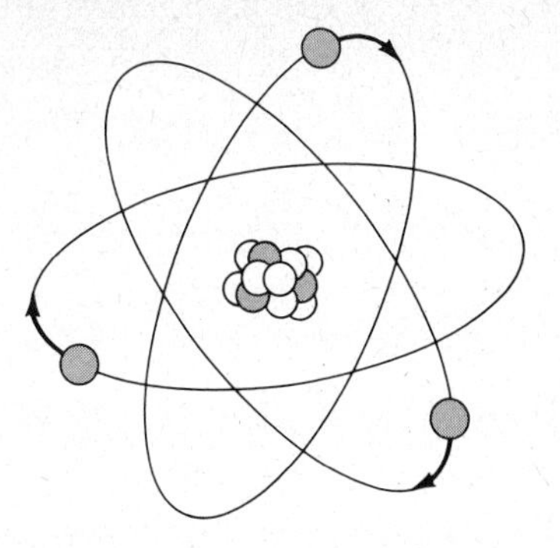

Figure 23.2 In an actual atom the number of electrons outside the nucleus equals the number of protons in it. Moreover, if drawn more to scale, the electron and the nucleus would be smaller than pinpricks in an atom this size.

A piece of metal or plastic or any other material that is large enough to be seen contains a tremendous number of atoms. As a typical example, a penny contains about 10^{23} copper atoms. All other pieces of substance about as large as a penny also contain about that many atoms. Although the atoms are tightly packed together in solids and liquids, they remain as separate units. Each has thermal energy. As a result, it vibrates back and forth rapidly as it tries to push its neighbors aside. In solids an atom seldom succeeds in pushing into a new position. In liquids the atoms are able to change their positions more often. Only in gases are the atoms able to move freely over large distances.

In many solids and liquids the atoms keep a strong hold on their electrons. The charges of the atoms are not free to move from place to place. They are held firmly by atoms.

> **Substances that have a negligible number of charges that are free to move about are called electric insulators or nonconductors.**

Typical insulators are most plastics, glass, wood, and oil.

Some atoms easily lose one or two of their electrons when they are packed together in solids and liquids. These substances are the metals. In them the atoms still are unable to move about easily. However, one or two electrons of each atom often escape. These electrons are then free to move about in the metal. As they move from place to place, they carry their negative charge along with them.

> **Substances that have many charges that are free to move about are called electric conductors.**

Metals are the most common conductors. However, other conductors do exist. For example, salt water contains ions that can carry charge from place to place. We shall discuss the subject of ionic conduction later.

Of great technical importance today is an in-between class of materials called semiconductors. As the name indicates, charge can move about in them but not very well.

23.5 Induced Charges

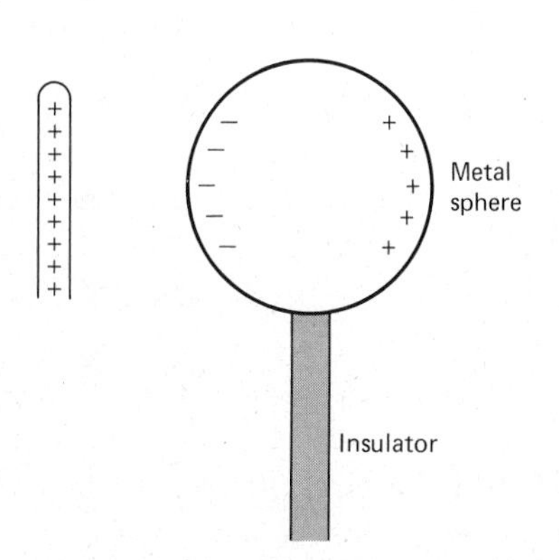

Figure 23.3 The positive rod induces negative and positive charges on the sphere. These induced charges are equal and opposite.

As we saw in the previous section, metals have electrons that are free to move. These free electrons can be caused to move within a metal. For example, suppose that a neutral metal sphere rests on top of an insulating pole as shown in Figure 23.3.

If a positively charged rod is brought close to it, the free electrons in the metal will be attracted to it. Some of them will move close to the rod. This produces an excess negative charge on one side of the sphere as shown. But these electrons left their original atoms behind on the other side of the sphere, and these atoms now have too few electrons. They are therefore positively charged. As a result, the opposite side of the sphere is positively charged as shown. In effect, positive charge was repelled by the rod and negative charge was attracted by it.

Of course, in this process no charge left or came onto the sphere. It must therefore still have zero net charge. The negative charges on one side must still balance the positive charges left on the other side. We call charges such as these *induced charges.* You will see many examples of induced charges as our study proceeds.

23.6 Coulomb's Law

We know that like charges repel and unlike charges attract, but we have not stated how large these forces are. The law that governs these forces was found by Coulomb and is called Coulomb's law. To understand what it is, refer to Figure 23.4.

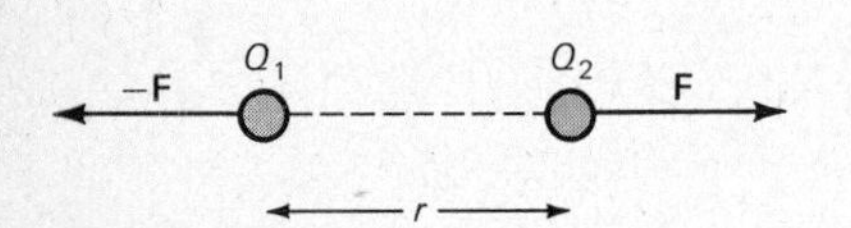

Figure 23.4 Coulomb's law gives the force on one point charge due to another point charge.

In Figure 23.4 are shown two charged balls taken to be so small that we refer to the charges as *point charges.* We represent the distance between them as r. (Because they are point charges, r is essentially the same between any two points on them.) Suppose that one ball has an excess charge Q_1 and the other has a charge Q_2. (This excess positive charge might have been produced by rubbing a positively charged rod against them.) Coulomb found by experiment that:

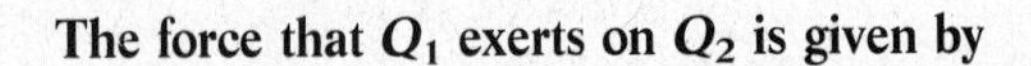

The force that Q_1 exerts on Q_2 is given by

$$F = (8.988 \times 10^9\ \mathrm{N \cdot m^2/C^2})\left(\frac{Q_1 Q_2}{r^2}\right) \qquad (23.1)$$

F is in newtons, Q_1 and Q_2 are in coulombs, and r is in meters.

This is *Coulomb's Law.*

Usually we shall approximate the constant in this equation by 9×10^9. It is often written as $1/4\pi\epsilon_0$, where $\epsilon_0 = 8.854 \times 10^{-12}\ \mathrm{C^2/N \cdot m^2}$ and is called the permittivity of free space.

As we see, the electrostatic force between charges is determined by an inverse square law. The force decreases in proportion to $1/r^2$. In this respect it is like the gravitational force. You will recall that Newton's law of gravitation also contains $1/r^2$. But gravitational and electrostatic forces are far different in magnitude, as we shall see in the example at the end of this section. The electrostatic force is strong, whereas the gravitational force is weak.

In Figure 23.4 the electrostatic force is drawn for Q_1 and Q_2 as like charges. They may be either both positive or both negative. In either case, the force is one of repulsion. Like charges repel each other. The law of action and reaction tells us that the forces on the two charges are equal and opposite. Coulomb's law applies to either charge.

If Q_1 and Q_2 are of opposite sign, Coulomb's law still gives the force on each. But we know that unlike charges attract. Therefore the forces in Figure 23.4 would be reversed in direction if the charges are of opposite sign.

EXAMPLE 23.1 Two metal spheres are 1.5 m apart. The diameter of each is 2.4 cm. Each sphere has a mass of 50 g and carries a charge of 5×10^{-8} C. (This is about the maximum charge such a sphere could hold in air.) (a) Find the Coulomb force on one sphere due to the other. (b) Find the gravitational force on one sphere due to the other.

Solution The situation is much like that shown in Figure 23.4, with $Q_1 = Q_2 = 5 \times 10^{-8}$ C, $r = 1.5$ m, and $M_1 = M_2 = 0.050$ kg. We have, using the units given and assuming the spheres to act like point charges,

$$\text{Coulomb force} = (9 \times 10^9\ \text{N}\cdot\text{m}^2/\text{C}^2)\frac{(5 \times 10^{-8}\ \text{C})(5 \times 10^{-8}\ \text{C})}{(1.5\ \text{m})^2}$$
$$= 1 \times 10^{-5}\ \text{N}$$

From Newton's law of gravitation, $F = GM_1M_2/r^2$,

$$\text{Gravitational force} = 6.67 \times 10^{-11}\frac{(0.050)(0.050)}{(1.5)^2}\ \text{N}$$
$$= 7.4 \times 10^{-14}\ \text{N}$$

As we see, the gravitational force is small in comparison to the electrostatic force in this case. The difference is much greater if we compare forces between two protons or two electrons.

23.7 Superposition in Electrostatics

Experiment shows that electrostatic forces can be superposed. This fact is stated in the *superposition principle.*

The resultant force on one charge due to all other charges is the vector sum of the forces due to each individual charge.

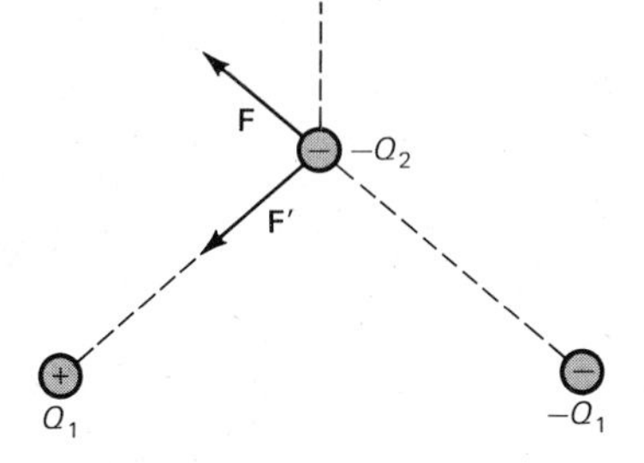

Figure 23.5 The resultant force on Q_2 is obtained by adding the vector forces **F** and **F′**. It will bisect the angle between **F** and **F′**.

To see what we mean by this, refer to Figure 23.5.

We see there a negative charge with magnitude Q_2. It experiences forces due to the other two charges: $-Q_1$ repels it with a force **F** while $+Q_1$ attracts it with a force **F′**. The resultant force on $-Q_2$ is found by taking the vector sum of **F** and **F′**. It will be directed toward the left in the figure.

We shall make frequent use of this superposition principle. It tells us that the Coulomb forces act independently of each other. As a result, we can consider the effect of each force separately. The total effect is then obtained by combining these individual effects.

23.8 Electric Field

In Chapter 19 we introduced the concept of the electric field. We stated there that

An electric field exists at a point if a tiny stationary positive test charge placed there experiences a force because of its charge. The strength and direction of the electric field is represented by the electric field vector *E*.

We now wish to investigate the electric field in more detail.

Suppose, as shown in Figure 23.6(a), we have a positive point charge $+Q$. We wish to know what the electric field near it looks like. To find out, we must imagine a tiny positive charge (called the *test charge*) to be placed at a point such as *A*, or *B*, or *C*, and so on. We then ask what sort of force the test charge experiences.

Because the charge Q is positive, the test charge (also positive) will be repelled. Therefore, the force on it at *A* will be in a direction away from *A* as shown. The repulsion force on the test charge at *B* or *C* is also as shown. In general, we see that the test charge will be repelled radially outward by the charge Q. To illustrate this fact, we draw the electric field lines shown in part (b) of the figure. (Of course, this diagram is only schematic. It should actually be three-dimensional.)

The electric field lines indicate the direction of the electric force on a positive test charge. The line of the force is tangent to the field line.

For example, we can see at once from Figure 23.6(b) the direction of the force at point *G*. If a positive test charge is placed there, it will feel a force

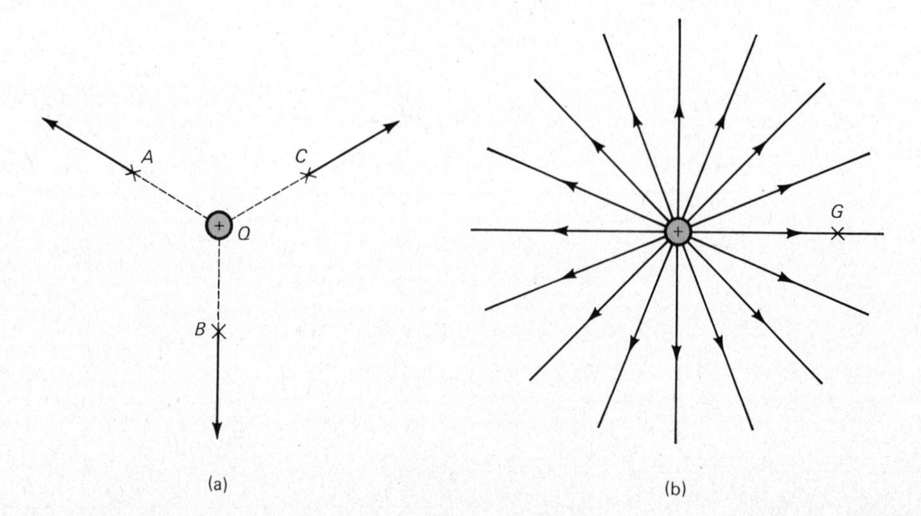

Figure 23.6 Electric field diagrams show the direction of the force on a positive test charge. They also tell us something about the magnitude of the force. What?

directed straight to the right. In fact, any positive charge placed at G will experience a force in that direction.

However, suppose that a negative charge is placed at A, B, C, or G. It will be attracted by the charge $+Q$. The arrows must be reversed in direction if they are to show the force on a negative charge. We therefore conclude

The force on a negative charge in an electric field is opposite in direction to the field line direction.

We can easily sketch the electric field for other situations. For example, study the diagrams in Figure 23.7. In (a) a positive test charge is attracted radially inward toward the charge $-Q$. The field lines are therefore drawn radially inward.

In Figure 23.7(b) we see the field due to two charges. At each point in it, the field lines show the direction of the resultant force on a positive test charge. For example, at point A the test charge experiences the two forces shown. Their resultant follows the field line. Check through the diagram in Figure 23.7(c) to make sure you understand the way it is drawn.

There are two other features we should notice about field lines. First, field lines originate on and come out of positive charges. They come into and end on negative charges. The reason for this is simple. If you place a positive test charge close to another charge, the force on it will be almost entirely due to that charge. The force will be repulsive at a positive charge. The field lines there must be outward. Just the opposite happens at a negative charge.

Second, field lines are closest together where the electric force is strongest. You will see that this is true in Figures 23.6 and 23.7. It can be proven to be true in all cases. If the field lines are properly drawn in three dimensions, then the force is proportional to the density of the field lines. This turns out to be true because of the fact that Coulomb's law is an inverse square law.

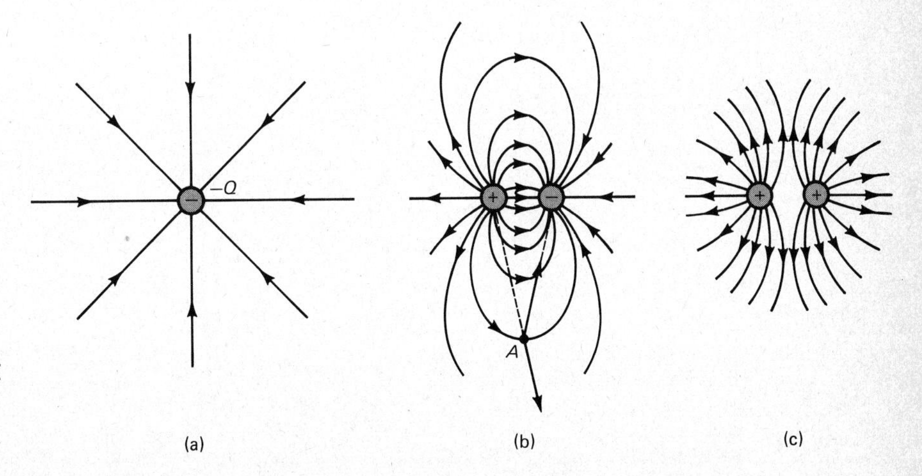

Figure 23.7 Typical field-line diagrams. Notice that the lines always come out of positive charges and go into negative charges.

23.9 Electric Field Intensity (or Strength)

To give the electric field quantitative meaning, we make use of the test charge concept.

A unit positive test charge is an imaginary charge defined as having a value of +1 C and as having the property that other charges experience no force from it.

This imaginary charge can be placed wherever we choose. The force that it experiences is then used to tell us the electric field intensity or strength.

The electric field intensity *E* at a point is as follows: It is the force experienced by a unit positive test charge placed at that point. The units of *E* are those of force per unit charge, newtons per coulomb. *E* is a vector that has the direction of the force.

To see the use of **E**, let us refer to Figure 23.8. We see there a region where the electric field is uniform (the lines are equally spaced) and directed toward the right. We are told that $E = 2000$ N/C. If the unit positive test charge of +1 C is placed in this region, it experiences a force of 2000 N toward the right. Remember, **E** is the force on a +1 C charge.

Suppose that any charge Q is placed in this region. A charge of +5 C, say, would experience a force five times larger than E. We therefore have the following general result.

When a charge *Q* is in a field *E*, it experiences a force

$$F = EQ \tag{23.2}$$

In the present case **E** is 2000 N/C directed toward the right. The force on a charge Q in this region is 2000 Q N toward the right. If the charge were $-Q$, the force would be made negative. This agrees with our previous result. Forces on negative charges are opposite in direction to the field lines.

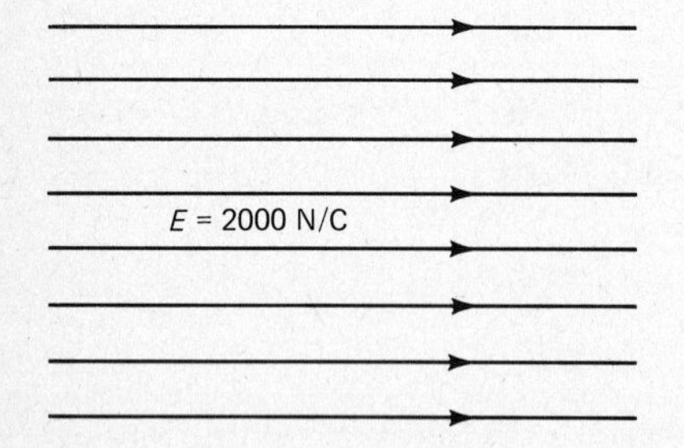

Figure 23.8 Because the field lines are equally spaced, the field is uniform in this region.

■ EXAMPLE 23.2 Find the electric field at point P in Figure 23.9 due to the $+5 \times 10^{-6}$ C point charge.

Solution Place an imaginary charge Q at point P and find the force on it due to the 5×10^{-6} C charge. From Coulomb's law we have

$$F = (9 \times 10^9\ \text{N}\cdot\text{m}^2/\text{C}^2)\frac{(5 \times 10^{-6}\ \text{C})Q}{(0.30\ \text{m})^2} = 5 \times 10^5 Q\ \text{N/C}$$

Since E is just F/Q, we have $E = 5 \times 10^5$ N/C.

The field is directed radially outward. If the charge had been -5×10^{-6} C, the field would have been numerically the same but directed radially inward. ■■

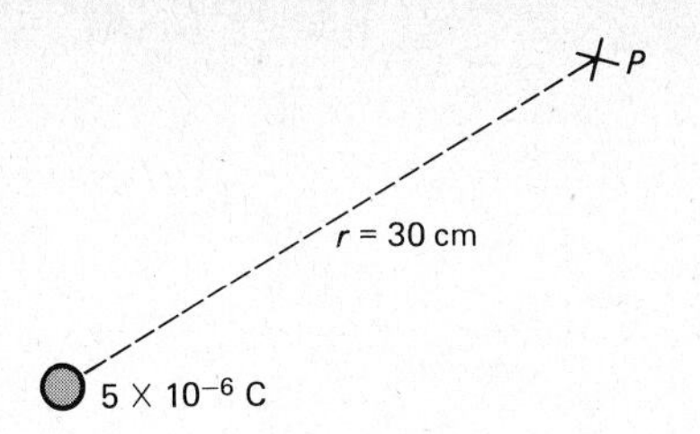

Figure 23.9 Find the electric field at P due to the 5×10^{-6}-C point charge.

23.10 Electric Field in and near Metals

There are three important facts concerning the electrostatic field in the case of metals.

Under electrostatic conditions (i.e., for charges at rest):

1. **The electric field is zero in the interior of metals (except in cavities in which there is a free charge).**
2. **The electric field just outside a metal is perpendicular to the metal surface.**
3. **All the excess charge on a metal is on the surfaces of the metal.**

Let us now see why these are true.

When a charged object is first brought near a piece of metal, the free electrons are induced to move here and there. We saw an example of this in Figure 23.3. This movement lasts only a moment. Soon, everything is at rest again except for the usual thermal motion. The electrons within the metal have reached static equilibrium.

If the free electrons are no longer moving, the unbalanced force on them must be zero. After all, they are free to move under any unbalanced force. If there is no unbalanced force on them, there must be no electric field where they are. You will recall that $F = QE$. In this case $Q = -e$, the electron charge. If $F = 0$, then E must be zero. We must therefore conclude that the electric field is zero inside, where the electrons are, in a metal. Under electrostatic conditions the electrons arrange on the metal in just such a way as to cause the field to be zero. Notice that our argument does not apply to cavities in the metal. It applies only to regions where free electrons exist.

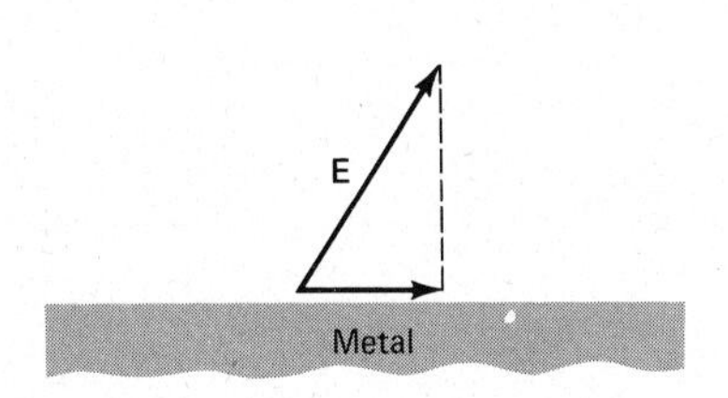

Figure 23.10 If **E** were directed as shown, its component along the surface would be nonzero. This cannot be true under electrostatic conditions. **E** must be perpendicular to the metal surface.

To show that the field is perpendicular to the metal surface, refer to Figure 23.10. Suppose (wrongly) as shown there, that the electric field is not perpendicular to the surface of the metal. The field would then have a component parallel to the surface. This component of the electric force would cause free electrons at the surface of the metal to move along the surface. But there is no such charge motion because we are assuming the situation to be electrostatic. Therefore, our original assumption must be wrong. The electric field can have no component parallel to the surface. Under electrostatic conditions the field lines must be perpendicular to the metal surface.

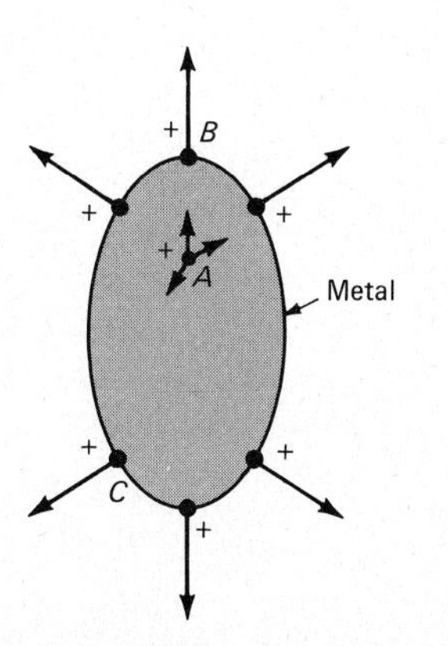

Figure 23.11 Only a few of the excess charges are shown. No excess charge can exist inside the solid metal at points such as A.

Suppose that we place an excess charge on a piece of metal. For example, a positively charged rod might have been rubbed on the object shown in Figure 23.11. We would like to know where on the metal object the excess charge is.

Our experience with electric fields tells us there would be no excess charge at A, a point in the solid metal. If there were, we should expect an electric field to exist close to it. But we know that the field there must be zero under electrostatic conditions. It therefore appears that the excess charge cannot exist inside the solid metal. Both experiment and theory show this to be correct.

The excess charge exists only on the surfaces of the metal. If we look at Figure 23.11, we see a curious fact about this. We have already learned that the field lines must be perpendicular to the metal surface. In addition, the field is zero in the metal, and so no field lines exist there. Therefore, the field must be as shown in the figure. Apparently the inward-directed electric field from the charges at B exactly cancels the inward-directed field of the charges at C. The field lines we have drawn are the result of all the charges on the metal. The charges arrange themselves on the surface so as to make the three facts stated above true.

23.11 Sketching Electric Field Diagrams

We now have considerable knowledge about electric fields and the lines we use to represent them. Let us apply it to a few situations. For example, suppose that a charged ball is held above a metal plate as shown in Figure 23.12(a). The symbol ⏚ tells us that the plate is attached to the earth. (We call this the "ground" and we say that the plate is "grounded.") Charges can run off and onto the plate from the ground.

The negative ball repels the free electrons on the plate. Some run off onto the ground. They leave behind an excess positive charge on the plate. The negative ball has induced a positive charge on the plate below it. To draw the field lines, we recall that

1. They come out of positive charges.
2. They end on negative charges.
3. They strike the metal perpendicularly.
4. They do not enter the metal.

With that in mind, the field lines can be sketched as shown.

In part (b) of Figure 23.12 we see a somewhat similar situation. Here the plate is replaced by a sharp needle. The positively charged ball induces negative charges to the tip of the needle. Because the field lines are very close together at

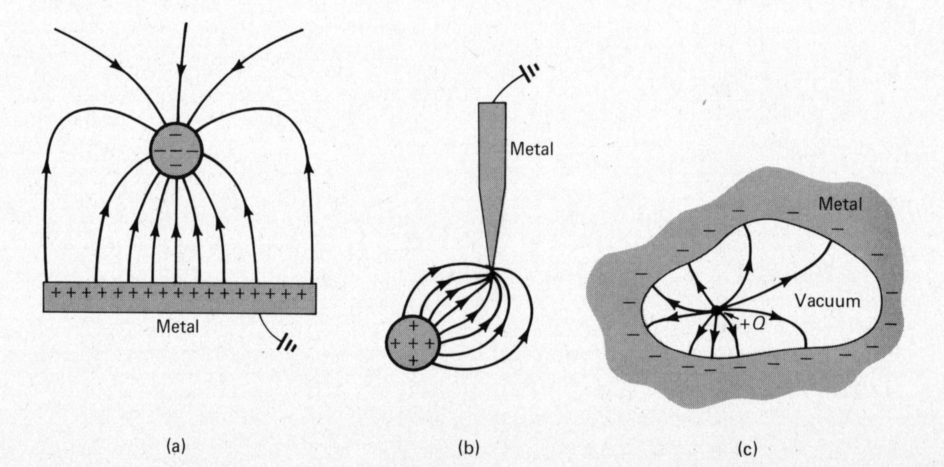

Figure 23.12 Typical electric field sketches. You should be able to justify how they are drawn.

the tip, the field must be very strong there. Sometimes in situations such as this, the field is strong enough to cause sparks to jump from the tip. A spectacular example of this is seen when lightning strikes between a highly charged cloud and a lightning rod.

Part (c) of Figure 23.12 shows a charge placed inside a cavity in a block of metal. The positive charge induces negative charges to appear on the surface of the metal cavity. Notice that all the field lines that leave $+Q$ end on the metal surface. We might guess from this that the induced charge is equal to Q in magnitude but is negative. This guess turns out to be correct. In fact, one can prove that equal numbers of field lines begin and end on equal charges.

23.12 Batteries and Potential Difference

We shall learn about the internal workings of a battery in the next chapter. For now we shall only discuss some of its effects. A simple situation involving a battery is shown in Figure 23.13. One terminal of the battery is the positive terminal. It is usually painted red or carries a symbol +. The other terminal, usually black or with the symbol −, is negative.

The negative terminal of a battery places a negative charge on metal objects fastened to it by wire. Positive charge appears on objects attached to the positive terminal. We can interpret what has happened in Figure 23.13 in the following way.

Negative charge has moved out of the negative terminal, through the metal wire, to the metal plate. This negative charge on the plate induces negative charge on the sphere to move through the wire back into the positive terminal of the battery. Batteries are constructed in such a way that equal charges move through the two terminals. As much charge flows out of one terminal as flows into the other.

Since the positive and negative charges attract each other, they arrange themselves on the sphere and plate approximately as shown. In addition, an electric field exists between the sphere and plate. This has important meaning, as we shall now see.

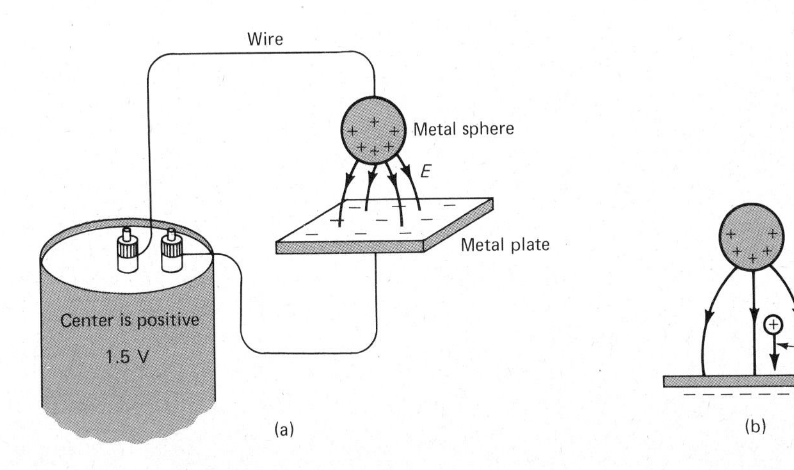

Figure 23.13 The battery places charges on the objects attached to its terminals.

The sphere and plate are drawn in cross section in Figure 23.13(b). If a ball with charge $+Q$ is placed between the sphere and the plate, it will feel a force in the direction of E. The force magnitude will be QE, where E is the electric field strength at that point. Obviously, one would have to exert a force and do work to pull the charged ball from the plate up to the sphere. (We ignore gravity, because we are assuming the ball to be very small.) Work has to be done against the repulsive force from the sphere and the attractive force from the plate.

The work done against electric forces in moving a +1 C test charge from a point *A* to a point *B* is called the potential difference from *A* to *B*. We represent it by the symbol *V*. It is measured in joules per coulomb, and this unit is called the volt, V.

In the figure the plate and sphere are attached to a 1.5-V battery. The potential difference between the plate and sphere is 1.5 V. Therefore, the work one would need do to carry a +1-C charge from the plate to the sphere is 1.5 J. If a +5-C charge, say, were carried from the plate to sphere, five times as much work would need to be done. Or for any charge $+Q$, an amount of work of $1.5Q$ J would have to be done. We therefore have the following important fact.

Suppose that a charge $+Q$ is carried from point *A* to point *B*. Suppose further that *V* is the potential difference (in volts) from *A* to *B*. Then the work done on *Q* against electric forces is

$$\text{Work} = QV \text{ joules} \tag{23.3}$$

We shall make much use of this fact.

EXAMPLE 23.3 How much work is done against electric forces as a proton is moved from the negative terminal of a 12-V battery to its positive terminal? Repeat for an electron. (See Figure 23.14.)

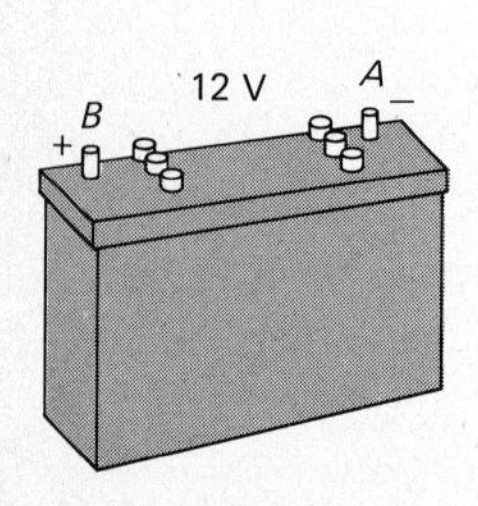

Figure 23.14 A proton will be attracted to the negative terminal and repelled by the positive. Work must be done to carry it from *A* to *B*. Negative work is done in carrying an electron from *A* to *B*. Why?

Solution The potential difference from *A*, the negative terminal, to *B*, the positive terminal, is 12 V. To move a charge q from *A* to *B* requires

$$\text{Work} = QV = (q)(12 \text{ V})$$

In our case $q = +1.6 \times 10^{-19}$ C, the charge on a proton. Therefore,

$$\begin{aligned}\text{Work on proton} &= (12 \text{ V})(1.6 \times 10^{-19} \text{ C}) \\ &= 1.9 \times 10^{-18} \text{ J}\end{aligned}$$

For an electron we have $q = -1.6 \times 10^{-19}$ C. Therefore,

$$\begin{aligned}\text{Work on electron} &= (12 \text{ V})(-1.6 \times 10^{-19} \text{ C}) \\ &= -1.9 \times 10^{-18} \text{ J}\end{aligned}$$

The work is negative because the electron is attracted to the positive terminal. It must be restrained from moving from A to B. This negative work is similar to the negative work that is done when a weight is lowered. In both cases the restraining force does negative work. ■■

23.13 Electric Potential Energy

In our study of mechanics we learned about gravitational potential energy. We saw that work and gravitational potential energy are closely related. A similar situation exists in electricity. When work is done against electric forces, a charge can be given potential energy. Let us now investigate the concept of electric potential energy.

In Figure 23.15 we see two metal plates connected to a battery. As explained in the previous section, the battery gives one plate a positive charge and the other an equal negative charge. The potential difference between them is the same as the potential difference of the battery. Let us say that the battery voltage is V volts. The potential difference between the plates is then V.

It is customary in electricity to give explanations in terms of positively charged particles. We shall therefore consider the behavior of a positive particle with charge $+q$. As we see in Figure 23.15, the particle will be repelled by plate B and attracted by plate A (like charges repel; unlike charges attract). If the charged particle is released between the two plates, it will start to move away from B and toward A because of the forces the charged plates exert on it. We say that "the particle will fall from plate B to plate A." This manner of speaking comes about in the following way.

Suppose that the charge $+q$ was at plate A. Because plate A attracts it, we need to pull on it to move it toward plate B. In fact, because B repels the particle, we need to pull it the whole way from A to B. The work we do on it was found in the last section. It is Vq, where V is the potential difference from A to B.

But what happens to the work we do on the particle in carrying it from A to B? It is stored as energy of the particle. In fact, the work will be given back to us if we move the particle back to A. The particle pulls us back. This is much like

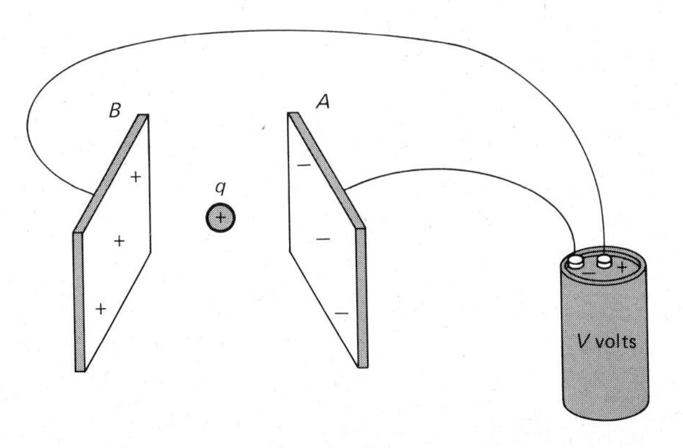

Figure 23.15 When the charge $+q$ is carried from plate A to plate B, its potential energy is increased by an amount Vq.

lifting and then lowering a weight. As we lift the weight, we store energy in it. As the weight is slowly lowered, it gives the energy back to its lifter.

As we see, the situation shown in Figure 23.15 is much like lifting a weight. We must pull the positive particle from A to B. But it will fall back to A if we release it. The work we do in carrying the charge from A to B is Vq. It appears as electric potential energy of the charge. When the charge is allowed to fall back to A, it gives back this potential energy. The electric potential energy is changed to work or to some other form of energy.

It is customary to speak of points of high and low potential. In Figure 23.15 plate B is at a higher potential than plate A. We say this because work has to be done to carry a positive charge from A to B. Using this way of speaking, the positive side of a battery is the high-potential (or just high) side of the battery. Work has to be done to carry a positive charge from the negative terminal of the battery to the positive.

The high side of a battery is its positive side. When a charge q is lifted through a potential difference V, it is given an electric potential energy Vq.

We shall soon see how practical use is made of this concept.

23.14 Interchange of Electric Potential Energy

Let us examine the situation in Figure 23.16. Two metal plates inside a vacuum tube are connected to a battery whose voltage is V volts. Notice the symbol used for a battery: $\overset{+}{-}\!|\!\vdash\!\overset{-}{-}$. (The + and − signs are sometimes omitted. You are expected to know that the longer line in the symbol represents the positive terminal.) The plates are charged as shown by the battery. The potential difference from A to B is the same as the battery voltage V. Because plate B is positive and plate A is negative, work must be done to lift a positive charge from A to B. Plate B is therefore at the higher potential.

Suppose that a particle with charge $+q$ and mass m is placed at plate B. It has a potential energy Vq relative to plate A. That much work must be done on it to lift it from A to B. That much energy will be given back as the charge $+q$ falls back from B to A.

If the positively charged particle is released at plate B, it will fall toward plate

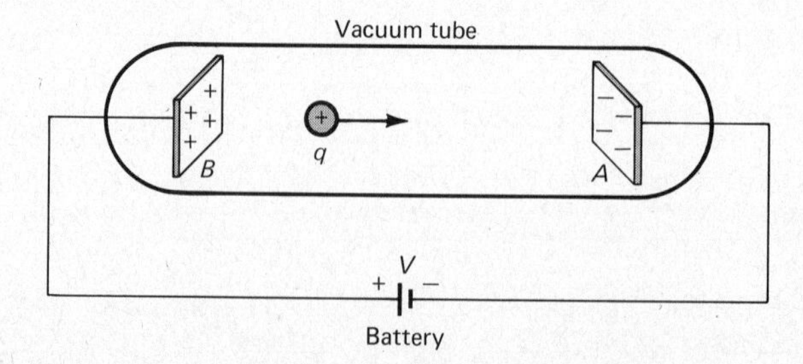

Figure 23.16 The charge q loses potential energy Vq as it falls from B to A. This lost energy appears as kinetic energy of the particle.

A. Because it is in vacuum, it will hit nothing. Of course, it will continue to speed up as it moves toward *A*. How fast is it going just before it strikes plate *A*?

We can answer this question easily if we notice that potential energy is changed to kinetic energy. At plate *B* the particle has a potential energy Vq. By the time it reaches *A*, all that potential energy has been lost. It has changed to kinetic energy of the particle. As an equation, this is

$$\text{Potential energy at } B = \text{kinetic energy at } A$$

We know that the potential energy is Vq and that the kinetic energy of the particle at *A* is $\frac{1}{2}mv_A^2$. Therefore, the equation becomes

$$Vq = \tfrac{1}{2}mv_A^2$$

As a numerical example, suppose that the particle is a proton and the battery voltage is 6 V. Then

$$V = 6\text{ V}$$
$$q = 1.6 \times 10^{-19}\text{ C}$$
$$m = 1.67 \times 10^{-27}\text{ kg}$$

Substituting these values and solving for v_A gives

$$v_A = 34{,}000\text{ m/s}$$

This is the speed of the proton just before it hits the plate at *A*.

EXAMPLE 23.4 Some nuclear accelerators accelerate the nuclei of helium atoms through large potential differences. In a typical low-energy device, the accelerating voltage might be 10×10^6 V. Find the kinetic energy of such a particle in both joules and electronvolts. Also find its speed. For the helium nucleus $m = 4 \times 1.67 \times 10^{-27}$ kg and $q = +2 \times 1.6 \times 10^{-19}$ C.

Solution When a charge q falls through a potential difference V, it acquires a kinetic energy Vq. Therefore,

$$\text{Kinetic energy} = (10 \times 10^6\text{ V})(3.2 \times 10^{-19}\text{ C}) = 3.2 \times 10^{-12}\text{ J}$$

Therefore, using the conversion 1 eV = 1.6×10^{-19} J (see Section 6.2),

$$\text{Kinetic energy} = (3.2 \times 10^{-12}\text{ J})\left(\frac{1\text{ eV}}{1.6 \times 10^{-19}\text{ J}}\right) = 20 \times 10^6\text{ eV} = 20\text{ MeV}$$

where MeV stands for million electronvolts.

To find the speed of the particle, we know that

$$\text{KE in joules} = \tfrac{1}{2}mv^2$$

Or

$$3.2 \times 10^{-12}\ \text{J} = \tfrac{1}{2}(4 \times 1.67 \times 10^{-27}\ \text{kg})v^2$$

From this we find that

$$v = 3.1 \times 10^7\ \text{m/s}$$

This speed is about one-tenth the speed of light. At speeds larger than this, the results of Einstein's theory of relativity must be used to obtain the correct speed. We shall see how this is done in Chapter 30. ■■

23.15 Acceleration of Negative Charges

The discussion so far has been concerned with positive particles. As we said earlier, positive charges are usually used in such discussions. The words we use, such as "high potential," refer to positive charges. Let us now see what to do when we deal with negative charges.

Suppose that a negative particle is placed between two charged plates as shown in Figure 23.17. This situation is identical to that in Figure 23.16 except that q is now negative. The negative charge will be repelled by plate A and attracted by plate B. Therefore, the charge will fall from A to B. Notice that the positive particle fell in the opposite direction. What is "downhill" electrically for a positive charge is "uphill" for a negative charge and vice versa.

As the charge falls from A to B, it will lose potential energy. This energy is equal to the work done in carrying the charge from B to A. Its magnitude is $|Vq|$. (The vertical lines on $|Vq|$ mean to give a positive sign to this quantity, i.e., to take its absolute value.) Therefore, the energy lost by the charge as it falls from A to B is $|Vq|$. This lost potential energy appears as kinetic energy of the particle. As we see, the concepts involved for a negative particle are the same as those when the particle is positive. The only difference is this:

Positively charged particles move by themselves from high to low potentials. Negatively charged particles fall from low to high potentials.

Let us now apply these concepts to a practical situation.

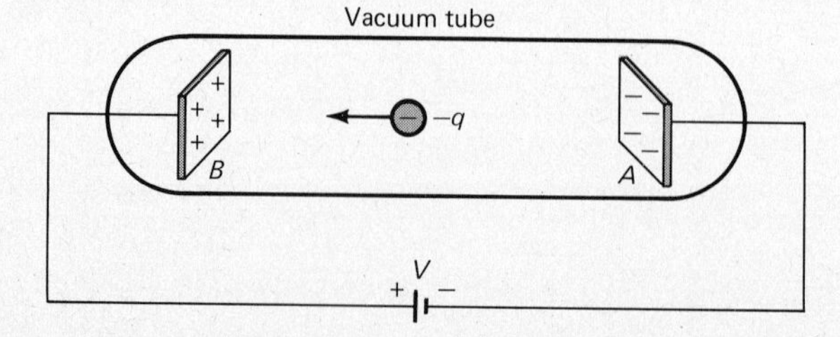

Figure 23.17 The charge $-q$ will fall from A to B. As it does, it loses a potential energy Vq.

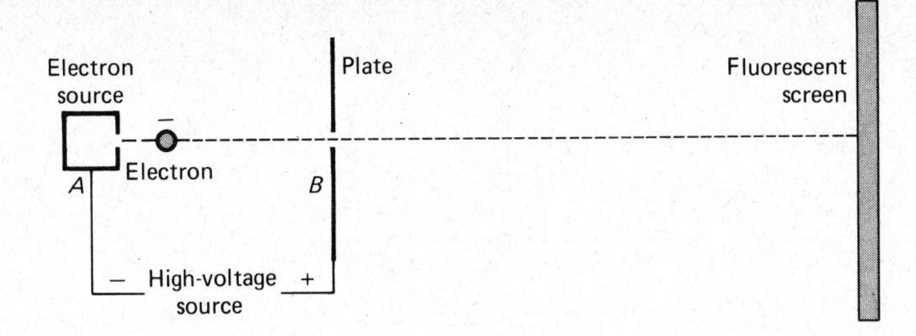

Figure 23.18 *B* is positive and *A* is negative. The electron experiences a force from *A* to *B*. The force causes it to speed up going from *A* to *B*.

23.16 Electron Beam in a Television Picture Tube

Many technical devices accelerate electrons by means of voltage difference. For example, in Figure 23.18 we see a highly simplified sketch of a television tube.[3] (The entire system is in vacuum so that collisions with air molecules do not occur.) A source of high voltage (let us say 10,000 V) is connected to *A* and *B*. The negative terminal is connected to the electron source. The positive terminal is connected to the metal plate at *B*.

An electron wanders out of the small hole in the source at *A*. It sees that the source is charged negative. The electron is therefore repelled by *A* because it, too, is negative. Moreover, it sees the plate at *B* is positive. The electron is therefore attracted by the plate.

As a result of these forces, the electron picks up speed as it moves from *A* to *B*. By the time it reaches *B*, it has fallen through a potential difference *V*. It has therefore gained kinetic energy equal to Ve, where e is the charge on the electron.

The fast-moving electron shoots through the hole in the plate at *B*. It then coasts along until it hits the fluorescent screen. This screen is simply the picture end of the TV tube. When the fast-moving electron strikes the fluorescent screen, a flash of light is given off. This light forms part of the picture we see on the TV screen.

Of course, this electron is not alone. Billions upon billions of electrons shoot down the tube each second. We call this stream of electrons the electron beam. The source and plate system that accelerates the electrons is called the electron gun. If you take apart a television or cathode ray tube, you will see that we have greatly simplified the construction of the gun. You will also see that other charged plates are used. These plates focus the beam and sweep it across the face of the tube. By means of them, the beam traces out a picture on the fluorescent screen.

Let us now find how fast the electron is going as it shoots through the hole in plate *B*. We shall assume that the electron has negligible kinetic energy as it wanders out from the source. It then falls through a potential difference $V =$ 10,000 V. This gives it a kinetic energy Ve, where $e = 1.6 \times 10^{-19}$ C. Therefore

$$\tfrac{1}{2}mv_B^2 = (10{,}000\ \text{V})(1.6 \times 10^{-19}\ \text{C})$$

[3]The cathode ray tube (CRT) on a computer is also much like this.

ELECTROSTATIC PRECIPITATORS

When most fuels are burned, an unburned residue is given off in the form of smoke. The smoke consists of tiny particles that do not settle out readily. These particles must be removed before the stack gases are vented to the atmosphere. One common way to do this is shown in the photo.

We see there two photos of the Kaiser Aluminum Plant in Baton Rouge, La. In (a) the electrostatic precipitator equipment is not turned on, but in (b) it is in operation. Clearly, the precipitator system is extremely effective.

In an electrostatic precipitator the stack gases pass through a wire grid system. High voltages on the grid cause the smoke particles to be subjected to intense electric fields. Free charges produced by the intense field cause the smoke particles to become charged. They are then pulled to electrodes of opposite polarity and precipitated out. The stack gases are therefore cleaned as the smoke passes through the electrostatic precipitator.

(a)

(b)

But $m = 9.1 \times 10^{-31}$ kg for an electron. Placing in this value, we solve and find that

$$v_B = 5.9 \times 10^7 \text{ m/s}$$

This speed is quite close to the speed of light, 3×10^8 m/s. In such cases the calculation should make use of Einstein's ideas about relativity. Our result must therefore be considered only approximate. The correct result is 5.8×10^7 m/s. As speeds come closer to the speed of light, the error increases rapidly. Notice that relativity becomes important in high-voltage electronic devices.

■ **EXAMPLE 23.5** An electron (charge $= -e$) is accelerated through V volts. What is its kinetic energy in electronvolts? Repeat for a helium nucleus (charge $= +2e$).

Solution The particle acquires a kinetic energy equal to Vq as it falls through a potential difference V. Therefore, for the electron

$$\text{KE of electron} = Ve \text{ joules}$$

where $e = 1.6 \times 10^{-19}$ C.

The conversion from joules to electronvolts is

$$1 \text{ eV} = 1.6 \times 10^{-19} \text{ J}$$
$$= e \text{ joules}$$

Therefore

$$\text{KE of electron} = \frac{Ve}{e} = V \text{ electronvolts}$$

Similarly, for the helium nucleus

$$\text{KE of helium nucleus} = V(2e) \text{ joules}$$
$$= 2V \text{ electronvolts}$$

The general rule we see exposed in this example is:

When a charge *n* times larger than the electron charge falls through a potential difference *V*, it acquires an energy of *nV* electronvolts.

This simple relation between V and energy makes the electronvolt a convenient energy unit. However, joules must always be used in our fundamental equations. ■■

■ **EXAMPLE 23.6** As shown in Figure 23.19, a proton originally has a speed of 2.00×10^5 m/s. It shoots through the holes in the two plates. Find its speed as it leaves the second plate.

Solution Plate B will be charged positive by the battery and A will be negative. Work must be done to carry a positive charge from A to B because of the charges on the plates. The electric field between the plates (recall that **E** is the force on a unit positive charge) is in the direction shown. Because B is at the high side of the battery, it is "uphill" electrically from A to B.

As the positive charge shoots between the plates, it is slowed down. Plate B repels it and plate A attracts it. As the positive charge moves from A to B it gains electric potential energy and loses an equal amount of KE. (This is similar to a ball thrown upward on the earth. It slows as it goes upward. Some of its kinetic energy is lost in doing work against gravity.)

As an equation, these ideas become

$$\text{Loss in KE} = \text{gain in electric PE}$$

from which

$$(\text{KE})_A - (\text{KE})_B = Vq$$

Or

$$\tfrac{1}{2}mv_A^2 - \tfrac{1}{2}mv_B^2 = (12 \text{ V})(1.6 \times 10^{-19} \text{ C})$$

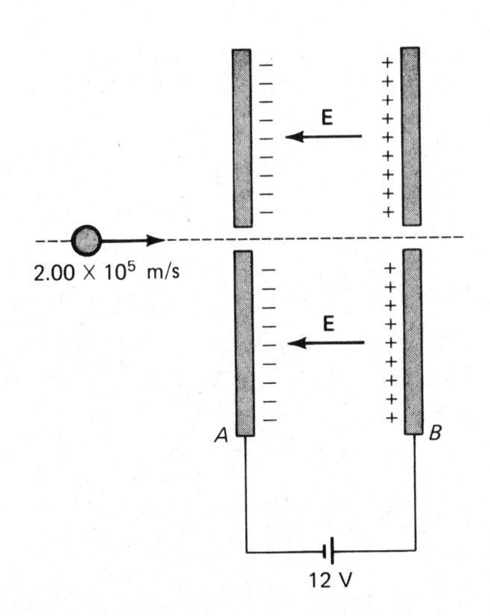

Figure 23.19 The proton is slowed by the electric field as it shoots between the plates.

Because the proton mass is 1.67×10^{-27} kg and $v_A = 2.00 \times 10^5$ m/s, we can solve for v_B. The result is

$$v_B = 1.94 \times 10^5 \text{ m/s}$$

Summary

There are two types of charge, positive and negative. All charges are integer multiples of the charge $e = 1.60219 \times 10^{-19}$ coulombs (C). The charge on the electron is $-e$, whereas the proton charge is $+e$.

Like charges repel each other; unlike charges attract each other. A point charge Q_1 exerts a force on a point charge Q_2 that is given by Coulomb's law:

$$F = (9 \times 10^9 \text{ N}\cdot\text{m}^2/\text{C}^2)\left(\frac{Q_1 Q_2}{r^2}\right)$$

The separation of the charges is r and is in meters. Q_1 and Q_2 are measured in coulombs, whereas F is in newtons. Often the constant 9×10^9 is written as $1/4\pi\epsilon_0$, where ϵ_0 is called the permittivity of free space.

The atom has a tiny inner core, the nucleus, that carries a positive charge. It is composed of protons (charge = e) and neutrons (charge = 0). The proton and neutron each have a mass of about 1.67×10^{-27} kg. Outside the nucleus are electrons. Each has a charge = $-e$ and a mass 9.1×10^{-31} kg. Because the atom is neutral, each atom has the same number of electrons as it has protons. The diameter of a typical atom is about 2×10^{-10} m.

Conductors of electricity have charges within them that can move freely. In metals the free charges are electrons. Each atom of the metal loses one or two electrons, and these are free to move through the metal. Electric insulators have negligible numbers of charges that can move freely. Semiconductors are on the borderline between conductors and insulators.

Coulomb forces can be superposed. The resultant electrostatic force on a charge is the vector sum of the individual forces exerted on it by charges in the vicinity.

An electric field exists at a point if a positive test charge placed there experiences a force of electric origin. The strength and direction of the electric field is represented by the electric field vector **E**.

Electric field lines indicate the direction of the electric force on a positive test charge. The field lines are tangent to the line of the force at any given point. These lines come out of positive charges and end on negative charges. Where the lines are closest together, the electric force is strongest.

The electric field strength (or intensity) **E** at a point is defined equal to the vector force per unit positive test charge placed at that point. A positive charge q coulombs placed in an electric field of strength **E** newtons per coulomb experiences a force given by

$$\mathbf{F} = q\mathbf{E} \text{ newtons}$$

If q is negative, the force is opposite in direction to **E.**

The electrostatic field is zero in the solid interior of metals. Field lines strike perpendicular to metal surfaces. They do not enter metals. When an excess charge is placed on a metal object, the charge comes to rest on the surfaces of the metal.

When a metal object is attached by a metal wire to a battery terminal, an excess charge is placed on the object by the battery. The positive terminal gives rise to positive charge; the negative terminal gives negative charge. If a negative charge flows out of the negative battery terminal, an equal negative charge must flow into the positive terminal.

The potential difference from a point A to a point B is defined as the work required to move a +1-C test charge from point A to point B. Its unit is joule per coulomb, which is renamed the volt (V). If the potential difference from A to B is V volts, the work done in carrying a charge Q from A to B is given by

$$\text{Work} = QV$$

A positive charge Q at a point A has an electric potential energy QV relative to a point B if the potential difference from B to A is V. In such cases A is said to be at a higher potential than B. Its potential is V volts higher than B. The high side of a battery is the positive terminal.

When a charge q falls freely through a potential difference V, its kinetic energy increases by an amount qV. Positive charges fall from high to low potential (+ to −). Negative charges fall from low to high (− to +).

When a charge n times larger than the electron charge falls through a potential difference V, it acquires an energy nV electronvolts. The conversion factor between joules and electronvolts is

$$1 \text{ electronvolt} = e \text{ joules}$$

where e is the fundamental charge magnitude.

Questions and Exercises

1. Many objects can be given static electric charge by rubbing. How could you determine the sign of the charge on a plastic pencil that has been rubbed with a woolen sweater?
2. Sparks can cause ether to explode. Years ago explosions used to occur in hospital operating rooms. Static electricity caused a spark, which led to an explosion. How might the static electricity be generated? What measures can be taken to decrease the chance of a spark?
3. Clothes dried in a rotating drier often cling together badly. What causes this? What factors determine how serious the problem will be? What can be done to decrease the problem?
4. Tear a tiny piece of paper from a sheet. Moisten it slightly by holding it in the palm of your closed hand. Now rub briskly a dry, plastic pencil or pen against a woolen sweater or some similar type of cloth. Bring the plastic close to the bit of paper as the paper lies on a table. Why does the paper jump and cling to the plastic? (Hint: Moist paper is slightly conducting.)
5. The dielectric strength of air is about 3×10^6 N/C, that is, a

spark will jump through air if the electric field in it is larger than this value. Why is a spark more likely to jump in part (b) of Figure 23.12 than in part (a)? Why are lightning rods made pointed?

6. Sketch the electric field outside a point charge. Sketch the electric field outside a metal sphere that carries the same charge Q spread uniformly on its surface. Using the fact that equal numbers of field lines come from equal charges, show that the following fact is reasonable: The field outside a uniformly charged sphere is the same as the field due to an equal charge placed at the position of its center.
7. A lone swimmer stands knee deep in water about 100 m from the shore of a large lake. Overhead a low thundercloud, which is highly charged, floats by. Sketch the electric field between swimmer and cloud assuming the cloud to be positive. Why is the swimmer in danger? What should you do (and not do) in a violent electrical storm? (Note: The human body, as well as the earth and most water, is slightly conducting.)
8. In an X-ray tube electrons are accelerated through perhaps 50,000 V. They strike a metal electrode and this electrode then gives off X rays. If the metal electrode is not properly cooled, it will get hot enough to melt. Where does the heat energy come from?
9. Metals are sometimes spray-coated by the use of electrostatic charge. The sprayer is attached to one terminal of a high voltage and the metal is attached to the other. Explain the principle of operation for this method.
10. The following quotation appeared in the "department of obscure information" section of a leading chemical magazine. "The earth's static electricity field increases by about 100 volts for every 3-foot increase in altitude in the area of Morgantown, W. Va." Correct it. Do you believe it? Could it be true sometimes?

Problems

1. How many electrons are there in a 1-C charge?
2. How large is the total charge in 1 kmol of protons? Of electrons?

*3. The atomic mass of copper is 63.5 kg/kmol and each atom has 29 electrons. (a) How many atoms are there in a 2.00-g copper coin? (b) How many electrons are there in the coin?

*4. Suppose the coin of Problem 3 carries a charge of 6.0×10^{-7} C. (This is a large charge for an object this size.) What percent of its electrons have been removed?

5. Two point charges have values $+5.0\ \mu C$ and $-7.0\ \mu C$. The 5-μC charge is at $x = 0$ whereas the -7-μC charge is at $x =$ 30 cm on the x axis. (a) Find the magnitude and direction of the force on the -7-μC charge. (b) Repeat for the 5-μC charge.
6. A $+25$-μC charge is at $x = 20$ cm and a $+30$-μC charge is at $x = 70$ cm on the x axis. Find the magnitude and direction of the force on the 25-μC charge. Repeat for the 30-μC charge.
7. In the hydrogen atom a single electron exists at a distance of about 0.053 nm from the nucleus. The nucleus is a single proton. Find the magnitude and direction of the force exerted by the nucleus on the electron.

*8. Radium nuclei are radioactive and throw out alpha particles. An alpha particle consists of two neutrons and two protons. The nucleus left behind has a charge of $+86e$. Find the magnitude and direction of the force exerted on the alpha particle by the nucleus when they are 10^{-13} m apart.

*9. Two small identical metal spheres carry charges of $+0.30\ \mu C$ and $-0.40\ \mu C$. (a) The spheres are 5.0 m apart. Find the force one exerts on the other. (b) The spheres are touched together and again separated to 5.0 m. What force does one now exert on the other?

*10. Two point charges are 3.0 m apart and each exerts an attractive force of 8.7×10^{-8} N on the other. One charge is twice the other in magnitude. Find the magnitudes of the two charges. Can you tell from these data which charge is negative?

*11. Two point charges exert a repulsive force of 250×10^{-10} N on each other when they are 4.0 m apart. (a) What force will one exert on the other when they are 3.0 m apart? (b) If the charges are Q_1 and Q_2 with $Q_2 = 7Q_1$, find the charges Q_1 and Q_2.

12. The following three charges exist on the x axis: $+5.0\ \mu C$ at $x = 0$, $-3.0\ \mu C$ at $x = 20$ cm, $-7.0\ \mu C$ at $x = 50$ cm. Find the magnitude and direction of the force on the -3.0-μC charge.
13. Repeat Problem 12 for the -7-μC charge.

**14. Two point charges are on the x axis. A 5.0-μC charge is at $x = 0$ and a -3.0-μC is at $x = 4.0$ m. Where can a third charge be placed so that the net force on it is zero? Why need we not specify the sign or magnitude of the third charge?

**15. Repeat Problem 14 for the case in which both charges are positive.

**16. Two identical balls, each with charge q, hang from a single point by two threads. The charges cause the balls to hang as

*Problems marked with an asterisk are not as easy as the unmarked ones.

**Problems marked with a double asterisk are somewhat more difficult than the average.

in the left part of Figure 23.1. The angle between the threads is 40° and the distance between the centers of the balls is 12 cm. Find the charge on each ball if the mass of each ball is 0.50 g.

17. Three equal 4.0-μC point charges are in the x, y plane at the following coordinates: $x = 0$, $y = 0$; $x = 150$ cm, $y = 0$; $x = 0$, $y = 300$ cm. Find the force on the charge at the origin.

*18. For the situation in Problem 17, find the force on the charge at $x = 150$ cm, $y = 0$.

19. Five identical 3.0-μC charges exist in a plane. Four are at the vertices of a square that has 7.0-m-long sides. The fifth is at the center of the square. Find the force on the charge at the center.

*20. For the situation in Problem 19, find the force on one of the corner charges.

*21. Repeat Problem 20 for the case in which the center charge is negative and the others are positive.

**22. The charges at the four corners of a 30-cm × 30-cm square are $+2.0\ \mu$C, $-3.0\ \mu$C, $-4.0\ \mu$C, and $+5.0\ \mu$C. The -3-μC charge is diagonally opposite the 5 μC. Find the magnitude of the force on the 5-μC charge because of the other three charges.

23. The electric field in a certain region is in the $+x$ direction. Its magnitude is 3000 N/C. What are the direction and the magnitude of the electric force on a charge q placed in this region if (a) $q = +5\ \mu$C and (b) $q = -5\ \mu$C?

24. In a certain region the electric field intensity is 2500 N/C in the $+x$ direction. (a) Find the magnitude and direction of the electric force experienced by a proton in this region. (b) Repeat for an electron.

25. In a certain region of space an electron experiences a force 3.0×10^{-16} N in the $+x$ direction. What are the magnitude and direction of **E** in that region?

26. A tiny sphere carries a charge of $+2.0 \times 10^{-6}$ C. (a) How large a force would it exert on a $+1$-C point charge placed 200 cm away? (b) What are the magnitude and the direction of the electric field strength 200 cm from the center of the sphere?

27. The value of **E** at a distance of 70 cm from a tiny charged sphere is 3500 N/C. Its direction is radially in toward the sphere. (a) Is the sphere's charge positive or negative? (b) How large a force would a $+1$-C charge experience 70 cm from the sphere? (c) What is the charge on the sphere?

*28. The electric field in a certain region is 800 N/C directed straight downward. A 2-g ball hangs from a thread in this region. Find the tension in the thread if the charge on the ball is (a) $+6\ \mu$C and (b) $-6\ \mu$C.

29. At what distance from a -5-μC point charge is the electric field 200 N/C? Is the electric field there directed radially inward or outward?

*30. Two point charges exist on the x axis. A $+3$-μC charge is at $x = 5$ cm and a $+3$-μC charge is at $x = -2$ cm. (a) What are the magnitude and the direction of the electric field at the coordinate origin? (b) Where is the electric field zero?

**31. Two charges are placed on the x axis. One is $-2.0\ \mu$C at $x = 0$. The other is $-3.0\ \mu$C at $x = 50$ cm. Where is the electric field due to these charges zero?

**32. Repeat Problem 31 for two positive charges.

**33. A 2.0-g ball hangs as a pendulum from a very thin thread. It hangs at an angle of 30° when placed in a horizontal field of 5.0×10^4 N/C. What is the charge on the ball?

*34. What is the electric field intensity just outside a uniformly charged 5.0-cm-radius metal sphere? The charge on the sphere is 2.5 μC. (Hint: Refer to Question 6 on page 508.)

*35. For the situation in Problem 22, what is the magnitude of the electric field at the center of the square?

*36. An electron is released from rest in a horizontal electric field of 3000 N/C directed eastward. (a) Find the acceleration of the electron. (b) What will be its speed after moving 50 cm?

37. Ten joules of work must be done against electric forces to carry a $+1$-C charge from point A to point B. (a) What is the potential difference between A and B? (b) How much work is needed to carry a proton from A to B?

38. How much work is required to carry a proton from the negative to the positive terminal of a 1.50-V battery? From the positive to the negative terminal?

39. Two metal plates are connected by wires to the terminals of a 12-V battery. (a) What is the potential difference between the plates? (b) How much work is done in carrying an electron from the positive to the negative plate?

40. How much kinetic energy does a proton acquire as it falls through a potential difference of 200 V?

41. A proton accelerates from rest as it falls through a potential difference. If the proton acquires a speed of 2×10^6 m/s, (a) how much kinetic energy did it acquire and (b) through how large a potential difference did it fall?

*42. An electron with unknown kinetic energy is brought to rest as it falls through a potential difference of 23 V. What was its kinetic energy?

*43. To light a 60-W light bulb, the power source must lift a charge of $\frac{1}{2}$ C through a potential difference of 120 V each second. How much work does the power source do each second?

*44. In a certain television picture tube a charge of 1×10^{-4} C falls through a potential difference of 20,000 V each second. How much electric potential energy is lost each second? How many calories of heat are generated each second as the charge strikes the end of the picture tube?

*45. In a certain X-ray machine a charge of 5×10^{-3} C falls through a potential difference of 4×10^4 V each second. (a) How much electric potential energy is lost each second? (b)

How many calories of heat are generated each second as the charge strikes and stops in the oppositely charged electrode?

46. In a certain vacuum tube electrons at rest are accelerated through a potential difference of 90 V. (a) Find the speed of an electron after it has been accelerated. What is the particle's energy in eV? (b) Repeat for a proton.
47. (a) Through how large a potential difference must an electron fall if it is to be going one-tenth the speed of light? What is the particle's energy in eV? (b) Repeat for a proton.
48. Radium atoms are radioactive and shoot out alpha particles. These are actually helium nuclei and have a charge of $2e$ and a mass $4 \times 1.67 \times 10^{-27}$ kg. The particles shoot out of the atom as though they had been accelerated from rest through a potential difference of 2.4×10^6 V. (a) How fast are the alpha particles going? (b) What is their energy in eV?

*49. An electron is moving at 2.5×10^6 m/s when it shoots from a positive plate toward a negative plate. What must be the potential difference between the plates if the electron is to have a speed of 1.5×10^6 m/s just before it hits the other plate?

*50. A proton is to be accelerated from a speed of 3.0×10^4 m/s to a speed of 7.0×10^4 m/s. Through how large a potential difference must it fall?

Bell Laboratories

CIRCUIT ELEMENTS

Performance Goals

When you finish this chapter, you should be able to

1. List several types of emf sources. Point out what major purpose they serve in electricity. Give two ways in which all emf sources are alike.
2. Give the fundamental relation between *C*, *Q*, and *V*. Draw a circuit to illustrate the meaning of these three quantities. Given two of the three quantities, compute the third and give its units.
3. State how the capacitance of a

Many practical applications of electricity involve circuits. We begin our study of circuit electricity in this chapter. It will be seen that sources of electromotive force provide the energy needed to maintain electric currents. The way in which resistors restrict the current flow will be discussed. Their action will be summarized in Ohm's law. In addition, this chapter will introduce the concepts of power loss, capacitance, series and parallel resistance combinations, resistivity, and several others. These concepts will be used repeatedly in the chapters that follow.

parallel-plate capacitor will be changed if A or d is changed by some known factor.

4. Find the equivalent capacitance for several capacitors in series and in parallel.
5. On a qualitative sketch of two close metal objects attached to the two terminals of a battery:
   a. Show the induced charges on each.
   b. Sketch the field lines between them.
   c. Sketch the equipotentials.
   d. Show what is meant by equipotential lines, planes, and volumes.
6. Sketch the field lines between closely spaced parallel plates that carry equal and opposite charges. Give the relation between E, V, and d for this situation. State the major restriction on it. Use it to find one quantity when the other two are known.
7. Explain what is meant by the statement that the electrostatic field is a conservative field.
8. State what data you would need to compute the electric current that flows in a wire. Given sufficient data concerning the charge flow, compute the current and give its units.
9. Compute how many electrons pass a point on a wire or beam in a given length of time when the current in the wire or beam is known. Find the current in a wire if the charge flow past a point is known.
10. State the direction of the current when the sign and the direction of motion of the charges are known.
11. Draw the circuit diagram for a

24.1 Sources of Electromotive Force

Electric systems require energy sources. The energy that heats a light bulb white-hot and the energy that drives a motor must come from somewhere. We call the devices that supply such energy *sources of electromotive force (emf)*.

Sources of emf supply energy to electric devices.

The name electromotive force is somewhat misleading. Sources of emf do not supply a force directly. Instead, they supply the energy by means of which work can be done. Typical sources of emf are batteries of all kinds and electric generators. For now we shall be concerned only with batteries. Electric generators will be discussed in Chapter 27. There are many kinds of batteries. Ordinary dry cells, wet cells (such as a car battery), and semiconductor-based solar cells are probably known to you. Fuel cells and thermoelectric emf sources are also widely known.

There are many complex devices that serve as sources of emf. A whole book would be needed to describe all of them. In spite of their variety, emf sources are all alike in two major respects. (1) Each provides two (or sometimes more) terminals between which a potential difference exists, and (2) when charge flows out from one terminal, an equal charge must flow into the other terminal. We shall see that these are the properties of a good battery that are of importance to us.

24.2 The Capacitor

Capacitors (also called condensers) are devices designed to hold charge. In our study of alternating currents in Chapter 28 we shall see why they are used in electronics. We shall discuss their design now, though, because they illustrate several principles that will be important to us in this chapter.

One of the most common types of capacitors consists of two metal plates close together, as diagramed in Figure 24.1(a). Part (c) shows a commercial version of this type. Each plate is a metal foil. A thin insulating sheet is used as a spacer between the plates. The flexible plates and spacer are folded into a tiny package sealed inside plastic. Two wires lead into the package and make contact with the two plates. We represent a capacitor by the symbol shown in part (b) of the figure.

A simple capacitor-battery circuit is shown in Figure 24.2. Notice the symbols in (b) used to represent the physical situation in (a). The source of emf, the battery, places equal but opposite charges on the two capacitor plates. (Recall that as much charge leaves one terminal as goes into the other terminal. This places a charge $+Q$ on the positive plate and a charge $-Q$ on the negative plate.) Let us call the magnitude of the charge on either plate Q.

Experiment shows that the charge Q on each of the capacitor plates is proportional to the potential difference of the battery. In fact, if the potential differ-

battery and resistor in series. Show on it the direction of the current. Show how an ammeter must be connected to measure I in it. Show how a voltmeter must be connected to measure the potential drop across the resistor. State the resistance of an ideal ammeter and of an ideal voltmeter.

12. State which is the high-potential (+) end of a resistor when the current flow direction is given.
13. Point out qualitatively how the potential varies from point to point in a circuit that consists of a battery connected across a series or parallel resistance combination. Tell how the currents in various parts of the circuit are related.
14. Explain how you could determine the resistance of a piece of wire or a metal cylinder if you were provided with a battery, an ideal voltmeter, and an ideal ammeter.
15. State Ohm's law and illustrate its use by means of an example devised by you.
16. State the relation between power, V, and I. Given a battery and resistance in series, state how much power is supplied to the circuit by the battery. Also state how much power is lost in the resistor. Where does this lost power go?
17. Give the current through and the resistance of a bulb when it is operated at its rated voltage. You are told that the bulb is stamped 75 W/120 V, for example.
18. Reduce a series or parallel resistor combination to its equivalent resistance.

ence between the two plates of a capacitor is V, then the charge Q on the capacitor is given by

$$Q = (\text{constant})V$$

This constant is a function only of the capacitor design. Its value depends on geometrical factors. As we would expect, the larger the area of the plates, the larger Q will be. The closer together the plates, the larger Q will be. It is given the symbol C and is called the *capacitance*.

We can see why it gets this name by solving for the constant in the equation above.

$$\text{Constant} = \frac{Q}{V}$$

or

$$\text{Capacitance} = C = \frac{Q}{V} \quad \text{or} \quad Q = CV \tag{24.1}$$

Notice that C measures the charge Q per potential difference V. It is a measure of the number of coulombs of charge the capacitor holds for each volt of potential difference. Although C has the units coulombs per volt, we give this unit a new name. It is called the farad, F, after Michael Faraday, a pioneer in experimental electricity. Most capacitors have values for C that are of the order of microfarads (μF) or smaller.

Both theory and experiment show that C for a parallel-plate capacitor is given by a simple formula. If the inside area of one plate is A and the separation of the plates is d, then

$$C = \frac{\epsilon_0 A}{d} \quad \text{(parallel plates only)} \tag{24.2}$$

The quantity ϵ_0 was introduced in the previous chapter. It is called the permittivity of free space and has the value 8.85×10^{-12} F/m. Notice the units for ϵ_0. Equation 24.2 shows that they are correct.

EXAMPLE 24.1 A metal plate 200 cm × 200 cm square is placed parallel to an identical plate and 0.50 mm away from it. Find the capacitance of this parallel-plate capacitor.

19. When given five or less resistors connected in series and parallel combinations, reduce them to a single equivalent resistor.
20. Write the relation between R, ρ, L, and A. Show the meaning of each symbol by means of a diagram. Given all but one of these quantities, find the unknown one.
21. When given an equation showing the temperature variation of R or ρ, be able to use it to find R or ρ at one temperature when the value is known at some other temperature.

Solution We are told that $d = 5 \times 10^{-4}$ m and $A = 4$ m^2. Substituting these values in Equation 24.2 gives

$$C = 7.1 \times 10^{-8} \text{ F} = 71 \text{ nF}$$

Notice how small this capacitance is. ■■

24.3 Capacitor Combinations

Manufacturers of capacitors make them with only certain values. When a capacitor with some other value is required, capacitor combinations must be used. There are two basic types of combinations. They are shown in Figure 24.3.

The capacitors in Figure 24.3(a) are said to be in parallel. Notice that the left plates of the capacitors are all attached to point A. In effect, the left plates have been combined to make a single plate of larger area. In the same way the right plates are combined at B. Because capacitance is proportional to plate area, it is no surprise that the combined capacitors act like a capacitor of value $C_1 + C_2 + C_3$. We say that the combined capacitors have an equivalent capacitance C_{eq} given by

$$C_{eq} = C_1 + C_2 + C_3 \qquad \text{(parallel)} \qquad (24.3)$$

Similar equations apply to any number of capacitors in parallel.

The capacitors shown in Figure 24.3(b) are said to be in series. In this case the equivalent capacitance is less than any of the three capacitances. It is given by

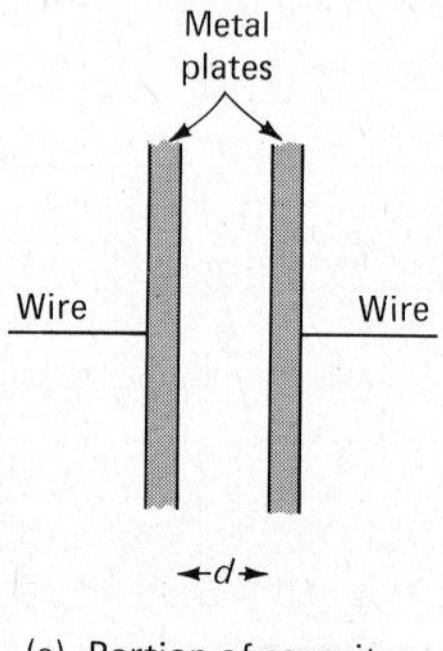

(a) Portion of capacitor

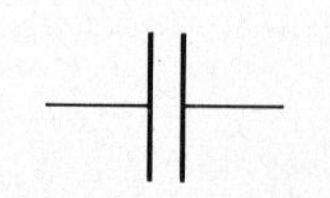
(b) Symbol for a capacitor

(c) Commercial capacitor broken open

Figure 24.1 The capacitor (or condenser).

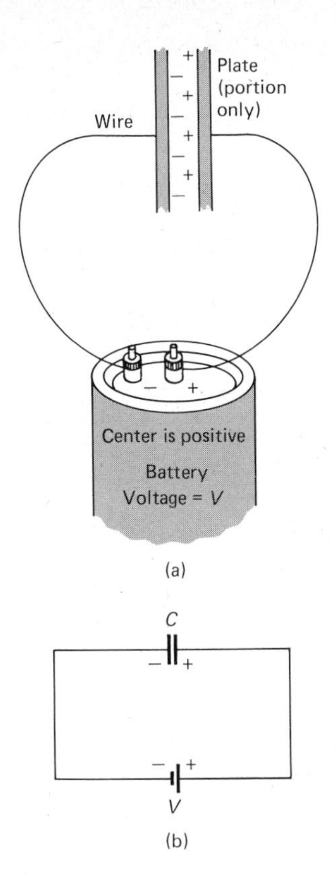

Figure 24.2 The circuit in (a) is represented by diagram in (b). Usually the + and − signs are omitted in the diagram.

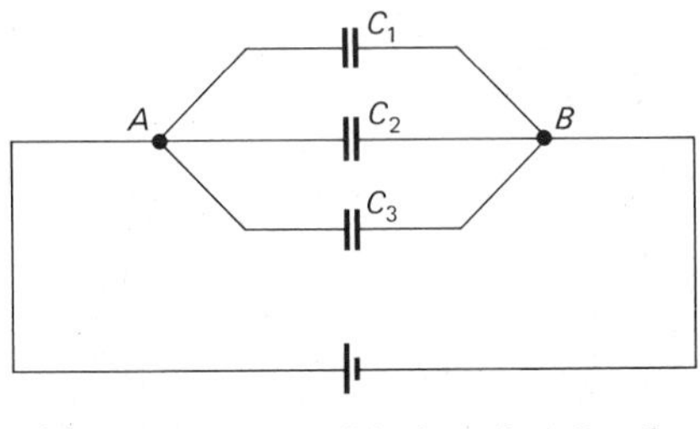

(a) Capacitors in parallel: $C_{eq} = C_1 + C_2 + C_3$

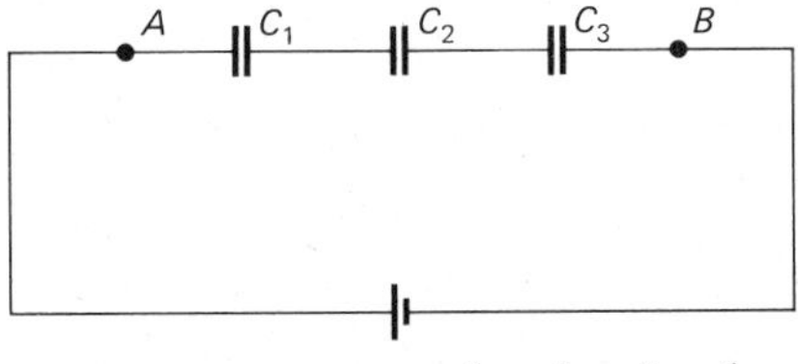

(b) Capacitors in series: $\frac{1}{C_{eq}} = \frac{1}{C_1} + \frac{1}{C_2} + \frac{1}{C_3}$

Figure 24.3 Equivalent capacitance.

$$\frac{1}{C_{eq}} = \frac{1}{C_1} + \frac{1}{C_2} + \frac{1}{C_3} \quad \text{(series)} \qquad (24.4)$$

with similar expressions applying to any number of capacitors in series.

EXAMPLE 24.2 You are given five capacitors with values 1, 2, 3, 4, and 5 μF. Find the equivalent capacitance if they are (a) in parallel and (b) in series.

Solution (a) For the capacitors in parallel,

$$C_{eq} = C_1 + C_2 + C_3 + C_4 + C_5 = 15\ \mu\text{F}$$

(b) For the capacitors in series,

$$\frac{1}{C_{eq}} = 1.00 + 0.50 + 0.33 + 0.25 + 0.20 = 2.28/\mu\text{F}$$
$$C_{eq} = 1/2.28 = 0.44\ \mu\text{F}$$

Notice, as stated previously, that the series equivalent is smaller than any of the single capacitors. ■■

24.4 Field Lines Between Parallel Plates

Refer to the parallel-plate capacitor shown in Figure 24.4(a). In most cases the plates would even be closer together and larger than shown there. You should be able to justify the way we have drawn the field lines in (a). Except for "fringing" at the edges of the plates, the electric field is uniform between them. In part (b) of the figure we show the central region of the plates enlarged.

Plate M carries a positive charge and plate N carries an equal negative charge. Far from the edges of the plates, the electric field is uniform. Let us call its intensity E. There is a simple relation between the potential difference between the plates, V, and the electric field between them, E.

To find this relation, recall that V is the work needed to carry a +1-C charge from plate N to plate M. But the force on a +1-C charge between the plates is constant and is simply E. As the charge is carried from N to M, the work done will be

$$\text{Work} = (\text{force})(\text{distance in direction of force}) = Ed$$

Because this is also equal to V, we have

$$V = Ed \quad \text{(parallel plates only)} \qquad (24.5)$$

This is a useful relation. But notice that it is restricted to E being constant. Usually a constant electric field is found only between parallel plates.

E is expressed as volts per meter (V/m) in Equation 24.5. Because a volt is a joule per coulomb and a joule is a newton-meter, we see that a volt per meter is the same as a newton per coulomb. Both units are used. However, it is most common to call the unit of E the volt per meter.

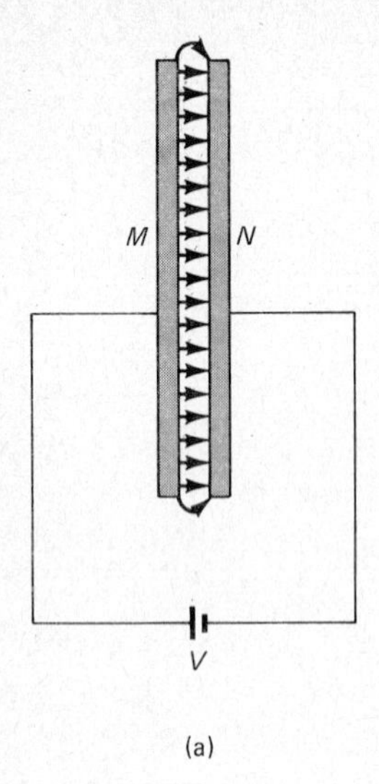

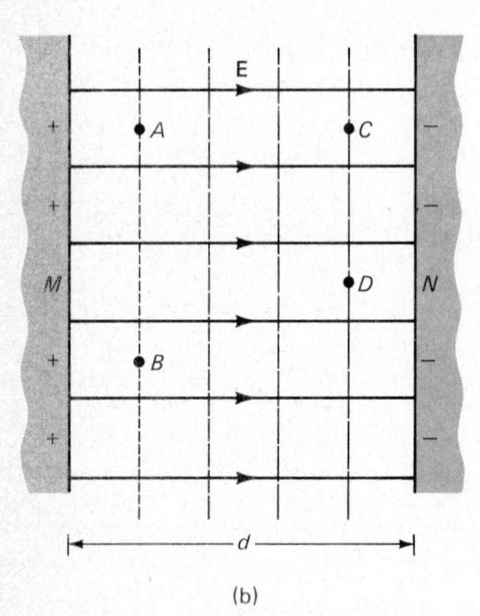

Figure 24.4 The field is uniform in the central region between parallel plates. What do the dashed lines in (b) represent? An infinite number of similar lines can be drawn.

▮ **EXAMPLE 24.3** Two large parallel metal plates are connected to a 12-V battery. If the gap between the plates is 0.20 cm, how large is the electric field, E, between the plates?

Solution For parallel plates Equation 24.5 applies.

$$V = Ed \qquad \text{or} \qquad E = \frac{V}{d}$$

Using $V = 12$ V and $d = 2.0 \times 10^{-3}$ m gives

$$E = \frac{12\ \text{V}}{2 \times 10^{-3}\ \text{m}} = 6000\ \text{V/m}$$

▮▮

24.5 Equipotentials and Field Lines

Let us refer again to part (b) of Figure 24.4. Work is needed to pull a positive test charge from plate N to M. But suppose that the charge is moved from A to B along the dashed line. The electric force on the charge is horizontal, but the distance moved is vertical. The force is perpendicular to the motion. There is no component of the force in the direction of motion. Therefore no work is done in moving a test charge from A to B. The potential difference is zero between any two points on this line. We call this line an *equipotential line.* Because the electric force is perpendicular to the equipotential line, we see that equipotential lines are perpendicular to electric field lines.

Equipotential lines are lines whose points are all at the same potential. Equipotential lines and field lines are perpendicular to one another.

We can imagine a plane through points A and B in Figure 24.4(b) which is parallel to the plates. The electric field lines are perpendicular to it. Therefore, no work is done in carrying a test charge from point to point on it. We conclude that the plane is an equipotential plane. All points on an equipotential plane are at the same electric potential.

We can also imagine similar planes located at each of the dashed lines in the figure. Each is an equipotential plane. In fact, there are an infinite number of equipotential planes that could be drawn parallel to the plates. In every case,

The electric field lines are always perpendicular to equipotential planes.

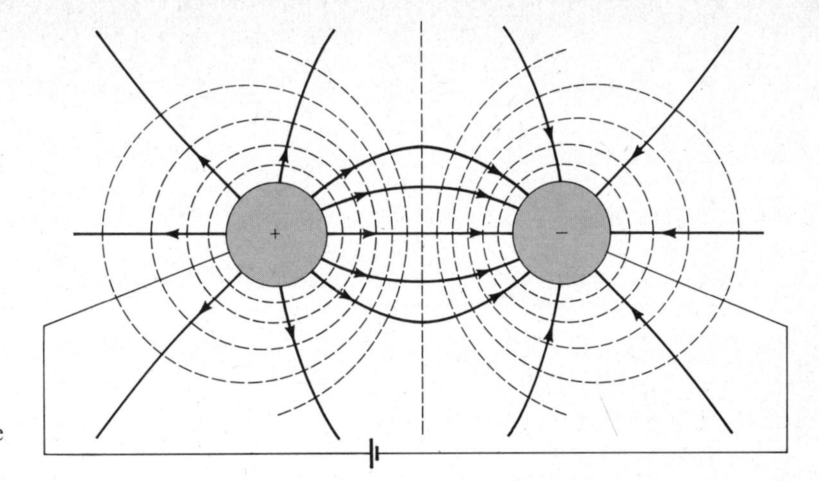

Figure 24.5 The field lines (solid) are perpendicular to the equipotentials. The battery is farther away than shown.

As another example, consider the oppositely charged metal spheres shown in Figure 24.5. The solid lines are field lines. The dashed lines are equipotentials. They are mutually perpendicular, as they must be. But we should notice something else in this figure. The surface of the metal is an equipotential surface. Indeed, as we shall now see, metal objects are equipotential solids.

Suppose that the two metal objects shown in Figure 24.6 are attached to a battery with voltage V. (These two objects act like plates of a capacitor, but we are not interested in that feature now.) This is an electrostatic situation. Therefore, the electric field in the metal is zero. As a result, points A, B, C, and D are all at the same potential. No work is required to carry a test charge from one to the other. We therefore conclude that the upper metal object and the wire leading to it are an equipotential volume.

Similarly, points E, F, and G are at the same potential. Hence the lower object and its wire are an equipotential volume. Of course, the battery that connects them ensures that a potential difference V exists between these equipotential volumes. The work required to carry a unit positive test charge from E to D is V joules per coulomb. Indeed, each point in the upper object is V volts higher in potential than each point in the lower object.

The concept of equipotentials reminds us of a similar situation in gravitation. We learned that all points at the same height from the earth are at the same gravitational potential. Gravitational potential energy depends only on height, mgh. Further, we learned that the work done against gravity in carrying a mass from one point to another is independent of the path. We called such a field and its force a conservative field and force.

A similar situation exists in electrostatics. Using the same methods as for gravitation, we can prove the following.

The electrostatic field is a conservative field. The work done in moving a test charge from one point to another is independent of the path.

We shall see that this has important application to electric circuits.

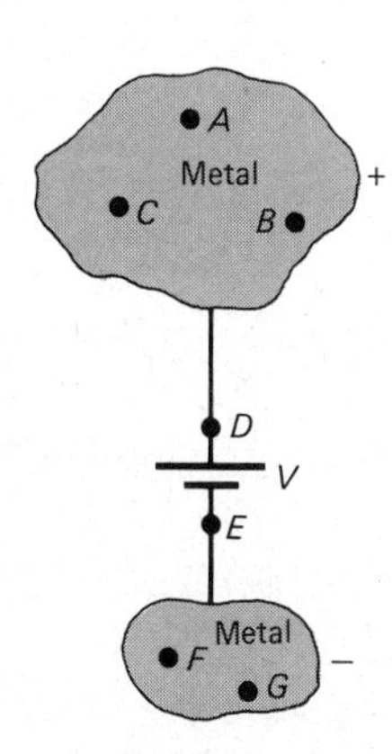

Figure 24.6 The solid metal objects are equal potential (equipotential) volumes. Every point in the upper one is V volts higher than every point in the lower one.

DIELECTRICS

Most capacitors have an insulating material called a *dielectric* separating the plates. Plastic film is commonly used for this purpose, but any nonconductor may be used. The dielectric prevents sparking between the plates since solid and liquid dielectrics have greater electric strength than air. In addition, the dielectric increases the capacitance of the capacitor by decreasing the field between the plates. Let us now see how it does this.

Although a dielectric contains no free charges, its atoms (and molecules) are polarizable. By this we mean that an impressed electric field distorts the atom so that its positive and negative charge are slightly separated. A typical situation is shown in the figure. Without the dielectric sheet the electric field between the plates is E_0. When the sheet is put in place, the atoms are distorted. The result is to induce charges on the two sides of the sheet as indicated. These induced charges are called *bound charges* because they are not free to move from their parent atoms.

The induced charges give rise to a reverse field E_i between the capacitor plates. Now the field is $E_0 - E_i$ rather than E_0. As you see, the dielectric reduces the electric field between the plates. Since $V = Ed$, this also causes the voltage difference between the plates to decrease. And, because the capacitance is given by $C = Q/V$, the insertion of the dielectric causes the capacitance to increase.

This increase in capacitance depends on how polarizable the dielectric is. Easily polarizable dielectrics give rise to much induced charge and large changes in capacitance. We define the *dielectric constant,* k_d, of the dielectric to be

$$k_d = \frac{\text{capacitance with dielectric}}{\text{capacitance without dielectric}}$$

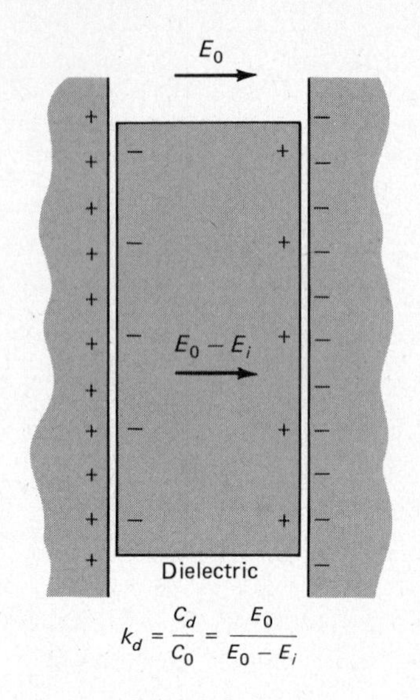

Most plastics and oils have k_d in the range from 2 to 5. But water has $k_d = 80$ and methyl alcohol has $k_d = 30$.

The dielectric constant is a measure of a material's ability to decrease the field due to a charge. In a capacitor the field is reduced by a factor $1/k_d$ if the charge on the capacitor remains the same. Similarly, when a point charge is immersed in a dielectric, the field due to the charge is reduced by a factor $1/k_d$. As a result, ions in water exert forces about 1/80 as large on each other as they do in air or vacuum. This is why ionic materials dissolve in water but not in liquids with much lower k_d. The electrostatic forces between the ions in water is small enough that oppositely charged ions can escape from each other.

24.6 Electric Current

Let us now consider the situation shown in Figure 24.7(a). There we see a wire connected between the two terminals of a battery.[1] This is no longer an electrostatic situation because now charges will move through the wire. They do so because there is an electric field in the wire. We can see this in the following way.

Suppose the potential difference of the battery is V. Then V joules of work must be done to carry a unit positive test charge from B to A by any path whatever. We could carry the test charge from B to A through the wire. Even so, the work we would do must still be V joules. There is an electric field in the wire and it tries to force positive charges to move from A to B.

[1] Do not make such a connection unless the wire has a high resistance.

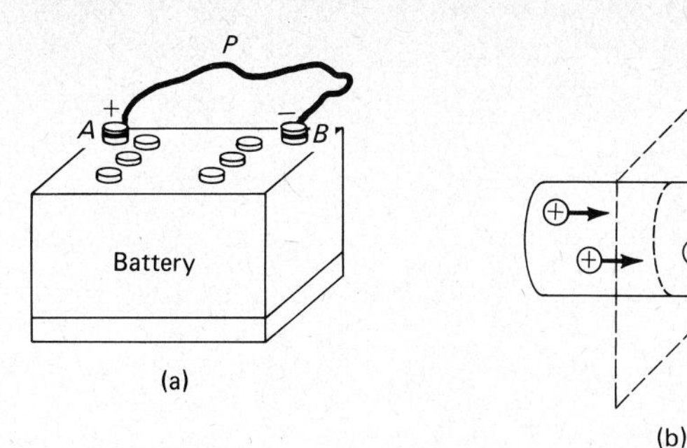

Figure 24.7 An electric field exists in the wire in (a). It causes an electric current to flow in the wire.

In the case of a metal wire, there are free electrons within the wire. Their number is large, of the order of the number of atoms in the wire. Forced on by this electric field in the wire, the electrons drift through the wire from B to A. We say that a current flows through the wire. If we follow an individual electron, it moves out of the battery at B, through the wire and point P, back into the battery at A, and through the battery back to point B. The path that the charge follows is called a *circuit.*

An electric circuit is a closed path along which charges can move.

The flow of charge that occurs in the wire is called a *current.* To define the term precisely, let us refer to part (b) of the figure. There we see positive charges flowing past point P in a circuit. Current is defined as having the following properties.

The magnitude of the electric current, *I*, at a given point in a circuit is the quantity of charge that passes the point each second.

The direction of current is taken opposite to the direction of negative charge flow or, if positive charge is moving, in the direction of positive charge flow.

The unit of current is the coulomb per second, which is called the ampere, A.

Once again we see the bias toward positive charges. It is traditional to take the current to flow in the direction in which positive charges move. If negative charges, such as electrons, are moving, the current direction is opposite to their motion direction. For example, in Figure 24.7(a), electrons move around the circuit from B to P to A to B. But these are negative charges. Therefore the current is defined to flow around the circuit in the opposite direction. The current flows from A to P to B to A.

Finally, we must extend our definition to changing currents. It is as follows:

If a charge ΔQ flows past a point in a circuit in a time Δt, then the current *I* is given by

$$I = \frac{\Delta Q}{\Delta t} \qquad (24.6)$$

The unit of *I* is amperes. The symbol for amperes is A.

▮ **EXAMPLE 24.4** In a television picture tube 5×10^{14} electrons shoot out of the electron gun in 1 s. How large is the current flowing in the tube? What is the direction of the current?

Solution The situation is much like that shown in Figure 23.18. We are told that 5×10^{14} electrons shoot from the gun in a time $\Delta t = 1$ s. Because each electron has a charge of magnitude 1.6×10^{-19} C, we have the following charge leaving the gun each second:

$$\Delta Q = (1.6 \times 10^{-19}\ \text{C})(5 \times 10^{14})$$
$$= 8.0 \times 10^{-5}\ \text{C}$$

Therefore, from the definition of current,

$$I = \frac{\Delta Q}{\Delta t} = \frac{8.0 \times 10^{-5}\ \text{C}}{1\ \text{s}} = 8.0 \times 10^{-5}\ \text{A}$$

The direction of the current is opposite to the direction of motion of the electrons. Why? ▮▮

24.7 Energy in a Circuit

We are all familiar with the flashlight bulb circuit shown in Figure 24.8(a). A battery causes current to flow through the bulb. This is similar to the situation in Figure 24.7(a) where a battery causes current to flow in a wire. A flashlight bulb is simply a filament (thin wire) sealed in a glass tube. As electrons move through the filament, they experience the equivalent of large friction forces. The resultant frictionlike energy losses cause the filament to become white hot. Of course the filament is made of a material that emphasizes this heating effect. The other wires of the circuit offer much less resistance to electron motion. There is therefore negligible heating in the connecting wires.

We can understand this circuit better if we consider the water circuit shown in Figure 24.8(b). There we see a water wheel driven by a water pump. The pump gives energy to the water as it lifts the water. This energy is (mostly) delivered to the water wheel as the water falls to its lower level. The water wheel itself can do useful work by driving machinery as it turns. In this system the energy given to the water by the pump results in work done by the water wheel. Notice that the water itself acts as a working fluid. It simply carries energy from the pump to the water wheel. The pump is the energy source.

In the flashlight circuit the battery is the energy source. The electrons in the wires simply carry energy from the battery to the light bulb. There the energy is

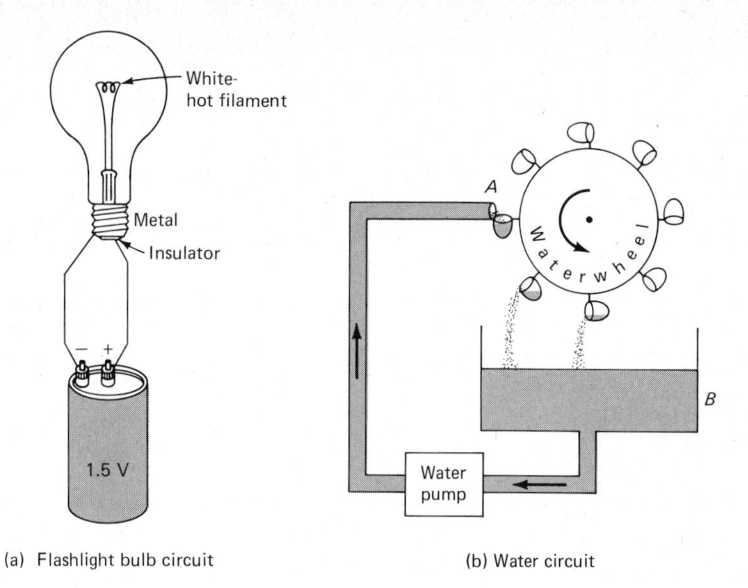

Figure 24.8 The water circuit in (b) is analogous to the electric circuit in (a).

released in the form of heat and light. The electrons, now energy depleted, return to the battery to obtain energy once again.

Or, put in another way, the battery gives the electrons electric potential energy by lifting them through the 1.5-V potential difference. They carry this energy to the filament by moving through the (nearly) equipotential connecting wires. At the filament they fall through a potential difference of 1.5 V and release their energy as heat and light. They then return to the battery through the nearly equipotential connecting wires. The battery then replenishes the energy of the electrons by lifting them once again through 1.5 V.

There is one other feature we can notice from the circuit in Figure 24.8. The connecting wires are assumed to be resistanceless. As a result, these wires do not heat. Moreover, the electrons lose no energy there and so these wires are equipotentials. This can never be perfectly true, of course. But,

As long as the heat generated in a circuit wire is negligible, the wire can be considered to be an equipotential volume.

24.8 Resistance and Ohm's Law

The flashlight circuit in Figure 24.8 is an example of a *series circuit*. There is only one path for the current to follow. All the current must pass through each portion of the circuit in turn.

We can summarize a series circuit in a circuit diagram such as Figure 24.9. The battery is shown by the symbol ⊣⊢ . (You know that its long side is the positive terminal.) The filament of the bulb offers resistance to current flow. Heat is generated there. We call it a *resistor* and represent it by the symbol —ʌʌʌ— on a circuit diagram.

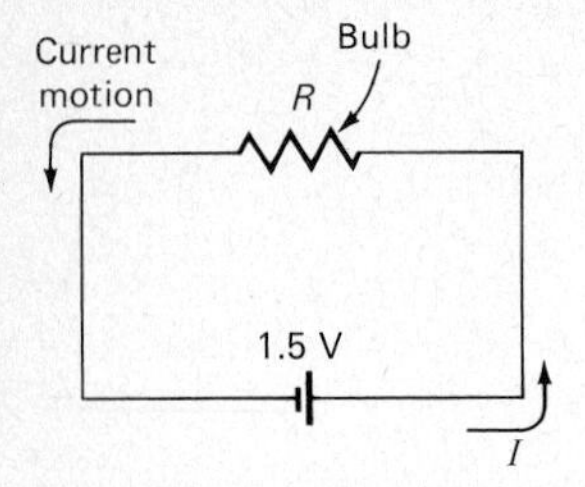

Figure 24.9 The circuit diagram for the circuit shown in Figure 24.8(a).

A resistor causes charges that flow through it to experience the equivalent of friction forces. Heat is generated in the resistor as current flows through it.

The direction of the current is also shown in Figure 24.9. Electrons flow out of the negative terminal, clockwise around the circuit, and into the positive terminal. Because they are negative charges, our definition of current insists the current flows in the opposite direction. As we see, current flows out of the positive battery terminal and back into the negative terminal. This is, of course, the direction the battery would cause positive charges to move if free positive charges existed in the circuit.

Let us now move on to Figure 24.10. This circuit is much like the previous one except for the symbol —Ⓐ— in it. We use this symbol to represent a device that measures current. It is called an ammeter. Notice that it is connected so that the current flows through it. Because we do not want the meter to disturb the circuit, a perfect ammeter should have zero resistance to current flow. In practice, a good ammeter will have very low resistance.

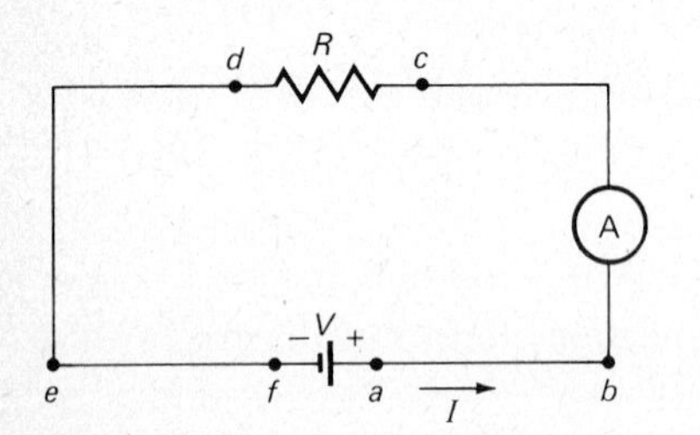

Figure 24.10 The current I is the same at points a, b, c, d, e, f. Must the ammeter be placed as shown? What should be the resistance of a good ammeter?

An ammeter measures the current *I* flowing through a wire. Its resistance to current flow should be very low.

Ammeter construction will be discussed in Chapter 26.

We can use the circuit of Figure 24.10 to assign a value to the resistance to flow of a resistor. To do so, we notice that the potential difference from c to d is the same as from a to f. This follows from the fact that the wires are equipotentials. The potential difference from one end of the resistor to the other is therefore V. (We say that the potential difference *across* the resistor is V.) This potential difference causes a current I to flow through it. In terms of V and I, our definition of resistance is as follows:

Let *V* be the potential difference (in volts) across a resistor *R* when a current *I* flows through the resistor. Then the resistance of the resistor is

$$R = \frac{V}{I} \quad \text{or} \quad V = IR \tag{24.7}$$

The units of *R*, which are volts per ampere, are given the name ohms. The symbol for this unit is Ω (Greek letter omega).

The relation 24.7 is often called Ohm's law, after Georg Simon Ohm, who first proposed it. Ohm also said that R remains constant for different values of V (battery voltage). This is often true, but sometimes it is not, particularly if the resistor becomes hot. We shall see later an example of how R varies with temperature.

EXAMPLE 24.5 When a 1.57-V battery is connected directly across a certain bulb, a current of 0.21 A flows through the bulb. What is the bulb's resistance under these conditions?

Solution According to Ohm's law, $V = IR$. In this case $V = 1.57$ V and $I = 0.21$ A. Solving,

$$R = \frac{1.57 \text{ V}}{0.21 \text{ A}} = 7.5 \text{ ohms} = 7.5\ \Omega$$

24.9 Simple Series Circuit

The circuit shown in Figure 24.11 is much like the one we have just discussed. Now, however, we show a voltmeter (—Ⓥ—) properly connected to it. Voltmeters are used to measure the potential difference between two points. In this case it is connected from one end of the resistor to the other. It therefore reads the potential difference across the resistor.

We do not want the voltmeter to disturb the circuit, and so no charge should flow through it. In order for this to be true, the voltmeter would have to have infinite resistance. Practical voltmeters have high resistance. For example, the resistance of a vacuum tube voltmeter is usually more than $10^6\ \Omega$. However, nonelectronic voltmeters have considerably lower resistance.

A voltmeter measures the potential difference between the two points to which it is connected. Its resistance should be very high.

We shall discuss voltmeter construction in Chapter 26.

As usual, in Figure 24.11 the current I flows out of the positive battery terminal. (Remember that I represents the flow of positive charge. It comes from the positive terminal.) We represent the emf of the battery (its voltage) by the symbol $\mathcal{E}$ (script E) because we want to use V for other quantities. Because point a is at the high (or positive) side of the battery, point a is at a higher potential than point d. The circuit from a to b is at the same high potential. The portion of the circuit from c to d is at the lowest potential.

Notice that the current flows through the resistor from b to c. It flows from the high side of the resistor to the low side. (This makes sense. Positive charges, which the current represents, will always fall from high to low potential if allowed to do so.) This is a general rule and an important one as we shall see.

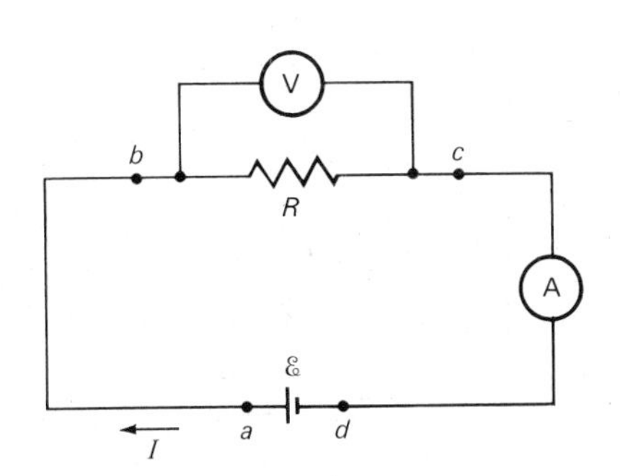

Figure 24.11 The emf (or voltage of this ideal battery) is represented by the symbol $\mathcal{E}$. Notice how the voltmeter is connected. If you had your choice, how large a resistance would you want it to have?

Current always flows through a resistor from its high-potential end to its low-potential end.

There is no similar rule for batteries. Usually the current flows out of the positive terminal of a battery and into its negative terminal. But if more than one source of emf exists in a circuit, one source may be charging the other. In that case the current would be going into the positive terminal of the battery that is being charged. The rule above, though, is *always* correct for resistors.

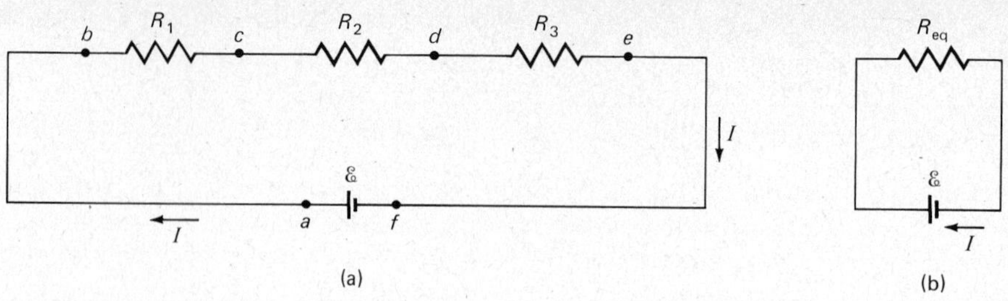

Figure 24.12 This is a series circuit. How large must the equivalent resistance, R_{eq}, in (b) be if the currents in (a) and (b) are to be the same?

Let us now look at the circuit in Figure 24.12(a). We call this a series circuit.

In a series circuit there is only one path that the moving charges (or current) can follow.

When the current leaves the battery at a, it has no choice but to follow the path *abcdefa*. Later we shall see circuits for which this is not the case.

The resistors in Figure 24.12(a) are said to be in series. By this we mean:

When the current must flow in succession through resistors R_1, R_2, R_3, and so on, the resistors are in series.

One might guess that the three resistors in Figure 24.12(a) could be replaced by a single resistor as shown in (b). This turns out to be true, as we shall now show.

24.10 Resistors in Series

We wish to find R_{eq} in part (b) of Figure 24.12 so that it will be equivalent to the three resistors in (a). In other words, for the same emf in (a) and (b), we want the current to be the same in either circuit. For the circuit in (b), Ohm's law tells us that

$$\mathscr{E} = IR_{eq}$$

What does Ohm's law tell us about the resistors in part (a) of the figure?

It tells us that the potential drop from b to c across resistor R_1 is

$$\text{Potential drop } b \rightarrow c = IR_1$$

We know this is a drop from b to c because of our general rule: Current always flows from high to low potential through a resistor. Similarly, the drops across R_2 and R_3 are

$$\text{Potential drop } c \rightarrow d = IR_2$$
$$\text{Potential drop } d \rightarrow e = IR_3$$

MINIATURIZED CIRCUITS

In the laboratory you will be using resistors, capacitors, and perhaps vacuum tubes and transistors. These circuit elements will be large enough to be easily handled. Until about 1960 nearly all circuits were composed of units such as these. Even today these same elements are in widespread use.

However, since 1960 we have learned how to decrease vastly the physical size of circuits. The invention of the transistor in 1948 spurred research in the area of solid state electronics. This research has paid off handsomely. Today it is possible to condense a circuit that contains thousands of elements into a tiny wafer (or "chip"). One such chip is shown here.

The pocket calculator and electronic computers are examples of the use of miniaturization. Prior to about 1960, such a device was not easily portable. Its cost was hundreds of times larger than that of today's equivalent. There are many other examples of the benefits of circuit miniaturization to technology. Indeed, these benefits extend into all areas of our society.

Still, however, conventional-size circuit elements are not outmoded. The tiny miniaturized circuits are not easy to experiment with or to change. They are practical only for mass production. Where circuit development and modification are concerned, conventional circuit elements must still be used.

This circuit chip, used in a watch, contains thousands of transistors in an area about 1 cm^2. (Courtesy of RCA Solid State Division)

We therefore have, for the total potential drop from b to e,

$$\text{Potential drop } b \rightarrow e = IR_1 + IR_2 + IR_3$$

But points a and b are at the same potential. And point e is at the same potential as point f. Therefore,

$$\text{Potential drop } b \rightarrow e = \text{potential drop } a \rightarrow f$$

As we see from the figure, the potential drop from a to f is simply the voltage of the battery, $\mathscr{E}$. Equating this to the drop from b to e gives

$$\mathscr{E} = IR_1 + IR_2 + IR_3$$

or

$$\mathscr{E} = I(R_1 + R_2 + R_3)$$

In the equivalent circuit of part (b) of Figure 24.12 we had

$$\mathscr{E} = IR_{eq}$$

Comparing these two equations, we see that R_{eq} is equivalent to $R_1 + R_2 + R_3$. We therefore conclude that

Several resistors in series are equivalent to a single resistor equal to the sum of the several resistors.

$$R_{eq} = R_1 + R_2 + R_3 + \cdots \qquad \text{(series)} \qquad (24.8)$$

24.11 Resistors in Parallel

The resistors shown in Figure 24.13(a) are said to be in parallel.

Several resistors are in parallel between two points if one end of each is connected directly to the one point and the other end of each is connected directly to the other point.

We wish to find the equivalent resistor, shown in part (b), that could replace them.

To analyze this situation, we need to notice what I does at point b in Figure 24.13(a). I represents a flow of positive charge. As the charges reach point b, they have a choice. They can follow a path through R_1 or through R_2 or through R_3. Some follow one path and others follow another. In other words, I splits into three parts, I_1, I_2, I_3. Because the current (charge) that flows into point b must flow out through these three paths, we have that

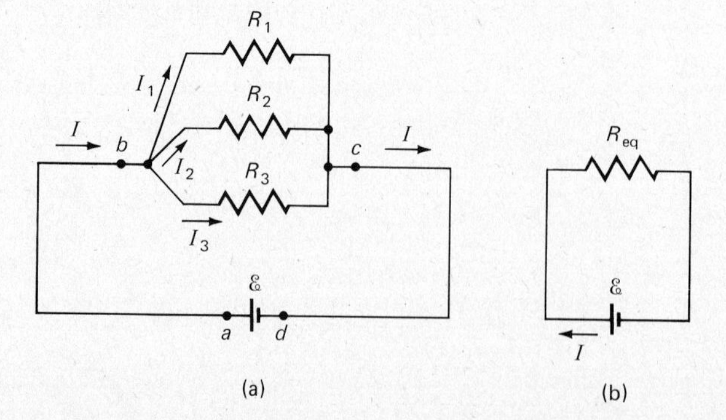

Figure 24.13 These resistors are in parallel. What is their equivalent resistance, R_{eq}?

$$I = I_1 + I_2 + I_3$$

But we also notice that the potential drop across each resistor is $\mathcal{E}$. This follows from the fact that the left end of each is connected directly by equipotential wires to the positive battery terminal. The right end of each is connected by equipotential wires to the negative battery terminal. Therefore, the potential drop across the battery, $\mathcal{E}$, is the same as across each of the resistors.

However, Ohm's law also tells us the potential drop across each resistor. It is I_1R_1 for R_1, it is I_2R_2 for R_2, and it is I_3R_3 for R_3. Because we have already concluded that each of these drops must be $\mathcal{E}$, we have

$$\mathcal{E} = I_1R_1 \quad \text{or} \quad I_1 = \frac{\mathcal{E}}{R_1}$$

$$\mathcal{E} = I_2R_2 \quad \text{or} \quad I_2 = \frac{\mathcal{E}}{R_2}$$

$$\mathcal{E} = I_3R_3 \quad \text{or} \quad I_3 = \frac{\mathcal{E}}{R_3}$$

We can now substitute these values in our previous equation.

$$I = I_1 + I_2 + I_3$$

$$I = \frac{\mathcal{E}}{R_1} + \frac{\mathcal{E}}{R_2} + \frac{\mathcal{E}}{R_3}$$

But we also had $\mathcal{E} = IR_{eq}$, from which $I = \mathcal{E}/R_{eq}$. Placing this value in for I gives, after canceling $\mathcal{E}$ from the whole equation,

$$\frac{1}{R_{eq}} = \frac{1}{R_1} + \frac{1}{R_2} + \frac{1}{R_3} \quad \text{(parallel)} \tag{24.9}$$

We therefore conclude that

Several resistors in parallel are equivalent to a single resistor. The reciprocal of the equivalent resistance is equal to the sum of the reciprocals of the several resistors.

■ **EXAMPLE 24.6** Find the resistor equivalent to those shown in Figure 24.14(a).

Solution The 3-Ω and 9-Ω resistors are in parallel. They are equivalent to R_1 given by

$$\frac{1}{R_1} = \frac{1}{3\ \Omega} + \frac{1}{9\ \Omega} \quad \text{or} \quad \frac{1}{R_1} = \frac{4}{9\ \Omega}$$

from which $R_1 = 2.25\ \Omega$. The parallel combination is replaced by a single resistor in (b) of the figure.

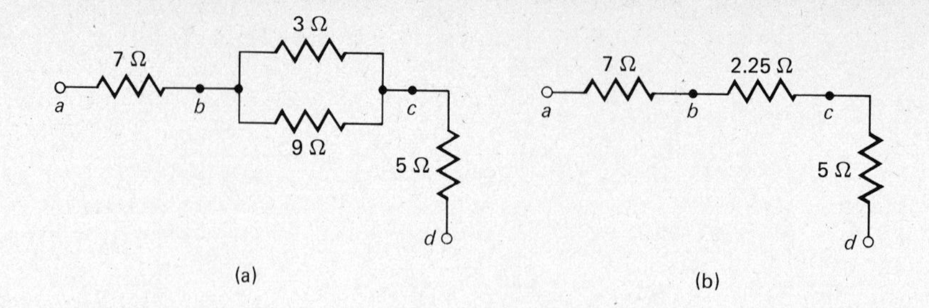

Figure 24.14 The resistor that could be used to replace those in (a) would have a resistance of 14.25Ω.

The resistors in (b) are in series. They are equivalent to a single resistor with value

$$\begin{aligned} R_{\text{eq}} &= 7\ \Omega + 2.25\ \Omega + 5\ \Omega \\ &= 14.25\ \Omega \end{aligned}$$

The combination in (a) can therefore be replaced by a single 14.25-Ω resistor. ▮▮

▮ **EXAMPLE 24.7** Find the equivalent resistance for the combination shown in Figure 24.15(a) from *a* to *d*.

Solution Notice that the 4-Ω and 5-Ω resistors are connected by equipotential wires to points *a* and *c*. They are therefore in parallel. Their equivalent is

$$\frac{1}{R_1} = \frac{1}{5\ \Omega} + \frac{1}{4\ \Omega} \qquad \text{or} \qquad R_1 = 2.22\ \Omega$$

We replace them in part (b) of the figure.

The 7-Ω and 2.22-Ω resistors are in series. They can be replaced by a 9.22-Ω resistor. This resistor is in parallel with the remaining two. Therefore,

$$\frac{1}{R_{\text{eq}}} = \frac{1}{9.22\ \Omega} + \frac{1}{8\ \Omega} + \frac{1}{6\ \Omega} = \frac{0.400}{\Omega}$$

from which

$$R_{\text{eq}} = 2.50\ \Omega$$

▮▮

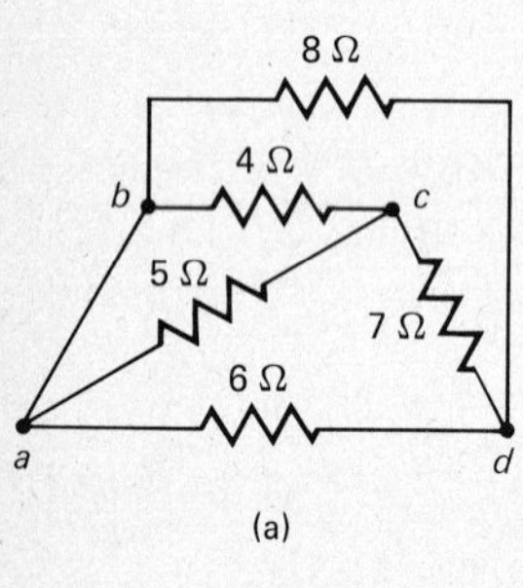

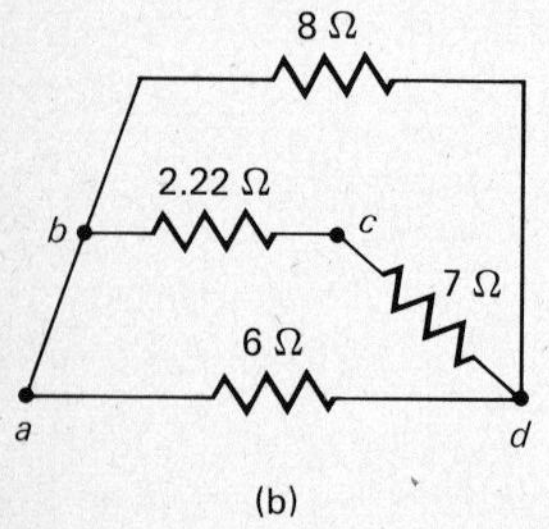

Figure 24.15 Find R_{eq} for the combination in (a).

24.12 Power in Steady Current Circuits

Most electric appliances and bulbs are rated in terms of their power requirements. To see what electric power is, we must recall the definition of power.

$$\text{Power} = \frac{\text{work done}}{\text{time taken}}$$

Let us now calculate the work done in electricity.

Suppose that a charge ΔQ is carried from one point to another. Suppose further that the magnitude of the potential difference between these two points is V. Then the work done in carrying the charge from the one point to the other is, from the definition of potential difference,

$$\text{Work} = (\Delta Q)V$$

If it took a time Δt to do this work, then the definition of power gives

$$\text{Power} = \frac{(\Delta Q)V}{\Delta t}$$

But in Equation 24.6, we defined the electric current to be

$$I = \frac{\Delta Q}{\Delta t}$$

Therefore, the power becomes

$$\text{Power} = VI \qquad (24.10)$$

If a positive charge (or the current) flows from high to low potential, this is a loss in power. That is what happens in resistors. The power loss gives rise to heat. Also, across a resistor, $V = IR$ and so the power loss there is I^2R.

The power lost as a current I flows through a potential drop V is VI. Because $V = IR$, the power loss in a resistor is I^2R.

EXAMPLE 24.8 A light bulb stamped 60 W/120 V means that the bulb will consume 60 W of power when the potential difference across its terminals is 120 V. Find the current that flows through the bulb when operated on 120 V. What is the bulb's resistance?

Solution The bulb is simply a resistor. Because $P = 60$ W and $V = 120$ V, we have from $P = VI$ that

$$I = \frac{60\text{ W}}{120\text{ V}} = \tfrac{1}{2}\text{ A}$$

To find the resistance of the bulb, we could use either $V = IR$ or $P = I^2R$. Using the power equation, we have

$$R = \frac{P}{I^2} = 240\ \Omega$$

We shall see in Section 24.14 that the resistance of a cold 60-W light bulb is much less than this. ▮▮

24.13 Resistivity

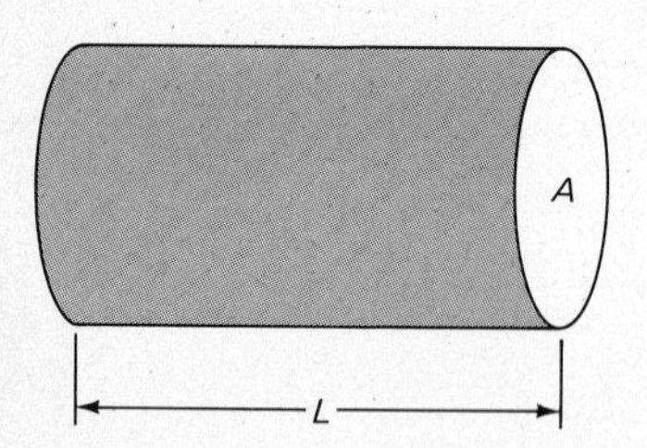

Figure 24.16 A uniform current is sent lengthwise through the cylinder. Its resistance is $R = V/I$. Knowing R, the resistivity ρ can be found from $R = \rho L/A$.

It is often necessary to state the amount of resistance a material shows to charge flow. For this purpose we define the resistivity of a material. It is represented by the symbol ρ (Greek letter rho). We can most easily define resistivity in terms of an experiment.

Suppose that the material in question is formed into a solid cylinder such as the one shown in Figure 24.16. (This might even be a wire.) When a potential difference V is maintained between its two ends, a current I flows through the cylinder. Ohm's law then tells us that the resistance of the cylinder is $R = V/I$. Therefore, R is easily determined for the cylinder. We need only measure I through it for a given potential drop across it.

We might suspect that R should increase as the length L of the cylinder is increased. Further, R should decrease as the cross-sectional area A increases. This is confirmed by experiment. One finds that

$$R = (\text{constant})\frac{L}{A}$$

The constant depends only on the material of the cylinder. For a good conductor, the constant is small. For a poor conductor, the constant is large.

We call the constant in this equation the resistivity of the material, ρ. We then have

$$R = \frac{\rho L}{A} \tag{24.11}$$

The units of ρ are ohm-meters ($\Omega \cdot \text{m}$) in the SI system. Equation 24.11 is not restricted to a cylinder, of course. It applies equally well to any solid that has a well-defined length and cross section. Typical values for ρ are given in Table 24.1.

Sometimes use is made of the electrical conductivity, σ, of a material. It is defined by the relation $\sigma = 1/\rho$. We do not list its values because they are readily obtained from ρ.

EXAMPLE 24.9 Find the resistance of 100 m of number 10 copper wire at 20°C. According to the American Wire Gauge scale, number 10 wire has a diameter of 2.59 mm.

Solution The resistivity of copper is $1.7 \times 10^{-8}\ \Omega \cdot \text{m}$. In the present case the wire has a cross-sectional area

$$A = \frac{\pi d^2}{4} = \frac{3.14(2.59 \times 10^{-3})^2}{4}\ \text{m}^2$$
$$= 5.27 \times 10^{-6}\ \text{m}^2$$

Using Equation 24.11, we find that

$$R = \frac{(1.7 \times 10^{-8}\ \Omega \cdot \mathrm{m})(100\ \mathrm{m})}{5.27 \times 10^{-6}\ \mathrm{m}^2} = 0.32\ \Omega$$

Notice the very low resistance of such a long piece of wire. It is for this reason that we can usually ignore the *IR* voltage drop along connecting wires. ▮▮

24.14 Variation of Resistance with Temperature

The resistance of materials changes with temperature. The resistance of metals usually increases with temperature. The resistance of semiconductors and insulators usually decreases with temperature.

Over a limited temperature range, the resistance R usually varies linearly with temperature t. That is

$$\frac{R - R_s}{R_s} = \alpha_s(t - t_s) \quad \text{or} \quad R = R_s + R_s\alpha_s(t - t_s) \quad (24.12)$$

In this expression t_s is a standard temperature. It is the temperature at which the resistor has a value R_s. The temperature coefficient of resistance, α_s, must also be appropriate to this temperature. In Table 24.1 we give α_s values for many materials. The reference temperature used is 20°C.

Because R is proportional to the resistivity ρ, we can write an equation similar to Equation 24.12. It is

$$\rho = \rho_s + \rho_s\alpha_s(t - t_s) \quad 24.12(a)$$

In it ρ is the resistivity at temperature t and ρ_s corresponds to t_s.

Table 24.1 ELECTRICAL RESISTIVITIES AND TEMPERATURE COEFFICIENTS AT 20°C

Substance	ρ ($\Omega \cdot$m)	α (per °C)
Silver	1.6×10^{-8}	3.8×10^{-3}
Copper	1.7×10^{-8}	3.9×10^{-3}
Aluminum	2.8×10^{-8}	3.9×10^{-3}
Tungsten	5.6×10^{-8}	4.5×10^{-3}
Iron	9.5×10^{-8}	5.0×10^{-3}
Mercury	95×10^{-8}	-0.9×10^{-3}
Graphite (carbon)	3500×10^{-8}	-0.5×10^{-3}
Ivory	2×10^{10}	—
Marble	10^{11}–10^{13}	—
Glass	10^{14}–10^{18}	—
Polyethylene	$>10^{17}$	—
Polystyrene	10^{19}–10^{23}	—

■ EXAMPLE 24.10 The white-hot tungsten filament of an incandescent lamp may have a temperature of about 1800°C. We found in Example 24.8 that the resistance of a 60-W/120-V bulb is about 240 Ω when operating. Find the resistance of the bulb at 20°C.

Solution We cannot give an exact solution because the temperature coefficient of resistivity equation does not apply over such a large temperature range. However, our result will be qualitatively correct at least. We are told that at $t = 1800°\text{C}$, $R = 240\ \Omega$. We know $\alpha_{20} = 4.5 \times 10^{-3}$ per °C for tungsten. Therefore, from Equation 24.12 we have

$$R = R_{20} + R_{20}\alpha_{20}(t - t_{20})$$

This becomes

$$240\ \Omega = R_{20} + R_{20}(4.5 \times 10^{-3}\ °\text{C})(1800 - 20°\text{C})$$

Solving gives $R_{20} = 27\ \Omega$. The bulb's resistance increases nearly tenfold as it heats white-hot. ■■

Summary

The devices that furnish electrical energy are called sources of emf. They are of great variety, although batteries and electric generators are most common. All emf sources provide at least two terminals. A potential difference exists between them. When charge flows out one terminal, an equal charge flows into the other.

Capacitors are devices that store charge. Often they consist of two parallel metal plates. The charge held by a capacitor for each volt potential difference between its terminals is called the capacitance, C, of the capacitor. In symbols, $C = Q/V$. The units of C are farads, F. Most capacitors have C values of microfarads or smaller.

The capacitance of a parallel-plate, air-filled capacitor is given by $C = \epsilon_0 A/d$. The plate separation is d and the area of either plate is A.

In a parallel-plate capacitor the electric field is uniform except near the edges. If the plate separation is d and the potential difference between the plates is V, then the electric field $E = V/d$. This relation is true only when the field is uniform. Two equivalent units for E are the volt per meter and the newton per coulomb.

Equipotential lines, surfaces, and volumes are regions in which all points are at the same potential. Electric field lines always strike perpendicular to equipotential lines and surfaces. Under electrostatic conditions, solid metal is always an equipotential volume.

The electrostatic field is a conservative field. The work done in moving a test charge from one point to another is independent of the path.

An electric circuit is a closed path along which charges can move.

If a charge ΔQ passes a point in a circuit in time Δt, then the magnitude of the current $I = \Delta Q/\Delta t$ amperes (A). The direction of the current is taken the same as the direction of positive charge flow. Or if negative charges are flowing, it is taken opposite to their direction of flow.

When charge flows through a resistor, electric potential energy is changed to heat. The resistance R of a resistor can be defined by the equation $R = V/I$. In this expression V is the potential difference across the resistor which causes a current I in it. The unit of R is called the ohm. Often the relation $V = IR$ is called Ohm's law. However, Ohm also implied that R is always constant. This is sometimes not true.

The connecting wires of a circuit are considered equipotentials if the IR drop (and accompanying heat developed) in them is negligible.

Ammeters measure I; they should have nearly zero resistance. Voltmeters measure V; they should have nearly infinite resistance.

Current always flows through a resistor from its high-potential end to its low end.

Several resistors in series are equivalent to a single resistor equal to the sum of the resistors. Several resistors in parallel are equivalent to a single resistor whose reciprocal is equal to the sum of the reciprocals of the several resistors.

The power lost as a current I flows through a potential drop V is VI. Because $V = IR$, the power loss in a resistor is I^2R.

The resistivity ρ of a material can be defined in terms of the resistance R of a wire whose length is L and whose cross-sectional area is A. Then $R = \rho L/A$. The units of ρ are ohm-meters.

Resistivity and resistance for metals usually increases with temperature. For insulators and semiconductors it usually decreases. Over a limited range, the variation of resistance with temperature can be represented by Equation 24.12.

Questions and Exercises

1. Potential differences exist between various parts of your body. Electrocardiograms, for example, measure them. Does this mean that the human body is an emf source?
2. Justify the following statement: Any two metal objects (not touching each other) attached by wires to the two terminals of a battery will act like a capacitor. What factors influence the capacitance of such a capacitor?
3. Suppose that you are given a complex circuit diagram showing batteries, resistors, and current directions. By looking at the diagram, how can you decide which is the high-potential (+) end of a certain battery? Of a resistor?
4. Explain this statement: Current is never lost in a circuit. What *is* lost as current flows in a circuit that contains a battery and resistances?
5. Give the physical justification for the following statement: Current is never used up in a resistor; the same amount of current flows out of a resistor as flows into it.
6. An electric field line cannot begin and end on the same conductor under electrostatic conditions. Why not?
7. There are two points in a given region of space, A and B. If $E = 0$ in this region, the potential difference between A and B is zero. But the potential difference between A and B can be zero even though $E \neq 0$. Explain.
8. What will happen if the two terminals of an ammeter are, by mistake, connected to the two terminals of a battery? Of a resistor through which current is flowing?
9. What will happen if a voltmeter is, by mistake, used in place of an ammeter in a simple series circuit consisting of a battery plus a resistance?
10. Justify the following statements: The value of R_{eq} for series resistors is always larger than the largest of the resistors; the value of R_{eq} for parallel resistors is always smaller than the smallest of the resistors. Do not do this by formula. Sketch a typical situation and use physical reasoning.
11. A large current flows at the first instant when an incandescent light bulb is turned on. Within a fraction of a second, the current becomes much smaller and remains small. Explain.

12. An ohmmeter consists basically of a battery and current-registering meter in series. The meter scale is calibrated to read the resistance placed across the terminals of the meter. Use an ohmmeter to measure the resistance through your body from hand to hand. Why does the value you obtain depend on how strongly you grasp the meter wires and how moist your hands are? The ohmmeter battery has low enough voltage so you will not feel a shock. *Do not attempt this with other sources of emf until you have studied the safety section in the next chapter!*

Problems

1. A potential difference of 25 V exists across a 0.75-μF capacitor. How large is the charge on the capacitor?
2. When a 0.35-μF capacitor is connected across a 12-V battery, how large a charge appears on it?
3. A cylindrical capacitor consists of two concentric metal pipes. When the potential difference between the two cylindrical plates is 90 V, the charge on either plate is 4.3×10^{-8} C. What is the capacitance of this capacitor?
4. A certain device consists of a metal sphere suspended above a flat metal plate. When the sphere and plate are connected to the two terminals of a 45-V battery, 5.0×10^{-10} C of charge flows onto the sphere. The charge on this "capacitor" is therefore 5×10^{-10} C. Find the capacitance of this device.
5. What must be the capacitance of a device that is to hold a charge of 2 μC when 1000 V is impressed across it?
6. Two metal plates are parallel and separated by a gap of 0.50 mm. What must be the area of each plate if the device is to have a capacitance of 2.0 μF?
7. In order to measure the motion of the end of a metal rod, its flat end is used as one plate of a parallel-plate capacitor. The other plate is a distance of 0.50 mm from it. The diameter of the plate area is 1.25 cm. Find the capacitance of this device.

*8. A spherical capacitor consists of two concentric metal spheres. Suppose there is a 0.5-mm gap between the outer surface of one sphere and the inner surface of the other. The outer radius of the inner sphere is 10 cm. Find the approximate capacitance of this capacitor.

*9. A cylindrical capacitor consists of two concentric metal cylinders. The cylinders are 20 cm long, they have a 0.10-mm gap between them, and the outer radius of the inner cylinder is 3 cm. What is the approximate capacitance of this capacitor?

10. Three capacitors (2 μF, 3 μF, and 4 μF) are placed in parallel. (a) What is their combined capacitance? (b) Repeat for the same capacitors placed in series.
11. What is the equivalent capacitance of 2-μF, 7-μF, and 14-μF capacitors when they are connected in (a) series and (b) parallel?

*12. How many different capacitors can you make from the following: 2 μF, 6 μF, and 12 μF? What are their values?

**13. The three capacitors given in Problem 12 are connected in parallel across a 12-V battery. (a) What is the charge on each capacitor? (b) Repeat for the capacitors in series.

14. Two large parallel metal plates are 0.135 cm apart. One is connected to the positive side of a 45-V battery while the other is connected to the negative side. (a) What is the electric field between the plates? (b) With the battery still connected, the plates are moved to a separation of 0.20 cm. What is the electric field between them now?
15. In a TV tube the beam shoots between two charged, closely spaced, parallel metal plates. The electric field between the plates deflects the beam across the TV screen. If the potential difference between the plates is 140 V, how large is E between them? The gap between the plates is 1.50 mm.
16. Sparking occurs in air when the electric field intensity exceeds about 3×10^6 V/m. Two parallel plates in air are 2.0 mm apart. How large a potential difference will cause an electric field of 3×10^6 V/m between them?

*17. The electric field between two parallel metal plates is 3×10^4 V/m. Consider two equipotential planes between the plates. We wish the potential difference between them to be 10 V. How far apart are these two equipotential planes?

*18. If a potential difference of 1.5 V is applied to the spherical capacitor described in Problem 8, what is the approximate electric field between the plates? Why is the answer only approximate?

*19. If a 12-V battery is connected to the plates of the cylindrical capacitor described in Problem 9, what is the approximate electric field between its plates? Why is the answer only approximate?

**20. Two square metal plates are opposite each other and are 2.00 cm apart. They are 25 cm on each side. A 6.0-V battery is connected to them so as to cause a uniform field between them. An electron moving at a speed of 3.0×10^7 m/s is shot between them along a straight line (initially) that is parallel to one edge of the plates. How far has the electron been

*Problems marked with an asterisk are not as easy as the unmarked ones.
**Problems marked with a double asterisk are somewhat more difficult than the average.

deflected from the straight line when it emerges from the region between the plates?

*21. From the fact that a volt is a joule per coulomb, prove that the units for E, the volt per meter and the newton per coulomb, are identical.

22. A certain wire has a current of 8 A flowing through it. How much charge flows through it in (a) 1 s and (b) 1 min?

23. The current through a certain lighted light bulb is 0.50 A. How much charge flows through the bulb each second? How many electrons flow through the bulb each second?

24. A 1600-W space heater has a current of 13.3 A flowing through it. How much charge flows through it each minute? How many electrons flow through it in one minute?

*25. A spark may jump from your hand to a metal object after static electricity has built up on your body. The charge on your body may be about 1×10^{-9} C. The current from your hand to the doorknob may last about 1×10^{-7} s. Assuming these values to be correct, find the average current that flows.

26. When a current of 2 A flows through a 6-Ω resistor, what is the potential difference between the ends of the resistor?

27. A certain radio resistor is marked 1.73×10^5 Ω. How much current will flow through it when it is connected across a 1.57-V battery?

28. To measure the resistance of a flashlight bulb, a 1.57-V battery, the bulb, and an ammeter are connected in a series circuit. The ammeter reads 0.34 A. What is the resistance of the bulb?

29. A certain ammeter has a resistance of 0.0020 Ω. By mistake a student connects it directly across a 1.50-V dry cell. (He thought it was a voltmeter and was using it to measure the cell voltage.) How much current flows through the meter before it burns out? (In a case like this the wires in the meter would become so hot that they would melt.)

30. What is the equivalent resistance of 4-Ω, 5-Ω, and 10-Ω resistors when they are all connected in (a) series and (b) parallel?

31. What is the equivalent resistance of 2-Ω, 6-Ω, and 12-Ω resistors when they are all connected in (a) series and (b) parallel?

32. What is the equivalent resistance of 3-Ω, 7-Ω, and 13-Ω resistors when they are all connected in (a) series and (b) parallel?

33. You are given two resistors. One is 3.0 Ω and the other is 5.0 Ω. By using them in different ways, you can obtain four different equivalent resistances. What are they?

*34. You are given three resistors. They have values of 3.0 Ω, 5.0 Ω, and 6.0 Ω. By using them in different ways, you can obtain 17 different equivalent resistances. What are they?

35. Find the equivalent resistance in Figure P24.1 between the points a and b.

36. Find the equivalent resistance in Figure P24.2 between the points a and b.

*37. Reduce the resistances between points a and b in Figure P24.3 to a single equivalent resistance.

**38. Reduce the resistors between a and b in Figure P24.4 to a single equivalent resistor.

*39. In Figure P24.1, a 12-V battery is connected from a to b. Find the current that flows in (a) the 3-Ω resistor and (b) the 4-Ω resistor.

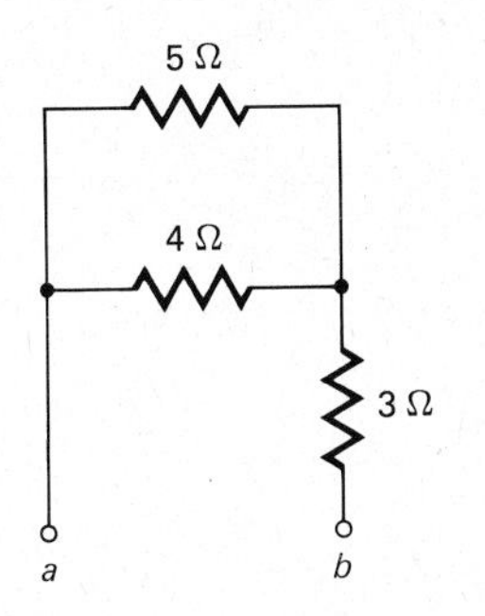

Figure P24.1

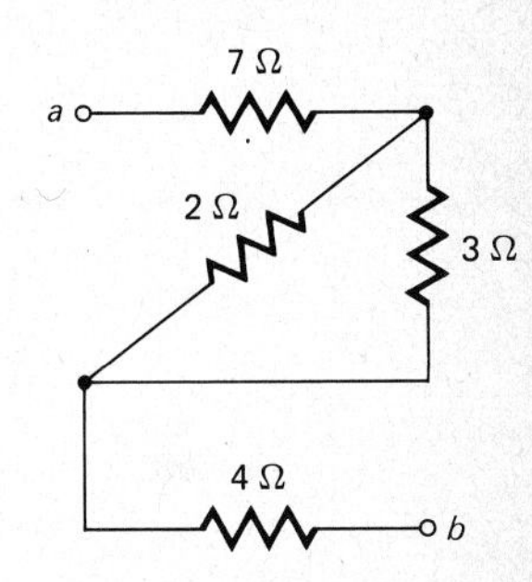

Figure P24.2

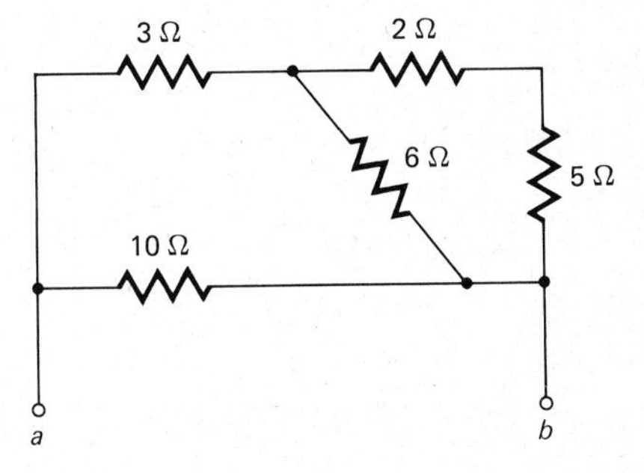

Figure P24.3

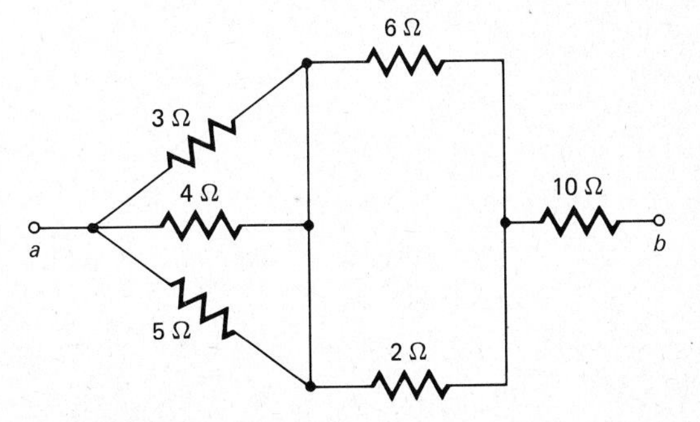

Figure P24.4

*40. When a certain battery is connected from a to b in Figure P24.1, the current through the 3-Ω resistor is 2.0 A. What is the potential difference of the battery?

**41. When a 20-V battery is connected from a to b in Figure P24.3, what is the current in each (a) the 10-Ω resistor and (b) the 2-Ω resistor.

42. How much current passes through a 40-W bulb when it is operating at its normal voltage, 120 V?

43. How much power is lost in a 7-Ω resistor when a current of 3 A flows through it?

44. How much power is lost in a resistor when a 1.5-V battery sends a current of 0.20 A through it?

45. In order for a 60-Ω resistor to dissipate 5 W, how much current would have to flow through it?

46. A certain 1650-W space heater is designed to operate on 120 V. How much current flows through it when it is operating? What is its resistance when operating?

47. (a) An electric clothes drier draws a current of 7.5 A when operating on 220 V. How much power does it use? (b) If it was redesigned to operate with the same power on 120 V, how much current would it draw?

*48. A certain electric motor has a power output of $\frac{1}{4}$ hp. Assume it to be 100 percent efficient. What is its power consumption in watts? How much current does it draw from a 100-V line?

*49. A flash heater for heating coffee water delivers 10,000 cal in 1 min. It operates on 120 V. (a) How much power is consumed by this heater? (b) How much current does it draw? (c) What is its resistance?

50. Number 18 wire has a diameter of 1.02 mm. What is the resistance (near 20°C) of 50 m of this size aluminum wire?

51. Number 10 Nichrome heater wire has a resistance of 0.213 Ω per meter length. Find the resistivity of Nichrome. (Number 10 wire has a diameter of 2.59 mm.)

52. A coil of copper wire has a resistance of 0.500 Ω at 20°C. What is its resistance at 40°C? At 10°C?

53. If the values given in Table 24.1 could be used over such a large range, at what absolute temperature would copper have zero resistivity? This does not happen in practice.

*54. A glass tube is 50 cm long and has an inner diameter of 0.40 cm. What would be the resistance at 20°C of a column of mercury that fills the tube? By what percentage would its resistance change for each Celsius degree of temperature change?

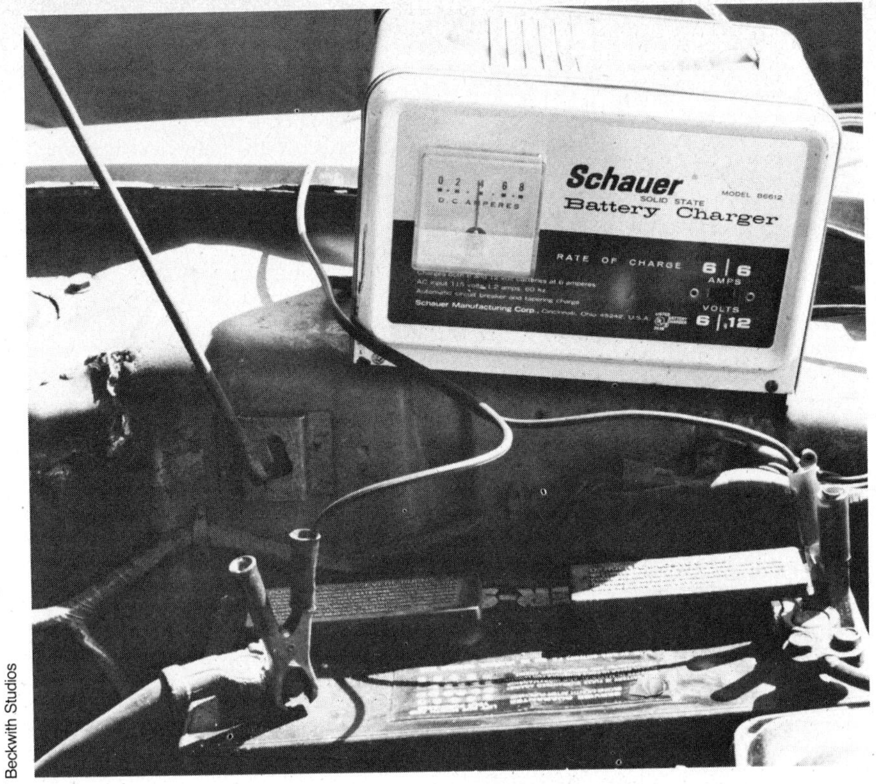

Beckwith Studios

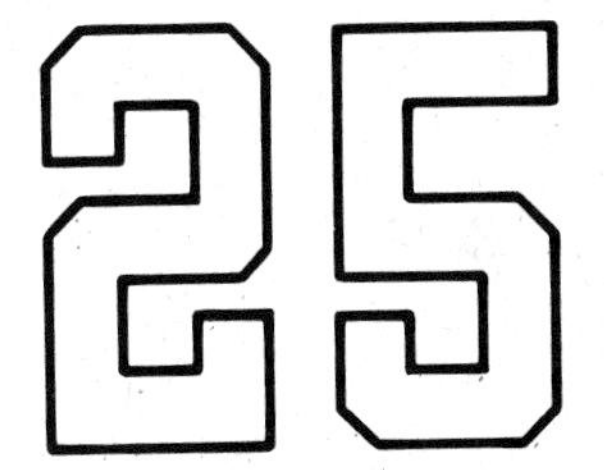

DIRECT CURRENT CIRCUITS

Performance Goals

When you finish this chapter, you should be able to

1. Assign currents to the wires of a given circuit and provide labels and assumed directions for them. Using these currents, write the point rule for any point in the circuit.
2. Assign currents to the wires in a given circuit and, using these assumed currents, state which is the high-potential end of each resistor.

In a direct current (dc) circuit the current flows steadily without reversing its direction of flow. This type of current flows from batteries. Direct current circuits are widely used in electronics and in many practical devices. We shall learn how to deal with dc circuits in this chapter.

Two rules, called Kirchhoff's rules, are presented as a logical method for analyzing all dc circuits. Later we shall see that they also apply to circuits with varying current.

In this chapter we shall learn about house circuits and aspects of safety connected with them. Other features and types of practical electric circuits will also be studied.

3. Write the independent loop equations for a given circuit.
4. Use Kirchhoff's rules to find the unknown features of a circuit.
5. In your own words, state the reasons behind the loop rule. Refer to the concept of conservative field in your explanation.
6. Draw a typical house circuit. Compute the current drawn by a bulb or heater of known power when connected to the circuit. Knowing the power of each appliance connected to the circuit, compute the current at each point in the circuit. Also point out the location and function of the fuse or circuit breaker.
7. Analyze a simple situation involving electrical safety. Point out actions that would be dangerous and explain why the danger exists.
8. Analyze the following situation. Three of the following are known for a battery or other emf source: emf, r, I, terminal p.d. Find the unknown quantity when the source is (a) connected across a known resistor and (b) connected across a known battery and is charging or discharging.
9. Sketch the circuit for a potentiometer measuring an unknown emf. Explain why the potentiometer is capable of measuring the emf.
10. Compute how much material plates out in a given length of time when the following are given: the plating current for a known plating solution, the ions that plate out, the ions' charge and atomic mass.

25.1 Kirchhoff's Point Rule

We saw in the last chapter that Ohm's law is useful in dealing with electric circuits. When a circuit contains resistors in series and parallel, we know how to simplify the resistances. Often circuits can be understood with these simple approaches. But there are other simple-looking circuits that cannot be understood using these methods alone. We shall see examples of such circuits as we proceed.

What we need is a general method that can be used on any circuit. Such a method exists. It is based on two rules, which are called Kirchhoff's rules. In this section we shall investigate one of these rules. It is called the point rule. We have already made use of it in the last chapter without naming it.

To see what the point rule is, let us refer to Figure 25.1. This is similar to a portion of Figure 24.13. A current I flows into point A of a circuit. We know that I is the number of coulombs of charge flowing into the point each second. Of course, this charge cannot be stored in the wire at A. As much charge must flow out each second as flows in. The charge flowing out each second is the sum $I_1 + I_2 + I_3$. Therefore, we see that

$$I = I_1 + I_2 + I_3$$

The same general reasoning can be used in any similar situation. As much current must flow out of a point in a circuit as flows in. This fact is summarized in *Kirchhoff's point rule:*

The sum of all the currents coming into a point must equal the sum of all the currents leaving the point.

We shall see that this simple rule is extremely useful in circuit analysis.

25.2 Kirchhoff's Loop (or Circuit) Rule

Kirchhoff's loop rule is based on the fact that an unchanging electric field is a conservative field. Let us review what this means. It means that the work done in carrying a test charge from A to B is independent of the path taken from A to B.

This has very important consequences if A and B coincide. Then, of course, no work is needed to carry the test charge from A to B. But if you refer to Figure 25.2, you will see a path that also leads from A to B. How much work is required to carry the test charge from A to B along this path? Because we are assuming a conservative field, the work done along any path between A and B is the same. It is zero in this case. We can therefore state that

The work done in carrying a unit positive test charge around a closed loop in an unchanging electric field is zero.

This has important meaning for circuits, as we shall soon see.

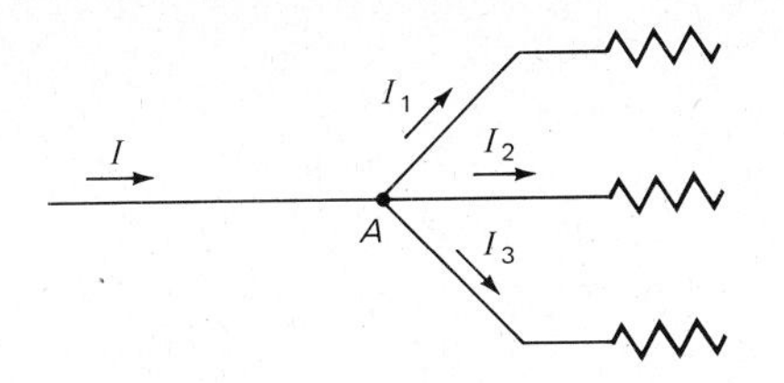

Figure 25.1 Kirchoff's point rule tells us that $I = I_1 + I_2 + I_3$.

Consider the circuit shown in Figure 25.3. The current flows in the direction shown. Because the current is steady, the electric field near the circuit is not changing. We are therefore concerned with an electrostatic field (i.e., one in which E does not vary with time). The conclusion we stated above about work around a closed path therefore applies. Before we see what it means for this circuit, let us review a few facts.

1. The potential difference between two points is the work done in carrying a unit positive test charge from the one to the other.
2. If the unit positive test charge is carried through a potential rise V, the work done is $+V$. If it is carried through a potential drop, the work done is $-V$.
3. Current flows from the high- to the low-potential end through a resistor. Going in the direction of the current, a potential drop of magnitude IR occurs in a resistor.
4. A potential rise of magnitude $\mathscr{E}$ occurs in going from the negative to positive terminal of an emf source.

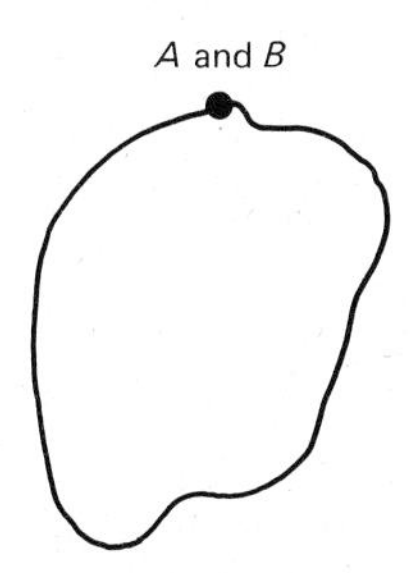

Figure 25.2 The work done in carrying a test charge around a closed path is zero if the field is not changing.

Now we are prepared to compute the work done as we carry a unit positive test charge through each portion of the circuit in Figure 25.3.

From items 1 and 2 we notice that work done on a unit positive test charge is equal to the potential rise through which we carry it. A potential rise results in positive work. A potential drop results in negative work. For example, suppose that we carry the test charge from 1 to 2 in Figure 25.3. We are carrying it through a potential drop IR_1 according to item 3. Therefore the work done is $-IR_1$. Or suppose that we carry the test charge from 8 to 9 through the emf $\mathscr{E}_2$. According to item 4, this is a potential rise of $\mathscr{E}_2$. The work done is therefore $+\mathscr{E}_2$. We can conclude the following from this discussion.

The work done in carrying a unit positive test charge around a closed dc circuit is zero. It is equal to the algebraic sum of the potential differences through which it is carried. Potential rises are taken positive. Potential drops are taken negative.

This statement leads at once to *Kirchhoff's loop* (or *circuit*) *rule:*

The algebraic sum of the potential changes around a closed dc circuit is zero.

In the next section we shall see how to use this rule.

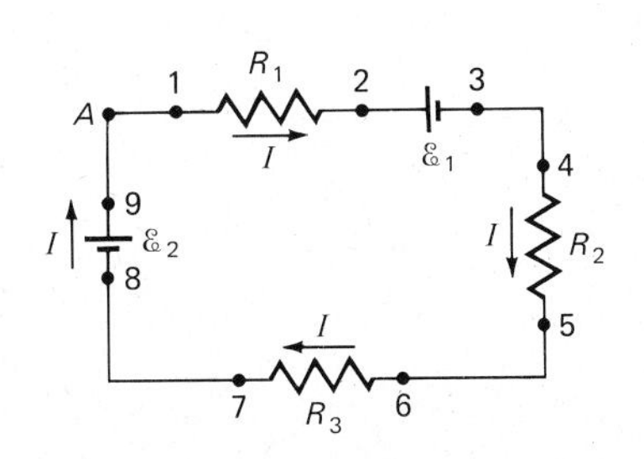

Figure 25.3 What does Kirchoff's loop rule tell us about this circuit?

25.3 Use of the Loop Rule

Consider the simple circuit shown in Figure 25.4. Our previous experience suggests to us that the current in it can be found using Ohm's law. We might guess that the two batteries are equivalent to a single 10-V battery. (Notice that

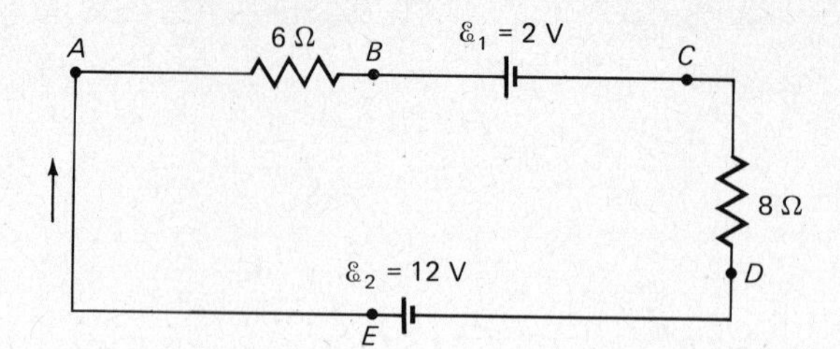

Figure 25.4 Because the 12-V battery has the larger emf, we expect the current to flow in the direction shown.

$\mathscr{E}_1$ wants the current to go counterclockwise around the circuit, but $\mathscr{E}_2$ wants it to go clockwise. We say that the batteries oppose each other.) In addition, the current should flow in the direction shown. Our guess would be that the 6-Ω and 8-Ω resistors act like a 14-Ω resistor. If all these guesses are correct, then Ohm's law tells us that the current in the circuit is 10 V/14 Ω = 0.71 A.

It turns out that this result is correct. But let us obtain it using the loop rule. To do so, we start at *A* and go around the circuit through *BCDE* and back to *A*. The loop rule tells us that the algebraic sum of the voltage changes should be zero as we go around the loop. Let us take them each in turn.

A → *B:* This is a voltage drop of $IR = 6I$ V. It is a drop because we are going through the resistor in the direction of the current.
B → *C:* This is a drop of 2 V because we are going from the positive to the negative side of the emf source.
C → *D:* This is a drop of $8I$ volts. It is a drop because we are going through the resistor in the direction of the current.
D → *E:* This is a rise of 12 V because we are going from the negative to the positive side of the battery.
E → *A:* There is no change in potential because the wire is an equipotential.

According to the loop rule, the algebraic sum of all these changes should be zero. Drops are to be taken negative. Rises are to be taken positive. We therefore have (all terms in volts)

$$-6I - 2 - 8I + 12 = 0$$

Solving for *I*, we have

$$-14I + 10 = 0$$
$$-14I = -10$$
$$I = 0.71 \text{ A}$$

This is the same result we obtained by the other method.

Before leaving the circuit, let us investigate two other features of the loop method. Suppose that we had gone around the circuit in the opposite direction. Then the situation would be as follows. (Be sure you understand the reason for the sign used.)

$A \to E$: change = 0
$E \to D$: change = −12 V
$D \to C$: change = +8I
$C \to B$: change = +2 V
$B \to A$: change = +6I

The loop rule is simply the sum of these changes equated to zero. Then, in volts,

$$-12 + 8I + 2 + 6I = 0$$

This gives the same value for I as before. We see that the loop rule applies no matter which direction we go around the loop.

As a final point, suppose that we are poor guessers. Then we might have guessed that I was in a direction opposite to that shown in Figure 25.4. Our voltage changes then would have been (check them and see)

$A \to B$: change = +6I V
$B \to C$: change = −2 V
$C \to D$: change = +8I V
$D \to E$: change = +12 V

The loop equation would then be (in volts)

$$6I - 2 + 8I + 12 = 0$$

Solving for I, we would have

$$14I + 10 = 0$$
$$I = -0.71 \text{ A}$$

Notice that the answer tells us that we guessed the wrong direction for I. The negative sign always results if the current direction is wrong. This is one of the beauties of Kirchhoff's rules. Even a poor guesser can use them successfully.

We shall now give several examples of the use of Kirchhoff's rules. Follow them through carefully. Be absolutely sure you understand the sign of each term in each equation. Signs are fundamental in the use of these rules.

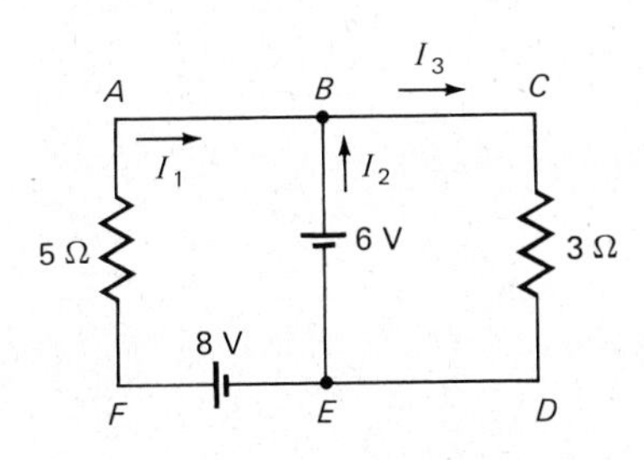

Figure 25.5 If the directions of I_1 and I_2 have been wrongly chosen, their numerical values will be negative. Notice that the two resistors are not in simple series or parallel.

EXAMPLE 25.1 Find the currents I_1, I_2, and I_3 in the circuit of Figure 25.5.

Solution We have assigned a direction to each current in the figure. If the direction is wrong, our answer will tell us so. Notice that the current everywhere in wire $EFAB$ is I_1, in wire BE it is I_2, and in $BCDE$ it is I_3.

Let us first write the point rule for point B.

Currents in = currents out

or

$$I_1 + I_2 = I_3$$

This equation has three unknowns.

Next let us write the loop equation for loop *ABEFA*. The potential changes are

$A \rightarrow B$: 0
$B \rightarrow E$: -6 V
$E \rightarrow F$: $+8$ V
$F \rightarrow A$: $-5I_1$ V

Be sure you understand the signs used. The sum of these changes must be zero.

$$-6 + 8 - 5I_1 = 0$$

This can be solved for I_1 to give

$$5I_1 = 2 \qquad \text{or} \qquad I_1 = 0.40 \text{ A}$$

Because I_1 is positive, its direction was chosen properly.

Now let us write the loop equation for loop *BCDEB*. The potential changes are

$B \rightarrow C$: 0
$C \rightarrow D$: $-3I_3$ V
$D \rightarrow E$: 0
$E \rightarrow B$: $+6$ V

Their sum must be zero.

$$-3I_3 + 6 = 0$$

From which

$$I_3 = 2.0 \text{ A}$$

The direction of I_3 was also chosen properly.

We can now return to equation (a) to find I_2. Substituting the values for I_3 and I_1, we find (in amperes) that

$$0.40 + I_2 = 2.0$$
$$I_2 = 1.60 \text{ A}$$

Our problem has been solved. ■■

You might wonder what would have happened if we had used loop *ABCDEFA*. We would have obtained a true equation. But it contains nothing new. The following rule applies.

A loop equation containing only potential changes already used in other loop equations is not a new equation. It can be obtained directly by algebra from the other loop equations.

For this reason a loop equation will not usually be useful unless it contains a potential change not used previously.

EXAMPLE 25.2 Refer to Figure 25.6. Suppose that originally no currents were labeled and all arrows were absent. The problem is to find the current in each wire.

Solution Our first task would be to label each wire with a current. We then assign a direction to each current. If the direction we assign is wrong, I will be found to be negative. Notice that I_1 flows through wire *BAF*. Also, I_4 and I_5 (or I_3) should not be considered to be the same.

Let us first write the point equation. At point B we have

$$\text{Current in} = \text{current out}$$
$$0 = I_1 + I_2 + I_3 \qquad \text{(a)}$$

This tells us at once that one of these currents has been assigned the wrong direction. But that does no harm.

For point E we have

$$I_2 + I_4 = I_5 \qquad \text{(b)}$$

At point F we see that

$$I_1 = I_4 + I_6 \qquad \text{(c)}$$

And at point D we see that

$$I_5 + I_6 + I_3 = 0 \qquad \text{(d)}$$

All these equations are not independent.

To write a loop equation, we note that loop *BEDCB* will give an equation with only one unknown. You should be able to show that the loop equation is (in volts)

$$+5 + 9I_3 - 8 = 0$$

From this, we find that

$$I_3 = \tfrac{1}{3}\text{ A}$$

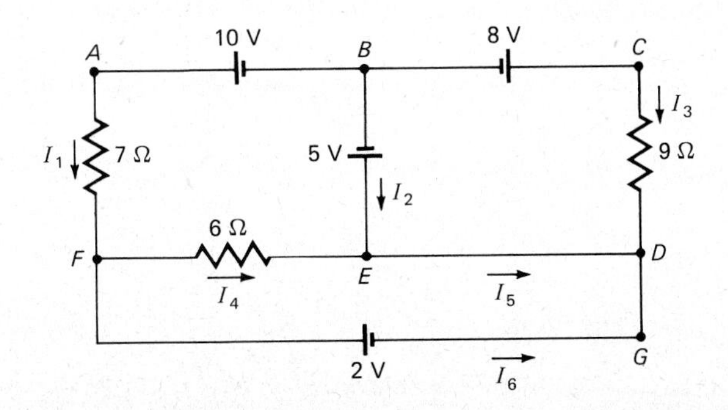

Figure 25.6 Find all six currents.

Let us now write the loop equation for loop *AFGDCBA*. It also is a nice loop to use, because I_3 is already known. We have, in volts,

$$-7I_1 - 2 + 9I_3 - 8 + 10 = 0$$

Since I_3 was found to be $\frac{1}{3}$ A, this equation becomes

$$-7I_1 + 3 = 0$$

And so we find that

$$I_1 = \tfrac{3}{7}\text{ A} = 0.429\text{ A}$$

Now let us follow loop *AFEBA*. This contains only one new unknown, I_4. Then the loop equation is, in volts,

$$-7I_1 - 6I_4 - 5 + 10 = 0$$

Since $I_1 = \frac{3}{7}$ A,

$$-3 - 6I_4 + 5 = 0$$

Solving gives

$$I_4 = \tfrac{1}{3}\text{ A} = 0.333\text{ A}$$

We have now covered all the wires except *ED*. But this wire contains no resistance or battery. It will therefore contribute no new term to our equations. No other independent loop equation is possible.

To proceed further, we return to equations (a), (b), (c), and (d). Because we know I_1 and I_3, we can use (a) to find I_2. We have, in amperes,

$$0 = 0.429 + I_2 + 0.333$$

From this,

$$I_2 = -0.762\text{ A}$$

Notice the negative sign. I_2 actually flows in a direction opposite to that shown.

Now, substituting in equation (b) gives

$$-0.762 + 0.333 = I_5$$

Notice that the negative sign is a part of I_2 and must be carried along. Solving for I_5 gives

$$I_5 = -0.43\text{ A}$$

It, too, flows in a direction opposite to that shown.

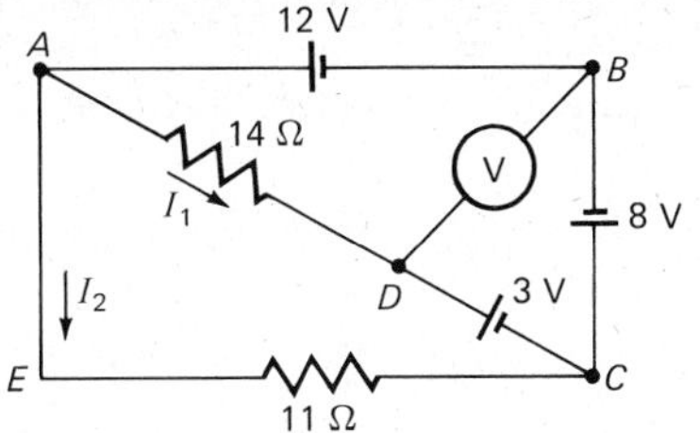

Figure 25.7 What does the voltmeter read?

We shall need I_6. It can be found from either equation (c) or (d). Each gives the result

$$I_6 = 0.096 \text{ A}$$

All the currents have now been found. ∎∎

∎ **EXAMPLE 25.3** In Figure 25.7 find the voltmeter reading and I_1 and I_2. Assume the voltmeter to be ideal.

Solution An ideal voltmeter has infinite resistance. No current flows through it. It has no effect on the circuit and can be ignored.

The voltmeter reads the potential difference from B to D. It is easy to find the potential difference in this case. To do so, start at B and go through C to D. Keep track of the potential rises and drops. Going from B to C, we have a rise of 8 V. And from C to D we have a further rise of 3 V. Therefore, point D is

$$8 \text{ V} + 3 \text{ V} = 11 \text{ V}$$

higher in potential than point B. The voltmeter will therefore read 11 V.

To find I_1, let us write the loop equation for $ADCBA$. It is, in volts,

$$-14I_1 - 3 - 8 + 12 = 0$$

From this,

$$I_1 = \tfrac{1}{14} \text{ A}$$

Similarly, for loop $AECBA$, we have

$$-11I_2 - 8 + 12 = 0$$

From this,

$$I_2 = \tfrac{4}{11} \text{ A}$$

We can check our value for the potential difference between B and D. Going by way of path BAD we have, in volts,

$$+12 - 14I_1$$

Placing in the value for I_1, this gives

$$+12 \text{ V} - 1 \text{ V} = 11 \text{ V}$$

This method also shows that point D is 11 V higher than point B. ∎∎

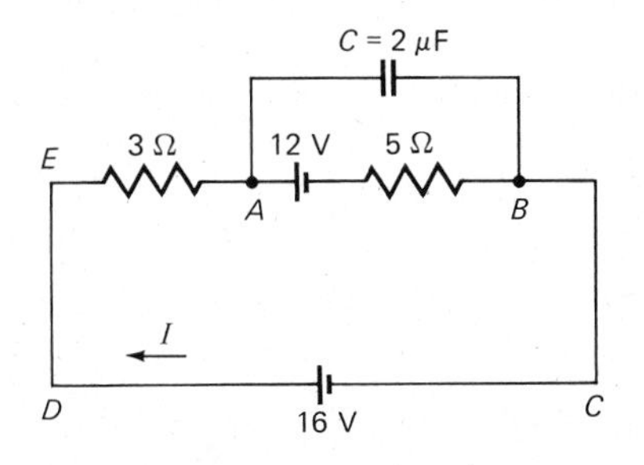

Figure 25.8 Under steady conditions how much current flows through the capacitor?

∎ **EXAMPLE 25.4** Find the charge on the capacitor in Figure 25.8.

Solution We shall see in a later chapter that it takes only a small fraction of a second for the capacitor to charge up. Once it is charged, no charge flows in the wires leading to it. The charged capacitor acts like an infinite resistance. It does not affect the circuit and so it can be ignored.

With that in mind, we know that the current I flows through $DEABCD$. Let us write the loop equation. Starting at D we have, in volts,

$$-3I - 12 - 5I + 16 = 0$$

Solving for I gives $I = 0.50$ A.

The potential across the capacitor is the same as the potential difference from A to B. (Its plates are connected to these two points by equipotential wires.) Going from A to B through the 12-V battery gives

$$-12\text{ V} - 5I\text{ V}$$

But $I = \frac{1}{2}$ A, and so this becomes

$$-12\text{ V} - 2.5\text{ V} = -14.5\text{ V}$$

In other words, point B is 14.5 V lower than point A. The potential difference across the capacitor is 14.5 V. The plate attached to A is the high (or positive) plate.

To find the charge on the capacitor, we recall from the last chapter that

$$C = \frac{Q}{V}$$

Therefore,

$$\begin{aligned} Q &= (2 \times 10^{-6}\text{ F})(14.5\text{ V}) \\ &= 2.9 \times 10^{-5}\text{ C} \end{aligned}$$

■■

25.4 House Circuits

The circuit of most concern to the average person is the house lighting circuit. It is a simple circuit, but because it is so widely used, it is of great importance. Let us consider a typical house and its electrical wiring.

Three wires are run to each house by the power company. One wire is grounded.[1] Its potential (or electrical level) is taken to be zero. One of the other two wires has a potential that is 110 V higher than ground. The third wire has a potential of 110 V lower than ground. [Actually these are alternating (ac) voltages. However, for our purposes they may be considered dc rather than ac. We shall discuss ac circuits in detail in Chapter 28.]

These three wires therefore furnish two different potential differences to the house. The potential difference between the ground wire and either of the other two wires is 110 V. But the potential difference between the two ungrounded wires is 110 + 110 or 220 V. As a result, there are two potential differences

[1] That is, one wire is connected to the moist earth by running a connecting wire from it to a water pipe (or metal post) buried in the earth.

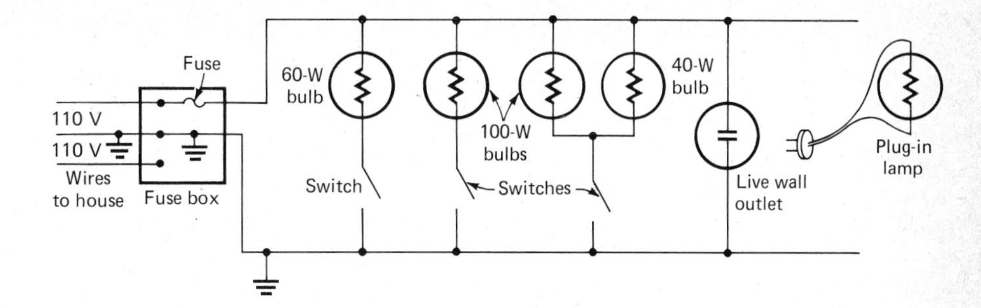

Figure 25.9 Most houses have many circuits such as this one. The fuse is often replaced by a circuit breaker.

available to a house: 110 V and 220 V. (These potential differences vary somewhat from place to place. They may range from 120 V to 100 V. In many parts of the world outside the United States, the two voltages are 220 V and 440 V.)

A schematic diagram of a house circuit is shown in Figure 25.9. The three wires coming to the house enter a central box (the circuit breaker or fuse box). A typical circuit, such as the one shown, services only a portion of the house. It consists of two wires. One is grounded (⏚) and the other is at a potential difference of 110 V from it. The second wire is called the high (or hot) wire.

Suppose that the 60-W bulb shown in the figure is a ceiling light controlled by a wall switch. One end of the bulb filament is connected to the high lead (or wire). The other is connected, through the switch, to the ground lead. When the switch is closed, current can flow from the high potential wire to the ground wire through the bulb. To light the bulb, one simply closes the switch. Then a potential difference of 110 V is placed across the bulb and it lights. Notice that the other lights (or appliances) connected to the line do not influence this bulb. This is always true when devices are connected in parallel as they are. (Can you explain that they are in parallel?)

A similar situation applies to the 100-W bulb on the left. It is also controlled by its own switch. But notice the 100-W and 40-W bulbs in combination, which are controlled by a single switch. When this switch is closed, the potential difference across each will be 110 V. (Why?) They go on and off together.

Finally, shown in Figure 25.9 is a live wall outlet. The potential difference between its two connectors is 110 V. What happens when the lamp shown is plugged in? Current can flow into the one prong of the plug, through the lamp bulb, and out the other prong. As a result, the lamp lights. Of course, most plug-in lamps have a switch built right into the lamp. How would we have to alter Figure 25.9 to show the lamp switch?

In Figure 25.10 we show a house circuit with many of its appliances operating. (We use 120 V for the line voltage so as to simplify the numbers.) To find the current being drawn from the 120-V line by any appliance, we make use of $P = VI$. For example, the 1200-W toaster draws a current given by

$$1200 \text{ W} = (120 \text{ V})I$$

which gives $I = 10$ A. Notice once again that, in a parallel system such as this, the appliances are independent of each other.

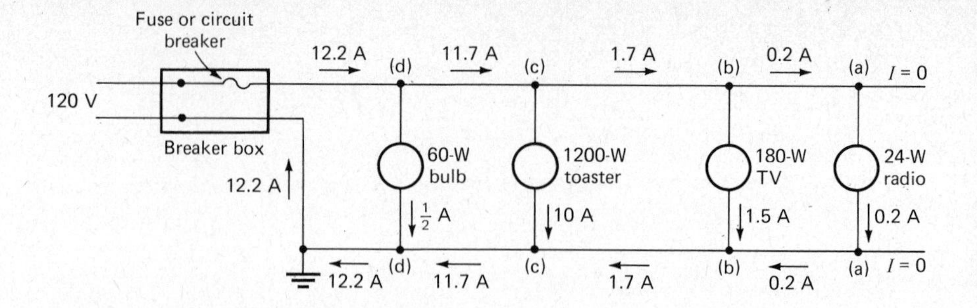

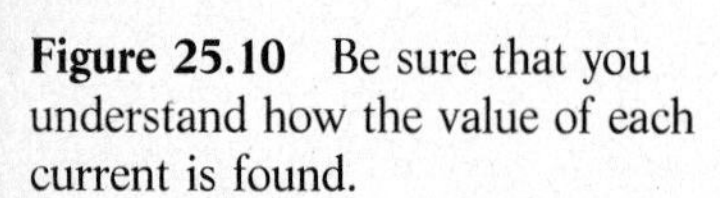

Figure 25.10 Be sure that you understand how the value of each current is found.

Let us now find the currents in the various wires. No current flows in the far-right portion of the circuit. (The ends of the wires act like plugs in a pipe.) You should be able to show that the current through the 24-W radio is 0.2 A. Then applying the point rule at either of the two points labeled (a) you will see that the current is 0.2 A in the upper and lower wires.

Going to points (b), we notice that the TV draws 1.5 A. Applying the point rule at either point (b) gives the current in the wires (c)–(b) to be 1.7 A. And so on, we can show the currents are as labeled. You should be able to find them in any circuit such as this.

Notice that the current flowing to this circuit from the breaker box is 12.2 A. This is quite a large current.[2] But most newer house wires can safely handle currents up to about 20 A. The wires in older houses are smaller in diameter and often are limited to currents below 15 A. Let us see why this limitation arises.

Usually we ignore the resistance of the connecting wires. Then the IR voltage drop along them is said to be zero because $R = 0$. But the wire really does have some resistance. For example, the resistance per meter length of house wire might be about 0.01 Ω per meter of length. When a current I flows through it, the power loss in it is I^2R. This power loss appears as heat. So the heat generated in a 1-m length of wire each second is I^2R, where R is about 0.01 Ω. For a current of 30 A this is about 9 J/s or about 2 cal of heat each second.

The wire has insulation on it, and this tends to hold the heat to the wire. As a result, the wire becomes hot. If it becomes hot enough, the insulation may melt or even burn. We see, then, that some way must be found to limit the current in the wires. As stated earlier, the maximum safe current in new houses is about 20 A.

To limit the current, either a fuse or circuit breaker is used. Notice where this device is placed in Figure 25.10. No wire will carry more current than flows through the fuse. If the current exceeds the safe current, the wire in the fuse melts. A gap then exists in the wire and no current can flow. We say that the fuse has "blown out." A circuit breaker serves the same purpose as a fuse. But it operates on a different principle. We shall discuss it in a later chapter.

Appliances such as electric stoves and clothes driers use a great deal of power. Because power = VI, we see that they require a large current. But the

[2] If we are just interested in this total current, it is easily found as follows. The appliances being run have a combined power of 60 + 1200 + 180 + 24 W = 1464 W. From $P = VI$, we see that $I = 1464\ \text{W}/120\ \text{V} = 12.2\ \text{A}$.

current is limited by the house wires. To avoid very high currents for such devices, they are often designed to operate on higher voltages. For example, a 3600-W drier operating on 120 V would require a current given by

$$3600\ \text{W} = (120\ \text{V})I \qquad \text{or} \qquad I = 30\ \text{A}$$

But if the drier is designed to operate on 240 V, then

$$I = \frac{3600\ \text{W}}{240\ \text{V}} = 15\ \text{A}$$

Therefore, using 240 V, there would be much less trouble with heating in the house wires. It is for this reason that devices that use a great deal of power are usually designed to operate on 240 V. What would happen if a 120-V appliance was, by mistake, connected to 240 V?

In most houses several circuits run from the central fuse box. Each circuit serves a different part of the house. Therefore, the current in each circuit need not be too large. If a fuse blows, only that one house circuit will be turned off. All of the others will still function normally.

25.5 Electrical Safety

Most electric appliances are designed in such a way that it is nearly impossible to use them unsafely. Even so, people sometimes manage to outsmart the designers and kill themselves. There are, of course, other ways in which accidents occur with electricity. Those who work with electricity as technicians must be on the alert constantly. In order to work safely with electricity, one should understand how electricity can cause damage to the body.

There are two major ways in which electricity damages the body. It can cause serious burns. It can also cause body cells to stop functioning properly. If the damaged cells are necessary to the operation of the heart or lungs, then these vital organs may stop working.

Even a small current through a muscle can cause the cells in it to undergo change. A person feels a shock when a current of about 0.001 A or larger flows through the body. At currents ten times larger, 0.01 A, the muscles of the hand can be paralyzed; the person is then unable to release the wire that is causing the shock. When a current of about 0.02 A or larger flows through the chest region, the muscles there become paralyzed. The person is then unable to breathe and, unless artificial respiration is given, the person will suffocate. Of course, the rescuer must first free the victim from the voltage source. A current of about 0.1 A through the heart region will shock the heart muscles into rapid, erratic contractions called ventricular fibrillation. Finally, currents of about 1 A through body tissue result in enough heating to cause serious burns.

As we see, the important quantity to control is current through the body cells. Voltage is important only because it can cause current to flow. The high-voltage

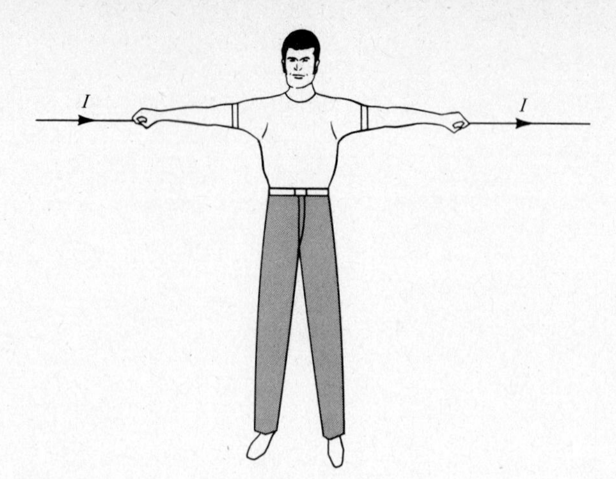

Figure 25.11 Some electronics technicians follow the adage, "Always keep one hand behind your back when probing a circuit." Why is that safer than the behavior shown?

shock you receive from electrostatic charge as you touch a metal door handle is very minor. A rather large current flows for a very short instant. But it lasts such an extremely short time that the cell damage is minor. The electrostatic charge on your body is very small.

In analyzing the safety of any situation, we must examine two features. First, the part of the body through which current goes is important. The person shown in Figure 25.11 is in great danger. The current flows in one hand and out the other. It flows through the chest region, and we have just seen that this is very dangerous. Paralysis of the chest muscles occurs at low currents. A much higher current could flow in through the hand and out the elbow above it without causing death.

Second, the part the body forms in the circuit is also important. In some situations the 120-V house circuit is almost certain to cause death. One of the two wires of the circuit is always grounded. It is therefore attached to the water pipes of the house. Suppose that a person is soaking in a bathtub. The bath water connects him well to the water pipes. If he accidentally touches the high-potential wire of the house circuit (as he might do by touching an exposed wire on an electrical heater or radio), current will flow through his whole body to the ground. Because of the large, efficient contact his body makes with the ground, the resistance of his body circuit is low. As a result, the current will be large and he is almost certain to be electrocuted.

Similar situations exist elsewhere. For example, if you accidentally touch an exposed wire while standing on the ground with wet feet, you are in far greater danger than if you are on a dry, insulating surface. The circuit through your body to the ground has a much higher resistance if your feet are dry. Similarly, if you are given a shock by touching a bare wire on a faulty appliance, the shock is greater if your other hand is touching the faucet on the sink or is in the dish-water.

As you can see from these examples, the danger from electrical shock can be eliminated by avoiding a current path through the body. When the voltage is greater than about 50 V, avoid touching any exposed metal portion of the circuit. If a high-voltage wire must be touched (for example, in case of a power line

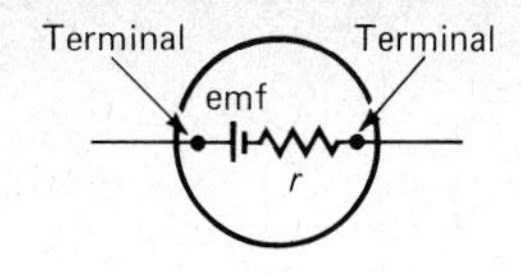

(a) The equivalent circuit for an emf source.

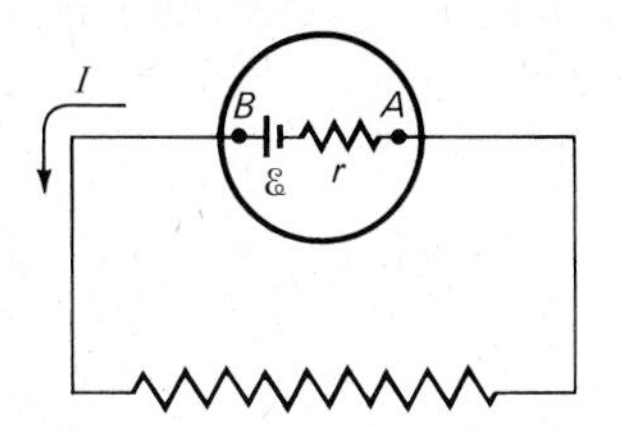

(b) The terminal p.d. is measured from A to B.

Figure 25.12 The terminal potential difference (p.d.) of an emf source is $\mathscr{E} - Ir$ when discharging. What is it when the source is being charged?

accident when help is not immediately available), use a dry stick or some other substantial piece of insulating material to move it. When in doubt about safety, avoid all contacts or close approaches to metal or to the wet earth. Above all, do not let your body become the connecting link between two objects that have widely different electric potentials.

25.6 Internal Resistance of emf Sources

You probably have noticed that the lights on an auto dim when you step on the starter. The starter requires a large amount of current from the battery. Apparently, the potential difference between the terminals of the battery decreases when a large current is drawn from it. Let us now see why this happens.

Each source of emf has within it the equivalent of a resistor. One should therefore picture a battery as shown in Figure 25.12(a). It acts like a pure emf in series with a resistor. Notice that the terminals of the battery are as shown. The pure emf and resistor are inside the battery between these terminals.

The potential difference (p.d.) between the two terminals of a battery is called the *terminal potential difference*. It is not necessarily equal to the emf of the battery. To see why not, let us look at part (b) of Figure 25.12.

The battery (or other emf source) shown there is discharging. Current is flowing from it in the direction the battery wishes it to flow. Let us say that the current is I as shown. Then the terminal potential difference (the change in potential going from A to B) is

$$-Ir + \mathscr{E}$$

Or, rewritten,

$$\text{Terminal p.d.} = \mathscr{E} - Ir$$

In other words, the potential difference across the battery is not equal to the emf $\mathscr{E}$. Instead, it is smaller by an amount Ir.

In previous sections of this text we have ignored Ir in comparison to $\mathscr{E}$. For a reasonably good battery from which I is not too large, $\mathscr{E}$ is usually much larger than Ir. Therefore, we are usually justified in using the emf $\mathscr{E}$ as the terminal potential of the battery. But when starting a car, for example, sometimes I is very large. Then $\mathscr{E} - Ir$ can no longer be approximated by $\mathscr{E}$. The terminal potential of the battery becomes less than its normal 12 V and the car lights dim.

When a battery becomes old, its internal resistance usually increases quite a bit. Then the terminal potential drops when even a small current flows from the battery. Quite often the battery is then unable to perform its function. The car's lights are always dim and the battery is unable to run the starter.

Other emf sources, by their nature, have high internal resistance. They do not make good batteries. For example, a potential difference exists between your foot and hand (or between two positions on your head). These potential differences are important for electrocardiograms and similar medical examinations.

Each acts like a pure emf source in series with a resistor. The internal resistance of the voltage source is very high, however. As a result, the measurement of these emf's on the body is difficult. Even a small current drawn from them alters the emf that we are trying to measure. In the next section we shall see one way for measuring emf's with high internal resistance.

EXAMPLE 25.5 Often in a house one notices that the lights dim for an instant when the motor on a dishwasher or washing machine kicks on. These devices draw quite large currents for the first instant when they start. The house power line acts like an emf source with internal resistance. In a certain house the line voltage drops from 110 V to 103 V when a current of 30 A is drawn from it. What is the equivalent internal resistance for the house line?

Solution The house line voltage is equivalent to the terminal p.d. of the battery shown in Figure 25.12(b). We are told that $\mathscr{E} = 110$ V. The terminal potential difference is, as before,

$$\mathscr{E} - Ir$$

When $I = 30$ A, then we have (in volts)

$$103 = 110 - (30)r$$

Solving, we find the equivalent internal resistance r to be 0.23 Ω. ∎∎

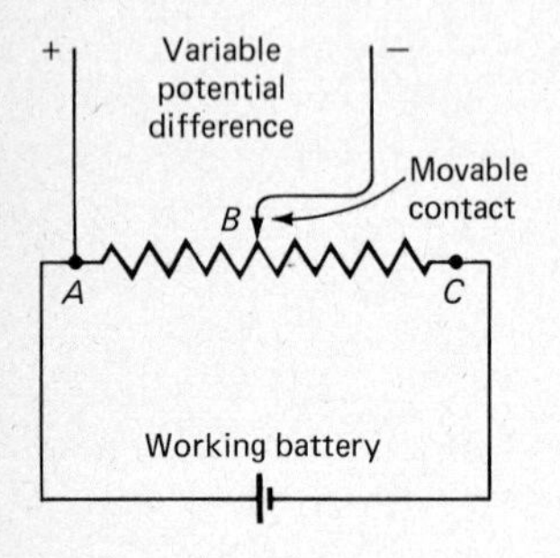

(a) Potential divider

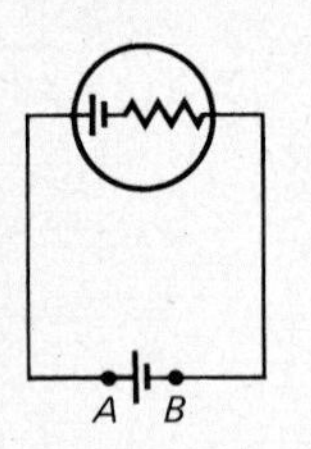

(b) Balancing batteries

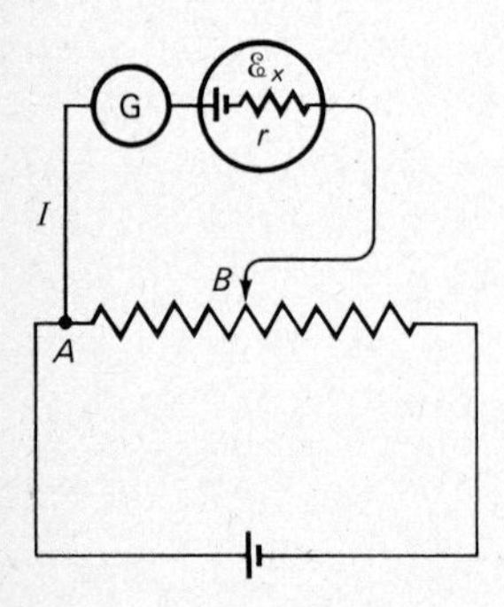

(c) Potentiometer

Figure 25.13 When no current flows through the galvanometer, then $\mathscr{E}_x$ is equal to the p.d. from A to B.

25.7 The Potentiometer

We saw in the last section that a need exists for a device to measure potential differences while drawing negligible current. Devices that do this fall into two general classes. One type of device uses electronics and is called an electrometer. For many purposes an electrometer is quite satisfactory. But when high accuracy is needed another device, called a potentiometer, is often used. It is this device that we shall now describe.

Part (a) of Figure 25.13 shows a simple device, called a *potential divider*, for obtaining a variable potential difference. The resistor from A to C has a potential drop across it equal to that for the working battery. But a sliding contact (B) can be moved along the resistor. Therefore, the potential difference between the wires going upward at A and B can be varied. When the slider is at A, the potential difference is zero. When B is moved to coincide with C, the potential difference is equal to that of the working battery. (Sometimes this resistor is called a potentiometer, but we reserve this name for the whole device to be described.)

Look now at part (b) of Figure 25.13. We see two batteries opposing each other. If their emf's are the same, they will exactly cancel. No current will then flow in the circuit. This fact is basic to the operation of a potentiometer.

In a potentiometer the lower battery in Figure 25.13 (b) is replaced by a potential divider. The situation is like that shown in part (c). In effect, the potential divider acts like a variable battery. Notice what happens when the potential difference from A to B equals the emf $\mathscr{E}_x$ of the cell at the top. The potential divider's voltage acts like the lower battery in part (b). It cancels $\mathscr{E}_x$ and so no current flows.

We see, then, that the galvanometer G (a very, very sensitive ammeter) will read zero when the potential difference from A to B equals $\mathscr{E}_x$. At that position of the slider, no current flows from the upper battery. Our circuit is therefore not influenced by any internal resistance the battery might have. Because I through r is zero, Ir is also zero.

In a potentiometer the potential divider is calibrated. A given position of the slider B corresponds to a known potential difference from A to B. In using the circuit in Figure 25.13(c), one simply moves B until no current flows through the galvanometer. At that position the potential difference from A to B (a known value) is equal to the unknown emf, $\mathscr{E}_x$. Therefore, the emf is determined. Again let us emphasize that this is the true emf of the source. We have ignored many practical considerations in our discussion. When you use a potentiometer in the laboratory, you will learn about them.

This general type of potentiometric circuit is widely used in chart recorders. A servomechanism moves point B until the zero-current position is found. The recorder pen is connected by levers to point B. It traces out the position of B on the unwinding chart paper of the recorder. As a result, the chart provides a record of the emf being measured.

25.8 Electrolysis and Electroplating

In industry it is frequently required to plate one metal onto another. For example, the chromium trim on autos is electroplated. A great deal of know-how is required to obtain a good surface in such an operation. But the basic idea behind electrolysis and electroplating is quite simple.

Suppose that you wish to plate silver onto a metal knife as shown in Figure 25.14. The knife is made negative by attaching it to the negative terminal of the battery. A piece of silver is used as the other electrode. It is made positive by the positive battery terminal. Both are placed in a slightly acid solution of silver nitrate, $AgNO_3$.

The silver nitrate ionizes in solution to form positive silver ions, Ag^+, and negative NO_3^- ions. Each ion has a charge of magnitude e. Attracted by the negative knife, the Ag^+ ions drift to it. Upon touching the knife, the Ag^+ ion combines with a free electron to form a silver atom. This atom then remains attached to the knife.

At the same time, some of the silver atoms in the other electrode dissolve into the solution. They enter as Ag^+ ions and, in the process, leave an electron behind on the silver electrode. This electron is then free to run through the battery and over to the knife. It can then combine with some other silver ion as the ion plates out onto the knife.[3]

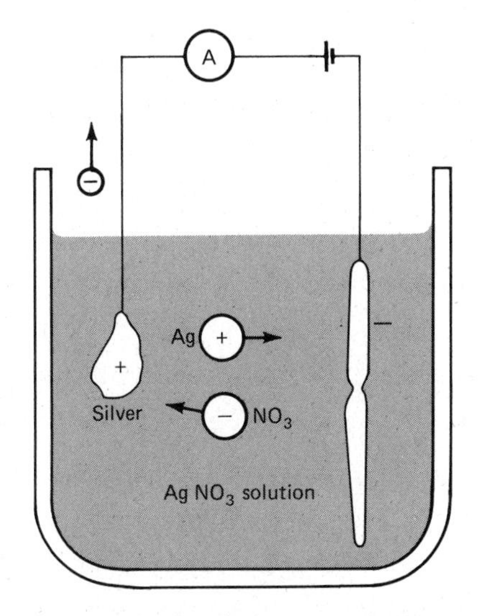

Figure 25.14 Silver ions leave the silver electrode and plate out on the negatively charged knife.

[3]The NO_3^- ions are attracted to the positive electrode, but they do not become neutralized there. The reasons for this are discussed in courses in physical chemistry.

As we see, a flow of electrons exists from the silver electrode through the battery to the knife. It exactly equals the current of silver ions through the solution. For every silver ion that crosses the solution, an electron flows from the silver through the battery to the knife to meet it. Thus a current flows counterclockwise around the circuit. Let us now compute how much silver is plated on the knife in a time t.

Suppose that a current I flows through the ammeter for a time t. Then we know that a charge $Q = It$ has been carried from the silver to the knife. But the charge carried across the electroplating cell by the silver ions in the time t has this same magnitude, It. Therefore

$$\text{Charge carried by silver ions in time } t = It$$

Now we know that each silver ion carried a charge q. Therefore, if n is the number of ions that crossed the cell in time t, we have

$$\text{Charge carried by silver ions in time } t = nq$$

If we equate these two expressions, we find that

$$nq = It$$

or

$$n = \frac{It}{q}$$

The chemists tell us that the atomic mass of a silver atom is $M =$ 108 kg/kmol. This means that each silver atom has a mass m_{atom} given by

$$m_{\text{atom}} = \frac{M}{N_A} = \frac{108 \text{ kg/kmol}}{6.02 \times 10^{26} \text{ atoms/kmol}} = 1.79 \times 10^{-25} \text{ kg}$$

where $N_A = 6.02 \times 10^{26}$ is Avogadro's number. This is the mass plated out on the knife by each of the n silver ions. Therefore, the total mass plated out in time t is

$$\text{Mass plated in time } t = m_{\text{atom}}n = (1.79 \times 10^{-25} \text{ kg})\left(\frac{It}{q}\right)$$

As a typical example, the current might be 0.20 A and flow for 15 min (900 s). Then

$$\text{Mass plated in 900 s} = (1.79 \times 10^{-25} \text{ kg})\left(\frac{0.20 \times 900}{1.60 \times 10^{-19}}\right)$$
$$= 2.0 \times 10^{-4} \text{ kg}$$

If the ions had been divalent, then each would have a carried a charge $q = 2e$. (For example, they might be Cu^{2+} ions when copper plating.) Then q in our equations above must be replaced by $2e$. In practice, there are a number of precautions one must take to obtain a nice, mirror surface. The reasons for some of them are quite involved. As a result, those who know the theory of electroplating appreciate the skill of those who exercise the art.

Summary

Kirchhoff's point rule: The sum of all the currents coming into a point must equal the sum of all the currents leaving the point.

Kirchhoff's loop rule: The sum of all the voltage changes around a closed loop is zero.

In using the loop rule, we take potential rises as positive changes. Potential drops are taken as negative changes.

The loop rule follows from the fact that the electrostatic field is a conservative field. As a result, the work done in carrying a unit positive test charge around a closed loop is zero. The loop need not be an actual circuit.

In house circuits the electric appliances are connected in parallel. The current drawn by each appliance can be found from its power rating by use of $P = VI$.

Even small currents through cells of the body can stop the cells from functioning. Currents through the chest region can cause the lungs and heart to stop operating. Very much larger currents are needed to cause serious burning of the body. To be safe, one must prevent the body from being a connecting link in an electrical circuit.

The terminal potential difference of an emf source is not always equal to the emf. It differs from it by the Ir drop in the internal resistance of the source. This is usually negligible for good batteries used at low currents. However, for old batteries and many other emf sources, r is large enough to be important.

A potentiometer measures potential difference without drawing appreciable current. It is therefore capable of measuring emf's. Its operation is based on balancing the unknown emf by a known potential difference.

In electrolysis and electroplating, ions carry charge through solution. Their charges are neutralized when they reach the electrode of opposite sign. The ions are then changed to neutral atoms and they then plate out or leave the solution.

Questions and Exercises

1. Kirchhoff's point rule is easily justified for points that are junctions of wires. But suppose that the "point" is actually a capacitor. Is the rule still true?
2. In applying the loop rule, we must know whether a given potential change is a drop or rise. How do we tell which is which in the case of a battery? Resistor? Capacitor?
3. In many houses a hall light, for example, can be turned on and off by two different switches. Draw a diagram showing how such a light is wired.
4. In some Christmas tree lights all the lights go off when one bulb is removed. In others the removal of one bulb does not affect the others. Draw a circuit diagram showing each type and explain why each behaves as it does.
5. The dome light in most autos comes on when the driver's

door is opened. Draw a diagram showing the car battery, dome light, and door switch. Locate each and explain where the connecting wires are in an actual auto.
6. Two motorists are using jumper cables to start one car from the battery of the other. They connect one wire and start to connect the other. A great deal of sparking occurs when they accidentally touch it to the bumper of one of the cars. Why? After connecting both wires to the batteries, the wires become very hot. What did they do wrong?
7. A safety rule one sometimes hears is the following: Always keep one hand behind your back while working on a live electronic circuit. Explain the idea behind this rule.
8. Electricians sometimes test a light plug with their fingers to see if it is live. How do they do this? Under what conditions could this test be very unsafe?
9. Birds run all over high-voltage lines and seem not to be bothered by the high voltage. Why aren't they electrocuted when they touch a bare wire?
10. A child is playing with a pair of wire cutter pliers. He decides to try them out on the wire leading to a lighted table lamp. What happens as he cuts the wire? Explain why the answer to this questions depends on several factors.
11. During power shortages power companies frequently lower the potential difference they supply to the public. Why does this result in a decrease of power use?
12. When a battery is being charged, its terminal potential difference is larger than its emf. Explain why.
13. Most present-day cars use 12-V batteries. Years ago they used 6-V batteries. Why was the change made?
14. Estimate the power consumed by a city near you. Which appliances in your house is it most important to use sparingly if you wish to conserve power?

Problems

1. The following are connected in a series circuit: 6.0-V battery, 9.0-V battery, 45-Ω resistor. Use the loop rule to find the current if the batteries are (a) aiding and (b) opposing.
2. A series circuit consists of an unknown resistance and two batteries opposing each other. The battery voltages are 12.0 V and 9.0 V. An ammeter measures the current to be 1.50 A. Use the loop rule to evaluate the resistor.
3. For the circuit shown in Figure P25.1, (a) write the point equation for point *A*. (b) Write the loop equation for loop *ABCDA*.
4. For the circuit shown in Figure P25.1, (a) write the point equation for point *B*. (b) Write the loop equation for loop *AEFBCDA*.

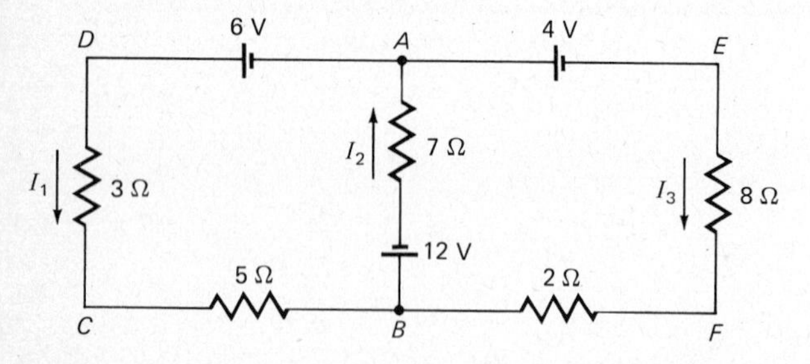

Figure P25.1

5. (a) Write the loop equation for loop *ABCDA* in Figure P25.2. (b) Write the point equation for point *A*. (c) Write the point equation for point *D*.

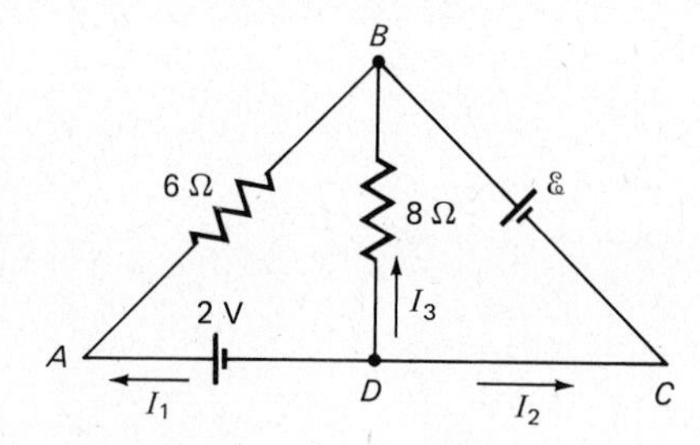

Figure P25.2

6. (a) Write the loop equation for loop *CBDC* in Figure P25.2. (b) Repeat for loop *BADB*. (c) Write the point equation for point *B*.
7. Find the magnitude and direction of I_1, I_2, and I_3 for the circuit shown in Figure P25.3.

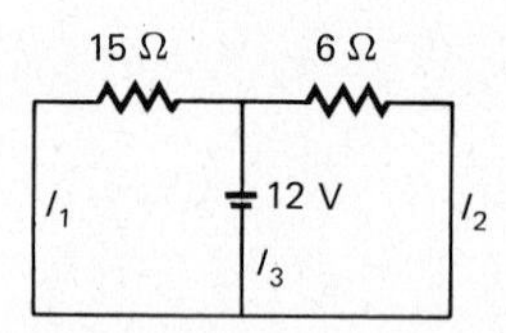

Figure P25.3

8. Find I_1, I_2, and I_3 in Figure P25.2. The battery emf is 5.0 V.
*9. In Figure P25.2, what must be the value of ℰ such that $I_1 = 0$? If ℰ has this value, what will be I_2 and I_3?
*10. Refer to Figure P25.2. A perfect voltmeter connected from *A*

*Problems marked with an asterisk are not as easy as the unmarked ones.

to B reads 1.2 V. Point A is at the higher potential. Find $\mathscr{E}$, I_1, I_2, and I_3.

*11. Refer to Figure P25.2. A perfect voltmeter connected from A to B reads 1.2 V. Point B is at the higher potential. Find $\mathscr{E}$, I_1, I_2, and I_3.

*12. Find the three currents in Figure P25.4.

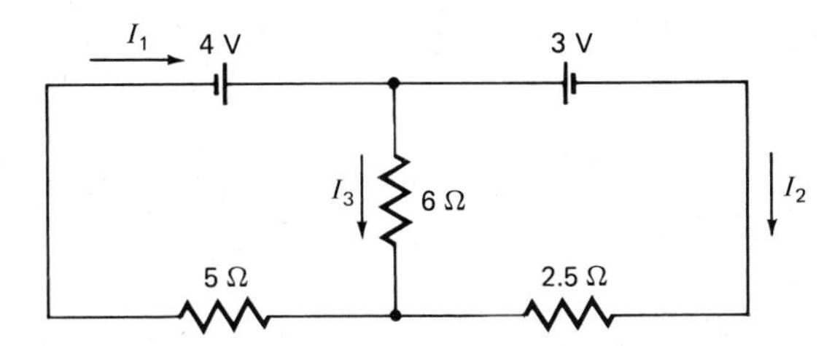

Figure P25.4

*13. If the 5-Ω resistor in Figure P25.4 is replaced by a 2.5-Ω resistor, what will the three currents be?

*14. Find I_1, I_2, and I_3 in the circuit of Figure P25.5. (Hint: It is usually best to reduce series and parallel resistance combinations if possible.)

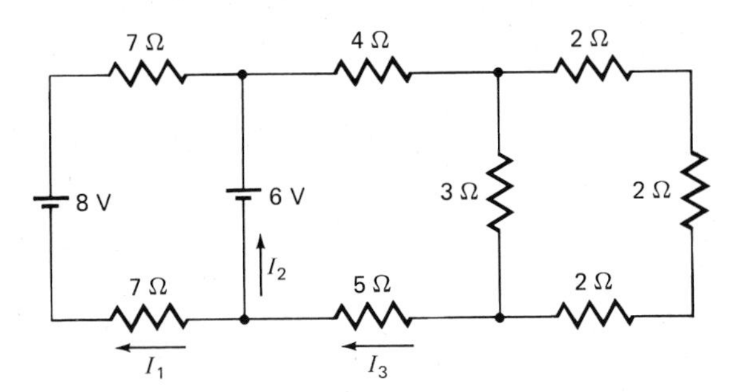

Figure P25.5

15. A certain electric clothes drier uses 2500 W of power. What current does it draw (a) from a 120-V line and (b) from a 240-V line? (c) Why are such appliances ordinarily designed to operate at the higher voltage?

16. A certain power-generating station has an output of 1.5 MW. How much current would have to flow from the station if it furnished its power at a voltage of (a) 120 V and (b) 100 kV?

17. The following are operating from a 120-V house line: a 60-W bulb, a 1200-W toaster, and a 100-W bulb. (a) How much current is being drawn by the toaster? (b) How much current is passing through the fuse for this line?

18. How many 75-W bulbs can be lit by a 120-V house line if the line is fused for 25 A?

19. Refer to Figure 25.9. Suppose all three switches are closed but nothing is plugged into the wall outlet. (a) How much current flows through the switch on the left? (b) Through the switch on the right? (c) How much current flows through the fuse?

20. A 120-V house circuit fused for 20 A has operating on it the following: two 60-W bulbs, a 1000-W heater, a 40-W radio, a 2.5-W clock. Will the fuse blow if an electric heater rated at 1200 W is plugged into it? How large a current is required for all but the second heater?

21. How many 75-W/120-V bulbs can be operated on a 20-A house circuit? Suppose that a 1200-W iron is operating at the same time. How many 75-W bulbs can then be used?

22. What is the terminal p.d. of a dry cell when the current being drawn from it is 3 A? The emf of the dry cell is 1.57 V and its internal resistance is 0.075 Ω.

23. An old dry cell has an emf of 1.57 V and an internal resistance of 0.20 Ω. What is its terminal p.d. when it is being (a) discharged at a current of 0.90 A and (b) charged at the same current?

24. An old dry cell that has an emf of 1.57 V is connected across a 2.00-Ω resistor. A current of 0.140 A flows from the battery. Find the terminal p.d. of the battery and its internal resistance.

*25. A certain new 12-V car battery has an internal resistance of 0.0095 Ω. On a cold day the starter of the car uses a current of 130 A. (a) By how much does the terminal p.d. of the battery decrease when the starter is used? (b) The cable connecting the starter to the battery has a resistance of 1.2×10^{-3} Ω. How many calories of heat are developed in it during the 5.0 s the starter is on?

*26. The performance of a battery is usually rated in ampere-hours. For example, a certain 1.35-V mercury cell has a rating of 3600 mA·h. This means, when fully charged, the battery can produce a current of 1 mA for 3600 h before it is completely discharged (or 2 mA for 1800 h, etc.). (a) How long can a current of 5.0 A be drawn from a fully charged 12-V battery which is rated at 90 A·h? (b) How much usable energy is stored in this 12-V battery?

**27. A certain long-life alkaline battery is connected across a 4.0-Ω resistor. Its terminal potential decreases with time as shown in Figure P25.6. (a) About how much usable electric energy is stored in this battery? (b) About what is its ampere-hour rating? (See Problem 26.)

*28. The voltage of a dry cell is 1.574 V when measured by a potentiometer. But a voltmeter that has a resistance of 2000 Ω reads 1.48 V when placed across it. (a) How much current does the voltmeter draw from the battery? (b) What is the internal resistance of the battery?

**Problems marked with a double asterisk are somewhat more difficult than the average.

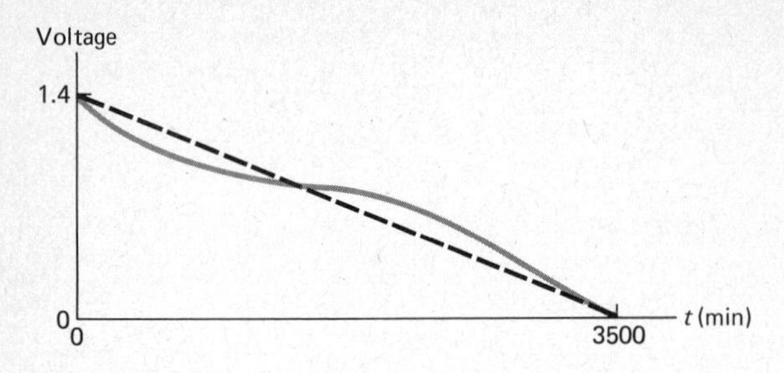

Figure P25.6

29. The quantity of charge 96,485 C is called the *faraday*. How many singly charged ions must plate out at an electrode to deposit one faraday of charge?

*30. Singly charged copper ions are plated out from solution by a current of 0.15 A flowing for 20 min. What mass of copper is plated out? For copper, $M = 63.54$ kg/kmol.

*31. When a current is passed through slightly acid water, hydrogen is freed at one electrode much like silver is freed in electroplating. The hydrogen ion has a charge $+e$ and an atomic mass $M = 1.00$ kg/kmol. How many hydrogen ions are changed to atoms and how much hydrogen is generated when a current of 2.0 A flows for 30 min?

*32. Refer to Problem 31. When the hydrogen is being generated at the negative electrode, oxygen is being generated at the positive electrode. (a) How many oxygen atoms are liberated by a current of 2.0 A flowing for 30 min? (b) What mass of oxygen is generated in this time? The charge on each oxygen ion is $-2e$ and the atomic mass of each is $M = 16.0$ kg/kmol.

*33. One way to measure the mass of silver atoms is by electroplating. The following data are taken for an experiment much like that described in Section 25.8:

Current = 0.45 A
Time of flow = 50 min
Mass plated out = 1.48 g

From these data, find the mass of a silver atom.

**34. To measure the atomic mass of a univalent metal atom, an electrolysis experiment is carried out. It is found that 0.475 g of metal is plated out by a current of 0.400 A in 30 min. Find the atomic mass of the metal atom. What atom is this?

**35. Refer to the circuit of Figure P25.7, Find I, I_1, and the charge on the 2-μF capacitor when the switch K is open.

**36. Repeat the previous problem for the case in which the switch K is closed.

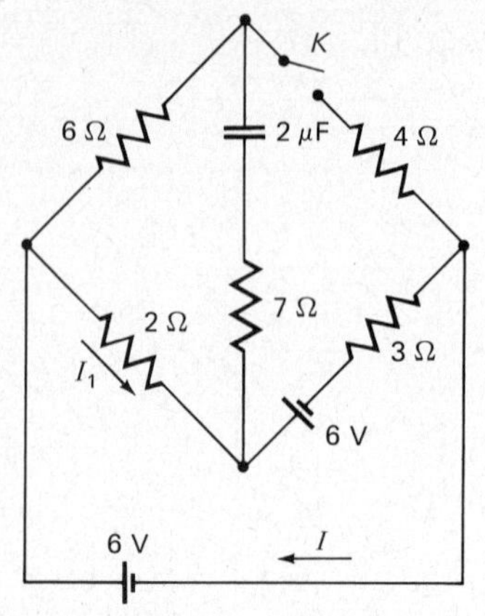

Figure P25.7

*37. The circuit shown in Figure P25.8 is called the Wheatstone bridge circuit. One adjusts the variable resistance, R_1, until no current flows through the galvanometer. The bridge is then said to be balanced. Show that the unknown resistance X is given by the following relation at balance:

$$X = R_1\left(\frac{R_2}{R_3}\right)$$

(Hint: Notice that at balance no current flows through the galvanometer. Write the loop equation for the two upper loops.)

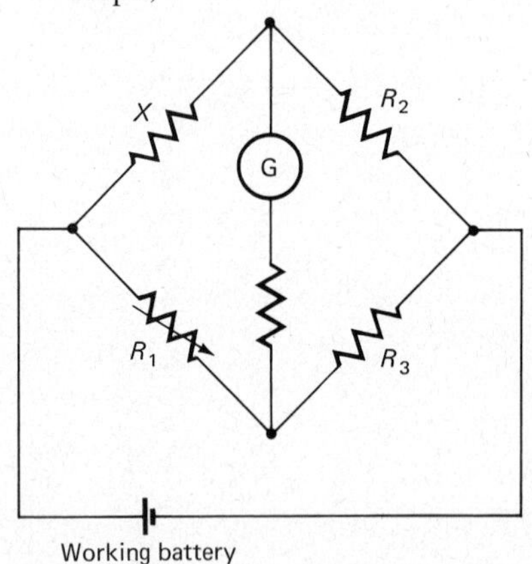

Figure P25.8

From PSSC Physics, D.C. Heath & Company, Lexington, Mass., 1981

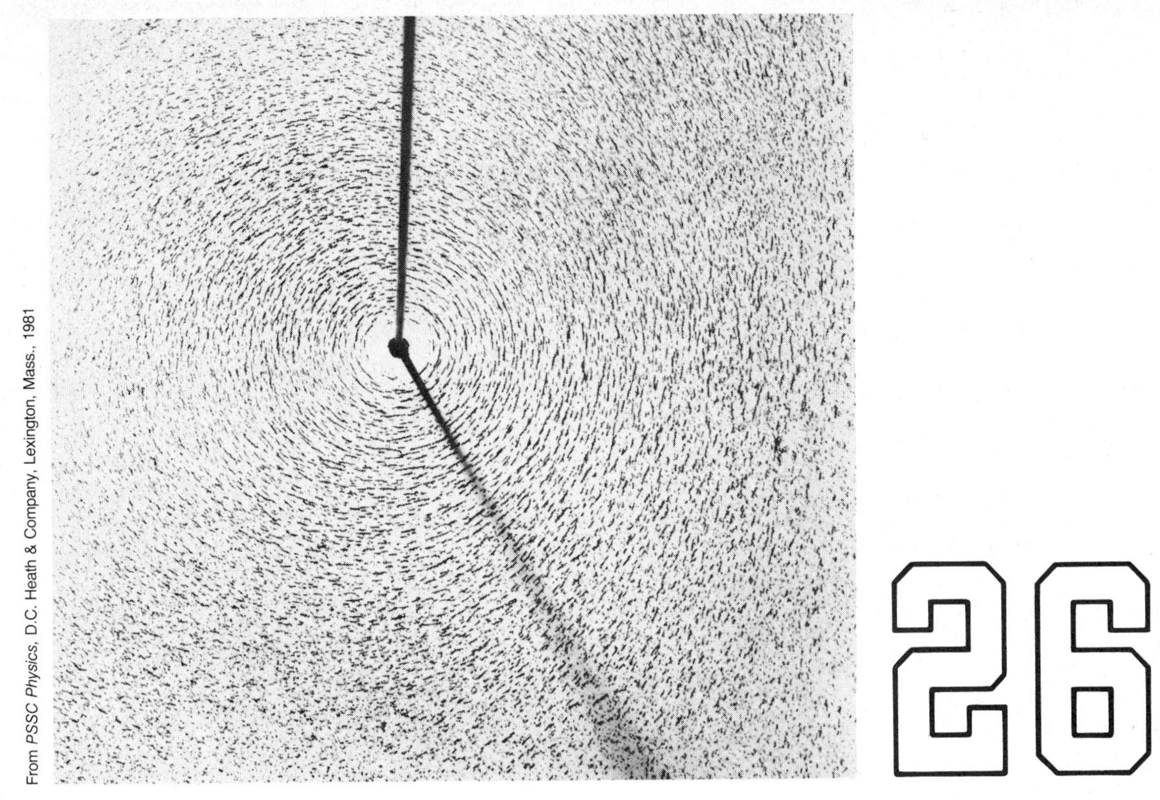

26 MAGNETISM

Performance Goals

When you finish this chapter, you should be able to do the following:

1. When given a compass and a magnet: Find which pole of the magnet is north; sketch the magnetic field of the magnet; list several materials the magnet will attract strongly; list several materials the magnet will not attract appreciably; explain why a chain of nails can be lifted by it.
2. Prepare a sketch showing the earth's magnetic field. Show the direction of the field lines, the

Magnetism is basic to the operation of electric motors, generators, and many other devices used in technology. In this chapter we shall learn about magnetic fields and the forces they cause on moving charges. We shall see that magnets are not the only sources of magnetic fields. All electric currents are accompanied by a magnetic field. By use of current-carrying coils, electromagnets can be made. We shall see that iron plays an important part in magnetism because it vastly increases the magnetic fields of currents. The force that a magnetic field exerts on moving charges provides the driving force for electric motors. These and similar topics are discussed in this chapter.

earth's rotation axis, and the north pole of the earth's equivalent magnet.

3. Find the magnitude and direction of the force on a given length of a current-carrying straight wire when placed in a given magnetic field.
4. Find the magnitude and direction of the force on a particle shooting through a magnetic field. Describe the path of the particle if it is moving in the direction of the field; if it is moving perpendicular to the field.
5. Describe quantitatively the path followed by a charged particle shot into a uniform magnetic field in a direction perpendicular to the field.
6. Give an order-of-magnitude value for the earth's magnetic field in both gauss and tesla.
7. Sketch the magnetic field for the following: straight wire, single loop of wire, flat coil of wire, solenoid. Show the directions of the current and **B**.
8. Explain what an electromagnet is. Given an iron rod and considerable wire, construct a simple electromagnet. Give five examples of practical uses of electromagnets.
9. Describe the torque on a given coil in a magnetic field and give the direction it will turn. Compute the torque on the coil when sufficient data are given.
10. Explain how a given meter movement can be made into an ammeter of a specified range. Repeat for a voltmeter.
11. Sketch the important elements of a single-coil dc motor. Explain the function of each element.

26.1 Magnets

Let us review some facts about magnets. As shown in Figure 26.1, an ordinary magnet has two poles. The north pole of one magnet repels the north pole of the other. South poles also repel each other. But a north pole is attracted by a south pole.

Like poles of magnets repel each other. Unlike poles attract.

Most materials are not affected much by a magnet, but materials that contain iron are attracted to it. For example, the iron nail in Figure 26.2 is attracted by the magnet. In (a) we see the nail as it was originally. It had no north or south pole. We say that it was *unmagnetized.* But as shown in (b), poles are induced on the nail as it is brought close to the magnet. As a result, it is attracted to the magnet. The nail then acts like a new magnet and attracts other nails as shown in (c). Most materials are not much attracted by magnets. Iron, nickel, cobalt, and their alloys are the only very common metals that are highly magnetic. They are called ferromagnetic materials.

Magnets can induce unmagnetized ferromagnetic materials to become magnets. As a result, a magnet attracts them.

One of the most commonly used magnets is a compass needle. The needle is actually a bar magnet. It is pivoted so it can rotate freely. The end of the compass needle that points north on earth is called the north pole of the needle magnet. This is shown in Figure 26.3(a). To find the type of pole on another magnet, we perform the experiment shown in part (b) of the figure. Because like poles repel, one can easily tell that the poles are as labeled in (b).

26.2 Plotting Magnetic Fields

A compass needle experiences a force when it is near a magnet. The poles of the compass needle experience forces because of the nearby magnet poles.

In a region where a pole of a magnet experiences a magnetic force, a magnetic field is said to exist.

The direction of the magnetic field is taken to be the same as the direction of the force on a north pole. This is also the direction that a compass needle points in the field.

A compass needle points in the direction of the magnetic field.

We can use a compass to sketch out a magnetic field. For example, the compasses in Figure 26.4(a) tell us how to sketch the field of the magnet. It is sketched in part (b) of the figure. Notice in particular that the compass needle points away from a north pole. It points toward a south pole. (Why?) As a result

12. Explain the words diamagnetic, paramagnetic, and ferromagnetic.
13. Explain the difference between an unmagnetized and a magnetized material. Explain what happens to the domains as the material is carried along the magnetization curve of Figure 26.26. Also explain the meaning of hysteresis in terms of them.
14. When given a description of the hysteresis properties of an iron, state whether it is more suitable for an electromagnet or a permanent magnet.

Magnetic field lines come out of north poles and enter south poles of magnets.

This is also shown in the magnetic field sketches shown in (c) and (d). You should be able to justify qualitatively the way the lines are drawn.

As you know, the earth acts like a huge magnet. A sketch of its magnetic field is shown in Figure 26.5. Because a compass needle points northward on the earth, the magnetic field must be in the direction shown. But the lines come out of north poles and go into south poles. Therefore, the earth's magnet has its north pole at the south end of the earth, whereas the earth's magnet has its south pole at the north end of the earth. As we see in the figure, the earth's rotation axis does not line up perfectly with the earth's magnet.

26.3 Magnetic Force on Currents

Magnetic fields exert a force on wires that carry current. This fact is of great importance in technology. It is fundamental to the operation of electric motors and many electric meters. Let us now investigate this type of force.

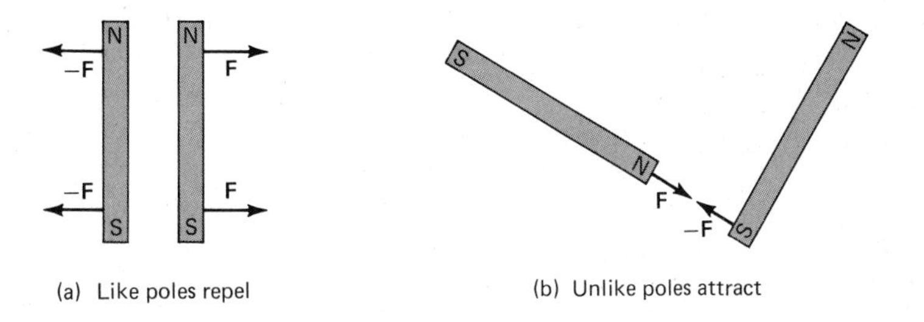

Figure 26.1 Like poles of magnets repel each other. Unlike poles attract.

Suppose that the experiment shown in Figure 26.6 is carried out. The wires from the battery carry a current through the magnetic field. Notice that the field lines skim across the tabletop on which the wires are lying. It is found that the wires experience forces. The directions of the forces are labeled in the diagram.

As we see, the wires that carry current along the field lines experience no force. But a force does exist on those wires that are perpendicular to the field lines. There is a simple rule for remembering the directions of the forces given in the figure. It is called the right-hand rule and it is illustrated in Figure 26.7.

Right-hand rule: The fingers of your right hand point in the direction of the field line. The thumb of the flat right hand is in the direction of current flow. The force on the wire is in the direction in which the palm of your hand is then ready to push.

Even if the wire is not perpendicular to the field lines, this rule can be used. Your hand must remain flat. But your thumb need not be perpendicular to your fingers.

The magnitude of the force depends on the angle between the field lines and

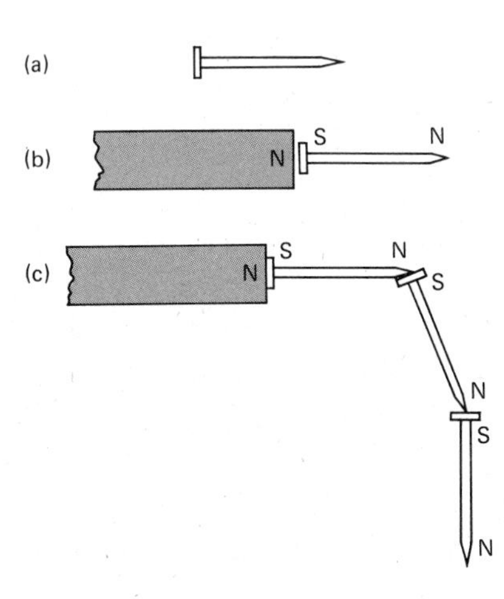

Figure 26.2 Ferromagnetic materials can be made to act like magnets by induction. As a result, magnets attract them.

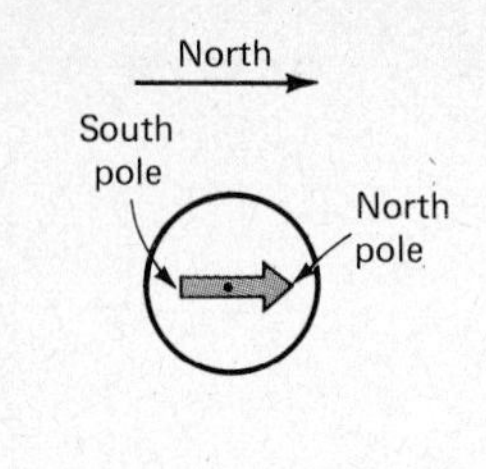

(a) Compass

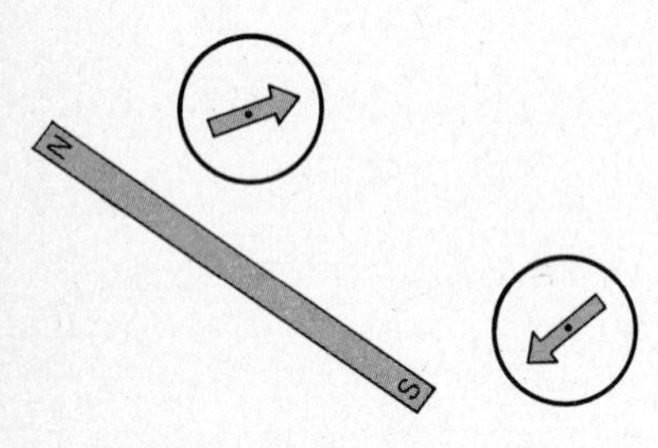

(b) Likes repel; unlikes attract

Figure 26.3 A compass can be used to label the poles on a magnet.

the current. This is shown in Figure 26.8. Maximum force occurs when the wire and field lines are perpendicular, that is, when $\theta = 90°$, as shown in Figure 26.8. Notice that θ is the angle between the wire and the field lines. When $\theta = 0$, the force on the wire is zero.

> **A wire carrying current along the field lines experiences no magnetic force.**

Experiment shows that, for all angles θ, the force is proportional to both I and $\sin\theta$. It is also proportional to the strength of the magnetic field. This quantity is defined in the next section.

26.4 Magnetic Flux Density

The strength of a magnetic field is represented by the symbol **B**. The quantity **B** goes under several names. It has two formal names. They are the *magnetic flux density* and the *magnetic induction.* But many people call it the *magnetic field strength.* **B** is a vector that points along the field lines. We define it precisely by means of an experiment.

The general idea behind the experiment used to define **B** is shown in Figure 26.9. As we have said, the direction of **B** is along the field lines. The magnitude of **B** is defined in terms of the force F on a length L of wire that carries a current I. As we pointed out in the previous section, F is proportional to I and $\sin\theta$; θ is the angle between **B** and the wire. F is also proportional to the strength of the magnetic field. Therefore,

$$F \sim ILB \sin\theta$$

We define **B** in such a way that the constant of proportionality in this expression is unity. Then by definition of **B**,

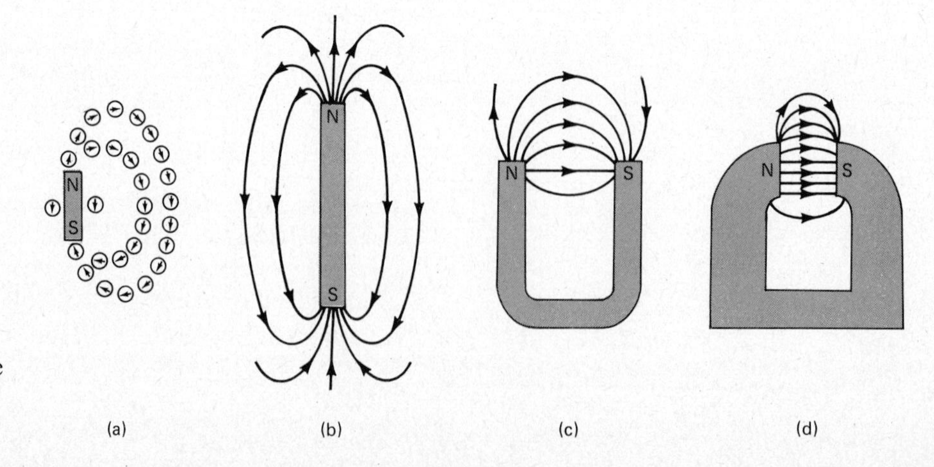

Figure 26.4 Magnetic field lines come out of the north pole and enter the south pole of a magnet.

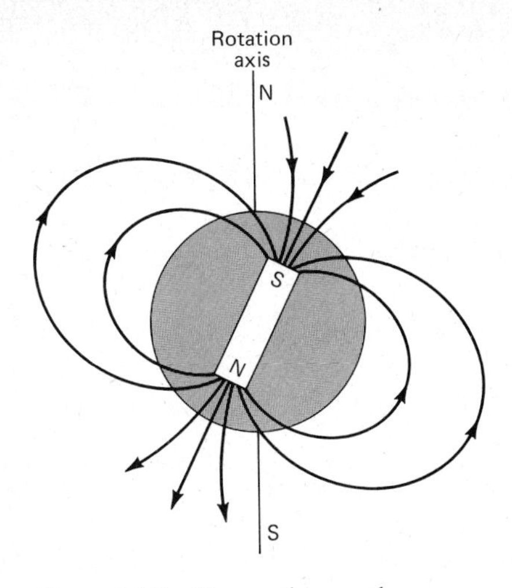

Figure 26.5 The earth's south magnetic pole is in the Arctic, near the North Pole of the earth.

$$F = ILB \sin \theta \qquad (26.1)$$

F is in newtons, I is in amperes, and L is in meters. The unit for **B** so defined has two equivalent names: the tesla (T) and the weber per square meter (Wb/m^2). Another unit used for **B** is called the gauss (G).

$$1 \text{ T} = 10^4 \text{ G} \qquad \text{(exactly)}$$

The gauss is not a member of the SI system of units. It should never be used in our equations.

Often it is convenient to notice that $B \sin \theta$ is an easily seen quantity. It is shown in the right-hand side of Figure 26.9. As we see, $B \sin \theta$ is the component of **B** that is perpendicular to the wire. We represent it by $B_{\perp}$. Then we have

The force on a length *L* of a wire carrying a current *I* in a magnetic field *B* is given by

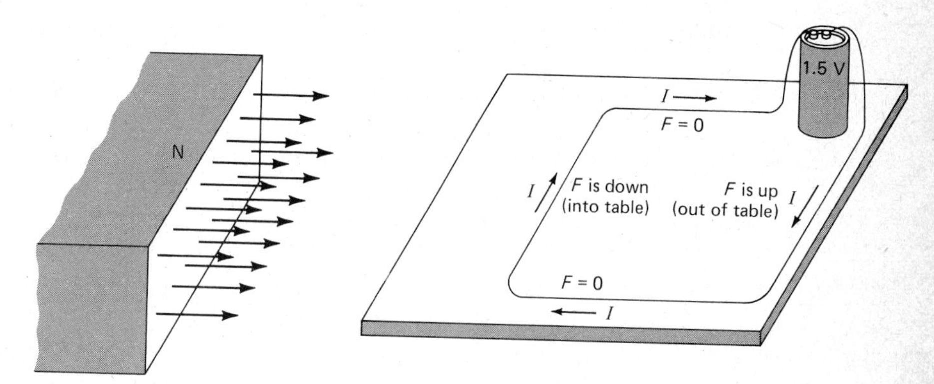

Figure 26.6 The circuit wires experience forces due to the magnetic field. Notice the directions of the forces.

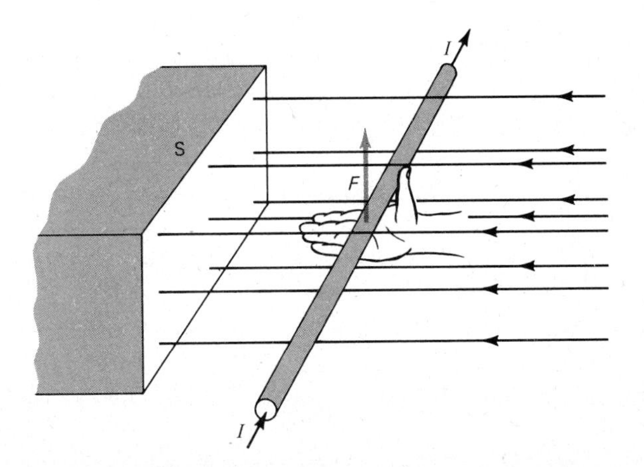

Figure 26.7 The right-hand rule tells us that, with the hand held flat, when the fingers point in the direction of the field and the thumb points in the direction of the current, the palm pushes in the direction of the force.

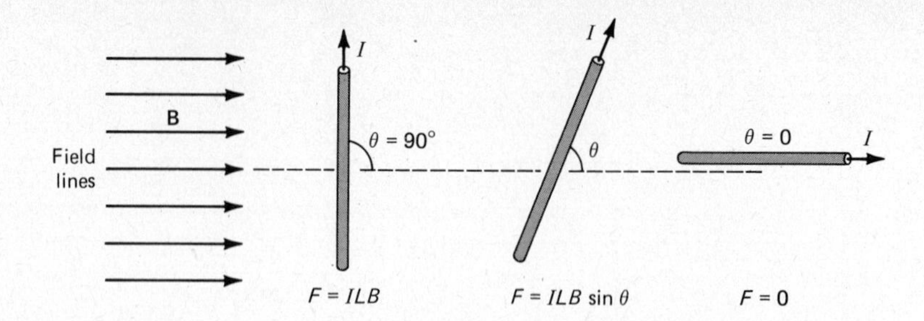

Figure 26.8 The force is greatest for $\theta = 90°$. It is zero when $\theta = 0°$. Can you show that the force on the wire is into the page?

$$F = ILB \sin \theta = ILB_{\perp}$$

The direction of the force is given by the right-hand rule.

▮ EXAMPLE 26.1 The earth's magnetic field in Colorado has an average value of 0.59 G. It is directed at an angle of 67° below the horizontal. (We say that its dip angle is 67°.) Find the force due to the earth's magnetic field on 5.0 m of an east-west wire carrying a westward current of 20 A. Repeat for a north-south wire.

Solution The situation is shown in Figure 26.10. In the first case of an east-west wire, shown in (a), **B** is perpendicular to the wire. Because $0.59 \text{ G} = 0.59 \times 10^{-4}$ T, we have

$$\begin{aligned} F &= ILB \sin \theta \\ &= (20)(5)(0.59 \times 10^{-4})(1) \text{ N} \\ &= 5.9 \times 10^{-3} \text{ N} \end{aligned}$$

The right-hand rule tells us the force on the wire in (a) is as shown. Check it and see. In situations such as this the exact direction of the force is most easily found in the following way.

The wire and the field line through it define a plane. The force is perpendicular to this plane. Its direction is given by the right-hand rule.

Once the plane has been found, only two directions are left to decide between.

The situation shown in (b) is for a current direction northward. As we see from the figure,

$$\begin{aligned} F &= ILB \sin \theta = ILB_{\perp} \\ &= (20)(5)(0.59 \times 10^{-4})(0.92) \text{ N} \\ &= 5.4 \times 10^{-3} \text{ N} \end{aligned}$$

The field line and wire define a plane, the plane of the page. **F** is therefore perpendicular to the page. Using the right-hand rule, we find that **F** is into the page. ▮▮

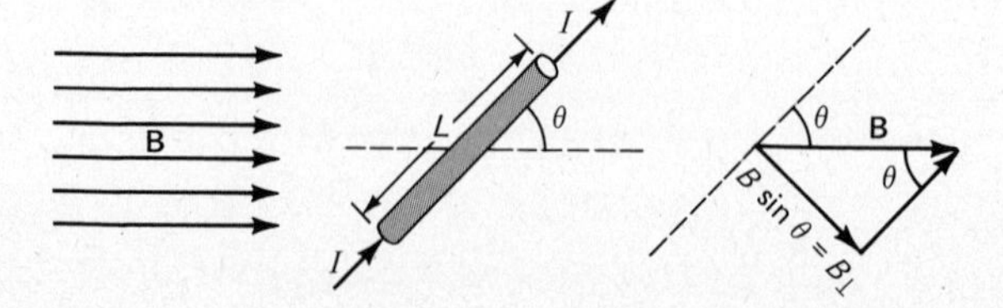

Figure 26.9 Let F be the force on the length L of wire. Then, by the definition of B, one has $F = ILB \sin \theta$. Notice that $B \sin \theta = B_{\perp}$. So $F = ILB_{\perp}$.

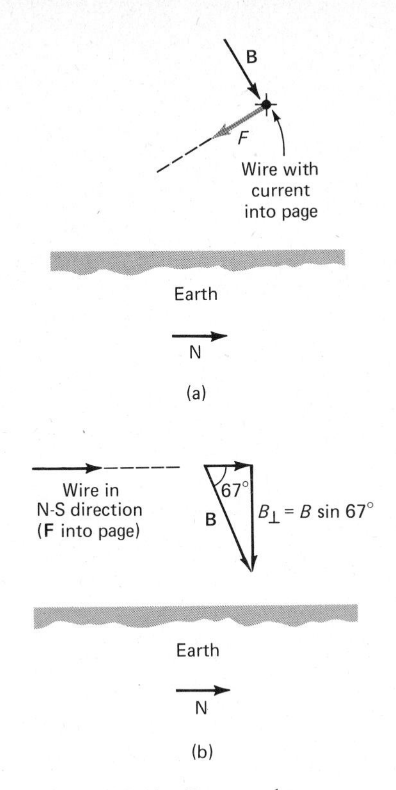

Figure 26.10 Be sure that you understand why the force is in the direction indicated in each case. Use the right-hand rule to find the direction.

26.5 Magnetic Deflection of Moving Charged Particles

An electric current consists of moving charges. As you might suspect, then, any charge that is moving through a magnetic field experiences a force. A simple experiment to show this is pictured in Figure 26.11. The path of the charged particles shooting from left to right is shown by a fluorescent screen. The tube is partly evacuated. A high voltage across it causes a corona discharge from a pointed needle at its left end.

Notice that in (a) the charged particles shoot straight through the tube when no magnetic field is present. But when a magnet is brought close in (b), the particle beam is deflected. As usual, the right-hand rule applies to the particle beam.

A beam of positive particles shoots through a magnetic field. The direction of the force on the particles is given by the right-hand rule. Fingers are in the direction of the field. Thumb is in the direction of the particle velocity.

If the particles carry a negative charge, the force direction is reversed. You can test your understanding of the rule by checking the force shown in part (c) of Figure 26.11.

It is easy to find the force on the moving charged particle. For the moving charges in a wire, we already know that

$$F = ILB \sin \theta$$

The current consists of charges q moving along the wire with speed v. If there are n charges per unit length of the wire, then the force per charge will be $F \div nL$ or

$$\text{Force on each charge} = \frac{F}{nL} = \frac{IB \sin \theta}{n}$$

We can simplify this by finding I in terms of the charge motion. In a time t all the free charges will move a distance vt along the wire. Hence all the charges in a length vt will pass through a given cross section of the wire in a time t. This number is nvt. If each carries a charge q, then the charge passing through the cross section in time t will be $(nvt)q$. Because current is charge per unit time, we find that

$$I = \frac{nvtq}{t} = nvq$$

Substituting this value for I, we find the force on each charge to be

$$\text{Force} = qvB \sin \theta \qquad (26.2)$$

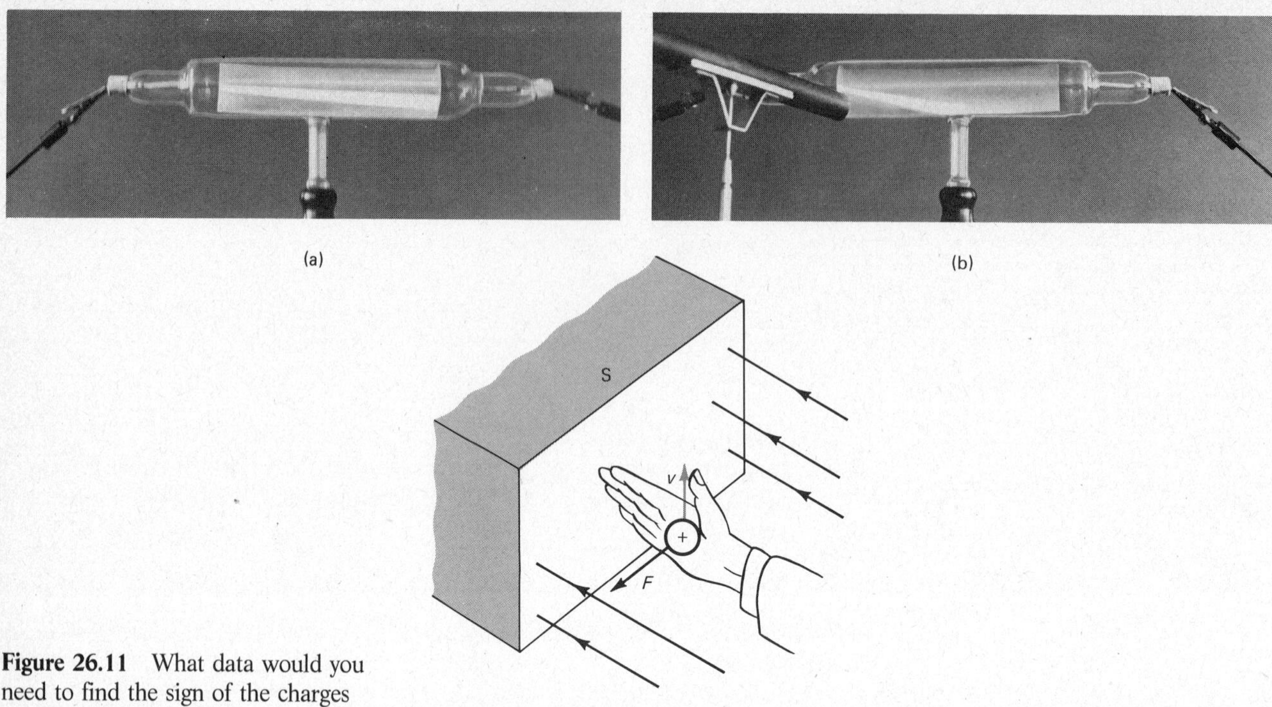

Figure 26.11 What data would you need to find the sign of the charges shooting down the tube?

A particle with charge q and velocity v is moving at an angle θ to a field B. The force on the particle is $qvB \sin \theta$. Its direction is given by the right-hand rule.

There is an important feature concerning this force that you should notice. The force on a moving charged particle is perpendicular to the direction of motion. Therefore, there is no force component in the direction of motion. As a result, this type of force cannot do work on the particle. In a field that is not varying with time,

The magnetic force on a moving charge does no work. It neither speeds up nor slows down the particle.

In spite of this, the force caused by the magnetic field does affect the particle. It causes its direction of motion to change.We saw this in Figure 26.11. Let us now investigate what type of motion results from this force.

26.6 Motion of Charges in a Magnetic Field

Suppose that a uniform magnetic field exists in a certain region of space. We picture the field in Figure 26.12. It is directed perpendicular to and into the page. (We represent a field into the page by crosses $\times$. This is to suggest the tail of the **B** vector arrow pointing away from you. A field out of the page is indi-

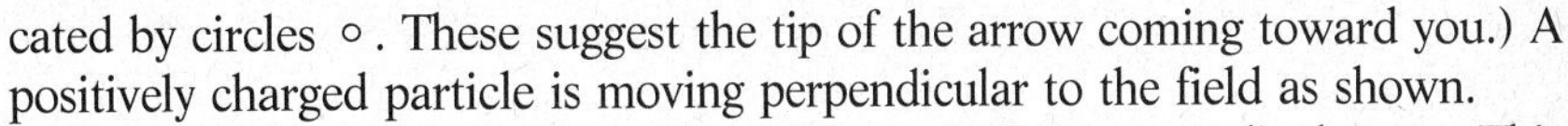

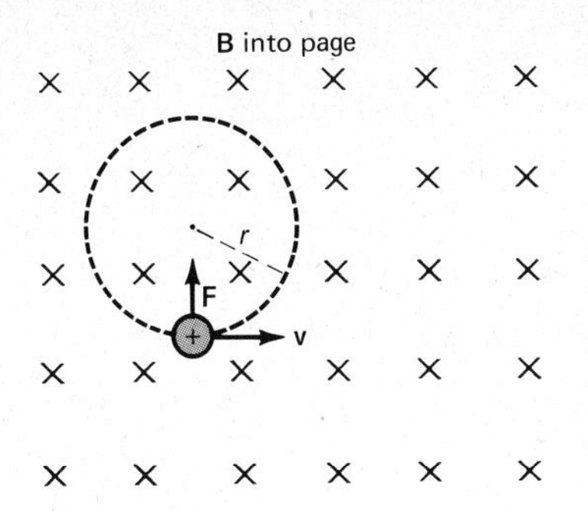

Figure 26.12 The charged particles will follow the circular path shown. Will their speed change?

cated by circles ∘. These suggest the tip of the arrow coming toward you.) A positively charged particle is moving perpendicular to the field as shown.

As we have said, the force on the moving particle is perpendicular to **v**. This situation is not new to us. The earth is held in its orbit about the sun by a force like this. A ball at the end of a string is caused to move in a circular path by a force of this type. In each case a steady force acts perpendicular to the direction of motion. The force causes the object to follow a circular path. Therefore, the charged particle of Figure 26.12 is caused to move in a circular path. Let us compute how large the circle will be.

The force F shown in Figure 26.12 supplies the needed centripetal force. You will recall that

$$\text{Centripetal force} = \frac{mv^2}{r}$$

for a particle of mass m and speed v moving in a circle of radius r.

In our case the centripetal force is being supplied by the magnetic field. That force is given by

$$\text{Magnetic field force} = qvB \sin 90°$$

because the particle is moving perpendicular to the field lines. Equating it to mv^2/r gives

$$\frac{mv^2}{r} = qvB(1)$$

This expression can be solved for r, the radius of the circular path. We then find that a particle (charge $= q$) moving with velocity **v** perpendicular to a field **B** follows a circle with radius

$$r = \frac{mv}{qB}$$

Notice that r is a measure of the momentum of the particle, mv. In atomic and nuclear physics this fact is used in measuring particle momentum. In industry and science this fact is used to measure the mass of atomic particles. This idea is illustrated in the following example.

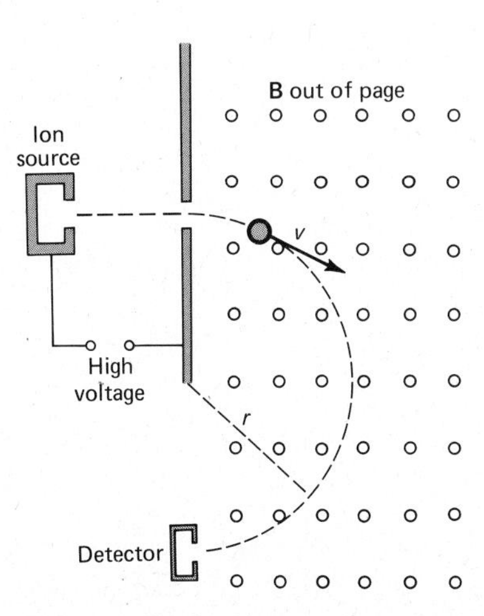

Figure 26.13 Is the particle positively or negatively charged?

■ **EXAMPLE 26.2** To analyze unknown gases and other substances, the atoms or molecules are ionized and then passed through a mass spectrometer. Ions from the ion source are deflected in a uniform magnetic field. This is shown schematically in Figure 26.13. The ions can be identified from their mass. Find the mass of the univalent ion shown in Figure 26.13 if $r = 0.400$ m, $B = 0.360$ T, and $v = 6.00 \times 10^5$ m/s.

Solution A univalent ion carries one unit of charge. Therefore $q = \pm e$. (Is the charge positive or negative in the figure?) Because the force qvB furnishes the centripetal force mv^2/r,

$$qvB = \frac{mv^2}{r}$$

Solving for m and substituting the given values, we have

$$m = \frac{qBr}{v}$$

$$m = \frac{(1.6 \times 10^{-19})(0.36)(0.40)}{6 \times 10^5} \text{kg}$$
$$= 3.84 \times 10^{-26} \text{ kg}$$

It is of interest to find the atomic mass, M, of this ion. To do so, we recall that there are Avogadro's number of atoms (or ions to good approximation) in 1 kmol (or atomic mass) of material. Therefore,

$$M = N_A m = (6 \times 10^{26} \text{ atoms/kmol})(3.84 \times 10^{-26} \text{ kg/atom})$$
$$= 23 \text{ kg/kmol}$$

Sodium atoms have an atomic mass of 23 kg/kmol. Therefore these are positive sodium ions. (Did your inspection of the figure tell you they were positive?) ▮▮

26.7 Sources of Magnetic Fields

Before turning our attention to the subject of motors, let us learn another way that magnetic fields can be obtained. A simple experiment can be done to show that magnets are not the only source of magnetic fields. The experiment is shown in Figure 26.14(a). Several compasses are set close to a wire. When the current I in the wire is zero, all the compasses point north. But when a large current flows, the needles circle the wire as shown.

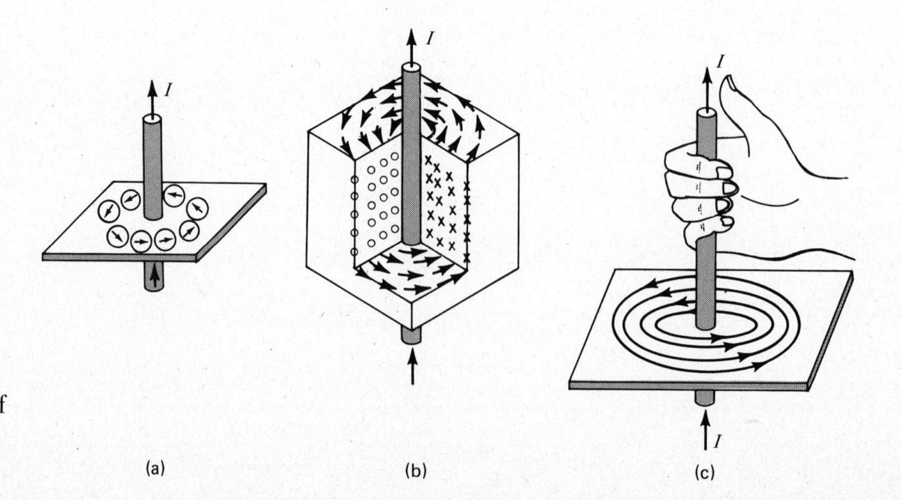

Figure 26.14 A current in a wire causes a magnetic field. The direction of the field can be found using the right-hand rule shown in (c).

The compass needles line up along the field lines. We therefore conclude that a wire carrying a current I generates a magnetic field. In the case of a long straight wire, the field is as shown in Figure 26.14(b). The field circles the wire; the field lines have no starting point and no end point.

A right-hand rule can be devised for determining the direction of the field lines near a wire. It is shown in Figure 26.14(c).

Grasp the wire with your right hand so that your thumb points in the direction of the current. Your fingers then circle the wire in the same sense (direction) as the field lines do.

Both experiment and theory show that B varies in a simple way outside a *long straight wire.* At a distance r from the axis of the wire,

$$B \text{ (long straight wire)} = \frac{\mu_0 I}{2\pi r}$$

All quantities must be in the SI system. The quantity μ_0 is exactly $4\pi \times 10^{-7}$ T·m/A. It is called the *permeability of free space.* We shall see later that it is important in other aspects of electromagnetism.

To see how large this magnetic field is, let us evaluate it for $r = 0.050$ m and $I = 20$ A. Substituting these values gives $B = 8.0 \times 10^{-5}$ T or 0.80 G. The field is comparable to the earth's magnetic field at this distance from the wire.

In order to obtain a stronger magnetic field, the wire can be made into a coil. This then gives rise to a magnetic field such as the one shown in Figure 26.15. If you use the right-hand rule (grasp a loop of the coil), you will see that the field direction is as shown. The field lines circle back upon themselves. They therefore have no beginning and no end. The equation for B at the center of the coil is

$$B = \frac{N\mu_0 I}{2a}$$

where N is the number of loops and a is the radius of the loop.

Very often it is desired to have a uniform magnetic field. This can be obtained in the center of a very long coil of wire such as the one shown in Figure 26.16. We call this type of coil a solenoid. (Use the right-hand rule by grasping any of the wires to see that the field is directed as shown.) Notice how the field lines circle through the solenoid. Here, too, they circle back on themselves. They have

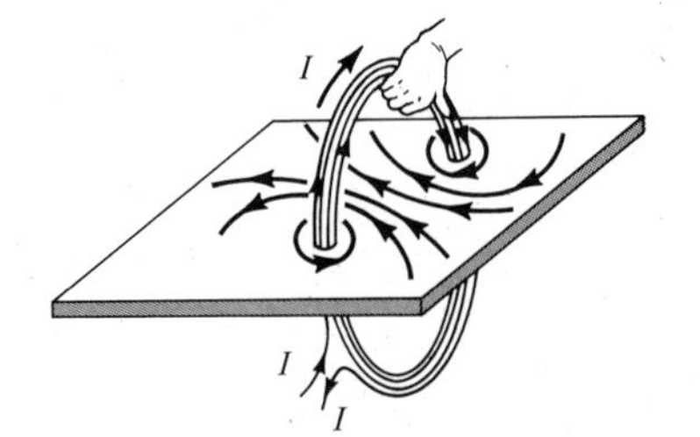

(a) Perspective view

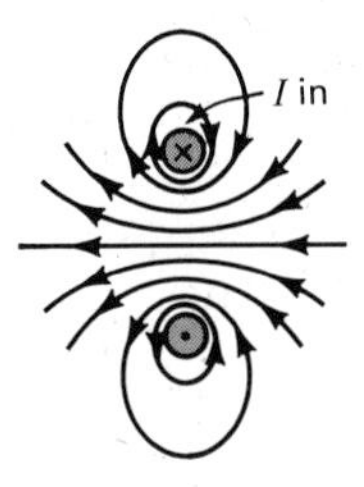

(b) Cross-sectional view

Figure 26.15 The magnetic field at the exact center of a coil (N loops and radius a) is given by $B = N\mu_0 I/2a$.

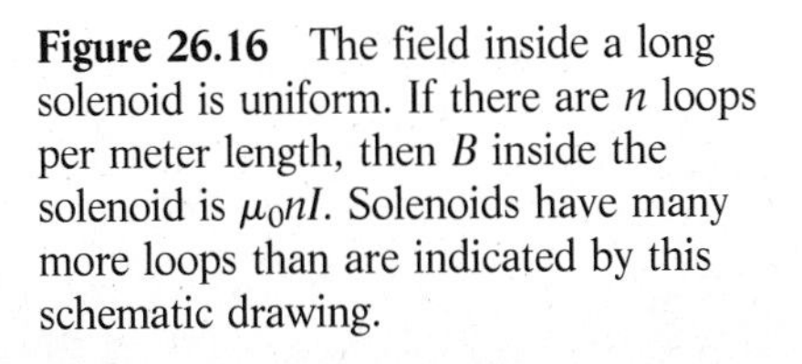

Figure 26.16 The field inside a long solenoid is uniform. If there are n loops per meter length, then B inside the solenoid is $\mu_0 nI$. Solenoids have many more loops than are indicated by this schematic drawing.

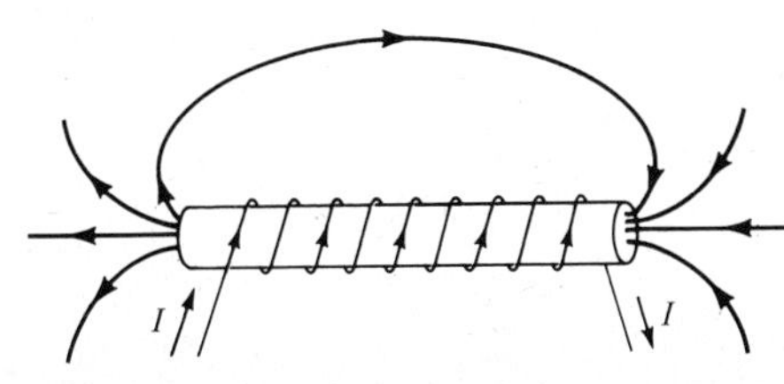

no beginning and no end. Inside the solenoid, the field is uniform and directed along the cylinder axis. The equation for B inside the solenoid is

$$B = \mu_0 n I$$

where n is the number of loops per meter of length.

26.8 Electromagnets

You may already have noticed a striking fact evident in Figure 26.16. The magnetic field of the solenoid looks very much like that of a bar magnet. Notice that it has a north end (left) and a south end (right). In fact, if the solenoid current and the number of loops on it are large, the solenoid will pick up iron nails. But an air-core solenoid such as this is not a strong magnet. It is usually too weak to be practical.

However, if the solenoid is wound on an iron core, the field is increased many times over. Typical iron-core fields are a few hundred times larger than those obtained with an air (or wood or plastic) core. (Remember that iron and its alloys are the only common metals that affect the magnetic field much.) Use is made of this fact in electromagnets.

A simple electromagnet is shown in Figure 26.17. It is a solenoid wound on an iron core. When no current flows through the coil, the core is "dead"; it has no magnetic field. But when the current is on, the coil and core act as a strong magnet. An electromagnet is of great value because it can be turned on and off. Also, all shapes of electromagnets can be made. One simply shapes the iron core and then winds a coil upon it. How would you construct a horseshoe electromagnet?

Many devices make use of electromagnets. The hammer on a door bell is pulled back by one. Electrically operated remote switches are frequently tripped by an electromagnet. One type of circuit breaker is activated by an electromagnet. When the current through the electromagnet becomes large, the magnet pulls a switch open. This acts like a fuse to open the circuit when the current in it becomes too large. There are many other uses of electromagnets. Can you list a few more?

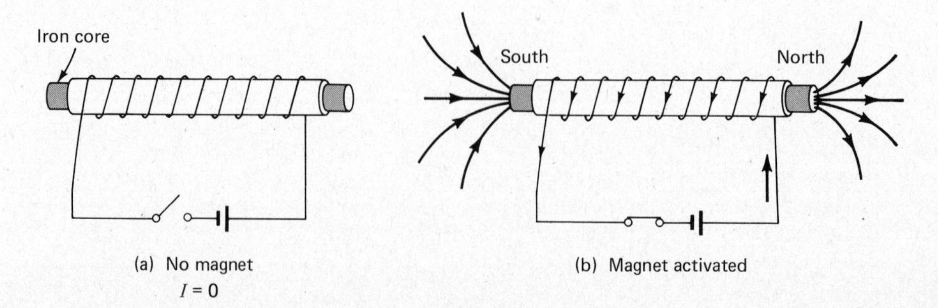

Figure 26.17 The iron core in the solenoid increases the strength of the electromagnet by a factor of hundreds.

26.9 Torque on a Current Loop

We have seen many times that a compass needle lines up in a magnetic field. It is simply a bar magnet. As shown in Figure 26.18(a), a bar magnet experiences a torque in a magnetic field. The field tries to line up the magnet. At equilibrium the axis of the magnet would lie along the field lines.

A similar situation exists for an electromagnet. Even a coil of wire acts like a magnet. It too tends to line up in the field, as shown in Figure 26.18(b). The torque on the magnet (and coil) is largest when the field lines are perpendicular to the axis of the magnet. In Figure 26.18(a) when $\theta = 90°$, the lever arms of the forces are largest. When $\theta = 0$, the forces are in line and they cause no turning effect. It can be shown that the torque is proportional to $\sin \theta$. In the case of a flat coil with N loops on it,

$$\text{Torque on coil} = AINB \sin \theta$$

where A is the end area of the coil. As you might expect, the torque is largest if N, A, and I for the coil are large. The quantity AIN is often called the *magnetic moment, M,* of the coil.

26.10 Moving Coil Meters

Many electric meters make use of the magnetic field torque on a current-carrying coil. Such a moving coil device is shown schematically in Figure 26.19. Notice that the large magnet furnishes the field in which the coil rotates. (In dc meters the magnet is usually a permanent magnet. In ac meters it is an electromagnet.) The current to be measured, I, flows through the coil. As we saw in the last section, the coil experiences a torque proportional to I. The coil rotates until this torque is balanced by the torque due to the restoring coil spring. Therefore, the rotation of the coil is a measure of I. This rotation is shown by the pointer needle, which moves along the meter scale.

Actual meters have design features not shown in Figure 26.19. The coil

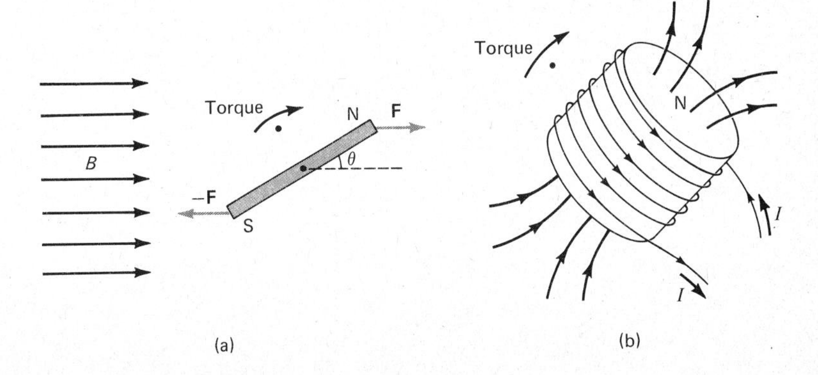

Figure 26.18 The magnetic field tries to line up the bar magnet and the loop with the field. At equilibrium a line from the south to the north pole of each will be along the field lines.

SUPERCONDUCTORS

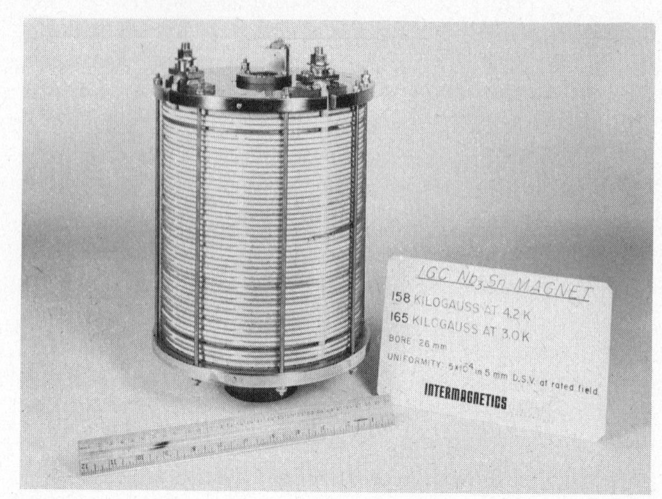

Courtesy of Intermagnetics General Corp.

As you know, the lowest temperature possible is −273°C or 0 K. Temperatures below about 10 K are obtained by use of liquid helium. This boils at 4.2 K under normal pressures. At still lower pressures its boiling point is lower than this. Liquid helium can therefore be used to cool substances to within a degree of 0 K.

Helium was first liquefied by Kammerlingh Onnes in 1908. He then used it to measure the electric resistance of substances at very low temperatures. To his amazement, he found that some materials had absolutely no resistance below a certain critical temperature. The critical temperatures for lead, mercury, tin, and indium are 7.2, 4.2, 3.7, and 3.4 K, respectively. It is now known that certain alloys have critical temperatures near 20 K. Below the critical temperature, these materials are called superconductors.

It must be emphasized that a superconductor has absolutely no electric resistance. When an induced emf causes a current to flow in a superconducting ring, the current continues even after the emf has become zero. Currents have been maintained for years in such a ring without an energy source. There is no doubt that the resistance is exactly zero.

For many years superconductivity found little practical use. However, recently new, higher-temperature superconducting materials have raised the useful temperature to near 20 K. Solenoids using superconducting wire, such as the one shown in the photo, are now fairly widely used. The one shown produces a magnetic field nearly 1000 times larger than a similar one that uses ordinary wire. This is only one of many examples of the use of superconducting wire in coils.

---

usually has an iron core. This greatly increases the torque on the coil. By proper shaping of the pole pieces, the field can be kept reasonably constant on the coil. This allows the meter scale to have uniform graduations.

26.11 The Ammeter

Ammeters and voltmeters are constructed using systems such as the one shown in Figure 26.19. This coil-magnet system is called the meter movement. A typical movement might have a resistance of 60 Ω (mostly the resistance of the coil). It would deflect full scale for a current through it of perhaps 5×10^{-4} A. Such a movement can be used to construct either an ammeter or voltmeter.

In the construction of an ammeter the movement is connected to the circuit shown in Figure 26.20(a). The whole circuit is contained inside the meter housing. A low resistance called a *shunt,* R_s, is placed in parallel with the movement. The value of R_s depends on the current range one wishes the ammeter to have.

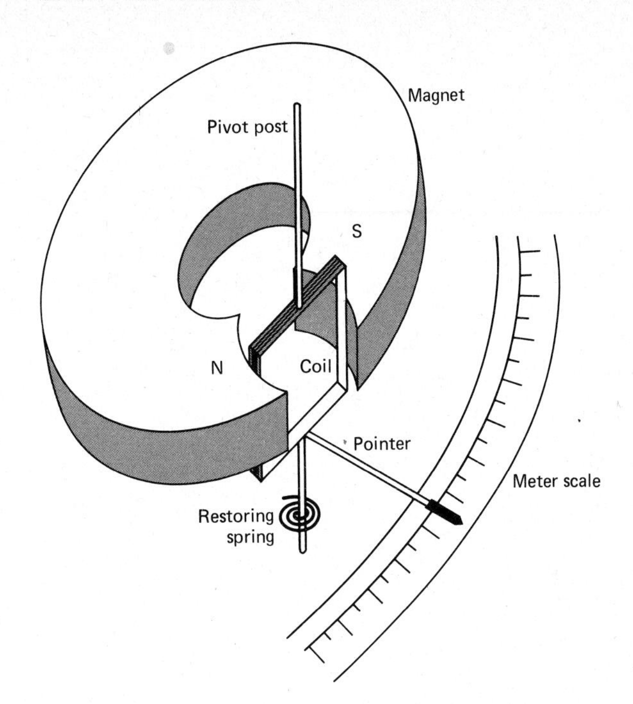

Figure 26.19 The coil rotates in the magnetic field until the torque due to the magnetic field is balanced by the torque due to the restoring force of the spring. In practice, the coil usually has an iron core. Why?

Suppose that you wish to construct a 2.0-A ammeter. The movement to be used has a resistance of 40 Ω and deflects full scale for a current of 5×10^{-4} A. Our ammeter must have a value for R_s such that 5×10^{-4} A flows through the movement when 2.0 A flows through the meter. The situation is shown in Figure 26.20(b).

As we see, 2.0 A comes into the meter. Of this, 5×10^{-4} A goes through the movement. The movement therefore deflects full scale. (Its reading would be 2 A.) The remainder of the current, 1.99995 A, goes through the shunt. To find R_s we simply write the loop equation for the movement and R_s. It is (in volts)

$$-(5 \times 10^{-4})(40) + (1.99995)R_s = 0$$

Solving gives

$$R_s = 0.0100\ \Omega$$

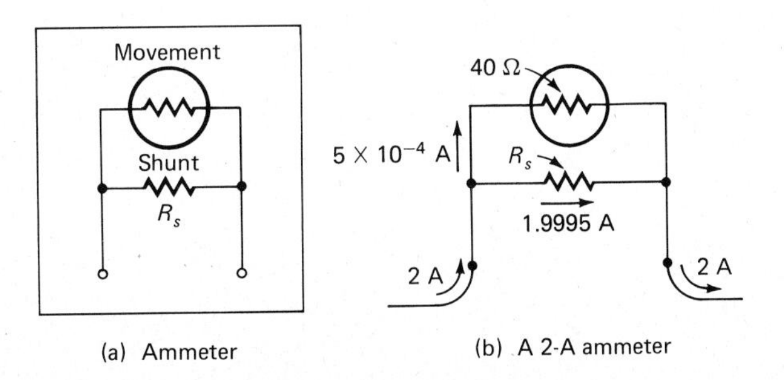

Figure 26.20 A single-scale ammeter. R_s is very small.

Notice how small the shunt resistance is. If you open the box of the ammeter, you can usually see it. It is just a short piece of fine wire.

You will recall that an ammeter should have very little resistance. Otherwise it would disturb the circuit into which it is connected. As you see, this meter will have a resistance slightly smaller than 0.01 Ω. (Remember that the equivalent of parallel resistors is smaller than any of the resistors.) It therefore satisfies our requirement for low resistance.

26.12 The Voltmeter

The movement used for a voltmeter is the same as that used for an ammeter. However, it is connected differently inside the meter box. A typical voltmeter circuit is shown in Figure 26.21(a). Notice that a large resistor, R_x, is placed in series with the movement. Let us see how large R_x must be in a typical situation.

Suppose that the movement deflects full scale for a current of 5×10^{-4} A and has a resistance of 40 Ω. We wish to construct a 25-V voltmeter (i.e., it deflects full scale for 25 V across its terminals). Therefore, as shown in part (b) of Figure 26.21, a current of 5×10^{-4} A must flow through the meter when a potential difference of 25 V is placed across its terminals.

The equivalent resistance of the meter is $R_x + 40\ \Omega$. Ohm's law, $V = IR$, then gives, in the usual units,

$$25 = (5 \times 10^{-4})(R_x + 40)$$
$$R_x = 50{,}000\ \Omega$$

Notice how large this resistance is. Why should a good voltmeter have a high resistance?

26.13 The dc Motor

We have seen in previous sections that a current-carrying coil experiences a torque in a magnetic field. This fact is basic to the operation of most electric motors. A simple dc motor is shown schematically in Figure 26.22. The shaft along AA' is the motor shaft.

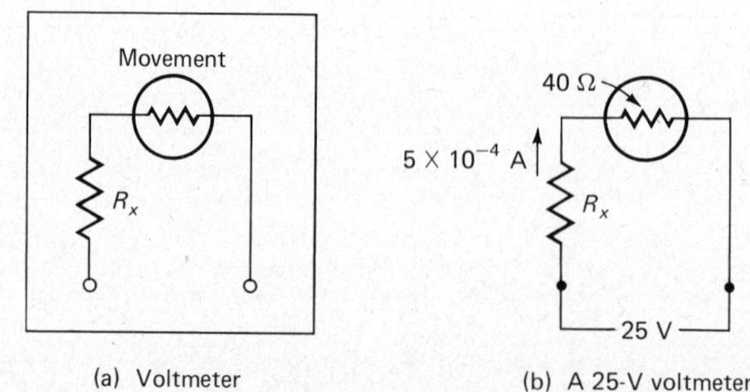

Figure 26.21 A single-scale voltmeter. R_x is very large.

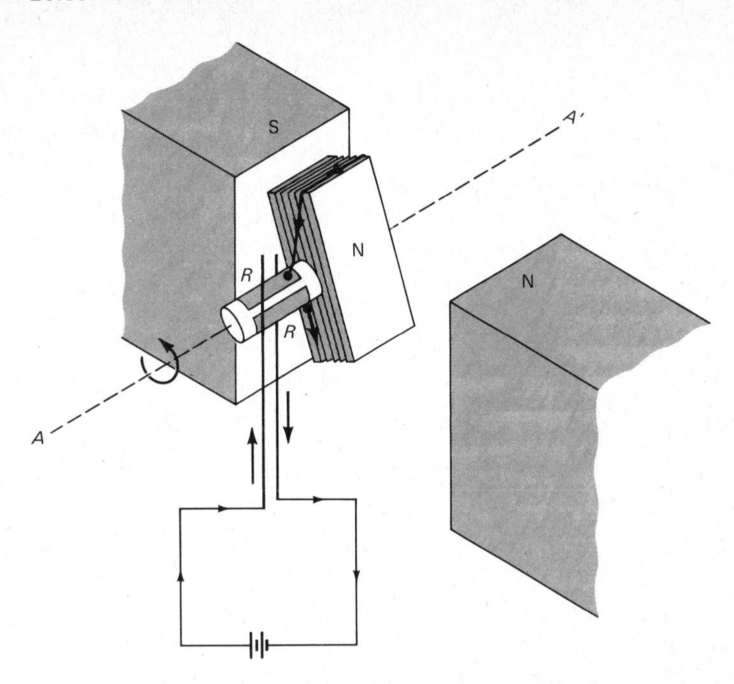

Figure 26.22 The dc motor. What is the purpose of the split rings, *R*?

As we see, a coil is pivoted so that it can rotate on an axis through *AA'*. This coil is called the *rotor* or *armature.* The coil is mounted between north and south poles of magnets. The poles cause a steady magnetic field to be impressed on the rotor coil. (The magnet poles may be from either permanent or electromagnets.)

In order to run the motor, a current is sent through the rotor coil as shown. The coil experiences a torque, which causes it to rotate. For example, consider the instant shown in Figure 26.22. The north pole of the rotor coil is repelled by the large magnet's north pole. A similar situation exists for the south pole. As a result, the torque causes the rotor to rotate in the direction indicated about *AA'* as axis.

Eventually, however, the coil will rotate until its north pole is next to the south pole of the large magnet and its south pole will be next to the north pole of the magnet. These attractions try to stop the coil from rotating further. If something was not done, the rotor would come to a stop.

To get around this difficulty, the current direction in the coil is reversed when the coil reaches this position. When the reversal has been made, the situation is again similar to that shown in Figure 26.22. The coil will continue to experience a torque in the proper direction to maintain its rotation. Let us now see how the current in the coil is reversed at just the proper instant.

Electrical connection to the coil is made through metal half-cylinders indicated by *R* in the figure. These are called split-ring commutators and act as slip rings. Contact is made to them by so-called brushes, which slide over the cylindrical surface. In the simple motor shown the brushes are simply pieces of spring steel. Usually the brushes are self-lubricating pieces of conducting material (often graphite) held in place by compressed springs.

Examine the arrangement of the split-ring commutator and brushes in Figure 26.22. You should notice that the current direction through the coil reverses after each half-rotation of the rotor. This in turn causes the torque acting on the rotor to continue the rotation in the same direction.

This simple motor has a serious drawback. The torque on the rotor is sometimes strong, but sometimes it is zero. When the rotor coil has its north-to-south line vertical in Figure 26.22, the torque is largest. When the coil's magnet is in line with the large magnet, the torque is zero. Practical motors need a more steady torque than this.

To obtain a steady torque, several coils are wound on the same armature. This is shown schematically in Figure 26.23. The commutator ring is now split into twice as many sections as there are coils in the armature. Contact is made with each coil as its plane rotates into a horizontal position. At that time, the coil experiences its maximum torque. As a result, the strong torque on the armature is maintained as the armature rotates.

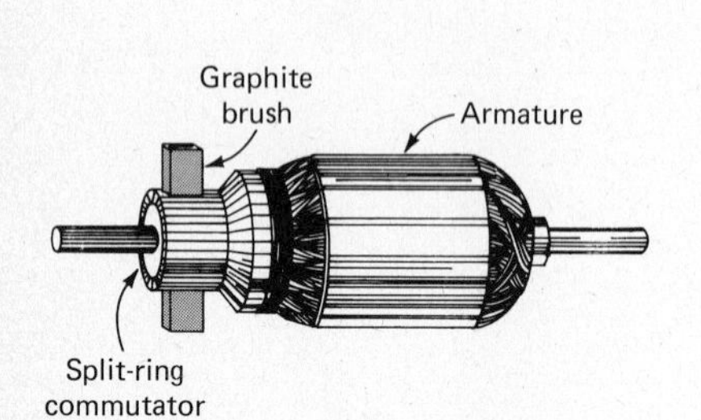

Figure 26.23 The armature in a practical motor consists of many coils wound at various angles.

26.14 Magnetic Properties of Materials

Most materials are not strongly attracted to magnets. Iron is the material we usually think of as being strongly attracted. There are two others of importance—cobalt and nickel. When a magnet is brought close to iron, nickel, cobalt, and their alloys, they become highly magnetized. They then act as magnets. This topic of induced magnetism was discussed in Section 26.1.

Let us refer to an experiment by which the magnetic properties of materials can be described. Suppose that one has a long, hollow solenoid as shown in Figure 26.24. When a current flows in its wires, the magnetic field inside it is B_v, let us say. The subscript v tells us that this is the field when the solenoid core is empty, that is, a vacuum.

When the vacuum core is replaced by a material, the field in the solenoid changes to some new value, B. We write

$$B = k_m B_v$$

The quantity k_m is called the *relative permeability* of the material of the core. For most materials k_m is extremely close to unity. This is just another way of saying that most materials are not much affected by steady magnetic fields. As typical examples, $k_m = 1.000021$ for solid aluminum and 0.999984 for solid lead.

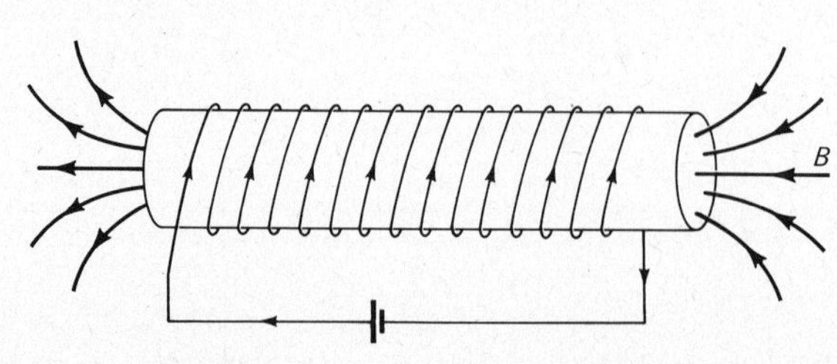

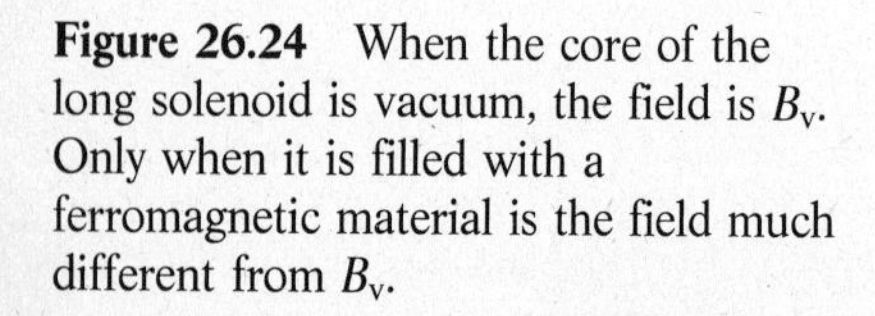

Figure 26.24 When the core of the long solenoid is vacuum, the field is B_v. Only when it is filled with a ferromagnetic material is the field much different from B_v.

Some materials increase the field slightly. For them k_m is slightly larger than unity. They are called *paramagnetic materials.* Other materials decrease the field slightly. For them k_m is slightly less than unity. They are called *diamagnetic materials.* The effects of both types are so small that they are usually ignored.

Unlike most materials, iron, nickel, and cobalt greatly change the magnetic field in which they are placed. When the core of the solenoid is filled with them (or their alloys), B is increased by a factor of hundreds. For these materials $k_m > 100$. They are called *ferromagnetic materials.* They have $k_m \gg 1$. Use is made of them wherever strong magnetic fields are required. We have already seen how they are used in electromagnets. In the next chapter we shall see other important uses of them.

26.15 Ferromagnetic Materials

As we have seen, ferromagnetic materials behave quite differently than most substances. They increase greatly any magnetic field in which they are placed. The reason for this is quite complicated. It involves the atomic details of these substances. Even so, we can understand a great deal about their behavior.

The atoms of ferromagnetic substances are very cooperative with each other. They pack together in such a way that they add their tiny atomic magnets together. (Many atoms act like bar magnets, but only in ferromagnetic substances do the atoms cooperate this way.) As a result, billions upon billions of cooperating atoms form a magnet within the solid substance. These magnetized regions are called *domains*. Typically they are only a fraction of a millimeter in largest dimension.

A domain is a tiny magnet formed by the cooperative action of many atoms in a ferromagnetic material.

Although the atoms wish to band together in this way, they are unable to do so if the temperature is too high. At high temperatures the atoms have a great deal of thermal energy. Their vibration breaks the domains apart. Each ferromagnetic substance has a maximum temperature at which domains can exist. This is called the *Curie temperature* (named after Madame Curie's husband). At temperatures above the Curie temperature, ferromagnetic substances lose their ferromagnetic properties. For iron, the Curie temperature is 770°C. It is 358° for nickel.

A bar of ordinary iron is usually not magnetized. Its atoms are arranged in domains and each domain acts like a tiny bar magnet, but the domains in the bar are not lined up with each other. This is shown in Figure 26.25. The arrow within each domain indicates the direction from its south pole to its north pole. The randomly oriented domains tend to cancel each other's effect. Therefore, the bar of iron acts as though it is unmagnetized.

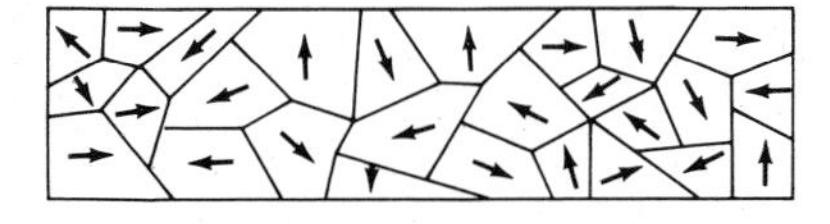

Figure 26.25 Each tiny domain acts like a magnet. The arrow shows the south-to-north direction in each. This diagram is only schematic. The domains are much smaller than shown.

But strange things happen to the domains if the iron is placed in an external magnetic field. These effects are shown clearly in Figure 26.26. There we see under a microscope a specially prepared piece of iron. It has been dusted with a

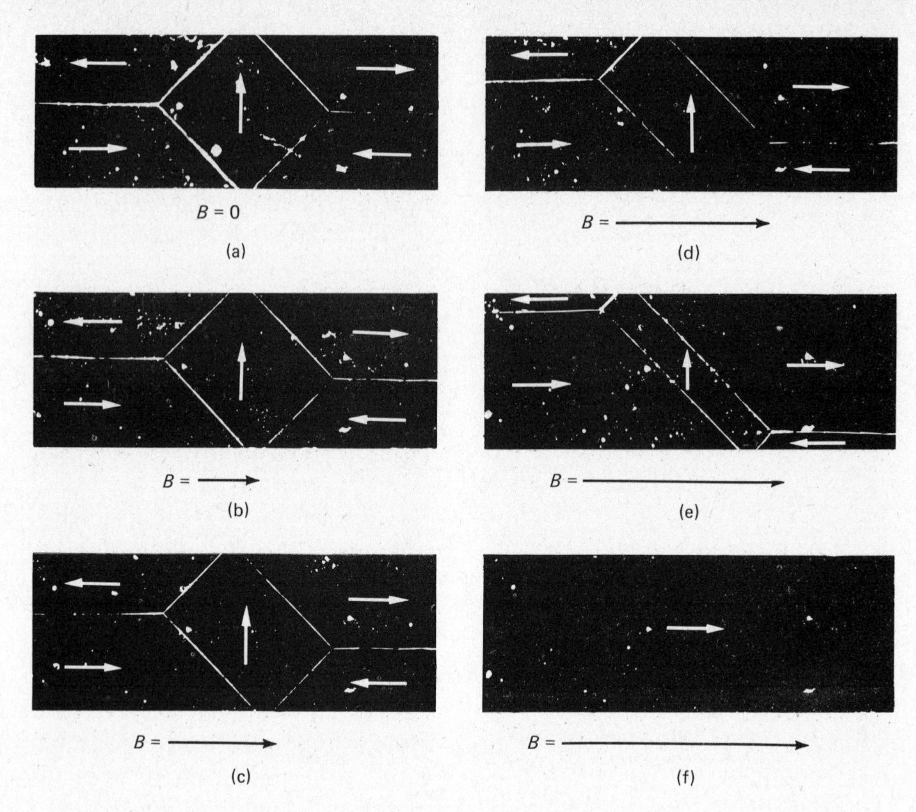

Figure 26.26 The photos show domain growth in an increasing rightward directed magnetic field. (Courtesy of General Motors Research Labs.)

powder that makes the domain walls visible. In (a) we see the material before it is placed in the external field. The magnetization in the various domains (shown by the arrows) tends to cancel.

Parts (b) through (f) of the figure show what happens as the iron is subjected to an increasing rightward-directed external field. Notice how the domains that are properly oriented in the field grow as the field increases. The domain walls move so as to allow more favorable orientation of the iron. Finally, at very high external fields the material is completely magnetized in a single direction, along the field, as shown in (f).

We can easily measure this effect for any type of iron or steel. The material is made into a cylindrical shape and used as the core for a solenoid. Then we compare the magnetic field within the solenoid to that which would exist without the core of iron. Call the field with the iron core B and that without B_V. We then increase the current in the coil so as to cause B_V and B to increase. The resultant values of B and B_V are plotted against each other in Figure 26.27. (Notice that the axes differ in scale; B is hundreds of times larger than B_V.) This graph is called the *magnetization curve* for the iron.

At low external fields (B_V) the field due to the iron increases greatly as the domains grow to increase the alignment. Eventually, though, the material is all aligned and the field reaches a nearly constant maximum value called the *saturation field.* In the case shown in the figure, this saturation field is about 1.8 T. At this point only about .001 of the total field is due to the current in the coil. All

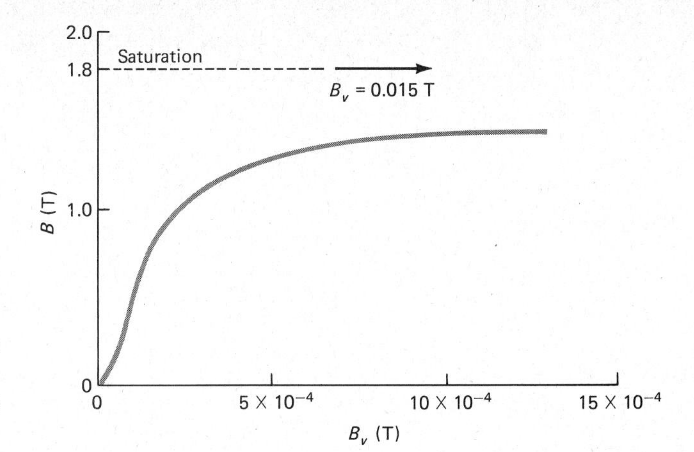

Figure 26.27 A long solenoid has a core of ordinary transformer steel. The field due to the wires of the solenoid (a vacuum-core solenoid) is B_V. The steel causes the actual field to be many times larger.

the rest is due to the magnetic effect of the iron. Clearly, iron increases magnetic fields tremendously.

An interesting effect occurs when the solenoid current is turned off. The domains in the ferromagnetic material do not become completely disordered. In so-called hard iron the domains remain lined up. The iron is therefore still magnetized and is then a permanent magnet. (Actually, to line them up in the first place, the iron would have to be heated.) But in soft iron the domains become largely disordered when the solenoid is turned off. Why would one want the core of an electromagnet to be made of soft iron, not hard iron?

This tendency of the domains to resist change leads to the phenomenon of *hysteresis.* Hysteresis is a lagging in the effect a magnetizing force has on a magnetic material. The magnetization of ferromagnetic materials depends on their past history. The magnetization is not necessarily zero when the magnetizing field is decreased to zero. The domains show a property very much like friction. They resist change. This causes heat loss in the ferromagnetic material as the domains are caused to line up or disalign.

Depending on the use to which a ferromagnetic material is put, it should have very definite magnetic properties. For some purposes it should have large hysteresis. For others it should show no hysteresis. There are other large features of its magnetic behavior that are also important. Metallurgists have designed ferromagnetic alloys of widely different properties. Each is tailored to a particular use in technology.

Summary

The north pole of a magnet is the one that points north on the earth. Like poles repel each other, and unlike poles attract.

Most materials are only slightly affected by steady magnetic fields, but iron, nickel, and cobalt (and their alloys) are ferromagnetic. They greatly increase magnetic fields in which they are placed. Magnets can induce unmagnetized ferromagnetic materials to become magnets. As a result, a magnet attracts them.

A compass needle is a bar magnet. In a region where a compass experiences a magnetic force, a magnetic field is said to exist. The compass needle points along the lines of the magnetic field. Magnetic field lines come out of the north pole of a magnet and enter the south pole.

The earth acts like a large magnet. Its equivalent bar magnet has its south pole in the Arctic and its north pole in the Antarctic.

A wire of length L carrying a current I at an angle θ to the magnetic field lines experiences a force

$$F = ILB \sin \theta$$

The quantity **B** measures the strength of the magnetic field. It is called the magnetic flux density or the magnetic induction. In the SI system its unit is the tesla (T), which is also called the weber per square meter (Wb/m^2). Another unit sometimes used for **B** is the gauss (G); 1 T = 10^4 G. The gauss should not be used in our equations. We define **B** to be a vector quantity. It is directed along the field lines.

The direction of the force experienced by a current in a magnetic field is given by the right-hand rule. The rule is illustrated in Figure 26.7. In using the rule, one notices that the current and a field line through it determine a plane. The force is perpendicular to this plane.

A positive charge q shoots with velocity **v** through a magnetic field **B**. It experiences a force

$$F = qvB \sin \theta$$

where θ is the angle between **v** and **B**. The direction of **F** is given by the right-hand rule with the direction of **v** replacing the current direction (see Figure 26.11). A negative charge experiences a force in the opposite direction.

A charged particle moving perpendicular to a uniform magnetic field follows a circular path. The circle is defined by equating the magnetic force, qvB, to the required centripetal force, mv^2/r.

When a current flows through a wire, a magnetic field circles the wire. The direction of the field can be found using a right-hand rule. The rule is illustrated in Figure 26.14. The magnetic field near a coil, or a solenoid, looks much like that of a bar magnet. An electromagnet can be made by winding a coil on an iron core. The iron greatly increases the magnetic field of the coil.

A coil that carries a current in a magnetic field experiences a torque. The torque is maximum when the field lines are tangent to the cross-sectional area defined by the coil. The torque is zero when the field lines are perpendicular to the cross-sectional area of the coil. This torque is used to rotate the moving element in motors and moving coil meters.

To make an ammeter from a meter movement, one places a low resistance shunt in parallel with the movement. To construct a voltmeter, one places a high resistance in series with the movement.

The relative magnetic permeability, k_m, of a material is defined in the following way. Suppose that the magnetic field in a very long solenoid is B_v when vacuum-filled. When the vacuum is replaced by a given material, the magnetic field changes to B. The field is changed by the factor $k_m = B/B_v$. For paramagnetic materials k_m is slightly larger than unity. For diamagnetic materials k_m is slightly less than unity. Only the ferromagnetic materials are not in one of these two classes. For ferromagnetic materials $k_m \gg 1$.

Ferromagnetic materials contain magnetically ordered regions called domains. To magnetize the material, the domains are altered so that they all take up the same orientation. Depending on the exact type of material, the domains show a hysteresis effect. Their orientation and size depend on the past history of the material.

Questions and Exercises

1. People often show the magnetic field near a magnet by sprinkling iron filings on a glass plate laid on the magnet. Explain why the iron filings show the field lines. Would brass filings or glass fibers work?
2. You are given a bar magnet. How could you determine which is its north pole? List as many ways as you can. What if it were a solenoid-type electromagnet run from a battery?
3. In Figure P26.1 is shown a magnet with some pieces of iron in the vicinity. Sketch the magnetic field in this region. Your sketch should show that "Magnetic field lines follow an iron path." Why do they do that? How would your sketch be changed if the iron pieces were replaced by aluminum pieces?

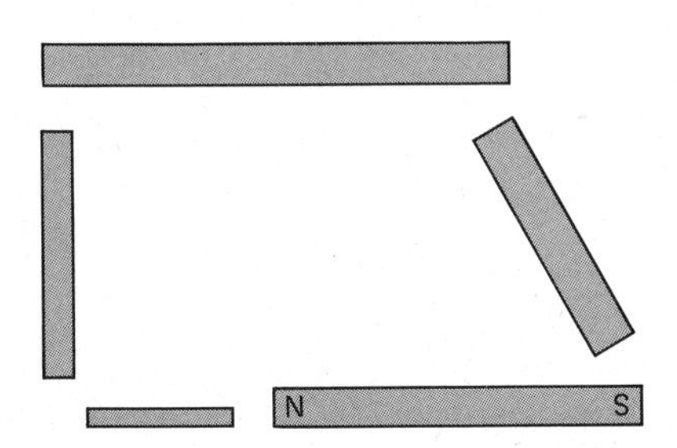

Figure P26.1

4. Refer to Figure P26.2. Describe the force experienced by the loop of current-carrying wire.
5. A straight wire carries a current along the axis of a solenoid. The solenoid also has a current flowing in it. Describe the force on the straight wire due to the magnetic field of the solenoid.

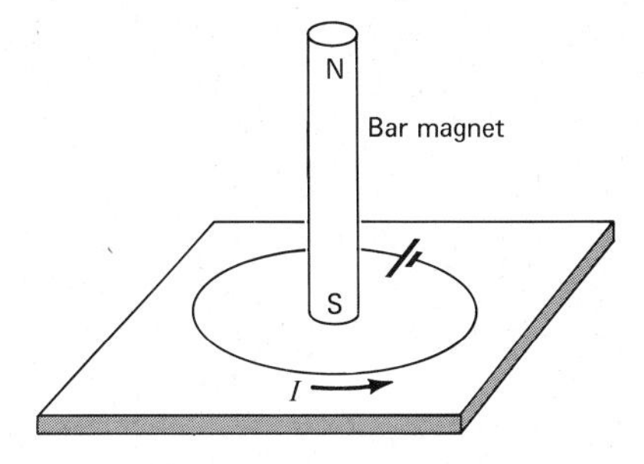

Figure P26.2

6. A charged particle is shot straight west, high above the earth along its magnetic equator. Could the earth's magnetic field cause the particle to circle the earth? Explain your answer.
7. Two parallel straight wires lie side by side on a tabletop. Equal currents are sent through them. Describe the force on each wire due to the other if the currents are (a) in the same direction and (b) in opposite directions.
8. A charged particle shoots toward the earth from distant space. It is moving along a radial line. If the particle is coming in toward the equator, it must have a high energy if it is to reach the earth's surface. But if it is headed in at one of the poles, it can easily reach the surface of the earth. Explain. Ignore collisions with molecules in the atmosphere.
9. Examine a loudspeaker and explain how the force that moves the diaphragm back and forth originates.
10. The unit of current, the ampere, can be defined in terms of forces and distances. Two parallel wires a given distance apart carry identical currents. How can the current be found by measuring the force on a given length of one wire due to the current in the other?

Problems

1. A 0.50-T magnetic field is directed vertically downward into a table top. On the table lies a wire that carries a current of 30 A straight westward. Find the force on a 50-cm length of the wire due to the field.
2. Suppose in Figure P26.3 that the magnetic field is 350 G and I is 20 A. Find the magnitude and direction of the force on a 30-cm length of the wire if the angle θ is (a) 90°, (b) 70°, and (c) 0°.
3. Refer to Figure P26.3. If $B = 250$ G and $I = 15$ A, find the magnitude and direction of the force on a 50-cm length of the wire if (a) $\theta = 90°$, (b) $\theta = 60°$, and (c) $\theta = 0°$.

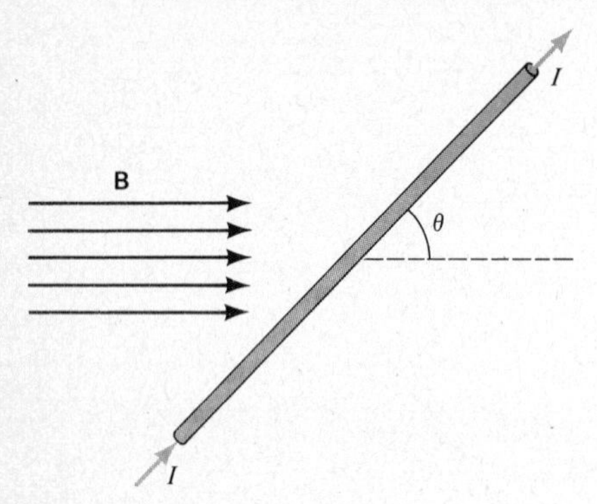

Figure P26.3

4. A horizontal wire carries a current of 20 A straight north. It experiences an eastward-directed force of 3×10^{-4} N for each 40 cm of its length. Assuming the force is due to a magnetic field, what are the magnitude and the direction of the magnetic field?
5. In Puerto Rico the horizontal component of the earth's magnetic field is 0.30 G. It is directed northward. The vertical component is 0.47 G and is directed downward. Find the magnitude and direction of the force on a 2.0-m length of vertical wire that is carrying a current of 17 A straight upward.
6. Repeat Problem 5 for a horizontal wire carrying a current southward.
7. In Ohio the earth's magnetic field is directed at an angle of 72° below the horizontal and is northward. Its value is 0.61 G. A horizontal, north-south wire there carries a current of 30 A northward. Find the magnitude and direction of the force on a 2.0-m length of the wire.
8. Repeat Problem 7 for a wire carrying the current vertically upward.
9. Suppose the wire in Figure P26.3 is replaced by a proton moving in the direction of the current with a speed of 5.0×10^6 m/s. The field has a value of 0.135 T. Find the magnitude and direction of the force on the proton if the angle θ is (a) 90°, (b) 50°, and (c) 0°.
10. Suppose the wire in Figure P26.3 is replaced by a charged particle moving in the direction of the current. Its speed is 5×10^6 m/s and $B = 250$ G. What are the magnitude and the direction of the force on the particle if $\theta = 40°$ and the particle is (a) an electron and (b) a proton?
11. A sodium ion ($q = +e$, $m = 23 \times 1.66 \times 10^{-27}$ kg) is moving with speed 3×10^4 m/s perpendicular to a magnetic field. What must be the magnitude of the field if the particle is to follow a circle with 20-cm radius?
12. A lithium ion ($m = 7 \times 1.66 \times 10^{-27}$ kg; $q = e$) is shot along the x axis in the $+x$ direction into a uniform magnetic field. The magnetic field is perpendicular to and into the x-y plane. (See Figure P26.4.) Its magnitude is 0.0350 T. (a) Will the ion describe a circular path above or below the x axis? (b) What will be the radius of the circle if the ion's speed is 4.8×10^5 m/s?

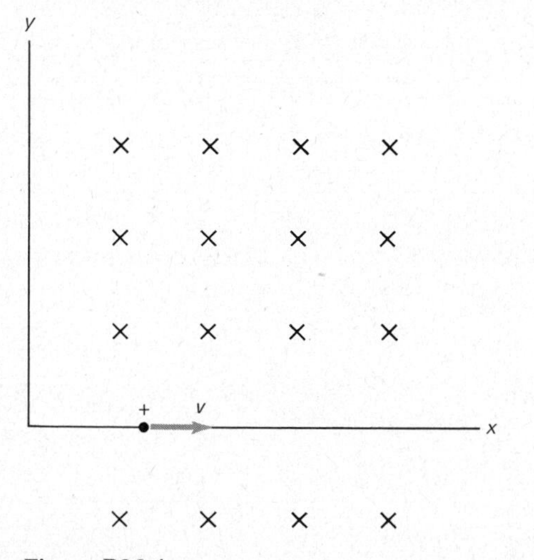

Figure P26.4

13. Repeat Problem 12 for a chlorine ion ($m = 35 \times 1.66 \times 10^{-27}$ kg; $q = -e$).
14. Each of the electrons in the beam of a TV tube has a momentum of 7.3×10^{-23} kg·m/s. If the beam is traveling perpen-

*Problems marked with an asterisk are not as easy as the unmarked ones.

dicular to the earth's magnetic field (0.75 G), how large is the radius of curvature of its path?

*15. The particle beam in a mass spectrometer consists of two types of chlorine ions. Both have the same charge and speed, but one has a mass of $35 \times 1.66 \times 10^{-27}$ kg and the other a mass of $37 \times 1.66 \times 10^{-27}$ kg. If the lighter one follows a path with $r = 9.00$ cm, what will be the radius of the path for the heavier one?

**16. A magnetic field is directed along the $+x$ axis with $B = 0.150$ T. A proton is shot into the field with speed 7×10^6 m/s at an angle of 20° to the $+x$ axis. The proton will follow a helical path. Find the radius and pitch of the helix.

17. The *velocity selector* is a device used to select charged particles of a known velocity from a beam of particles. As shown in Figure P26.5, the beam is shot into uniform crossed electric and magnetic fields. (E** is as shown and **B** is into the page.) Show that only those particles that have $v = E/B$ will follow a straight line path through the region.

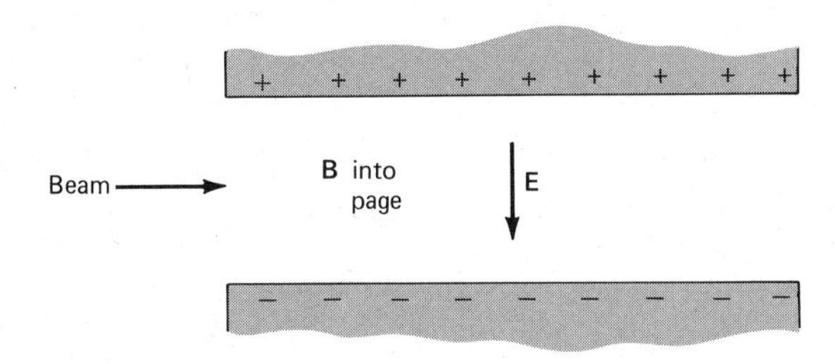

Figure P26.5

18. What is the magnitude of B outside a long straight wire at a point 5 cm from the axis of the wire? The current in the wire is 25 A.

19. How large a current must flow in a long straight wire to produce a field of 6 G at a distance of 2 cm from the axis of the wire?

*20. Two parallel wires lie 5.0 cm apart on a tabletop. They carry currents of 20 A in the same direction. (a) Find B at the position of one wire due to the other. (b) Find the force on one wire (per meter) due to the other. Is it attractive or repulsive?

*21. For the situation described in Problem 20, find the magnetic field midway between the two wires (a) if the current directions are the same and (b) if they are opposite.

22. A flat coil of wire has 200 loops. The radius of the coil is 3 cm. Find the field at the center of the coil when the current in it is 5 A.

23. How large a current must flow in the coil of Problem 22 to produce a magnetic field of 0.01 T at its center?

24. A long solenoid is wound on a wooden rod. It has 5000 loops on its 40-cm length. How large is the magnetic field in the solenoid when a current of 2 A flows in it?

25. How large a current must flow in the solenoid of Problem 24 if the field inside the solenoid is to be 200 G?

26. A flat circular coil of 50 loops has a radius of 3 cm. It is in a magnetic field of 500 G. How large is the torque on the coil when the current in it is 8 A and the axis of the coil makes an angle θ to the field lines with (a) $\theta = 0°$, (b) $\theta = 30°$, and (c) $\theta = 90°$?

27. A flat coil has a radius of 20.0 cm and has 50 loops of wire. It is suspended with its plane vertical and in a north-south direction. A current of 15 A flows in the coil. Find the torque on it due to the earth's field, which is 0.65 G and directed northward with a dip angle of 70°.

28. Repeat Problem 27 for the same coil lying on a horizontal tabletop.

29. The coil in Figure P26.6 ($N = 200$, $A = 3.0$ cm^2) is in the field of a magnet ($B = 0.60$ T). The current in the coil is 0.25 A. It flows as shown in the portion of the coil closest to the viewer. (a) Will the coil turn clockwise? (b) How large a torque acts on it in the position shown?

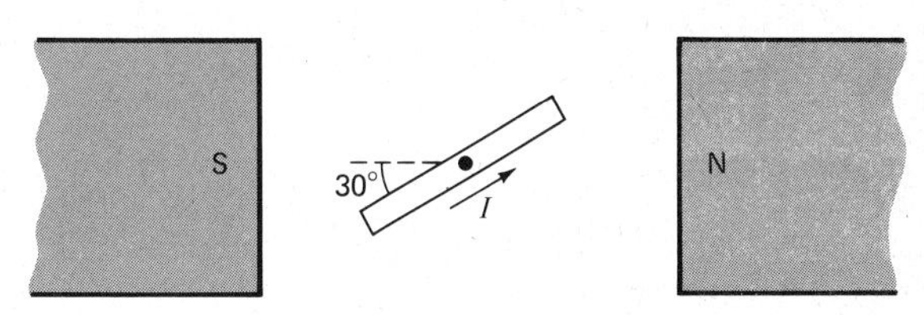

Figure P26.6

*30. In a certain motor the armature coil has an area of 12.0 cm^2. There are 50 loops of wire on it and the current through it is 2.0 A. A magnetic field of 0.85 T exists parallel to the plane of the coil. On the axle of the motor is wound a thin string with a mass M hanging from it. What is the largest value M can have if the motor is to be able to lift it? The axle radius is 0.70 cm.

31. A certain meter movement has a resistance of 60 Ω and deflects full scale for a current of 0.00100 A. How large a shunt resistance is needed to make this into a 0.50-A ammeter?

32. Using the same meter movement as in Problem 31, how large a series resistor is needed to make it into a 2.5-V voltmeter?

33. A certain meter movement has a resistance of 80 Ω and deflects full scale for a voltage of 2 mV across its terminals. What shunt resistance would be required to make this into a 1.50-A ammeter?

**Problems marked with a double asterisk are somewhat more difficult than the average.

34. How could you make a 12-V voltmeter from the movement in Problem 33?

**35. Refer to Figure P26.7. The meter movement has $R = 50\ \Omega$ and deflects full scale for a current of 0.200 mA. Find R_1 and R_2 to make this into the multirange meter shown.

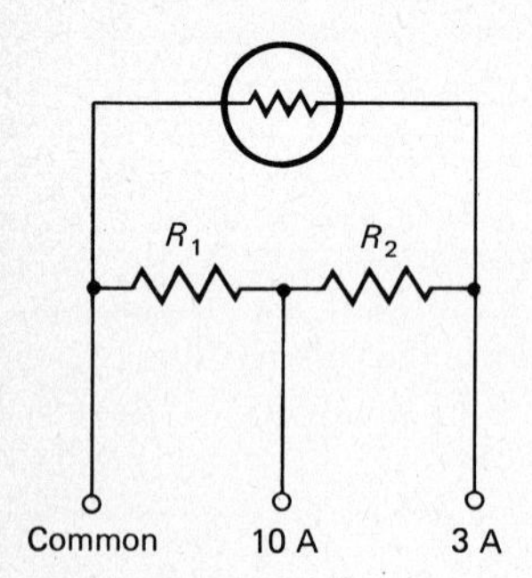

Figure P26.7

**36. Refer to Figure P26.8. The meter movement has $R = 50\ \Omega$ and deflects full scale for a current of 0.25 mA. Find R_1 and R_2 so that the voltmeter will have the ranges indicated.

37. A long solenoid has a field within it of 20 G when air-filled. What will the field be if its core is made of iron that has a relative permeability of 500?

38. The field within a long solenoid has a value of 8.0 G when its core is air. When an iron core is slipped into it, the field becomes 0.12 T. What is the relative permeability of the iron?

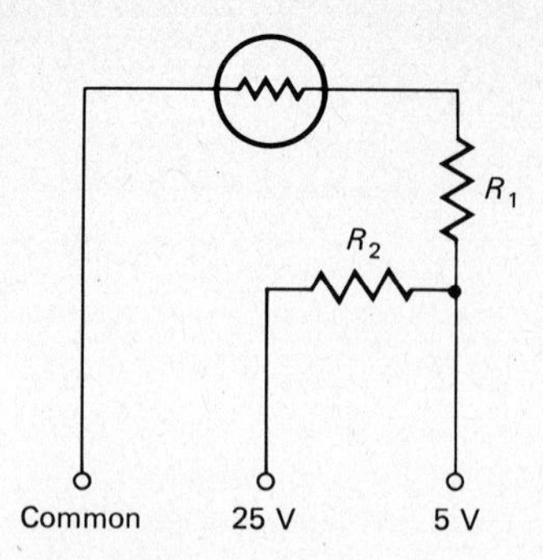

Figure P26.8

*39. A long, hollow solenoid has 13,000 loops of wire on its total 2.0-m length. The current in it is 7.0 A. (a) Find B inside the empty solenoid. (b) Find B inside it if its core is the transformer steel shown in Figure 26.27. (c) What is k_m for this steel under these conditions? (d) Repeat the problem for a solenoid with 130 loops of wire on its 2.0-m length.

40. For the transformer steel shown in Figure 26.27, what are the values of k_m at the following values of B_v: 1×10^{-4} T, 2.5×10^{-4} T, 5×10^{-4} T, 10×10^{-4} T, 150×10^{-4} T?

An electric generator station.

Consolidated Edison

27 INDUCED EMF'S

Performance Goals

When you finish this chapter, you should be able to

1. State how the current in a secondary coil behaves as the current in the primary coil is changed. Assume that the two coils are wound on the same iron core.
2. State how the emf induced in a coil by a nearby magnet varies as the magnet is moved from place to place.
3. Compute the flux through a given coil in a known magnetic field. Describe in words and

When a simple coil of wire is rotated in a magnetic field or when a magnetic field through a coil is changed, the coil acts as though it had an emf source in it. We call such an emf an induced emf. Induced emf's are extremely important in modern-day electricity. They are basic to the operation of electric generators, transformers, and motors. In this chapter we shall learn about this very fundamental effect and its application to technology. Other applications will be encountered in the following chapters.

with a graph how the flux is changing as a function of time if the coil is rotating in the field.
4. State Faraday's law in words and in equation form. Explain how Lenz's law is related to it.
5. When the flux through a loop is changing at a constant, known rate, compute the induced emf in the loop.
6. Describe qualitatively the induced emf in a coil when a graph showing the variation of flux through it with time is given.
7. Make a rough sketch of an ac generator and describe its principle of operation. Sketch a graph of the output voltage of the generator. State how the induced emf varies with N, B, and f.
8. Explain how the mechanical energy input to a generator is changed to electric energy output.
9. Sketch a simple transformer. Describe the basic idea behind its operation. Given the values of N_p and N_s, find the ratio of input to output voltage.
10. Explain why power companies transmit electric power at the highest voltage that is practical.
11. By use of a sketch, explain what is meant by a counter emf and how it originates. In your explanation state why the current drawn by a motor depends on its loading and rotation rate.
12. Calculate the back emf of a motor if you are given the resistance and operating current and voltage for it.
13. Describe eddy currents. Explain why they are often to be

27.1 Induced emf's in Coupled Coils

A simple but important experiment is shown in Figure 27.1. Two coils are wound on the same iron core. The coils are insulated from each other. Current cannot flow from one coil to the other. One coil is connected through a switch to a battery. The other coil is connected to a galvanometer, but no battery exists in its circuit. We call the circuit that contains the battery and its coil the *primary circuit* and *primary coil.* The other circuit and coil are called the *secondary.*

No current flows in the secondary circuit because it has no power source. But at the exact instant that the switch is closed in the primary, the galvanometer deflects, and then, almost at once, it returns back to its normal reading of zero. It is as though a battery existed for that instant in the secondary. After the current has flowed for some time in the primary, the switch is opened and the same effect appears again. At the instant the switch is pulled open, the galvanometer suddenly deflects. But this time it deflects in the opposite direction. Again, the deflection lasts only an instant before the galvanometer returns to a reading of zero.

We show the galvanometer behavior in the lower part of Figure 27.1. The figure is not to scale. Usually the current pulses (shown by the galvanometer deflections) last only a fraction of a second. Even so, the essential features are clear.

1. The current in the secondary lasts only an instant. It exists only when a change is being made in the current in the primary.
2. The current in the secondary has different directions. It depends on whether the primary current is increasing or decreasing.

Let us now go on to a different experiment, which gives more information about this effect.

27.2 Induced emf's and Moving Magnets

In Figure 27.2 we again see a coil in series with a galvanometer. It need not be wound on an iron core, but an iron core makes the effect stronger. Notice what happens when a magnet is near the coil. Parts (a) and (c) show what happens if the magnet is at rest. The galvanometer reads zero. No current flows in the coil. But an entirely different situation occurs if the magnet is moving. As shown in parts (b), (d), and (e), a current flows in the coil when the magnet is moving.

Notice another thing, too. The current direction depends on which way the magnet is moving. This can be seen by comparing parts (b) and (d). Also, the current direction depends on which pole of the magnet is approaching the coil. You can see this by comparing parts (b) and (e).

What can we conclude from these experiments? In both cases the galvanometer deflects only while a change is being made. The experiment in Figure 27.1 shows that the effect occurs only when the current in the primary is changing. Here we see that the effect occurs only when the magnet is moving.

avoided. State how eddy currents are minimized in transformers and similar devices. Give an example of a situation where eddy currents prove useful.

14. Explain what is meant by motional emf. Given a rod or wire moving through a magnetic field in a simple way, compute the potential difference between its ends.

The galvanometer circuit shows an induced battery effect. For an instant the circuit appears to have a battery in it. But any source of emf would cause the same result. We therefore say that "an emf is induced in the coil."

The induced emf can be described in terms of changing magnetic fields. In both experiments the magnetic field through the coil is changing in some way whenever the induced emf exists. In Figure 27.2 the emf exists only when the magnet is moving. At that time the field near the coil is either increasing or decreasing. When the field is not changing, no induced emf exists.

A similar situation exists in Figure 27.1. When the current in the primary coil is zero, no magnetic field exists. But when the switch is closed, the primary becomes an electromagnet. A magnetic field then exists near the secondary coil. As we see, the induced emf is zero unless the magnetic field is changing. It exists only when the electromagnet is being turned on or off. In the next two sections we shall learn how to describe this situation in a precise way.

27.3 Magnetic Flux

Michael Faraday (1791–1867) was the first to unify the experiments relating to induced emf's. He was a careful and ingenious experimenter, who was largely self-educated. He noticed that the important quantity in all these experiments is the magnetic flux. Let us now see what is meant by this term.

To begin with, we shall make a rule about how to draw magnetic field lines. Suppose that the magnetic field is as shown in Figure 27.3(a). Its value is B tesla. We erect an area A with surface perpendicular to the field lines. Our rule then tells us how many field lines to draw through this area A. The rule is as follows.

An area *A* has its surface perpendicular to *B*. We agree to draw *BA* field lines through this area.

If the area A is 1 m^2, then the number of field lines through it is equal in value to B. For example, in Figure 27.3(b) the number of lines through the unit area is

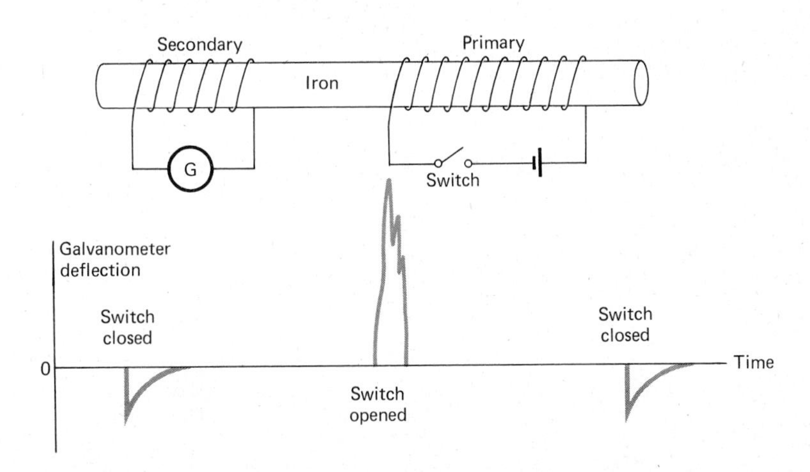

Figure 27.1 The galvanometer deflection lasts only a fraction of a second in each case. Therefore the graph is not to scale.

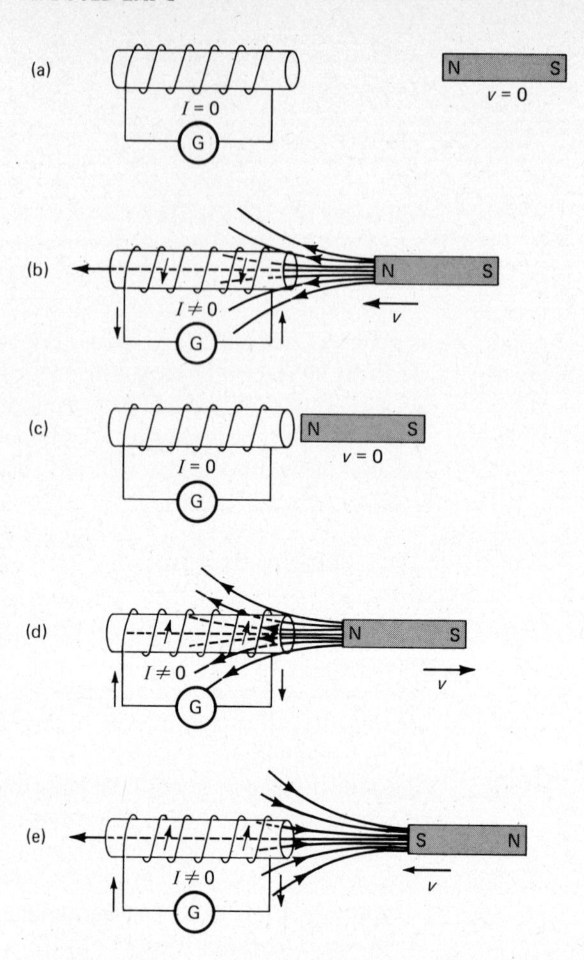

Figure 27.2 The current in the coil exists only when the magnetic field through it is changing.

16. Therefore $B = 16$ T. Usually we shall not draw our field line diagrams this accurately, but for quantitative work we shall make use of the rule. It can be restated as follows.

A number of field lines equal to *B* goes through unit area perpendicular to *B*.

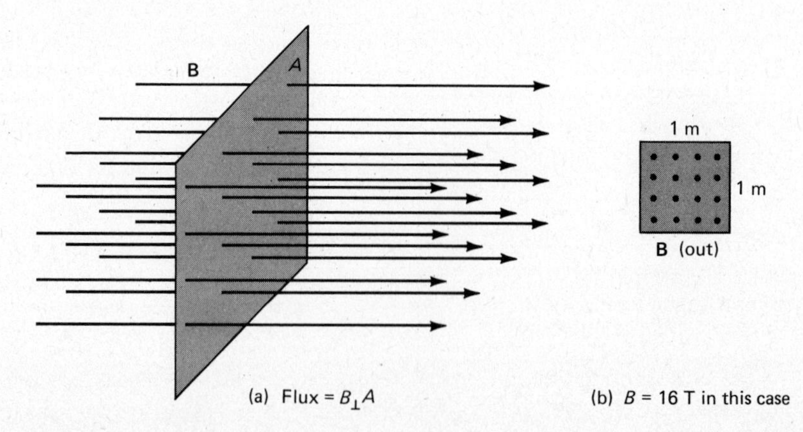

Figure 27.3 The number of field lines through an area A perpendicular to **B** is BA.

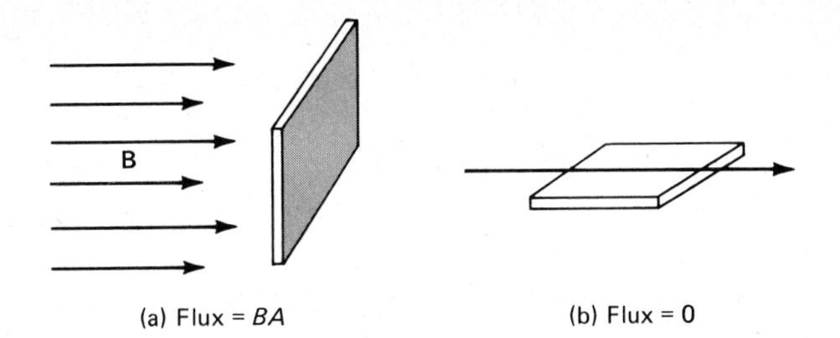

Figure 27.4 The flux through an area depends on the angle between the surface and the field lines.

We are now able to define what we mean by *flux*. The definition is stated in terms of the field lines that pass through an area.

The flux, Φ, through an area is the number of field lines that pass through that area.

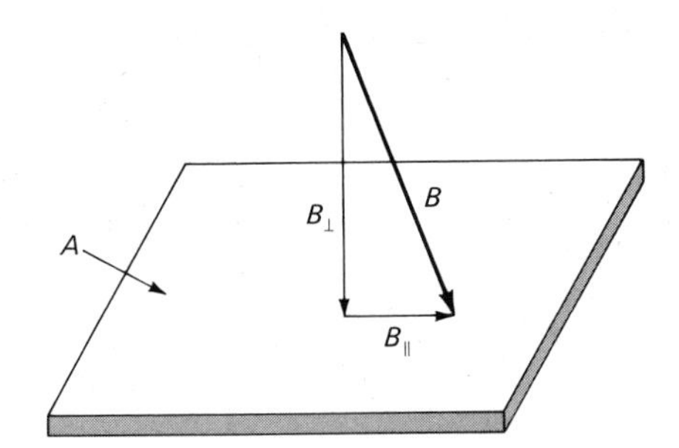

Figure 27.5 The flux through the area A is $B_\perp A$.

We use the symbol Φ (Greek letter phi) to represent flux. Let us apply the definition to Figure 27.3. The number of lines passing through the area in (a) is 16. Therefore, the flux through that area is 16. In part (b) 16 lines pass through the area. Therefore, the flux through it is also 16.

The flux through an area can be found even if the area is not perpendicular to the field lines. In Figure 27.4(a) the flux is simply BA as before. In (b) no flux lines pass through the area and so $\Phi = 0$. Let us now find the value for Φ when the area is oriented at an arbitrary angle to the field as in Figure 27.5. We can see at once how to deal with this case if we split the field into its two component fields $B_\perp$ and $B_\parallel$. Notice that the only flux lines going through the area are due to $B_\perp$, the component of B perpendicular to the area. Hence we see at once that the flux through the area is due entirely to $B_\perp$ and is $B_\perp A$.

A flat area A is in a magnetic field B. The flux Φ through the area A is given by

$$\Phi = B_\perp A \tag{27.1}$$

This is the equation we shall use to compute flux. Can you show that the units of Φ are webers or the equivalent tesla-square meters?

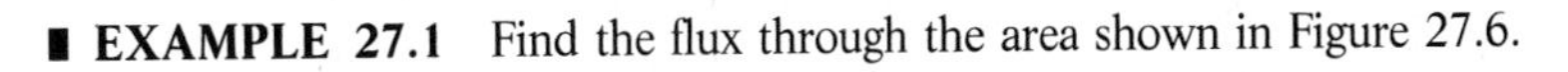

■ **EXAMPLE 27.1** Find the flux through the area shown in Figure 27.6.

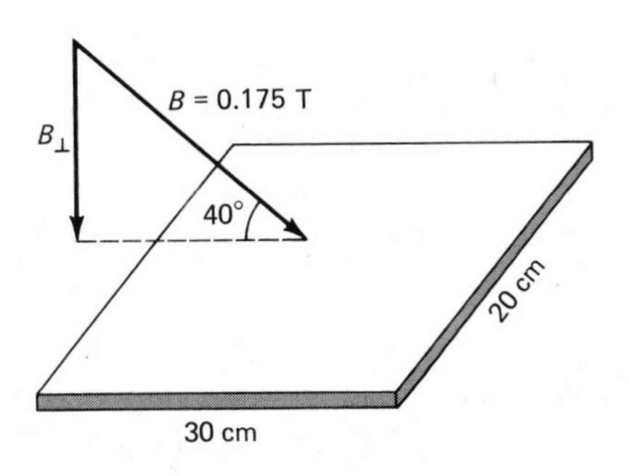

Figure 27.6 The flux through the area is $B_\perp A$.

Solution From the figure we see that $B = 0.175$ T. But

$$\begin{aligned} B_\perp &= B \sin 40^\circ \\ &= 0.112 \text{ T} \end{aligned}$$

Because

$$\Phi = B_\perp A$$

with $A = (0.20\ \text{m})(0.30\ \text{m}) = 0.060\ \text{m}^2$, we have

$$\Phi = (0.112\ \text{T})(0.060\ \text{m}^2) = 0.067\ \text{T}\cdot\text{m}^2$$

27.4 Faraday's Law and Induced emf

Faraday noticed that an induced emf occurs in a coil whenever the flux is changing through a coil. For example, look once again at Figure 27.2. Notice that the number of field lines passing through the coil (the flux) changes whenever the magnet is moved. During the time the flux is changing, an induced emf exists.

Similarly, in Figure 27.1 when the primary is turned on, the flux through the secondary is large. It is zero when the primary is turned off. Each time the flux changes in the secondary, an induced emf exists.

Whenever the flux through a coil is changing, an induced emf exists in the coil.

Faraday went one step further. He measured the magnitude of the emf caused when the flux changed by an amount $\Delta\Phi$ in a time Δt. His measurements showed the following.

If the flux through a coil with N loops is changing at the rate $\Delta\Phi/\Delta t$, then the induced emf in the coil is given by

$$\textbf{emf} = -N\frac{\Delta\Phi}{\Delta t} \qquad \textbf{(27.2)}$$

This equation is called Faraday's law. We shall discuss the meaning of the negative sign in the next section.

EXAMPLE 27.2 In Figure 27.7 the field increases steadily from zero to 0.25 T in a time of 3.6 s. The coil has 150 loops of wire and a cross-sectional area of 0.75 cm^2. Find the induced emf in the coil. Also, how large a current flows in the coil circuit if its resistance is 5.0 Ω?

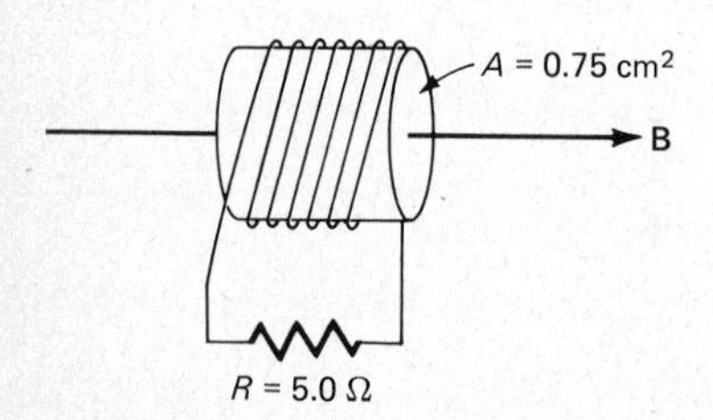

Figure 27.7 B is increasing and there are really 150 loops of wire on the coil.

Solution We know that $A = 0.75 \times 10^{-4}\ \text{m}^2$ and $N = 150$. Further, Φ changes from zero to $(0.25\ \text{T})A$, so $\Delta\Phi = (0.25\ \text{T})A$. This occurs in a time $\Delta t = 3.6$ s. Using Faraday's law, we have

$$\text{emf} = -N\frac{\Delta\Phi}{\Delta t} = -(150)\frac{(0.25)(0.75 \times 10^{-4})}{3.6}\text{V}$$
$$= -7.8 \times 10^{-4}\ \text{V}$$

The emf is small in this case because the flux is changing so slowly. In most practical applications the flux changes much more rapidly. We shall soon examine the practical use of this effect. The effect would have been hundreds of times larger if the core of the coil had been iron. Why?

The current in the circuit is

$$I = \frac{\text{emf}}{R} = \frac{7.8 \times 10^{-4}}{5.0}\ \text{A} = 1.56 \times 10^{-4}\ \text{A}$$

■■

27.5 Direction of Induced emf

Until now we have said very little about the direction of the induced emf. We saw in Figures 27.1 and 27.2 that the induced current is not always in the same direction. The rule that tells us the direction of the induced emf is called *Lenz's law.*

When the flux through a coil is changing, an emf is induced in the coil. The current caused by the induced emf generates a flux through the coil. The induced flux will be in such a direction as to cancel the original change in flux.

This may be condensed to say:

The induced emf opposes the change in flux.

The opposing nature of the induced emf is represented by the negative sign in Equation 27.2. When Φ is increasing, the induced emf tries to decrease Φ. When $\Delta\Phi$ is negative, the induced emf tries to increase Φ. Let us look at a few examples that illustrate this point.

In Figure 27.8(a) the magnet is moving to the left. As a result, B (and the flux) increases to the left through the coil. The induced emf will try to cancel this change. It causes current to flow in the direction shown. This current generates a flux toward the right. The induced flux is directed so as to cancel the change due to the moving magnet.

Examine the other parts of Figure 27.8. In each case Lenz's law tells us the current direction in the coil. Notice that the coil in (c) is wound differently from those in (b) and (a), but Lenz's law still tells us the direction. Also, look at Figure 27.1. Can you show that the induced currents given there are correct? If you understand Lenz's law, then these and similar examples are easily figured out.

There is another feature of Figure 27.8 worth noticing. In part (a) the coil acts like an electromagnet. Its right end is its north pole. This induced north pole repels the approaching magnet. In this way, also, it tries to stop the change. In part (b) the coil has its south pole on the right-hand end. It opposes the approach of the magnet and thereby tries to stop the change. In part (c) the right-hand end is a south pole. It tries to attract the north pole of the magnet, which is moving away. Again the induced current is trying to stop the change.

The opposing nature of induced emf's is a necessary result of energy conser-

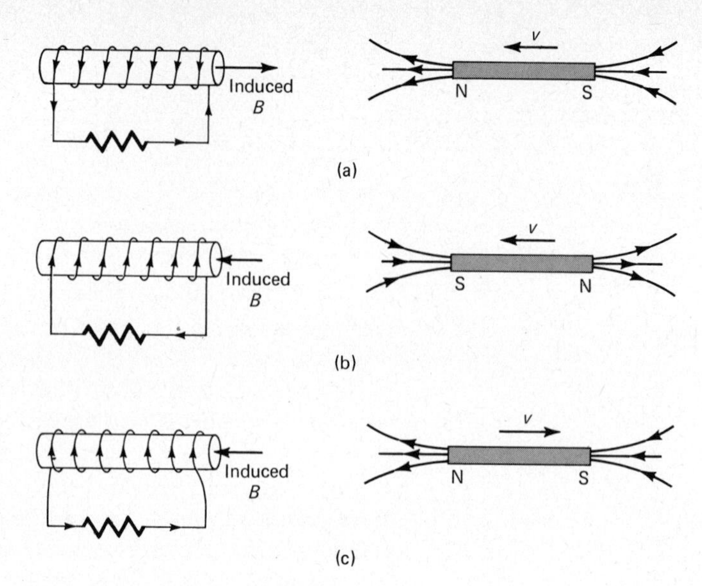

Figure 27.8 You should be able to tell in which direction the induced current will flow under similar conditions. Notice that the coil in (c) is wound differently from the one in (b).

vation. Suppose that the induced emf caused a flux that aided, rather than opposed, the changing flux. Then the induced flux would induce still another emf. This would aid the first emf and would induce still another emf, and so on. All these emf's would be aiding each other, and so an infinite emf would be induced.

Or we can view this situation in another way. Refer to Figure 27.8 to see how work and energy are related to induced emf's. For example, consider the situation shown in (a). The induced emf causes the coil to act like a magnet. Its right-hand end is a north pole and so it opposes the approach of the nearby bar magnet. We must do work to push the magnet to the left against the repulsion effect of the coil. This work gives rise to the heat generated in the resistor by the induced current. As we see, the following changes occur.

1. The approaching bar magnet induces an emf in the coil.
2. The induced emf causes a current.
3. The induced current causes a magnetic field, which opposes the motion of the bar magnet.
4. Work must therefore be done to move the magnet.
5. The induced current in the coil and resistor generates heat energy.
6. The heat energy generated equals the work done in moving the magnet.

All of this follows from the fact that the induced emf *opposes* the change. If it aided rather than opposed the change, no work would be needed to move the bar magnet. Indeed, the induced current and field would exert forces on the magnet to cause it to move more rapidly. In addition, heat would be generated by the induced current as it flows through the resistor. As we see, in this case heat energy would be created without external work being done. This contradicts the law of conservation of energy.

If you examine all situations involving induced emf's, you will see this same effect. The induced emf must oppose the change or energy could not be conserved. We see from this that Lenz's law and the opposing nature of induced emf's are a necessary consequence of the law of energy conservation.

27.6 The ac Generator

Perhaps the two most important applications of Faraday's law are the transformer and the alternating current (ac) generator. In this section we shall discuss the generator. The transformer will be the topic of a later section.

The simplest electric generator consists of a coil rotating in a magnetic field as shown in Figure 27.9. An external energy source rotates the coil in the magnetic field. Notice that the flux through the coil changes as the coil rotates. When the surface of the coil is horizontal, the flux through it is zero. When the plane of the coil is vertical, the flux through it is maximum. Faraday's law tells us at once that this changing flux results in an induced emf in the coil.

The two ends of the wire of the coil run to slip rings. Brushes slide along the rings. The output wires running from the brushes have a potential difference equal to the emf induced in the coil. Let us now see what sort of emf this generator provides.

According to Faraday's law, the induced emf in the coil is given by

$$\text{emf} = -N\frac{\Delta\Phi}{\Delta t}$$

where N is the number of loops on the coil. We therefore need to know how Φ, the flux through the coil, behaves. To learn about Φ, refer to Figure 27.10(a).

From Figure 27.10(a) we see that $B_\perp$ is $B \sin\theta$. Because $\Phi = B_\perp A$, we have

$$\Phi = AB\sin\theta$$

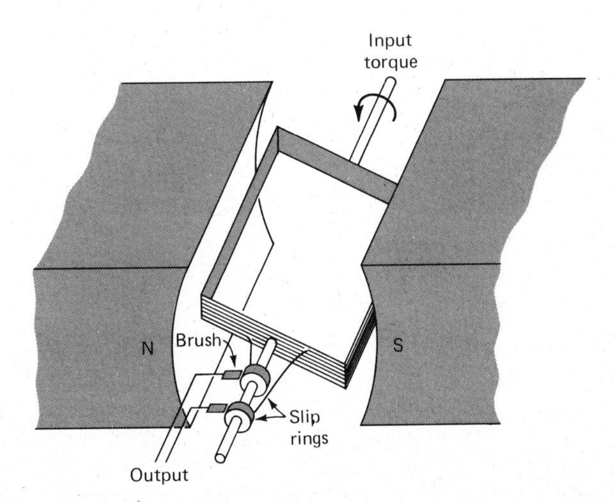

Figure 27.9 A simple ac generator. An external energy source rotates the coil with uniform angular speed ω.

You can easily check this by noticing the following. When $\theta = 0$, the lines skim the loop and so $\Phi = 0$. This checks, because $\sin \theta = 0$ when $\theta = 0$.

But the loop is rotating with angular speed ω (or $2\pi f$). We make use of the rotational motion equation, $\theta = \omega t$. Then we have

$$\Phi = AB \sin \omega t = AB \sin(2\pi f t)$$

In other words, the flux follows a sine curve as shown in Figure 27.10(b).

To find the induced emf, we need $\Delta\Phi/\Delta t$. This is simply the slope of the Φ versus t curve, the curve in (b). If we look at this curve, we can easily sketch its slope. At first the slope is positive and large. The slope then decreases to zero at the crest of the curve. Then the slope becomes negative, and so on. As we see, the slope's graph is as shown in Figure 27.10(c). This curve looks like a cosine curve. It is easily shown, using calculus, for example, that our guess is correct. We therefore have the following important result.

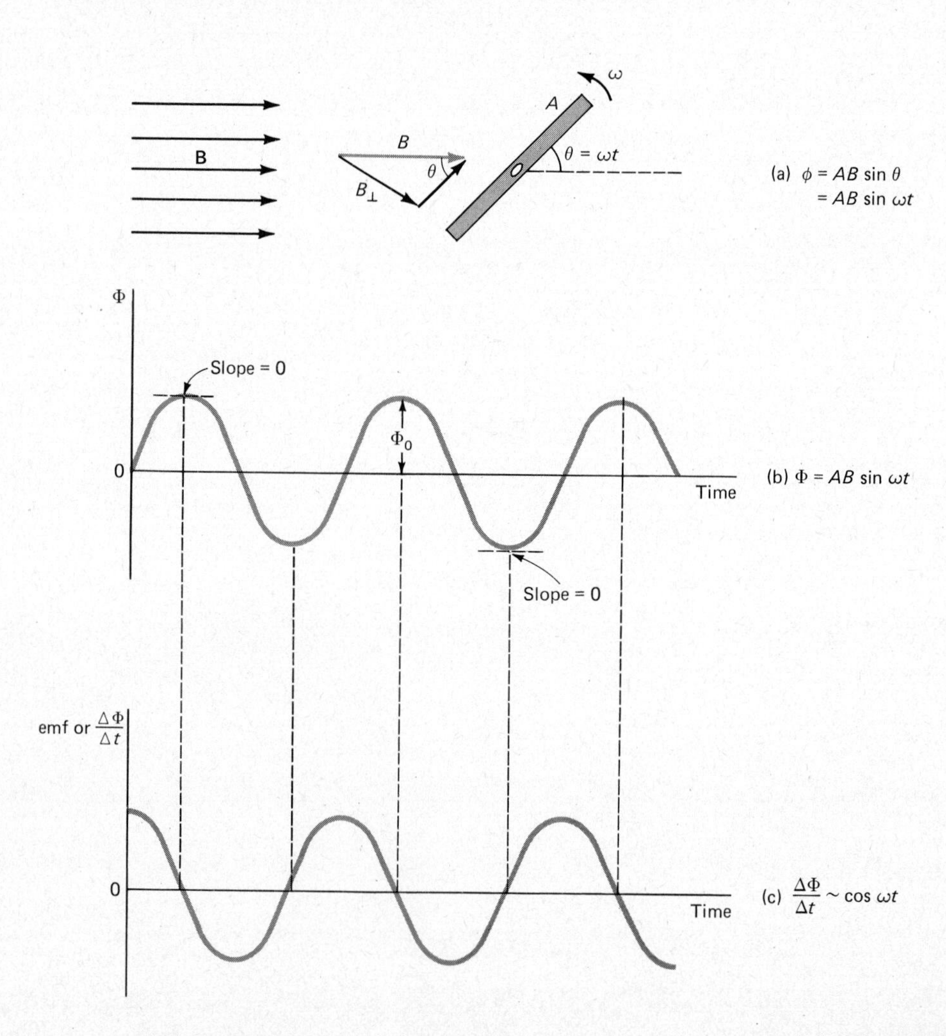

Figure 27.10 The rotating coil causes ϕ through it to vary as shown in (b). Because $\Delta\phi/\Delta t$ is the slope of the ϕ graph, this quantity varies in the way shown in (c).

If a quantity Φ plotted against t is a sine curve, then $\Delta\Phi/\Delta t$ follows a cosine curve.

Because the emf induced in the coil is proportional to $\Delta\Phi/\Delta t$, we see that the graph of $\Delta\Phi/\Delta t$ is also a graph of the induced emf. Therefore,

A flat coil rotating in a uniform magnetic field has an alternating emf generated in it. The emf has a sinusoidal (or cosinusoidal) form. Its frequency is the same as the rotation frequency of the coil. We call this type of potential difference an ac voltage.

The induced emf is given by (see Example 27.6)

$$\text{emf} = -NAB\omega \cos(\omega t)$$

Of course, $\omega = 2\pi f$, where f is the rotation frequency of the coil. Because $\cos\theta = \sin(\theta + 90°)$, the emf can be written as either a sine function or a cosine function.

As we would expect, the emf is large when the number of loops N on the coil is large. It also increases with B and A. This is also to be expected. However, the factor $\omega = 2\pi f$ may surprise you. It should not, however. After all, the larger f is, the faster the coil rotates. Because the flux changes fastest when f is large, the emf will also be large for high rotation frequencies.

Power companies rotate their generator coils at a precise rate. Usually the alternating voltage they obtain from the generator has a frequency of 60 Hz. Indeed, they maintain this frequency so precisely that electric clocks are geared to it.

27.7 Energy Balance in the ac Generator

As you probably know, the power station usually uses steam or water power to turn the generator coil. At first you might think that the friction in the bearings accounts for most of the work needed. But this cannot be true. The law of conservation of energy tells us otherwise. We obtain our electric energy from the power station generators. This energy must be supplied to the generator by whatever turns the coil. In comparison, the work done against friction in the bearings is negligible.

To see how energy interchange occurs in the generator, refer to Figure 27.9. As soon as use is made of the generator output, current flows in the output wires. But the same current also flows in the wires of the coil. These current-carrying wires are in the magnetic field of the magnet. They therefore experience a force. The force is in such a direction as to stop the rotation of the coil. Examine Figure 27.11 and show that the induced current flows as indicated. Does the torque on the current try to stop the rotation?

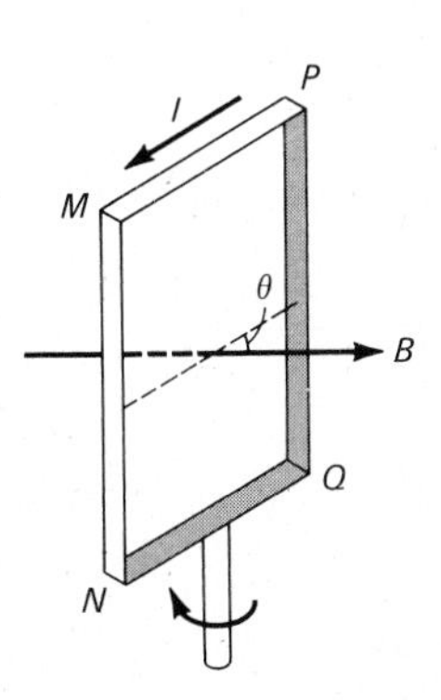

Figure 27.11 If θ is less than 90°, show that the induced current flows as indicated. Describe the torque on the loop due to the field.

We see, then, that the generator coil becomes hard to rotate when a large current is drawn. When the generator is supplying no current, only the friction in the bearings opposes the coil's motion. But when power is supplied by the generator, a current must flow from it. This causes forces that try to slow the coil. The turbine or other turning agent must supply energy to the generator to keep the coil turning. We can summarize the effect as follows.

The electric power drawn from the generator is provided to the generator by the energy source that rotates the coil. A retarding torque proportional to the current drawn from the generator acts on the coil. Mechanical energy must be applied to the coil to overcome this retarding torque.

The law of conservation of energy tells us at once that

Mechanical energy input = electric energy output + energy losses (friction)

In most generators the energy loss term is quite small. Therefore, the electric generator is an efficient method for changing mechanical to electric energy. Even so, the overall efficiency of a power station is far from 100 percent. The difficulty lies in the production of the mechanical energy. For example, the generation of mechanical energy from coal using steam is a relatively low-efficiency process.

■ **EXAMPLE 27.3** A small ac generator has a coil (200 loops) of area = 40 cm^2 that rotates at a frequency of 60 Hz in a field $B = 0.35$ T. Find an expression for the ac voltage furnished by the generator.

Solution We had in such a case that

$$\text{emf} = NBA(2\pi f)\cos(2\pi ft)$$

In our case $N = 200$, $B = 0.35$ T, $A = 40 \times 10^{-4}\ \text{m}^2$, and $f = 60$ Hz. Placing these values in the equation gives

$$\text{emf} = 106\cos(120\pi t)\ \text{volts}$$

This generator provides an ac voltage that varies between +106 V and −106 V. ■■

27.8 The Transformer

Alternating currents and voltages make it possible to use transformers. These are devices that change (or transform) one potential difference to another. For example, the television picture tube requires a potential difference of about 20,000 V. A transformer is used to change the 110-V line voltage to the required 20,000 V. As another example, a door bell requires a potential difference of perhaps 9 V. Use is made of a transformer to obtain this potential difference from the 110-V line. Let us see what a transformer is and how it works.

A schematic diagram of a typical transformer is shown in Figure 27.12. It consists of two coils wound on an iron yoke (or core). The primary coil has N_p loops on it. It is connected to the ac power source. Notice the symbol –⊝–

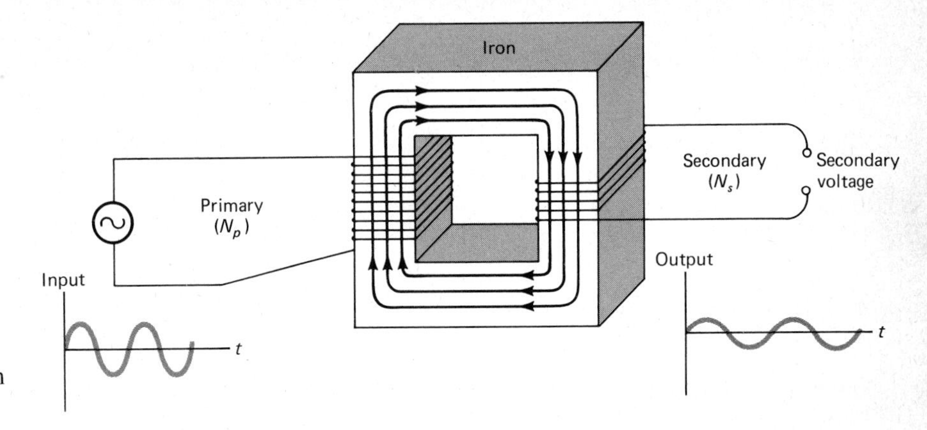

Figure 27.12 A transformer. The flux (or field lines) follows the iron yoke. Notice that the flux is the same through both the secondary and the primary.

used for an ac voltage source. The wave in the circle reminds us that the voltage is alternating. An alternating current flows through the primary because of the ac power source.

The secondary coil has no power source in its circuit. We call the number of loops on it N_S. Let us find the emf induced in the secondary because of the alternating emf (or voltage source) in the primary.

The flux from the primary coil is due to the ac voltage source in the primary. But magnetic field lines follow through iron, as was pointed out in the last chapter. In Figure 27.12 the field lines follow the iron through both coils. The same amount of flux threads through the secondary as goes through the primary.

As the sinusoidal voltage in the primary varies, the flux also varies. This varying flux passes through the secondary coil. There, it induces an emf in the secondary coil. The emf is given by Faraday's law to be

$$\text{emf in secondary} = (\text{emf})_s = -N_S \frac{\Delta\Phi}{\Delta t}$$

Because the same flux threads the primary and secondary, $\Delta\Phi/\Delta t$ is the same for each. The emf driving the primary is given by

$$\text{emf driving primary} = (\text{emf})_p = -N_p \frac{\Delta\Phi}{\Delta t}$$

provided the primary has small resistance.

Let us now divide one of these equations by the other. (Remember that equals divided by equals are equal.) Then

$$\frac{(\text{emf})_S}{(\text{emf})_p} = \frac{N_S}{N_p} \qquad (27.3)$$

In other words,

> **The output voltage of a transformer is in the same ratio to the input voltage as N_s is to N_p.**

We see from this how to transform 110 V to a much higher voltage. We can use a transformer for which $N_s/N_p \gg 1$. This is called a step-up transformer. If we want to reduce the voltage to a lower value, we need $N_s/N_p \ll 1$.

We shall discuss the relation between the currents in the primary and secondary in the next chapter (Section 28.13).

■ **EXAMPLE 27.4** In a TV set the 110-V, 60-Hz line voltage is to be raised to 20,000 V. What type of transformer is needed?

Solution We clearly need a step-up transformer. We want

$$\frac{(\text{emf})_s}{(\text{emf})_p} = \frac{20{,}000}{110} = 182$$

Therefore, the ratio of the number of loops on the secondary to the number on the primary is 182. The transformer must have

$$\frac{N_s}{N_p} = 182$$

■■

27.9 Power Transmission

Transformers have another important duty. They make it practical to transmit electrical energy over large distances.

Suppose that a city of 200,000 people is to be supplied with electricity from a distant hydroelectric plant. We can estimate the power consumption of the city. Let us say that each average person consumes 200 W. (This includes all uses of power in the city.) Then the total power used by the city is

$$\text{Power used} = \left(200\ \frac{\text{W}}{\text{person}}\right)(200{,}000\ \text{persons})$$

$$= 4 \times 10^7\ \text{W}$$

Suppose that this power was carried to the city by 120-V dc lines. Then the current in the lines to the city can be found from

$$P = VI$$

or

$$I = \frac{40 \times 10^7\ \text{W}}{120\ \text{V}} = 3.3 \times 10^5\ \text{A}$$

To carry this huge current without melting, a wire with a huge diameter would be required. Although this is not completely impossible, it would be inefficient. However, since $P = VI$, if power is transmitted at high voltage (say 5×10^5 V), then only a small I is needed. For example, in the case of our assumed city, at $V = 5 \times 10^5$ V,

$$I = \frac{4 \times 10^7 \text{ W}}{5 \times 10^5 \text{ V}} = 80 \text{ A}$$

This is a much more easily managed current. Power companies are currently doing research to design even higher-voltage power lines than this.

Of course, voltages of the order of 10^5 V are dangerous. They must be vastly lowered for power transmission throughout a city. Therefore, power companies have transformer stations that lower the power line voltage. These lower voltages are then carried to local areas. There, still other transformers lower the voltage to the values furnished to houses in the area. As we see, transformers are indeed important in our society.

27.10 Back emf of Motors

If you take a motor apart, you will see that the armature wires are quite large. A good motor has very little resistance. After all, we do not want to use the motor as a heater. We want it to change the input electrical energy to output mechanical energy. Resistance in the coils is therefore kept to a minimum.

Because a typical motor may draw only 1 A from 120 V, you might conclude from $V = IR$ that the motor's resistance is 120 Ω. This is not true, however. There is another feature of the motor that keeps the current low. It is the back emf of the motor. We shall now see what is meant by this term.

Refer back to the motor shown in Figure 26.22. Notice that it consists basically of a coil that rotates in a magnetic field. But this is exactly what a generator is. As the motor armature rotates in the field, it acts like the coil of a generator. There is, therefore, an emf generated in the motor coil. The direction of this emf is opposite to the voltage driving the motor. We call it a *back emf* or *counter emf.*

Every motor is also a generator. Its back emf opposes the voltage source that runs the motor.

When we consider the electric circuit of a motor, we should draw it with a back emf as is done in Figure 27.13. The effect of the back emf is to decrease the current in the circuit. If the resistance of the motor is r, then the current through it is given by (from Ohm's law)

$$I = \frac{\text{driving voltage} - \text{back emf}}{r}$$

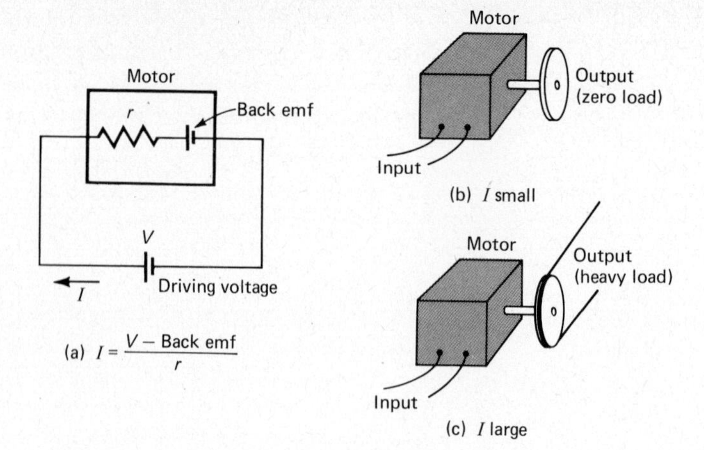

Figure 27.13 The back emf decreases as the load slows the motor.

Usually r is only a few ohms. Because the driving voltage is usually 120 V, it is clear that the back emf in a motor must nearly equal the driving voltage.

The back emf is generated by the rotating coil of the motor. The faster the coil rotates, the larger the back emf will be. When running under low load, the motor turns rapidly. The back emf nearly equals the driving voltage. As a result, the motor draws only a small current. However, the motor draws a much larger current under heavy load. It is then turning more slowly. As a result, the back emf is less, and the driving voltage causes a large current to flow.

A similar situation exists when a motor first is turned on. Initially the back emf is zero and so a large initial current flows. But as the motor builds up speed, the back emf increases. The current drawn by the motor rapidly drops to its normal operating level.

EXAMPLE 27.5 With an ohmmeter the resistance of a particular motor is read to be 1.50 Ω when disconnected. When running on 120 V, the motor normally draws 6.0 A. Find the normal back emf of the motor.

Solution From $V = IR$, we know that

$$\text{Driving voltage} - \text{back emf} = Ir$$

where r is the motor resistance. The ohmmeter reading tells us that $r = 1.50\ \Omega$. Therefore,

$$120\text{ V} - \text{back emf} = (6\text{ A})(1.50\ \Omega)$$

from which

$$\text{Back emf} = 111\text{ V}$$

27.11 Eddy Currents and Induction Heating

All ac devices are surrounded by varying magnetic fields. The current flowing to and in them is alternating. As a result, the magnetic field due to the current is constantly changing. This changing magnetic field induces emf's in the metal objects in the region. These emf's cause currents to flow in the metal.

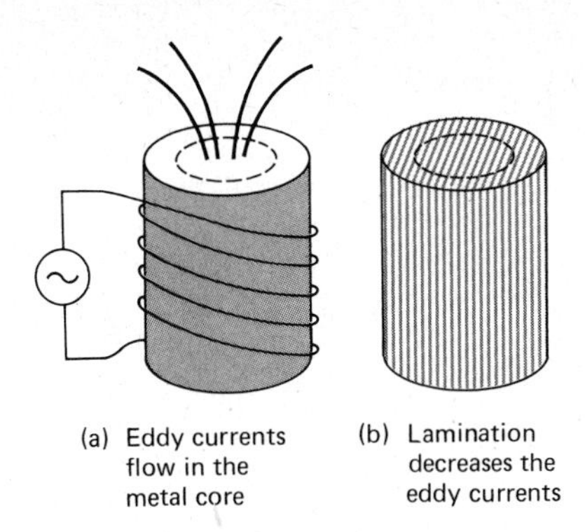

Figure 27.14 The changing flux through the metal core causes eddy currents to flow in it.

Induced currents, called eddy currents, flow in metal objects that are in a changing magnetic field.

Let us now examine this phenomenon.

Consider a coil of wire wound on a metal core as shown in Figure 27.14(a). (This might be part of a transformer.) The ac voltage source causes a flux, which is directed alternately upward and downward. This changing flux causes emf's to be induced.

Consider the core to be made of tightly packed cylindrical metal shells. The dashed circle drawn on the top of the core circle represents the end of such a cylindrical shell. Because of the changing flux, an emf is induced in the cylindrical shell. The emf will be alternately clockwise and counterclockwise around the circle. As a result, circular currents will flow back and forth within the metal core. These currents are eddy currents.

Most often eddy currents are not wanted. They cause heating of the core and loss of energy. To reduce them, the cores of transformers and other metal ac machinery are laminated, that is, the core is slit into thin slices as shown in Figure 27.14(b). Each slice is insulated from the other. Because of these insulating barriers, current can no longer flow around the circle. Eddy currents are therefore much reduced.

Even so, tiny eddy currents still flow in the thin laminations. They generate magnetic fields and thereby exert forces on each other. These forces alternate at twice the power-line frequency and cause the laminations to vibrate. If you listen to a transformer, you can often hear this vibration. The effect is very noticeable in devices where the laminations have become loose from each other.

Sometimes eddy currents are desirable. For example, metals can be melted by placing them in rapidly varying magnetic fields. The induced eddy currents generate the heat needed to melt the metal. This widely used process is called *induction heating.*

27.12 Motional emf's

In certain cases an easily pictured explanation can be given for Faraday's law. One such case is shown in Figure 27.15. Let us first compute the induced emf in the loop.

The area of the loop is increasing as the bar moves to the right with speed v. In a time Δt the bar rolls a distance $v\ \Delta t$. Therefore, in this same time the area of the loop increases by an amount $(v\ \Delta t)d$, where d is the width of the loop. But the increase in flux through the loop is simply this increase in area multiplied by B. We have

$$\Delta\Phi = B(v\ \Delta t)d$$

To find the induced emf in the loop, we use Faraday's law. We have

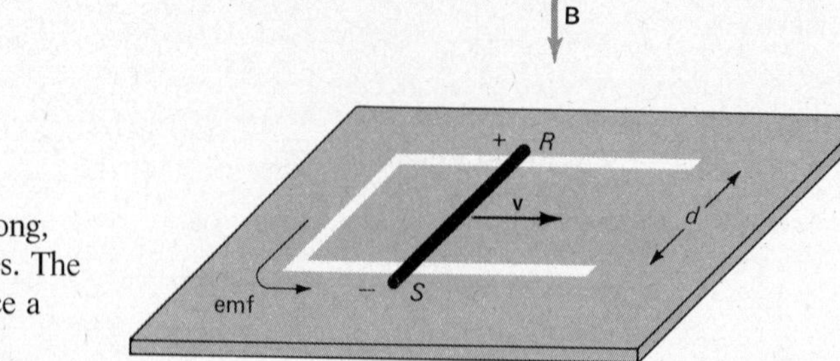

Figure 27.15 As the bar rolls along, the flux through the loop increases. The free electrons in the bar experience a force toward the viewer.

$$\begin{aligned}\text{emf} &= -N\frac{\Delta\Phi}{\Delta t}\\ &= -(1)\frac{Bvd\,\Delta t}{\Delta t}\\ &= -Bvd\end{aligned}$$

You should be able to use Lenz's law to find the direction of the induced emf. Its direction is shown in the figure.

There is another way of looking at this situation. In Chapter 26 we learned that a charge q moving with velocity **v** through a magnetic field experiences a force.

$$F = qvB \sin \theta$$

where θ is the angle between **v** and **B**. Each charge in the moving rod experiences this force. Our right-hand rule tells us that the force on a positive charge in the rod is away from the viewer. Notice that this force would cause a counterclockwise current to flow in the loop. This agrees with the current direction caused by the emf we calculated before.

But, in addition, this new way of looking at things can tell us how large the emf is. You will recall that the electric field **E** is defined as the electric force on unit positive test charge. We have just seen that a charge q in the moving rod experiences a force $F = qvB \sin \theta$. (In the present case $\theta = 90°$ and $\sin \theta = 1$, but because we want a general result, we shall retain θ.) Therefore, each charge in the rod seems to experience an electric field directed from S to R:

$$E = \frac{F}{q} = vB \sin \theta$$

Let us summarize this result.

A charge moving with velocity v through a magnetic field B acts as though it is in an electric field given by $vB \sin \theta$, where θ is the angle between v and B.

We can now ask what the potential difference is between one end of the bar and the other, from R to S. By definition, the potential difference is the work done in carrying a unit positive test charge from R to S. The force needed to carry the unit positive charge has a magnitude E. We have just found it to be $vB \sin \theta$. The distance from one end of the rod to the other is d. The work done is the

$$\text{Potential difference from } R \text{ to } S = (vB \sin \theta)d$$

In our case $\theta = 90°$ and $\sin \theta = 1$. Therefore,

$$\text{Potential difference from } R \text{ to } S = Bvd$$

This is the same result we found for the emf induced in the loop! Apparently the induced emf is due to the forces on the moving charges of the rod.

Let us review what we have found. A metal wire or rod or bar of length d is moving with velocity **v** perpendicular to the bar. It is in a magnetic field which is at an angle θ to the velocity. Because of the motion, there exists a potential difference between the two ends of the bar. We call this induced potential difference a *motional emf.* It is given by the equation

$$\text{Motional emf} = Bvd \sin \theta \qquad (27.4)$$

EXAMPLE 27.6 Use the idea of motional emf to compute the induced emf in the rotating coil of Figure 27.16. The coil, $PQRS$, has N loops.

Solution Notice first that wires SP and QR are not cutting through the field lines. They are simply skimming over the lines of flux. No emf is generated in them, and so their effect is zero.

Let us next look at wire RS. The motional emf induced in it is, from Equation 27.4,

$$\text{emf in } SR = Bvd \sin \theta$$

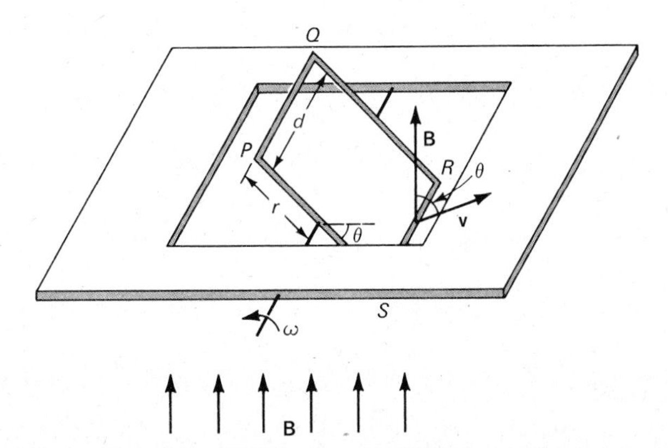

Figure 27.16 As the coil turns, the wires PQ and RS cut the field lines. An induced emf therefore appears in the coil.

MHD POWER GENERATORS

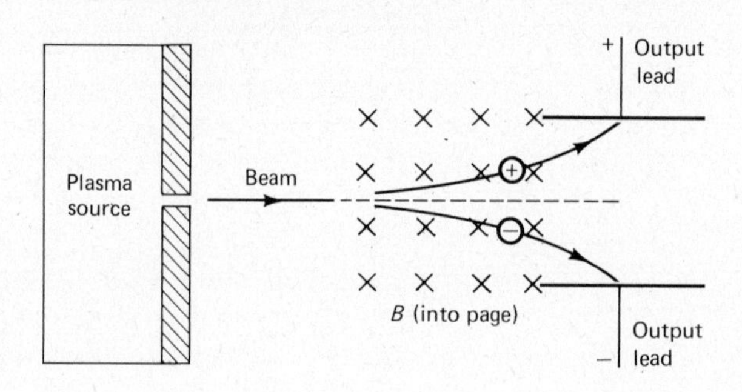

Most of our electricity is generated by steam-driven turbine generators. The steam is obtained using the heat from the burning of coal and oil or from the fission reaction in nuclear reactors. All such processes are limited in efficiency according to the laws of thermodynamics. You will recall from Chapter 16 that the efficiency increases with the temperature of the heat source. Unfortunately, steam turbine systems are quite restricted by the temperature range within which the use of steam is practical. For this reason other methods of generating electricity are of interest.

One of the most promising alternatives is based on the use of a plasma. A plasma is a highly ionized gas containing essentially no net charge. Plasmas can be generated efficiently by the use of nuclear reactors. It is possible to build a device that generates a voltage (and power) from a plasma beam. The process involves the interaction of the beam with a magnetic field. We call the device that uses this process an MHD (magnetohydrodynamic) generator.

A simple MHD generator is shown in the figure. The power source emits a neutral beam of charged particles, a plasma beam. The beam passes through a magnetic field, which causes the charged particles to deflect. You will recall that no work is done by the magnetic field in this process. A permanent magnet could be used. As shown, the positive particles strike the top plate and the negative particles strike the lower one. This gives rise to a potential difference between the two plates.

The two charged plates are the electric power source. Current can be drawn from them. Of course, the maximum current drawn cannot exceed the charge furnished to the plates each second by the beam. The voltage generated by the device would be regulated by the beam intensity and the capacitance of the circuit connected to the plates. As you see, the energy in the extremely hot plasma beam can be converted directly to electrical energy. Because of the high temperature of the beam, the system has a high thermodynamic efficiency.

At the present time MHD generators are under development. Practical generators for space vehicles are available. However, MHD generators will become most important when (and if) nuclear fusion power sources become practical. These types of power sources are discussed in Chapter 30.

You should be able to show, using the right-hand rule, that it is directed toward S from R. Similarly, the emf generated in wire PQ is

$$\text{emf in } PQ = Bvd \sin \theta$$

Its direction is from P to Q. Both emf's are directed around the loop from $RSPQR$. They therefore aid each other. The emf induced in one loop of the coil is therefore

$$\text{emf in one loop} = 2Bvd \sin \theta$$

We can put this in better form if we recall the relation between tangential speed and angular speed. The quantity v is the tangential speed of a point on a circle. (The point R, for example, describes a circle as the coil rotates.) If the angular speed of the coil is ω, then we know that

$$v = \omega r$$

where r is shown in the figure.

Substitution in the emf equation gives

$$\text{emf in loop} = B(2rd)\omega \sin \theta$$

where we have grouped the factors in a special way. But $2rd$ is simply the area of the loop. Therefore,

$$\text{emf in loop} = BA\omega \sin \theta$$

Because the coil has N loops on it, the coil's emf will be N times this large. We then find that

$$\text{emf in coil} = NBA\omega \sin \theta$$

Because $\theta = \omega t$, we see that the emf in the coil varies sinusoidally with time. This confirms the result we stated in Section 27.6. ■■

Summary

In a region where the magnetic field is **B**, the number of field lines that pass through unit area perpendicular to the field lines is B. We call the number of field lines passing through an area A the flux through the area. It is represented by Φ. If $B_\perp$ is the component of **B** perpendicular to the surface area **A**, then $\Phi = B_\perp A$.

During the instant that the flux is changing through a coil, an induced emf exists in the coil. Suppose that the coil has N loops. The flux through it is changing at the rate $\Delta\Phi/\Delta t$. Then Faraday's law gives the induced emf to be

$$\text{emf} = -N\frac{\Delta\Phi}{\Delta t}$$

The negative sign represents Lenz's law. This law states that the induced emf is in such a direction as to oppose the change in flux. When the flux through a coil has a steady value, there is no induced emf in the coil.

An alternating current (ac) generator consists basically of a coil rotating in a magnetic field. The coil rotates with frequency $\omega = 2\pi f$. As it does so, the flux through it keeps changing. This causes an emf to be generated in the coil. The emf varies sinusoidally with frequency f. Its value is (for a coil of N loops and area A)

$$\text{emf} = -2\pi NABf \cos(2\pi ft)$$

When current is drawn from the generator, retarding forces exist on the rotating coil. As a result, more work must be done to keep the coil rotating when the generator is developing more current. In this way mechanical energy is changed to electrical energy.

A transformer consists of two coils, a primary and a secondary. The ac current in the primary induces an emf in the secondary. We have

$$\frac{\text{Primary voltage}}{\text{Secondary voltage}} = \frac{N_p}{N_s}$$

A step-up transformer changes a low input voltage to a high output voltage. Transformers will not operate on dc.

Electrical power is transmitted at high voltages over large distances. This decreases the required current and therefore reduces I^2R losses in the power lines. Transformers are then used to reduce the voltage to usable levels.

The rotating coil of a motor acts as a generator. A back (or counter) emf is generated in it. This emf opposes the voltage source that is running the motor and thereby keeps the current through the motor low.

Eddy currents are induced to flow in metal portions of ac machinery and transformers by the changing magnetic field. For this reason the core and similar massive parts are laminated so as to reduce eddy currents. Induction heating uses eddy currents to produce the heat.

A metal bar or wire of length d is moving with velocity **v** in a direction perpendicular to the bar or wire. The velocity makes an angle θ with the field **B** through which it is moving. There is induced a potential difference $vBd \sin \theta$ between the two ends of the bar because of this motion. We call this a motional emf.

Questions and Exercises

1. Suppose you have available a bar magnet, a galvanometer, and a large amount of fine wire. Describe an experiment you could do to show your little sister the basics of induced emf.
2. A flat coil of wire lies on a tabletop. The table has a hole in it near the center of the coil. A bar magnet is held high above the table with its north pole pointed downward. It is then dropped down through the hole. Sketch a graph showing how the induced emf in the coil varies as the magnet falls through the center of the coil.
3. In sensitive electronic circuits stray induced emf's are very undesirable. To avoid induced emf's, the input and output wires to the various circuit elements are often twisted together. This helps eliminate stray induced emf's in two ways. What are they? Stray induced emf's are usually more noticeable in high-frequency circuits than in those that operate at lower frequencies. Why?
4. A transducer is a device for changing one type of energy into another type of energy. One way of constructing a transducer is shown in Figure P27.1. Shown there is a magnetic-type microphone. When sound waves strike the diaphragm, they cause it to move back and forth. Explain how this motion is changed into an electric signal (i.e., voltage).

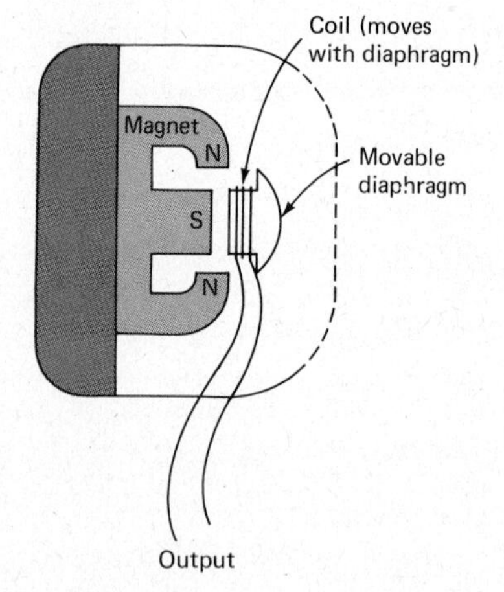

Figure P27.1

5. A long, straight wire is connected through a resistor and switch to a battery. It lies on a table and near it lies a flat coil. Sketch the situation for each of several relative positions of the coil and wire. In each case, what is the direction of the induced emf in the coil when the switch is just closed? Just pulled open?
6. A straight bar-type electromagnet is operated using ac current. At the end of the magnet is held a metal ring cut from the end of a metal pipe. The ring is placed concentric to the bar. Why does the metal ring become very hot? Why doesn't the ring heat appreciably if it is cut through on one side by a saw?
7. In some houses the lights dim for an instant when the refrigerator or washing machine motor turns on. Explain why this happens.
8. The back emf of a dc motor helps to keep the speed of the motor constant as the load is changed. Explain why.
9. The metal ring and rod shown in Figure P27.2 swing freely as a pendulum. However, if the pole of a bar magnet is placed so that the ring swings in front of it as shown, the pendulum is highly damped and comes to rest almost at once. Explain.
10. The motor shaft of a motor that uses permanent magnets will continue to rotate for some time after the power source is turned off. This happens if the motor is not loaded and if the motor circuit is open so no current can flow through it. But a different situation exists if the ends of the motor leads are connected together. Then the motor slows very rapidly. Explain why this difference occurs.

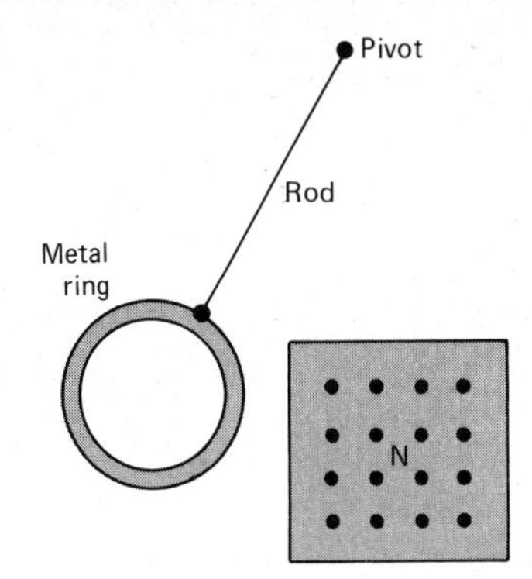

Figure P27.2

11. A long piece of copper pipe is held with its axis vertical. Into the upper end is dropped a strong bar magnet. Describe the forces that act on the magnet as it falls. Why does it soon reach a constant velocity?
12. What will happen to an ordinary door bell transformer if you try to run it using a 12-V battery? The same thing will happen to a dc motor if you hold its shaft so it cannot rotate. Explain why.

Problems

1. A flat, rectangular board is 20 cm $\times$ 30 cm. It is placed in a 0.150-T magnetic field in such a way that a perpendicular to the board's surface makes an angle θ to the field. Find the flux through the board for θ values of (a) 0°, (b) 90°, (c) 40°.
2. The magnetic field in a certain solenoid is 20 G. If the cross-sectional area of the solenoid is 1.5 cm^2, how much flux threads through the solenoid?
3. The earth's magnetic field in Tennessee has an average value of 0.58×10^{-4} T. It is directed at an angle of 67° below the horizontal. Find the flux through a circular coil ($r = 20$ cm) lying on a tabletop.
4. In Arizona the average magnetic field of the earth is 0.52 G. It is directed at an angle of 59° below the horizontal and nearly straight north. Find the flux through a metal window frame (80 cm $\times$ 120 cm) that is part of the north wall of a room.

*5. A certain solenoid is wound on a 1.8-cm-diameter iron cylinder. There are 300 loops on the 25-cm-long solenoid. When the current through the wire is 0.40 A, the relative permeability of the iron is 150. How much flux then threads the solenoid?

6. A flat loop of wire has an area of 40 cm^2. It is in a region where $B = 200$ G and is directed along the x axis. Call θ the angle between the axis of the loop and the x axis. What is the change in flux through the loop as θ is changed from (a) 0° to 60° and (b) 30° to 40°?
7. Refer to the situation described in Problem 6. (a) Suppose the change in part (a) takes 0.5 s. How large an average emf is induced in the loop in this time? (b) Repeat for the situation in part (b).
8. A 200-loop flat coil with a 3-cm radius lies flat on a table. When a vertical magnetic field of 0.150 T is suddenly turned on, an average emf of 80 V is induced in the coil. How long does it take for the field to rise to its final value?
9. Two coils are wound on a bar of iron. The secondary coil has 500 loops of wire on it. Each loop has a radius of 0.40 cm. As the switch is turned on in the primary, the magnetic field increases from 0 to 0.62 T in 0.030 s. Find the average emf induced in the secondary during this time.

*10. A man's wedding band has a resistance of 6.0×10^{-6} Ω. It is slipped over a solenoid (radius = 0.70 cm) in which B changes from zero to 0.31 T in a time of 0.075 s. Find the average emf and current in the ring as the solenoid field changes in this way.

11. At a certain place the earth's magnetic field is 0.65 T and directed at an angle of 60° below the horizontal. A young woman places her ringlike bracelet (r = 3.0 cm) on a table and slides it across the table at a speed of 40 cm/s. How large is the induced emf in the bracelet? Would your answer be different if the bracelet were rolled in a straight line without wobbling?

12. In an effort to measure the earth's magnetic field, a boy performs the following experiment. He connects the two ends of a horizontal loop (area = 0.80 m^2) to a sensitive voltmeter. While carrying the loop and voltmeter, he runs along the flat earth at a speed of 5.0 m/s. If the vertical component of the earth's field there is about 0.48 G, how large a voltage should the voltmeter read? Would your answer be different if the loop flopped around as he ran?

*13. As shown in Figure P27.3, a loop of wire is being pulled out of a uniform magnetic field (B = 0.25 T). (a) At what rate is flux changing through the loop? (b) Find the induced emf in the loop. Is it clockwise or counterclockwise? (c) What is the current in the loop if R = 0.020 Ω for it?

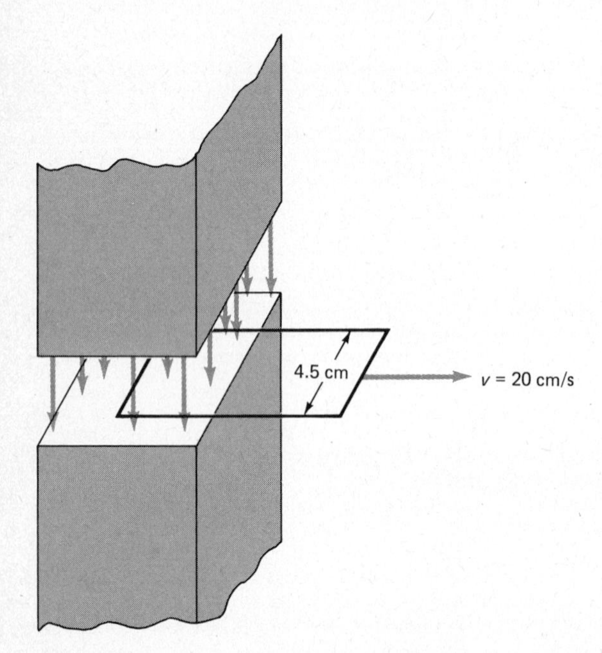

Figure P27.3

*14. The induced emf in the loop of Figure P27.3 at the instant shown is 1.73×10^{-3} V. How large is the uniform magnetic field shown?

*15. As a lecture demonstration, a teacher mounts a coil on an axis that is aligned east-west. For the coil, A = 0.75 m^2 and N = 50. The teacher rotates the coil at a rate of 3.0 rev/s. If the earth's magnetic field there is 0.65 G, what is the maximum value of the induced voltage in the coil?

*16. A rectangular coil that has 70 loops on it is 5 cm × 2 cm. It rotates about an axis that is perpendicular to a magnetic field of 0.80 T. How fast, in revolutions per second, must the coil rotate if the maximum induced emf in it is to be 110 V? Assume the geometry is similar to that shown in Figure 27.9.

17. A certain transformer used in a radio changes the 120-V line voltage to 9.0 V. (a) What is the turns ratio, N_p/N_s, for this transformer? (b) By mistake it is connected into the circuit backwards. About what output voltage does it deliver before everything burns out?

18. A neon sign transformer is designed to change 120 V ac to 15,000 V ac. (a) What is the turns ratio, N_p/N_s, for this transformer? (b) If the transformer was connected up backwards (120 V to the secondary), what voltage would appear across the primary?

19. A homemade transformer is constructed by winding a 100-loop solenoid on an iron rod. This is to be the primary. On top of this are wound 2000 loops of fine wire. This is the secondary. (a) Is this a step-up or a step-down transformer? (b) For an input voltage of 120 V ac, what will be the output voltage?

*20. An industrial electric heater is to provide 10,000 cal of heat each second. On what voltage should the heater be designed to operate if it is to draw a current of 200 A?

21. The dc resistance of a certain motor is 5.7 Ω. When connected to 120 V dc, it draws 2.0 A. (a) How large is the counter emf of the motor? (b) How much current flows through the motor the instant it is turned on?

22. The dc resistance of a certain motor is 4.2 Ω. When operating at rated speed on 120 V, it draws 1.6 A. (a) What is the back emf when the motor is running at its rated speed? (b) If the motor is suddenly loaded down so that it can no longer turn, how much current would it draw?

*23. A certain $\frac{1}{4}$-hp motor has a resistance of 0.50 Ω. Assume it to have nearly 100 percent efficiency. (a) How much current does it draw on 110 V when its output is $\frac{1}{4}$ hp? (b) What is its back emf? [Note: If you want to make this a ** problem, consider the effect of the internal resistance when doing part (a).]

*24. Large electric motors may take 30 s after being turned on to get up to their operating speed. One such motor has a resistance of 0.80 Ω and draws 7.0 A on 220 V. (a) How large a resistance (the starting resistance) must be placed in series with the motor when it is first turned on if the current is not to exceed 15 A when first turned on? (b) This starting resistance is later removed, of course. What is the back emf of this motor when operating at normal speed?

25. What is the potential difference between the ends of a 70-cm-long straight wire that is moving sideways perpendicular to a magnetic field? Its speed is 15 m/s and $B = 0.27$ T.

26. A 120-cm-long horizontal rod oriented east-west is falling toward the earth with a speed of 20 m/s. The horizontal component of the earth's field there is directed northward and is 0.60 G. (a) What is the potential difference between the ends of the rod? (b) Which end of the rod is positive?

27. An airplane whose wing span (tip to tip) is 20 m flies horizontal and due north. Its speed is 150 m/s. The earth's magnetic field there is northward and has a downward vertical component of 0.46 G. (a) Find the potential difference between the wing tips. (b) Is the right-hand wing tip (as seen by the pilot) positive or negative?

28. Repeat Problem 27 for the case in which the plane is flying west.

*29. In order to decrease the drain on his car battery, an ingenious young man proposes the following. "Mount a 1.50-m wire across the width of the car. When the car runs along the road, an emf is induced in the wire. Run this emf to the lights and light them that way." (a) Compute how fast the car must be going north to generate 12 V in the 1.5-m wire. Assume that the vertical component of the earth's field is 0.50 G. (b) Assuming that 12 V could be generated, would the lights light?

**30. The metal rod shown in Figure P27.4 rolls down the incline as indicated. It is part of the rectangular loop circuit shown. A vertical magnetic field of B is present. (a) Find the induced emf in the rod when its speed is v. (b) If the resistance of the loop is R, what is the current in the loop? (c) Is the current clockwise or counterclockwise? (d) How large is the force that acts on the rod because of the current in the magnetic field? (e) Does this force tend to speed or stop the rod?

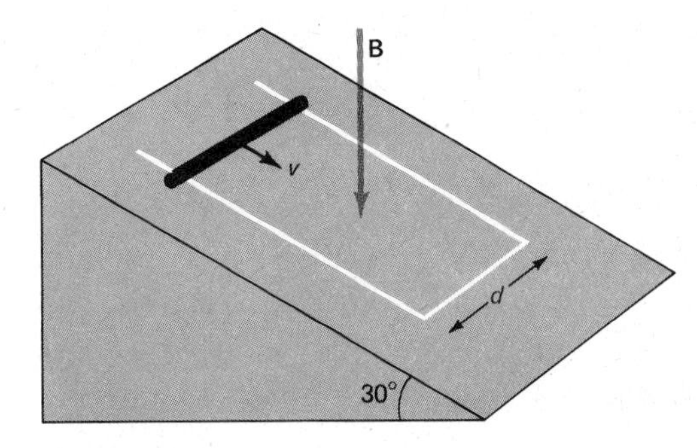

Figure P27.4

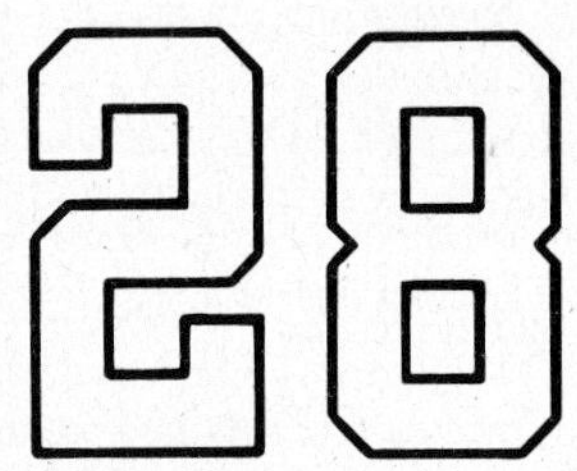

28 TIME-VARYING CURRENTS AND FIELDS

Performance Goals

When you finish this chapter, you should be able to

1. Define in your own words the following terms for a sinusoidal voltage or current: peak value, average value, rms value, and effective value. Where possible, relate the values to each other.
2. Given an ac voltage or current as read by the usual ac meter,

Most applications of electricity make use of alternating currents and voltages. This is the type of voltage furnished by the power company. Much of what we learned about dc circuits also applies to ac circuits. However, the effects of coils and capacitors are much different in ac applications. In this chapter we shall see how resistors, capacitors, and inductance coils behave under ac conditions. We shall also show how oscillating charges give rise to em waves.

sketch the time variation of the wave and give its amplitude. Show the period of the wave on the sketch.
3. Sketch v and i on the same graph for an ac voltage applied to a resistance. Given two of the following (V, I, and R), find the third quantity.
4. Sketch q and i as a function of t during the charging of a capacitor by a battery. Using the sketch, explain what is meant by the time constant. Given sufficient data, compute the time constant. Explain the meaning of the time constant for discharge of the capacitor through a resistor.
5. Given two of the following (V, X_C, I), find the third quantity for an ac voltage applied to a capacitor. State the phase relationship between v and i.
6. Define X_C and give its units. Explain how the impeding effect of a capacitor depends on frequency.
7. Explain why an inductance coil can be of considerable effect in an ac circuit, but not in a dc circuit. Define the quantity L and give its units.
8. Give the meaning of X_L and justify qualitatively its frequency dependence. Given two of the following (V, I, X_L), find the third for an ac voltage across an inductor. State the phase relation between v and i for an inductor.
9. On one graph sketch i, v_R, v_C, and v_L as a function of time for an RCL series circuit. Use the sketch to explain why the voltmeter reading across the combination is not equal to the

28.1 Alternating Current and Voltage

A varying current or potential difference that repeats the same pattern over and over again is called an alternating current or voltage. Typical alternating voltages are shown in Figure 28.1. (We call them ac voltages.) All these waveforms have widespread use in electronics. Only the sinusoidal waveform is furnished by the power company. It is also the waveform most often obtained at the output terminals of electronic oscillators. With proper electronic circuits the sinusoidal form can be changed to many other waveforms.

We shall be concerned only with sinusoidal waves. Those of you who do more advanced work in electronics will learn about the uses of other waveforms. But outside of electronics, sinusoidal ac voltages are by far the most important.

Consider the ac voltage shown in Figure 28.2(a). The way it is drawn, we see that it is a sine wave voltage. (Note that $v = 0$ when $t = 0$. It is a sine function because $\sin 0 = 0$.) We can write its equation as

$$v = v_0 \sin(2\pi f t) \tag{28.1}$$

It is customary to use small letters v and i to represent the varying voltage and current. The quantity v_0 is the amplitude of the voltage.

The amplitude of the voltage is 20 V in this case, as you can see in the figure. To find the period of the oscillating voltage, notice that it takes 0.4 s for one complete cycle. Therefore, the period $T = 0.4$ s. Because the frequency $f = 1/T$, we have

$$f = \frac{1}{0.4 \text{ s}} = 2.5 \text{ Hz}$$

If we substitute these values for the amplitude v_0 and the frequency f we have

$$v = 20 \sin(5\pi t) \text{ volts}$$

An alternating current is shown in Figure 28.2(b). This is a cosine curve. You should be able to show that its equation is

$$i = i_0 \cos(2\pi f t)$$

or

$$i = 3.0 \cos(5\pi t) \text{ amperes}$$

Its frequency is 2.5 Hz and its amplitude is 3.0 A.

It usually makes little difference whether an ac voltage or current is a sine or a cosine wave. As you see in Figure 28.2, the ac voltage has a sine waveform simply because $t = 0$ at the instant v was increasing through zero. If the timing clock had been started 0.10 s later, $t = 0$ would have occurred at a voltage peak.

sum of the individual voltages. Given the voltage reading across each element, compute the voltage across the combination.

10. Define the quantity Z for a series RCL circuit. Given a typical series RCL circuit, compute the current through it. Knowing the current, compute the voltage across each of its elements.
11. State, without any data given, the ac power loss in an ideal inductor and an ideal capacitor. Also, compute the power loss in a resistor when the minimum sufficient data are given.
12. Define each of the quantities in $P = VI \cos \Phi$. State the value of the power factor for each of the following: resistor, ideal capacitor, ideal inductor. Also compute the power factor for a series RCL circuit when R, C, L, and f are known.
13. Sketch a graph showing the variation of I as a function of frequency in an LC circuit. Show what happens to the curve if the circuit has resistance. Using $V = IZ$, point out what has happened to Z at the resonance frequency. Use this fact to find the relation between the resonance frequency and LC.
14. Sketch a circuit that uses a diode for rectification. Explain the operation of the circuit. Sketch a graph showing the input and output current.
15. Explain in words and with sketches how two reversing charges give rise to an electric field wave.
16. Explain in words and with sketches how two reversing

The wave would then have been a cosine wave. Usually we don't care when the timing clock was started. Therefore, usually it makes no difference if v or i is a sine or a cosine function. The important fact is that they alternate in a sinusoidal (or cosinusoidal) way.

28.2 ac Meter Readings

An ordinary dc voltmeter or ammeter will read zero if it is used to measure ac. The reason is simple: dc meters are usually designed to read the average voltage or current. But ac voltages and currents are positive as much as they are negative. Their average value is therefore zero, and this is what the dc meter reads.

Because an ac current flows first one way, then the reverse, it is useless for some purposes. It cannot be used for electrolysis because it will constantly alternate electrodes. During the negative part of the cycle the material plated out will be removed. There are also many electronic applications where ac is not suitable.

However, heating and many other applications can use ac and dc equally well. If you examine these situations, you will find that the use depends on i^2 (not i) and v^2 (not v). For example, the heat generated in a resistor is i^2R. Because these applications depend on the square of i and v, and because $(+i)^2 = (-i)^2$, it makes no difference whether i is plus or minus. Alternating current is therefore quite usable in these applications.

We cannot measure an ac current or voltage in terms of its average value. That value is always zero. Instead, we measure the effective or root mean square values of v and i. These terms are explained in the discussion that follows.

The heating effect of all types of current is given by the familiar power loss equation:

$$\text{Power loss} = i^2R$$

In the case of an ac current, such as the one shown in Figure 28.2(b), i varies continuously with time. Because the current varies so rapidly (60 Hz for house current), we are usually interested only in the average power loss. It will be

$$\text{Average power loss} = R(\text{average value of } i^2)$$

The important quantity is the average value of the square of i.

This is true in most applications of ac current. The important quantity is the average value of the current squared, or $(i^2)_{av}$. Similarly, the average value of the voltage squared, $(v^2)_{av}$, is usually the important quantity.

Most ac current meters are so made that they read $\sqrt{(i^2)_{av}}$, which we represent by I. Similarly, ac voltmeters read $\sqrt{(v^2)_{av}}$, and we represent this quantity by V. Because these are values of the square root of the average (or mean) squared quantity, they are often called the *root mean square* (rms) current and voltage. Or, alternatively, they are called the *effective* values.

charges give rise to a magnetic field wave.

17. Sketch a diagram showing an em wave traveling out along the *x* axis. Show both its electric and magnetic portions. Indicate the wavelength of the wave. State how fast the wave travels through vacuum.

18. Given a list of several types of em waves, arrange them in order of decreasing frequency (or wavelength). Given the wavelength in vacuum of an em wave (or its frequency), compute its frequency (or wavelength). Further, state whether the wave is radio and radar, infrared, visible, ultraviolet, or X ray.

19. Explain how straight wire and loop antennas detect radio waves.

20. Explain what is happening within a radio when we tune it to select the waves from one particular station.

Most ac meters are calibrated to read $\sqrt{(i^2)_{av}}$ or $\sqrt{(v^2)_{av}}$. These readings are represented by *I* and *V*. We call them the rms (or effective) values.

If we refer to Figure 28.3, we can see an important feature of the rms values of I and V. As shown, $(i^2)_{av}$ is $\frac{1}{2}i_0^2$. Therefore, I, which is equal to $\sqrt{(i^2)_{av}}$, is given by

$$I = \sqrt{\tfrac{1}{2}i_0^2} = \frac{i_0}{\sqrt{2}}$$
$$\simeq 0.707 i_0$$

Similarly, the rms voltage is given by

$$V = \frac{v_0}{\sqrt{2}} \simeq 0.707 v_0$$

Of course, these relations between the rms values (i.e., meter readings) and the wave amplitudes are generally true only for sinusoidal waves.

Let us summarize what we have said about ac meters.

Most ac meters read the effective or rms values of sinusoidal waves. We designate these readings as *V* and *I*. They are related to the amplitudes of the waves, v_0 and i_0, through the relation

$$\mathbf{V = \frac{v_0}{\sqrt{2}} \quad \text{and} \quad I = \frac{i_0}{\sqrt{2}}}$$

As you will see, most of our discussion will be in terms of the readings V and I.

EXAMPLE 28.1 With an ac voltmeter the potential difference of the house line is read to be 117 V. Its frequency is known to be 60 Hz. Write the equation for this voltage wave.

Solution The meter reads rms voltage. Therefore, because $V = 0.707 v_0$, we have

$$v_0 = \frac{117}{0.707}\text{ V} = 165\text{ V}$$

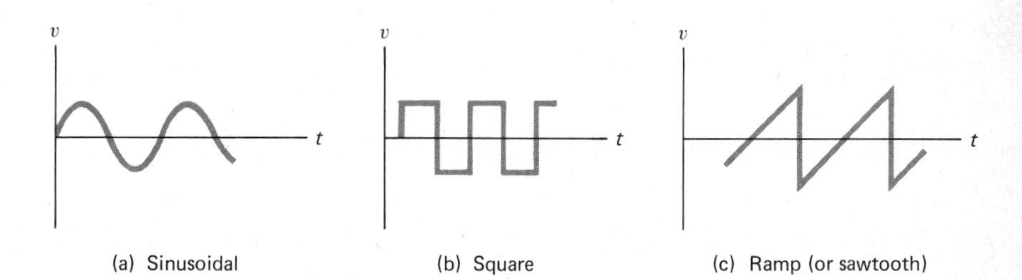

Figure 28.1 Typical ac waveforms.

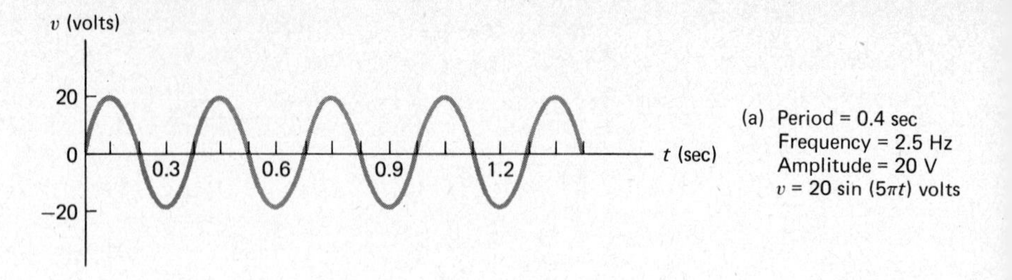

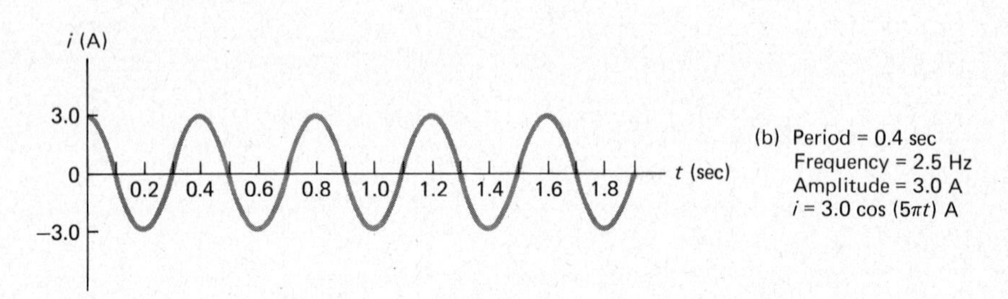

Figure 28.2 Whether a wave is a sine or cosine function depends on when the zero of time is taken. The voltage wave in (a) would have been a cosine form if the clock had been started 0.10 s later.

As we see, v_0, the peak voltage, is 165 V. The general waveform is

$$v = v_0 \sin(2\pi f t)$$

where we use the sine form for convenience. Using the values $v_0 = 165$ V and $f = 60$ Hz, this becomes

$$v = 165 \sin(120\pi t) \text{ volts}$$

Notice that the so-called line voltage is 0.707 times the peak voltage. The ac voltmeter reads (165)(0.707) = 117 V in the present case. ■■

28.3 Resistance in ac Circuits

A simple ac circuit is shown in Figure 28.4(a). The symbol —⊖— represents the ac voltage source. If this is the power line in a house, then $f = 60$ Hz and $V = 120$ V for it. At any instant the current through the resistor is i and the voltage across it is v. Ohm's law then tells us that

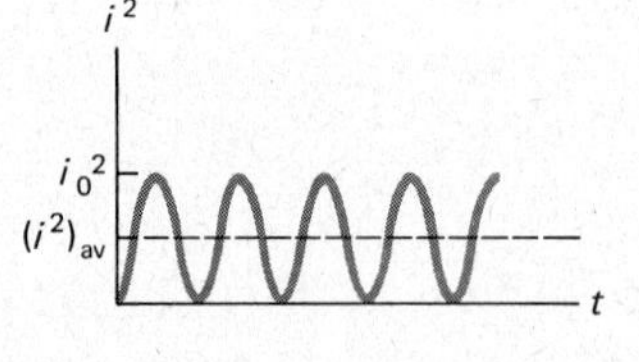

(a) $I = \sqrt{(i^2)_{av}} = \sqrt{i_0^2/2}$

(b) $I = i_0/\sqrt{2} = 0.707 i_0$

Figure 28.3 Most ac meters read the effective (or rms) values, $I = i_0/\sqrt{2}$ and $V = v_0/\sqrt{2}$.

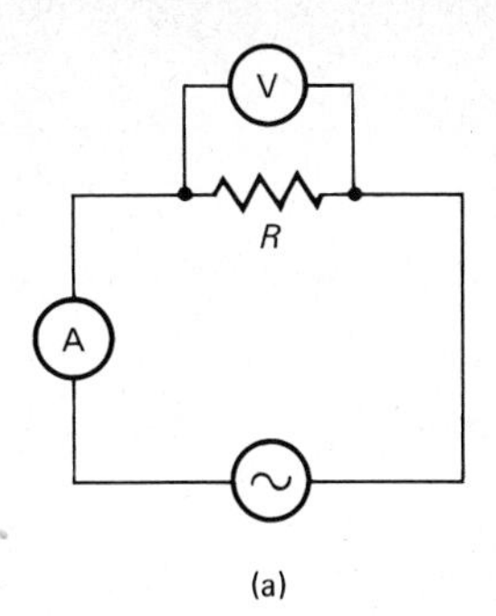

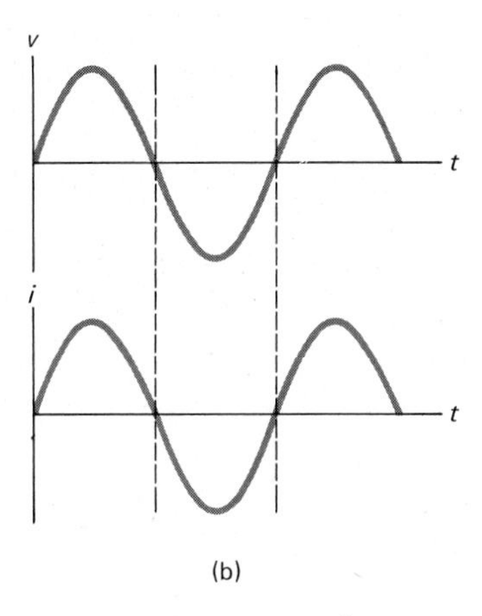

Figure 28.4 In a simple resistor v and i are in phase.

$$v = iR$$

Let us see what the ac voltmeter and ammeter would read in this circuit.

The voltmeter reads $\sqrt{(v^2)_{av}} = V$ and the ammeter reads $\sqrt{(i^2)_{av}} = I$. We can find these values by squaring both sides of $v = iR$ and averaging over one cycle. Doing this, we have

$$(v^2)_{av} = (i^2)_{av}R^2$$

But $(v^2)_{av} = V^2$ and $(i^2)_{av} = I^2$, so this becomes

$$V^2 = I^2R^2$$

from which

$$V = IR$$

So we see that the usual Ohm's law equation applies to ac as well as dc circuits.

In an ac circuit the potential difference across a resistor and the current through it are related by

$$\boldsymbol{V = IR} \tag{28.2}$$

where V and I are the rms meter readings.

To complete the story for this simple circuit, we shall notice two other features. The first feature is shown in Figure 28.4(b). We see there plots of the voltage v across the resistor and the current i through the resistor. As you would expect, the current and voltage are in step. Each reaches its maximum at the same time.

For a resistor v and i are in phase.

We shall see later that this fact is of importance.

The second feature concerns the power loss in the resistor. At any instant $P = vi$, as we proved in Chapter 24. This can also be written as $P = i^2R$. To find the average power loss in the resistor, we need the average of i^2R. But the average of i^2 is simply I^2. Therefore we find the following.

The average power loss in a resistor is given by I^2R. Because $V = IR$, we may rewrite this as VI.

As we see, the average power loss in a resistor obeys the same relation we found for dc circuits. However, the rms values for V and I must be used.

■ EXAMPLE 28.2 A 75-W bulb is operated on its rated 120-V, 60-Hz power source. Find the current drawn by the bulb.

Solution We know that the average power is given by VI. Therefore, because 75 W is the average power,

$$I = \frac{75\text{ W}}{120\text{ V}} = 0.625\text{ A}$$

■■

28.4 Time Constant of *RC* Circuit

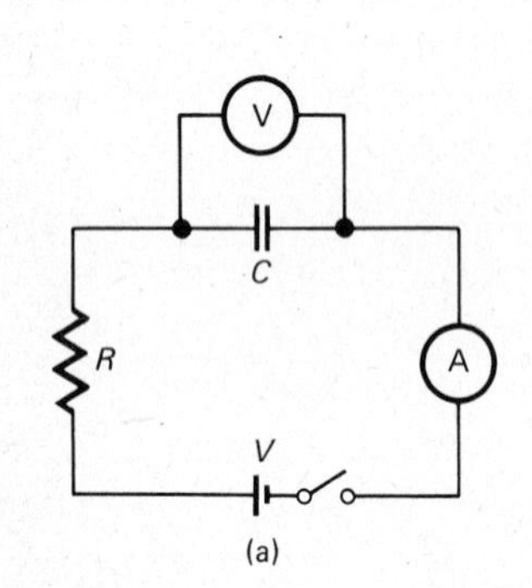

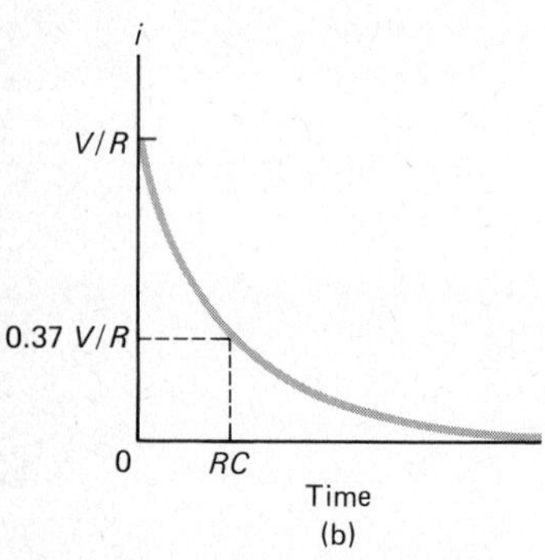

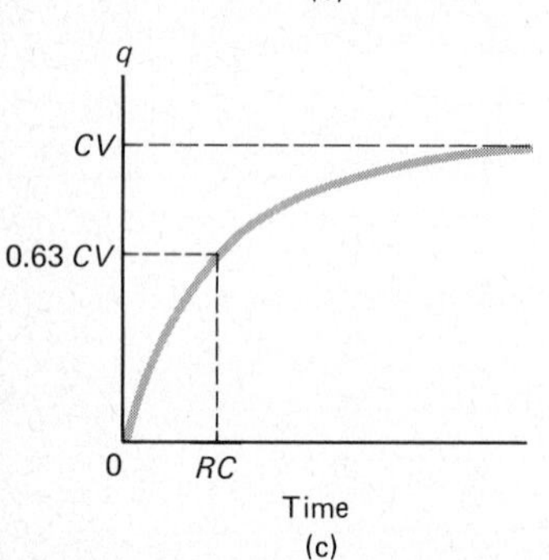

Figure 28.5 When the switch is first closed, the current is limited only by the resistor. At $t = 0$, $i = V/R$. At very long times, $i \to 0$ and the charge on the capacitor is CV. The time constant, $t = RC$, is a convenient measure of the charging time.

Capacitances are frequently used in ac circuits. We shall first consider a simple but important case. It is shown in Figure 28.5(a). A resistor, capacitor, battery, and switch are in series. Let us assume that the switch is open and the capacitor has no charge on it. What happens if the switch is now closed?

The qualitative answer to this question is quite easy to see. After the switch is closed, current will flow and the capacitor will charge. This will continue until the potential difference across the capacitor equals the voltage of the battery. Then the capacitor will have its maximum charge and the current in the circuit must stop. If we were to guess how the current behaved, we might guess it to start large and then slowly drop to zero. This is actually the way it behaves, as shown in Figure 28.5(b). Notice in (c) how the charge q on the capacitor starts at zero and rises to a maximum value.

At the instant the switch is closed, only the resistor holds the current back. As a result, the current at that instant is large. But as time goes on, a charge builds up on the capacitor. The charged capacitor also tends to hold the current back. Therefore, the current decreases as the capacitor charges.

The voltage drop across the capacitor was found in Chapter 24 to be given by $v = q/C$. When q gets large enough, $v = V$, the battery voltage. Then the battery can place no more charge on the capacitor and the current must stop. Notice how q varies with time in Figure 28.5(c). Compare its variation to the current.

As we see from Figure 28.5, there is no sharp point in time when the current stops. These curves are actually exponential functions, and they have no distinct end point. Therefore we use another way to describe how long it takes for a capacitor to charge. We describe its behavior in terms of the time constant, RC.

At a time $t = RC$ after the switch is closed, the capacitor is about 63 percent charged. The current in the circuit has dropped to about 37 percent of its initial value. Theory shows that these numbers apply to any RC circuit. As you would expect, the larger R is, the longer it takes to charge the capacitor. (After all, R keeps the current small.) In addition, the larger C is, the longer it takes to charge the capacitor. This is because the final charge on the capacitor is large for large C. It is not surprising, then, that the time constant is given by RC.

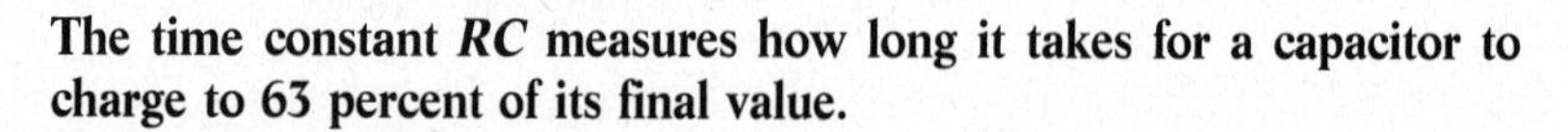
The time constant *RC* measures how long it takes for a capacitor to charge to 63 percent of its final value.

If you trace through the units of RC, namely, ohm-farad, you will find that they are equivalent to seconds. Therefore, RC is a time in seconds.

When a charged capacitor of value C is connected across a resistor R and allowed to discharge, its original charge Q_0 will be reduced to $0.37Q_0$ in one time constant, RC. In a time of n constants, the charge remaining on the capacitor will be $(0.37)^n Q_0$. For example, after five time constants have passed, the charge will have decreased to $(0.37)^5 = 0.0069$ of the original charge.

EXAMPLE 28.3 In many applications of high voltages, a high potential difference is placed across a capacitor. For example, in a TV set a dc potential of about 20,000 V is used in the TV tube. In the set a capacitor is also charged to this potential. For safety the charge on the capacitor is allowed to leak off slowly through a "bleeder" resistor connected to its two terminals. Suppose that the capacitor has $C = 2\ \mu F$ and the bleeder resistor is $10^6\ \Omega$. About how long would it take, after the TV is turned off, for the capacitor to discharge?

Solution The capacitor discharges through the bleeder. Therefore, for this circuit

$$RC = (10^6\ \Omega)(2 \times 10^{-6}\ F) = 2\ s$$

In this length of time the capacitor will be about 63 percent discharged. If one waits about ten times this long, the charge remaining on the capacitor should be harmless. ■■

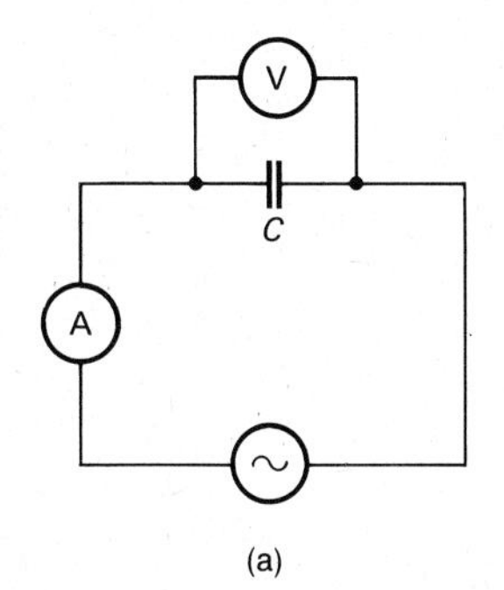

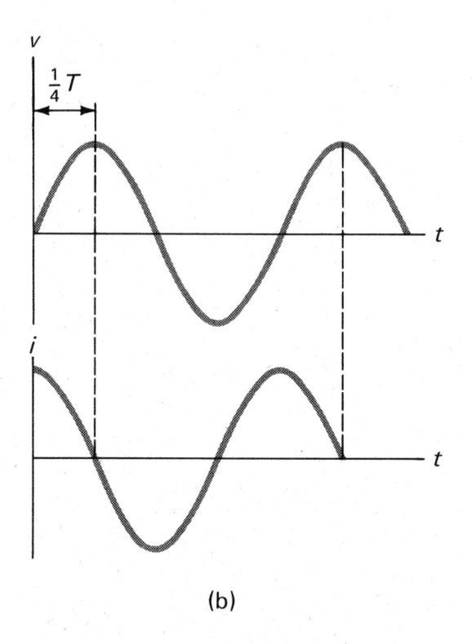

Figure 28.6 The voltage across a capacitor reaches its maximum $\frac{1}{4}$ of a cycle (90°, $\pi/2$) later than the current. Their phase difference is 90°.

28.5 Capacitance in an ac Circuit

Let us turn now to the simple circuit shown in Figure 28.6(a). A capacitor is connected directly across an ac power source. The power source provides the ac potential difference shown in the upper part of Figure 28.6(b). It is our purpose to find how the current in this circuit behaves.

Both experiment and theory show that the current follows the curve given in the lower part of Figure 28.6(b). The current and voltage are out of phase. The current is maximum when the source voltage is zero. We might have guessed that something like this would happen. After all, recall what we saw in the previous section. The current was greatest when the charge (and voltage) on the capacitor was zero. This also happens when an ac power source is used. In Figure 28.6(b), the voltage curve lags $\frac{1}{4}$ of a period behind the current curve.

The voltage lags $\frac{1}{4}$ of a cycle behind the current in a capacitance circuit.[1] The phase difference between current and voltage is 90°.

Compare this with the situation in a resistance circuit. There the voltage and current are in phase. Is there some relation similar to $V = IR$ that applies to the capacitance circuit? There is. Let us now see what it is.

Clearly, the capacitor opposes the flow of current. For example, in a dc circuit the capacitor stops the current completely. We call this stopping effect of the

[1] Some people use the word ICE to remember this: I in a C circuit is ahead of E (voltage).

capacitor the *impedance* or the *capacitive reactance* of the capacitor. The symbol X_C is used for this quantity. It is given in terms of the capacitance C and frequency of the voltage source f as

$$X_C = \frac{1}{2\pi f C}$$

Its unit is the ohm.

What we are saying can be stated the following way in terms of the voltmeter and ammeter readings.

When an ac voltage of frequency f is applied to a capacitance C, then

$$V = IX_C \tag{28.3}$$

where the capacitive reactance (or impedance)

$$X_C = \frac{1}{2\pi f C}$$

In other words, an Ohm's law relation applies to a capacitance. In it R is replaced by $X_C = 1/2\pi fC$.

We can easily justify the way f and C enter into the impedance of a capacitor. The higher the frequency, the less time the capacitor has to charge up. At large values of f the capacitor does not nearly become fully charged. It therefore offers little impedance to current flow. However, at low frequencies we are approaching dc conditions. Then the impeding effect of the capacitor is large. From this type of reasoning we see that it seems sensible that $X_C \sim 1/f$.

The fact that $X_C \sim 1/C$ is also easy to understand. If the value of C was near zero, the capacitor could hold no charge. The current flowing to it would be very small. As a result, when C is small, X_C should be very large. This, too, agrees with our expression for X_C.

■ **EXAMPLE 28.4** A 2-μF capacitor is connected directly across the 120-V, 60-Hz power lines. Find the current that flows to it. Repeat for the case in which the power source has a frequency of 6000 Hz.

Solution We know that $V = IX_C$. In the first case

$$X_C = \frac{1}{2\pi(60)(2 \times 10^{-6})}\ \Omega$$
$$= 1330\ \Omega$$

Then we find, from $V = IX_C$, that

$$I = \frac{120}{1330}\ A = 0.090\ A$$

In the second case the procedure above gives

$$X_C = 13.3\ \Omega$$

and so I is

$$I = 9.0\ \text{A}$$

Notice that, as we saw previously, the impeding effect of a capacitor decreases with increasing frequency. ■■

28.6 Self-Inductance

In addition to resistance and capacitance there is a third circuit element that is important in ac circuits. It is called self-inductance. A self-inductance is usually a coil of wire, often with a ferromagnetic core. As its name implies, it makes use of an emf it induces in itself.

For example, consider the coil in Figure 28.7(a). Before the switch is closed, the flux through it is zero. When the switch is first closed, a current starts to flow through the coil. As a result, flux appears in the coil. This increasing flux in the coil induces an emf in the coil. The coil induces an emf in itself. Lenz's law tells us that this emf is directed so as to oppose the change in flux and current through the coil.

When a current is changing in a coil, a self-induced emf exists in the coil. The induced emf opposes the change in current.

We can see how the current in the coil of Figure 28.7(a) will change as the switch is first closed. If there were no coil, the current would rise at once to a value $I = V/R$. Only the resistor would limit the current. But with the coil present, as the current tries to increase from its original zero value, there will be an emf induced in the coil. This induced emf opposes the increase in current. Therefore, the current does not rise at once to its final value, V/R. Instead, the current rises rather slowly to this value, as shown in Figure 28.7(c).

Every coil induces an emf in itself as the current through it changes. We

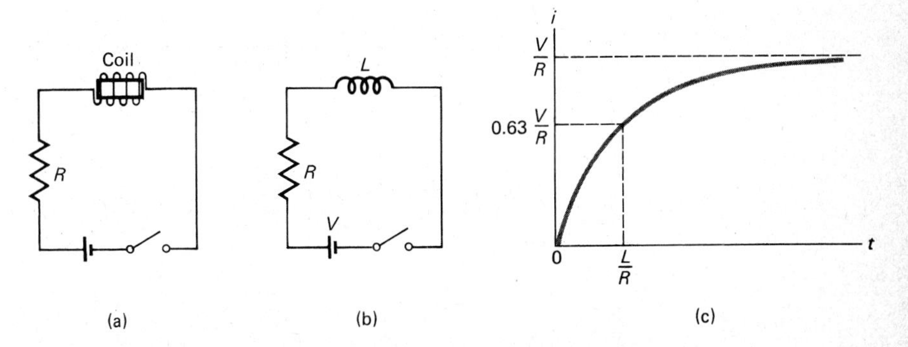

Figure 28.7 The induced emf in the self-inductance causes the current to rise less rapidly than it would otherwise.

measure this property of a coil in terms of what is called the *self-inductance* of the coil. We represent this quantity by the symbol L.

The self-inductance of a coil is defined in the following way. Suppose that the current through the coil is changing at a rate $\Delta i/\Delta t$. This changing current causes a changing flux and therefore an induced emf. We define L, the self-inductance of the coil, by the equation

$$\text{emf} = -L\frac{\Delta i}{\Delta t} \tag{28.4}$$

Because the emf opposes the change in current, the negative sign appears in the equation. The unit of self-inductance is the henry (H). A transformer-size coil with an iron core might have an inductance near 1 H.

Now let us return to the circuit of Figure 28.7(a). We show its schematic diagram in (b). Notice the symbol used for an inductor ⅏. Do not confuse it with the resistance symbol.

We define a time constant for an RL circuit just as we did for an RC circuit. The definition is shown in Figure 28.7(c). In a time equal to L/R, the current rises to 63 percent of its final value. This time, L/R, is called the time constant. Can you justify the way L and R appear in it? Can you show that its units are seconds?

▮ **EXAMPLE 28.5** A steady current of 5.0 A is flowing through a transformer coil that has a self-inductance of 2.0 H. The current is suddenly stopped in 0.01 s by opening a switch. About how large an emf is induced in the coil?

Solution From the definition of self-inductance,

$$\text{emf} = -L\frac{\Delta i}{\Delta t}$$

We approximate $\Delta i/\Delta t$ by setting $\Delta i = 5$ A, and $\Delta t = 0.01$ s. Then

$$\text{emf} = -(2.0)\frac{5.0}{0.01}\text{ V}$$

$$= -1000 \text{ V}$$

Notice how large this induced voltage is. These large voltages often cause sparking to occur at the switch as it is opened. Care must be taken to avoid injury due to high voltages generated in this way. ▮▮

28.7 Inductance in an ac Circuit

Suppose that an ac voltage source is connected across an inductor as shown in Figure 28.8(a). The sinusoidal voltage causes a sinusoidal current to flow. But the induced emf in the inductor opposes the change in the current. It has the net

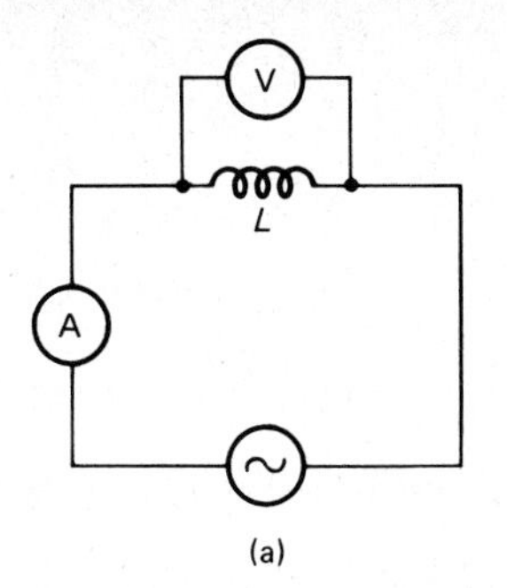

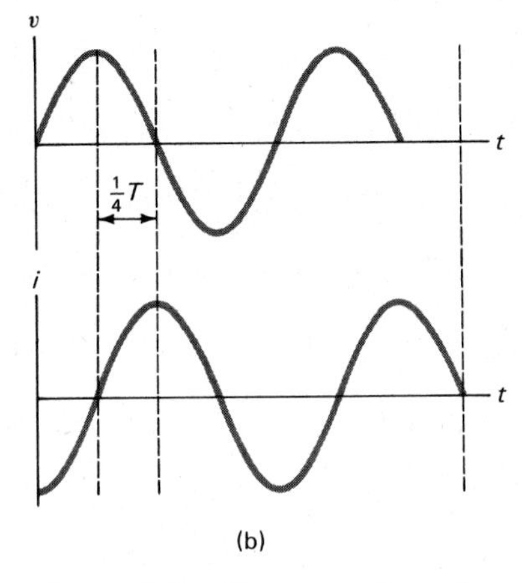

Figure 28.8 The current through an inductor reaches its maximum $\frac{1}{4}$ of a cycle later than the ac voltage that causes it. The current lags 90° behind the voltage.

effect of delaying the current in the circuit. This delay is shown in part (b) of the figure.

As we see, the current is delayed by $\frac{1}{4}$ of a period. The current wave is 90° (or $\frac{1}{4}$ of a cycle) behind the voltage wave. This is much like what happened with the capacitor. But in that case the current wave was ahead of the voltage wave. Let us state this finding.

The current lags $\frac{1}{4}$ of a cycle behind the voltage in an inductance circuit.[2] The phase difference between current and inductance is 90°.

An inductor opposes the changing current in an ac circuit. Like R and C, it shows an impedance effect, called the *inductive reactance.* We represent this quantity by the symbol X_L. It is related to the current I and the voltage V across an inductor by

$$V = IX_L$$

The unit of X_L is the ohm.

Here, too, we see that an Ohm's law relation applies. In it the inductive reactance replaces the usual resistance factor. It can be shown that the impeding effect of the inductor is given by

$$X_L = 2\pi fL$$

Let us summarize these facts.

When an ac voltage V with frequency f is applied to an inductor L, then

$$V = IX_L \qquad (28.5)$$

where the inductive reactance (or impedance)

$$X_L = 2\pi fL$$

If f is in hertz and L is in henrys, then X_L is in ohms.

It is easy to see why X_L depends on frequency. The induced emf is proportional to the rate of change of the current. But the change rate is large for high frequencies. As a result, the impeding effect of the inductor is greatest at high frequencies. At very low frequencies nearly dc conditions apply. Then the change is so slow that induced emf's are very small. The impedance effect of the inductor is small at low frequencies.

■ **EXAMPLE 28.6** The primary of a certain radio transformer has a dc resistance of 2 Ω and an inductance of 0.15 H. Find X_L for it when operated at 60 Hz. Repeat for 6000 Hz.

[2] Some people use the word ELI to remember this: E (voltage) in an L circuit is ahead of I.

Solution At 60 Hz we have

$$X_L = 2\pi(60)(0.15)\ \Omega$$
$$= 57\ \Omega$$

Notice that the impedance due to the self-inductance is large compared to the impeding effect of the coil's resistance. This is even more evident at 6000 Hz. Then

$$X_L = 2\pi(6000)(0.15)\ \Omega$$
$$= 5700\ \Omega$$

■■

28.8 *R, C,* and *L* in a Series ac Circuit

In previous sections we have examined the effect of *R, C,* and *L* separately. We now wish to consider the general series circuit shown in Figure 28.9(a). Because it is a series circuit, the current everywhere is the same. The current wave is represented by the top curve in Figure 28.9(b).

We learned in previous sections the following facts about the voltage waves.

1. v_R is in phase with i.
2. v_C is $\frac{1}{4}$ of a cycle behind i.
3. v_L is $\frac{1}{4}$ of a cycle ahead of i.

Using these facts, we can draw the graphs for the voltages across *R, C,* and *L*. These also are shown in Figure 28.9(b).

If you examine these graphs, you will notice a striking fact.

In a series circuit v_C and v_L always have opposite signs. They are 180° out of phase. Therefore, they tend to cancel each other.

In view of this, we encounter a peculiar feature about ac circuits. It has to do with the voltmeter readings across each element and how they add. Let us see what it is.

You will recall that Kirchhoff's loop rule states that the sum of the voltage changes around a closed circuit loop is zero. This rule is correct even for ac circuits such as the one shown in Figure 28.9(a). (At microwave frequencies this rule must be used with caution, however. In that case the swiftly changing flux near the circuit induces appreciable emf's in the circuit.) It is therefore true that the algebraic sum of the voltage changes around the circuit is zero *at any instant.* But ac meters do not read the voltage at a particular instant. Instead they read an average value, namely, $\sqrt{(v^2)_{av}}$. Moreover, the voltage they read is always positive. For this reason ac meters give readings that can never add to zero; the readings are all positive, and so no negative voltage changes are read by them. For this reason Kirchhoff's loop rule cannot be written in terms of the sum of ac voltmeter readings.

The voltage reading across two elements in an ac circuit is not always equal to the sum of the voltages across each of the two elements.

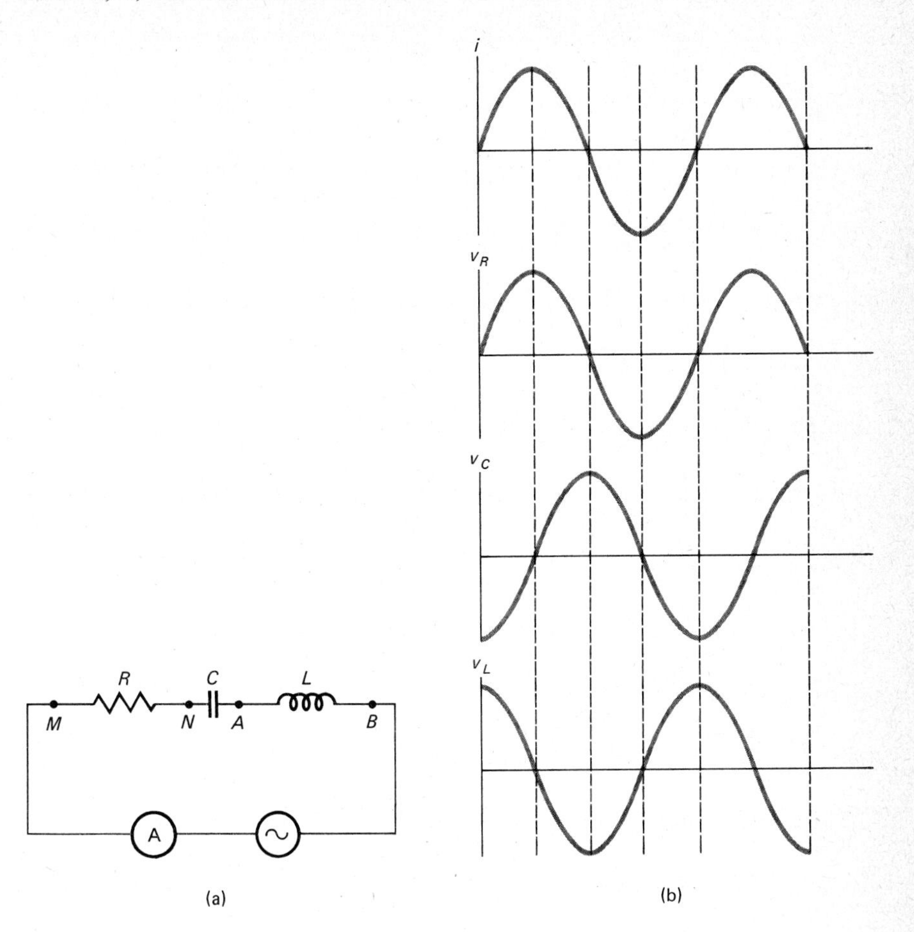

Figure 28.9 Notice how the voltages across the inductor and capacitor tend to cancel each other. Even though $v_{\text{source}} = v_R + v_C + v_L$, we cannot write $V_{\text{source}} = V_R + V_C + V_L$. Why?

We shall now find out how the voltage differences behave in situations such as this.

A voltmeter connected from N to A in Figure 28.9(a) will read the rms voltage across the capacitor, V_C. This value is 0.707 of the amplitude of the v_C wave shown in (b). Similarly, a voltmeter placed from A to B across the inductor will read V_L. This reading is 0.707 of the amplitude of the v_L wave. What will a voltmeter read when placed across both the capacitor and inductor from N to B?

At any instant the voltage difference from N to B will be the value of v_C at that instant added to the value of v_L at that instant. But we notice from part (b) of Figure 28.9 that when v_L is positive, v_C is negative, and vice versa. Therefore, these two voltages tend to cancel each other. The following result is therefore not surprising.

In an ac circuit a voltmeter placed across an inductor and capacitor in series reads a magnitude $|V_L - V_C|$.

The same voltmeter reads V_L when placed across the inductor alone and V_C when placed across the capacitor.

When the meter is connected across all three elements, the resistance as well as C and L, the situation is more complicated. As we see from Figure 28.9(b), the voltage across the resistor, v_R, sometimes adds and sometimes subtracts from v_L and v_C. Although the total effect of these waves can be found using reasonably simple trigonometry, we shall simply state what the voltmeter will read.

In a series ac circuit, if V_L, V_C, and V_R are the rms voltages across the three individual elements, then the voltage V across the RCL combination is given by

$$V^2 = V_R{}^2 + (V_L - V_C)^2 \tag{28.6}$$

As an example of the use of Equation 28.6, consider the circuit in Figure 28.9(a). Suppose that an ac voltmeter placed across the individual elements gives readings $V_R = 20$ V, $V_L = 60$ V, and $V_C = 90$ V. The equation tells us that the voltmeter should read the following when connected directly across the voltage source:

$$\begin{aligned} V &= \sqrt{V_R{}^2 + (V_L - V_C)^2} \\ &= \sqrt{400 + 900} \text{ V} \\ &= 36 \text{ V} \end{aligned}$$

Notice how different this is from the sum of the individual voltages.

There is an easy way to visualize Equation 28.6. The equation looks very much like the Pythagorean theorem for a right triangle, namely,

$$(\text{Hypotenuse})^2 = (\text{side } A)^2 + (\text{side } B)^2$$

To make use of this likeness, we construct a right triangle as shown in Figure 28.10(a). We take the two sides of the triangle to be V_R and $(V_L - V_C)$. Then the value of V is simply equal to the hypotenuse of this triangle. We have

$$V = \sqrt{V_R{}^2 + (V_L - V_C)^2}$$

As we see, Equation 28.6 is represented by the triangle shown.

Another important aspect of this triangle representation has to do with the angle ϕ. This angle turns out to be the phase angle between the current through the circuit and the voltage across it. For example, if the circuit contained no L and C, then $V_L - V_C$ would be zero. In that case the angle ϕ in the figure is zero. This agrees with our previous discussion, in which we found the current and voltage to be in phase for a simple resistor.

Or if $R = 0$ and there was no capacitor in the circuit, then $V_R = 0$ and $V_C = 0$. In that case the triangle of Figure 28.10(a) would have $\phi = 90°$. This also agrees with our previous result; in a simple inductance circuit the current and voltage are 90° out of phase.

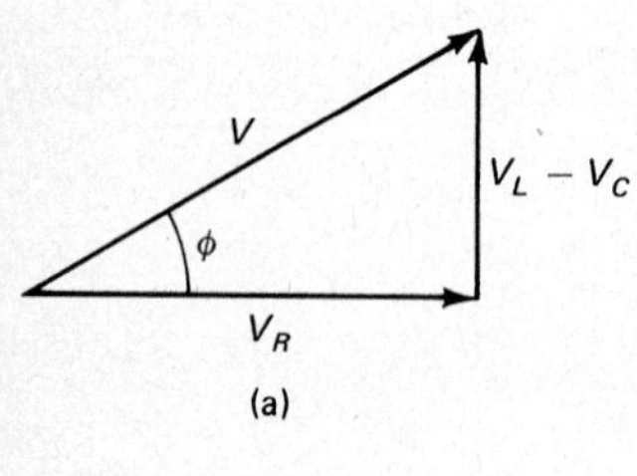

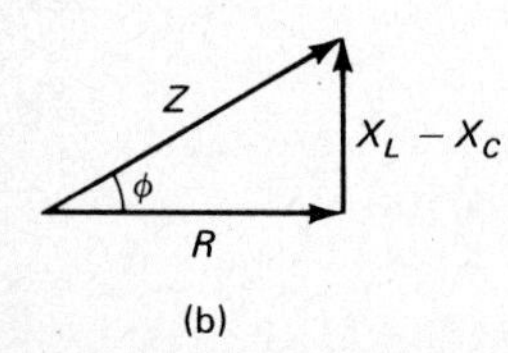

Figure 28.10 The triangle in (a) represents the fact that $V^2 = V_R{}^2 + (V_L - V_C)^2$ in a series RCL circuit. What does the triangle in (b) represent?

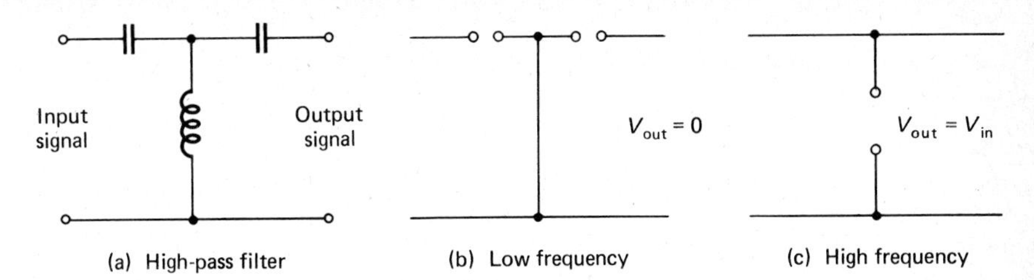

Figure 28.11 This simple filter passes high frequencies, but not low frequencies.

EXAMPLE 28.7 There are alternating 60-Hz magnetic fields nearly everywhere because of the power lines. A sensitive electronic circuit often shows a 60-Hz induced emf because of these varying fields. A filter is used to eliminate this unwanted voltage wave. Consider the high-pass filter shown in Figure 28.11(a). Explain why it passes high-frequency voltages but not low-frequency voltages.

Explanation You will recall that at low frequencies a capacitor has a high impedance, but an inductor has a low impedance. As a result, at very low frequencies a capacitor acts almost like a gap in the wire; its impedance is nearly infinite. But under these same conditions, an inductor acts almost like a wire that has zero impedance. With this in mind, we can represent the circuit in (a) by the circuit in (b) at very low frequencies. Because the capacitors act like gaps in the wire, the filter does not allow low-frequency signals to pass through it.

At high frequencies a capacitor offers little impedance to current flow, but the impedance of an inductor is large. It will act like a gap in the wire. As a result, the filter behaves as in (c) at high frequencies. As we see, it acts like two wires and so it allows high-frequency signals to pass through. ▮▮

EXAMPLE 28.8 A certain series ac circuit has *R*, *C*, and *L* connected across a 120-V power source. The voltmeter reading across the inductor is 205 V. The voltmeter reading across the capacitor is also 205 V. What will the voltmeter read across the resistor?

Solution This circuit is the same as the one in Figure 28.9. Notice that the voltmeter readings do not add directly. For example, $V_L + V_C = 410$ V, whereas the power source voltage is only 120 V. To relate the voltages in a series ac circuit, we must use Equation 29.6,

$$V^2 = V_R^2 + (V_L - V_C)^2$$

Substitution of the known values gives, in volts,

$$(120)^2 = V_R^2 + (205 - 205)^2$$
$$V_R = 120 \text{ V}$$

Can you show that this result also follows from the triangle of Figure 28.10? ▮▮

28.9 Impedance in *RCL* Series Circuit

We have seen that Ohm's law equations apply to the resistor, the capacitor, and the inductor in ac circuits. A similar relation applies to the combination of *R*, *C*, and *L* in series.

Equation 28.6 applies to the voltage V across R, L, and C in series.

$$V^2 = V_R{}^2 + (V_L - V_C)^2$$

If we divide this whole equation by I^2, we obtain

$$\frac{V^2}{I^2} = \frac{V_R{}^2}{I^2} + \frac{(V_L - V_C)^2}{I^2}$$

But we know that $V_R/I = R$, $V_L/I = X_L$, and $V_C/I = X_C$. Substitution gives

$$\left(\frac{V}{I}\right)^2 = R^2 + (X_L - X_C)^2$$

After taking square roots of both sides and multiplying by I, this becomes

$$V = I\sqrt{R^2 + (X_L - X_C)^2}$$

Notice that here, too, we have an Ohm's law form. But now R is replaced by the square root factor. We call this the *impedance* of the series circuit and represent it by the symbol Z.

In a series *RCL* circuit the voltage *V* across the circuit and current *I* through it are related by

$$\mathbf{V = IZ} \tag{28.7}$$

The quantity *Z* is the impedance and is given by

$$\mathbf{Z = \sqrt{R^2 + (X_L - X_C)^2}}$$

Its unit is the ohm.

We see that Z contains much of our previous information. If $X_L = X_C$, then $X_L - X_C = 0$ and $Z = R$. As another example, for a pure resistance, $Z = R$. Similarly, $Z = X_L$ for an inductor and $Z = X_C$ for a capacitor. For any series connection of R, C, and L, we see that $V = IZ$ applies.

The quantity Z can also be shown in a right triangle. Because $Z^2 = R^2 + (X_L - X_C)^2$, we see that Z corresponds to the hypotenuse whereas R and $(X_L - X_C)$ are the sides. This representation is also shown in Figure 28.10. The angle ϕ is the same in these two triangles. Can you prove this fact?

▮ **EXAMPLE 28.9** A 0.025-H inductor whose resistance is 5.0 Ω is connected across a 120-V, 60-Hz power source. Find the current that flows through it. Repeat for a frequency of 6000 Hz.

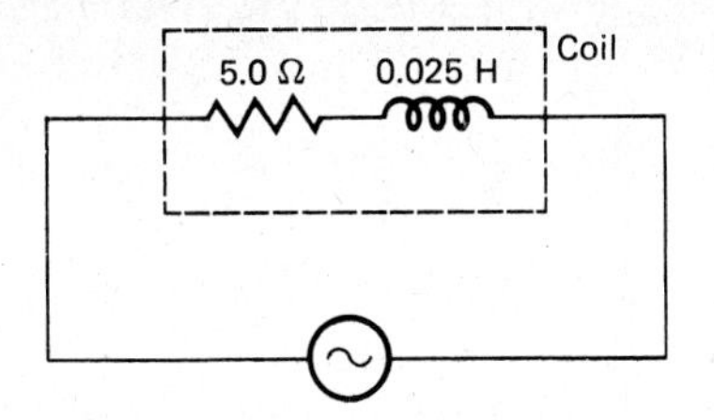

Figure 28.12 The resistance of an inductance coil acts like a series resistor.

Solution We can picture the coil and its resistance to be in series as shown in Figure 28.12. We are told that $R = 5.0\ \Omega$. To find X_L we use

$$X_L = 2\pi f L$$

which is, in the first case,

$$X_L = 2\pi(60)(0.025)\ \Omega = 9.4\ \Omega$$

The impedance of the coil is

$$Z = \sqrt{R^2 + X_L^2}$$

because X_C is zero (i.e., there is no capacitor). Placing in the values gives

$$Z = \sqrt{25 + 89}\ \Omega = 10.7\ \Omega$$

Using $V = IZ$ gives

$$I = \frac{120}{10.7}\ \text{A} = 11.3\ \text{A}$$

At the higher frequency, R still is $5.0\ \Omega$. But now X_L turns out to be $940\ \Omega$. Notice how much larger the impedance of the inductor is at this higher frequency. Using the same procedure as before, we find that $Z = 940\ \Omega$. The resistance effect is negligible in comparison to the effect of the inductance. To find I in this case, we use $V = IZ$ to give $I = 0.128$ A. This example shows clearly that an inductor can be used to keep currents low at high frequencies. Its blocking effect is much less at low frequencies. ■■

■ **EXAMPLE 28.10** A 0.50-μF capacitor, a 20-Ω resistor, and a 0.050-H inductor are connected in series across a 25-V, 1000-Hz power source. Find the voltmeter readings across the capacitance and inductance.

Solution We shall use $V = IZ$ to find I. To use it, we need R, X_L, and X_C. They are

$$R = 20\ \Omega$$
$$X_L = 2\pi(1000)(0.050)\ \Omega = 314\ \Omega$$
$$X_C = \frac{1}{2\pi(1000)(0.5 \times 10^{-6})}\ \Omega = 318\ \Omega$$

For future reference, notice that X_L and X_C are nearly equal in this case.

We now need Z. It is given by

$$Z = \sqrt{(20)^2 + (314 - 318)^2}\ \Omega$$
$$= \sqrt{400 + 16}\ \Omega = 20.4\ \Omega$$

Notice how little effect the combined X_L and X_C have. This is true even though both have impedances much larger than R.

$$I = \frac{V}{Z} = \frac{25}{20.4}\ \text{A} = 1.23\ \text{A}$$

Clearly the resistance of the circuit is what limits I in this particular case.

We can find the voltages across the inductor and capacitor by use of

$$V_L = IX_L \qquad \text{and} \qquad V_C = IX_C$$

We then have

$$V_L = (1.23)(314)\ \text{V} = 385\ \text{V}$$

and

$$V_C = (1.23)(318)\ \text{V} = 391\ \text{V}$$

Notice how large the voltages are across these elements. They are much larger than the source voltage, 25 V. But we also notice that the voltage across the series combination of L and C is $V_L - V_C$, which is -6 V. Their two voltages tend to cancel each other. ■■

28.10 Resonance in *RCL* Series Circuits

In the previous example we saw a situation where X_L was nearly equal to X_C. The capacitance and inductance nearly canceled each other out. If X_L had exactly equaled X_C, then they would have canceled exactly. This is an interesting and important case, as we shall now see.

Both $X_L = 2\pi fL$ and $X_C = 1/2\pi fC$ are dependent on frequency. Notice that X_L increases with frequency whereas X_C decreases with frequency. Because of this it is possible to find a frequency at which $X_L = X_C$. Let us see what happens at this special frequency.

Consider the circuit shown in Figure 28.13(a). The current that flows in it is

$$I = \frac{V}{Z} = \frac{V}{\sqrt{R^2 + (X_L - X_C)^2}}$$

Let us suppose for a moment that R is negligibly small. Then this equation becomes

$$I = \frac{V}{\sqrt{(X_L - X_C)^2}} = \frac{V}{|X_L - X_C|}$$

If the power source is a variable frequency oscillator, then we can measure I in this circuit as a function of f. At very low frequencies X_C is very large and X_L is very small. I is therefore small. At very high frequencies the reverse is true; X_C is very small and X_L is very large. But I is again very small.

However, at some intermediate frequency, $X_L = X_C$. Then if the resistance is

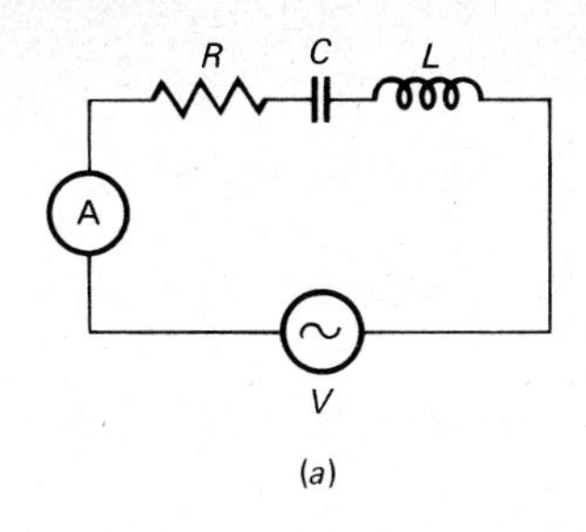

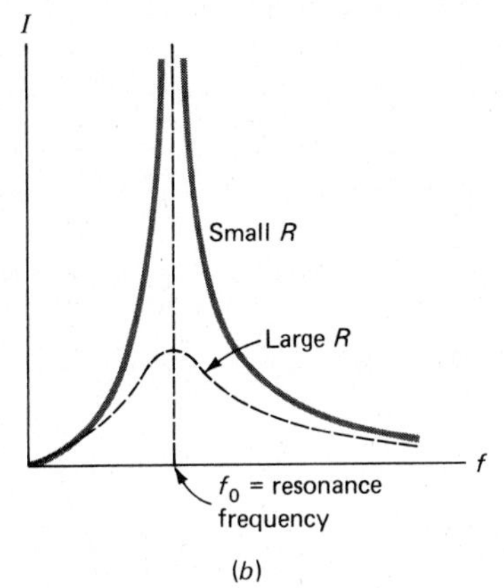

Figure 28.13 As the oscillator frequency is varied, the current in the circuit changes. The sharpness of the resonance is decreased by large circuit resistance.

negligible, I would become infinite. This variation of I with frequency is shown in Figure 28.13(b). Of course, R is never zero. Therefore I follows a curve more like the broken curve shown in the figure.

The frequency at which the current in an LC circuit reaches its maximum value is called the resonance frequency. It occurs when $X_C = X_L$ and is given by

$$f_0 = \frac{1}{2\pi}\frac{1}{\sqrt{LC}} \qquad \textbf{(28.8)}$$

Equation 28.8 for the resonance frequency is easy to obtain. It is the frequency at which $X_L = X_C$, that is, when

$$2\pi f_0 L = \frac{1}{2\pi f_0 C}$$

Solving this equation for f_0 gives the result in Equation 28.8.

The resonating circuit is of great importance in electronics. As we shall see later, it is used to tune radios and TVs. Its great advantage is that it responds strongly to one particular frequency. Suppose that the oscillator in Figure 28.13 is actually a power input source that has many frequencies all existing at once. If one of these frequencies is the resonance frequency, then the circuit responds strongly to it. The other frequencies have little effect. A resonant circuit can select one signal of a definite frequency from a jumble of other signals.

■ **EXAMPLE 28.11** A student wishes to construct a circuit that will resonate on the 120-V, 60-Hz house line. He has available a 0.25-H coil that has a resistance of 3.0 Ω. What value capacitor must he connect in series with it? How large a current will the resonating circuit draw on the house line?

Solution At resonance, $X_L = X_C$. We know that $f = 60$ Hz and $L = 0.25$ H, and so we can find C. Equating X_L to X_C gives

$$2\pi f L = \frac{1}{2\pi f C}$$

Placing in the known values and solving for C gives 28 μF. He is unlikely to find a capacitor this large that will withstand high voltages. (They exist, but they are not widely used.)

If he does find one, the current in his circuit is, from $V = IZ$,

$$I = \frac{120 \text{ V}}{\sqrt{R^2 + (X_L - X_C)^2}} = \frac{120 \text{ V}}{R}$$

Remember, at resonance, $X_L = X_C$. Because $R = 3.0\ \Omega$, we obtain $I = 40$ A. Such a large current would have blown a house fuse, so it is probably lucky that the capacitor is not readily available. ■■

28.11 Power Loss in ac Circuits

No power loss occurs in an ideal capacitor or inductor. This important fact can be seen by referring to the curves of Figure 28.14. These are taken from Figure 28.6. They show how the current and voltage waves are related in the case of a capacitor.

At any instant the power being used in the capacitor is vi. This quantity is plotted in the lower part of the figure. During part of the cycle the capacitor is using power. But during other parts of the cycle the power is negative. The capacitor is giving the power back to the energy source. The power curve is negative as much as it is positive. Therefore, the capacitor uses zero net power. No power loss occurs in an ideal capacitor.

You will recall that i and v are also 90° out of phase for an ideal inductor. The same reasoning shows that the power loss there is also zero during a cycle. We can therefore conclude that no power loss occurs in an ideal inductor.

In the case of a resistor, v and i are in phase. The power loss there is VI, which is the same as I^2R. The power loss in a resistor is I^2R.

We can summarize all these results in terms of a quantity called the *power factor.* Let us designate the phase angle between v and i by ϕ. This is the same angle ϕ we encountered in Figure 28.10. As we know, ϕ is 90° for an inductor

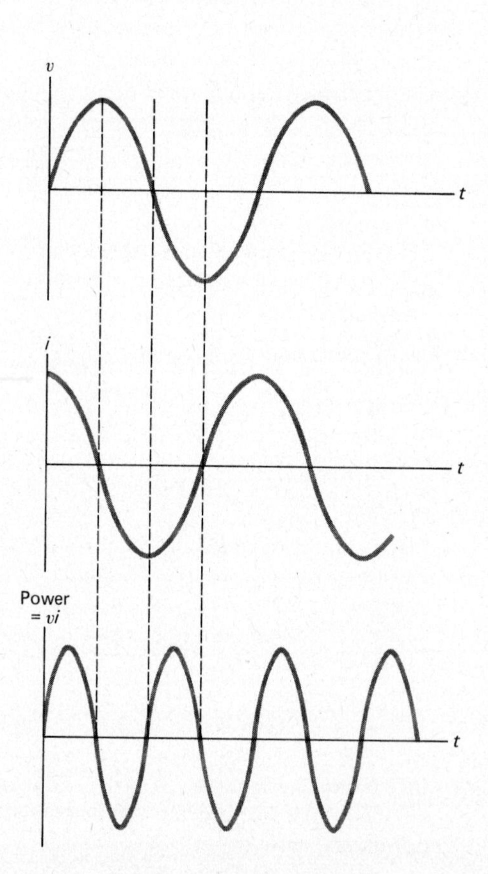

Figure 28.14 For an ideal capacitor, v and i are 90° out of phase. The instantaneous power, vi, is therefore negative as much as it is positive. As a result, the capacitor consumes zero average power.

and also for a capacitor. (Strictly speaking, ϕ is defined in such a way that it is $-90°$ in this case, as we can easily see by referring back to Figure 28.10. But we shall not be concerned with this detail here.) In the case of a resistor, $\phi = 0$. We call $\cos \phi$ the power factor.

The power consumed in a portion of an ac circuit is given by

$$\textbf{Power} = VI \cos \phi \qquad \textbf{(28.9)}$$

The quantity $\cos \phi$ is called the power factor for that portion of the circuit.

Because $\phi = 0$ for a resistor, $\cos \phi = 1$ for a resistor. The power loss in it is VI. In the case of an ideal capacitor or inductor, $\cos \phi = 0$ because $\phi = \pm 90°$. The power loss in these cases is zero.

Equation 28.9 applies to all portions of an ac circuit. For example, a resistor, capacitor, and inductor might be connected in series. Call V the voltage across the combination and let I be the current through it. Then the power loss in this case is still $VI \cos \phi$. As we have said, the phase angle ϕ turns out to be the same as the angle ϕ shown in Figure 28.10. From that figure we see that

$$\text{Power factor} = \cos \phi = \frac{R}{\sqrt{R^2 + (X_L - X_C)^2}} \qquad (28.10)$$

$$= \frac{V_R}{\sqrt{V_R{}^2 + (V_L - V_C)^2}}$$

We can easily see that this checks our previous results. In any case where $R = 0$, Equation 28.10 shows that the power factor is zero.

EXAMPLE 28.12 An inductance coil is never ideal. It always has some resistance. Also, any eddy current heating of the coil's core makes the coil act like it has resistance. (This resistance varies with frequency because the eddy current heating increases with frequency.) Suppose that a particular coil has $L = 0.20$ H and has a resistance of 3.0 Ω at 1000 Hz. Find the power loss in the coil when operated on a 15-V, 1000-Hz power source.

Solution As we did in Figure 28.12, we consider the coil to be an ideal inductor in series with a resistor. The inductive reactance is

$$X_L = 2\pi(1000)(0.20)\ \Omega = 1260\ \Omega$$

We then have

$$Z = \sqrt{R^2 + X_L{}^2} \simeq 1260\ \Omega$$

Notice in this case that the impedance is essentially all due to the inductor. The current in the coil is given by

$$I = \frac{V}{Z} = \frac{15}{1260}\ \text{A}$$
$$= 0.0120\ \text{A}$$

Method 1: We can find the power in two different ways. Because all the power loss occurs in the resistance, we have

$$\text{Power} = I^2R = (0.0120)^2(3)\ \text{W}$$
$$= 4.3 \times 10^{-4}\ \text{W}$$

Method 2: The power factor is

$$\cos\phi = \frac{R}{Z} = \frac{3}{1260} = 2.38 \times 10^{-3}$$

Then

$$\text{Power} = VI\cos\phi$$
$$= (15)(0.0120)(2.4 \times 10^{-3})\ \text{W}$$
$$= 4.3 \times 10^{-4}\ \text{W}$$

Both methods give the same result. ▪▪

28.12 Conversion of ac Voltage to dc Voltage

Because the power company furnishes us with ac voltage, the need often arises to change this to dc. This conversion is carried out by means of a device called a rectifier. There are three different types of rectifiers in widespread use. Each allows current to flow in only one direction.

1. The vacuum tube rectifier makes use of the flow of electrons from a hot filament to a cold metal plate. Electrons boil out of the hot filament into the vacuum. Because they cannot leave the cold plate, electrons can flow in only one direction through the device. Current flows only when the plate is positive. (See Figure 28.15.)
2. The gas tube rectifier is much like the vacuum tube rectifier. However, the tube contains a nonreactive gas. Electrons passing through the tube ionize the gas atoms. The current through the tube is greatly increased by the ions.
3. The semiconductor diode rectifier makes use of two crystalline semiconducting materials. One material is called n-type. The other is called p-type. Conventional current can only flow from p-type to n-type at the junction between the two materials. As a result, a p-n junction acts as a rectifier of current. We will describe it more fully in Chapter 30.

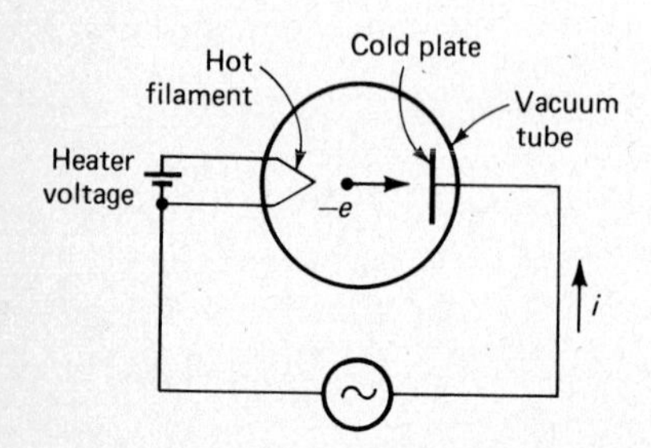

Figure 28.15 Electrons boil out of the filament and move to the plate when the plate is positive. No reverse flow can occur. Therefore, i in the circuit flows only in the direction shown.

The p-n junction rectifier (or diode) is usually simplest to use. For this reason it has received widespread application since its development. We shall use it to

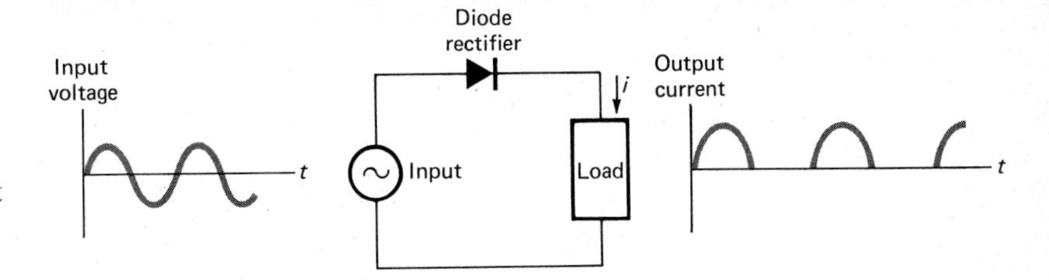

Figure 28.16 The diode passes current in only one direction. It therefore rectifies the current in the circuit.

illustrate how a rectifier does its job, but the discussion applies equally well to the other types.

We show in Figure 28.16 how a rectifier can be used to provide dc current from an ac source. The box labeled "Load" might be a resistor, an electroplating cell, a battery being charged, or any other use of dc current. Notice the symbol used for the diode (—▶|—). Most diodes have this symbol stamped on them. The diode conducts current in the direction of the arrow in the symbol, but not in the reverse direction.

As seen in Figure 28.16, the current to the load is dc. The diode allows current to flow in only one direction. Of course, the current pulsates. No current at all flows when the source tries to produce current in the wrong direction.

Often the pulsating current must be smoothed out for a particular application. This can be done by several techniques. Some smoothing can be done by using two diodes. A full-wave rectifier using two diodes is shown in Figure 28.17. Notice where the load is placed. The current through it is shown at the right in the figure. You should convince yourself that one of the diodes passes current no matter which end of the power source is positive. Often the resistor is replaced by an inductance coil that has a wire connected to its midpoint (a center-tap coil).

By placing inductors and capacitors in the load circuit, one can smooth the dc output further. The capacitor-inductor system used to do this is called a smoothing filter.[3] A typical, not too smooth, output is shown in Figure 28.18. Frequently the smoothness of the dc output is specified in terms of the *ripple*. This quantity is defined as the ratio of the current (or voltage) variation to the maximum current (or voltage). A good dc power supply might well have a ripple of less than 0.01 percent (i.e., 1×10^{-4}).

[3] A rather crude smoothing filter can be made by connecting a large capacitance across the load. It tends to maintain a constant charge if the pulsation time is short compared to the effective RC time constant. As a result, the voltage across the load is held relatively constant.

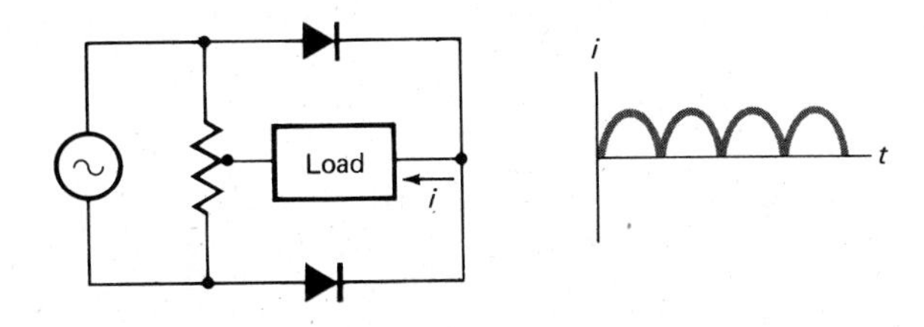

Figure 28.17 A full-wave rectifier gives a smoother dc output than a single diode does.

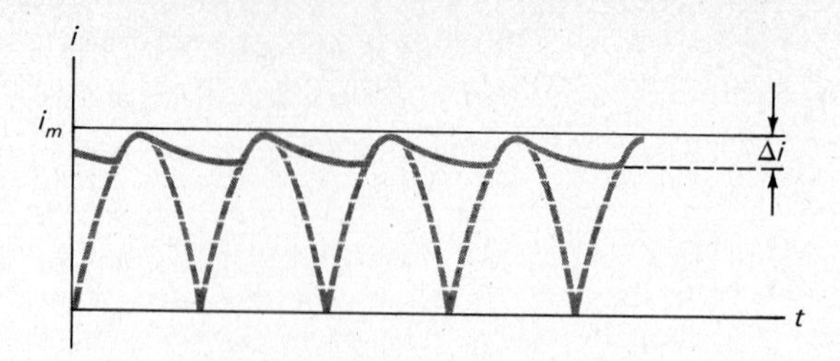

Figure 28.18 The output from a rectifier system can be smoothed further by use of capacitors and inductors. The ratio $\Delta i/i_m$ is called the ripple of the wave.

28.13 Transformer Currents

In the last chapter we found that the following relation applies to a transformer: $V_s = V_p(N_s/N_p)$. It states that the induced voltage in the secondary is proportional to the turns ratio, N_s/N_p. There is no simple general relation that relates the current in the secondary (I_s) to the current in the primary (I_p). Let us see why.

The law of conservation of energy tells us that the average input power to the primary must equal the average output power from the secondary provided losses in the transformer are negligible. We can then write

$$V_pI_p \cos \phi_p = V_sI_s \cos \phi_s$$

where ϕ_p and ϕ_s are the phase angles for the primary and secondary, respectively. If we could set $\phi_p = \phi_s$, then we could write this as

$$I_s = I_p\left(\frac{V_p}{V_s}\right) = I_p\left(\frac{N_p}{N_s}\right)$$

But this equation is obviously wrong. For example, if the resistance of the secondary is infinite, then we know that $I_s = 0$. But the primary coil still draws current from the power source, and so $I_p \neq 0$. As we see, I_s does not equal $I_p(N_p/N_s)$ in this case. The difficulty occurs, of course, because we made the unjustified step of setting $\phi_p = \phi_s$.

Another approach would be to discuss instantaneous power, vi. Then we might try to write $v_pi_p = v_si_s$. But this too is wrong. It neglects the fact that part of the input power is stored in the transformer coils.

In spite of these complications, there is one special case that can be treated easily. Suppose that the current in the primary is negligible when no current is being drawn from the secondary (i.e., when the resistance of the secondary is infinite). For such a transformer a simple relation applies when enough current is being drawn by the secondary so that the primary current is large. In that case it is clear that negligible energy is being stored and consumed in the transformer coils. Then we can write, for the instantaneous values,

$$v_pi_p = v_si_s$$

But both v_p and v_s are proportional to $\Delta\phi/\Delta t$. They are therefore in phase.

Because they are in phase, so must i_p and i_s be in phase. As a result, the relation above is true for the rms values as well as for the instantaneous values. We thus find that in this special case

$$I_s = I_p\left(\frac{V_p}{V_s}\right) = I_p\left(\frac{N_p}{N_s}\right)$$

Again we must caution that this relation is correct only if $I_p \cong 0$ when no current is being drawn from the secondary. It tells us that a step-up transformer steps up the voltage and steps down the current.

28.14 Origin of Radio Waves

We have mentioned that stray magnetic fields have important effects. When changing, these fields lead to eddy currents in nearby pieces of metal and induced emf's in nearby circuits. Because ac currents produce continuously changing fields, this effect is quite noticeable for them. The higher the frequency of the alternating current, the faster the fields will be changing. Therefore, stray induced emf's are most easily observed at very high frequencies.

In effect what we have here is a changing current in one circuit inducing emf's in a distant circuit. One can conceive of this effect being used to transmit signals over large distances. Indeed, this is basically what happens in radio wave transmission as we will now see.

During our study of electrostatics we learned that a charge causes an electric field. As an example, the positive and negative charges in Figure 28.19 cause the field shown. We call two equal and opposite charges such as this a *dipole*. This type of charge combination is important to us because it is the simplest charge combination used in radio transmitting antennas. In this section we shall see how a charged radio antenna causes a radio wave.

All sources of waves send out a signal. For example, the situation shown in Figure 28.20(a) is familiar to all of us. A small stone falls into water. It causes a wave on the water surface. This wave travels out along the surface and might hit a bug sitting at *A*. When the wave strikes the bug, the signal (the wave) tells it that something fell in the water. Notice that the signal takes time to reach the bug. The wave on the water surface tells the bug what happened earlier at the wave source.

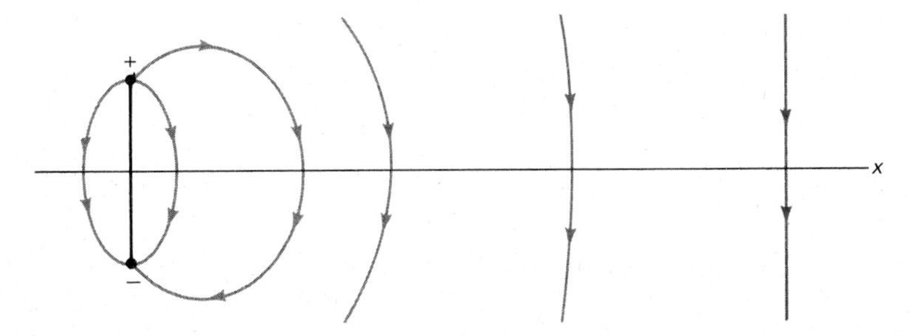

Figure 28.19 The charged dipole antenna blankets the surrounding region with an electric field.

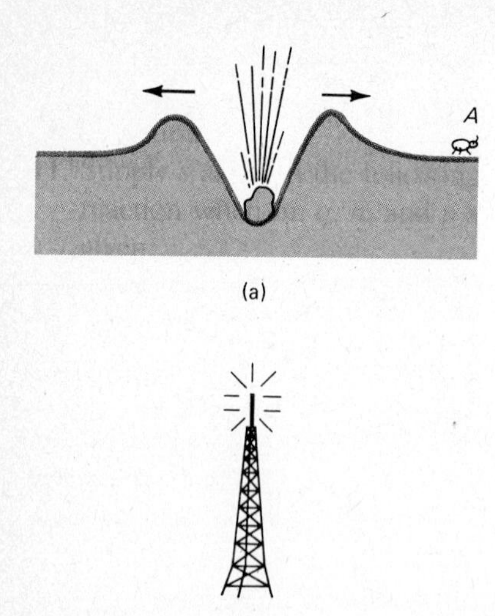

Figure 28.20 The water wave caused by the stone in (a) will soon reach the bug at *A*. It takes time for the wave generated at the source to reach distant observers. The wave sent from the radio transmitter in (b) also takes time to travel out from the station.

A similar situation exists with the electric signal sent out by the radio transmitter shown in Figure 28.20(b). At the transmitter the antenna acts as the wave source. We can picture the basic element of the antenna to be a charge dipole. At a certain instant the antenna might be charged as shown in Figure 28.19. Notice the direction in which the electric field points in the region close to the charged antenna.

However, the situation in Figure 28.19 will not exist for long. The antenna is charged by use of an electric oscillator. The oscillator provides ac to the antenna. As a result, the charges on the antenna reverse with the same frequency as the oscillator. Each radio station is assigned a frequency at which to operate. Ordinary radio stations have frequencies in the range of about 0.5×10^6 to 1.5×10^6 Hz. The charge on the transmitting antenna is reversed about a million times each second.

Each time the charge on the antenna reverses sign, the electric field in Figure 28.19 must reverse direction. This constantly reversing electric field is a changing signal that rushes out from the transmitter. It moves out over the earth to tell people miles away what is happening at the transmitting antenna.

In order to illustrate this, we have tried to represent the situation in Figure 28.21(a). We see there the electric field that has proceeded out along the earth from the transmitter. The upward-directed field was produced when the bottom of the antenna was positive. At the instant shown in (a), the bottom of the antenna is positive because we see that the field close to it is directed upward. In part (b) of the figure, we have plotted the electric field as a function of position from the antenna. Notice how this plot corresponds to the field shown in (a).

Of course, Figure 28.21 shows the situation at a certain instant. As time passes, the signal (or field wave) will travel outward from the antenna toward the right in the figure. Just as a water wave travels out along the water surface from a vibrator, an electric field wave travels out from the antenna. But the electric field wave is only part of the wave sent out by the antenna. In the next section we shall learn about the magnetic field wave that accompanies it.

28.15 Magnetic Field Wave

Let us look at the antenna in Figure 28.22(a). The radio station oscillator causes the top end of the antenna continually to reverse its charge. To do this, charge

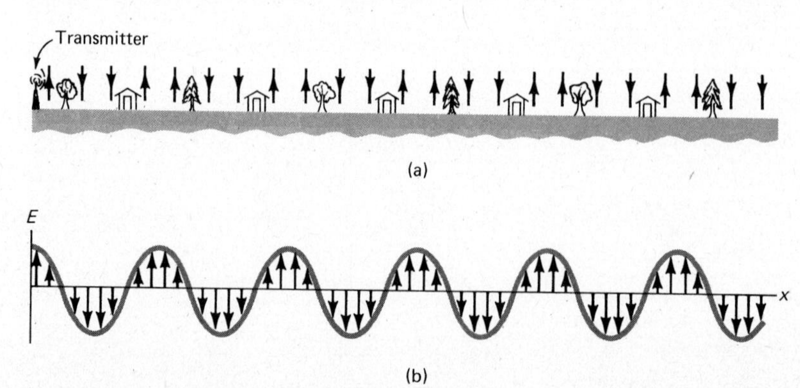

Figure 28.21 An electric field wave blankets the earth for miles from the radio station. It moves to the right with the speed of em radiation.

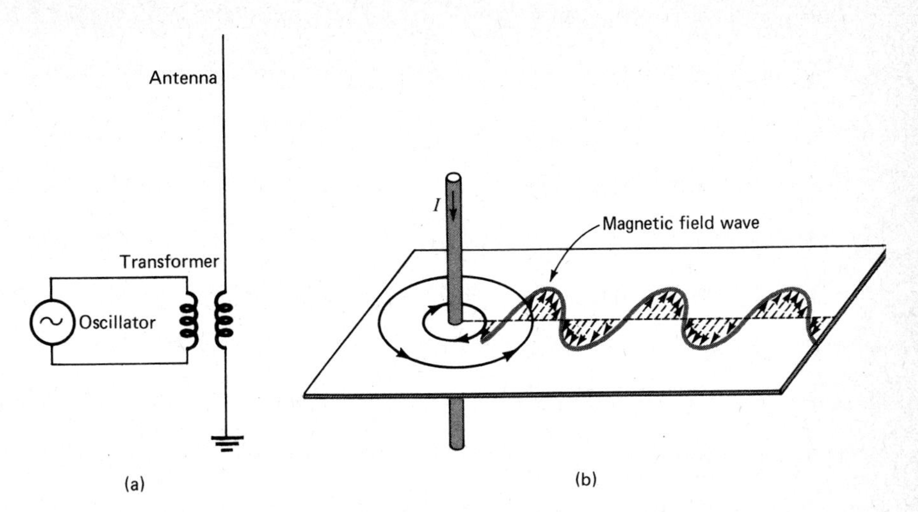

Figure 28.22 As charge rushes up and down the antenna, a magnetic field wave is sent out.

must run up and down the antenna wire. This means that an ac current flows in the antenna wire.

We know that a current generates a magnetic field. As shown in Figure 28.22(b), the magnetic field circles the wire. Because the current keeps reversing, the *B* field also reverses. As with the *E* field wave generated by the changing charge on the antenna, a *B* field wave is generated by the current in it. Notice the orientation of the *B* field wave. The vectors of the *B* field circle the antenna wire. But the *E* field vectors are perpendicular to them. As a result, the electric field wave is perpendicular to the magnetic field wave. This is shown in Figure 28.23. Both waves travel out along the earth, the *x* axis. In the case shown the **E** vectors are vertical, whereas the **B** vectors are horizontal.

The combination of the electric field and magnetic field is called an electromagnetic (em) wave. In it B and E are mutually perpendicular. They are in phase with each other.

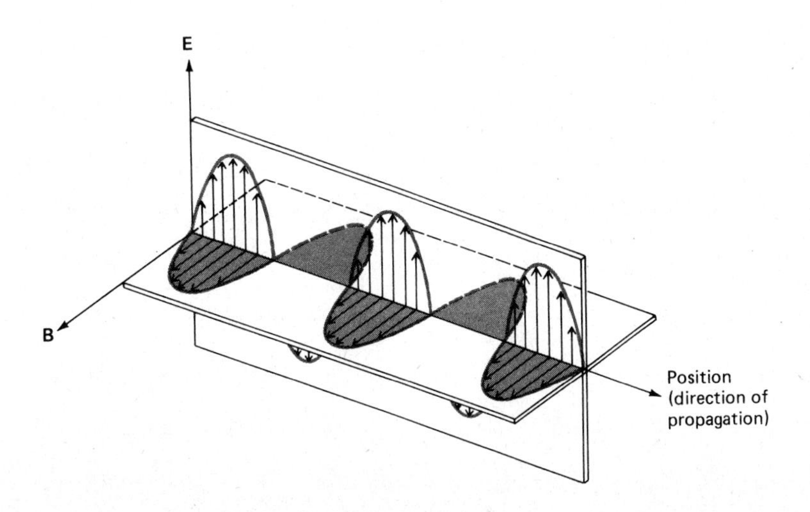

Figure 28.23 The electromagnetic wave consists of an electric field perpendicular to a magnetic field.

28.16 Speed and Nature of em Waves

We pointed out in Chapter 19 that James Clerk Maxwell predicted many facts about radio waves long before the first radio transmitter was built. He made use of Faraday's law for induced emf's together with the other laws of electricity we have studied. His work was based on the equations that represent these laws. By means of these equations, he was able to predict how fast radio waves should travel. He, and the rest of the scientists of the time, were amazed by his result. The radio waves he predicted could exist should travel with a speed that was already known. They should travel with the speed of light.

Electromagnetic waves travel with the speed of light. In vacuum this speed is 2.9979×10^8 m/s and is represented by the letter *c*.

Later, when radio waves could be produced, they were found to have a speed c, the speed predicted by Maxwell. We now know that all the types of radiation (or waves) listed in Figure 19.5 are em waves. They all have the same basic nature. Each consists of an E wave and a B wave, which travel through space together.

In spite of their basic similarity, em waves differ greatly in one respect. They interact with matter in different ways depending on their frequency. This fact influences how we can detect each type of wave. For example, a radio easily detects the em waves from a nearby station, but it cannot detect a strong beam of light. In the next section we shall discuss how radio and television waves can be detected.

EXAMPLE 28.13 A certain radio station is assigned a frequency of 1150 kHz. What is the wavelength of its radio waves?

Solution The speed of em waves in air is nearly the same as in vacuum. Therefore, $v \simeq c = 3 \times 10^8$ m/s. We know that, for any wave,

$$\lambda = v\left(\frac{1}{f}\right)$$

In this case $f = 1.15 \times 10^6$ Hz, so

$$\lambda = (3 \times 10^8 \text{ m/s})\left(\frac{1}{1.15 \times 10^6 \text{ s}^{-1}}\right) = 261 \text{ m}$$

28.17 Reception of Radio and Television Waves

There are two basic ways to detect a radio wave. One way uses the electric field portion of the wave and the other uses the magnetic field portion.

The idea behind the first method is shown in the top portion of Figure 28.24. As we saw in Example 28.13, radio waves are many meters long. TV waves are also several meters long. The waves in Figure 28.24 are therefore not drawn to

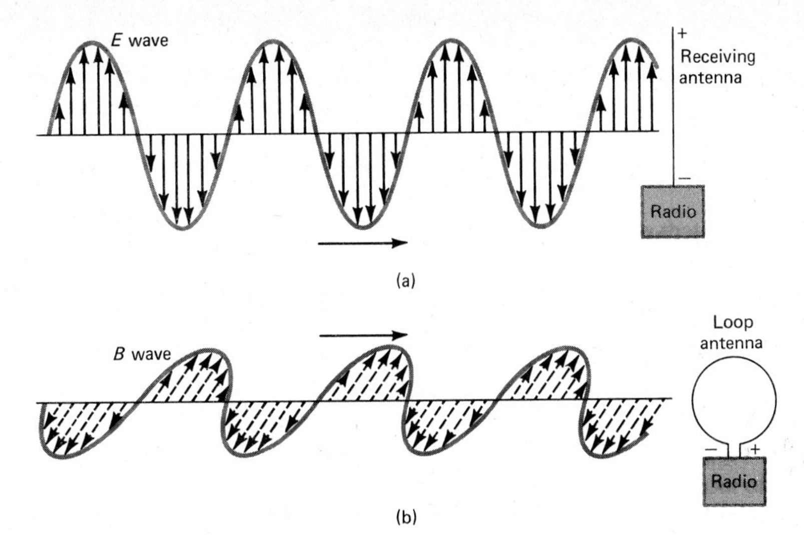

Figure 28.24 The radio's antenna acts like a voltage source because the radio wave passing by it induces potential differences in it.

scale. The wave passes over the receiving antenna in (a). This antenna is a straight wire or metal rod. The E field of the wave causes charges in the wire to move. At the instant shown in the figure, $\mathbf{E}$ has made the top of the antenna positive and the bottom negative. As time goes on, the changing E wave will cause charge to rush up and down the receiving antenna. This charge motion is detected by the radio in a way we shall soon describe.

The second way in which radio waves are detected is shown in Figure 28.24(b). This method is used in nearly all portable radios. It makes use of the magnetic part of the em wave. Notice that the B wave, as it sweeps by the radio, will cause a changing flux through the loop antenna (actually a coil wound on a ferromagnetic core). This changing flux induces an emf and charge motion in the loop. The radio is designed to detect this small induced current. Let us now see how the radio makes use of the tiny currents that flow in the antenna.

The voltages and currents induced in the antenna can be used as an ac power source. For example, the voltage induced in the loop antenna acts like an ac oscillator. The antenna voltage source is made part of an inductance-capacitance circuit as shown in Figure 28.25. You will recall that such a circuit responds strongly to one frequency, its resonance frequency, given by $f = 1/2\pi\sqrt{LC}$. Signals of other frequencies cause little response.

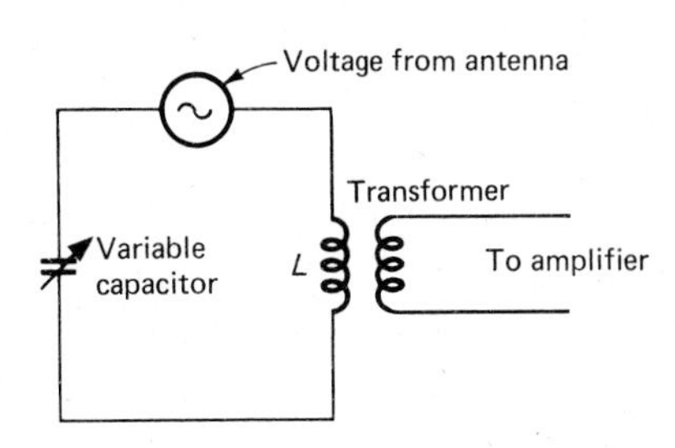

Figure 28.25 The antenna drives the resonant LC circuit. By tuning the variable capacitor, one selects the frequency (station) to which the circuit repsonds strongly.

In most places on earth radio waves from many radio stations exist. The radio antenna responds to all of them at the same time. But the resonant circuit selects only one of these frequencies to which it responds strongly. As a result, the resonance frequency of the resonant circuit determines which station the radio will receive. When you tune a radio (or TV set), you are simply adjusting the frequency of the resonance circuit. This is usually done by changing the capacitance of a variable capacitor.

The resonant circuit greatly increases the voltage or current provided to it by the antenna. In turn, the current in the resonant circuit is used to induce an emf in another circuit. This circuit, an amplifier, magnifies the response still more.

Additional circuits modify the response still further. At last, the resultant voltage is fed to the radio loudspeaker, and the loudspeaker emits a sound that represents the signal originally given to the radio station transmitter.

■ **EXAMPLE 28.14** Refer to Figure 28.24(b). What will happen to the radio response as the radio is rotated about a vertical axis?

Solution When positioned as shown, maximum flux goes through the loop antenna. (Recall that $\Phi = B_{\perp}A$. In this case **B** is perpendicular to the loop area.) As a result, the flux changes will be the largest possible. The radio response will therefore be large.

Suppose that the radio antenna is rotated so its plane is perpendicular to the page. Now no flux goes through the loop. (The lines of **B** simply skim by the loop.) As a result, the induced emf is zero. The radio response will be negligible.

You can easily observe this effect with a portable radio. Tune it to a station that is not too close. Now orient the radio in various directions. Notice how the response varies with orientation. ■■

28.18 Difficulties in Generating and Receiving Very Short Waves

There is a wide range of radio waves. The simplest AM radio can receive waves with $0.5 \leq f \leq 1.5$ MHz. Many radios also have shortwave bands. These bands have resonant circuits that can be adjusted to higher frequencies (i.e., shorter wavelengths). The most expensive radios have several frequency bands. They can respond to a large range of em wave frequencies.

The longest-wavelength radio wave possible approaches infinity. For example, a charged ball at the end of a pendulum string emits an em wave as it vibrates back and forth. Its wavelength would be of the order of 10^9 m (of the order of 10^6 miles). But em waves such as these are difficult to detect. Usually their energy is low (for reasons to be pointed out later). More important, the electric circuit for their detection must be very large. The LC resonance circuit for such low-frequency waves requires values for L and C that are far larger than commonly available.

Radio waves much shorter than normal broadcast wavelengths are in widespread use. The frequencies of VHF (very high frequency) television waves range from 39 to 80 MHz. The UHF (ultra high frequency) television frequencies range from 80 to 800 MHz. It is interesting to note that for $f = 800 \times 10^6$ Hz, the corresponding wavelength is 37.5 cm. Commercial radio and TV waves range from less than 1 m to hundreds of meters in wavelength.

Difficulties appear when we try to generate and detect waves of extremely high frequency. At such high frequencies the values of L and C required for the necessary resonant circuit are very small. The capacitance between two of the wires of the circuit is nearly as large as the capacitance needed. In addition, even a single loop of wire in the circuit has an inductance that is no longer negligible. For these reasons, the construction of circuits at extremely high frequencies requires considerable skill and attention to details.

Figure 28.26 A microwave circuit. (Courtesy of Raytheon Co.)

Frequencies of about 600 MHz and higher give rise to radar (or micro) waves. They are simply extremely high-frequency radio-type waves. Radar waves often have wavelengths in the 1-to-10-cm range. For such short waves wires and coils are no longer usable. A coil 2 cm long no longer obeys the usual circuit rules for a 1-cm wave. The electric and magnetic fields in it vary in a complicated way because the wire is much longer than the wave. Radar-length waves require entirely different types of circuits.

Because they are so short, radar waves begin to behave like light waves. They can be carried from place to place through highly polished pipes. But, in addition, they are long enough so that tubes can be made in which the waves resonate. (For example, a 1-cm-long radar wave will resonate in a polished metal tube much like a 1-cm sound wave would. This is true even though the em radar wave is entirely different in nature from the sound wave.) The resonant circuit for a radar wave is no ordinary circuit. It consists of a metal cavity in which the wave resonates. In fact, short-wavelength radar and microwave equipment looks like plumbing. It is composed of pipes (called waveguides) and boxes (called cavities). A typical microwave circuit is shown in Figure 28.26.

At wavelengths shorter than about 0.3 cm, the handling of em waves becomes even more difficult. The necessary waveguides and cavities become too small to be manageable. Other equipment difficulties also arise. For these reasons we reach the limit of present-day mastery of em waves. We are unable to control precisely the generation of such extremely short em waves. They can be generated, of course. They are long-wavelength infrared waves. But our ability to generate and use them is limited. These waves are too short to be handled using radar techniques, yet they are too long to be handled easily using the methods we use for light waves.

Summary

A sinusoidal current or voltage is an ac current or voltage. The equation of a typical voltage is $v = v_0 \sin(2\pi ft)$. We use small letters, v and i, to represent varying currents and voltages. The amplitude of the voltage is v_0. This voltage is often called the peak voltage.

Most ac meters read effective values of v and i. These are also called the root mean square (rms) values. They are represented by V and I. It is found that $V = v_0/\sqrt{2}$ and $I = i_0/\sqrt{2}$. The factor $1/\sqrt{2} = 0.707$.

The average value of an ac voltage or current is zero. As a result, ac voltage is not usable for applications such as electroplating.

When an ac voltage V is applied across a resistor R, the current through the resistor is given by an Ohm's law form, $V = IR$. The ac current through the resistor is in phase with the voltage across it.

When an ac voltage V is applied across a capacitor C, the current in the wire to the capacitor is given by an Ohm's law form, $V = IX_C$. The quantity X_C is called the capacitive reactance (or the impedance of the capacitor). It is given by $X_C = 1/2\pi fC$, and its unit is the ohm. This shows that the impeding effect of the capacitor is large at low frequencies and small at high frequencies. The ac current to the capacitor is 90° out of phase with the voltage across it. The current is $\frac{1}{4}$ of a cycle ahead of the voltage.

The time taken for a capacitor to charge through a resistor to 63 percent of its final value under dc conditions is RC. This quantity is called the time constant. It is also the time taken for the capacitor to discharge 63 percent through a resistor R.

A changing current through a coil induces an emf in the coil. The induced emf opposes the change in current. This property of a coil is called self-inductance. Its magnitude is represented by L, defined as follows: emf $= -L\ \Delta i/\Delta t$. The unit of L, the self-inductance, is the henry, H.

When an ac voltage V is applied across an ideal inductor L, the current through the inductor is given by an Ohm's law form, $V = IX_L$. The quantity X_L is called the inductive reactance of the coil (or the impedance of the coil). It is given by $X_L = 2\pi fL$, and its unit is the ohm. This shows that the impeding effect of an inductor is low at low frequencies and high at high frequencies. The ac current through the inductor is 90° out of phase with the voltage. The voltage is $\frac{1}{4}$ of a cycle ahead of the current.

When an ac voltage is applied to a series RCL circuit, the current and voltage are related through $V = IZ$, where Z is called the impedance. The impedance $Z = \sqrt{R^2 + (X_L - X_C)^2}$ and is measured in ohms. In such a circuit the current lags behind the voltage by a phase angle ϕ, where $\tan \phi = (X_L - X_C)/R$.

In an ac series RCL circuit, the voltage across C is always opposite in sign to that across L. Voltages read by voltmeters do not add directly in ac circuits. The voltage across the combination is given by $V^2 = V_R{}^2 + (V_L - V_C)^2$.

Resonance occurs in an ac series circuit at the frequency for which $X_L = X_C$. The resonance frequency $f_0 = (1/2\pi)\sqrt{1/LC}$. At that frequency Z is a minimum and the current is a maximum. The less resistance the circuit has, the sharper the resonance is.

No average power loss occurs in an inductor or capacitor. The power loss in a resistor is VI, which is the same as I^2R. In general, the power loss in a circuit is $VI \cos \phi$. The voltage across the circuit is V and the current through it is I. We call the quantity $\cos \phi$ the power factor. The angle ϕ is the phase angle between V and I. For ideal inductors and capacitors the power factor is zero. It is unity for a resistor.

Diodes can be used to rectify ac. They conduct current in only one direction, converting ac to dc voltage. The pulsating current and voltage generated by a rectifier can be smoothed using smoothing filters. The lack of smoothness is measured in terms of the ripple. A small ripple means a smooth dc voltage or current.

Radio waves are sent out by periodic changing of charge on an antenna. The frequency of the wave is the same as the frequency of charge oscillation. In ordinary broadcast radio, the frequency ranges from about 0.5 to 1.5 MHz.

An em wave consists of two component waves, an electric field wave perpendicular to a magnetic field wave. The two waves are in phase. Because these waves involve only fields, they can travel through vacuum.

Radio waves as well as all other electromagnetic (em) waves travel with speed $c = 3.00 \times 10^8$ m/s through vacuum. Their wavelengths in vacuum are found from $\lambda = c/f$. Radio waves have wavelength values of hundreds of meters. TV frequencies are higher than those for broadcast radio. Microwaves and radar waves have still higher frequencies. Their wavelengths range down to about a centimeter.

Radio and TV waves are detected by a receiving antenna. The oscillating electric field wave induces a voltage in a straight-wire antenna. Or, the oscillating magnetic field wave induces a voltage in a loop antenna. This voltage then drives a resonant circuit. The radio then responds to the station whose frequency matches the frequency of the resonant circuit.

Radar waves of very short wavelength are generated in circuits that contain hollow tubes to guide the waves. Resonant circuits are replaced by resonant cavities for these waves. For still shorter waves the generators and detectors are of molecular and atomic size. We use atoms and molecules to generate and detect them.

Questions and Exercises

1. Fluorescent lights flicker on and off 120 times each second. You can sometimes see this effect when objects are moving rapidly in the light of a fluorescent bulb. Why is this effect not easily noticed in the case of an incandescent lamp bulb?
2. List as many applications as you can where dc must be used instead of ac. What major advantages does ac have over dc?
3. A dc voltmeter is connected across the terminals of a variable-frequency oscillator. Explain what the meter will show as the frequency is slowly increased from zero.
4. A variable-frequency, constant-voltage power source is placed across a circuit element. The current is measured as the frequency is changed from zero to extremely high frequency. Sketch on one graph the i versus f curve if the circuit element is a resistor, a capacitor, and an inductor.
5. A so-called low-pass filter is shown in Figure P28.1. Explain why this device passes low-frequency, but not high-frequency signals.

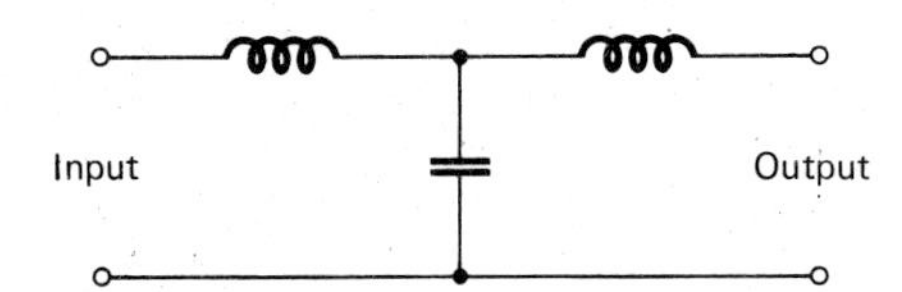

Figure P28.1

6. A large solenoid is connected in series with a light bulb and a power source. The iron core of the solenoid can be moved in and out of the solenoid. Explain what will happen to the light bulb when the core is slowly moved into the solenoid if (a) the power source is 120 V dc; (b) the power source is 120 V, 60 Hz ac. A device such as this is sometimes used as a light dimmer.
7. A resistance coil with no self-inductance is sometimes desirable. How can a solenoidal coil be made so that its self-inductance is essentially zero?
8. A transformer is designed to be used on 120 V, 60 Hz. Why is the transformer likely to burn out if it is connected to 120 V dc?
9. A student connects in series an inductance coil, a capacitor, and a 15-V, variable-frequency power supply. The capacitor is capable of withstanding a voltage of 200 V. But at a certain frequency the capacitor burns out (that is, the insulation between its plates breaks down because of too high a voltage). Explain what has probably happened.
10. You have probably noticed that the power company has a large number of transformers on poles in your area. These transformers run day and night, and the power company makes no attempt to turn them off when no one is using them. Why don't they? (Hint: The dc resistance of a good transformer is very low.)
11. Electromagnetic waves from the sun reach the earth even though the space between us and the sun is vacuum. Sound waves emitted by the sun cannot reach the earth. Why this difference?
12. The following story is told. Near a very powerful radio transmitter used by the U.S. government during wartime, an interesting effect was observed. Sparks were seen to jump from place to place on an old nearby farm fence. How could such sparks be explained?

13. A certain radio station has its antenna oriented vertically. How should a straight-wire receiving antenna be oriented to obtain maximum reception? Repeat for a loop-receiving antenna.
14. In the movies at least, the good guys are able to locate the bad guys' radio transmitter in the following way. The good guys travel in a mobile unit. On top of the unit is a loop antenna that can be rotated. Explain the basis of the method.
15. Figure P28.2 is a sketch of a bridge-type rectifier circuit. Explain how it rectifies the current to the load. Sketch graphs of the input and output current.

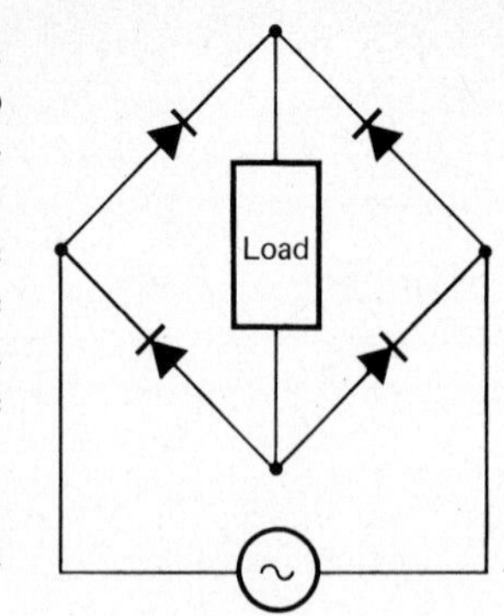

Figure P28.2

Problems

1. For the voltage curve shown in Figure P28.3, find the following: (a) maximum or peak voltage, (b) rms voltage, (c) period, (d) frequency.
2. A sinusoidal current makes 1000 cycles each second and has an effective value of 5 A. Find (a) peak current, (b) period, (c) equation for the current.
3. A current is given by $i = 2 \cos (40t)$ A. Find (a) peak current, (b) effective current, (c) rms current, (d) frequency.
4. Write the equation for the voltage shown in Figure P28.3.

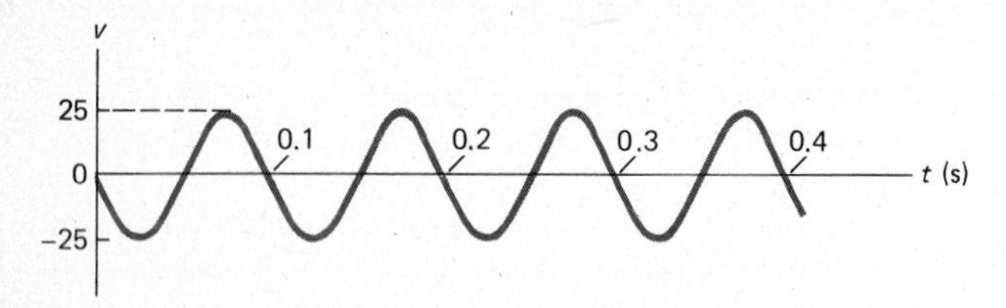

Figure P28.3

5. The current flowing through a 5-Ω resistor has an rms value of 3 A. (a) How much power is lost in the resistor? (b) How much heat (in calories) is produced in the resistor in 1 min?
6. The sinusoidal current flowing through a 12-Ω resistor has a peak value of 4 A. (a) How much power is lost in the resistor? (b) How many calories of heat are produced in the resistor in 8 s?
7. A current $i = 5 \sin (20t)$ amperes flows through a 15-Ω resistor. How much power is lost in the resistor?
8. A 20-Ω resistor is connected directly across the 120-V ac power line. (a) What is the maximum voltage across the resistor? (b) What will an ordinary ac voltmeter read across the resistor? (c) What will an ordinary ac ammeter read in series with the resistor? (d) What is the maximum current through the resistor?
9. A 30-Ω resistor is connected in series with an ac 60-Hz power source and an ordinary ac ammeter. An ordinary ac voltmeter is connected across the resistor. It reads 80 V. Find the following: (a) current read by ammeter, (b) rms voltage of power source, (c) peak voltage of power source, (d) peak current in circuit.
10. In a certain radio circuit, a 2×10^6-Ω resistor is connected in series with a 0.020-μF capacitor and a 70-V dc power source. How long does it take for the source to charge the capacitor to 63 percent of its maximum charge? What is the charge on the capacitor at this time?
11. An uncharged 2-μF capacitor is connected in series with a switch, a 5×10^6-Ω resistor, and a 12-V battery. Find (a) the current in the circuit just after the switch is closed, (b) the time constant, (c) the final charge on the capacitor, (d) the charge on the capacitor after one time constant, (e) the current after one time constant.
12. A 0.075-μF capacitor has been charged to a voltage of 12 V. It is then removed from the power source. Its two terminals are held in the two hands of a person whose resistance between hands is 40,000 Ω. How long does it take the capacitor to lose 63 percent of its charge? What will be the charge on it then?

*13. A capacitor with original charge Q_0 is discharged through a resistor. (a) What fraction of Q_0 will be left on the capacitor after four time constants have passed? (b) What then will be the voltage across the capacitor in terms of the original voltage V_0?

**14. Starting from the fact that the units of RC are ohm-farad, prove that its units are seconds.

*Problems marked with an asterisk are not as easy as the unmarked ones.
**Problems marked with a double asterisk are somewhat more difficult than the average.

15. What is the impedance of a 2-μF capacitor when used at a frequency of (a) 10 Hz and (b) 10^5 Hz?
16. When connected across a 40-V-rms, 60-Hz source, a capacitor draws a current of 4.0 mA. (a) What is the impedance of the capacitor? (b) What is its capacitance?
17. A 70-V power source and a 0.30-μF capacitor are connected in series. Find the current in the circuit if the power source frequency is (a) 100 Hz and (b) 10,000 Hz.
18. An oscillator is connected directly across a 0.25-μF capacitor. What must be the oscillator voltage if the current in the circuit is to be 0.50 A and the oscillator frequency is (a) 100 Hz and (b) 10,000 Hz?
**19. Starting from the fact that the unit of the time constant of an *LR* circuit is henrys per ohm, show that its unit is seconds.
20. Find the inductive reactance of a 5-mH coil when used at a frequency of (a) 10 Hz and (c) 100 kHz.
21. When connected across a 20-V (rms), 1000-Hz power source, a coil with negligible resistance draws a current of 2 mA. Find (a) the impedance and (b) the inductance of the coil.
22. A certain essentially resistanceless coil passes a current of 20 mA at a frequency of 1000 Hz. How much current would it pass at the same voltage at a frequency of (a) 1 Hz and (b) 1 MHz?
23. The primary of a transformer used in a radio has an inductance of 0.243 H and negligible resistance. (a) How much current does it draw on 120 V, 60 Hz? (b) Repeat for 120 V, 6000 Hz.
*24. A certain air-core solenoid has an inductance of 1.38×10^{-4} H. (a) How much current does it draw from a 50-V, 100-Hz source? (b) Repeat for a 50-V, 10,000-Hz source. (c) If the core is now filled with laminated iron, what will be the coil's inductance? Assume the relative permeability of the iron to be 300.
*25. A 2-μF capacitor in series with a 300-Ω resistor is connected across a 50-V (rms), 400-Hz power source. Find (a) X_C, (b) impedance of the series combination, (c) current in the circuit, (d) voltage across R, (e) voltage across C.
*26. A series circuit consists of a 0.50-μF capacitor, a 200-Ω resistor, and a 1000-Hz, 30-V power source. Find (a) X_C, (b) impedance of the *RC* combination, (c) current in the circuit, (d) voltage across R, (e) voltage across C.
*27. A series circuit consists of a 3-mH coil having negligible resistance, a 200-Ω resistor, and a 10-kHz, 50-V power source. Find (a) X_L, (b) impedance of the *RL* combination, (c) current in the circuit, (d) voltage across R, (e) voltage across L.
*28. The following are placed in series: a 40-Ω resistor, a 2×10^{-6}-F capacitor, a 0.150-H inductor, and a 70-V, 300-Hz power source. Find (a) current in the circuit, (b) V across R, (c) V across C, (d) V across L, (e) V across the *LC* combination, (f) V across the *RC* combination.
*29. A series circuit consists of a 30-Ω resistor, a 0.027-H inductor, a 1.60×10^{-6}-F capacitor, and a 40-V, 1000-Hz power source. Find (a) current in the circuit, (b) V across R, (c) V across C, (d) V across L, (e) V across the *LC* combination, (f) V across the *RL* combination.
30. A 5.0-Ω resistor is connected in series with a 0.20-μF capacitor and a 0.35-H inductor. (a) What must be the driving frequency of a 40-V power source if this circuit is to be at resonance? (b) How much current will it then draw from the power source?
31. It is desired to design a circuit that has a resonance frequency of 60 Hz. A coil with an inductance of 0.70 H and a resistance of 4.5 Ω is available. (a) What value capacitor must be connected in series with it? (b) If this combination is connected across 120 V, 60 Hz, how much current will it draw?
*32. A coil is found to draw a current of 2.10 A when connected across a 12.0-V dc power source. When connected directly across a 60-Hz, 115-V line, it draws a current of 14.0 A. Assuming that its dc and ac resistance are the same, find the inductive reactance of the coil at 60 Hz.
*33. An air-core solenoid has an ohmmeter reading of 2.70 Ω. When placed in series with a 2.0-μF capacitor, it resonates to a frequency of 14,700 Hz. The power source has a voltage of 17.0 V. Find the inductance of the coil and the voltage across it at resonance.
34. A 50-Ω resistor and a 2-μF capacitor are placed in series across an ac power source. The current in the circuit is 2 A. How much average power is lost in (a) the capacitor and (b) the resistor? (c) How much power does the source furnish to the circuit?
35. For the circuit of Problem 34, suppose the power source frequency is 1 kHz. Find (a) X_C, (b) the impedance of the combination, (c) the power factor for the circuit.
36. A series circuit consists of a 200-Ω resistor, 0.5-μF capacitor, and 1000-Hz power source. Find the power factor for the circuit.
*37. In a series *RCL* circuit the voltages are: $V_R = 30$ V, $V_C = 20$ V, $V_L = 50$ V. The current in the circuit is 3 A. Find (a) the power factor for the circuit, (b) source voltage, (c) average output power of the source.
*38. The primary coil of a transformer has a resistance of 0.55 Ω and an inductance of 1.70 H. It is connected to a 60-Hz, 440-V line. Find (a) the current drawn, (b) the power factor, (c) the power consumed.
*39. A coil used to smooth rectified voltage (called a choke coil) has a resistance of 150 Ω and an inductance of 10 H. (a) How much current does this coil draw when connected across 60 Hz, 120 V? (b) What is the power factor for it? (c) How much power does it consume?
**40. Resonance frequency is proportional to $1/\sqrt{LC}$. Show that the unit, (henry $\times$ farad)$^{1/2}$, is equivalent to seconds.

*41. A certain coil has a resistance of 145 Ω and an inductance of 0.027 H. (a) Find its impedance at 80,000 Hz. (b) What value capacitor must be placed in series with it to make its power factor unity?

**42. In a series *RLC* circuit the voltages across the various circuit elements are: $V_R = 23$ V, $V_L = 45$ V, $V_C = 30$ V. (a) From these data find the voltage of the power source. (b) If the current in the circuit is 0.50 A, find R, X_L, and X_C. (c) What is the power factor for this circuit?

**43. In an ideal transformer the primary coil would have $R = 0$. When no power is being drawn from the secondary, the power factor for this ideal coil is zero. The current to it would be given by $V_P = IX_L$. Suppose now that power P is drawn from the secondary. This power must be furnished to the primary. As a result, the ideal transformer now appears as though its primary coil has a resistance r. Because r is negligible compared to X_L, the current I to the primary will still be nearly V_p/X_L. Show that the apparent resistance r of this ideal transformer is related to the power P drawn from the secondary by

$$r = P\left(\frac{X_L}{V_P}\right)^2$$

Also show that the power factor for the transformer input is

$$\cos \phi = \frac{PX_L}{V_P^2}$$

What importance does this result have for Question 10?

44. A common microwave apparatus uses 3.4-cm-wavelength microwaves. What is the frequency of these waves?

45. What frequency should a radio station have if its waves are to have a wavelength of 200 m?

46. A student wishes to design a resonant circuit that will resonate to 2.00×10^{10}-Hz microwaves. A 1.2×10^{-9}-F capacitor will be used. What is the inductance of the coil needed? For comparison purposes, the inductance of a 1-cm-diameter circular loop is of the order of 10^{-8} H.

47. A student wishes to design a resonant circuit that will resonate to 2.0-cm microwaves. A 0.50-mH coil will be used. (a) What is to be the resonance frequency of the circuit? (b) What value capacitor should be used? For comparison purposes, the capacitance of two parallel 10-cm-long wires with 1 mm diameter and 1 cm apart is about 10^{-12} F.

American Institute of Physics

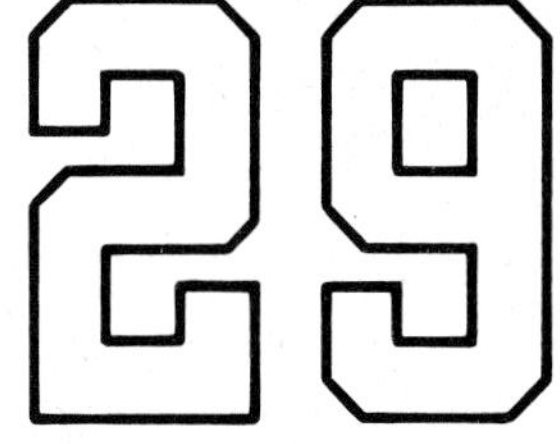

PHYSICS OF THE ATOM

Performance Goals

When you finish this chapter, you should be able to

1. State the two assumed facts upon which Einstein based his theory of relativity.
2. Write an equation that relates the mass of an object to its rest mass and its speed. Sketch the relation in a m versus v graph.
3. Compute the mass of a particle at speed v if its rest mass is known.
4. Explain in words what is meant

Shortly after 1900, a series of scientific discoveries began that has revolutionized our world. These discoveries have led to an understanding of the atom and its nucleus. In 1927 the people most influential in this scientific revolution gathered at the famous Solvay Conference. Their photo, shown here, provides us with the faces that accompany the famous names we shall study in this chapter.

by the relations $\Delta E = \Delta mc^2$ and $\text{KE} = (m - m_0)c^2$. Use these relations for simple computations.

5. Sketch the apparatus for an experiment to illustrate the photoelectric effect. State what happens to the current as a function of wavelength. Explain why the wave concept of em radiation has difficulty in explaining this experiment.
6. In your own words, explain the concept of a light beam as a stream of photons. In your explanation give the speed of the photons and the relation between photon energy and wavelength.
7. Explain what is meant by the de Broglie wavelength of a particle.
8. Sketch a diagram of the Balmer series for the hydrogen emission spectrum.
9. Explain why energy must be given to a hydrogen atom if its electron is to be moved from the ground state to an excited state. Point out that only certain excited states for the atom are possible. Sketch qualitatively the energy-level diagram for hydrogen. Explain the meaning of the lowest level and of the zero level.
10. Calculate the photon wavelength emitted as an atom falls from one given level to another. If the energy-level diagram for hydrogen is given, show which transitions result in the Balmer series and the Lyman series. Use the diagram to explain how the Lyman series limit and ionization energy are related.
11. Find the wavelengths of em radiation that hydrogen atoms will emit when electrons (or

29.1 Einstein and Relativity

In 1905 Albert Einstein presented the first part of his famous theory of relativity. Only 26 at the time, he had set out to find the consequences of two puzzling experimental facts. Extensive measurements by others had indicated the following to be true.

1. Accurate measurements of the speed of light in vacuum give the value $c = 2.998 \times 10^8$ m/s no matter how the light source or measurer are moving.
2. We can measure velocities only relative to some object or marker. We can say only that something is at rest relative to some other thing. It makes no sense to state that an object is at rest unless reference is made to some other object.

Einstein found that the consequences of these two facts are simple and astonishing. Assuming the truth of the two statements, he concluded that:

1. Neither an object nor any form of energy can be accelerated to a speed faster than c, the speed of light in vacuum.
2. The mass of an object increases as its speed increases.
3. Mass is a form of energy in the sense that mass and energy can be interchanged.
4. Suppose that any type of timing device (a clock) is moving at high speed past a person. The person's measurements will show that the clock is ticking slower than when it was at rest relative to the person.
5. Suppose that an object is moving at high speed past a person. The person will measure the object to be shortened along its line of motion.

Of all these results, the first three are of most importance to us.[1]

We have already mentioned conclusion 1 in previous chapters. When electrons, for example, are accelerated to high speeds, their behavior is not normal. In particular, when v becomes close to c, our usual equations no longer apply. No matter how hard we try, an electron cannot be accelerated to a speed c or greater. In fact, no object or particle can be accelerated to a speed as large as the speed of light in vacuum.

The reason for conclusion 1 is provided by conclusion 2. At very high speeds the mass of an object increases rapidly with speed. This can be demonstrated in the case of electrons. To do so, one shoots electrons of known speed into a magnetic field as shown in Figure 29.1. The electron then follows a circular path. As usual, the centripetal force, mv^2/r, must be furnished by the magnetic field force, qvB. Equating the two gives

$$\frac{mv^2}{r} = qvB$$

[1] You can find a simple discussion of Einstein's theory in F. Bueche, *Principles of Physics,* 4th ed., (New York, McGraw-Hill, 1982), p. 674.

photons) of known energy collide with them. The energy-level diagram for hydrogen will be given.

12. State (and sketch) qualitatively what happens to an atom's energy levels as the atoms are packed into a solid. Use the sketch to show why a continuous spectrum is emitted by a red-hot solid.

13. Explain qualitatively why atoms with large nuclear charges can be caused to give off X-ray line spectra.

14. Sketch a simple X-ray tube and its circuit. Explain why the tube gives off a continuous spectrum of X rays as well as a line spectrum.

15. Calculate the shortest possible wavelength emitted from an X-ray tube operated at a known voltage.

16. Explain why hydrogen atoms absorb only the wavelength of the Lyman series.

17. Point out three major differences between a laser beam and an ordinary light beam. Explain what feature of the laser causes each difference.

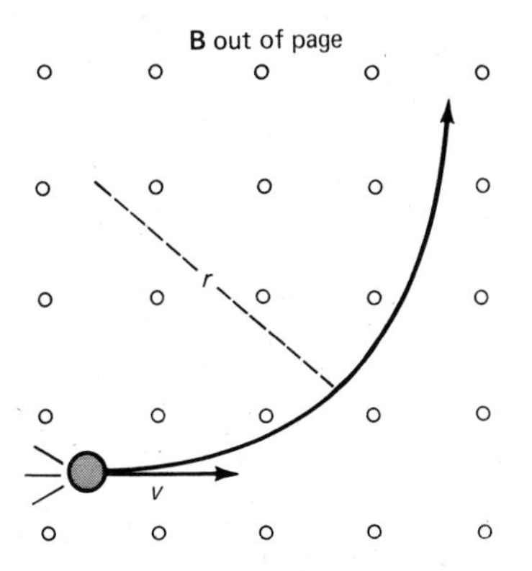

Figure 29.1 The mass of the electron can be found from the radius of its circular path.

From this

$$m = \frac{qB}{v}r$$

As we see, the mass of the electron can be measured.[2] The results of measuring m as a function of speed are shown in Figure 29.2. We call the mass of the particle when it is at rest the *rest mass* of the particle. It is denoted by m_0. In Figure 29.2 we see that the electron's mass is close to m_0 for low speeds. But when v becomes close to c, the mass increases rapidly with v. Einstein predicted that $m \to \infty$ as v gets very close to c. Measurements have confirmed this fact for m/m_0 values in excess of several thousand. We believe that any and all objects show this same behavior.

The apparent mass of an object increases without limit as its speed approaches the speed of light in vacuum.

Einstein showed that the mass m increases in the following way:

$$m = \frac{m_0}{\sqrt{1 - (v/c)^2}} \qquad (29.1)$$

Notice that when $v = 0$, then $m = m_0$, the rest mass. But when $v = c$, the denominator of Equation 29.1 becomes zero. Then $m \to \infty$. The curve shown in Figure 29.2 is a plot of Equation 29.1.

EXAMPLE 29.1 How fast must an object be moving if $m/m_0 = 1.010$?

Solution Equation 29.1 tells us that

$$\frac{m}{m_0} = \frac{1}{\sqrt{1 - (v/c)^2}}$$

In our case this becomes

$$1.010 = \frac{1}{\sqrt{1 - (v/c)^2}}$$

Squaring both sides of this equation gives

$$1.020 = \frac{1}{1 - (v/c)^2}$$

[2] Actually, it is the momentum that is measured directly. We had $mv = qBr$ with B and r being the quantities measured in the experiment. Since m as such is not measured, it is perhaps better to refer to m as the *apparent* mass.

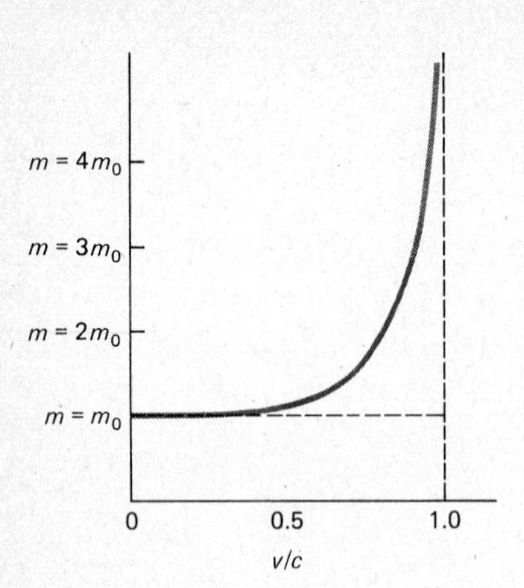

Figure 29.2 The mass of an object approaches infinity as its speed approaches c.

After multiplying each side by $1 - (v/c)^2$, this equation becomes

$$1.020 - 1.020\left(\frac{v}{c}\right)^2 = 1$$

or

$$\left(\frac{v}{c}\right)^2 = \frac{0.020}{1.020}$$

and so

$$v = 0.14c = 4.2 \times 10^7 \text{ m/s}$$

We therefore conclude that an object must have a speed 14 percent of the speed of light before its mass has increased by 1 percent. Clearly, relativistic effects become important only at very high speed. ■■

29.2 Einstein's Mass-Energy Relation

We saw in the last section that an object's mass increases as its speed approaches the speed of light. As we learned many chapters ago, the mass of an object is a measure of its inertia. Therefore, as the mass increases, the force needed to cause a given acceleration increases. When the object's speed is close to the speed of light, a nearly infinite force is needed because the object's mass is nearly infinite. Because infinite forces are not available, it is impossible to accelerate an object to the speed of light. This gives us a reason for the conclusion that no object can exceed the speed c.

There is another way of looking at this effect. The accelerating force does work on the object. This work gives rise to kinetic energy at low speeds. But at speeds close to c, the force causes only a small acceleration even though it does a great deal of work. Where does this work go?

Einstein answered this question by showing that $\frac{1}{2}mv^2$ is the correct equation for kinetic energy only under certain limiting conditions. At low speeds, where $m = m_0$, then $\frac{1}{2}mv^2$ is correct. But at high speeds, and at all speeds, the correct relation is

$$\text{Kinetic energy} = (m - m_0)c^2 \qquad (29.2)$$

If you substitute for m from Equation 29.1, then the usual form $\frac{1}{2}m_0v^2$ can be obtained. But it is obtained only in the limiting case of $v/c \rightarrow 0$.

The kinetic energy of an object with rest mass m_0 is $(m - m_0)c^2$. At low speeds this reduces to $\frac{1}{2}m_0v^2$.

We see from this that the increase in kinetic energy of an object gives rise to an increase in its mass. This is true even for a car moving at 20 m/s along the road. (In that case $m/m_0 = 1 + 2 \times 10^{-15}$. The increase in mass is far too small to measure.)

Einstein showed even more than this. His theory predicts (and experiment has confirmed) that any change in energy results in a change of mass. It does not matter whether the energy is potential, kinetic, heat, chemical, or any other kind. The following is true.

When the energy of an object changes by an amount ΔE, the mass of the object changes by an amount Δm. They are related by the mass-energy relation:

$$\Delta E = (\Delta m)c^2 \qquad (29.3)$$

You will notice that Equation 29.2 is a special case. In it Δm is $m - m_0$ and ΔE is the total kinetic energy of the object. Equation 29.3 is more powerful than Equation 29.2. It applies to any kind of energy. Indeed, it says that mass and energy are convertible from one to the other. In fact, mass can be thought of as being one form of energy.

■ **EXAMPLE 29.2** A kilogram of gasoline (about 1.5 liters) yields an energy of 4.8×10^7 J when burned. Compare this to the mass energy of 1 kg of a substance.

Solution According to Einstein, the mass energy is given by

$$\Delta E = (\Delta m)c^2$$

or

$$\Delta E = (1 \text{ kg})(3.0 \times 10^8 \text{ m/s})^2 = 9.0 \times 10^{16} \text{ J}$$

The mass energy of 1 kg is the following number of times larger than the combustion energy of a like mass of gasoline:

$$\frac{\text{mass energy}}{\text{combustion energy}} = \frac{9.0 \times 10^{16} \text{ J}}{4.8 \times 10^7 \text{ J}} = 1.89 \times 10^9$$

This huge ratio is responsible for the importance of nuclear energy as we shall see in Chapter 31. ■■

29.3 Photoelectric Effect

In the same year that he presented his theory of relativity, Einstein made another discovery. He found that a beam of light sometimes acts like a beam of particles. Let us see how he was led to this conclusion.

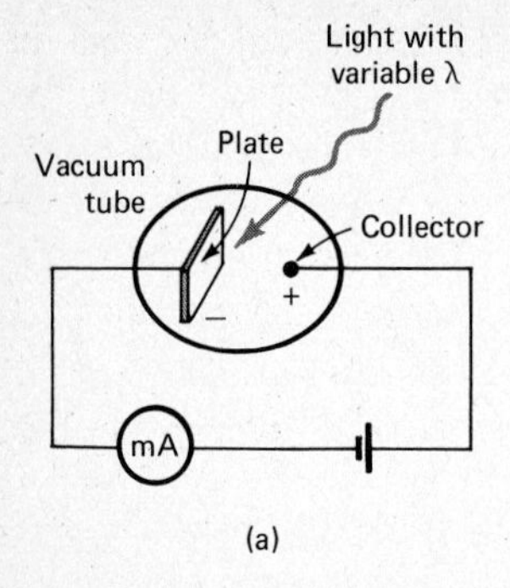

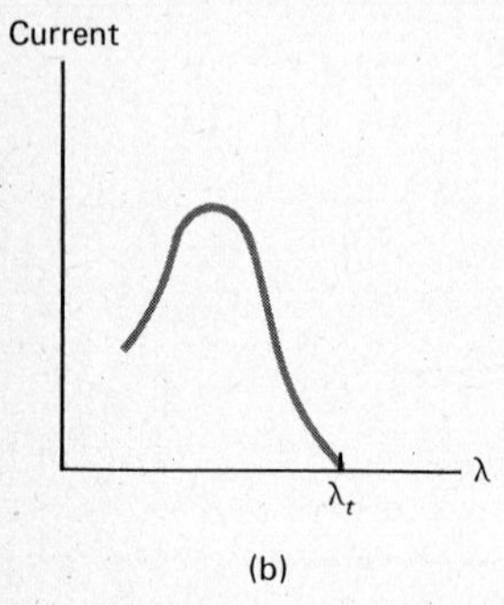

Figure 29.3 The light ejects electrons from the plate provided that λ is shorter than λ_t.

Einstein was trying to explain the photoelectric effect. This effect had been discovered in 1888 by Heinrich Hertz. It can be illustrated with the apparatus sketched in Figure 29.3(a). A metal plate and collecting wire act as two electrodes in a vacuum tube. Because of the vacuum gap between the plate and collector, no current can flow through the tube. The milliammeter in the circuit therefore reads zero *provided that the tube is in darkness.*

However, when light of proper wavelength is shined on the plate, a current is found to flow. Current flows only if the wavelength is shorter than a certain threshold wavelength, λ_t. For example, blue light might cause a current to flow, while red light might not. The value of λ_t depends on the material from which the plate is made. We can see this behavior more clearly in Figure 29.3(b) where the dependence of current on wavelength is shown for a typical material. As you see there, current does not flow if the wavelength is greater than λ_t. Current is found to flow at once if light with wavelength less than λ_t is used.

Many people had tried to explain the photoelectric effect, but with little success. It is apparent that the light beam gives energy to the electrons in the plate and throws them out of the plate. But even the weakest beam with wavelength less than λ_t knocks out electrons at once. Even very strong beams with wavelength greater than λ_t cannot knock electrons from the plate unless the plate is heated red hot. This latter effect is simply thermionic emission, the boiling of electrons from the plate. It is quite different from photoelectric emission where electrons are emitted immediately by cold materials.

Einstein recognized that somehow the energy in a light beam must be localized, not spread out over the whole beam. He therefore theorized that a light beam consists of little pieces of energy shooting through space with the speed of light. We call these pieces of energy *photons* or *light quanta.*

Suppose light does consist of photons. When a photon strikes a surface, it may give all its energy to an electron. If the electron now has enough energy, it will escape from the surface and become free. The situation is sketched in Figure 29.4. The photon has an energy E_{ph}. It is customary to call the energy needed to tear an electron loose from a material the *work function* energy. We represent it by E_{WF}. Therefore, the emitted electron will have an energy

$$\tfrac{1}{2}mv^2 = E_{ph} - E_{WF} \qquad (29.4)$$

If the photon's energy is less than the work function energy, the photon cannot release the electron. We therefore begin to understand why some wavelengths of light eject electrons and others do not. The photon energy must be at least E_{WF} if an electron is to be freed. To proceed further, we must analyze experimental data for the effect.

Experiments had shown that the kinetic energy ($\tfrac{1}{2}mv^2$) of the emitted electrons varies with the wavelength of the incident light. The results showed that, for wavelengths in meters,

$$\tfrac{1}{2}mv^2 = \frac{1.989 \times 10^{-25}\ \mathrm{J\cdot m}}{\lambda} - E_{WF}$$

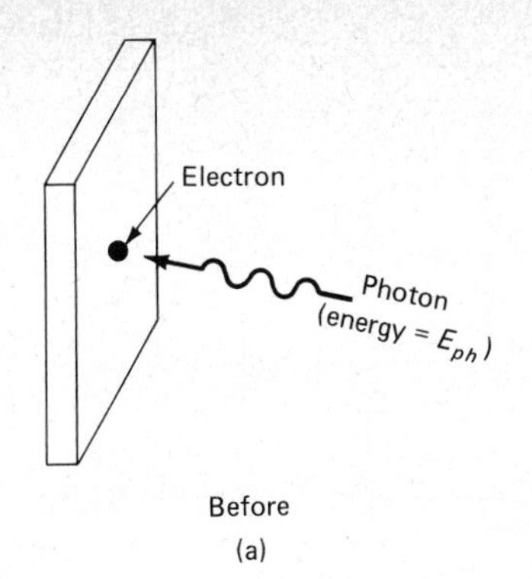

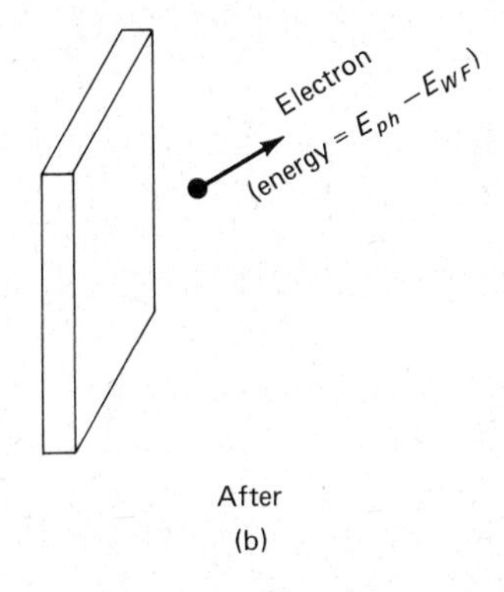

Figure 29.4 The photon gives its energy to the electron. To escape from the plate, the electron must expend the work function energy, E_{WF}.

This relation corresponds well with the equation we just reasoned out from the idea of photons. In fact, the photon energy term E_{ph} in Equation 29.4 is just 1.989×10^{-25} J·m/λ. We thus find that

$$\text{Photon energy} = E_{ph} = \frac{1.989 \times 10^{-25}\ \text{J}\cdot\text{m}}{\lambda}$$

It is customary to write this in a somewhat different way. Using the speed of light $c = 3 \times 10^8$ m/s, we have

$$E_{ph} = \frac{(6.63 \times 10^{-34}\ \text{J}\cdot\text{s})(c)}{\lambda}$$

The constant 6.63×10^{-34} J·s is called *Planck's constant* and is represented by h. Thus we have that the energy of a photon is

$$\text{Photon energy} = \frac{hc}{\lambda} = hf \qquad (29.5)$$

where Planck's constant $h = 6.63 \times 10^{-34}$ J·s. The form of the equation involving f, the frequency of the light, is obtained by remembering that frequency is related to wavelength through $f = c/\lambda$.

Let us review what the photoelectric effect led Einstein to conclude about light and, indeed, about all em radiation.

> **A beam of em radiation with wavelength λ and frequency *f* consists of a stream of photons. Each photon is a tiny piece of energy that travels with speed *c* along the beam. The energy of each photon is *hc*/λ or *hf*.**

Notice that the photon energy increases with decreasing wavelength. Blue-light photons have more energy than red-light photons. Moreover, X-ray photons should have very high energy because λ is so small for them. To interpret the threshold wavelength in the photoelectric effect, we remember that the photon must have at least the work function energy to free an electron. Photons with $\lambda > \lambda_t$ do not have enough energy, while those with $\lambda < \lambda_t$ do.

Sometimes it is helpful to try to make a picture of a photon in our mind. Perhaps a photon is something like the wave pulse shown in Figure 29.5. It is a burst of radiation. If the radiation has short wavelength, then the photon has high energy. If the radiation has long wavelength, the photon's energy is lower. But we should enter a word of caution. We are unable to prove whether the picture in Figure 29.5 is right or wrong. It is simply a mental aid for those who insist on picturing a photon.

Today we know that Einstein's concept is correct. Electromagnetic radiation has two faces. When traveling through space, it behaves like a wave. Like all

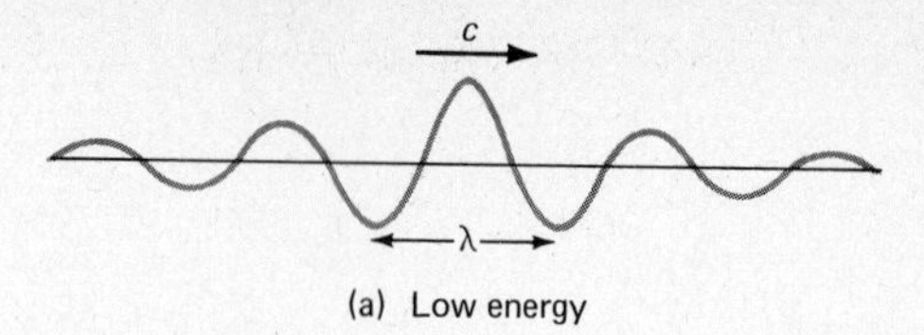

c

λ

(b) High energy

Figure 29.5 The shorter the wavelength of the pulse, the higher is the photon energy.

waves, em waves can interfere with each other and be refracted. But em radiation behaves like a beam of energy pulses, photons, when it interacts with matter. The energy of each photon is determined by the wavelength (or frequency) of the beam. Each photon has an energy hc/λ (or hf). As we proceed with our study, you will see that the concept of photons provides us with the key to the understanding of light emission from atoms and molecules.

You might be puzzled by the fact that photons travel with the speed of light. Does not Equation 29.1 predict the photon to have infinite mass? The answer to this is "No." The reason is as follows.

The photon does indeed have a speed $v = c$. But its rest mass is zero. A photon at rest cannot exist. Its speed is always c. If $m_0 = 0$ and $v/c = 1$, then Equation 29.1 gives $m = 0/0$. You may recall from your mathematics courses that this is an indeterminant form. In simple words, this means 0/0 can have any value. Therefore Equation 29.1 tells us nothing about the photon. It simply confirms that a particle with zero rest mass can (and must) have a speed c.

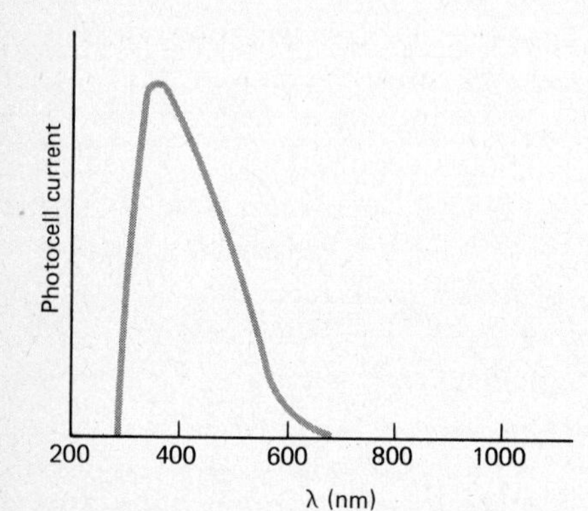

Figure 29.6 In the photoelectric response of a commerial S-4 photocell, the cutoff at small λ results because the glass tube does not transmit ultraviolet light.

▮ EXAMPLE 29.3 The wavelength response of a photocell in widespread use is shown in Figure 29.6. (This is called an S-4 surface and its response is nearly the same as that of the human eye.) Estimate the work function for this surface.

Solution From the graph, λ_t is about 700 nm. This wavelength photon has just enough energy to overcome the work function and free an electron. Therefore,

$$E_{WF} = \frac{hc}{\lambda_t} = \frac{(6.63 \times 10^{-34}\ \text{J}\cdot\text{s})(3 \times 10^8\ \text{m/s})}{700 \times 10^{-9}\ \text{m}}$$
$$= 2.8 \times 10^{-19}\ \text{J} = 1.8\ \text{eV}$$

The latter form is obtained by recalling that $1\ \text{eV} = 1.6 \times 10^{-19}\ \text{J}$. ▮▮

▮ EXAMPLE 29.4 Find the energy of a photon in the sodium arc's yellow light, $\lambda = 589$ nm. Compare this to the energy needed to tear two carbon atoms apart in a molecule that composes human skin. (Take this latter energy to be about 3.5 eV.)

Solution The photon energy is hc/λ. In our case $\lambda = 589 \times 10^{-9}$ m. Therefore,

$$\text{Photon energy} = \frac{(6.63 \times 10^{-34}\ \text{J}\cdot\text{s})(3 \times 10^{8}\ \text{m/s})}{589 \times 10^{-9}\ \text{m}}$$
$$= 3.38 \times 10^{-19}\ \text{J}$$

Since $1\ \text{eV} = 1.6 \times 10^{-19}$ J,

$$\text{Photon energy} = (3.38 \times 10^{-19}\ \text{J})\left(\frac{1\ \text{eV}}{1.6 \times 10^{-19}\ \text{J}}\right)$$
$$= 2.1\ \text{eV}$$

It was stated in the problem that 3.5 eV is needed to tear a carbon-carbon bond apart in a skin molecule. Therefore, yellow light photons do not have enough energy to tear the molecules of the skin apart. Yellow light therefore does not damage human skin. But ultraviolet light photons (λ much shorter) have enough energy to do this. In fact, ultraviolet light causes sunburn. Its photons tear the skin molecules apart on impact. Sunlight causes sunburn because of the short wavelength (ultraviolet) radiation present in it. ■■

29.4 de Broglie Waves

Einstein's discovery that light waves sometimes behave like particles (photons) tempts one to ask if particles might sometimes behave like waves. It turns out that particles do indeed have some wavelike properties. This discovery was made in 1923. In that year Louis de Broglie presented a theory that suggested all objects have an associated wavelength. We will find it by a simplified method based on analogy.

We can guess the wavelength of a particle's wave by analogy to the photon. We have

$$\text{Photon energy} = \frac{hc}{\lambda} \quad \text{or} \quad \lambda_{\text{photon}} = \frac{hc}{\text{photon energy}}$$

According to relativity, energy is related to mass through $\Delta E = (\Delta m)c^2$. Although the photon has no rest mass, it does have the mass equivalent of its energy. Calling this mass m_p, we have

$$\text{Photon energy} = m_p c^2$$

If we substitute this value in the above relation for λ_{photon}, we obtain

$$\lambda_{\text{photon}} = \frac{hc}{m_p c^2} = \frac{h}{m_p c} = \frac{h}{\text{photon momentum}}$$

because m_pc is mv, the momentum for the photon. If a particle has a wavelength associated with it, perhaps by analogy it is

$$\lambda_{\text{particle}} = \frac{h}{\text{particle momentum}} = \frac{h}{mv}$$

This proposed wavelength for a particle is called the *de Broglie wavelength.*

The de Broglie wavelength for a particle of mass *m* and speed *v* is

$$\lambda_{\text{particle}} = \frac{h}{mv} \tag{29.6}$$

Of course, this guess has no value unless there are experimental results to confirm it.

Waves undergo interference and diffraction. To prove that an object has wave properties, we must show it gives rise to these wave effects. Let us see what wavelength should be associated with a particle. Then we may be able to design an interference experiment to show its wave nature.

A baseball moving at a few meters per second has, from Equation 29.6, a wavelength of about 1×10^{-33} m. This wavelength is far too small to produce practical interference and diffraction effects. It would be impossible to observe wave properties of a baseball even if the de Broglie wavelength applies to it.

An electron, on the other hand, has a much smaller mass and larger de Broglie wavelength. After being accelerated through 100 V, an electron has a speed of 6×10^6 m/s. Equation 29.6 gives its de Broglie wavelength to be 0.12 nm. If we are to observe interference of electrons, we must use a thin film about 0.1 nm thick or two slits about that far apart. Because atoms have a diameter of a few tenths of a nanometer, interference effects should be noticed as an electron beam is reflected off the planes in a crystal. The effect would be much like that for X-ray diffraction, a topic we discussed in Section 22.9.

The first experiments that showed clearly the wave properties of electrons were published in 1927 by Davisson and Germer. They used an arrangement somewhat like that shown in Figure 29.7. A beam of electrons was reflected off a crystal. As shown in (b), different reflected beams occur as the main beam strikes the atomic layers in the crystal. Suppose the electrons have a de Broglie wavelength λ. Then if beam 2 travels λ further than beam 1, beams 1 and 2 will reinforce. In fact, all of the reflected beams will reinforce in that case. Hence a strong reflected beam will occur at this particular value of θ.

Davisson and Germer found that a crystal *does* reflect an electron beam strongly only at certain angles. The angles of strong reflection are exactly those predicted for the de Broglie waves of the electrons. We therefore have striking confirmation of the wave nature of electrons. Many later experiments have shown that de Broglie's supposition is correct.

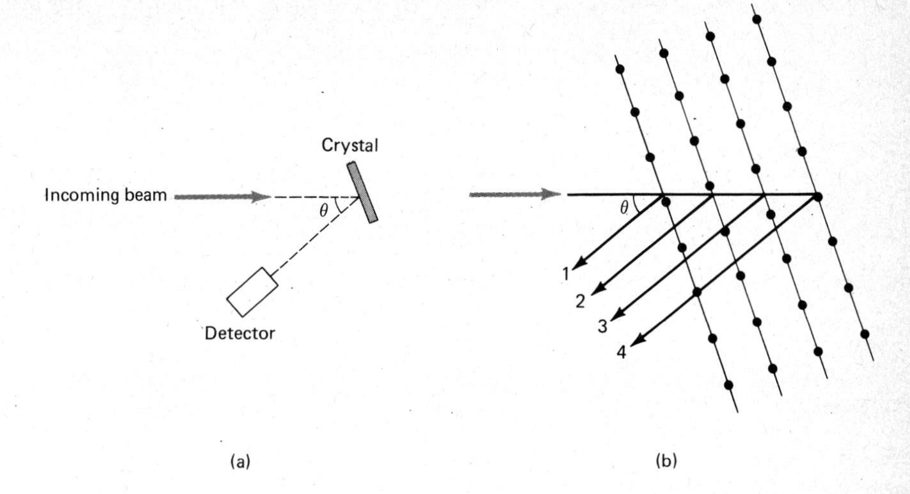

Figure 29.7 The electron de Broglie waves are reflected from the crystal planes. If the path lengths of the reflected beams differ by integer multiples of λ, a strong reflection will be observed.

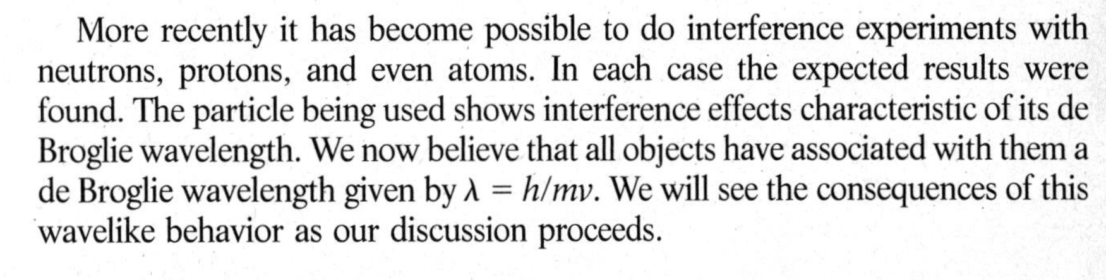

More recently it has become possible to do interference experiments with neutrons, protons, and even atoms. In each case the expected results were found. The particle being used shows interference effects characteristic of its de Broglie wavelength. We now believe that all objects have associated with them a de Broglie wavelength given by $\lambda = h/mv$. We will see the consequences of this wavelike behavior as our discussion proceeds.

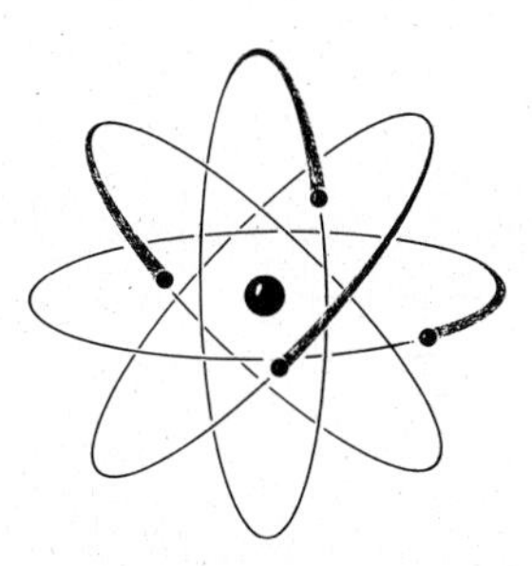

Figure 29.8 One concept of an atom.

29.5 Hydrogen Atom Orbits

De Broglie's discovery of the wave nature of the electron provided a valuable clue for understanding the structure of the atom. It had been known for some time that the atom contains a positively charged nucleus at its center. Electrons were thought to circle the nucleus as shown schematically in Figure 29.8. The nucleus contains a positive charge equal to the combined charges of the negative electrons. Much as the planets are held in orbit about the sun by the gravitational force, the electrons are held in orbit by the electric attraction of the positive nucleus for the negative electrons. This is shown in Figure 29.9 for the hydrogen atom, an atom that has only one electron.

In 1913 Neils Bohr had presented a somewhat successful theory for the hydrogen atom. He assumed that the single hydrogen electron is held in orbit by the Coulomb force. Therefore, for an electron of mass m and charge e that orbits at a radius r with speed v,

$$\text{Centripetal force} = \text{Coulomb force}$$

or

$$\frac{mv^2}{r} = k\frac{e^2}{r^2} \tag{29.7}$$

where $k = 9 \times 10^9\ \text{N}\cdot\text{m}^2/\text{C}^2$ (the Coulomb law constant).

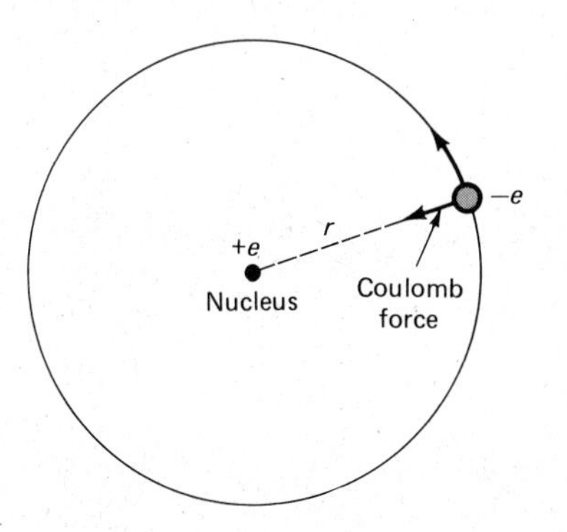

Figure 29.9 The Coulomb force supplies the centripetal force required to hold the electron in orbit.

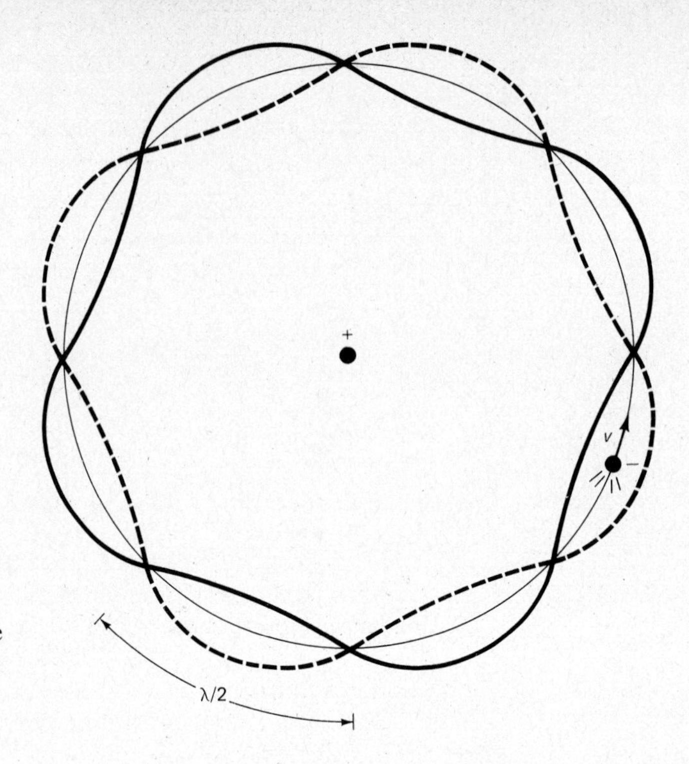

Figure 29.10 An electron will resonate in a circular orbit only if the orbit is a whole number of de Broglie wavelengths in circumference.

But to obtain agreement with experiment, Bohr had to assume that the electron could circle in only certain special orbits. He could give no reason for the orbits he chose except that they worked. However, once we know that electrons have wave properties, it becomes clear why certain orbits are special.

We see in Figure 29.10 an electron circling an orbit that has a circumference of a special length. If you trace the electron's de Broglie wave around the circle, you will see the orbit is 4 λ long. As the electron circles the orbit over and over again, the de Broglie wave will reinforce itself. The wave is in resonance with itself on an orbit of this size.

Similarly, the de Broglie wave will resonate with itself when the orbit is λ long, or 2 λ, or 3 λ, and so on. In other words, the electron wave resonates within the atom only when the orbit circumference is $n\lambda$, where n is an integer. It is therefore reasonable to assume that the atom will have its electron circling in one of these so-called stable orbits.

Let us try to compute the radii of these stable orbits. We will designate them as $r_1, r_2, r_3, \ldots, r_n$, where r_1 will be the smallest orbit. For these stable orbits we have

$$\text{Orbit circumference} = n \cdot (\text{electron wavelength})$$

or

$$2\pi r_n = n \frac{h}{mv} \tag{29.8}$$

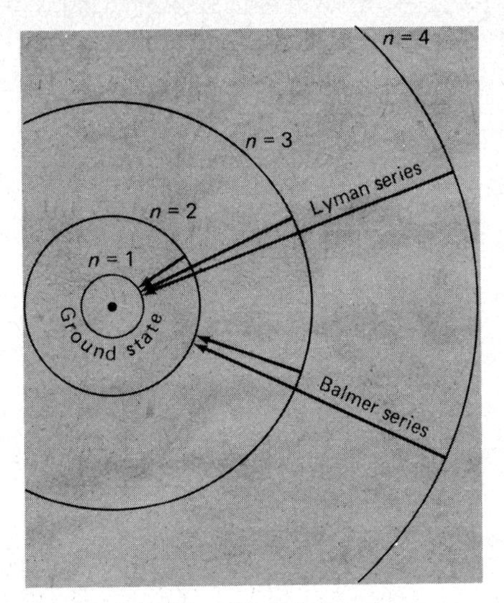

Figure 29.11 The hydrogen atom orbits.

where $n = 1, 2, 3, \ldots$, any integer. (Bohr actually postulated this relation ten years earlier, but he had no clear justification for it.)

This equation contains two unknowns, r_n and v, the electron's speed. But Equation 29.7 contains these same two unknowns if we recognize that r there is the same as r_n. Solving these two equations simultaneously gives r_n and v to be

$$r_n = \frac{n^2h^2}{4\pi^2ke^2m} \quad \text{and} \quad v = \frac{2\pi ke^2}{nh}$$

If you evaluate these for the smallest orbit, r_1, you will find that $r_1 = 5.3 \times 10^{-11}$ m and $v = 2 \times 10^6$ m/s. The value for r is reasonable because we know the hydrogen atom is about 10^{-10} m in diameter. The value found for the electron speed is also reasonable. For future reference we should notice that

$$r_n = (5.3 \times 10^{-11}\text{ m}) \times n^2$$

A diagram for these orbits is shown in Figure 29.11. (The additional data given in the figure will be discussed later.) The smallest orbit occurs for $n = 1$. Orbits for which n becomes very large, near infinity, have such large radii that the electron is nearly free from the atom.

An electron circling in any of these orbits has two kinds of energy. It has kinetic energy because it is in motion. It also has electrical potential energy. By custom, the electron's potential energy is taken to be zero when it is infinitely far from the nucleus, that is, when it is free from the atom. Because the nucleus attracts the electron, the electron loses potential energy as it comes closer to the nucleus. Hence the potential energy of the electron is negative. (This is like the gravitational potential energy of a ball in a room when the ceiling is taken as the zero level for potential energy. As long as the ball is in the room, its potential energy is negative.)

When we compute these two energies and take their sum, we obtain the energy of an electron in the nth orbit to be

$$E_n = -\frac{2\pi^2k^2e^4m}{n^2h^2} \tag{29.9}$$

Upon evaluation and changing the energy in joules to electronvolts, this becomes

$$E_n = -\frac{13.6}{n^2}\text{ eV} \tag{29.10}$$

The energy is negative simply because of the way we chose the zero of potential energy for the electron.

EXAMPLE 29.5 How much energy must be given to a hydrogen atom electron to pull it out to the fifth orbit if it is originally in the first orbit? Compare this to the energy needed to tear the electron completely loose from the atom.

Solution The energy we wish to find is

$$\text{(Energy in 5th orbit)} - \text{(energy in 1st orbit)}$$

Or, in symbols,

$$\text{Energy needed} = E_5 - E_1$$

Using the values of E_n from Equation 29.10, this becomes

$$\text{Energy needed} = \left(-\frac{13.6}{25}\right) - \left(-\frac{13.6}{1}\right) = -13.6\left(\frac{1}{25} - 1\right)$$
$$= 13.1 \text{ eV}$$

We wish to compare this with the energy difference $E_\infty - E_1$ because $n \rightarrow \infty$ represents $r \rightarrow \infty$. Using $E_n = -13.6/n^2$,

$$E_\infty = -\frac{13.6}{\infty} = 0$$

and so the energy requested, the so-called *ionization energy* of the atom, is

$$\text{Ionization energy} = E_\infty - E_1 = 0 - \left(-\frac{13.6}{1}\right) = 13.6 \text{ eV}$$

As you see, nearly as much energy is needed to pull the electron out to the fifth orbit as is needed to pull the electron completely loose. This fact is easily seen from an energy level diagram, a topic discussed in the next section. ■■

29.6 Hydrogen Energy-Level Diagram

A convenient way to show the energy levels of atoms, molecules, and solids is the *energy-level diagram*. To become familiar with these diagrams, let us construct one for a ball of mass m suspended in a room as shown in Figure 29.12.

The gravitational PE of the ball depends on where we take the zero for PE. Let us be nonconventional and take zero PE to be at the ceiling. Then, when the ball is in the room, it will have negative PE because it is below the zero level. For example, when the ball is on the table, its PE is $-mgh_t$. (In this system height is measured down from the ceiling; h_t is the height from ceiling to tabletop.) When it is at the level of the bulb, its PE is $-mgh_b$. The lowest energy level possible for the ball is on the floor, $-mgh_f$. We call this the *ground state energy*.

We can show the energy levels of the ball when it is in the room by the diagram in Figure 29.12(b). The diagram consists of a vertical energy scale with

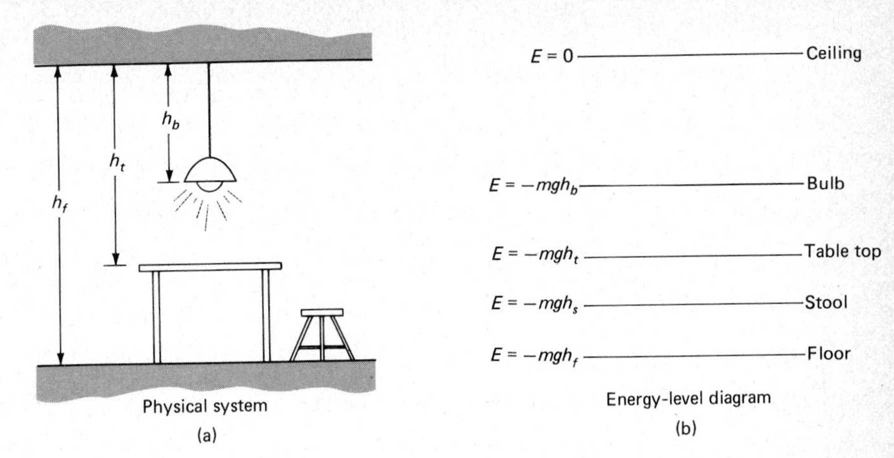

Figure 29.12 Part (b) is a diagram of the potential energy levels for a mass m placed in the physical situation shown in (a). Notice that the energy levels are negative because of our choice of $E = 0$ at the ceiling.

horizontal lines drawn at the appropriate energy levels. At the top is the $E = 0$ level, representing the ball at the ceiling. The lowest level is $E = -mgh_f$, representing the ball at the floor. Lines are drawn in a similar way to show the energy of the ball at the other levels in the room. As we see, an energy-level diagram simply summarizes the energies of the system.

In the case of the hydrogen atom, the energies the electron can have are given by $E_n = -13.6/n^2$ eV. An atom that has its electron in the $n = 1$ orbit, the ground (or lowest) energy state, has an energy of -13.6 eV. The energy is negative because we chose zero to be when the electron is loose from the atom. If an electron is in the $n = 2$ orbit, the energy of the atom is $-13.6/2^2$ eV or -3.4 eV. In the $n = 3$ orbit, the energy of the atom is $-13.6/3^2$ eV or -1.5 eV. And so on for the other orbits. Finally, when $n \to \infty$, the energy is zero because

$$E_\infty = -\frac{13.6}{\infty} = 0 \text{ eV}$$

This is for an infinitely large electron orbit, that is, when the electron is free from the atom.

These hydrogen atom energy levels are summarized in the energy-level diagram in Figure 29.13. Only the levels for $n = 1$ to $n = 8$ are shown. All the levels for $n = 9$ to $n = \infty$ are crowded between the $n = 8$ level and the $n = \infty$ level. As you see, the levels corresponding to large orbital radii have nearly the same energy, very close to zero. (For example, $E_9 = -0.17$ eV.) The diagram makes it easy to see the relative energies of the various atomic energy states.

We should notice three important features of this system. First, its energies are *quantized*. The atom at rest can have only those energies shown in the diagram. These are the only energies with which the electron wave can resonate in the atom. The atom can have an energy of E_1 or E_2 or E_3, and so on. It cannot have an in-between energy of -10 eV, for example. No such resonance energy exists. The atom will not be found with this energy.

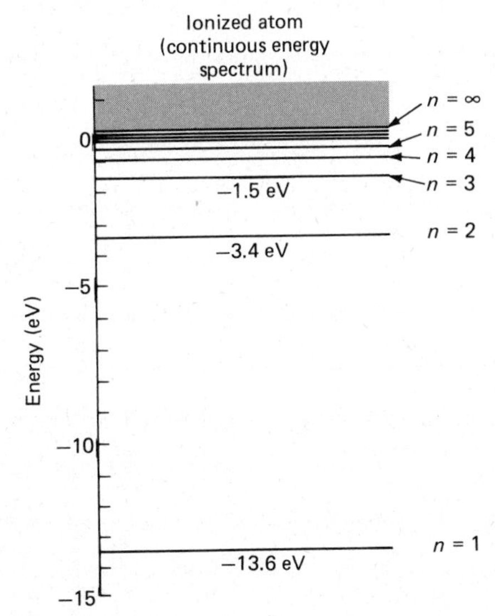

Figure 29.13 Energy-level diagram for hydrogen. There are an infinite number of levels (not shown) between the $n = 8$ and $n = \infty$ levels. Level $n = 1$ is the ground state; $n = \infty$ is the level at which ionization occurs.

Second, the $n = 1$ level is the ground state for the atom. This corresponds to the lowest energy the atom can have. Just as a physical system such as a ball falls to its state of lowest energy when allowed to do so, the atom also falls to its lowest energy level. Except at high temperatures, the usual hydrogen atom is in its $n = 1$ state. The electron is in its smallest orbit.

Third, energies above $E = 0$ in the diagram correspond to ionized atoms. When the electron is infinitely far from the nucleus and at rest, the energy of the atom is zero, $E_\infty = 0$. But if the electron is not only free, but also moving, then the energy of the system will be still larger and so $E > 0$. After ionization, the electron's KE can be anything. Hence the energy is not quantized for energies above $E = 0$. As shown in the diagram, the levels above $E = 0$ blend into a continuous energy region called the *continuum*.

29.7 Atomic Excitation

Hydrogen atoms on the earth are usually in the $n = 1$ state, the ground state. It is easy to see why this is true. The energy of the $n = 1$ state is -13.6 eV, while that of the $n = 2$ state is -3.4 eV, a difference of 10.2 eV. Thermodynamics tells us systems tend to fall to their lowest energy state, the $n = 1$ state in this case. After a hydrogen atom does so, it requires 10.2 eV of energy to lift the electron to the $n = 2$ orbit.

But 10.2 eV is a large amount of energy. An atom's translational kinetic energy due to thermal motion is only about 0.04 eV at room temperature. Therefore, collisions between atoms are not likely to be energetic enough to excite an atom, that is, to throw it from the $n = 1$ state to a higher state. Visible light photons have far less than 10 eV energies. They have no effect when incident on hydrogen atoms. They do not have enough energy to excite a hydrogen atom.

Suppose, however, that a hydrogen atom is placed in a high voltage discharge tube. In such a tube there will be many electrons shooting around with energies much larger than 10 eV. When one of these collides with an atom, it can excite the atom to states with large values of n. Indeed, many hydrogen atoms would be ionized by the discharge. We see, then, that high voltages can be used to excite atoms.

Another method makes use of high-energy photons. A 10-eV photon has a wavelength of 124 nm, a wavelength in the border region between ultraviolet and X rays. We see then that a beam of X-ray photons striking hydrogen atoms can excite the atoms by collision. In fact, photons with energies higher than 13.6 eV will be able to ionize the hydrogen atoms they strike. But visible light photons have much smaller energies. They cause no effect when incident on hydrogen atoms. They do not have enough energy to excite an atom and so they are not absorbed by hydrogen gas.

Of course, if you heat hydrogen atoms to a high temperature, their thermal energies become very large. Then they may excite one another upon collision. Usually this mechanism is not important until the gas is at thousand-degree temperatures. At these temperatures the fastest moving atoms have enough energy to cause excitation.

▮ EXAMPLE 29.6 What is the longest wavelength em radiation that can ionize unexcited hydrogen atoms?

Solution It requires 13.6 eV to lift an electron from the ground state to the $n = \infty$ level. Therefore, a photon must have an energy of 13.6 eV to ionize hydrogen. The energy of a photon is hc/λ. Converting electron volts, we have

$$(13.6 \text{ eV})\left(1.60 \times 10^{-19} \frac{\text{J}}{\text{eV}}\right) = 2.18 \times 10^{-18} \text{ J}$$

Then, using $h = 6.63 \times 10^{-34}$ J·s and $c = 3 \times 10^{8}$ m/s, we have

$$\frac{hc}{\lambda} = \frac{(6.63 \times 10^{-34} \text{ J·s})(3 \times 10^{8} \text{ m/s})}{\lambda} = 2.18 \times 10^{-18} \text{ J}$$

$$\lambda = 91 \text{ nm}$$

This is in the very short ultraviolet wavelength region. ▮▮

29.8 Hydrogen Emission Spectrum

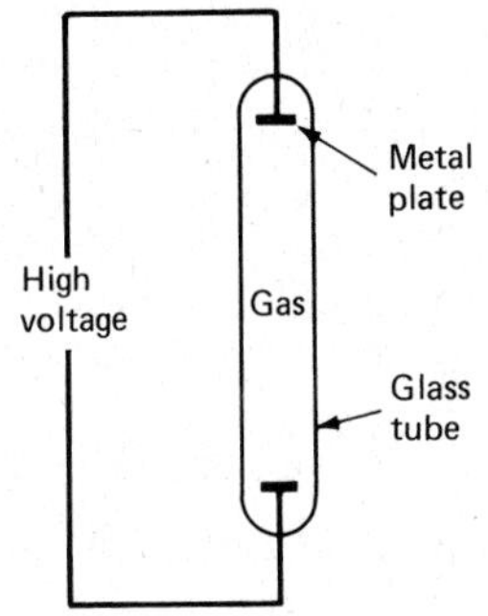

Figure 29.14 When a high voltage is applied to the low-pressure gas, a glow discharge occurs in the tube.

Hydrogen gas can be caused to emit light by exciting the atoms. We can excite the gas atoms by use of the discharge tube, photon bombardment, or heating, but the discharge tube method is most common. These tubes are much like the familiar neon lights (see Figure 29.14). The gas in the tube is at low pressure. A high voltage applied to electrodes at the ends of the tube subjects the tube to a large electric field. This causes high-energy electrons to shoot through the tube. The gas atoms are bombarded by these electrons and become excited or even ionized.

Let us see what happens to these highly excited atoms. They are unstable in the excited state and, if allowed to do so, they will return to the ground state. For example, consider the hydrogen atom in Figure 29.15. Its electron is in the $n = 4$ state. The electron will eventually fall to a smaller orbit. In doing so, it loses energy. What happens to this energy?

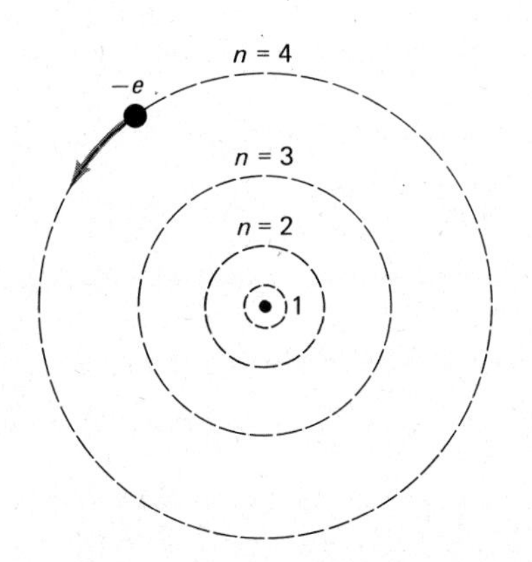

Figure 29.15 The electron can fall to any of the lower energy states, emitting a photon to carry away the energy.

An isolated atom can lose energy in only one way; it must throw out a photon. When the hydrogen electron falls to the $n = 2$ state, for example, it must throw out a photon with energy given by $E_4 - E_2$. We know that $E_n = -13.6/n^2$ eV and so this energy would be

$$E_4 - E_2 = \left(\frac{-13.6}{4^2}\right) - \left(\frac{-13.6}{2^2}\right) = -0.85 - (-3.40)$$
$$= -0.85 + 3.40 = 2.55 \text{ eV}$$

Equating this to the energy of the emitted photon gives

$$\text{Photon energy} = E_4 - E_2$$

$$\frac{hc}{\lambda} = (2.55 \text{ eV})(1.60 \times 10^{-19} \text{ J/eV})$$

Solving for λ gives it to be 486 nm, a wavelength in the blue region of the visible spectrum.

Of course, the atom need not have fallen to the $n = 2$ state. It could fall to the $n = 3$ state or the $n = 1$ state. The energy lost in each case is different. Therefore, different wavelength photons would result from each transition. Moreover, once the electron has fallen to any but the $n = 1$ orbit, it will undergo additional transitions until it reaches the $n = 1$ state.

It follows, then, that excited hydrogen atoms will emit many wavelengths of light. As the electron in an atom falls to a lower state, it will emit a photon characteristic of the two levels between which it falls. Long before Einstein had discovered the photon, scientists had measured the wavelengths of light emitted by hydrogen atoms. Let us now learn about the spectrum they observed and see how it can be explained in terms of our understanding of the atom.

29.9 Hydrogen Spectral Series

To study the light given off by excited hydrogen atoms, we can use either a prism or a diffraction grating spectrometer. A prism spectrometer and the wavelengths emitted in the visible and near ultraviolet are shown in Figure 29.16. The spectrum emitted in this region is shown in more detail in Figure 29.17. (See Figure 19.6 for the visible lines shown in color.)

This series of spectral lines was shown by Balmer to be representable by a simple formula.

$$\frac{1}{\lambda} = R\left(\frac{1}{2^2} - \frac{1}{n^2}\right) \qquad n = 3, 4, \ldots$$

In this formula $R = 1.0974 \times 10^7 \text{ m}^{-1}$ and n is any integer larger than 2. For example, when $n = 3$, the formula gives $\lambda = 656$ nm. This is the longest wavelength line in the series, called the *Balmer series.* For $n = 4$, the formula gives

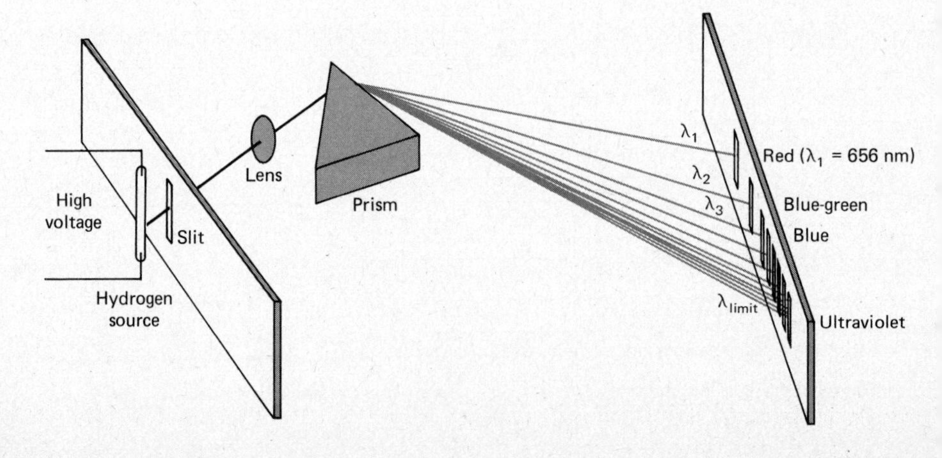

Figure 29.16 The high voltage causes a gas discharge in the hydrogen gas tube. (The tube acts like a neon sign, but in this case it is more correctly a hydrogen sign.) Light coming from the tube consists of a series of spectral lines.

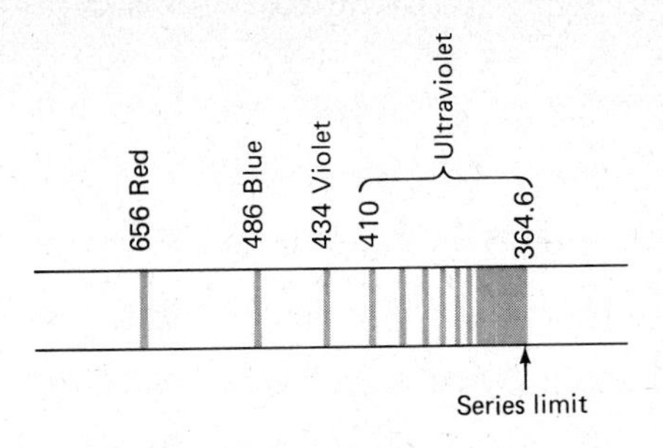

Figure 29.17 Excited hydrogen atoms emit the spectral lines of the Balmer series. The wavelengths are given in nanometers (nm).

$\lambda = 486$ nm, the next line in the series shown in Figure 29.17. Similarly, each value of n gives a line in the series. Finally, for $n = \infty$, the result is 365 nm, the shortest wavelength or *series limit* for the series.

As technology advanced to allow infrared and ultraviolet measurements, other hydrogen spectral series were found. The three series of shortest wavelength are shown in Figure 29.18. Each fits a formula similar to the Balmer formula. It is found that

$$\text{Lyman series:} \quad \frac{1}{\lambda} = R\left(\frac{1}{1^2} - \frac{1}{n^2}\right) \qquad n = 2, 3, \ldots$$

$$\text{Paschen series:} \quad \frac{1}{\lambda} = R\left(\frac{1}{3^2} - \frac{1}{n^2}\right) \qquad n = 4, 5, \ldots$$

Other series similar to these are found still further in the infrared.

We can understand these series easily in terms of an energy-level diagram. The hydrogen diagram in Figure 29.13 is redrawn and modified in Figure 29.19. Recall that the atom (or its electron) can exist only in the energy states indicated by the horizontal lines. In a highly excited gas there are some atoms in each state. As the atoms fall to lower states, they emit photons. The wavelength of the emitted photon varies *inversely* with the energy difference between the two levels. Large energy difference means small emitted wavelength.

Look at the transition arrows labeled Lyman series. These represent the atom falling to the $n = 1$ (or ground) state from higher states. We know that $E_n = -13.6/n^2$ eV for the hydrogen atom. Therefore, the energy lost by the atom as it falls from the nth state to the $n = 1$ state is

$$E_n - E_1 = 13.6\left(\frac{1}{1^2} - \frac{1}{n^2}\right) \text{ eV} \qquad n = 2, 3, \ldots$$

This energy is emitted as a photon.

We can find the wavelength of the emitted photon by equating the above energy difference (in joules) to hc/λ. We have

$$\frac{1}{\lambda} = \left(\frac{13.6 \text{ eV}}{hc}\right)(1.60 \times 10^{-19} \text{ J/eV})\left(\frac{1}{1^2} - \frac{1}{n^2}\right)$$

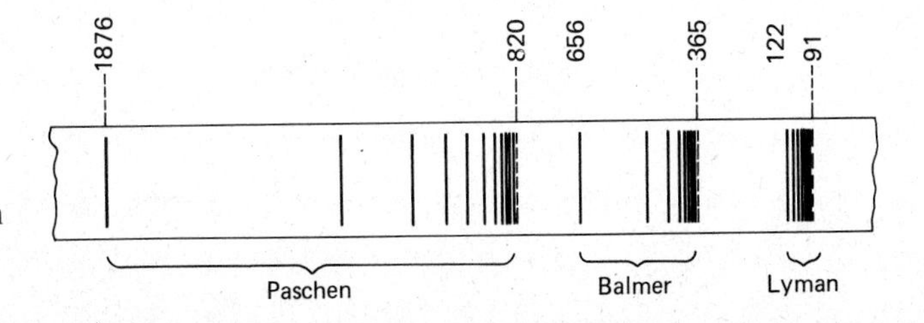

Figure 29.18 Wavelengths are given in nm for the three shortest wavelength series emitted by hydrogen.

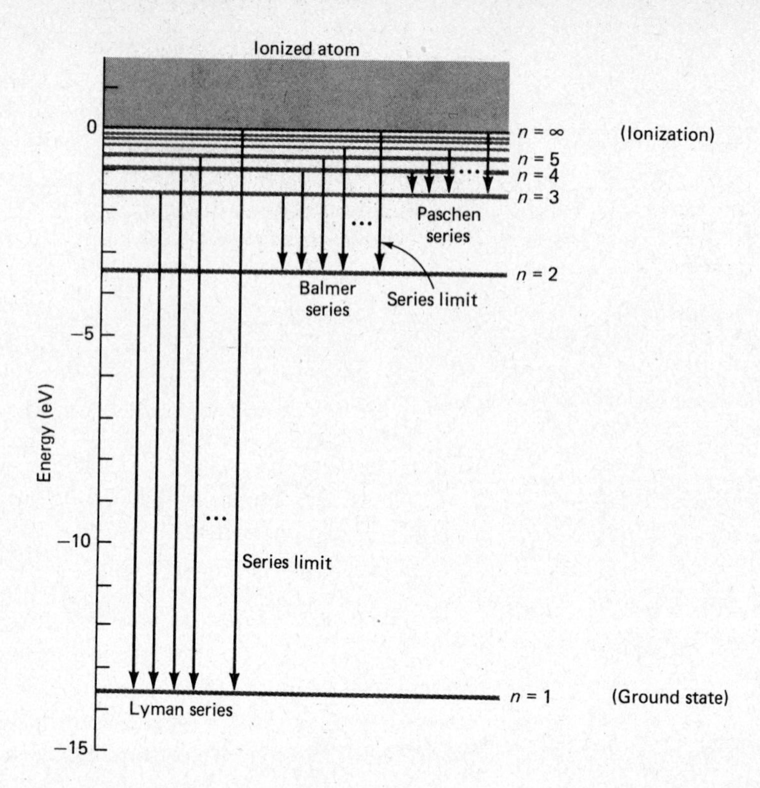

Figure 29.19 Transitions are shown easily on an energy-level diagram.

Upon evaluation this becomes

$$\frac{1}{\lambda} = 1.0974 \times 10^7 \left(\frac{1}{1^2} - \frac{1}{n^2}\right)\text{m}^{-1} \qquad n = 2, 3, \ldots$$

Notice that this is the same formula as the one that represents the Lyman series of spectral lines. We therefore conclude that

The Lyman series is emitted as the atom falls from higher states to the $n = 1$ state.

In a similar way we can show that transitions to the $n = 2$ state give rise to the Balmer series of spectral lines. Transitions to the $n = 3$ level cause the Paschen series. Clearly, the energy-level diagram provides a convenient way for interpreting the emission spectrum of hydrogen.

We can see another important feature of the spectral series from the diagram. Notice that the energy levels at high values of n are closely spaced. Hence the transition arrow for a transition starting at $n = 10$ has nearly the same length as one starting at $n = 1000$. Therefore, the emitted photons for all these transitions have about the same energy and wavelength. It is for this reason that the spectral lines all blur together as we go to higher n values in the series. Finally, the longest arrow occurs for the $n = \infty$ transition. This transition gives rise to the shortest wavelength, the series limit, shown for the Balmer series in Figure 29.17.

Let us now summarize what we have found out about the hydrogen atom.

1. The atom exists in only certain definite states.
2. Each state corresponds to a certain energy. These energies are shown by the energy-level diagrams of Figures 29.13 and 29.19.
3. When the electron is infinitely far from the atom (i.e., the atom is ionized), the energy of this state is defined to be zero.
4. The ground state of the atom has the lowest possible energy. Unexcited atoms are in the ground state.
5. When the electron falls from one level to a lower one, the atom changes from one state to another. The emitted photon has energy equal to the difference in energy between the two states.
6. The photon energy is related to the wavelength of the emitted spectral line through

$$\text{Photon energy} = \frac{hc}{\lambda}$$

EXAMPLE 29.7 Using the diagram of Figure 29.19, find the wavelength of the second line of the Lyman series.

Solution This line corresponds to the next to longest line of the series. Therefore, it results from the transition shown as the second transition from the left. It starts at a level with energy of about -1.6 eV (the $n = 3$ level). It ends at the -13.6-eV level. Therefore, the emitted photon has an energy of

$$13.6 \text{ eV} - 1.6 \text{ eV} = 12.0 \text{ eV}$$

Using

$$\text{Photon energy} = \frac{hc}{\lambda}$$

we have

$$\lambda = \frac{hc}{\text{photon energy}}$$

We must express the photon energy in joules, not electronvolts, so this becomes

$$\lambda = \frac{(6.63 \times 10^{-34} \text{ J}\cdot\text{s})(3 \times 10^{8} \text{ m/s})}{(12.0 \text{ eV})(1.6 \times 10^{-19} \text{ J/eV})} = 104 \times 10^{-9} \text{ m}$$

This is the wavelength of the second line of the Lyman series.

EXAMPLE 29.8 Find the series limit wavelength for the Balmer series.

Solution The line corresponds to the transition from $n = \infty$ to $n = 2$. Therefore

$$\text{Photon energy} = 13.6\left(\frac{1}{2^2} - \frac{1}{\infty^2}\right) = 13.6\left(\frac{1}{4} - 0\right) = 3.4 \text{ eV}$$

Then, from photon energy $= hc/\lambda$,

$$\lambda = \frac{hc}{(3.4)(1.60 \times 10^{-19})} = 365 \text{ nm}$$

29.10 Deficiencies of the Orbital Model

The idea of electron wave resonance has been very successful in finding atomic energy levels. But the model of precise orbits we have used in our explanation is not exact. We can best understand the difficulty by comparing electron de Broglie waves to situations involving light waves.

In light interference experiments we find the positions of maximum light intensity by finding the positions of reinforcement. This is also what we have done using de Broglie waves for reinforcement within the atom. The places where the de Broglie waves reinforce are the places where the electron is found to be. But this is not strictly true. Just as light exists elsewhere than at the maxima of the light interference pattern, the electron is found elsewhere than at the positions of complete reinforcement. Hence the orbits we have computed are far sharper than the actual positions where the electron may be.

The inaccuracies of our simple calculation of orbits can be removed by use of *wave mechanics* (or *quantum mechanics*). This is a method of calculation that relies heavily on the wave nature of electrons and objects in general. By means of it, the electron in the atom is found to stray far from the orbits we have pictured. Even so, the atomic states have the same energies as we have calculated. But the de Broglie wave resonances and the possible positions of the electron in the atom are far more complex than simple circular orbits around the nucleus.

29.11 Spectra and Energy Levels of Other Atoms

Hydrogen has the simplest of all spectra. Even helium, with only two electrons, has a considerably more complicated spectrum. As the number of electrons in an atom increases, the number of lines in the spectrum increases greatly. This trend is indicated by the spectra in Figure 19.6. However, there are many fainter lines that do not show in the photo. For example, the spectrum of calcium (Ca) has nearly 100 lines in this region, many more than appear in the photo.

Each type of atom has its own unique spectral lines. For this reason industry makes use of spectra to identify atoms. As an example, suppose that a certain impurity is suspected in a metal. The metal is vaporized in an electric arc and the light emitted from the arc is examined. If the spectral lines for the impurity are present, then the impurity is known to be present.

The energy-level diagrams of atoms that contain many electrons are also very complicated. In spite of this, most of the energy levels have been found for many of them. Knowledge of these energy-level diagrams has made it possible to predict many useful properties of these atoms. This is especially true in the use of materials for lasers and solid state electronics.

29.12 Spectra and Energy Levels of Solids

We have been discussing the energy levels and spectra of isolated atoms. We saw that these levels are separated. As an atom changes from one level to a lower one, it emits a photon. This photon has an energy equal to the difference between the two levels. Because the energy levels are separated, so too are the wavelengths of the emitted photons. The emitted radiation is therefore a line spectrum.

But in solids (and liquids) the situation is somewhat different. The electrons near the outer edge of the atom are disturbed by the neighboring atoms. As a result, their energies are also disturbed. This causes the higher energy levels (those just below the zero level) to broaden out. They become wide, overlapping energy bands. The situation is shown schematically in Figure 29.20.

Notice that the higher energy levels broaden into bands when the atoms are packed in a solid. As a result, these higher states do not have separate energies. The energies mix together to form a wide band of continuous energy levels. This means that a whole range of photon energies (and wavelengths) results for transitions from one state to the other. The original sharp spectral lines emitted by the separated atoms no longer exist. The lines have broadened and merged

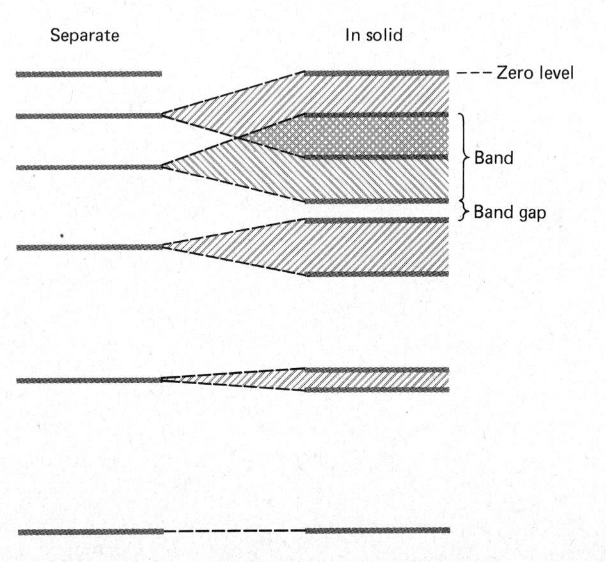

Figure 29.20 When atoms are packed together in a solid, their higher energy levels widen into overlapping bands. Transitions involving these bands result in a continuous spectrum of wavelengths. Transitions involving only the deeper levels still give rise to a line spectrum.

together. They form a continuous band of color, as shown in the top line of the color plate in Figure 19.6.

As we see, the excited atoms in a solid emit a *continuous spectrum*. It is for this reason that a white-hot solid emits a continuous band of color. The energy levels have broadened. The spectral lines have also broadened. They no longer exist as separate lines. Instead, they form a continuous spectrum of emitted wavelengths.

29.13 X-Ray Emission

Suppose that the nucleus of the hydrogen atom had a charge of $+50e$ instead of $+1e$. Then the Coulomb force on its electron would be increased by a factor of 50. As a result, a much larger energy would be required to knock the electron out of the ground state. The ground state would have a negative energy of many hundreds of electronvolts rather than -13.6 eV. We would therefore expect that the energy difference between energy levels would be much larger. Because of this, the photons emitted during transitions would have much higher energies and shorter wavelengths. They would be in the X-ray region.

This type of reasoning can be extended to atoms much more complicated than hydrogen. Atoms such as iron, copper, tin, tungsten, and so on have nuclear charges much larger than $+1e$. As a result, the innermost electrons of these atoms are tightly held in the ground state. Very large energies are needed to excite them. When they fall back to the ground state, photons of very high energy are emitted. These photons have wavelengths in the X-ray region.

Of course, these atoms can also emit much longer wavelengths. The energy levels close to the zero level are much closer together than the lower levels. Transitions between these levels can give rise to photons of much longer wavelength. In solids and liquids these longer emitted wavelengths are part of a continuous spectrum. As pointed out before, these longer-wavelength transitions give rise to the glow given off by red-hot objects.

However, the innermost electrons of the atom are not much influenced by neighboring atoms. The neighboring atoms are held away from them by the

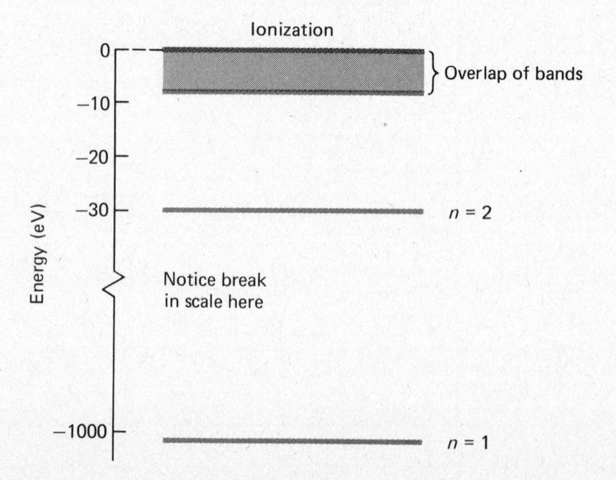

Figure 29.21 On the energy-level diagram for sodium metal, notice that the $n = 1$ level is near -1000 eV.

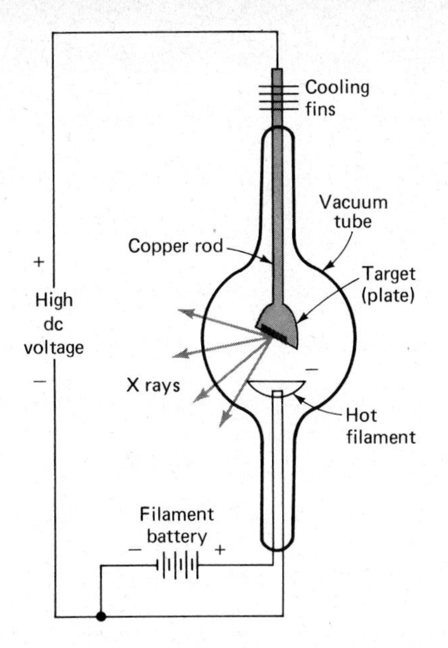

Figure 29.22 Electrons are accelerated from the filament to the plate across the X-ray tube. When they strike the plate, X rays are emitted.

outer electrons of the atom. Because of this, the widely spaced, lowest energy levels of the atom are not broadened much. This was shown in Figure 29.20. It is also shown in Figure 29.21, where we see the energy levels for solid sodium. Notice the huge difference in energy between the $n = 1$ and $n = 2$ levels. This difference is even larger for most atoms. (Sodium has a nuclear charge of $+11e$.)

To generate X rays, use is made of an X-ray tube such as that shown in Figure 29.22. It is simply a sturdy vacuum tube with two electrodes. A hot filament emits electrons. They are accelerated across the tube through a potential difference of perhaps 40,000 V. (Their energies are then 40,000 eV.) When they strike the target, or plate, the bombarding electrons are stopped. This sudden deceleration causes em radiation to be emitted. Because the deceleration can occur in many ways, this radiation (called bremsstrahlung or stopping radiation) is a continuous spectrum of X-ray wavelengths.

But, in addition, the impact with the target knocks electrons out of the atoms of the target. As these electrons fall back to the lowest energy levels of the atom, a line spectrum of X rays is also emitted. Therefore, the radiation from an X-ray tube consists of a line spectrum on top of a continuous spectrum. A typical example of this is shown in Figure 29.23. For the case shown there 35,000-eV electrons strike a molybdenum target.

As you know, X rays are highly penetrating. The photons have very high energies. When they collide with molecules, they can tear the molecules apart. Exposure to X rays can burn the skin, producing an effect similar to sunburn. The skin molecules have been torn apart. Deeper within the body, the X rays can also destroy body cells by breaking their molecules apart. Use is made of this fact to destroy cancer cells by X-radiation. In such applications great care must be taken or serious damage may result to the cells necessary for the life of the individual being treated. Also, care must be used not to damage gene cells in those of childbearing age. Damaged gene cells can cause defects in children later born to the patient. We will return to this subject in Chapter 31.

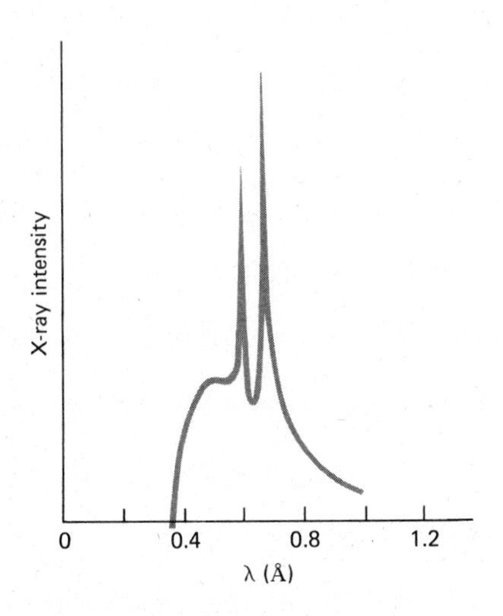

Figure 29.23 When 35,000-eV electrons strike a molybdenum target, the X rays shown are emitted. (1 Å = 0.1 nm.)

■ **EXAMPLE 29.9** What is the shortest X-ray wavelength that can be emitted from an X-ray tube operating at 35,000 V? Check your answer by referring to Figure 29.23.

Solution The bombarding electrons have an energy of 35,000 eV. If one of these is stopped by a single collision with an atom of the target, it is sometimes possible for all its energy to be radiated as one photon. Such a photon has the maximum possible energy. To find its wavelength, we use

$$\text{Photon energy} = \frac{hc}{\lambda}$$

In the present case the energy is $3.5 \times 10^4 \times 1.60 \times 10^{-19}$ J. Substitution of this value gives

$$\lambda = 0.36 \times 10^{-10}\ \text{m} = 0.036\ \text{nm} = 0.36\ \text{Å}$$

The shortest X-ray wavelength in Figure 29.23 does indeed have this wavelength. ■■

29.14 Absorption of em Radiation

Consider the situation shown in Figure 29.24(a). The incident beam of em radiation consists of a continuous spectrum. All wavelengths are present in it. The beam passes through a tube containing a gas of hydrogen atoms. We wish to know which wavelengths of radiation are absorbed by the gas atoms.

Experiment shows that only certain wavelengths are absorbed strongly. All the rest pass through the gas without being stopped. The spectrum of the radiation after it passes through the tube is shown in part (c) of Figure 29.24. For comparison, the spectrum of the original beam is shown in (b). Notice that the hydrogen atoms absorb strongly only certain wavelengths. These wavelengths are the same as the wavelengths of the Lyman series. The reason for this is easily understood.

Originally the hydrogen atoms are in the ground state. As the beam of radiation passes through the gas, the atoms are struck by photons of the beam. But the electron in an atom cannot do just anything. It can only be knocked out to the $n = 2$ level or the $n = 3$ level, and so on. The atom cannot exist with the electron between these levels. The electron can therefore absorb only very special energies.

If the photon has just enough energy to knock the electron to the $n = 2$ level, then the photon can be absorbed. But a photon with less energy than this cannot be absorbed. If it gave its energy to the electron, the electron would end up somewhere between the $n = 1$ and $n = 2$ levels. Since the electron cannot exist there, such a photon cannot be absorbed.

The only photons that can be absorbed are those that have energies equal to the transition energies shown by the arrows in Figure 29.25. But these are the same energies as those emitted as the excited atom falls to the ground state. In Figure 29.19 we saw that these gave rise to the Lyman series. As a result, we see that unexcited hydrogen atoms absorb only certain wavelengths. These wavelengths are those of the Lyman series.

Similar situations exist for other atoms. They can absorb only certain wavelengths of radiation. These wavelengths are determined by the energy-level structure of the atom. Like the wavelengths of the emission spectrum, the wavelengths of the absorption spectrum can be used to identify the atoms present in a material. The absorption spectra of atoms, liquids, and solutions are widely used to identify atoms and molecules.

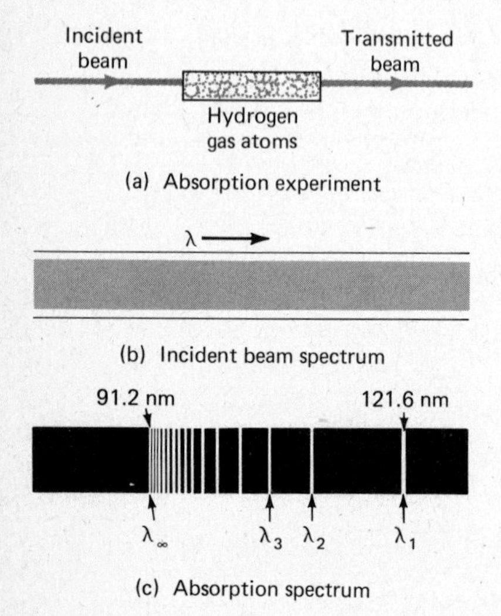

Figure 29.24 Radiation with wavelengths of the Lyman series is strongly absorbed.

■ **EXAMPLE 29.10** When a continuous spectrum is shined through a gas composed of vaporized sodium atoms, the wavelength 589 nm is strongly absorbed. (This is also the wavelength of the yellow light given off by a sodium vapor lamp.) What is the difference in energy between the starting state and the end state for the transition?

Solution A photon energy hc/λ with $\lambda = 589 \times 10^{-9}$ m changes the atom from one state to another. Therefore,

$$\text{Difference in energy} = \frac{hc}{\lambda}$$

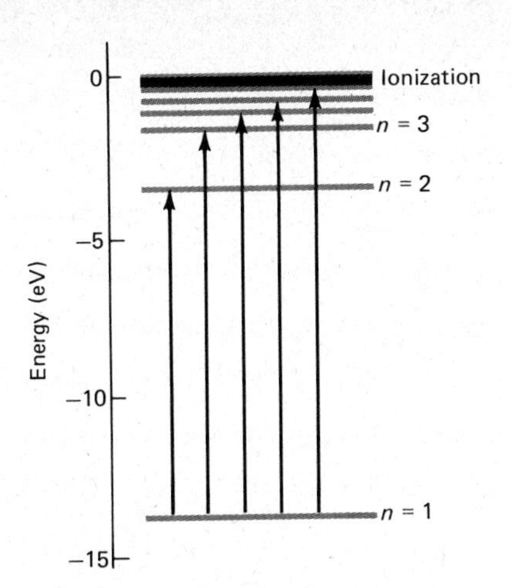

Figure 29.25 Photons having the energies shown by the transition arrows can be absorbed by hydrogen atoms.

Placing in the values gives

$$\text{Difference in energy} = 3.38 \times 10^{-19}\ \text{J}$$
$$= 2.1\ \text{eV}$$

■■

29.15 The Laser

Nearly always, atoms in a light source act independently from one another. Their photons are emitted at random times. As a result, the light waves emitted by the various atoms are not usually in phase. Sometimes they cancel and sometimes they aid each other. The beam of light from the source is far weaker than it would be if all the waves were in phase with each other.

There is one type of light source in which the atoms all emit their light waves in phase. It is called a laser. (The name comes from the words "*L*ight *A*mplification by *S*timulated *E*mission of *R*adiation.") Lasers produce an intense, narrow, pencillike beam of parallel light. Many types of lasers exist. However, they all operate in the following basic way.

1. An outside agent (an electrical discharge, for example) excites atoms within the laser.
2. By proper selection of the laser material, the majority of certain atoms in the material are excited to metastable states. A metastable state is one in which the atom will exist for some time. The excited atoms remain excited for an appreciable time.
3. One excited atom falls to a lower state and emits a photon. This photon's wave travels past the other excited atoms and triggers them to fall to the lower state. In so doing, they emit photons with waves in step with the original photon wave.
4. The laser material is in the form of a rod or tube. As shown in Figure 29.26(a), the ends of the laser material are closed by accurately parallel mirrors. The beam travels back and forth many times between these mirrors. On each trip it causes more excited atoms to emit waves in phase with it. As a result, a very intense, coherent beam of in-phase waves exists in the laser tube or rod.
5. Only those waves that travel accurately along the axis of the tube will remain in the tube after many trips up and down it. Therefore, the beam is almost entirely composed of waves going in exactly the same direction. Its rays are accurately parallel.
6. One mirror at the end of the laser leaks light slightly. A small fraction of the beam passes through it. This last portion of the beam emerges from the laser. It is what we call the laser beam.

As you see, the laser beam is very intense because the waves are all in phase. Each excited atom sends out the same wavelength wave. Therefore, the laser beam has a single, highly precise wavelength. The waves are identical and in step. They are therefore coherent. Because the beam reflects many times be-

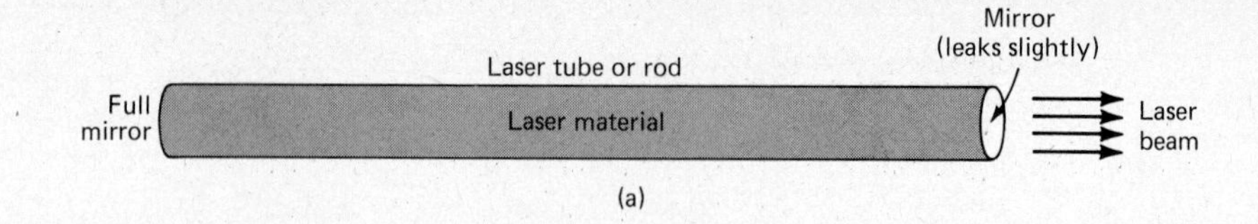

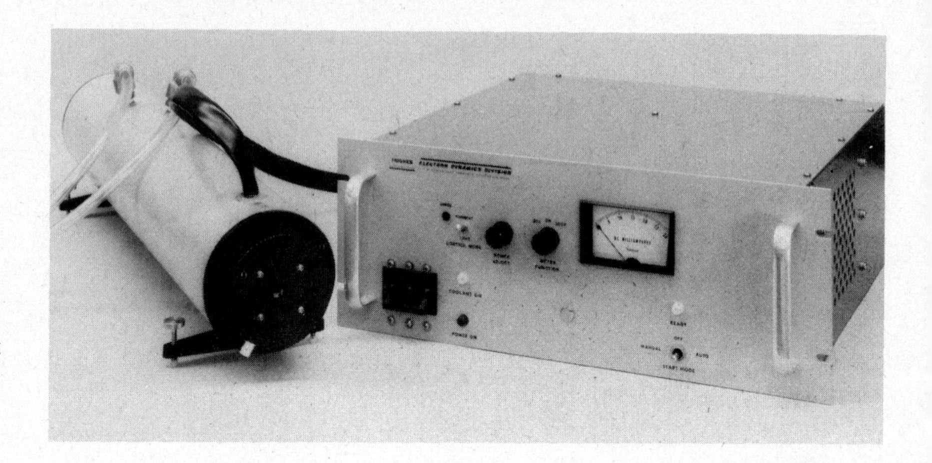

Figure 29.26 The laser produces an intense, coherent beam of parallel light with a precise wavelength. (Photo courtesy of Hughes Aircraft Co., Industrial Products Division)

tween the parallel mirrors, its rays are accurately parallel. The beam retains its pencillike quality over distances of many miles.

Because of these unique properties, lasers have become widely used. Their high intensity makes them useful for such diverse applications as drilling precise holes through even the hardest materials and welding a loosened retina in place in an eye. Their high coherence and precise wavelength has led to the use of laser light in communications and signal transmission. Their pencillike quality has led to their use in surveying and equipment alignment. The list of present and probable future uses of lasers is far too long to be included here. One of its possible future uses, to ignite the nuclear fusion reaction, will be mentioned in Chapter 31. This use, if successful, could influence the course of civilization in the future.

Summary

Einstein's special theory of relativity shows, among other things, that

1. No material object can be accelerated to a speed faster than the speed of light in vacuum, c.
2. The mass of an object increases as its speed increases. Near the speed c, the mass increases rapidly with speed and $m \to \infty$ as $v \to c$. The mass of the object when $v = 0$ is called the rest mass, m_0. Einstein found that $m = m_0/\sqrt{1 - (v/c)^2}$.

3. Mass and energy are interchangeable. When an object undergoes a change ΔE in energy, its mass changes by Δm, given by $\Delta E = (\Delta m)c^2$. The kinetic energy of an object is equal to $(m - m_0)c^2$. At low speeds this reduces to $\frac{1}{2} m_0 v^2$.

In the photoelectric effect em radiation causes electrons to be emitted from the surface of a material. There is a threshold wavelength for each material. Electrons are emitted only when λ is smaller than this threshold wavelength.

We now accept Einstein's explanation of the photoelectric effect. He pointed out that a beam of light behaves like a stream of little pieces of energy. We call these energy pulses photons. The energy of each photon in a beam of light with wavelength λ (or frequency f) is hc/λ (or hf). The quantity h is equal to 6.63×10^{-34} J·s and is called Planck's constant. The photons travel along the beam with the speed of light. Short-wavelength em radiation has higher-energy photons than long-wavelength radiation does.

The de Broglie wavelength of a particle of mass m traveling with speed v is h/mv, where h is Planck's constant. Particles undergo interference and diffraction effects as predicted for a wave with this wavelength.

Atoms can take on only certain energies. The lowest energy is for the ground state of the atom. An atom normally exists in this state. When the atom is struck by a particle or photon, it may be thrown to an excited state. Each atom has its own set of excited states. Each state has a definite energy. The atom can exist only in one of these states.

The em spectrum emitted by an atom depends on its possible energy states. These energies that an atom can take on are summarized in energy-level diagrams. The photons emitted by an atom have energies equal to the energy differences between these levels.

The spectrum of hydrogen consists of several series of spectral lines. When the excited atom falls to the ground ($n = 1$) state, the Lyman series of lines is emitted. The Balmer series is emitted when it falls to the $n = 2$ state. The Paschen series is emitted when it falls to the $n = 3$ state. The Lyman series is in the ultraviolet, the Balmer in the visible and ultraviolet, and the Paschen in the infrared.

An atom that has lost an electron is said to be ionized. The work required to ionize an atom is called the ionization energy of the atom.

When atoms are packed together in a solid, the highest energy levels spread into overlapping bands. As a result, photons of all energies can be emitted. The visible spectrum of a solid is therefore continuous. For this reason red-hot objects emit continuous spectra.

The lowest energy levels in most atoms are at negative energies of hundreds and thousands of electronvolts. Photons emitted for transitions to these levels have wavelengths in the X-ray region.

In an X-ray tube a high-energy electron beam strikes a target. The sudden deceleration of the electrons causes a continuous X-ray spectrum to be emitted. In addition, the impacts cause the target atoms to be excited. As they fall back to the lowest energy states, a line spectrum of X rays is also emitted.

Photons can be absorbed by an atom if their energy is exactly correct. It must correspond to the energy needed to throw the atom from the ground state to an excited state. Therefore, the wavelengths absorbed by atoms (the absorption spectrum) make up a line spectrum.

The laser produces an intense, accurately monochromatic, highly coherent beam. The beam consists of parallel rays, and so its pencillike quality persists over long distances.

Questions and Exercises

1. Explain why, qualitatively, you would expect the current in the photocell of Figure 29.3 to vary linearly with the light intensity.
2. Describe an experiment you might do to determine the work function of a surface. Assume you have a variable wavelength light source available. How could you measure the KE of the fastest electrons emitted from the surface in a phototube?
3. A student takes home a test tube of hydrogen gas he has just made in chemistry lab. He goes into a dark closet with the tube and expects to see the gas emit the reddish blue glow of the Balmer series. Instead he only sees darkness. What went wrong?
4. Using the photon concept, suggest reasons why one would expect the following generalization to be true: The possible damage to humans caused by em radiation increases with decreasing wavelength.
5. Most doctors are hesitant to prescribe an X-ray of the gastrointestinal tract for a pregnant woman. An arm or head X-ray is prescribed with less hesitation. Why? Would similar attitudes extend to young men and to women who are not pregnant?
6. Hydrogen atoms are bombarded by electrons that have fallen through 12.8 V. Which lines of the Lyman and Balmer series will be emitted by the atoms?
7. Ordinary hydrogen atoms absorb only wavelengths of the Lyman series. Extremely hot hydrogen gas absorbs wavelengths of the Balmer series also. Why this difference? This forms the basis for one method for measuring the temperature of very hot gases.
8. The spectrum of the sun is continuous. However, sunlight reaching the earth is missing several wavelengths in the visible. These are called the Fraunhofer lines. Explain their origin. There are many helium atoms between the sun and earth. The existence of helium was first learned from these absorption lines in the sun's spectrum.
9. An iron alloy is suspected to contain about 0.5 percent chromium. Devise a method for checking this. Assume that fairly complete chemical and physical laboratories are available to you. Repeat for the case in which a tank of benzene is suspected of having a slight quantity of purple dye in it.
10. As a rough generalization, the ability of a material to stop X rays is proportional to the number of electrons per unit volume of the material. Why is lead a better shield against X rays than aluminum is? How well would you expect plastics to act as an X-ray shield?
11. A glass tube contains two electrodes at its ends and is filled with air. The electrical resistance across the tube is essentially infinite because the gas is nonconducting. However, when a strong beam of X rays shines into the tube, the gas becomes conducting. Explain why. Would a strong beam of visible light cause the same effect?

Problems

1. Find the ratio m/m_0 for an object at the following speeds: (a) $0.10c$, (b) $0.30c$, (c) $0.70c$, (d) $0.90c$.
2. How fast (relative to c) must an object be moving so that $m = 10m_0$?
*3. How much energy, in electronvolts, must an electron have if $m = 10m_0$ for it? Through how large a potential difference must it be accelerated to obtain this energy?
*4. A proton is accelerated through a potential difference of 5×10^6 V. Find m/m_0 for it.
*5. In present-day nuclear accelerators protons can be accelerated through 10^9 V. (a) What is the ratio m/m_0 for such a proton? (b) How fast, in comparison to c, is the proton moving?
*6. A car moving at 20 m/s will have $m/m_0 = 1 + 2 \times 10^{-15}$.

*Problems marked with an asterisk are not as easy as the unmarked ones.

Show that this is true. [Hint: For small values of x, $(1 - x)^a$ is approximately equal to $1 - ax$.]

7. An object with a rest mass of $m_0 = 3.0$ kg is lifted a distance of 5.0 m. (a) By how much does this increase its mass? (b) What fraction of m_0 is this?

**8. Show that for $v \ll c$, the relation $(m - m_0)c^2$ reduces to $m_0v^2/2$.

9. A certain radio station radiates 5000 W of power at a frequency of 1.2×10^6 Hz. (a) What energy photons does the station radiate? (b) How many photons does it send out each second?

10. A certain helium-neon laser sends out a beam of red light with $\lambda = 6328 \times 10^{-10}$ m. The beam has a power of 1.00 mW. (a) What is the energy of each photon in the beam? Give your answer in joules and electronvolts. (b) How many photons are emitted each second?

*11. It is known that em radiation with λ = 1240 nm has 1.00-eV photons. Because photon energy is inversely proportional to wavelength, other photon energies (or wavelengths) can be found by inverse proportion. (a) Show that the photons in a 1240-nm beam have energies of 1.00 eV. (b) What is the wavelength of a 3.50-eV photon? (c) What do we call em radiation that has λ = 1240 nm? (d) What type contains 3.50-eV photons?

12. The work function of tungsten is 4.5 eV. What is the photoelectric threshold wavelength for tungsten? (Remember that the work function is the least possible energy that can tear an electron loose from the material.) What type of radiation is this?

13. The photoelectric threshold wavelength for pure sodium metal is 546 nm. How much energy (in electronvolts) is needed to tear an electron away from this metal?

14. Sunburn occurs when photons in sunlight strike skin molecules and break them apart. (In a burn caused by a hot object, enough thermal energy is given to the skin molecules so that they vibrate widely enough to rupture.) To break a chemical bond in flesh, an energy of about 3.5 eV is needed. (a) What wavelength photons have this energy? (b) Which causes sunburn—visible, infrared, or ultraviolet light?

*15. The photoelectric threshold for pure sodium metal is 546 nm. Suppose 480-nm light strikes sodium metal. (a) How much energy will the fastest electron emitted from the metal have? (b) What will be its speed?

16. A particle moving at a speed of 5×10^6 m/s has how large a de Broglie wavelength if the particle is (a) an electron and (b) a proton?

17. A particle is accelerated through 200 V. Find its speed and de Broglie wavelength if it is (a) an electron and (b) a proton.

18. The yellow light given off by a sodium lamp has a wavelength of about 589 nm. It is emitted as the atom falls from one energy state to another. What is the energy difference between these two states?

19. When the fourth line of the Balmer series (λ = 410 nm) is emitted, the atom falls from the $n = 6$ to the $n = 2$ state. What is the difference in energy between these two states?

20. What is the wavelength of the third line in the Lyman series?

*21. Hydrogen atoms are usually found in the ground state, that is, the electron is in the $n = 1$ level. What wavelength light striking the atom will excite it to the $n = 2$ level?

*22. Refer to Figure 19.6 and the second spectrum shown. The dark lines are at the positions of wavelengths that are highly absorbed by the outer layers of the sun and the material between the sun and the earth. (a) What is the starting state and the final state of the hydrogen atom that gives rise to the line labeled H on the right? (b) Repeat for the line labeled H on the left.

23. When one electron is torn away from a helium atom, the result is a helium ion. This ion has one electron and a nuclear charge of $+2e$. Therefore, except for the nuclear charge, the ion is much like a hydrogen atom. The energy-level diagram for a helium ion (Figure P29.1) is similar to the diagram for

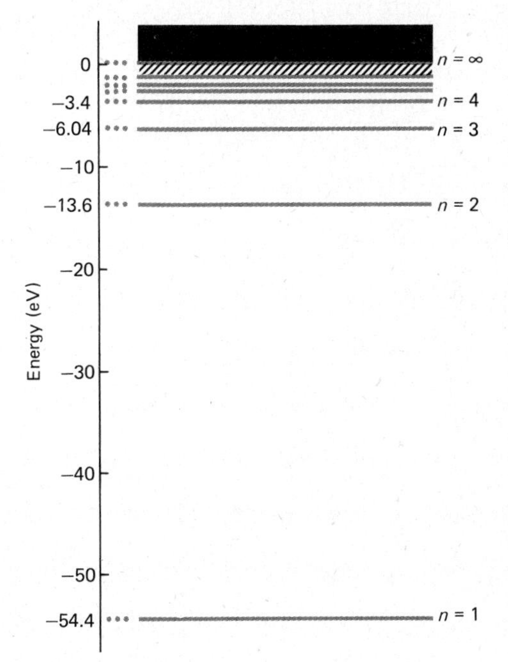

Figure P29.1 Energy-level diagram for singly ionized helium.

**Problems marked with a double asterisk are somewhat more difficult than the average.

hydrogen. Using it, find the wavelength of the first line of the Lyman series for the helium ion.

24. Refer to the energy-level diagram for ionized helium in Figure P29.1. Using it, find the first line of the Balmer series for this ion. What is the series limit of the Balmer series?
25. Refer to the energy-level diagram for a singly ionized helium atom in Figure P29.1. The ion is normally in the $n = 1$ or ground state. (a) How much energy is needed to remove the second electron from the ion? (b) What is the wavelength of the Lyman series limit for the ion?

*26. A continuous spectrum of em radiation passes through a gas composed of singly ionized helium atoms. The energy-level diagram for these ions is shown in Figure P29.1. Normally the ions are in the $n = 1$ or ground state. Give the two longest wavelengths that the gas will strongly absorb.

27. A nitrogen molecule (N_2) can be pictured as two balls (the two atoms) at the two ends of a spring (the bond between the atoms). The molecule can vibrate with an alternate stretching and compressing of the spring. Two ordinary balls at the ends of a spring can be considered to vibrate with any energy. But when the balls are of atomic size, it turns out that only certain energies of vibration are possible. The lowest possible vibration energy is 0.147 eV, as shown in the energy-level diagram of Figure P29.2. It can also vibrate with other energies, as shown. Notice that the spacing between these lowest energy levels is 0.293 eV. (a) What wavelength radiation does the N_2 molecule emit as it falls from one level to the next lowest level? (b) What is this type of em radiation called?
28. Refer to the description given in Problem 27 for the energy-level diagram of Figure P29.2. (a) What wavelength of radiation will nitrogen molecules absorb as they are lifted from one vibrational state to the next higher one? (b) In what portion of the em spectrum is this? (c) Infrared absorption and emission spectra are used to identify molecules. How can N_2 molecules be identified in this way?
29. In a certain TV set the electron beam is accelerated through a potential difference of 20,000 V. What is the shortest-wavelength X ray emitted as the beam strikes the end of the picture tube? Fortunately, very little of this radiation can penetrate 0.5 cm of glass. Even thinner sheets of metal will effectively stop the X-radiation. Longer-wavelength X rays are much more easily stopped.

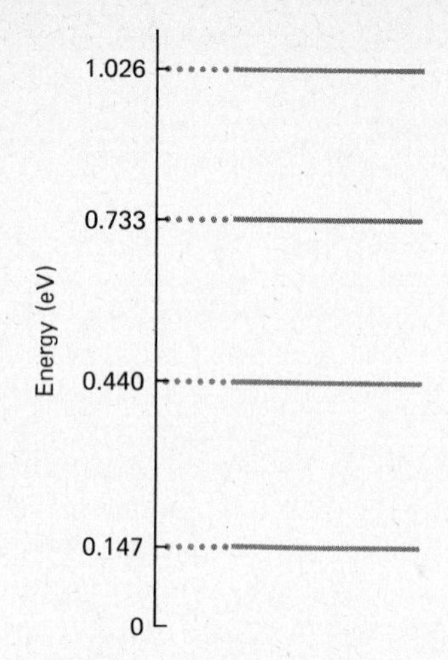

Figure P29.2 The nitrogen molecule can vibrate with only certain energies, the lowest of which are shown in the diagram.

30. Some hospitals use a device called a betatron for radiation therapy. A typical betatron accelerates electrons to an energy of 25 MeV. These electrons strike a target material and cause X rays to be emitted. What is the shortest-wavelength X ray emitted from the device? After passing through about 10 cm of lead, such a beam is reduced to about 0.003 of its original intensity. About 100 cm of water would be required for this same loss in intensity.

**31. In a certain spectral range an atom gives rise to the following spectral series: 72.9, 54.0, 48.2, 45.6 nm, with the series limit at 40.5 nm. Find a set of energy levels within the atom that could give rise to this series.

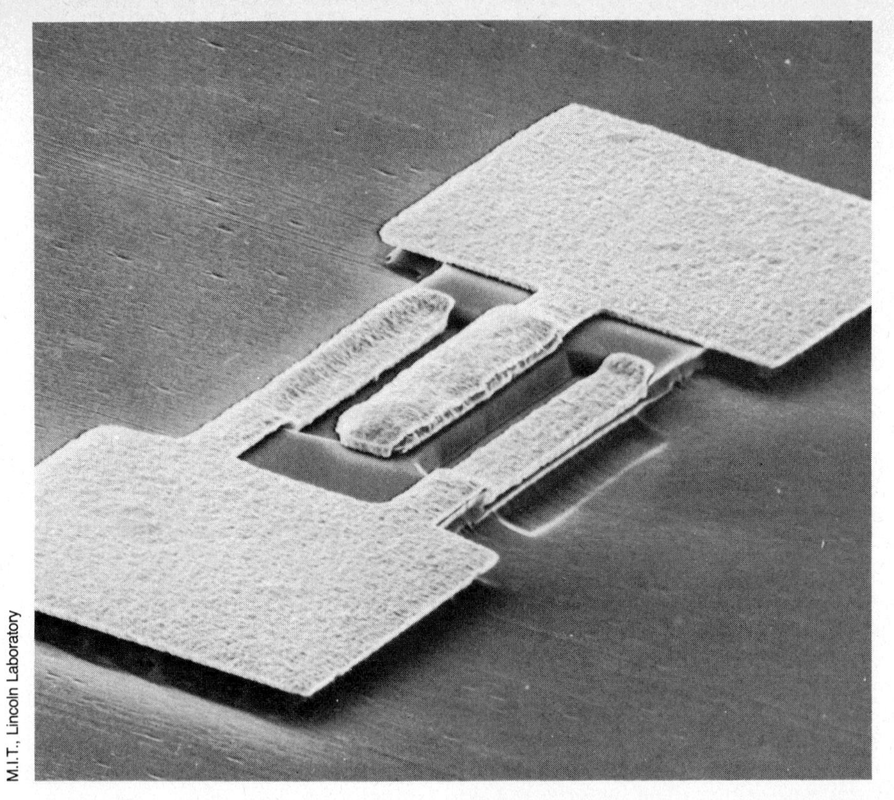

M.I.T., Lincoln Laboratory

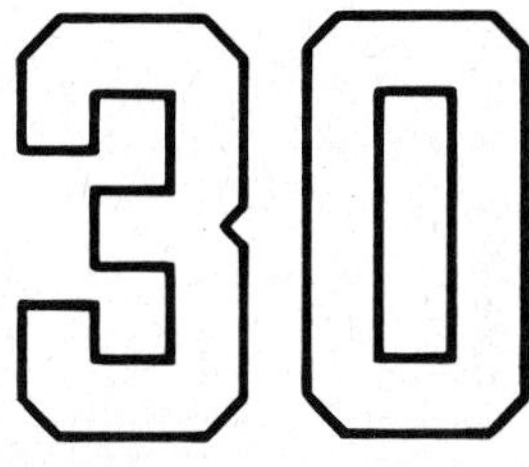

30 SOLID STATE ELECTRONICS

Performance Goals

When you finish this chapter, you should be able to

1. Define the following:
   a. Intrinsic semiconductor
   b. *n*-type semiconductor
   c. *p*-type semiconductor
   d. Hole
   e. Semiconductor diode
   f. Depletion region
   g. Forward and reverse bias
   h. Zener diode
   i. LED
   j. Photodiode

The field of electronics entered a period of rapid change in the late 1940s with the development of the transistor. This device blazed the way for the wide use of solid state electronics. Within a short time the vacuum tube circuits of previous years were largely replaced by semiconductor devices. As our mastery of semiconductor electronics grew, a revolution in practical electronics occurred. This revolution has led to the wide use of the computers and electronic control devices that we see in all branches of modern technology.

k. Solar cell
l. Transistor
m. Alpha of a transistor
n. IC

2. Sketch the energy-band diagram for a typical conductor and a typical nonconductor. Use it to explain their behavior.
3. Compare the energy-band diagrams for a nonconductor, an intrinsic semiconductor, an *n*- and a *p*-type semiconductor.
4. Explain which dopants give *n*-type semiconductors and which give *p*-type.
5. Explain why a semiconductor diode conducts current better in one direction than in the other.
6. Describe how the depletion region arises and why it behaves in some ways like a capacitor.
7. Give a typical use of a Zener diode in a circuit.
8. Explain what an LED is and how it can be used to display numbers.
9. Explain how exposure to light gives rise to a voltage in a solar cell.
10. Compare the currents in the emitter, collector, and base wires of a transistor and explain why the base current is usually very small.
11. Explain how a transistor can be used for current, voltage, and power amplification.

30.1 Conductors and Nonconductors

The chemical properties of an atom are controlled by its outer electrons. We picture the electrons of an atom to exist in shells, or layers. The innermost electrons are tightly bound to the nucleus. They are not easily torn from the atom and so they do not influence its chemical behavior. Only the electrons in the outermost electronic shell can be freed with energies of a few electronvolts or less. We call these outer shell electrons *valence electrons.* Solid state electronics is concerned mainly with them.

When atoms are packed together in a solid, the behavior of the valence electrons determines the electrical properties of the material. In the case of metals the few valence electrons of each atom escape from the parent atom. As a result, a block of metal acts like a box with an electron gas within it. The valence electrons, essentially free, travel through the entire metal block much like gas molecules in a box would do. Under the influence of an applied electric field, the free electrons move in the way we have discussed in our study of electricity. Materials such as these are good conductors of electricity. Currents flow readily in them.

Many other materials, however, do not conduct electricity. For example, diamond, which is pure crystalline carbon, is a nonconductor. In a material such as this, the valence electrons are still firmly held by the atoms even in the solid. Because there are no free electrons, an applied electric field cannot cause a current to flow. The material is a nonconductor.

But to show you that the situation is not simple, consider the case of graphite. It, too, is pure carbon, but unlike diamond it conducts electricity and so is a conductor. Apparently, the crystalline form in which the atoms are arranged influences their electrical behavior. It turns out that the graphite crystal has a layered structure and electrons can move more or less freely within a layer. Graphite crystals conduct electricity along the layer direction, but they are poor conductors in the perpendicular direction.

Most materials we think of as nonconductors, unlike diamond, consist of more than one type of atom. Plastics, for example, usually contain carbon, hydrogen, and oxygen atoms. Glass contains silicon and oxygen plus other atoms. What can we say about the valence electrons in these materials? No matter how complicated the atomic and molecular structure, the valence electrons are still held firmly to the atoms. These materials have no free electrons.

30.2 Energy-Level Diagrams for Solids

Because of the wide diversity of materials, it proves convenient to use energy-level diagrams to describe the behavior of their electrons. As you recall from the last chapter, such diagrams show the allowed energies of the electrons. In the last chapter we spoke only of the electrons in a single atom. Now, however, we will use such diagrams to describe the electrons in the multitude of atoms that make up the solid.

We saw in Section 29.12 that the energy levels of the atoms change as the atoms are brought together to form a solid. The higher energy levels widen into bands. For example, Figure 30.1(a) shows the energy levels of an isolated sodium atom. When the atoms are joined to form a solid piece of sodium metal, the levels become as shown in Figure 30.1(b). The $n = 3$ level has widened into a band that extends all the way to the $E = 0$ level. In other words, the solid has a continuous band of allowed electron energies extending from about -9 eV to freedom, $E = 0$. However, the $n = 2$ level is about the same in the solid as for the isolated atom. Let us now examine this situation in more detail.

The sodium atom has eleven electrons. The $n = 1$ level can hold only two electrons. So two of the eleven electrons are in it. The $n = 2$ level can hold only eight electrons. It therefore accounts for eight more of the eleven electrons. The eleventh electron is left over and must be placed in the $n = 3$ level. It is sodium's single valence electron. Because only about 5 eV is required to lift it to the $E = 0$ level and freedom, this electron is easily removed. Hence sodium atoms ionize readily and are quite reactive chemically.

In the solid the situation is the same for the $n = 1$ and $n = 2$ electrons. These ten electrons are tightly bound to the atom and their energies are virtually unchanged. However, the single $n = 3$ electron, the valence electron, is free from its parent atom, although it is still confined to the solid. The electron is no longer confined to a definite, fixed energy level. The $n = 3$ level has widened into a band. Therefore, the free valence electron can take on any energy above level B in Figure 30.1(b). As a result, currents can be set up in the solid by means of electric fields that accelerate the free electrons.

There is another important point we should mention about the electrons in an energy band. Only a certain number of electrons can exist in each atomic energy level. There are only two $n = 1$ electrons, eight $n = 2$ electrons, and so on. Similarly, the electrons in a solid cannot all have the same energy. What this means in the present case is that not all the $n = 3$ electrons can fall to level B, the bottom of the band. At absolute zero of temperature, when the electrons all have their lowest possible energy, the band is filled with electrons from level B

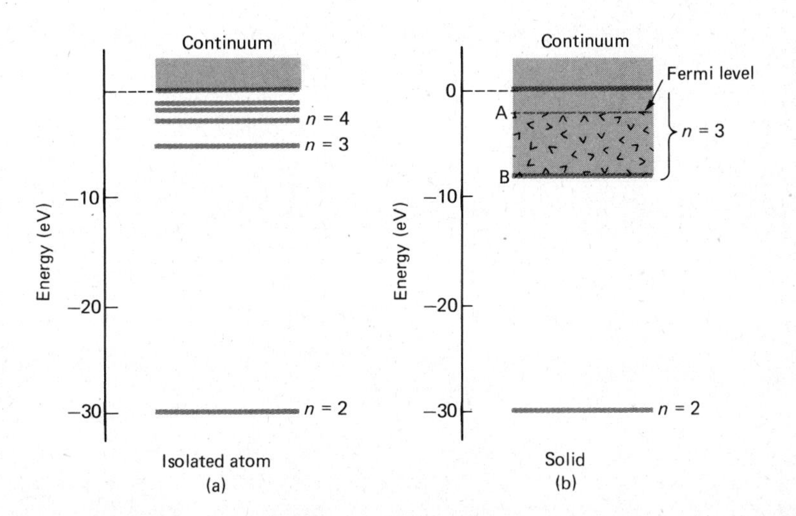

Figure 30.1 The energy-level diagram for isolated sodium atoms is shown in (a) and for a solid in (b). Levels 7 to ∞ are not shown, nor is the $n = 1$ level at about -80 eV.

to level A. But above A, the band is empty. We call the energy at level A the *Fermi energy*, E_F, of the system. It is the highest electron energy for a system in which the electrons have their lowest possible energy.

Suppose sodium metal is heated above absolute zero. Then some of the electrons near level A in Fig. 30.1(b) will obtain thermal energy. They thus spill over into the unoccupied levels in the region above level A. It is only at absolute zero that level A is a sharp boundary between filled and vacant levels. Thermal energy blurs this boundary at higher temperatures. But because the average thermal energy at 300 K is less than 0.04 eV, the blurring would be scarcely noticeable on the scale of this diagram.

30.3 Intrinsic Semiconductors

The example we have just given, sodium, is typical of conductors. Let us now consider a nonconductor. For simplicity, let us consider the atom that contains one less electron than sodium. It is neon. As you know, this atom is nonreactive because it does not have an easily freed valence electron. It has eight $n = 2$ electrons and none in the $n = 3$ state.

Although neon liquifies only at low temperatures, let us consider a solid composed of it. Its energy diagram would not differ much from Figure 30.1(b). But in this case there would be no electrons in the $n = 3$ band. We call the $n = 3$ band the *conduction band.* The $n = 2$ level would be completely filled. We show the situation on a larger scale in Figure 30.2.

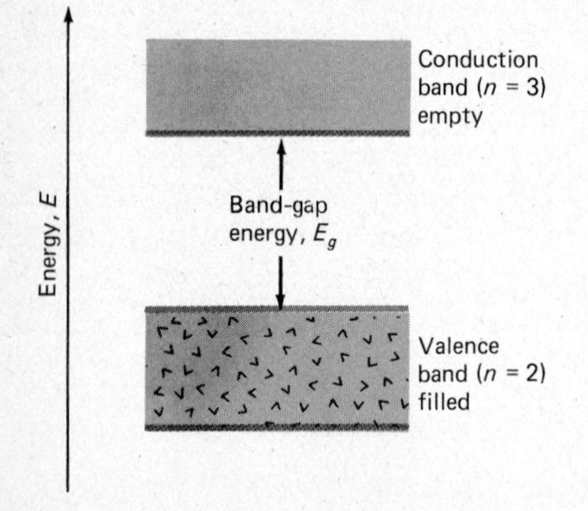

Figure 30.2 If the band-gap energy is a few electronvolts, this material is a nonconductor.

The $n = 2$ level has widened slightly into a band. It contains the eight $n = 2$ valence electrons for the neon atom. We call this the *valence band.* This band is completely filled just as the $n = 2$ level was filled for the isolated atoms. Moreover, these electrons are tightly held by each atom even in the solid. To free an electron, the electron must be lifted to the $n = 3$ band, the conduction band. In other words, to obtain a free electron for conduction through the solid, an energy equal to the band-gap energy, E_g, must be furnished to an electron in the valence band. But E_g is much larger than 0.04 eV, ordinary thermal energies. Hence, essentially no electrons can be excited to the conduction band by thermal motion. Lacking electrons in the conduction band, the solid has no free electrons and is therefore a nonconductor.

All nonconductors have energy-level diagrams similar to this. They have a filled valence band and an empty conduction band. The energy gap between the bands is too large for thermal motion to excite electrons from the filled to the empty band. Impressed electric fields are unable to give small energy to the electrons because the only allowed energies are at much higher energies, namely, an energy E_g higher. As a result, there are essentially no free electrons in these materials and they are electrical nonconductors.

However, certain materials have a much smaller band-gap energy. For example, in pure silicon it is 1.1 eV and in germanium it is 0.7 eV. These materials have filled valence bands and empty conduction bands at absolute zero. They are therefore nonconductors. But at room temperature a few electrons are excited from the valence band to the conduction band by thermal energy. As a result, these materials show some conductivity at moderate temperatures. Because their conductivity is much smaller than that of metals, we term them semiconductors. Moreover, the behavior is characteristic of (or intrinsic to) the pure material and so we call them *intrinsic semiconductors.*

The conductivity of these materials increases with increasing temperature. The higher the temperature, the more electrons will be thrown into the conduction band. This effect is often pronounced. Conductivities may increase by a factor of ten when the temperature is increased from 0°C to 40°C.

The temperature sensitivity of semiconductors is utilized in devices called *thermistors.* They are essentially semiconductor resistors. As the resistor is heated, its resistance changes greatly with temperature. Its resistance can be used as a measure of temperature. Moreover, because they are electrical in nature, they are easily incorporated in electric circuitry. Thermistors are used in temperature control devices, in safety switches to prevent overheating, and as short time lag thermometers.

Another feature we should mention is shown in Figure 30.3. This is Figure 30.2 redrawn to show schematically what happens when an electron is excited from the valence band to the conduction band. The excited electron is free to roam throughout the conduction band. But the electron leaves behind a vacancy in the otherwise filled valence band. We call this a *hole.* As we shall see later, the hole is capable of carrying current just like the free electron.

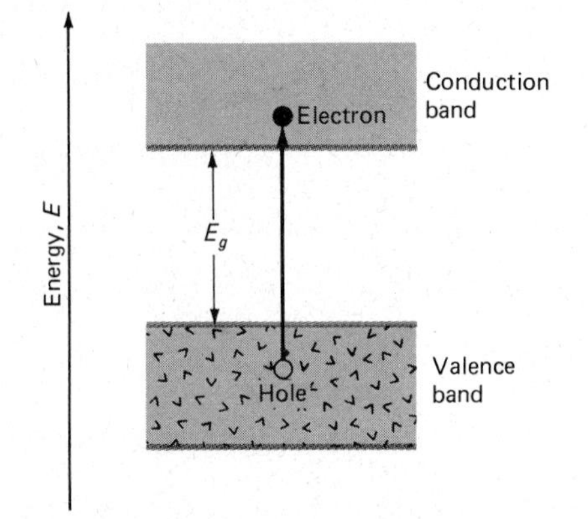

Figure 30.3 The electron excited from the valence band to the conduction band leaves behind a hole. It, too, can carry current.

EXAMPLE 30.1 Find the average kinetic energy and speed of a free electron at 300 K.

Solution Equation 12.7 gives the relation between the KE and temperature of a gas molecule of mass m_0 as

$$\tfrac{1}{2}m_0v^2 = \tfrac{3}{2}kT$$

Applying this to an electron gas gives

$$\begin{aligned} \text{KE} = \tfrac{1}{2}m_0v^2 &= (\tfrac{3}{2})(1.38 \times 10^{-23}\ \text{J/K})(300\ \text{K}) \\ &= 6.2 \times 10^{-21}\ \text{J} = 0.039\ \text{eV} \end{aligned}$$

Solving for v gives

$$v = \sqrt{\frac{6.2 \times 10^{-21}\ \text{J}}{\tfrac{1}{2}(9.1 \times 10^{-31}\ \text{kg})}} = 1.2 \times 10^5\ \text{m/s}$$

30.4 *n*-Type Semiconductors

Absolutely pure intrinsic semiconductors are difficult to produce commercially. Moreover, E_g, the gap energy, is usually too large for widespread utilization of these materials. For these reasons other types of semiconductors have been designed and manufactured.

Most commercial semiconductor devices are made using the atoms of silicon

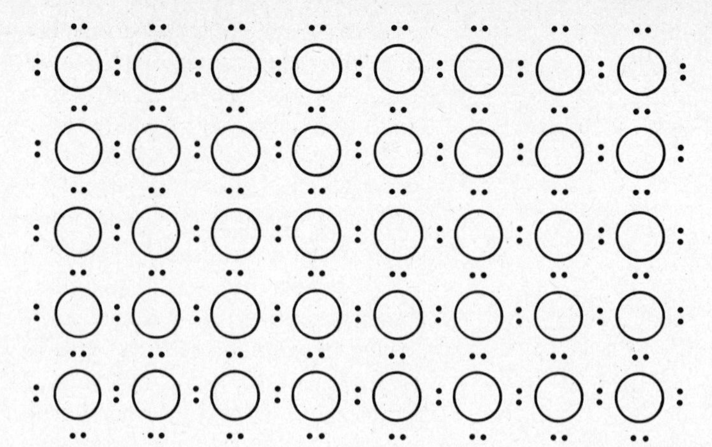

Figure 30.4 In the lattice each atom of silicon (or germanium) shares electrons to achieve a stable, filled outer shell.

or germanium. These atoms, as well as those of carbon, have four electrons in their outer atomic shell. For the shell to be complete, four more electrons are needed. When these atoms are crystallized as a solid, they can form a diamondlike lattice. In it the atoms share electrons, as indicated schematically in Figure 30.4, in such a way that, in effect, they have eight electrons in their outer shells. This configuration is stable, and the electrons are held rather tightly in it.

The energy needed to free one of the electrons from a silicon atom in the pure crystal is 1.09 eV, that is to say, E_g in Figure 30.3 is 1.09 eV. (For carbon in diamond it is 7 eV. For germanium it is 0.72 eV.) Because this energy is large compared to thermal energy, scarcely any electrons are knocked free from their parent atoms. The intrinsic conductivity of pure silicon is therefore low. At room temperature the conductivity of silicon is only a factor of about 10^{-11} that of copper.

To make silicon a higher conductivity semiconductor, it is "doped" with impurity atoms. By adding only a fraction of a percentage of the proper impurity atom, the conductivity can be increased by many powers of ten. To produce a so-called *n*-type semiconductor, the impurity atom is so chosen that it adds free electrons to the crystal lattice. In the case of silicon the impurity atoms frequently added are phosphorus, arsenic, and antimony. These are all atoms that have a valence of five, that is to say, they have five electrons in their outer shell. Let us illustrate their effect by using arsenic as an example.

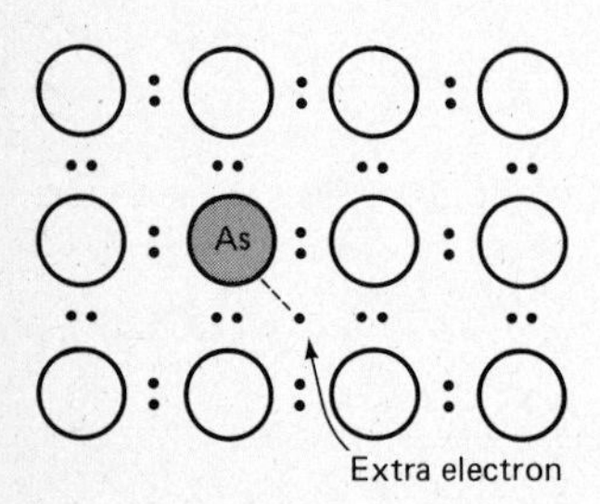

Figure 30.5 The arsenic atom contributes an extra electron to the silicon lattice.

An arsenic (As) atom is much like a silicon atom. When arsenic is used to dope silicon, the arsenic atom (called the dopant atom) fits nicely into the silicon crystal lattice, as shown in Figure 30.5. Because the arsenic atom has five outer electrons, it has one more electron than is needed by the perfect silicon lattice. As a result, this extra electron is held loosely in the lattice and escapes easily under thermal motion. It is then free to wander through the crystal. In this way free electrons are added to the crystal by the impurity atoms, and the crystal becomes a much better electrical conductor.

The charge carriers in the doped crystal of Figure 30.5 are electrons, negative particles. We call such a semiconductor an *n-type semiconductor* because the charge carriers are negative.

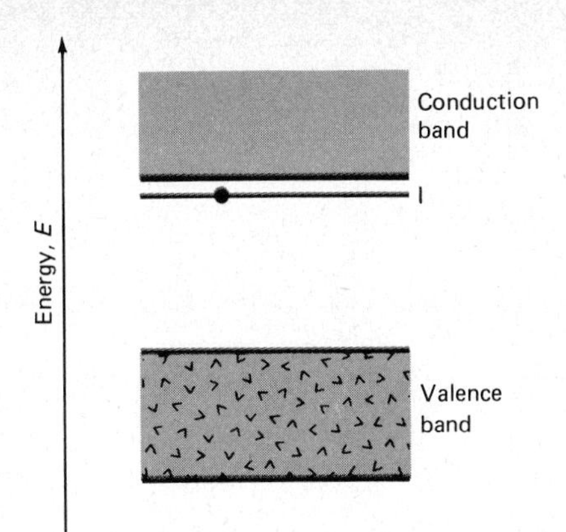

Figure 30.6 In an *n*-type semiconductor the impurity level, I, is close to the conduction band. Thermal energy easily excites its electron to the conduction band.

The addition of the dopant changes the energy level diagram for the system. The new energy-level diagram is shown in Figure 30.6. Because the extra electron supplied by the dopant atom is easily freed, its energy level is only slightly below the conduction band. Therefore the arsenic atom contributes the impurity level labeled *I*. The electron in it, the nearly free extra arsenic electron, is easily excited through the small gap between *I* and the conduction band. For this reason, the extra arsenic electron is normally excited to the conduction band except at the very lowest temperatures.

30.5 *p*-Type Semiconductors

An entirely different type of semiconductor is formed when a trivalent impurity atom is added to the silicon lattice. The impurity shown in Figure 30.7 is gallium (Ga); aluminum, boron, and indium are also used. These atoms have only three valence electrons. Because the lattice requires four valence electrons to be complete, there is an electron vacancy.

As you might guess, only a small energy is required to cause one of the nearby lattice electrons to move over into the vacancy next to the gallium atom. The result is shown in Figure 30.7(b). A vacant electron position still exists in the lattice, but it is no longer at the site of the gallium atom. As before, the vacancy left by an electron is called a *hole*. (A little thought will convince you that this is the same as the definition we gave earlier.) A nearby electron will soon fill this vacancy, so that in effect the hole wanders about throughout the crystal lattice as indicated in part (c) of the figure.

It is obvious that charge motion occurs in the crystal as the hole moves from place to place. But there is a more subtle point we should also notice. The moving hole acts like a positive charge moving through the crystal. Each atom of the crystal is originally electrically neutral. When an electron from another atom moves into the hole, it leaves its parent atom charged positively, that is, the presence of the hole means that the atom has a positive charge. As the hole moves from atom to atom, it in effect carries a positive charge with it. The holes act like movable (free) positive charges in the crystal. As a result, current can flow in a semiconductor doped with trivalent impurity atoms. We call this type of semiconductor a *p-type semiconductor* to indicate that the charge carriers are positively charged holes.

Figure 30.7 The electron vacancy can wander away from the impurity atom. It becomes a mobile hole within the lattice.

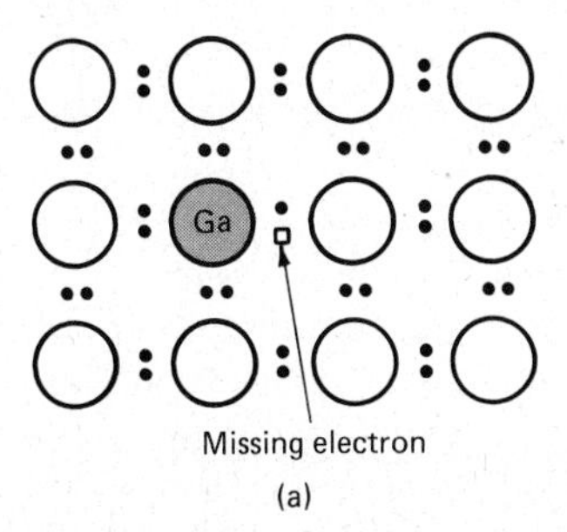

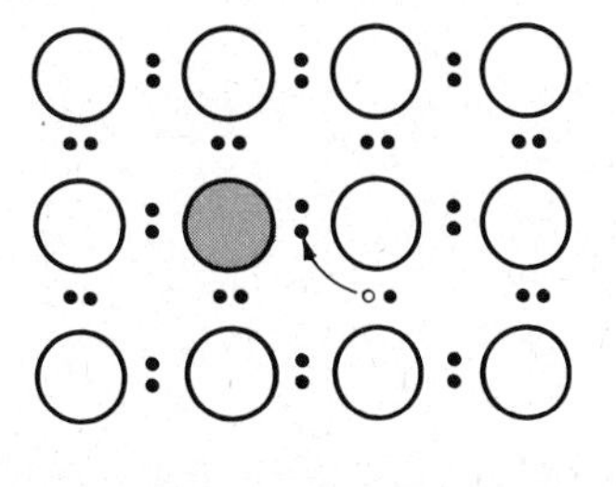

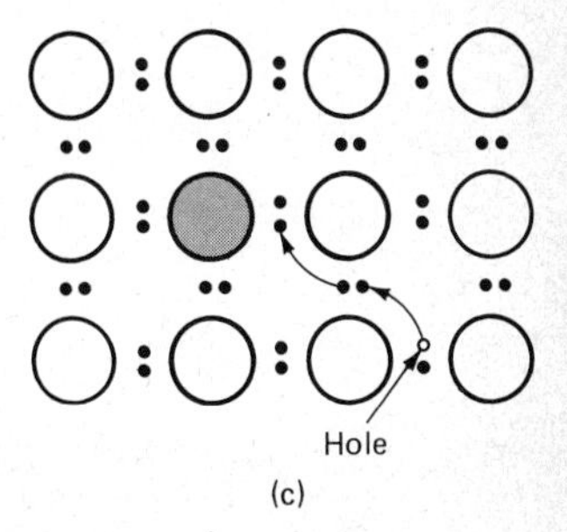

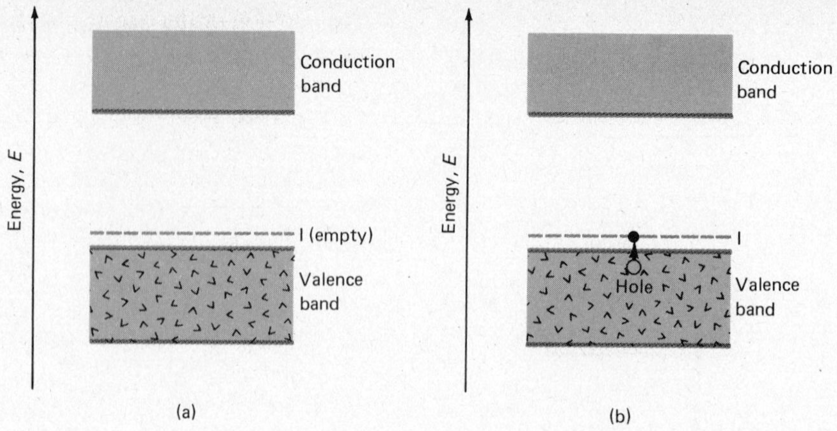

Figure 30.8 In a p-type material the impurity level is close to the valence band.

We can summarize our discussion in the energy-level diagrams in Figure 30.8. The vacant site furnished by the gallium atom contributes the vacant impurity level I. Notice how close the impurity level is to the valence band below it. This reflects the fact that only a small energy is required to cause one of the shared electrons to move from its parent atom to the electron vacancy. Because the empty impurity level I is so close to the valence band, thermal energy easily throws an electron from the valence band to the impurity level. This leaves behind a hole in the valence band. This hole acts as a free positive charge that is able to move throughout the crystal.

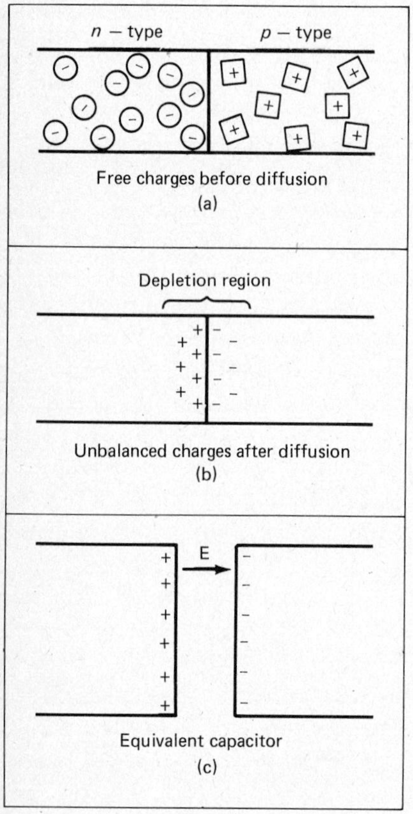

Figure 30.9 A charge double layer exists at a p-n junction.

30.6 The Semiconductor Diode

The p-type and n-type semiconductors we have just described are the materials used in solid state electronics. Let us begin our study of their function by discussing the solid state diode rectifier. (We discussed the use of this device in circuits in Chapter 28.)

Suppose an n-type semiconductor is fused to a p-type semiconductor to form what is called a p-n junction. As shown in Figure 30.9(a), the free charges are electrons (negative) in the n-type semiconductor and holes (positive) in the p-type. Remember that these are the only charges that can move freely in these materials.

A word of caution should be given concerning the representation shown in Figure 30.9(a). One must remember that the two materials are electrically neutral and so the free charge shown is not a net charge on the material. The charges shown are only those that are free to move. They are compensated by an equal bound charge of opposite sign.

Except for the tiny amount of impurity, the n- and p-type materials are identical. Hence it is possible for some of the free electrons to wander out of the n-type material and into the p-type. (We say that they have *diffused* into the material.) Similarly, some of the holes diffuse into the n-type material. This causes a profound change in the junction. Before diffusion each material is electrically neutral. But after diffusion the electrons have carried a net negative charge to the *p*-type material, and the holes have carried a net positive charge to the *n*-type material.

The result is shown in Figure 30.9(b). A charge *double layer* exists at the boundary. This double layer is analogous to the two plates of a capacitor as shown in (c). Notice that the electric field at the junction opposes further diffusion. Positive holes are repelled to the right and negative free electrons are repelled to the left. As a result, diffusion stops after the double layer has acquired enough charge. Thus an equilibrium electric field is set up at the junction. Of course, the left side of the double layer region has lost electrons, and the right side has lost holes. We therefore refer to this region as the *depletion region*. Typically it is about 10^{-5} cm thick. (The diagram exaggerates its width.)

Let us see what happens when we impress a voltage on the junction. One possibility is shown in Figure 30.10(a). The polarity of the battery is such that the electrons are forced to the right at the junction, while the holes are caused to move to the left. This is the direction in which the charges diffuse if there is no battery present, and so it is a direction of easy charge motion. The electrons and holes wish to move to the junction where they can combine with each other. Unlike the situation in Figure 30.9, the connection to the battery prevents net charge from building up on the semiconductors. Therefore, a current will flow continuously. The battery furnishes electrons to the *n*-type material and removes from the *p*-type those that have moved across the junction into the *p*-type material. As we see, for a battery polarity as shown, called *forward bias*, current flows readily through the diode.

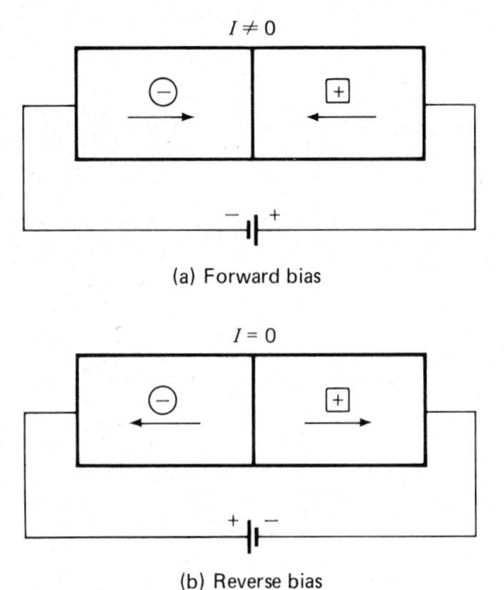

Figure 30.10 The *p-n* junction acts as a rectifier.

In Figure 30.10(b) we see what happens when the battery is reversed, the *reverse bias*. Now the battery tries to pull the electrons and holes apart. In so doing, it leaves oppositely charged regions on the two sides of the junction. This causes a potential difference between the two sides of the junction. In a fraction of a second this potential becomes equal to the battery voltage, and so the current stops. As we see, under reverse-bias conditions, the diode does not pass current.[1] The diode can therefore be used as a rectifier, as discussed in Chapter 28.

We can summarize our discussion in terms of the *characteristic curve* for a diode. It is the curve relating current I through the diode to the voltage V across it. The curve for a typical diode is shown in Figure 30.11. As we see, the diode is a good current rectifier. Let us now look at a few other uses of the *p-n* junction.

30.7 The Varactor

As stated earlier, two oppositely charged regions occur on the two sides of a *p-n* junction. They are at the two sides of the depletion region. (A similar situation occurs when the junction is reverse biased.) These two charged regions act like the two plates of a capacitor. Hence the junction has a definite capacitance

[1] A small reverse current does flow. It is due to motion of charges other than the free electrons and holes we have been discussing.

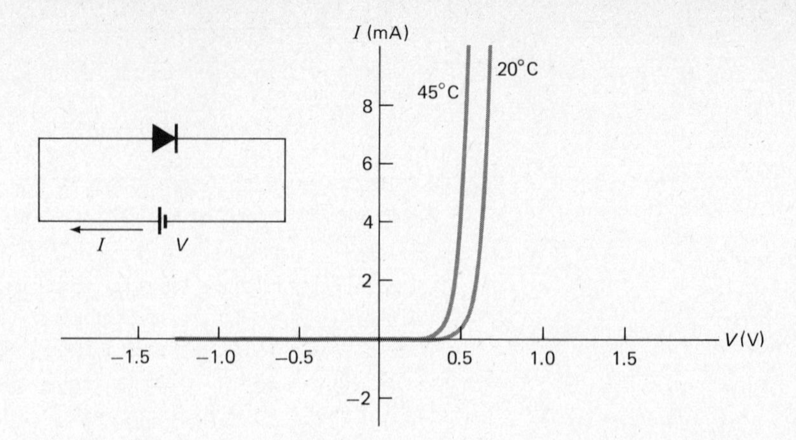

Figure 30.11 The voltage-current curve for a silicon diode. (A reverse current of about 10^{-12} A occurs, but it is too small to be shown.)

whose magnitude depends on the width of the depletion region. But this width can be changed by impressing a reverse-bias voltage on the junction. We therefore have a means for obtaining a capacitor whose magnitude can be controlled by an impressed voltage. Junctions made for this purpose are called *varactors* or *varicaps.* Their capacitance decreases with increasing reverse bias because the width of the depletion region (i.e., the distance between plates) increases with reverse bias.

30.8 The Zener Diode

Figure 30.11 for the current through a diode only applies if the reverse voltage is not too high. Very little current flows through the diode at low reverse voltages. But if the reverse voltage is made large enough, the diode acts as though it is breaking down, as shown in Figure 30.12. The high reverse bias causes electrons to be torn loose from the parent atoms in the lattice. Moreover, accelerated by the high voltage, freed electrons strike other atoms and free still more electrons. This action causes an avalanche of charge to flow through the crystal. The voltage at which this breakdown occurs is determined by the construction and material of the diode. Diodes with breakdown voltages ranging from a few volts upward can be purchased. These diodes are called *Zener diodes,* after the person who first explained the basis for their behavior.

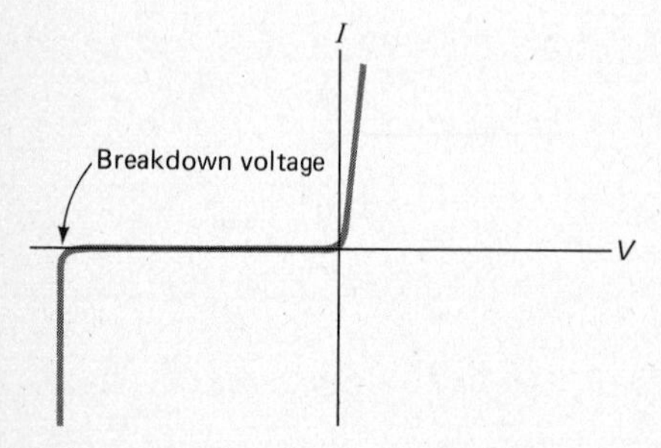

Figure 30.12 In the Zener diode breakdown occurs at a well-defined reverse bias.

Zener diodes are useful in voltage control applications. These uses are based on the fact that the current changes extremely rapidly with voltage in the breakdown region. We can see its basic operation by referring to Figure 30.13. (Notice the symbol used for the Zener diode.) A continuously increasing input voltage is converted to a voltage that reaches a constant value. When the input voltage exceeds the breakdown voltage, the Zener diode still maintains the breakdown voltage across the output. The resistor is needed to supply the difference in voltage drop between the source and the output. If it were not there, the current through the diode would become extremely large (trying to supply the needed voltage drop), and the diode would burn out. There are many other circuits in which the Zener diode can be used.

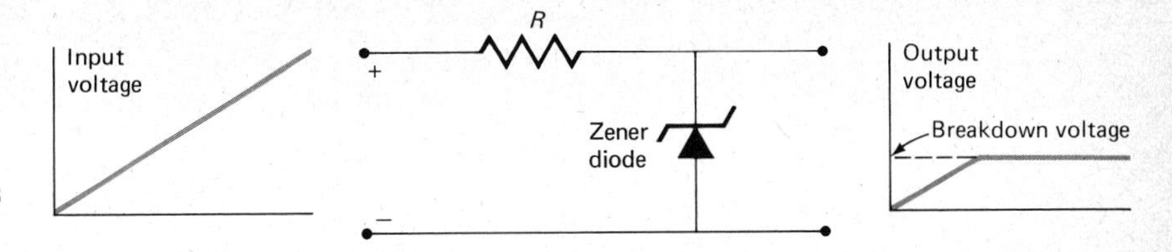

Figure 30.13 The Zener diode causes the output voltage to remain constant even though the input voltage continues to rise.

30.9 The Light-Emitting Diode (LED)

It is possible to construct a *p-n* junction that emits light as current passes through the diode in the forward direction. You will recall from Figure 30.10(a) that electrons combine with holes at the junction when a forward current flows. The electron loses energy as it "falls" into the hole. In most cases this lost energy appears as thermal energy in the crystal. However, by using special dopant materials in selected crystals, the lost energy is emitted as a light photon. As a result, these diodes emit light when a current flows through them. They are called *light-emitting diodes,* LEDs.

LEDs are widely used to display numbers on calculators and digital devices. Because an LED lights when a forward voltage is applied to it, the LED can be turned on and off by simple circuitry. Seven LEDs are arranged as shown in Figure 30.14(a) in a typical single-digit display. When selected LEDs are lighted, numerals may be reproduced as shown in (b) and (c).

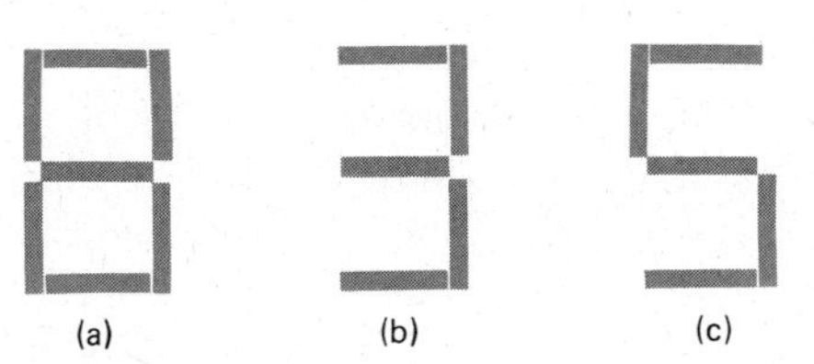

Figure 30.14 Seven LEDs are used in the array shown in (a) to display numerals.

30.10 Light-Induced Conductivity

You will recall that in the photoelectric effect an electron is torn loose from a material by a photon. Photons with enough energy can also tear electrons loose from free atoms. This is called the atomic photoelectric effect. A similar effect can occur within a semiconductor.

Pure semiconductors have low conductivity. However, if sufficiently energetic photons bombard a semiconductor, its conductivity increases roughly in proportion to the radiation intensity. The photons strike atoms within the material and thereby free electrons from their parent atoms.[2] These freed electrons are available for conduction, and so the conductivity of the material is increased. Use of this effect is made in *photoresistors,* semiconductor resistors whose resistance can be altered by shining light on them.

Such light-induced conductivity is not readily utilized in doped semiconductors. Difficulty arises because the number of free electrons and holes produced by the light is usually much smaller than the number produced by doping. Hence, except for very high light intensities, the light produces little change in conductivity of the material.

[2]That is to say, the photons eject electrons from the valence band and throw them to the conduction band.

This difficulty is overcome in the *photodiode*. In this device a *p-n* junction is used under reverse bias conditions. No current flows through the diode when the junction is in darkness. But when light is shined on the depletion layer at the junction, free electrons and holes are formed. If you liken the depletion region to the region between the plates of a charged capacitor, you will readily recognize the effect that occurs. The light, in effect, injects charges into the depletion region, the region between the equivalent "capacitor plates" at the junction. The freed electrons and holes will move through the region under the action of the electric field between the "plates." As a result, current flows through the junction. Therefore, the current that flows through the junction is produced by the incident light. Hence the current is a direct measure of the light intensity.

A somewhat similar situation occurs in the *solar cell.* This device is usually a thin layer of *p*-type material on top of an *n*-type material. Light passes through the thin layer and strikes the junction. Let us consider what happens under open-circuit conditions.

As in the photodiode, the light injects free electrons and holes into the depletion region. Here, too, the boundaries of the region act like charged capacitor plates. These plates collect the free electrons and holes from the depletion region until a new diffusion-controlled equilibrium is reached. But this new charge collected by the plates causes the potential difference between the plates to change. This change in potential appears as a voltage difference between the two materials of the solar cell. Under open-circuit conditions the voltage so generated is close to 1 V. A typical solar cell in sunlight can produce a current of a few tenths of an ampere per square centimeter at a voltage of perhaps 0.7 V. Present-day solar cells have an efficiency that is less than 20 percent. However, much research is going on to raise cell efficiency.

30.11 The Transistor

There are many applications in electronics that require electric signals to be amplified. For example, in a radio the weak input signal from the antenna is amplified until it is capable of driving a loudspeaker. In applications such as this, devices more complex than diodes are required. Let us now learn what a transistor is and how it can be used to amplify an electric signal.

A *transistor* consists of a layer of *p*-type material sandwiched between two layers of *n*-type material (an *n-p-n* transistor) or a layer of *n*-type material sandwiched between two layers of *p*-type material (a *p-n-p* transistor). These two types are shown in Figure 30.15. The way they are represented in a circuit diagram is shown below each type. Notice the different directions of the arrows on the emitter in the two cases. The arrow shows the conventional current direction through the device in normal operation. We shall also need to know that the emitter and collector regions are much more highly doped than is the base region.

Let us consider the *n-p-n* transistor in Figure 30.16. If you check the battery polarity, you will see that junction 1 is forward biased, while 2 is reverse biased. Current flows readily through junction 1, but only with great difficulty through

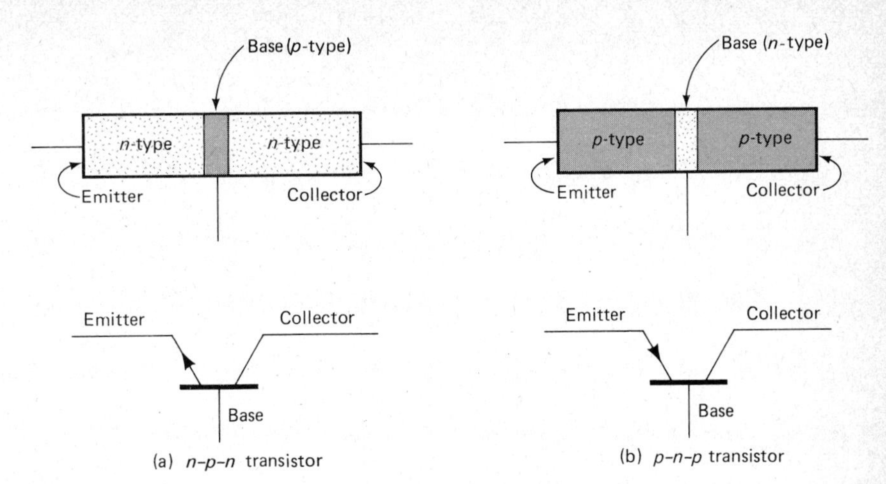

Figure 30.15 Two types of transistors.

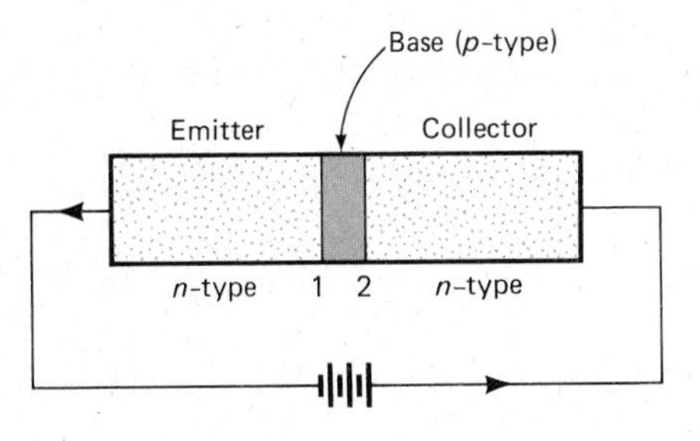

Figure 30.16 Junction 1 is forward biased, while junction 2 is reverse biased.

junction 2. Therefore, the voltage drop across 2, namely V_c, is much larger than the drop across 1, V_e.

An electron entering the emitter from the attached wire moves freely through the *n*-type material of the emitter. It easily crosses through the forward-biased junction at 1. As it travels through the *p*-type base, the electron may encounter a hole and combine with it. But this effect is kept to a minimum by two means. The base is lightly doped and therefore contains few holes; in addition, the base is thin so the distance from 1 to 2 is small.

After passing through the base, the electron reaches junction 2. Since this junction is reverse biased, holes in the *p*-type base are forced to the left by the battery. But we are now concerned with the uncommon situation of a free electron in the *p*-type base. The electron is attracted to the right by the positive side of the battery. As a result, junction 2 does not present a barrier to the movement of the electron. We therefore conclude that, for the situation shown in Figure 30.16, electrons move easily to the right through both junctions.

If the transistor is to be useful, connection to the base must be made. A typical circuit is shown in Figure 30.17. Because the base region is so narrow, most of the electrons entering it from the emitter are collected by the collector. As a result, the current I_b in the base wire is much smaller than the emitter current I_e and the collector current I_c. Therefore, $I_e \simeq I_c \gg I_b$.

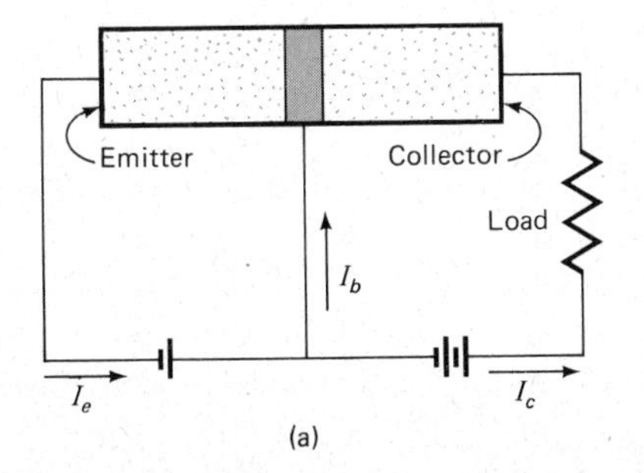

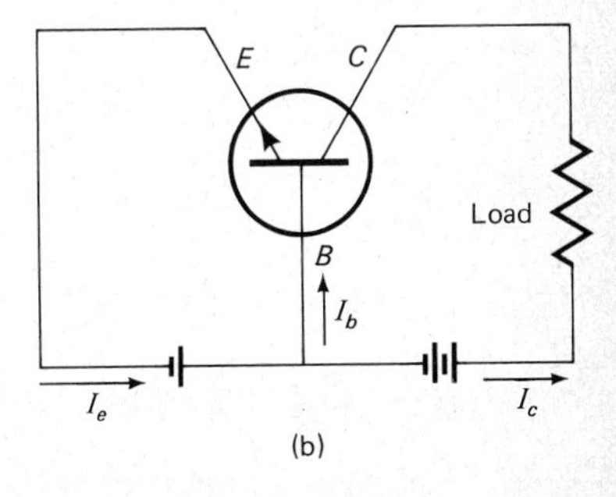

Figure 30.17 In this circuit $I_e \approx I_c >>> I_b$.

The current drawn from the base can be used to control the emitter and collector currents. To see why this is true, notice the following fact. In Figure 30.16 electrons enter the left side of the emitter and travel through it into the base. Some of the electrons are captured by the holes in the base, and the base becomes negatively charged. Unless this excess charge is drained off by a connection to the base, the excess charge will stop further electron flow to the base and to the collector. Hence if I_b is zero in Figure 30.17, the currents I_e and I_c would drop to zero unless the battery voltage is very high. However, only a small base current is needed to drain off the excess charge on the base. We therefore see that a small base current can control much larger currents through the emitter and collector.

Transistor manufacturers usually specify the ratio I_c/I_e for their transistors. The ratio is called alpha, α.

$$\alpha = \frac{I_c}{I_e}$$

In most cases alpha ranges from about 0.95 to 0.99. The values of I_e and I_c are nearly the same. By applying Kirchhoff's point rule to Figure 30.17, we have

$$I_b = I_e - I_c$$

from which, because $I_c/I_e = \alpha$,

$$I_b = I_e(1 - \alpha)$$

For an alpha value of 0.98, only 2 percent of the current is diverted out the base connection.

30.12 Amplification

As we have seen, the current flowing from the base of a transistor is much smaller than the emitter and collector currents. This fact allows us to use the transistor as an amplifier, as we shall now see. Consider the circuit shown in Figure 30.18. A sinusoidal input current with rms value I_b is provided to the base wire as shown.[3] We are interested in the sinusoidal output current, I_c, furnished to the load, R_L.

We know that

$$I_b = I_e(1 - \alpha)$$

Since $I_e = I_c/\alpha$, this equation becomes

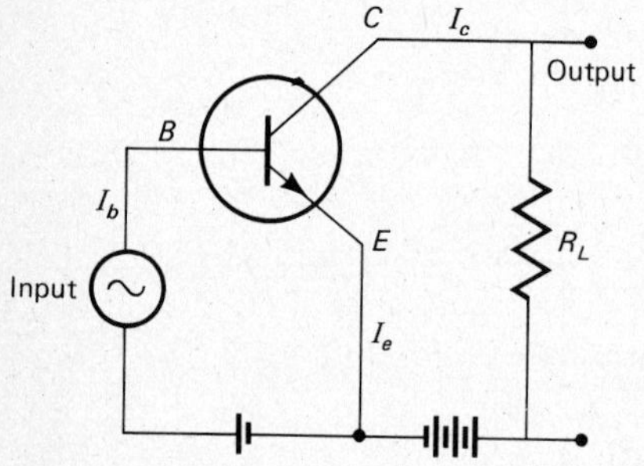

Figure 30.18 Common-emitter amplifier circuit.

[3] We shall assume this sinusoidal current to be small compared to the direct current caused by the bias batteries.

$$I_b = I_c \frac{1 - \alpha}{\alpha}$$

Solving for the ratio I_c/I_b gives

$$\frac{I_c}{I_b} = \frac{\alpha}{1 - \alpha}$$

Notice that the output current I_c is very much larger than the input current I_b. For an alpha of 0.98, the so-called *current gain* is

$$\text{Current gain} = \frac{\alpha}{1 - \alpha} = \frac{0.98}{0.02} = 49$$

The transistor circuit has greatly amplified the input current.

We are also concerned with the output voltage (the sinusoidal voltage drop across R_L) as compared to the input voltage (the voltage of the input source). Let us suppose that the sinusoidal input voltage is small enough so that it is much smaller than the forward bias on the emitter connection. Because the input current will be in the forward direction through the emitter, it will encounter small impedance. Hence the input voltage need not be large to produce sizable currents.

However, the output voltage across R_L is simply I_cR_L. Because R_L can be quite large, the output voltage can also be made large.[4] As a result, the output voltage can be made much larger than the input voltage, and so voltage amplification occurs.

Because power = IV, the fact that current and voltage amplification occurs leads to power amplification as well. For a typical transistor used in the circuit of Figure 30.18, the power can be amplified by a factor of several hundred.

The circuit shown in Figure 30.18 is called a *common-emitter amplifier* circuit. It derives its name from the fact that the base wire and collector wire meet at the emitter wire; they have the emitter wire in common. Other connections are also possible, as we shall see in the exercises at the end of this chapter.

30.13 Integrated Circuits (ICs)

Individual transistors can be made very small, and so electronic equipment using solid state devices can be made quite compact. As solid state electronics replaced vacuum tubes in radios and television sets, considerable savings were made in space and energy requirements. Moreover, complex devices such as computers were also greatly simplified with solid state circuitry. Even so, these

[4] A practical limit on R_L exists. If it is large compared to the effective resistance of the reverse-biased collector junction, then the performance of the transistor is affected. However, the effective resistance of a reverse-biased junction is very large.

circuits were still too bulky for many applications we now take for granted. Today's pocket calculators and electronic watches are examples of systems that depended on the next important advance in solid state electronics, the invention and use of *integrated circuits,* ICs.

An integrated circuit is a circuit contained entirely within a single block (or chip) of semiconductor. A single IC chip may contain thousands of circuit elements. Its size might be 2 mm × 2 mm × 0.5 mm. By combining ICs, complex circuits can be packaged in a very small volume. ICs have made it possible to construct such widely different devices as electronic watches, compact missile guidance systems, heart pacemakers, and affordable personal computers. Let us now see how circuit elements in ICs are produced.

In a typical case one starts with a thin wafer of pure silicon. By a proper treatment of its surface, various circuit elements can be made on and in the wafer. For example, Figure 30.19 shows the steps by which an n-type region is made.

1. The surface of the wafer is oxidized. The thin oxide surface layer acts as a shield for the silicon below.
2. A thin layer called a *photoresist* is applied. This material, when exposed to ultraviolet (UV) light, becomes insoluble in most acids.
3. On top of the photoresist is placed a mask, a pattern for the device to be constructed. It is a photographic image on film. It is obtained by taking a picture of the pattern required. Opaque areas of the film prevent exposure of the photoresist below them. The pattern has been reduced to a very small size by photographic reduction.

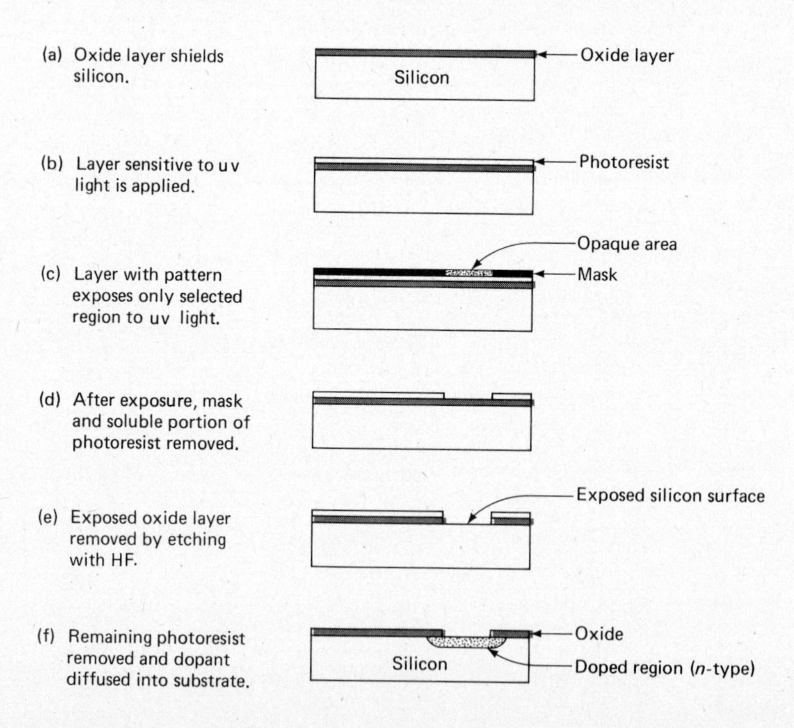

Figure 30.19 Preparation of a small portion of an integrated circuit involves the steps shown.

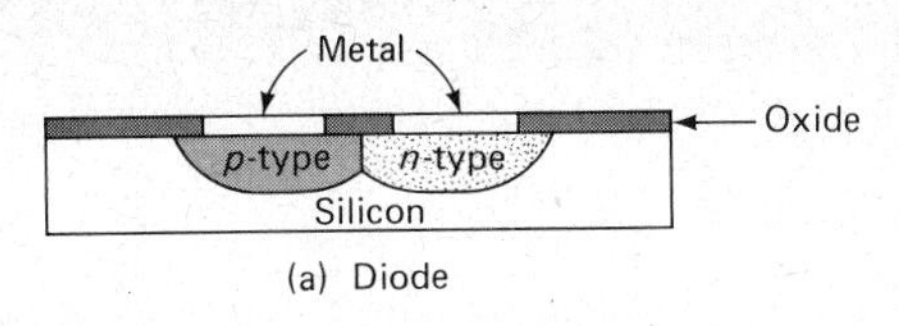

(a) Diode

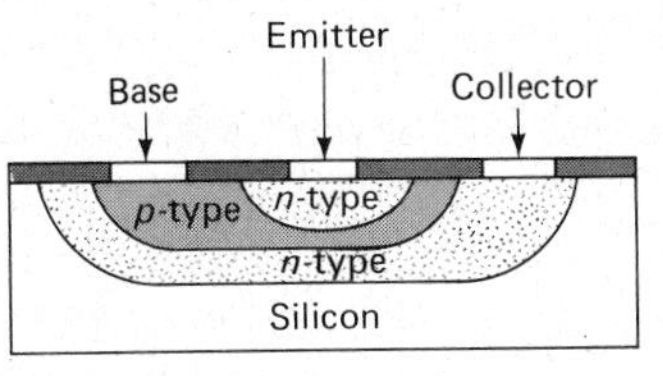

(b) *n-p-n* transistor

Figure 30.20 IC circuit elements.

4. Ultraviolet light is shined onto the mask. Those areas of the photoresist below the transparent portions of the mask are thus rendered insoluble.
5. The mask is removed and the soluble portion of the photoresist layer is dissolved.
6. The silicon oxide layer beneath these regions is removed by dissolving it in hydrofluoric (HF) acid. The photoresist protects the remainder of the oxide surface during this process.
7. The remaining photoresist layer is stripped off. An *n*-type dopant is diffused into the exposed portion of the silicon wafer. As a result, an *n*-type region is formed in the wafer.

Procedures similar to this are used to prepare many other types of doping patterns. These may be combined in such a way as to form diodes, transistors, and other devices. For example, Figure 30.20 shows a diode and a transistor.

Many other types of solid state electronic components can be made. Much research effort is being expended to produce new and novel components. As they become available, still greater advances will be made in solid state electronic circuitry. We can foresee continued rapid advances in the fields of computers, controls, and solid state electronics in general.

Summary

In solids the higher energy levels widen into bands. If the conduction band contains electrons, the substance is a conductor. If it is empty and the valence band is filled, the substance is an insulator. In intrinsic semiconductors, the band gap is small enough that thermal energies excite electrons from the valence band to the conduction band.

By adding atoms with valence five to silicon and germanium, an *n*-type semiconductor is made. When trivalent atoms are used, the material becomes a *p*-type semiconductor. The major current carriers are free electrons in *n*-type and positive holes in *p*-type semiconductors.

A semiconductor diode consists of a *p*-type material fused to an *n*-type material. Conventional current flows easily from *p*-type to *n*-type through the junction, but not in the reverse direction.

The Zener diode makes use of the fact that a diode breaks down if the reverse bias is sufficiently large. These diodes are often used for voltage regulation.

In light-emitting diodes, LEDs, the diode is so made that light is emitted when current passes through the diode. Photons are given off in the junction as electrons and holes combine while the current is flowing. LEDs are frequently used in digital displays.

When light strikes a semiconductor, free electrons and holes can be produced. Use of light-induced conductivity is made in photoresistors. The solar cell's voltage is also generated by the process. A voltage arises because of charge separation that occurs in the depletion region.

A transistor consists of a narrow base region sandwiched between an emitter and a collector. If the base is *p*-type, then the emitter and collector are *n*-type, and vice versa. In operation the collector current and emitter current are about equal, while the base current is much smaller than those two. A small change in the base current can cause a large change in the collector current. As a result, the transistor can be used to amplify a signal.

Integrated circuits, ICs, consist of many circuit elements contained in a tiny chip of semiconducting material. ICs make it possible to construct highly complex circuits in a small volume. This has made possible the many sophisticated electronic devices that are so common in present-day technology.

Questions and Exercises

1. The resistivities of metals usually increase slowly with temperature, but the resistivities of many semiconductors decrease rapidly with temperature. Why do semiconductors behave in this way?
2. A silicon crystal contains *n*-type impurity that gives an impurity energy level 0.010 eV below the conduction band. Sketch a rough curve showing how the conductivity of this material varies with temperature from 0 K to about 800 K.
3. Refer to Figure 30.1(b) and explain what is meant by the Fermi level. In terms of it, explain why thermionic emission (the emission of electrons by a hot solid) occurs only at high temperatures.
4. What would you expect *n*-type doping to do to the photoelectric threshold wavelength for silicon? What about *p*-type doping? Does your answer depend on how sensitive your measurements are?
5. When a *p*-type material is fused to an *n*-type, why don't the free electrons in the *n*-type run into the *p*-type and neutralize the positive holes? After all, don't the positive holes attract the electrons?
6. Equal, evenly mixed, very small amounts of gallium and arsenic dopant atoms are placed in the same silicon crystal. What would you predict for the electrical properties of this crystal?
7. Figure P30.1 is a rough representation of the depletion region near a *p-n* junction. The charges shown represent net charge in the regions. Describe how the charges shown arise and what they represent (a) if there is no voltage impressed on the junction and (b) if the junction is reverse biased. (c) In the case of reverse bias explain how the region will change as the bias voltage is increased. (d) Describe the electric field in the region depicted in the figure. (e) Describe what will happen to electrons released by high-energy photon radiation passing through the region. Relate this to the photodiode and solar cell.

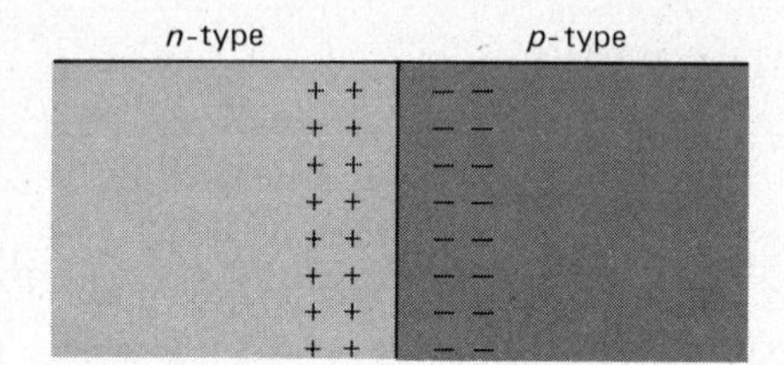

Figure P30.1

8. Explain why the diode characteristic curve shown in Figure 30.11 varies with temperature as indicated.
9. A Zener diode is connected in series with a 5000-Ω resistor and a variable 0–50-V dc power supply. Describe how the current through the diode and the voltage across it will vary as the voltage source increases slowly from 0 to 50 V. Assume the diode is in (a) the forward bias direction and (b) the reverse bias direction.
10. Discuss the difficulties that might be encountered in devising an array of LEDs that could be used to represent the capital letters of the alphabet.
11. A *p-n-p* transistor is connected in series with a battery in the following way: plus side of battery to the emitter and negative side of battery to the collector. The base wire is not connected. Compare the voltage between the emitter wire and the base wire to that between the collector wire and the base wire. Repeat for an *n-p-n* transistor.

Problems

1. The value of E_g for a diamond crystal is about 7.0 eV. It is 1.09 eV for silicon. What wavelengths of em radiation will cause conduction in diamond? In silicon? What does this have to do with the fact that diamond is transparent, while silicon is much less so?
2. A certain *n*-type semiconductor has its impurity level at

0.25 eV below the conduction band. What wavelengths of em radiation will cause the substance to be a conductor at absolute zero?

3. An electron falls from the bottom of the conduction band to the top of the valence band in silicon. If the energy lost in the process is emitted as a photon, what wavelength radiation is emitted?
4. What wavelength em radiation is required to excite a valence band electron in germanium to the conduction band. Will this wavelength cause photoelectric emission from germanium?
5. A sound wave of frequency f carries energy through a crystal by vibrations of the crystal lattice. Quantum mechanics tells us that this wave energy comes in packets called *phonons.* The energy of a phonon is hf, where h is Planck's constant. What would be the wavelength of the sound wave that corresponds to a phonon energy of 0.20 eV in a crystal where the speed of sound is 5000 m/s?
6. The depletion region in a diode is of the order of 1×10^{-5} cm thick. Suppose a varactor, in effect, has plates this far apart and with area of 2.0 mm^2. What will be the capacitance of the varactor?
7. An LED emits light of 520 nm wavelength as the holes and electrons combine at the junction. Estimate the energy gap for this semiconductor.
8. Integrated circuits use thin strips of semiconductor to carry currents. What is the resistance of a strip 5.0 μm wide, 2.0 μm thick, and 12 μm long? Assume the resistivity of the material to be 0.20 $\Omega \cdot$m.
9. The emitter current is 20.0 mA in a transistor that has an alpha of 0.985. What are the collector and base currents?
10. The base current for a transistor with an alpha of 0.970 is increased by 10 μA. How much does the collector current change? The emitter current? Assume the connection is as shown in Figure 30.18.
11. When a transistor is connected as shown in Figure P30.2, it forms a common-collector-type amplifier circuit. In this circuit the current gain is I_e/I_b. Show that it is $1/(1 - \alpha)$.

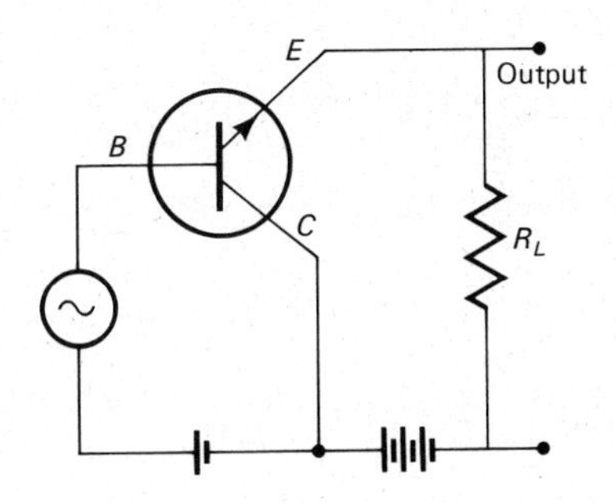

Figure P30.2

Stanford University

Linear accelerator facility at Stanford.

31 THE ATOMIC NUCLEUS: NUCLEAR ENERGY

Performance Goals

When you finish this chapter, you should be able to

1. Given Z and A for an isotope, sketch the atom so as to show the positions of electrons and nucleus, give order-of-magnitude diameters of the atom and nucleus, give the charge of the nucleus in units of e, give the

Nuclear technology has grown greatly since its birth in the early 1940s. Today it is important in many branches of science extending from medicine to nuclear energy. As time goes on, the use of nuclear reactors will become increasingly vital to civilization. In many less noticeable applications, radioactivity and other aspects of nuclear behavior are widely used. These uses will expand as the years go by. For this reason an increasing number of technicians and scientists will need to know the basics of nuclear behavior. This chapter introduces the major topics of nuclear science and technology.

approximate mass of the nucleus in atomic mass units, give the number of neutrons and/or protons in the nucleus.
2. Describe the neutron and proton by giving the charge and approximate mass of each. The mass should be given in both atomic mass units and kilograms.
3. Explain what is meant by the nuclear force and give its approximate range.
4. Explain what isotopes are by giving two imaginary isotopes and stating how they differ. Explain why chemists consider them to be the same element.
5. Given the symbol for an isotope such as ${}^{A}_{Z}X$, state its nuclear mass and charge. Also state how many protons or neutrons exist in it.
6. Given the masses of the proton and neutron together with the mass and charge of the nucleus of an isotope, find the mass defect for this isotope, find the energy required to tear its nucleons apart, give the binding energy of the nucleus.
7. Sketch roughly the mass defect per nucleon versus Z graph. Using it, point out why nuclear fission and nuclear fusion of certain nuclei can result in the release of energy.
8. State the equivalent energy (in megaelectronvolts) of 1 u of mass.
9. Sketch how the quantity of radioactive substance decreases with time. Give the fraction of material unchanged after any specified number of half-lives has passed.

31.1 The Atomic Nucleus

The nucleus is a tiny object at the very center of the atom. A typical atom has a diameter on the order of 10^{-10} m. The nucleus at its center is less than 1/10,000 as large. A typical nucleus has a diameter of about 10^{-15} m.

We frequently picture the nucleus as shown in Figure 31.1. It contains two types of particles, the neutron and the proton. Both types of particles are also called nucleons.

The neutron has no charge. Its mass is 1.675×10^{-27} kg. The proton has nearly the same mass as the neutron, 1.673×10^{-27} kg. But the proton has a charge $+e$, where e is the charge quantum, 1.602×10^{-19} C.

The picture of the nucleus shown in Figure 31.1 should not be taken too seriously. It is only schematic in nature.

What holds a nucleus together? The protons repel each other because they carry like charges. Clearly, electric forces do not hold the neutrons and protons together. The gravitational force is much smaller than the Coulomb force, so its effect is too small to be of importance. We need yet a third type of force to hold the nuclear particles together. It is called the *nuclear force.*

The nuclear force is a strong, short-range force. It is an attractive force between nucleons. There is little difference in the nuclear force between two protons, two neutrons, or a neutron and a proton. The nucleons behave much the same as far as the nuclear force is concerned.

Unlike the Coulomb and gravitational forces, the nuclear force is not a $1/r^2$ type of force. It decreases rapidly for particle separations larger than 1×10^{-15} m and is essentially zero for separations larger than 5×10^{-15} m. Two nucleons attract each other strongly when they are in near contact, but they exert little force on each other when they are separated by a distance greater than one nucleon diameter. In a nucleus each nucleon feels strong forces due to the nucleons that are its closest neighbors, but nucleons in the rest of the nucleus exert only small nuclear forces on it.

It is instructive to compare the potential energy between two protons at various separation distances. At large separations the Coulomb repulsion predominates. Work is required to push the like charges toward each other. Hence their potential energy increases as their separation distance decreases as shown in Figure 31.2. But when their separation is about 5×10^{-15} m, the nuclear attractive force begins to be felt. This force draws the particles together and therefore causes the potential energy to decrease. When the energy becomes negative at small separations, the particles have become bound together by the nuclear force.

Because the nuclear force is so strong, the nucleons are held together tightly in an essentially spherical structure. We can define an approximate radius R for the nucleus. It is given by experiment to be

10. Define half-life, decay constant, decay law, activity, becquerel, and curie.
11. Supply s and r in the following reaction when m, q, n, and p are given:

$$^{m}_{n}X \rightarrow {}^{q}_{p}Y + {}^{r}_{s}Z + T$$

You are told whether T is an alpha or beta particle, a gamma ray, or a neutron.
12. Explain how the age of a piece of wood can be determined using radiocarbon dating.
13. Explain why a uranium nuclear-type reactor must use uranium 235 rather than natural uranium. In your explanation point out the essential difference (and its reason) between a fission reactor and a fission bomb. Why cannot either be made using only 1 g of uranium 235?
14. Explain the source of energy given off in the fission reactor. Using a sketch, explain in general terms how the fission-released energy is used to generate electrical power.
15. Give at least two uses of a nuclear reactor besides the generation of power.
16. Describe what is meant by the fusion reaction. Point out the difficulty that has prevented us from using it as a controlled power source. In spite of this difficulty, explain why the reaction is occurring continuously in the sun and other stars.

$$R = (1.2 \times 10^{-15})A^{1/3}\text{m} \qquad (31.1)$$

where the number of nucleons in the nucleus is A.

EXAMPLE 31.1 Find the mass density of a typical nucleus.

Solution Each nucleon has a mass of about 1.67×10^{-27} kg. The mass of a nucleus that contains A nucleons is therefore $A\ (1.67 \times 10^{-27}\ \text{kg})$. The nuclear volume is $4\pi R^3/3$ and so

$$\text{Density} = \frac{\text{mass}}{\text{volume}} = \frac{A\ (1.67 \times 10^{-27}\ \text{kg})}{(4\pi/3)(1.2 \times 10^{-15}\ \text{m})^3\ A}$$
$$= 2.3 \times 10^{17}\ \text{kg/m}^3$$

Because A cancels, all nuclei have about this same extremely high density.

31.2 Atomic Number and Mass

Hundreds of years ago scientists began to look for a systematic way to list the elements. The most successful method was found by the Russian chemist Mendeleev (1834–1907). He prepared a chart that grouped the chemical elements (or atoms) in a way we call the *periodic table*. A modern form of the periodic table is shown in Figure 31.3.

You will notice that each element in the periodic table is assigned a number. Hydrogen (H) is 1, helium (He) is 2, sodium (Na) is 11, and so on. These numbers are referred to as the *atomic number* for each element, and we represent it by the symbol Z. We know the following important fact.

The atomic number, Z, of an element gives the number of protons in the nucleus of each atom. Because each proton carries a charge $+e$, the charge on the nucleus is $+Ze$.

For example, gold (Au) has an atomic number $Z = 79$. The nucleus of a gold atom therefore has 79 protons in it and has a charge $+79e$.

The periodic table also shows the atomic masses as determined by the chemists. We express the mass in a special unit called the *atomic mass unit*, represented by the symbol u.

The atomic mass unit, u, is related to the kilogram through $1\ \text{u} = 1.6606 \times 10^{-27}$ kg

In terms of this unit the proton and neutron have masses of about 1.01 u.[1] The electron mass is only about 0.00055 u. Because the mass of the electrons in an

[1] More exact values are $m_{\text{neutron}} = 1.008665$ u and $m_{\text{proton}} = 1.007276$ u.

Figure 31.1 The nucleus is composed of protons (light) and neutrons (dark).

atom is so small, the mass of the atom and the mass of the nucleus are nearly the same.

As we shall see later, the atomic mass listed in the periodic table is an average of several values in most cases. The mass is given in atomic mass units and is shown below the symbol for each element. For example, the atomic mass of potassium (K) is 39.102 u.

EXAMPLE 31.2 The average atomic mass of mercury (Hg) is 200.59 u. Find the average *nuclear* mass for the Hg atom.

Solution The atomic mass includes the electrons as well as the nucleus. From the periodic table we see that the atomic number for Hg is 80. Its nucleus has a charge of $+80e$. Because the atom is electrically neutral, it must possess 80 electrons to balance this charge. Each electron has a mass of 0.00055 u. Therefore,

$$\text{Mass of electrons in atom} = (80)(0.00055\ \text{u}) = 0.044\ \text{u}$$

The average atomic mass was given as 200.59 u. Therefore,

$$\textbf{Mass of nucleus} = 200.59\ \text{u} - 0.044\ \text{u} = 200.55\ \text{u}$$

Notice that the electron mass is only a small portion of the total.

31.3 Isotopes

The mass spectrometer makes it possible to measure with great accuracy the masses of single atoms. The principle of operation of this device is shown in

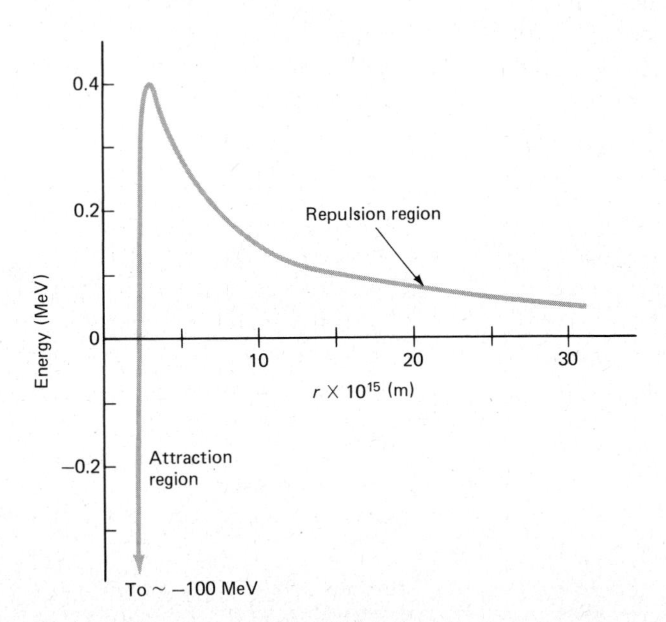

Figure 31.2 The nuclear attraction force between two protons overpowers the Coulomb repulsion when the proton separation is a few femtometers. (femto = 10^{-15}, one-quadrillionth).

Group		I	II	III	IV	V	VI	VII	VIII	O
Period	Series									
1	1	1 H 1.00797								2 He 4.003
2	2	3 Li 6.939	4 Be 9.012	5 B 10.81	6 C 12.011	7 N 14.007	8 O 15.9994	9 F 19.00		10 Ne 20.183
3	3	11 Na 22.990	12 Mg 24.31	13 A1 26.98	14 Si 28.09	15 P 30.974	16 S 32.064	17 Cl 35.453		18 Ar 39.948
4	4	19 K 39.102	20 Ca 40.08	21 Sc 44.96	22 Ti 47.90	23 V 50.94	24 Cr 52.00	25 Mn 54.94	26 Fe 55.85 27 Co 58.93 28 Ni 58.71	
	5	29 Cu 63.54	30 Zn 65.37	31 Ga 69.72	32 Ge 72.59	33 As 74.92	34 Se 78.96	35 Br 70.909		36 Kr 83.80
5	6	37 Rb 85.47	38 Sr 87.62	39 Y 88.905	40 Zr 91.22	41 Nb 92.91	42 Mo 95.94	43 Tc [98]	44 Ru 101.1 45 Rh 102.905 46 Pd 106.4	
	7	47 Ag 107.870	48 Cd 112.40	49 In 114.82	50 Sn 118.69	51 Sb 121.75	52 Te 127.60	53 I 126.90		54 Xe 131.30
6	8	55 Cs 132.905	56 Ba 137.34	57-71 Lanthanide series*	72 Hf 178.49	73 Ta 180.95	74 W 183.85	75 Re 186.2	76 Os 190.2 77 Ir 192.2 78 Pt 195.09	
	9	79 Au 196.97	80 Hg 200.59	81 T1 204.37	82 Pb 207.19	83 Bi 208.98	84 Po [210]	85 At [210]		86 Rn [222]
7	10	87 Fr [223]	88 Ra [226]	89-103 Actinide series†						

Lanthanide series	57 La 138.91	58 Ce 140.12	59 Pr 140.91	60 Nd 144.24	61 Pm [147]	62 Sm 150.35	63 Eu 152.0	64 Gd 157.25	65 Tb 158.92	66 Dy 162.50	67 Ho 164.93	68 Er 167.26	69 Tm 168.93	70 Yb 173.04	71 Lu 174.97
†*Actinide series*	89 Ac [227]	90 Th 232.04	91 Pa [231]	92 U 238.03	93 Np [237]	94 Pu [242]	95 Am [243]	96 Cm [247]	97 Bk [247]	98 Cf [251]	99 E [254]	100 Fm [253]	101 Md [256]	102 No [254]	103 Lw [257]

Figure 31.3 Periodic table of the elements. Masses include electrons. The isotope carbon 12 is assigned a mass of exactly 12 u. For unstable nuclei the mass of the most stable isotope is shown in brackets. See also Appendix 3.

Figure 31.4. The atoms are vaporized and ionized in the ion source. Those ions that wander out through the hole in the source are accelerated through a known voltage, V. Their velocities as they travel through the second slit can be found from

$$\tfrac{1}{2}mv^2 = Vq$$

In this expression q is the charge and m is the mass of the ion.

The fast-moving ions then enter a region perpendicular to a known magnetic field B. As we learned in Chapter 26, the ions will travel in a circle of radius r for which

$$\frac{mv^2}{r} = qvB$$

We can replace v in this equation by the value found from the previous equation. Then after solving for m, we find that

$$m = \frac{qB^2r^2}{2V}$$

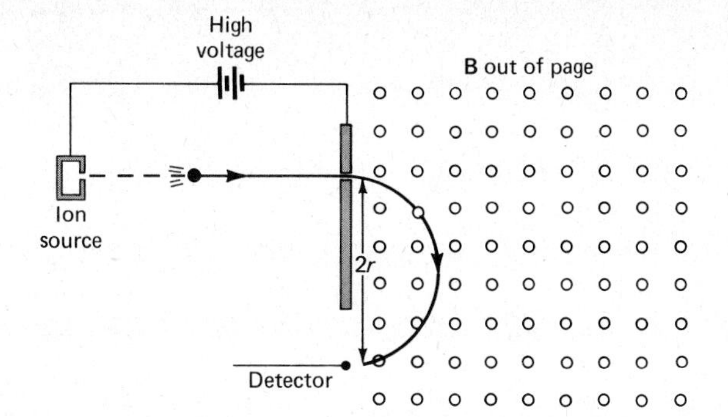

Figure 31.4 The radius of the circular path is a measure of the ion's mass. The path within the mass spectrometer is in vacuum.

Therefore, the mass m of the ion can be found from the measured values of B, r, and V. Notice that the larger the mass of the particle, the larger will be the radius of its circular path.

Use of the mass spectrometer produced interesting results. The chemists provided material of extremely high purity, but often in the mass spectrometer the pure material would act like two or more separate materials. For example, pure lithium was found to consist of two types of atoms. One had a mass of 6.015 u and the other had a mass of 7.016 u. There was about 8 percent of the first type and 92 percent of the second. And yet the chemists knew the material to be better than 99.99 percent pure lithium. How can we explain this?

In order to understand these results, we must understand how the chemists classify the elements. Atoms are classified according to the chemical reactions they undergo. For example, all lithium atoms combine explosively with water. But these reactions are determined by the number of electrons in the atom and their energy levels. As we have pointed out before, the number of electrons and the atom's energy levels depend on the charge of the nucleus. They do not depend seriously on the mass of the nucleus.

With this in mind, the lithium experiment can be interpreted as follows. The lithium atom has an atomic number $Z = 3$. It therefore has three electrons outside the nucleus. To compensate for these, the nucleus must have three protons. The remainder of the nuclear mass is furnished by its neutrons.

When the chemists supply pure lithium, they are supplying atoms of the same known nuclear charge. Each lithium atom has a nuclear charge of $+3e$. These are furnished by the three protons within the nucleus. But the chemists cannot be sure of the number of neutrons in each nucleus. The neutrons are not important for chemical reactions.

We therefore conclude that chemically pure lithium consists of atoms with three electrons each. The nucleus of each atom contains three protons. But the mass spectrometer tells us that the number of neutrons is not the same in the nuclei of all the atoms. Because the masses of the protons and neutrons are each very close to 1.00 u, we interpret the measured masses as follows.

About 8 percent of the atoms have a nuclear mass of 6 u. Each nucleus contains three protons. They supply a mass 3 u to the nucleus. The remaining mass must be due to three neutrons. These atoms, then, have nuclei that contain three protons and three neutrons.

The other 92 percent of the atoms have a nuclear mass of 7 u. Of this, 3 u is furnished by the three protons. But there must be four neutrons in the nucleus to make up the extra 4 u. Therefore, this type of lithium nucleus contains three protons and four neutrons. Both types behave the same chemically, because they both have the same nuclear charge.

Nuclei of the same chemical element have the same nuclear charge. They contain the same number of protons. Those nuclei that contain different numbers of neutrons, but the same number of protons, are called *isotopes* of the element.

In our example lithium was found to contain two isotopes, one with nuclear mass 6 u and the other with mass 7 u. The average value found for these is 6.939 u. This is the mass value given in the periodic table.

31.4 Symbols Used for Isotopes

All the elements in the periodic table have been studied by using the mass spectrometer. The masses of the isotopes of the elements have been measured and their percentages (abundances) have been found. If we examine a listing of the isotopes, we see that the mass of each isotope is close to being an integer. The largest differences from an integer are 119.902 (for tin) and 255.09 (for mendelevium). Even these are close to the integers 120 and 255.

The integer closest to the mass (in atomic mass units) of an isotope is called the mass number of the isotope. It is represented by *A*.

The mass number of an isotope is easily interpreted. Recall that the nucleus is composed of protons and neutrons. Both of these particles have a mass close to 1.00 u. As a result, the number of nucleons (i.e., protons + neutrons) in a nucleus should be close to the nuclear mass in atomic mass units. Therefore,

The mass number, *A*, of an isotope is equal to the number of nucleons for that isotope.

Suppose that the mass number, A, and the atomic number, Z, for an isotope are given. We can then find the number of protons and neutrons in the nucleus. The number of protons is equal to Z. Because there are A nucleons in the nucleus, there must be $A - Z$ neutrons.

In a nucleus with mass number A and atomic number Z, there are Z protons and $A - Z$ neutrons.

Let us now consider an element whose symbol is X, for example. The element has its own atomic number Z. One of its isotopes has a mass number A. We then agree to represent this isotope of the element as follows:

The symbol ^A_ZX represents an isotope with nuclear charge *Z* and mass number *A*.

We should emphasize that all isotopes of a given element have the same value for Z. For example, all atoms of cobalt have an atomic number $Z = 27$. The isotope of cobalt found on earth has a mass number of 58. We represent it as $^{58}_{27}$Co and refer to it as cobalt 58. There is another isotope of cobalt that is radioactive and is manufactured in nuclear reactors. It is widely used in technical work as a radiation source. It is cobalt 60 and its symbol is $^{60}_{27}$Co.

■ **EXAMPLE 31.3** The material used in the first nuclear bomb was uranium 235. Natural uranium contains only 0.715 percent of this isotope. Uranium is element number 92 in the periodic table. How many neutrons and protons are in the nucleus of uranium 235?

Solution For this isotope the given data tell us that $Z = 92$ and $A = 235$. There are therefore 92 protons in each nucleus. The number of neutrons is $A - Z$, which is 143 in this case. The symbol for this isotope is $^{235}_{92}$U. ■■

31.5 Mass Defect

From what we have learned thus far, we might expect that the mass of a nucleus would be equal to the sum of the masses of its protons and neutrons. This turns out to be wrong. The highly accurate mass spectrometer measurements show that a slight discrepancy occurs. Let us give an example of it.

The helium isotope $^4_2$He consists of two protons and two neutrons. The combined mass of two protons and two neutrons is

$$\begin{aligned} \text{Mass of 2 protons} &= 2 \times 1.007277 \text{ u} \\ \text{Mass of 2 neutrons} &= 2 \times 1.008665 \text{ u} \\ \text{Sum} &= 4.031884 \text{ u} \end{aligned}$$

However, the mass of $^4_2$He is measured to be

$$\text{Mass of } ^4\text{He nucleus} = 4.001506 \text{ u}$$

The helium nucleus has a mass of 4.0015 u, but the mass of the protons and neutrons that compose it is 4.0319 u. Apparently mass is lost as neutrons and protons are joined together to form a nucleus. In this case 0.0304 u of mass is lost. This general situation turns out to be true for all nuclei.

All nuclei have a mass that is smaller than the sum of the masses of its separated nucleons. This difference in mass is called the mass defect.

It is usually most convenient to deal with the mass defect per nucleon. This quantity is the mass defect divided by the number of nucleons in the nucleus. For the case of helium 4,

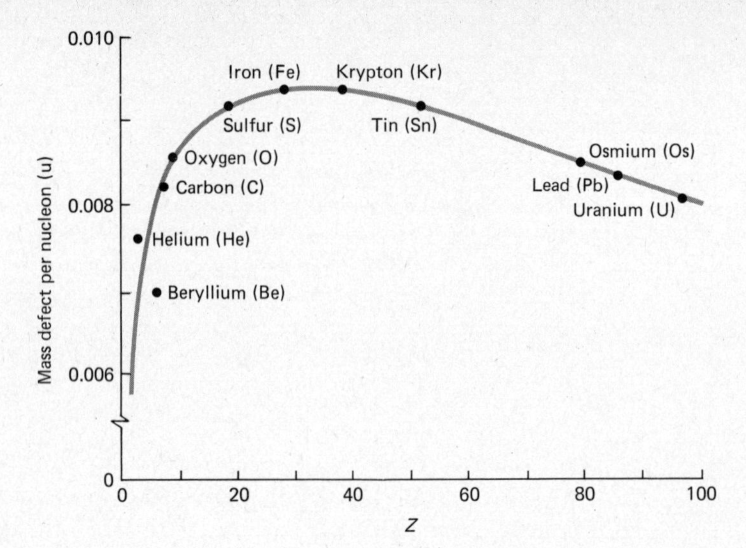

Figure 31.5 Mass loss per nucleon as the nucleons are combined into nuclei.

$$\text{Mass defect per nucleon} = \frac{0.0304\text{ u}}{4} = 0.0076\text{ u}$$

Typical values for this quantity are shown in Figure 31.5. Notice that the mass loss per nucleon is greatest for those elements near $Z \simeq 30$.

This loss in mass as protons and neutrons are joined to form a nucleus is of great technical importance. It is basic to nuclear power generators. Both the fission nuclear reactor and the dreamed-of fusion power source are based on it. We shall see in the next section how this mass loss can be understood in terms of Einstein's mass-energy formula.

31.6 Nuclear Binding Energy

The mass defect can be explained in terms of Einstein's theory of relativity. As you will recall, Einstein predicted that mass and energy could be interchanged. The relation between these two quantities is $\Delta E = (\Delta m)c^2$. Let us now see how this relation can be used to provide information about nuclei.

When two protons and two neutrons are brought together to form a helium nucleus, the mass loss is 0.0304 u. Therefore,

$$\begin{aligned} \Delta m &= (0.0304\text{ u})\left(1.66 \times 10^{-27}\frac{\text{kg}}{\text{u}}\right) \\ &= 5.0 \times 10^{-29}\text{ kg} \end{aligned}$$

The energy loss equivalent to this mass loss is

$$\begin{aligned} \text{Energy} &= (\Delta m)(c^2) = (5.0 \times 10^{-29}\text{ kg})(3 \times 10^8\text{ m/s})^2 \\ &= 4.5 \times 10^{-12}\text{ J} \end{aligned}$$

We can better understand the magnitude of this energy if we change it to electronvolts. Then

$$\text{Energy} = (4.5 \times 10^{-12}\ \text{J})\left(\frac{1}{1.6 \times 10^{-19}}\ \frac{\text{eV}}{\text{J}}\right)$$
$$= 28\ \text{MeV}$$

This energy is equivalent to that obtained by an electron as it falls through 28 million volts. As we see, a large quantity of energy is lost as a nucleus is formed. We shall learn later that the proposed fusion nuclear power source is based on this fact.

Now, however, we wish to examine the reverse situation. What must be done to tear a nucleus apart into its separate nucleons? The situation in the case of the helium nucleus is shown in Figure 31.6. The separated nucleons have a mass Δm larger than the original nucleus had. Therefore, an energy $(\Delta m)c^2$ must be added to the helium nucleus to tear it apart into the neutrons and protons that compose it.

To tear a nucleus into its separated nucleons, mass must be created.

This mass must be created from energy according to the mass-energy relation. When the nucleons are pulled apart from each other, work must be done. This work furnishes energy to the nucleons. The energy thus furnished is converted to mass.

We see, then, that energy must be furnished if a nucleus is to be torn apart into its separate neutrons and protons. This added energy is needed to create the extra mass of the separated particles.

The energy needed to tear apart a nucleus into its separate nucleons is called the binding energy of the nucleus.

Notice that this energy is needed to create the extra mass of the separated particles.

We can relate the binding energy of a nucleus to the mass defect for the nucleus. The mass defect is the mass lost as the neutrons and protons are put together in the nucleus. This same amount of mass must be created when the nucleus is torn apart. As a result, the binding energy of a nucleus is proportional

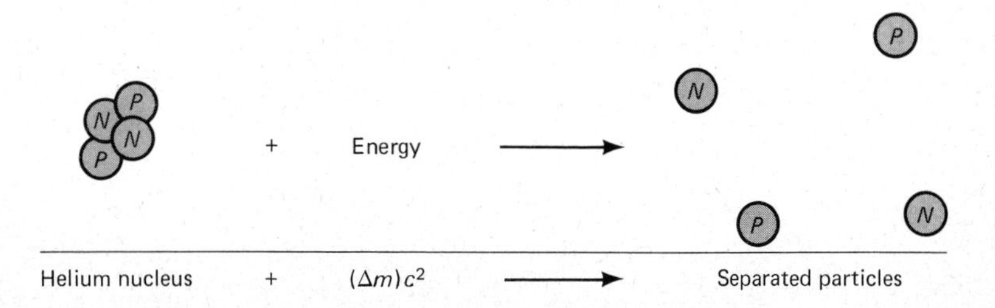

Figure 31.6 Energy is needed to tear a nucleus apart. If Δm is the mass defect for the nucleus, then the energy required is $(\Delta m)c^2$.

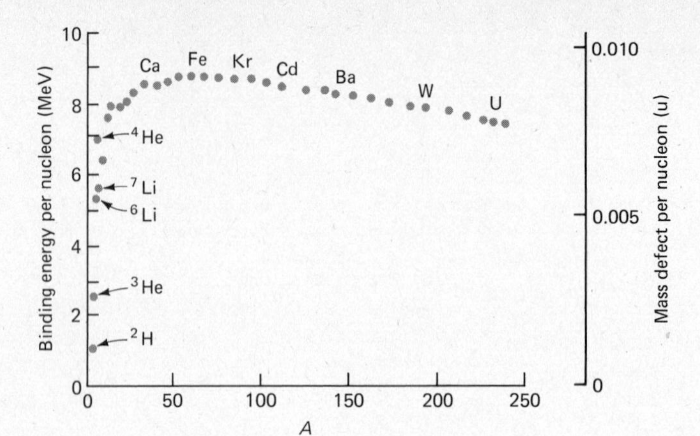

Figure 31.7 Those nuclei for which the binding energy per nucleon is largest are the most stable.

to the mass defect for the nucleus. The more mass lost when the nucleus was first formed, the more energy must be furnished to tear the nucleus apart.

We can gain important information from a graph of binding energy per nucleon as a function of atomic mass number. This graph is shown in Figure 31.7. Because binding energy and mass defect are proportional, this graph looks much like that for the mass defect in Figure 31.5.

Notice what a high binding energy per nucleon means. When the binding energy per nucleon is large, the particles are very strongly held to the nucleus. Therefore, the most stable nuclei are those for which the binding energy per nucleon is largest. From the graph the very smallest and very largest nuclei are least stable. The nuclei close to $A = 70$ are the most stable.

■ **EXAMPLE 31.4** How much energy is released as 1 kg of neutrons and 1 kg of protons are changed into 2 kg of helium 4 nuclei? (This reaction goes on in the sun and gives off tremendous heat.)

Solution We found that 28 MeV of energy is released when one ^{4}He nucleus is formed. The mass of a ^{4}He nucleus is 4.0 u or $4 \times 1.66 \times 10^{-27}$ kg. In 2 kg of helium the number of ^{4}He nuclei is:

$$\text{Number} = \frac{2\ \text{kg}}{\text{mass of He nucleus}}$$
$$= \frac{2\ \text{kg}}{4 \times 1.66 \times 10^{-27}\ \text{kg}} = 3.0 \times 10^{26}$$

The total energy released when these are formed is $(28\ \text{MeV}) \times (3 \times 10^{26})$. This follows because 28 MeV is released when one nucleus is formed. Therefore,

$$\text{Energy released} = 8.4 \times 10^{33}\ \text{eV}$$
$$= 1.3 \times 10^{15}\ \text{J}$$

For comparison purposes, when 1 kg of carbon is burned in oxygen, about 3×10^7 J of heat is released. Chemical reactions furnish on the order of 10^{-7} times less energy than the joining together of protons and neutrons does. ■■

■ **EXAMPLE 31.5** Show that 1 u of mass is equivalent to 931 MeV of energy.

Solution We simply substitute in the mass-energy relation, $\Delta E = (\Delta m)c^2$:

$$\begin{aligned}\text{Energy} &= \left(1\text{ u} \times 1.6605 \times 10^{-27}\frac{\text{kg}}{\text{u}}\right)(2.998 \times 10^8\text{ m/s})^2 \\ &= 1.492 \times 10^{-10}\text{ J} \\ &= 931\text{ MeV}\end{aligned}$$

This is a convenient factor to remember:

1 u is equivalent to 931 MeV of energy. ■■

31.7 Radioactivity

The Coulomb repulsion forces between the charged protons in a nucleus try to tear the nucleus apart. The nuclear attractive force overcomes this repulsion in the stable elements found in nature. Even so, the balance between forces is rather delicate. Only the isotopes found in nature are stable. There were probably many more isotopes in existence on the earth over 4 billion years ago when the solar system was first formed. But they were unstable and have disintegrated since then.

There are still a few unstable isotopes found in nature. They disintegrate so slowly that they have not completely disappeared. All the elements with Z larger than 83 are unstable. They are slowly disappearing from the earth. Among them are radium, thorium, and uranium.

Other unstable isotopes are formed artificially. Large nuclear reactors produce many unstable nuclei. We shall see how this is done in a later section. Unstable isotopes are often formed in nuclear laboratories by bombardment. Also, high-energy particles from space, called cosmic rays, strike the earth. They collide with stable nuclei and change them to unstable forms.

All unstable nuclei, no matter what their origin, are said to be radioactive. As time goes on they undergo change. These changes are spontaneous. They happen even though no outside disturbance acts on the nuclei.

Radioactive nuclei found in nature undergo change in three major ways (see Figure 31.8).

1. The nucleus may emit gamma rays (γ rays). This radiation is equivalent to short-wavelength X rays.
2. The nucleus may throw out an alpha particle (α particle). This particle consists of two protons and two neutrons. It is simply a helium 4 nucleus.
3. The nucleus may throw out a beta particle (β particle). This particle is an ordinary electron.

Each of these methods causes the nucleus to change in a different way.

When a gamma ray is thrown out, neither the mass nor the charge of the

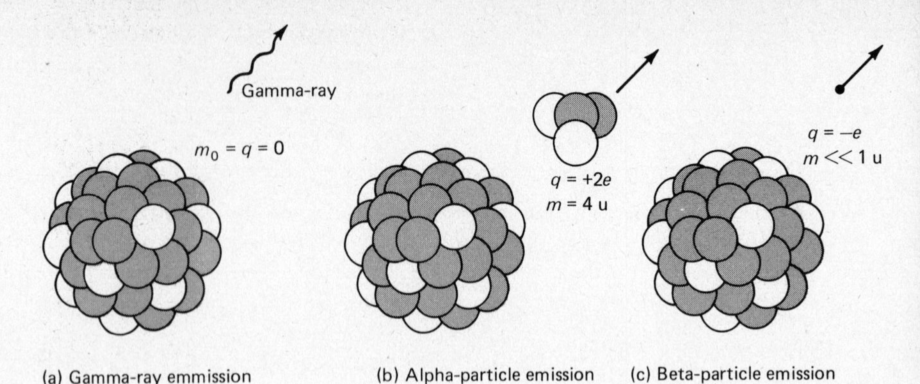

Figure 31.8 Radioactive elements found in nature undergo change mainly by three types of emission.

nucleus changes.[2] The gamma ray is a photon. It simply carries energy away from the nucleus. When the nucleus emits a gamma ray, it falls from an excited state to a lower state. We write the emission of a gamma ray as a reaction in the following way. Suppose that the original isotope is ${}^{A}_{Z}\text{X}$. Then the reaction is

$${}^{A}_{Z}\text{X} \rightarrow {}^{A}_{Z}\text{X} + \gamma$$

Notice that the isotope retains the same charge (Z) and mass (A).

When a nucleus emits an alpha particle, both its mass and charge change. The alpha particle is ${}^{4}_{2}\text{He}$. When it is thrown out, the nucleus loses 2 units of charge and 4 units of mass. We can write the reaction as

$${}^{A}_{Z}\text{X} \rightarrow {}^{A-4}_{Z-2}\text{Y} + {}^{4}_{2}\alpha$$

The products are a new element and an alpha particle. As we see, the new element has an atomic number $Z - 2$ instead of Z. This particular isotope has a mass number of $A - 4$.

It is surprising perhaps that a nucleus should emit a beta particle, an electron. The nucleus contains no electrons. However, as a mental aid we can think of the neutron as being a proton plus an electron. As a result, the end result of beta-particle emission is that a neutron within the nucleus is replaced by a proton. We can write this reaction as

$${}^{A}_{Z}\text{X} \rightarrow {}_{Z+1}{}^{A}\text{Q} + {}^{0}_{-1}\beta$$

The beta particle, an electron, has a mass much less than 1 u. We assign it a mass number zero.

As a check when writing nuclear reactions, you should notice the following:

1. The sum of the atomic numbers on the two sides of the arrow must be equal. For example, in the last reaction, $Z = (Z + 1) + (-1)$. (This follows from the fact that charge is conserved.)

[2]The mass changes by an amount $<<<$ 1 u.

2. The sum of the mass numbers on the two sides of the arrow must be equal. For example, in the last reaction, $A = A + 0$.

Let us now illustrate these reactions by some examples.

EXAMPLE 31.6 The radium nucleus $^{226}_{88}\text{Ra}$ emits an alpha particle. Write the reaction.

Solution We have

$$^{226}_{88}\text{Ra} \rightarrow {}^{222}_{86}? + {}^{4}_{2}\alpha$$

The product is represented by a question mark. Looking in the periodic table, we see that element 86 is radon (Rn). Therefore the reaction is

$$^{226}_{88}\text{Ra} \rightarrow {}^{222}_{86}\text{Rn} + {}^{4}_{2}\alpha$$

Radon is a radioactive gas. Its chemical properties are much like those of neon. ■■

EXAMPLE 31.7 Cobalt 60 emits a gamma ray. Write the reaction.

Solution Z for cobalt is 27. We have

$$^{60}_{27}\text{Co} \rightarrow {}^{60}_{27}\text{Co} + {}^{0}_{0}\gamma$$

■■

EXAMPLE 31.8 Bismuth 210 ($^{210}_{83}\text{Bi}$) is radioactive and emits a beta particle. Write the reaction.

Solution The reaction is

$$^{210}_{83}\text{Bi} \rightarrow {}^{210}_{84}? + {}^{0}_{-1}\beta$$

Element 84 is polonium (Po), so this becomes

$$^{210}_{83}\text{Bi} \rightarrow {}^{210}_{84}\text{Po} + {}^{0}_{-1}\beta$$

■■

31.8 Half-Life of Radioactive Substances

Even a tiny piece of substance contains billions upon billions of atoms. If the substance is radioactive, the nuclei of the atoms undergo change, one by one. We need some way to state how fast the change takes place. To do this, we define the half-life, $T_{1/2}$, of the substance.

The half-life, $T_{1/2}$, of a radioactive substance is a time. It is the time taken for half of the nuclei of the substance to undergo change.

For example, we saw in the last section that radium changes to radon by emitting an alpha particle. If you are given 1 mg of pure radium, half of it will change to radon in a time of 1620 years. The half-life of radium is 1620 years.

Half-lives vary widely. The half-life of uranium 238, the most common isotope, is 4.5×10^9 years. But some of its other isotopes have half-lives of only a few minutes. Cobalt 60 has a half-life of 5.25 years. Some isotopes formed in nuclear reactors have a half-life of only a fraction of a second.

In spite of the difference in half-lives, all radioactive substances undergo decay (change) in the same way with time. They decay according to the laws of probability. To illustrate what we mean, consider an isotope for which $T_{1/2} = 1$ day. Suppose that there are N nuclei of the substance at time $t = 0$. After 1 day, there will be $\frac{1}{2}N$ nuclei left. The rest have undergone change. During the second day, half of these $\frac{1}{2}N$ will undergo change. So when $t = 2$ days, the number left will be $\frac{1}{4}N$. Similarly, at the end of 3 days, there will be $\frac{1}{2}(\frac{1}{4}N)$ or $\frac{1}{8}N$ left. And so on. A graph of this behavior, the decay curve, is shown in Figure 31.9.

The decay curve never reaches zero. No matter how long the substance has existed, only half of it will decay in the next half-life. It is for this reason that some uranium 238 is still in existence on the earth. Although the universe is about 12×10^9 years old, only half of what remains is lost during each half-life. We shall see in Section 31.10 why substances such as radon (half-life = 3.8 days) are still found in nature.

■ **EXAMPLE 31.9** Cobalt 60 is provided to industry by agencies that operate research-type nuclear reactors. It is used as a source of gamma rays (X rays). Its half-life is 5.25 years. About what fraction of the original material will still exist after about 26 years?

Solution Twenty-six years is about five half-lives. We can make the following table:

Time in half-lives →	0	1	2	3	4	5
Amount left →	1	0.5	0.25	0.125	0.0625	0.0313

Therefore, after 26 years only about 3 percent of the original material is left. The radiation source will have lost about 97 percent of its intensity. ■■

31.9 The Decay Law

The decay curve in Figure 31.9 shows the number N of particles remaining undecayed at time t after there had been an original number N_0. To emphasize that the number of particles remaining is decreased by half during each half-life, we have redrawn the decay curve in Figure 31.10. Notice that the curve will never reach zero because half of something is still something.

The relation between N and N_0 can be represented in equation form as well. The curve of Figure 31.10 is given by

$$N = N_0 e^{-\lambda t} \tag{31.2}$$

We call this the *exponential decay law*. It receives its name from the fact that $e^{-\lambda t}$ is called an exponential function in mathematics. It will be sufficient for us to know that this equation, when plotted, gives the graph shown in Figure 31.10.

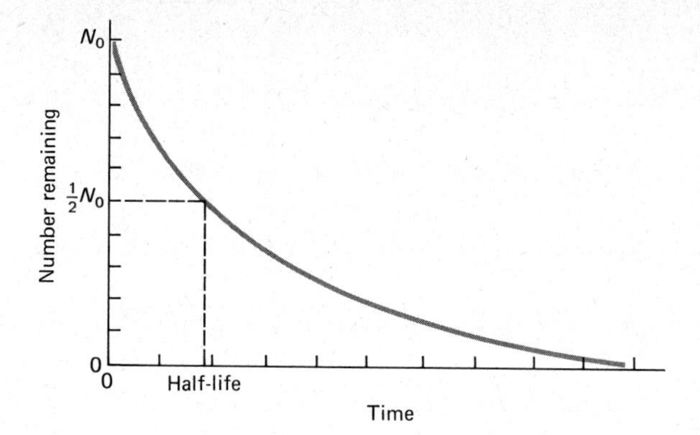

Figure 31.9 All pure radioactive substances follow the decay curve shown. It is an exponential function and approaches zero only at $t \to \infty$.

The quantity λ in Equation 31.2 is called the *decay* or *disintegration constant.* It is related to the half-life through

$$\lambda = \frac{0.693}{T_{1/2}} \tag{31.3}$$

It has the following important use. If a radioactive sample contains N nuclei with decay constant λ, then the number ΔN of nuclei that will decay in time Δt is given by

$$\Delta N = \lambda N \, \Delta t \tag{31.4}$$

The decay law, Equation 31.2 can be put in a more usable form by taking logarithms of each side. Upon simplification, one finds that

$$2.303 \log_{10}\left(\frac{N_0}{N}\right) = \lambda t$$

Let us now use these relations in an example.

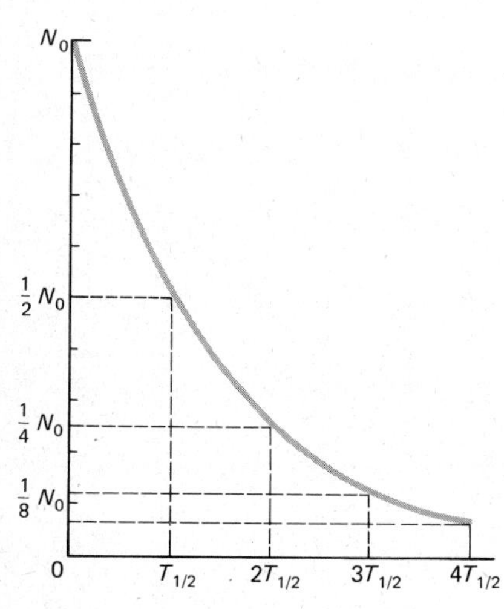

Figure 31.10 The height of the decay curve (the number of particles remaining) decreases by half in each half-life.

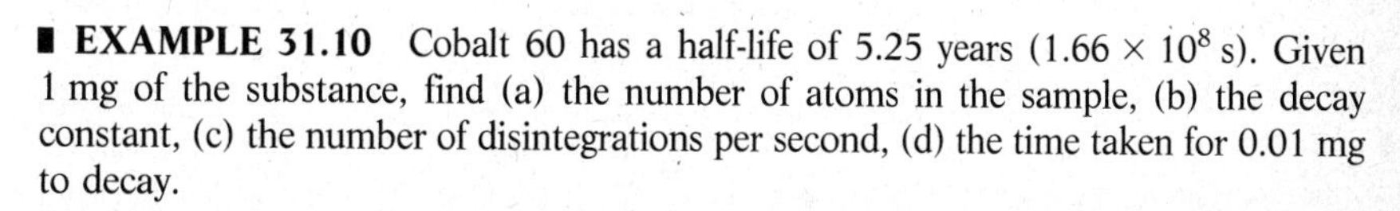

■ **EXAMPLE 31.10** Cobalt 60 has a half-life of 5.25 years (1.66×10^8 s). Given 1 mg of the substance, find (a) the number of atoms in the sample, (b) the decay constant, (c) the number of disintegrations per second, (d) the time taken for 0.01 mg to decay.

Solution (a) The mass of a cobalt 60 atom is about 60 u.[3] This is $60 \times 1.66 \times 10^{-27}$ kg, or 10×10^{-26} kg. In 1×10^{-6} kg (i.e., in 1 mg) there are the following number of atoms.

$$N_0 = \frac{\text{mass}}{\text{mass/atom}} = \frac{1 \times 10^{-6}\ \text{kg}}{10 \times 10^{-26}\ \text{kg/atom}} = 1 \times 10^{19}\ \text{atoms}$$

[3]The exact value, found in tables that list the masses of the isotopes, is 59.93381 u.

(b) We know that

$$\lambda = \frac{0.693}{T_{1/2}} = \frac{0.693}{1.66 \times 10^8 \text{ s}} = 4.2 \times 10^{-9} \text{ s}^{-1}$$

(c) At the start, $N = N_0 = 1 \times 10^{19}$ atoms. So we have

$$\Delta N = \lambda N \, \Delta t = (4.2 \times 10^{-9} \text{ s}^{-1})(1 \times 10^{19} \text{ atoms}) \, \Delta t$$

from which

$$\frac{\Delta N}{\Delta t} = 4.2 \times 10^{10} \text{ atoms/s}$$

(d) We wish to find t when $N/N_0 = 0.99$. Using

$$2.303 \log\left(\frac{N_0}{N}\right) = \lambda t$$

gives

$$2.303 \log(1.010) = (4.2 \times 10^{-9} \text{ s}^{-1})t$$

or

$$2.303(0.0043) = (4.2 \times 10^{-9} \text{ s}^{-1})t$$

from which $t = 2.4 \times 10^6$ s. ■■

31.10 Radioactive Series

Many isotopes originally present in the universe have long since decayed to negligible amounts. If it were not for the long half-lives of uranium 238 (4.5×10^9 years), uranium 235 (7.1×10^8 years), and thorium 232 (1.4×10^{10} years), the highest Z element found on earth would be lead ($Z = 82$). These three isotopes are still found on earth. Each decays slowly enough that appreciable amounts of them are still left. Let us see what happens as one of them decays.

When uranium 238 decays, it emits an alpha particle, and the $^{238}_{92}U$ nucleus is changed to thorium 234. But this nucleus is also unstable and decays by emission of a beta particle, forming another unstable nucleus. The decay progresses through what is called a radioactive series, as shown in Figure 31.11. Finally, the decay results in lead 206. This is a stable nucleus, and the series then ends.

In a similar way, the uranium 235 series eventually ends in lead 207. The thorium 232 series finally ends in lead 208. These are the three radioactive series found in nature. All the elements in the series above lead are unstable with fairly short half-lives. As a result, they would long since have disappeared if they were not being formed continuously by the series.

We should also notice that most of the helium that exists on earth is a result of these radioactive series. Alpha particles (helium nuclei) are emitted during

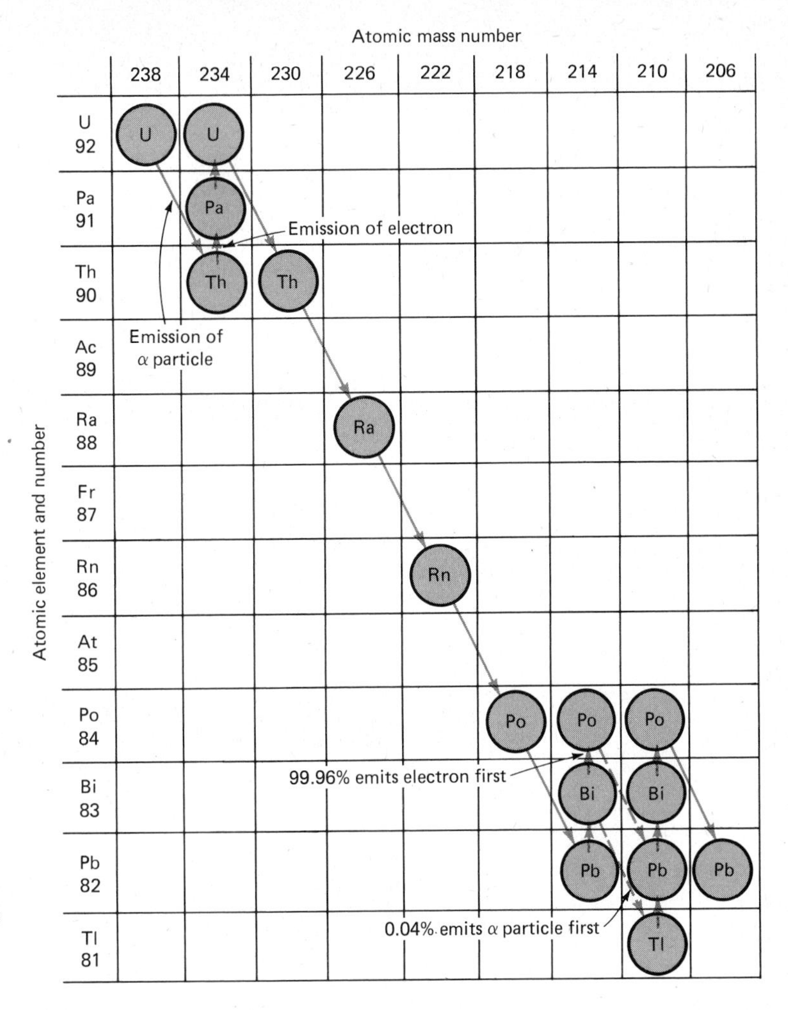

Figure 31.11 The uranium 238 series is typical of the three radioactive series found in nature. They are named for the parent nucleus-uranium 238, uranium 235, and thorium 232.

the decay processes. These particles are slowed by collision and pick up electrons to become helium atoms. Because helium is a chemically nonreactive gas, it eventually enters the atmosphere. It is such a light atom that it then escapes from the earth. Most of the helium found on earth is the result of radioactive decay processes within the earth.

■ **EXAMPLE 31.11** Rocks are found with about 50-50 mixtures of uranium 238 and lead 206. The half-life of uranium 238 is about 4.5 billion years. Use this fact to estimate the age of these rocks.

Solution The lead 206 mixed with uranium 238 almost certainly comes from the decay of the uranium 238. Because they are about equal in quantity, one estimates that about half the original uranium 238 has decayed. The time taken for this would be about 4.5 billion years. Therefore it appears that the rock solidified about 4.5 billion years ago. Data such as these lead us to believe that the earth was formed about 4.5×10^9 years ago. ■■

EXAMPLE 31.12 The isotope of carbon, carbon 14, is formed at the upper edge of the atmosphere as a result of bombardment by high-energy particles from space, cosmic rays. This isotope has a half-life of 5730 years. The isotope mixes with the normal carbon 12 on earth and becomes part of the plants and animals on earth. Before atomic bomb testing began, the ratio of ^{14}C to ^{12}C in plant life was 1.5×10^{-12}. Suppose that an old piece of wood has a ^{14}C to ^{12}C ratio of $\frac{1}{4} \times 1.5 \times 10^{-12}$. How old is the piece of wood?

Solution Originally, when the wood first died, it had a ^{14}C to ^{12}C ratio of about 1.5×10^{-12}. As time went on, the ^{14}C decayed. Now there is only one-fourth of it left. Therefore, a time equal to two half-lives must have gone by since the wood died. The age of the wood is therefore two ^{14}C half-lives, or about 11,500 years old. This method is called *radiocarbon dating*. Similar techniques are sometimes used with other isotopes. ■■

EXAMPLE 31.13 An old piece of wood has a ratio of ^{14}C to ^{12}C that is only 0.70 as large as for a new piece of wood. Find the age of the wood.

Solution Radioactive decay has occurred long enough that $N/N_0 = 0.70$. The half-life of ^{14}C is 5730 years, and so

$$\lambda = \frac{0.693}{T_{1/2}} = \frac{0.693}{5730 \text{ years}} = 1.21 \times 10^{-4} \text{ years}^{-1}$$

Then, using

$$2.303 \log\left(\frac{N_0}{N}\right) = \lambda t$$

gives

$$2.303 \log(1.43) = (1.21 \times 10^{-4} \text{ year}^{-1})t$$

or

$$(2.303)(0.155) = (1.21 \times 10^{-4} \text{ year}^{-1})t$$

Solving for t gives the age of the wood to be

$$t = 2950 \text{ years}$$

■■

31.11 Properties of Radiation

Before proceeding further with our discussion of radioactivity, let us discuss the radiation radioactive substances emit. We will be concerned with the way each radiation behaves as it shoots through material.

Alpha Particles Alpha particles are helium nuclei, two protons joined with two neutrons. Their charge is $+2e$. Because of their large charge and large size, they easily ionize atoms that they strike. Only rarely do they strike the tiny nucleus of an atom. Most of their energy is lost in interactions with atomic electrons. They lose about 30 eV during each electron encounter. Because alpha particles usually have energies of several MeV, they will ionize nearly a million atoms before coming to rest.

When passing through matter, an alpha particle behaves like a cannon ball shooting through a room filled with ping-pong balls. The massive particle has no difficulty in pushing the electrons aside. As it tears electrons loose from atoms and molecules, it causes great damage. It leaves a line of ionized atoms and broken molecules along its path.

Because an alpha particle loses its energy bit by bit in millions of tiny collisions, its path is straight. Particles of the same initial energy travel almost exactly the same distance before stopping. We call this distance the *range* of the particle. A 1-MeV alpha particle has a range of about 0.5 cm in air, 0.003 cm in flesh, and 0.0003 cm in aluminum. The distances are approximately doubled for a 2 MeV particle. As you see, alpha particles are easily stopped. Hence it is easy to shield oneself from them. But danger becomes greater if alpha emitters are swallowed. Then they may localize in the body and cause serious damage.

Protons Protons pass through mater much like alpha particles do, but because of their smaller charge and mass, they cause less ionization per unit path length. As a result, a proton usually has a larger range than an alpha particle of the same energy. A 1-MeV proton's range in air is about ten times larger than that for an alpha particle of the same energy.

Neutrons Because neutrons have no charge, they are much more penetrating than protons, even though their masses are the same. This points out that energy loss is mainly the result of ionization. Charged particles can dislodge an electron from an atom by simply passing by. The Coulomb force acts as a collision force of one particle on the other. But this is not possible for neutrons because they have no charge. They are best stopped by direct collisions with light nuclei. Remember that in an elastic head-on collision between two equal mass particles, the particles simply interchange their velocities.

Beta Particles Beta particles are electrons. Because of their tiny mass, they are hard to stop, but easy to deflect. A 1-MeV electron has a range of perhaps 5 m in air. Electrons with energy this low have a less definite range than larger particles do. Their momentum is so low that they deflect easily, thus following a nonlinear path.

Gamma Rays Like all photons, gamma rays have neither charge nor rest mass. They lose energy in entirely different ways than do charged particles. At energies less than about 1 MeV, they lose energy chiefly by the atomic photoelectric effect. They eject an electron from an atom upon collision. This effect is so large for very soft X rays (or gamma rays) that they are stopped by the skin of a person. (A "soft" X ray is one that has an energy below about 1 keV.) But 30,000-eV X rays can pass through a person's arm. High energy gamma rays are very penetrating.

In the photoelectric effect the photon loses all its energy to the electron and thus disappears at that point. Therefore a gamma-ray beam does not have a definite range. Photons are lost from the beam all along its path, and the beam loses intensity as it travels through matter. The penetrating power is characterized by the distance the beam goes until its intensity is reduced by one-half. This distance or thickness is called the *half-value layer.* For a 1-MeV photon, it is about 10 cm in flesh and 0.9 cm in lead. Because lead has more electrons per unit volume than most materials, lead is the most effective common shield against gamma and X rays.

EXAMPLE 31.14 The half-value layer for 10,000-eV X rays in lead is 0.00076 cm. Electrons striking a TV screen generate X rays in this range. About what thickness of lead is required to reduce the beam from the tube to 0.01 of its original value?

Solution Each 0.00076 cm decreases the beam intensity by $\frac{1}{2}$. Because 0.01 is about $(\frac{1}{2})^7$, we see that the required thickness is about

$$7\ (0.00076\ \text{cm}) = 0.0053\ \text{cm}$$

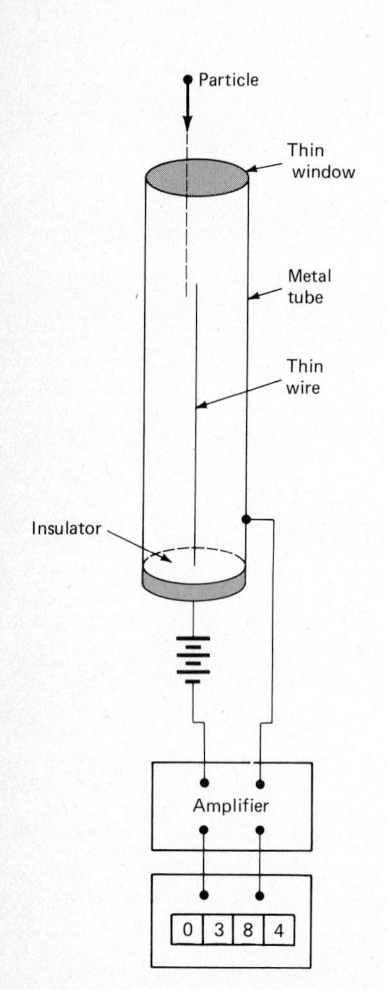

Figure 31.12 The Geiger counter. A proportional counter is similar in design.

31.12 Radiation Detectors

As we saw in the last section, all common radiation except neutrons produces ions. This fact is basic to the operation of many detectors of radiation. In this section we will describe a few of the most widely used detectors.

Geiger Counter The Geiger counter is sketched in Figure 31.12. The metal tube is closed at one end by a very thin film. Within the tube is a special gas at reduced pressure. A potential difference of perhaps 1000 V exists between the metal tube and the thin central wire.

When a charged particle shoots through the tube, it produces positive ions along its path. The ions migrate toward the negative electrode. The corresponding freed electrons accelerate toward the positive wire. They produce further free electrons by collisions and, as a result, an electrical discharge occurs in the tube. This discharge lasts only an instant. The discharge causes a current pulse to pass through the amplifier. A counter records each pulse that the amplifier feeds to it. Thus each particle that enters the Geiger tube is counted.

Charged particles such as electrons are counted easily by the Geiger counter. Gamma rays and neutrons can also be detected by use of special techniques. Usually it is more convenient to use other devices for detecting them.

Proportional Counter The proportional counter is similar to the Geiger counter, but it operates at smaller voltages and no discharge occurs in the tube. Instead, each current pulse has a magnitude proportional to the ionization caused by the incoming radiation. Because the number of ions is proportional to the energy of the particle, the pulse magnitude is a measure of particle energy. Using proper circuitry, the energies of the incoming particles can be measured.

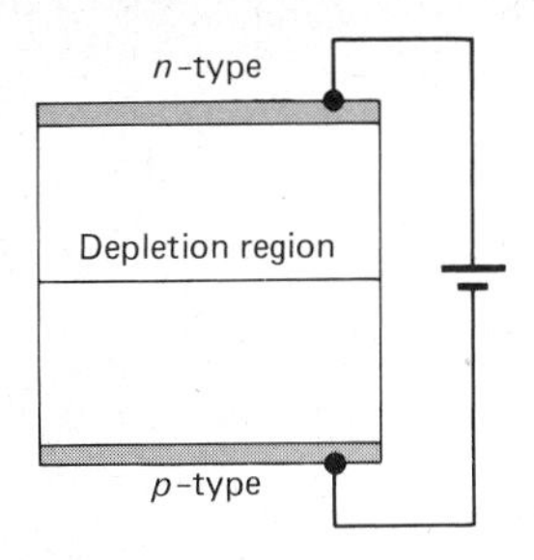

Figure 31.13 An ionizing particle passing through the depletion region causes a current pulse.

Scintillation Counter When an electron strikes the fluorescent material on the screen of a television tube, a flash of light is given off. This same effect can be used to count ionizing particles and gamma rays. A block of transparent fluorescent material, called the *scintillator,* is used as a detector. The block is cemented to the face of a phototube (or, in practice, a photomultiplier tube). A particle or photon entering the fluorescent block causes a pulse of light. The light is channeled into the phototube and a current pulse is generated. These pulses are counted electronically, thereby counting the incident particles. Because the light pulse is proportional to the energy of the particle, particle energies can be measured as well.

Semiconductor Diode Consider the reverse-bias semiconductor diode shown in Figure 31.13. No current flows under these conditions. However, if an ionizing particle or photon shoots through the junction, excess holes and free electrons are generated. You will recall from Chapter 30 that a high electric field exists in the depletion region under reverse-bias conditions. The excess holes and electrons move in opposite directions because of this field. As a result, a pulse of current occurs for each particle that passes through the junction. These detectors are extremely fast and versatile. They are widely used.

Figure 31.14 This bubble chamber photo shows two antiprotons entering at the bottom. They are annihilated by protons at the site of the four-pronged stars. (Lawrence Berkeley Laboratory, University of California)

Cloud, Bubble, and Spark Chambers These devices give rise to a visible trace of the particle's path. The path is then usually photographed. In the cloud chamber the particle generates ions in a supersaturated vapor. Vapor droplets form preferentially on the ions and so the droplets trace out the path. Similarly, in a bubble chamber bubbles in a superheated liquid form along the path. A spark chamber consists of many parallel plates with large electric fields between them. Sparks jump along the ionized path and thus trace out the path. See Figure 31.14 for a typical bubble chamber photo.

31.13 Activity of Radioactive Samples

A convenient way to describe a radioactive sample is to state how many decays occur in it each second. We call this quantity the *activity* of the sample.

The activity of a radioactive sample is the number of decays that occur in unit time in the sample.

In equation form,

$$\text{Activity} = \frac{\Delta N}{\Delta t} = \lambda N \qquad (31.5)$$

where use has been made of Equation 31.4. The activity of a sample is a measure of the radioactive strength of the sample; it is a measure of the radiation that the sample emits.

The SI unit of activity is the becquerel (Bq), after Henri Becquerel, who discovered radioactivity in 1896. Another unit commonly used for activity is the curie (Ci), after Marie Curie, who discovered radium in 1898.

In 1 Bq of a substance, 1 nucleus decays per second. In 1 Ci of a substance, 3.70×10^{10} nuclei decay per second.

$$\mathbf{1\ Ci = 3.7 \times 10^{10}\ Bq}$$

The curie is a large unit. Most radiation sources are in the millicurie range. It was originally intended that the activity of 1 g of radium be exactly 1 Ci. But more precise measurements of the half-life of radium showed that the curie as defined was a little larger than the activity of 1 g of radium. Typical radium sources used in medical work contain only a fraction of a curie. However, cobalt 60 gamma-ray sources used for cancer therapy and in industry often have activities of several thousand curies.

■ **EXAMPLE 31.15** What is the activity of 1 g of radium? The half-life of radium is 5.1×10^{10} s and its atomic mass number is 226.

Solution The activity is equal to λN, where

$$\lambda = \frac{0.693}{5.1 \times 10^{10}\ \text{s}} = 1.36 \times 10^{-11}\ \text{s}^{-1}$$

The mass of one atom is about

$$(226\ \text{u/atom}) \times (1.66 \times 10^{-27}\ \text{kg/u}) = 3.75 \times 10^{-25}\ \text{kg/atom}$$

Therefore N is given by

$$N = \frac{\text{mass}}{\text{mass/atom}} = \frac{1 \times 10^{-3}\ \text{kg}}{3.75 \times 10^{-25}\ \text{kg/atom}} = 2.67 \times 10^{21}\ \text{atoms}$$

We then have that

$$\begin{aligned}\text{Activity} &= \lambda N = (1.36 \times 10^{-11}\ \text{s}^{-1})(2.67 \times 10^{21}\ \text{atoms}) \\ &= 3.63 \times 10^{10}\ \text{decays/s} = 3.63 \times 10^{10}\ \text{Bq} \\ &= 0.98\ \text{Ci}\end{aligned}$$

■■

31.14 Absorbed Radiation Dose

When high-energy radiation strikes atoms and molecules, it causes ions to be formed. It does this by knocking electrons loose from atoms. Often the molecule of which the atom is a part is destroyed in the process. In many applications of radiation, the effects of radiation are proportional to the amount of energy absorbed from the beam. For this reason a unit is needed to measure the total amount of radiation absorbed by a material exposed to the radiation beam.

The common unit of absorbed radiation is called the rad.

When 1 kg of material absorbs 10^{-2} J of radiation, the absorbed dose is 1 rad. For example, if your body receives a dose of 0.10 rads (100 millirads), each kilogram of your body absorbs $0.10 \times 0.01 = 10^{-3}$ J of radiation energy on the average. It does not matter whether the energy comes from a beam of X rays, high-energy protons, or whatever. If 0.01 J of beam energy is absorbed per kilogram, the radiation dose is 1 rad.

The SI unit for absorbed radiation is the gray (Gy).

One gray of radiation is equivalent to 1 J of radiant energy absorbed per kilogram.

From the definitions, we see that 1 Gy = 100 rads. Although the gray is the preferred unit, the rad is frequently used.

31.15 Biologically Equivalent Dose

Your body is damaged less by 1 rad of X rays than it is by 1 rad of high-energy protons. Even though both would deposit 0.01 J of energy per kilogram of your body, the damage done to your body would differ in the two cases. We need, then, another unit to describe the effect of radiation on biological tissue.

To set up such a unit, we use what is called the *relative biological effectiveness* (RBE) of radiation. A beam of 200-keV X rays is taken as standard for this comparison. It is assigned an RBE of unity. Approximate values of RBE for various types of radiation beams are shown in Table 31.1.

We see that protons are about two times as damaging as 200-keV X rays. For them, RBE = 2. A 1-rad dose of protons causes the same damage as a 2-rad dose of 200-keV X rays. To account for this, we define a unit that describes biological damage resulting from a beam. This unit we call the *rem*. It is defined in the following way.

Table 31.1 APPROXIMATE RBE VALUES (BASED ON 200-keV X RAYS HAVING RBE = 1)

Radiation	RBE
5-MeV gamma rays	0.5
1-MeV gamma rays	0.7
200-keV gamma rays	1.0
Electrons	1.0
Protons	2.0
Neutrons	2–10
Alpha particles	10–20

$$\text{Biologically equivalent dose in rem} = (\text{dose in rad}) \times (\text{RBE})$$

For example, a 1-rad dose of 200-keV X rays gives a biologically equivalent dose of 1 rem because the RBE is unity in this case. A 1-rad dose of protons gives a biologically equivalent dose of 2 rem. A 1-rad dose of cobalt 60 gamma rays that have an energy of about 1.3 MeV gives a biologically equivalent dose of 0.7 rem.

31.16 Radiation Safety

Exposure to radiation should be kept to a minimum. When a high-energy particle or photon strikes the body, it has the ability to damage a molecule of the body. When the particle strikes a chemical bond within a molecule, the bond will be broken if the particle has enough energy. For example, ultraviolet photons in sunlight damage the molecules of the skin and cause sunburn. More penetrating radiation such as X rays and fast particles can damage molecules deeper within the body. Use of this fact is made when high-energy X rays (or gamma rays) are used to treat cancer. The radiation is focused onto the tumor cells and destroys them.

Because radiation can destroy and damage cells, it can harm the body. In addition to the effects of direct cell damage, other bad effects also may occur. The damaged cells may live but, as they multiply, form mutant cells that are different from the original cell. These mutant cells can cause cancer. If the mutant cells are part of the reproductive system, they can cause birth defects in future children.

It is impossible to escape all radiation. On the average, we all experience a yearly dose of about 40 millirems from the radioactive atoms in the soil, rocks, and wood around us. This type of radiation varies widely from place to place. On the Atlantic Coast of the United States it is about 20 millirems, while it is about 90 millirems near the Rocky Mountains.

Other inescapable radiation is the result of cosmic rays. These highly energetic particles from outer space bombard the earth. A few particles from cosmic rays pass through your body each minute. Because of them, the average person receives a yearly dose of about 45 millirems. The dose rate varies with altitude above the earth, because the atmosphere tends to protect us. At sea level it is about 40 millirems per year, while it is about 150 millirems per year at high altitudes in the Rocky Mountains.

It is estimated that the average dose received by a person due to medical X rays is about 70 millirems per year. In addition, traces of radioactivity in our food and water supplies contribute a dose of about 20 millirems per year.

We see from this that the average person receives a radiation dose of about 170 millirems per year from relatively inescapable sources. It is estimated that this radiation dose causes about 6000 cancer deaths per year in the United States. This is about 0.2 percent of the total number of yearly deaths due to cancer. We should view such estimates with caution, however, because they are based on extrapolation from data at much higher doses. To put these numbers in perspective, we should notice that this percentage is at least a factor of 100 times smaller than cancer deaths caused by smoking.

Workers in medical and nuclear facilities are protected by regulations that limit allowed exposure to radiation. It is recommended that no worker be subject to a dose of more than 3 rems in 3 months, with a yearly limit of 5 rems. The reason for the time limitation has to do with the fact that the body is able to repair itself. Although a dose of 100 rems in 1 day will cause serious illness, the same dose over a period of years is much less noticeable.

Considerable controversy exists concerning the effects of small radiation doses. It is well to relate newspaper accounts to the 170-millirem-per-year figure given for unavoidable exposure. One can also gain perspective by recalling that the upper limit for safe nuclear work is 5 rems per year. But, in the final analysis, our objective should be to reduce radiation exposure wherever it is feasible to do so. The first 100 millirems due to the environment is unavoidable. Additional radiation exposure is ours to choose.

31.17 Nuclear Fission

In late 1938 Otto Hahn, together with Lise Meitner and Fritz Strassmann, made a discovery that was to affect the world greatly. They found that uranium 235 nuclei could be caused to explode. The situation is shown in Figure 31.15.

Suppose that a free neutron wanders near to a uranium 235 nucleus. The nucleus is found to capture the slowly moving neutron. When the neutron becomes a part of the nucleus, the nucleus is no longer stable. It suddenly splits apart, as shown in Figure 31.15. The nucleus splits into two or more nuclei of smaller mass, that is, into the nuclei of elements near the center of the periodic table. A typical reaction might be:

$$ {}_0^1n + {}_{92}^{235}\mathrm{U} \rightarrow {}_{92}^{236}\mathrm{U} \rightarrow {}_{56}^{141}\mathrm{Ba} + {}_{36}^{92}\mathrm{Kr} + 3{}_0^1n $$

where ${}_0^1n$ is a neutron. Many other sets of products are also found. The nucleus splits apart in many different ways. On the average, about 2.5 neutrons are produced in the fission of each uranium 235 nucleus.

When a heavy nucleus splits apart into two or more nuclei of moderate mass, the nucleus is said to have undergone nuclear fission.

About 90 different products of uranium 235 fission have been found. These, in turn, are highly unstable. (For example, krypton 92 is far from the closest stable form of krypton, namely, krypton 86.) The fission fragments decay to more stable nuclei. As a result, a large quantity of radiation is given off before the products reach their final state.

Uranium 235 is the only isotope available in quantity in nature that undergoes fission easily.[4] It requires only a slow neutron to cause the process. But

[4] Many other nuclei can be torn apart by bombarding them with high-energy particles, but such reactions consume too much energy to be useful in technology.

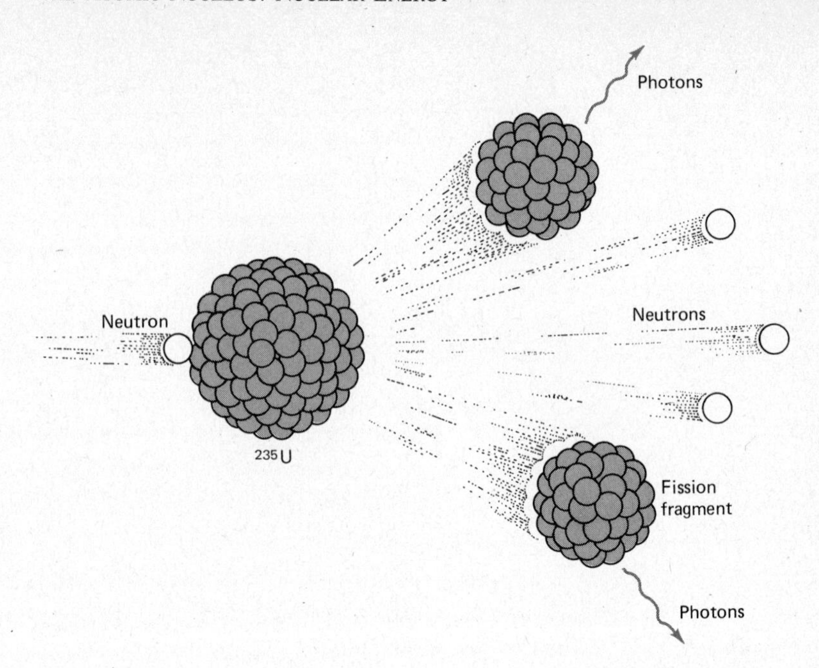

Figure 31.15 When a slowly moving neutron is captured by a uranium 235 nucleus, the nucleus undergoes fission.

uranium 235 is itself not found on earth in large quantities. Natural uranium contains about 99.3 percent uranium 238 with only about 0.7 percent uranium 235. The separation techniques of chemistry can separate only elements, not isotopes of the same elements. Therefore, extracting uranium 235 from natural uranium is a difficult task. Huge plants have been constructed to separate the isotopes by diffusion processes. The cost of the plants and the electricity to operate them is so great that the government must finance them.

Notice that the fission reaction changes a high-Z nucleus to nuclei near the center of the periodic table. The mass defect curve of Figure 31.5 points out that the product nucleons have less mass than the nucleons in uranium 235. We therefore conclude the following:

When a high-Z nucleus splits, mass is destroyed. The fission process releases energy.

This released energy is mainly kinetic energy of the fission products. But in addition, gamma rays and beta particles from radioactive decay processes also carry away energy. Eventually this energy is mostly lost through collisions. It is thereby changed to random molecular motion, which we then call heat energy.

The energy released in uranium 235 fission is extremely large. For example, the fission of 1 kg of uranium 235 generates about as much energy as the burning of 10^7 kg of coal or 500,000 gal of gasoline.[5] As we see, the fission

[5] Because the density of uranium is 18.7 g/cm^3, a comparison of volumes is even more striking. One cubic centimeter of ^{235}U generates about as much energy as a cube of coal that is 10 m on each edge.

process can be used as an energy source. We shall see how this is done in the next section, where nuclear reactors are discussed.

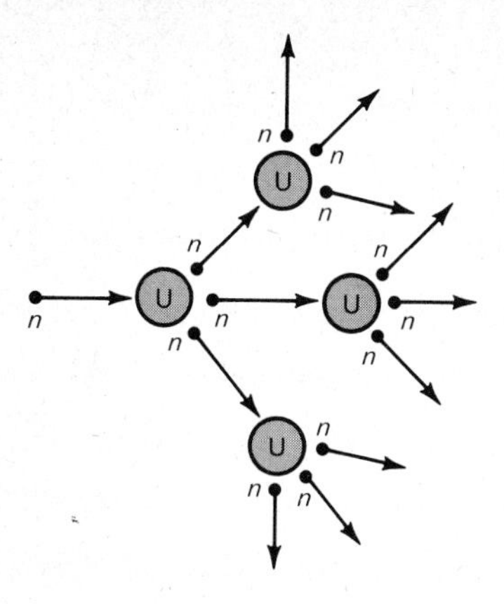

Figure 31.16 After two steps in the chain reaction, the single initial neutron has been increased to nine.

31.18 Nuclear Reactors

Two facts about the uranium 235 fission reaction make it practical as an energy source. First, because slow-moving neutrons can cause the reaction, essentially no input energy is needed to cause the reaction. Second, because each fission process generates about 2.5 neutrons, the needed neutrons are furnished by the reaction itself.

Because the reaction produces more neutrons than it requires, a chain reaction is possible. Such a reaction is shown schematically in Figure 31.16. We assume there that each fission process generates three neutrons. Notice that the number of neutrons multiplies rapidly. Suppose that each fission requires only 0.01 s. Then after 1 s, about 10^{23} kg of uranium would have been consumed! Obviously, the chain reaction is capable of releasing tremendous amounts of energy in a very short time.

Of course, in our example we made the favorable assumption that each neutron generated caused a further reaction. This is not true in general. The fraction of neutrons that cause fissions to occur depends on several factors. One of the most important is the size and shape of the original piece of uranium. If the material is in the form of a thin sheet, most of the neutrons will be lost through the surface, as shown in Figure 31.17(a). Even if the material is spherical as in (b), some neutrons will escape from the surface. Only when the sphere is made very large will the number of neutrons lost be a small fraction of the total.

There is a critical size for the uranium if the reaction is to continue. If the uranium mass is too small, then most of the neutrons will escape and the reac-

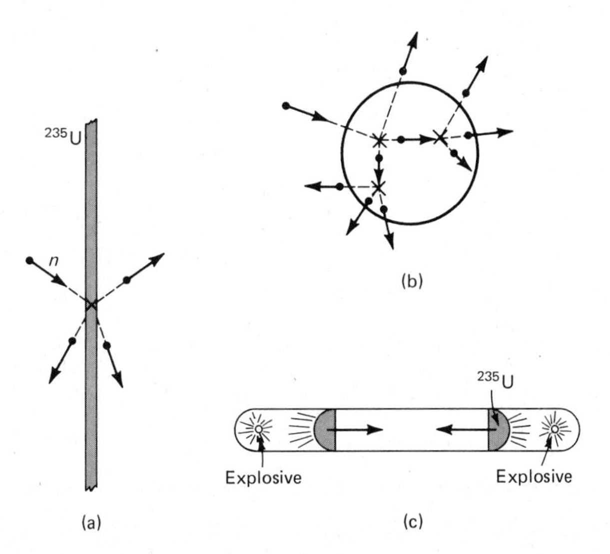

Figure 31.17 A critical size is necessary if the chain reaction is to continue and grow.

tion will stop. If the mass of uranium is extremely large, nearly all the neutrons will cause reaction and so much energy will be released in such a short time that an explosion will occur. There is a dividing line between these two cases. If the sphere is of just a critical size, then one neutron from each fission will cause a further fission. The reaction will continue on at a constant rate.

At critical size the fission reaction continues at a constant rate.

The critical size for uranium 235 is about the size of a baseball.

If uranium 235 is available in quantity, it is easy to make a simple bomb. Part (c) of Figure 31.17 shows one way to do this. Two pieces of uranium 235 slightly smaller than the critical size are placed in the far ends of a tube. They are suddenly forced together by some mechanism such as the one shown. The combined pieces are then larger than critical size. An explosion will then result. In practice, many refinements are necessary if the explosion is to be efficient and large.

Nuclear reactors are designed to maintain the fission reaction close to the critical condition. A schematic diagram of a reactor system is shown in Figure 31.18. An operational power reactor is shown in Figure 31.19.

Basically, the reactor core consists of the fuel (^{235}U in our case) contained in special sealed cylinders. These fuel cylinders are submerged in a material called the moderator. The moderator is used to slow and reflect the neutrons generated by the fission reaction. Typical moderators are water, heavy water, and carbon. Heavy water is made by replacing the hydrogen isotope ^1H in H_2O by the isotope ^2H. This isotope is called deuterium.

The rate of reaction is controlled by several control rods that can be lowered into the reactor. They usually consist of a material such as cadmium, which absorbs neutrons. To increase the rate of reaction, the rods are moved out a little. To decrease or stop the reaction, the rods are moved farther into the reactor core. In steady operation the reactor is at its critical state.

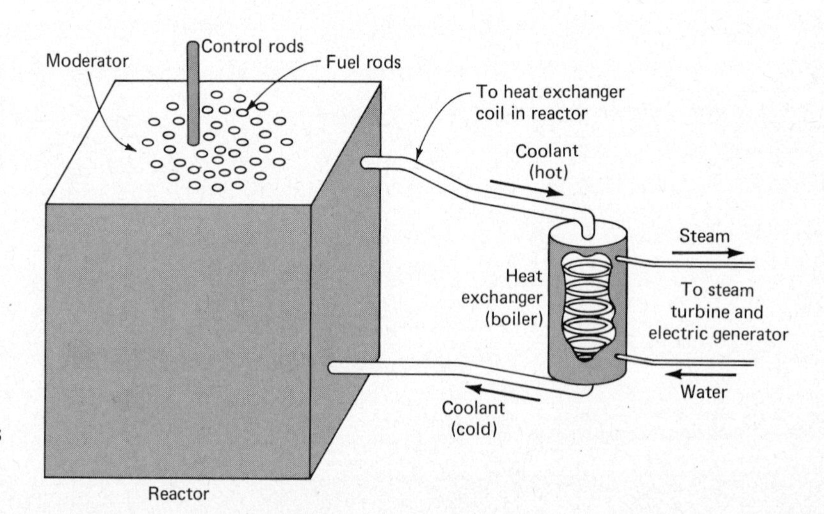

Figure 31.18 This diagram of a nuclear reactor system is only schematic. Notice that the reactor plays the part of the furnace in an ordinary coal-fired power plant.

Figure 31.19 The Palisades nuclear power plant on Lake Michigan has an output power of 700 MW. (Courtesy of Consumers Power Co.)

As we said earlier, the energy from the fission reaction eventually ends as heat. The reactor therefore acts like a furnace whose fuel is uranium 235 rather than coal, oil, or gas. The heat is removed by means of heat exchanger coils within the reactor and a fluid circulated into them. Because this fluid is subject to severe radiation within the core, it demands precautions not needed in conventional power plants.

The hot fluid circulated through the reactor is then brought to an external heat exchanger. There it is used to generate steam. This hot steam is then used as in any power plant. It drives steam turbines, which, in turn, drive electric generators.

31.19 Using the Radiation of the Reactor

Most of the products of the fission reaction are highly unstable. As a result, tremendous amounts of radioactivity exist within the reactor core. There are vast quantities of high-energy neutrons, gamma rays, and beta particles as well as other particles. Any material placed within the reactor will undergo strong bombardment by these various radiations.

One use of the reactor radiation is to produce plutonium 239, ${}^{239}_{94}\text{Pu}$. This isotope is not found in quantity on earth, and it has a half-life of 24,000 years. But ${}^{239}\text{Pu}$ has the property that it undergoes fission much like ${}^{235}\text{U}$. To produce plutonium, one places ${}^{238}\text{U}$ in the reactor core. The following reaction then occurs:

$$ {}^{238}_{92}\text{U} + {}^{1}_{0}n \rightarrow {}^{239}_{92}\text{U} $$

The ^{239}U is unstable (half-life = 24 min) and changes to neptunium 239 (^{239}Np) with the emission of a beta particle. But the ^{239}Np also decays by beta-particle emission (half-life = 2.4 days). The end product is ^{239}Pu.

Of course, the plutonium is still mixed with some ^{238}U. But these are different elements and can be separated chemically. In a *breeder reactor* more plutonium is generated than the ^{235}U used to generate it. As a result, we no longer have to rely on ^{235}U for nuclear fuel. The equally usable material, ^{239}Pu, can be prepared from the much more abundant ^{238}U.

Hundreds of useful isotopes have been made using radiation from reactors. Many of these are used in medicine. For example, iodine 131 has a half-life of 8 days and emits beta particles and gamma rays. When humans eat foods containing this isotope, it tends to accumulate in the thyroid gland. The gland then becomes radioactive and can be studied from outside the body. Further, when radiation of the thyroid is prescribed, the radiation can be carried directly to the site by this isotope.

There are too many uses of reactor-made isotopes for us to describe them all here. We have already mentioned cobalt 60. It furnishes gamma rays with energies of about 1.2 MeV. As a result, cobalt 60 can often be used in place of a 1-MV X-ray machine. Cobalt 60 and other isotopes produced in reactors are used in many engineering applications. Uses have also been found in nearly all branches of science and technology. Even if the reactor had no utility as a power source, it would still be of great value in the production of isotopes.

31.20 Nuclear Fusion

The fission power source is based on the fact that nuclei with high atomic number have more mass per nucleon than do nuclei with medium atomic number. As a result, the splitting of high-Z nuclei causes a loss in mass. This lost mass appears as energy.

A similar situation occurs at the low-atomic-number end of the periodic table. Mass is lost as nuclei such as helium, lithium, and carbon are formed from protons and neutrons. In Section 31.6 we showed that 28 MeV of energy is released as a helium 4 nucleus is formed. We can therefore imagine a power source that gives off energy as small nuclei are fused together to form larger nuclei.

There is one great difficulty with this reaction. In the fission reaction the nucleus splits almost by itself. But in a fusion reaction protons (or positive nuclei) must be forced together against their Coulomb repulsion for each other. A great amount of work must be done to get the nuclei close enough so that the nuclear force will be strongly felt. Once the particles are close together, the nuclear force will pull them together and hold them there, but a great amount of work is needed to push them together in the first place.

In order to get two protons close enough together to react, they must be shot at each other with energies of about 0.10 MeV. This can be done using high-voltage machines. However, the energy needed to operate such a device far exceeds the energy given off in the fusion reaction. There appears to be only one practical way to use fusion for power generation. It requires that the nuclei to be fused must be heated to temperatures of about 10^7 °C. At these high temperatures the nuclei have tremendous kinetic energy. Their collisions bring them close enough together to fuse.

The fusion reaction actually does occur in the sun and in the stars. For example, in the sun the temperature is about 10^8 °C. At that high temperature the electrons have been stripped from all atoms. The sun consists of a fluid (called a plasma) composed of protons, neutrons, nuclei of atoms, other particles, and radiation. At these high temperatures the particles have large thermal energy. They collide with energies great enough to overcome the Coulomb repulsion. In certain cases they fuse together and give off energy.

The sun's energy is generated primarily by two types of fusion reaction. Both reactions consist of several steps. One of them can be summarized as follows:

$$4\,{}^1_1\mathrm{H} \rightarrow {}^4_2\mathrm{He} + 2 \text{ positrons} + \text{energy}$$

(A positron is much like the electron. It has the same mass as the electron, but its charge is $+e$). The other fusion reaction going on in the sun fuses helium 4 nuclei together to form carbon 12 nuclei. This reaction is thought to be the less important of the two, although in stars considerably hotter than our sun it may be more important. In any case, the energy of the sun is obtained from fusion reactions. The reactions furnish enough energy to keep the sun hot as well as to furnish heat to the solar system and earth.

In principle, at least, a fusion-type power source can be constructed on earth. We have a limitless supply of protons (i.e., hydrogen nuclei) from the water of the oceans. Therefore, fuel would be no problem. However, after years of research no one has yet found a practical, nonexplosive way for carrying out the fusion reaction.

A practical system must heat the fuel (protons or other light nuclei) to a temperature of about 10^8 °C. The reactants must be held at this temperature as fusion proceeds. A means must be found to make use of the energy given off during the reaction without cooling the reactants. Because all containers vaporize at these temperatures, the problem becomes apparent.

Several approaches are being tried in current research. Some make use of magnetic fields to hold the plasma (the hot gas composed of charged particles). By proper design the moving charges are continuously deflected by the magnetic field. No container walls are touched by the particles. The high temperatures needed to start the reaction may be produced by intense laser beams or by huge electrical discharges.

At the present time the problems of ignition and confinement have not been fully solved. It is true that the fusion reaction is carried out in the H-bomb. There the ignition is done by exploding a fission-type bomb. The intense heat generated by the fission bomb ignites the fusion reaction in the H-bomb. Obviously, though, the confinement problem is impossible when this form of ignition is used.

Work is still progressing to harness the fusion reaction. Its potential is great. No lack of fuel would exist for billions of years. Unlike the nuclear fission reactor, radioactive waste products are not an extremely serious problem. It remains to be seen whether we can design a system to provide useful energy by nuclear fusion.

Summary

The nucleus is a tiny object at the center of the atom. Its diameter, about 10^{-15} m, is only about 1/10,000 as large as the diameter of the atom. In the nucleus are found neutrons and protons. Each has a mass of about 1.67×10^{-27} kg. The neutron has no charge. The charge on the proton is $+1e$.

Nucleons, that is, protons and neutrons, are held together in the nucleus by the nuclear force. This force is very strong at distances up to about 1×10^{-15} m. It is essentially zero for separations greater than 5×10^{-15} m.

The number of protons in a nucleus is equal to the atomic number (Z) for that nucleus. The total number of protons and neutrons in a nucleus is equal to the atomic mass number (A). It follows that the number of neutrons in a nucleus is given by $A - Z$.

Each neutral atom has Z electrons to compensate for its Z protons in the nucleus. Atoms that have the same Z are members of the same element. They all behave alike in chemical reactions.

In nature one often finds atoms of the same element that have different mass numbers. Their nuclei contain the same number of protons, so Z is the same for each, but the nuclei contain different numbers of neutrons. Nuclei that contain the same number of protons but different numbers of neutrons are called isotopes of the element. An isotope of an element X is represented by the symbol $^{A}_{Z}X$.

The atomic mass unit (u) is equal to 1.6606×10^{-27} kg. In terms of it the proton and neutron masses are about 1.01 u. The electron mass is 0.00055 u.

When neutrons and protons are combined to form nuclei, they lose mass. This lost mass is called the mass defect of the nucleus. The mass defect per nucleon is least for nuclei that have very small Z or very large Z.

The mass defect of a nucleus is a measure of the binding energy of the nucleons within it. To tear the nucleons apart, mass equal to the mass defect must be created. The energy equivalent to this mass must be furnished. When the nucleons were first joined to form the nucleus, this same energy must have been given off.

Nuclei that are radioactive undergo spontaneous change. All nuclei with Z larger than 83 are radioactive. The radioactive substances found in nature emit gamma rays (X rays), beta particles (electrons), and alpha particles (helium nuclei). Radioactive isotopes decay in the same way with time. In a time interval of one half-life, $T_{1/2}$, only half of the material present at the start of the interval will remain unchanged.

Three radioactive series are found in nature. Each starts with an isotope of high Z that has a half-life greater than about 10^9 years. Many radioactive isotopes found on earth exist only because they are being formed continuously by these series. Each series ends in an isotope of lead.

The activity of a radioactive sample is $\Delta N/\Delta t$, the number of atoms that decay in unit time. It is related to the decay constant λ through $\Delta N/\Delta t = \lambda N$. The decay constant is related to half-life by $T_{1/2} = 0.693/\lambda$. The SI unit of activity is the becquerel (Bq); 1 Bq = 1 decay/s. The curie (Ci) is also used; 1 Ci = 3.7×10^{10} Bq.

Radiation dose is measured in rads. A 1-rad dose is equivalent to 10^{-2} J of radiation energy absorbed per kilogram. A unit called the rem is also used. It is a measure of the biological damage caused by the radiation.

When a nucleus with high Z splits apart into two or more nuclei of moderate Z, the nucleus is said to have undergone fission. The fission process releases energy and a large quantity of radiation.

Fission of uranium 235 can be caused by a slow neutron. In the process more neutrons are generated. These new neutrons can cause further fission reactions. Under certain conditions a chain reaction results in a rapid buildup of the number of nuclei undergoing fission. Nuclear reactors make use of the fission reaction to generate energy.

When small Z nuclei are fused together in the fusion reaction, large amounts of energy are released. To cause fusion, the Coulomb force must be overcome. This requires extremely high temperatures. The fusion process furnishes the energy of the sun and the stars. On earth a useful, controlled fusion reaction system is still being sought.

Questions and Exercises

1. The isotope strontium 90, $^{90}_{38}Sr$, is produced in explosions of nuclear bombs. Its half-life is 28 years. It is taken up by growing plants and enters our bodies with food. How many protons exist in an atom of this substance? Electrons? Neutrons?
2. Can an isotope $^{61}_{25}X$ be separated by chemical means from the isotope $^{59}_{25}Y$? From the isotope $^{61}_{26}Q$?
3. The half-life of radium is only about 1620 years. Even so, there is still a considerable quantity of radium found in rocks on earth. Why hasn't it all long since disappeared?
4. A tiny hole is accidentally left in a supposedly sealed tube containing radium. It is stored in a cupboard for a few months. Later it is found that the whole cupboard is slightly radioactive. How could the radioactivity have spread in this way? Radium itself is a solid.
5. Deuterium (heavy hydrogen) is the isotope ^2_1H. Its atomic mass is 2.0141 u. The atomic mass of ^1_1H is 1.0078 u and the mass of a neutron is 1.00867 u. What do you predict about the stability of the deuteron, the deuterium nucleus?
6. Alpha particles are often emitted by high-Z radioactive elements. These particles appear to be especially stable. How could you predict this from a consideration of the mass defect curve?
7. How does the nucleus of an isotope change when it emits an alpha particle? A beta particle? A gamma ray?
8. The isotope neptunium 237, $^{237}_{93}Np$, is made in nuclear reactors. It decays by emitting in succession the following particles: $\alpha, \beta, \alpha, \alpha$. By the time these emissions have taken place, what are the atomic number and mass number for the nucleus?
9. Ten million years from now the amount of radon on the earth probably will be about the same as it is now. Explain why. (Radon, Rn, has a half-life of about 3.8 days. It is a member of the series given in Figure 31.11.)
10. You sometimes hear the saying "Mass cannot be created and it cannot be destroyed." Is the saying correct?
11. It is sometimes said that every one of our sources of energy is the result of nuclear energy. Are fuels such as coal and oil the result of nuclear energy? How about hydroelectric power? Can you think of any that are not?
12. After a nuclear bomb has been tested in the atmosphere, added radioactivity is noticed in the air in various parts of the world. Close to the test area, the rise is immediate. Thousands of miles away, the rise occurs some days later. Explain how this added radioactivity arises.
13. Alpha particles with energy of 1 MeV can travel a distance of

about 0.5 cm in air before stopping. The distance they can travel in flesh is about 3×10^{-3} cm and in aluminum about 3×10^{-4} cm. What type of injury would you expect to occur to a person exposed to a strong beam of 1-MeV alpha particles? What type of shielding is needed against such a beam?

14. Repeat Question 13 for 1-MeV beta particles. Their range in air, flesh, and aluminum is about 400 cm, 0.5 cm, 0.15 cm, respectively.
15. Repeat Question 13 for 1-MeV gamma rays. Such a beam is reduced to half its intensity by the following approximate thicknesses of air, flesh, aluminum, and lead: 5000 cm, 10 cm, 4 cm, and 1 cm.

Problems

1. Iodine 131 is radioactive. It is used in biology and medicine to trace the path followed by iodine that enters the body. Its symbol is $^{131}_{53}I$. How many electrons exist in its atom? Protons? Neutrons?
2. Tumor tissue is difficult to detect by X rays because it does not differ much from normal flesh. However, radioactive technetium 99 tends to localize in tumors. A patient who has received an injection of $^{99}_{43}Tc$ will show high radioactivity in the region of a tumor. How many electrons exist in an atom of $^{99}_{43}Tc$? Protons? Neutrons?
3. How many protons, neutrons, and electrons does each of the following atoms contain: (a) lithium 6, (b) copper 64, and (c) gold 197?
4. How many electrons, protons, and neutrons does each of the following atoms contain: (a) cobalt 60, (b) chlorine 35, (c) carbon 14?
5. The atomic mass of the cobalt 59 atoms found in nature is 58.93 u. (a) What is the mass of the atom in kilograms? (b) Repeat for the carbon 12 atom that has a mass of exactly 12 u. (c) What is the mass of a cobalt 59 nucleus?
6. Find the mass in kilograms of a helium 4 atom. Its atomic mass is 4.0015 u. What is the mass of its nucleus?

*7. What is the approximate radius of the cobalt 60 nucleus? What is its approximate mass density?

*8. What is the approximate radius of the lead 208 nucleus? What is its approximate mass density?

*9. The average atomic mass of oxygen is 16 u. How many oxygen atoms are there in 16 kg of oxygen? (Do not use Avogadro's number in your calculation.)

*10. Gold has an average atomic mass of 197 u. How many gold atoms are there in 1 kg of gold.

11. In a certain mass spectrometer the lithium ion of mass 6.015 u travels in a circle with radius = 4.8 cm. At what radius will the 7.016-u isotope of lithium be found? Assume the ions to have the same speed.
12. As stated in the text, the abundances of lithium isotopes in nature are as follows: 8 percent have 6.015-u mass and 92 percent have 7.016-u mass. Show that the average atomic mass for lithium on earth should be 6.94 u. [Hint: Average mass value = $\sum_n$ (fraction)$_n$(mass)$_n$.]
13. Boron as found in nature consists of two isotopes; 80.22 percent have atomic mass 11.0093 u and 19.78 percent have mass 10.0129 u. Find the atomic mass the chemists list for this element. See Problem 12 for a hint.
14. Refer to Figure 31.5. (a) What is the mass defect per nucleon for carbon 12, the most common form of carbon found on earth? (b) How much larger than the mass of the nucleus is the mass of the separated protons and neutrons? (c) Using these data, together with the masses of the proton and neutron, what is the predicted mass of the carbon 12 atom?
15. The mass of the neutron is 1.008665 u. The mass of a hydrogen atom (1 proton + 1 electron) is 1.007825 u. (a) Find the mass defect for the isotope $^{7}_{3}Li$, which has an atomic mass of 7.016005 u. (b) What is the mass defect per nucleon?
16. Using the data of Problem 15, find the mass defect for $^{13}_{6}C$, which has an atomic mass of 13.003354. What is the mass defect per nucleon?
17. (a) How much energy is needed to separate all the nucleons in $^{120}_{50}Sn$ which has an atomic mass of 119.90220 u? (b) What is the binding energy per nucleon for this material?
18. The atomic mass for the isotope $^{14}_{7}N$ is 14.003074 u. (a) How much energy is required to separate this nucleus into its nucleons? (b) What is the binding energy per nucleon in this case?
19. Refer to Figure 31.5. Using the data shown there for iron, find the binding energy per nucleon. Check you answer by reference to Figure 31.7.

*20. From a consideration of energy alone, could the following decay occur?

$$^{1}_{0}n \rightarrow {}^{0}_{-1}\beta + {}^{1}_{1}H$$

In nature a free neutron decays in the way shown with a half-life of 720 s. How much energy is released in the process? (The atomic mass of $^{1}_{1}H$ is 1.007825 u and includes the mass of one electron.)

*Problems marked with an asterisk are not as easy as the unmarked ones.

21. Complete the following reactions. Use the periodic table to find the symbol for the isotope.

$$^{231}_{91}Pa \rightarrow \alpha +$$
$$^{240}_{92}U \rightarrow {}_{-1}\beta +$$
$$^{99}_{43}Tc \rightarrow \gamma +$$

22. Complete the following reactions. Use the periodic table to find the symbol for the isotope.

$$^{60}_{27}Co \rightarrow \gamma +$$
$$^{66}_{28}Ni \rightarrow {}_{-1}\beta +$$
$$^{158}_{72}Hf \rightarrow \alpha +$$

23. Cobalt 60 nuclei emit gamma rays of about 1.2-MeV energy. (a) How much mass does a nucleus lose (in atomic mass units) as it emits such a gamma ray? (b) What fraction of its total mass is this?

*24. Show that, from an energy standpoint, the following decay process is not possible: $^{16}_{8}O \rightarrow {}^{12}_{6}C + \alpha$. The mass of ^{16}O atoms is 15.9949 u.

25. Bismuth 212 has a half-life of about 6.0 minutes. About what fraction of the original material would be present after one hour?

26. The half-life of iodine 131 is 8.0 days. How much of an original 0.050-g sample would be left after about 1 month?

27. A certain radioactive substance loses half of its radioactivity in 27 h. (a) What is its half-life? (b) How long will it take to reduce its radioactivity to one-eighth of its original value? (c) To one thirty-second?

*28. Radioactive calcium 45 is used in medicine because it localizes in bone. It has a half-life of 152 days. (a) How long does it take for the activity of a sample to decrease to 75 percent of its original value? (b) To 30 percent?

*29. You have a vial of iodine 131 that has an activity of 2.0 μCi. What will be its activity after 45 days? The half-life of this isotope is 8.1 days.

**30. What is the activity of 1.0 mg of iodine 131? Its half-life is 8.1 days. Give your answer in both becquerels and curies.

**31. What is the activity of 3×10^{-4} g of iron 53? The half-life of this isotope is 46.3 days.

32. A cobalt 60 source is used for X-raying castings. When first obtained, its activity was 3.0 Ci. What will be its activity after about 16 years? The half-life of cobalt 60 is 5.3 years.

*33. The following data were taken to obtain the age of a campsite used by ancient peoples. Wood chips near the campfire site have only about 1/30 as much ^{14}C as present-day wood does. About how old is the campsite? The half-life of ^{14}C is 5730 years.

*34. The power output of a certain uranium 235 nuclear reactor system is to be 5×10^6 W. Assume that the efficiency of the reactor is 20 percent. (a) How much mass will the reactor lose in 1 h? (b) How much uranium 235 will undergo fission each hour? (Assume that uranium 235 loses 0.10 percent of its mass per fission process.)

*35. In very hot stars the "triple alpha" fusion reaction occurs. It is, overall, $3{}^{4}_{2}He \rightarrow {}^{12}_{6}C$. How many megaelectronvolts are given off as each $^{12}_{6}C$ nucleus is formed? The mass of ^{12}C is exactly 12 u.

36. A certain substance decreases the intensity of a beam of X rays by a factor of 16 for a thickness of 0.73 cm. What is the half-value layer for this material?

37. By what factor is the intensity of a gamma ray beam decreased as the beam goes through a thickness of 2.80 cm of a substance whose half-value layer is 0.40 cm for this radiation?

** Problems marked with a double asterisk are somewhat more difficult than the average.

APPENDIXES

Appendix 1
Mathematics Review

There are certain mathematical operations that technicians use over and over again. They are summarized here. In addition to them you should be able to use the pocket calculator.

A. Addition The order in which ordinary numbers are added is not important. For example, $5 + 3$ and $3 + 5$ are both 8. We sometimes represent numbers by letters. In that case our addition rule becomes $a + b = b + a$.

B. Subtraction You know that $9 - 5 = 4$ and $5 - 9 = -4$. It is also true that $-9 + 5 = 5 - 9 = -4$. The general rule is $a - b = -b + a$.

To subtract a negative number, we simply change its sign and add it. For example,

$$12 - (-8) = 12 + 8 = 20$$

and

$$a - (-b) = a + b$$

C. Multiplication The order in which common numbers are multiplied is not important. For example, $3 \times 5 = 5 \times 3 = 15$, and $ab = ba$. The rules concerning signs are as follows.

$$\text{(Plus)} \times \text{(plus)} = \text{plus} \rightarrow (2)(6) = 12 \rightarrow (a)(b) = ab$$
$$\text{(Plus)} \times \text{(minus)} = \text{minus} \rightarrow (2)(-6) = -12 \rightarrow (a)(-b) = -ab$$
$$\text{(Minus)} \times \text{(minus)} = \text{plus} \rightarrow (-2)(-6) = 12 \rightarrow (-a)(-b) = ab$$

D. Division The sign rules for division are as follows:

$$\frac{\text{(Plus)}}{\text{(Plus)}} = \text{plus} \rightarrow \frac{12}{3} = 4 \rightarrow a \div b = \frac{a}{b}$$

$$\frac{\text{(Plus)}}{\text{(Minus)}} = \frac{\text{(minus)}}{\text{(plus)}} = \text{minus} \rightarrow \frac{+12}{-3} = -4 \rightarrow \frac{a}{-b} = -\frac{a}{b}$$

$$\rightarrow \frac{-12}{+3} = -4 \rightarrow \frac{-a}{b} = -\frac{a}{b}$$

$$\frac{\text{(Minus)}}{\text{(Minus)}} = \text{plus} \rightarrow \frac{-12}{-3} = 4 \rightarrow \frac{-a}{-b} = \frac{a}{b}$$

E. Parentheses The following rules apply to quantities contained within brackets or parentheses.

$$(a + b) = (b + a) = a + b$$
$$-(a + b) = -a - b$$
$$c(a + b) = ca + cb$$
$$(c + d)(a + b) = c(a + b) + d(a + b) = ca + cb + da + db$$
$$\frac{(a + b)}{c} = \frac{a}{c} + \frac{b}{c}$$

But notice!

$$\frac{a + b}{c + d} \quad \textit{is not} \quad \frac{a}{c} + \frac{b}{d}$$

F. Fractions In a fraction such as a/b, a is the numerator and b is the denominator. Numerical fractions are evaluated by division. For example,

$$\frac{1}{2} = 0.500, \qquad \frac{210}{3} = 70, \qquad \text{and} \qquad \frac{71}{140} = 0.507$$

The denominator of a fraction may be a fraction. In that case the following rule is often useful.

To divide a number by a fraction, invert the fraction and multiply by it.

For example,

$$\frac{3}{\frac{2}{3}} = 3 \cdot \frac{3}{2} = \frac{9}{2} \qquad \text{and} \qquad \frac{a}{b/c} = a \cdot \frac{c}{b} = \frac{ca}{b}$$

When we multiply both the numerator and denominator by the same number, we are in effect multiplying by unity. A similar situation exists when we divide both numerator and denominator by the same number. The general rule is

The value of a fraction is not changed by multiplying (or dividing) both its numerator and denominator by the same number.

For example:

$$\frac{a}{b} = \frac{ac}{bc}$$

$$\frac{ac}{bc} = \frac{ac \div c}{bc \div c} = \frac{a}{b}$$

$$\frac{(a + b)c}{(e + f)c} = \frac{(a + b)c \div c}{(e + f)c \div c} = \frac{(a + b)}{(e + f)}$$

In the last two examples we say that we have "canceled" c from the numerator and denominator. Similarly,

$$\frac{3a}{3b} = \frac{a}{b}, \qquad \frac{3a + 3b}{12e + 9f} = \frac{a + b}{4e + 3f}, \qquad \frac{ac + bc}{ce} = \frac{a + b}{e}$$

But not

$$\frac{3a + b}{3e} = \cancel{\frac{a + b}{e}}$$

Each term in the numerator and denominator must be treated the same way.

G. Equations It is often necessary to *solve an equation*. For example, suppose we are told that

$$5x + 7 = 3(1 - x)$$

and we wish to find x. Of course, our rules for parentheses tell us that we can write this equation as

$$5x + 7 = 3 - 3x$$

In addition, there are other rules that help us solve equations. They are as follows.

Rule 1. Quantities equal to the same quantity are equal to each other. If

$$a + b = c \qquad \text{and} \qquad x + y + z = c$$

then

$$a + b = x + y + z$$

Rule 2. Both sides of an equation can be multiplied by the same quantity without invalidating the equation. If

$$\tfrac{1}{7}c = a + b$$

then

$$7 \times \tfrac{1}{7}c = 7 \times (a + b)$$

or

$$c = 7a + 7b$$

Rule 3. Both sides of an equation can be divided by the same quantity. If

$$16c = 3a + 2b$$

then

$$\frac{16c}{16} = \frac{3a + 2b}{16}$$

or

$$c = \frac{3a}{16} + \frac{2b}{16}$$

which can be written as

$$c = \frac{3}{16}a + \frac{1}{8}b$$

Rule 4. An equation of the general form

$$\frac{a}{b} = \frac{c}{d}$$

can be cross multiplied, that is,

$$\frac{a}{b} \times \frac{c}{d} \quad \text{gives} \quad ad = bc$$

The same result is achieved by multiplying the numerators of both sides of the equation by bd, giving

$$\frac{abd}{b} = \frac{cbd}{d}$$

which gives, after canceling,

$$ad = cb$$

This is the same result we found by cross multiplying.

Rule 5. We can square both sides of an equation.

$$3x = 4 \qquad \text{is equivalent to} \qquad 9x^2 = 16$$

We can use rule 2 to multiply one side of the equation by $3x$ and the other by 4 because $3x = 4$.

Rule 6. The reverse of rule 5 is also permitted, that is, we can take the square root of both sides of an equation. Then

$$9x^2 = 16 \qquad \text{becomes} \qquad 3x = 4$$

Rule 7. An isolated quantity can be moved from one side of an equation to the other, or transposed, provided that we change its sign. Thus

$$3 + x = 16$$

can be written

$$x = 16 - 3 \qquad \text{or} \qquad x = 13$$

This rule simply involves subtracting the same quantity (3 in this case) from both sides of the equation.

H. Exponents In a quantity such as a^c, we call c the exponent of a. Its meaning is shown by the following examples.

$$a^3 = a \cdot a \cdot a \qquad \text{or} \qquad 5^3 = 5 \cdot 5 \cdot 5 = 125$$

$$a^{-1} = \frac{1}{a} \qquad \text{or} \qquad 5^{-1} = \frac{1}{5} = 0.20$$

$$a^{-3} = \frac{1}{a^3} = \frac{1}{a \cdot a \cdot a}$$

There are certain rules involving exponents that one must use frequently. They are

Rule 1. $a^n \cdot a^m = a^{n+m} \rightarrow 2^3 \cdot 2^2 = 2^5 = 32$ (Notice that both numbers raised to the two exponents must be the same, a in this case.)

Rule 2. $a^{-n} = \dfrac{1}{a^n} \rightarrow 5^{-3} = \dfrac{1}{5^3} = \dfrac{1}{125}$

Rule 3. $\dfrac{1}{a^{-n}} = a^n \rightarrow \dfrac{1}{5^{-3}} = 5^3 = 125$

Rule 4. $(a^n)^m = a^{nm} \rightarrow (3^2)^3 = 3^6 = 729$

Rule 5. $(a^n)^{1/2} = \sqrt{a^n} = a^{n/2} \rightarrow (5^6)^{1/2} = 5^3 = 125$

Rule 6. $a^0 = 1 \rightarrow (5)^2(5)^{-2} = (5)^0 = 1$

Rule 7. $\dfrac{a^n}{a^m} = a^{n-m} \rightarrow \dfrac{5^6}{5^2} = 5^4$

Rule 8. $\dfrac{a^n}{a^{-m}} = a^{n+m} \rightarrow \dfrac{5^6}{5^{-2}} = 5^8$

Rule 9. $(ab)^n = a^n b^n \rightarrow (3 \cdot 5)^2 = 3^2 \cdot 5^2 = 225$

I. Scientific Notation (Powers of Ten) See Section 1.9 of the text.

J. Trigonometry See Section 1.4 of the text.

K. Some Useful Formulas

Circumference of a circle = $2\pi r$
Area of a circle = πr^2
Surface area of a sphere = $4\pi r^2$
Volume of a sphere = $\frac{4}{3}\pi r^3$
Area of a rectangle = ab
Volume of a right circular cylinder = $(\pi r^2)L$

Appendix 2
Trigonometric Functions

Angle (deg)	Sine	Cosine	Tangent	Angle (deg)	Sine	Cosine	Tangent
0°	0.000	1.000	0.000				
1°	.018	1.000	.018	46°	0.719	0.695	1.036
2°	.035	0.999	.035	47°	.731	.682	1.072
3°	.052	.999	.052	48°	.743	.669	1.111
4°	.070	.998	.070	49°	.755	.656	1.150
5°	.087	.996	.088	50°	.766	.643	1.192
6°	.105	.995	.105	51°	.777	.629	1.235
7°	.122	.993	.123	52°	.788	.616	1.280
8°	.139	.990	.141	53°	.799	.602	1.327
9°	.156	.988	.158	54°	.809	.588	1.376
10°	.174	.985	.176	55°	.819	.574	1.428
11°	.191	.982	.194	56°	.829	.559	1.483
12°	.208	.978	.213	57°	.839	.545	1.540
13°	.225	.974	.231	58°	.848	.530	1.600
14°	.242	.970	.249	59°	.857	.515	1.664
15°	.259	.966	.268	60°	.866	.500	1.732
16°	.276	.961	.287	61°	.875	.485	1.804
17°	.292	.956	.306	62°	.883	.470	1.881
18°	.309	.951	.325	63°	.891	.454	1.963
19°	.326	.946	.344	64°	.899	.438	2.050
20°	.342	.940	.364	65°	.906	.423	2.145
21°	.358	.934	.384	66°	.914	.407	2.246
22°	.375	.927	.404	67°	.921	.391	2.356
23°	.391	.921	.425	68°	.927	.375	2.475
24°	.407	.914	.445	69°	.934	.358	2.605
25°	.423	.906	.466	70°	.940	.342	2.747
26°	.438	.899	.488	71°	.946	.326	2.904
27°	.454	.891	.510	72°	.951	.309	3.078
28°	.470	.883	.532	73°	.956	.292	3.271
29°	.485	.875	.554	74°	.961	.276	3.487
30°	.500	.866	.577	75°	.966	.259	3.732
31°	.515	.857	.601	76°	.970	.242	4.011
32°	.530	.848	.625	77°	.974	.225	4.331
33°	.545	.839	.649	78°	.978	.208	4.705
34°	.559	.829	.675	79°	.982	.191	5.145
35°	.574	.819	.700	80°	.985	.174	5.671
36°	.588	.809	.727	81°	.988	.156	6.314
37°	.602	.799	.754	82°	.990	.139	7.115
38°	.616	.788	.781	83°	.993	.122	8.144
39°	.629	.777	.810	84°	.995	.105	9.514
40°	.643	.766	.839	85°	.996	.087	11.43
41°	.658	.755	.869	86°	.998	.070	14.30
42°	.669	.743	.900	87°	.999	.052	19.08
43°	.682	.731	.933	88°	.999	.035	28.64
44°	.695	.719	.966	89°	1.000	.018	57.29
45°	.707	.707	1.000	90°	1.000	.000	∞

Appendix 3
Table of the Elements

The masses listed are based on ${}^{12}_{6}C = 12u$. A value in parentheses is the mass number of the most stable (long-lived) of the known isotopes.

Element	Symbol	Atomic Number Z	Average Atomic Mass
Actinium	Ac	89	(227)
Aluminum	Al	13	26.9815
Americium	Am	95	(243)
Antimony	Sb	51	121.75
Argon	Ar	18	39.948
Arsenic	As	33	74.9216
Astatine	At	85	(210)
Barium	Ba	56	137.34
Berkelium	Bk	97	(247)
Beryllium	Be	4	9.0122
Bismuth	Bi	83	208.980
Boron	B	5	10.811
Bromine	Br	35	79.904
Cadmium	Cd	48	112.40
Calcium	Ca	20	40.08
Californium	Cf	98	(251)
Carbon	C	6	12.01115
Cerium	Ce	58	140.12
Cesium	Cs	55	132.905
Chlorine	Cl	17	35.453
Chromium	Cr	24	51.996
Cobalt	Co	27	58.9332
Copper	Cu	29	63.546
Curium	Cm	96	(247)
Dysprosium	Dy	66	162.50
Einsteinium	Es	99	(254)
Erbium	Er	68	167.26
Europium	Eu	63	151.96
Fermium	Fm	100	(257)
Fluorine	F	9	18.9984
Francium	Fr	87	(223)
Gadolinium	Gd	64	157.25
Gallium	Ga	31	69.72
Germanium	Ge	32	72.59
Gold	Au	79	196.967
Hafnium	Hf	72	178.49
Helium	He	2	4.0026
Holmium	Ho	67	164.930
Hydrogen	H	1	1.00797
Indium	In	49	114.82
Iodine	I	53	126.9044
Iridium	Ir	77	192.2
Iron	Fe	26	55.847
Krypton	Kr	36	83.80
Lanthanum	La	57	138.91

Table of the Elements *(continued)*

Element	Symbol	Atomic Number Z	Average Atomic Mass
Lawrencium	Lr	103	(257)
Lead	Pb	82	207.19
Lithium	Li	3	6.939
Lutetium	Lu	71	174.97
Magnesium	Mg	12	24.312
Manganese	Mn	25	54.9380
Mendelevium	Md	101	(256)
Mercury	Hg	80	200.59
Molybdenum	Mo	42	95.94
Neodymium	Nd	60	144.24
Neon	Ne	10	20.183
Neptunium	Np	93	(237)
Nickel	Ni	28	58.71
Niobium	Nb	41	92.906
Nitrogen	N	7	14.0067
Nobelium	No	102	(254)
Osmium	Os	76	190.2
Oxygen	O	8	15.9994
Palladium	Pd	46	106.4
Phosphorus	P	15	30.9738
Platinum	Pt	78	195.09
Plutonium	Pu	94	(244)
Polonium	Po	84	(209)
Potassium	K	19	39.102
Praseodymium	Pr	59	140.907
Promethium	Pm	61	(145)
Protactinium	Pa	91	(231)
Radium	Ra	88	(226)
Radon	Rn	86	222
Rhenium	Re	75	186.2
Rhodium	Rh	45	102.905
Rubidium	Rb	37	85.47
Ruthenium	Ru	44	101.07
Samarium	Sm	62	150.35
Scandium	Sc	21	44.956
Selenium	Se	34	78.96
Silicon	Si	14	28.086
Silver	Ag	47	107.868
Sodium	Na	11	22.9898
Strontium	Sr	38	87.62
Sulfur	S	16	32.064
Tantalum	Ta	73	180.948
Technetium	Tc	43	(97)
Tellurium	Te	52	127.60
Terbium	Tb	65	158.924
Thallium	Tl	81	204.37
Thorium	Th	90	232.0381
Thulium	Tm	69	168.934
Tin	Sn	50	118.69
Titanium	Ti	22	47.90
Tungsten	W	74	183.85

Table of the Elements *(continued)*

Element	Symbol	Atomic Number Z	Average Atomic Mass
Uranium	U	92	238.03
Vanadium	V	23	50.942
Xenon	Xe	54	131.30
Ytterbium	Yb	70	173.04
Yttrium	Y	39	88.905
Zinc	Zn	30	65.37
Zirconium	Zr	40	91.22

Appendix 4
Periodic Table of the Elements

The values listed are based on ${}^{12}_{6}C = 12$ u exactly. For artificially produced elements, the approximate atomic weight of the most stable isotope is given in brackets.

Period	Series	Group I	Group II	Group III	Group IV	Group V	Group VI	Group VII	Group VIII	Group 0
1	1	1 H 1.00797								2 He 4.003
2	2	3 Li 6.939	4 Be 9.012	5 B 10.81	6 C 12.011	7 N 14.007	8 O 15.9994	9 F 19.00		10 Ne 20.183
3	3	11 Na 22.990	12 Mg 24.31	13 Al 26.98	14 Si 28.09	15 P 30.974	16 S 32.064	17 Cl 35.453		18 Ar 39.948
4	4	19 K 39.102	20 Ca 40.08	21 Sc 44.96	22 Ti 47.90	23 V 50.94	24 Cr 52.00	25 Mn 54.94	26 Fe 55.85, 27 Co 58.93, 28 Ni 58.71	
4	5	29 Cu 63.54	30 Zn 65.37	31 Ga 69.72	32 Ge 72.59	33 As 74.92	34 Se 78.96	35 Br 70.909		36 Kr 83.80
5	6	37 Rb 85.47	38 Sr 87.62	39 Y 88.905	40 Zr 91.22	41 Nb 92.91	42 Mo 95.94	43 Tc [98]	44 Ru 101.1, 45 Th 102.905, 46 Pd 106.4	
5	7	47 Ag 107.870	48 Cd 112.40	49 In 114.82	50 Sn 118.69	51 Sb 121.75	52 Te 127.60	53 I 126.90		54 Xe 131.30
6	8	55 Cs 132.905	56 Ba 137.34	57-71 Lanthanide series*	72 Hf 178.49	73 Ta 180.95	74 W 183.85	75 Re 186.2	76 Os 190.2, 77 Jr 192.2, 78 Pt 195.09	
6	9	79 Au 196.97	80 Hg 200.59	81 Tl 204.37	82 Pb 207.19	83 Bi 208.98	84 Po [210]	85 At [210]		86 Rn [222]
7	10	87 Fr [223]	88 Ra [226]	89-103 Actinide series†						

* *Lanthanide series*	57 La 138.91	58 Ce 140.12	59 Pr 140.91	60 Nd 144.24	61 Pm [147]	62 Sm 150.35	63 Eu 152.0	64 Gd 157.25	65 Tb 158.92	66 Dy 162.50	67 Ho 164.93	68 Er 167.26	69 Tm 168.93	70 Yb 173.04	71 Lu 174.97
† *Actinide series*	89 Ac [227]	90 Th 232.04	91 Pa [231]	92 U 238.03	93 Np [237]	94 Pu [242]	95 Am [243]	96 Cm [247]	97 Bk [247]	98 Cf [251]	99 E [254]	100 Fm [253]	101 Md [256]	102 No [254]	103 Lw [257]

ANSWERS TO THE ODD-NUMBERED PROBLEMS

Chapter 1

1. 10.0 mm at 49°
3. 17 km at 29°
5. (a) 64 km, (b) 51°
7. $s_x = -10$ m, $s_y = 17.3$ m
9. (a) 55, 95, (b) 35, −61, (c) −173, 100
11. (a) 3.6 m, (b) 0.556, (c) 0.833, (d) 0.667
13. (a) 4.00 m, (b) 6.93 m
15. (a) 2.44 m, (b) 1.40 m
17. (a) 3.92 m, (b) 2.52 m
19. $s_x = 40$ m, $s_y = 30$ m
21. (a) −23.9 m, 65.8 m, (b) −15 m, −26 m, (c) 3 m, −5.2 m
23. 58 m at 149°
25. 7.5 m
27. 26.6°, 63.4°
29. 85 N at 45°
31. 17 km at 29°
33. (a) 1.73 km, (b) −26°
35. 16.6 N at 25°
37. (a) 100 km/h, (b) 173 km/h
39. (a) 25 m east, (b) 5 m west, (c) 5 m east
41. (a) 28.3 m at 45°, (b) 28.3 m at −45°
43. 320 km/h east
45. 240 km/h at −17°
47. (a) 5.79×10^{-4}, (b) 3.6×10^{-3}, (c) 7.49×10^{3}, (d) 2.0001×10^{4}
49. (a) 672,300, (b) 242, (c) 0.00036, (d) 0.00365
51. (a) 17.4×10^{-2}, (b) 75.47×10^{4}, (c) 19.5×10^{-8}, (d) 0.187×10^{-5}
53. (a) 7.5×10^{7}, (b) 7.8×10^{-5}, (c) 6.6×10^{-6}, (d) 16.4×10^{-4}
55. (a) 0.41, (b) 125, (c) 2.14, (d) 0.169

Chapter 2

1. $F_x = -30$ N, $F_y = 20$ N
3. 45 N at 90°
5. 60 lb, 150 lb
7. $W_2 = 600$ N, $T_1 = T_2 = T_3 = 300$ N, $T_4 = 600$ N
9. (a) $W_m + 2W_g$, (b) $2W_g$, (c) equal and opposite
11. (a) $F_x = -40$ N, $F_y = -70$ N, (b) 80.6 N, (c) 240°
13. 33.3 N, 26.7 N
15. $T_1 = 39.6$ N, $T_2 = 21.9$ N
17. 4020 N
19. 15.3 N
21. 800 N
23. (a) $-0.453F_1L$, (b) $0.25F_2L$, (c) 0, (d) $F_2 = 1.81F_1$
25. $8F$
27. $P = 145$ lb, $F = 115$ lb
29. 64.3 cm
31. 65.2 cm from A
33. 257 N, 197 N, 115 N
35. $W_2 = 1.0$ N, $W_3 = 0.50$ N
37. $P = 63$ N, $H = 63$ N, $V = 340$ N
39. (a) Proof, (b) No
41. 52.8 N and 39.2 N
43. 70 N, 166 N
45. 29.7°

Chapter 3

1. 7.95 h
3. 6020 km/h
5. 155 mi/h, 0
7. 1.61×10^{6} cm
9. 99 km/h
11. 3.0 cm/s in x direction, yes
13. 4 m/s in x direction, 0, 2.0 m/s in $-x$ direction
15. 3.5 m/s, 0; −3.5 m/s
17. (a) 1.8 m/s, (b) −0.2 m/s
19. (a) 14.0 m/s, (b) 10.0 m/s
21. 10.8 s
23. 2.5 m/s^2
25. −4.0 m/s^2
27. (a) 3.3 s, (b) 3.5 s
29. (a) 5.0 m/s^2, (b) 4.0 s
31. (a) 0.80 m/s^2, 14 m/s, (b) 2.0 ft/s^2, 40 ft/s
33. (a) 0.133 m/s^2, (b) −0.20 m/s^2, (c) 375 m
35. 466 m, 68.5 m/s, no
37. (a) 1.63 s, 16.0 m/s, (b) 1.58 s, 51 ft/s
39. (a) 19.8 m/s, (b) 2.02 s
41. (a) 14.7 m/s, (b) 11.0 m
43. 4.7 m/s
45. 34.2 m/s upward
47. (a) 25.0 m/s, (b) 3.2 s, (c) 0.12 m/s upward
49. 9.6 m
51. 880 m
53. 29 m and 6.4 m
55. 14.2 m/s
57. 7.1 m/s

Chapter 4

1. (a) 29.4 N, (b) 61 kg
3. 0.125 N

5. (a) 6 kg, (b) 6.67 m/s^2
7. (a) 5.0 m/s^2, (b) 9.4 ft/s^2
9. (a) 19.6 N, (b) 3.2 N
11. (a) 4.08 kg, (b) 68.6 N, (c) 1.55 slugs, (d) 644 lb
13. (a) 6 N, (b) 0.62 lb
15. (a) 28 N, (b) 2.33 N
17. 24.3 m/s^2
19. (a) 2 m/s^2, (b) 1800 N, (c) 6 ft/s^2, 522 lb
21. 10,000 m/s^2, 200 N
23. (a) 1.2×10^5 N, (b) 37,300 lb
25. (a) 9.3 m/s^2, (b) 20.6 ft/s^2
27. (a) 2.14 m/s^2, (b) 1.50 N
29. 2.67 m/s^2, 8.0 N
31. 0.049 m/s^2, 5.7 s
33. 3.75×10^{-9} N, 1.57×10^{10}
35. 1.68×10^5 N, 3.78×10^4 lb, 1.96×10^5 N vs. 1.68×10^5 N
37. (a) $0.80W_e$, (b) 7.8 m/s^2
39. (a) 196 N, (b) 210 N
41. (a) 50.5 N, (b) 47.5 N
43. (a) 9.8 N, (b) 11.0 N, (c) 8.6 N, (d) 4.9 N, 5.5 N, 4.3 N

Chapter 5

1. 300 N
3. (a) 98 N, 148 N, 48 N, 141 N, 73 N, (b) 40 lb, 70 lb, 10 lb, 66 lb, 25 lb
5. 390 N, 290 N
7. 0.38, 0.32
9. 640 N, 480 N
11. (a) 30 N, (b) 24 lb
13. 1.19, 0.84
15. 3680 N, 2710 N
17. 0.51
19. 41.5 N
21. 59 N
23. 31.5 N
25. (a) −0.53 m/s^2, (b) 0.267 N, (c) 0.054
27. (a) 1.56 m/s^2, 16.5 N; (b) 0, 19.6 N
29. 0.66 m/s^2, 25.4 N

Chapter 6

1. (a) 15 J, (b) −10 J, (c) 0, (d) 1 kJ, (e) −1 kJ
3. (a) 20 kJ, (b) 15,000 ft·lb
5. 118 N, 590 J
7. (a) 580 J, (b) 800 ft·lb
9. (a) 78 J, −24 J, (b) 50 ft·lb, −12.5 ft·lb
11. (a) 588 J, (b) −588 J, (c) 0
13. 3.92 J
15. (a) 384 J, (b) 1.2 m, (c) 365 J, (d) 9.5 N
17. (a) 2.62 hp, (b) 1.82 hp
19. 1.52 m/s
21. 1210 N
23. 4.0×10^5 J, 2.25×10^5 J, 310 ft·lb
25. 225 J, 310 ft·lb
27. 560 N, 190 lb
29. 3.7 m/s
31. 2.4 m/s
33. 7.92 m/s
35. 1590 N
37. 1.77 m/s, 0.16 m
39. 0.98 N
41. 0.0272 N
43. (a) 2.01 s, (b) 2.05 s
45. 497 s
47. (a) 111 hp, (b) 0.54

Chapter 7

1. 36, 16.7, 0.46
3. 36, 720 N
5. W, $\frac{1}{2}W$, $2W$, 1.0, 2.0, 0.50
7. 1490 N
9. 10
11. 250 N
13. 23.5 N
15. 8, 5, 0.625
17.

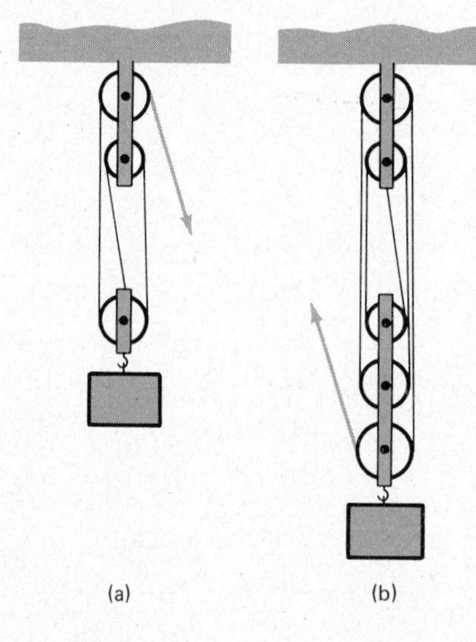

19. 8
21. 6.7, 4.0
23. 144 cm^2
25. 346 N
27. 15,600, 5200, 1730
29. (a) 10.8 m, (b) 187 J, (c) 17.3 N
31. Proof
33. (a) 10, (b) 2, (c) 20

Chapter 8

1. (a) 0.30 kg·m/s in x direction, (b) 0.0466 slug·ft/s in y direction
3. (a) 45,000 kg·m/s, (b) 8 kg·m/s, (c) 9300 slug·ft/s
5. (a) 0.079 kg·m/s down, (b) 0.069 kg·m/s up, (c) 0.148 kg·m/s up
7. 0.60 kg·m/s
9. 15 N·s
11. −0.100 kg·m/s, $-x$
13. 3440 N
15. (a) 7300 N, (b) 1360 lb
17. (a) 4×10^{-4} s, (b) 20,000 N
19. 5.66 m/s
21. 2.3 m/s
23. 2.7×10^5 m/s
25. 3850 kg
27. Going west at 0.67 m/s
29. 2.5 m/s, no
31. −0.33 m/s, 4.67 m/s
33. (a) 1.98 m/s, (b) 6.6 ft/s
35. 5.4 km/h, 11 km/h
37. $v_x = 2.67$ m/s, $v_y = 1.33$ m/s
39. 8.9 cm
41. Proof

Chapter 9

1. (a) 0.10 rev, 0.628 rad, (b) 26.9 deg, 0.075 rev, (c) 785 deg, 13.7 rad
3. (a) 18,000 deg/s, 50 rev/s, 314 rad/s, (b) 0.0722 deg/s, 2×10^{-4} rev/s, 1.26×10^{-3} rad/s (c) 44.9 deg/s, 0.125 rev/s, 0.783 rad/s.
5. (a) 57.3 deg/s^2, 0.159 rev/s^2, 1.0 rad/s^2, (b) 1440 deg/s^2, 4 rev/s^2, 25.1 rad/s^2, (c) 0.0154 deg/s^2, 4.29×10^{-5} rev/s^2, 2.69×10^{-4} rad/s^2
7. 18.2 rad/s
9. 1.05 rad/s^2
11. 125 rev
13. 20.7 rad/s
15. (a) 30 s, (b) 0.10 rev/s^2
17. 0.236 m/s
19. 38.2
21. 2.65 rev/s
23. (a) 2.67 rad/s^2, (b) 1.06 rev
25. 2.81 N
27. 0.062
29. 2.72

31. 871 N
33. (a) 16.4 N, (b) 0.69 N
35. 8.85 m/s
37. 1.38 km/s
39. Proof

Chapter 10

1. (a) 0.045 kg·m^2, (b) 30 cm, (c) 0.047 slug·ft^2, 1 ft
3. (a) 0.016 kg·m^2, (b) 0.040 slug·ft^2
5. Proof
7. (a) 8000 N·m, (b) 3.14×10^4 N·m
9. (a) 2.0 N·m, (b) 0.73 kg·m^2
11. (a) 0.178 rad/s^2, (b) 0.59 s
13. 16.7 N
15. (a) 14.7 rad/s^2, (b) zero
17. (a) 0.0116 kg·m^2, (b) 0.29 N
19. 0.317 m/s^2
21. 2.5 J
23. 90 J, 224 J, 314 J
25. 0.64 m
27. 5.3 m/s
29. 2.02 rev/s
31. 9.7×10^{-3} N·m
33. 59.7 N
35. 0.236 kg·m^2/s
37. 8.0 h

Chapter 11

1. No
3. (a) 6.37×10^7 N/m^2, (b) 3.18×10^{-4}, (c) 0.064 cm
5. 3.1×10^{11} N/m^2
7. (a) 440 N, (b) 0.2 percent
9. 1.04×10^6 N/m^2
11. (a) 3.45 rev/s, (b) 3.36 rev/s
13. (a) 0.150 kg, (b) 18.7
15. 891 kg/m^3
17. 5280 m
19. 2.74×10^5 Pa
21. 0.22 MPa
23. 2.2×10^8 Pa, 22.4 km
25. (a) 66.6 kPa, (b) 0.84 N
27. 0.857
29. 690 kPa
31. 98.6 kPa
33. 84 kPa
35. (29.4 N)(A_b/A_t)
37. (a) 770 N, (b) 161 lb
39. (a) 21.3 cm^3, (b) 1.46 g/cm^3
41. 0.042 m^3

Chapter 12

1. −38°C, 234 K, 422°R; 673.9°F, 630 K, 1134°R
3. −40°
5. 5.55 mol
7. 2.99×10^{-26} kg
9. (a) 2.5×10^{24}, (b) 1.0×10^{26} atoms
11. (a) 0.0537 kmol, (b) 0.42 kmol, (c) 7.9×10^{-5} kmol
13. 1.03×10^{24} molecules
15. −450°F
17. 14,000 cm^3
19. (a) 218 kPa, (b) 4.36 mm^3
21. 0.076
23. 6.09×10^6 Pa
25. 1.31×10^5 Pa
27. 112 kg
29. 2.5×10^5 m/s
31. 1.2 mm
33. (a) zero, (b) 4.0×10^{-3} cm
35. 253 cm^3
37. 100.0090 cm^3
39. 1.27 cm^3
41. (a) 5×10^{10}, (b) about 10^{-6} cm Hg
43. Proof

Chapter 13

1. (a) 4000 cal, (b) 8860 Btu
3. (a) 74,700 cal, (b) 239 Btu
5. (a) 1.60×10^5 J, (b) 38 kcal
7. 427 m
9. 0.21°C
11. 148 g
13. 280°C
15. (a) 64°C, (b) 39.6°C
17. 295 cal
19. 2650 cal
21. 771 cal
23. 4080 cal
25. 8.2 g
27. −11.3°C
29. 10.4°C
31. (a) 3.25×10^5 cal, (b) 2800 m
33. (a) 58 kcal/mol, (b) 530 kcal/mol

Chapter 14

1. (a) 125.5 J, (b) −251 J
3. −4700 J
5. −2260 J
7. 168 J, 300 kPa, 600 K
9. 308°C
11. (a) 1.35 J, (b) zero, (c) −1.05 J
13. (a) 200 J, (b) zero
15. (a) 900 J, (b) zero
17. 38 J
19. 4.0 J
21. 244 J
23. (a) 2.2 J, (b) 4.9 J, (c) no
25. 218 kg
27. (a) *DC*, (b) *AD* and *BA*, (c) *DC* and *CB*
29. (a) *AD*, (b) *CB* and *BA*, (c) *AD* and *CD*
31. (a) −120 J, (b) −120 J
33. 19.57°C
35. 43 percent
37. (a) 250 J, (b) 750 J
39. (a) 0.49, (b) 244 cal, (c) 124 cal

Chapter 15

1. 5.8×10^6 J/h
3. 0.0185 cal/(cm)(s)(°C)
5. 2.6 Btu/s, 2.7 kW
7. 83 cm
9. (a) 0.0133 cal/s, (b) 0.0133 cal/s, (c) −32°C, (d) −34°C
11. 8.5°C
13. 253 g
15. 2400 Btu
17. 1.74
19. 3.1×10^4 cal
21. 5740 K
23. (a) 26.5, (b) 430 cal/s
25. 2.65 tons
27. 3.7 hp
29. (a) 361 cal, (b) 1.51
31. 7.7 kg
33. 49 percent

Chapter 16

1. 2.0 cm, 8 s, 0.125 Hz
3. (a) 2.43 s, (b) 0.41 Hz
5. 0.196 cm
7. (a) 14.7 m/s^2, (b) 0, (c) 9.8 m/s^2
9. (a) 0, (b) 2.0 m/s^2, (c) 1.5 m/s^2
11. (a) 133 N/m, (b) 167 mJ
13. (a) 0.28 m/s, (b) 0, (c) 0.187 m/s
15. Proof
17. 28 m/s
19. 0.333 s, 3 cm, 14.2 N/m
21. (a) 4×10^4 N/m, (b) 8.5 Hz, (c) yes
23. $2\pi Af$, $(2\pi f)^2 A$
25. 7.0 s
27. $k_1k_2/(k_1 + k_2)$, $k_1 + k_2$

29. 2.23 Hz
31. 0.58 m

Chapter 17

1. 1.5×10^9 Hz
3. 294 m
5. (a) 0.4 cm, (b) 10 cm, (c) 20 Hz
7. (a) 125 Hz, (b) 0.10 mm, (c) 2.0 cm
9. (a) 1.20 m, (b) 48 m/s
11. 21.4 Hz, 42.9 Hz, 64.3 Hz, 85.7 Hz
13. (a) 0.125 Hz, (b) 0.25 Hz
15. 0.214 Hz, 0.429 Hz, 0.643 Hz
17. (a) 2 m, (b) 400 m/s, (c) 66.7 Hz
19. (a) 120 Hz, (b) 60 Hz
21. 180 Hz, 540 Hz, 900 Hz
23. 42.5 Hz, 127.5 Hz, 212.5 Hz, 297.5 Hz
25. 335 m/s
27. 1390 Hz
29. 866 m/s
31. 153 m/s, 95.6n Hz
33. 3920 m/s
35. 1111 Hz
37. 3333 Hz
39. (a) 4900 m/s, (b) 2.0×10^{11} N/m^2

Chapter 18

1. About 1.7 cm to 17 m
3. 3.0×10^{-5} m
5. 14,600 m
7. 349.8 m/s
9. 260.2 Hz
11. 308 m/s
13. 2.25×10^9 N/m^2
15. 830 m
17. 5.25×10^{-6} J
19. 1.56×10^{-4} J
21. Proof
23. (a) 0.108 J, (b) 2.6×10^{-5} Cal
25. (a) 50 dB, (b) 150 dB, (c) −20 dB
27. 58 dB
29. 2.0×10^{-4} W/m^2
31. 3.16×10^{-3} W/m^2, 0.36 W
33. 425 Hz, 378 Hz
35. 1040 Hz, 777 Hz
37. 999.3 Hz and 1000.7 Hz
39. 2.1 beats/s
41. Proof

Chapter 19

1. 4.74×10^{14} Hz
3. 5.1×10^{14} Hz, 442 nm
5. 2.67 cm
7. 175 cm, 325 cm
9. 955 W/m^2
11. (a) 8.37 W, (b) 5.6×10^5 W/m^2
13. 754 lm
15. 1.40×10^{-6} sr
17. 503 lm
19. 1630 lm
21. 8960 W/m^2
23. 15.6 lm/W
25. 50,000 lx
27. 5000 lx
29. 31 lx
31. 50 cm
33. 270 cd
35. 3.6 mW

Chapter 20

1. (a) 6.0 m, (b) 1.75 m, (c) 100 cm
3. 40°
5. $(1.5 + 3n)$ meters behind each mirror for n an integer
7. (a) 40 cm, 0.67 cm, real and inverted, (b) 48 cm, 1.20 cm, real and inverted, (c) 120 cm, 6.0 cm, real and inverted, (d) −60 cm, 6.0 cm, virtual and erect
9. (a) −24 cm, 0.4 cm, (b) −21.8 cm, 0.545 cm, (c) −17.1 cm, 0.86 cm, (d) −12 cm, 1.2 cm; all images virtual and erect
11. (a) 60 cm, 1.5 cm, real and inverted, (b) 80 cm, 3 cm, real and inverted, (c) −24 cm, 4.8 cm, virtual and erect
13. (a) −30 cm, 0.75 cm, virtual and erect, (b) −26.7 cm, 1.0 cm, virtual and erect, (c) −10.9 cm, 2.18 cm, virtual and erect
15. (a) 4.0 cm, (b) inverted
17. Proof
19. (a) $p = 36$ cm, (b) $q = -180$ cm
21. (a) −5.5 cm, (b) −6.2 cm, (c) 0.27 in (a), 0.176 in (b)
23. 1.899×10^8 m/s
25. 1.94×10^8 m/s, 380 nm, 5.1×10^{14} Hz
27. (a) 45°, (b) 27.1°
29. 57°
31. 1.329
33. (a) 37.3°, (b) 70°
35. Proof
37. 48.8°

Chapter 21

1. (a) converging, (b) 12 cm
3. (a) 80 cm, yes, no, (b) 200 cm, yes, no, (c) −40 cm, no, yes
5. (a) 80 cm, yes, no, 5.0 cm, (b) 200 cm, yes, no, 20 cm, (c) −40 cm, no, yes, 10 cm
7. 58.6 cm
9. (a) −46 cm, no, yes, (b) −37.5 cm, no, yes, (c) −20 cm, no, yes
11. (a) −46 cm, no, yes, 0.23, (b) −37.5 cm, no, yes, 0.375, (c) −20 cm, no, yes, 0.67
13. 16 cm
15. (a) 100 cm, (b) 400 cm
17. (a) 120 cm, (b) −40 cm
19. Either 55 cm or 145 cm from object
21. Proof
23. On the first lens, 5 cm long, virtual, inverted
25. 4.6 cm
27. 12 cm in front of diverging lens, 1.2 cm, no, no
29. Proof
31. 21 cm in front of eyepiece, 3.4, upright image

Chapter 22

1. (a) loud sound, (b) weak sound, (c) loud, (d) weak
3. 0, 0.70, 1.40, 2.10, 2.80, 3.50, 4.20, 4.90 m
5. 2.80 m, 0.93 m, 0.56 m, 0.40 m
7. (a) 0, (b) 546 nm, (c) 2 × 546 nm, (d) 3 × 546 nm
9. (a) 2.65 cm, (b) 2.21 cm
11. (a) 594 nm, (b) no
13. 9.16×10^{-7} m, (b) 1.09×10^4 lines/cm
15. 0.0424°
17. 1.66°
19. (a) yes, (b) 0.0021°
21. 4.1 m
23. (a) 316 nm, (b) 238 nm, (c) 214 nm
25. $n(\lambda_{vac}/1.33) = d \sin \theta$, where λ_{vac} is wavelength in vacuúm
27. 110 nm, 330 nm, 550 nm, 770 nm
29. 137 nm, 410 nm, 683 nm, 956 nm
31. 203 nm
33. (a) 201 nm, (b) 401 nm
35. 1.000292

Chapter 23

1. 6.3×10^{18}
3. (a) 1.9×10^{22}, (b) 5.5×10^{23}
5. (a) 3.5 N in $-x$ direction, (b) 3.5 N in $+x$ direction
7. 8.2×10^{-8} N radially inward

9. (a) 4.3×10^{-5} N, (b) 9.0×10^{-7} N
11. (a) 4.4×10^{-8} N, (b) 2.52×10^{-9} C, 1.76×10^{-8} C
13. 0.84 N in $+x$ direction
15. $x = 2.25$ m
17. 0.066 N at 194°
19. Zero
21. 0.15×10^{-3} N inward along diagonal
23. (a) 0.0150 N in $+x$ direction, (b) 0.0150 N in $-x$ direction
25. 1875 N/C in $-x$ direction
27. (a) negative, (b) 3500 N, (c) -1.91×10^{-7} C
29. 15 m, inward
31. $x = 22.5$ cm
33. 2.26×10^{-7} C
35. 2.0×10^{6} N/C
37. (a) 10 V, (b) 1.6×10^{-18} J
39. (a) 12 V, (b) 1.92×10^{-18} J
41. (a) 3.34×10^{-15} J, (b) 20,900 V
43. 60 J
45. (a) 200 J, (b) 48 cal
47. (a) 2560 V, 4.7×10^{6} V, (b) 2560 eV, 4.7 MeV
49. 11.4 V

Chapter 24

1. 1.88×10^{-5} C
3. 480 pF
5. 2.0 nF
7. 2.2 pF
9. 3.34 nF
11. (a) 1.40 μF, (b) 23 μF
13. (a) 24, 72, and 144 μC, (b) 16 μC
15. 93 kV/m
17. 0.33 mm
19. 120 kV/m
21. Proof
23. 0.50 C, 3.1×10^{18}
25. 0.010 A
27. 9.1×10^{-6} A
29. 750 A
31. (a) 20 Ω, (b) 1.333 Ω
33. 1.875, 3.0, 5.0, and 8.0 ohms
35. 5.22 Ω
37. 3.84 Ω
39. (a) 2.30 A, (b) 1.275 A
41. (a) 2.0 A, (b) 1.48 A
43. 63 W
45. 289 mA
47. (a) 1650 W, (b) 13.75 A
49. (a) 697 W, (b) 5.8 A, (c) 20.7 Ω
51. 1.12×10^{-6} Ω·m
53. 37 K

Chapter 25

1. (a) $\frac{1}{3}$ A, (b) $\frac{1}{15}$ A
3. (a) $I_2 = I_1 + I_3$, (b) $7I_2 + 12 + 5I_1 + 3I_1 - 6 = 0$
5. (a) $-6I_1 - \mathscr{E} + 2 = 0$, (b) $I_1 = I_1$, (c) $0 = I_1 + I_2 + I_3$
7. $I_1 = 0.80$ A, $I_2 = 2.0$ A, $I_3 = 2.80$ A
9. 2.0 V, $I_2 = 0.25$ A, $I_3 = -0.25$ A
11. $\mathscr{E} = 3.2$ V, $I_1 = -0.20$ A, $I_2 = 0.60$ A, $I_3 = -0.40$ A
13. $I_1 = 0.44$ A, $I_2 = -0.041$ A, $I_3 = 0.483$ A
15. (a) 20.8 A, (b) 10.4 A
17. (a) 10 A, (b) 11.3 A
19. (a) 0.545 A, (b) 1.27 A, (c) 2.73 A
21. 32, 16
23. (a) 1.39 V, (b) 1.75 V
25. (a) 1.24 V, (b) 24.2 cal
27. (a) 32 kJ, (b) 13 A·h
29. 6.02×10^{23}
31. 2.25×10^{22}, 3.7×10^{-5} kg
33. 1.75×10^{-25} kg
35. $I = I_1 = 0$, $Q = 0$
37. Proof

Chapter 26

1. 7.5 N south
3. (a) 0.1875 N, into the page, (b) 0.162 N, into the page, (c) 0
5. 1.02×10^{-3} N westward
7. 3.5×10^{-3} N westward
9. (a) 1.08×10^{-13} N, into the page, (b) 8.3×10^{-14} N, into the page, (c) 0
11. 0.0358 T
13. (a) below, (b) 5.0 m
15. 9.51 cm
17. Proof
19. 60 A
21. (a) 0, (b) 3.2×10^{-4} T perpendicular to tabletop
23. 2.39 A
25. 1.27 A
27. 6.1×10^{-3} N·m
29. (a) no, (b) 7.8×10^{-3} N·m
31. 0.120 Ω
33. 1.33 mΩ
35. $R_1 = 1.00$ mΩ, $R_2 = 2.33$ mΩ
37. 1.00 T
39. (a) 0.057 T, (b) 1.8 T, (c) 31, (d) 5.7×10^{-4} T, 1.3 T, 2300

Chapter 27

1. (a) 9.0×10^{-3} Wb, (b) 0, (c) 6.9×10^{-3} Wb
3. 6.7×10^{-6} Wb
5. 2.3×10^{-5} Wb
7. (a) 8.0×10^{-5} V, (b) 1.6×10^{-5} V
9. 0.52 V
11. zero, no
13. (a) 2.25×10^{-3} Wb/s, (b) 2.25 mV c.w., (c) 0.1125 A
15. 0.046 V
17. (a) 13.3, (b) 1600 V
19. (a) step-up, (b) 2400 V
21. (a) 108.6 V, (b) 21 A
23. (a) 1.70 A, (b) 109.2 V
25. 2.84 V
27. (a) 138 mV, (b) negative
29. (a) 160,000 m/s, (b) no

Chapter 28

1. (a) 25 V, (b) 17.7 V, (c) 0.10 s, (d) 10 Hz
3. (a) 2 A, (b) 1.41 A, (c) 1.41 A, (d) 6.37 Hz
5. (a) 45 W, (b) 645 cal
7. 187.5 W
9. (a) 2.67 A, (b) 80 V, (c) 113 V, (d) 3.78 A
11. (a) 2.4 μA, (b) 10 s, (c) 24 μC, (d) 15.2 μC, (e) 0.888 μA
13. (a) 0.0183, (b) $0.0183V_0$
15. (a) 7960 Ω, (b) 0.796 Ω
17. (a) 0.0132 A, (b) 1.32 A
19. Proof
21. (a) 10,000 Ω, (b) 1.59 H
23. (a) 1.31 A, (b) 0.0131 A
25. (a) 199 Ω, (b) 360 Ω, (c) 0.139 A, (d) 41.7 V, (e) 27.7 V
27. (a) 188 Ω, (b) 275 Ω, (c) 0.182 A, (d) 36.4 V, (e) 34.2 V
29. (a) 0.525 A, (b) 15.8 V, (c) 52.2 V, (d) 89.3 V, (e) 37 V, (f) 91 V
31. (a) 10 μF, (b) 27 A
33. 5.87×10^{-5} H, 38 V
35. (a) 79.6 Ω, (b) 94.0 Ω, (c) 0.53
37. (a) 0.71, (b) 42.4 V, (c) 90 W
39. (a) 31.8 mA, (b) 0.040, (c) 0.152 W
41. (a) 13,600 Ω, (b) 0.147 nF
43. Proof
45. 1.5×10^{6} Hz
47. (a) 1.5×10^{10} Hz, (b) 2.3×10^{-19} F

Chapter 29

1. (a) 1.0050, (b) 1.048, (c) 1.400, (d) 2.29
3. 4.61 MeV, 4.61×10^{6} V

5. (a) 2.065, (b) $v/c = 0.875$
7. (a) 1.6×10^{-15} kg, (b) 5.4×10^{-16}
9. (a) 7.95×10^{-28} J, (b) 6.3×10^{30}
11. (a) proof, (b) 354 nm, (c) infrared, (d) ultraviolet
13. 2.27 eV
15. (a) 4.96×10^{-20} J, (b) 3.3×10^{5} m/s
17. (a) 8.4×10^{6} m/s, 8.7×10^{-11} m, (b) 1.96×10^{5} m/s, 2.0×10^{-12} m
19. 3.02 eV
21. 122 nm
23. 30.4 nm
25. (a) 54.4 eV, (b) 22.8 nm
27. (a) 4230 nm, (b) infrared
29. 0.062 nm (0.62 Å)
31. −3.4, −4.9, −7.7, −13.6, −30.6 eV

Chapter 30

1. 177 nm, 1140 nm
3. 1140 nm
5. 1.03×10^{-10} m
7. 0.42 eV
9. $I_c = 19.7$ mA, $I_b = 0.3$ A
11. Proof

Chapter 31

1. 53, 53, 78
3. (a) 3, 6, 3, (b) 29, 35, 29, (c) 79, 118, 79
5. (a) 9.786×10^{-26} kg, (b) 1.99×10^{-26} kg, (c) 9.784×10^{-26} kg
7. 4.7×10^{-15} m, 2.3×10^{17} kg/m^3
9. 6.02×10^{26}
11. 5.6 cm
13. 10.812 u
15. (a) 0.042 u, (b) 0.0060 u
17. (a) 1020 MeV, (b) 8.5 MeV/nucleon
19. 8.7 MeV
21. $^{227}_{89}Ac$, $^{240}_{93}Np$, $^{99}_{43}Tc$
23. (a) 2.13×10^{-30} kg, (b) 2.1×10^{-5}
25. 9.8×10^{-4}
27. (a) 27 h, (b) 81 h, (c) 135 h
29. 42.5×10^{-9} Ci
31. 5.9×10^{11} Bq (16 Ci)
33. 29,000 years
35. 7.3 MeV
37. 0.0078

INDEX

A

B

C

D

T

U

V

W

X

Y

Z